D0217876

BIOLOGY

Life On Earth

ABOUT THE COVER PHOTOGRAPH

In the clear tropical waters of the Red Sea a cup coral, also called tube coral, of the genus *Tubastrea* trails its tentacles in the current. Like a predatory flower, it sits silently waiting to snare shrimp, juvenile fish, or other morsels brought by the tides. Radial symmetry gives this cnidarian the ability to capture food from any direction. The yellow speckles studding its tentacles are clusters of microscopic, poisoned darts that fire in response to the touch and taste of potential prey. Much of the body of *Tubastrea* resembles a water-filled balloon. When disturbed, it rapidly deflates and retreats into a limestone cup, as seen below the extended individuals. (Photograph by Christopher Newbert.)

BIOLOGY

Life On Earth

SECOND EDITION

Gerald *Teresa*

Audesirk & Audesirk

UNIVERSITY OF COLORADO
AT DENVER

Paula Nicholas

BIOLOGICAL ILLUSTRATOR

Macmillan Publishing Company
NEW YORK

Collier Macmillan Publishers
LONDON

To Our New Daughter
and
To Her Grandparents

Printed in the United States of America

Macmillan Publishing Company
866 Third Avenue, New York, New York 10022

Collier Macmillan Canada, Inc.

Library of Congress Cataloging in Publication Data

Audesirk, Gerald.
Biology : life on earth / Gerald Audesirk and Teresa Audesirk ; Paula Nicholas, biological illustrator. — 2nd ed.
p. cm.
Includes bibliographies and index.
ISBN 0-02-305070-5
1. Biology. I. Audesirk, Teresa. II. Title.
QH308.2.A93 1989
574--dc19 88-39363
CIP

Printing: 3 4 5 6 7 8 Year: 0 1 2 3 4 5 6 7 8

Acknowledgments and credits for all photographs appear on pages *P-1–P-5*, which constitute an extension of the copyright page.

Preface

Textbooks designed for nonmajors courses in biology should be a breed apart from majors texts, because the students are very different. For many nonmajors, this course may well be their only formal encounter with biology during their college careers. Most enroll in a biology course to fulfill a general education requirement. Of these, many choose biology out of a genuine interest in learning more about living things, but many others take biology as an alternative to chemistry or physics. These students are usually unprepared for, and often intensely uninterested in, detailed chemical and biophysical explanations. Nevertheless modern biology, to an ever-increasing extent, incorporates large amounts of biochemistry and biophysics, and understanding the functioning of living organisms often requires understanding the interactions among the molecules of which they are made.

To make matters still more challenging for instructor and author alike, many colleges and universities offer combined courses for both majors and nonmajors. How does one offer the depth and breadth of coverage required for the pre-health student and the prospective professional ecologist or molecular biologist, while keeping the business or English major from sinking in a sea of incomprehensible detail?

No text can hope to meet all of these needs completely. However, we have learned much from our own experience in teaching general biology to thousands of majors and nonmajors, from writing the first edition of this text, and from receiving many helpful suggestions from users, and as a result we have tried to organize this second edition according to a few important principles.

First, a general biology text must capture and hold the students' interest. Users and reviewers have responded enthusiastically to our writing style, calling it lively, friendly, and interesting. In this revision, we have tried to retain the comfortable style of the first edition. We have also tightened up the writing, cutting down on overly long sentences, elaborate examples, and excessively detailed explanations that may confuse the student. As a result, this edition is shorter and easier to read, although we have not omitted any important topics.

Second, any text, but especially one designed for nonmajors and mixed courses, must begin with basic principles and work up from there. Ideally, the text should be flexible enough, and organized well enough, so that instructors can choose the level of material that they wish their students to master. Some instructors, especially in one-semester nonmajors courses, may wish to assign only the essential principles of difficult topics. Others, especially in two-semester mixed courses, will want to assign both the principles and the anatomical, biophysical, and biochemical details. In many parts of this edition, we have explicitly written difficult material at two levels, to serve these two audiences. Subjects such as photosynthesis, respiration, DNA replication, transcription, translation, the Hardy-Weinberg principle, and kidney function are first explained as a "take-home lesson," describing the essential facts and mechanisms that all biology students, at any level, must know. Only then do we fill in the details, either in a subsequent section of the text or in special boxed figures. Some topics, such as immunology, have grown enormously complex over the past few years. We have deliberately omitted some of the more complicated aspects of immunology, restricting our explanations to information that we feel the beginning student can understand.

Third, people learn best when they have mental "hooks" on which to hang new information. Wherever possible, we integrate new concepts in biology with everyday experience (see, for example, the Reflection at the end of Chapter 7). We also offer unusual examples in the hope that they will help to fire the student's imagination. We think that this blend of the familiar and the exotic gives a big assist to learning.

FEATURES

Second editions of texts are always outgrowths from the first, and ours is no exception. Users of the first edition will recognize many parts of this second edition, including what one reviewer called a "relentless" use of evolution as an organizing theme; thought-provoking Essays and Reflections; and many similarities in descriptions and explanations where users, reviewers, and we all agreed that they worked well. However, there are major changes too. For one, we have departed from a longstanding tradition among nonmajors biology texts: *the second edition is shorter than the first*. Nonmajors texts have undergone creeping, or rather galloping, increases in complexity and length. The average nonmajors text has 800 pages or more these days. Besides that, increasing amounts of material are often stuffed into tables to take up less room. Our own first edition, at 850+ pages, was a typical example of length overkill. By tightening up the writing and simplifying some of the explanations, we have reduced the length without sacrificing coverage of any major subjects. Whether used for a full-year or a one-semester course, this edition will be welcomed by students who have been submerged beneath an ever-mounting wave of textbook length and complexity.

Other new features of this edition include:

- *Chapter outlines and outlined chapter summaries*
- *Pronouncing end-of-chapter glossaries as student study aids*
- *Full color art throughout*
- *Simplified and expanded explanations of basic chemistry*
- *A box on interpreting chemical structures and formulas*
- *Multilevel explanations of complex materials such as photosynthesis, respiration, and protein synthesis*

Some of these changes, as well as valuable features retained from the first edition, are explained more fully below.

Organizational Themes

We have organized the text by using two main themes: *evolution* and *adaptation to the environment*. As Theodosius Dobzhansky so aptly phrased it, "Nothing in biology makes sense, except in the light of evolution." Life on Earth has been forged in the crucible of evolution, and any modern biology text must reflect that fact. Our second organizational theme, adaptation to the environment, flows naturally from the first—organisms are adapted to the environment because that environment places selective pressures upon them that help to shape their evolution. Nevertheless, organisms are not perfect, and not all biological structures are optimal solutions brought about by natural selection. Genetic drift, chance catastrophes both ancient and modern, and the inevitable requirement that evolutionary change must build upon pre-existing structures have led to many unusual, unlikely, and, from an engineering standpoint, inefficient structures. In the words of Sydney Brenner, "Anything that is produced by evolution is bound to be a bit of a mess," and we have tried to show that, too.

Chapter Sequence

One difficulty for biology teachers and textbook writers is that there is no "correct" sequence of topics in biology. Virtually every subject would be more understandable if everything else preceded it. Further, an introductory course cannot cover all topics equally well, and different instructors may wish to emphasize different areas. In one-semester courses, many instructors may leave out entire sections of the text. We have taught nonmajors biology using both the "traditional" levels-of-organization sequence and an ecology/evolution-first sequence, and didn't find a clear advantage either way. In any sequence, teacher and writer must try to overcome the students' lack of knowledge of the topics that haven't yet been covered. How does one relate biochemistry and genetics to evolution and ecological niches if these subjects won't be covered for several weeks? Or how does one discuss the carbon cycle if photosynthesis is put off until near the end of the semester?

Some texts claim that their chapters and units are completely self-contained, and can be shuffled into whatever sequence the instructor wishes. Being teachers ourselves, we know that complete interchangeability is impossible. However, there *are* features that can be incorporated into a text to maximize flexibility. We have begun that flexibility in the first chapter. In addition to the usual sections on the nature of life and the scientific method, we have two other important features. First, a Gallery of Life briefly introduces the five kingdoms of life and provides illustrations of a few examples of each kingdom. Therefore, instructors can fairly assume that their students have at least a passing acquaintance with the living world, and so they can employ examples of real organisms in lecture. Second, we provide a brief, understandable explanation of the principles of evolution. This will enable instructors to refer to the evolutionary significance of many topics even though they are covered before evolution is formally taught. A thorough discussion of evolution is, of course, presented in Unit III. The student will not have to refer back to Chapter 1 to fill in any material.

Sequence flexibility is also enhanced by selective repetition. For example, learning about plant adaptations for life on land is essential to understanding the plants' original conquest of dry land, plant diversity, plant physiology, and ecology. We have not hesitated to repeat important concepts—not in tiresome detail, but with enough thoroughness to make them understandable separately in each presentation. Such selective repetition not only aids sequence flexibility, but is a form of emphasis, strengthening the student's comprehension of important points.

Artwork

This new edition features a tremendous increase in *full-color art*, made possible by the generosity and foresight of the Macmillan Publishing Company and the marvelous talents of fellow Coloradan Paula Nicholas. In addition to her outstanding artistic skills, Paula is herself a biologist, with a master's degree in botany from Colorado State University. Her dedication to combining accuracy with simplicity and beauty is evident throughout this text. Our delight in working with Paula was tempered only by our sadness at the death of Carol Dufficy, the artist for our first edition, who was killed last year in an auto accident.

Learning Aids

New to this edition are *pronouncing glossaries at the end of each chapter*. All biology texts have end-of-text glossaries, but these suffer from two major drawbacks. First, most are too limited in coverage to be of real value as a student study aid. Second, owing to a combination of general inertia and disappointment at failing to find several of the first terms they try to look up, most students make little use of end-of-text glossaries. In this text, each chapter has its own glossary. Virtually all boldfaced terms are defined in the glossary of the chapter in which they are used or in the glossary of a previous chapter. Definitions of particularly important terms, such as "gene," are often repeated in several glossaries. A complete glossary is found at the end of the text as well.

Also new to this edition are *chapter outlines*. Each chapter begins with an outline consisting of the first two levels of section headings, so that the student can glean some idea of where the chapter is going. The *chapter summaries* are organized by using the first level of headings, so that the student can easily tell what part of the chapter the summary sections describe.

We have continued to use Essays and Reflections—many new to this edition—to pique student interest and to provide a glimpse into research techniques, unusual aspects of biology, and the everyday relevance of seemingly esoteric topics.

Each chapter ends with a set of study questions and suggested readings. The study questions are mostly a straightforward review of the chapter material, designed for the students to test their recall and comprehension. Thought-provoking discussion questions are included in both the Study Guide and the Instructor's Manual. The suggested readings have been revised and updated to include readings on the latest research findings. However, we have resisted the temptation to throw out the old and bring in the new for the sole purpose of having lots of 1987 and 1988 references. Most nonmajors, and even many majors for that matter, are simply not ready for current *Science* and *Nature* articles. Even some of the more recent *Scientific American* articles, particularly in immunology, are beyond the level of the beginning student. We have tried to limit the articles that we suggest to ones that a student might actually understand and enjoy.

The Supplements Package

A comprehensive set of supplements are available, each coordinated closely with the text and with one another. The supplements were prepared by a team of biologists at Virginia Commonwealth University and Clemson University.

The *Study Guide* by Joseph P. Chinnici, Virginia Commonwealth University, contains an outline review of each chapter plus review questions with answers.

The *Instructor's Manual* by Margaret L. May, Virginia Commonwealth University, contains an overview of each chapter, teaching suggestions, thought questions for written assignments or discussion, sources of films and computer software, and answers to the Study Questions in the text.

The *Test Bank* by Gail C. Turner, Virginia Commonwealth University, consists of over 1700 objective test items (about 45 per chapter). The Test Bank is available in *computerized format*, in both IBM- and Apple-compatible versions.

The *Lab Manual*, by William H. Leonard of Clemson University, is a revision of Leonard's popular *Laboratory Investigations in Biology*. It has been completely reorganized to follow the sequence of topics covered in the text.

Teaching Aids include approximately 100 *color overhead transparencies* and *color slides*, both with presentation notes. Enlarged *transparency masters* of another 100 diagrams, in black and white, allow the instructor to produce readable transparencies of other diagrams from the text, the Study Guide, and the Instructor's Manual.

ACKNOWLEDGMENTS

This text, like all textbooks, is not the sole work of its authors. We have benefited enormously from the opinions and expert advice of numerous colleagues at the University of Missouri and the University of Colorado at Denver. Many users and reviewers have offered helpful criticisms and comments, especially Daniel Chiras, Joseph P. Chinnici, Sandra Winicur, Hendrick J. Ketellapper, Gary L. Meecker and David Thorndill. A complete listing of reviewers is given below.

Usually unacknowledged, but perhaps most valuable of all, are the thousands of students whom we have taught over the years. Students, by the questions they ask and by their performance on tests, cast the deciding vote on whether an explanation is really helpful and understandable. Student feedback constantly refines our teaching, and therefore also our writing. Although they often remain unaware of it, our students are actually partners with us in a combined teaching/learning endeavor, and their role is really much more important than ours.

Despite all of this good advice from colleagues and students, we undoubtedly have slipped up now and then, and we remain solely responsible for errors and ineffective explanations. In cases where we have failed to follow the advice of users and reviewers, we ask their indulgence and request their continued gentle remonstrance.

Finally, we wish to thank the staff at Macmillan Publishing Company for their unstinting support and effort on our behalf. We are particularly indebted to Kristin Watts and Amado Toribio, who kept track of what seemed like thousands of photographs, permissions, and communications; Yvonne Gerin, the Sherlock Holmes of photo researchers, who pursued attractive, interesting, and pedagogically essential photos across the globe; Gerald Wilkie, who designed an entirely new look for our second edition; and Ed Neve and Pam Kennedy, who shepherded photos, artwork, and raw text back and forth among all of the elements of the production team, until a finished book finally emerged. Most of all, we thank Senior Editor Bob Rogers whose enthusiam and energy proved invaluable to everyone involved in the second edition. Bob became not only the guiding force behind the entire project, but a friend as well.

As authors, we now have an extended classroom that reaches across most of North America. We hope that these colleagues and students will find this edition to be both enjoyable and valuable as a tool to understanding biology. As we wrote in Chapter 1, biology can be much more than just another course to take and another set of facts to memorize. Maybe we're biased, but what can be more fascinating than learning about life on Earth?

Gerry and Terry Audesirk
Golden, Colorado
July, 1988

Reviewers and Feedback Sources

SECOND EDITION REVIEWERS

William F. Burke, *University of Hawaii*
Joseph P. Chinnici, *Virginia Commonwealth University*
Dan Chiras, *University of Colorado, Denver*
Mary U. Connell, *Appalachian State University*
Joyce Corban, *Wright State University*
David J. Cotter, *Georgia College*
John P. Harley, *Eastern Kentucky University*
Michael D. Hudgins, *Alabama State University*
Hendrick J. Ketellapper, *University of California, Davis*
Margaret May, *Virginia Commonwealth University*
Gary L. Meeker, *California St. Univ.–Sacramento*
Thoyd Melton, *North Carolina State University*
Glendon R. Miller, *Wichita State University*
Russell V. Skavaril, *Ohio State University*
David Thorndill, *Essex Community College*
Gail Turner, *Virginia Commonwealth University*
P. Kelly Williams, *University of Dayton*
Sandra Winicur, *Indiana University at South Bend*

FIRST EDITION REVIEWERS

G. D. Aumann, *University of Houston*
Vernon Avilia, *San Diego State University*
J. Wesley Bahorick, *Kutztown State University*
Gerald Bergtrom, *University of Wisconsin*
Raymond Bower, *University of Arkansas*
Bob Coburn, *Middlesex Community College*
Martin Cohen, *University of Hartford*
David J. Cotter, *Georgia College*
Douglas M. Deardon, *University of Minnesota*
Katherine J. Denniston, *Towson State University*
Ronald J. Downey, *Ohio University*
Wayne Elmore, *Marshall University*
Nancy Eyster-Smith *Bentley College*
Gerald Farr, *Southwest Texas State University*
Douglas Fratianne, *Ohio State University*
Michael Gaines, *University of Kansas*
David Glenn-Lewin, *Iowa State University*
Margaret Green, *Broward Community College*
Alexander Henderson, *Millersville University*
Laura Mays Hoopes, *Occidental College*
Arnold Karpoff, *University of Louisville*
William H. Leonard, *Clemson University*
William Lowen, *Suffolk Community College*
Steele R. Lunt, *University of Nebraska–Omaha*
C. O. Patterson, *Texas A & M University*
Harry Peery, *Tompkins-Courtland Community College*
Bernard Possident, *Skidmore College*
Robert Robbins, *Michigan State University*
Albert Ruesink, *Indiana University*
Alan Schoenherr, *Fullerton College*
Eldon Sutton, *University of Texas at Austin*
Professor Tobiessen, *Union College*
Richard Tolman, *Brigham Young University*
Nancy Wade, *Old Dominion University*
Jerry Wermuth, *Purdue University–Calumet*
Jacob Wiebers, *Purdue University*
Chris Wolfe, *North Virginia Community College*

Brief Contents

Detailed Contents

Unit I
The Life of a Cell

Unit II
Inheritance

12 HUMAN GENETICS 216

Unit III Evolution

13 EVOLUTION: ORIGIN OF SPE... 236

14 THE PROCESSES AND RESULTS OF EVOLUTION 256

15 THE HISTORY OF LIFE ON ...H 279

16 TAXONOMY: IMPOSING ORDER ON DIVERSITY 299

17 THE DIVERSITY OF LIFE. I. MICROORGANISMS 302

Unit IV
Anatomy and Physiology of Plants and Animals

Unit V
Ecology

Essays and Reflections

1
Introduction

Figure 1-1 *Viewed from the distance of the moon, the astonishing thing about the earth, catching the breath, is that it is alive. The photographs show the dry, pounded surface of the moon in the foreground, dead as an old bone. Aloft, floating free beneath the moist, gleaming surface of bright blue sky, is the rising earth, the only exuberant thing in this part of the cosmos.*

—Lewis Thomas

On your way to class tomorrow morning, take the time to observe the astonishing array of creatures that live even in a place as domesticated as a city or a college campus. Of course, there are grass, bushes, and trees; dogs, cats, and sparrows; and lots of people. But if you look more closely, you may also encounter insects and earthworms, a spider spinning its web, and mushrooms in the shade beneath a bush. And in your mind's eye, picture all the things that are too small to see, such as the bacteria on your teeth that give you "morning mouth" when you wake up.

Why is there such an astounding diversity of living things? How do they interact with one another? In what ways are bacteria, plants, and people alike, and what makes them so different? What processes must occur in the body of each organism for it to live? These are a few of the questions biologists ask in their quest to understand life on Earth, and that we invite you to ask too, as we explore biology together. For although we are part of the web of life, we humans are more than just another strand; only we can contribute that particular brand of exuberance that comes from understanding the nature of the Earth and its inhabitants, and appreciating the beauty of it all. We are life's way of understanding itself.

THE CHARACTERISTICS OF LIVING THINGS

To study the nature of life on Earth, we should begin at the beginning: *What is life?* If you look up "life" in a dictionary, you will find definitions like "the quality that distinguishes a vital and functioning being from a dead body," but you won't find out what that "quality" is, and for good reason. We all have an intuitive understanding of what it means to be alive—a mouse or a tree is alive, whereas a rock is not. Nevertheless, defining "life" is difficult, because living things are so diverse, and nonliving matter is sometimes so lifelike, that virtually any criterion that we use to define life will also be met by something in the inanimate world. We can, however, describe some of the characteristics of living things that, by and large, are not shared by nonliving objects. *Living organisms* (1) *have complex, organized structures;* (2) *actively maintain their complex structures, a process called homeostasis;* (3) *grow;* (4) *acquire energy and materials from the environment and convert them into new forms;* (5) *respond to their environment;* (6) *reproduce;* and (7) *taken as a whole, have the capacity to evolve.*

Figure 1-2 *Living organisms are complex and organized compared to nonliving matter.* **(a)** *Each crystal of table salt, sodium chloride, is an exact cube, showing great organization but minimal complexity.* **(b)** *The water and dissolved materials in the ocean represent nonliving complexity, but there is very little organization.* **(c)** *Living things have both complexity and organization. The water flea,* Daphnia pulex, *is only 1 millimeter long, yet it has legs, mouth, digestive tract, gonads, eyes of a sort, and even a rather impressive brain considering its size.*

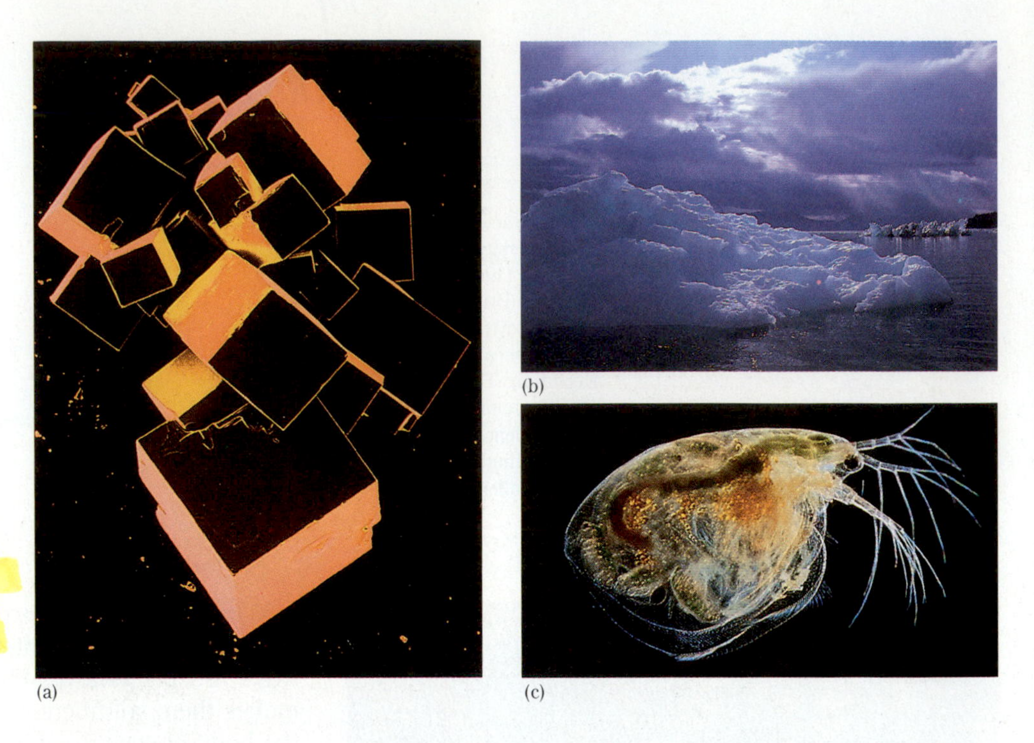

COMPLEXITY AND ORGANIZATION. Compared to nonliving matter of similar size, living organisms are highly complex and organized. A crystal of table salt (Fig. 1-2a), for example, consists of just two elements, sodium and chlorine, arrayed in a precise cubical arrangement: organized but simple. The oceans (Fig. 1-2b) contain some atoms of all the naturally occurring elements, but these atoms are randomly distributed: complex but not organized. In contrast, even the tiny water flea (Fig. 1-2c) contains dozens of different elements, linked together in thousands of specific combinations of atoms called molecules. These molecules are further organized into ever larger and more complex assemblies to form structures such as eyes, legs, digestive tract, and brain.

The complementary principles of complexity and organization reveal that the living world consists of a hierarchy of structures, each based on the one below it and providing the foundation for the one above (Fig. 1-3). An **atom** is the smallest particle of an element that retains the properties of that element. If it were possible to cut up a diamond, which is pure carbon, into ever-smaller pieces, each would still be carbon, until we had separated the diamond into individual carbon atoms: any further division would produce isolated subatomic particles that would no longer be carbon. Atoms may combine in specific ways to form assemblies called **molecules;** for example, one carbon atom can combine with two oxygen atoms to form carbon dioxide. Although many simple molecules form spontaneously, really large and complex molecules are manufactured only by living things.

Just as an atom is the smallest unit of an element, the **cell** is the smallest unit of life (Fig. 1-4). Taken at its simplest, a cell consists of the genes that contain the information needed to control the life of the cell, subcellular structures called **organelles** that act as miniature chemical factories to use the information in the genes and keep the cell alive, and a thin membrane that separates the cell from the outside world. Some organisms, mostly microscopic, consist of just one cell, but larger ones are com-

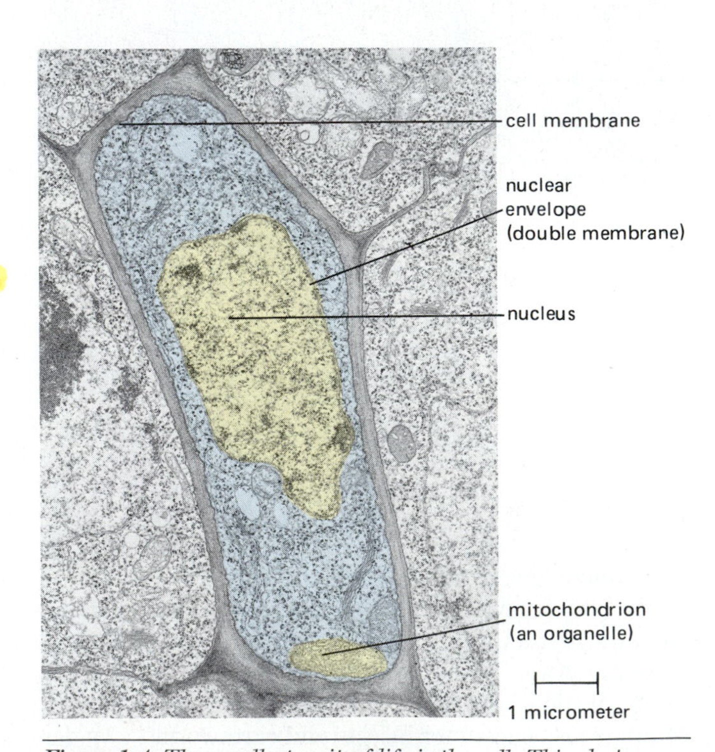

Figure 1-4 *The smallest unit of life is the cell. This electron micrograph clearly shows the membrane that surrounds the cell, separating it from its environment; the nucleus that contains the cell's genes; and many other specialized structures, called organelles, that perform particular functions in the life of the cell.*

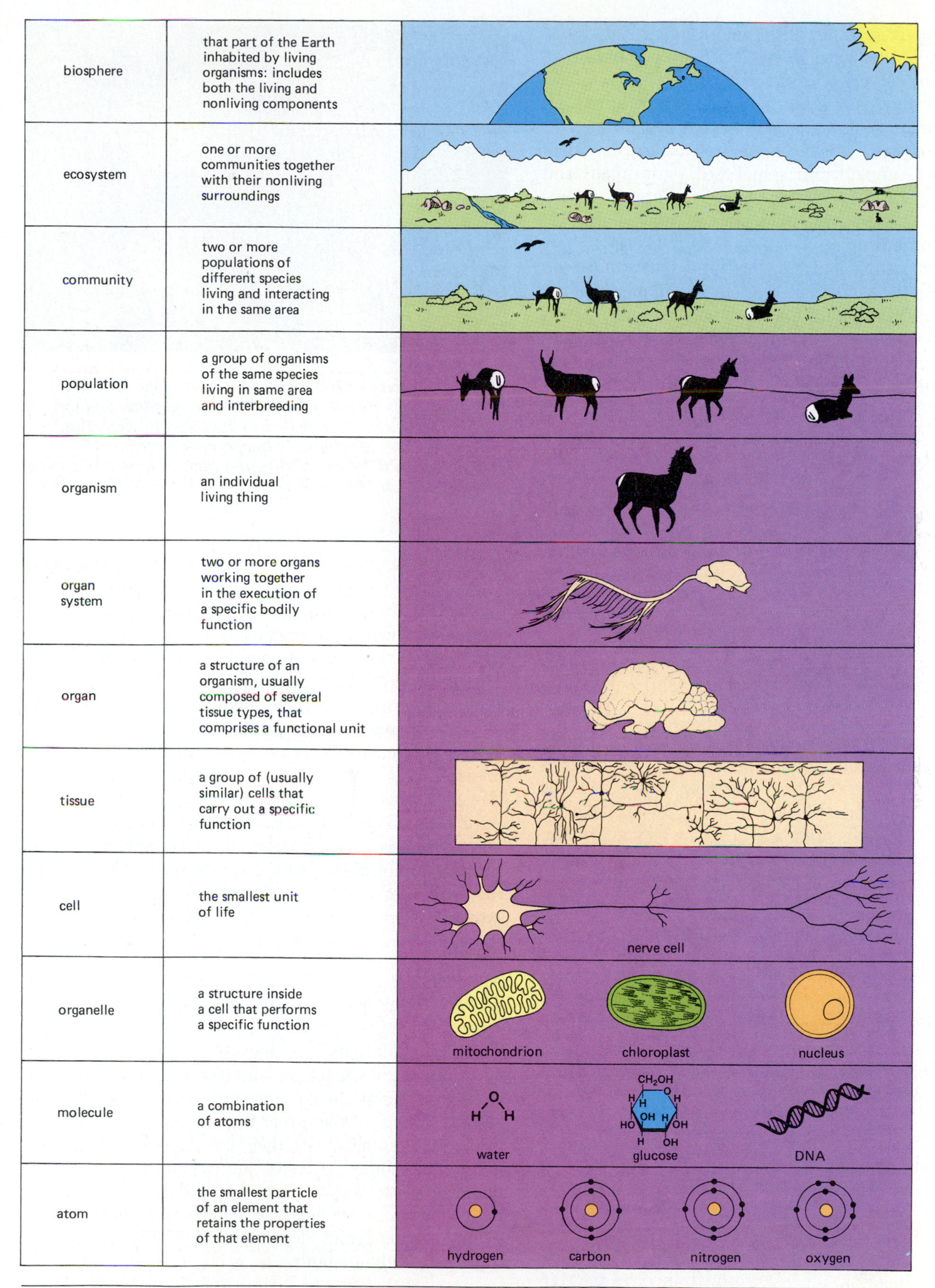

Figure 1-3 *Levels of organization of matter on Earth. Only the simplest levels of organization are possible without life. At each level, interactions among objects allow the development of the next-higher level of organization.*

posed of many cells. In these multicellular organisms, cells of similar type form **tissues,** such as nervous or connective tissue. A variety of tissue types (e.g., nervous and connective) may combine to make up a structural unit, called an **organ** (e.g., the brain). Several organs that collectively perform a single function are called an **organ system;** for example, brain, spinal cord, sense organs, and nerves form the nervous system. All the organ systems functioning cooperatively comprise an individual living thing, the **organism.**

The organization and complexity of living things do not end with the individual. A group of organisms of the same species that live in the same area and may potentially breed with one another is called a **population.** Populations of several different species living and interacting in the same area comprise a **community.** Many communities, plus their nonliving surroundings, make up an **ecosystem.** Finally, the entire surface of the Earth inhabited by living things is called the **biosphere.**

We will use this pattern of organization of life on Earth as the framework for organizing this text, beginning with atoms and molecules, moving on to cells, organs, and organisms, and concluding with the study of the interactions among organisms that are the focus of ecology.

HOMEOSTASIS. Complex, organized structures are not easy to maintain. Whether we consider the molecules of your body or the clothes in your closet, organization tends to disintegrate into chaos unless energy is used to sustain it (we will explore this tendency more fully in Chapter 6). To stay alive, organisms must keep the conditions within their bodies fairly constant, a process called **homeostasis** (derived from Greek words meaning "to stand the same"). Among warm-blooded animals, for example, vital organs such as the brain and heart are kept at a warm, constant temperature despite wide fluctuations in environmental temperature. This is accomplished by a variety of automatic mechanisms, including sweating during hot weather and burning up more food when it's cold, and by behaviors such as basking in the sun or even, in the case of people, adjusting the thermostat in the room.

Not everything "stands the same" throughout an organism's life. Major changes occur, such as growth and reproduction, but these are not really failures of homeostasis. Rather, they are usually specific, genetically programmed parts of the life cycle.

GROWTH. At some time in its life cycle, every living thing becomes larger, that is, it grows. This is obvious for plants, birds, and mammals, which all start out very small and undergo tremendous growth during their lives. Even single-celled bacteria, however, are small when they are first formed and grow to about double their original size before they divide. In all cases, growth involves the conversion of materials acquired from the environment into the specific molecules of the organism's own body.

Figure 1-5 Living organisms acquire energy and materials from their environment: the plants in this meadow capture energy from the sun and materials from the air, water, and soil, while the grazing moose acquires both energy and materials from the plants. Without a continual input of energy from the sun, all life would cease.

ACQUISITION AND USE OF ENERGY AND MATERIALS. To achieve homeostasis and to grow, organisms need energy and nutrients (Fig. 1-5). **Nutrients**—the atoms and molecules of which all organisms are made—may be acquired from the air, water, soil, or other living things. These nutrients must then be converted into the molecules of the organism's own body. **Energy**—the ability to do work, including carrying out chemical reactions, growing leaves in the spring, or contracting a muscle—is obtained in one of two basic ways. Plants and some single-celled organisms capture the energy of sunlight and store it in energy-rich sugar molecules, a process called **photosynthesis.** In contrast, most bacteria, fungi, and animals cannot photosynthesize, and must eat the energy-rich molecules contained in the bodies of other organisms instead. As you might guess, the original sources of these energy-rich molecules are photosynthetic organisms.

RESPONSIVENESS. Living organisms perceive and respond to stimuli in their internal and external environments. When you feel hungry, you are perceiving internal stimuli, including the emptiness of your stomach and low levels of sugars and fats in your blood. You then respond to external stimuli by choosing appropriate objects to eat, such as a piece of pie rather than the plate and fork.

Not only animals, with their elaborate nervous systems and mobile bodies, perceive and respond to stimuli. The plants on your windowsill grow toward the light, and even the bacteria in your intestine manufacture a different set of digestive enzymes depending on whether you drink milk, eat candy, or both.

REPRODUCTION. Living organisms are not immortal; sooner or later, everything dies. The continuity of life on

Figure 1-6 Living organisms reproduce: as it grows, this baby orangutan will resemble, but not be identical to, its parents. The similarity and variability of offspring are crucial to the evolution of life.

Earth occurs because organisms reproduce, giving rise to others of the same type (Fig. 1-6). The diversity of life happens in part because the offspring are usually somewhat different from their parents. The combination of small but significant variation within overall similarity is an important component of our final property of life, the capacity to evolve.

THE CAPACITY TO EVOLVE. During its lifetime, an individual organism maintains homeostasis. This tendency toward individual constancy stands in stark contrast to the capacity of groups of organisms to change over time. Although the genetic makeup of a single organism remains essentially the same over its lifetime, the genetic composition of the species as a whole changes over many lifetimes, that is, the species **evolves.** Some of the changes that occur in a species, such as the development of larger and harder hooves in horses (see Fig. 1-13), help individual organisms to cope with the rigors of their environment. Such changes are called **adaptations.** We will return to the concepts of evolution and adaptation later in this chapter.

THE DIVERSITY OF LIFE

Although all living things share these general characteristics, evolution has brought forth an amazing variety of life forms. We will discuss the classification and structures of organisms in detail in Chapters 16 through 19. In the meantime, as you read this text and listen to your instructor, some of the organisms mentioned may be unfamiliar to you. Therefore, we present a brief description of the major categories of living organisms and a photographic essay, "Gallery of Life," to introduce you to the diversity of life on Earth.

Living organisms are grouped into five major categories, called **kingdoms:** Monera, Protista, Fungi, Plantae, and Animalia. Although biologists think that the groupings reflect evolutionary relationships among organisms, we must classify living things according to the characteristics we can see and measure today. There are exceptions to any simple set of criteria used to define the kingdoms, but three characteristics are particularly useful: cell type, the number of cells in each organism, and the mode of acquiring energy.

CELL TYPE. There are two fundamentally different kinds of cells, called **prokaryotic** and **eukaryotic.** The word "karyotic" refers to the **nucleus** of the cell: a membrane-enclosed sac containing the genetic material (Fig. 1-4). "Eu" means "true" in Greek, and eukaryotic cells are recognized by the presence of a membrane-enclosed nucleus. Eukaryotic cells are larger than prokaryotic cells and contain a variety of other organelles, many surrounded by membranes. Prokaryotic cells do not have a nucleus; their genes reside in the cytoplasm of the cell. They are small, only 1 or 2 micrometers in length, and lack membrane-bound organelles. "Pro" means "before" in Greek, and prokaryotic cells are thought to be the most primitive cells, having evolved before eukaryotic cells. As we will see in Chapter 15, eukaryotic cells probably evolved from prokaryotic cells. All the organisms of the kingdom Monera have prokaryotic cells, while the cells of the other four kingdoms are eukaryotic.

CELL NUMBER. Organisms of the kingdoms Monera and Protista are usually single-celled, or **unicellular,** although a few live in strands or mats of cells with little communication, cooperation, or organization among cells. Fungi, plants, and animals are **multicellular,** with the life of both individual cells and entire organisms completely dependent on intimate cooperation among cells.

ENERGY ACQUISITION. All organisms need energy to live. Photosynthetic organisms capture energy from sunlight and store it in molecules such as sugars and fats. These organisms, including plants, some monerans, and some protists, are therefore called **autotrophs,** meaning "self-feeders." Organisms that cannot photosynthesize must acquire energy prepackaged in the molecules of the bodies of other organisms; hence these are called **heterotrophs,** meaning "other-feeders." Many monerans and protists, and all animals and fungi, are heterotrophs. Heterotrophs differ in the size of the food they eat. Some, such as bacteria and fungi, absorb individual food molecules; others, including most animals, eat whole chunks of food and break them down to molecules in their digestive tracts.

Gallery of Life

Kingdom Monera

Key Characteristics: prokaryotic; unicellular; some autotrophic (photosynthetic), some heterotrophic (absorptive).

The Monera (bacteria) are unicellular, prokaryotic organisms. Some, the cyanobacteria, can photosynthesize, but most cannot. Almost all the nonphotosynthetic bacteria absorb their food molecule by molecule from their surroundings. Most moneran cells are surrounded by a thick, rigid cell wall. In a few, whiplike appendages protrude through the cell wall and enable the cell to move.

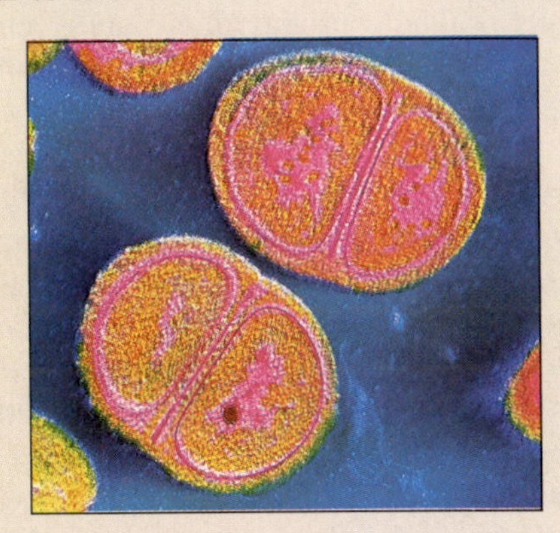

A color-enhanced electron micrograph of dividing bacteria.

Kingdom Protista

Key Characteristics: eukaryotic; unicellular, some autotrophic (photosynthetic), some heterotrophic (either absorptive or ingestive).

Protists are large, single eukaryotic cells. Some, such as *Euglena* and the diatoms, can photosynthesize and are therefore autotrophic, but most are heterotrophic. Some, such as *Amoeba,* can ingest food particles nearly as large as they are, whereas others, such as the malaria parasite, are limited to absorbing individual food molecules. Most protists, like the *Paramecium* illustrated here, are mobile, moving with cilia or flagella.

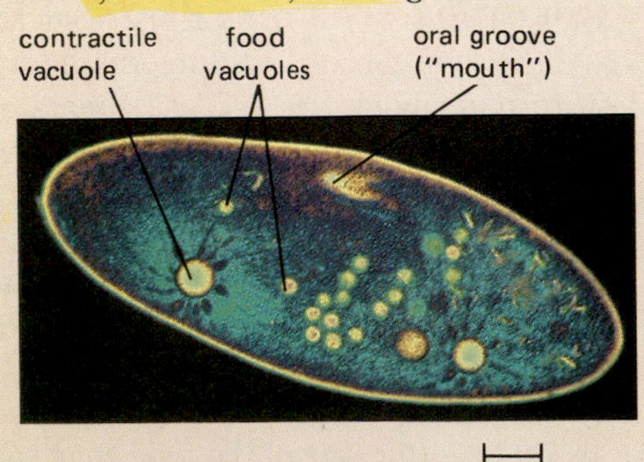

A light micrograph of Paramecium *shows how complex single eukaryotic cells can be.*

An exotic mushroom from Peru.

Kingdom Fungi

Key Characteristics: eukaryotic, mostly multicellular, heterotrophic (absorptive).

Most fungi are multicellular, often with distinct cell types specialized for feeding or reproduction. The mushroom, for example, consists of a maze of fine underground feeding threads that provide nutrients and energy for the aboveground reproductive stalk and cap. Fungi usually secrete digestive enzymes outside their bodies into their food, usually the dead bodies or wastes of plants and animals. These enzymes break down large food particles into separate molecules that can be absorbed by the fungal cells. Most fungi cannot move.

Kingdom Plantae

Key Characteristics: eukaryotic, multicellular, autotrophic (photosynthetic).

Plants are multicellular, nonmotile eukaryotes with considerable specialization of cell types. The plant kingdom is divided into smaller categories, called di-

The giant kelp Macrocystis *forms dense underwater forests along the southern California coast.*

Gallery of Life

visions, that show different degrees of cell specialization and consequently different habitat requirements.

Algae. The three divisions of algae (Rhodophyta, the reds; Phaeophyta, the browns; and Chlorophyta, the greens) all live in water, either fresh or marine, and show the least amount of specialization of cells into tissues and organs. Although a few, such as the giant kelps, can grow fairly large, most are quite small.

Mosses. The mosses and their relatives comprise the division Bryophyta. These are all land plants, but lack true roots for obtaining water and efficient conducting systems for transporting water throughout their bodies. Consequently, all bryophytes are very small and must live in moist places.

Ferns grow in lush profusion on damp forest floors.

Reproduction in mosses. The capsule is ejecting spores that will drift about on the wind. With luck, a few may land in damp places capable of supporting a new moss plant.

Flowering plants, such as these black-eyed susans, owe much of their success to a mutually beneficial relationship with insects, in which the flower provides food for the insect, while the insect fertilizes the flower.

Vascular plants. The vascular plants include ferns, conifers, and flowering plants. These are called vascular plants because they contain efficient vessels that transport water and nutrients between roots and above ground structures. Accordingly, they can inhabit drier habitats than can the mosses. Ferns, however, are still restricted to moist places, because they release swimming sperm that thrash their way through a film of water on the forest floor to reach the egg. Conifers and flowering plants enclose their sperm in pollen grains that are carried by wind or animals to their destinations, and consequently they are better adapted for living on land.

Kingdom Animalia

Key Characteristics: eukaryotic, multicellular, heterotrophic (most are ingestive).

Animals are multicellular eukaryotes that typically ingest their food in fairly large pieces. Animal bodies contain a wide variety of tissues and organs, composed of specialized cell types. Nearly all animals are mobile, at least at some stage in their life cycles. Animals live in both fresh and salt water, and virtually every type of habitat on land, from deserts to the surfaces of glaciers. The animal kingdom is subdivided into units called phyla (singular, phylum; equivalent to divisions of plants).

Sponges. Sponges (phylum Porifera) are the simplest animals, usually composed of only a few cell types, with minimal coordination among cells. All sponges are aquatic, feeding by pulling water into their

Gallery of Life

bodies through pores ("porifera" means "pore bearer") and filtering out minute plants and animals for food.

Sponges draw water into their bodies through tiny pores, filter out their prey, and eject the water out through the large holes visible here.

Coelenterates. The jellyfish, hydras, anemones, and their relatives (phylum Cnidaria) are exclusively aquatic; most are marine, but a few inhabit fresh water. All coelenterates bear stinging tentacles with which they capture prey and bring it to the mouth for ingestion. Since coelenterates have only one opening to their digestive tracts, indigestible remnants of a meal have to exit through the mouth.

Jellyfish swim constantly by pumping their contractile bells.

Annelids. Segmented worms (phylum Annelida), including earthworms, leeches, and exquisitely colored marine tube worms, can be found in both aquatic and terrestrial habitats. All, however, require at least damp surroundings, since their moist skins lose water quickly in dry conditions. Annelids, and all the phyla that follow, have digestive tracts with two openings, a mouth and an anus. Potentially, at least, feeding can be continuous, as food enters the mouth and wastes exit via the anus, with digestion proceeding in between.

Annelids are segmented worms, clearly demonstrated by this colorful marine worm that bears a pair of legs on each segment.

Molluscs. The phylum Mollusca includes animals of very diverse body form, including snails, clams, scallops, and octopuses. Most molluscs inhabit marine or freshwater environments, although a few snails and slugs live on land. The terrestrial molluscs, like the worms, are not very waterproof, and hence are largely restricted to damp habitats.

The octopus is the brainiest of the invertebrates, with learning abilities almost on a par with the rat.

Arthropods. All the members of the phylum Arthropoda are covered with a stiff suit of armor called an exoskeleton. Like armor, the appendages of the arthropod exoskeleton have joints to allow movement ("arthropoda" means "joint-footed"). In terms of both the number of species and the number of individuals, arthropods are the most successful of the animals, as you will recognize by a short list of representatives: spiders, scorpions, centipedes, millipedes, insects, crabs, shrimp, and lobsters. Arthropods live on land, in the sea, and in fresh water. The weighty exoskeleton, however, limits arthropod size, especially on land; not

Gallery of Life

surprisingly, the largest arthropods inhabit the sea, where the water can support some of the weight of the exoskeleton.

A fearsome-looking wolf spider awaits its insect prey, which it will kill with its poisoned fangs.

Echinoderms. The starfish, sea urchins, sea cucumbers, and sand dollars comprise the phylum Echinodermata. The phylum name means "spiny skin," which is especially obvious in sea urchins; however, if you touch a starfish, you will find that the name applies here too. Echinoderms are exclusively marine.

The long spines of the red sea urchin deter most potential predators.

Chordates. The familiar vertebrates are members of the phylum Chordata: fish, amphibians, reptiles, birds, and mammals. Chordates are found in freshwater, saltwater, and terrestrial habitats. All chordates have an internal skeleton of bone or cartilage, covered over with muscles and skin. Internal skeletons are stronger per unit weight than are the external skeletons of arthropods, and not surprisingly all large land animals are chordates.

A clownfish cuddles up to its "home" anemone, unharmed by the stinging tentacles that protect the fish from predators.

The bright colors of this tropical frog are a warning: Don't try to eat me, I'm poisonous!

The size and pizzazz of the dewlap hanging from this lizard's neck advertises its strength and virility to other lizards.

BIOLOGY: THE SCIENCE OF LIFE

Biology is a science, and its principles and methods are the same as those of any other science. In fact, a basic tenet of modern biology is that living things obey the same laws of physics and chemistry that govern inanimate matter.

Scientific Principles

All intellectual endeavors begin with a set of assumptions that usually remain unproven. Euclidean geometry, for example, assumes that parallel lines never intersect, no matter how far they extend. All scientific inquiry, including biology, is based on a small set of assumptions that we might call *scientific principles: causality, uniformity in space and time,* and *common perception.*

CAUSALITY. Historically, two approaches have been taken to the study of life and other natural phenomena. The first assumes that some events happen through the intervention of supernatural forces beyond our understanding. To the ancient Greeks, gods and goddesses were the immediate causes of powerful and unpredictable events: Zeus hurled thunderbolts from the sky, while Poseidon made earthquakes and storms at sea. By the time of Socrates, Plato, and Hippocrates, this view was being questioned. A striking example is offered by physicians studying diseases of the human body, including epilepsy. As you probably know, attacks of *grand mal* epilepsy cause convulsions and unconsciousness. Until relatively recent times, epilepsy was commonly thought to be a visitation from the gods. The ancient Greeks, in fact, called epilepsy "the sacred disease." Although he did not understand how epilepsy produces its symptoms, a Greek physician living about the time of Hippocrates firmly rejected the notion of supernatural causes of disease: "It seems to me that the disease called sacred has a cause, just as other diseases have. Men think it divine merely because they do not understand it. But if they called everything divine that they did not understand, there would be no end of divine things! . . . In Nature all things are alike in this, in that they can be traced to preceding causes."

In contrast to what we might call "supernatural causality," science adheres to the principle of **natural causality:** all Earthly events "can be traced to preceding [natural] causes." Today we realize that epilepsy is an organic disease of the brain, in which groups of nerve cells fire uncontrollably. No one knows exactly *why* they fire in such a catastrophic way, but scientists are confident that there is a natural cause, and that sooner or later they will discover it.

Natural causality is the very essence of science. Our confidence in this principle rests on the steady procession of causes that we have discovered for events formerly regarded as supernatural: lightning, earthquakes, epilepsy, leprosy, volcanoes, the movement of the sun across the sky, and getting pregnant, to name but a few.

The principle of causality has an important corollary: the evidence we gather about the causes of natural events is not deliberately distorted to fool us. If apples fall from trees at rates that conform to Newton's laws of gravity, this is evidence that the laws are correct. This corollary may seem foolish, yet not so very long ago people argued that fossils are not evidence of evolution, but were placed in the Earth by God as a test of our faith. If we cannot trust the evidence provided by the universe, the entire enterprise of science is futile.

UNIFORMITY IN SPACE AND TIME. A second fundamental principle of science is that natural laws do not change with time or distance. The laws of gravity, for example, are the same today as they were a billion years ago or will be a billion years hence. We also assume that these laws will hold just as well in Moscow as in New York, or in the spiral galaxy in Andromeda, for that matter. Although we cannot prove it conclusively, this principle seems to be valid. The calculations required to send the *Voyager* spacecraft to Saturn, for example, were based on this principle, with obvious success.

Uniformity in space and time is especially vital to biology, since many events of great importance to biology happened before humans were around to observe them, such as the evolution of today's diversity of living things. Some people believe that all the different types of living organisms were individually created in the past by the direct intervention of God. As scientists, we freely admit that we cannot disprove this idea. However, creationism is contrary both to natural causality and uniformity in time. The overwhelming success of science in explaining natural events through natural causes has led almost all scientists to reject creationism.

COMMON PERCEPTION. A final assumption of science is that we all enjoy *common perception:* all human beings perceive natural events through their senses in fundamentally the same way. If an elephant were to walk down your dormitory hall, all the students who happen to be looking at the hall would see the elephant, with obvious exceptions such as blind people. Further, the different senses would agree: the elephant could be seen, heard, felt, and smelled.

Although causality and uniformity in space and time are fundamental to almost all human activities, common perception is, to some extent, a peculiarly scientific principle. Value systems, such as those involved in the appreciation of art, poetry, and music, do not assume common perception. We may perceive the colors on the canvas in a similar way (the scientific aspect of art) but we do not perceive the aesthetic value of the painting identically (the humanistic aspect of art). Moral values also differ radically among people, often due to their culture or religious beliefs. Since value systems are subjective, not objective, science cannot solve certain types of problems, such as the morality of abortion.

Scientific Method

Given these assumptions, how do biologists study the workings of life? Ideally, biology and the other sciences use the *scientific method,* which consists of four interrelated operations: observation, hypothesis, experiment, and conclusion. All scientific inquiry begins with an *observation* of the world. Then in a flash of insight, a dream, or more likely after long, hard thought, a *hypothesis* is formulated, that a certain preceding cause produces the observed event. To be useful, a hypothesis must be able to predict the outcome of further observations or *experiments.* Carrying out these experiments either supports or refutes the hypothesis, and a *conclusion* is drawn about its validity.

Although there can be no science without initial observations and imaginative, testable hypotheses, probably the most difficult and time-consuming part of the scientific method is the experiment. The simplest form of experiment tests the hypothesis that a single preceding event is the cause of a single subsequent observation. As an example, consider the experiments of the Italian physician Francesco Redi (1621–1697) as he investigated why maggots appear in spoiled meat. Before Redi, this was considered evidence of **spontaneous generation,** the production of living organisms from nonliving matter. Redi observed that flies swarm around fresh meat, and that maggots appear in meat left out for a few days. Unlike the millions of other people who had probably seen this happen, Redi formed a testable hypothesis. He proposed that the flies lay eggs that hatch into maggots, and predicted that if flies could be kept away from fresh meat, maggots would never appear.

This is the essence of Redi's experiment (Fig. 1-7). Many different factors might influence the appearance of maggots, including not only the presence of flies, but also temperature, time of year, and whether the experiment was conducted indoors or outdoors. These potentially important factors are called **variables.** Redi wanted to test just one variable, the access of flies to the meat. Therefore, he took two clean jars and filled them with similar pieces of meat. He left one jar open (the "control" jar) and covered the other with gauze to keep out flies (the "experimental" jar). He did his best to keep all the other variables the same for the experimental and control jars (e.g., type of jar, type of meat, and temperature).

After a few days, maggots swarmed over the meat in the open jar, but no maggots appeared in the covered jar, although Redi reported that the meat nevertheless became "putrid and stinking." (Bacteria, of course, passed through the gauze and spoiled the meat.) Redi concluded that his hypothesis was correct, and that meat did not spontaneously generate maggots. Only through controlled experiments could this age-old superstition be laid to rest.

It is important to keep in mind the limitations of the scientific method. In particular, scientists can seldom be sure that they have controlled all the variables other than the one they are trying to study. For example, while Redi's gauze certainly kept flies away from the meat, the gauze may also have kept something else away that was the real source of maggots. Or perhaps gauze has some kind of "anti-maggot" effect, such as an insecticide. While Redi's experiment supported his hypothesis, it did not preclude all other possible hypotheses. Therefore, *scientific conclusions must always remain tentative,* and be subject to revision if new observations or experiments demand it.

SCIENTIFIC THEORIES. As science progresses, certain hypotheses are tested over and over again. To take a trivial example, each time an apple falls *down* to Earth rather than *up* into the sky, the hypothesis of gravitational attraction between masses is supported. When many independent observations all support a hypothesis, and the hypothesis is of general importance in explaining a large number of phenomena, the hypothesis is elevated to the status of a **theory.** In scientific terminology, a theory is a hypothesis that has been supported by so many cases that few scientists seriously doubt its validity. Thus a scientific theory is similar to what we would call a law or fact in common English usage. However, scientists must be open-minded about even their most cherished theories, and be willing to discard them if new observations render them obsolete.

Probably the foremost theory in biology is evolution. Since its formulation by Charles Darwin and Alfred Wallace in the mid-nineteenth century, the theory of evolution has been supported by fossil finds, geological studies, radioactive dating of rocks, genetics, biochemistry, and breeding experiments. People who refer to evolution as "just a theory" profoundly misunderstand what scientists mean by that term.

Real Science

To many people, science is cloaked in cold objectivity, as unemotional people calculatingly apply the scientific method to unravel the mysteries of nature. However, scientists are real people. They often make mistakes in their research and are driven by the same ambitions, pride, and fears as other people. As you will read in Chapter 10, ambition played an important role in the discovery of the structure of DNA by James Watson and Francis Crick.

The scientific method is not a formula that can be applied mindlessly with a Nobel Prize as the end result. Accidents, lucky guesses, controversies with competing

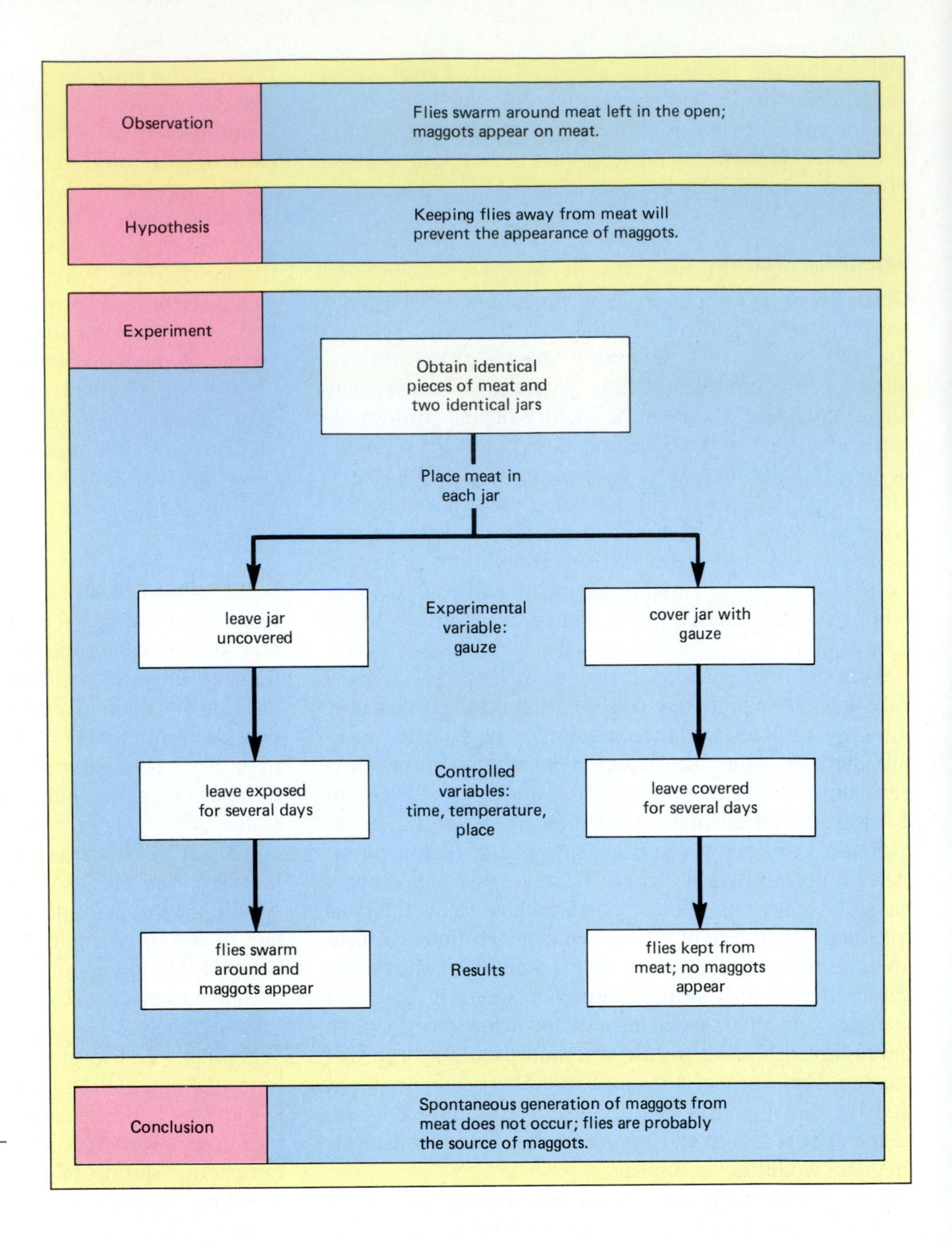

Figure 1-7 The experiments of Francesco Redi.

scientists, and of course, the intellectual powers of individual scientists contribute greatly to scientific advances. To illustrate what we might call "real science," let's consider two actual cases.

Chance and the Prepared Mind. When microbiologists study bacteria, one of the most important requirements is to have a pure culture, that is, a culture free from contamination by other bacteria, molds, and so on. Only by studying a single type at a time can we learn about the properties of that particular bacterium. Consequently, at the first sign of contamination, a culture is usually thrown out, often with appropriate mutterings about sloppy technique. On one such occasion, however, in the late 1920s, Alexander Fleming turned a ruined culture into one of the greatest medical advances in history.

One of Fleming's bacterial cultures became contaminated with a patch of a mold called *Penicillium.* Before throwing out the culture dish, Fleming noticed that *no bacteria were growing near the mold* (Fig. 1-8). Why not? Fleming hypothesized that perhaps the mold releases a substance that kills off bacteria growing nearby. To test this hypothesis, Fleming grew some pure *Penicillium* in a liquid medium, filtered out the mold, and then applied the remaining liquid to an uncontaminated bacterial culture. Sure enough, the bacteria were killed. Further research into these mold extracts resulted in the production of the first antibiotic, penicillin.

Figure 1-8 *Penicillin diffuses outward from a penicillin-soaked disk of paper, creating a pronounced bald spot in the "lawn" of bacteria on this petri dish.*

Fleming's experiments are a classic in scientific methodology, proceeding from observation to hypothesis to experimental tests of the hypothesis. However, scientific method alone would have been useless without the serendipitous combination of accident and acumen. Had Fleming been a "perfect" microbiologist, he would not have had any contaminated cultures. Had he not been acquainted with the previous research of other scientists, he would not have known how to attack the problem. Had he been less observant or less brilliant, the contamination would have been just another spoiled culture dish. Instead, it was the beginning of antibiotic therapy for bacterial diseases. As Louis Pasteur said, "Chance favors the prepared mind."

THE VISION OF BEES. Few scientific discoveries occur in a single bound, as Fleming's experiments may suggest. More often, a researcher will perform an experiment and draw a conclusion from it. Further thought about the problem, or perhaps the objections of his colleagues, will require further experiments. A case in point is the investigation of color vision in bees by Karl von Frisch.

Flowers are often brightly colored and are visited by bees collecting nectar and pollen. Perhaps bees locate flowers by their colors; if so, then bees must have color vision. To study color discrimination in bees, von Frisch began by placing honey on colored paper, for instance blue (Fig. 1-9a). Bees found the honey, lapped some up, went back to the hive to fill up a comb, and returned to the paper for more. After a while, von Frisch took away the honeyed paper and replaced it with two empty papers, one red and one blue. The bees swarmed around the blue paper (Fig. 1-9b). Does this experiment show that bees have color vision and use it to discriminate between red and blue?

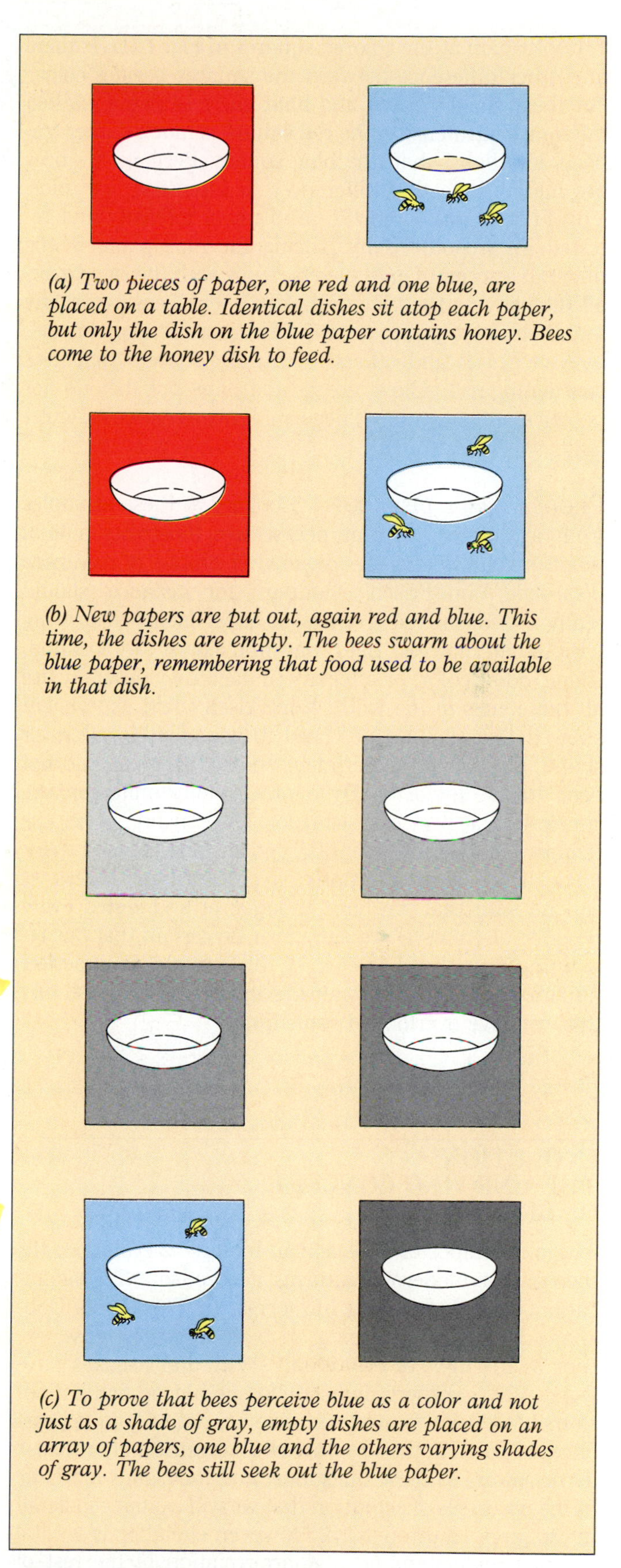

(a) Two pieces of paper, one red and one blue, are placed on a table. Identical dishes sit atop each paper, but only the dish on the blue paper contains honey. Bees come to the honey dish to feed.

(b) New papers are put out, again red and blue. This time, the dishes are empty. The bees swarm about the blue paper, remembering that food used to be available in that dish.

(c) To prove that bees perceive blue as a color and not just as a shade of gray, empty dishes are placed on an array of papers, one blue and the others varying shades of gray. The bees still seek out the blue paper.

Figure 1-9 *The experiments of Karl von Frisch, demonstrating color vision in bees.*

Look again at the colored squares in Fig. 1-9a. Is there any other difference between the squares besides color? For these shades of red and blue, there is also a marked difference in intensity: the red square is much darker (reflects less light) than the blue square. Perhaps the bees are merely detecting differences in intensity, not color. Von Frisch considered the problem of intensity and retested his hypothesis by placing the blue paper in the midst of a series of gray papers, some with the same overall intensity as the blue (Fig. 1-9c). The bees unerringly found the blue paper anyway. Many more experiments, by von Frisch and others, eventually showed that bees can distinguish yellow, green, and blue, but they do not perceive red as a color. They can, however, see ultraviolet light, which we cannot.

PROOF VERSUS DISPROOF. Looking at these examples from a slightly different perspective illustrates a final point: *scientific advances occur by a process of successive disproofs.* Von Frisch's findings, for instance, should really be considered a series of experiments disproving various alternatives. Bees may find flowers by location, shape, odor, color, or intensity (in fact, they probably use all these cues in the wild). Von Frisch's first experiment showed that they do not need to use odor, location, or shape. His second experiment showed that they do not need to use intensity. By a process of elimination, we conclude that bees can see color. Once again, we see the tentative nature of scientific conclusions. Each experiment disproves an alternative hypothesis, and unless we can be sure that we have thought of and tested every possible alternative, we can never be certain that our remaining hypothesis is correct. In a world of imperfect human minds, we must always entertain the possibility that we have overlooked something.

EVOLUTION: THE UNIFYING CONCEPT OF BIOLOGY

The most important concept in biology is evolution: the theory that modern organisms descended, with modification, from preexisting life forms. In Theodosius Dobzhansky's words, "Nothing in biology makes sense, except in the light of evolution." Why don't snakes have legs? Why are there dinosaur fossils but no living dinosaurs? Why are monkeys so like us, not only in appearance, but even in the structure of their genes and proteins? The answers to these questions, and thousands more, lie in the processes of evolution that we will examine in detail in Chapters 13 through 15. However, evolution is so vital to your understanding and appreciation of the rest of biology that we must take a brief look at its important principles before going further.

The Mechanisms of Evolution

In the mid-nineteenth century two English naturalists, Charles Darwin and Alfred Russel Wallace, formulated the theory of evolution that still forms the basis of our modern understanding. Evolution arises as a consequence of three natural processes: *genetic variation* among members of a population; *inheritance* of those variations from parent to offspring; and *natural selection,* the survival and enhanced reproduction of organisms with favorable variations.

VARIATION. Look around at your classmates and notice how different they are. People vary in pigmentation, body build, height, and the shape of their noses, just to name a few obvious features. Although some of this variation is due to differences in environment and life-styles, it is also strongly influenced by our genes: most of us could pump iron for the rest of our lives and never develop a body like Arnold Schwartzenegger's.

Where does genetic variation come from? Genes are the blueprints of life, providing the instructions needed to build an organism. Occasionally, genes suffer accidents: perhaps radiation strikes a gene just so, altering the blueprint. These accidents are called **mutations.** An organism with a mutation will develop characteristics that are a little different from the original. Due to mutations, many of which occurred millions of years ago and have been passed from parent to offspring through countless generations (see below), members of the same species are often slightly different from one another.

INHERITANCE. The genetic makeup of offspring is inherited from their parents. Mutations are rare, and the genes we inherit are mostly identical to those possessed by our parents. Therefore, a gene providing instructions for, say, bigger teeth in beavers will usually be passed on unchanged from parent to offspring. These offspring will be able to cut down trees more efficiently, build bigger dams and lodges, and eat more bark than "ordinary" beavers.

NATURAL SELECTION. Organisms that best meet the challenges of their environment will survive to leave the most offspring. The offspring will inherit genes that made their parents successful. Natural selection will thus preserve genes that help organisms to flourish in their environment, and discard the rest. Our hypothetical buck-toothed beavers will probably obtain more food and better shelter than their smaller-toothed relatives, and will probably raise more offspring as a result. The offspring will inherit their parents' genes for larger teeth. Over time, less-well-endowed beavers will become increasingly scarce, and after many generations all beavers will have large teeth. Structures, physiological processes, or behaviors that aid survival and reproduction in a particular environment are called **adaptations.** Most of the fea-

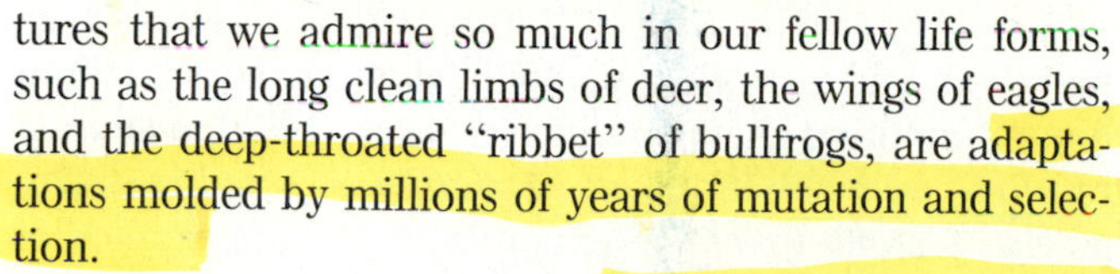

Figure 1-10 *A leopard shark, common in Pacific waters off the coast of southern California, displays the features that have characterized sharks for millions of years: sleek shape; long, powerful tail; acute sense of smell; and sharp rows of teeth.*

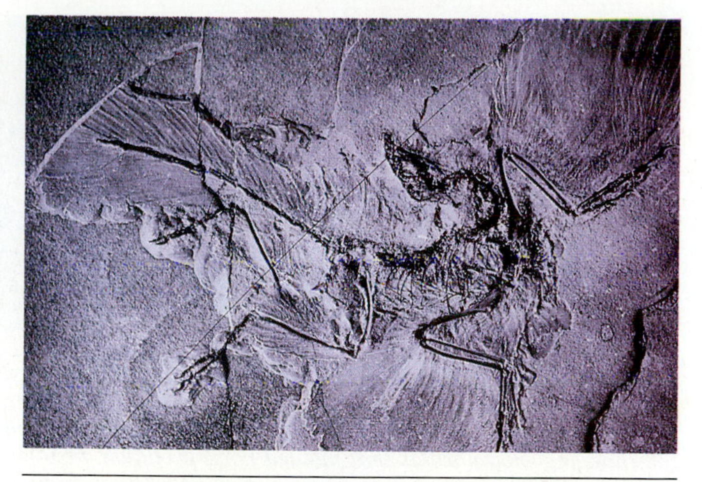

Figure 1-12 *A possible "missing link" between reptiles and birds:* Archaeopteryx, *a reptile that, although it probably couldn't fly, had feathers and fairly substantial wings.*

tures that we admire so much in our fellow life forms, such as the long clean limbs of deer, the wings of eagles, and the deep-throated "ribbet" of bullfrogs, are adaptations molded by millions of years of mutation and selection.

In the long run, however, natural selection is unpredictable: what helps an organism to survive today may be a liability tomorrow. If environments change—for example, as ice ages come and go—then the genetic makeup that best adapts organisms to their environment will also change. New mutations may arise that increase the fitness of an organism in the altered environment, and these mutations in their turn will spread throughout the population.

Over millennia, *the interplay of environment, variation, and selection results in evolution: the modification of the genetic makeup of species.* In constant environments, some well-adapted forms persist relatively unchanged. For example, sharks (Fig. 1-10) have retained essentially the same body form and probably behaviors for tens of millions of years, as their sleek shape, acute sense of smell, and formidable teeth have made them predators *par excellence.* In changing environments, some species fail to adapt and become extinct. The petrified forests of Arizona (Fig. 1-11) are the remnants of giant trees that could not withstand the conversion of tropical lowland to high desert. Other species, through mutations at the right times and places, change to meet new challenges, giving rise to new and different life forms, epitomized by protobirds such as *Archaeopteryx* (Fig. 1-12).

Figure 1-11 *Logs in Petrified Forest National Park. The petrified trunks littering the ground are remnants of vast forests of trees, unable to survive in desert conditions.*

Figure 1-13 An artist's rendering of the evolution of the horse, starting from Hyracotherium *(also called* Eohippus*) and leading through several intermediate forms to the modern horse,* Equus. *The changing backgrounds through which the horses run depict the changing environments that molded horse evolution, from forests through open woodlands to hard, flat plains. See Chapter 13 for more on horse evolution.*

The Evidence of Evolution

How do we know that evolution occurs? The major evidence for evolution, presented more completely in Chapter 13, is threefold. First, in many instances *fossils record gradual changes in organisms over millions of years,* just as would be expected if one form evolved from preceding forms. Fossil horses are a familiar example of sequential changes over time (Fig. 1-13). Second, *related modern organisms show remarkable similarities in physiology and structure,* such as the arrangement of bones in the wings of birds and bats, the flippers of seals and porpoises, the arms of humans, and the legs of dogs and sheep (Fig. 1-14). Such similarities are what one would expect if all these organisms descended from a common ancestor, as mutation and selection wrought slow changes in limb bones, adapting them to different uses. Third, *evolution resulting from selection can be readily observed in action at the present time,* for example in the appearance of pesticide resistance among insects (Fig. 1-15).

Although evolutionary biologists continue to debate the exact mechanisms whereby evolution occurs, the hypothesis of evolution has become a biological theory, one that has passed innumerable scientific challenges with flying colors. Virtually all biologists regard evolution as a fact of life. As with all scientific conclusions, evolutionary theory is subject to modification and refinement if new evi-

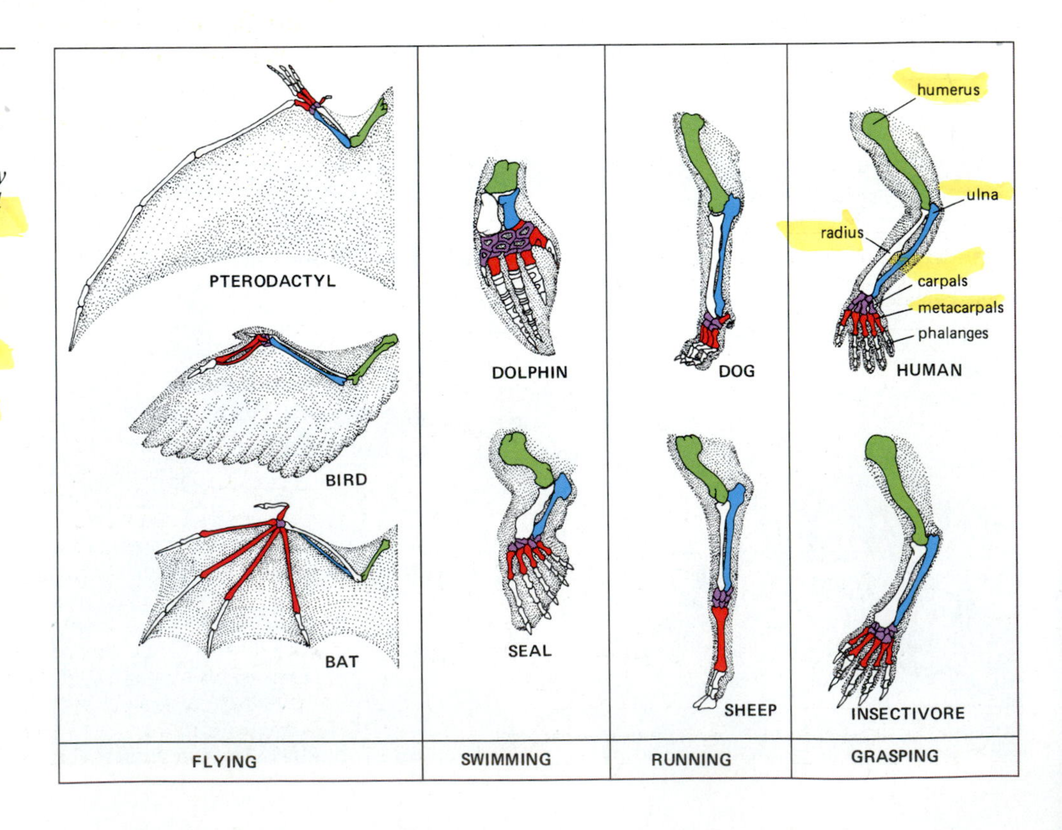

Figure 1-14 Similarities in the arrangement of bones in the forelimbs of different vertebrates. Even though these limbs have radically different functions, they are all built on the same general plan. This would be expected if all evolved from a common, though distant, ancestor. The limbs were not designed from scratch to perform a specific function; rather, mutation and natural selection worked on structures that were already available.

dence should appear, but no one expects the essential framework to fail.

THE CONTINUING JOURNEY

Some people regard science as a "dehumanizing" activity, feeling that too deep an understanding of the world robs us of vision and awe. Nothing could be further from the truth, as we repeatedly discover anew in our own lives. A few years ago, we visited the Denver Botanic Gardens in spring. We watched a bee foraging at a spike of lupines, flowers that we had seen hundreds of times, but never really looked at closely before. Lupines have a complicated structure, with two petals on the lower half of the flower enclosing the pollen-laden male reproductive structures (stamens) and sticky female pollen-capturing stigma (Fig. 1-16). In young lupine flowers, the weight of a bee pushing on these petals compresses the stamens, squirting pollen all over the bee's abdomen, just as your finger pushing the lever of the Colgate pump squirts toothpaste onto your brush. In older flowers, the stigma protrudes through the lower petals, and when a pollen-dusted bee visits, it usually leaves behind a few grains of pollen.

Did our newfound insights into the functioning of lupine flowers detract from our appreciation of the gardens?

A field is sprayed with DDT. Most of the insects are killed, but a few possess a random, formerly useless mutation which confers resistance to DDT. Only these resistant insects survive and reproduce.

The next year, most of the insects are descendants of resistant mutants, and are themselves resistant. When the field is sprayed again, far fewer insects are killed by the DDT.

Eventually, all insects are DDT resistant: evolution has occurred.

Figure 1-15 *Human activities have triggered some cases of extremely rapid evolutionary change. In the years since DDT was first introduced in the 1940s, over 400 species of insects have evolved resistance to its effects. Some points to note: (1) A few insects possessed mutations conferring DDT resistance* before DDT was ever invented. *The mutations occurred completely by chance and were not caused by exposure to DDT. (2) The mutation for resistance to DDT is itself neither good nor bad, but only acquired usefulness after humans invented and used DDT. (3) "Artificial selection" by DDT eliminated nonresistant insects, allowing the resistant ones to increase. Environmental conditions (here human farmers using DDT) are the guiding force behind evolutionary change.*

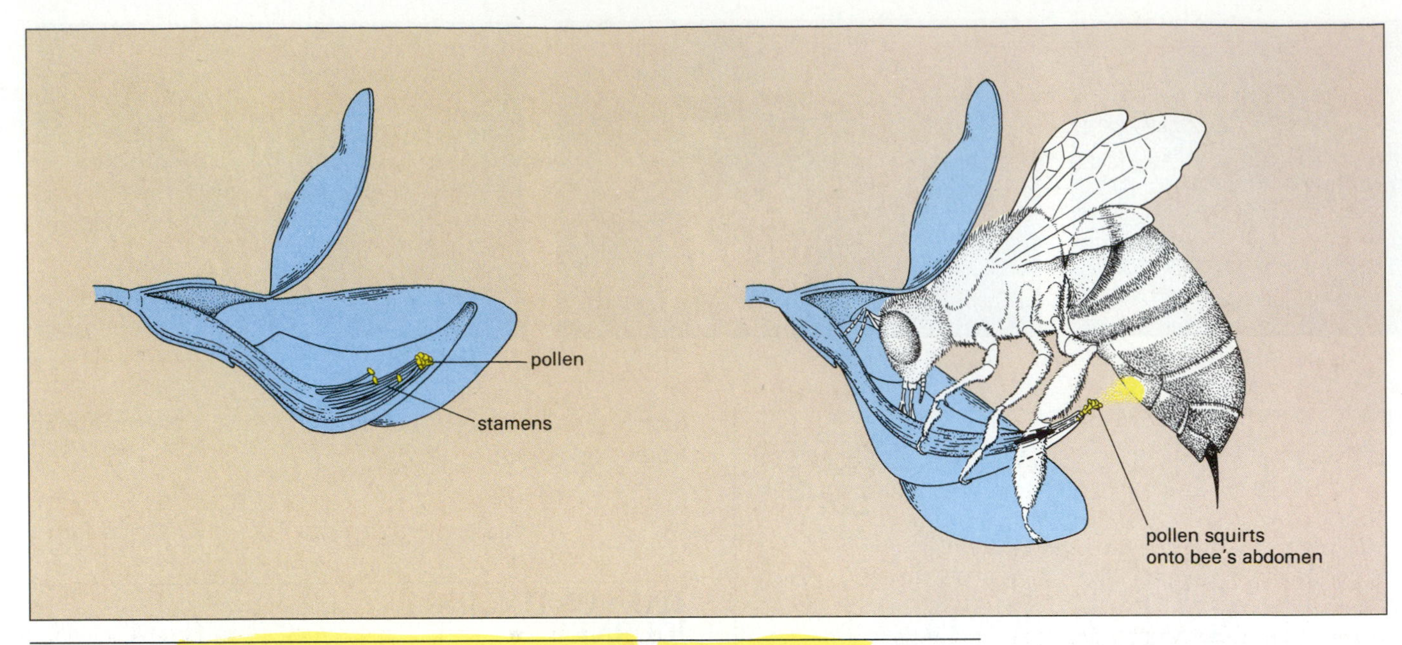

Figure 1-16 Lupines, like many members of the pea family, have complex flowers. The reproductive structures are enclosed within the lower petals. In young lupine flowers, the lower petals form a tube within which the stamens fit snugly like the plunger of a bicycle pump. The stamens shed pollen within the tube. When the weight of a foraging bee pushes on the lower petals, the stamens are thrust forward, squirting pollen out the end all over her abdomen. Some pollen adheres to the abdomen and may come off on the sticky, pollen-receiving stigma of the next flower visited by the bee.

Far from it. Rather, we now looked on lupines with new delight, understanding something of the interplay of form and function, bee and flower, that shaped the evolution of the lupine.

A few months later we ventured atop Hurricane Ridge in Olympic National Park, where the alpine meadows fairly burst with color in August (Fig. 1-17). As we crouched beside a wild lupine, an elderly man stopped to ask what we were looking at so intently. He listened with interest as we explained the structure to him, and then went off to another patch of lupines to watch the bees foraging. He, too, felt the increased sense of wonder that comes with understanding.

We try to convey that dual sense of understanding and wonder to you throughout this book. We also try to emphasize that biology is not a completed work but an exploration that we have really just begun. Lewis Thomas, physician and natural philosopher *extraordinaire,* has stated our situation in his usual elegant way: "The only solid piece of scientific truth about which I feel totally confident is that we are profoundly ignorant about nature. Indeed, I regard this as the major discovery of the past hundred years of biology . . . but we are making a beginning."

We cannot urge you strongly enough, even if you are not contemplating a career in biology, to join in the journey of biological discoveries throughout your life. Don't look upon biology as just another course to take, just another set of facts to memorize. Biology can be much more than that. It can be a pathway to a new understanding of yourself and the living Earth around you.

Figure 1-17 Wild lupines and subalpine fir on Hurricane Ridge in Olympic National Park. Thousands of people visit Hurricane Ridge each summer to gaze in awe at Mt. Olympus, but few bother to investigate the wonders at their feet.

SUMMARY OF KEY CONCEPTS

The Characteristics of Living Things

Living organisms usually possess the following characteristics: complexity and organization of structure; homeostasis; growth; the ability to acquire energy and materials from the environment; responsiveness; reproduction; and the capacity to evolve.

The Diversity of Life

Living organisms can be grouped into five major categories, called kingdoms: Monera, Protista, Fungi, Plantae, and Animalia. We classify organisms into kingdoms based on the type of cell the organism possesses, the number of cells per organism, and the mode of acquiring energy.

Cell type: Eukaryotic cells have their genetic material enclosed within a membrane-bound nucleus. Prokaryotic cells do not have a nucleus.

Cell number: Organisms may consist of a single cell (unicellular) or many cells bound together and working cooperatively (multicellular).

Energy acquisition: Autotrophic organisms obtain energy from inanimate sources, usually converting sunlight energy into energy-rich molecules through photosynthesis. Heterotrophic organisms obtain energy by eating energy-rich molecules (food) synthesized in the bodies of other organisms. The food may be eaten in large chunks (ingestion) or may be absorbed molecule by molecule from the environment (absorption).

The five kingdoms can be recognized by the following characteristics.

Monera: prokaryotic, unicellular, autotrophic or heterotrophic (absorptive).

Protista: eukaryotic, unicellular, autotrophic or heterotrophic (ingestive or absorptive).

Fungi: eukaryotic, multicellular, heterotrophic (absorptive).

Plantae: eukaryotic, multicellular, autotrophic.

Animalia: eukaryotic, multicellular, heterotrophic (ingestive).

Biology: The Science of Life

Biology is based on the scientific principles of natural causality, uniformity in space and time, and common perception. These principles are assumptions that cannot be directly proven, but that are validated by experience.

Knowledge in biology is acquired through the application of the scientific method. First, an observation of the world is made. A hypothesis is formulated, which suggests a natural cause for the observation. The hypothesis is used to predict the outcome of further observations or experiments. A conclusion is then drawn about the validity of the hypothesis.

Evolution: The Unifying Concept of Biology

Evolution is the theory that modern organisms descended, with modification, from preexisting life forms. Evolution occurs as a consequence of (1) genetic variation among members of a population, (2) inheritance of those variations by offspring, and (3) natural selection of the variations that best adapt the organism to its environment.

Three important types of evidence supporting the theory of evolution are: (1) fossils recording gradual changes over time from ancestral to modern forms; (2) similarities in anatomy, biochemistry, and physiology among modern organisms; and (3) present-day evolutionary changes in organisms occurring from natural or artificial selection.

GLOSSARY

Adaptation a specific structure, physiological mechanism, or behavior that promotes the survival and/or reproduction of an organism.

Atom the smallest particle of an element that retains the properties of the element.

Autotroph (aw'-tō-trōf) an organism that can manufacture all its high-energy organic molecules (e.g., sugars, proteins) from simple inorganic molecules (such as carbon dioxide, water, and minerals), using a nonliving energy source (usually sunlight); an organism that does not have to consume organic molecules as food.

Biosphere (bī'-ō-sfēr) that part of the Earth inhabited by living organisms; includes both the living and nonliving components.

Causality the scientific principle that natural events occur as a result of preceding natural causes.

Cell the smallest unit of life, consisting, at a minimum, of an outer membrane enclosing a watery medium containing organic molecules, including genetic material composed of DNA.

Community two or more populations of different species living and interacting in the same area.

Ecosystem (ēk'-ō-sis-tem) one or more communities together with their nonliving surroundings.

Energy the ability to do work.

Eukaryotic (ū-kar-ē-ot'-ik) referring to the type of cell characteristic of members of the kingdoms Protista, Fungi, Plantae, and Animalia. Eukaryotic cells have their genetic material enclosed within a membrane-bound nucleus, contain other membrane-bound organelles, and are usually larger than prokaryotic cells.

Evolution the descent of modern organisms from preexisting life forms; strictly speaking, any change in the overall genetic composition of a population of organisms from one generation to the next.

Gene the physical unit of inheritance; part of a DNA molecule.

Heterotroph (het'-er-ō-trōf) an organism that cannot use inanimate energy sources (such as sunlight) to synthesize all its energy-rich organic molecules; hence heterotrophs must acquire energy-rich organic molecules manufactured by other living organisms as food.

Homeostasis (hō-mē-ō-stā'-sis) maintenance of a stable internal environment in the face of changes in the external environment.

Hypothesis (hī-poth'-e-sis) in science, a supposition based on previous observations, which is offered as an explanation for an event, and used as the basis for further observations or experiments.

Inheritance the transmission of inborn characteristics from parent to offspring.

Kingdom the most inclusive category in the classification of organisms. The five kingdoms are Monera, Protista, Fungi, Plantae, and Animalia.

Membrane in cells, a thin sheet of lipids and proteins that surrounds the cell or its organelles, separating them from their surroundings.

Molecule a relatively stable combination of atoms.

Mutation a change in a gene; usually refers to a genetic change that is significant enough to change the appearance or function of the organism.

Natural selection the differential survival and/or reproduction of organisms due to environmental forces, resulting in the preservation of favorable adaptations. Usually, natural selection refers specifically to differential survival or reproduction based on genetic differences among individuals.

Nutrients the atoms and molecules that living organisms need to acquire in their diets.

Organ a structure of an organism (e.g., intestine), usually composed of several tissue types, that is organized into a functional unit.

Organelle (or-ga-nel') a structure inside a cell that performs a specific function.

Organism (or'-ga-niz-em) an individual living thing.

Organ system two or more organs working in together in the execution of a specific bodily function (e.g., digestive tract).

Population: a group of organisms of the same species living in the same area and interbreeding.

Prokaryotic (prō-kar-ē-ot'-ic) referring to the type of cell characteristic of members of the kingdom Monera (bacteria). Prokaryotic cells do not have their genetic material enclosed within a membrane-bound nucleus, lack other membrane-bound organelles, and are usually smaller and simpler in structure than eukaryotic cells.

Spontaneous generation the proposal that living organisms can arise from nonliving matter.

Theory in science, an explanation for natural events that is based on a large number of observations and is in accord with scientific principles, especially causality.

Tissue a group of (usually similar) cells that together carry out a specific function.

Variable a condition, particularly in a scientific experiment, that is subject to change.

STUDY QUESTIONS

1. What are seven characteristics that can distinguish living from nonliving things? Which would be particularly applicable in distinguishing a living organism from one that has just died? For each characteristic, name a nonliving thing that comes close to possessing it.
2. Name the five kingdoms of living organisms. For each, list the type of cell, whether its members are unicellular or multicellular, and their method of acquiring energy.
3. Define homeostasis. Why must organisms continually acquire energy and materials from the external environment to maintain homeostasis?
4. Distinguish between prokaryotic and eukaryotic cells.
5. Distinguish between autotrophic and heterotrophic organisms, and give an example of each. Which kingdoms are composed exclusively of heterotrophic organisms?
6. Describe the scientific method. In what ways do you use the scientific method in everyday life, for instance in finding out why your car won't start?
7. What is evolution? Briefly describe how evolution occurs.

SUGGESTED READINGS

Attenborough, D. *Life on Earth.* Boston: Little, Brown and Company, 1979. Based on the BBC television series of the same name, Attenborough's book provides a fascinating look at the living creatures with which we share our world, with particular emphasis on their evolution.

Discover. New York: Family Media, Inc. A monthly newsmagazine of science, *Discover* provides up-to-the-minute coverage of events in all of science, as well as articles on leading scientific personalities.

Science News. Washington, D.C.: Science Service, Inc. A weekly magazine offering brief summaries of current discoveries in science.

Scientific American. San Francisco: W. H. Freeman and Company, Publishers. This magazine of science has been fascinating readers for well over a century. Each month a half-dozen articles explore topics from black holes to genetic engineering.

Thomas, L. *The Lives of a Cell* and *The Medusa and the Snail.* New York: Bantam Books, 1979 and 1980. Thomas is a physician, scientist, and philosopher. There is probably no more eloquent spokesman for science as an exciting, invigorating, human activity.

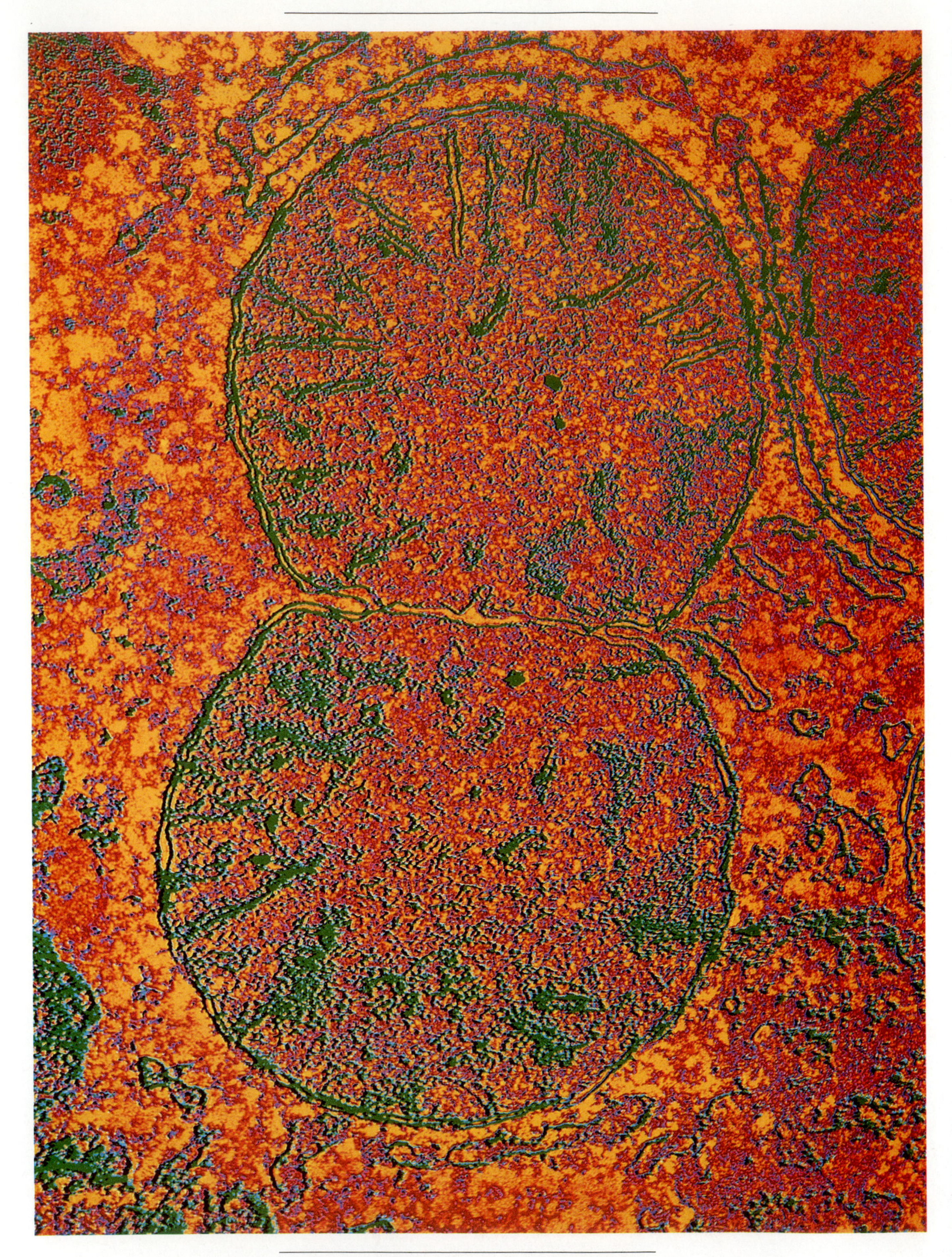

A color-enhanced electron micrograph of a dividing mitochondrion.

Unit I
The Life of a Cell

2 The Chemistry of Life. I. Atoms and Molecules

A few centuries ago, physics, chemistry, and biology were very separate sciences. Physicists studied the movements of large objects here on Earth and the planets about the sun; chemists sought to define the properties of matter, and if possible, to transmute base metals into gold; and biologists collected and classified plants and animals. The last 200 years, however, have witnessed the steady merging of the three sciences. Physicists have discovered a strange world of subatomic particles and forces that determine the properties of atoms; chemists study physics so that they can understand how atoms interact to form molecules; and biologists learn organic chemistry to help them study the large, complex molecules that form the bodies of living things.

In many instances, the study of life on Earth ultimately involves the study of the molecules of which living organisms are composed. How does photosynthesis convert the energy of sunlight into the energy of sugar molecules? What is the structure of the cell membrane, and how does it function in controlling the movement of materials into and out of the cell? How do muscles contract? How do the nerve cells in your brain communicate with one another? What causes cancer? To understand the answers to these questions, you must first learn a little about energy and matter, the properties of atoms, and how atoms interact with one another to form molecules.

MATTER AND ENERGY

Technically speaking, matter and energy are interchangeable, as expressed by Albert Einstein's famous equation, $E = mc^2$. However, for the chemical reactions that occur within living organisms, we can treat matter and energy as being quite distinct from one another: **matter** is the physical material of the universe, while **energy** is the capacity to do work, usually manifested by moving pieces of matter around.

FORMS OF ENERGY. Energy can exist in several forms, and may be converted from one form to another. The two major categories of energy are kinetic energy and potential energy. **Kinetic energy** is energy of movement. This includes not only movement of large objects, but also electrical energy (movement of electrons within a wire) and heat (movement of atoms and molecules). **Potential energy** is "stored" energy that can be released as kinetic energy under the right conditions. A pine cone has potential energy due to its position up in a tree. Its potential energy is converted to kinetic energy of motion when it falls on your head. The food you eat has chemical potential energy that is converted to the kinetic energy of movement and heat when you run (Fig. 2-1).

Figure 2-1 When 20,000 people run the New York Marathon each year, they do more than just make the bridges quake. They also collectively convert over 50,000,000 kilocalories of food into movement, heat, and sore muscles—that's equivalent to burning off over 14,000 pounds of fat or eating about 30,000 pounds of chocolate cake.

Transfers of energy regulate all interactions among bits of matter. In this chapter we examine the structure of matter, from atoms to molecules. In Chapter 3 we will focus on the molecules that make up the bodies of living organisms. In Chapter 6 we will explain how the flow of energy among atoms and molecules determines the chemical reactions that occur within cells.

THE STRUCTURE OF MATTER

The matter of the universe consists of 92 naturally occurring **elements.** An element, such as carbon or oxygen, is a substance with specific properties that can neither be broken down to simpler substances nor converted to different substances by ordinary chemical reactions. In contrast, a **compound** such as carbon dioxide is a substance composed of two or more elements. Compounds can be broken down into their component elements (e.g., carbon and oxygen) by chemical means. The elements that comprise most of the substance of the universe, the Earth, and living things are listed in Table 2-1.

Atomic Structure

If you took a diamond (a form of carbon) and cut it up into pieces, each piece would still be carbon. Finer and finer divisions would eventually result in a pile of carbon **atoms,** the smallest possible particles of the element carbon. Atoms themselves, however, are composed of two parts: a central **nucleus** and one or more outer **orbitals** (Fig. 2-2). Within the nucleus and orbitals reside **subatomic particles.** The nucleus contains heavy, positively charged **protons** and equally heavy but uncharged **neu-**

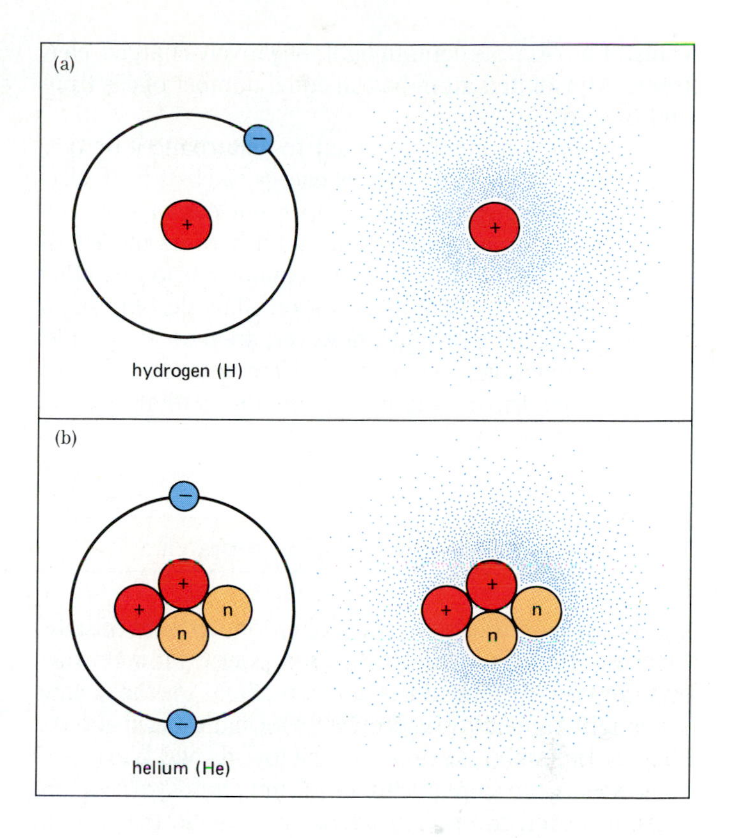

Figure 2-2 Structural representations of the two smallest atoms, hydrogen **(a)** *and helium* **(b).** *On the left are shown "planetary" models, in which the electrons are imagined as miniature planets, circling about the nucleus in precisely defined orbits. On the right are more accurate "electron cloud" models. Electrons may be found almost anywhere, but most of the time they stay within the stippled region, which represents the electron orbital. The denser the dots are, the higher the probability that electrons will be found at that place. Although electron cloud models of atoms are more accurate, planetary models are much easier to visualize and understand, and we will use them most of the time.*

Table 2-1 **Common Elements Important in Living Organisms**

Element	Symbol	Atomic Number[a]	Percent in Universe[b]	Percent in Earth[b]	Percent in Human Body[b]
Hydrogen	H	1	91	0.14	9.5
Helium	He	2	9	Trace	Trace
Carbon	C	6	0.02	0.03	18.5
Nitrogen	N	7	0.04	Trace	3.3
Oxygen	O	8	0.06	47	65
Sodium	Na	11	Trace	2.8	0.2
Magnesium	Mg	12	Trace	2.1	0.1
Phosphorus	P	15	Trace	0.07	1.0
Sulfur	S	16	Trace	0.03	0.3
Chlorine	Cl	17	Trace	0.01	0.2
Potassium	K	19	Trace	2.6	0.4
Calcium	Ca	20	Trace	3.6	1.5
Iron	Fe	26	Trace	5.0	Trace

[a]Atomic number = number of protons in the nucleus.
[b]Approximate percentage of atoms of this element, by weight, in the universe, in the Earth's crust, and in the human body.

trons. The orbitals contain light, negatively charged **electrons.** An isolated atom has an equal number of electrons and protons.

Nuclei are extremely resistant to disturbances by outside forces. Ordinary sources of energy, such as heat, electricity, or light, cannot break up a nucleus or even noticeably alter its structure. The stability of its nucleus is the reason why a carbon atom remains carbon whether it is part of a diamond, carbon dioxide, or sugar molecule. Electron orbitals, however, are more malleable: chemical reactions occur when electrons are removed from, added to, or shared among the orbitals of two or more atoms.

The Nucleus. The number of protons in the nucleus, called the **atomic number,** is characteristic of each element. For example, every hydrogen atom has one proton in its nucleus, every carbon atom has six, and every oxygen atom has eight. Different atoms of a given element may, however, have different numbers of neutrons. Atoms with the same number of protons (i.e., atoms of the same element) but with different numbers of neutrons are called **isotopes.** All the isotopes of an element are virtually identical in their chemical reactions, but may vary in their physical properties. In particular, the nuclei of some isotopes spontaneously disintegrate, releasing radioactivity as they do. Radioactive isotopes are extremely useful as

The Use of Isotopes in Biological Research

As you read this book you will encounter a lot of statements that may cause you to wonder: "How do they know *that*?" How do we know that DNA is the genetic material of cells (Chapter 10)? How do we measure the ages of fossils (Chapter 13)? How do we know that sugars made in plant leaves during photosynthesis are transported to other parts of the plant in the phloem tubes (Chapter 21)? These discoveries, and many more, have been possible only through the use of isotopes.

Atomic nuclei contain protons and neutrons. Every atom of a particular element has the same number of protons, but the number of neutrons may vary. Neutrons do not affect the chemical reactivity of an atom very much, but they do make their presence felt in other ways. First, neutrons add to the mass of an atom. Sophisticated instruments can detect the difference in mass between a chlorine atom with 17 protons and 18 neutrons and one with 17 protons and 19 neutrons. Second, the ratio of neutrons to protons helps to determine how stable a nucleus is. Nuclei with "too many" neutrons break apart spontaneously, often emitting radioactive particles in the process. Instruments such as Geiger counters can detect the radioactivity, and therefore detect the presence of radioactive isotopes.

Let's look at the use of radioactive isotopes in solving a simple biological problem. As you know, oysters have shells. The shells are made of calcium carbonate, secreted by the oyster as it grows. A basic biological question, which might be interesting to oyster "farmers" as well, is where the calcium comes from: does the oyster use calcium dissolved in the water, or calcium from its food, or both?

The most common isotope of calcium has 20 protons and 20 neutrons, and is not radioactive. The combined mass of protons and neutrons is 40, and this isotope is often called ^{40}Ca (pronounced "calcium 40"). A very rare isotope of calcium, ^{45}Ca, has 25 neutrons and is radioactive. By special chemical means, we can obtain calcium samples that have a lot of ^{45}Ca. Using ^{45}Ca, we can design an experiment to find out the source of calcium in oyster shells.

We might set up two aquaria, each seeded with identical numbers of baby oysters. All environmental conditions in the two tanks, such as temperature and type of food, are kept the same, except for which isotope of calcium is used in the food and water. In one tank, the water contains only ^{40}Ca, while the food given to the oysters includes radioactive ^{45}Ca. In the other tank, the conditions are reversed: the water has radioactive ^{45}Ca and the food has only nonradioactive ^{40}Ca. After a few weeks, the oysters in each tank are harvested. By comparing the amount of radioactivity in the shells, we can determine the relative contribution of dissolved calcium versus food calcium in shell synthesis. We find that the shells of oysters raised in tanks with dissolved ^{45}Ca are extremely radioactive, whereas the shells of oysters raised in tanks with ^{45}Ca in their food are not very radioactive. Therefore, oysters extract most of the calcium for their shells from the water, not from their food.

The use of isotopes in biological research is an example of the interplay among physics, chemistry, and biology that is so crucial to modern biological research. Most of the advances of the last 25 years, including molecular genetics, new therapies for cancer and heart disease, and our understanding of the flow of nutrients in ecosystems, have occurred because biologists recognize the importance of the other sciences in their own research.

"labels" in studying biological processes (see the essay, "The Use of Isotopes in Biological Research").

Electron orbitals, shells, and energy levels. At one time, it was thought that electrons were like tiny planets, circling the nucleus in well-defined orbits. However, electrons are really much harder to pin down than that. Rather than orbiting at a precise distance from the nucleus, electrons are found in a region of space called an **orbital** (Fig. 2-2). As you may know, particles with the same charge (e.g., both negative) repel one another, while particles with opposite charges attract one another. Therefore, you might expect that the negatively charged electrons would be pulled into the positively charged nucleus. This does not happen because the electrons zip about at high speed. Their energy of movement keeps them from collapsing toward the nucleus, analogous to the energy of movement that keeps the Earth from crashing into the sun, despite enormous gravitational pull.

Because of their mutual electrical repulsion, only two electrons can fit into a single orbital, and only a few orbitals can fit at a given distance from the nucleus. Large atoms accommodate many electrons by having layers of orbitals, called **electron shells,** at increasing distances from the nucleus (Fig. 2-3). Electrons in shells closest to the nucleus have the least energy and those farthest away have the most. For this reason, electron shells are often called **energy levels.**

The electron shell closest to the nucleus, with the lowest energy level, has only one orbital and can hold only two electrons. The second shell has four orbitals, and can hold up to eight electrons. The electrons in an atom normally occupy the shells closest to the nucleus. Thus a carbon atom, with six electrons, has two electrons in the first shell, closest to the nucleus, and four electrons in its second shell.

Why do we speak of energy *levels* and *shells*? Because orbitals are found only at specific distances from the nucleus—distances that correspond to discrete energy levels. Most people aren't used to thinking about energy occurring in steps, but perhaps an analogy with gravity can help. Picture yourself ascending a flight of stairs. You must spend a certain amount of kinetic energy to climb each step, and no intermediate amount will suffice. In

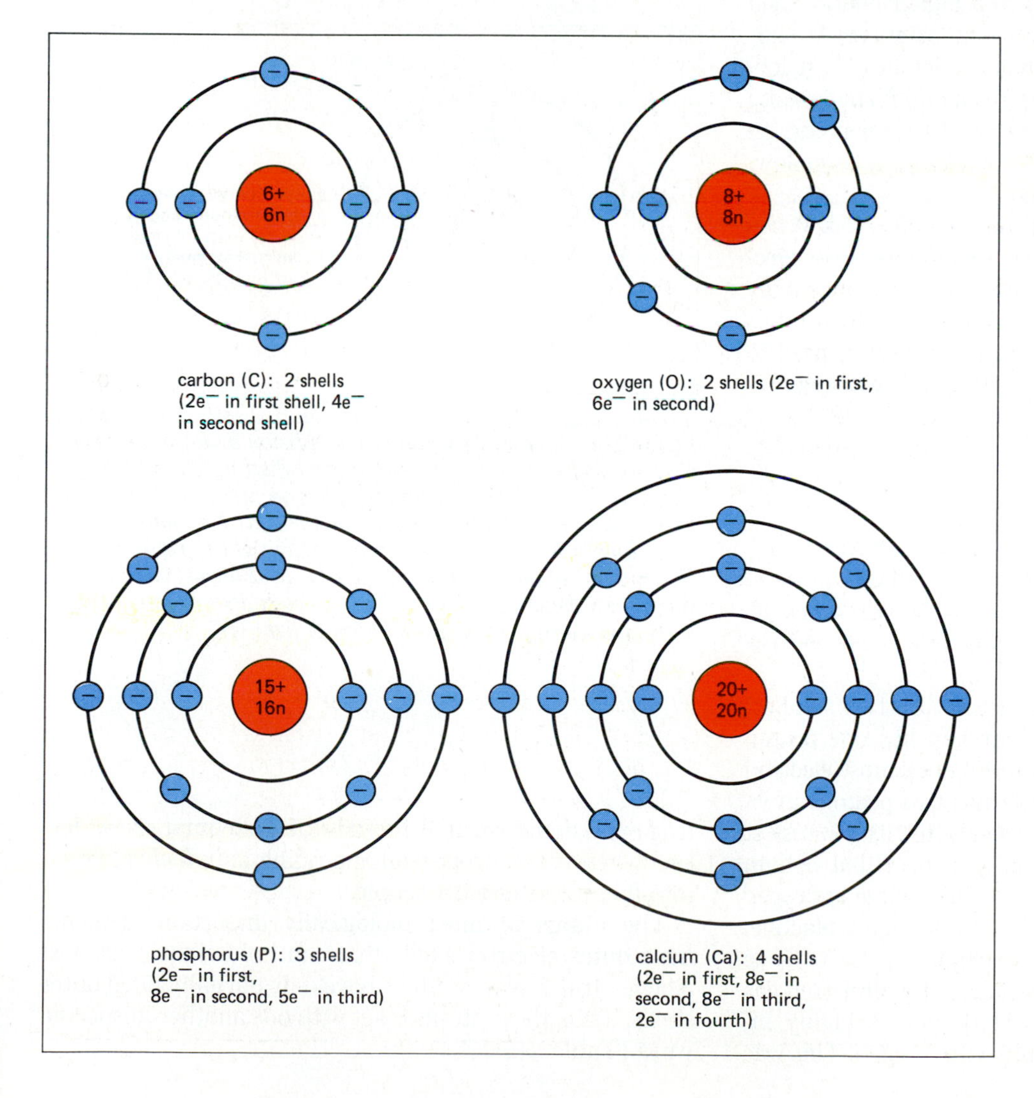

Figure 2-3 *Electron shells in atoms. Most biologically important atoms have at least two shells of electrons. The first shell, closest to the nucleus, can hold two electrons, while the next three shells usually hold a maximum of eight electrons.*

other words, you cannot lift your foot half a step and stand between steps. The energy used in climbing each step is stored as the potential energy of a higher position relative to the bottom of the stairs. This potential energy can be transformed back into kinetic energy by jumping back down the steps: the higher up you start, the harder you land.

Energy levels in atoms are analogous to the staircase (Fig. 2-4). It takes energy to force an electron out of a low-energy shell into a higher shell. Since an electron cannot occupy an intermediate position between shells, a certain amount of energy must be imparted to an electron to boost it to the next shell. Usually, the electron spontaneously returns to the lower energy level, releasing energy once again, often in a different form, such as light or heat. Such "energy level hopping" occurs in fluorescent light bulbs. Electric current flows through the gas that fills the hollow bulb. The electrical energy causes electrons in the atoms of the gas to jump to higher energy levels, from which they soon return to their original low level. When they do, they give off light waves of a particular energy, corresponding to a specific color (wavelength). These cycles of jumping to a higher-energy shell and back down again continue as long as energy is supplied to the bulb. Each element has particular energy levels to which its electrons can jump back and forth, lending characteristic colors to fluorescent bulbs, depending on which gas they are filled with: a sodium lamp is yellow, mercury is violet, and neon is red.

Energy-level hops have biological relevance, too. When plants photosynthesize, the energy of sunlight kicks electrons in chlorophyll molecules out of their low-energy shells. When the energized electrons return to orbitals with lower energy levels, some of their energy is used to synthesize sugar from carbon dioxide and water, as we describe in Chapter 7.

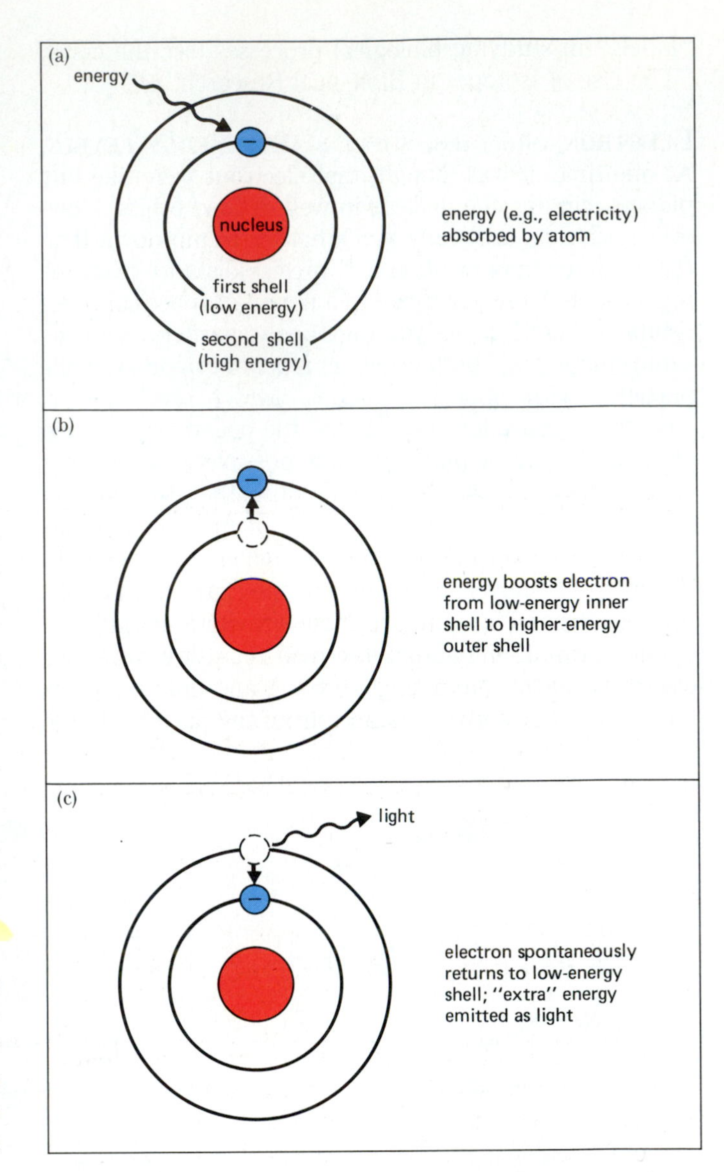

Figure 2-4 *Energy absorption and emission by an atom. The electron shells of an atom are found at discrete energy levels. If energy (e.g., electrical energy) enters an atom* **(a)**, *only specific amounts can be absorbed by the electrons, namely those amounts of energy that are exactly right to raise an electron out of one shell to a new shell at a higher energy level* **(b)**. *When the electron drops back down to its "resting" shell, it releases energy, often in the form of light* **(c)**.

Atomic Reactivity

Most atoms can react with other atoms to form molecules. The basic principle of atomic reactivity is this: *an atom is stable (i.e., will not react with other atoms) when its outermost electron shell is either completely full or completely empty.*

To illustrate this principle, consider the two smallest atoms, hydrogen and helium. Hydrogen has one proton in its nucleus, and one electron in its outermost electron shell, which can hold two. Helium has two protons in its nucleus, and two electrons completely fill its outermost electron shell. Therefore, we would suspect that helium atoms, with a full shell, should be stable, while hydrogen atoms, with a half-empty shell, should be highly reactive. Our deductions are correct, as is shown by the use of these gases in blimps. Helium, being stable, is the preferred gas with which to inflate a blimp. Hydrogen is highly inflammable, that is, it reacts readily with oxygen. This was tragically demonstrated in 1937 at Lakehurst, New Jersey, when the hydrogen-filled dirigible *Hindenburg* burst into flames, killing 36 people.

The atoms of most biologically important elements have outer electron shells that can hold eight electrons. The neutral, isolated atoms have only partially filled outer shells. Thus these atoms react with one another in specific ways (Table 2-2).

Table 2-2 **Bonding Patterns of Atoms Commonly Found in Biological Molecules**

Atom	Capacity of Outer Electron Shell	Electrons in Outer Shell	Number of Covalent Bonds Usually Formed	Common Bonding Patterns
Hydrogen	2	1	1	—H
Carbon	8	4	4	—C— —C= =C= —C≡
Nitrogen	8	5	3	—N— —N= N≡
Oxygen	8	6	2	—O— O=
Phosphorus	8	5	5	—P=
Sulfur	8	6	2	—S—

MOLECULAR STRUCTURE

An atom with an outermost electron shell that is partially full can gain stability by losing electrons (emptying the shell completely), gaining electrons (filling the shell), or sharing electrons with another atom (with both atoms behaving as if they had full shells). Losing, gaining, and sharing electrons result in forces called **chemical bonds** that hold atoms together in molecules.

Ionic Bonds

Some atoms, such as sodium, have an outermost electron shell containing only one electron (Fig. 2-5a). Other atoms, such as chlorine, have an outer shell that lacks only one electron to be completely full. Sodium can therefore achieve stability by losing the electron from its outer shell, leaving that shell empty, while chlorine can fill its outer shell by gaining an electron. This is precisely what each of them does to form sodium chloride: sodium loses an electron, becoming positively charged, and chlorine picks up the electron, becoming negatively charged (Fig. 2-5b). Atoms that have lost or gained electrons, and hence are charged, are called **ions.**

Since oppositely charged particles attract one another, sodium ions and chloride ions tend to stay near one another, forming crystals of salt (Fig. 2-5c). (To a chemist, the word "salt" refers not only to sodium chloride, but is a general term applied to any substance composed of positively and negatively charged ions.) This electrical attraction between oppositely charged ions, holding them together in salt crystals, is called an **ionic bond.**

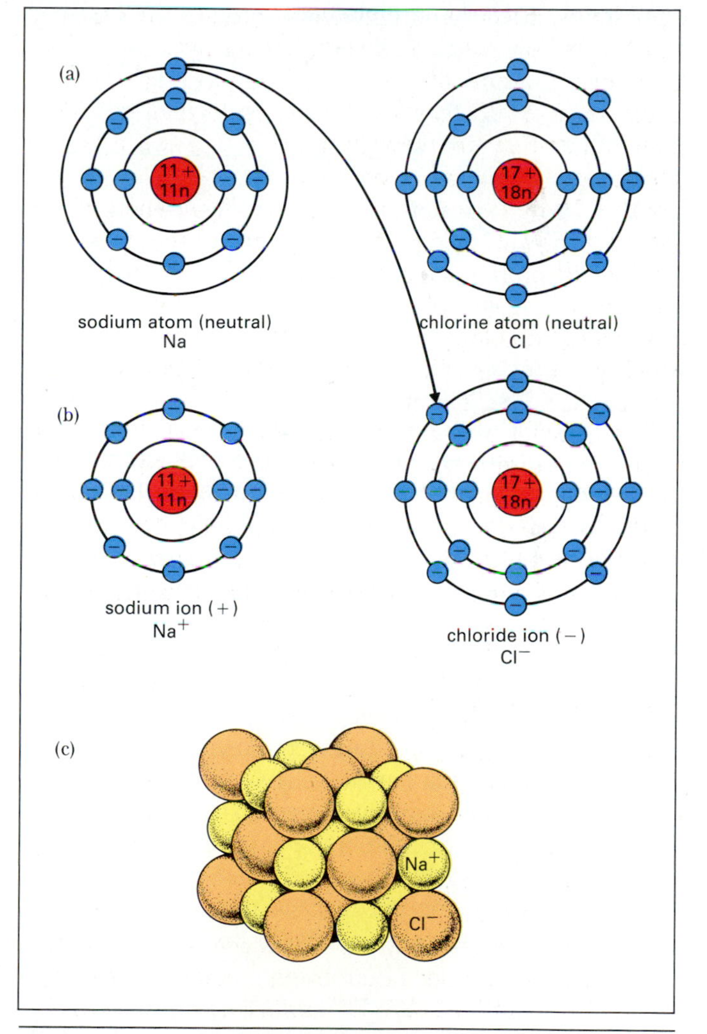

Figure 2-5 The formation of ions and ionic bonds. **(a)** *Sodium has only one electron in its outer electron shell, while chlorine has seven (lacking only one electron to fill its outer shell).* **(b)** *Sodium can achieve stability by donating an electron to chlorine, so that both have empty or full outer shells. Sodium therefore becomes a positively charged ion, and chlorine a negatively charged ion.* **(c)** *Since oppositely charged particles attract one another, the resulting sodium and chloride ions nestle closely together in a crystal of salt, NaCl.*

Covalent Bonds

An atom with a partially full electron shell can gain stability by *sharing electrons with another atom in a* **covalent bond.** To understand what this means, consider the hydrogen atom, which has one electron in a shell built for two. A hydrogen atom can become reasonably stable if it shares its single electron with another hydrogen atom,

Interpreting Chemical Structures and Formulas

Many of you have never taken a chemistry course, and therefore you may not know how to interpret drawings and formulas of molecules. In this figure we introduce you to the ways in which simple inorganic molecules are illustrated.

Subatomic structure cannot be seen even by the most sophisticated scientific instruments. However, most people understand and remember something most easily if they can picture it. Consequently, physicists have developed visual models of atoms and molecules. Different types of models serve different intellectual purposes.

Using water as an example, let's look at five common ways of representing molecules:

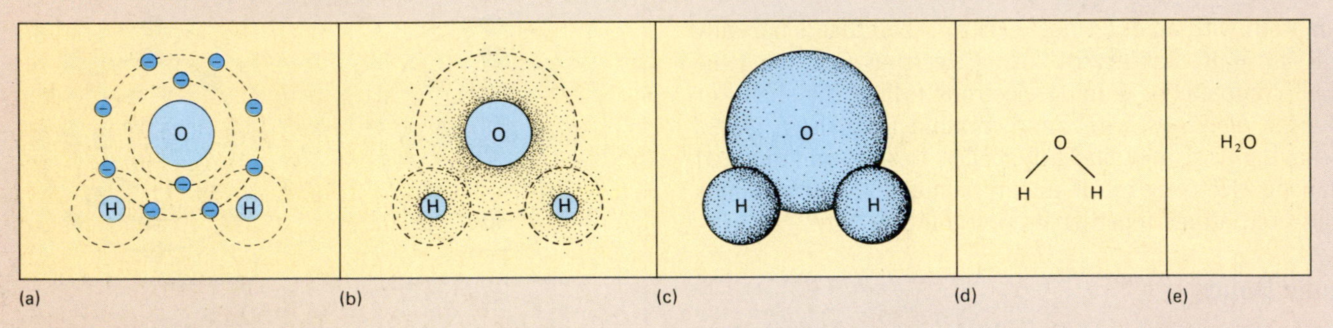

One of the first, and still very useful, ways to visualize atomic structure is the planetary model (a). The atomic nucleus is thought of as the center of a "solar system" with the electrons orbiting about it, much as the planets orbit around the sun. When you are trying to figure out how two atoms bond together to make a molecule, planetary models are convenient, because they allow you to count electrons and see what arrangements result in filled electron shells. For example, oxygen has six electrons in its outer shell; eight electrons would fill the shell. Hydrogen has one electron in its outer shell, with two electrons needed to fill the shell. The planetary model shows that filled orbitals occur when two hydrogens each share one electron with a single oxygen atom.

Planetary models, however, are inaccurate: electrons do not revolve about the nucleus in well-defined orbits. In addition, planetary models do not show the three-dimensional structure of atoms and molecules. Electron cloud models (b) are the most accurate atomic portraits, because electrons roam about over a relatively large volume of space. The density of dots in an electron cloud model corresponds to the likelihood that an electron will be found at any given place at any given time. However, except for their accuracy, electron cloud models are not very useful, partly because they are tedious to draw. It is easier to visualize the three-dimensional structure of a molecule with a space-filling or ball model (c). You can think of the ball as representing the "normal" outer edge of the electron cloud of an atom. Ball models portray atoms within a molecule or even groups of molecules in reasonably accurate spatial relationships (see Fig. 2-8).

Really quick representations of molecules can be drawn using only the atomic symbol for the individual atoms, with lines for the bonds joining them (d). Single bonds are shown as dashes and double bonds as double dashes. Finally, if the structure of the molecule does not need to be shown at all, the molecule can be represented by its chemical formula (e). The subscript number after each atomic symbol is the number of atoms of each element in the molecule; for example, "H_2O" means "two hydrogens and one oxygen make up water."

Large molecules, including virtually all organic molecules, are often drawn using even further shortcuts, such as leaving out carbon and hydrogen atoms. These abbreviated portraits of organic molecules are explained in Figs. 3-2 and 3-9 in the next chapter.

forming a molecule of hydrogen gas, H_2 (Fig. 2-6a). Since the two hydrogen atoms are identical, neither can capture the other's electron. Both electrons spend part of the time in the shell of both hydrogen atoms, with the result that each hydrogen behaves almost as if it had two electrons in its shell. The stability of such a covalent bond depends on the atoms involved. The bond in a hydrogen molecule is not very strong. If an oxygen atom comes by, stronger and more stable covalent bonds will be formed by electron sharing between oxygen and hydrogen atoms (Fig. 2-6b). This is what happened to the *Hindenburg:* weak bonds between hydrogen atoms were replaced by strong bonds

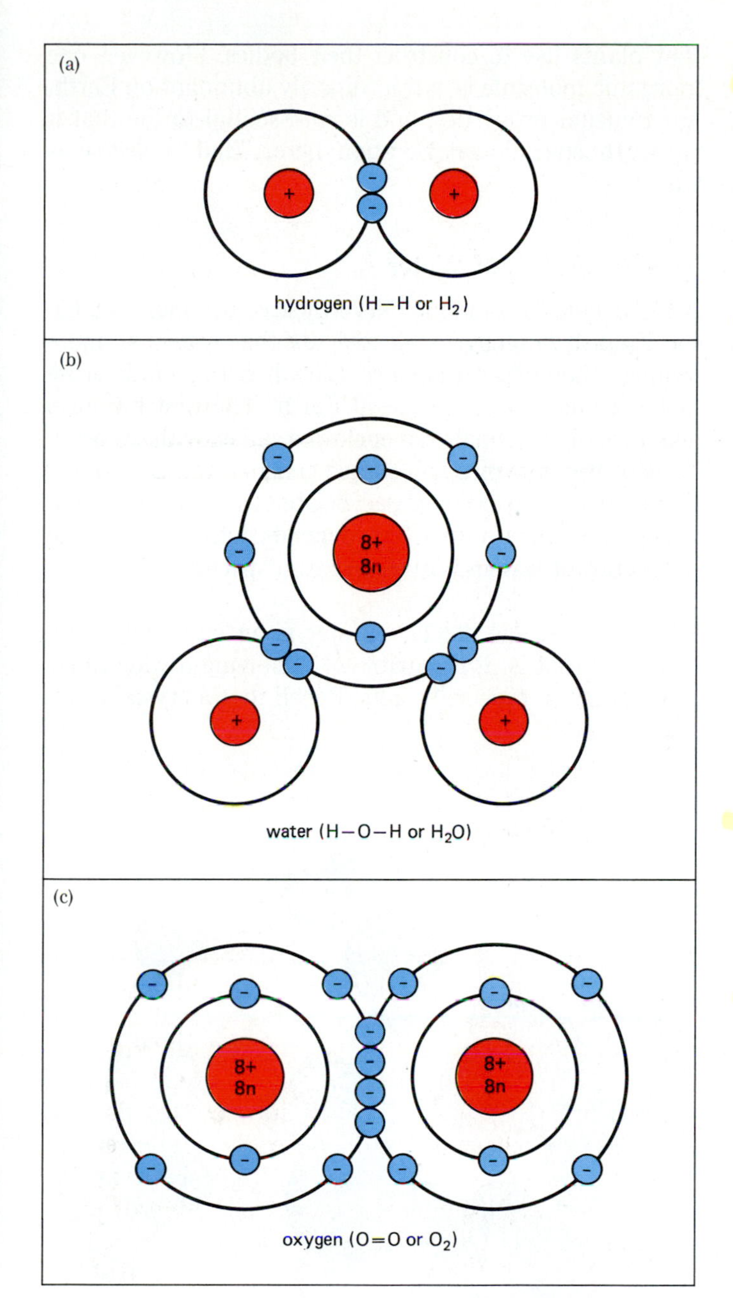

Figure 2-6 In covalent bonds, electrons are shared between atoms. **(a)** *In hydrogen gas, one electron from each hydrogen atom is shared, forming a single covalent bond. The resulting bond is usually represented by a single dash connecting the atomic symbols, and the molecule of hydrogen gas may be represented as H—H or H_2.* **(b)** *Oxygen lacks two electrons to fill its outer shell, so oxygen can make two single bonds, one with each of two hydrogen atoms to form water (H—O—H or H_2O).* **(c)** *In oxygen gas, two oxygen atoms share four electrons, making a double bond, usually represented by a double dash or equal sign (O═O or O_2).*

between hydrogen and oxygen atoms, with explosive results.

As this example suggests, all covalent bonds are not created equal. Some covalent bonds, such as those in water (H_2O) and carbon dioxide (CO_2), are extremely stable. A molecule with stable bonds between its atoms can be considered to be a "low-energy molecule." Other bonds, such as those in hydrogen gas, oxygen gas, or gasoline, are less stable and come apart more easily. Molecules with such unstable bonds are "high-energy molecules." When a chemical reaction occurs in which less stable bonds are broken and more stable bonds are formed (such as burning gasoline with oxygen to form carbon dioxide and water), energy is released (see Chapter 6).

Biological molecules are largely composed of atoms joined by covalent bonds. Hydrogen, carbon, oxygen, nitrogen, phosphorus, and sulfur are the most common atoms found in biological molecules. Except for hydrogen, each of these atoms lacks two or more electrons to fill its outermost electron shell, and consequently can share electrons with two or more other atoms. Hydrogen can form a covalent bond with one other atom, oxygen and sulfur with two, nitrogen with three, and carbon and phosphorus with up to four other atoms (Table 2-2). This diversity of bonding arrangements permits biological molecules to be constructed in almost infinite variety and complexity.

Sometimes, an atom will share two pairs of electrons with another atom, forming a **double covalent bond** (each atom contributes two electrons; Fig. 2-6c). If atoms share three pairs of electrons, a **triple covalent bond** is formed. Double and triple bonds create further variety in the shapes and functions of biological molecules.

POLAR COVALENT BONDS. In hydrogen gas, the shared electrons spend equal time in the vicinity of each of the two hydrogen nuclei. Therefore, not only is the molecule as a whole electrically neutral, but each hydrogen "end" of the molecule is electrically neutral too. Such an electrically symmetrical bond is called **nonpolar.** In other molecules, one nucleus may pull more strongly on the electrons than the other nucleus does. This forms a **polar covalent bond.** Although the molecule as a whole is electrically neutral, it has charged parts: the atom that attracts the electrons more strongly will have a slightly negative charge (the negative pole of the molecule), while the other has a slightly positive charge (the positive pole). In water, for example, oxygen attracts electrons more strongly than hydrogen does, so that the oxygen end of a water molecule is negative, while each hydrogen is positive (Fig. 2-7).

Hydrogen Bonds

A glass of water contains countless water molecules, each carrying a slight negative charge on its oxygen atom, and a slight positive charge on each of its two hydrogen atoms. When two water molecules get close together, the oppositely charged parts of the molecules attract each other. The electrical attraction between polar parts of molecules

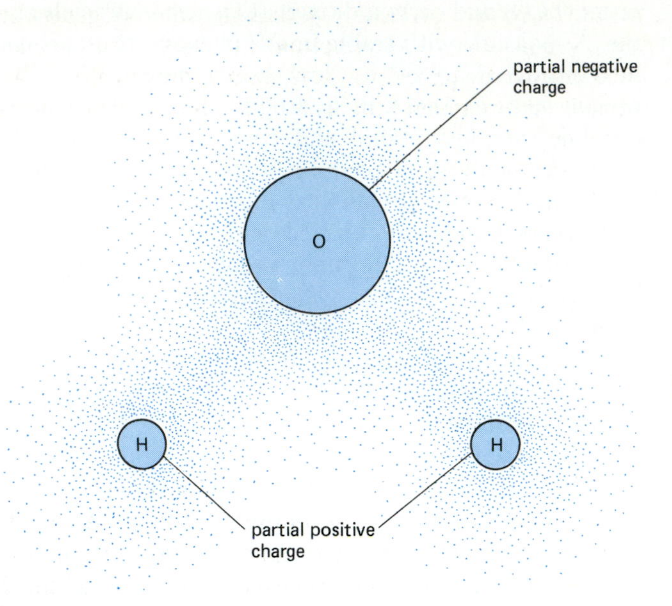

Figure 2-7 The electrons in polar covalent bonds are unequally shared. In water, the oxygen atom attracts the electrons more strongly than do the hydrogens. The oxygen atom therefore has a partial negative charge while the two hydrogens are left somewhat deprived of electrons and thus are partially positive.

is called a **hydrogen bond** (Fig. 2-8). Hydrogen bonds hold liquid water in a slightly organized way, producing some unique properties that are essential to life on Earth (see the next section).

Other molecules also form hydrogen bonds. In biological molecules, both nitrogen and oxygen atoms attract electrons more strongly than hydrogen atoms do, becoming slightly negative themselves and leaving the hydrogen slightly positive. As we shall see in Chapter 3, hydrogen bonds play crucial roles in shaping the three-dimensional structures of proteins and nucleic acids.

IMPORTANT INORGANIC MOLECULES

Chemists classify molecules as organic or inorganic. The word "organic" originally signified that these molecules could only be manufactured within living organisms. Today, however, organic chemists can synthesize many of these molecules in test tubes. Thus the modern definition of an **organic molecule** is: any molecule that contains both carbon and hydrogen. Most organic molecules are large, with complex structures. **Inorganic molecules** include carbon dioxide and all molecules lacking carbon.

Many inorganic molecules are important to living organisms, including, for instance, minerals in the soil and carbon dioxide in the air, which are the raw materials that plants use to construct their bodies. However, one inorganic molecule is extraordinarily abundant on Earth, has unusual properties, and is so essential to life that it is worth special consideration here. That molecule is water.

The Properties of Water

All molecules are unique. Nevertheless, we might stretch the English language a bit and say that water is "more unique" than other molecules. Life almost certainly arose in the waters of the primeval Earth. The first life form was probably a small sac enclosing water with an array of dissolved enzymes and simple genetic material. Living organisms are still composed of about 60 to 90 percent water, and all life on Earth depends intimately on the properties of water. Why is water so special?

WATER AS A SOLVENT. Water is an extremely good **solvent;** that is, it is capable of dissolving a wide range of substances, especially salts. Recall that a crystal of ta-

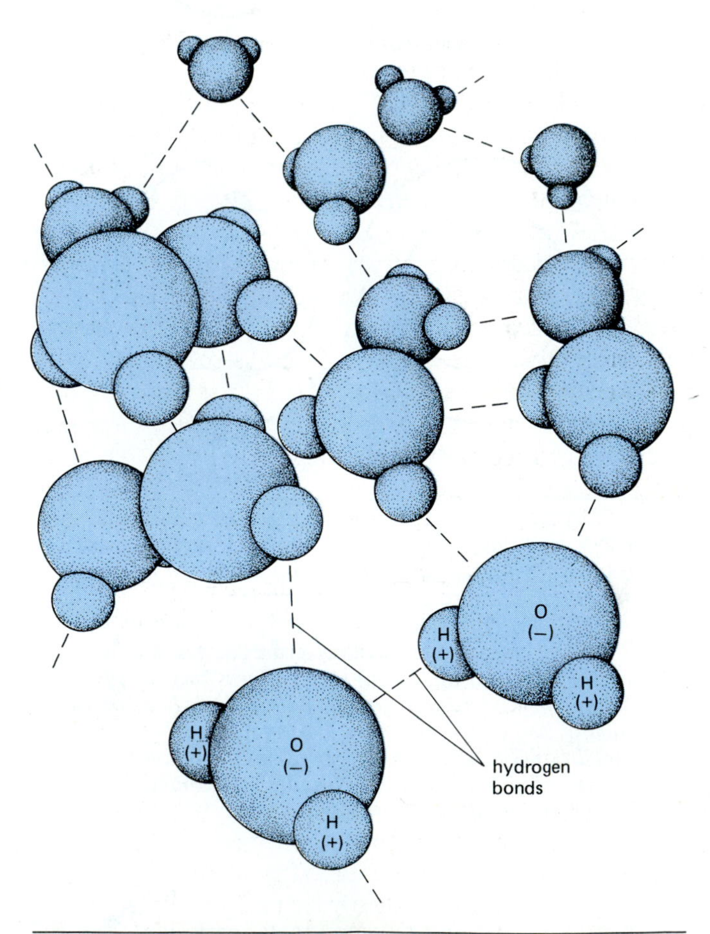

Figure 2-8 The partial charges on different parts of water molecules produce weak attractive forces called hydrogen bonds (dashed lines) between the hydrogens of one water molecule and the oxygens of other molecules.

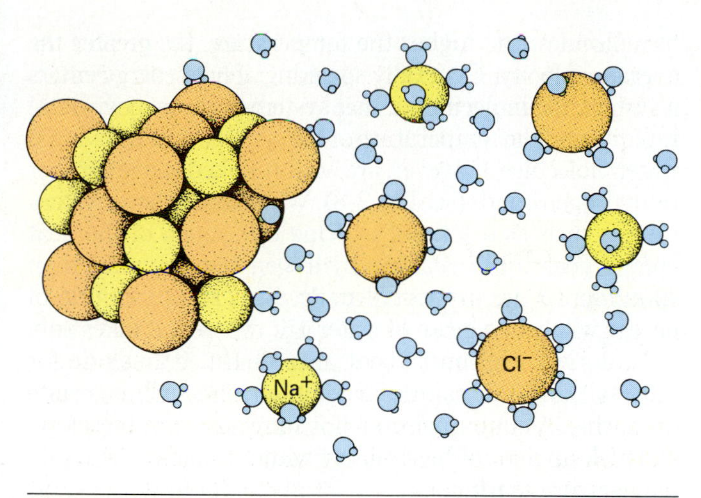

Figure 2-9 The polarity of water molecules allows water to dissolve polar or charged substances. When a salt crystal is dropped into water, the water molecules worm their way around the outsides of the sodium and chloride ions, surrounding them with oppositely charged ends of the water molecules. Thus insulated from the attractiveness of other molecules of salt, the ions float away, and the whole crystal gradually dissolves.

ble salt is held together by the electrical attraction between positively charged sodium ions and negatively charged chloride ions. Since water is a polar molecule, it has positive and negative ends. If a salt crystal is dropped into water, the positively charged hydrogen ends of water molecules will be attracted to and surround the negatively charged chloride ions, while the negatively charged oxygen ends of water molecules will surround the positively charged sodium ions. As water molecules enclose the sodium and chloride ions, the ions separate from the crystal and drift away in the water—the salt dissolves (Fig. 2-9).

Water dissolves polar molecules in a similar fashion, as its positive and negative poles are attracted to oppositely charged regions of dissolving molecules. Many biological molecules, such as sugars and amino acids, are either charged or polar, and hence dissolve readily in water. Water also dissolves gases such as oxygen and carbon dioxide. Other liquids can dissolve some of these substances, but not all of them. Alcohol, for example, dissolves some sugars and proteins but not salts. By dissolving such a wide variety of molecules, the watery cytoplasm of a cell provides a suitable environment for the myriad chemical reactions essential to life on Earth.

WATER AS A REACTANT. Water also enters into many of the chemical reactions that occur in cells. The oxygen that green plants release into the air is derived from water during photosynthesis. When your body manufactures a protein, fat, nucleic acid, or sugar, it produces water in the process, and conversely, when you digest proteins, fats, and sugars in the foods you eat, water is used in the reactions.

IONIZATION OF WATER. Although water is generally regarded as a stable compound, individual water molecules constantly gain, lose, and swap hydrogen atoms. As a result, at any given time a small fraction of water molecules are ionized:

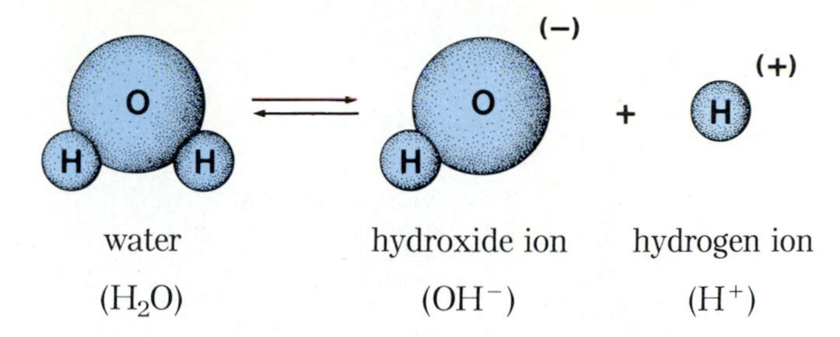

Pure water has equal numbers of hydrogen ions and hydroxide ions, but in many solutions the concentrations of hydrogen and hydroxide ions are not the same. If the concentration of H^+ ions exceeds the concentration of OH^- ions, the solution is an **acid;** if the concentration of OH^- ions is greater, the solution is a **base.** For example, if we add hydrochloric acid (HCl) to pure water, almost all the HCl molecules separate into H^+ and Cl^- ions. Thus there are far more H^+ ions than OH^- ions, and the solution is acidic. If we add sodium hydroxide (NaOH) to water, the NaOH molecules separate into Na^+ and OH^-. The resulting solution, with more OH^- ions than H^+ ions, is basic.

The degree of acidity is expressed on the **pH scale** (Fig. 2-10), in which neutrality (equal numbers of H^+ and OH^- ions) is assigned the number 7. Acids have a pH below 7, while bases have a pH above 7.

The ionization of water is crucial to life. The conformation of proteins depends to a considerable extent on pH. If the pH becomes too acidic or too basic, many proteins stop functioning. Even minor changes in blood pH, for example, severely impair the ability of hemoglobin to carry oxygen.

WATER AS A TEMPERATURE MODERATOR. Organisms can survive only within a limited temperature range. For one thing, organisms depend on protein enzymes that guide the chemical reactions essential to life. Enzymes are damaged by high temperatures, becoming nonfunctional at temperatures well below boiling. It is also important that the water within living cells remain liquid. The smaller molecules upon which the enzymes work, such as sugars, salts, and amino acids, come into contact with the enzymes by diffusing through liquid water. If everything were frozen in place, the chemical reactions of the cell would cease. Further, the spearlike ice crystals formed as water freezes can rupture a cell.

Water has three properties that moderate temperatures inside cells. First, water has a high **specific heat,** which is the energy needed to raise the temperature of a gram of a given substance by 1°C. Temperature measures the velocity

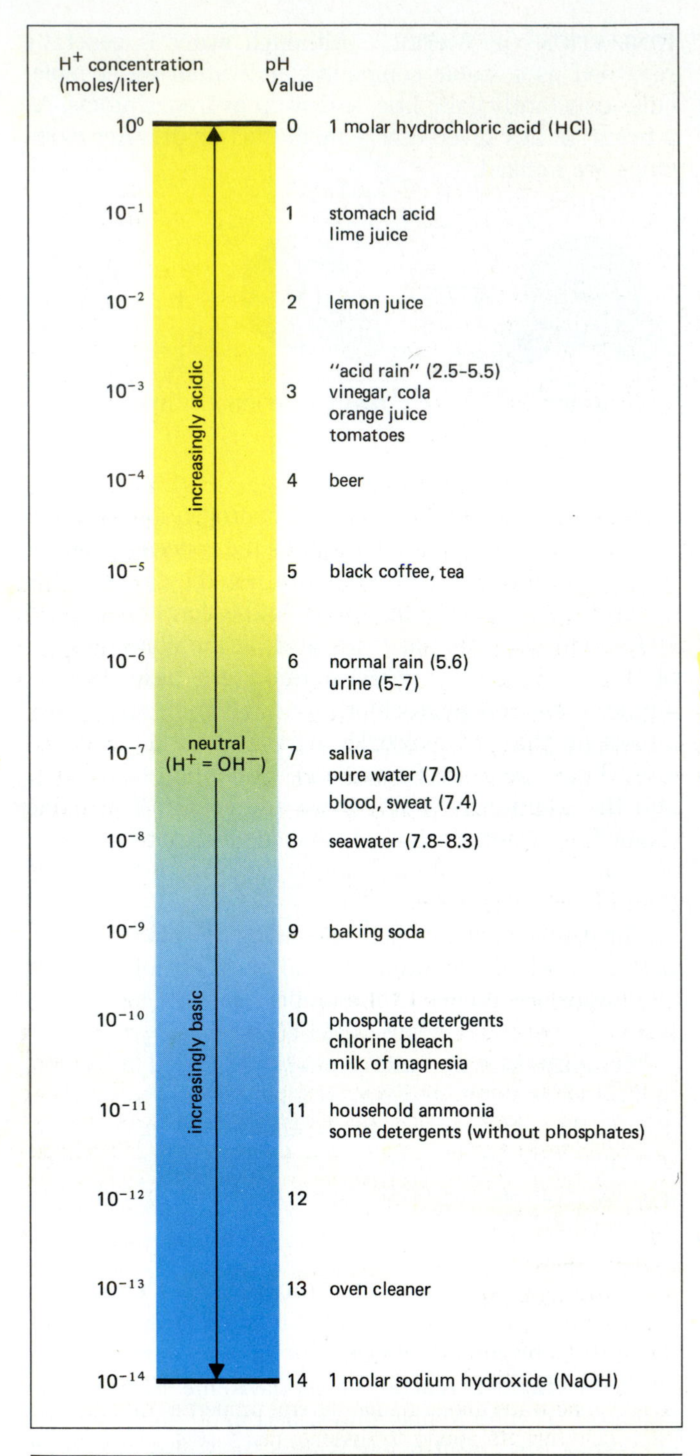

Figure 2-10 The pH scale expresses the concentration of hydrogen ions in a solution on a scale of 0 (very acidic) to 14 (very basic). The pH scale is logarithmic: each unit of change in pH represents a tenfold change in the concentration of acid or base. Lemon juice, for example, is about ten times more acidic than orange juice, and the severe acid rains in the northeastern United States are almost 1,000 times more acidic than normal rainfall. Except for the insides of your stomach, nearly all the fluids in your body are finely adjusted to a pH of 7.4; a pH of 7.2, although it may not sound like much of a change, is enormously more acidic than normal.

of molecules: the higher the temperature, the greater the average velocity. Generally speaking, if heat energy enters a system, the molecules of that system begin to move more rapidly, and the temperature of the system rises. Individual water molecules, however, are weakly linked to one another by hydrogen bonds (see Fig. 2-8). When heat enters a watery system such as a lake or a living cell, much of the heat energy goes into breaking hydrogen bonds rather than speeding up the individual molecules. Thus 1 calorie of energy will heat a gram of water 1°C, whereas it takes only 0.6 calorie per gram to heat alcohol 1°C, 0.2 calorie for table salt, and 0.02 calorie for common rocks such as granite or marble. A sunbather on a hot summer's day, therefore, can pick up a lot of heat energy without sending her body temperature soaring.

Water moderates high temperatures through its great **heat of vaporization,** the heat required to convert liquid to vapor at a constant temperature. Water has one of the highest heats of vaporization known, 539 calories per gram. Again, this is due to the hydrogen bonds interconnecting individual water molecules. For a water molecule to evaporate, it must move fast enough to break all the hydrogen bonds holding it to the other water molecules in the solution. Only the fastest-moving water molecules, carrying the most energy, can break their hydrogen bonds and escape into the air as water vapor. The remaining liquid is cooler for the loss of these high-energy (and high-temperature) molecules. If our sunbather's body temperature begins to rise, she perspires, covering her body with a film of water. Heat energy is transferred from her skin to the water. Evaporating just 1 gram of water cools 539 grams of her body 1°C, so a great loss of heat can occur without much loss of water.

Finally, water moderates low temperatures through its high **heat of fusion,** the energy that must be removed from the molecules of a liquid before they form the precise arrangement of a crystal. Very low temperatures are required for a long period of time before a sizable body of water turns to ice, which protects living organisms from the damaging effects of freezing.

WATER AS A SOLID. Water, of course, will become a solid after prolonged exposure to temperatures below its freezing point. But even here the unique properties of water come into play, to the advantage of living things: ice floats. Most liquids become more dense when they solidify, and the solid sinks. Ice, however, is less dense than liquid water, so when a pond or lake starts to freeze in winter, the ice stays on top, forming an insulating layer that delays the freezing of the rest of the water. If ice sank, ponds and lakes in much of North America would freeze solid during the winter, eliminating fish and most other animal life.

Water also behaves in solidlike ways even when it is actually still liquid. This is also due to the presence of hydrogen bonds between water molecules, which causes **cohesion,** the tendency for a substance to stick together

(a)

(b)

Figure 2-11 **(a)** *Cohesion between water molecules allows water striders to skate across the surface of still waters.* **(b)** *In giant redwoods, cohesion holds water molecules together in continuous strands from the roots to the topmost leaves hundreds of feet above the ground.*

in one piece. Cohesion in water at the surface of a lake or pond produces a solidlike effect called **surface tension.** In general, things that are more dense than water sink. However, because the water molecules at the surface of a pond cohere to one another, the surface film acts almost as a solid, supporting relatively dense objects such as fallen leaves, water striders (Fig. 2-11a), and for a painful split-second, the body of an inept human diver belly-whopping into a pool.

A more important role of cohesion occurs in the life of land plants (Fig. 2-11b). A plant absorbs water through its roots. How does the water reach the aboveground parts, especially if the plant is a 100-meter-tall redwood? As we shall see in Chapter 21, the water is pulled up by the leaves. Water fills tiny tubes interconnecting leaves, stem, and roots. Water molecules evaporating from the leaves pull water up the tubes, much like a rope being pulled up from the top. The system works because the hydrogen bonds interconnecting water molecules are stronger than the weight of the water in the tubes, even a hundred meters worth, so the water "rope" does not break. Without the cohesion of water, there would be no land plants, and therefore no land animals, including people.

SUMMARY OF KEY CONCEPTS

Matter and Energy

Matter is the physical material of the universe. Energy is the capacity to do work. Energy can exist in several forms that may be converted from one to another. The two major categories are kinetic energy (the energy of motion) and potential energy (stored energy).

The Structure of Matter

All matter is made of particles, called atoms, that are composed of a central nucleus containing protons and neutrons and outer orbitals containing electrons.. Every atom of a given element has the same number of protons, which is different from the number found in the atoms of any other element.

Electron orbitals are found at specific distances from the nucleus, called electron shells. Each shell can contain a fixed maximum number of electrons. The chemical reactivity of an atom depends on the number of electrons in its outermost electron shell; an atom is most stable, and therefore least reactive, when its outermost shell is either completely full or completely empty.

Molecular Structure

Atoms may combine to form molecules. The forces holding atoms together in molecules are called chemical bonds. There are two principal types of chemical bond, called ionic and covalent bonds. If one atom fills its outermost shell by acquiring electrons, while another atom empties its shell by losing electrons, this results in negatively and positively charged particles called ions. Ionic bonds are electrical attractions between charged ions, holding them together in crystals. Covalent bonds involve the sharing of electrons by two atoms, in which neither atom completely gains nor loses an electron. A third type of interaction, called a hydrogen bond, is the attraction between slightly charged regions of different molecules or distant parts of a large molecule.

Important Inorganic Molecules

The most important inorganic molecule for life on Earth is water. The water molecule has several properties that facilitate life, including its ability to dissolve many substances, to participate in chemical reactions, and to maintain a fairly stable temperature in the face of wide temperature fluctuations in the environment.

GLOSSARY

Acid a substance that releases hydrogen ions (H^+) into solution; a solution with a pH of less than 7.

Atom the smallest particle of an element that retains the properties of the element; composed of a central nucleus containing protons and neutrons, and outer orbitals containing electrons.

Atomic number the number of protons in the nuclei of all atoms of a particular element.

Base a substance that releases hydroxide ions (OH^-) into a solution or that is capable of combining with and neutralizing hydrogen ions, producing a solution with a pH greater than 7.

Calorie the amount of heat needed to raise the temperature of 1 gram of water 1°C.

Chemical bond the force of attraction between neighboring atoms that holds them together in a molecule.

Cohesion the tendency of a substance to hold together.

Compound a substance composed of two or more elements that can be broken into its constituent elements by chemical means.

Covalent bond (kō-vā′-lent) a chemical bond between atoms in which electrons are shared.

Double bond a covalent bond that occurs when two atoms share two pairs of electrons.

Electron a subatomic particle, found in the orbitals outside the nucleus of an atom, bearing a unit of negative charge and very little mass.

Electron shell all the electron orbitals at a given distance from the nucleus of an atom.

Element a substance that cannot be broken down to a simpler substance by ordinary chemical means.

Energy the capacity to do work.

Energy level a particular amount of energy characteristic of a given electron shell in an atom.

Heat of fusion the energy that must be removed from a substance to transform it from a liquid into a solid at a constant temperature.

Heat of vaporization the energy that must be supplied to a substance to transform it from a liquid into a gas at a constant temperature.

Hydrogen bond the weak attraction between a hydrogen atom bearing a partial positive charge (due to polar covalent bonding with another atom) and another atom, usually oxygen or nitrogen, bearing a partial negative charge. Hydrogen bonds may form between atoms of a single molecule or of different molecules.

Inorganic referring to any molecule that does not contain both carbon and hydrogen.

Ion (ī′-on): an atom or molecule that has either an excess of electrons (and hence is negatively charged) or has lost electrons (and is positively charged).

Ionic bond a chemical bond formed by the electrical attraction between positively and negatively charged ions.

Kinetic energy (kin-et′-ik): the energy of movement.

Matter the material of which the universe is made.

Molecule a particle composed of one or more atoms, held together by chemical bonds. A molecule is the smallest particle of a compound that displays all the properties of that compound.

Neutron a subatomic particle found in the nuclei of atoms, bearing no charge and having mass approximately equal to that of a proton.

Nonpolar covalent bond a covalent bond with equal sharing of electrons.

Nucleus the central region of an atom, composed of protons and neutrons.

Orbital the region of an atom, outside the nucleus, in which an electron is likely to be found.

Organic referring to a molecule that contains both carbon and hydrogen.

pH a scale with values from 0 to 14, used for measuring the relative acidity of a solution. At pH 7 a solution is neutral, pH 0 to 7 is acidic, and pH 7 to 14 is basic. Each unit on the pH scale represents a tenfold change in the concentration of hydrogen ions.

Polar covalent bond a covalent bond with unequal sharing of electrons, so that one atom is relatively negative, while the other is relatively positive.

Potential energy "stored" energy, usually chemical energy or energy of position within a gravitational field.

Proton a subatomic particle found in the nuclei of atoms, bearing a unit of positive charge and a relatively large mass roughly equal to the mass of the neutron.

Radioactive pertaining to an atom with an unstable nucleus that disintegrates spontaneously with the emission of radiation.

Single bond a covalent bond that occurs when two atoms share a single pair of electrons.

Solvent a liquid that is capable of dissolving (uniformly dispersing) other substances in itself.

Specific heat the amount of energy required to raise the temperature of 1 gram of a substance 1°C.

Subatomic particles the particles of which atoms are made: electrons, protons, and neutrons.

Surface tension the property of a liquid to resist penetration by objects at its interface with the air, due to cohesion between molecules of the liquid.

Triple bond a covalent bond that occurs when two atoms share three pairs of electrons.

STUDY QUESTIONS

1. Describe the structure of an atom.
2. Distinguish between atoms and molecules; elements and compounds; protons and electrons.
3. What factors influence the reactivity of an atom?
4. Describe how ionic bonds are formed.
5. Describe how covalent bonds are formed. What determines the number of covalent bonds that an atom can form?
6. What is a hydrogen bond? How does it influence the structure of water and of biological molecules?
7. List the elements commonly found in biological molecules.

SUGGESTED READINGS

Dickerson, R., and Geis, I. *Chemistry, Matter, and the Universe.* Menlo Park, Calif.: W. A. Benjamin, Inc., 1976. An introduction to basic chemical principles, written for the nonchemistry major.

Miller, G. *Chemistry: A Basic Introduction,* 4th ed. Belmont, Calif.: Wadsworth Publishing Company, Inc., 1987. A good introductory chemistry text.

3
The Chemistry of Life. II. Biological Molecules

Living things contain an amazing variety of molecules, from the proteins of horn and hoof to the carbohydrates of sugars and tree trunks. Organic molecules, although diverse in form and function, all owe their versatility to the carbon atom. A carbon atom has four electrons in an outermost shell, with room for eight. Therefore, carbon atoms achieve stability by sharing four electrons with other atoms, forming up to four single covalent bonds or smaller numbers of double or triple bonds. Molecules with many carbon atoms can assume complex shapes, including chains, branches, and rings.

Organic molecules are much more than just complicated skeletons of carbon atoms, however. Attached to the carbon backbone are groups of atoms, called **functional groups,** that determine the characteristics, solubility, and chemical reactivity of the molecules. The common functional groups found in organic molecules are diagrammed in Table 3-1.

SYNTHESIZING ORGANIC MOLECULES: A MODULAR APPROACH

In principle, there are two ways to manufacture a large, complex molecule: one could synthesize the molecule atom by atom according to an extremely detailed blueprint, or one could take preassembled smaller molecules and hook them together. Just as trains are made by coupling engines, boxcars, coal cars, and cabooses together, life on Earth also takes the modular approach. Small molecules (e.g., amino acids) are used as **subunits** with which to synthesize longer molecules (e.g., proteins), like cars in a train (Fig. 3-1a). Also like a train, which has standard connectors on each car so that they can all attach to one another, biological molecules almost always use the same type of chemical reaction to join subunits to one another.

Living organisms synthesize long organic molecules through **condensation reactions,** so called because a hy-

Table 3-1 **Important Functional Groups in Biological Molecules**

Group	Structure	Properties	Types of Molecules
Hydrogen	—H	Polar or nonpolar, depending on what atom hydrogen is bonded to; involved in condensation and hydrolysis	Almost all organic molecules
Hydroxyl	—OH	Polar; involved in condensation and hydrolysis	Carbohydrates, nucleic acids, alcohols, some acids and steroids
Carboxylic acid (carboxyl)	—C(=O)—O—H	Acid; negatively charged when H^+ dissociates; involved in peptide bonds	Amino acids, fatty acids
Amino	—N(—H)—H	Basic; may bond an additional H^+, becoming positively charged; involved in peptide bonds	Amino acids, nucleic acids
Phosphate	—O—P(=O)(—O—H)—O—H	Acid; up to two negative charges when H^+ dissociates; links nucleotides in nucleic acids	Nucleic acids, phospholipids

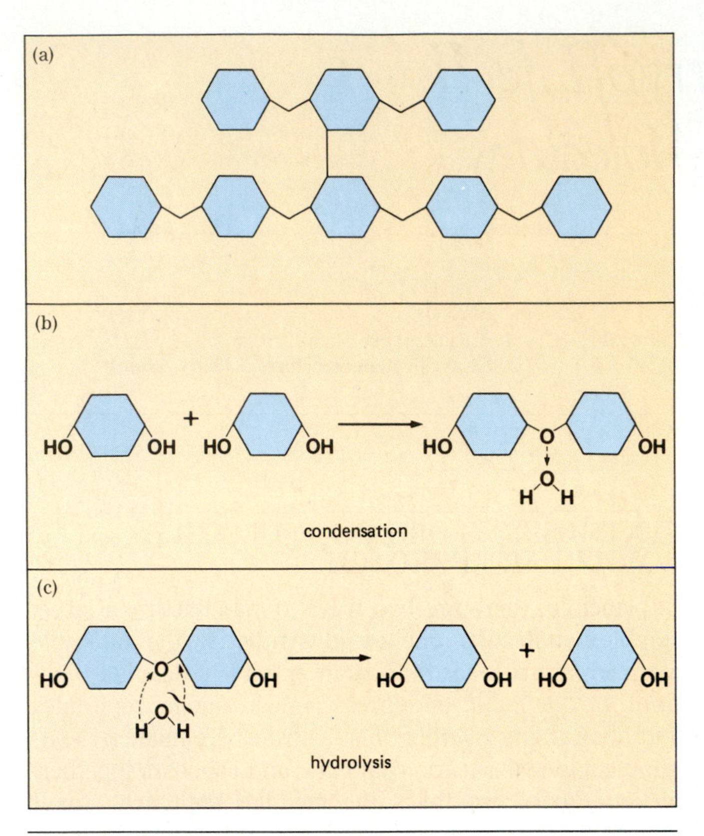

Figure 3-1 Synthesis and breakdown of organic molecules.
(a) *A typical organic molecule is composed of similar or identical subunits linked together.*
(b) *In a condensation reaction, two subunits are joined by a covalent bond. Simultaneously, a hydroxyl group is removed from one subunit and combines with a hydrogen removed from a second subunit to form water.*
(c) *Hydrolysis is the reverse of condensation. Hydrogen and hydroxyl from water are added to the subunits as the large organic molecule is broken apart into its subunits.*

drogen, removed from one subunit, and a hydroxyl, removed from a second subunit, "condense" to form a molecule of water as the subunits are joined by a covalent bond (Fig. 3-1b). The reverse reaction, called **hydrolysis** (literally, "to break apart with water"), can break the molecule apart into individual subunits again (Fig. 3-1c). During hydrolysis, water is added back, a hydrogen to one subunit and a hydroxyl to the other.

THE PRINCIPAL TYPES OF BIOLOGICAL MOLECULES

Although the bodies of living things often include thousands of different organic molecules, nearly all fall into one of four categories: carbohydrates, lipids, proteins, or nucleic acids.

Carbohydrates

All **carbohydrates** are either small, water-soluble sugars (glucose, fructose) or chains made by stringing sugar subunits together (starch, cellulose). If a carbohydrate consists of just one sugar molecule, it is called a **monosaccharide** (meaning "single sugar" in Greek). When two or more monosaccharides are linked together, they form a **disaccharide** ("two sugars") or a **polysaccharide** ("many sugars").

Sugars usually have a backbone of three to seven carbon atoms (Fig. 3-2a, left). Most of the carbon atoms have both a hydrogen (—H) and a hydroxyl group (—OH) attached to them. Therefore, the general formula for a sugar is $(CH_2O)_n$, where n is the number of carbons in the backbone. When they are dissolved in water, such as in the cytoplasm of a cell, the carbon backbone of a sugar usually "circles up" into a ring (Fig. 3-2a, right). It is in this ring form that sugars link together to make disaccharides and polysaccharides (see Fig. 3-3).

Most carbohydrates, at least small ones, are water soluble. As with water molecules, the O—H bond in a hydroxyl group is polar, because oxygen attracts electrons more strongly than hydrogen does. Hydrogen bonds between water molecules and the polar hydroxyl groups keep the carbohydrate in solution.

MONOSACCHARIDES. Glucose (Fig. 3-2a) is the most common monosaccharide in living organisms and is the subunit of which most polysaccharides are made. Glucose has six carbons and hence has the chemical formula $(CH_2O)_6$, or $C_6H_{12}O_6$. Plants use the energy of sunlight to synthesize glucose from carbon dioxide and water. Glucose molecules, either singly or as subunits of disaccharides or polysaccharides, carry the sun's energy to parts of the plant that cannot photosynthesize, such as its roots, and store energy for the entire plant during the night.

Many organisms also synthesize other monosaccharides that have the same chemical formula as glucose, but with slightly different structures (Fig. 3-2b). These include fructose (the "corn sugar" of the food industry) and galactose (part of lactose, or "milk sugar"). Other important monosaccharides, such as ribose and deoxyribose, have five carbons (Fig. 3-2c). Ribose and deoxyribose are parts of the genetic molecules of RNA and DNA, respectively.

DISACCHARIDES AND POLYSACCHARIDES. Monosaccharides, especially glucose and its relatives, have a short life span in a cell. Most are either metabolized, freeing their chemical energy for use in driving needed cellular reactions, or are linked together in chains of varying length to form disaccharides and polysaccharides (Fig. 3-3). Disaccharides are often used for short-term energy storage or for transport, especially in plants. Common disaccharides include sucrose (table sugar: glucose plus

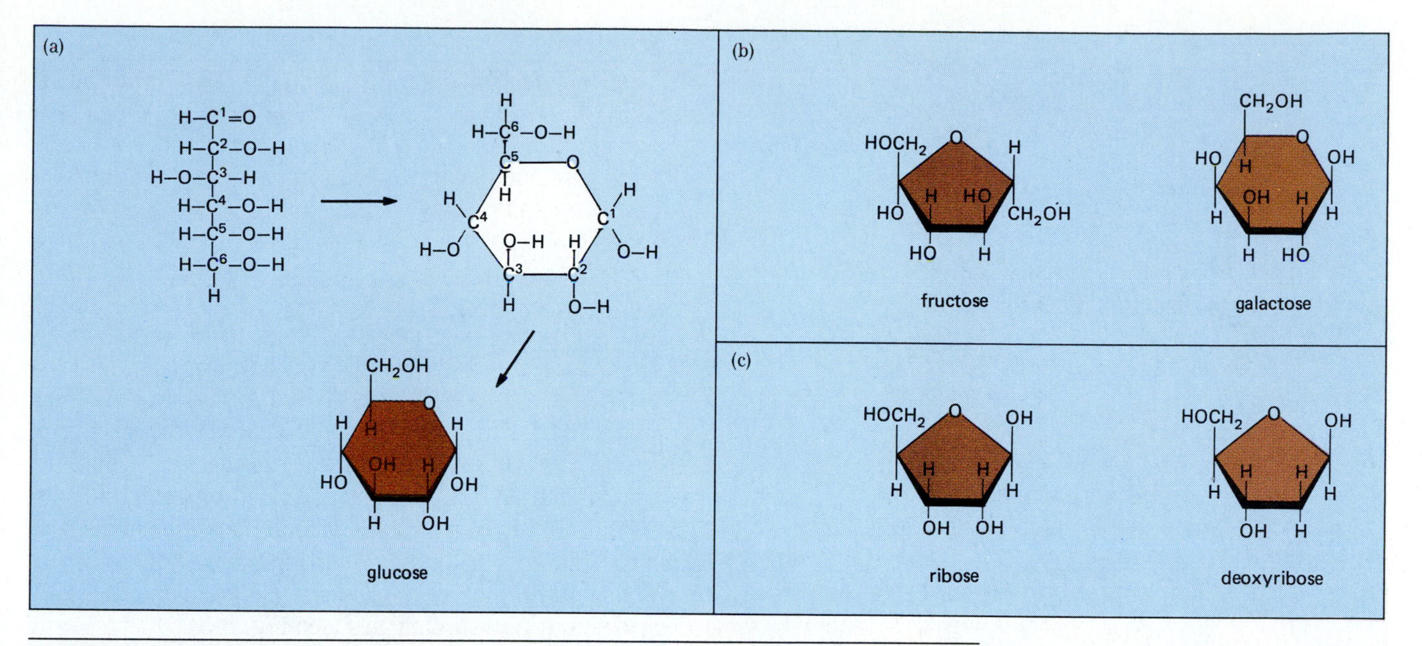

Figure 3-2 Monosaccharide structure.
(a) *The most common monosaccharides have "backbones" of either five or six carbon atoms (chain diagram of glucose, upper left). When dissolved in water, however, the chain bends upon itself to form a ring (upper right). These ring forms are usually drawn on paper as if you were looking at the edge of the polygon (bottom). The thick edge projects out of the paper toward you, the thin edge recedes behind the paper, and the —H, —OH, and —CH_2OH groups are perpendicular to the ring, in the plane of the paper. For convenience, carbon atoms at the corners of the polygons are omitted, but other atoms (such as oxygen) are usually shown.*
(b) *Fructose and galactose have the same atomic composition as glucose, but a different structure.*
(c) *Ribose and deoxyribose are five-carbon monosaccharides that form parts of nucleic acids.*

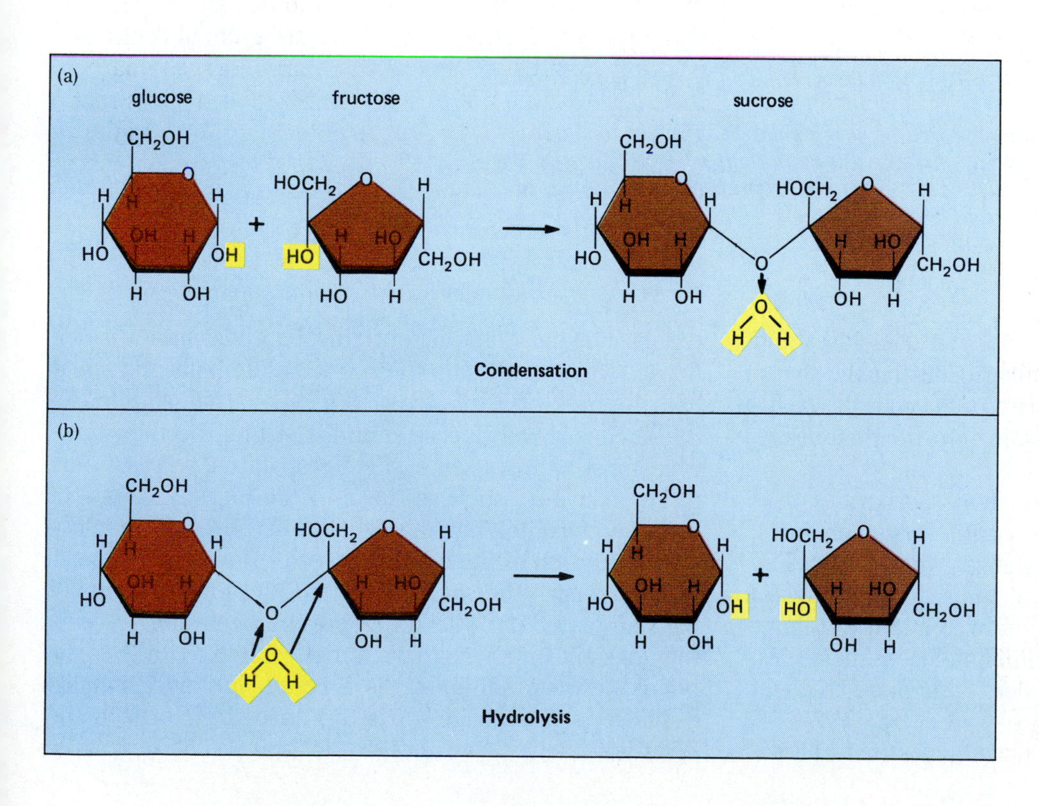

Figure 3-3 Synthesis and breakdown of a disaccharide.
(a) *The disaccharide sucrose is synthesized by a condensation reaction, in which a hydrogen (—H) is removed from glucose and a hydroxyl group (—OH) is removed from fructose, leaving the two monosaccharide rings joined by single bonds to the remaining oxygen atom.*
(b) *Hydrolysis of sucrose is just the reverse of its synthesis, as water is split and added back to the monosaccharides.*

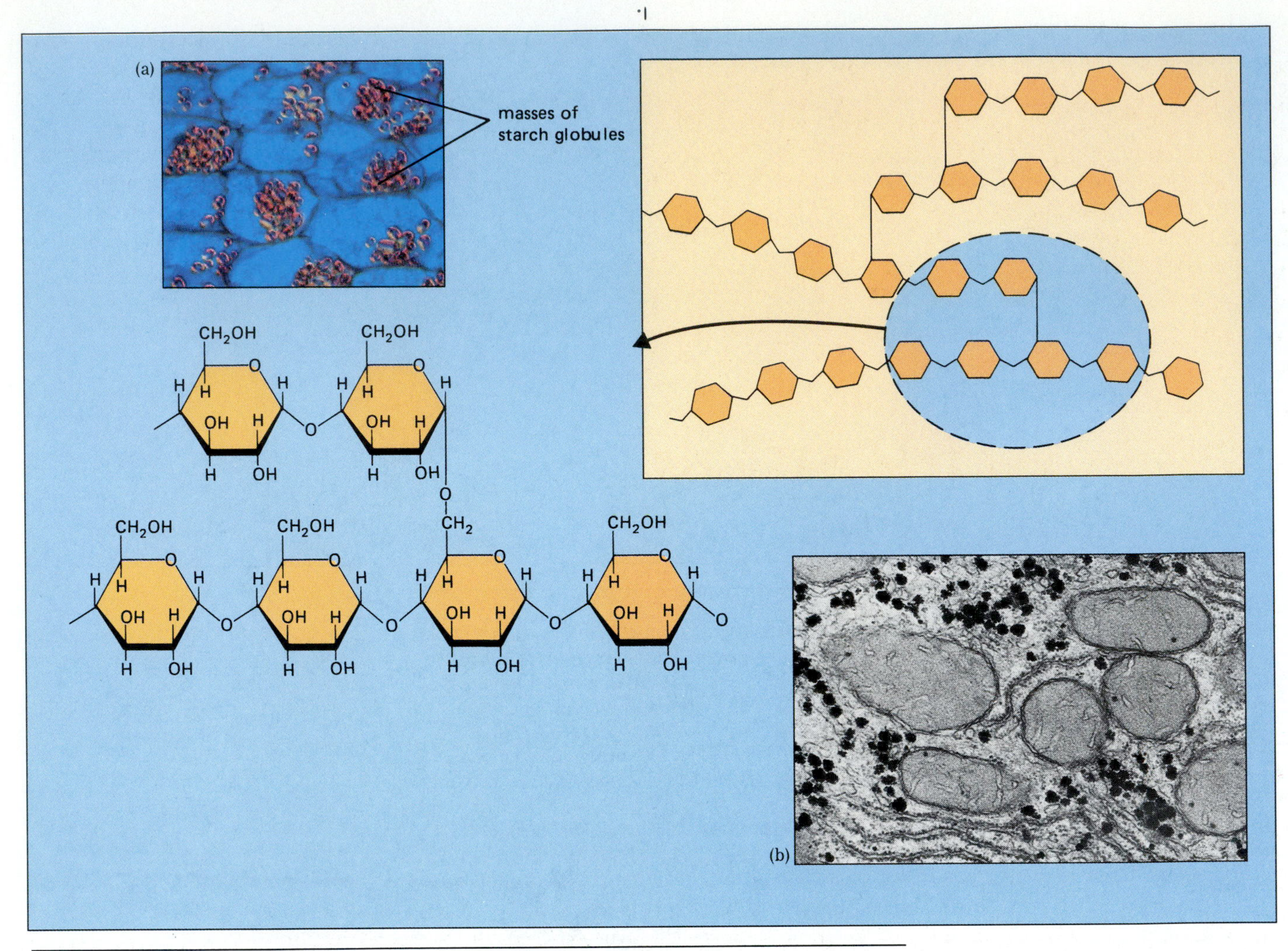

Figure 3-4 Both plants and animals use polysaccharides composed of glucose subunits for energy storage.
(a) *Most plants synthesize starch, which commonly occurs as branched chains of up to half a million glucose subunits. Starch forms water-insoluble globules such as these which make up most of the bulk of a potato.*
(b) *The animal storage polysaccharide, glycogen, is very similar in structure to starch, except that branches occur more frequently, making it a more compact molecule, and the total number of glucose subunits is smaller. This electron micrograph shows granules of glycogen in a salamander's liver.*

fructose), lactose (milk sugar: glucose plus galactose), and maltose (glucose plus glucose, formed during the digestion of starch). When the organism needs energy, the disaccharides are broken apart again into their monosaccharide subunits by hydrolysis.

For long-term storage of energy, monosaccharides, usually glucose, are joined together into polysaccharides, forming **starch** (in plants) or **glycogen** (in animals; Fig. 3-4). Starch may occur as coiled, unbranched chains of up to 1000 glucose subunits, or, more commonly, as huge branched chains of up to half a million glucose molecules. Glycogen is usually much smaller, with more frequent branching.

Many organisms use polysaccharides as structural materials. The most familiar structural polysaccharide is **cellulose,** which makes up the cell walls of plants and about half the bulk of a tree trunk (Fig. 3-5). Like starch, cellulose consists of glucose subunits strung together; however, most animals can easily digest starch, whereas only a few microbes, such as those in the digestive tracts of cows and termites, can digest cellulose. Why? In cellulose the orientation of the bonds between subunits is different, so that every other glucose is "upside down" (compare Fig. 3-5 with 3-4). This difference in orientation prevents the digestive enzymes of animals from getting at the bonds between subunits. As a result, for most animals cellulose is "roughage," passing unscathed through the digestive tract.

Polysaccharides are also the starting point for the synthesis of many other important molecules. The hard outer coverings (exoskeletons) of insects, crabs, and spiders are made of **chitin,** a polysaccaride in which the individual glucose subunits have been chemically modified (Fig. 3-6). Interestingly, chitin also stiffens the cell walls of many fungi. Bacterial cell walls contain still other types of modified polysaccharides, as do the lubricating fluids in our joints and the transparent corneas of our eyes.

Many nonstructural molecules also incorporate carbohydrates as essential components. Perhaps the most important examples are the nucleic acids (discussed later), the carriers of hereditary information in all organisms. Other molecules that are part carbohydrate include some components of mucus, some hormones, and many molecules in cell membranes, including "identification molecules" such as those on red blood cells that determine blood type.

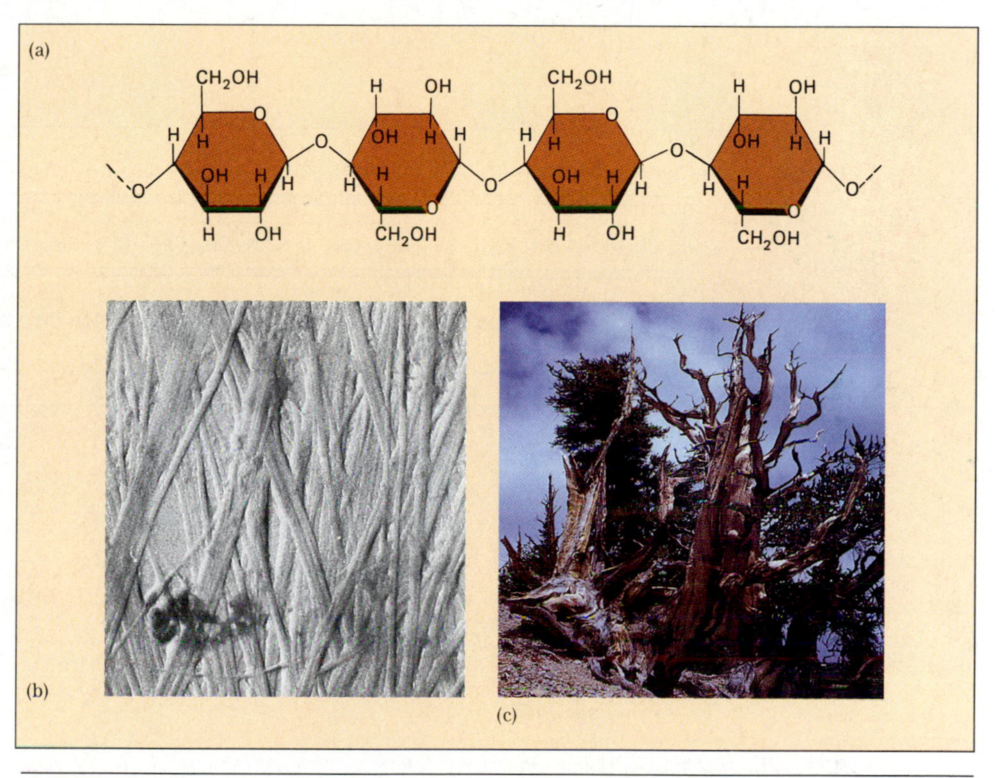

Figure 3-5 *Cellulose structure and function.* **(a)** *Cellulose, like starch, is composed of glucose subunits, but the orientation of the bond between subunits is different (compare with Fig. 3-4a), so that every other glucose molecule is "upside down." Unlike starch and glycogen, cellulose has great structural strength, due partly to the difference in bonding and partly to the arrangement of cellulose fibers.* **(b)** *Plant cells often lay down cellulose fibers in layers that run at angles to each other, resulting in resistance to tearing in both directions.* **(c)** *The final product can be incredibly tough, as this 3000-year-old bristlecone pine in California's White Mountains testifies.*

Figure 3-6 **(a)** *Chitin has the same "alternating upside down" bonding of glucose molecules as cellulose, with the difference that the glucose subunits are modified by replacing of one of the hydroxyl groups with a nitrogen-containing side group (yellow).* **(b)** *Tough, slightly flexible chitin supports the otherwise soft bodies of arthropods and fungi.*

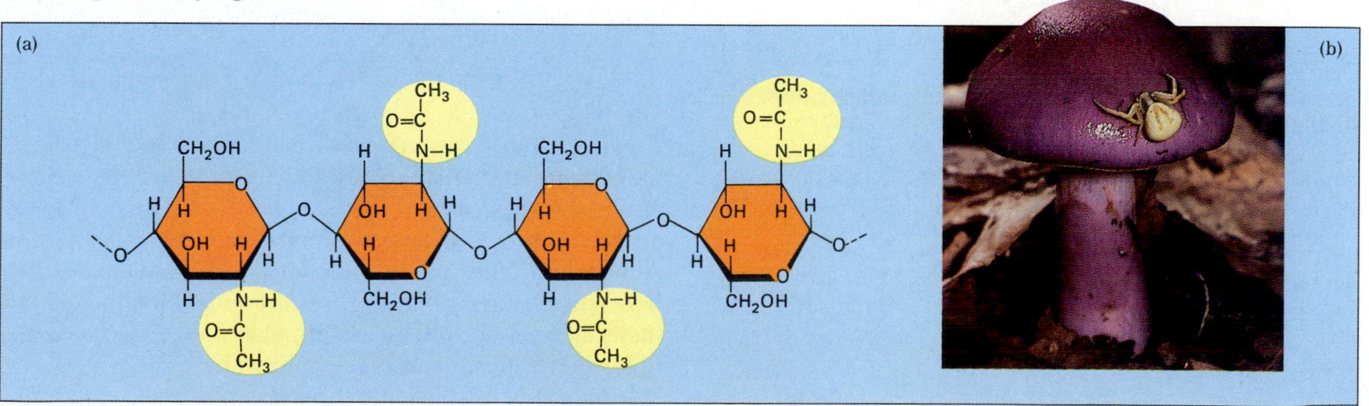

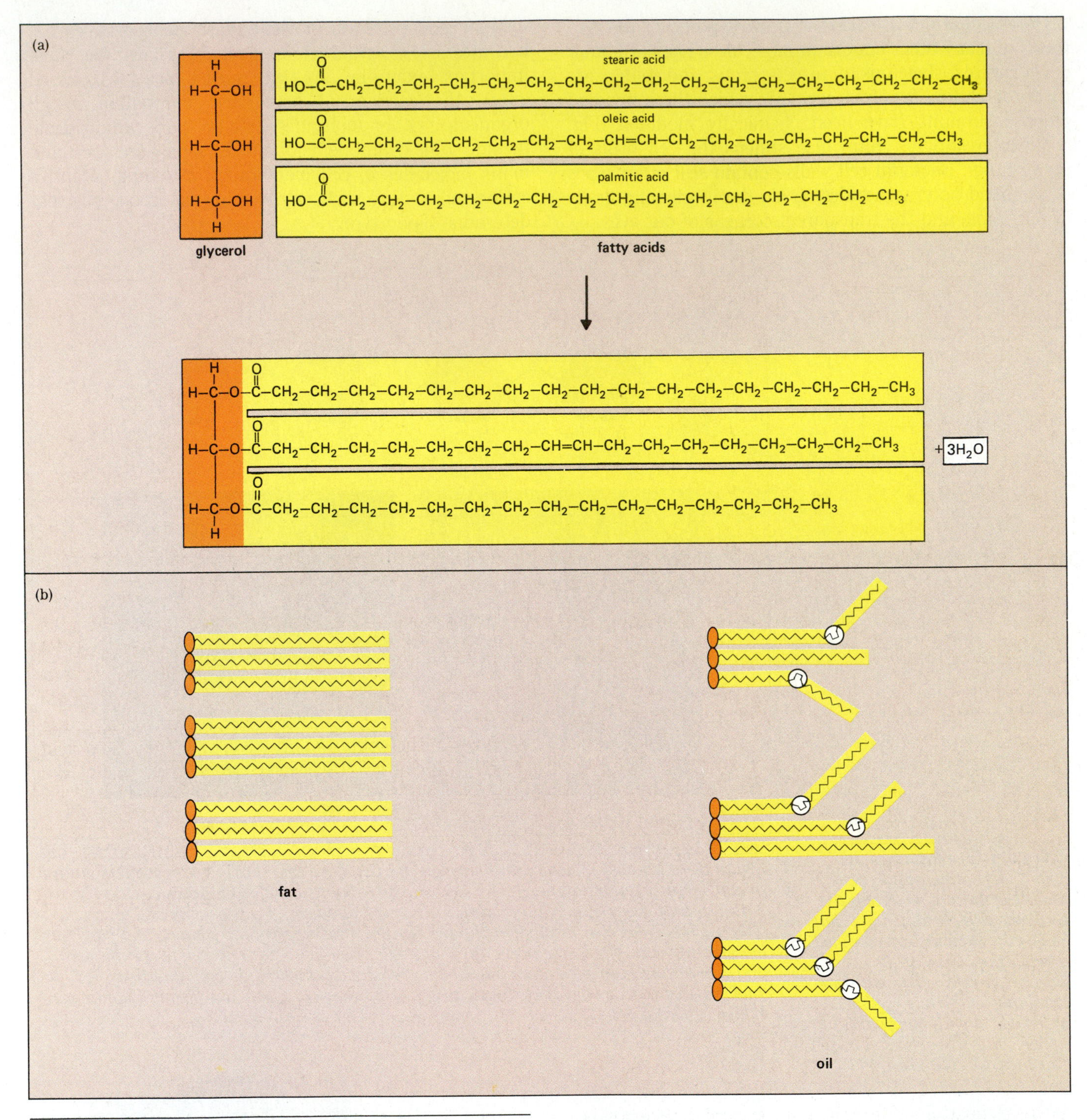

Figure 3-7 Fats and oils.
(a) *Fats and oils are synthesized by a condensation reaction linking three fatty acids to a single glycerol molecule.*
(b) *If the fatty acids are saturated, with all single bonds between carbons, the chains pack closely together, forming a fat that is solid at room temperature (left). Unsaturated fatty acids have some double bonds between carbons, which cause kinks in the chains (right, white circles). The kinks prevent close packing and produce an oil that is liquid at room temperature.*

Lipids

Lipids include a diverse assortment of molecules, whose unifying characteristic is insolubility in water. As we examine the structures of lipids, notice that lipids have few charged regions, hydroxyl groups, or other polar parts. Almost all the bonds in lipids are nonpolar carbon–carbon or carbon–hydrogen bonds. Nonpolar lipid molecules cannot interact with polar water molecules, so lipids generally do not dissolve in water.

Lipids are classified into three groups: (1) oils, fats, and

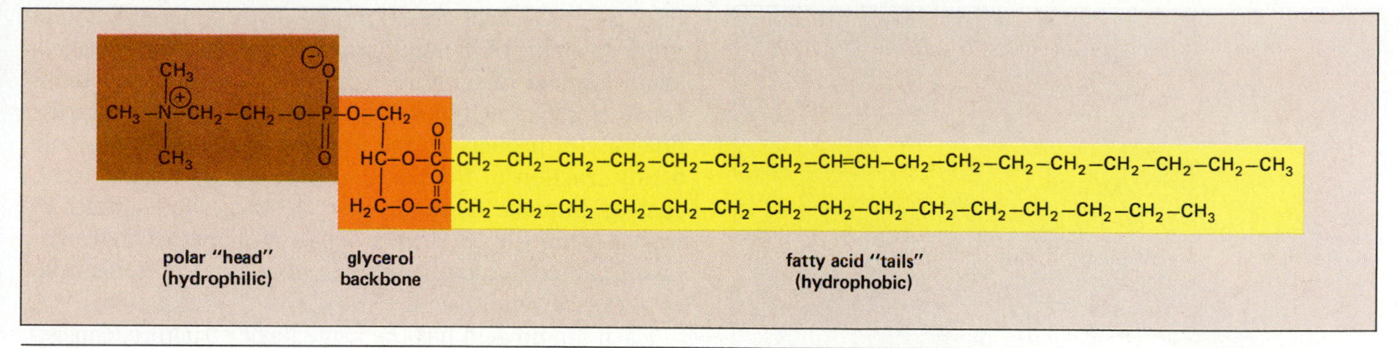

Figure 3-8 *Phospholipids are similar to fats or oils, except that only two fatty acid "tails" are attached to the glycerol backbone. The third position on the glycerol is occupied by a polar "head" composed of a phosphate group (PO_4) to which is attached a second, often nitrogen-containing group.*

waxes, which are similar in structure and contain only carbon, hydrogen, and oxygen; (2) phospholipids, structurally akin to oils but also containing phosphorus and nitrogen; and (3) the fused-ring family of steroids.

OILS, FATS, AND WAXES. These compounds are related in two ways: they contain only carbon, hydrogen, and oxygen, and they usually do not have ring structures. **Fats** and **oils** are formed by condensation reactions from one molecule of **glycerol** (a short three-carbon molecule with one hydroxyl group per carbon) and three molecules of **fatty acids** [long chains of carbon and hydrogen with a carboxylic acid group (—COOH) at one end (Fig. 3-7a)]. Fats and oils have a high concentration of chemical energy, about 9300 calories per gram (compared to 4100 for sugars or proteins). They are used for semipermanent energy storage, for example, in bears that feast during summer and fall, putting on fat to tide them over during their winter hibernation.

The difference between a fat (solid at room temperature) and an oil (liquid at room temperature) lies in their fatty acids. Fats have fatty acids with all single bonds in their carbon chains. Hydrogens occupy all the other bond positions on the carbons. The resulting fatty acid (e.g., stearic acid in Fig. 3-7a) is called **saturated** because it is "saturated" with hydrogens; that is, it has as many hydrogens as possible. If there are one or more double bonds between carbons, and consequently fewer hydrogens, the fatty acid is called **unsaturated** (oleic acid in Fig. 3-7a). Oils have mostly unsaturated fatty acids. The difference in solidity between fats and oils is due to the "kinkiness" of double bonds, which keeps unsaturated fatty acids from nestling closely together the way that saturated fatty acids can (Fig. 3-7b). An oil can be converted to a fat by breaking the double bonds between carbons, replacing them with single bonds, and adding hydrogens to the remaining bond positions. This is the "hydrogenated oil" in the ingredients on a box of margarine.

Waxes are similar to fats and oils except that the fatty acids are linked to large, long-chained alcohols instead of glycerol. Waxes form a waterproof coating over the leaves and stems of land plants. Animals also synthesize waxes, as waterproofing for mammalian fur and insect exoskeletons, and in a few cases, to build elaborate structures such as beehives.

PHOSPHOLIPIDS. The cell membrane that separates the inside of a cell from the outside world contains several types of **phospholipids.** These are similar to oils, except that one of the fatty acids is replaced by a phosphate group with a short, polar, often nitrogen-containing group attached to the end (Fig. 3-8). Unlike the fatty acid "tails," which are insoluble in water, the phosphate–nitrogen "heads" are polar or charged, and are water soluble. Thus phospholipids have two contradictory ends: heads that are **hydrophilic** (Greek for "water-loving") and tails that are **hydrophobic** ("water-fearing"). As you will see in the next chapter, this dual nature of phospholipids is crucial to the structure of the cell membrane.

STEROIDS. Except for being insoluble in water, **steroids** are unlike the other lipids. All steroids have similar structures, composed of four fused rings with various functional groups protruding from them (Fig. 3-9), and all are synthesized from cholesterol. Besides being the precursor for other steroids, cholesterol itself occurs in most cell membranes. Despite their structural similarity, steroids serve a wide variety of functions. Common steroids include the vertebrate male and female sex hormones, salt-regulating hormones, bile "detergents" for emulsifying fats consumed in the diet, and insect molting hormones.

Proteins

Proteins perform a wide variety of functions. Protein catalysts called **enzymes** guide almost all the chemical re-

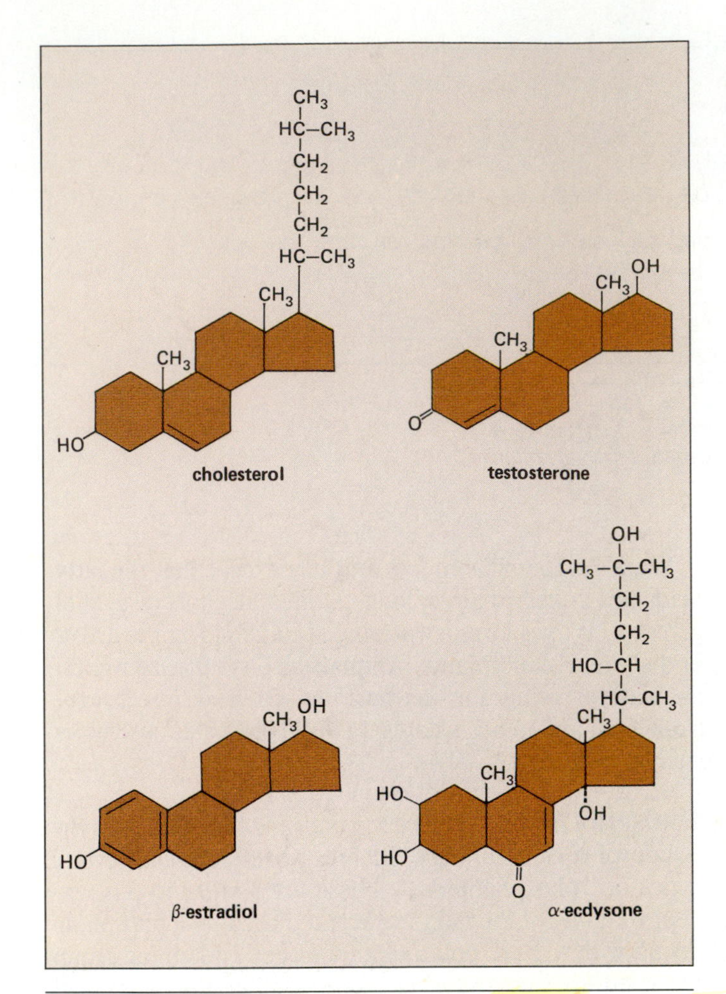

Figure 3-9 Steroids are synthesized from cholesterol, and all have almost the same molecular structure (colored polygons; note that both the carbon atoms at the corners, and the hydrogen atoms, have been omitted from these drawings). Differences in functional groups attached to the polygons cause great differences in physiological functioning, including the male sex hormone testosterone, the female sex hormone estradiol (a type of estrogen), and the insect molting hormone, ecdysone.

actions that occur inside cells (see Chapter 6). Since each enzyme assists only one or a few specific reactions, leaving others untouched, cells usually contain hundreds of different enzymes. Other proteins are used for structural purposes, such as elastin, which gives skin its elasticity; the tough proteins of hair, horns, and claws; and the silk of spider webs and silkmoth cocoons (Fig. 3-10). Still other types of proteins are used for energy and material storage (albumin in eggs, casein in milk), transport (hemoglobin to carry oxygen in the blood), and cell movement (contractile proteins in muscle). Hormones (insulin, growth hormone), antibodies, and many poisons (rattlesnake venom) are also proteins.

AMINO ACIDS AND PROTEIN SYNTHESIS. Proteins are chains of smaller subunit molecules called **amino acids** (Fig. 3-11a). There are 20 amino acids commonly found in the proteins of living organisms. This may seem like a small number of subunits from which to construct thousands of different types of proteins. In fact, though, the diversity of proteins in the organisms alive today is only a tiny fraction of the possible proteins that *could* exist. Just as humans have invented a staggering number of words using the 26 letters of the alphabet, living organisms can construct incredible numbers of proteins from 20 different amino acids.

Each amino acid has the same basic structure, consisting of a central carbon bonded to four different functional groups: a nitrogen-containing amino group (—NH_2); a carboxylic acid group (—COOH); a hydrogen; and a variable group (usually denoted by the letter "R"). The R group differs among the amino acids and gives each its distinctive properties (Fig. 3-11b). The 20 amino acids can be grouped into a few functional types, based on the nature of the R group: (1) hydrophilic (with acidic, basic, or polar R groups), (2) hydrophobic (large, nonpolar R groups), (3) ambivalent (small, uncharged R groups, not very hydrophilic or hydrophobic), and (4) the sulfur-containing cysteine, two of which can link protein chains together via bonds called *disulfide bridges* between the

Figure 3-10 Common structural proteins include hair **(a)**, *horn* **(b)**, *and spider web silk* **(c)**.

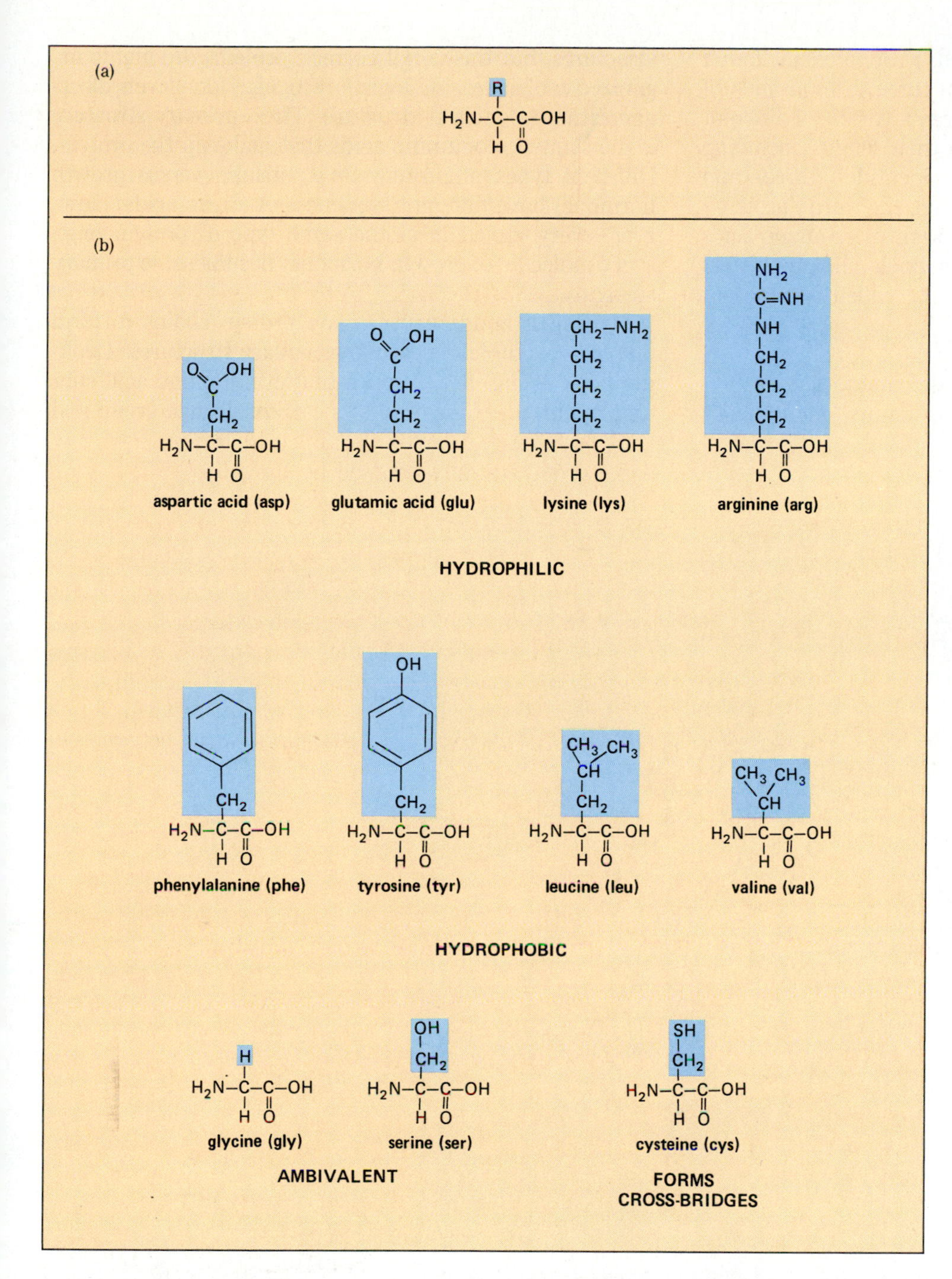

Figure 3-11 Amino acid structure.
(a) *Amino acids all contain a central carbon to which four different side groups are attached. Three of these groups are the same for every amino acid: a hydrogen atom, a carboxylic acid group (—COOH), and an amino group (—NH_2). The fourth group (R) is variable, and gives each amino acid its unique characteristics.*
(b) *Most amino acids may be classified according to the variable R group as hydrophilic, hydrophobic, or ambivalent (usually, with small R groups that have little effect on the water solubility of the amino acid). Cysteine stands in a class by itself. Two cysteines in distant parts of a protein molecule can form a covalent bond between their sulfur atoms, making a "cross-bridge" that brings the cysteines very close together and bends the protein chain.*

sulfurs (see the essay, "Protein Structure"). The different functional properties of proteins are caused by the different R groups of their amino acids.

Like lipids and polysaccharides, proteins are synthesized by a condensation reaction. The nitrogen of the amino group of one amino acid is joined to the carbon of the carboxylic acid group of another amino acid by a single covalent bond (Fig. 3-12). This bond is called a **peptide bond,** and the resulting chain of two amino acids is called a **peptide.** More amino acids are added, one by

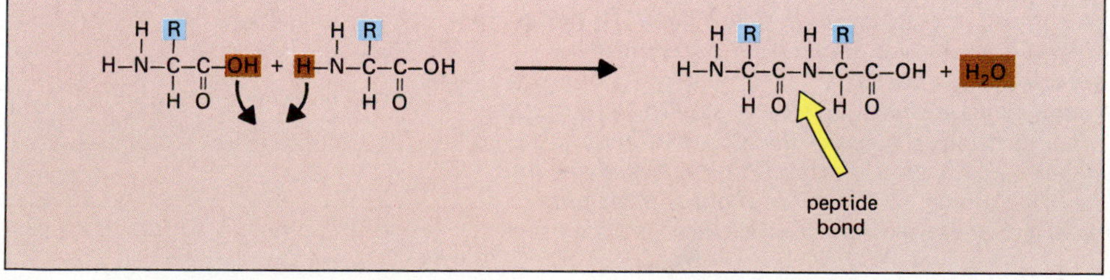

peptide bond

Figure 3-12 In protein synthesis, a condensation reaction joins the carbon of the carboxylic acid group of one amino acid to the nitrogen of the amino group of a second amino acid. The resulting covalent bond is called a peptide bond.

one, until the protein is completed. Protein chains found in living cells vary in length from three to thousands of amino acids. Biochemists have given a variety of different names to proteins, depending on their length, including "dipeptide" (two amino acids), "tripeptide" (three), and "polypeptide" (three or more). Often, the word "protein" is reserved for long peptides, say 50 or more amino acids in length. For simplicity, in this book all amino acid chains, regardless of length, will be called peptides or proteins.

PROTEIN STRUCTURE. The phrase "amino acid chains" may evoke images of proteins as floppy, monotonous structures, but this is not correct: proteins are highly organized molecules. Biologists recognize four levels of organization to protein structure. The **primary structure** is the sequence of amino acids that make up the protein. Different types of proteins (e.g., insulin versus growth hormone) have different sequences of amino acids. However, every molecule of the same type of protein (e.g., every molecule of growth hormone) has the same primary structure.

Hydrogen bonds cause many protein chains to form one of two simple, repeating **secondary structures.** Looking back at Fig. 3-12, notice that every amino acid subunit retains a —C=O from its carboxylic acid group and

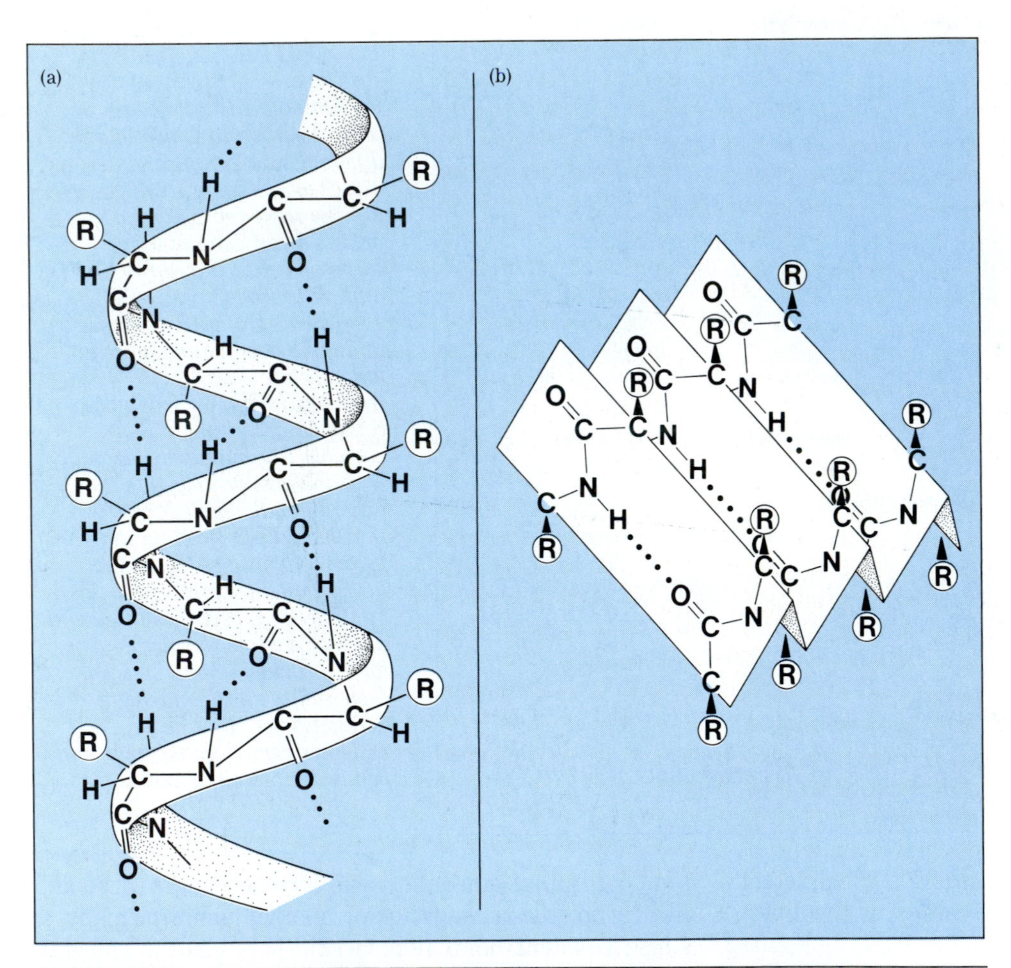

Figure 3-13 The two most common secondary structures of proteins are the helix and the pleated sheet.
(a) *In a helix, hydrogen bonds (dotted lines) form between the acid group oxygen of one amino acid and the amino group hydrogen of the third amino acid "up" the helix. These hydrogen bonds hold the protein chain in a spiral configuration, in which about $3\frac{1}{2}$ amino acids make up each turn of the helix (for simplicity, only 3 amino acids per turn are shown). The variable R groups project outward from the helix.*
(b) *In the pleated sheet arrangement, several peptide chains lie side by side (running zigzag* across *the sheets left to right). Hydrogen bonds between peptides (lengthwise* along *the sheets top to bottom) hold each sheet together. The R groups project alternately above and below the sheet. Despite its accordion-like appearance, each peptide chain is in its fully extended state and cannot easily be stretched. For this reason, pleated sheet proteins such as silk are strong but not elastic.*

an —N—H from its amino group. Since oxygen attracts electrons more strongly than carbon, the oxygen is relatively negative. Similarly, nitrogen attracts electrons more strongly than hydrogen, leaving the hydrogen relatively positive. Therefore, hydrogen bonds can form between the —C=O and —N—H groups of a protein chain. Many proteins, such as the hair protein keratin, have a coiled, Slinky-toy-like shape called a **helix,** in which hydrogen bonds hold together the turns of the coils (Fig. 3-13a). Some proteins, such as silk, are composed of many protein chains lying side by side, with hydrogen bonds holding adjacent chains in a **pleated sheet** arrangement (Fig. 3-13b).

It is rare for an entire protein to be a simple helix or pleated sheet. Most proteins assume complex three-dimensional **tertiary structures** (Fig. 3-14). Disulfide bridges may bring otherwise distant parts of a protein close together. Internal stresses, due to the particular amino acids present, may contort a protein too. For example, amino acids with very large R groups, such as phenylalanine (see Fig. 3-11b), are too bulky to fit side by side in a simple helix. As a result, the helix bends.

Perhaps the most important influence on the tertiary structure of a protein is its cellular environment, specifically whether the protein is dissolved in the water of the cytoplasm, in the lipids of the membranes, or half in one and half in the other. Hydrophobic amino acids orient themselves as far away from water as possible, while hydrophilic amino acids may form hydrogen bonds with nearby water molecules. Therefore, a protein chain dissolved in water folds into an irregular glob, with its hydrophilic amino acids facing the outside watery environment and its hydrophobic ones clustered in the center of the molecule.

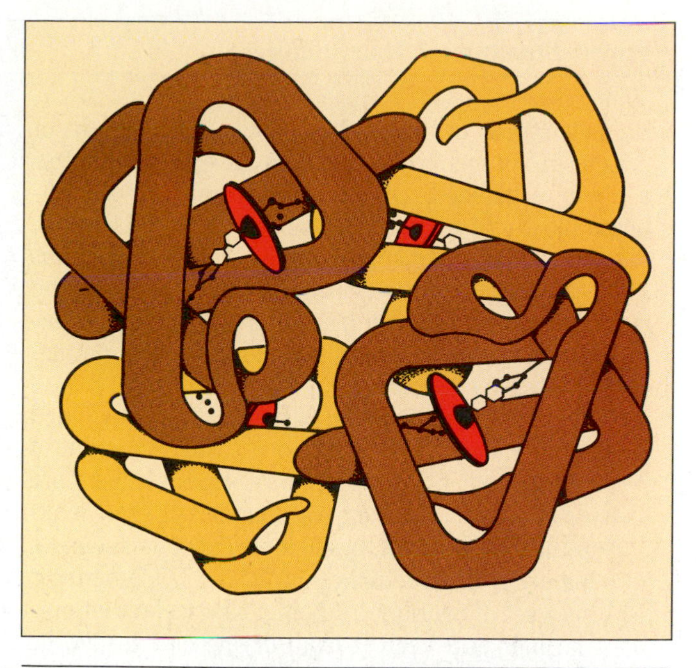

Figure 3-15 *A single hemoglobin molecule consists of two pairs of very similar subunits, each of which closely resembles myoglobin (note the similarity in structure between the hemoglobin subunit on the lower right and the myoglobin molecule shown in Fig. 3-14). Each subunit contains one heme group, which can bind one oxygen molecule, so the entire hemoglobin molecule can hold four oxygen molecules at once.*

Figure 3-14 *Myoglobin, an oxygen storage protein in muscles, has a complex tertiary structure consisting of a series of helical segments (straight "tubes" in the drawing) connected by bends and kinks. In the midst of the myoglobin molecule sits a flat, ringed structure called a heme group (red) containing an iron atom that binds oxygen.*

Besides assuming complex three-dimensional structures of their own, peptide chains often join with other peptides to form complexes referred to as the **quaternary structure** of a protein. The best studied such "super-protein" is hemoglobin (Fig. 3-15), the oxygen-carrying pigment in red blood cells. A single molecule of hemoglobin consists of two pairs of very similar peptide chains, held together by hydrogen bonds. Although no one knows precisely why hemoglobin has four chains, it is known that a single chain binds oxygen more tightly than the four-chain molecule does. This would be great for picking up oxygen in the lungs, but since the single chains would not release the oxygen again under normal conditions, the rest of the body's cells would die from lack of oxygen. Interactions among the four chains seem to cause hemoglobin to bind oxygen tightly enough to acquire oxygen in the lungs and still be able to give it up again to the body tissues.

Protein Structure: A Hairy Subject

A single strand of human hair, about 20 millionths of a meter in diameter and not even alive, is nonetheless a highly organized, complex structure. Hair is composed mostly of a single, helical protein called keratin. (There are also pigments that give hair its color, but these won't concern us here.) If we look closely at the structure of hair, we can learn a great deal about biological molecules, chemical bonds, and why human hair behaves as it does.

In the molecular structure of hair, nature anticipated the technique of sailors who made strong ropes out of weak individual fibers of hemp: a series of fibers are twisted about one another, with bundles of fibers making up the final product. A single strand of hair consists of a hierarchy of structures (Fig. E3-1). The outermost layer is a set of overlapping shinglelike scales that protect the hair and keep it from drying out. Inside the hair lie closely packed, dead cells, each filled with long

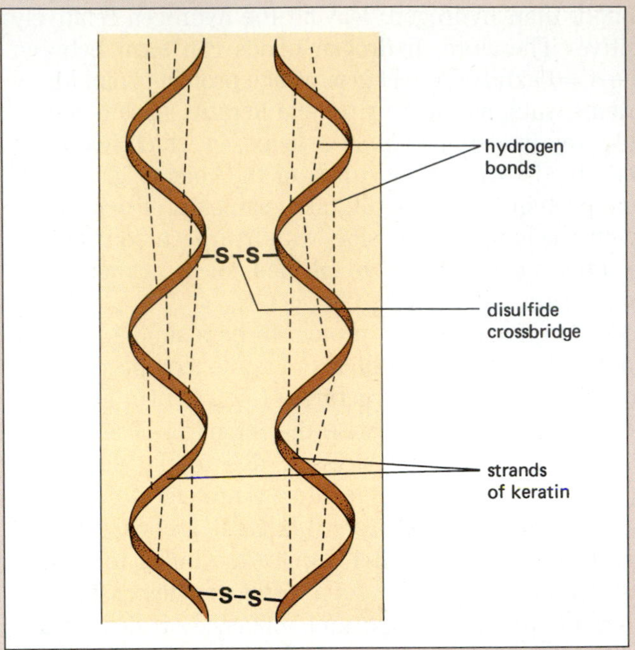

Figure E3-2. Hydrogen bonds and disulfide bridges between cysteines impart strength and elasticity to individual hairs.

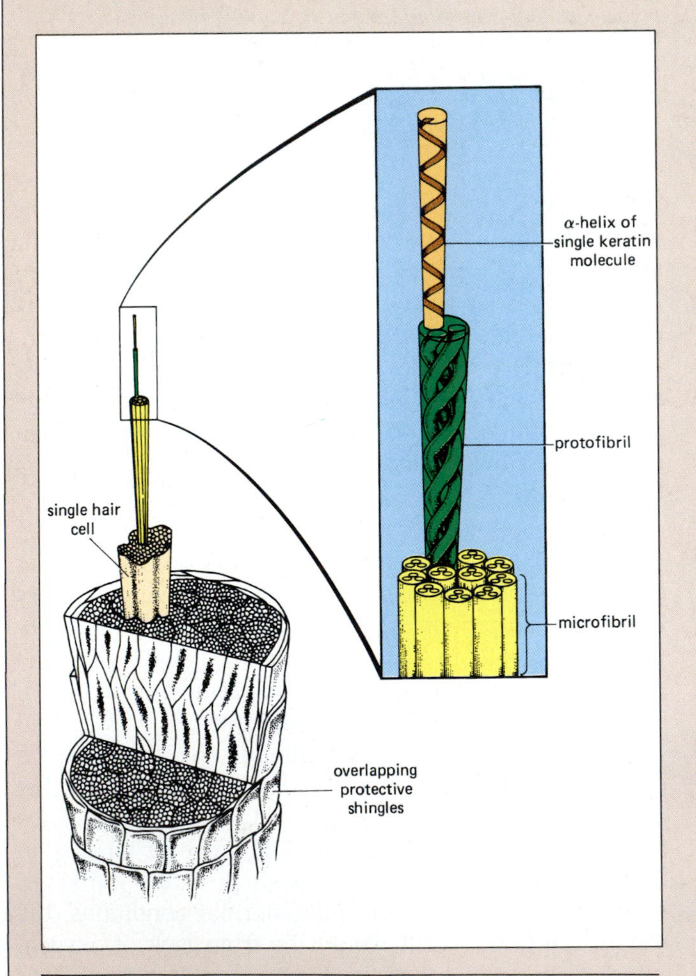

Figure E3-1. The microscopic organization of a single hair is one of bundles of fibers within further bundles of fibers.

strands running parallel to the axis of the hair. These strands in turn are composed of thinner strands called microfibrils, embedded in a protein matrix. Each microfibril is a bundle of protofibrils, and each protofibril is composed of three or more helical keratin molecules twisted together. As a hair grows, living cells in the hair follicle embedded in the skin whip out new keratin at the rate of 10 turns of the protein helix every second.

If you pluck a hair from your head, you will notice that it is rather strong: it smarts a bit to break off a healthy hair. Hair gets its strength from three types of chemical bonds. First, the individual molecules of keratin are held in their helical shape by many hydrogen bonds (Fig. E3-2). Before a hair will break, all the hydrogens bonds of all the keratin molecules in one cross-sectional plane of the strand must be broken to allow the helix to be stretched out to its maximal extent. Second, each molecule is cross-linked to neighboring keratins by disulfide bridges between cysteines. Some of these bridges have to separate as the hair stretches. Finally, at least one peptide bond in each keratin molecule must break before the strand as a whole breaks.

Hair is also fairly stiff. The stiffness arises from hydrogen bonds within the individual helices of keratin and from disulfide bridges holding neighboring keratin

Protein Structure: A Hairy Subject

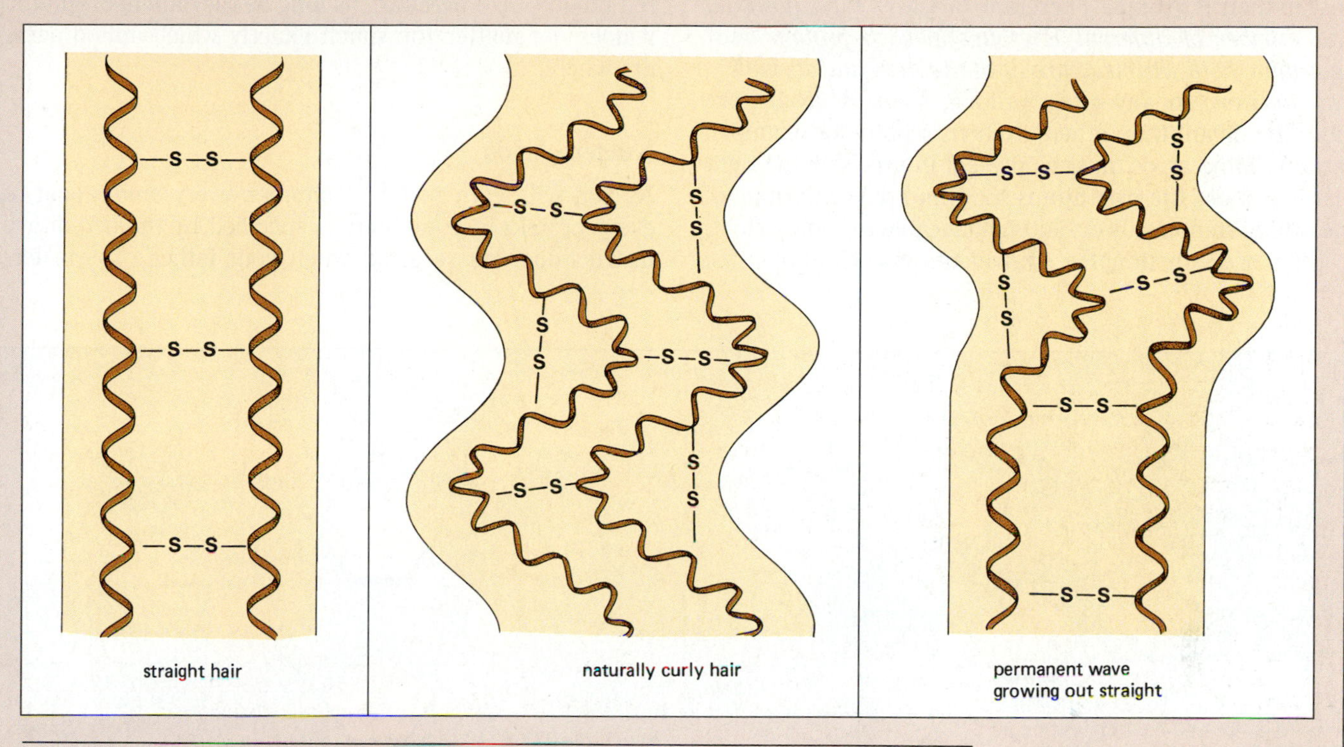

Figure E3-3. If the disulfide bridges join individual keratin molecules in a hair at the same level, the hair will be fairly straight. If disulfide bridges connect different levels within or between molecules, the hair will have a natural curl. The same effect can be obtained with a permanent wave that breaks and reforms the disulfide bridges of naturally straight hair. The new hair growing out will be straight again.

molecules together. When hair gets wet, however, the hydrogen bonds between turns of the helices are replaced by hydrogen bonds between the amino acids and the water molecules surrounding them, so the helices collapse. Wet hair is therefore very limp. If wet hair is rolled onto curlers and allowed to dry, the hydrogen bonds reform in slightly different places, holding the hair in a curve. The slightest moisture, even humid air, allows these hydrogen bonds to rearrange into their natural configuration, and the hair straightens out again.

Pull gently and you will discover still another property of hair. It stretches and then springs back into shape when you release the tension. When the hair stretches, many of the hydrogen bonds within each keratin helix are broken, allowing the helix to be extended. Most of the disulfide bonds between different levels of the neighboring helices, on the other hand, are distorted by stretching but do not break. When tension is released, these disulfide bridges contract again, returning the hair to its normal length.

Finally, each hair has a characteristic shape: it may be straight, but more likely it is wavy to some extent, perhaps even coiled like a spring. The curliness of hair is genetically specified, and biochemically is determined by the arrangement of disulfide bridges (Fig. E3-3). Curly hair has disulfide bonds crosslinking the various keratin molecules at *different levels,* while straight hair has bridges mostly at the *same level.* When straight hair is given a "permanent wave," two lotions are applied. The first lotion breaks disulfide bonds between neighboring helices. The hair is then rolled tightly onto curlers, and a second solution, which re-forms the bridges, is applied. The new disulfide bridges connect helices at different levels, holding the strands of hair in a curl. These new bridges are more or less permanent, and genetically straight hair can be transformed into biochemically curly hair. The new hair that grows in, of course, will have the genetically determined arrangement of bridges, and will not be curly.

Protein function All proteins are synthesized with identical peptide bonds linking amino acids that differ only in their R groups. Therefore, *the exact type, position, and number of different R groups in each protein must determine both the structure of the protein and its biological function.* In any given protein, some R groups are more important than others. In hemoglobin, for example, certain amino acids must be present in precisely the right places to hold the iron atoms that bind oxygen. Many of the other amino acids are interchangeable to some extent, if they are functionally equivalent. For example, the amino acids on the outside of a hemoglobin molecule mostly serve to keep it dissolved in the cytoplasm of the red blood cell. Therefore, as long as they are hydrophilic, it does not matter too much exactly which amino acids are where.

Nucleic Acids

As you will learn in later chapters, every amino acid of every protein in your body is specified by the hereditary instructions you received when your father's sperm fer-

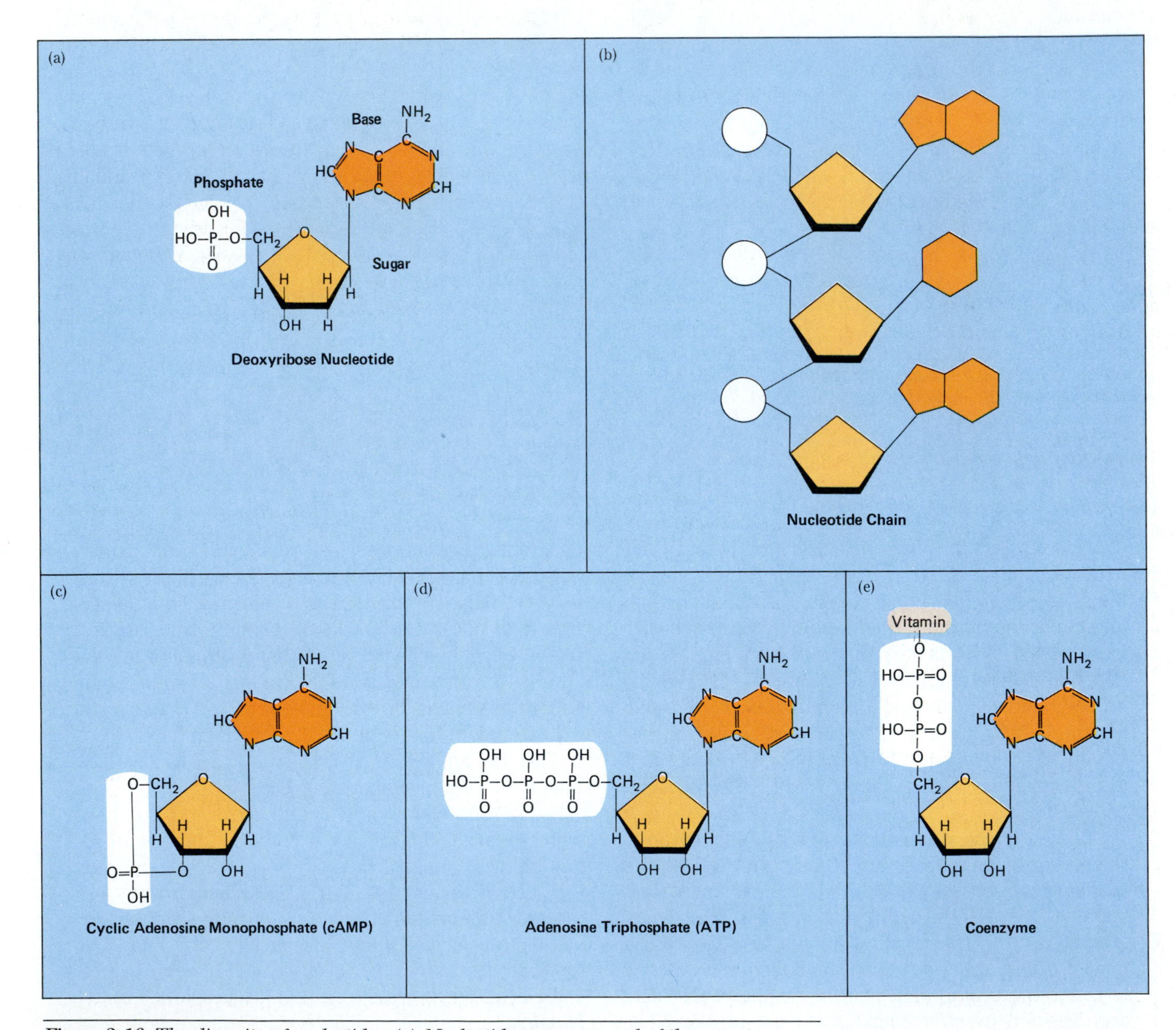

Figure 3-16 The diversity of nucleotides. **(a)** *Nucleotides are composed of three parts: a five-carbon sugar, either ribose or deoxyribose, a phosphate group, and a nitrogen-containing base that differs among nucleotides (see Chapter 10). Nucleotides may link together to form chains* **(b)***, as in DNA and RNA. Single modified nucleotides function in intracellular communication and regulation (***c:** *cyclic AMP), energy transfer (***d:** *ATP), and metabolism (***e:** *coenzymes).*

tilized your mother's egg. In fact, that is precisely what genes are: a set of instructions spelling out the amino acid sequences of your body's proteins. Genes are composed of the fourth major category of biological molecule, **nucleic acids.** Nucleic acids are chains of similar but not identical subunits called **nucleotides.** There are two distinct classes of nucleotides, the **ribose** nucleotides and the **deoxyribose** nucleotides. All nucleotides have a three-part structure: a five-carbon *sugar* (ribose or deoxyribose), a *phosphate group,* and a nitrogen-containing *base* that differs among nucleotides (Fig. 3-16a). Because the bases vary, there are four different types of ribose nucleotides and four types of deoxyribose nucleotides (see Chapters 10 and 11).

Nucleotides may be strung together in long chains, with the phosphate of one nucleotide covalently bonded to the sugar of another (Fig. 3-16b). Deoxyribose nucleotides form chains millions of units long called **deoxyribonucleic acid,** or **DNA.** DNA is found in the chromosomes of all living things, and its sequence of nucleotides, like the dots and dashes of a biological Morse code, spells out the genetic information needed to construct the proteins of the organism. Chains of ribose nucleotides, called **ribonucleic acid,** or **RNA,** are copied from the central repository of DNA in the nucleus of each cell. RNA carries DNA's genetic code into the cytoplasm and directs the synthesis of proteins.

OTHER NUCLEOTIDES. Not all nucleotides are part of DNA or RNA molecules. Some exist separately in the cell or occur as parts of other molecules. The **cyclic nucleotides** (Fig. 3-16c) are intracellular messengers that carry information from the cell membrane to other molecules in the cell. These messenger molecules are synthesized when certain hormones come in contact with the cell membrane. The messengers then activate essential reactions in the cell cytoplasm or nucleus. Other nucleotides have extra phosphate groups. These diphosphate and triphosphate nucleotides, such as **adenosine triphosphate** (ATP) (Fig. 3-16d), are unstable, energetic molecules that carry energy from place to place within a cell. They capture energy where it is produced (during photosynthesis, for example) and give it up to drive energy-demanding reactions elsewhere (e.g., to synthesize a sugar). Finally, certain nucleotides assist enzymes in their action: these are called **coenzymes,** and usually consist of a nucleotide combined with a vitamin (Fig. 3-16e). You will learn more about energy carrier nucleotides and coenzymes in Chapter 7, when we discuss energy production and use in the cell.

SUMMARY OF KEY CONCEPTS

Synthesizing Organic Molecules: A Modular Approach

Most large biological molecules are synthesized by linking together many smaller subunit molecules. Chains of subunits are connected by covalent bonds through condensation reactions; the chains may be broken apart again by hydrolysis reactions.

The Principal Types of Biological Molecules

The most important organic molecules are carbohydrates, lipids, proteins, and nucleic acids.

Carbohydrates include sugars, starches, and cellulose. Sugars (monosaccharides and disaccharides) are used for temporary storage of energy and for the construction of other molecules. Starches and glycogen are polysaccharides that serve for long-term energy storage in plants and animals, respectively. Cellulose and related polysaccharides form cell walls of bacteria, fungi, plants, and some protists.

Lipids are water-insoluble molecules of diverse chemical structure, and include oils, fats, waxes, phospholipids, and steroids. Lipids are used for energy storage (fats and oils), as waterproofing for the outer layers of land organisms (waxes), as the principal component of cell membranes (phospholipids), and as hormones (steroids).

Proteins are chains of amino acids. The function of each protein is determined by the sequence of amino acids in the chain. Proteins may be enzymes (biological catalysts), structural molecules (hair, horn), hormones (insulin), or transport molecules (hemoglobin).

Nucleic acids are chains of nucleotides. Each nucleotide is composed of a phosphate group, a sugar group, and a nitrogen-containing base. The two types of nucleic acids are deoxyribonucleic acid (DNA) and ribonucleic acid (RNA). Other nucleotides include energy carrier molecules (ATP), intracellular messengers (cyclic AMP), and coenzymes.

GLOSSARY

Adenosine triphosphate (a-den′-ō-sēn trī-fos′-fāt; ATP) a nucleotide with three phosphate groups that serves as an energy carrier molecule in cells.

Amino acid the individual subunit of which proteins are made, composed of a central carbon atom to which is bonded an amino group (—NH_2), a carboxylic acid group (—COOH), a hydrogen atom, and a variable group of atoms denoted by the letter R.

Carbohydrate a compound composed of carbon, hydrogen, and oxygen, with the chemical formula $(CH_2O)_n$; includes sugars and starches.

Cellulose an insoluble carbohydrate composed of glucose subunits; forms the cell wall of plants.

Chitin (kī′-tin) a compound found in the cell walls of fungi and the exoskeletons of arthropods, composed of chains of nitrogen-containing, modified glucose molecules.

Coenzyme (kō-en′-zīm) an organic molecule that assists enzymes in their actions.

Condensation a chemical reaction in which two molecules are joined by a covalent bond, with the simultaneous removal of a hydrogen from one molecule and a hydroxyl group from the other, forming water.

Cyclic nucleotide (sik′-lik nū′-klē-ō-tīd) a nucleotide in which the phosphate group is bonded to the sugar at two points, forming a ring. Cyclic nucleotides serve as intracellular messengers.

Deoxyribonucleic acid (dē-ox-ē-rī-bō-nū-klā′-ik; DNA) a molecule composed of deoxyribose nucleotides; the genetic information of all living cells.

Disaccharide (dī-sak′-a-rīd) a carbohydrate formed by the covalent bonding of two monosaccharides.

Enzyme a protein molecule that speeds up specific chemical reactions but which is not itself used up or permanently altered; a protein catalyst.

Fat a lipid composed of three saturated fatty acids covalently bonded to glycerol; fats are solid at room temperature.

Fatty acid an organic molecule composed of a long chain of carbon atoms, with a carboxylic acid (—COOH) group at one end. Fatty acids may be saturated (all single bonds between the carbon atoms) or unsaturated (one or more double bonds between carbon atoms).

Functional group one of several groups of atoms commonly found in organic molecules, including hydrogen, hydroxyl, amino, carboxyl, and phosphate groups.

Glucose the most common monosaccharide, with the molecular formula $C_6H_{12}O_6$. Most polysaccharides, including cellulose, starch, and glycogen, are made of glucose subunits covalently bonded together.

Glycerol (glis′-er-ol) a three-carbon alcohol to which fatty acids are covalently bonded to make fats and oils.

Glycogen (glī′-kō-gen) a polysaccharide composed of branched chains of glucose subunits, used as a carbohydrate storage molecule in animals.

Helix (hē′-licks) a spiral structure similar to a corkscrew or a spiral staircase; a type of secondary structure of a protein.

Hydrolysis (hī-drol′-i-sis) the chemical reaction that breaks a covalent bond through the addition of hydrogen to the atom forming one side of the original bond, and a hydroxyl group to the atom on the other side.

Hydrophilic (hī-drō-fil′-ik) pertaining to a substance that is attracted to and usually dissolves in water. Such molecules either are ions or have polar parts.

Hydrophobic (hī-drō-fō′-bik) pertaining to a substance that is insoluble in water; usually uncharged and lacking any polar parts.

Lactose (lak′-tōs) a disaccharide composed of glucose and galactose; found in mammalian milk.

Lipid (li′pid) one of a number of water-insoluble organic molecules, containing large regions composed solely of carbon and hydrogen. Lipids include oils, fats, waxes, phospholipids, and steroids.

Maltose (mal′-tōs) a disaccharide composed of two glucose molecules.

Monosaccharide (mo-nō-sak′-a-rīd) the basic molecular unit of all carbohydrates, composed of a backbone of carbon atoms to which are bonded hydrogen and hydroxyl groups.

Nucleic acid (nū-klā′-ik) an organic molecule composed of nucleotide subunits. The two common types of nucleic acids are ribonucleic acids (abbreviated RNA) and deoxyribonucleic acids (DNA).

Nucleotide (nū′-klē-ō-tīd) an organic molecule composed of a phosphate group, a five-carbon monosaccharide (ribose or deoxyribose), and a nitrogen-containing base.

Oil a lipid composed of three fatty acids, some of which are unsaturated, covalently bonded to a molecule of glycerol. Oils are liquid at room temperature.

Peptide (pep′-tīd) a chain composed of two or more amino acids linked together by peptide bonds.

Phospholipid (fos-fō-li′-pid) a lipid consisting of glycerol to which two fatty acids and one phosphate group are bonded. The phosphate group bears another group of atoms, often containing nitrogen, and usually either polar or bearing an electrical charge.

Pleated sheet a type of secondary structure of a protein. In a pleated sheet, protein chains lie side by side, held to one another by hydrogen bonds.

Polysaccharide (pol-ē-sak′-a-rīd) a large carbohydrate molecule composed of branched or unbranched chains of repeating monosaccharide subunits, usually glucose or modified glucose molecules. Polysaccharides include starches, cellulose, and glycogen.

Primary structure: the amino acid sequence of a protein.

Protein an organic molecule composed of one or more chains of amino acids.

Quaternary structure (kwat′-er-nā-rē) the complex three-dimensional structure of a protein that is composed of more than one peptide chain.

Ribonucleic acid (rī-bō-nū-klā′-ik; RNA) a molecule composed of ribose nucleotides; transfers hereditary instructions from the nucleus to the cytoplasm; also the genetic material of some viruses.

Saturated referring to a fatty acid with as many hydrogen atoms as possible bonded to the carbon backbone; a fatty acid with no double bonds in its carbon backbone.

Secondary structure a repeated, regular structure assumed by protein chains, held together by hydrogen bonds; usually either a helix or a pleated sheet.

Starch a polysaccharide composed of branched or unbranched chains of glucose molecules, used by plants as a carbohydrate storage molecule.

Steroid a lipid composed of four fused rings of carbon atoms to which functional groups are attached.

Sucrose a disaccharide composed of glucose and fructose.

Tertiary structure (ter′-shē-ār-ē) the complex three-dimensional structure of a single peptide chain. The tertiary structure is held in place by disulfide bonds between cysteine amino acids, by attraction and repulsion among amino acid side groups, and by interaction between the cellular environment (water or lipids) and the amino acid side groups of the protein.

Unsaturated referring to a fatty acid with fewer than the maximum number of hydrogen atoms bonded to its carbon backbone; a fatty acid with one or more double bonds in its carbon backbone.

Vitamin an organic substance that cannot be synthesized by an organism but that is required in trace amounts.

Wax a lipid composed of fatty acids covalently bonded to long-chain alcohols.

STUDY QUESTIONS

1. Which elements are commonly found in biological molecules?
2. List the four principal types of biological molecules and give an example of each.
3. Describe condensation and hydrolysis reactions.
4. Distinguish among the following: monosaccharide, disaccharide, and polysaccharide. Give two examples of each, and their functions.
5. Describe the molecular structures of fats, oils, and steroids. Give one example of each, and its function.
6. Draw the structure of an amino acid. What determines the properties of each amino acid? What are the four general types of amino acids?
7. Describe the synthesis of a protein from amino acids.
8. Define the primary, secondary, tertiary, and quaternary structures of a protein.
9. Draw the structure of an individual nucleotide and a nucleic acid, and label the parts (use boxes to represent the various parts).

SUGGESTED READINGS

Doolittle, R. F. "Proteins." *Scientific American,* October 1985. Proteins are the molecular tools that, directly or indirectly, carry out almost all the functions of a cell.

Sharon, N. "Carbohydrates." *Scientific American,* November 1980 (Offprint No. 1483). Carbohydrates, although all quite similar in structure, perform a variety of roles in the life of a cell.

Stryer, L. *Biochemistry,* 3rd ed. New York: W. H. Freeman and Company, Publishers, 1988. The diagrams of different types of biological molecules, and how they function, are particularly effective.

4 Cells: The Units of Life. I. Interactions with the External Environment

Before the invention of the microscope, biologists often speculated about how the bodies of plants and animals are constructed. Dissections of animals and human cadavers revealed that animals are composed of various parts, such as muscle, brain, and liver, that are very unlike one another in appearance. Plants look so completely different from animals that it seemed unlikely that the two could have very much in common.

Then in 1665 Robert Hooke peered through a primitive microscope at an "exceeding thin . . . piece of Cork" and discovered that cork, which is actually the bark of certain trees, is made up of "a great many little Boxes" (Fig. 4-1). The "boxes" of cork, being long dead, were empty, but later observations of living plants showed that "these cells [are] fill'd with juices." Hampered by the poor optics of the microscopes of his day, Hooke could not see many details of plant cell structure.

Over the next 150 years, biologists noted that many plants are composed of cells, a task made easy by the thick cell wall that outlines every plant cell. Animal cells, however, escaped notice until the 1830s, when Theodor Schwann saw that cartilage contains cells, each with a thin but well-defined cell boundary. Soon, better microscopes and new types of stains revealed that other animal tissues are also cellular. This led to the formulation of the first two principles of the **cell theory:**

1. Every living organism is made up of one or more cells.
2. Cells are the functional units of living organisms.

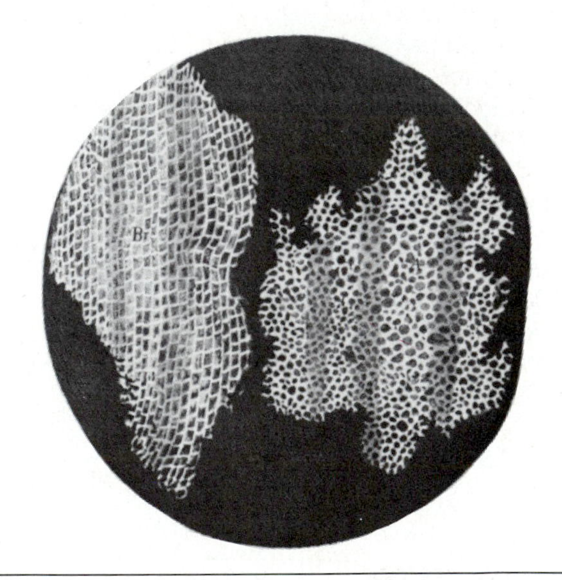

Figure 4-1 *Robert Hooke's drawings of the cells of cork, as seen with an early light microscope. The cells of cork are not living, and only the cell walls remain to outline the cell.*

During the 1840s and 1850s, biologists saw that cells could enlarge and divide to produce two new cells. As far as they could tell, every cell originated from the division of another cell. In 1855, Rudolf Virchow proclaimed the third principle of cell theory:

3. All cells arise from preexisting cells.

What exactly *is* a cell, and why are all living organisms composed of cells? *A* **cell** *is the smallest unit of life.* All

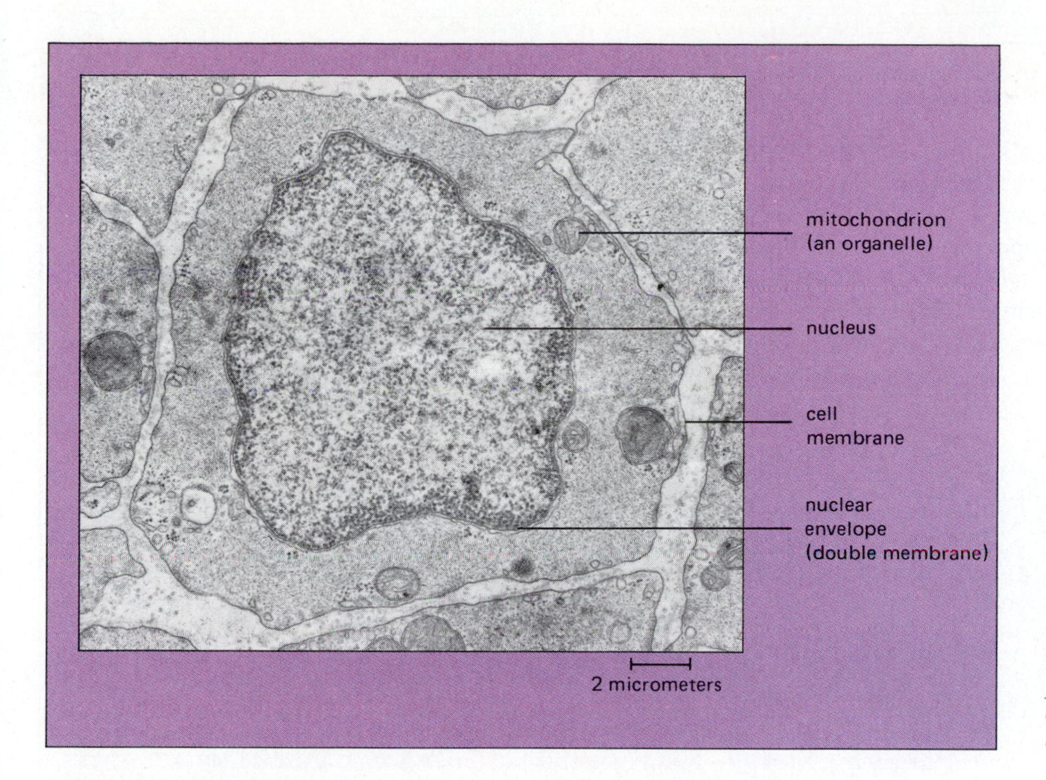

Figure 4-2 An animal cell as seen with an electron microscope. Note the cell membrane (thin line surrounding cell), cytoplasm (material enclosed by the cell membrane), and the nucleus containing the genetic material. The function of the various particles within the cell, called organelles, is the topic of the next chapter.

cells have at least three components (Fig. 4-2): a **membrane** that separates the inside of the cell from the outside world and regulates traffic into and out of the cell; **cytoplasm,** the various substances found inside the cell membrane, including water, salts, sugars, enzymes, and often particles called organelles; and **genetic material** composed of deoxyribonucleic acid (DNA), that encodes the information needed to construct the cell and direct its activities.

Generally speaking, cells are small, ranging from about 1 to 100 micrometers (millionths of a meter) in diameter. Living organisms may consist of just one (bacteria, protists) or aggregates of interacting, cooperating cells (fungi, plants, and animals; see the photographic essay, "Viewing the Cell").

CELL FUNCTIONS AND LIMITATIONS

Why are all living things composed of cells? From the essential components of cells listed above, you can probably guess why microscopic organisms are cellular: the cell membrane separates the complex life processes in the cytoplasm from the chaos of the outside world, while the genetic material is the data bank that directs those life processes.

What may be less obvious is why large organisms consist of myriads of cells rather than just one large cell. To answer that question, we must consider two physical factors that limit cell size. First, transfer of information from the genetic material to other parts of the cell may be limited by distance. Second, exchange of nutrients and wastes with the external environment may be limited both by the distance from the center of the cell to its surface and by the surface area of the cell.

The genetic material (DNA) usually resides in a localized region of a cell. In eukaryotes, this region, called the nucleus, is separated from the rest of the cytoplasm by a double membrane. The information encoded in the DNA must be transferred to other regions of the cell to control uptake of materials, growth, and reproduction. These instructions take the form of large molecules of ribonucleic acid (RNA). RNA is copied from DNA and moves to other parts of the cell, where its information is used to synthesize proteins. (Information transfer within cells will be examined in detail in Chapter 11.) If a cell becomes too large, RNA movement may be too slow for adequate control over the activities of distant regions of the cell. Some cells, notably some of the large protists such as *Paramecium,* get around this limitation by having many copies of the DNA in different parts of the cell. However, in extremely large cells, coordinating dozens of copies of DNA might be difficult.

The second size constraint, exchange with the environment, is even more difficult to circumvent. Acquiring nutrients and eliminating wastes both occur through the cell membrane. This would cause two major problems for very large cells (Fig. 4-3). First, as a cell becomes larger its innermost regions become farther removed from the membrane. Both nutrients and wastes move into, through, and out of cells by **diffusion,** the net transport of molecules from places of high concentration to places

Viewing the Cell: A Gallery of Microscopic Images

Collaboration among biologists, physicists, and engineers have resulted in a variety of microscopes that allow us to examine structures too small to see with the naked eye. *Light microscopes* use lenses, usually of glass, to focus and magnify light rays that either pass through or bounce off a specimen. Light microscopes provide a wide range of images, depending on the method of illumination of the specimen and whether or not it has been stained (Fig. E4-1a–e). The wavelengths of visible light limit the resolving power of light microscopes to about 1 micrometer.

Electron microscopes use beams of electrons instead of light. The negatively charged electrons are focused by magnetic fields rather than by conventional lenses. *Transmission electron microscopes* (TEM) pass electrons through a thin specimen and can reveal minute subcellular structures (Fig. E4-1f), including organelles and cell membranes (Fig. E4-2). *Scanning electron mi-*

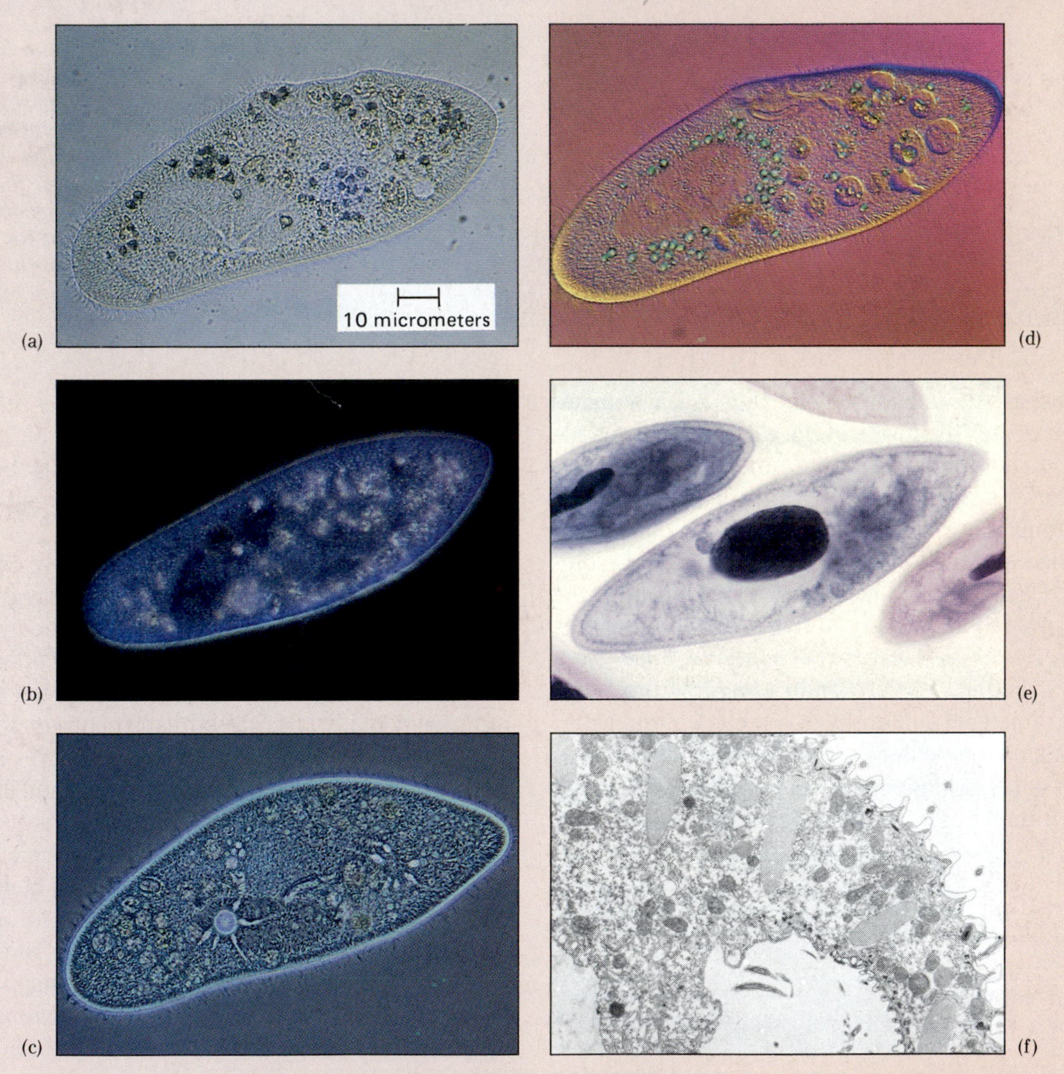

Figure E4-1. *Images of the protozoan* Paramecium *seen through different types of microscopes.* ***(a through d)*** *Living* Paramecia *seen with* ***(a)*** *conventional brightfield,* ***(b)*** *darkfield,* ***(c)*** *phase contrast, and* ***(d)*** *differential interference contrast illumination.* ***(e)*** *This* Paramecium *has been stained to reveal some subcellular structures, and photographed with a brightfield microscope.* ***(f)*** *A transmission electron micrograph of part of a* Paramecium. *Note the increased detail compared to the specimens photographed through a light microscope.*

Viewing the Cell: A Gallery of Microscopic Images

croscopes (SEM) bounce electrons off specimens that have been coated with metals, and provide three-dimensional images. SEMs can be used to view structures ranging in size from entire insects (Fig. E4-3) down to cells (Fig. E4-4) and even parts of cells (Fig. E4-5).

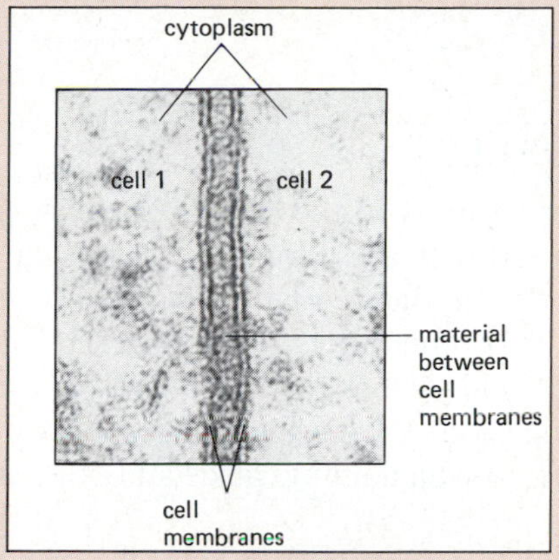

Figure E4-2. Transmission electron microscopes can magnify much more than light microscopes. This photo shows a pair of cell membranes that have been cut perpendicularly (like cutting a sheet of paper with scissors and looking at the paper edge on). In micrographs such as this one, membranes appear as a pair of dark lines (the phospholipid heads) separated by a thin layer of paler material (the phospholipid tails). A cell membrane is only about 7 billionths of a meter in thickness.

Figure E4-3. Scanning electron microscopes provide images that look three-dimensional, such as this ant carrying a computer chip.

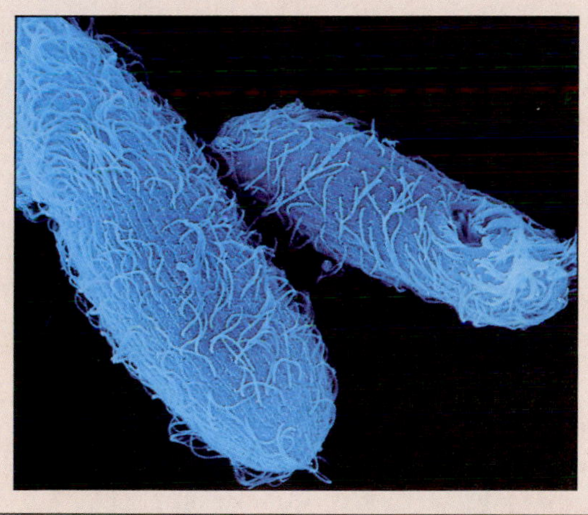

Figure E4-4. An SEM photograph of Paramecium *shows the pattern of hair-like cilia that cover the surface of the cell.*

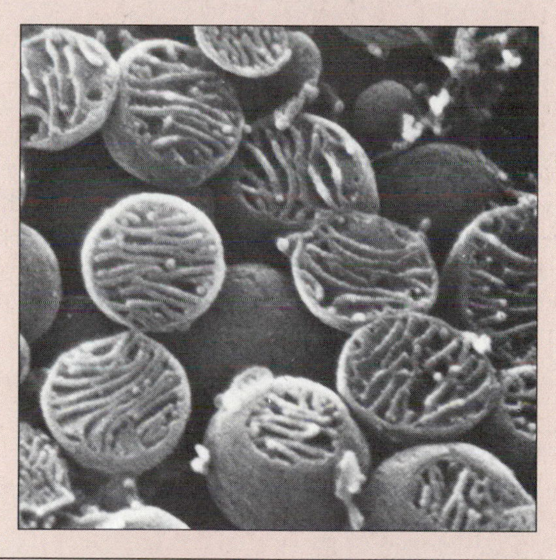

Figure E4-5. An SEM photograph of the inside of a cell; the oval structures with folds of membrane inside are mitochondria, important organelles found in all eukaryotic cells, which we will discuss in Chapter 5.

of low concentration (we will discuss diffusion more thoroughly in a moment). Relying on diffusion alone, it would take oxygen molecules over *200 days* to reach the center of a cell 20 centimeters in diameter (about the thickness of your chest). Clearly, not much life could go on inside a cell that large! The second difficulty arises from geometry: as a cell enlarges, its volume increases more rapidly than its surface area (from high school days, you may remember the equations $V = \frac{4}{3}\pi r^3$ and $A = 4\pi r^2$; see Fig. 4-3). A cell that doubles its radius therefore becomes eight times greater in volume but only four times greater in surface area. In general, as a cell increases in volume, more chemical reactions occur. Thus more nutrients and oxygen are needed, and more waste products must be eliminated, all through the cell surface. In a very large cell the surface area of membrane would be too small to

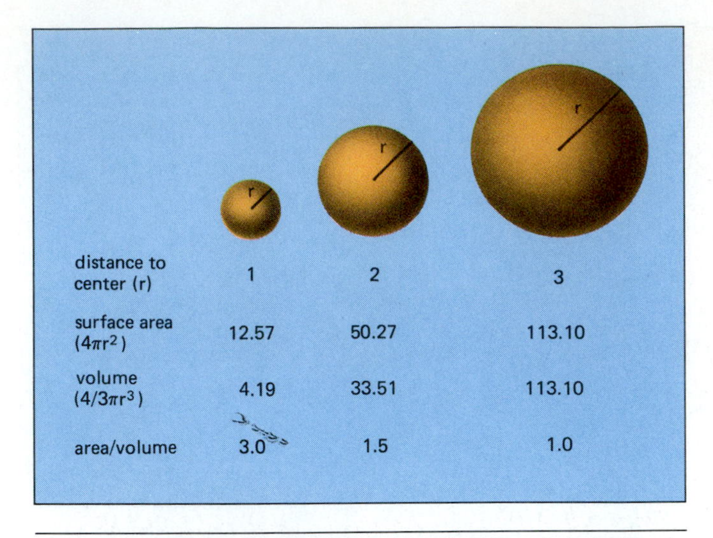

Figure 4-3 Geometrical considerations limit the size of cells. As a cell enlarges, the distance from the center of the cell to the outside world increases. Further, the volume increases much more rapidly than the surface area. As these spherical cells illustrate, doubling the radius halves the surface/volume ratio. Thus each unit volume of cytoplasm within the cell has only half the membrane area available to exchange nutrients and wastes with the external environment.

keep up with the cell's metabolic needs. Large organisms, therefore, consist of numerous cells bound together, with special provisions made for internal cells to contact the external environment and/or exchange materials with it via respiratory and circulatory systems.

From the bacteria in your large intestine quietly dining on the remnants of your lunch, to the cheetah chasing a gazelle on the African plains, the activities of living organisms are possible only because of the activities of the cells of which they are composed. We will explore the structures and functions of cells, working from the outside in. In this chapter we examine the outermost structures of cells, the cell wall that supports and protects many cells and the cell membrane that regulates the interactions of cells with their external environment. The internal organization of cells is the subject of Chapter 5.

CELL WALLS

The outer surfaces of the cells of bacteria, plants, fungi, and some protists are covered with stiff, nonliving coatings called **cell walls.** Plant cell walls are composed of **cellulose** and other polysaccharides (see Fig. 3-5), while fungal cell walls are made of the modified polysaccharide **chitin** (Fig. 3-6). Bacterial cell walls have a chitinlike framework to which amino acids and other molecules are bound.

Cell walls are produced by the cells they surround. In plants, a newly formed cell secretes sticky polysaccharides such as pectin (the ingredient that congeals grape juice into jelly). These polysaccharides glue adjacent cells together, forming the *primary cell wall* (Fig. 4-4). Many plant cells then secrete cellulose and other polysaccharides through the cell membrane (but beneath the primary wall), forming a *secondary cell wall.* In some plant cells, the secondary wall may become thicker than the rest of the cell.

Cell walls support and protect otherwise fragile cells. For example, cell walls allow plants and mushrooms to stand erect on land without the skeleton required by animals. Tree trunks are the ultimate cell wall, so to speak, being almost entirely composed of cellulose and other ma-

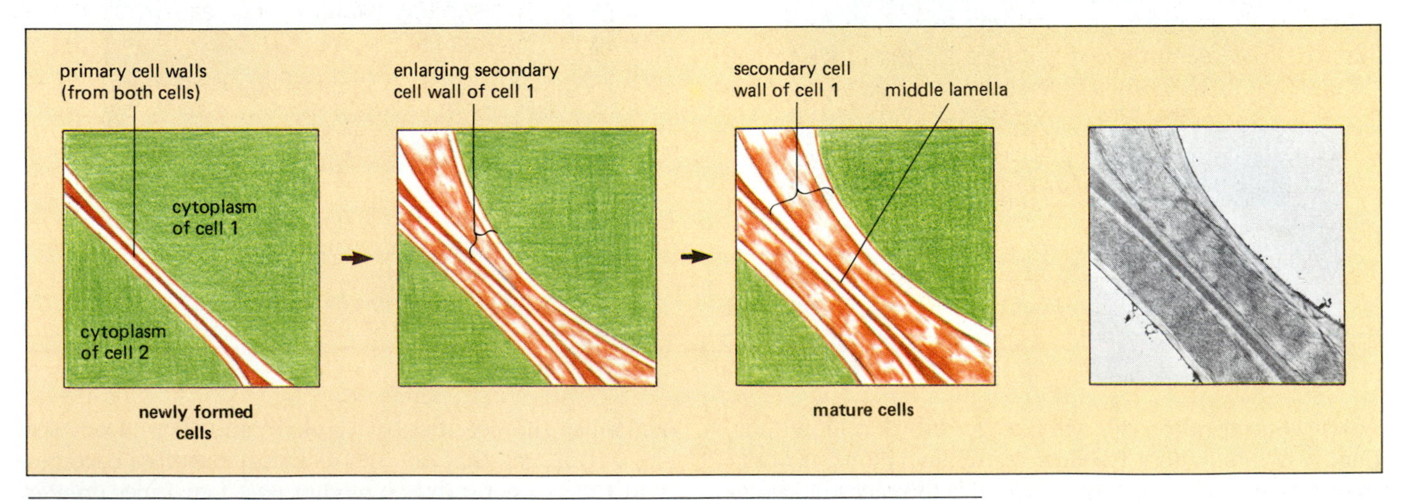

Figure 4-4 A close-up view of plant cell walls. When it is first formed by cell division, a plant cell secretes a primary cell wall of pectin and other polysaccharides, just outside the cell membrane. The primary cell walls of adjacent cells fuse together to form a middle lamella between the cells. Later the cells may secrete cellulose and other carbohydrates beneath the primary wall, forming secondary cell walls, pushing the primary cell wall and middle lamella farther away from the cell membranes.

terials laid down over the years, and capable of supporting impressive loads.

Although strong, cell walls are usually porous, permitting easy passage of materials (otherwise, the cell within would soon die). The structure that really governs the interactions between a cell and its external environment is the cell membrane.

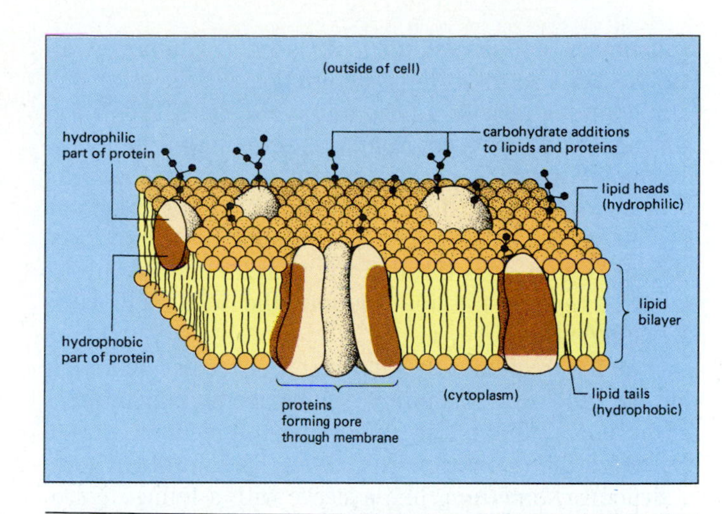

Figure 4-5 *According to the fluid mosaic model, the cell membrane is a bilayer of phospholipids in which are embedded various proteins. The orientation of a membrane protein is determined by its amino acids. Hydrophobic amino acids (brown) interact with the nonpolar lipid tails in the middle of the bilayer. Hydrophilic amino acids (tan) interact with the polar phospholipid heads and water, and stick out of the membrane on one side or the other. Some proteins extend entirely across the membrane, forming pores through which water-soluble molecules can enter or leave the cells. Many proteins and lipids have carbohydrates attached to them, especially on the side facing the exterior of the cell.*

CELL MEMBRANES

Cell membranes serve three major functions. As we have already noted, life processes within the cell must be separated from the outside world. Precious nutrients, enzymes, and DNA must not be allowed to leak uncontrollably out of the cytoplasm. Therefore, *the cell membrane must isolate the cytoplasm from the external environment.* However, to paraphrase the poet John Donne, "no cell is an island, sufficient unto itself." Every cell must acquire nutrients and eliminate wastes. The second function of the cell membrane, then, is to *regulate the flow of materials into and out of the cell.* Finally, cells constantly send messages to, and receive messages from, other cells. Thus the third function of the cell membrane is to *communicate with other cells.* As we shall see, cell membranes are composed of phospholipids, proteins, and protein–carbohydrate complexes called glycoproteins. As a general rule, *isolation is a function of the phospholipids, while regulation and communication are functions of the proteins and glycoproteins.*

Cell Membrane Structure

To understand how cell membranes can simultaneously isolate, regulate, and communicate, we must first understand their structure. According to the **fluid mosaic model,** cell membranes consist of a "sea" of phospholipids (the fluid) in which "icebergs" of protein (the mosaic) are embedded (Fig. 4-5). As you learned in Chapter 3, a phospholipid consists of two very different parts, a hydrophilic head and a pair of hydrophobic tails:

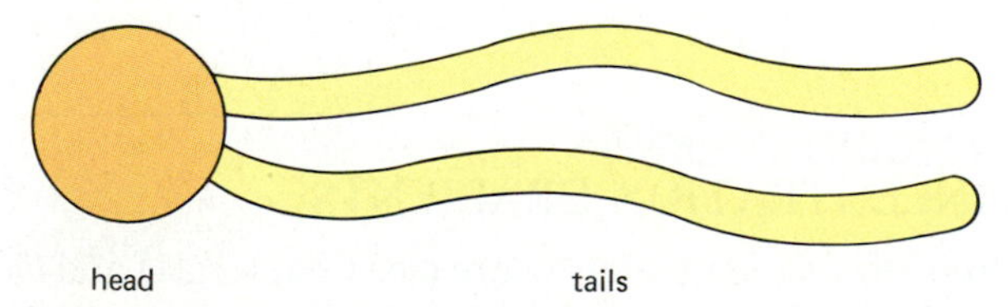

All living cells are surrounded by water, whether the pond in which an *Amoeba* spends its life or the extracellular fluid that bathes the cells of animals. The cell cytoplasm is also mostly water. Cell membranes therefore separate a watery cytoplasm from a watery external environment. Under these conditions, phospholipids spontaneously arrange themselves in a double layer called a *lipid bilayer* (Fig. 4-5). The hydrophilic heads face the cytoplasm or the extracellular fluid, while the hydrophobic tails hide inside the bilayer. Since individual phospholipid molecules are not bonded to one another, this double layer is quite fluid, with individual molecules moving about easily.

Embedded within the lipid bilayer are many proteins of different sorts (Fig. 4-5). Remember that some amino acids are hydrophilic, while others are hydrophobic. Membrane proteins are held in the lipid bilayer by interactions between hydrophobic amino acids and the hydrophobic tails of the phospholipids. Hydrophilic parts of the proteins protrude either into the cytoplasm or into the extracellular fluid. As you might expect from our analogy of protein icebergs in a lipid sea, many membrane proteins can move about within the lipid bilayer. Some membrane proteins, however, are anchored in place by a network of protein strands in the cytoplasm called the cytoskeleton (see Chapter 5).

Cells also contain internal membranes that enclose the nucleus and many other organelles. These membranes are similar in structure to the cell membrane, except that their specific combinations of lipids and proteins vary according to the function of the organelle.

Cell Membrane Function

The combination of lipid and protein in the cell membrane governs which materials can enter and leave a cell.

Substances that are soluble in oil, such as vitamins A, D, and E, can also dissolve in the lipid part of the cell membrane and can easily diffuse into a cell. Such substances are rare in the cell's environment, since lipid-soluble, hydrophobic molecules do not readily dissolve in the water of the extracellular fluid. Water-soluble materials, such as salts, amino acids, and sugars, cannot dissolve in lipids and so cannot pass through the bilayer; that is why we said that the phospholipids perform the isolation function of the membrane.

Proteins regulate the movement of most materials through the cell membrane. Some membrane proteins form "pores" that allow small water-soluble molecules to penetrate the membrane (Fig. 4-5). These proteins form a structure something like a sleeve with a lining: hydrophobic amino acids (the outer material of the sleeve) anchor the protein in the lipid bilayer while hydrophilic amino acids form the inside of the pore (the lining of the sleeve). Every cell membrane sports a large assortment of protein pores, each lined with specific amino acids that allow certain molecules to pass through. Some of these protein pores are open all the time, while others have "gates" that open and close depending on the requirements of the cell (such gated pores are essential to the electrical activity of nerve cells, as we shall see in Chapter 29). Other proteins that span the cell membrane have special sites that can grab onto specific molecules, such as glucose or sodium ions, and use cellular energy to pump these molecules into or out of the cell.

Some of the proteins that stick out into the extracellular fluid enable the cell to communicate with other cells. Many of these proteins have carbohydrates attached to their ends, forming **glycoproteins.** Some, such as the glycoproteins on red blood cells that determine blood type, serve as identification molecules, saying, in effect, "This is a red blood cell of the human being Mary Gomez." Other glycoproteins may bind hormones such as insulin and trigger a metabolic change in the cell. This allows hormone-producing cells in one part of the body to elicit responses from distant cells in other parts of the body.

TRANSPORT ACROSS CELL MEMBRANES

The Movement of Particles in Fluids

A cell consists of a fluid cytoplasm separated by a cell membrane from a fluid environment. Therefore, before we discuss how materials move across cell membranes, we should briefly consider the behavior of particles that are part of a fluid.

Particles in a fluid (a gas or liquid) are in constant motion. In general, the *movement of individual particles is random,* as they collide with one another and rebound, but the *net movement of particles may be very nonrandom.* For example, if you were to drift with the wind in a hot-air balloon, and if you could see them, the air molecules all around you would appear to be moving randomly. As you could tell by watching the ground, however, the entire mass of air would be moving along, driven by the pressure of the wind. *Unless living things intervene, particles in a fluid undergo net movement in response to gradients, that is, differences in concentration, pressure, or electrical charge.* Particles move from regions of high concentration to low concentration (e.g., sugar dissolving in tea) or from high pressure to low pressure (e.g., air flowing out of a bicycle pump when you depress the piston). Charged particles move in response to electrical gradients, toward unlike charges or away from like charges. The gradient of concentration, pressure, or electrical potential provides the energy that drives the net movement of the particles. By analogy with gravity, it is customary to refer to such movements as going "down" the gradient.

Movement Across Cell Membranes

Materials may move across cell membranes by passive or active processes (Table 4-1). In **passive transport,** substances move *down* gradients of concentration, pressure, or electrical charge. The gradient provides the energy that drives movement and controls the direction of transport across the cell membrane, either in or out. The membrane acts like a filter, with the lipids and protein pores regulating which molecules can cross, but not influencing their direction of transport.

All living cells, however, need to acquire some substances *against* gradients of concentration, pressure, or electrical charge. This is accomplished by **energy-requiring transport** processes in which chemical energy from cellular metabolism drives the movement of particles across the cell membrane. Specific transport proteins in the membrane control the direction of movement. (A helpful analogy is to consider what happens when you ride a bike. If you don't pedal, you can only go downhill; if you expend enough energy pedaling, you can go wherever you wish.)

TRANSPORT DOWN CONCENTRATION GRADIENTS

Although gradients of pressure (see Chapter 21) and electrical potential (Chapter 29) are important in regulating the movement of certain substances in living organisms, most materials that move passively across cell membranes do so in response to concentration gradients.

Diffusion

Diffusion is the net movement of particles in a fluid from regions of high concentration to regions of low concen-

Table 4-1 **Transport Across Cell Membranes**

Passive transport:	movement of substances across a cell membrane, going down a gradient of concentration, pressure, or electrical charge. Does not require the expenditure of energy by the cell.
Simple diffusion:	diffusion of water, dissolved gases, or lipid-soluble molecules through the phospholipid bilayer of a membrane.
Osmosis:	diffusion of water across a differentially permeable membrane, that is, a membrane that is more permeable to water than to dissolved solutes.
Facilitated diffusion:	diffusion of (usually water-soluble) molecules through a membrane, assisted by membrane proteins.
Energy-requiring transport:	movement of substances across a cell membrane, usually against a concentration gradient, using cellular energy.
Active transport:	movement of individual small molecules or ions through membrane-spanning proteins, using cellular energy, usually ATP.
Endocytosis:	movement of large particles, including large molecules or entire microorganisms, into a cell by a process in which the cell membrane engulfs extracellular material, forming membrane-bound sacs that enter the cytoplasm.
Exocytosis:	movement of materials out of a cell by enclosing the material in a membranous sac that moves to the cell surface, fuses with the membrane, and opens to the outside, allowing its contents to diffuse away.

tration, driven by a concentration gradient. To see how concentration gradients can cause molecules to move from one place to another, let's consider a simple example, placing a drop of dye in a glass of water (Fig. 4-6). With time, the drop will seem to become larger and paler, until eventually, even without stirring, the entire glass of water will be uniformly, faintly colored.

Why? At the border of the drop, dye molecules are moving in all directions. Those that stay within the drop do not change its composition, which was all dye to begin with, but some, simply due to random motion, go out into the water. Thus the *net movement* of dye molecules is from the region of high dye concentration (the drop) into the region of low concentration (the water). The same thing happens with water molecules. Random motion causes some to enter the drop, and the net movement of water is from the high water concentration outside the drop into the low water concentration inside the drop. To the outside observer, the dye molecules that moved beyond the original borders of the drop have made the drop

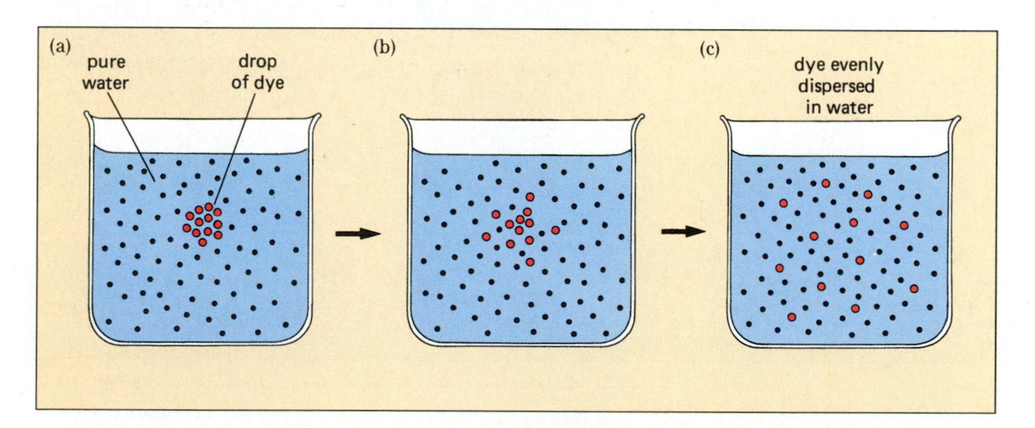

Figure 4-6 Diffusion of a dye in water. **(a)** *A drop of pure dye (red dots) is suspended in a glass of pure water (black dots in blue background).* **(b)** *Although individual molecules move at random, concentration gradients cause diffusion of dye molecules into the water, and water molecules into the drop of dye.* **(c)** *Eventually, dye and water are both evenly dispersed. Individual molecules still move, but there is no longer any concentration gradient, and therefore no further diffusion.*

grow larger; the water molecules that invaded the drop diluted the dye, making the drop paler. At first, when the drop is pure dye and the water is pure water, there is a very steep concentration gradient, and the dye diffuses rapidly. As the concentration differences lessen, the dye diffuses more and more slowly. However, as long as the concentration of dye within the expanding drop is greater than the concentration of dye in the rest of the glass, the net movement of dye will be from drop to water, until the dye becomes uniformly dispersed in the water. With no concentration gradient of either dye or water, diffusion stops.

As you can appreciate from this imaginary experiment, diffusion cannot move molecules rapidly over long distances. Although the drop of dye immediately begins to diffuse into the water, uniform dispersion may take many minutes. As we described earlier, the slow rate of diffusion over long distances is one of the reasons why cells have to be small.

In summary, then, (1) *diffusion requires a difference in concentration;* (2) *the net movement of particles is down a gradient from high to low concentration;* (3) *the rate of diffusion depends on the gradient: the steeper the gradient, the faster the rate of diffusion;* (4) *diffusion cannot rapidly move molecules over long distances;* but (5) *if no other processes intervene, diffusion will continue until the concentration gradient is eliminated.*

DIFFUSION ACROSS CELL MEMBRANES. Many molecules cross cell membranes by diffusion, in response to concentration differences between the cytoplasm and the extracellular fluid. However, cell membranes are **differentially permeable;** that is, they allow some molecules to pass through, or **permeate,** more easily than others. Water, dissolved gases such as oxygen and carbon dioxide, and lipid-soluble molecules such as vitamin A easily diffuse across the lipid bilayer. This process is called **simple diffusion** (Fig. 4-7a).

In contrast, most water-soluble molecules cannot cross the bilayer. These molecules can diffuse across only with

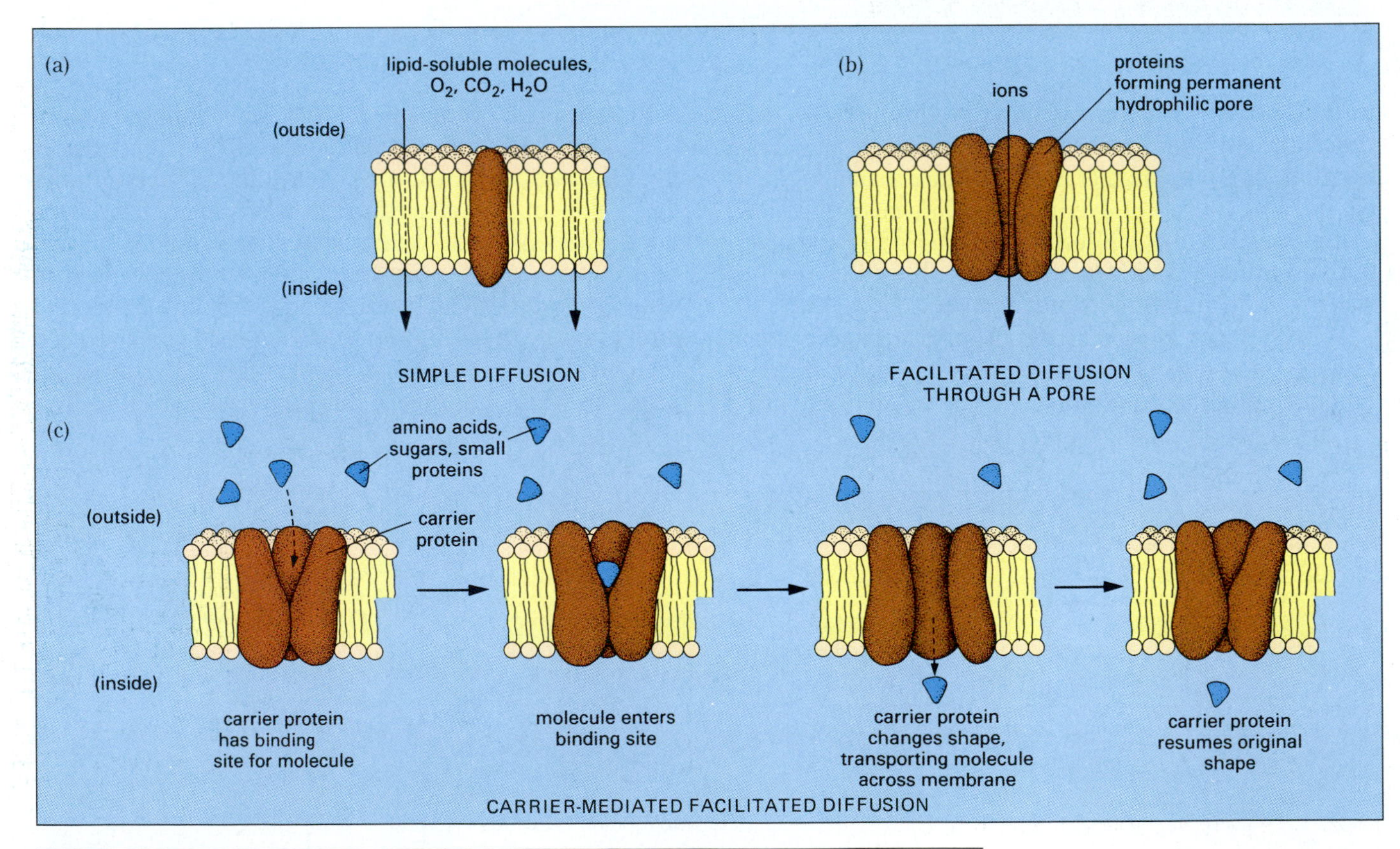

Figure 4-7 *The cell membrane regulates diffusion into and out of the cell.*
(a) *Simple diffusion: Water, gases such as oxygen and carbon dioxide, and lipid-soluble molecules can diffuse through the lipids.*
(b) *Facilitated diffusion through pores: Water-soluble molecules cannot pass through the lipid bilayer. Protein pores allow some water-soluble molecules to penetrate the cell.*
(c) *Carrier-mediated facilitated diffusion: Carrier proteins may bind specific molecules and, as a result, change their shape, passing the molecule through the middle of the protein to the other side of the membrane.*

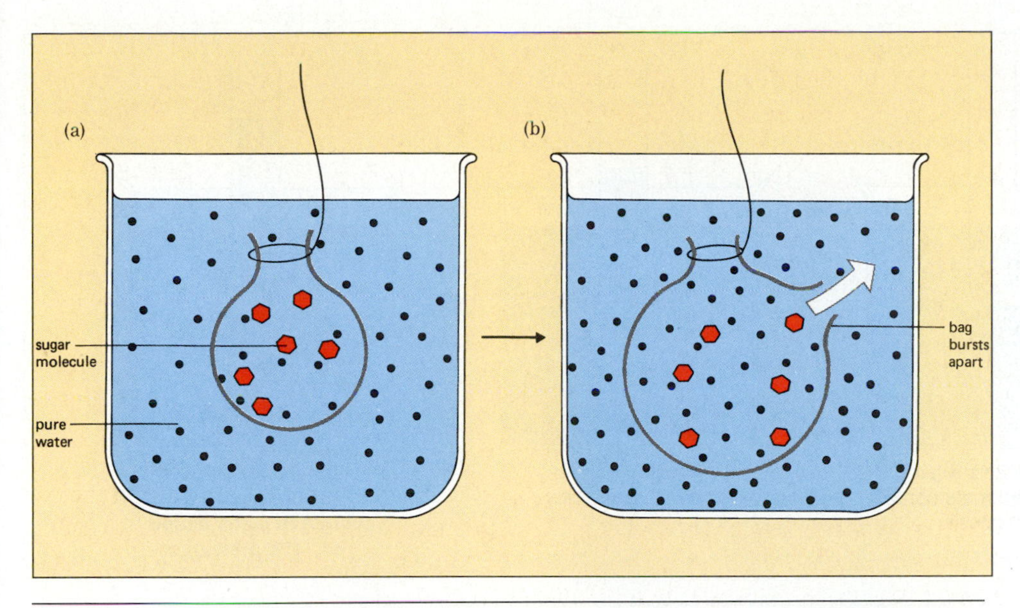

Figure 4-8 Osmosis of water through a differentially permeable membrane. The membrane is permeable to water (black dots) but not to larger molecules such as sugar (red hexagons). **(a)** *If a bag made of such a membrane is filled with a sugar solution and suspended in pure water, water will diffuse down its concentration gradient from the high concentration of pure water outside the bag to the lower concentration of water inside. The bag swells up as water enters.* **(b)** *If the bag is weak enough, the increasing water pressure will cause it to burst.*

the aid of membrane-spanning proteins. This process is called **facilitated diffusion** (Fig. 4-7b). In some cases of facilitated diffusion, membrane proteins form pores with hydrophilic linings. In effect, then, the lipid bilayer is perforated with holes through which certain water-soluble molecules can diffuse. Many ions, such as potassium and sodium ions, diffuse through protein pores. For other water-soluble molecules, especially larger ones, membrane-spanning proteins act as **carriers** (Fig. 4-7c). These proteins have groups of amino acids that bind specific molecules in the cytoplasm or extracellular fluid. Binding triggers a change in the shape of the protein carrier, opening a temporary channel in the middle of the protein, through which the molecule passes across the membrane. Cell membranes contain relatively few pores or carriers through which any given molecule can pass. Therefore, molecules that cross the membrane by facilitated diffusion usually do so more slowly than those that can cross by simple diffusion through the lipid bilayer.

It is important to remember that molecules moving by both simple and facilitated diffusion can go in either direction across the membrane, depending on the concentration gradient. For example, the concentration of oxygen in the water of a rushing mountain brook is very high. A protist drifting in the stream uses up oxygen in its metabolism, so the oxygen concentration in its cytoplasm is extremely low. Therefore, oxygen diffuses rapidly into the cell down a steep concentration gradient. Just the opposite happens with carbon dioxide. The protist produces CO_2, raising the concentration in its cytoplasm, while the stream water has very little CO_2. The concentration gradient for CO_2 promotes rapid diffusion out of the cell.

Osmosis

Water, like any other molecule, moves by diffusion from high water concentrations to low water concentrations. However, the diffusion of water across differentially permeable membranes has such dramatic consequences that it has been given a special name: **osmosis.** Let's investigate osmosis in another thought experiment.

A very simple kind of differentially permeable membrane consists of an impervious sheet perforated with tiny pores. The pores allow water molecules to pass through, but not larger molecules such as sugar. Suppose that we make a bag out of such a membrane, fill it with a sugar solution, tie off the top, and place the bag in a glass of pure water. The bag will rapidly swell up, and if it is weak enough, it will burst (Fig. 4-8).

If you could visualize individual molecules, you would see that the concentration of water molecules is lower in the sugar solution than in the pure water outside the bag. In fact, the sugar dilutes the water! (You can prove this for yourself by measuring out a liter of water and dissolving a kilogram of sugar in it: the volume of the resulting solution will be greater than 1 liter.) Even more important, hydrogen bonds form between water molecules and the polar hydroxyl groups on the sugar mole-

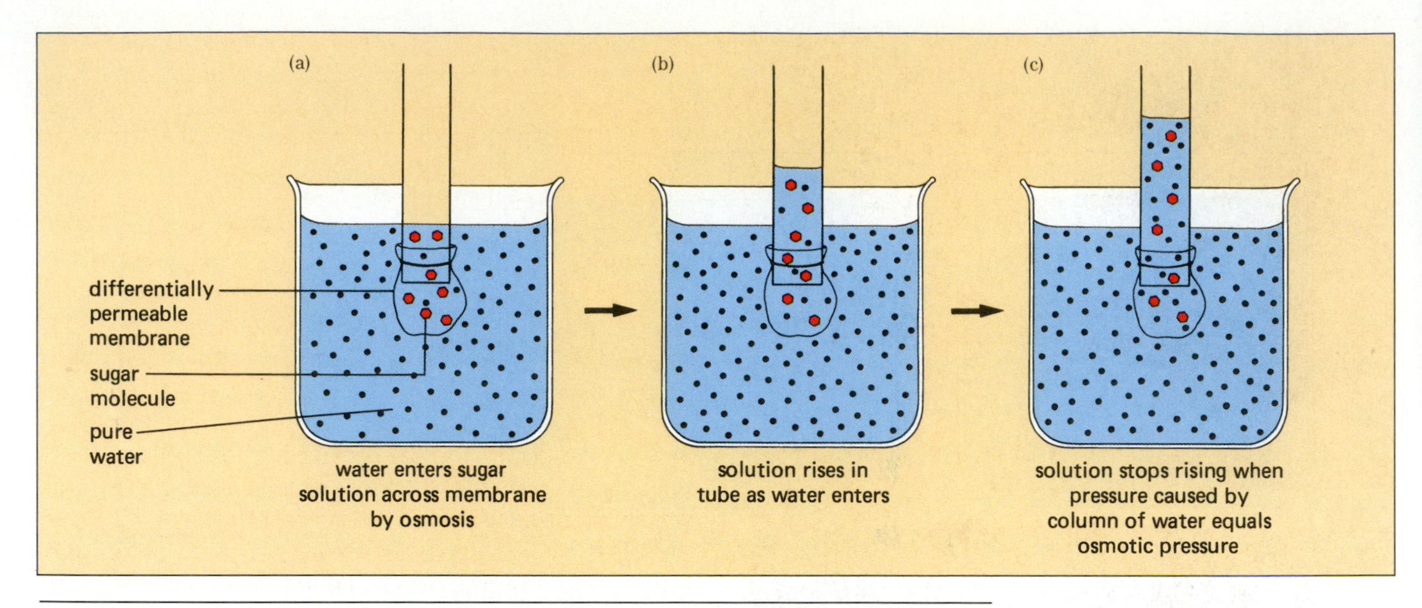

Figure 4-9 Osmotic pressure. **(a)** *A membrane permeable to water (black dots) but not sugar (red hexagons) is fastened across the bottom of a glass tube. A sugar solution is placed in the tube and lowered into a glass of pure water until the level of the sugar solution is even with the surface of the water.* **(b)** *Water moves by osmosis across the membrane into the tube. As a result, the solution rises within the tube.* **(c)** *When the pressure of the column of solution in the tube forces water out through the membrane as fast as osmosis causes water to enter, the column stops rising. The resulting hydrostatic pressure equals the osmotic pressure of the solution.*

cules. These water molecules tend to remain clustered around the sugars and are not free to diffuse through the pores in the membrane. Therefore, the concentration of "free" water molecules is lower inside the bag than in the pure water outside.

Since sugar molecules cannot fit through the pores, whereas free water molecules can, only water can move down its concentration gradient from the pure water outside to the sugar solution inside the bag. The bag swells up as water enters. The sugar cannot escape at all, so the water concentration inside the bag is always lower than in the pure water outside. Water continues to enter the bag until it bursts.

Now let's redo our experiment, but with one change. Suppose that we tie a piece of the same differentially permeable membrane across the end of a glass tube, suspend the tube in pure water, and put the sugar solution in the tube (Fig. 4-9). Water will diffuse across the membrane from the pure water into the solution in the tube. Since the upper end of the tube is open, as the water enters, it will raise the level of the solution in the tube. Eventually, the level will stop rising. Why? As the solution in the tube rises above the level of water in the glass, its weight applies a pressure to the membrane across the bottom of the tube. Therefore, two opposing gradients are set up: a concentration gradient moving water *into* the tube, and a pressure gradient pushing water *out of* the tube. When the two gradients are equal, no further net movement of water will occur across the membrane. The lower the concentration of free water molecules in the tube (i.e., the higher the sugar concentration), the greater will be the tendency for water to move across the membrane and into the tube. The physical pressure that exactly balances the osmosis of water due to the concentration difference between a solution and pure water is defined as the **osmotic pressure** of the solution. Osmotic pressure can be used to measure of the concentration of free water molecules in the solution.

As we have emphasized, osmosis is driven by *water concentration differences between solutions.* Therefore, osmosis does not depend on the *type* of dissolved molecules in a solution, only on their *concentration.* We can produce the same osmotic pressure by adding sucrose, glucose, amino acids, sodium ions, or a mixture of all of these to a solution, as long as the concentration of all types of nonpermeable particles stays the same.

Osmosis across cell membranes. Most cell membranes are highly permeable to water. Since all cells contain appreciable amounts of dissolved salts, proteins, sugars, and so on, the flow of water across the cell membrane depends on the concentration of water in the liquid that bathes the cells. The extracellular fluids of animals are usually **isotonic** ("having the same strength") to the insides of the body cells: that is, the concentrations of water and dissolved particles outside the cells are the

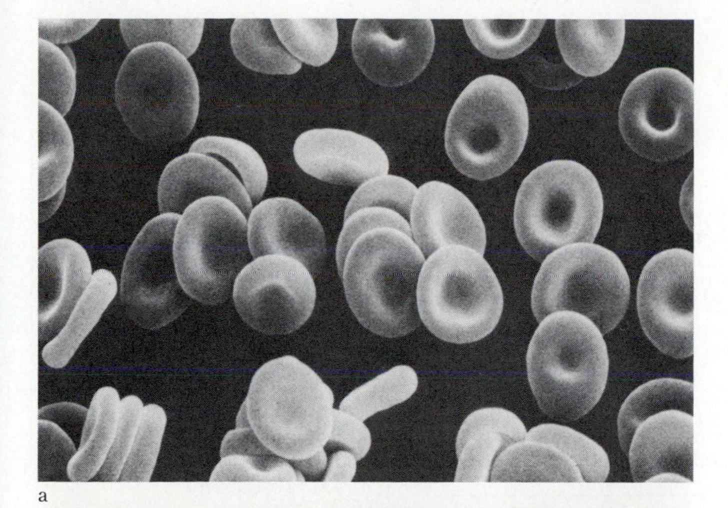
a

b

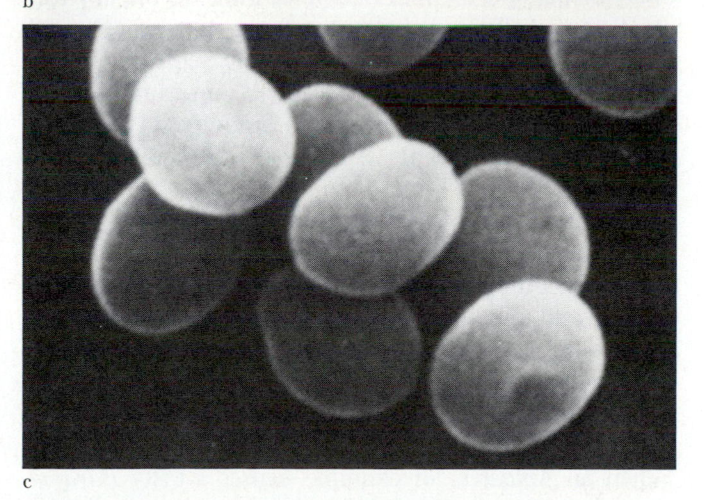
c

Figure 4-10 The effects of osmosis on animal cells. Red blood cells are normally suspended in the fluid environment of the blood and lack the ability to regulate water flow across their cell membranes. **(a)** *If red blood cells are immersed in an isotonic salt solution, which has the same concentration of dissolved substances as the blood, there is no net movement of water across the cell membrane, and the red blood cells will keep their characteristic dimpled disk shape.* **(b)** *A hypertonic solution, with too much salt, will cause water to leave the cells, shrivelling them up.* **(c)** *A hypotonic solution, without enough salt, will cause water entry, and the cells will swell.*

same as those inside, so there is no tendency for water either to enter or leave the cells. Note that the *types* of dissolved particles are seldom the same inside and outside the cells, but the *total concentration* of all particles is equal.

If cells, say red blood cells, are taken out of the body and immersed in salt solutions of varying concentrations, the effects of the differential permeability of the cell membrane to water and dissolved particles become dramatically apparent (Fig. 4-10). If the solution has a higher salt concentration than the cytoplasm (i.e., if the solution has a lower water concentration), water will leave the cells by osmosis. The cells will shrivel up until the concentrations of water inside and outside become equal. Such a solution is called **hypertonic** ("having greater strength"). Conversely, if the solution has little or no salt, water will enter the cells, causing them to swell up. If the solution has little enough salt, the cells will burst. Solutions that cause water to enter cells by osmosis are called **hypotonic** ("having lesser strength").

These are not merely academic considerations. At least some cells of all organisms are in contact, not with well-regulated body fluids, but with the external environment. In many cases the environment is not isotonic to the cells. Freshwater protozoans, for example, constantly battle the tendency for water to move into their cells. Although the cell membrane of *Paramecium,* for example, is only about 1 percent as permeable to water as red blood cell membranes are, water does enter. Consequently, paramecia must expend energy pumping water back out again (see Fig. 5-12 in the next chapter). Although there is just as large a gradient for water to enter a freshwater algal cell, the alga has the advantage of a cell wall surrounding it. As water enters the algal cell, pressure builds up but is withstood by the cell wall. Soon, the pressure within the cell forces water out as fast as it diffuses in, so the alga does not have to spend energy in actively expelling water.

ENERGY-REQUIRING TRANSPORT ACROSS CELL MEMBRANES

All cells need to move some materials "uphill" across their cell membranes, against diffusion gradients. For example, every cell requires some nutrients that are extremely scarce in its environment. Since these nutrients may be less concentrated in the environment than in the cell cytoplasm, the cell would lose, not gain, these nutrients by diffusion. Other substances, such as sodium and calcium ions in your brain cells, must be maintained at much lower concentrations inside the cells than in the extracellular fluid. When these ions diffuse into the cells, they must be pumped out again against their concentration gradients. Finally, many cells acquire or expel some

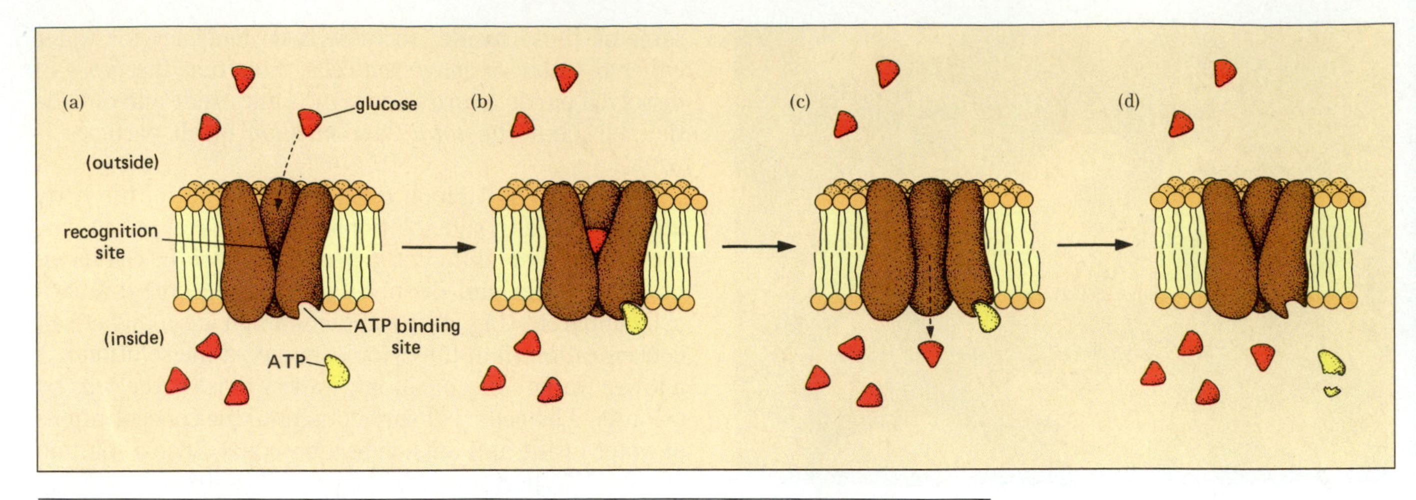

Figure 4-11 Active transport uses cellular energy to move molecules across the cell membrane, often against a concentration gradient. **(a)** *A transport protein (brown) has a recognition site for the molecule to be transported (glucose in this example) and an ATP binding site.* **(b)** *The transport protein binds ATP and glucose.* **(c)** *Energy from ATP changes the shape of the transport protein and moves the glucose molecule across the membrane.* **(d)** *The carrier releases the glucose and the remnants of the ATP, and resumes its original configuration.*

substances, such as whole bacteria or large proteins, that are too large to diffuse across a membrane regardless of concentration gradients. Cells have evolved several processes that use cellular energy to move materials into or out of the cell.

Active Transport

In **active transport,** specific membrane proteins use cellular energy to move individual molecules across the cell membrane, usually against their concentration gradients (Fig. 4-11). Active transport proteins span the membrane and have two specific groupings of amino acids, called **active sites.** One active site recognizes a particular molecule, say a sugar, and binds it. The second site binds an energy-carrier molecule, usually adenosine triphosphate (ATP; see Chapter 3 for the structure of ATP and Chapter 6 for a description of energy use in cells). The ATP donates energy to the protein, causing it to change shape and move the sugar molecule across the membrane.

Endocytosis and Exocytosis

Cells can also acquire particles, especially large proteins or entire microorganisms such as bacteria, by a process called **endocytosis** (Greek for "into the cell"). During endocytosis, the cell membrane engulfs the particle and pinches off a membranous sac called a **vesicle,** with the particle inside, into the cytoplasm (Figs. 4-12 and 4-13). Two types of endocytosis can be distinguished, based on the size of the particle acquired and the method of acquisition.

In **pinocytosis** ("cell drinking"), a very small patch of membrane dimples inward and buds off into the cytoplasm as a tiny vesicle (Fig. 4-12). As you can see by its name, biologists formerly thought that pinocytosis is merely the acquisition of fluid from outside the cell. This is usually not the case. Most often, pinocytosis is a method of acquiring specific nutrients from the external environment, especially molecules too large to pass through membrane pores. How does the cell do this? Certain membrane proteins, called **receptors,** protrude into the extracellular fluid, each bearing a binding site for a particular nutrient molecule. If the right molecule contacts a receptor protein, it attaches to the binding site. The protein/nutrient complex then moves sideways within the lipid bilayer to special places where contractile proteins pull the membrane inward, creating dimples. The dimples deepen and pinch off vesicles into the cytoplasm. The vesicles carry both the bound molecules and some of the extracellular fluid into the cell.

Phagocytosis ("cell eating") is used to pick up large particles, including whole microorganisms (Fig. 4-13). When an *Amoeba,* for example, senses a tasty *Paramecium,* it produces extensions of its surface membrane, called **pseudopodia** (Latin for "false foot"; sing., **pseudopod**). The pseudopodia surround the luckless *Paramecium,* their ends fuse, and the prey is carried into the interior of the *Amoeba* for digestion.

The reverse of endocytosis, called **exocytosis** (Greek for "out of the cell"), is often used by cells to dispose of unwanted materials, such as the waste products of digestion, or to secrete materials, such as hormones, into the extracellular fluid (Fig. 4-14). During exocytosis, a vesicle

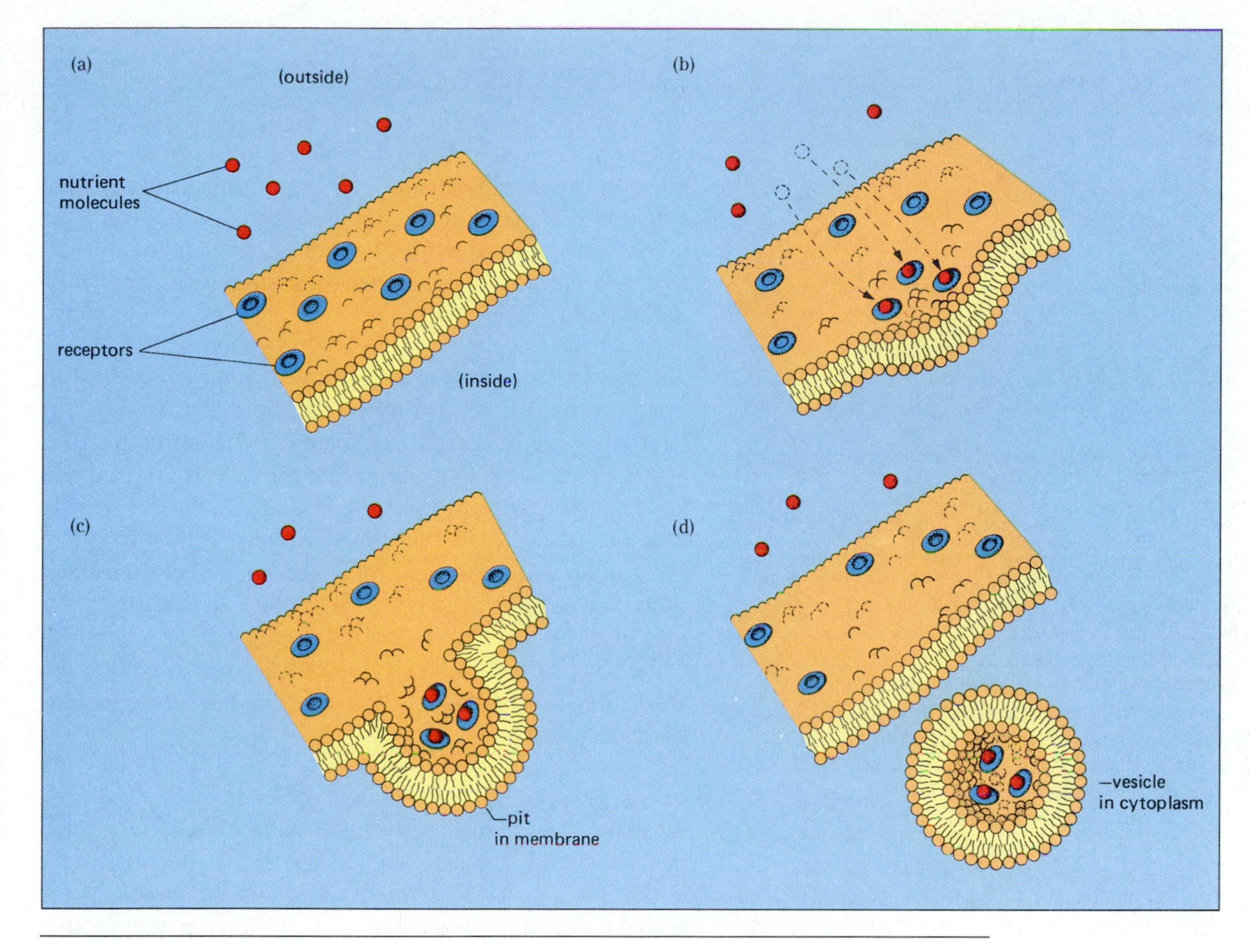

Figure 4-12 *A model for receptor-mediated pinocytosis.* **(a)** *The cell membrane bears receptor proteins that can bind extracellular molecules (e.g., nutrients).* **(b)** *The receptor proteins bind the nutrients, and migrate along the fluid lipid bilayer of the membrane to dimpling sites.* **(c)** *The membrane dimples inward to form a pit, carrying the receptor/captured molecule complexes with it.* **(d)** *The end of the membrane pit buds off a vesicle into the cytoplasm of the cell.*

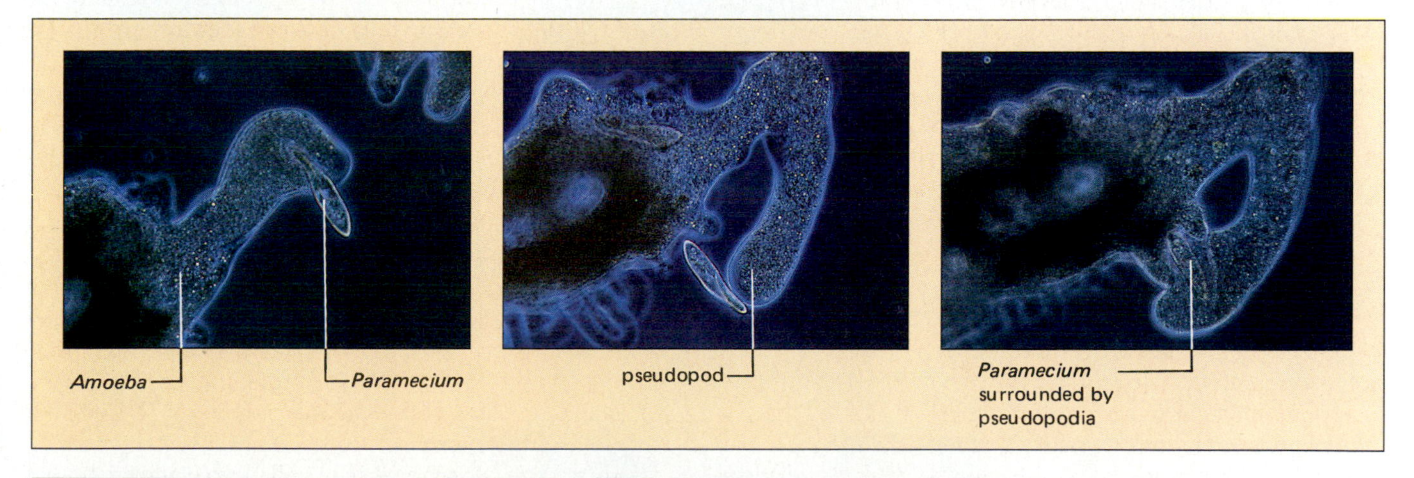

Figure 4-13 *Phagocytosis. An* Amoeba *senses a* Paramecium *nearby and sends out a pseudopod that encircles the hapless prey and rejoins the main cell body, trapping the* Paramecium. *As in pinocytosis, the prey enters the* Amoeba *enclosed within a vesicle made of cell membrane. Fusion of the food vesicle with a digestive organelle spells the end of the* Paramecium *but continued life for the* Amoeba.

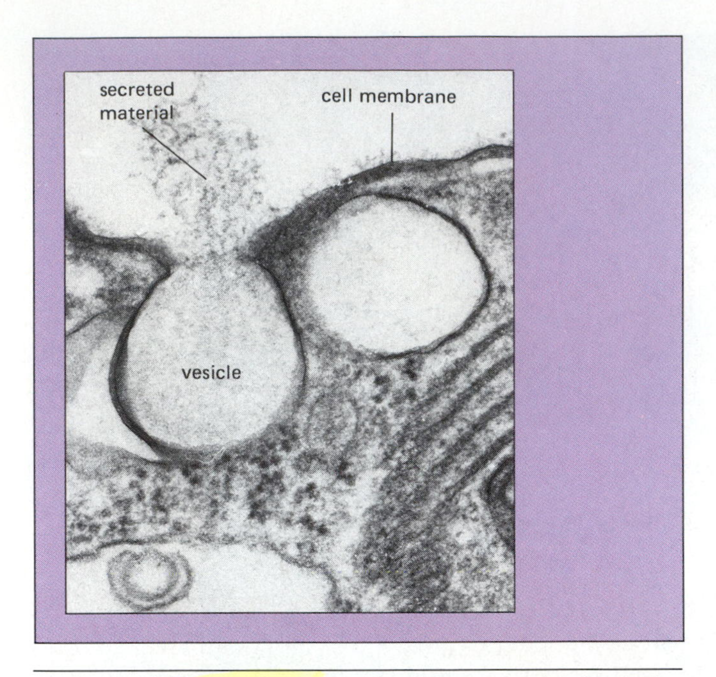

Figure 4-14 Exocytosis operates by the reverse of endocytosis, as the material to be ejected from the cell is encapsulated into a membrane-bound vesicle which moves to the cell membrane and fuses with it. As the vesicle opens to the outside, the material within leaves by diffusion.

created by organelles within the cell (see Chapter 5) moves to the cell surface, where the membrane of the vesicle fuses with the cell membrane. The vesicle opens to the extracellular fluid and its contents diffuse out.

CELL CONNECTIONS AND COMMUNICATION

In multicellular organisms, cell membranes also function in holding clusters of cells together and in providing avenues through which cells can communicate with their neighbors. Depending on the organism and the cell type, one of four types of connections may occur between cells: desmosomes, tight junctions, gap junctions, and plasmodesmata.

ADHESION. Animals, as you know, tend to be flexible, mobile organisms. Many of an animal's tissues are stretched, compressed, and bent as the animal moves about. If the skin, intestines, stomach, urinary bladder, and other organs are not to tear apart under the stresses of movement, their cells must adhere firmly together. Such animal tissues have junctions called **desmosomes** that hold adjacent cells together (Fig. 4-15). In a desmosome, the membranes of adjacent cells are glued together by proteins and carbohydrates. Protein strands extend into the interior of each cell, further strengthening the attachment.

LEAK-PROOFING. The body contains many tubes or sacs that must hold their contents without leaking: a leaky urinary bladder would spell disaster for the rest of the body. The spaces between the cells lining such sacs are sealed with **tight junctions** (Fig. 4-16). The membranes of adjacent cells nearly fuse along a series of ridges, effectively forming waterproof gaskets between cells. Con-

Figure 4-15 Cells lining the small intestine are firmly attached to one another by desmosomes.

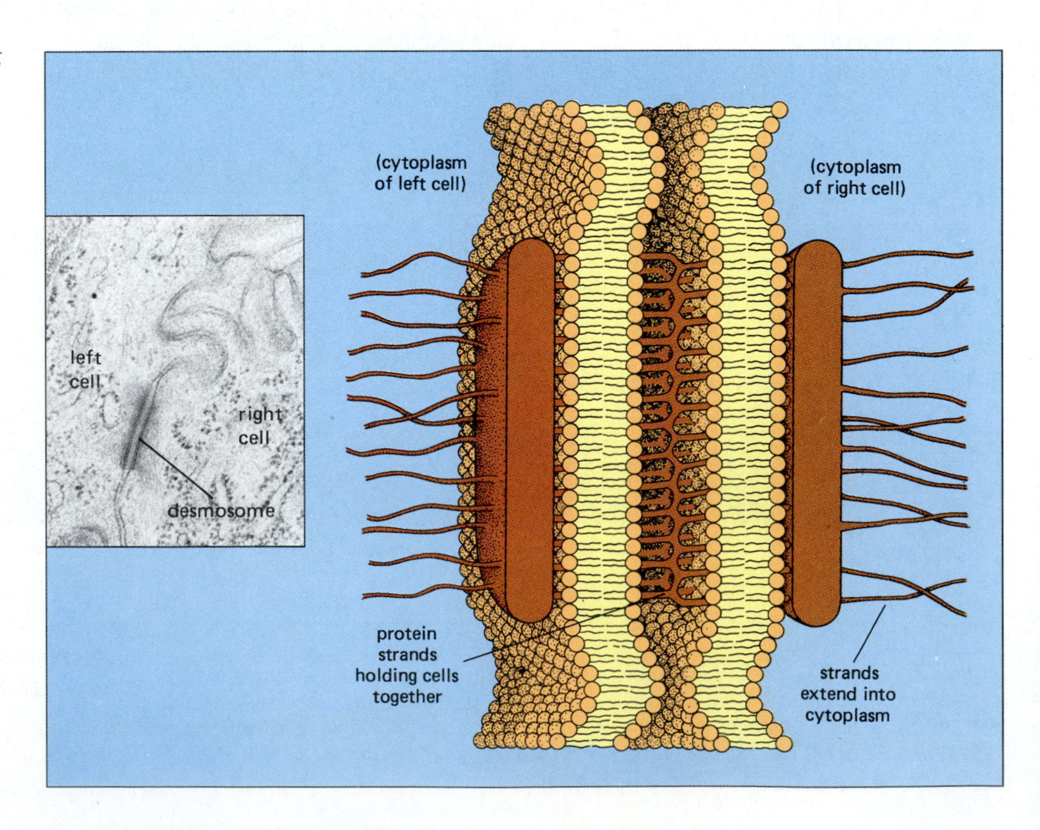

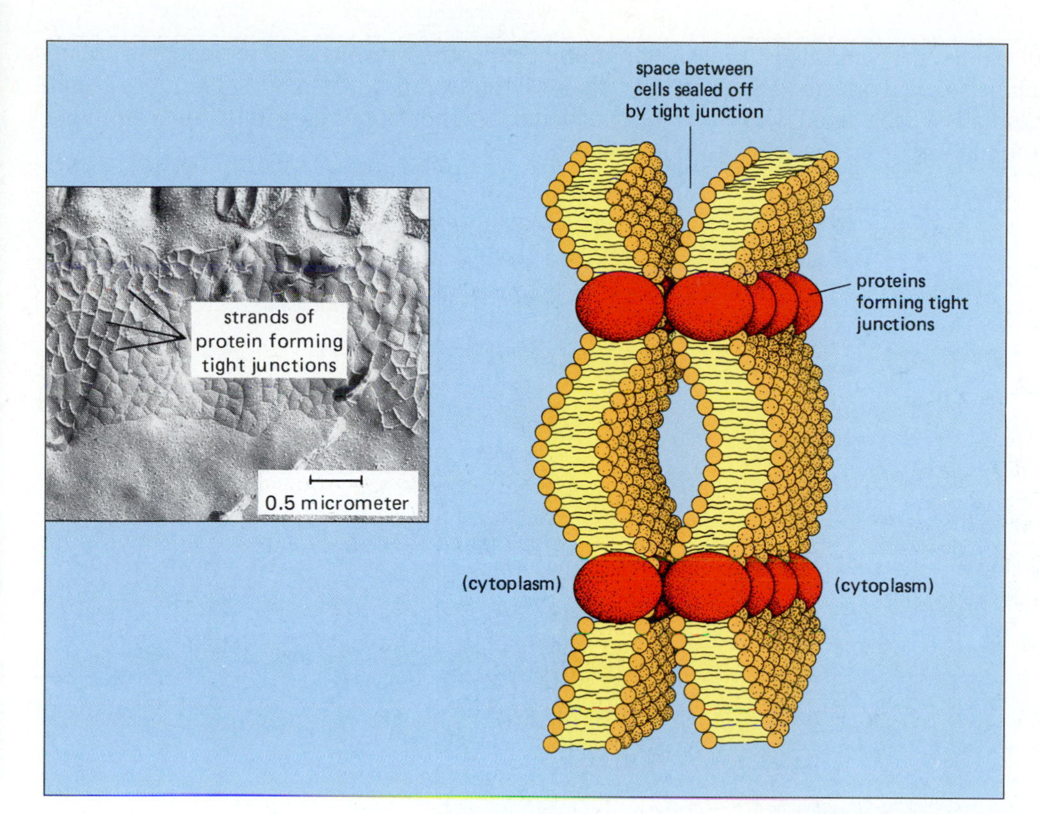

Figure 4-16 *Leakage between cells is prevented by close-fitting tight junctions.*

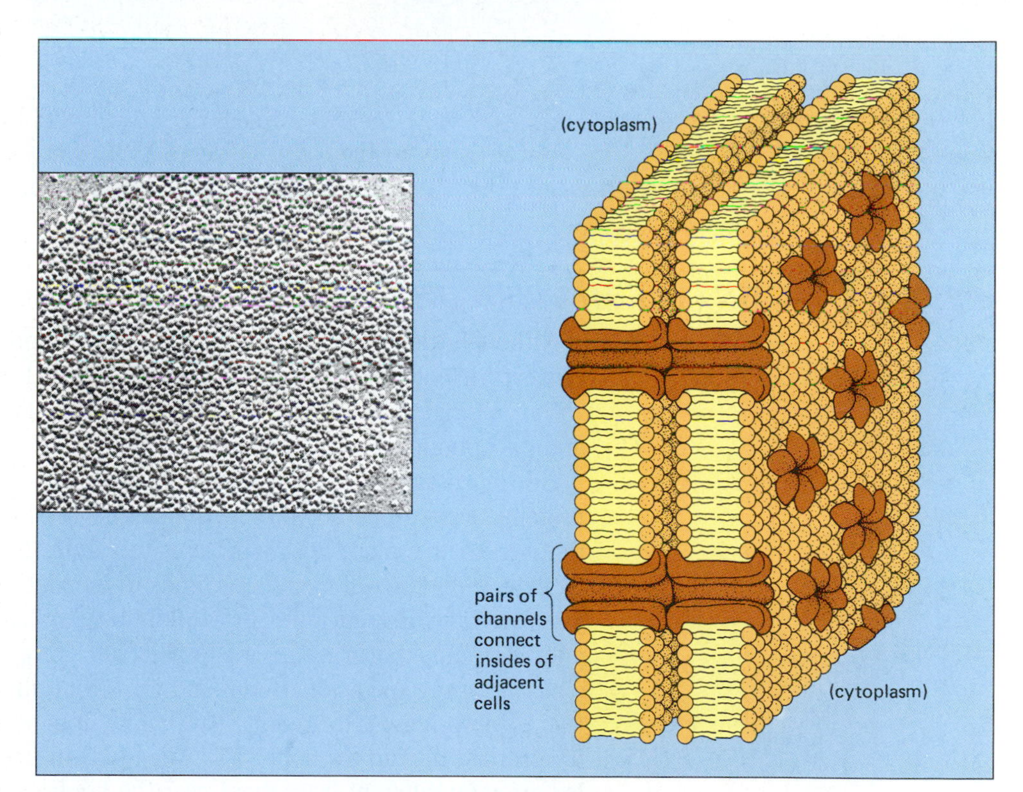

Figure 4-17 *Gap junctions contain cell-to-cell channels that interconnect the cytoplasm of adjacent cells.*

tinuous tight junctions sealing each cell to its neighbors keep molecules from escaping between cells.

CELL-TO-CELL COMMUNICATION. Multicellular organisms must coordinate the actions of their component cells. Many animal cells, including some brain cells, most gland cells, and every cell of very young embryos, communicate through protein channels directly connecting the insides of adjacent cells (Fig. 4-17). These cell-to-cell channels are clustered in specialized regions called **gap junctions.** Hormones, nutrients, ions, and even electrical signals can pass through the channels at gap junctions.

Virtually all the living cells of plants are connected to one another by much larger channels called **plasmodesmata** (Fig. 4-18). Each plasmodesma is a tube, lined with cell membrane, that penetrates the cell wall from the cytoplasm of one cell to the cytoplasm of its neighbor. Many plant cells have thousands of plasmodesmata. As a result, water, nutrients, and hormones pass quite freely from one cell to another.

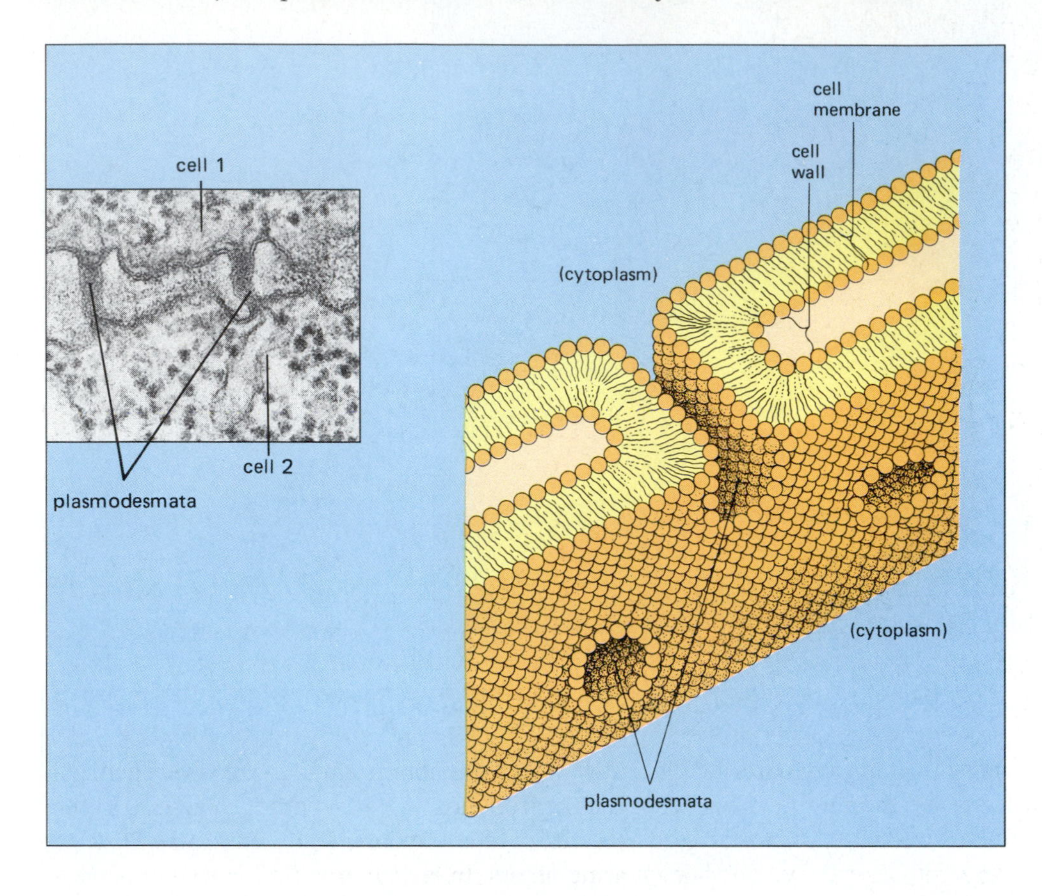

Figure 4-18 *Plant cells are widely interconnected by rather large cell-to-cell pores called plasmodesmata.*

REFLECTIONS ON CELL MEMBRANES

Although the membranes of all cells have a similar structure, there is tremendous diversity in membrane function from organism to organism, and from cell to cell within a single organism. This diversity arises largely from the different proteins and phospholipids in the membrane.

Our discussion of membranes emphasized the unique functions of the membrane proteins. Consequently, you may think that the lipids are just a waterproof matrix that serves as a place for the proteins to reside. This isn't quite true, as we can see by examining the cell membrane lipids in the legs of caribou (Fig. 4-19). During the long arctic winters, temperatures plummet far below freezing. Keeping their legs and feet really warm would waste precious energy, and caribou have evolved specialized arrangements of arteries and veins that allow the temperature of their lower legs to drop almost to freezing (0°C). The upper legs and main trunk of the body, in contrast, remain at about 37°C. The lipids in the membranes of cells in the upper leg are very different from those near the hoofs. Why?

Remember, the membrane of a cell needs to be somewhat fluid. For example, protein receptors that gather materials from the extracellular fluid must move through the lipids to sites of pinocytosis. Now the fluidity of a membrane is a function of the fatty acid tails of its phospholipids: unsaturated fatty acids remain fluid at lower temperatures than saturated ones (Chapter 3). Caribou legs have a gradient of lipid types in their cell membranes. The membranes of cells near the

Figure 4-19 *A caribou browses on the Alaskan tundra. The lipid composition of the membranes in the cells of its legs varies with distance from the trunk. Unsaturated phospholipids predominate near the hoof, while a higher proportion of saturated lipids is found near the trunk.*

chilly hoof have lots of unsaturated fatty acids, while the membranes of cells near the warmer trunk have more saturated ones. This arrangement gives the cell membranes throughout the leg the proper fluidity despite great differences in temperature.

As important as the phospholipids are, the membrane proteins probably play the major roles in determining cell function and in governing the interactions between a cell and its neighbors. Every nerve cell in your body, for instance, has membrane proteins essential for producing electrical signals and conducting them along the nerves to various parts of the body. Other membrane proteins receive chemical messages from neighboring nerve cells or from hormones and other chemicals in the blood. Each cell in the brain has a specific set of membrane proteins, allowing it to respond to some stimuli while ignoring others. In fact, your ability to read this page depends on the proteins residing in the membranes of your brain cells.

As we progress through this book, we shall return many times to the concepts of membrane structure presented in this chapter. Diversity in membrane lipids and proteins is the key to understanding not only the isolated cell, but entire organs, which function as they do largely because of the properties of the membranes of their component cells.

SUMMARY OF KEY CONCEPTS

Cell Functions and Limitations

The essentials of modern cell theory can be summarized in three statements: (1) all living organisms are made up of cells; (2) cells are the functional units of living organisms; and (3) all cells arise from pre-existing cells. Cells consist of at least three components: a cell membrane that separates the cell contents from the environment; cytoplasm, the material within the membrane; and genetic material, DNA.

Cell Walls

Moneran, plant, fungal, and some protist cells are surrounded by a rigid cell wall outside the cell membrane. The cell wall protects and supports the cell.

Cell Membranes

The cell membrane has three major functions: (1) isolate the cytoplasm from the external environment; (2) regulate the flow of materials into and out of the cell: and (3) communicate with other cells. The membrane consists of a bilayer of phospholipids in which a variety of proteins is embedded. Molecules that are soluble in oil can pass through the lipid portion of the membrane. Most water-soluble molecules pass through the membrane only with the assistance of proteins.

Transport Across Cell Membranes

Unless living things intervene, particles in a fluid move in response to gradients of concentration, pressure, or electrical charge. In cells, passive transport is the movement of substances across the cell membrane down gradients, usually of concentration. Cellular energy is not used. Energy-requiring transport processes move substances across the membrane, usually against concentration gradients, using cellular energy.

Transport Down Concentration Gradients

Diffusion is the movement of particles from regions of higher concentration to regions of lower concentration. Many substances diffuse across cell membranes. In simple diffusion, water, dissolved gases, and lipid-soluble molecules diffuse through the lipid bilayer. In facilitated diffusion, water-soluble molecules cross the membrane through protein pores or with the assistance of protein carriers. In both cases, molecules move down their concentration gradients, and cellular energy is not required.

Osmosis is the diffusion of water across a differentially permeable membrane, down its concentration gradient. Again, cellular energy is not used.

Energy-Requiring Transport Across Cell Membranes

In active transport, protein carriers in the membrane use cellular energy to drive the movement of molecules across the cell membrane, usually against concentration gradients. Large molecules (e.g., proteins), particles of food, and microorganisms may be acquired by endocytosis, in which the cell membrane surrounds small volumes of extracellular fluid containing the desired particles and forms a vesicle that pinches off the inside of the membrane and enters the cytoplasm.

The secretion of substances such as hormones from the cell is accomplished by exocytosis. The molecules to be voided are enclosed in a vesicle that moves to the cell surface, fuses with the membrane, and empties its contents outside the cell.

Cell Connections and Communication

Cells may be connected to one another by a variety of junctions. Desmosomes attach cells firmly to one another, preventing tearing of a tissue during movement or stress. Tight junctions seal off the spaces between adjacent cells, leak-proofing organs such as the urinary bladder. Gap junctions in animals and plasmodesmata in plants are locations in which the cytoplasm of two adjacent cells is interconnected by pores through adjoining cell membranes.

GLOSSARY

Active site a group of amino acids on a protein that binds specific molecules and/or performs specific functions, such as catalyzing a chemical reaction.

Active transport the movement of molecules across a cell membrane through the use of cellular energy, usually against a concentration gradient.

Carrier a protein in a cell membrane that binds specific molecules and facilitates their transport across the membrane.

Cell the basic unit of life, consisting of an outer cell membrane surrounding cytoplasm and genetic material.

Cell membrane the outer membrane of a cell, composed of a bilayer of phospholipids in which proteins are embedded.

Cell theory a theory stating that all living things are composed of cells, cells are the functional units of living things, and all cells arise from preexisting cells.

Cell wall a layer of material, usually made up of cellulose or cellulose-like materials, found outside the cell membrane of plants, fungi, bacteria, and plantlike protists.

Concentration gradient a difference in concentration of a substance between two parts of a system or across a barrier such as a cell membrane.

Cytoplasm (sī′-tō-plazm) the material contained within the cell membrane but outside the nucleus.

Desmosome (dez′-mō-sōm) a strong cell-to-cell junction that functions in attaching cells to one another.

Differential permeability the property of a membrane by which some substances can permeate more readily than other substances.

Diffusion the net movement of particles from a region of high concentration to a region of low concentration, driven by the concentration gradient.

Endocytosis (en-dō-sī-tō′-sis) the movement of material into a cell by a process in which the cell membrane engulfs extracellular material, forming membrane-bound sacs that enter the cell interior.

Exocytosis (ex-ō-sī-tō′-sis) the movement of material out of a cell by a process whereby intracellular material is enclosed within a membrane-bound sac that moves to the cell membrane and fuses with it, releasing the material outside the cell.

Facilitated diffusion diffusion of molecules across a cell membrane, assisted by protein pores or carriers embedded in the cell membrane.

Fluid mosaic a model of cell membranes. According to this model, the membrane is composed of a double layer of phospholipids in which a variety of proteins is embedded. The lipid bilayer is a somewhat fluid matrix that allows movement of proteins within it.

Gap junction a type of cell-to-cell junction in animals, in which channels connect the interiors of adjacent cells.

Glycoprotein a protein to which a carbohydrate is attached.

Hypertonic (hī-per-ton′-ik) referring to a solution that has a higher concentration of dissolved particles (and therefore a lower water concentration) than a cell.

Hypotonic (hī-pō-ton′-ik) referring to a solution that has a lower concentration of dissolved particles (and therefore a higher water concentration) than a cell.

Isotonic (ī-sō-ton′-ik) referring to a solution that has the same concentration of dissolved particles (and therefore the same water concentration) as a cell.

Osmosis (oz-mō′-sis) the diffusion of water across a differentially permeable membrane, usually down a concentration gradient of free water molecules. Water moves into the solution that has a lower water concentration from the solution with the higher water concentration.

Osmotic pressure a measure of the tendency of water to move from a solution with a lower concentration of water into one with a higher concentration of water; the physical pressure which must be applied to a solution to prevent water movement into it from pure water.

Passive transport movement of materials across a cell membrane down a gradient of concentration, pressure, or electrical charge, and not using cellular energy.

Permeate (per′-mē-āt) to pass through, as through pores in a membrane.

Phagocytosis (fā-gō-sī-tō′-sis) a type of endocytosis in which extensions of a cell membrane engulf extracellular particles and transport them into the interior of the cell.

Pinocytosis (pī-nō-sī-tō′-sis) a type of endocytosis in which part of the cell membrane, when contacted by appropriate extracellular substances, forms a vesicle that enters the cytoplasm, carrying materials into the cell.

Plasmodesma (plaz-mō-dez′-ma; pl. plasmodesmata) a cell-to-cell junction in plants that connects the interiors of adjacent cells.

Pseudopod (sū′-dō-pod) a temporary extension of the cell membrane used for locomotion or phagocytosis in certain cells such as the protist *Amoeba* or white blood cells of vertebrates.

Receptor a protein or glycoprotein, located on a membrane or in the cytoplasm of a cell, that recognizes and binds to specific molecules. Binding by receptors often triggers a response by a cell, such as pinocytosis, protein synthesis, or exocytosis.

Simple diffusion diffusion of water, dissolved gases, or lipid-soluble molecules through the phospholipid bilayer of a cell membrane.

Tight junction a type of cell-to-cell junction in animals that prevents the movement of materials through the spaces between cells.

Vesicle (ves′-i-kul) a membrane-bound sac, usually found within a cell.

STUDY QUESTIONS

1. List and explain the principles of cell theory.
2. Describe the physical constraints that prohibit the development of very large cells.
3. Name two chemical compounds in cell walls and give an example of an organism that synthesizes each.
4. Describe the process of cell wall formation in plant cells.
5. Describe and diagram the structure of a cell membrane. What are the two principal types of molecules in cell membranes? What are the three main functions of cell membranes?
6. Define diffusion and osmosis.
7. What types of molecules diffuse through cell membranes, and through which parts of the membrane do they pass?
8. Define hypotonic, hypertonic, and isotonic. What would be the fate of an animal cell immersed in each of the three types of solution?
9. Describe the following types of transport processes: simple diffusion, facilitated diffusion, active transport, pinocytosis, phagocytosis, and exocytosis.
10. Name four types of cell-to-cell junctions and the function of each. Which allow communication between the interiors of adjacent cells?

SUGGESTED READINGS

Bretscher, M. S. "The Molecules of the Cell Membrane." *Scientific American,* October 1985. Beautifully illustrated, this article explores the structure and function of cell membranes, with special attention to cell junctions.

Fawcett, D. *The Cell: An Atlas of Fine Structure.* Philadelphia: W. B. Saunders Company, 1966. An excellent collection of light and electron micrographs of cell structures.

Kessel, R., and Shih, C. *Scanning Electron Microscopy in Biology: A Student's Atlas of Biological Organization.* New York: Springer-Verlag New York, Inc., 1974. The scanning electron microscope has opened new vistas into the three-dimensional structure of organisms, tissues, and cells, nowhere more evident than in this beautiful collection.

Luria, S. E. "Colicins and the Energetics of Cell Membranes." *Scientific American,* December 1975. Active transport is studied by observing the effects of molecules that inhibit various phases of the process.

5
Cells: The Units of Life. II. Internal Organization

As you learned in Chapter 4, all cells possess a cell membrane composed of phospholipids and proteins. Further, all cells regulate the flow of materials across their cell membranes in fundamentally similar ways. Finally, the hereditary material of all living cells is DNA. Despite these fundamental similarities, cells vary tremendously in size, shape, and internal organization.

The most fundamental differences among cells occur between the **prokaryotes** [monerans (bacteria); Fig. 5-1a], which probably resemble the earliest cells to evolve on Earth, and the **eukaryotes** (protists, plants, animals, and fungi; Fig. 5-1b). The most obvious difference is the presence of discrete, membrane-bound compartments, called **organelles,** within the cytoplasm of eukaryotic cells. Although some prokaryotic cells have membranes

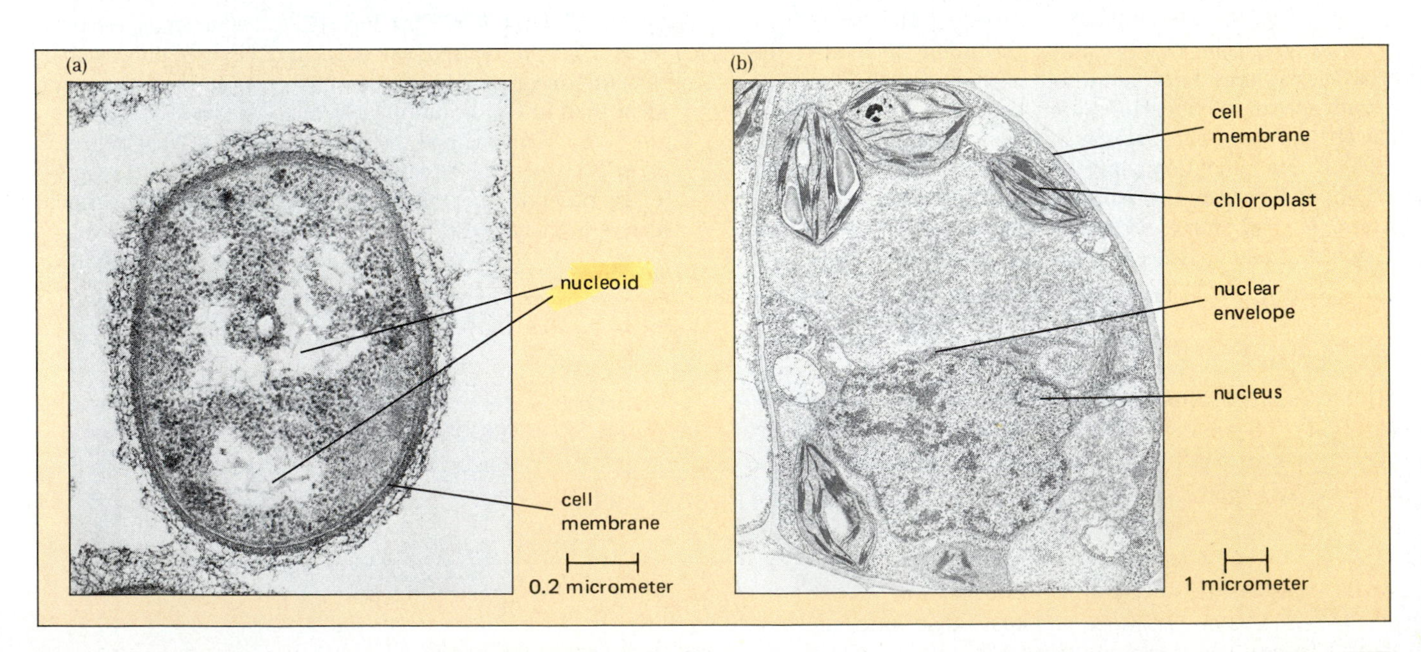

Figure 5-1 Electron micrographs of a prokaryotic cell **(a)** *and a eukaryotic cell* **(b)**. *Both have a cell membrane enclosing cytoplasm. However, the eukaryotic cell has many membrane-bounded organelles, including a nucleus containing DNA. The pale area in the prokaryotic cell, the nucleoid, is the location of the cell's DNA, but it is not separated from the rest of the cytoplasm by a membrane.*

Table 5-1 **Cell Structures, Their Functions, and Their Distribution in Living Cells**

Structure	Function	Prokaryotes[a]	Eukaryotes: Plants	Eukaryotes: Animals
Cell surface				
Cell wall	Protect, support cell	Present	Present	Absent
Cell membrane	Isolate cell contents from environment; regulate movement of materials into and out of cell; communicate with other cells	Present	Present	Present
Organization of genetic material				
Genetic material	Encode information needed to construct cell and control cellular activity	DNA	DNA	DNA
Chromosomes	Contain and control use of DNA	Single, circular, no proteins	Many, linear, with proteins	Many, linear, with proteins
Nucleus	Membrane-bound container for chromosomes	Absent	Present	Present
Nuclear membrane	Enclose nucleus; regulate movement of materials into and out of nucleus	Absent	Present	Present
Nucleolus	Synthesize ribosomes	Absent	Present	Present
Cytoplasmic structures				
Mitochondrion	Produce energy by aerobic metabolism	Absent	Present	Present
Chloroplast	Perform photosynthesis	Absent	Present	Absent
Ribosome	Provide site of protein synthesis	Present	Present	Present
Endoplasmic reticulum	Synthesize many proteins and lipids	Absent	Present	Present
Golgi complex	Modify and package proteins and lipids; synthesize carbohydrates	Absent	Present	Present
Lysosome	Contain intracellular digestive enzymes	Absent	Present	Present
Plastid	Store food, pigments	Absent	Present	Absent
Central vacuole	Contain water and wastes; provide turgor pressure to support cell	Absent	Present	Absent
Other vesicles and vacuoles	Contain food obtained through phagocytosis; contain secretory products	Absent	Present (some)	Present
Cytoskeleton	Give shape and support to cell; position and move cell parts	Absent	Present	Present
Centriole	Synthesize microtubules of cilia and flagella; may produce spindle in animal cells	Absent	Absent (in most)	Present
Cilia and flagella	Move cell through fluid or fluid past cell surface	Absent[b]	Absent (in most)	Present

[a]Many structures are listed as "absent" in prokaryotes; however, their functions are often essential to the life of any cell. In prokaryotes, these functions still occur, but not in discrete structures. For example, prokaryotes synthesize digestive enzymes on ribosomes attached to the cell membrane and immediately secrete the enzymes, where they digest food outside the cell. Therefore, although endoplasmic reticulum, secretory vesicles, and lysosomes are not found in prokaryotes, the functions of these organelles are still carried out.
[b]Many prokaryotes have structures called flagella, but these are not made of microtubules and move in a fundamentally different way than eukaryotic cilia or flagella do.

within the cytoplasm, these are derived from and probably connected to the cell membrane, and do not form discrete organelles. Eukaryotic cells have a variety of organelles, each performing specific functions such as synthesis of organic molecules, digestion, or energy production. The most conspicuous organelle is the **nucleus,** a membrane-bound sac containing the cell's genetic material. ("Eukaryote," in fact, means "true nucleus" in Greek.) In contrast, although the DNA of prokaryotes is localized within a specific region called the **nucleoid,** it is not enclosed within membranes; hence the name "prokaryotic" ("before the nucleus"). As a reference and study guide, Table 5-1 lists the major characteristics of prokaryotic and eukaryotic cells, including brief descriptions of the functions of the various organelles.

THE ORGANIZATION OF EUKARYOTIC CELLS

Conceptually, cell biologists divide the material within a eukaryotic cell into the membrane-bound **nucleus,** which contains the hereditary material, and the **cytoplasm,** which includes everything else. The cytoplasm in turn is composed of several types of organelles, occupying as

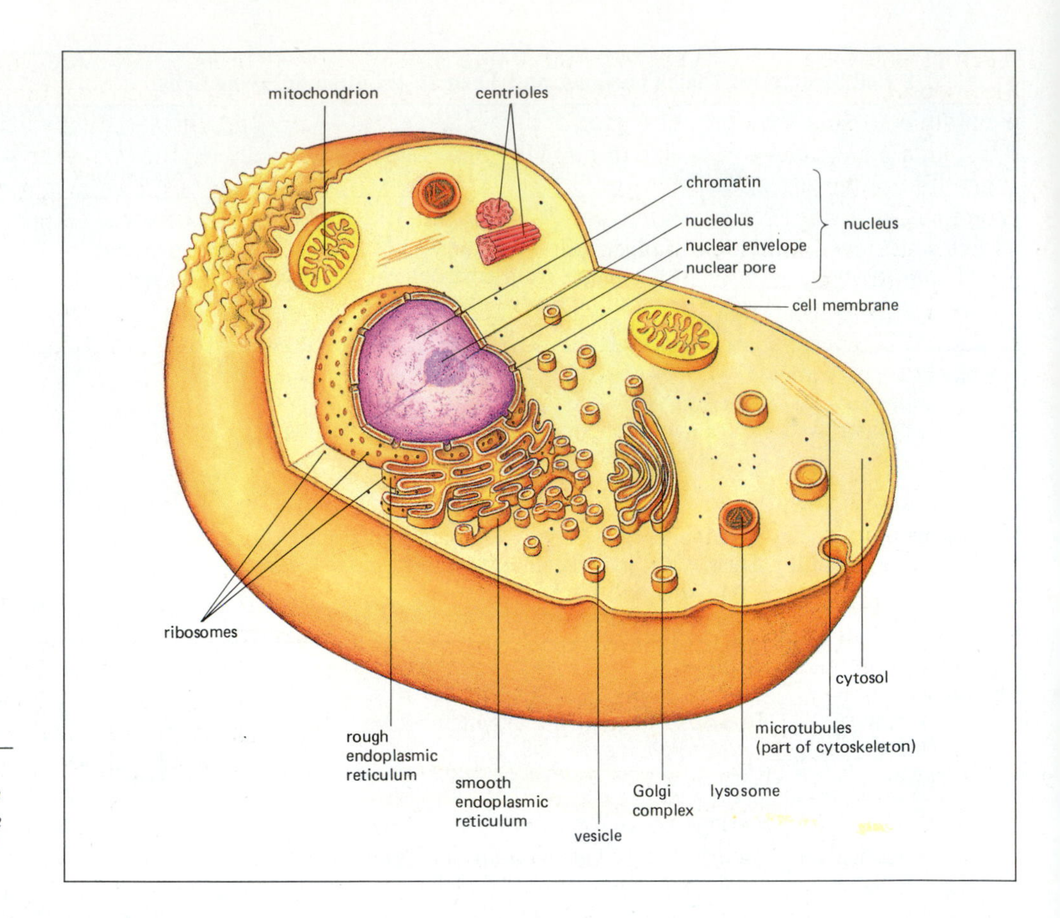

Figure 5-2 *An illustration of a "typical" animal cell. The entire range of organelles depicted here seldom occurs in a single animal cell.*

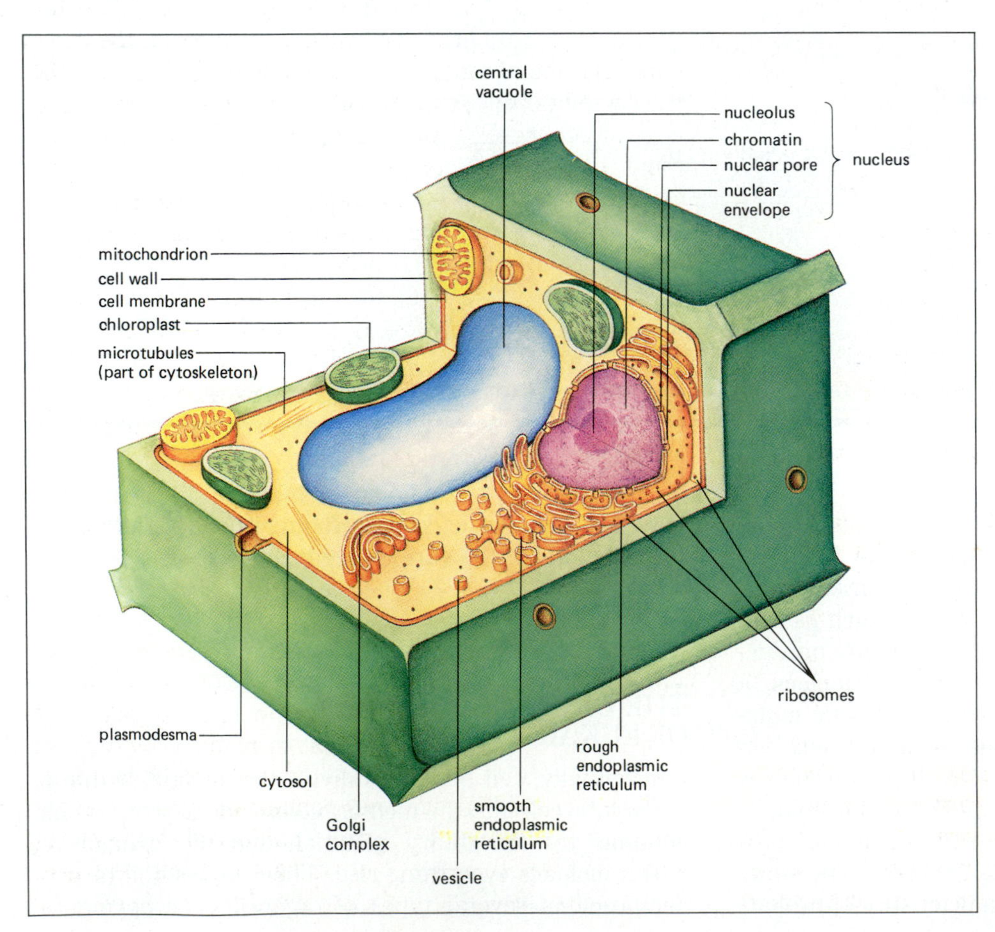

Figure 5-3 *An illustration of the major features of a "typical" plant cell. Not all of these structures occur in every plant cell. Compare this drawing with that of an animal cell in Figure 5-2.*

much as half the volume of the cell, and a fluid matrix, the **cytosol,** in which the organelles reside. The cytosol is an aqueous solution of salts, sugars, amino acids, proteins, fatty acids, nucleotides, and other materials. Giving shape and organization to the cytoplasm is a network of protein fibers called the **cytoskeleton.** Many of the organelles and even individual molecules of the cytoplasm are thought to be attached to the cytoskeleton.

Eukaryotic cells, however, are not all alike. Indeed, striking differences occur among eukaryotic cells, especially between plant and animal cells. Figures 5-2 and 5-3 illustrate the structures that are found in animal and plant cells, although few individual cells contain all the features shown in the drawings. You may want to refer back to these illustrations as we describe the subcellular structures in more detail. In contrast to animal cells, plant cells (1) are surrounded by a cell wall outside the cell membrane; (2) contain *plastids,* organelles that are usually involved in photosynthesis (chloroplasts) or storage of food or pigments; (3) contain one or more large *vacuoles* that store water, pigments, and wastes; and (4) usually lack centrioles, lysosomes, cilia, and flagella.

In the following pages, we introduce you to the major components of eukaryotic cells. We emphasize the union of *structure and function* that is crucial to understanding biology. In studying cellular anatomy, remember that each structure originally evolved, and persists today, because it carries out a specific function that is essential to the survival and reproduction of the cell.

THE STRUCTURE AND FUNCTION OF EUKARYOTIC ORGANELLES

Energy: Mitochondria and Chloroplasts

Every cell has prodigious energy needs: to manufacture materials, to pick things up from the environment and throw other things out, to move, and to reproduce. As Lewis Thomas put it: "There are structures squirming inside each of our cells that provide all the energy for living." These are the **mitochondria** and **chloroplasts,** "essentially foreign creatures" thought to have evolved from bacteria that took up residence long ago within a fortunate eukaryotic cell (see the essay, "The 'Essentially Foreign Creatures' Within Us").

Mitochondria and chloroplasts are similar to each other in many ways (Figs. 5-4 and 5-5). Both are usually oblong in shape, about 1 to 5 micrometers long, and are surrounded by a double membrane. Both have enzyme assemblies that synthesize the energy-carrier molecule ATP, although the systems are used in a very different manner (Chapter 7). Finally, both have many characteristics, including their own DNA, which seem to be remnants of their probable evolution from free-living organisms (see the Essay). However, they also have many differences, corresponding to their vastly different roles in cells: chloroplasts capture the energy of sunlight during photosynthesis and store it in sugar, whereas mitochondria convert the energy of sugar into ATP for use by the cell.

MITOCHONDRIA. Almost all eukaryotic cells have *mitochondria,* often called the "powerhouses of the cell." Well deserving of their nickname, mitochondria extract energy from food molecules and store it in the high-energy bonds of ATP. As you will see in Chapter 7, various amounts of energy can be released from a food molecule, depending on how it is metabolized. The breakdown of food molecules begins in the cytosol, but the cytosol lacks the enzymes needed to use oxygen in food metabolism. This **anaerobic** (without oxygen) metabolism does not convert very much food energy into ATP energy. Mitochondria are the only places in a cell where oxygen can be used in food metabolism. This **aerobic** metabolism is much more effective than anaerobic metabolism: about 19 times more ATP is generated by aerobic metabolism in the mitochondria than by anaerobic metabolism in the cytosol.

Mitochondria are round, oval, or tubular sacs made of a pair of membranes (Fig. 5-4). Although the outer mitochondrial membrane is smooth, the inner membrane loops back and forth to form deep folds called **cristae** (singular, **crista,** meaning "crest"). As a result, the mitochondrial membranes enclose two fluid-filled spaces, the **intermembrane compartment** between the inner and outer membranes, and the **matrix,** or inner compartment, within the inner membrane. Some of the reactions of food metabolism occur in the fluid matrix contained within the inner membrane, while the rest are conducted by a series of enzymes attached to the membranes of the cristae. We will examine the function of mitochondria in more detail in Chapter 7.

CHLOROPLASTS. Chloroplasts are found only in plants and certain protists, notably the unicellular algae. Like mitochondria, chloroplasts have outer and inner membranes, but in chloroplasts there is very little space between the membranes. The inner membrane encloses a semifluid material called the **stroma.** Embedded within the stroma are interconnected stacks of hollow, disk-shaped sacs (Fig. 5-5). The individual sacs are called **thylakoids,** while a stack of sacs is a **granum** (plural **grana**). During chloroplast development, thylakoids probably arise as invaginations of the inner membrane that bud off into the stroma, so that in mature chloroplasts the thylakoids are not connected to the inner membrane.

The thylakoid membranes contain the green pigment molecule **chlorophyll.** During photosynthesis, chlorophyll captures the energy of sunlight and transfers it to other molecules in the thylakoid membranes. These mole-

The "Essentially Foreign Creatures" Within Us: The Evolution of Chloroplasts and Mitochondria

In the nineteenth century, microscopists noted that the chloroplasts of some red algae have a fairly simple structure. Instead of the precise stacks of thylakoids typically found in land plants, these chloroplasts are filled with ribbons of membrane (Fig. E5-1a). In fact, the chloroplasts are similar in both size and structure to certain photosynthetic bacteria (Fig. E5-1b). As biochemistry and microscopy advanced, biologists realized that both chloroplasts and mitochondria resemble prokaryotes and are very different from other eukaryotic organelles.

First, chloroplasts and mitochondria contain a single circular chromosome composed of DNA without any proteins. Prokaryotic cells also have circular chromosomes of DNA without proteins. Second, chloroplasts and mitochondria contain their own ribosomes and synthesize proteins. The ribosomes resemble prokaryotic ribosomes more closely than they resemble the ribosomes found in the rest of the cell. Third, chloroplasts and mitochondria can grow, duplicate their DNA, and reproduce (Fig. E5-2). Fourth, they apparently cannot be manufactured by the cell. During cell division, occasionally one of the daughter cells receives all the chloroplasts or mitochondria, and the other receives none. When this happens, the cell that lacks these organelles never gets them back again. Finally, mitochondria and chloroplasts are surrounded by two separate membranes, whereas most other organelles are bounded by one. The molecular composition of the inner membrane resembles that of a prokaryotic membrane, while the outer membrane resembles a eukaryotic membrane.

Why should chloroplasts and mitochondria resemble prokaryotic cells so strongly? Many years ago, a few biologists suggested that they resemble prokaryotes because *chloroplasts and mitochondria might be the descendants of prokaryotic cells that took up residence within the cytoplasm of protoeukaryotic cells.* More recently, this **endosymbiotic theory** has been persuasively argued by Lynn Margulis, and today most biologists agree that mitochondria and chloroplasts are, in Lewis Thomas's words, "essentially foreign creatures" within eukaryotic cells.

How could free-living bacteria wind up as organelles

Figure E5-2. This series of micrographs illustrates division of mitochondria. A membranous partition (arrow) grows inward from one side of the mitochondrion, eventually reaching the other side and producing two daughter mitochondria from the parent. Not visible in these micrographs is the mitochondrial DNA, which duplicates before division, with one copy being incorporated into each daughter organelle.

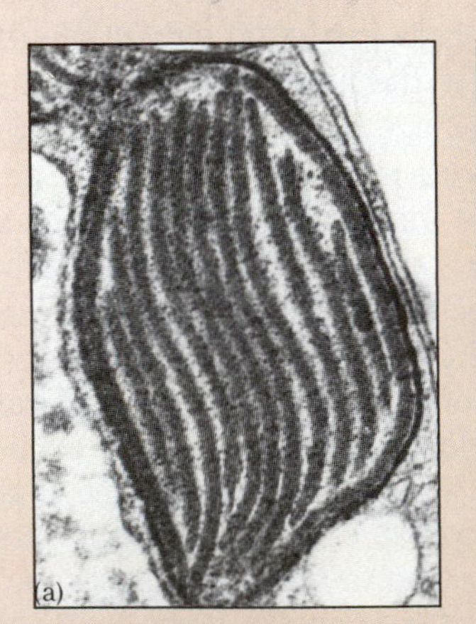

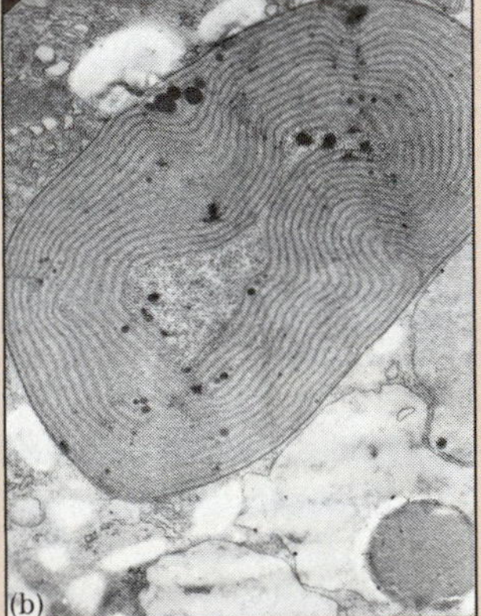

Figure E5-1. The chloroplast of a red alga **(a)** *compared to an entire cyanobacterium* **(b)**.

The "Essentially Foreign Creatures" Within Us: The Evolution of Chloroplasts and Mitochondria

of eukaryotic cells? Let's look first at a hypothetical scenario for the origin of mitochondria. Many single-celled organisms feed by phagocytosis. They engulf entire prey organisms in extensions of the cell membrane, "swallowing" the prey in a vesicle of membrane. Usually, the prey is doomed, as the vesicle fuses with enzyme-containing lysosomes and the prey inside is digested. According to the endosymbiotic theory, over a billion years ago the ancestors of eukaryotic cells had evolved, and some of them fed by phagocytosis. However, these cells lacked the enzymes needed to use oxygen in the metabolism of food molecules. As you will see in Chapter 7, such a cell can only extract a small amount of usable energy as ATP from its food. Some bacteria, however, had evolved these enzymes, and these aerobic bacteria could produce much more ATP from the same amount of food.

Bacteria were probably favorite prey of protoeukaryotic cells, and occasionally some bacteria were captured as prey but not digested. The bacteria took up residence in the cytoplasm of their new host. In most cases this must have harmed the host, for host and bacterium would compete for the same nutrients. The waste products of the bacterium may also have been poisonous. The host cell would probably have died—done in, as it were, by an indigestible bit of food.

However, an aerobic bacterium in this situation would be in prokaryotic paradise. The cytoplasm of an anaerobic host cell would be rich in half-digested food molecules. Awash in food, the bacterium would probably generate large amounts of surplus ATP, some of which might pass into the host cytoplasm and could be used by the host. If the bacterium did *not* kill the host, the relationship would be beneficial for both. The host–bacterium combination would survive better than either type could by itself. The "mitochondrion–bacterium" multiplied inside its host, and when the host cell divided, each daughter cell contained a few of these newfound powerhouses. The aerobic, truly eukaryotic cell—ancestor to all protists, plants, animals, and fungi—had been born.

Later, one of these new mitochondrion-containing cells may have similarly captured a photosynthetic cyanobacterium. It, too, found a congenial home in its host's cytoplasm. It photosynthesized, grew, and multiplied. Surplus sugars synthesized by the protochloroplast may have passed to the host, providing some of the energy needed for the host to live and grow. The mitochondria already present in the cell metabolized these sugars, providing plenty of energy. The plantlike protists and all true plants would be the descendants of this cell.

If mitochondria and chloroplasts descended from bacteria, this would explain why they have bacterialike DNA and ribosomes, and can grow and duplicate themselves. The endosymbiotic theory also neatly explains the nature of the two membranes. Remember, during phagocytosis a cell surrounds its prey with a vesicle made from its cell membrane. A live bacterium has its own cell membrane. Therefore, the newly engulfed protomitochondrion or protochloroplast would have two membranes, an outer one from the eukaryotic predator and an inner one from the bacterium (Fig. E5-3).

Needless to say, if chloroplasts and mitochondria arose from such predation/parasitisms, it was a very long time ago, and evolution has not stood still since. Chloroplasts and mitochondria are now fully incorporated into the lives of eukaryotic cells and cannot survive by themselves. Indeed, many of the proteins found in chloroplasts and mitochondria are not encoded by the organelle DNA, but are specified by the nuclear DNA of the cell. This should not surprise us, because for over a billion years, natural selection has operated not on the "host" alone, nor on the "parasite" alone, but on the functional whole of the entire eukaryotic cell.

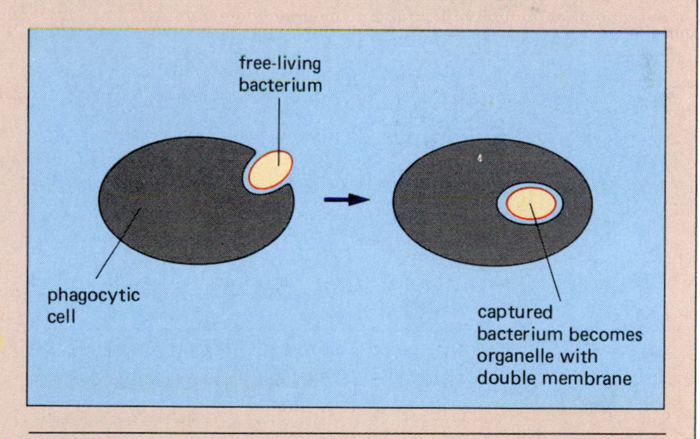

Figure E5-3. *A model for the incorporation of bacteria into the cytoplasm of host cells, and the origin of the double membrane surrounding mitochondria and chloroplasts.*

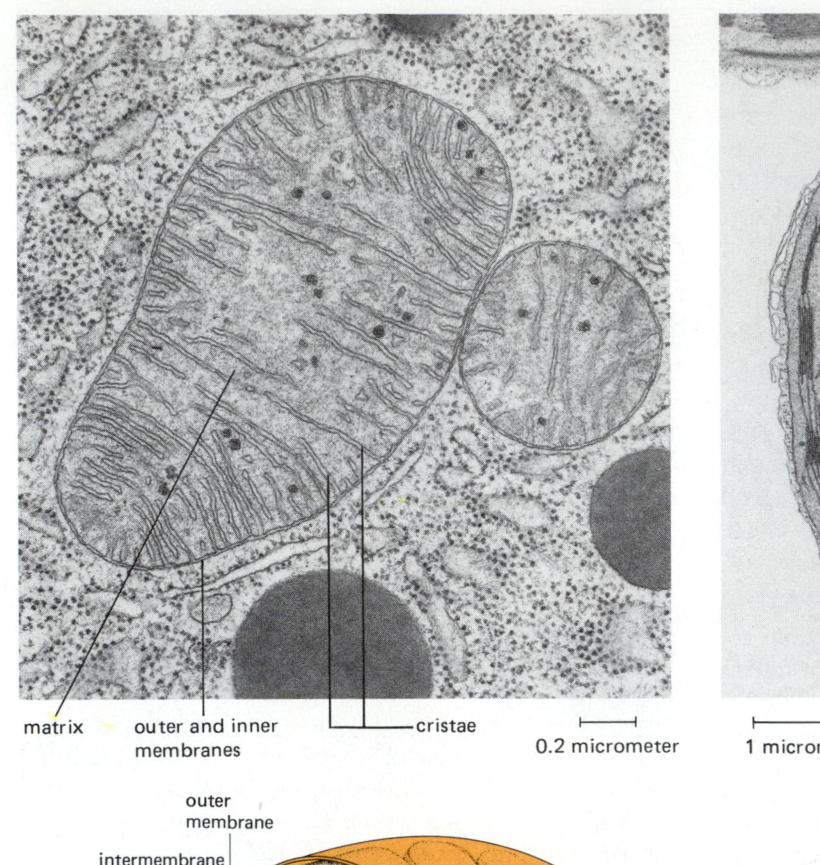

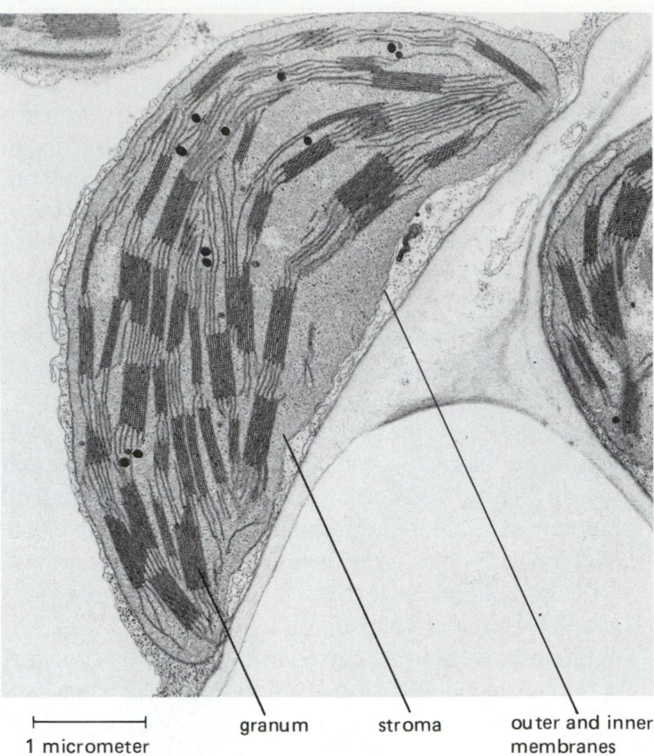

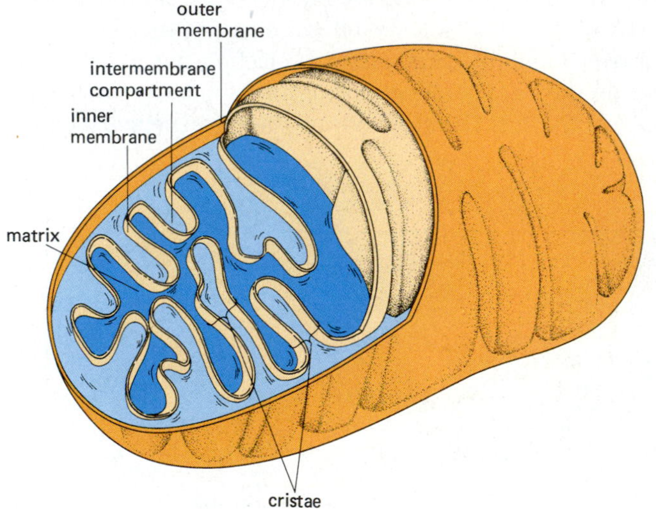

Figure 5-4 *Mitochondria consist of a pair of membranes enclosing two fluid compartments. Mitochondria are the site of aerobic metabolism described in Chapter 7.*

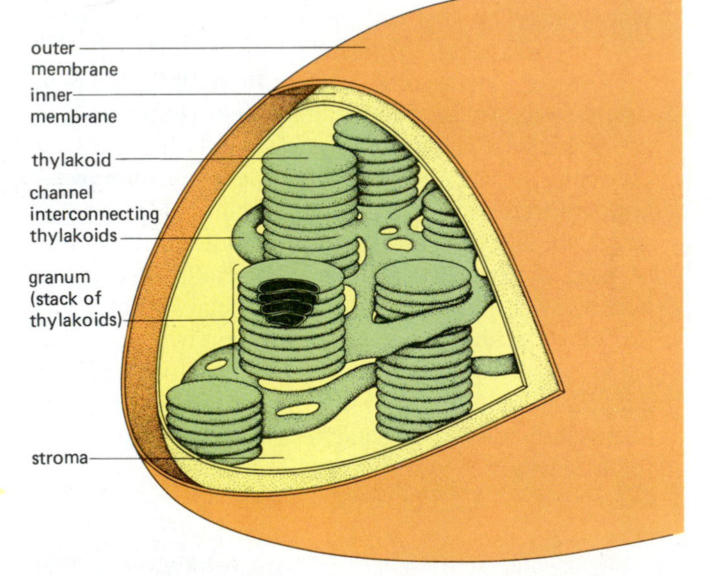

Figure 5-5 *Chloroplasts are surrounded by a double membrane, although the inner membrane is not usually visible in electron micrographs. Enclosed by the inner membrane is the semifluid stroma, in which are embedded stacks of sacs collectively referred to as grana. The individual sacs of the grana are called thylakoids.*

cules in turn transfer the energy to ATP and other energy-carrier molecules. The energy carriers diffuse into the stroma, where their energy is used to drive the synthesis of sugar from carbon dioxide and water.

Manufacture: Ribosomes, Endoplasmic Reticulum, and Golgi Complex

Every cell synthesizes many different molecules, to replace structures that break down and wear out, to grow new structures, and to reproduce. Many cells also export materials into the environment, including such molecules as hormones, digestive enzymes, and milk proteins. Most

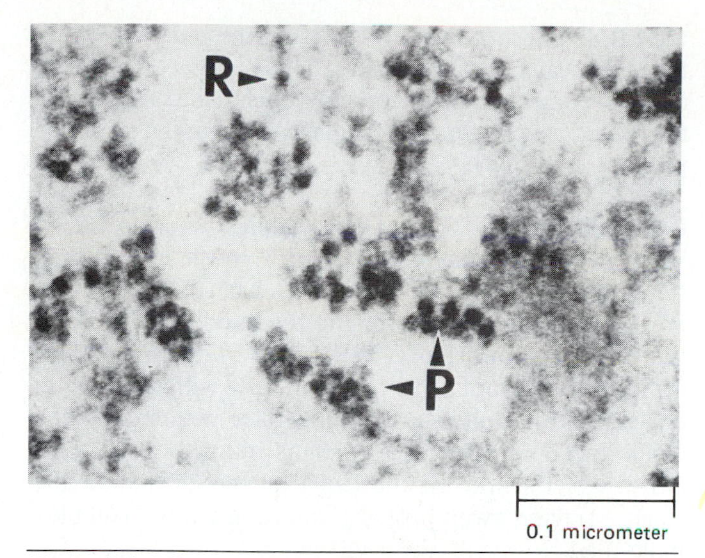

Figure 5-6 *Ribosomes may be found free in the cytoplasm either singly (R) or in groups called polyribosomes (P). Free ribosomes synthesize proteins that will be used within the cell. Other ribosomes are attached to the endoplasmic reticulum (see Fig. 5-7).*

of these molecules are manufactured in the cytoplasm. Some, including most carbohydrates, are synthesized by enzymes in the cytosol. Other molecules, particularly proteins and lipids, are synthesized in or on specific organelles: ribosomes, endoplasmic reticulum, and Golgi complex.

RIBOSOMES. In electron micrographs, ribosomes appear as dark granules, either distributed in the cytosol (Fig. 5-6) or clustered along the membranes of the nuclear envelope and the endoplasmic reticulum (Fig. 5-7). A **ribosome** is a small particle composed of RNA and protein that serves as a "workbench" for the synthesis of proteins (Chapter 11). Like many human workshops, ribosomes are nonspecific, and any ribosome can be used to synthesize any of the hundreds of proteins made by a cell.

The location of ribosomes within a cell seems to be related to the function of the cell. Cells that retain most of the proteins they manufacture have mostly "free" ri-

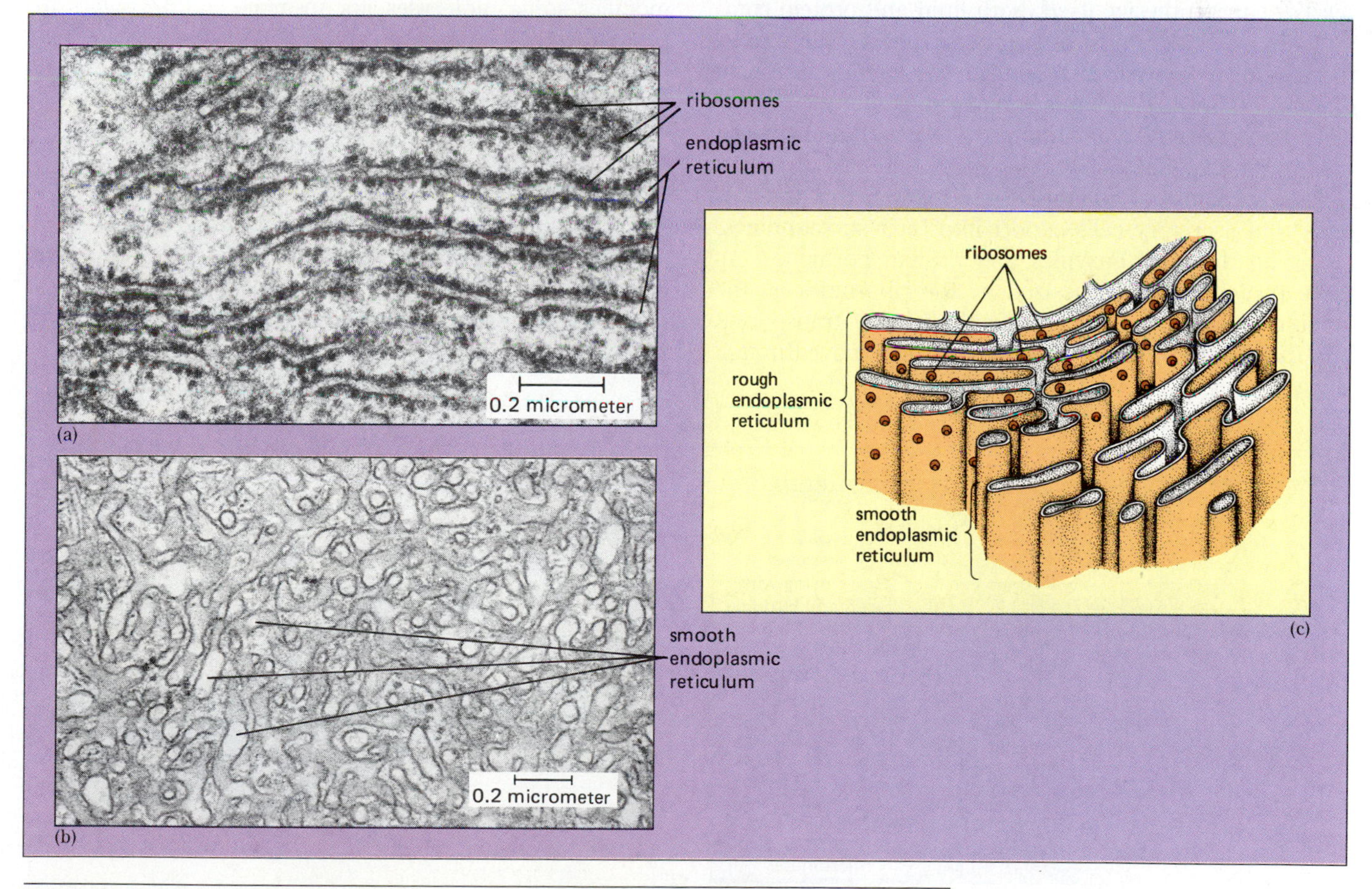

Figure 5-7 *There are two types of endoplasmic reticulum, rough ER, coated with ribosomes* **(a)**, *and smooth ER without ribosomes* **(b)**. *Although the ER looks like a series of tubes and sacs in electron micrographs, it is actually a maze of folded sheets and interdigitating channels* **(c)**. *In many cells the rough and smooth ER are thought to be continuous, as depicted in the drawing. Ribosomes stud the cytoplasmic face (tan) of the rough ER membrane.*

bosomes, not attached to membranes. On the other hand, cells that secrete proteins, such as the cells of the pancreas that produce digestive enzymes, have huge numbers of ribosomes lining an enormous endoplasmic reticulum that practically fills the cytoplasm.

Endoplasmic reticulum. The **endoplasmic reticulum** (ER for short) is a series of interconnected membranous tubes and channels in the cytoplasm (Fig. 5-7). Most eukaryotic cells have two forms of ER, rough and smooth. Numerous ribosomes stud the outsides of **rough endoplasmic reticulum,** while **smooth ER** lacks ribosomes.

The different structures of smooth and rough ER reflect different functions. *Protein enzymes embedded in the membranes of the smooth ER are the major site of lipid synthesis,* including the phospholipids of the ER and other membranes. In some cells the smooth ER synthesizes other types of lipid as well, such as the steroid hormones testosterone and estrogen produced by the gonads of mammals. *The ribosomes on the outside of rough ER synthesize proteins, including membrane proteins.* Therefore the ER can synthesize itself, both lipid and protein components. Although most of the membrane synthesized in the ER forms new or replacement ER membrane, some of it moves inward to replace nuclear membrane or outward to form the Golgi complex, lysosomes, and the cell membrane (see Fig. 5-10).

The ribosomes on rough ER also manufacture the proteins that secretory cells export into their surroundings, including digestive enzymes and protein hormones. As these proteins are synthesized by the ribosomes on the outside of the ER, they are simultaneously transported into the channels inside. The proteins then move through the ER and accumulate in pockets at the ends, especially the ends near Golgi complexes. These pockets then "bud off," forming membrane-bound vesicles that migrate to the Golgi, where final packaging and processing of the proteins take place.

Golgi complex. The **Golgi complex** is a specialized set of membrane channels derived from the ER. In fact, the Golgi looks very much like a stack of smooth ER that has been stepped on, flattening the middle and making the ends bulge out (Fig. 5-8). In reality, vesicles that bud off from the endoplasmic reticulum fuse with the sacs on one side of the Golgi stack, adding their membrane to the Golgi and emptying their contents into the Golgi sacs. Other vesicles bud off the Golgi on the far side of the stack, carrying away proteins, lipids, and other complex molecules.

The Golgi complex performs three major functions. (1) It separates out proteins and lipids received from the ER according to their destinations; for example, the Golgi separates digestive enzymes that are bound for lysosomes from hormones that will be secreted from the cell. (2) It modifies some molecules, for instance adding sugars to proteins to make glycoproteins. (3) It packages these materials into vesicles that are then transported to other parts of the cell or to the cell membrane for exocytosis.

Digestion: Lysosomes

Some of the proteins manufactured in the endoplasmic reticulum and sent to the Golgi are intracellular digestive enzymes that can break down proteins, fats, and carbohydrates into their component subunits. In the Golgi, these enzymes are packaged into membrane-bound vesi-

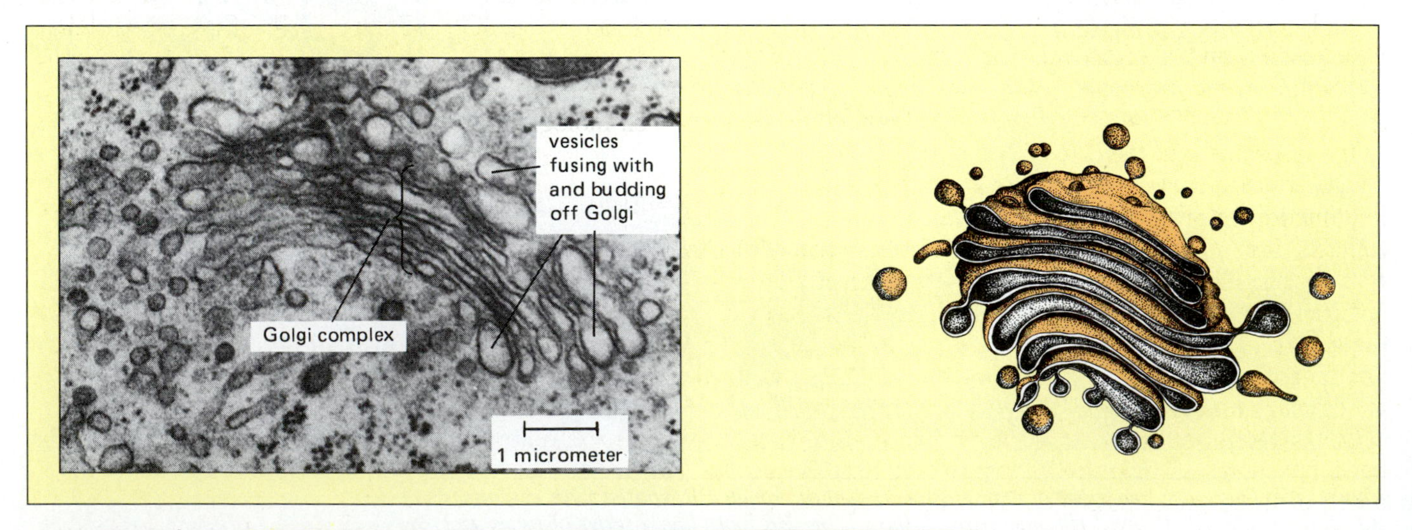

Figure 5-8 The Golgi complex is a stack of flat, membranous sacs derived from the endoplasmic reticulum. Vesicles constantly bud off and fuse with the Golgi and ER, transporting material in several directions: from the ER to the Golgi and back again, and from the Golgi to cell membrane, lysosomes, and vesicles.

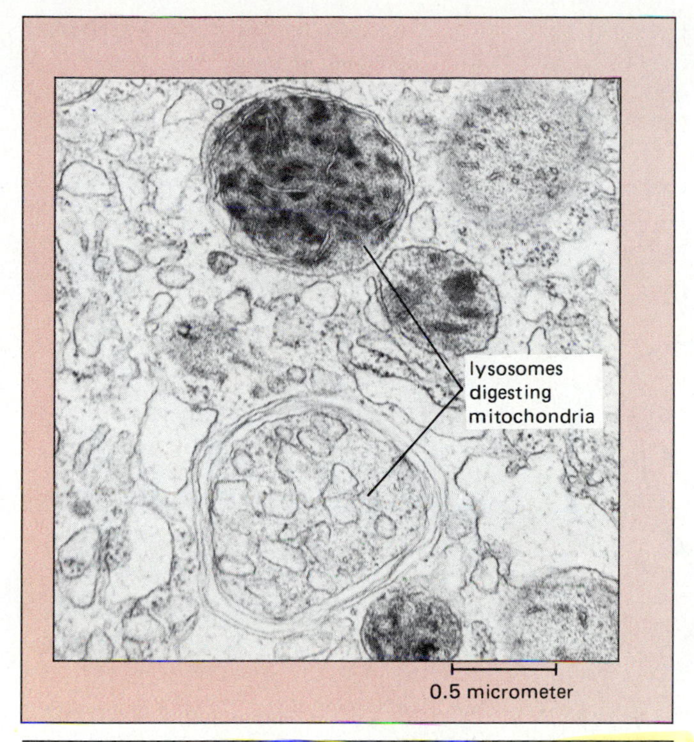

Figure 5-9 Lysosomes are enzyme-filled vesicles that bud off the Golgi complex. These enzymes digest food and worn-out organelles.

cles called **lysosomes** (Fig. 5-9). The major function of lysosomes is to digest food particles, from individual proteins to complete microorganisms. As we saw in Chapter 4, many cells "eat" by phagocytosis, engulfing extracellular particles with extensions of the cell membrane. The food particles are then moved into the cytoplasm, enclosed within membranous sacs now called **food vacuoles.** By an unknown process, lysosomes recognize these food vacuoles and fuse with them. The contents of the two vesicles mix, and the lysosomal enzymes digest the food into amino acids, monosaccharides, fatty acids, and other small molecules. These simple molecules then diffuse out of the lysosome into the cytosol to nourish the cell.

Lysosomes also assist the cell by digesting defective or malfunctioning organelles, such as mitochondria or chloroplasts. It is not known how the cell identifies organelles that have outlived their usefulness, but apparently they are enclosed in vesicles made of membrane from the endoplasmic reticulum. These vesicles fuse with lysosomes, and digestive enzymes within the lysosome enable the cell to recycle valuable materials from the defunct organelles.

One problem, of course, with being able to digest proteins, fats, and carbohydrates is that the cell itself is composed of these same molecules. That is one reason why lysosomes are membrane bound: the membrane isolates the potentially deadly enzymes from the rest of the cell. Even so, the enzymes chew away at the lysosomal membrane, and the cell constantly expends energy to keep the membrane in good repair. Nevertheless, occasionally a lysosome breaks open. Its enzymes digest ribosomes, mitochondria, cell membrane, and other vital parts, killing the cell.

The Flow of Membrane Through the Cell

The nuclear envelope, rough and smooth ER, Golgi complex, lysosomes, food vacuoles, and the cell membrane all form an interrelated membrane system. Membrane is synthesized in the ER and flows back and forth among these structures in an orderly way. As an example, let's look at the movement of materials destined for inclusion in the cell membrane (Fig. 5-10). The ER synthesizes the phospholipids and proteins that make up the cell membrane, and simultaneously synthesizes proteins and lipids that belong in the ER itself. How are these sorted out? The ER buds off a vesicle whose membrane is a mixture of ER and cell membrane components. The vesicle fuses with the Golgi complex, adding its "mixed membrane" to the Golgi membrane. In the Golgi, ER and cell membrane components are separated. ER material is removed as "recycle" vesicles that pinch off and return to the ER. Cell membrane material continues on through the Golgi, where it may be modified, for example by adding sugars to make glycoproteins or glycolipids. Eventually, the cell membrane material becomes an "outward bound" vesicle that buds off the far side of the Golgi and moves to the cell surface. The vesicle fuses with the cell membrane, replenishing and enlarging the membrane.

The Golgi complex processes and packages all membrane-enclosed materials produced by the cell. Many of the vesicles pinched off from the Golgi contain secretory products (e.g., hormones) that are released outside the cell by exocytosis (see Fig. 4-14). Lysosomes contain digestive enzymes and often fuse with food vacuoles to carry out intracellular digestion. How these diverse materials are recognized, purified, modified properly, separated out, and individually packaged remains a challenge to cell biologists.

Storage and Elimination: Plastids and Vacuoles

If a cell finds itself in a favorable environment, some of the molecules that it synthesizes and some of the food that it acquires and digests will be more than it needs at the moment. Cells have evolved organelles in which to store such valuable, if temporarily surplus, molecules. Other materials, such as the waste products of digestion, must be eliminated, and still other organelles serve this function.

Plastids. Besides mitochondria and nuclei, plant cells have other organelles with double outer membranes. These are the **plastids;** we have already met the most important plastid, the chloroplast. Other types of plastids

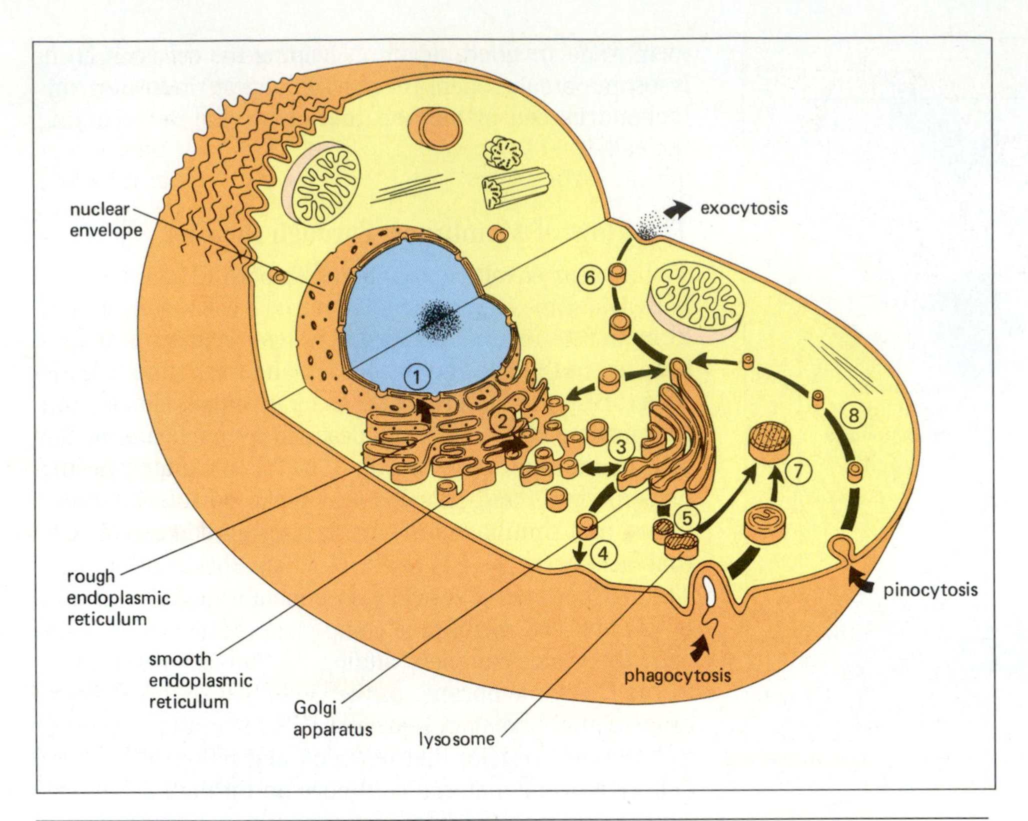

Figure 5-10 The flow of membrane and protein within a cell. Membrane, regardless of its destination, is synthesized by the endoplasmic reticulum. (1) Some of this membrane moves inward to form new nuclear membrane, while most of it moves outward to form (2) smooth ER and (3) Golgi membrane. From the Golgi, membrane moves to form (4) new cell membrane and (5) the membranes that surround other cell organelles such as lysosomes. Many proteins are synthesized in the rough ER and move through the smooth ER to the Golgi. Here the proteins are sorted according to function; some remain as parts of the Golgi membrane, some are returned to the ER, some are packaged in vesicles bound for the cell membrane where they will be (6) secreted from the cell, and some are packaged in lysosomes. (7) Lysosomes may fuse with phagocytic food vacuoles for intracellular digestion of food particles. (8) Cell membrane brought into the cell through pinocytosis is recycled through the Golgi complex.

are used as storage containers for various types of molecules, including pigments such as the carotenoids that give ripe fruits their rich yellow, orange, and red colors. Especially important, particularly for perennial plants, are plastids that store photosynthetic products from the summer for use during the following winter and spring. Plants usually convert the sugars made during photosynthesis into starch and store it in plastids (Fig. 5-11). Potatoes, for example, are masses of cells stuffed with starch-filled plastids.

VACUOLES. Most cells contain sacs, called **vacuoles,** that are bounded by a single membrane. We have just seen that food vacuoles are at least temporary features of phagocytic cells.

Freshwater protozoans such as *Paramecium* often contain complex **contractile vacuoles** composed of collecting ducts, a central reservoir, and a tube leading to a pore in the cell membrane (Fig. 5-12). Since fresh water is hypotonic to the cytosol of protozoans, water constantly enters the cell by osmosis (see Chapter 4). The increasing volume of water might soon burst open the fragile protozoan. However, water is pumped into the collecting ducts and drains into the central reservoir. When the reservoir is full, it contracts, squirting the water up the exit tube and out through the pore in the cell membrane.

Three-fourths or more of the volume of many plant cells is occupied by a huge, water-filled **central vacuole** (see Fig. 5-3). Dissolved salts within the vacuole promote water entry by osmosis, tending to cause the vacuole to swell up. The pressure of the water within the vacuole, called **turgor pressure,** pushes the cytosol up against the

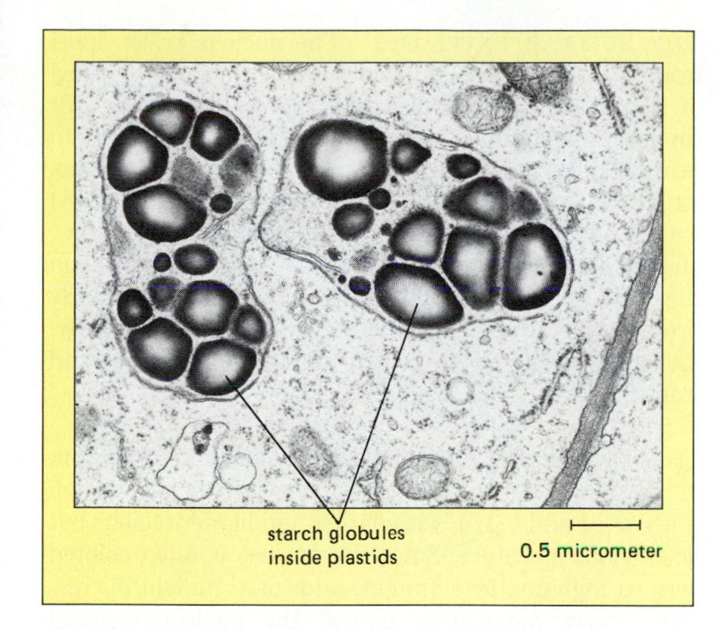

Figure 5-11 *Plastids are organelles, found in the cells of plants and plantlike protists, that are surrounded by a double membrane. Chloroplasts are the most familiar plastids; other types store various materials, such as the starch filling these plastids in potato cells.*

wall with considerable force, helping the cell to keep its shape (Fig. 5-13). Turgor pressure, in fact, provides support for many nonwoody plants. If you fail to water your houseplants, the central vacuoles lose water and the cytoplasm shrinks away from the cell walls. Just as a balloon goes limp when its air leaks out, the plant droops as its cells lose turgor pressure.

Plant vacuoles serve other purposes besides support. They provide a dump for hazardous wastes, which plant cells often cannot excrete. Some plant cells store extremely poisonous substances such as sulfuric acid in their vacuoles. These poisons may deter animals from munching on the otherwise tasty leaves. Vacuoles may also store sugars and amino acids not immediately needed by the cell. Finally, the colors of many flowers are due to blue or purple pigments in the central vacuoles of their cells.

Control: The Nucleus

Deoxyribonucleic acid, or **DNA,** is the genetic material of all living cells. A cell's DNA is the repository of information needed to construct the cell and direct its life processes. Just as a blueprint is used selectively, one part

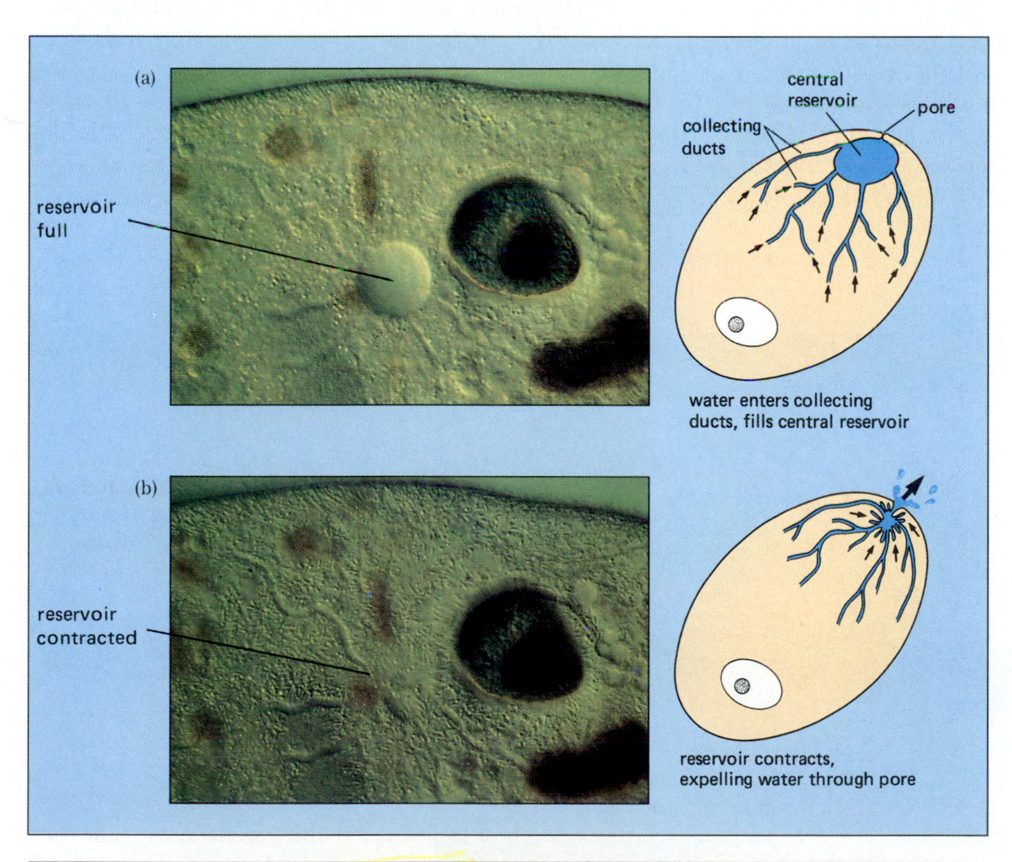

Figure 5-12 *Many freshwater protists contain contractile vacuoles.* **(a)** *Water constantly leaks into the cell, where it is taken up by collecting ducts and drains into the central reservoir of the vacuole.* **(b)** *When full, the reservoir contracts, expelling the water outside the cell through a pore in the cell membrane.*

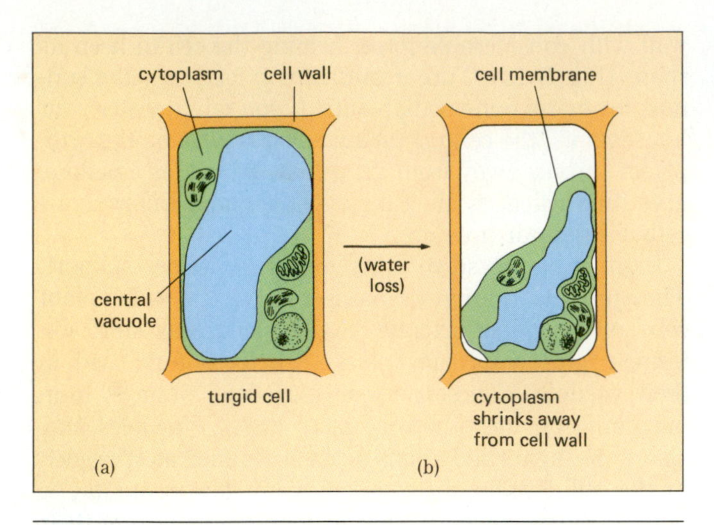

Figure 5-13 The role of turgor pressure in plant cells. **(a)** *Water pressure within the central vacuole of plant cells pushes the cytoplasm up against the cell wall, helping the cell maintain its shape.* **(b)** *When a plant wilts, the central vacuole shrinks, the cytoplasm pulls away from the cell wall, and the whole cell may become soft and shrunken.*

at a time, to build a house, the hereditary information in DNA is used selectively by the cell, depending on its stage of development and its environmental conditions. In eukaryotic cells, the DNA is housed within the nucleus.

The nucleus consists of three readily distinguishable components: a granular-looking material called **chromatin,** a darker region called the **nucleolus,** and the **nuclear envelope** separating the nuclear material from the cytoplasm (Fig. 5-14).

THE NUCLEAR ENVELOPE. The nucleus is set apart from the rest of the cell by the **nuclear envelope** formed of two membranes riddled with pores (Fig. 5-15). Although in some microscopic images the pores appear to be open holes in the envelope, in reality they are complex structures made up of protein and ribonucleic acid (RNA) with a small channel through the middle. Water, ions, and small molecules such as ATP can pass freely through the pores, but the passage of large molecules, particularly proteins and RNA, is probably regulated. Consequently, the pores may help to control information flow to and from the DNA.

CHROMATIN: THE HEREDITARY MATERIAL. The bulk of the nucleus appears granular in an electron micrograph (Figs. 5-14 and 5-15), with darker and lighter regions but no obvious structure. Since the nucleus is often colored very strongly by the common stains used in light microscopy, early histologists named the nuclear material **chromatin,** meaning merely "a colored substance." We have since learned that chromatin consists of DNA complexed with proteins. Although you cannot see them in ordinary electron micrographs, eukaryotic DNA and its associated proteins form long strands called **chromosomes** ("colored bodies"). When cells divide, each chromosome coils upon itself, becoming thicker and shorter. The resulting "condensed" chromosomes are easily visible even in light microscopes (Fig. 5-16).

Cellular processes are governed by the information encoded in DNA. Since the DNA stays in the nucleus, while most of the chemical reactions that it controls occur in the cytoplasm, *information molecules must be exchanged between the nucleus and the cytoplasm.* Genetic infor-

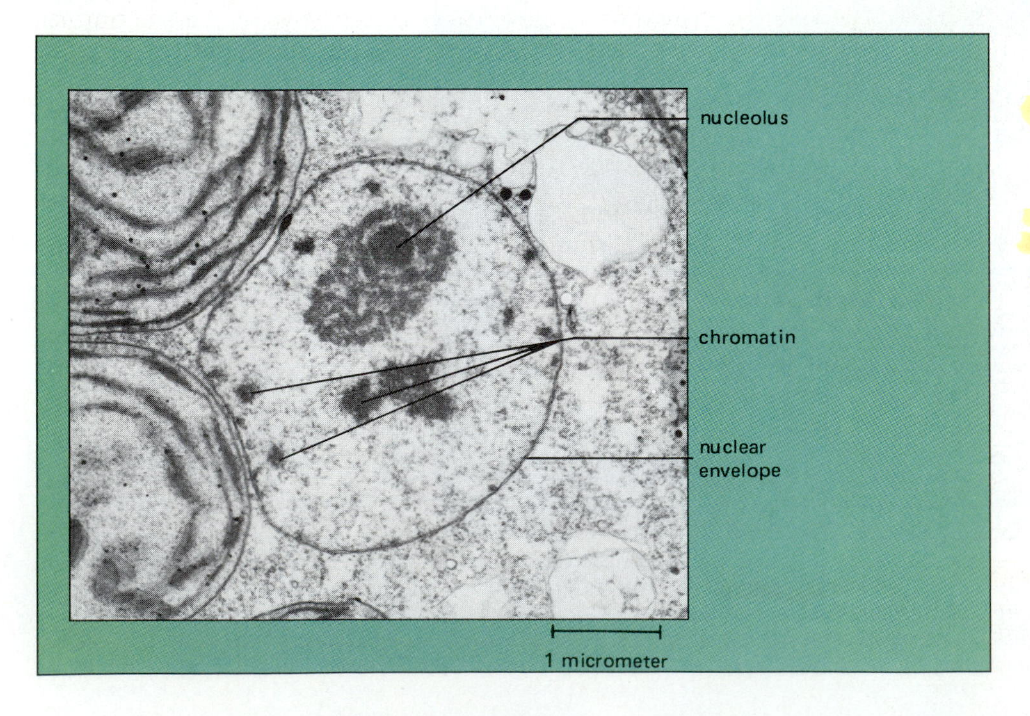

Figure 5-14 The eukaryotic nucleus consists of an outer double membrane, termed the nuclear envelope, enclosing the genetic material, DNA. The DNA is complexed with proteins to form chromatin. One region of the chromatin, the nucleolus, stains much more darkly than the rest of the nucleus.

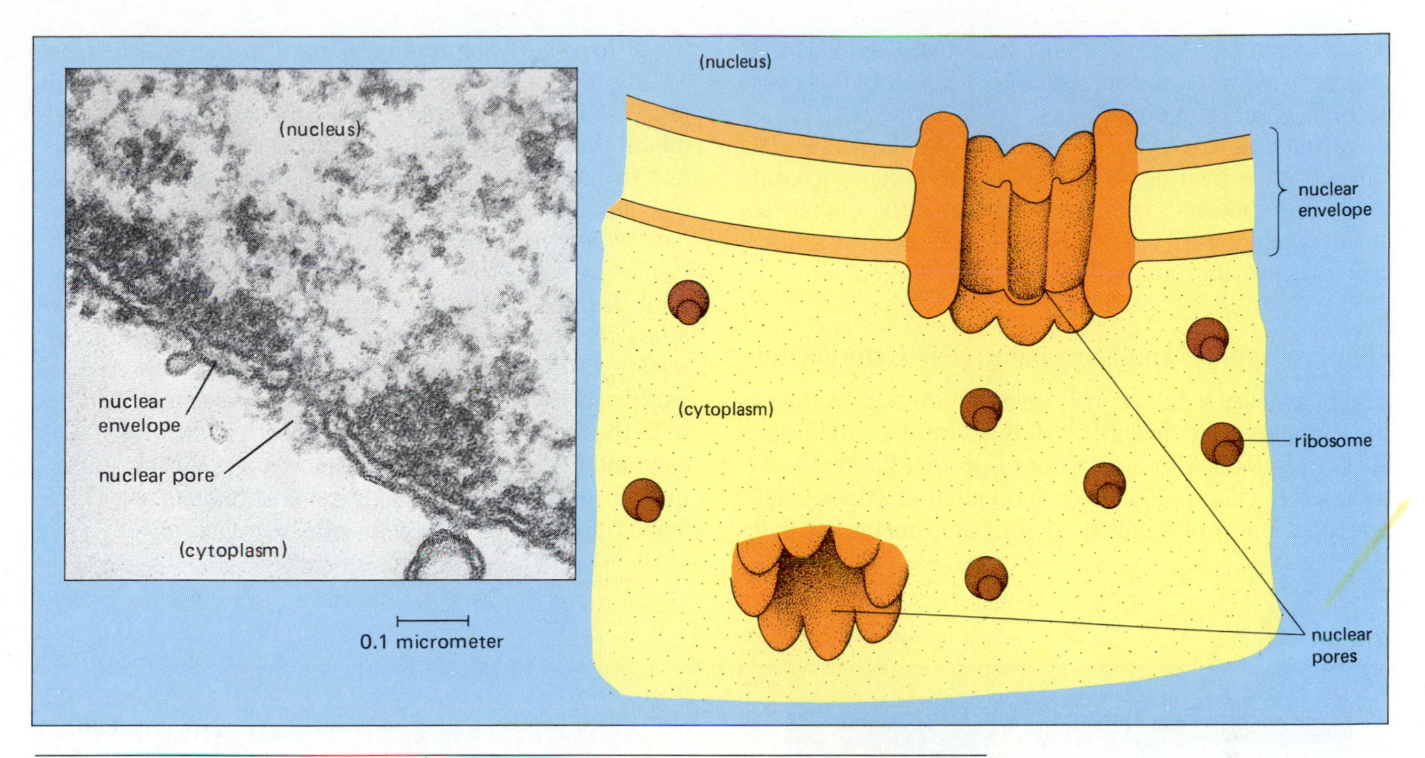

Figure 5-15 *A transmission electron micrograph of the nuclear envelope clearly shows that it is composed of a double membrane that is fused together around the lips of the nuclear pores. Micrographs such as this one can only hint at the structure of the pore, which is shown more clearly in the accompanying drawing.*

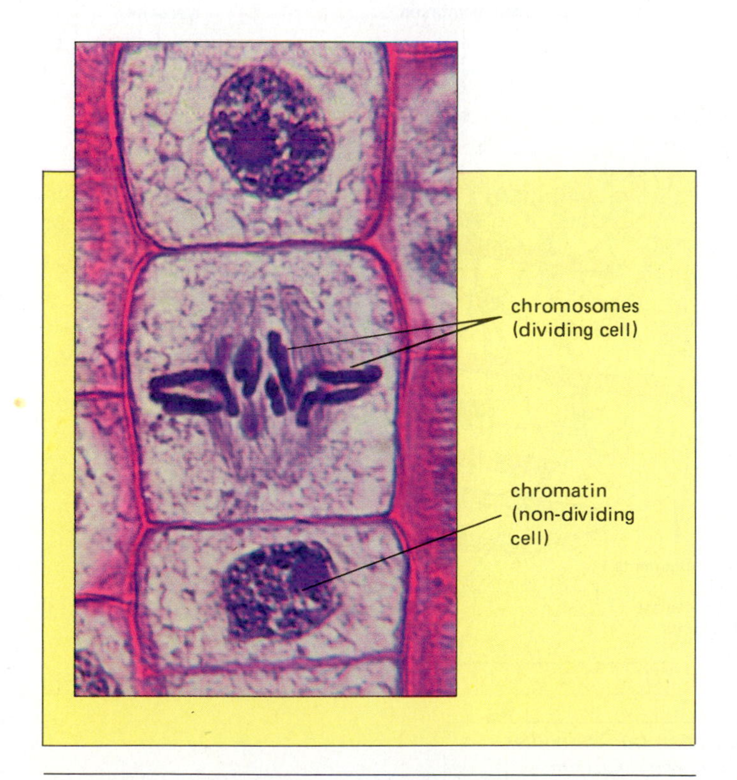

Figure 5-16 *Chromosomes, seen here in a light micrograph of a dividing cell in an onion root tip, are the same material as the chromatin seen in nondividing cells (DNA and proteins) but in a more compact state.*

mation is copied from DNA into molecules of RNA, which move through the pores of the nuclear envelope into the cytoplasm. Ribosomes in the cytoplasm use the information in the RNA molecules to direct protein synthesis. Other molecules, especially proteins, pass from the cytoplasm into the nucleus and regulate information transfer from DNA to RNA, depending on what is happening in the cytoplasm and the extracellular environment. We will take a closer look at these processes in Chapter 11.

THE NUCLEOLUS. Most eukaryotic nuclei have one or more darkly staining regions called **nucleoli** ("little nuclei", Fig. 5-14), which are the sites of ribosome synthesis. Ribosomes consist of two subunits, each composed of proteins and special types of RNA molecules called ribosomal RNA. Eukaryotic DNA usually contains many genes that encode the information for ribosomal RNA. These genes, often located on different chromosomes (ten in humans), cluster together in the nucleolus, where they direct synthesis of ribosomal RNA. Like all proteins, ribosomal proteins are synthesized in the cytoplasm, and they pass from the cytoplasm through the nuclear envelope to the nucleolus. Here the ribosomal RNA and proteins are assembled into ribosome subunits. Completed subunits then pass through the nuclear pores into the cytoplasm. *A nucleolus, then, consists of DNA (the genes for ribosomal*

RNA), ribosomal RNA, proteins, and ribosomes in various stages of synthesis, and stains darkly because of this dense accumulation of material.

During cell division, ribosome synthesis slows. The chromosomes bearing the ribosomal RNA genes separate as the chromosomes condense. Consequently, the nucleolus disappears until after cell division is complete and the cell resumes ribosome synthesis.

Shape, Support, and Movement: The Cytoskeleton

The organelles we have just described do not float about the cytoplasm haphazardly. Rather, most of the organelles are attached to a network of protein fibers, the **cytoskeleton** (Fig. 5-17). Even individual enzymes, which are often parts of detailed, step-by-step metabolic pathways, may be fastened in sequence to the cytoskeleton, so that molecules can be passed from one to the next for processing. Several types of protein fibers, including thick **microtubules** [about 25 nanometers (billionths of a meter) in diameter], medium-sized **intermediate filaments** (10 nanometers), and thin **microfilaments** (5 to 7 nanometers), contribute to the cytoskeleton (Table 5-2).

The cytoskeleton performs several crucial functions. First, for cells without cell walls, *the cytoskeleton determines the shape of the cell* (Fig. 5-17a). Second, *the assembly, disassembly, and sliding of microfilaments and microtubules causes cell movement.* Cell movement includes both the familiar locomotion of amoebas and white blood cells and the migration and shape changes that occur during the development of multicellular organisms. Third, *microtubules and microfilaments move organelles from*

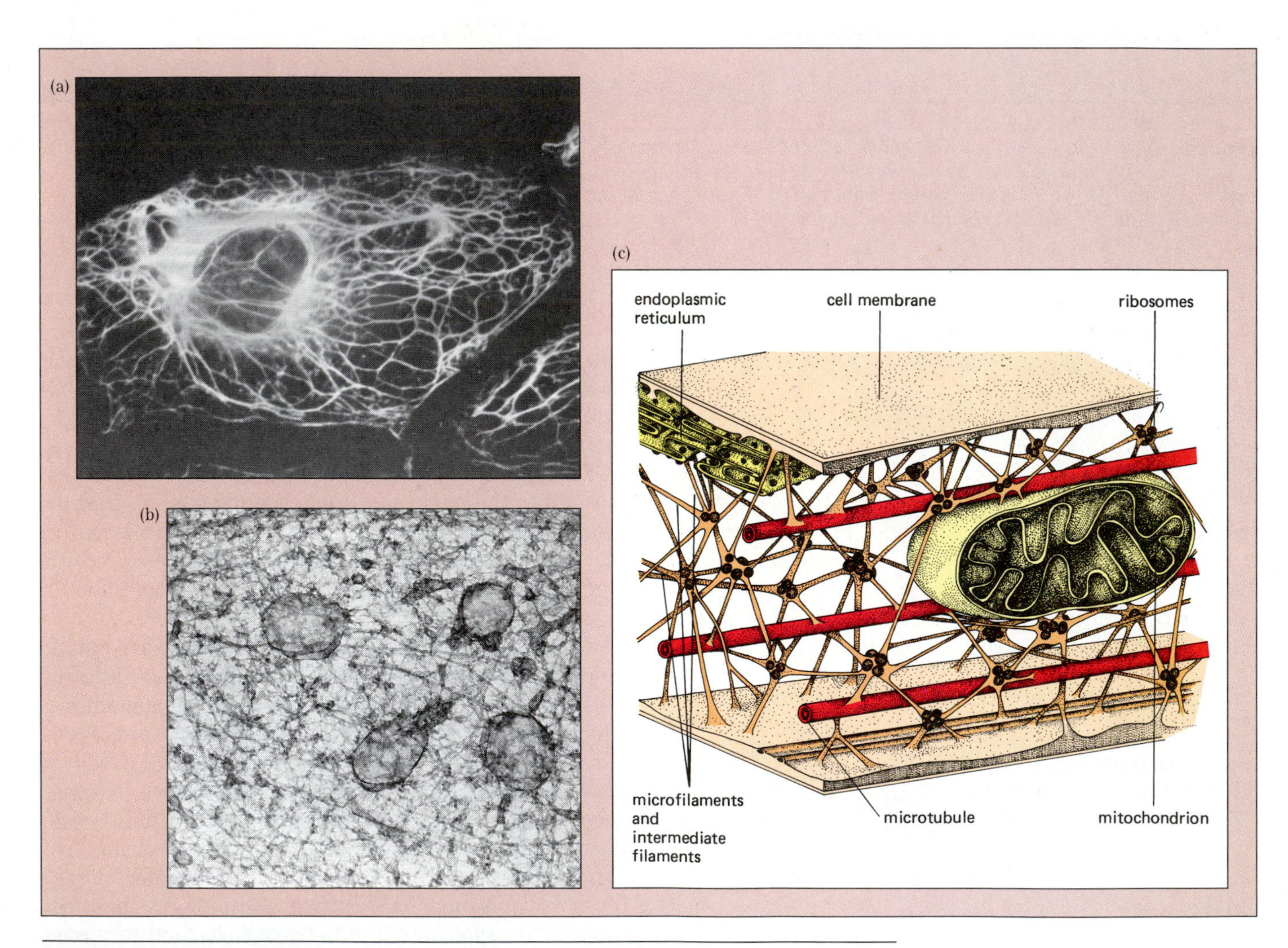

Figure 5-17 *Cells are given shape and organization by the cytoskeleton.* **(a)** *Antibodies bound to microtubules outline part of the cytoskeleton.* **(b)** *High-resolution electron microscopy reveals the maze of filaments and fibers that comprises the cytoskeleton.* **(c)** *An artist's sketch of the cytoskeleton. Many organelles are probably attached to filaments and tubules, which also reinforce the cell membrane. Chemical reactions, movement of cell parts, and acquisition and release of materials are all coordinated by the cytoskeleton.*

Table 5-2 Components of the Cytoskeleton

	Microtubules	Microfilaments	Intermediate Filaments
Structure	Hollow tubes about 25 nm in diameter, may be more than 50 μm in length	Solid strands about 7 nm in diameter, may be several cm long (muscle cells)	Hollow tubes about 8 to 10 nm in diameter, 10 to 100 μm in length
Protein	Tubulin	Actin (most) and/or myosin	At least five different proteins
Function	Major part of cytoskeleton; movement of chromosomes during cell division; movement of organelles within cytoplasm; movement of cilia and flagella	Major part of cytoskeleton; muscle contraction; changes in cell shape, including division of cytoplasm in dividing animal cells; cytoplasmic streaming; movement of pseudopodia	Major part of cytoskeleton; maintenance of cell shape; attachment of microfilaments in muscle cells; support of nerve cell processes (axons)

place to place within a cell. For example, microfilaments attach to vesicles formed during pinocytosis (see Chapter 4) and pull the vesicles into the cell. The vesicles budded off the ER and Golgi complex are probably guided by the cytoskeleton as well. Fourth, during eukaryotic cell division, *microtubules move chromosomes into the daughter cells.* Finally, *division in animal cells results from the contraction of a ring of microfilaments that pinch the parent cell around the middle, producing two new daughter cells.*

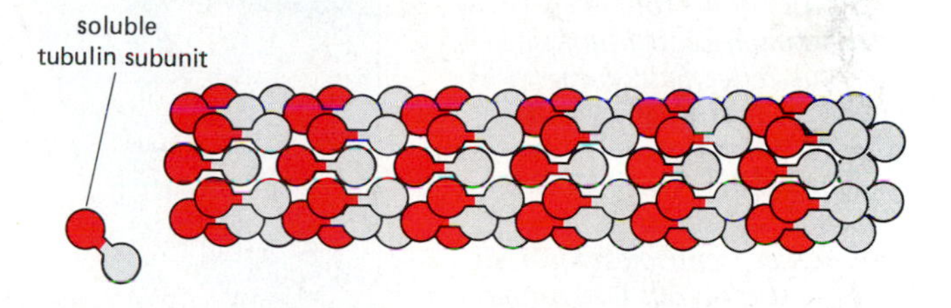

Figure 5-18 Microtubules are assembled in the cell from subunits of the protein tubulin that are dissolved in the cytoplasm.

MICROFILAMENTS. Microfilaments are strands that consist mostly of the protein actin, sometimes in association with a second protein, myosin. Myosin has small extensions that can grab onto actin and flex, sliding the actin filament past the myosin, much like a sailor pulling up an anchor line hand over hand. The most familiar function of actin and myosin is muscle contraction (see Chapter 30), but actin and myosin contribute to shape changes and organelle movement in most, if not all, eukaryotic cells.

MICROTUBULES. Microtubules are hollow cylinders made from many dumbbell-shaped subunits of the protein tubulin (Fig. 5-18). Individual tubulin subunits are available dissolved in the cytoplasm and can be used to construct microtubules when needed by the cell. Some microtubules are relatively permanent features, such as those that make up the cilia and flagella by which many cells move (see below). Many microtubules, however, are transitory, appearing and disappearing as required by the cell.

Both temporary and permanent microtubules are generated by **microtubule organizing centers** in various locations within the cell. All eukaryotic cells have a microtubule organizing center near the nucleus. During cell division, this center produces a football-shaped array of microtubules, called the **spindle apparatus,** that moves the chromosomes into the two daughter cells (see Chapter 8). In animal cells, a prominent pair of **centrioles** is found at the microtubule organizing center near the nucleus (Fig. 5-19). Centrioles are short, barrel-shaped rings of microtubules. When a cell divides, the centrioles are duplicated, and one pair moves into each daughter cell. It is not clear if the centrioles are an active part of the microtubule organizing center or are merely located there and go along for the ride during cell division. In ciliated cells, the centrioles multiply and the "offspring centrioles" move to the surface of the cell. Here they become anchored just beneath the cell membrane and give rise to the microtubules of the cilia. Due to their location at the base of the cilia, these centrioles are often called **basal bodies.**

Microtubules cause cellular movement in two fundamentally different ways. First, microtubules can slide past one another, like escalators moving in opposite directions. Any cellular objects attached to the microtubules also

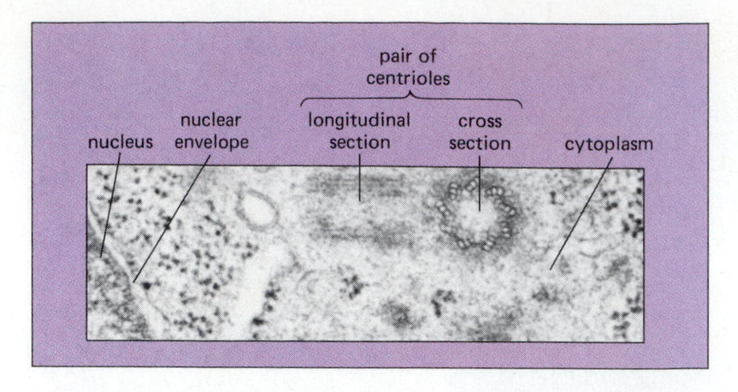

Figure 5-19 *In animal cells, a pair of centrioles is found at the microtubule organizing center near the nucleus. The centrioles, always lying at right angles to one another, are composed of nine short triplets of microtubules. A centriole is identical to the basal body of a cilium or flagellum (see Fig. 5-20).*

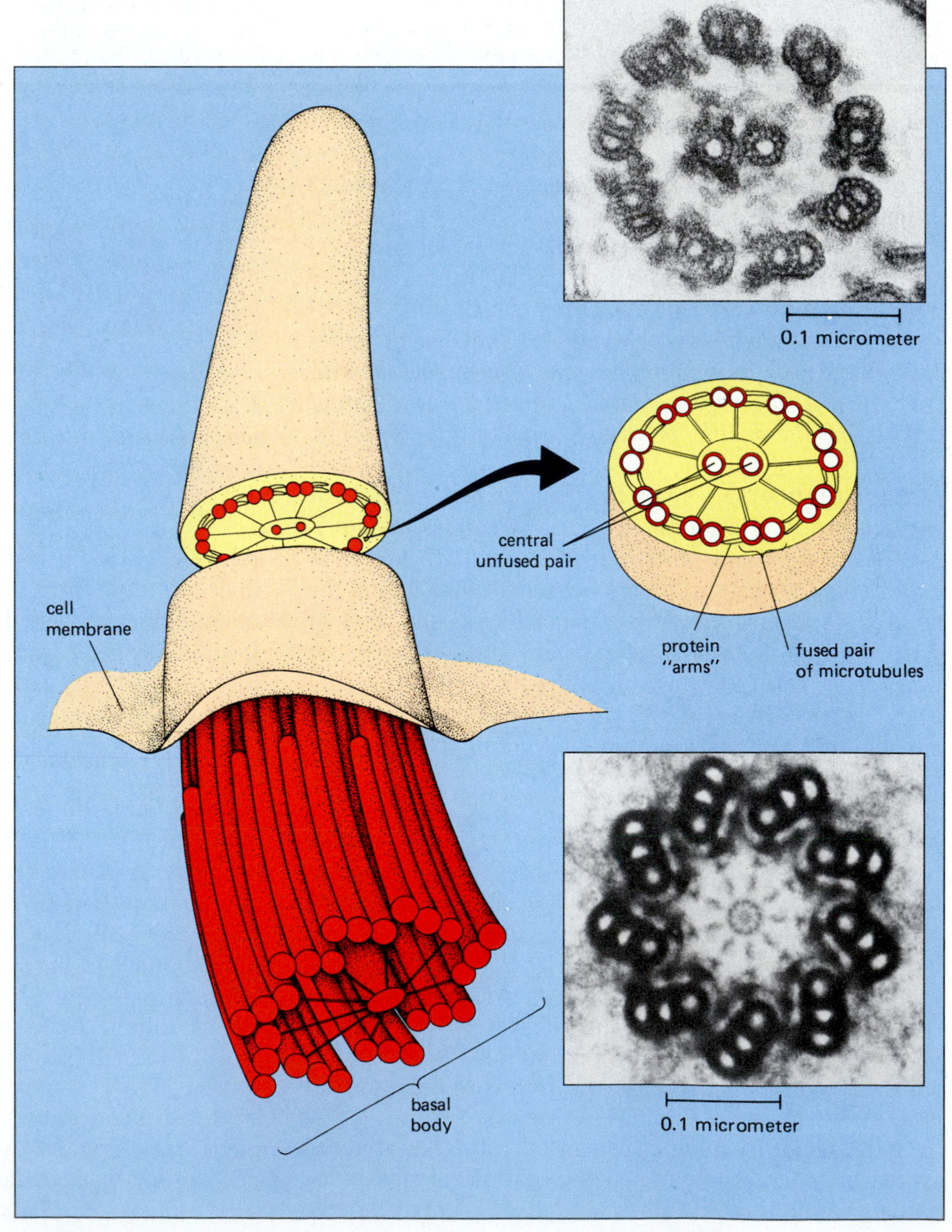

Figure 5-20. *Both cilia and flagella contain microtubules arranged in an outer ring of nine fused pairs of microtubules surrounding a central unfused pair. The outer pairs have extensions that interact with adjacent pairs to provide the force for bending (see Fig. 5-21). Cilia and flagella arise from basal bodies (centrioles) located just beneath the cell membrane. Basal bodies have nine fused triplets, from which the fused pairs arise, and an indistinct central hub. It is not known how the basal body organizes the somewhat different structure of the cilium or flagellum.*

move apart in opposite directions. If one microtubule is fixed in place, for example to the cell membrane, the second microtubule and its passenger organelles will move away to another part of the cell. Microtubule sliding is also responsible for the whiplike action of cilia and flagella (see below). The second mechanism of microtubule movement is, at first glance, slightly bizarre. Tubulin subunits can be added to the ends of a microtubule, making it longer, or removed from the ends, making it shorter, or added at one end and removed from the other, effectively moving the microtubule along in one direction. Assembly and disassembly of microtubules contribute to changes in the overall shapes of cells, for example when nerve cells grow processes from your spinal cord out to the muscles of your legs.

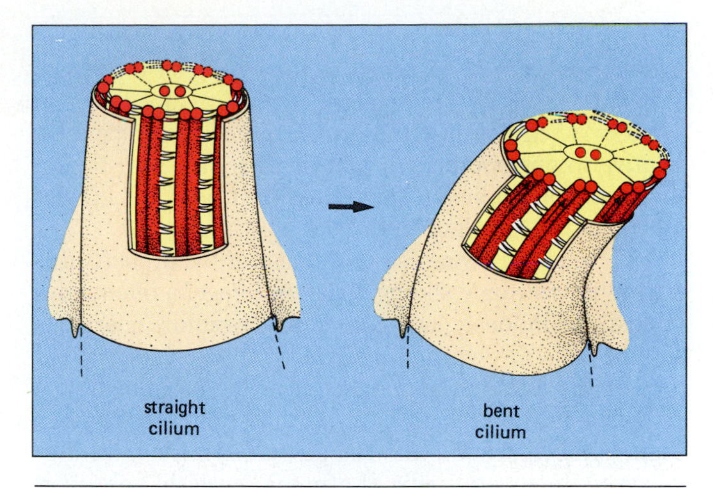

Figure 5-21 Cilia and flagella bend when the arms of one fused pair of microtubules temporarily attach to the shaft of an adjacent pair and flex, "walking" along the shaft. If the microtubules were free, this would result in one pair sliding past the other pair. Since the microtubules are anchored in place by the basal body, the sliding causes the cilium or flagellum to bend.

MICROTUBULAR APPENDAGES: CILIA AND FLAGELLA. Both cilia and flagella are whiplike extensions of the cell membrane. Each cilium and flagellum contains a ring of nine fused pairs of microtubules, with an unfused pair of microtubules in the center of the ring (the so-called "9 + 2" arrangement; Fig. 5-20). This pattern of microtubules is produced by a **basal body** (centriole) located just beneath the cell membrane. A basal body consists of a ring of nine triplets of short microtubules. Two of the members of each triplet give rise to the pairs of microtubules in the cilium or flagellum.

Figure 5-21 diagrams how cilia and flagella bend. Tiny protein "arms" project out from each pair of microtubules in the outer ring. These arms attach to the neighboring pair of microtubules and flex, thereby moving the first pair along relative to the second. However, the basal body firmly anchors the "bottom" of all the microtubules in the entire cilium or flagellum. Therefore, adjacent microtubules can slide past one another only if the whole cilium

Figure 5-22 Characteristic patterns of movement of cilia and flagella.
(a) *Cilia usually "row" along, providing a force of movement parallel to the cell membrane, just as oars provide movement parallel to the sides of a rowboat.*
(b) *Flagella often "scull" with a continuous bending movement proceeding from the base to the tip. This provides a force of movement perpendicular to the cell membrane. In this way a flagellum attached to a sperm can move the sperm straight ahead.*

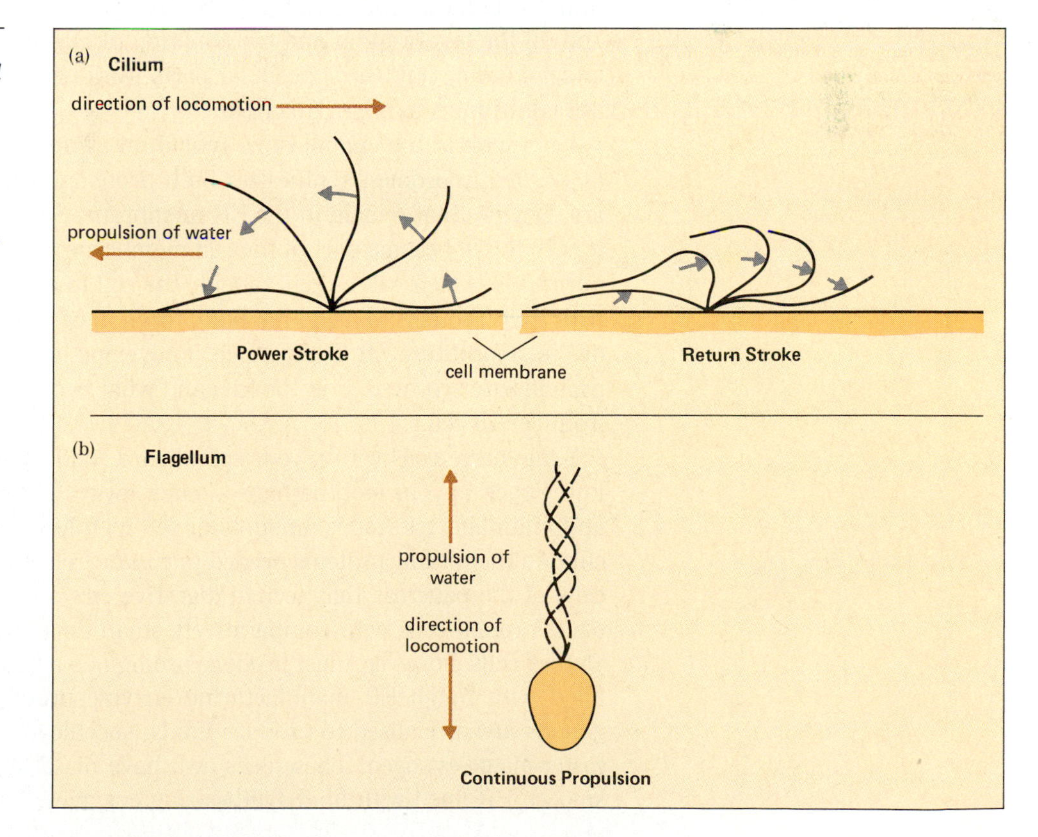

or flagellum bends. (Think of a paintbrush: Bristles of equal length are all attached to the handle of the brush. When the bristles are straight, their tips are lined up evenly. When you flex the brush, the bristles slide a bit relative to one another, so that the tips of the bristles on the inside of the bend extend past the tips of the bristles on the outside of the bend.)

ATP energy powers the movement of the protein arms during microtubule sliding. Cilia and flagella often move almost continuously, and consequently require enormous supplies of ATP, which is generated by mitochondria that are usually found in abundance near the basal bodies.

The main differences between cilia and flagella lie in their number, length, and the direction of the force they generate. In general, **cilia** (Latin for "eyelash") are short (about 10 to 25 micrometers in length), numerous, and provide force in a direction *parallel* to the cell membrane. This is accomplished through a fairly stiff "rowing" motion during the power stroke, with most of the bending occurring at the base of the cilium, and a flexible return stroke (Fig. 5-22a). **Flagella** (Latin for "whip") are long (50 to 75 micrometers), usually few in number, and provide force *perpendicular* to the cell membrane. Flagella undulate with a continuous bending motion, without distinct power and return strokes (Fig. 5-22b).

Unicellular organisms such as *Paramecium* and *Euglena* use cilia or flagella to move about. Animal sperm also rely on flagella for movement. In multicellular animals, cilia are occasionally used for locomotion, moving the animal through a fluid. Many small aquatic invertebrates, for example, swim by the coordinated beating of rows of cilia, like the oars on a Roman slave galley. More often, however, the organism stays put and its cilia move fluids and suspended particles past a surface. Ciliated cells line such diverse structures as the gills of oysters (moving food- and oxygen-rich water), the oviducts of female mammals (moving the eggs along from the ovary to the uterus), and the respiratory tracts of most land vertebrates (clearing debris and microorganisms from the trachea and lungs).

Prokaryotic cells may also bear slender protrusions that undulate or spin, thereby enabling the cell to move about. However, the "flagella" of prokaryotes do not contain microtubules, and have no evolutionary relationship to eukaryotic flagella or cilia.

REFLECTIONS ON CELLULAR DIVERSITY

Every cell in a multicellular organism is highly specialized, down to the level of its organelles, to perform specific tasks. Later in the book, when we discuss plant and animal physiology, you will see that the functioning of each organ, whether brain or kidney, root or leaf, depends both on the specialization of individual cells and on the organization and coordination of all the cells within the organ. Only by understanding how the subcellular parts work can we hope to understand how each cell contributes to organ function.

Let's look at the endoplasmic reticulum as an example. The ER, along with its associated ribosomes, synthesizes both proteins and lipids. Some of those proteins are enzymes that remain in the ER membrane. Some are pores, carriers, and receptors that will become part of the cell membrane. Others are export proteins such as hormones, which will be sent outside the cell to other parts of the body, or even, as with milk proteins, outside the body to another organism. Similarly, lipids may be destined for future ER, Golgi, nuclear envelope, or cell membrane. Which lipids and proteins are synthesized in the ER, and what becomes of them, varies tremendously from cell to cell.

Some nerve cells within your spinal cord send a long fiber all the way from your lower back to your foot, perhaps a meter away. The ER of these cells manufactures and maintains enormous amounts of cell membrane. Further, this membrane must contain the specific proteins needed to conduct electrical signals along the fiber. The cells of the pancreas that secrete digestive enzymes, on the other hand, are more-or-less round cells with comparatively small amounts of cell membrane. These secretory cells, however, must produce prodigious amounts of enzymes, and are nearly filled with rough ER manufacturing enzyme proteins. The secretory cells of the gonads are specialized to produce lipids, specifically steroid hormones such as testosterone or estrogen. These cells also have massive amounts of ER, but here it is smooth ER lined with lipid-synthesizing enzymes.

Cells differ with respect to other organelles as well. In plants, for example, most leaf cells are packed with chloroplasts, whereas root cells lack chloroplasts entirely. Although all cells have many features in common, the subcellular specializations of cells yield a variety of functions, allowing the construction of marvelously complex multicellular organisms.

SUMMARY OF KEY CONCEPTS

The Organization of Eukaryotic Cells

All cells are either prokaryotic or eukaryotic. Prokaryotic cells lack membrane-bound organelles. Eukaryotic cells have several types of membrane-bound organelles, including a nucleus containing genetic material. In eukaryotic cells, the cytoplasm includes all the material within the cell membrane but outside the nucleus. The cytoplasm, in turn, consists of a fluid matrix, the cytosol, and the organelles.

The Structure and Function of Eukaryotic Organelles

All eukaryotic cells contain mitochondria, organelles that use oxygen to complete the metabolism of food molecules, capturing much of their energy as ATP. Cells of plants and some protists contain chloroplasts, which capture the energy of sunlight during photosynthesis, enabling the cells to manufacture organic molecules, particularly sugars, from simple inorganic molecules. Both mitochondria and chloroplasts probably originated from bacteria that were captured by protoeukaryotic cells over a billion years ago and have become incorporated into the normal functioning of the eukaryotic cell.

Ribosomes, which are particles composed of RNA and protein, are the sites of protein synthesis. Ribosomes may be found free in the cytoplasm or attached to the membranous channels of the endoplasmic reticulum (ER). ER with ribosomes, called rough ER, manufactures many cellular proteins. ER without ribosomes, called smooth ER, manufactures lipids. The ER is the site of all membrane synthesis within the cell. The Golgi complex is a series of membranous sacs derived from the ER. The Golgi processes and modifies materials synthesized in the rough and smooth ER. Some substances in the Golgi are packaged into vesicles for transport elsewhere in the cell. Lysosomes are vesicles budded off the Golgi that contain digestive enzymes, which digest food particles and defective organelles.

Plastids are organelles with double membranes found in plant cells, including both chloroplasts and storage plastids containing pigments or starch. Many cells contain sacs bounded by a single membrane, called vacuoles. Plant cells have a large central vacuole that stores water, wastes, and some pigments. Animal and protist cells may contain small vacuoles, including food vacuoles. Certain freshwater protists have contractile vacuoles that expel water that enters the cells by osmosis.

Genetic material (DNA) is contained within the nucleus, which is bounded by the double membrane of the nuclear envelope. Pores in the nuclear envelope regulate the movement of molecules between nucleus and cytoplasm. The genetic material of eukaryotic cells is organized into linear strands called chromosomes, which consist of DNA and proteins. The nucleolus consists of the genes that code for ribosome synthesis, together with ribosomal RNA and protein.

The cytoskeleton organizes and gives shape to the cytoplasm of eukaryotic cells. The cytoskeleton is made of several types of proteins that form strands and filaments within the cytoplasm. Some proteins of the cytoskeleton, notably the microfilaments and microtubules, also contribute to movement of cell parts. Cilia and flagella are whiplike extensions of the cell membrane that contain microtubules in a characteristic 9 + 2 pattern. Cilia and flagella bend through sliding movements of the microtubules.

GLOSSARY

Aerobic using oxygen.

Anaerobic not using oxygen.

Basal body the organelle, structurally identical to a centriole, that gives rise to the microtubules of cilia and flagella.

Central vacuole a large, membrane-bound organelle, containing mostly water and dissolved substances, that occupies most of the volume of mature plant cells.

Centriole (sen′-trē-ōl) in animal cells, a microtubule-containing structure found at the microtubule organizing center and the base of each cilium and flagellum. Gives rise to the microtubules of cilia and flagella, and may be involved in spindle formation during cell division.

Chlorophyll (klor′-ō-fil) a green pigment found in chloroplasts that captures light energy during photosynthesis.

Chloroplast (klor′-ō-plast) the organelle of plants and plantlike protists that is the site of photosynthesis; surrounded by a double membrane and containing an extensive internal membrane system bearing chlorophyll.

Chromatin (krō′-ma-tin) the complex of DNA and proteins that makes up the chromosomes of eukaryotic cells.

Chromosome (krō′-mō-sōm) threadlike bodies that contain the genetic material, DNA; eukaryotic chromosomes also contain proteins bound to the DNA.

Cilium (sil′-ē-um; pl. cilia) a short, hairlike projection from the surface of certain eukaryotic cells, containing microtubules in a 9 + 2 arrangement. Movement of cilia may propel cells through a fluid medium or move fluids over a stationary surface layer of cells.

Contractile vacuole a membrane-bound organelle found in certain protists that takes up water from the cell, contracts, and expels the water outside the cell via a pore in the cell membrane.

Crista (kris′-ta; pl. cristae) a fold in the inner membrane of a mitochondrion.

Cytoplasm (sī′-tō-plazm) the material within a cell, exclusive of the nucleus.

Cytoskeleton a network of protein fibers in the cytoplasm that gives shape to a cell, holds and moves organelles, and is often involved in cell movement.

Cytosol (sī′-tō-sol) the fluid part of the cytoplasm.

Deoxyribonucleic acid (DNA) the molecules that are the genetic material of all living cells.

Endoplasmic reticulum (en-dō-plaz′-mik re-tik′-ū-lum) a system of membranous channels within eukaryotic cells; the site of most protein and lipid synthesis.

Eukaryotic (ū-kar-ē-ot′-ik) referring to cells of organisms of the kingdoms Protista, Fungi, Plantae, and Animalia. Eukaryotic cells have their genetic material enclosed within a membrane-bound nucleus and contain other membrane-bound organelles.

Flagellum (fla-jel′-um; pl. flagella) a long, hairlike extension of the cell membrane. In eukaryotic cells, contains microtubules arranged in a 9 + 2 pattern. Movement of flagella propel some cells through fluid media.

Food vacuole a membranous sac containing food obtained by phagocytosis.

Golgi complex (gōl′-jē) a stack of membranous sacs found in most eukaryotic cells, which is the site of processing and separation of membrane components and secretory materials.

Granum (gra′-num; pl. grana) in chloroplasts, a stack of thylakoids.

Intermediate filament part of the cytoskeleton of eukaryotic cells, probably functioning mainly for support.

Intermembrane compartment the fluid-filled space between the inner and outer membranes of a mitochondrion.

Lysosome (lī′-sō-sōm) a membrane-bound organelle containing intracellular digestive enzymes.

Matrix the fluid contained within the inner membrane of a mitochondrion.

Microfilament part of the cytoskeleton of eukaryotic cells, composed of the proteins actin and (sometimes) myosin; functions in movement of cell organelles and in locomotion by pseudopodia.

Microtubule a hollow cylindrical strand found in eukaryotic cells, composed of the protein tubulin; part of the cytoskeleton used in movement of cell organelles, cell growth, and construction of cilia and flagella.

Microtubule organizing center a region of a eukaryotic cell at which tubulin is assembled into microtubules.

Mitochondrion (mī-tō-kon′-drē-un) an organelle bounded by two membranes that is the site of the reactions of aerobic metabolism.

Nuclear envelope the double membrane system surrounding the nucleus of eukaryotic cells. The outer membrane is often continuous with the endoplasmic reticulum.

Nucleoid (nū′-klē-oyd) the location of the genetic material in prokaryotic cells; not membrane enclosed.

Nucleolus (nū-klē′-ō-lus) the region of the eukaryotic nucleus engaged in ribosome synthesis, consisting of the genes encoding ribosomal RNA, newly synthesized ribosomal RNA, and ribosomal proteins.

Nucleus the membrane-bound organelle of eukaryotic cells that contains the cell's genetic material.

Organelle (or-ga-nel′) a structure found in the cytoplasm of eukaryotic cells that performs a specific function; sometimes used to refer specifically to membrane-bound structures such as the nucleus or endoplasmic reticulum.

Plastid (plas′-tid) in plant cells, an organelle bounded by two membranes that may be involved in photosynthesis (chloroplasts), pigment storage, or food storage.

Prokaryotic (prō-kar-ē-ot′-ik) referring to cells of the kingdom Monera. Prokaryotic cells do not have their genetic material enclosed within a membrane-bound nucleus and also lack other membrane-bound organelles.

Ribosome (rī′-bō-sōm) a particle composed of RNA and protein that is the site of protein synthesis in both eukaryotic and prokaryotic cells.

Rough endoplasmic reticulum endoplasmic reticulum lined on the outside with ribosomes.

Smooth endoplasmic reticulum endoplasmic reticulum without ribosomes.

Spindle apparatus a structure found in dividing eukaryotic cells, composed of microtubules, which appears to guide the movement of chromosomes.

Stroma (strō′-ma) the semifluid material of chloroplasts in which the grana are embedded.

Thylakoid (thī′-la-koyd) a membranous sac within a chloroplast; the thylakoid membranes contain chlorophyll.

Turgor pressure pressure developed within a cell (especially the central vacuole of plant cells) as a result of osmotic water entry.

Vacuole (vak′-ū-ōl) a large vesicle consisting of a single membrane enclosing a space. *See* Central vacuole; Food vacuole; Contractile vacuole.

Vesicle (ves′-i-kul) a small membrane-bound sac within the cytoplasm.

STUDY QUESTIONS

1. Diagram "typical" prokaryotic and eukaryotic cells, and describe their important similarities and differences.
2. Which organelles are common to both plant and animal cells, and which are unique to each?
3. Define nucleus, cytoplasm, and cytosol.
4. Describe the cytoskeleton of a eukaryotic cell.
5. Describe the structure and function of the nuclear membrane.
6. Define chromatin, chromosome, and DNA.
7. What is the nucleolus? Why does it disappear during cell division?
8. What are the functions of mitochondria and chloroplasts? What is the evidence that these originated as captured prokaryotic cells?
9. What is the function of ribosomes? Where in the cell are they typically found? What are the differences in the function of ribosomes in these different cellular locations?
10. Describe the structure and function of the endoplasmic reticulum and Golgi apparatus.
11. How are lysosomes formed? What is their function?
12. Diagram the structure of cilia and flagella.

SUGGESTED READINGS

Becker, W. M. *The World of the Cell.* Menlo Park, Calif.: The Benjamin-Cummings Publishing Co., 1986. Clear, easy to read, with magnificent drawings of cell structures.

DeDuve, C. *A Guided Tour of the Living Cell.* New York: Scientific American Books, 1984. Profusely illustrated and entertainingly written.

Dustin, P. "Microtubules." *Scientific American,* August 1980. Construction and action of microtubules in cell division, cilia, and flagella.

Fawcett, D. *The Cell: An Atlas of Fine Structure.* Philadelphia: W. B. Saunders Company, 1966. An excellent collection of light and electron micrographs of cell structures.

Kessel, R., and Shih, C. *Scanning Electron Microscopy in Biology: A Student's Atlas of Biological Organization.* New York: Springer-Verlag New York, Inc., 1974. The scanning electron microscope has opened new vistas of comprehension of the three-dimensional structure of organisms, tissues, and cells, nowhere more evident than in this beautiful collection.

Lazarides, E., and Revel, J. P. "The Molecular Basis of Cell Movement." *Scientific American,* May 1979 (Offprint No. 1427). The functioning of the cytoskeletal proteins in diverse types of cell movement.

Margulis, L. "Symbiosis and Evolution." *Scientific American,* August 1971 (Offprint No. 1230). Evidence for the hypothesis of the evolution of mitochondria and chloroplasts as captured bacteria, by the leading proponent of the theory.

Rothman, J. "The Compartmental Organization of the Golgi Apparatus." *Scientific American,* September 1985 (Offprint No. 1563). Although they look homogeneous in electron micrographs, the sacs of the Golgi actually form three biochemically distinct processing units.

Weber, K., and Osborn, M. "The Molecules of the Cell Matrix." *Scientific American,* October 1985. New techniques in biochemistry and microscopy are used to probe the structure and function of the cytoskeleton.

6
Energy Flow and Biochemical Reactions

Disorder spreads through the universe, and life alone battles against it.

—G. Evelyn Hutchinson

A lynx chases; a hare flees (Fig. 6-1). These behaviors are possible only because of the coordinated action of billions of individual cells. The cells of predator and prey churn with activity, not only during the chase, but in the quiet times too: taking in nutrients and secreting hormones, contracting muscles, and synthesizing proteins and lipids, to name just a few. These activities all require **energy,** which can be defined simply as *the capacity to do work,* including synthesizing molecules, moving things around, and generating heat and light. In this chapter we examine the physical laws that govern energy flow in the universe, how energy flow in turn governs chemical reactions, and how the pathways that energy takes in a cell are controlled by the molecules of the cell itself. In Chapter 7 we focus on photosynthesis, the chief "port of entry" for energy into the biosphere, and glycolysis and cellular respiration, the most important pathways for energy release.

Figure 6-1 A lynx chases a hare across a snow-covered meadow in Montana. Both predator and prey convert the chemical energy of their food molecules into the kinetic energy of the chase.

ENERGY FLOW IN THE UNIVERSE

All interactions among pieces of matter are influenced by the flow of energy among them, which is governed by the **laws of thermodynamics.** These laws deal with "isolated systems," which are any parts of the universe that cannot exchange either matter or energy with any other parts. The universe as a whole is the largest isolated system. No small part of the universe is truly isolated from all possible exchange with every other part, but the concept of an isolated system is a useful abstraction in dealing with energy flow.

In discussing how energy flow governs events within an isolated system, we need to know two things: (1) *how much energy a system has,* and (2) *how useful the energy is.* The quantity and usefulness of energy are the subjects of the two laws of thermodynamics (Fig. 6-2).

The First Law of Thermodynamics

The **first law of thermodynamics** is often called the law of conservation of energy. It states that *within any isolated system, energy can neither be created nor destroyed,*

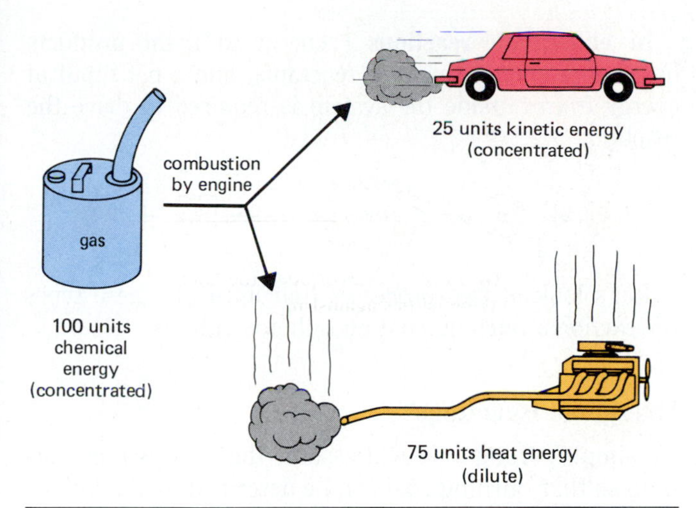

Figure 6-2 Thermodynamics of automobiles. According to the first law, the quantity of energy originally present in the gasoline equals the energy in the resulting products: the molecules in the exhaust, the moving car, and heat. The form of the energy changes from chemical energy into heat or energy of movement. According to the second law, some of the energy in the products are in a less concentrated, less usable form than the chemical energy of the gasoline. In the case of cars, about 75 percent of the chemical energy of the gas is given off as waste heat to the environment, contributing to random movement of molecules in the air.

although it can be changed in form (e.g., from chemical energy to heat energy). In other words, within an isolated system *the total quantity of energy remains constant.* To use a familiar example, let's see how the first law applies to driving your car (Fig. 6-2). To a first approximation, we can consider that your car (with a full tank of gas), the road, and the surrounding air constitute an isolated system. When you drive your car, you convert the chemical energy of gasoline into kinetic energy of movement and heat energy. The total amount of energy in the gasoline is the same as the total amount of kinetic energy plus heat.

An important principle of energy conversion is that *energy always flows "downhill," from places with a high concentration of energy to places with a low concentration of energy.* This is the principle behind engines. As you may know, temperature is a measure of the speed of movement of molecules. The burning gasoline in your car's engine consists of molecules moving at extremely high speeds: a high concentration of energy. The cooler air outside the engine consists of molecules moving at much lower speeds: a low concentration of energy. The molecules in the engine hit the piston harder than the air molecules outside do, so the piston moves upward, moving the car along. Work is done. When the engine is turned off, it cools down. The molecules on both sides of the piston move at the same speed, so the piston stays still. No work is done.

The Second Law of Thermodynamics

The **second law of thermodynamics** states that *any change in an isolated system causes the quantity of concentrated, useful energy to decrease.* To continue our example of a car engine, the heat of the burning gasoline moves the piston, which moves the car. However, the kinetic energy of the moving car is much less than the chemical energy that was originally contained in the gasoline (Fig. 6-2). Where is the "missing" energy? Feel the engine and the exhaust: they are both hot. The burning gas not only moved the piston, it also heated up the engine, the exhaust system, and the air around the car. The friction of tires on pavement heated the road up a little, too. So, as the first law dictates, there really isn't any missing energy: the original quantity of energy in the gasoline still exists, only in different forms. However, some of the energy (the heat given off) merely increased the random movement of molecules in the engine block, the road, and the air. Such random movements are not useful: you can't very well gather up the energy found in the warmed-up road and use it to drive your car a few more blocks.

The second law also tells us something about the organization of matter. Regions of concentrated energy are usually also regions of great orderliness. The eight carbons in a single molecule of gasoline have a much more orderly arrangement than do the eight separate, randomly moving molecules of carbon dioxide formed during combustion. Therefore, another way of looking at the second law is that *all processes in an isolated system result in an increase in randomness and disorder.* This increase in randomness, disorder, and low-level energy is called **entropy.**

Eventually, several billion years from now, all the energy in the universe will be evenly dispersed as heat, and all the matter will be randomly distributed in small molecules. Without differences in energy, no further work will be possible. Life, therefore, will cease. (This rather gloomy prospect is explored in a marvelous science fiction short story by Isaac Asimov entitled "The Last Question": definitely recommended reading.)

Entropy and Life. If you stop to think about the second law of thermodynamics, you may wonder how life can exist at all. If chemical reactions, including those inside living cells, cause the amount of "dilute," unusable energy to increase, and if matter tends toward increasing randomness and disorder, how can organisms accumulate the concentrated energy and precisely ordered molecules that characterize living things? The answer is that the second law applies only to isolated systems, and *living organisms are not isolated systems.* Nuclear reactions in the sun produce concentrated energy (sunlight) at the expense of vast increases in entropy. Living things on Earth use that concentrated energy to synthesize complex molecules and maintain orderly structures (Fig. 6-3). Small

(a) *A plant is placed in an insulated, light-tight box. Without any energy input, the plant dies. Its organized structure and complex molecules break down into simpler, less organized structures and molecules: entropy increases.*

(b) *The isolated system now includes the entire solar system. Nuclear reactions in the sun liberate energy as light and heat. The entropy of the sun increases. Some of the light energy is captured by the plant, which uses it to maintain its organized structure and to grow, that is, to create even more organization. The entropy of the plant decreases. The entropy of the sun increases much more rapidly than the entropy of the plant decreases, so the overall entropy of the solar system increases.*

Figure 6-3 *Life does not violate the second law of thermodynamics. Although entropy must increase within an isolated system as a whole, parts of the system may decrease in entropy.*

pockets of decreasing entropy on Earth do not violate the second law, because the entropy of the solar system as a whole constantly increases.

THERMODYNAMICS AND CHEMICAL REACTIONS

A chemical reaction begins with one set of substances, called the **reactants,** and converts them into another set, the **products.** Thermodynamically, all chemical reactions fall into one of two categories: exergonic and endergonic. In **exergonic reactions** (Greek for "energy out"), the reactants have more energy than the products. Consequently, energy is released by the reaction:

1 + 2 → 3 + 4 + energy

In **endergonic reactions** ("energy in"), the products have more energy than the reactants, and a net input of energy from outside the system is required to drive the reaction:

A + B + energy → C + D

Let's look at two processes that illustrate these types of reactions: burning coal and photosynthesis.

Exergonic Reactions

To simplify things, we will assume that coal is pure carbon, so that burning coal can be described by the following equation:

$$C + O_2 \rightarrow CO_2 + \text{energy}$$

This reaction illustrates two important concepts (Fig. 6-4a). First, molecules of carbon and oxygen contain much more energy than molecules of carbon dioxide do, so the reaction releases energy. *This release of energy allows exergonic reactions to occur without a net input of energy.* In our particular example, coal fires, once started, burn by themselves until the coal is used up: you don't have to keep the fire going by adding outside energy from a flame thrower. [It may be helpful if you think of exergonic reactions as being "downhill," both in the graphical sense shown in Fig. 6-4a and in the everyday sense that cars and bicycles can coast downhill without engines or legs propelling them, and in fact acquire energy (go faster) as they do.]

Although burning coal releases energy, a lump of coal does not burst into flames by itself. This leads to the second point: *all chemical reactions require an initial input of energy, called the* **activation energy,** *to get them started.* Chemical reactions require activation energy because atoms and molecules are surrounded by a cloud of negatively charged electrons. For two molecules to react with each other, their electron clouds must be forced together, despite their electrical repulsion. This, of course, requires energy. The usual source of activation energy is kinetic energy of movement. Only molecules moving with sufficient speed can collide hard enough to force their electron clouds to mingle and react. Since molecules move faster at higher temperatures, most reactions occur more readily and more rapidly at high temperatures. This is what setting coal on fire does: atoms of carbon move faster due to the heat, some vaporizing completely off the lump of coal. These energetic carbon atoms collide with oxygen molecules and react to form carbon dioxide, releasing heat. This released heat causes more carbon atoms to vaporize and collide with oxygen molecules, producing more CO_2 and releasing more heat, thus sustaining the reaction until the coal is consumed.

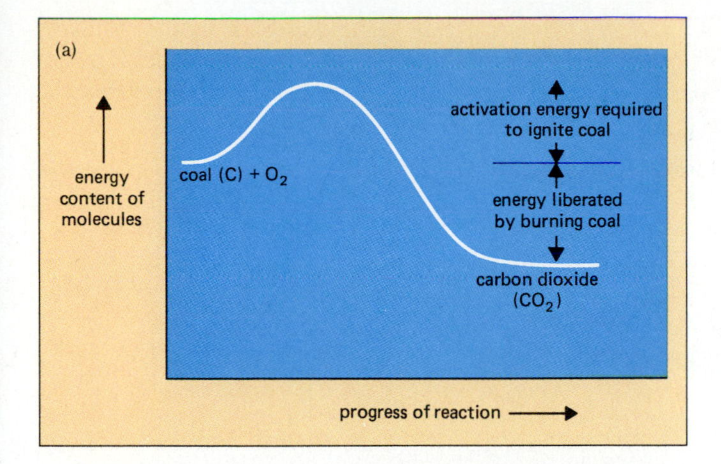

(a) *An exergonic ("downhill") reaction proceeds from high-energy reactants such as coal to low-energy products such as CO_2. The energy difference between the chemical bonds of the reactants and products is released into the environment as heat. To get the reaction started, however, an initial input of energy, the activation energy, is required.*

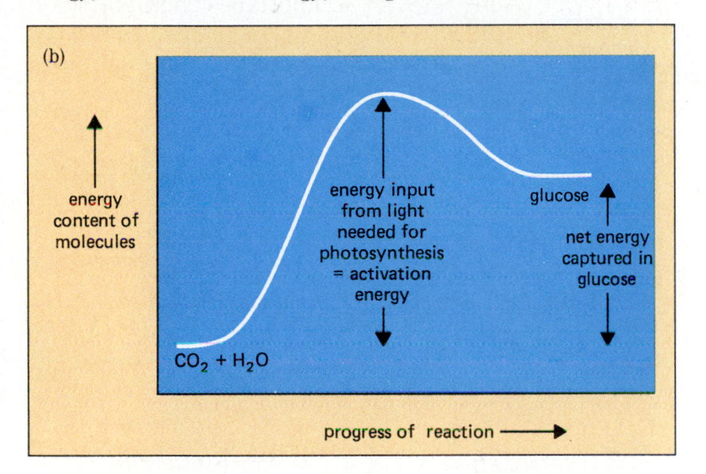

(b) *An endergonic ("uphill") reaction, such as photosynthesis, proceeds from low-energy reactants such as CO_2 and H_2O to high-energy products such as glucose, and therefore requires a net input of energy.*

Figure 6-4. Energy relations in exergonic and endergonic reactions.

Endergonic Reactions

Many reactions in living systems occur in which the products contain more energy than the reactants. These reactions must be driven through an input of energy into the low-energy reactants (Fig. 6-4b). As we will see in the next chapter, photosynthesis in green plants takes low-energy water and carbon dioxide and produces high-energy sugar:

$$6\ CO_2 + 6\ H_2O + \text{energy} \rightarrow C_6H_{12}O_6\ (\text{glucose}) + 6\ O_2$$

This reaction requires energy, which plants obtain from sunlight. We might call endergonic reactions "uphill," in the same sense that vehicles require energy inputs to go uphill.

Coupled Reactions

If you have been following carefully, about this time you may be saying, "Wait a minute. Photosynthesis is an endergonic reaction in an isolated plant, all right, but what if you include the sun in the system?" You are exactly right. Photosynthesis is no exception to the rule that all reactions are driven by energy flowing from high concentration to low concentration. Photosynthesis is an example of a **coupled reaction,** in which *an exergonic reaction provides the energy needed by an endergonic reaction:*

1. Nuclear fusion of hydrogen in the sun → helium + large energy.
2. $6\ CO_2 + 6\ H_2O + \text{small energy} \rightarrow C_6H_{12}O_6 + 6\ O_2$.

The reactions occurring in the sun liberate energy as light. The sunlight captured by a plant possesses much more energy than is needed to drive photosynthesis. Therefore, the overall process, if we include the sun, is exergonic.

Living organisms are chemists *par excellence,* constantly using the energy given off by exergonic reactions (metabolism of food, either eaten or manufactured during photosynthesis) to drive essential endergonic reactions (synthesis of complex molecules, brain activity, or movement; Fig. 6-5).

The exergonic and endergonic halves of coupled reactions often occur in different places, so there must be some way to transfer the energy from the exergonic reaction to the endergonic reaction. In photosynthesis, sunlight carries energy from exergonic reactions in the sun to the endergonic reactions in plants. In coupled reactions occurring within cells, energy is usually transferred from place to place by **energy carrier** molecules, of which the most common is **adenosine triphosphate,** or **ATP.** We will examine the synthesis and use of ATP later in this chapter.

Chemical Equilibria

So far, we have discussed chemical reactions as if they always proceed in one direction, from reactants to products, and then are finished. This is generally not the case. *Chemical reactions are reversible, proceeding spontaneously in one direction but capable of being driven in the reverse direction under the right conditions.* Furthermore, all reactions, if left to themselves, eventually reach a steady state called a **chemical equilibrium,** in which the reaction proceeds at an equal rate in both directions.

To understand the reversibility of chemical reactions and the nature of chemical equilibria, let's look at a reaction of vital importance to all of us: the combination of oxygen with hemoglobin in the blood. Hemoglobin, as

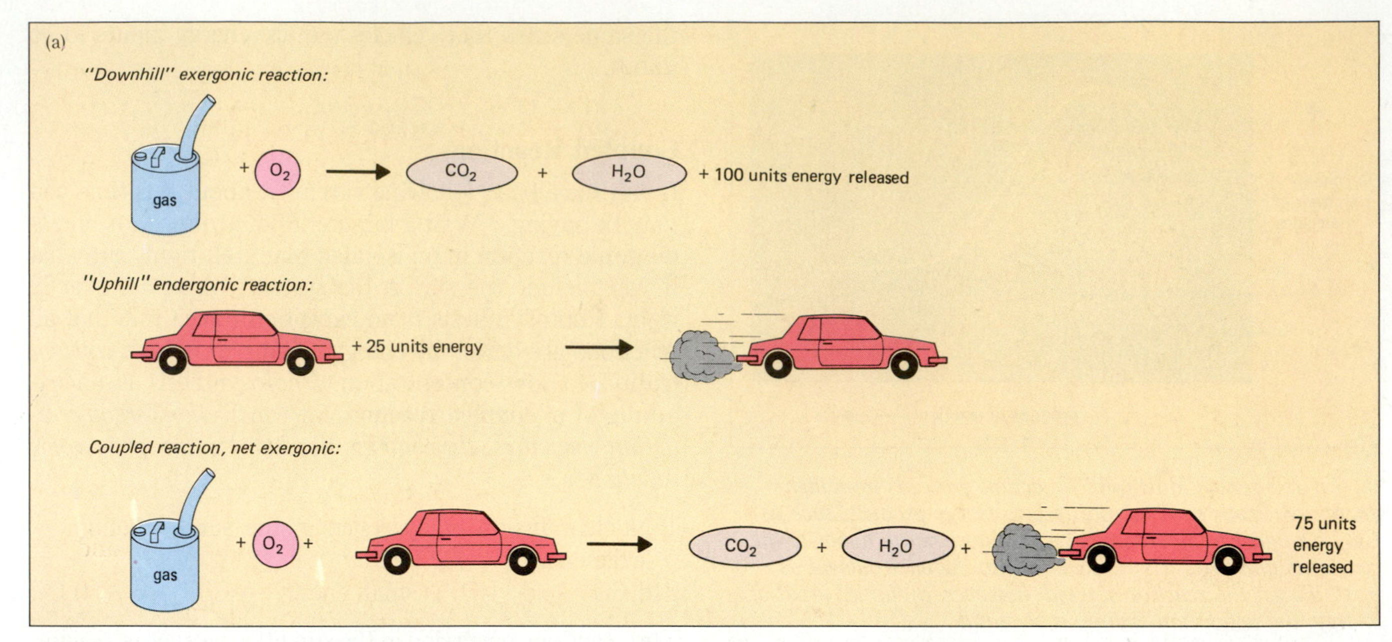

(a) *Burning gasoline is an exergonic reaction that can be coupled to the endergonic reaction of moving a car. According to the second law of thermodynamics, all reactions are inefficient, so the exergonic reaction must actually release more energy than the endergonic reaction requires. The excess energy is given off to the environment as heat. The overall reaction is exergonic.*

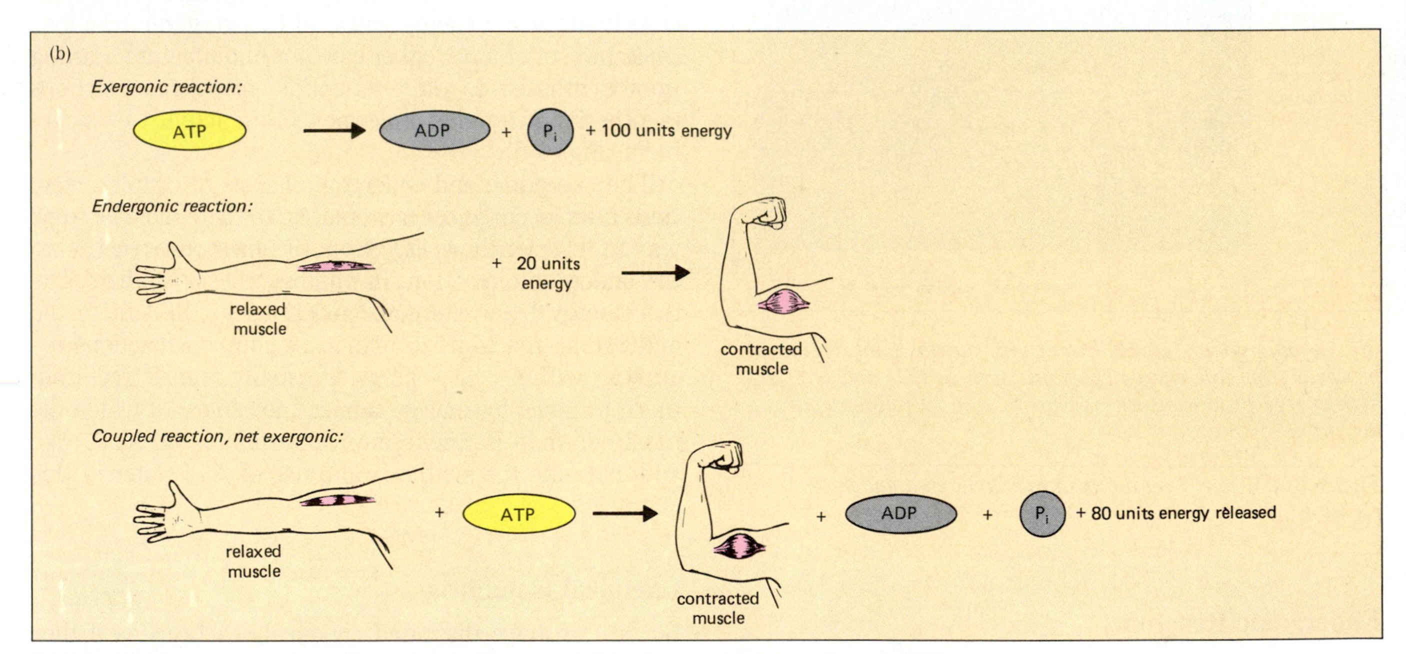

(b) *Muscle movement is an endergonic reaction coupled to the exergonic reaction of ATP hydrolysis. The energy released by ATP hydrolysis exceeds the energy put into muscle contraction, and the overall reaction is exergonic.*

Figure 6-5 Coupled reactions.

you will recall from Chapter 3, is the pigment molecule in your red blood cells. As your blood circulates through your lungs, the hemoglobin picks up oxygen. When the blood passes through the other organs of the body, the oxygen is given off again. Why?

The iron atom held by each hemoglobin subunit can reversibly bind one oxygen molecule. During normal breathing, the concentration of oxygen in the air in your lungs is 15 to 20 percent. When blood enters the lungs, hardly any hemoglobin molecules have oxygen bound to them (you'll see why in a minute). Therefore, oxygen binds to the hemoglobin, with the initial reaction proceeding from reactants (oxygen and hemoglobin) to products (oxyhemoglobin):

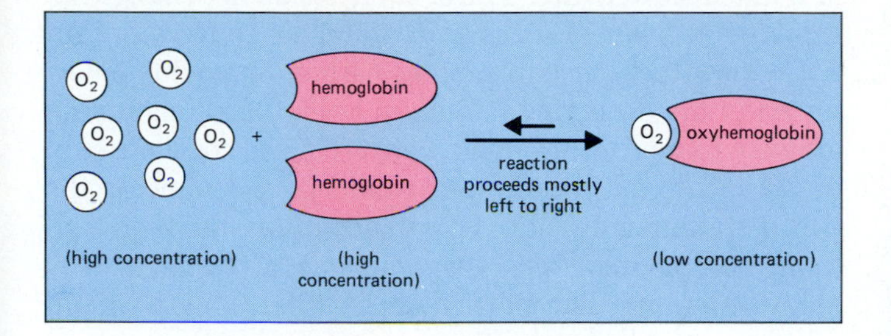

This binding, however, is reversible. Some oxygen molecules leave the hemoglobin again and reenter the air in the lung. Initially, this happens rarely, because very few of the hemoglobin molecules have oxygen bound to them. As more and more hemoglobin molecules become oxygenated, a few give up their oxygens. Finally, an equilibrium is established: just as many oxygen molecules leave the air and bind to hemoglobin as leave the hemoglobin and reenter the air:

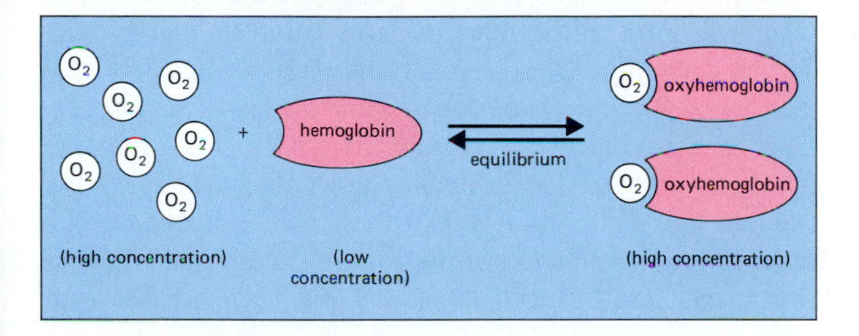

At equilibrium, we cannot really speak of products and reactants anymore, since the reaction proceeds in both directions at the same rate.

Chemical equilibria are affected by the concentrations of the molecules involved. At the high concentration of oxygen in newly breathed air, the equilibrium point occurs with most hemoglobin molecules having oxygen bound to them. In body tissues, say in an exercising muscle, chemical reactions in the mitochondria consume oxygen, so the tissue concentration of oxygen is very low. When oxygenated blood enters the muscle, oxygen molecules will sporadically leave the hemoglobin molecules, just as they have always done. However, there are no free oxygen molecules in the muscle to replace them, so the overall reaction shifts direction:

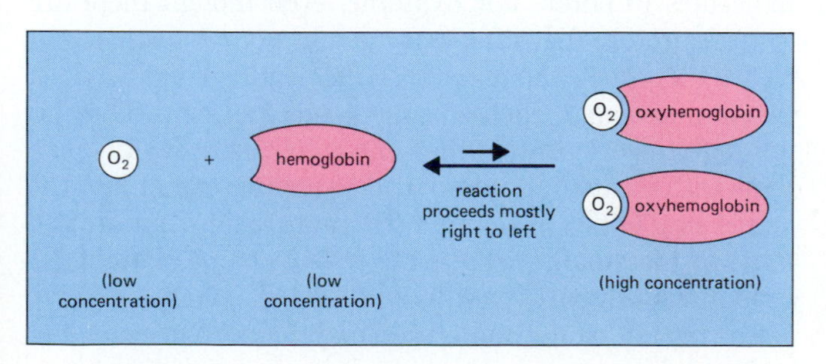

Since the muscle uses up oxygen as fast as the hemoglobin releases it, the oxygen concentration in the muscle remains low. Almost all the oxygen leaves the hemoglobin and enters the muscle cells. A new equilibrium is established, this time with almost all the hemoglobin in the deoxygenated state:

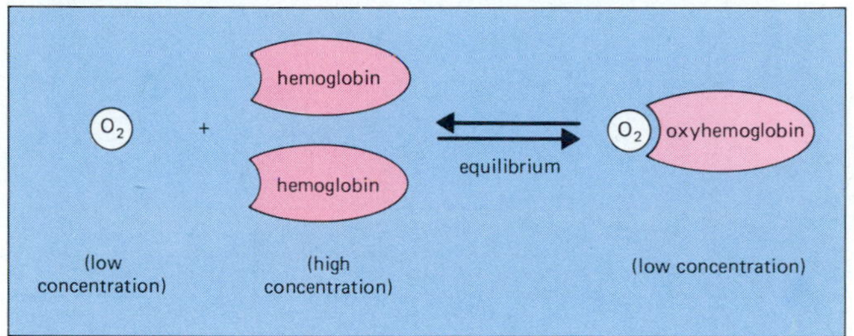

Eventually, the hemoglobin returns to the lungs, and the cycle repeats.

The hemoglobin–oxygen reaction illustrates an important principle of reactions occurring in living organisms. *By adding and removing reactants and products, cells drive reversible reactions back and forth as required by the metabolic demands of the organism.*

CONTROLLING REACTIONS WITHIN THE CELL

The reactions occurring within cells are governed by the same laws of thermodynamics that control any other reactions. However, living cells finely tune their biochemistry in three ways: (1) *They regulate their chemical reactions with protein catalysts called* **enzymes;** (2) *they synthesize energy carrier molecules that store energy, transport it from place to place, and regulate its flow;* and (3) *they couple reactions together, driving energy-requiring endergonic reactions with the energy released by exergonic reactions.*

ACTIVATION ENERGY AND REACTION RATES. The laws of thermodynamics tell us that energy-releasing reactions can occur spontaneously, but they do not say how fast they will occur. In general, the rate of a reaction is determined by its activation energy, which can be thought of as a measure of the velocity with which reactants must collide to force their electron clouds together. Therefore, the rate of most reactions is limited by the velocity of reactant molecules—the faster they travel, the faster the reaction goes. Most reactions can be accelerated by raising the temperature, which increases the velocity of molecules. Without high temperatures, however, many substances would be almost immortal, were it not for the enzymes found in living organisms. The sugar molecules in a candy bar, for example, contain large amounts of energy, but the activation energy required to cause sugar

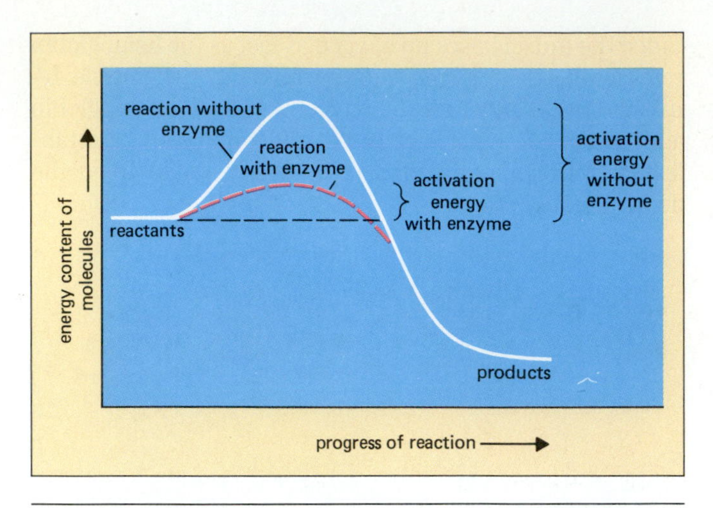

Figure 6-6 Catalysts speed up the rate of reactions by lowering the activation energy.

to react with oxygen is enormous. As any cook knows, you can boil a sugar solution for hours and hardly any sugar breaks down. However, the enzymes in your body allow you to metabolize sugar at far lower temperatures.

Catalysts

Catalysts are molecules that speed up the rate of a reaction without themselves being used up or permanently altered. *Catalysts speed up reactions by reducing the activation energy* (Fig. 6-6).

As an example of catalytic action, let's consider the catalytic converters placed on cars to reduce pollution. When gasoline is completely burned, the final products are carbon dioxide and water:

$$2\ C_8H_{18}\ (\text{octane}) + 25\ O_2 \rightarrow 16\ CO_2 + 18\ H_2O$$

However, flaws in the combustion process generate other substances, including poisonous carbon monoxide, CO. Left to itself, carbon monoxide will slowly react with oxygen in the air to form carbon dioxide:

$$2\ CO + O_2 \rightarrow 2\ CO_2$$

At equilibrium, this reaction results in air with a minuscule concentration of carbon monoxide. In large cities, however, so many cars emit so much CO that the spontaneous reaction of CO with O_2 cannot keep up, and unhealthy levels of carbon monoxide accumulate. Enter the catalytic converter. The platinum catalysts in the converter hasten the conversion of CO to CO_2, thereby reducing air pollution.

Catalytic converters illustrate four important principles about catalysts: First, as we already noted, *catalysts speed up reactions.* Second, *catalysts cannot cause energetically unfavorable reactions to occur;* catalysts only facilitate reactions that would occur spontaneously anyway, although at a much slower rate. Third, *catalysts do not change the equilibrium point of a reaction.* Catalytic converters cannot reduce the concentration of carbon monoxide below the naturally occurring minimum level. If fresh mountain air were to be run through a converter, the concentration of CO in the air coming out would be the same as the concentration going in. Fourth, *catalysts are not consumed in the reactions they promote.* No matter how many reactions they participate in, the catalysts themselves are not permanently changed. Converters on cars have to be replaced, not because of the numbers of CO molecules they have oxidized to CO_2, but because they become poisoned by trace materials in gasoline, or by a few tankfuls of leaded gas.

Enzymes: Biological Catalysts

Enzymes are protein catalysts synthesized by living organisms. These biological catalysts therefore possess the four features listed above. However, enzymes have two additional attributes that set them apart from inorganic catalysts. First, although inorganic catalysts can usually accelerate many different reactions, *enzymes are very specific, promoting only one or at most a few chemical reactions.* In most cases, an enzyme catalyzes a single reaction involving one or two specific molecules, while leaving even quite similar molecules untouched. You may recall from Chapter 3, for example, that animals have enzymes that break apart starch molecules but leave cellulose intact, despite the fact that starch and cellulose are both composed of glucose subunits. The second distinctive trait of enzymes is that *their activity can be regulated, that is, enhanced or suppressed, often by the very molecules whose reactions they promote.*

ENZYME STRUCTURE AND FUNCTION. Why are enzymes specific, and how are they regulated? Enzyme function is intimately related to enzyme structure. Enzymes are proteins with complex three-dimensional shapes. Each enzyme has a dimple or groove, called the **active site,** into which reactant molecules can enter (Fig. 6-7). In enzyme-catalyzed reactions, the reactants are usually called the **substrates** of the enzyme.

The active site of each enzyme has a distinctive shape and distribution of electric charge that allows only certain molecules to enter. For example, even though there are several protein-digesting enzymes in your intestine that cleave the same chemical bond (the peptide bond between two amino acids), each enzyme is quite specific. They do not digest every protein they encounter because only proteins with particular amino acid sequences can fit into the active site. Other proteins, with amino acids that are too large or too small, or have the wrong charge, cannot get in and consequently cannot be digested.

Conversely, some molecules may be able to enter the

Figure 6-7 Enzyme action.

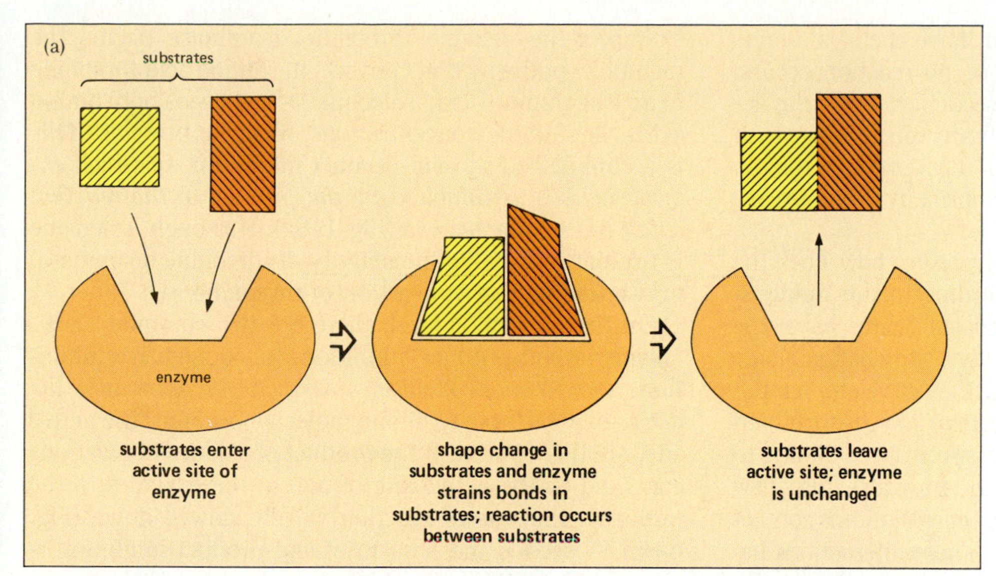

(a) *Both substrate molecules and the active sites of enzymes have specific shapes and electrical charges. According to the induced-fit model of enzyme action, the binding of substrates to the active site alters the shape of both. This strains the chemical bonds in the substrates, making them more reactive.*

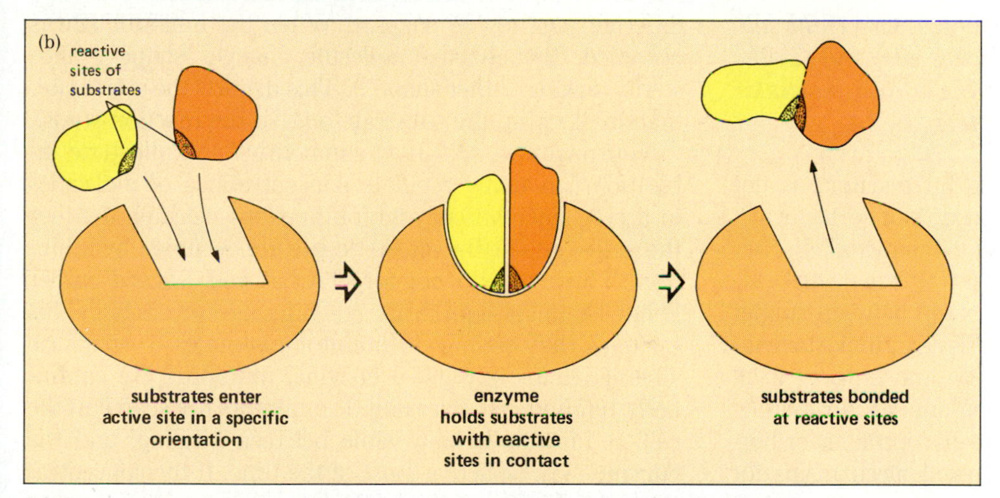

(b) *Large substrate molecules can usually react only at specific sites. Binding to the active site of an enzyme may bring these sites into close contact, promoting chemical recombinations.*

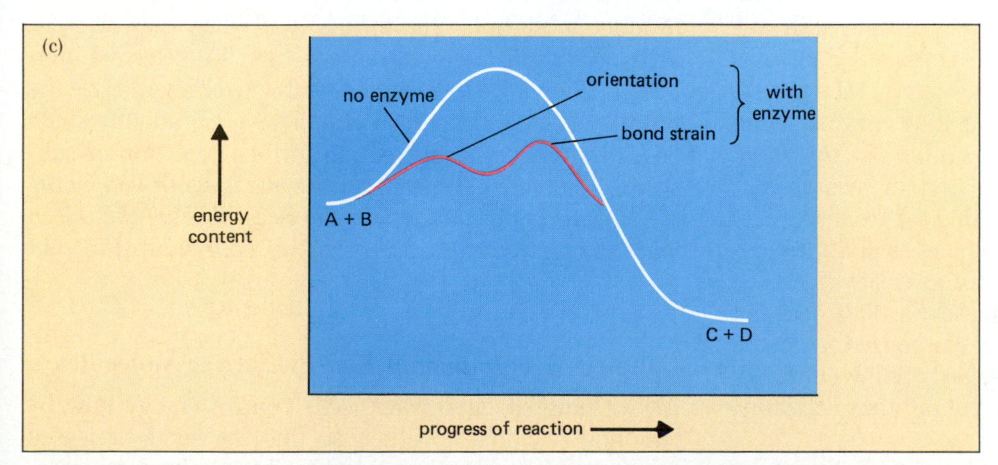

(c) *Substrate orientation and bond strain act as mini-reactions with low activation energy, allowing enzyme-catalyzed reactions to bypass an otherwise high activation energy requirement.*

active site of an enzyme but do not have chemical bonds upon which the enzyme can act, so no reaction occurs. Many poisons, including some insecticides, enter the active site of enzymes essential in brain functioning or in respiration and never leave again. The enzymes remain plugged up, and the reactions they normally promote cannot occur.

Once a substrate enters the active site, how does the enzyme promote a reaction? According to the **induced-fit** model of enzyme action, when substrates enter the active site, both substrate and active site change shape (Fig 6-7a). Specific amino acids of the active site temporarily bind to parts of the substrates. This distorts and weakens the chemical bonds in the substrates. Binding to the active site may also move the substrates into new positions, orienting them so that the correct atoms contact one another and react (Fig. 6-7b). These interactions between substrate and enzyme are like mini-reactions with very low activation energy, allowing the overall reaction to bypass its otherwise high activation energy barrier (Fig. 6-7c).

For an enzyme to catalyze a reaction, then, (1) *the substrate must be able to enter the active site,* and (2) *the substrate and enzyme must have the correct molecular composition to interact with each other.*

Enzyme regulation. Speeding up reactions is not always desirable. For example, you do not want to metabolize every glucose molecule you eat immediately after every meal. For one thing, you might starve to death overnight if you could not store any energy between supper and breakfast. What's more, glucose is an important ingredient in essential body chemicals, and burning it all up would mean there would not be any left for synthesizing molecules such as membrane glycoproteins or hormones. Fortunately, cells have evolved mechanisms for regulating enzyme activity.

One of the simplest ways of regulating enzyme-assisted reactions is to *limit the amount of enzyme present in the cell.* Obviously, if there is no enzyme to begin with, the reaction will not occur. As we will see in Chapter 11, cells can regulate the synthesis of enzymes to meet their changing needs. A second method of regulating enzyme action is to *synthesize the enzyme initially in an inactive form, and activate it only when needed.* Certain cells in your stomach, pancreas, and small intestine, for example, produce enzymes that digest food molecules such as proteins and lipids. These enzymes could just as easily digest the proteins and lipids of the cells themselves. This does not happen, because the enzymes are synthesized in an inactive form, with the active site blocked off. In your digestive tract, the interfering parts are cut off, thus activating the enzymes (see Chapter 24).

More flexible enzyme regulation can be achieved by *temporarily activating and inactivating enzymes, depending on the conditions in the cell at any given time.* For example, the enzyme threonine deaminase begins the metabolic pathway that converts the amino acid threonine to another amino acid, isoleucine. A cell needs both amino acids, in suitable concentrations, to make proteins. This is accomplished by **end-product inhibition,** in which *enzyme activity is inhibited by the product of the reaction catalyzed by the enzyme* (Fig. 6-8a). If enough isoleucine is present, it inhibits the activity of threonine deaminase, preventing further conversion of threonine.

On the molecular level, there are two common mechanisms of end-product inhibition. In **competitive inhibition,** *two or more molecules compete for entry into the active site.* Obviously, if one molecule occupies the active site, another cannot. If the product of a series of reactions can bind to the active site of one of the enzymes in the pathway, the rate of reaction will be slowed down (Fig. 6-8b). A second mechanism of end-product inhibition is **allosteric inhibition,** in which *enzyme action is blocked by molecules binding to the enzyme someplace away from the active site* (Fig. 6-8c). Many enzymes have both an active site that catalyzes the reaction and an inhibitor site on a different part of the enzyme. When the inhibitor site is occupied, the enzyme molecule changes shape ("allosteric" means "other shape"). This distorts the active site, thereby keeping the substrate out, so the reaction stops.

You might wonder how competitive and allosteric inhibition can really regulate concentrations of molecules in a cell, since an enzyme totally stops working if either the active site or the allosteric site are occupied by inhibitors. However, do not forget what you learned earlier about chemical equilibria: reactions are reversible. This includes the binding of inhibitor molecules to enzymes. Therefore, for any given enzyme, inhibitors are continually binding and releasing. If inhibitors are scarce in the cell, it may be a long while between binding, and the enzyme may be active most of the time. If the concentration of inhibitors is high, then as soon as one inhibitor leaves the enzyme, another is likely to take its place, and the enzyme will be inactive most of the time. In end-product inhibition, this means that when there is very little product in the cell, there is very little enzyme inhibition. Therefore, the enzymes will actively synthesize the product. As the product accumulates, it gradually shuts down the enzymes, preventing further depletion of substrate. Thus *the activity of an enzyme is controlled by the concentration of inhibitors in the cell, allowing the cell to maintain optimal concentrations of both substrates and products.*

Coupled Reactions and Energy Carrier Molecules

As we pointed out earlier, cells control energy flow by coupling reactions together, so that the energy released by exergonic reactions is used to drive endergonic reactions. Energy transfer is the role of energy carrier molecules. Energy carriers work something like rechargeable

Figure 6-8 Enzyme regulation.

end product inhibition

(1) A → B, enzyme, inhibition

(2) W → X → Y → Z, enzyme 1, enzyme 2, enzyme 3, inhibition

(a) *In end-product inhibition, enzyme activity may be inhibited by the product of the reaction catalyzed by the enzyme (1) or by the product of a later reaction in the metabolic pathway (2).*

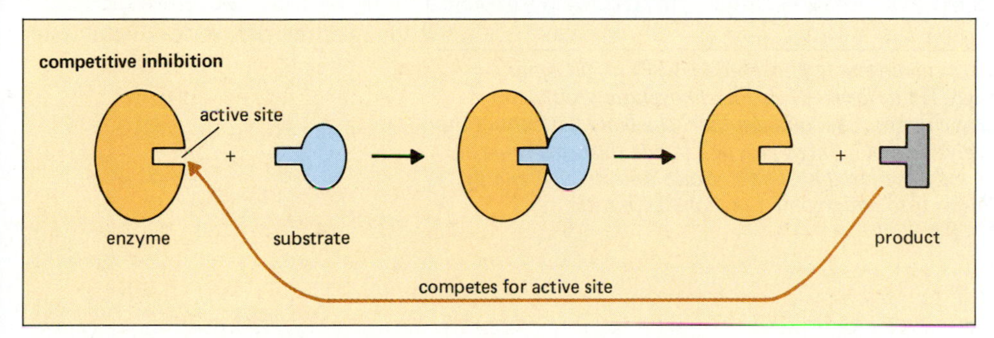

(b) *In competitive inhibition, product and substrate compete for entry to the active site: whenever the product occupies the active site, the substrate cannot get in. Therefore, as product accumulates, the reaction slows down. Other molecules, including regulatory molecules produced by the cell, may also compete with substrates for entry to the active site.*

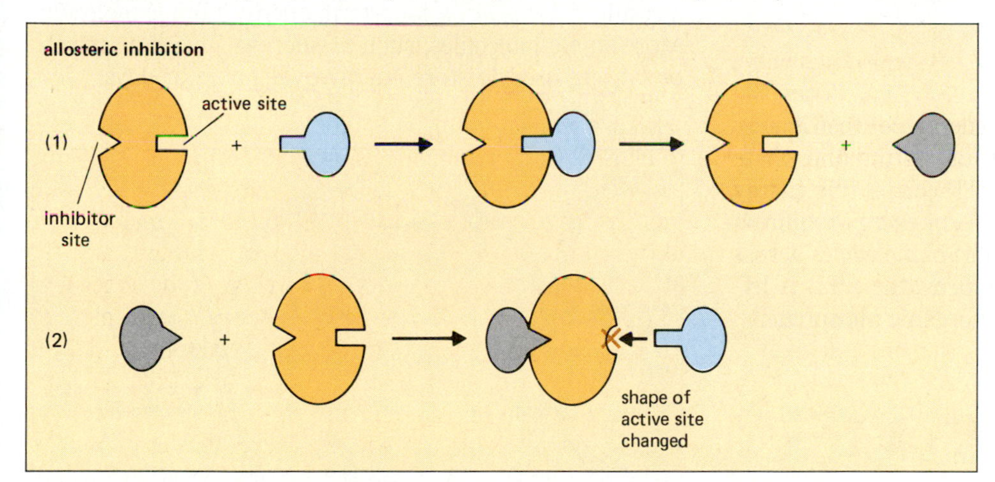

(c) *In allosteric inhibition, the product (or other regulatory molecule) binds to a site on the enzyme other than the active site. This alters the shape or charge of the active site, preventing the substrate from entering.*

batteries, picking up an energy charge at an exergonic reaction, moving to another location in the cell, and releasing the energy again to drive an endergonic reaction.

ATP. The most common energy carrier in living cells is **adenosine triphosphate,** or **ATP.** As you learned in Chapter 3, ATP is composed of a nitrogen-containing base, adenine; a sugar, ribose; and three phosphate groups (Fig. 6-9). Most of the chemical bonds in ATP are "ordinary" covalent bonds, but the bonds joining the last two phosphate groups to the rest of the molecule are special bonds sometimes called "high-energy" bonds. A high-energy bond requires a lot of energy to form and is fairly unstable, being easily broken to give up its energy again. Under most circumstances, only the last high-energy bond of ATP is used by the cell to carry energy.

Energy released in the cell through glucose metabolism is used to drive the synthesis of ATP from adenosine

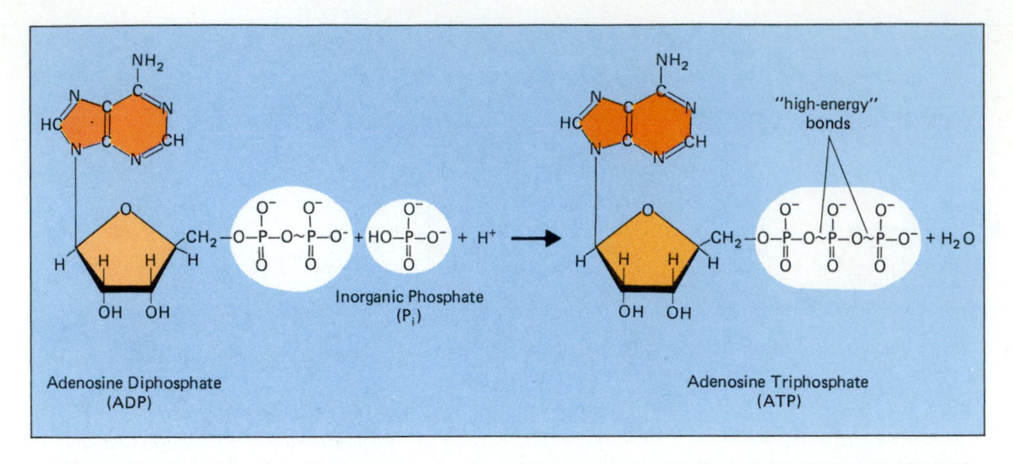

Figure 6-9 *The synthesis and structure of adenosine triphosphate (ATP). A phosphate group bonds to adenosine diphosphate (ADP) to form ATP. The phosphate groups are highly negatively charged. Since negative charges repel one another, the bonds attaching the last two phosphates of ATP are relatively unstable, "high-energy" bonds that can be easily broken, releasing energy. In most cases, only the last phosphate group and its high-energy bond are used to carry energy and transfer it to endergonic reactions within the cell.*

diphosphate (ADP) and inorganic phosphate (P_i):

ADP + P_i + energy ⟶ ATP + H_2O

ATP synthesis requires energy, and some of that energy is stored in the high-energy bond linking the phosphate to the rest of the ATP molecule. ATP carries this energy to various sites in the cell that perform energy-requiring reactions, such as the active transport of molecules across the cell membrane. When this happens, the ATP is hydrolyzed once again to ADP and inorganic phosphate:

ATP + H_2O ⟶ ADP + P_i + energy

The energy released by ATP hydrolysis may be used to drive the energy-requiring reaction. The ADP and P_i are usually recycled back to energy-generating reactions that resynthesize ATP.

Energy use in a cell involves pairs of coupled reactions linked by ATP (Fig. 6-10). In the first coupled reaction, energy release drives ATP synthesis; in the second, ATP hydrolysis drives an endergonic reaction. Each coupled reaction is mediated by a specific enzyme that positions the molecules properly and ensures that the ATP energy is channeled properly. Enzyme-catalyzed coupled reactions are the key to all the synthetic activities of a cell, providing the energy and the specificity necessary to construct the different types of molecules needed by the cell.

So many ATP-driven reactions occur in living cells that the lifetime of any given ATP molecule is very short. As a result, ATP is *not* a long-term energy storage molecule. More stable molecules, such as sucrose, glycogen, starch, or fat, are used to store energy over hours or days.

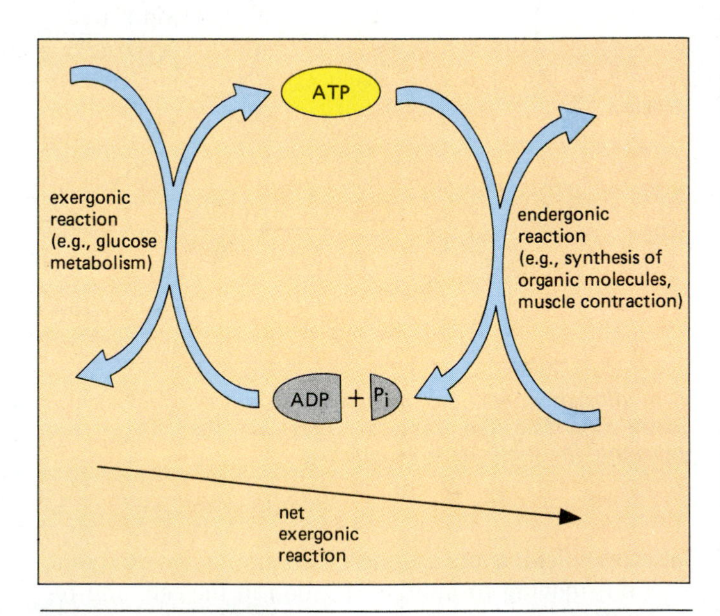

Figure 6-10 *Coupled reactions within living cells. Exergonic reactions (e.g., metabolism of food) drive the endergonic reaction of ATP synthesis from ADP. The ATP molecule moves to another part of the cell, where ATP hydrolysis liberates some of this energy again to drive essential endergonic reactions. The ADP and inorganic phosphate are recycled back to the exergonic reactions, where they will be converted to ATP once again.*

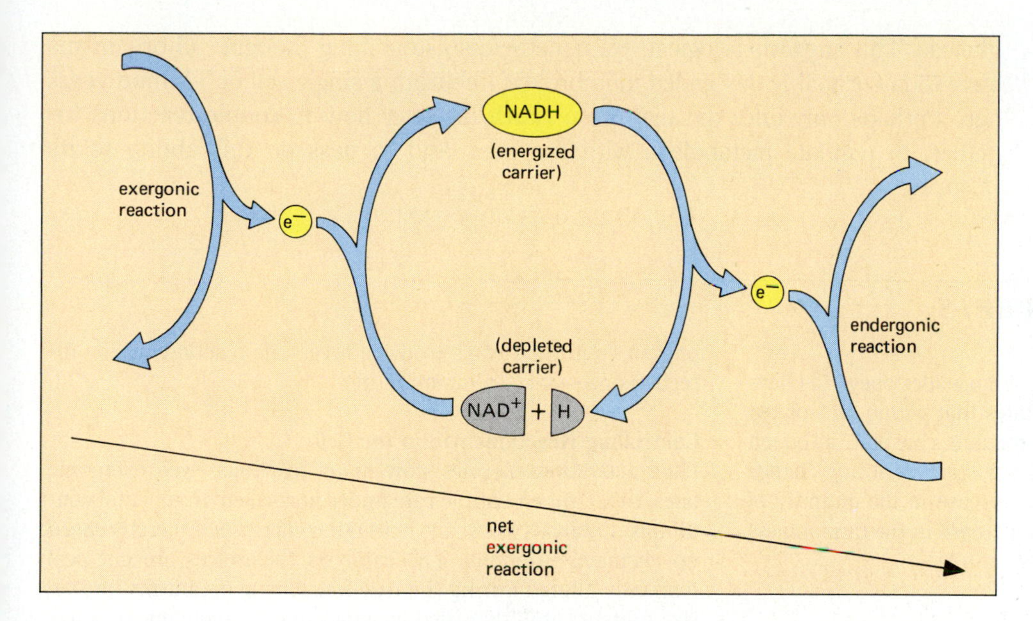

Figure 6-11 *Electron carrier molecules such as nicotinamide adenine dinucleotide (NAD^+) pick up electrons generated by exergonic reactions and hold them in high-energy outer electron shells. Hydrogen atoms are often picked up simultaneously. The electron is then deposited, energy and all, with another molecule to drive an endergonic reaction.*

OTHER ENERGY CARRIERS. Energy may also be transported around a cell by other carrier molecules. In some exergonic reactions, including both glucose metabolism and the light-capture stage of photosynthesis, energy is transferred to electrons. These energetic electrons (sometimes along with hydrogens) may be captured by **electron carriers** (Fig. 6-11), such as nicotinamide adenine dinucleotide (NAD^+) and its relative, flavin adenine dinucleotide (FAD). Loaded electron carriers then donate the electrons, along with their energy, to other molecules. We will see more about electron carriers and their role in cellular metabolism in Chapter 7.

REFLECTIONS ON ENZYMES AND THE EVOLUTION OF LIFE

The reactions that occur inside your body often go excruciatingly slowly if left to themselves or need carefully regulated inputs of energy, or both. Enzymes are essential if these reactions are to proceed smoothly and in proper sequence, or indeed often if they are to occur at all. In fact, it seems certain that life cannot exist without enzymes. The first living organisms, whatever their other characteristics, probably had some molecules that catalyzed vital reactions.

As we shall see in Chapter 11, every amino acid of every enzyme is specified by an organism's genes. One of the most striking findings of the last two decades of biology is the similarity of many enzymes across the spectrum of living organisms. For example, the enzymes that catalyze the early steps of glucose metabolism are very similar in every organism, from bacteria and fungi to higher plants and people.

How could such complex proteins have arisen in the earliest cells, and why should they have remained so similar over billions of years? The origin of enzymes may remain forever speculative, since it happened so long ago and left no fossil record. Almost certainly, early protein catalysts were much smaller, much less specific, and much less efficient than modern enzymes. But once chance mutation had stumbled upon a protein configuration that effectively catalyzed an important reaction, natural selection would strongly oppose too much tinkering with it. Certain doom awaited the organism unfortunate enough to suffer major changes in an essential enzyme.

Diverse living organisms have many biochemical features in common: all use ATP as an energy carrier and all utilize nearly the same pool of nucleotides and amino acids to synthesize their genes and proteins, respectively. Even large molecules such as enzymes remain similar after eons of evolution. We just do not find organisms with genes made of lipids or with strikingly different enzymes that break down glucose. This suggests a logical, yet nonetheless astonishing thought: all organisms alive today probably descended in unbroken line from a single cell of 3 billion years ago. Perhaps only once did matter on Earth discover how to couple reactions together, to regulate metabolism with enzymes, and to pass on this ability to its offspring.

SUMMARY OF KEY CONCEPTS

Energy Flow in the Universe

The flow of energy among atoms and molecules obeys the laws of thermodynamics. The first law states that within an isolated system the total amount of energy remains constant, although it may change in form. The second law states that any change within an isolated system causes a decrease in the quantity of concentrated, useful energy, and an increase in the randomness and disorder.

Thermodynamics and Chemical Reactions

Chemical reactions fall into two categories. In exergonic reactions, the product molecules have less energy than the reactant molecules, so the reaction releases energy. In endergonic reactions, the products have more energy than the reactants, so the reaction requires an input of energy. Exergonic reactions can occur spontaneously, but all reactions, including exergonic ones, require an initial input of energy, called the activation energy, to overcome electrical repulsions between reactant molecules. Exergonic and endergonic reactions may be coupled together, so that the energy liberated by an exergonic reaction drives the endergonic reaction. Living organisms couple exergonic reactions such as light-energy capture or sugar metabolism with endergonic reactions such as the synthesis of organic molecules.

Both exergonic and endergonic reactions are reversible and can proceed in either direction, given suitable inputs of products, reactants, and energy. Living organisms provide these inputs in controlled ways to drive reversible reactions in the directions necessary to maintain life.

Controlling Reactions within the Cell

High activation energies slow many reactions, even exergonic ones, to an imperceptible rate under normal environmental conditions. Catalysts lower the activation energy and thereby speed up chemical reactions. The catalysts themselves are not permanently altered during the reaction. Living organisms synthesize protein catalysts called enzymes that promote one or a few specific reactions. The reactants temporarily bind to the active site of the enzyme, which strains their original chemical bonds and makes it easier to form the new chemical bonds of the products. Enzyme action is regulated in three ways: (1) by altering the rate of enzyme synthesis, (2) by activating previously inactive enzymes, and (3) by inhibiting enzyme activity. One common form of enzyme inhibition is end-product inhibition, in which the end product of a metabolic pathway inhibits the activity of one of the enzymes that mediates an earlier step. As products accumulate, they automatically slow down their own rate of synthesis.

Energy released by chemical reactions within a cell is captured and transported about the cell by energy carrier molecules such as ATP and electron carrier molecules. These molecules are the major means whereby cells couple exergonic and endergonic reactions occurring at different places within the cell.

GLOSSARY

Activation energy in a chemical reaction, the energy needed to force the electron clouds of reactants together, prior to the formation of products.

Active site the region of an enzyme molecule that binds substrates and performs the catalytic function of the enzyme.

Adenosine triphosphate (a-den′-ō-sēn trī-fos′-fāt; ATP) a molecule composed of ribose sugar, adenine, and three phosphate groups. The last two phosphate groups are attached by "high-energy bonds" that require considerable energy to form and release that energy again when broken. ATP serves as the major energy carrier in cells.

Allosteric inhibition (al-ō-ster′-ik) enzyme regulation in which an inhibitor molecule binds to an enzyme at a site away from the active site, changing the shape or charge of the active site, so that it can no longer bind substrate molecules.

Catalyst (cat′-a-list) a substance that speeds up a chemical reaction without itself being permanently changed in the process. Catalysts lower the activation energy of a reaction.

Chemical equilibrium the condition in which the "forward" reaction of reactants to products proceeds at the same rate as the "backward" reaction from products to reactants, so that no net change in chemical composition occurs.

Competitive inhibition in enzyme-catalyzed reactions, a condition in which two molecules (at least one a substrate for the enzyme) compete for entry into the active site of the enzyme, thus slowing down the rate of reaction.

Coupled reactions a pair of reactions, one exergonic and one endergonic, that are linked together so that the energy produced by the exergonic reaction provides the energy needed to drive the endergonic reaction.

Electron carrier a molecule that can reversibly gain and lose electrons. Electron carriers generally accept high-energy electrons produced during an exergonic reaction and donate the electrons to acceptor molecules that use the energy to drive endergonic reactions.

Endergonic (en-der-gon′-ik) pertaining to a chemical reaction that requires an input of energy to proceed; an "uphill" reaction.

End-product inhibition in enzyme-mediated chemical reactions, the condition in which the product of a reaction inhibits one or more of the enzymes involved in synthesizing the product.

Energy the capacity to do work.

Energy carrier a molecule that stores energy in "high-energy" chemical bonds and releases the energy again to drive coupled endergonic reactions. ATP is the most common energy carrier in cells.

Entropy (en′-trō-pē) a measure of the amount of randomness and disorder in a system.

Enzyme (en′-zīm) a protein catalyst that speeds up the rate of specific biological reactions.

Exergonic (ex-er-gon′-ik) pertaining to a chemical reaction that releases energy; a "downhill" reaction.

Induced fit a theory of enzyme activity proposing that binding of substrates to an enzyme active site changes the shape or charge of both the substrates and the active site.

Product an atom or molecule resulting from a chemical reaction.

Reactant an atom or molecule that is used up in a chemical reaction to form a product.

Substrate the atoms or molecules that are the reactants for an enzyme-catalyzed chemical reaction.

STUDY QUESTIONS

1. State the two laws of thermodynamics. Define an isolated system in which they apply.
2. Why don't living organisms violate the second law of thermodynamics? What is the ultimate energy source for life?
3. Define endergonic and exergonic reactions, and give an example of each.
4. What are coupled reactions? How do living organisms use coupled reactions to sustain life?
5. What is activation energy? How do catalysts affect activation energy? How does this change the rate of reactions?
6. What is a chemical equilibrium? Why are reversible reactions important in living organisms?
7. What is an enzyme? In what ways are enzymes similar to, and different from, inorganic catalysts?
8. Describe the structure and function of enzymes. How is enzyme activity regulated?
9. What is an energy carrier? What is the most common energy carrier in living cells? How does it couple exergonic and endergonic reactions together?

SUGGESTED READINGS

Baker, J. J. W., and Allen, G. E. *Matter, Energy, and Life,* 3rd ed. Reading, Mass.: Addison-Wesley Publishing Co., Inc., 1974. A discussion of energetics from a biological point of view.

Dickerson, R. E. "Cytochrome *c* and the Evolution of Energy Metabolism." *Scientific American,* March 1980 (Offprint No. 1464). Cytochrome *c* is an electron carrier in mitochondria, and one of the best studied proteins. Dickerson points out the evolutionary heritage of this molecule in diverse organisms, and also discusses its role in energy acquisition by cells. See also Dickerson's earlier article on cytochrome *c,* "The Structure and History of an Ancient Protein," *Scientific American,* April 1972.

Koshland, Jr., D. E. "Protein Shape and Biological Control." *Scientific American,* October 1973 (Offprint No. 1280). Enzyme function is intimately related to structure. Koshland discusses enzyme specificity and regulation in terms of its protein structure.

Stryer, L. *Biochemistry,* 3rd ed. New York: W. H. Freeman and Company, Publishers, 1988. This text is especially good in its discussion of enzyme structure and function.

7
Energy Use in Cells

With few exceptions, the flow of energy through life on today's Earth begins with the sun (Fig. 7-1). This was not always so. About $4\frac{1}{2}$ billion years ago, the Earth formed as chunks of matter collided and fused, transforming their energy of movement into heat. Storms and volcanic eruptions released still more energy on the newly formed planet, but no organisms existed to harness the enormous energy fluxes, nor is it likely that any could have withstood the violence of that time. Energy-rich organic molecules formed, their synthesis driven by heat and lightning (see Chapter 15). As the Earth cooled and calmed, living cells arose, feeding on the soup provided by the earlier chemical cauldron. However, the cells gradually consumed the organic molecules, and the soup thinned. Sources of organic energy became scarce.

Figure 7-1 Almost all of the energy available to life on Earth comes from the sun. Photosynthesis by plants traps a small fraction of the sunlight energy striking the Earth, and nearly all other life forms obtain their energy, directly or indirectly, from plants.

All the while, another source of energy flowed to the Earth: the light of the sun. Through chance changes in their molecules, some fortunate cells acquired the ability to capture the energy of sunlight and use it to synthesize organic molecules, such as glucose, that are rich in chemical energy—**photosynthesis** had arisen. These cells prospered, and the seas became filled with their progeny.

Evolution continued, and several types of photosynthesis evolved. The most common type released oxygen as a by-product. Gradually, so much oxygen was released that it began to accumulate in the atmosphere. Free oxygen is an extremely reactive chemical, and it would have been dangerous to early life forms, breaking down their hard-won organic molecules. But this very reactivity also provided an opportunity. When cells metabolize glucose without oxygen, a process called **glycolysis,** the products are not much lower in energy than the reactants, and therefore not much energy is released (Fig. 7-2, solid line). This is true even for photosynthetic organisms: without oxygen, only a small fraction of the light energy captured by photosynthesis and stored in the chemical energy of glucose molecules can be extracted again to drive essential cellular reactions.

When glucose is broken down in the presence of oxygen, however, the products are extremely low-energy molecules of carbon dioxide and water (Fig. 7-2, dashed line). If a cell could control these reactions, it could obtain vast new supplies of useful energy. Eventually, cells evolved the enzymes that enabled them to do just that. The advantages of this new process, called **cellular respiration,** were enormous. Cellular respiration can extract about 19 times more energy from each food molecule than glycolysis can. Cells that could respire would grow faster and reproduce more rapidly than cells that relied on glycolysis alone.

The complementary reactions of photosynthesis and cellular respiration are perhaps 2 billion years old, and together they drive the flow of energy and the cycling of carbon through individual organisms and ecosystems (Fig. 7-3).

The molecular events of photosynthesis, glycolysis, and respiration are very complex. Therefore, we will divide

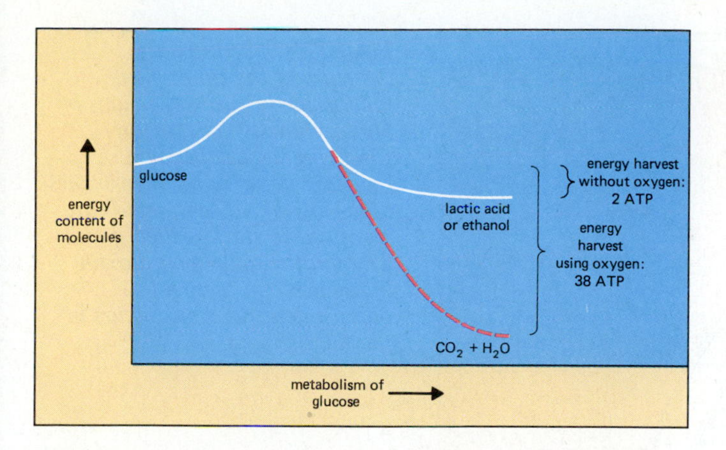

Figure 7-2 A chemical energy diagram for glucose metabolism, with and without oxygen. Without oxygen, glucose cannot be broken down further than lactic acid or ethanol, which still retain most of the chemical energy originally present in the glucose. The use of oxygen allows glucose to be broken down completely to carbon dioxide and water. The extremely low energy levels of carbon dioxide and water permit a rich energy harvest by the cell.

our descriptions into two parts. In the main body of the text and its accompanying illustrations, we examine each of these processes in terms that minimize the molecular and biochemical details. Special illustrations set in tan boxes and titled "A Closer Look at . . . " describe the molecular mechanisms in more depth.

CAPTURING LIGHT ENERGY: PHOTOSYNTHESIS

Photosynthesis *uses the energy of sunlight to convert low-energy reactants, carbon dioxide and water, into high-energy products, glucose and oxygen:*

$$6\,CO_2 + 6\,H_2O + \text{energy} \rightarrow C_6H_{12}O_6 + 6\,O_2$$

In plant and algal cells, photosynthesis occurs within the chloroplasts. Let's begin, then, by reviewing the structure of the chloroplast and how this structure relates to the events of photosynthesis.

The Chloroplast: Site of Photosynthesis

Chloroplasts are organelles found in plant and algal cells that consist of a double outer membrane enclosing a semi-

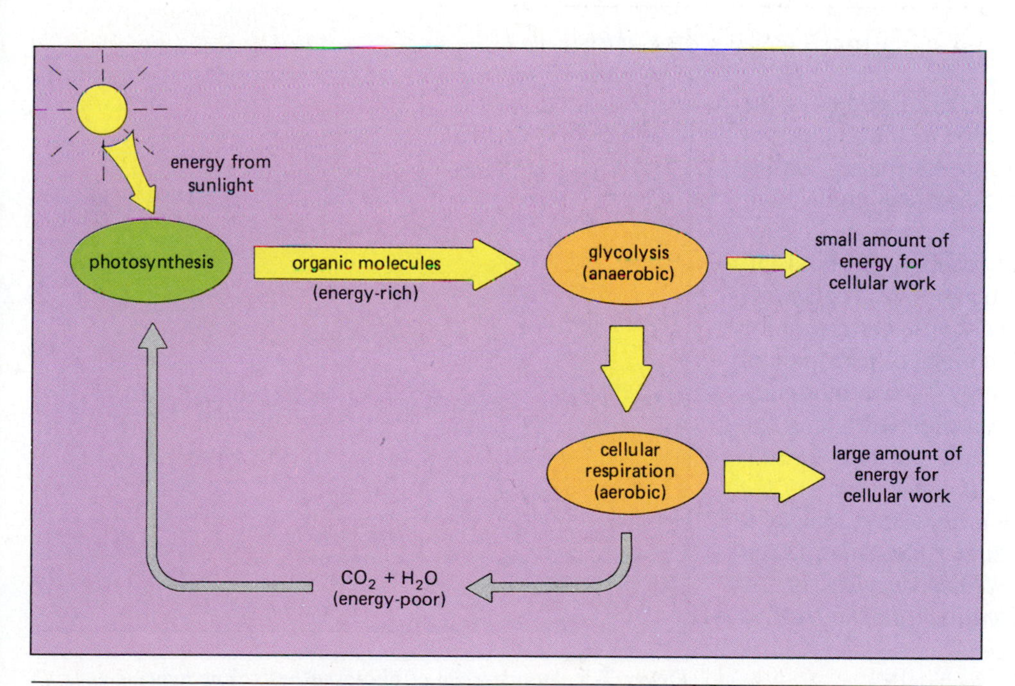

Figure 7-3 Carbon cycles through the environment in the mirror-image reactions of photosynthesis and cellular respiration. Chloroplasts in green plants use the energy of sunlight to synthesize high-energy carbon compounds such as glucose from low-energy molecules of water and carbon dioxide. Plants themselves, and other organisms that eat plants or one another, extract energy from these organic molecules by glycolysis and cellular respiration, yielding water and carbon dioxide once again. This energy, in turn, drives all the reactions of life.

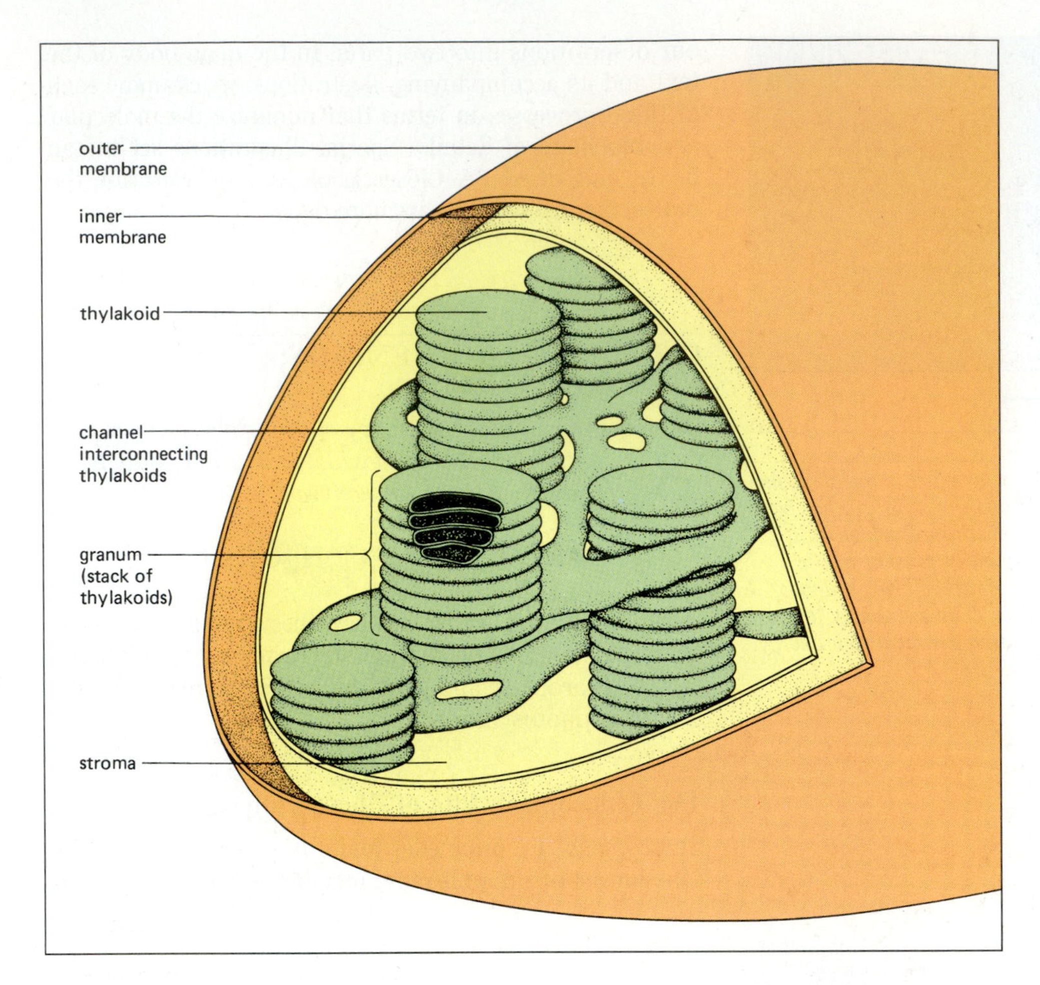

Figure 7-4 The intricate structure of the chloroplast reflects the intricate reactions of photosynthesis. The energy-capturing light-dependent reactions occur in the membranes of the thylakoid sacs, which are arranged in stacks called grana. The fixation of carbon dioxide into glucose in the light-independent reactions occurs in the fluid stroma.

fluid medium, the **stroma** (Fig. 7-4). Embedded in the stroma are disk-shaped, interconnected membranous sacs called **thylakoids.** In most chloroplasts, the thylakoids are piled atop one another in stacks called **grana** (singular, **granum**).

The seemingly simple chemical reaction of photosynthesis actually involves dozens of enzymes catalyzing dozens of individual intermediate reactions. Conceptually, however, photosynthesis can be thought of as a pair of reactions coupled together by energy carrier molecules (Fig. 7-5). Each reaction occurs in a different site in the chloroplast. (1) *In the light-dependent reactions, chlorophyll and other molecules bound in the membranes of the thylakoids capture sunlight energy and convert some of it to the chemical energy of energy carrier molecules.* (2) *In the light-independent reactions, soluble enzymes in the stroma use the chemical energy contained in the carriers to power the synthesis of glucose.*

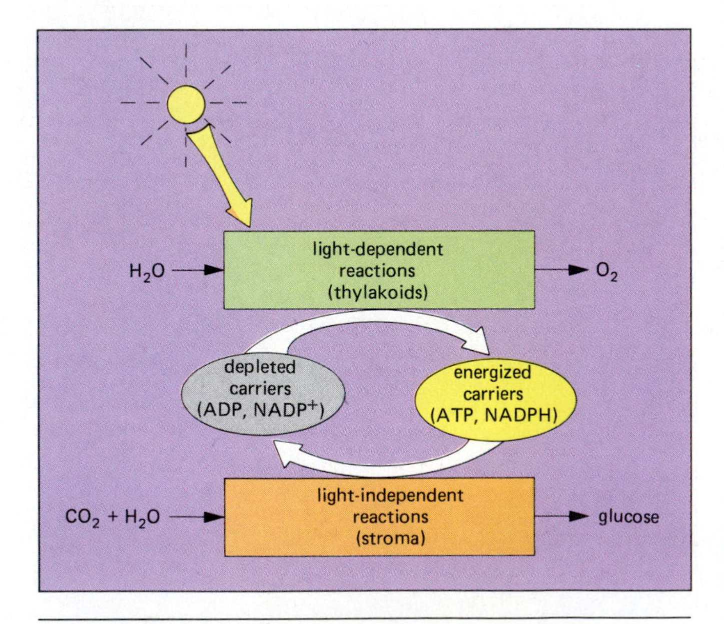

Figure 7-5 An overview of photosynthesis. The light-dependent reactions capture light energy and convert it to the high-energy chemical bonds of ATP and energetic electrons of NADPH (yellow). The light-independent reactions use the energy of ATP and NADPH to synthesize glucose from carbon dioxide and water. In the process, ADP and $NADP^+$ (gray) are regenerated and pass back to the light-dependent reactions to be recharged.

Converting Light to Chemical Energy: The Light-Dependent Reactions

The first steps in photosynthesis, the light-dependent reactions, convert the energy of sunlight into the chemical energy of two different carrier molecules: the familiar energy carrier ATP and the electron carrier nicotinamide adenine dinucleotide phosphate (NADPH).

LIGHT, PIGMENTS, AND PHOTOSYNTHESIS. The electromagnetic radiation emitted by the sun covers a wide spectrum, from short wavelength gamma rays, through ultraviolet, visible, and infrared light, to very long wavelength radio waves (Fig. 7-6a). As you probably know, light and the other types of radiation are composed of individual packets of energy, called **photons.** The energy of a photon corresponds to its wavelength: short wavelength photons are very energetic, while longer wavelength photons have lower energies. Visible light, although only a small portion of the spectrum, contains the wavelengths with energies that are most useful to living organisms.

The color of an object is determined by the wavelengths of visible light that it absorbs and the wavelengths that

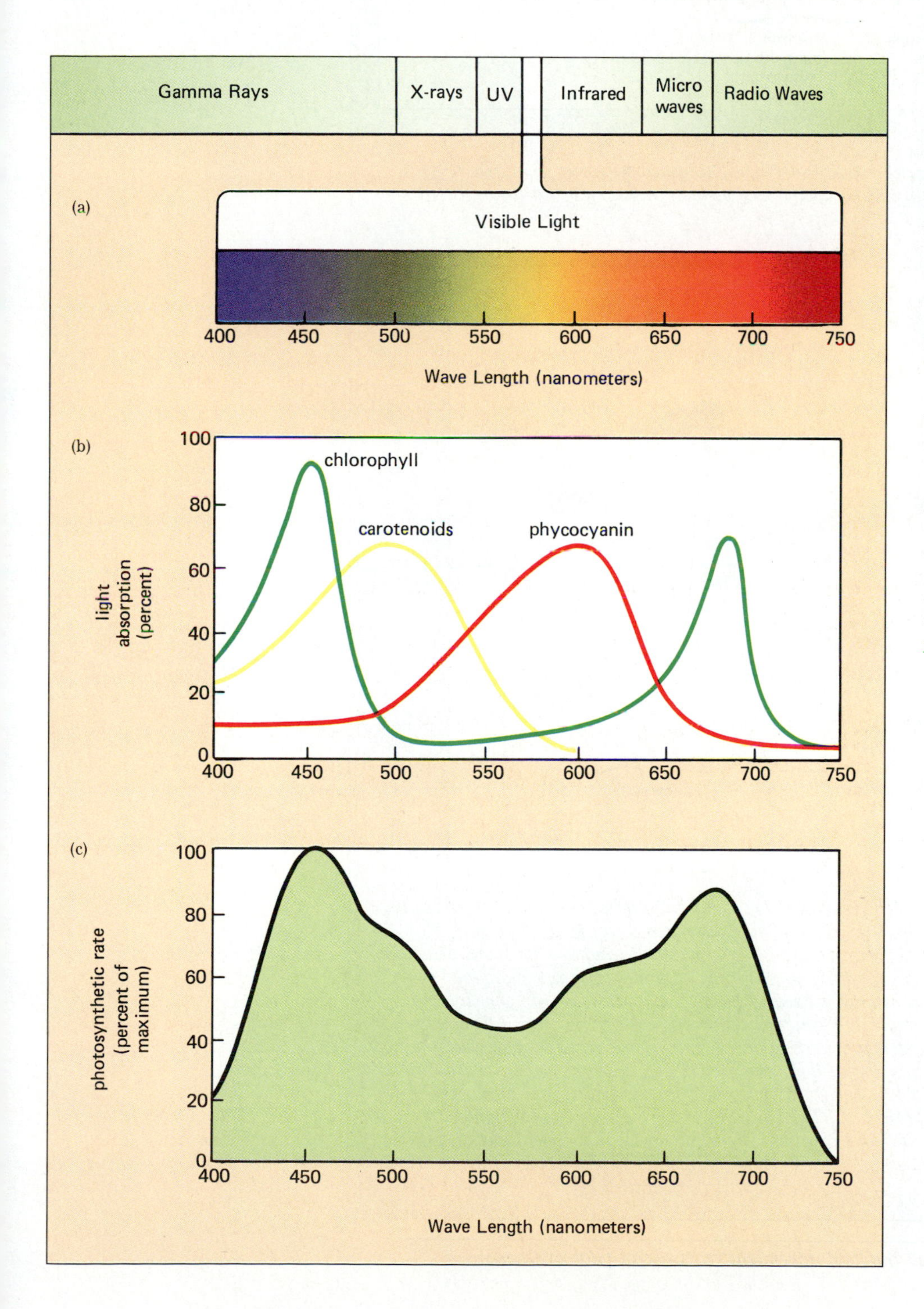

Figure 7-6 Light, chloroplast pigments, and photosynthesis. **(a)** *Visible light, a small part of the electromagnetic spectrum, consists of different wavelengths that correspond to the colors of the rainbow.* **(b)** *The first step in photosynthesis is the absorption of light by pigment molecules in the thylakoid membranes of chloroplasts. Different types of pigments selectively absorb certain colors; the height of each line represents the ability of each pigment to absorb different colors of light. Chlorophyll (green line) strongly absorbs violet, blue, and red light and therefore looks green. The other pigments in chloroplasts absorb other colors of light.* **(c)** *Photosynthesis is driven to some extent by all colors of light, due to the absorption of light by the different thylakoid pigments.*

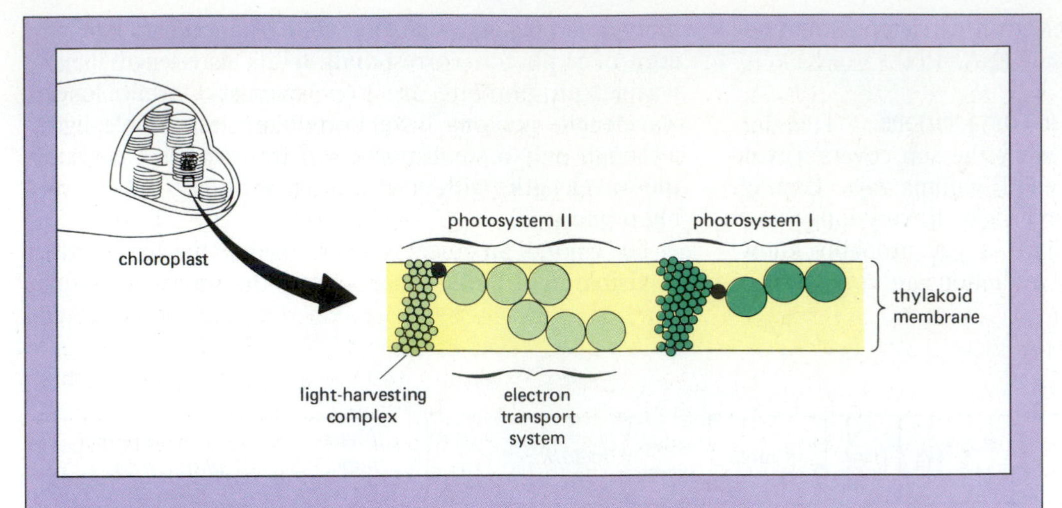

(a) *The thylakoid membranes contain many copies of photosystems I and II. Each photosystem consists of a light-harvesting complex of pigment molecules and an adjacent electron transport system.*

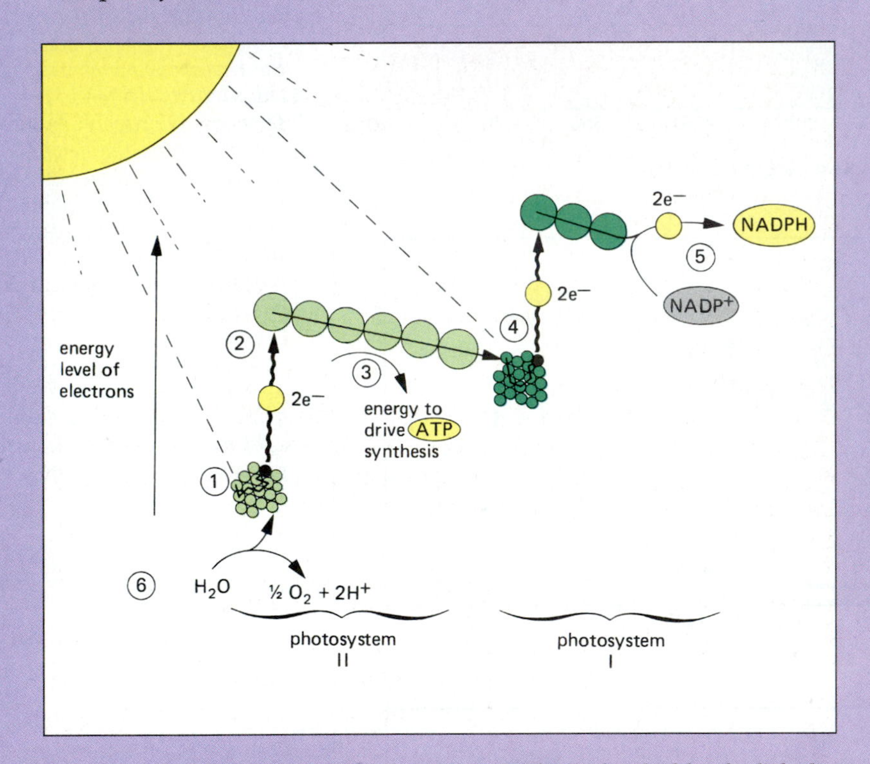

(b) *A summary of the light-dependent reactions. (1) Light is absorbed by the light-harvesting complex of photosystem II (light green), and the energy is passed to the reaction center chlorophyll. (2) This energy boosts electrons to high energy levels, and the electrons pass to the adjacent electron transport system. (3) The transport system passes the energetic electrons along, and some of their energy is used to pump hydrogen ions into the thylakoid interior. The hydrogen ion "battery" thus generated can drive ATP synthesis (see boxed Figure 7-8 for details). (4) Light strikes photosystem I (dark green), causing it to emit electrons to a second electron transport system. These electrons are replaced by those coming from the transport system of photosystem II. (5) The energetic electrons from photosystem I are captured in molecules of NADPH. (6) The electrons lost from the reaction center of photosystem II are replaced by electrons obtained from splitting water. This reaction also releases oxygen.*

Figure 7-7 *Thylakoid structure and the light-dependent reactions of photosynthesis.*

it transmits or reflects. **Chlorophyll,** the key light-capturing molecule in thylakoid membranes, absorbs blue and red light, but reflects green, and therefore appears green to human eyes (Fig. 7-6b). Thylakoids also contain other pigment molecules, called accessory pigments. **Carotenoids** absorb blue and green light, and therefore appear yellow or red, while **phycocyanins** absorb green and yellow, and therefore appear blue or purple. These accessory pigments capture light energy and transfer it to chlorophyll. Therefore, photosynthesis is driven to some extent by nearly all wavelengths of light (Fig. 7-6c).

THE LIGHT-DEPENDENT REACTIONS. In the thylakoid membranes, chlorophyll, accessory pigment molecules, and electron carrier molecules form highly organized assemblies called **photosystems.** Each thylakoid contains thousands of copies of two different types of photosystems, named **photosystem I** and **photosystem II** (Fig. 7-7a). Each consists of two major parts, a **light-harvesting complex** and an **electron transport system.**

The light-harvesting complex is composed of about 300 chlorophyll and accessory pigment molecules. These molecules absorb light and pass the energy to a specific chlorophyll molecule called the **reaction center.** By analogy with TV reception, the light-absorbing pigments are called antenna molecules, since they gather energy and transmit it to the energy-processing reaction center. The reaction center chlorophyll is located adjacent to the electron transport system, which is a linked series of electron carriers embedded in the thylakoid membrane. When the reaction center chlorophyll receives energy from the antenna molecules, one of its electrons absorbs the energy, leaves the chlorophyll, and jumps over to the electron transport system. This energetic electron passes along from one carrier to another. At some of the transfers, the electron releases energy that drives reactions resulting in the synthesis of ATP or NADPH. With this overall scheme in mind, let's look a little more closely at the actual sequence of events in the light-dependent reactions (Fig. 7-7b).

For historical reasons, the photosystems are numbered "backwards," and the usual process of light-energy capture is most easily understood by starting with photosystem II. The light-dependent reactions begin when a photon of light is absorbed by an antenna molecule in photosystem II. The photon's energy passes from molecule to molecule until it reaches the reaction center, where it boosts an electron completely out of the chlorophyll molecule. The first electron carrier of the adjacent electron transport system instantly accepts this energized electron. The electron passes from carrier to carrier, releasing energy as it goes. Some of the energy is used to charge up a hydrogen ion "battery" within the thylakoid that powers the synthesis of ATP (see boxed Figure 7-8, "A Closer Look at ATP Synthesis in Chloroplasts").

Meanwhile, light rays have also been striking the light-harvesting complex of photosystem I, ejecting an electron from its reaction center chlorophyll. This electron jumps to photosystem I's electron transport system. Photosystem I's reaction center chlorophyll immediately obtains a replacement for its lost electron from the last electron carrier in photosystem II's electron transport system. Photosystem I's high-energy electron moves through its electron transport system to the "empty" electron carrier $NADP^+$. Each $NADP^+$ molecule picks up two energetic electrons and one hydrogen ion, forming the energized carrier NADPH. $NADP^+$ and NADPH are both water-soluble molecules that dissolve in the chloroplast stroma.

Overall, electrons flow from the reaction center of photosystem II, through the photosystem II electron transport system, to the reaction center of photosystem I, through the photosystem I electron transport system, and on to NADPH. To sustain this one-way flow of electrons, photosystem II's reaction center must be continually supplied with new electrons to replace the ones it gives up. These replacement electrons come from water. In a poorly understood series of reactions, photosystem II's reaction center chlorophyll attracts electrons from water molecules within the thylakoid compartment, causing the water molecules to split apart:

$$H_2O \rightarrow \frac{1}{2}O_2 + 2\,H^+ + 2\,e^-$$

For every two photons captured by photosystem II, two electrons are boosted out of the reaction center chlorophylls and are replaced by the two electrons obtained by splitting one water molecule. As water molecules are split, the liberated oxygen atoms combine to form a molecule of oxygen gas, O_2. The oxygen may be used directly by the plant in its own respiration or it may be given off to the atmosphere.

Summary of the Light-Dependent Reactions:

The light-dependent reactions begin with (1) the absorption of light by light-harvesting complexes. The light (2) energizes electrons that are ejected from the reaction center of the complex and (3) are transferred to an adjacent electron transport system. As the electrons pass through the transport system, they release energy. (4) Some of the energy is used to create a hydrogen ion battery that drives ATP synthesis, some is captured as NADPH, and some is used to split water, replacing the ejected electrons and generating oxygen as a byproduct.

Fixing Chemical Energy as Glucose: The Light-Independent Reactions

The ATP and NADPH synthesized in the light-dependent reactions diffuse into the stroma, where they drive the synthesis of glucose from carbon dioxide and water.

Figure 7-8. A Closer Look at ATP Synthesis in Chloroplasts: Chemiosmosis

In the electron transport system of photosystem II, energetic electrons are passed from carrier to carrier. Until about 20 years ago, it was thought that ATP synthesis was directly coupled to these electron transfers: where the exergonic steps were large enough, the energy given up by the electrons drove ATP synthesis. More recent data show that ATP synthesis in chloroplasts is not *a simple coupled reaction. The electron transfers do not directly drive ATP synthesis; rather, the energy released during the transfers is used to charge up a hydrogen ion "battery" in the thylakoids. In a completely separate reaction, the energy stored in the thylakoid battery then powers ATP synthesis. Let's follow these reactions in some detail.*

In the first step of the light-dependent reactions of photosynthesis, a photon strikes the light-harvesting complex of photosystem II, and the energy is absorbed by an antenna pigment molecule. The energy hops around from molecule to molecule within the complex until it reaches the reaction center chlorophyll. Here an electron absorbs the energy and is ejected completely out of the chlorophyll molecule. Within a billionth of a second, the electron is captured by the first electron carrier of the adjacent electron transport system.

The electrons pass from carrier to carrier, losing energy as they go. The exergonic reaction of electron movement is coupled to the endergonic reaction of active transport of hydrogen ions across the thylakoid membrane from the stroma into the thylakoid compartment:

This raises the concentration of hydrogen ions (and therefore also positive charge) inside the thylakoid, creating a hydrogen ion "battery" of sorts. The thylakoid membrane is impermeable to hydrogen ions, except at specific protein pores that are coupled to ATP-synthesizing enzymes. When hydrogen ions flow through these pores, down their gradients of charge and concentration, the energy released drives the synthesis of ATP:

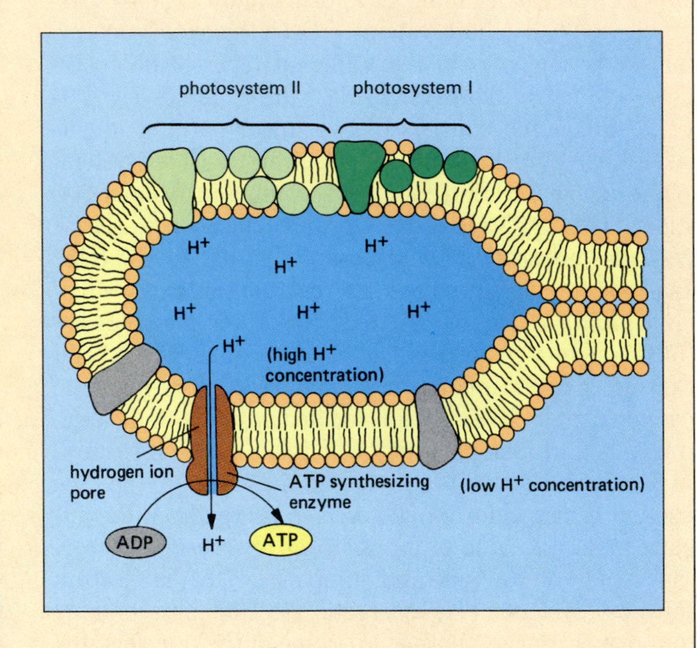

Exactly how this works is not completely understood, but perhaps we can get a feeling for the process by pursuing the battery analogy further. In a flashlight battery, the negative pole (the flat end of the battery) tends to release electrons, and the positive pole (knobbed end) tends to absorb electrons and transfer them to chemical reactions going on inside the battery. The positive and negative poles, however, are insulated from each other. Electrons can flow from negative to positive poles only if wires connect the two. The flow of electrons can do work, such as lighting a bulb or running a motor. The chloroplast thylakoid sac operates in an analogous way. Hydrogen ions in the thylakoid interior diffuse down their concentration and charge gradient out into the stroma, but can only do so through the ATP-synthesized enzyme pores. This flow of hydrogen ions can do work, namely drive ATP synthesis. This mechanism of ATP generation was first proposed in 1961 by Peter Mitchell, who called it ***chemiosmosis.***

Chemiosmosis has been shown to be the mechanism of ATP generation in chloroplasts, mitochondria (see Figure 7-20), and bacteria. For his brilliant hypothesis, Mitchell was awarded a Nobel Prize in 1978.

These reactions are termed the light-independent reactions, since they can occur independently of light if ATP and NADPH are available.

Carbon dioxide capture and glucose synthesis occur in a set of reactions known as the **Calvin-Benson cycle** (after its discoverers) or the **C_3 (three-carbon) cycle,** since some of the important molecules in the cycle have three carbon atoms in them (Fig. 7-9). The C_3 cycle requires (1) CO_2 (usually from the air); (2) a CO_2-capturing sugar, ribulose bisphosphate; (3) enzymes to catalyze all the reactions; and (4) energy in the form of ATP and NADPH.

The C_3 cycle begins (and ends) with a five-carbon sugar, ribulose bisphosphate (RuBP). RuBP combines with CO_2 to form an extremely unstable six-carbon compound. This compound spontaneously reacts with water to form two three-carbon molecules of phosphoglyceric acid (PGA), which gives the C_3 cycle its name. Capturing CO_2 in PGA is called **carbon fixation,** since it "fixes" gaseous CO_2 into a relatively stable organic molecule. Energy donated by ATP and NADPH is then used to convert PGA to phosphoglyceraldehyde (PGAL).

PGAL is a versatile molecule and can take one of several paths. Two PGAL molecules (three carbons each) may combine, becoming one molecule of glucose (six carbons). PGAL may also be used in the synthesis of lipids or amino acids. Alternatively, through a complex series of reactions requiring ATP energy, ten molecules of PGAL (10 × 3 carbons) can regenerate six molecules of RuBP (6 × 5 carbons). Overall, to fix six molecules of CO_2 as one molecule of glucose, six molecules of RuBP enter and are regenerated in each "turn" of the C_3 cycle (Fig. 7-9).

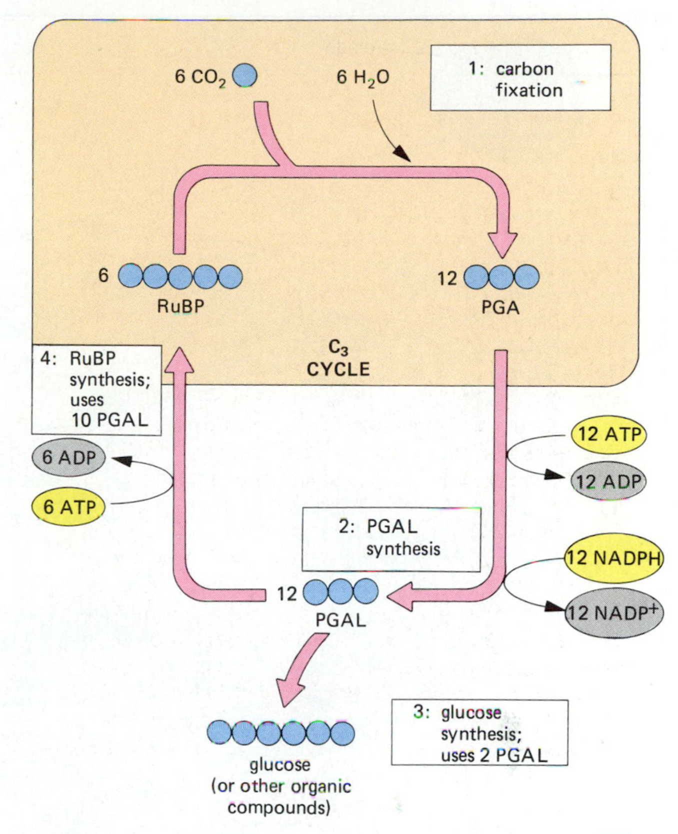

Figure 7-9 The C_3 cycle of carbon fixation. Carbon atoms are colored blue, depleted carrier molecules ADP and $NADP^+$ are gray, and energized carriers ATP and NADPH are yellow. (1) Six molecules of ribulose bisphosphate react with six molecules of CO_2 and six molecules of H_2O to form 12 molecules of phosphoglyceric acid (PGA). This is carbon fixation, the capture of carbon from CO_2 into organic molecules. (2) The energy of 12 ATPs and the electrons and hydrogens of 12 NADPHs are used to convert the 12 phosphoglyceric acids to 12 phosphoglyceraldehydes (PGAL). (3) Two of the phosphoglyceraldehydes are further processed into glucose or other organic molecules such as glycerol, fatty acids, or the carbon skeleton of amino acids, depending on the needs of the plant. (4) Energy from six ATPs is used to rearrange ten PGALs into six ribulose bisphosphates, completing one turn of the C_3 cycle.

Summary of the Light-Independent Reactions:

For the synthesis of one molecule of glucose: six molecules of CO_2 are captured by six molecules of RuBP. A series of reactions driven by energy from ATP and NADPH produces 12 molecules of PGAL. Two PGAL molecules join to become one molecule of glucose. Further ATP energy is used to regenerate the six RuBP molecules from ten PGAL molecules.

The Relation Between the Light-Dependent and Light-Independent Reactions

As Fig. 7-4 illustrates, the light-dependent and light-independent reactions are closely coordinated. The light-dependent reactions in the thylakoids use light energy to "charge up" the energy carrier molecules ATP and NADPH. These energized carriers move to the stroma, where their energy is used to drive glucose synthesis. The depleted carriers, $NADP^+$ and ADP, then return to the light-dependent reactions for recharging.

Water, CO_2, and the C_4 Pathway

Photosynthesis requires light and CO_2. Either of these may limit the rate of photosynthesis; that is, photosynthesis cannot occur in the dark, regardless of CO_2 levels, and conversely carbon fixation cannot occur without a supply of CO_2, regardless of the light intensity. An ideal leaf therefore should have a large surface area to intercept sunlight and be very porous to allow lots of CO_2 to enter the leaf from the air. For land plants, however, being porous to CO_2 also means being porous to water vapor, which evaporates out of the leaf. Broadleaf plants have evolved leaves that are a compromise between obtaining adequate CO_2 supplies and reducing water loss: large,

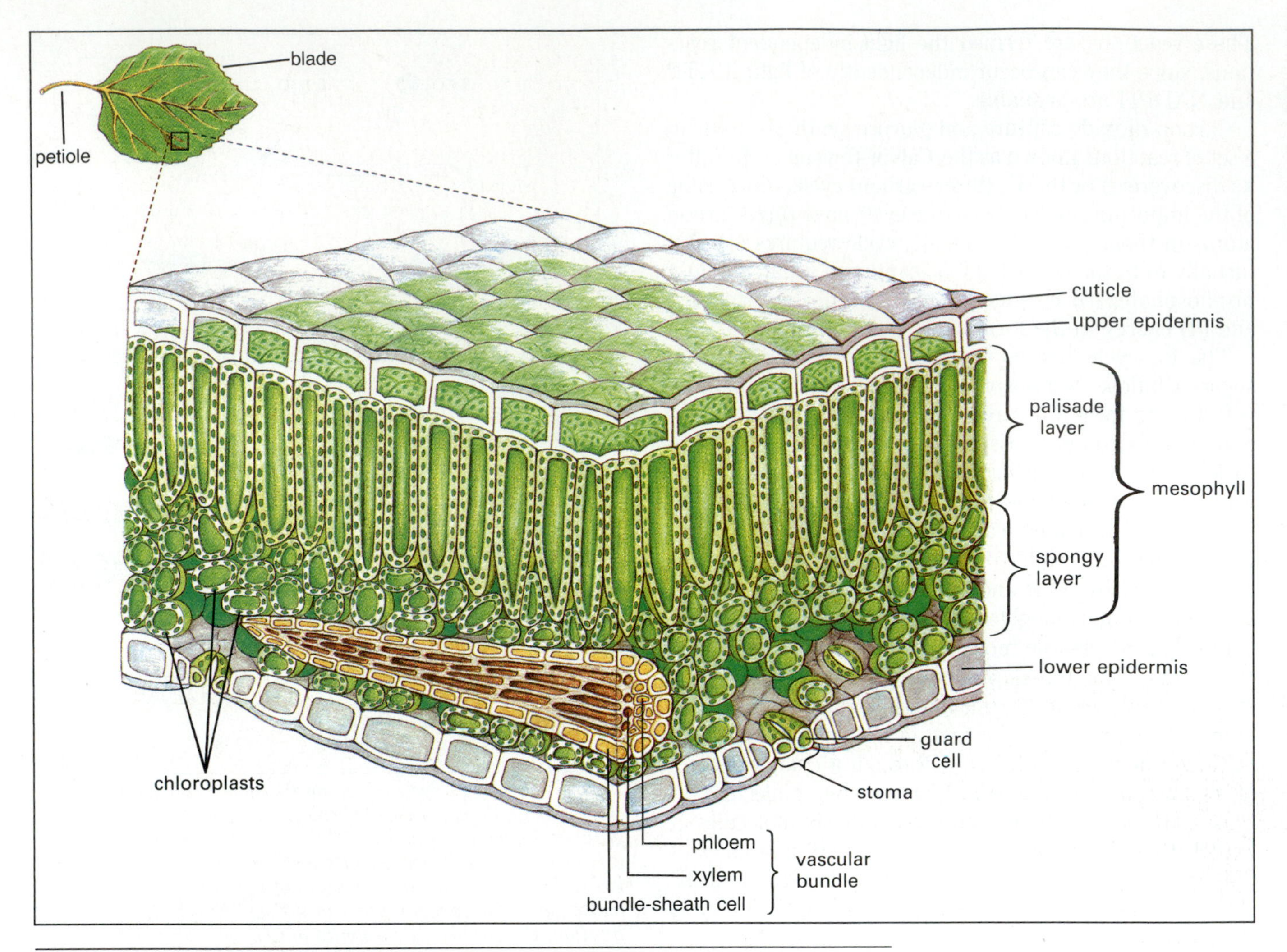

Figure 7-10 *Leaf structure in a land plant. In most leaves, the mesophyll cells contain chloroplasts and carry out most of the leaf's photosynthesis. The epidermal cells lack chloroplasts and are quite transparent, allowing sunlight to penetrate to the mesophyll cells. The epidermal cells secrete an airtight, waterproof cuticle that covers the leaf surface except where the epidermis is perforated by the adjustable pores of the stomata. Almost all gas exchange between the interior of the leaf and the atmosphere occurs through the stomata. The vascular bundles (veins) carry water, minerals, and sugars between the leaf and the stem.*

waterproof leaves, with adjustable pores called **stomata** (singular, **stoma**) that admit carbon dioxide (Fig. 7-10; see Chapter 21 for more details of leaf structure).

CO_2 can enter the leaf only when the stomata are open. Since the atmosphere is only about 0.03 percent CO_2, diffusion of CO_2 through open stomata is slow. Open stomata also allow water to evaporate out of the leaf. If the plant is in danger of drying out, the stomata stay closed, even though there may not be enough CO_2 in the leaves for photosynthesis.

During hot and/or dry weather, the stomata will seldom open, and CO_2 supplies will seldom be replenished. The rate of photosynthesis will therefore depend on how efficiently CO_2 can be fixed in the first step of the C_3 cycle, that is, how low the CO_2 concentration can drop within the leaf before carbon fixation ceases. As it happens, the RuBP + CO_2 reaction is not very efficient, and the enzyme that catalyzes it is not very selective. Oxygen, which comprises about 20 percent of the atmosphere, is a competitive inhibitor of the enzyme (see Chapter 6). Therefore, at low CO_2 concentrations, carbon fixation grinds to a halt.

Some plants, especially tropical and desert ones, have evolved a way to circumvent this problem (Fig. 7-11). The mesophyll cells in the leaves of some plants have a different CO_2 capture molecule, phosphoenolpyruvate (PEP). CO_2 reacts with PEP to form a four-carbon molecule of oxaloacetic acid. This **four-carbon** or **C_4 pathway** tolerates low CO_2 and high O_2 concentrations. Oxaloacetic acid is shuttled into the bundle-sheath cells sur-

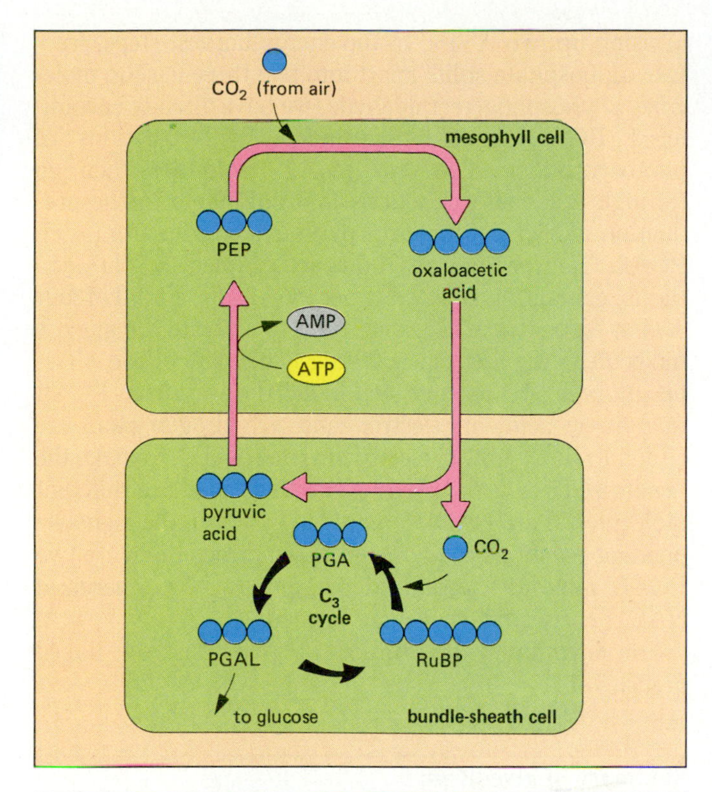

Figure 7-11 In C_4 plants, carbon fixation is a cooperative process involving two types of leaf cells, the mesophyll cells and the bundle-sheath cells surrounding the veins (see Fig. 7-10). In the mesophyll cells, CO_2 reacts with phosphoenolpyruvic acid (PEP) to form a four-carbon compound, oxaloacetic acid. This molecule is shuttled into the bundle-sheath cells. There it breaks down into CO_2 and pyruvic acid. The CO_2 feeds into the C_3 cycle to produce glucose. The pyruvic acid returns to the mesophyll cells, where ATP is used to regenerate PEP.

rounding the leaf veins, where it breaks down, releasing CO_2 again. This creates a high CO_2 concentration in the bundle-sheath cells, allowing the RuBP + CO_2 reaction to fix carbon in the C_3 cycle. The remnant of the shuttle molecule returns to the mesophyll cells, where ATP energy is used to regenerate the PEP capture molecule.

The C_4 pathway is advantageous when lots of light and little water are available. The C_4 pathway is not *always* better, because regenerating PEP uses up ATP from the light-dependent reactions. If water is plentiful (so the stomata can stay open, letting in lots of CO_2) or light is weak (making ATP hard to come by), the straightforward C_3 carbon fixation pathway would be more efficient. However, a plant with the enzymes of the C_4 pathway always uses the C_4 method of carbon fixation, regardless of environmental conditions. Therefore, C_3 plants tend to have the advantage in cool, wet, cloudy climates, where their exclusive use of the C_3 pathway is more energy efficient. C_4 plants are favored in deserts and in midsummer in temperate climates, where light energy is plentiful but water is scarce.

ENERGY RELEASE: GLYCOLYSIS AND CELLULAR RESPIRATION

The end products of photosynthesis are organic molecules such as glucose. Although glucose contains a lot of chemical energy, that energy is not in a form the cell can readily use. As we saw in Chapter 6, the high-energy bonds of ATP provide the energy for most reactions that occur within living cells. *Therefore, both photosynthetic organisms such as plants and heterotrophic organisms such as animals and fungi must convert the energy of glucose into the readily accessible energy of ATP.*

Figure 7-12 provides an overview of glucose metabolism in eukaryotic cells. The first stage, **glycolysis,** does not depend on the availability of oxygen, and proceeds in exactly the same way under both aerobic (with oxygen) and anaerobic (without oxygen) conditions. Glycolysis

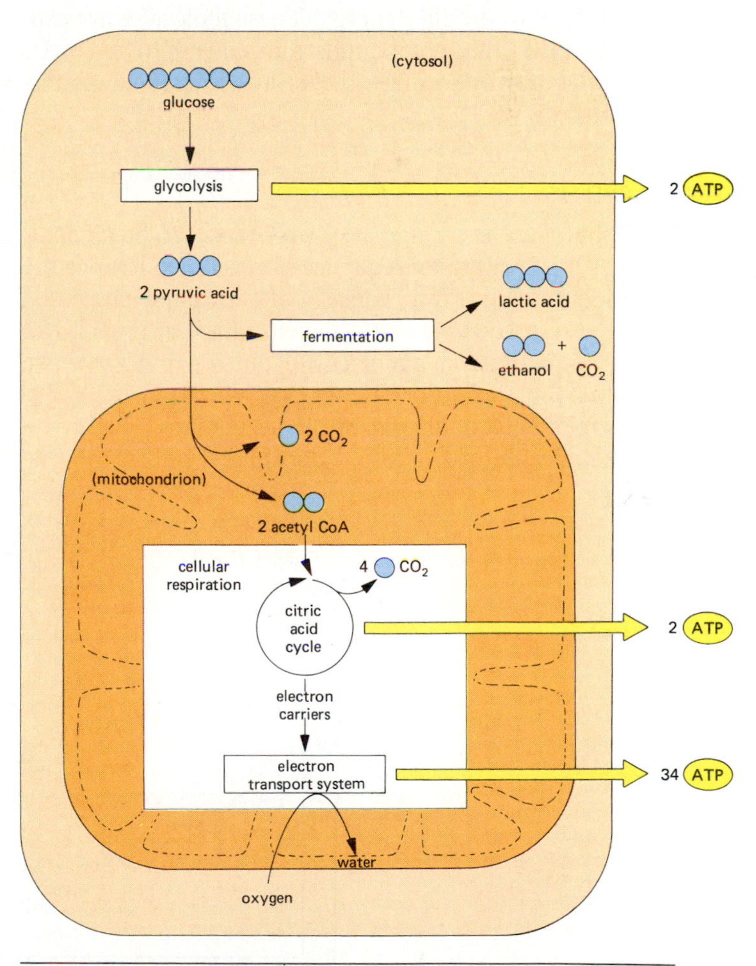

Figure 7-12 A summary of glucose metabolism. Use this diagram as a reference as we progress through the reactions of glycolysis (in the cell cytosol) and cellular respiration (in the mitochondria). Note that the breakdown of glucose occurs in stages, with various amounts of energy harvested as ATP along the way. The vast majority of the ATP is produced in the mitochondria, justifying their nickname of "powerhouse of the cell."

splits apart a single glucose molecule (a six-carbon sugar) into two three-carbon molecules of pyruvic acid. This releases a small fraction of the chemical energy stored in the glucose, some of which is used to generate two ATP molecules. Under anaerobic conditions, the pyruvic acid is usually converted by **fermentation** into lactic acid or ethanol. Fermentation does not produce any more ATP energy. Both glycolysis and fermentation occur in the cytosol. If oxygen is available, the pyruvic acid produced by glycolysis enters the mitochondria, where **cellular respiration** breaks it down completely to carbon dioxide and water, generating an additional 36 ATP molecules. Chemically, the complete metabolism of glucose through glycolysis and cellular respiration is the reverse of photosynthesis:

$$C_6H_{12}O_6 + 6\,O_2 \rightarrow 6\,CO_2 + 6\,H_2O + \text{energy}$$

Cells can also use other organic molecules, such as proteins or fats, to produce energy. These molecules are usually converted into compounds that enter into glycolysis or cellular respiration (see the essay, "Metabolic Transformations").

Glycolysis

Like photosynthesis, glycolysis (in Greek, "to break apart a sweet") is a complex sequence of reactions. Reduced to its essentials, glycolysis consists of two major processes: (1) glucose activation and (2) energy harvest (Fig. 7-13). In glucose activation, a molecule of glucose undergoes two reactions that each use ATP energy. These reactions convert a relatively stable glucose molecule into a highly reactive molecule of fructose diphosphate, but at the cost of using up two ATPs. In the energy harvest steps, fructose diphosphate splits apart into two three-carbon molecules of phosphoglyceraldehyde (we have already encountered PGAL in the light-independent reactions of photosynthesis). The two PGAL molecules then go through a series of reactions that culminate in the production of two molecules of pyruvic acid, one from each PGAL. Two of these reactions are coupled to ATP synthesis, generating two ATPs per PGAL, for a total of four ATPs. Since two ATPs were used to activate the glucose molecule in the first place, there is a net gain of two ATPs per glucose. At another step along the way from PGAL to pyruvic acid, an electron and a hydrogen atom are added to the "empty" electron carrier NAD^+ to make the "energized" carrier NADH (note that these are different molecules than $NADP^+$ and NADPH in the light-dependent reactions of photosynthesis). Since there are two PGAL molecules produced per glucose, two energized NADH carrier molecules are formed.

For a complete description of glycolysis, see boxed Figure 7-14.

Summary of glycolysis:

Each molecule of glucose is broken down to two molecules of pyruvic acid. During these reactions, two ATPs and two NADH electron carriers are formed.

The electrons carried in NADH are highly energetic, but their energy can only be used to synthesize ATP when oxygen is available (see the section "Cellular Respiration" below). Under anaerobic conditions (under which life, and probably glycolysis, evolved), NADH production is not a

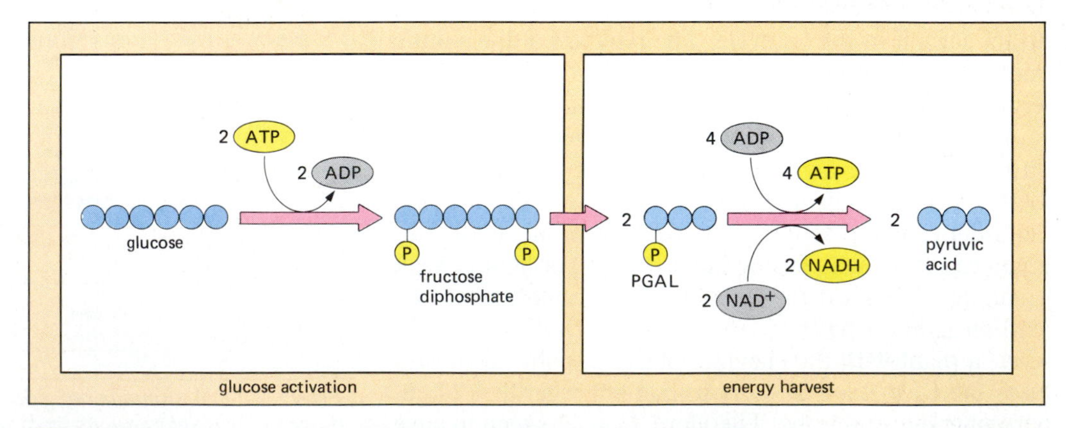

Figure 7-13 *The essentials of glycolysis. (1) Glucose activation: The energy of two ATP molecules is used to convert glucose to the highly reactive fructose diphosphate. Fructose diphosphate splits into two smaller, but still reactive, molecules of phosphoglyceraldehyde (PGAL). (2) Energy harvest: The PGAL molecules both go through a series of reactions that generate four ATPs and two NADH molecules. Therefore, glycolysis results in a net harvest of two ATP molecules per glucose molecule.*

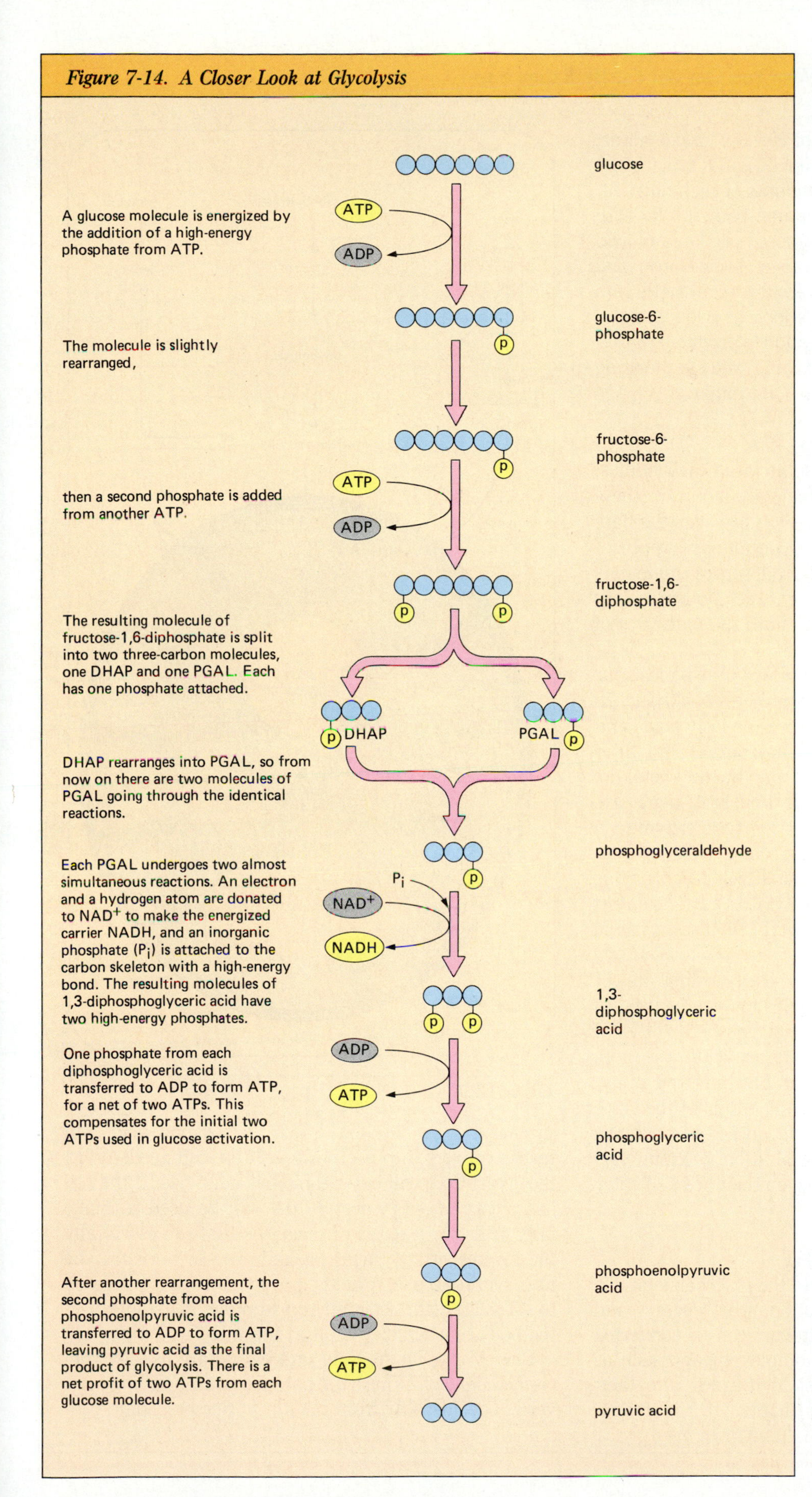

The detailed reactions of glycolysis. Carbon atoms are colored blue; depleted carrier molecules of ADP and NAD^+ are gray; energized carriers ATP and NADH, and phosphate groups attached to the carbons are yellow.

Metabolic Transformations—Why You Can Get Fat By Eating Sugar

As you know, humans do not live by bread alone, much less by glucose alone. Nor does the typical diet contain exactly the required amounts of each nutrient. Accordingly, the cells of the human body seethe with biochemical reactions, synthesizing one amino acid from another, making fats from carbohydrates, and channeling surplus organic molecules of all types into energy storage or release. Let's look at two examples of these molecular interconversions: producing ATP from fats and proteins during fasting, and synthesizing fats from sugars during surplus food intake.

Metabolizing Fats and Proteins

Even the leanest people have some fat in their bodies. During fasting or starvation, the body mobilizes these energy reserves for ATP synthesis, since even the bare maintenance of life requires a continuous supply of ATP, while seeking out new food sources demands even more energy expenditure. Fat metabolism is very straightforward, flowing directly into the pathways of glucose metabolism.

Chapter 3 illustrated the structure of a typical fat: three fatty acids connected to a glycerol backbone. In fat metabolism, the bonds between the fatty acids and glycerol are hydrolyzed. The glycerol part of a fat, after activation by ATP, feeds directly into the middle of the glycolysis pathway (Fig. E7-1). The fatty acids are transported into the mitochondria, where enzymes in the inner membrane and matrix chop them up into two-carbon acetyl groups. These attach to coenzyme A to form acetylcoenzyme A (acetyl-CoA), which enters the citric acid cycle just like acetyl-CoA from pyruvic acid does.

In cases of severe starvation, or of feeding almost exclusively on protein, amino acids too can be used to produce energy. Some amino acids can be readily cut down to pyruvic acid. Other amino acids are more complicated in structure, but can be converted to pyruvic acid, acetyl-CoA, or the compounds of the citric acid cycle. These molecules then proceed through the remaining stages of cellular respiration, yielding various amounts of ATP depending on their point of entry into the pathway.

Converting Carbohydrates to Fat

Just as fats can be funneled into the glucose metabolic pathway for energy production, the sugars and starches in cornflakes or candy bars can be converted into fats for energy storage. Complex sugars, such as starches and sucrose, are first hydrolyzed into their monosaccharide subunits (see Chapter 3). The monosaccharides are broken down to pyruvic acid and thence converted to acetyl-CoA. If the cell needs ATP, the acetyl-CoA will enter the citric acid cyle. If the cell has plenty of ATP, acetyl-CoA will be used to make fatty acids via a series of reactions that are essentially the reverse of fatty acid breakdown. In humans, the liver synthesizes fatty acids, but fat storage is relegated to fat cells, with their all too familiar distribution in the body, particularly around the waist and hips.

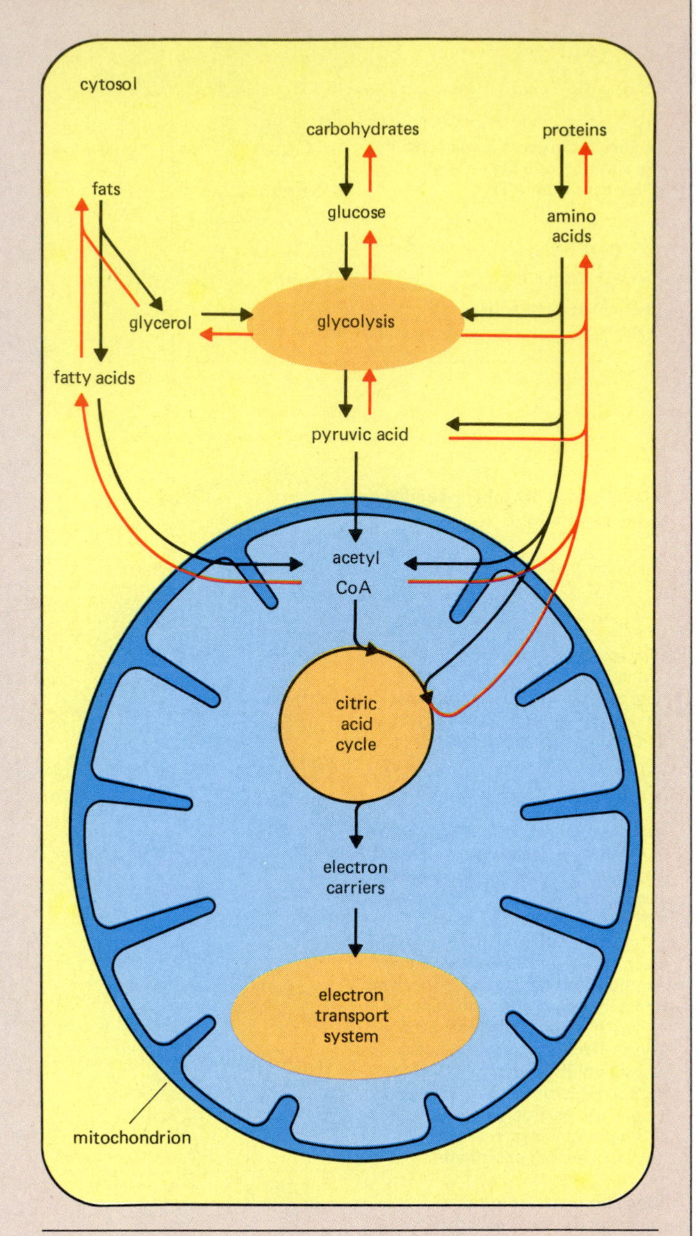

Figure E7.1

Energy use, fat storage, and eating are usually precisely balanced. Where the balance point lies, however, varies considerably from person to person. Some peo-

Metabolic Transformations—Why You Can Get Fat By Eating Sugar

ple seem to eat nearly continuously without ever storing much fat, probably because their metabolic rates are high or adjust upward to compensate for excess food intake. Others crave high-calorie foods even when they have a lot of fat stored already. As every chronic dieter knows, many people seem to have bodies that are "biologically set" to have large fat deposits. Constant restraint can slim most people down, but as soon as the discipline is relaxed, the fat once again accumulates. Before the fashionably thin become too smug about their svelte condition, they should remember that overeating during times of easy food availability is highly adaptive behavior from an evolutionary point of view. If hard times come, the plump may easily (if hungrily) survive while the trim succumb to starvation. Only recently have many societies enjoyed regular agricultural surpluses, in which the drive to eat can lead to obesity.

method of energy capture; it is actually a way of getting rid of hydrogens and electrons during the metabolism of glucose to pyruvic acid. This works fine except for the fact that NAD^+ is used up as it accepts electrons and hydrogens to become NADH. If a cell could not dispose of the electrons and hydrogens, and regenerate NAD^+, glycolysis would have to stop once the supply of NAD^+ was exhausted.

Fermentation

Fermentation solves this problem by allowing pyruvic acid to accept electrons and hydrogens from NADH, thereby regenerating NAD^+ for use in further glycolysis. There are two main types of fermentation, one converting pyruvic acid to lactic acid, the other converting it to carbon dioxide and ethanol (Fig. 7-12).

Lactic acid fermentation. When you exercise vigorously, your muscles need lots of ATP as an energy source (Fig. 7-15a). Even though cellular respiration generates much more ATP than glycolysis alone, *cellular* respiration is limited by *organismal* respiration (breathing): you may not be able to get enough oxygen out of the air, into your blood, and delivered to your muscles to keep cellular respiration going at the necessary pace. When this happens, your muscles do not immediately stop working. Instead, glycolysis continues for a while, providing its meager two ATP molecules per glucose, and both

(a)

(b)

Figure 7-15 Fermentation. **(a)** *A sprinter at the end of a race. The runner's respiratory and circulatory systems cannot supply oxygen to his leg muscles fast enough to keep up with the demand for energy, so glycolysis and lactic acid fermentation must provide the energy. Panting after the race brings in the oxygen needed to remove the lactic acid through cellular respiration.*
(b) *A newly opened bottle of champagne. The bubbles are CO_2 trapped in the bottle by fermenting yeast.*

pyruvic acid and NADH pile up. To regenerate NAD^+, muscle cells convert pyruvic acid molecules to lactic acid, using electrons and hydrogens from NADH:

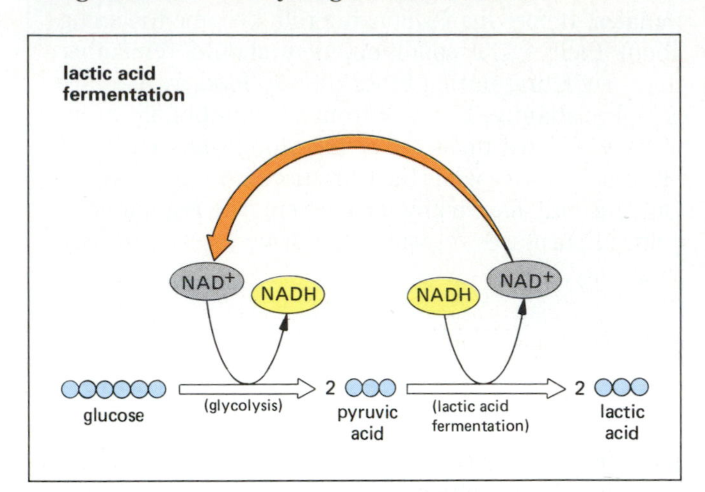

The regenerated NAD^+ can again accept electrons during glycolysis, and energy production can continue. Lactic acid, however, is toxic in high concentration. Eventually, the muscles suffer from lactic acid poisoning, and fatigue sets in. After exercise ceases, the lactic acid is reconverted to pyruvic acid, which is broken down through cellular respiration to carbon dioxide and water. Interestingly enough, this reconversion occurs not in the muscle cells, which lack the necessary enzymes, but in the liver.

A variety of microorganisms also use lactic acid fermentation, including the bacteria that produce yogurt, sour cream, and cheese.

ALCOHOLIC FERMENTATION. Many microorganisms use another process to regenerate NAD^+ under anaerobic conditions: alcoholic fermentation. This series of reactions produces ethanol and CO_2 from pyruvic acid, using hydrogens and electrons from NADH:

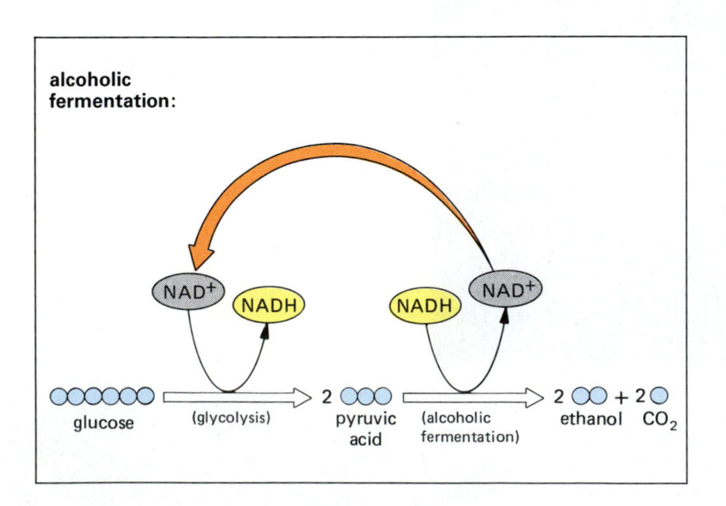

Alcoholic fermentation is a somewhat chancy proposition for the microbe, because the alcohol it generates is poisonous. If the alcohol concentration becomes too great, the microbe will die. Most yeasts, for example, die if the alcohol concentration exceeds 12 percent, although certain strains developed for wine production do not succumb until 18 percent. (Under natural conditions, of course, yeasts are unlikely to find themselves confined in a small volume of water, with lots of sugar for energy production and no oxygen. Selective pressure for alcohol tolerance was probably very slight before humans developed a taste for wine.) Sparkling wines, such as champagne, are bottled while the yeasts are still alive and fermenting, trapping both the alcohol and the CO_2. When the cork is removed, the pressurized CO_2 is released, sometimes rather explosively (Fig. 7-15b).

Cellular Respiration

In eukaryotic cells, cellular respiration occurs in the mitochondria (Fig. 7-16). Recall that a mitochondrion has two membranes, producing two compartments: an inner compartment enclosed by the inner membrane and containing the fluid **matrix,** and an **intermembrane compartment** between the two membranes. Two very different processes occur in mitochondria. (1) *In the matrix, pyruvic acid is converted into carbon dioxide, with much of its chemical energy captured in a few ATP molecules and many electron carriers. (2) A second series of reactions involving the inner membrane and the intermembrane compartment uses the energetic electrons caught by the electron carriers to synthesize most of the ATP produced during cellular respiration.*

THE MATRIX REACTIONS. The two pyruvic acid molecules formed from each glucose molecule during glycolysis pass through both mitochondrial membranes and into the matrix. Here each pyruvic acid reacts with a molecule called coenzyme A (Fig. 7-17). Each pyruvic acid is split into CO_2 and a two-carbon molecule called an acetyl group, which immediately attaches to coenzyme A to form the complex acetylcoenzyme A (acetyl-CoA). During this reaction, an energetic electron and a hydrogen are transferred to NAD^+, forming NADH.

The two acetyl-CoA molecules then enter a cyclic pathway known as the **Krebs cycle,** after its discoverer Hans Krebs, or the **citric acid cycle,** after the first product in the reaction sequence (Fig. 7-17). Each acetyl-CoA molecule combines briefly with a molecule of oxaloacetic acid.

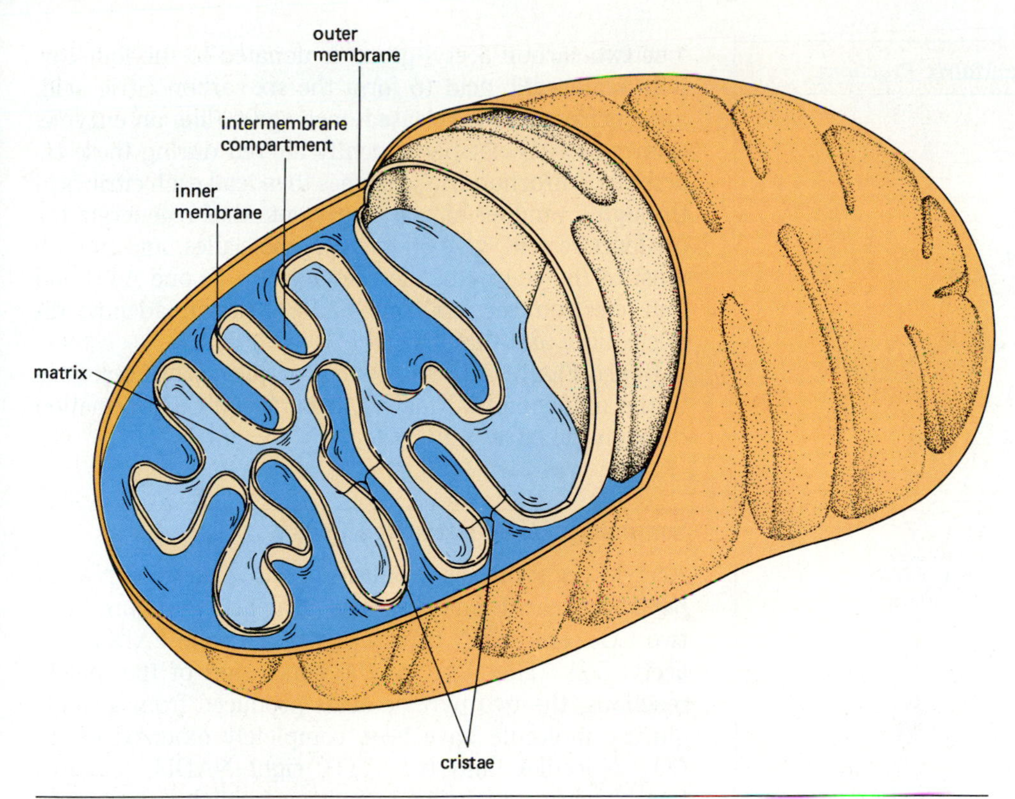

Figure 7-16 *The structure of a mitochondrion, like the structure of a chloroplast, reflects the compartmentalized reactions that occur there. The reactions of the citric acid cycle, in which pyruvic acid is broken down to CO_2 and electrons are transferred to carriers, occur in the fluid matrix contained within the inner membrane. The electron transport system, which then uses the energy of the captured electrons to pump hydrogen ions into the intermembrane compartment, resides in the inner membrane. As in chloroplasts, these hydrogen ions are used to drive ATP synthesis.*

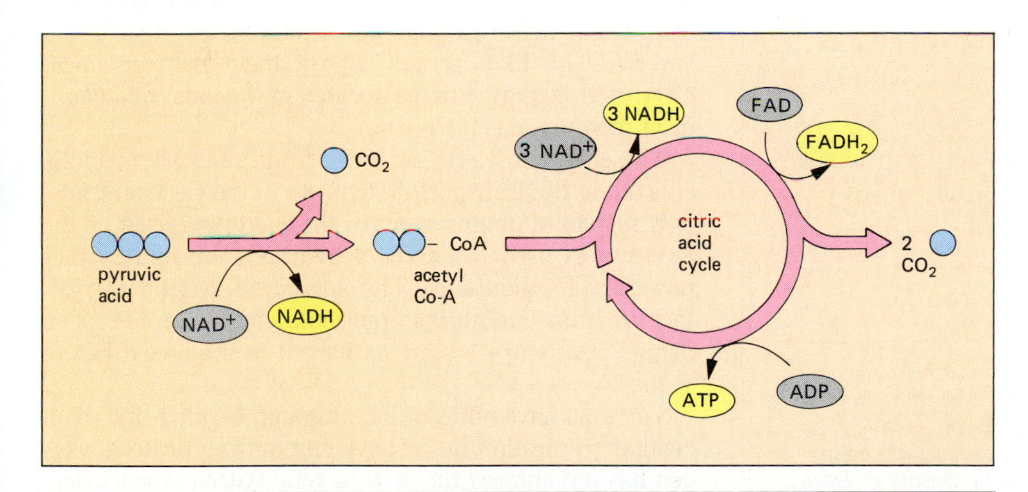

Figure 7-17 *The essentials of the metabolic reactions occurring in the mitochondrial matrix. Pyruvic acid reacts with coenzyme A to form CO_2 and acetyl-CoA. During this reaction, an energetic electron is added to NAD^+ to form NADH. The acetyl-CoA enters the citric acid cycle, where the acetyl group is broken down to two molecules of CO_2. The reactions of the citric acid cycle form one ATP, three NADH, and one $FADH_2$ per acetyl-CoA. Since each glucose molecule yields two pyruvic acids, the total energy harvest in the matrix is 2 ATP, eight NADH, and two $FADH_2$.*

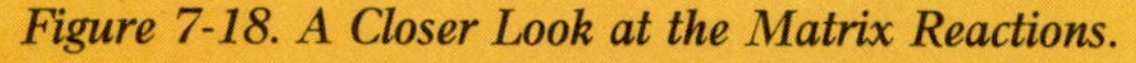

Figure 7-18. A Closer Look at the Matrix Reactions.

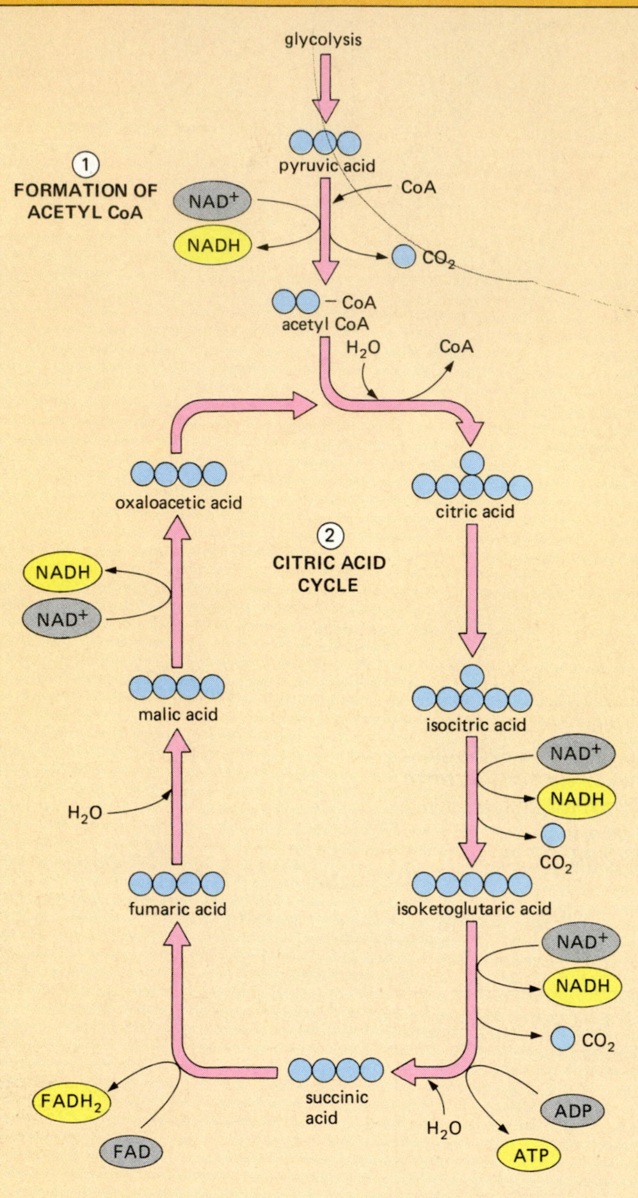

Color coding as in boxed Figure 7-14. Note that there are two pyruvic acids produced during glycolysis from each glucose molecule, so that each of these reactions occurs twice during the metabolism of a single glucose molecule. (1) Pyruvic acid is split to CO_2 and an acetyl group, which attaches to coenzyme A to form acetyl-CoA; simultaneously, NAD^+ receives an electron and a hydrogen atom to make NADH. (2) Acetyl-CoA donates its acetyl group to oxaloacetic acid to make citric acid. As citric acid proceeds through the remaining reactions, two more molecules of CO_2 are given off, accounting for all three carbons of the original pyruvic acid. Three more molecules of NADH, one $FADH_2$, and one ATP are formed per acetyl CoA.

The two-carbon acetyl group is donated to the four-carbon oxaloacetic acid to form the six-carbon citric acid, and coenzyme A is released once again (like an enzyme, coenzyme A is not permanently altered during these reactions). Mitochondrial enzymes then lead each citric acid through a number of rearrangements that regenerate the oxaloacetic acid, give off two CO_2 molecules, and capture most of the energy of the acetyl group as one ATP and four electron carriers, one $FADH_2$ (flavin adenine dinucleotide) and three NADH.

Boxed Figure 7-18 shows the complete set of reactions in the mitochondrial matrix, from acetyl-CoA formation through the citric acid cycle.

Summary of the Matrix Reactions

Acetyl-CoA synthesis produces one CO_2 and one NADH per pyruvic acid molecule. The citric acid cycle produces two CO_2, one ATP, three NADH, and one $FADH_2$ per acetyl-CoA. Therefore, at the conclusion of the matrix reactions, the two pyruvic acids produced from a single glucose molecule have been completely oxidized to six CO_2 molecules, and two ATP, eight NADH, and two $FADH_2$ electron carriers have been formed.

THE INNER MEMBRANE REACTIONS. At this point the cell has produced only four ATP molecules from the original glucose molecule: two during glycolysis and two during the citric acid cycle. The cell has, however, captured many energetic electrons in carrier molecules: two NADH during glycolysis, and eight more NADH and two $FADH_2$ from the matrix reactions, for a total of ten NADH and two $FADH_2$. The carriers deposit their electrons in an **electron transport system** located in the inner mitochondrial membrane (Fig. 7-19).

The energetic electrons move from molecule to molecule along the transport system. As in the electron transport system of photosystem II, the energy released by the electrons is used to charge a hydrogen ion battery that powers ATP synthesis. The energized electron carriers formed from one glucose molecule yield 34 ATPs. The details of this process are explained in the boxed Figure 7-20.

You may have noticed that although we have described cellular respiration as an oxygen-requiring process, oxygen has not entered the picture yet. Oxygen accepts electrons from the end of the electron transport system, two electrons (and two hydrogen ions) combining with one oxygen atom to form water. This clears out the transport system, leaving it ready to run more electrons through. Without oxygen, the electrons "pile up" in the transport system, with the result that the hydrogen ion battery is not charged up, and therefore ATP synthesis stops.

The complete breakdown of pyruvic acid to carbon dioxide and water in the mitochondria results in an im-

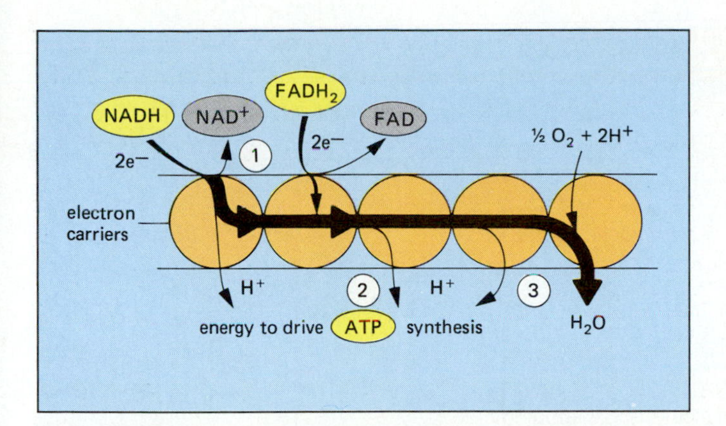

Figure 7-19 The electron transport system of mitochondria. (1) The electron carrier molecules NADH and $FADH_2$ deposit their energetic electrons with the carriers of the transport system located in the inner membrane. (2) The electrons move from carrier to carrier within the transport system. Some of their energy is used to pump hydrogen ions across the inner membrane from the matrix into the intermembrane compartment. This charges up a hydrogen ion battery that can be used to drive ATP synthesis (for details, see boxed Figure 7-20). (3) At the end of the electron transport system, the energy-depleted electrons combine with oxygen and hydrogen ions in the matrix to form water.

Figure 7-20. A Closer Look at ATP Synthesis in Mitochondria

ATP synthesis in mitochondria is essentially the same as ATP synthesis in chloroplasts. The inner membrane of a mitochondrion has an electron transport system that functions similarly to the one in photosystem II of a thylakoid. Further, the intermembrane compartment between the outer and inner membranes of a mitochondrion is analogous to the interior of a thylakoid.

Anatomically, the arrangement in mitochondria looks like this:

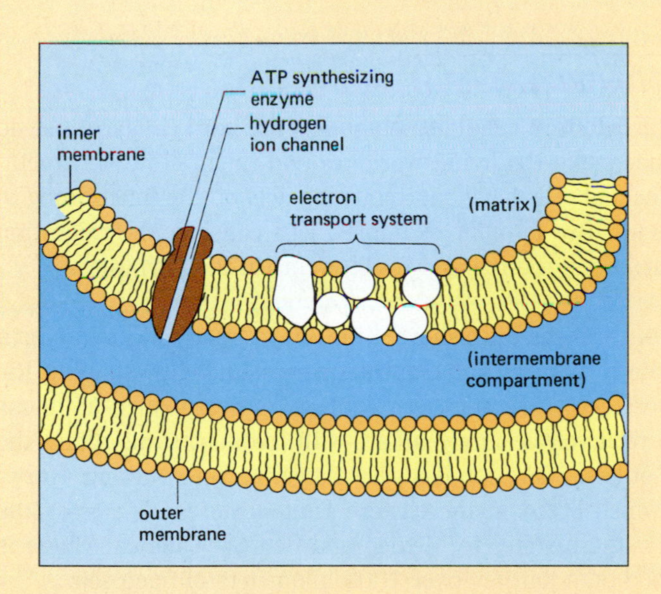

The electron carriers formed during glycolysis and the citric acid cycle, NADH and $FADH_2$, deposit their electrons with the electron transport system. As they pass through the electron transport system, the electrons provide the energy to pump hydrogen ions across the inner membrane, from the matrix to the intermembrane compartment:

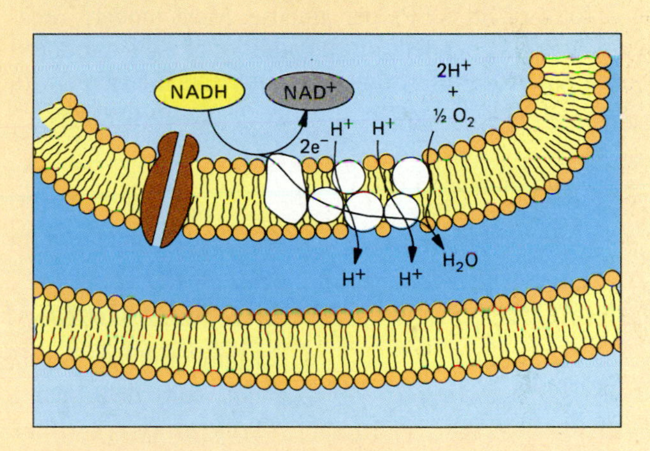

As in chloroplasts, the inner membrane is only permeable to hydrogen at special pores that are coupled with ATP-synthesizing enzymes. As the hydrogen ion battery discharges through the pores, the released energy drives ATP synthesis:

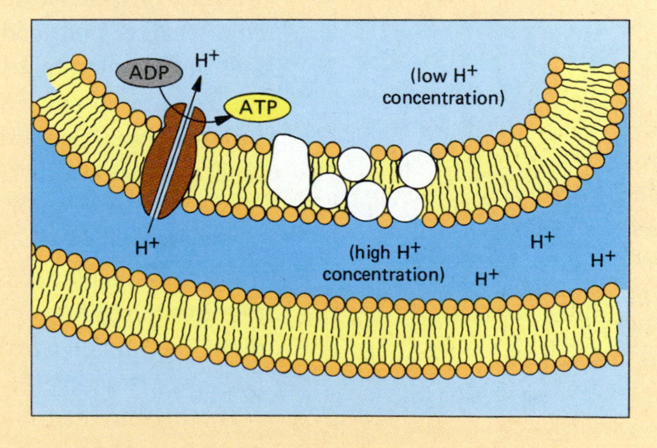

Table 7-1 **Summary of Glycolysis and Cellular Respiration**

Process	Location	Reactions	Electron Carriers Formed	ATP Yield
Glycolysis	Cytosol	Glucose broken down to 2 pyruvic acids	2 NADH	2 ATP
Acetyl-CoA formation	Matrix of mitochondrion	Each pyruvic acid combined with coenzyme A to form acetyl-CoA and CO_2	2 NADH	
Citric acid cycle	Matrix of mitochondrion	Acetyl group of acetyl-CoA metabolized to 2 CO_2	6 NADH, 2 $FADH_2$	2 ATP
Electron transport	Inner membrane, intermembrane compartment	Energy of electrons from NADH, $FADH_2$ used to pump H^+ ions into intermembrane compartment, H^+ gradient used to synthesize ATP: 3 ATP per NADH, 2 ATP per $FADH_2$		34 ATP

pressive gain in ATP molecules for the cell (Table 7-1). Whereas glycolysis produces only two ATP molecules per glucose, the mitochondrial reactions add 36 more ATPs, for a total of 38 ATPs per glucose. Most modern organisms are so dependent on the additional ATP produced by cellular respiration that anything that interferes with it, such as lack of oxygen, quickly results in death.

Summary of the Inner Membrane Reactions:

Electrons from the electron carriers NADH and $FADH_2$ enter the electron transport system of the inner mitochondrial membrane. Here their energy is used to charge a hydrogen ion battery. The battery drives the synthesis of 34 molecules of ATP. At the end of the electron transport system, two electrons combine with one oxygen atom and two hydrogen ions to form water.

REFLECTIONS ON CELLULAR METABOLISM AND ORGANISMAL FUNCTION

Many students feel that the details of cellular metabolism are hard to learn and do not really help them to understand the living world around them. However, metabolic processes within individual cells have enormous impacts on the functioning of entire organisms. To take just two familiar examples, let's consider the migration of ruby-throated hummingbirds in the fall and track events in the Olympics.

Hummingbird migration. As the days shorten in August, many birds in North America prepare to migrate to Central and South America, thereby escaping the cold weather and food shortages of the northern winter. For some, such as ducks and geese, migration is a relatively leisurely affair, with frequent stops along the way for rest and food. For others, migration is a very dangerous undertaking. Ruby-throated hummingbirds, which breed in the eastern United States, fly across the Gulf of Mexico from the southern states to Mexico and Central America. There is nowhere to stop and no food on a journey over 1000 kilometers of open sea.

This presents real difficulties for a migrating hummingbird. With its short wings, a hummer could not fly at all if it weighed too much. Yet flying 1000 kilometers requires a lot of energy that must be stored in the bird's body. Hummers solve this dilemma through a twofold strategy of storing the highest-energy molecules possible, fat, and extracting the maximum usable energy during flight.

A ruby-throated hummingbird weighs 2 to 3 grams prior to putting on fat for migration. It adds as much as 2 grams of fat in late summer, nearly doubling its weight. As you recall from Chapter 3, fats contain over twice as much energy per unit weight as proteins or carbohydrates. If a hummer had to store 4 or 5 grams of glycogen or protein, it would be too heavy to lift off.

Even so, the hummer must still squeeze every ATP possible out of each fat molecule. The hummer that just makes it to Guatemala on 2 grams of fat using cellular respiration would collapse almost within sight of the Gulf coast if it used lactic acid fermentation instead. A hummingbird's flight muscles are packed with mitochondria, and its respiratory system is exquisitely designed to extract oxygen out of the air (Chapter 26), so that even during strenuous flight cellular respiration never slows down.

THE OLYMPICS. Humans, like hummingbirds, must regulate energy reserves and energy use. Why is the average speed of the 5000-meter run in the Olympics slower than that of the 100-meter dash? It is not because the distance runners could not, at any given time, run faster; the finishing kick at the end of the race testifies to that. The reason is that at top speed, their leg muscles consume ATP faster than their lungs can extract oxygen from the air to keep cellular respiration going. Glycolysis and lactic acid fermentation can keep the muscles functional for a short time, but soon the toxic effects of lactic acid buildup cause fatigue and cramps. While runners may be able to do a 100-meter dash anaerobically, 5000 meters is out of the question. Distance runners must therefore pace themselves, using cellular respiration to power their muscles for most of the race and saving the anaerobic sprint for the finish.

Marathon runners face somewhat the same dilemma that hummingbirds do. Nobody ever saw a fat marathoner: it's too much work to lug a lot of weight around for 26 miles. Nevertheless, a marathon may require 3000 kilocalories of energy, and would require even more if cellular respiration did not provide all the ATP. Marathoners train 50 or 100 miles a week, not so much to build up their leg muscles (which are usually pretty stringy) as to build up the capacity of their respiratory and circulatory systems to deliver enough oxygen to their muscles.

As you can see, life on Earth depends on obtaining, storing, and using energy efficiently. Your knowledge of the principles of cellular metabolism should enable you to appreciate the energy-related adaptations of living organisms more fully.

SUMMARY OF KEY CONCEPTS

Capturing Light Energy: Photosynthesis

Photosynthesis uses the energy of sunlight to convert low-energy inorganic molecules of CO_2 and H_2O into high-energy organic molecules such as glucose. Photosynthesis occurs in two steps: the light-dependent and the light-independent reactions. In the light-dependent reactions, light excites electrons in chlorophyll molecules and transfers the energetic electrons to electron transport systems. Some of the energy is used to pump hydrogen ions into the thylakoid sacs, generating a hydrogen ion battery that powers ATP synthesis. Some of the energy, in the form of energetic electrons, is added to electron carrier molecules of $NADP^+$ to make the highly energetic carrier NADPH. Some of the energy is used to split water, generating oxygen as a by-product. In the light-independent reactions, ATP and NADPH provide the energy that drives the synthesis of glucose from CO_2 and H_2O. The depleted carriers, ADP and $NADP^+$, return to the light-dependent reactions for recharging.

Energy Release: Glycolysis and Cellular Respiration

Cells produce usable energy by breaking down glucose to lower-energy compounds and capturing some of the released energy as ATP. In glycolysis, glucose is metabolized in the cytosol to two molecules of pyruvic acid, generating two ATP molecules and two energetic NADH electron carriers. In the absence of oxygen, pyruvic acid is converted by fermentation to lactic acid or ethanol and CO_2, which regenerates NAD^+. This allows glycolysis to proceed but produces no additional ATP. If oxygen is available, the pyruvic acids are metabolized through cellular respiration in the mitochondria. Reactions in the matrix of the mitochondria complete the breakdown of pyruvic acid to CO_2, generating two ATP, eight NADH, and two $FADH_2$ molecules from the two pyruvic acids produced by glycolysis from a single glucose molecule. The electron carriers NADH and $FADH_2$ donate their energetic electrons to an electron transport system in the inner mitochondrial membrane, where their energy is used to pump hydrogen ions from the matrix into the intermembrane compartment, generating a hydrogen ion battery. This hydrogen ion battery drives ATP synthesis, producing 34 ATPs per glucose. Thus a total of 38 ATPs are produced from a single glucose molecule: 2 from glycolysis, 2 from the matrix reactions, and 34 from the electron transport system. At the end of the electron transport system, the electrons combine with oxygen atoms and hydrogen ions to form water. The use of oxygen empties the transport system of energy-depleted electrons, allowing new electrons to run through.

GLOSSARY

C_3 cycle the cyclic series of reactions whereby carbon dioxide is fixed into carbohydrates during the light-independent reactions of photosynthesis. Also called the Calvin–Benson cycle.

C_4 pathway the series of reactions in certain plants that fixes carbon dioxide into organic acids for later use in the C_3 cycle of photosynthesis.

Calvin–Benson cycle *see* C_3 cycle.

Carbon fixation the initial steps in the C_3 cycle, in which carbon dioxide reacts with ribulose bisphosphate to form a stable organic molecule.

Carotenoid (ka-rot′-en-oyd) a family of pigments, usually yellow, orange, or red, found in chloroplasts of plants and serving as accessory light-gathering molecules in thylakoid photosystems.

Cellular respiration the oxygen-requiring reactions occurring in mitochondria that break down the end products of glycolysis into carbon dioxide and water, while capturing large amounts of energy as ATP.

Chlorophyll (klor′-ō-fil) a green pigment that acts as the primary light-trapping molecule for photosynthesis.

Citric acid cycle a cyclic series of reactions in which the pyruvic acid produced by glycolysis is broken down to CO_2, accompanied by the formation of ATP and electron carriers. Occurs in the matrix of mitochondria.

Electron transport system a series of molecules found in the inner membrane of mitochondria and the thylakoid membranes of chloroplasts, which extract energy from electrons, which is used to generate ATP or other energetic molecules.

Fermentation anaerobic reactions that convert the pyruvic acid produced by glycolysis into lactic acid or alcohol and CO_2.

Glycolysis (glī-kol′-i-sis) anaerobic reactions carried out in the cytosol that break glucose down into two molecules of pyruvic acid, producing two ATP molecules. Glycolysis does not require oxygen, but can proceed when oxygen is present.

Granum (pl. grana) in chloroplasts, a stack of thylakoids.

Krebs cycle the citric acid cycle (in honor of Hans Krebs, who discovered many of its biochemical details).

Light-dependent reactions the first step of photosynthesis, in which the energy of light is captured as ATP and NADPH; occurs in thylakoids of chloroplasts.

Light-harvesting complex in photosystems, the assembly of pigment molecules (chlorophyll and often carotenoids or phycocyanins) that absorb light energy and transfer the energy to electrons.

Light-independent reactions the second stage of photosynthesis, in which the energy obtained by the light-dependent reactions is used to fix carbon dioxide into carbohydrates; occurs in the stroma of chloroplasts.

Matrix the fluid contained within the inner membrane of mitochondria.

Photon (fō′-ton) the smallest unit of light.

Photosynthesis the series of chemical reactions in which the energy of light is used to synthesize high-energy organic molecules, usually carbohydrates, from low-energy inorganic molecules, usually carbon dioxide and water.

Photosystem in thylakoid membranes, a light-harvesting complex and its associated electron transport system.

Phycocyanin (fī-kō-sī′-a-nin) a bluish pigment found in the membranes of chloroplasts and used as an accessory light-gathering molecule in thylakoid photosystems.

Reaction center in the light-harvesting complex of a photosystem, the chlorophyll molecule to which light energy is transferred by the antenna pigments. The captured energy ejects an electron from the reaction center chlorophyll, and the electron is transferred to the linked electron transport system.

Stoma (pl. stomata; stō′-ma) adjustable openings in plant leaves. Most gas exchange between leaves and the air occurs through the stomata.

Stroma (strō′-ma) the semifluid medium of chloroplasts, in which the membranous grana are embedded.

Thylakoid (thī′-la-koyd) a disk-shaped, membranous sac found in chloroplasts, the membranes of which contain the photosystems and the ATP-synthesizing enzymes used in the light-dependent reactions of photosynthesis.

STUDY QUESTIONS

1. Define photosynthesis. What is the overall chemical equation for photosynthesis? Is photosynthesis an exergonic or endergonic process?
2. Diagram and describe the structure of the chloroplast.
3. Briefly describe the light-dependent and light-independent reactions of photosynthesis. In which part of the chloroplast does each occur?
4. What is the difference between carbon fixation in C_3 and C_4 plants? Under what conditions does each method of carbon fixation work most effectively?
5. What is the overall chemical reaction for the complete metabolism of glucose via cellular respiration? Is this an endergonic or an exergonic process?
6. Briefly describe glycolysis and cellular respiration. In what part of the cell does each occur? What conditions are necessary if cellular respiration is to occur? What is the overall

energy harvest (in terms of ATP molecules generated per glucose molecule) for each?

7. Briefly describe fermentation. Why is fermentation essential if glycolysis is to continue under anaerobic conditions?
8. Diagram and describe the structure of the mitochondrion. In which parts do the various reactions of cellular respiration occur?
9. Describe the citric acid cycle in general terms. In what form is most of the energy harvested?
10. Describe the mitochondrial electron transport system. What is the energy of the electrons used for?

SUGGESTED READINGS

Bjorkman, O., and Berry, J. "High-Efficiency Photosynthesis." *Scientific American,* October 1973.

Govindjee, G. and Govindjee, R. "The Primary Events of Photosynthesis." *Scientific American,* December 1974 (Offprint No. 1310).

Hinkle, P. C., and McCarthy, R. E. "How Cells Make ATP." *Scientific American,* March 1978 (Offprint No. 1383).

Miller, K. R. "The Photosynthetic Membrane." *Scientific American,* October 1979 (Offprint No. 1448).

Stryer, L. *Biochemistry,* 3rd ed. New York: W. H. Freeman and Company, Publishers. 1988. Good presentation on the processes of photosynthesis and respiration.

Although all Emperor penguins may look alike to the human eye, they are not identical. This penguin father has passed on a unique set of genes to his chick. The genes endow the chick with characteristics that allow the father to recognize his own offspring in the midst of thousands born each year in the colony.

Unit II
Inheritance

8 Cellular Reproduction and the Life Cycles of Organisms

"All cells come from cells." With these words, Rudolf Virchow captured the crucial importance of cellular reproduction for both unicellular and multicellular organisms. If all living organisms consist of one or more cells, and if all cells are descended from preexisting cells, it follows that cellular reproduction is absolutely essential for the continued existence of life on Earth. In this chapter we will study two forms of reproduction at the cellular level. First, we will examine cell division, the process by which an individual cell divides into two nearly identical, though smaller, copies of itself. Cell division is the mechanism whereby unicellular organisms reproduce and the bodies of multicellular organisms grow. Second, we describe a specialized form of cell division called meiosis, which is the basis of sexual reproduction. In animals, the daughter cells produced through meiosis are the sex cells—sperm or egg—that unite during sexual reproduction to give rise to a new generation.

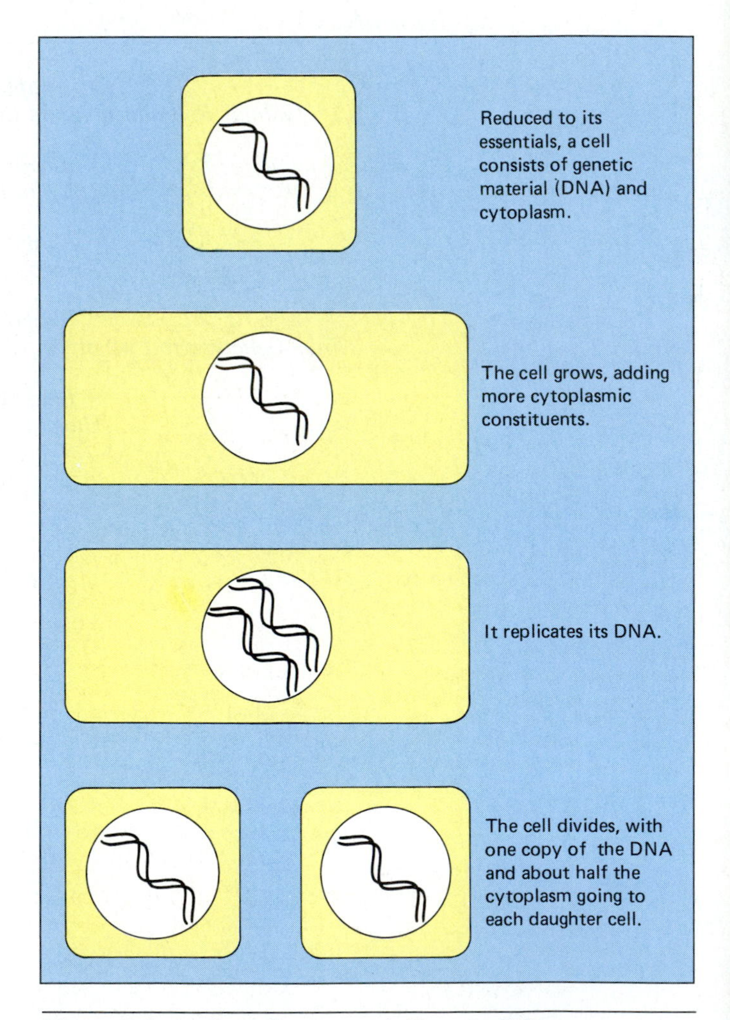

Figure 8-1 *A schematic representation of cellular reproduction.*

ESSENTIALS OF CELLULAR REPRODUCTION

Every organism, and each individual cell of every organism, must pass on to its offspring two essential requirements for life: (1) *hereditary information to direct life processes*, and (2) *other materials needed by the offspring to survive and utilize its hereditary information* (Fig. 8-1).

The hereditary information of all living cells is **deoxyribonucleic acid, DNA**. Like many large biological molecules, a molecule of DNA consists of a long chain of smaller subunits. DNA has four different types of subunits, called nucleotides. Like a biological Morse code, the sequence of nucleotides in DNA encodes a message. Segments of DNA a few hundred to a few thousand nu-

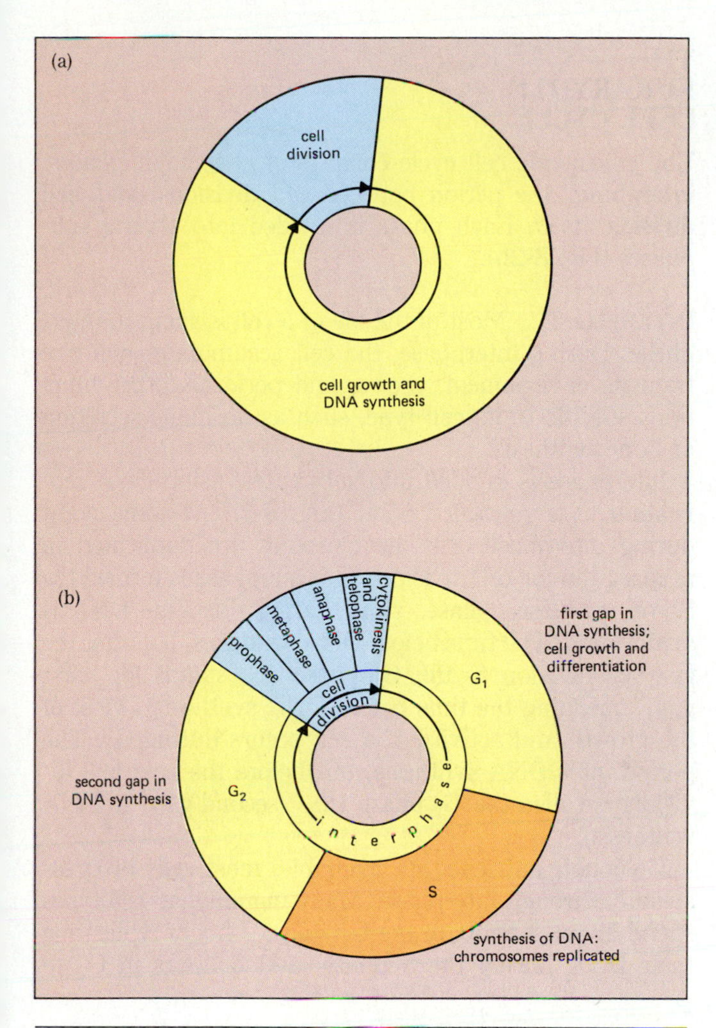

Figure 8-2 Prokaroytic and eukaryotic cell cycles.
(a) *In prokaryotic cells, growth of the cell and DNA replication occur simultaneously. When the DNA has finished replicating, the cell divides.*
(b) *The eukaryotic cell cycle consists of two major phases, interphase and cell division. Each is divided into several subphases, as described in the text.*

cleotides long are the units of inheritance—the **genes** — that specify the enzymes and other proteins needed to build a cell and carry out its metabolic activities. We will see in Chapters 10 and 11 how DNA encodes genetic instructions and how a cell regulates which genes it uses at any given time.

For any cell to survive, it must have a *complete set* of genetic instructions. Therefore, when a cell divides, it cannot simply split its set of genes in half and give each daughter cell half a set. Rather, the cell must first *duplicate its DNA*, much like making a photocopy of an instruction manual. Each daughter cell then receives a complete "DNA manual" containing all the genes.

Even a full set of DNA, however, is useless without materials with which to work. Each newly formed cell must receive the molecules it needs to read its genetic instructions and to keep it alive long enough to acquire new materials from the environment and to process them into new cellular components. Furthermore, as we described in Chapter 5, a cell cannot synthesize either mitochondria or chloroplasts from scratch: these organelles arise only by division of previously existing mitochondria and chloroplasts. Usually, when a cell divides, its cytoplasm is divided about equally between the two daughter cells. This simple mechanism normally provides both daughter cells with all the organelles, nutrients, enzymes, and other molecules they need.

The complete sequence of activities of a cell from one cell division to the next constitutes the **cell cycle** (Fig. 8-2). Newly formed cells usually acquire nutrients from their environment, synthesize more of their own materials, and grow. After a variable amount of time, depending on the organism, the type of cell, and the nutrients available, the cell divides. This general description applies both to eukaryotic and to prokaryotic cells. However, prokaryotic cells are structurally and functionally very different from eukaryotic cells, and accordingly their cell cycle differs in many respects. We therefore discuss prokaryotic and eukaryotic cycles separately.

THE PROKARYOTIC CELL CYCLE

For most of its cell cycle, a prokaryotic cell absorbs nutrients, synthesizes more of its own molecules, and grows. During this time, it also slowly replicates its DNA, which forms a single circular **chromosome** (Fig. 8-3). Shortly after DNA replication is finished, the cell divides.

You will recall from Chapter 4 that a prokaryotic cell does not have a membrane-bound nucleus. With only a single chromosome and no nucleus, prokaryotic cells divide by the fairly simple mechanism, called **fission,** diagrammed in Figure 8-4.

1. Prior to DNA replication, one point on the chromosome is attached to the cell membrane.
2. The chromosome replicates (by a process we will describe in Chapter 10) and the resulting pair of identical chromosomes attach to the cell membrane at nearby, but distinct, points.
3. New cell membrane is added between the two attachment points, pushing them apart.

4 and 5. As the two chromosomes move toward opposite ends of the cell, the cell membrane around the middle of the cell grows inward.

6. Finally, two new daughter cells are formed.

Each cell thus receives one of the replicated chromosomes and about half the cytoplasm. With a complete set of genes and enough materials to work with, the two daughter cells begin to grow, starting the cycle over again.

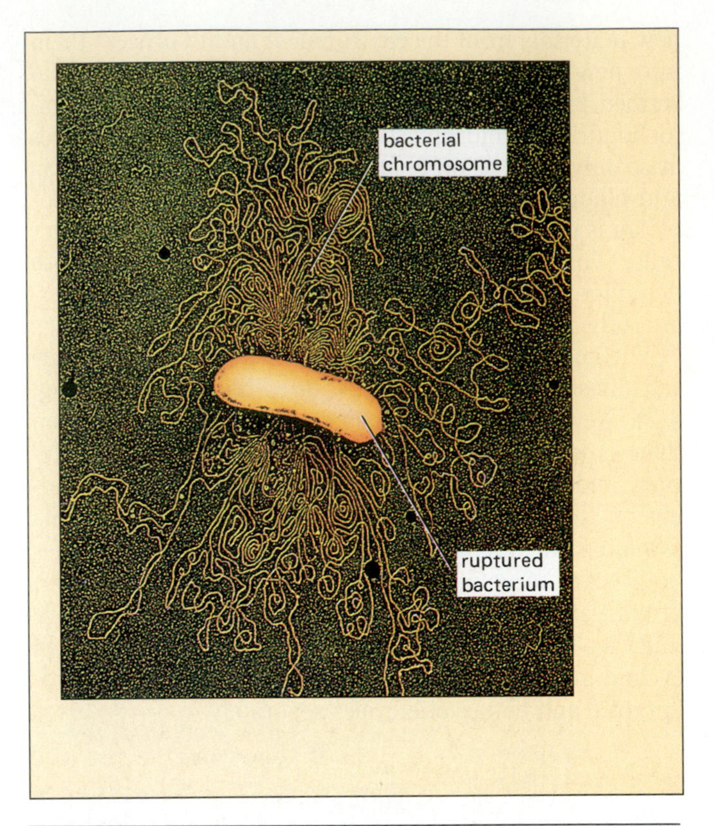

Figure 8-3 *This bacterial cell (yellow central oblong) is about 1 × 3 micrometers in size. The cell was ruptured osmotically, freeing its chromosome. If you look carefully, you will see that there are no free ends: the chromosome is a circular molecule of DNA 1 or 2 millimeters in circumference.*

THE EUKARYOTIC CELL CYCLE

The eukaryotic cell cycle consists of two major phases: **interphase,** the period between cell divisions, and **cell division** itself. Each phase is divided into several subphases (Fig. 8-2b).

Interphase. Most of the life of a cell is spent in interphase. During interphase, the cell accumulates material from its environment, grows, and performs other functions specific to its cell type, such as hormone secretion or bone synthesis.

Interphase is divided into subphases, using DNA replication as a reference point (Fig. 8-2). At some point during interphase, the chromosomes are duplicated in preparation for cell division. This subphase is termed the **S,** or **synthesis, phase,** since during this time DNA is synthesized. The time before DNA synthesis, but after the last cell division, is the **G_1 phase** (G_1 stands for "first gap," meaning the time before DNA synthesis). Most of the growth and activity of a cell occurs during G_1. The period after DNA synthesis, but before the next cell division, is called the **G_2 phase** (the "second gap" in DNA synthesis).

If enough nutrients are available, most cells progress steadily through interphase. Many mammalian cell types spend about 5 hours in the G_1 stage, 7 hours replicating their DNA during the S phase, and 3 hours in G_2 to

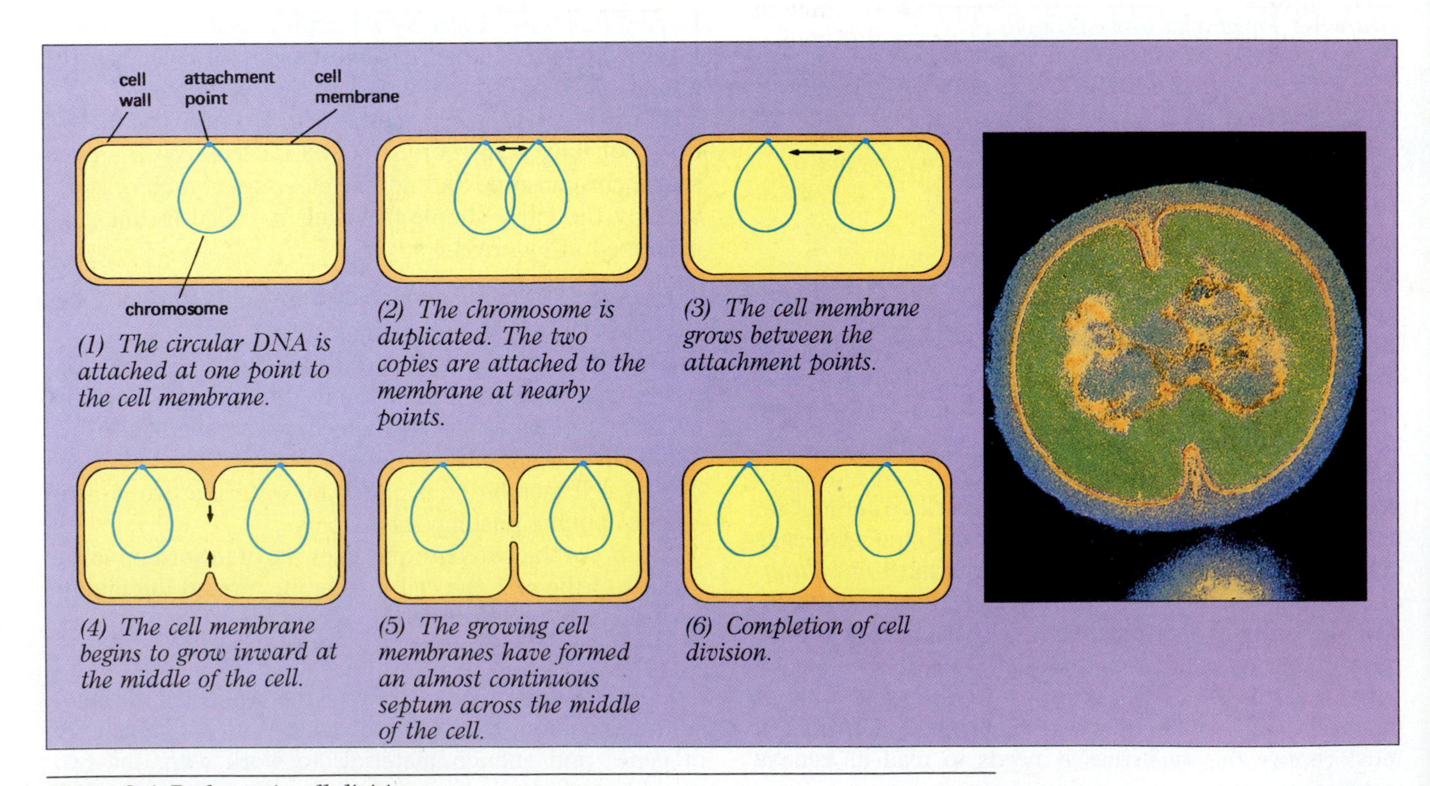

Figure 8-4 *Prokaryotic cell division.*

prepare for division. Cell division itself takes about 1 hour. Exceptions occur mainly in the G_1 phase. The early cell divisions of an embryo, for instance, occur in rapid succession with virtually no G_1 phase at all and therefore almost no growth between divisions (see Chapter 32). In contrast, nerve cells in the adult brain no longer divide, but remain in the G_1 phase of the cell cycle for life.

CELL DIVISION. Cell division consists of two potentially independent events, **nuclear division** and **cytoplasmic division.** During nuclear division (called **mitosis**), identical copies of the DNA (which was previously duplicated during interphase) are packaged in two new nuclei. During cytoplasmic division (called **cytokinesis**), the cytoplasm is split into two daughter cells, with each cell receiving one of the newly formed nuclei and roughly equal amounts of cytoplasm. Therefore, the two daughter cells are essentially identical to each other, both genetically and cytoplasmically. They are also genetically identical to the parent cell.

Not all cells adhere to these scheme. Some cells, including vertebrate muscle cells and some fungi, undergo mitosis without cytokinesis, which produces single cells with many nuclei.

As we mentioned already, a multicellular organism grows from a single fertilized egg by repeated cell divisions. Therefore, with a few exceptions such as muscle cells and sex cells, *every cell of a multicellular organism is genetically identical.*

Much of biology, reduced to its essentials, is the study of cellular activities during interphase, which result in such diverse phenomena as photosynthesis, muscle movement, and thought. Most of the rest of this book is devoted to these topics. Here we focus on the events of cell division.

CELL DIVISION IN EUKARYOTES

Eukaryotic cells contain much more DNA than prokaryotic cells do. The circular chromosome of a bacterium is 1 or 2 millimeters in circumference, whereas the total length of all the chromosomes in a human cell, for example, is about *two thousand* millimeters. Eukaryotic cells package these enormous amounts of DNA, along with a roughly equal amount of protein, into many separate chromosomes. Even so, eukaryotic chromosomes may be much longer than prokaryotic chromosomes. Human cells have 46 chromosomes, so the average length of a human chromosome is over 40 millimeters.

The complicated events of eukaryotic cell division are largely an evolutionary solution for sorting out a large number of long chromosomes. To help you understand the mechanisms of eukaryotic cell division, we will begin by taking a closer look at the structure of the eukaryotic chromosome.

The Eukaryotic Chromosome

Under the light microscope, interphase nuclei appear as large, darkly staining bodies without much internal structure. Early microscopists named the material in the nucleus **chromatin,** from the Greek word for color. During cell division, dark threadlike structures form where the nucleus used to be. The microscopists called these threads **chromosomes,** meaning "colored bodies." With the advent of the electron microscope and advances in molecular biology, we now know that chromatin and chromosomes are actually the same thing (DNA complexed with proteins), but in different stages of condensation. Each eukaryotic chromosome is a single long molecule of DNA, continuous from end to end, complexed with special proteins called histones. During most of its life (i.e., interphase), a cell must be able to read the information contained in the chromosomes, which can be done only when the chromosomes are extended (see Chapter 11). In this extended state, chromosomes are too thin to be visible in light microscopes, or even in ordinary electron microscopes. During cell division, however, the chromosomes must be sorted out and moved into the daughter nuclei. Just as sewing thread is easier to organize when it is wound onto spools, sorting and transporting chromosomes is easier if they are condensed and shortened (Fig. 8-5).

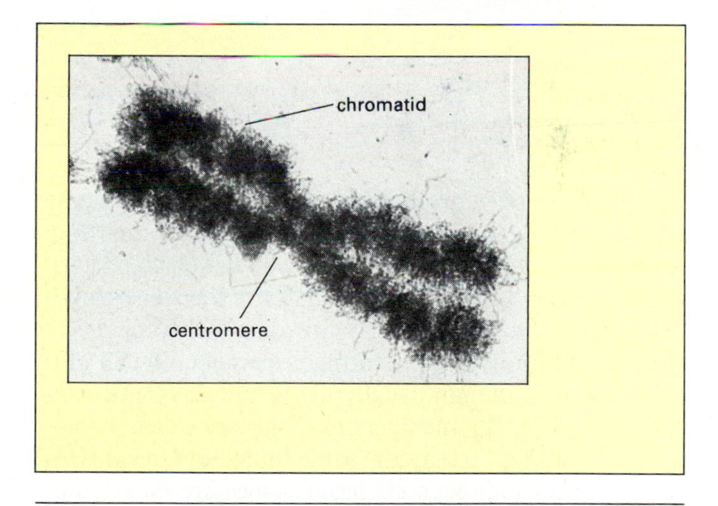

Figure 8-5 An electron micrograph of a eukaryotic chromosome as seen during cell division. The DNA and protein of the chromosome, which are thin and extended during interphase, have coiled up into the thick, short structure seen here. The chromosome has been duplicated prior to condensation; the two strands, called chromatids, are attached at the centromere. To get a sense of the degree of compaction, look at the "fuzz" around the edges: these are loops of the chromosome, about as thin as the whole chromosome would look during interphase. The condensed chromosome is about 10 micrometers long, while the same chromosome during interphase would relax into a thin strand over 20,000 micrometers long.

The nomenclature used by biologists to describe chromosomes during cell division often troubles beginning students. However, you will find it difficult to understand cell division without understanding the terminology, so follow carefully. For simplicity, the drawings that follow show chromosomes in a condensed state.

Most chromosomes consist of two arms that extend out from a specialized region called the **centromere** (meaning "middle body"):

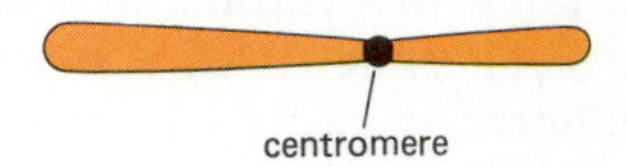

During interphase, chromosomes are duplicated, but the two copies remain attached at the centromere. *As long as they are attached to one another, the copies are called* **sister chromatids.** The entire structure (two sister chromatids attached to one centromere) is still considered to be a single chromosome:

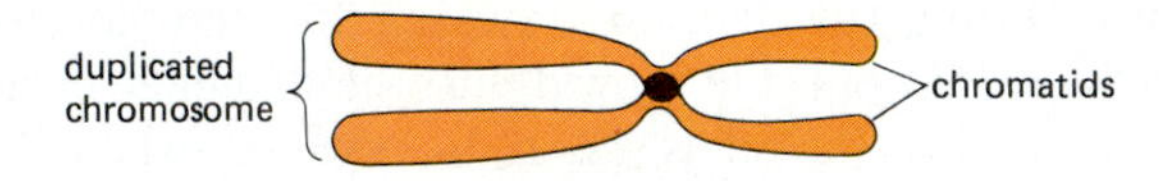

Figure 8-5 therefore shows a single chromosome consisting of two chromatids.

During mitosis, the two sister chromatids separate. *When the chromatids separate, each chromatid becomes an independent chromosome:*

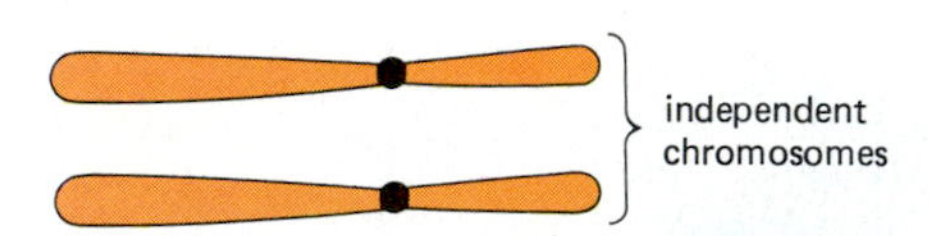

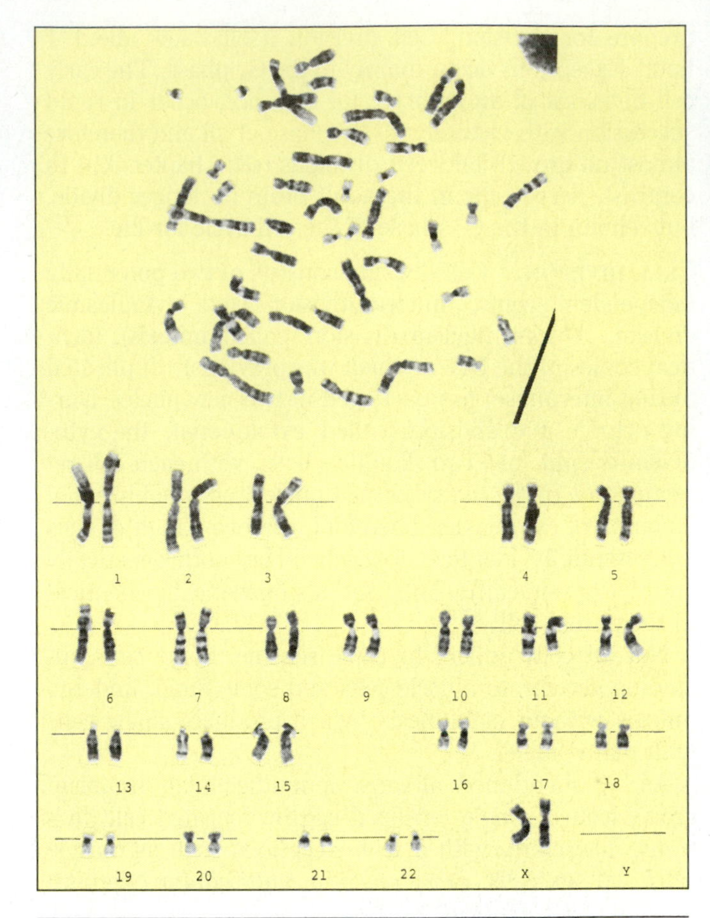

Figure 8-6 The complete set of chromosomes for a human female. The chromosomes of a dividing cell were stained and photographed (top). Pictures of the individual chromosomes were cut out and arranged according to size. Note that the chromosomes occur in pairs, which are similar in both size and staining patterns. If these chromosomes were from a male, there would be one X and one Y chromosome. In humans, the Y chromosome is much smaller than the X chromosome.

The condensed chromosomes of each species assume characteristic shapes, sizes, and staining patterns (Fig. 8-6). In the nonreproductive cells of many organisms, there are *pairs of chromosomes* that are the same length and stain in the same pattern. Breeding and biochemical experiments show that the chromosomes of each pair also have similar, although usually not identical, genetic content. Therefore, the members of a pair are called **homologues,** meaning "to say the same thing" in Greek. Cells with pairs of homologous chromosomes are called **diploid.** The 46 chromosomes of the nonreproductive cells of human beings occur as 23 pairs of homologues.

Not all cells have pairs of homologous chromosomes. At some point in the life cycle of all sexually reproducing organisms, cells are produced that have only one copy of each type of chromosome: these cells are called **haploid.** In animals, these haploid cells are the sex cells, or **gametes** (sperm and egg). Gametes cannot live independently; male and female gametes fuse together to re-form a diploid cell, the fertilized egg or **zygote.** As you may have already surmised, one chromosome of each pair is inherited from each parent. We shall see how this happens in the second half of this chapter.

Some organisms have more than two homologues of each type of chromosome in their cells, and are called **polyploid.** In biological shorthand, *the number of different types of chromosomes* is termed the **haploid number,** and is designated n. For humans, $n = 23$; for other organisms, n ranges from 1 to several hundred. Diploid cells, with two homologues of each type, are $2n$; polyploid cells may be $3n$ (triploid), $4n$ (tetraploid), and so on.

Mitosis

As we outlined earlier, eukaryotic cell division consists of mitosis (nuclear division) and cytokinesis (cytoplasmic division). These two processes usually occur together, but it is simpler to consider them separately. Before we begin our discussion of mitosis, remember that the chromo-

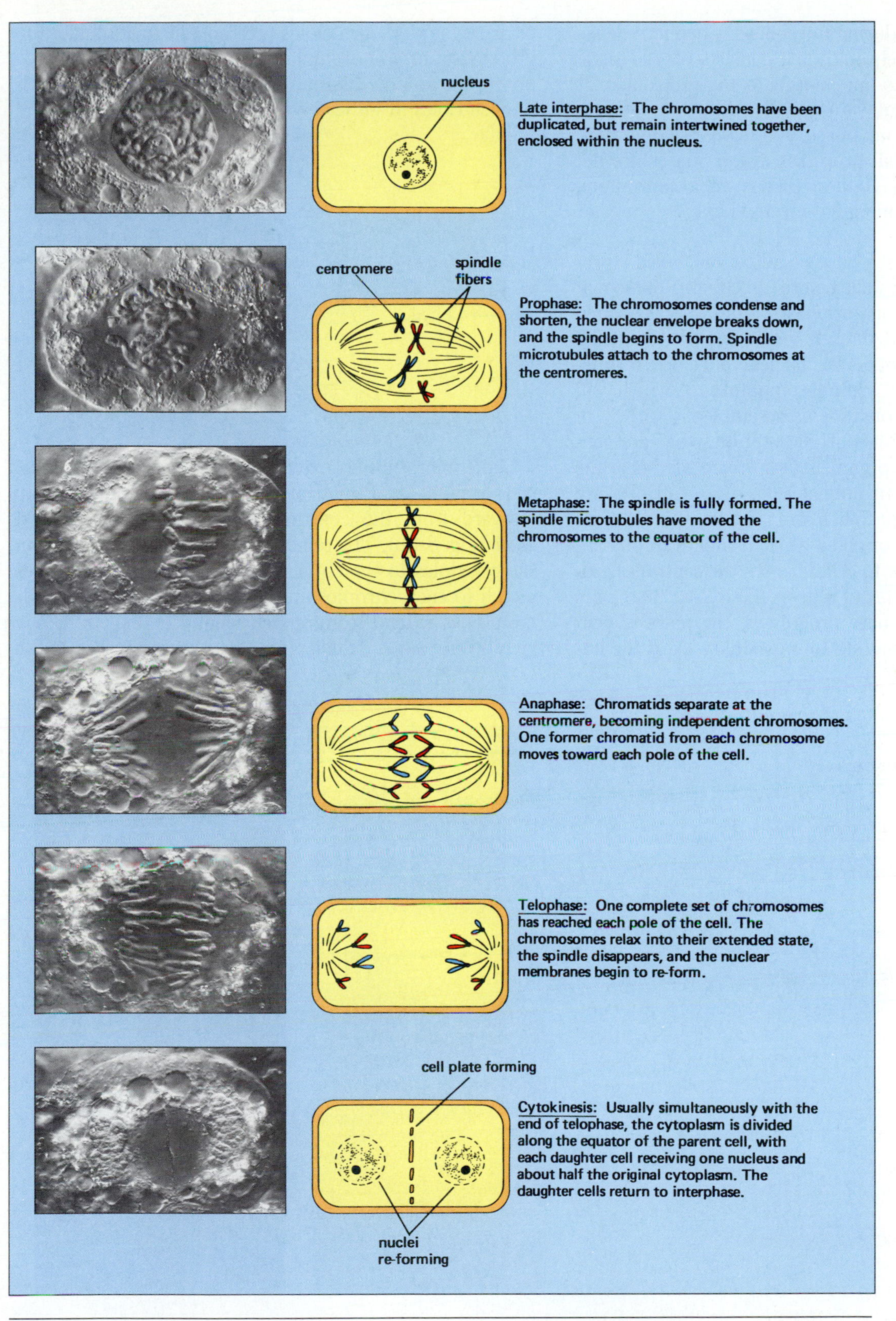

Figure 8-7 *Mitosis in a living plant cell, seen by a special optical system that makes the chromosomes appear as three-dimensional threads.*

somes are duplicated during interphase. Therefore, when mitosis begins, each chromosome consists of two sister chromatids attached to one another at the centromere.

For convenience, mitosis is divided into four phases: prophase, metaphase, anaphase, and telophase (Fig. 8-7). As with most biological processes, these phases are not really discrete events. Rather, they form a continuum, each phase gradually merging into the next.

PROPHASE. The first phase of mitosis is called **prophase** (meaning "the stage before" in Greek). Three major events occur during prophase. First, the replicated chromosomes coil and thicken, becoming visible in the light microscope; second, the nuclear envelope disintegrates; and third, the **spindle apparatus,** which will separate the sister chromatids, is assembled.

Spindles are assemblies of microtubules that are generated by microtubule organizing centers located near the nucleus (Chapter 5). In animal cells, the centrioles are located here and appear to help organize the spindle. Centrioles occur in pairs, oriented at right angles to one another (Fig. 8-8a). Prior to cell division, the centrioles replicate, so that at the start of mitosis the cell has two pairs of centrioles. As prophase progresses, the pairs of centrioles separate and migrate to opposite sides of the nuclear region. The centrioles are thought to stimulate the production of spindle microtubules. By the end of prophase, the spindle assumes a shape like a football, with the centrioles at either end and the chromosomes in the middle:

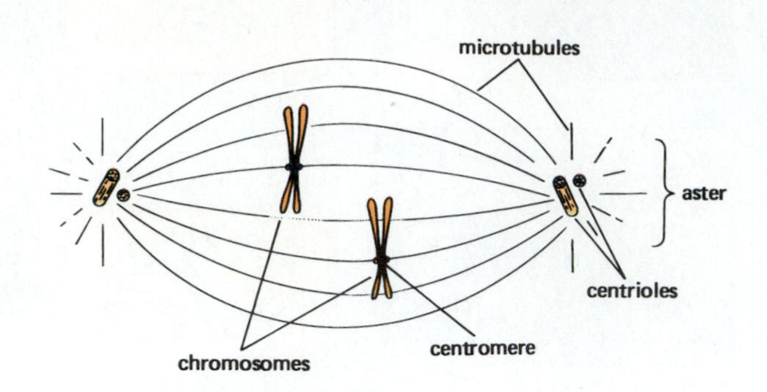

Some microtubules extend completely through the nuclear region from centriole to centriole. Others radiate outward from the centrioles, forming the star-shaped **aster** (Fig. 8-8b). The remaining microtubules reach from the centrioles to the chromosomes. Several microtubules attach to each chromosome at its centromere. Some microtubules extend from one chromatid to one end of the spindle and some extend from the other chromatid to the

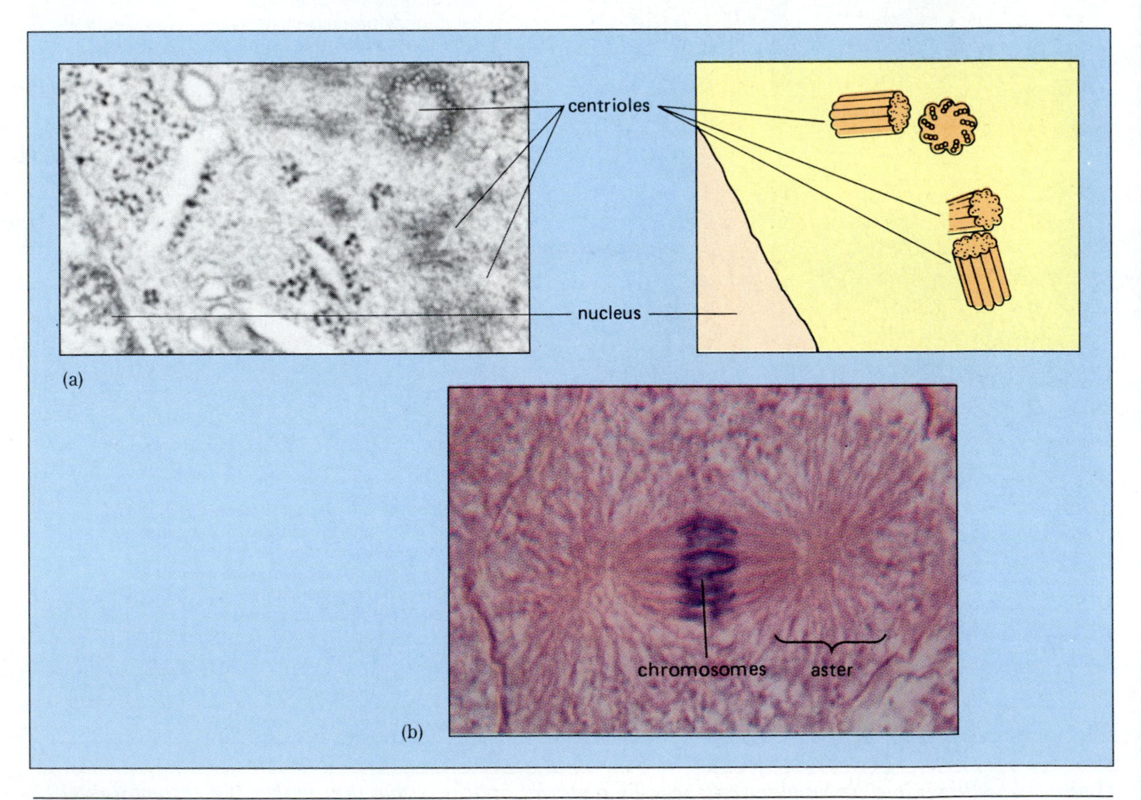

Figure 8-8 **(a)** *An electron micrograph of centrioles in an animal cell. The centrioles have been duplicated, just prior to cell division. The centrioles of each pair lie at right angles to each other. During mitosis, the two pairs migrate to opposite sides of the nuclear region.*
(b) *Asters in embryonic whitefish cells. The centrioles, which are not visible at this magnification, lie in the center of the asters.*

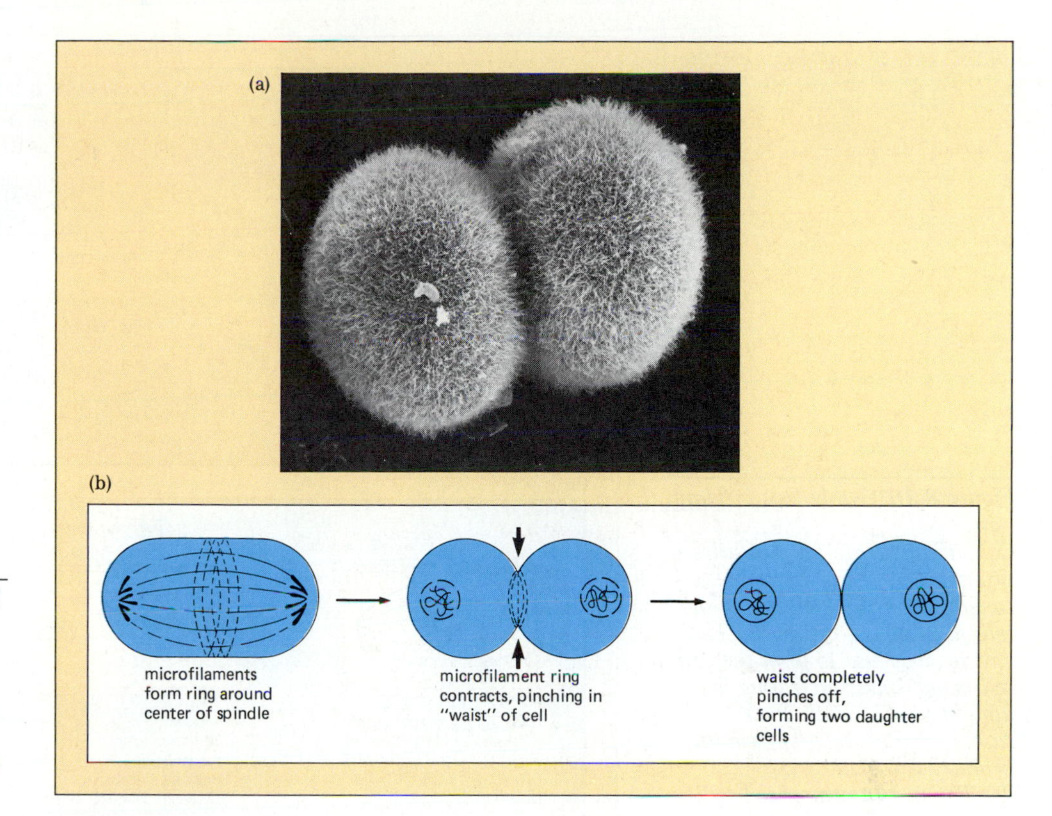

Figure 8-9 Cytokinesis in animal cells.
(a) *A ring of microfilaments just beneath the cell membrane contracts around the equator of the cell, pinching it in two.*
(b) *The mechanism of cytokinesis in animal cells.*

opposite end of the spindle. Plant cells form spindles that closely resemble those of animal cells, except that plant cells do not have centrioles and do not form asters. Prophase ends when the nuclear envelope has disappeared entirely, the spindle has completely formed, and all the chromosomes are attached to spindle microtubules.

METAPHASE. During **metaphase** (the "middle stage"), the spindle microtubules align the chromosomes along the equator of the cell. The sister chromatids of each chromosome face opposite poles of the cell, with microtubules running from each centromere to the appropriate end of the spindle.

ANAPHASE. At the beginning of **anaphase,** the centromere of each chromosome divides, and the sister chromatids separate into two independent chromosomes. The chromosomes move toward opposite poles of the cell. Two forces probably contribute to chromosome separation. First, each chromosome travels poleward by moving along the microtubles that connect it to a pole. In 1988, researchers discovered that the centromere, much like a microscopic Pakman, appears to chew its way poleward along the microtubule, discarding tubulin subunits as it goes. Second, the poles themselves move apart, pulling their spindle microtubules and attached chromosomes with them.

Note that one sister chromatid of each chromosome moves toward each pole of the cell. Since the sister chromatids are identical copies of the original chromosomes, the two clusters of chromosomes that form each contain one copy of every chromosome.

TELOPHASE. When the chromosomes reach the poles of the spindle, **telophase** (the "end stage") has begun. The spindle disintegrates, a nuclear envelope begins to form around each group of chromosomes, and the chromosomes themselves become thin and extended once again. During telophase of most cells, cytokinesis occurs, enclosing each nucleus in its own separate cell.

An important point to note here is that mitosis works equally well in haploid, diploid, or polyploid cells, whether there are few or many chromosomes. In every case, all the chromosomes are replicated, and one copy of each ends up in each daughter cell.

Cytokinesis

In most cells, the division of the cytoplasm into nearly equal halves begins during telophase. In animal cells, microfilaments composed of actin and myosin form rings around the equator of the cell, surrounding the remnants of the spindle (Fig. 8-9). The microfilaments are attached to the cell membrane. During cytokinesis, the rings contract and pull in the equator of the cell, much like pulling the drawstring around the waist of a pair of sweatpants. Eventually, the "waist" contracts down to nothing, dividing the cytoplasm into two new daughter cells.

Cytokinesis in plant cells is quite different, probably

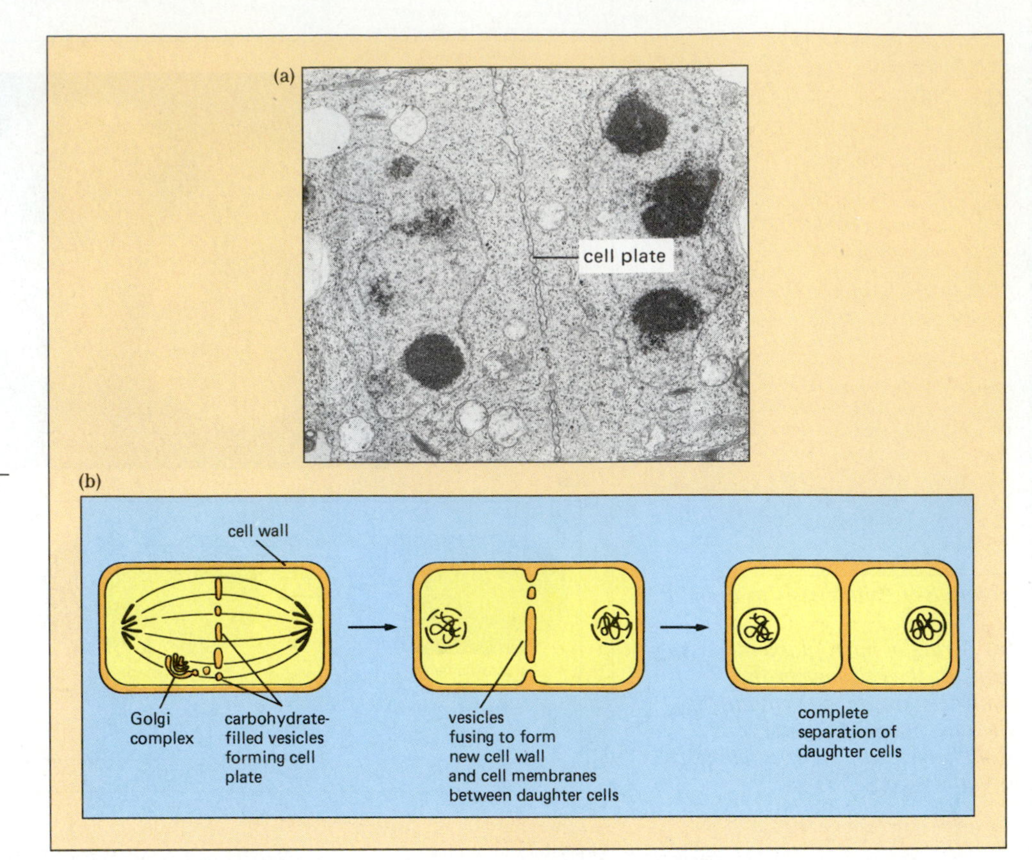

Figure 8-10 *Cytokinesis in plant cells.*
(a) *Carbohydrate-filled vesicles produced by the Golgi complex congregate at the equator of the cell, forming the cell plate. The vesicles will fuse to form the two cell membranes separating the daughter cells, while their carbohydrate contents form the primary cell wall.*
(b) *The mechanism of cell plate formation and cytokinesis in plant cells.*

Figure 8-11 *Modes of asexual reproduction.* **(a)** *In unicellular microorganisms, such as the* Paramecium *shown here, cell division produces two new, independent organisms.*
(b) Hydra, *a freshwater relative of jellyfish and anemones, grows a miniature replica of itself (a bud) protruding from its side. When fully developed, the bud falls off and assumes independent life.*
(c) *The trees in aspen groves are genetically identical, each tree growing as shoots from the roots of a single ancestral tree. This photo shows three groves near Aspen, Colorado. The genetic identity within a grove, and genetic difference between groves, is shown by their leaves in fall. One entire grove is still green, the second has turned bright gold, while the third has already lost its leaves.*

because the stiff cell wall makes it impossible to divide one cell into two by pinching at the waist. Instead, the Golgi complex buds off carbohydrate-filled vesicles that line up along the equator of the cell between the two nuclei (Fig. 8-10). The vesicles fuse together, producing a hollow, pancake-shaped structure, the **cell plate.** When enough vesicles have fused, the edges of the cell plate merge with the original cell membrane around the circumference of the cell. Seen as if you were facing the dividing cell in Figure 8-10, the left membrane of the cell plate becomes part of the cell membrane of the left daughter cell, while the right membrane of the cell plate becomes the cell membrane of the right daughter cell. The carbohydrate formerly contained in the vesicles remains between the cell membranes as the beginning of the primary cell wall.

Cell Division and Asexual Reproduction

Through cell division and subsequent diversification of cells, a single cell, usually a fertilized egg, generates the multicellular bodies of plants and animals. Cell division is also the basis of **asexual reproduction,** in which offspring are formed from a single parent without the necessity of uniting male and female gametes. This is the normal mode of reproduction for many unicellular organisms, such as *Paramecium, Euglena,* and yeasts (Fig. 8-11a).

Many multicellular organisms can also reproduce asexually. Through cell division, small replicas of the parent are grown. A *Hydra,* for example, can reproduce by growing a miniature new *Hydra* as a bud (Fig. 8-11b). Eventually, the bud separates from its parent, going off to live independently. Since mitosis produces genetically identical cells, these offspring are genetically identical to their parents. The beautiful aspen groves of Colorado and New Mexico (Fig. 8-11c) develop asexually from shoots growing up from the root system of a single parent tree. The entire grove, although appearing as a population of separate trees to the admiring visitor, is actually a clone of genetically identical individuals, interconnected by their root systems.

MEIOSIS AND SEXUAL REPRODUCTION

Asexually produced offspring are genetically identical to their parents. This may be an advantage if the parent organism is well adapted to its environment, if that environment never changes, and if the organism and its offspring never move to a new location. In changing environments, however, offspring that are carbon copies of the parent may not survive. If the offspring varied somewhat, on the other hand, perhaps some of them would be well adapted to a new or changing environment.

Early in the history of life on Earth, organisms evolved a way to produce offspring that are similar but not identical to their parents: **sexual reproduction.** In sexual reproduction, two haploid sex cells, or **gametes,** usually from two different parents, fuse together to form a diploid cell, the **zygote.** The zygote then develops into the offspring organism. Each gamete contains genetic instructions from one parent. Therefore, insofar as the two parents have different genes, their offspring will receive a combination of genes that is unique to itself and different from either parent. As we shall see, sexual reproduction offers virtually unlimited possibilities for producing genetically unique offspring.

Although the details vary widely from organism to organism, there are three features typical of sexual reproduction in multicellular eukaryotes: (1) *The organisms engaging in sexual reproduction have diploid cells with pairs of homologous chromosomes at some stage in their life cycle.* (2) At some point, *the homologues are separated from one another through a special cell division process called* **meiosis,** *which produces haploid cells.* In animals, these haploid cells are gametes (i.e., sperm or egg). In plants and many fungi, the haploid cells are spores that undergo mitosis to produce a multicellular, haploid body. This haploid form then produces sex cells through further mitosis. (3) *During fertilization, the haploid gametes fuse to form a diploid cell once again (the fertilized egg), with one copy of each homologous chromosome donated by each parent.*

The key to sexual reproduction is meiosis, the production of haploid cells with unpaired chromosomes. If organisms produced sex cells with the full diploid complement of chromosomes, their offspring would be tetraploid ($4n$), having received $2n$ chromosomes from its father and $2n$ from its mother. The next generation would become $8n$, the next $16n$, and so on. Only by separating homologous chromosomes in meiosis and fusing the resulting haploid cells can each new generation remain diploid:

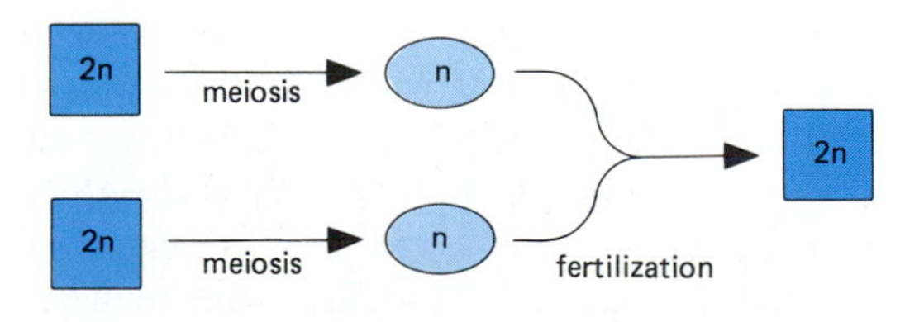

Meiosis

The word **meiosis** comes from a Greek word meaning "to diminish," and that is just what meiosis does: it reduces the number of chromosomes by half. Meiosis is not, however, a random division. In meiosis, *each daughter cell receives one member of each pair of homologous chromosomes.*

Meiosis I

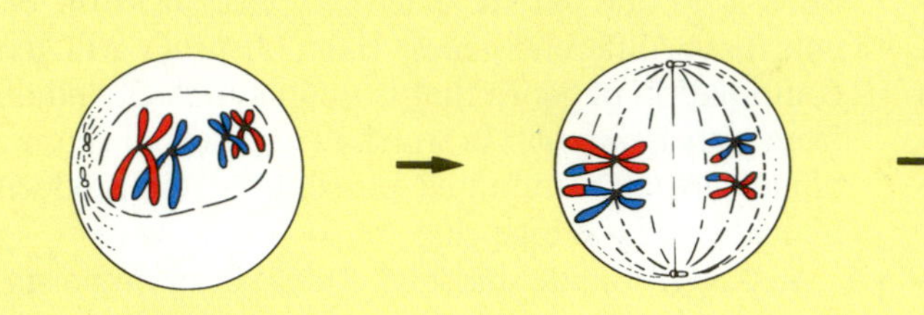

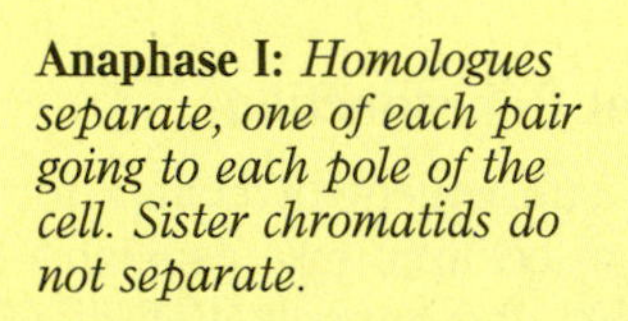

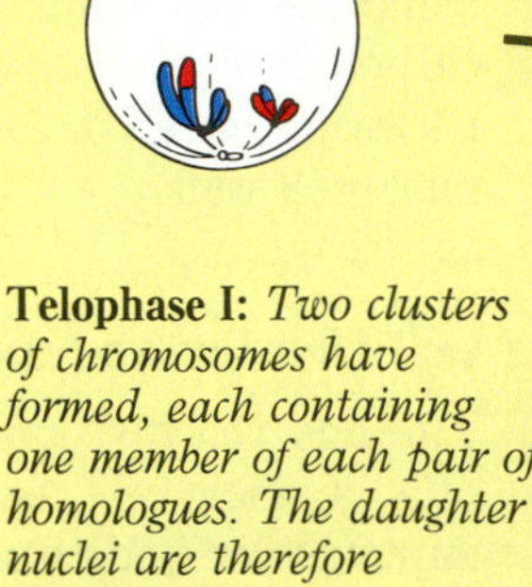

Prophase I: *Chromosomes thicken and condense. Homologous chromosomes come together in pairs, and chiasmata occur, as chromatids of homologues exchange parts. The nuclear membrane disintegrates, and the spindle forms.*

Metaphase I: *Paired homologous chromosomes line up along the equator of the cell. One homologue of each pair faces each pole of the cell, with both chromatids of a given homologue becoming attached to microtubules leading to the same pole.*

Anaphase I: *Homologues separate, one of each pair going to each pole of the cell. Sister chromatids do not separate.*

Telophase I: *Two clusters of chromosomes have formed, each containing one member of each pair of homologues. The daughter nuclei are therefore haploid. Cytokinesis often occurs during telophase I. There is usually little or no interphase between meiosis I and meiosis II.*

Figure 8-12 *Meiosis is the process by which the homologous chromosomes of a diploid cell are separated, producing four haploid daughter cells. In these diagrams, two pairs of homologous chromosomes are shown, large and small. The blue chromosomes are from one*

Meiosis involves two nuclear divisions, termed meiosis I and meiosis II. In brief, the chromosomes are duplicated during interphase prior to meiosis I. Sister chromatids remain attached at the centromere. *During* **meiosis I,** *homologous chromosomes are separated, with each daughter nucleus I receiving one homologue of each pair of chromosomes.* Therefore, the daughter nuclei are haploid. The sister chromatids do *not* split, so each chromosome still consists of two joined chromatids at the end of meiosis I. The chromosomes do not duplicate again between meiosis I and meiosis II. *In* **meiosis II,** *the sister chromatids split into two independent chromosomes, with one going into each daughter nucleus II.* Therefore, the reduction in chromosome number from diploid to haploid occurs during meiosis I. Meiosis II is similar to mitosis occurring in a haploid cell, with the number of chromosomes remaining the same. Each nuclear division is usually accompanied by cytokinesis. Therefore, a cell that undergoes both meiosis I and II produces a total of four haploid cells. The phases of meiosis are given the same names as the roughly equivalent phases in mitosis, although, as we shall see, there are important differences, especially in meiosis I (Fig. 8-12).

MEIOSIS I. During **prophase I,** the nuclear envelope disintegrates and the duplicated chromosomes condense, as they do in prophase of mitosis. Unlike mitosis, however, early in prophase I of meiosis the homologous chromosomes pair up with one another, side by side. Often the arms of the two chromosomes intertwine, forming crosses or **chiasmata** (singular, **chiasma;** Fig. 8-13). As we will see in the next chapter, at least some of these chiasmata are sites of exchange of corresponding segments of DNA between chromosomes. As the chromosomes continue to shorten and thicken, the chiasmata disappear. The chromosomes, still paired, attach to spindle microtubules and move toward the equator of the cell.

When the paired chromosomes reach the equator, the cell enters **metaphase I.** Two striking differences between meiosis I and mitosis become evident. (1) In mitosis the chromosomes line up individually along the equator of the cell. In meiosis, the chromosomes are aligned *in pairs* along the equator. (2) In mitosis, the two sister chromatids of each chromosome attach to spindle fibers extending toward opposite poles of the cell (see p. 140). Therefore, during anaphase the spindle microtubules pull the chromatids apart, dragging one to each pole. *In meiosis*

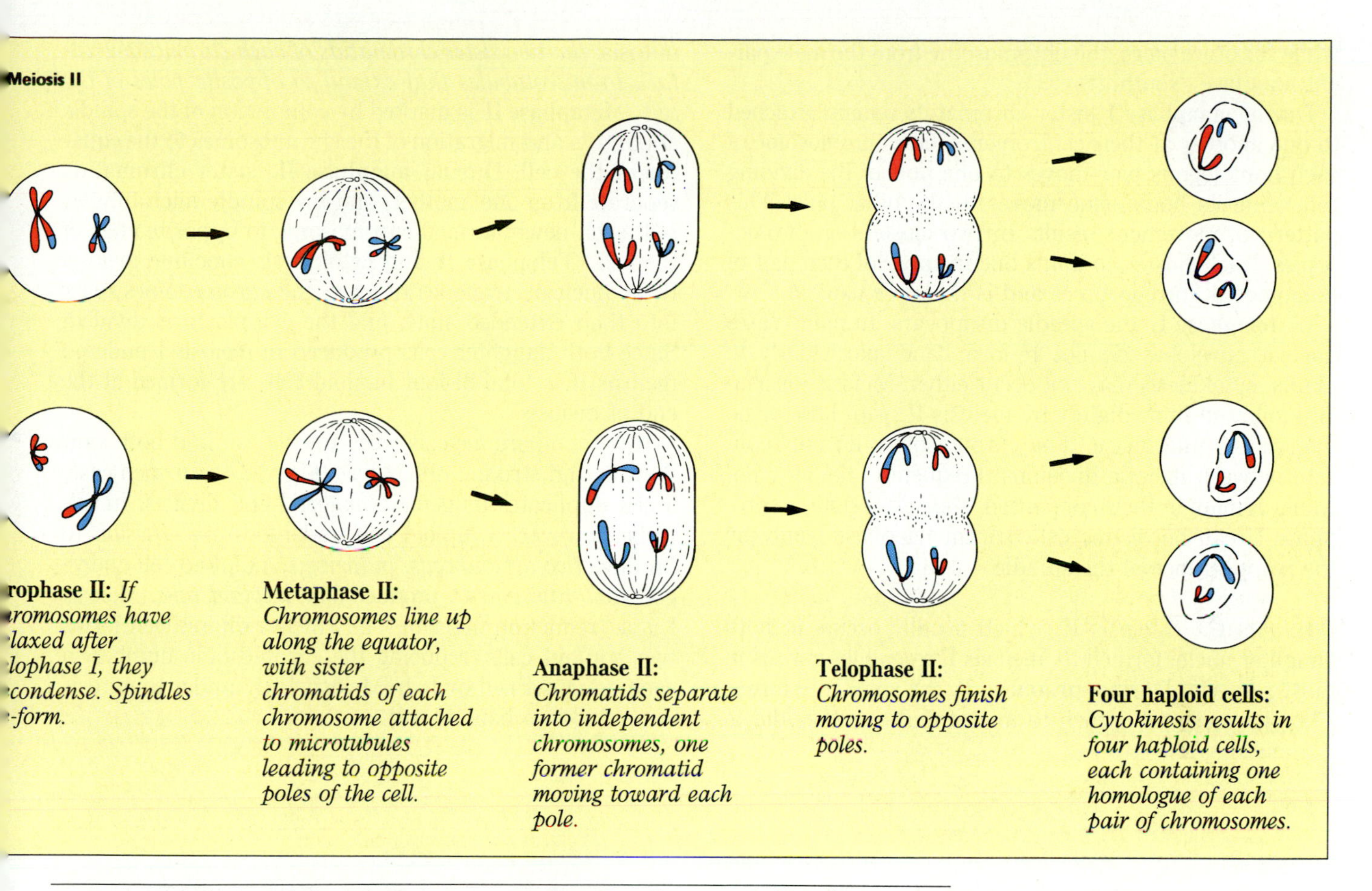

parent (e.g., father), and the red chromosomes are from the other parent. A random assortment of chromosomes from both parents may end up in the haploid daughter cells as a result of meiosis.

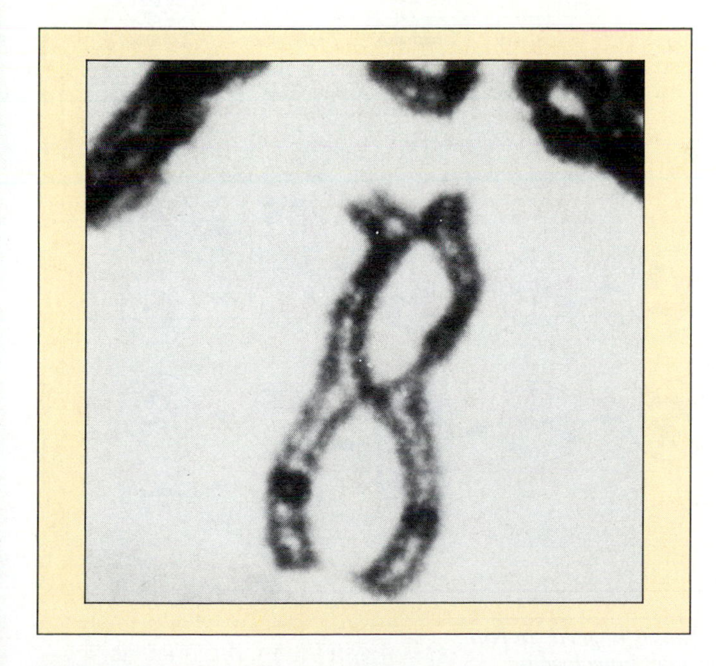

Figure 8-13 Chiasmata in homologous chromosomes during prophase I in a salamander testis. Chiasmata are sites of potential interchange of DNA between homologues.

I, however, both chromatids of one homologous chromosome attach to spindle microtubules extending toward one pole of the cell, while both chromatids of its paired chromosome attach to spindle fibers leading to the opposite pole:

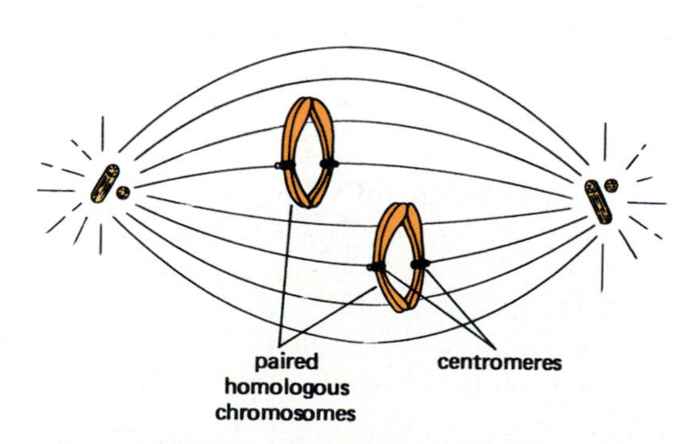

Movement of the spindle microtubules will thus separate the paired homologous chromosomes but leave the sister chromatids attached together at the centromere.

Which member of a pair of chromosomes faces which pole of the cell is random. For some pairs, the chromosome derived from the male parent may face "north," while for other pairs, the chromosome from the male parent may face "south."

During **anaphase I** sister chromatids remain attached to one another at their centromeres. One chromosome of each homologous pair moves to one pole of the dividing cell while its homologue moves to the other pole. This pattern of movement results in two clusters of chromosomes. Each cluster contains one member of each pair of homologous chromosomes and is therefore haploid.

In **telophase I,** the spindle disappears. In many cases nuclear envelopes do not re-form, and, particularly in plants, cytokinesis may not occur either. Meiosis I is usually followed immediately by meiosis II, with little or no intervening interphase. The chromosomes do not replicate between meiotic divisions. Frequently, the chromosomes remain in their condensed, shortened state, so prophase II can be distinguished from telophase I only by the reappearance of the spindle.

Meiosis II. Meiosis II, which usually occurs in both daughter nuclei formed by meiosis I, resembles mitosis in most respects. During **prophase II** the spindle re-forms. The chromosomes attach to microtubules as they did in mitosis: *the two sister chromatids of each chromosome attach to microtubules that extend to opposite poles of the cell.* **Metaphase II** is marked by completion of the spindle apparatus and migration of the chromosomes to the equator of the cell. During **anaphase II,** sister chromatids separate from one another, and the spindle microtubules pull each newly formed chromosome to opposite ends of the cell. **Telophase II** and cytokinesis conclude meiosis II, as nuclear envelopes re-form, the chromosomes relax into their extended state, and the cytoplasm is divided. Since both daughter cells produced in meiosis I undergo meiosis II, a total of four haploid cells are formed at the end of meiosis.

If we compare meiosis with mitosis, we find both similarities and striking differences (Fig. 8-14). In each case, a cell duplicates its genetic material and divides. In mitosis, however, a diploid cell undergoes *one division* to produce *two diploid cells;* in meiosis, a diploid cell undergoes *two divisions* to produce *four haploid cells.* Meiosis I is a "reduction division," in which a diploid cell forms two haploid cells, reducing the chromosome number in each daughter cell by half. Meiosis II is similar to mitosis occurring in a haploid cell.

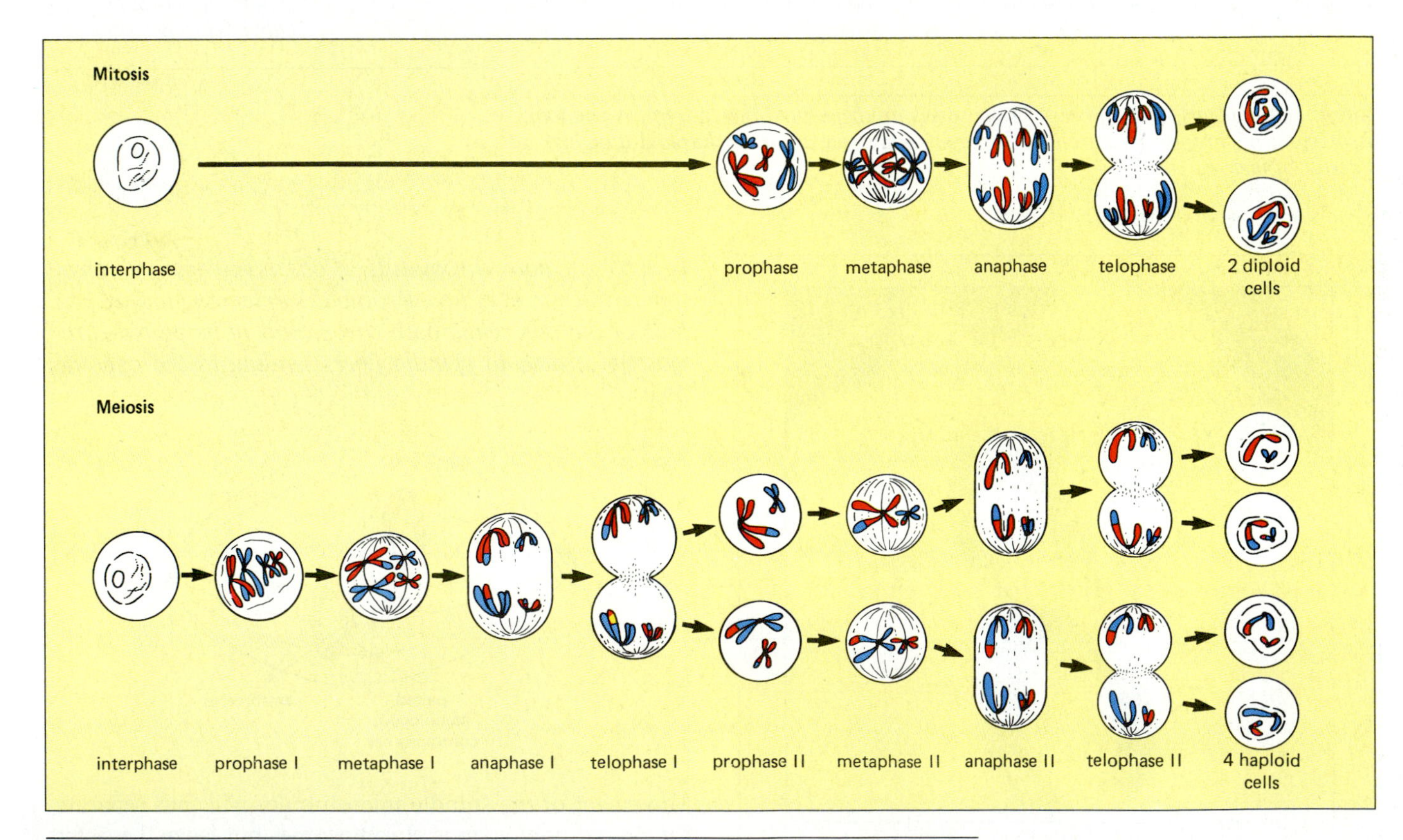

Figure 8-14 A comparison of mitosis and meiosis, with comparable phases aligned. In both, chromosomes are duplicated during interphase. Meiosis I, with pairing of homologous chromosomes, formation of chiasmata and exchange of chromosome parts, and separation of homologues to form haploid daughter nuclei, has no counterpart in mitosis. Meiosis II, however, is similar in many respects to mitosis occurring in a haploid cell.

MITOSIS, MEIOSIS, AND EUKARYOTIC LIFE CYCLES

Most eukaryotic organisms have life cycles in which mitosis and meiosis both play crucial roles (Fig. 8-15). All life cycles have a common overall pattern. (1) A diploid organism produces haploid cells through meiosis. (2) Two haploid cells fuse, bringing together genes from different parent organisms and endowing the resulting diploid cell with new gene combinations. (3) At some point, meiosis occurs again, re-creating haploid cells. (4) At some time in the life cycle, mitosis of either haploid or diploid cells, or both, results in the growth of multicellular bodies and/or asexual reproduction. The seemingly vast differences between the life cycles of, say, ferns and humans are due to variations in three aspects: (1) the interval between meiosis and the fusion of haploid cells; (2) at what points in the life cycle mitosis and meiosis occur; and (3) the relative proportions of the life cycle spent in the diploid and haploid states. These aspects of life cycles are interrelated, and we can conveniently label life cycles according to whether diploid or haploid stages predominate.

Haploid life cycles. Some eukaryotes, such as the unicellular alga *Chlamydomonas,* spend most of their life cycles in the haploid state, with unpaired chromosomes (Fig. 8-16). Asexual reproduction by mitosis produces a population of identical, haploid cells. Under certain environmental conditions, specialized "sexual" haploid cells

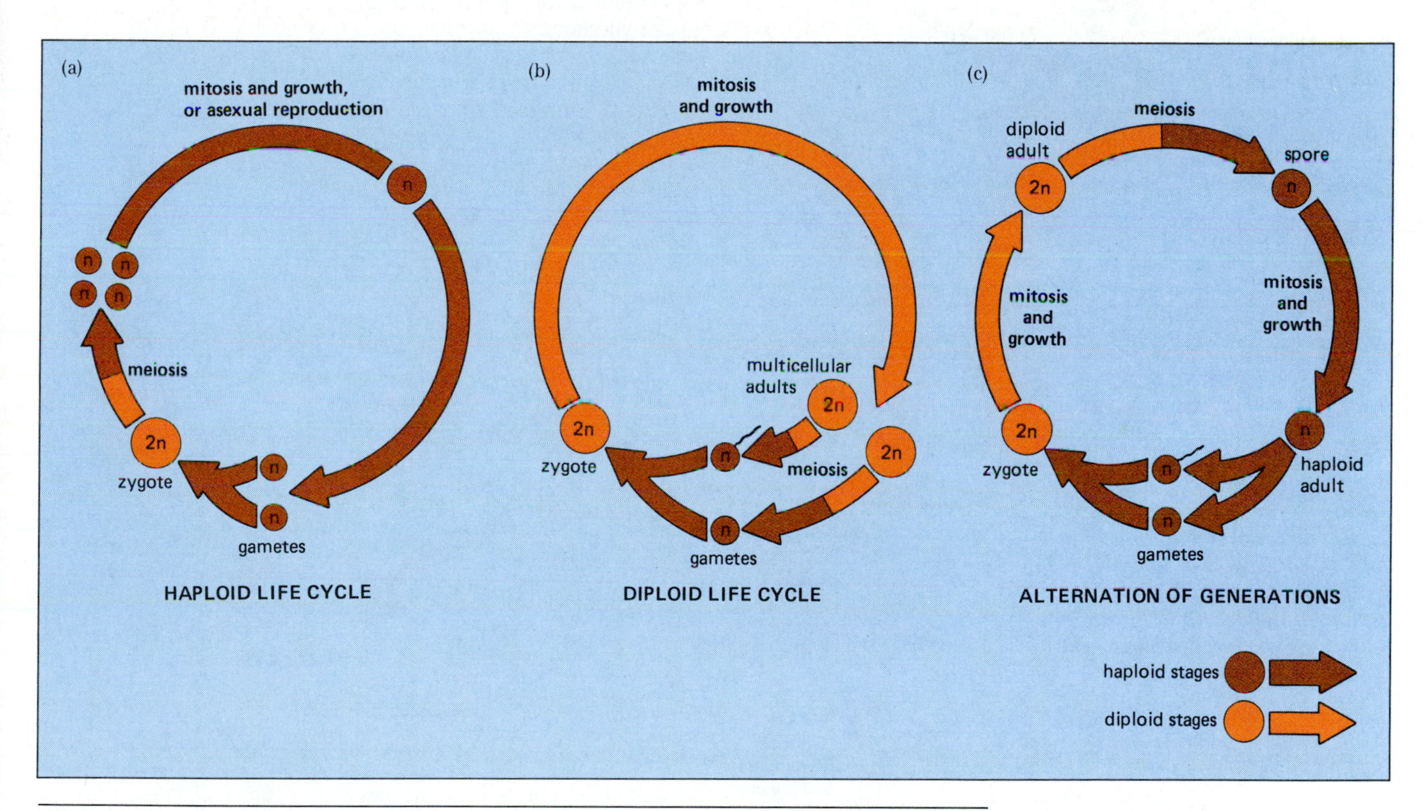

Figure 8-15 Schematic diagrams of the three major types of eukaryotic life cycles. The lengths of the arrows correspond roughly to the proportion of the life cycle spent in each stage.

(a) Haploid life cycle *(many protists, algae, and fungi): Most of life cycle is spent with haploid cells. Mitosis of haploid cells results in either asexual reproduction of unicellular organisms or growth of multicellular organisms. At some point, specialized reproductive haploid cells fuse, and the resulting diploid cell almost immediately undergoes meiosis.*

(b) Diploid life cycle *(animals): Most of the life cycle is spent with diploid cells. When haploid cells are produced, they immediately fuse to form a diploid zygote. Mitosis of the zygote produces multicellular bodies composed of diploid cells.*

(c) Alternation of generations *(plants): Both multicellular diploid and multicellular haploid stages occur. Cells of the diploid stage undergo meiosis to form spores. The spores then undergo mitosis to produce a multicellular haploid body. Later, specialization of haploid cells produces gametes, which fuse to form a zygote. Through mitosis, the zygote develops into a multicellular diploid form.*

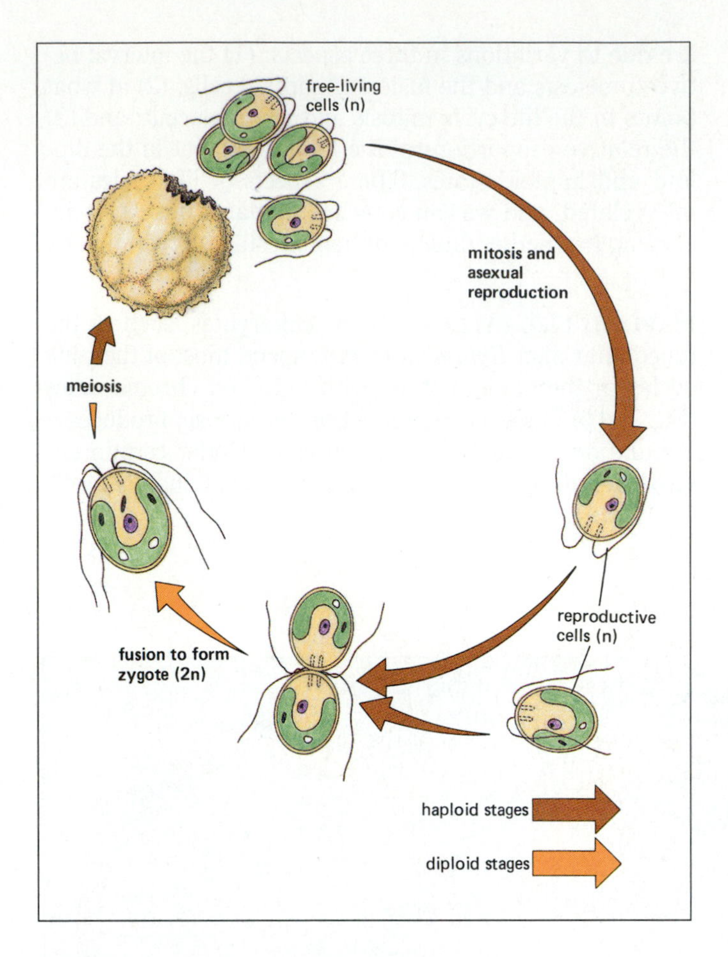

Figure 8-16 The life cycle of the unicellular alga Chlamydomonas. Chlamydomonas *multiplies asexually by mitosis of haploid cells. When nutrients are scarce, specialized haploid cells (usually of different strains) fuse to form a diploid cell. Meiosis then immediately produces four haploid cells, usually with different genetic composition than either of the original parental strains.*

Figure 8-17 The human life cycle. Through meiosis, the two sexes produce different types of gametes, sperm in males and eggs in females, that fuse to form a diploid zygote. Mitosis and specialization of cells produce an embryo, child, and ultimately sexually mature adults. The haploid state lasts only a few hours to a few days, whereas the diploid state may survive for a century.

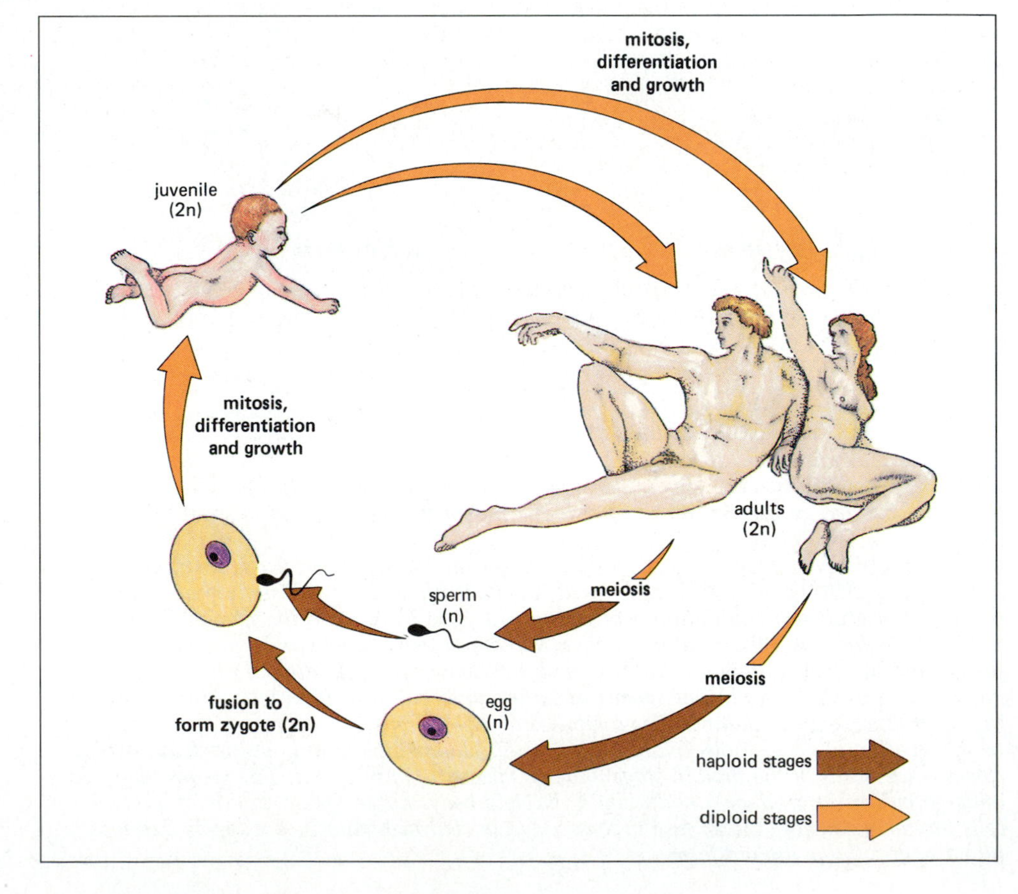

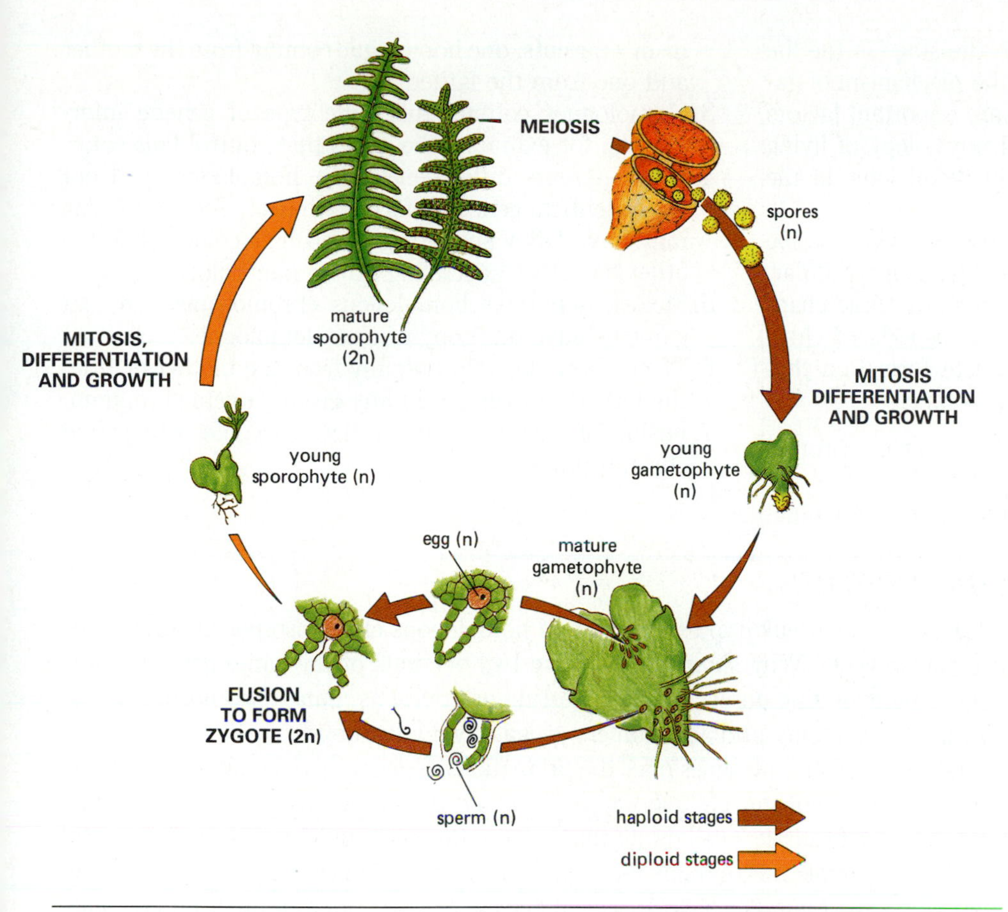

Figure 8-18 The life cycle of a fern is representative of alternation of generations. The diploid stage is called the sporophyte; this is the fern usually seen in the woods. Specialized cells in the sporophyte undergo meiosis, producing haploid spores. Spores differ from gametes in that spores undergo mitosis and grow into a multicellular haploid body (the gametophyte). In ferns, the haploid stage is inconspicuous, a small, flat plant growing in moist places on the forest floor. "Gonads" form on the haploid stage, producing sperm and eggs. Sperm and eggs fuse to make a diploid zygote. Through mitosis, the zygote develops into a new multicellular diploid body.

are produced. Two of these sexual haploid cells fuse together, forming a diploid cell. This cell immediately undergoes meiosis, producing haploid cells again. Mitosis nevers occurs in diploid cells.

Diploid life cycles. Most animals have life cycles that are just the reverse of the *Chlamydomonas* cycle. Virtually the entire animal life cycle is spent in the diploid state (Fig. 8-17). Haploid gametes (sperm in males, eggs in females) are formed through meiosis. These fuse to form a diploid fertilized egg, the zygote. Growth and development of the zygote into the adult organism is a result of mitosis in diploid cells.

Alternation of generation life cycles. The life cycles of most plants include both multicellular diploid and multicellular haploid stages. In the typical pattern (Fig. 8-18), a multicellular diploid body gives rise to haploid cells, called **spores,** through meiosis. The spores then undergo mitosis to create a multicellular haploid stage (the "haploid generation"). At some point, certain haploid cells differentiate into haploid gametes. Two gametes then may fuse to form a diploid zygote. The zygote grows by mitosis into a diploid multicellular body (the "diploid generation"). In "primitive" plants, such as ferns, both the haploid and diploid stages are free-living, independent plants. The flowering plants, however, have reduced the haploid stage to a minimum, represented only by the pollen grain and a small cluster of cells in the ovary of the flower.

CHROMOSOMES, MEIOSIS, AND INHERITANCE

As you can see from these life-cycle summaries, the details of sexual reproduction differ tremendously among the various forms of life on Earth. Both the stage in the life cycle in which meiosis occurs, and the mechanism of gamete formation, vary widely. These are important factors in understanding the evolution and physiology of living organisms, and will be discussed in detail later in the book.

In the remaining chapters of this unit, we will describe the mechanisms of inheritance on the organismic, cellular, and molecular levels. In preparing to study these chapters, you should concentrate on those aspects of chromosome structure, meiosis, and gamete formation that are essential for your understanding of genetics:

1. Diploid organisms have pairs of homologous chromosomes.
2. The paired chromosomes are derived from an organism's parents, one homologue coming from the mother and one from the father.
3. Homologues contain the same type of genetic information, for example, the genes that control hair color.
4. If the parents differ genetically, homologues will not have identical genetic information (e.g., one homologue may have DNA specifying blond hair color, while the other has DNA specifying brown hair color).
5. Meiosis separates homologous chromosomes, so that gametes have one copy of each homologue.
6. Which homologue (originally from the mother or from the father) is included in any given gamete is random.
7. Fusion of gametes forms a diploid zygote with paired chromosomes.

REFLECTIONS ON PAIRED CHROMOSOMES

All multicellular eukaryotes have paired homologous chromosomes at some stage in their life cycle. Why should a cell have two versions of the same genetic information, necessitating duplicating and sorting out twice as many chromosomes during mitosis? Diploidy probably confers at least two advantages.

First, pairs of chromosomes offer the protection of having two copies of each gene, one on each homologue. If one copy should be damaged—by radiation or toxic chemicals, for example—its counterpart on the homologous chromosome may still provide the genetic information the organism needs to survive. As we shall see in Chapter 12, most humans have dozens of defective genes, but still function normally because of intact genes on homologous chromosomes.

A second, perhaps more important advantage of paired chromosomes and meiosis is the production of variable offspring through **sexual recombination:** if the two gametes that form an offspring come from genetically different parents, the offspring will have a combination of genes that differs from either parent. As we pointed out earlier, environments change and organisms move about, so that what is a "perfect" set of genes for the parent may not be so great for the offspring. Therefore, it is an evolutionary advantage to produce offspring that are at least somewhat different from the parents and from each other.

How do meiosis and the fusion of haploid gametes generate genetic variation among offspring? Consider the common fruit fly, *Drosophila melanogaster. Drosophila* has four pairs of chromosomes, for a diploid number of eight. Let us suppose that the members of each pair are slightly different from one another. During gamete formation, the pairs are separated, with one member of each pair entering each gamete. Which chromosome of each pair of homologues a particular gamete receives is random. When fertilization occurs, the four chromosomes of the sperm join with the four chromosomes of the egg to produce a zygote with four pairs again. How many different combinations of chromosomes can the gametes have, and how many unique offspring with different gene combinations can be formed through fertilization? A single fly can produce 16 different types of sperm or egg. [You can work this out by assigning numbers and letters to the chromosomes (e.g., 1a, 1b, 2a, 2b, 3a, 3b, 4a, and 4b) and grouping them into all possible combinations of gametes, each receiving either the "a" or "b" chromosome of each number; you will find 16 different combinations. A simpler mathematical way is to calculate the number of homologues of any type of chromosome (2) raised to the power of the number of types of chromosomes (4 in *Drosophila*): $2^4 = 16$.] A single pair of flies can produce

16 × 16 = 256 unique offspring. A single human being, with 23 pairs of chromosomes, can produce 2^{23} or about 8 million different gametes. One man and one woman could theoretically people the Earth with 64 trillion unique offspring!

SUMMARY OF KEY CONCEPTS

Essentials of Cellular Reproduction

Every cell must begin life with (1) a complete set of genetic information (DNA) and (2) the materials necessary to survive and utilize the genetic information. Cellular reproduction involves duplication of the DNA, with one copy packaged into each daughter cell, and cytoplamic division, with roughly half the parental cytoplasm passing into each daughter cell.

The Prokaryotic Cell Cycle

The DNA of a prokaryotic cell is in the form of a single circular chromosome, attached at one point to the cell membrane. When a prokaryotic cell reproduces, it duplicates the chromosome, moves the two resulting copies to opposite ends of the cell, and divides the cell in two.

The Eukaryotic Cell Cycle

The eukaryotic cell cycle consists of two major phases, interphase and cell division. During interphase, the cell takes in nutrients, grows, and duplicates its chromosomes. Cell division consists of two processes, nuclear division (mitosis) and cytoplasmic division (cytokinesis). Mitosis parcels out one copy of each chromosome into two separate nuclei, and cytokinesis subsequently encloses each nucleus in a separate cell.

Cell Division in Eukaryotes

The chromosomes of eukaryotic cells consist of both DNA and protein. During interphase, the chromosomes are in an extended form, accessible for use in reading their genetic instructions. During cell division, the chromosomes condense to form short, thick structures. At some point in all eukaryotic life cycles, cells contain pairs of chromosomes called homologues. Homologues have virtually identical appearance and very similar genetic information. Cells with pairs of homologous chromosomes are diploid. Those with only a single member of each pair are haploid, and those with more than two copies of each homologous chromosome are polyploid.

The chromosomes are duplicated during interphase, prior to mitosis. The two identical replicas, called chromatids, are attached to one another at the centromere. During mitosis, the chromatids separate, one going to each daughter nucleus. The two daughter nuclei thus receive identical genetic information. Mitosis consists of four phases:

Prophase: the chromosomes condense, the nuclear envelope breaks down, and the spindle microtubules form. Some of the spindle microtubules attach to the chromosomes at the centromeres. The spindle microtubules will move the chromosomes to opposite ends of the cell later in mitosis.

Metaphase: the chromosomes move to the equator of the cell.

Anaphase: the two chromatids of each chromosome separate, and the spindle microtubules move one chromatid of each chromosome to opposite ends of the cell.

Telophase: the chromosomes relax into their extended state and nuclear envelopes reform around each new daughter nucleus.

Cytokinesis usually divides the cytoplasm into roughly equal halves. In animal cells, the cell membrane is pinched in along the equator by a ring of microfilaments, separating the cytoplasm into two daughter cells, each containing one of the daughter nuclei. In plant cells, new cell membrane forms along the equator by fusion of vesicles produced by the Golgi complex.

Meiosis and Sexual Reproduction

Sexual reproduction usually combines genetic contributions from two parental organisms. At some point in the life cycle of all sexually reproducing organisms, diploid cells undergo meiosis, which separates homologous chromosomes and produces haploid cells with only one homologue from each pair. These haploid cells or their descendants fuse to form diploid cells that receive one homologue of each pair from each parent, reestablishing pairs of homologous chromosomes.

During meiosis, a diploid cell undergoes two specialized cell divisions, meiosis I and meiosis II, to produce four haploid daughter cells. During interphase before meiosis, chromosomes are duplicated. The cell then proceeds into meiosis I.

Meiosis I: During prophase I, homologous chromosomes, each consisting of two chromatids, pair up together, sometimes exchanging parts. During metaphase I, paired homologues move together to the equator of the cell, one member of each pair facing opposite ends of the cell. Homologous chromosomes separate during anaphase I, and two nuclei form in telophase I. Each daughter nucleus receives only one member of each pair of homologues and is therefore haploid.

Meiosis II: Meiosis II occurs in both of the daughter nuclei and resembles mitosis in a haploid cell. The chromosomes move to the equator of the cell. The two chromatids separate and are moved to opposite ends of the cell. This second division produces four haploid nuclei. Cell membranes form around the nuclei to yield four haploid cells.

Meiosis separates pairs of homologous chromosomes in diploid cells to form unpaired chromosomes in haploid cells. Homologous chromosomes contain similar, but usually slightly different, genetic information. The separation of homologues during meiosis I is random: either member of each pair of homologues may end up in either daughter nucleus. Therefore, the haploid cells produced by meiosis may differ in their genetic composition.

Mitosis, Meiosis, and Eukaryotic Life Cycles

All eukaryotic life cycles are similar in two respects: (1) sexual reproduction combines haploid gametes to form a diploid cell, and (2) at some point in the life cycle diploid cells undergo meiosis to produce haploid cells once again. Three general types of life cycle can be distinguished, based on the proportion of the life cycle spent in diploid or haploid stages.

GLOSSARY

Anaphase (an′-a-fāz) The stage of mitosis and meiosis II in which the sister chromatids of each chromosome separate from one another and are moved to opposite poles of the cell. In meiosis I, the stage in which homologous chromosomes are separated.

Asexual reproduction reproduction that does not involve the fusion of haploid gametes.

Aster during cell division in animals and some protists, a star-shaped array of microtubules extending in all directions outward from the centrioles.

Cell cycle the sequence of events in the life of a cell, from one division to the next.

Cell division in eukaryotes, the process of reproduction of single cells, usually into two identical daughter cells, by mitosis accompanied by cytokinesis.

Cell plate in plant cell division, a series of vesicles that fuse to form the new cell membranes and cell wall separating the daughter cells.

Centromere (sen′-trō-mēr) the region of a replicated chromosome at which the sister chromatids are held together.

Chiasma (kī-as′-ma; pl. chiasmata) during prophase I of meiosis, a point at which a chromatid of one chromosome crosses with a chromatid of the homologous chromosome. Exchange of chromosomal material between chromosomes takes place at a chiasma.

Chromatid (krō′-ma-tid) one of the two identical strands of DNA and protein forming a replicated chromosome. The two sister chromatids are joined at the centromere.

Chromatin (krō′-ma-tin) the complex of DNA and proteins that makes up eukaryotic chromosomes.

Chromosome (krō′-mō-sōme) in eukaryotes, a linear strand composed of DNA and protein, found in the nucleus of a cell, containing the genes; in prokaryotes, a circular strand composed solely of DNA.

Cytokinesis (sī-tō-ki-nē′-sis) division of the cytoplasm and organelles into two daughter cells during cell division. Usually cytokinesis occurs during telophase of mitosis or meiosis.

Diploid (dip′-loyd) referring to a cell with pairs of homologous chromosomes.

Fertilization the fusion of male and female haploid gametes to form a zygote.

Gamete (gam′-ēt) a haploid sex cell formed in sexually reproducing organisms.

Gene a unit of inheritance; the segment of DNA that carries the genetic information required to produce a specific trait.

Haploid (hap′-loyd) referring to a cell that has only one member of each pair of homologous chromosomes.

Homologue (hō′-mō-log) a chromosome that is similar in appearance and genetic information to another chromosome with which it pairs during meiosis. Also called homologous chromosome.

Interphase the stage of the cell cycle between cell divisions. During interphase, chromosomes are replicated, and other cell functions occur, such as growth, movement, and acquisition of nutrients.

Meiosis (mī-ō′-sis) a type of cell division found in eukaryotic organisms, in which a diploid cell divides twice to produce four haploid cells.

Metaphase (met′-a-fāz) a stage in mitosis and meiosis in which the nuclear membrane has completely disappeared, the chromosomes are lined up along the equator of the cell, and each chromosome is attached to microtubules of the fully formed spindle.

Mitosis (mī-tō′-sis) a type of nuclear division found in eukaryotic cells. Chromosomes are duplicated during interphase before mitosis. During mitosis, one copy of each chromosome moves into each of two daughter nuclei. The daughter nuclei are therefore genetically identical to each other.

Polyploid (pol′-ē-ployd) having more than two homologous chromosomes of each type.

Prophase (prō′-fāz) the first stage of mitosis or meiosis, in which the chromosomes first become visible in the light microscope as thickened, condensed threads, the nuclear membrane disintegrates, and the spindle forms. In meiosis I, the homologous chromosomes pair up and exchange parts at chiasmata.

Sexual recombination during sexual reproduction, the formation of offspring with (usually different) genetic contributions from two parental organisms.

Sexual reproduction a form of reproduction in which genetic material from two parental organisms is combined. In eukaryotes, two haploid gametes fuse to form a diploid zygote.

Spindle a football-shaped array of microtubules that moves the chromosomes to opposite poles of a cell during anaphase of meiosis and mitosis.

Telophase (tel′-o-fāz) the last stage of meiosis and mitosis, in which a nuclear membrane re-forms around each new daughter nucleus, the spindle disappears, and the chromosomes relax from their condensed form.

Zygote (zī′-gōt) in sexual reproduction, a diploid cell formed by the fusion of two haploid cells.

STUDY QUESTIONS

1. Diagram and describe the prokaryotic cell cycle.
2. Diagram and describe the eukaryotic cell cycle. Name the various phases and briefly describe the events that occur during each.
3. Define mitosis and cytokinesis. Do these two processes always occur together? Give an example.
4. Diagram the stages of mitosis. How does mitosis ensure that each daughter nucleus receives a complete set of chromosomes?
5. Define homologous chromosome, centromere, chromatid, diploid, haploid, and polyploid.
6. Describe the process of cytokinesis in animal and plant cells.
7. Diagram the events of meiosis. At what stage do haploid cells first appear? When do chiasmata appear? What occurs at chiasmata?
8. In what ways are mitosis and meiosis similar? In what ways are they different?

9. Draw a generalized diagram of each of the three types of life cycles of eukaryotic organisms, and give an example of an organism using each type. Label the haploid and diploid stages, and indicate at which point in the cycle meiosis occurs.

SUGGESTED READINGS

Mazia, D. "The Cell Cycle." *Scientific American*, January 1974. A clearly illustrated description of the essentials of cell division.

Wolfe, S. *Introduction to Cell Biology*. Belmont, Calif.: Wadsworth Publishing Company, Inc., 1983. A good overview of mitosis, meiosis, and the structure of chromosomes.

9
Principles of Inheritance

In the mingling of the seed, sometimes the woman, with sudden force, overpowers the man, and then the children, born of maternal seed, will resemble more the mother; but if from paternal seed, the father. The children you see resembling both their parents, having the features of both, have been created from father's body and mother's blood, when the seeds course through the bodies excited by Venus, in harmony of mutual passion, breathing as one, with neither conquering and neither being conquered.

On the Nature of Things
Lucretius (96–55 B.C.)

Speculation about reproduction and inheritance goes back thousands of years. As the passage from Lucretius quoted above indicates, fanciful explanations, with no evidence to support them, were once widely accepted. However, the ancients were not totally ignorant about reproduction. People have known for millennia that most animals and plants reproduce sexually, the offspring arising from the mating of male and female parents. Further, organisms mate with other organisms of the same kind, and the resulting offspring are also of the same kind as the parents. Nevertheless, offspring are almost never exactly identical to either parent.

Any hypotheses about how organisms pass their traits on to their offspring must explain at least this short list of observations:

1. All living things arise from preexisting living things.
2. The offspring of two parents usually resemble the parents and each other (Fig. 9-1).
3. Many traits in the offspring are not exactly like those of either parent.

Historically, philosophers and scientists have devised many hypotheses to explain our first two observations. Both Hippocrates (ca. 460–377 B.C.) and Charles Darwin (1809–1882) believed that each part of the body gives off tiny particles that travel to the gonads. During intercourse, the father's particles merge with the mother's to form the offspring. Aristotle (384–322 B.C.) favored a fluid rather than particles, but the general idea was similar. After the invention of the microscope, when sperm and eggs were seen for the first time, two new schools of thought arose. One held that an entire tiny person is contained in each sperm (and that tiny person has even smaller sperm with yet smaller people in them, with still smaller sperm . . .), while the other believed that the egg contains the tiny person with tiny eggs inside.

Figure 9-1 Two members of the Hapsburg royal family of Spain, Philip IV (left) and his brother Carlos (right). Note the similarities in facial features, especially the prominent "Hapsburg chin."

In the nineteenth century, many scientists felt that although the physical mechanisms of inheritance were uncertain, the resulting patterns of inheritance could be explained by **blending.** This hypothesis held that parents pass on their traits to their offspring, where they blend to produce intermediate traits. At first glance, blending seems quite reasonable. For example, if a tall man marries a short woman, most of the sons are not as tall as the father, and most of the daughters are not as short as the mother. Superficially, it appears as if the tallness of the father and the shortness of the mother mix together in the children. When carried to its logical conclusion, however, blending is unsatisfactory: surely over thousands of generations all individuals of any given species would become identical to one another, with all variation "blended out." However, everyone can see that members of the same species differ in appearance and behavior. Further, if blending *is* the major principle of inheritance, can the theory of evolution by natural selection among variable individuals possibly be correct? When he published the *Origin of Species* in 1859, Darwin recognized that the notions about inheritance then prevalent could not explain the inheritance of variation required by his theory of evolution, but he couldn't offer a better suggestion.

Enter Gregor Mendel.

Figure 9-2 *A portrait of Gregor Mendel, painted about the time of his pioneering genetics experiments.*

GREGOR MENDEL AND THE ORIGIN OF GENETICS

Gregor Mendel (Fig. 9-2) was a monk in the monastery of St. Thomas in Brunn (now Brno, Czechoslovakia). Before settling down in the monastery, Mendel tried his hand at several pursuits, including health care and high school teaching. In an effort to earn his teaching certificate, Mendel attended the University of Vienna for two years, where he studied botany and mathematics, among other subjects. This training proved crucial to his later experiments that were the foundation for the modern science of genetics. At St. Thomas, Mendel carried out both his monastic duties and an elegant series of experiments on inheritance in the common edible pea.

Mendel was by no means the first person to study inheritance. However, as Mendel himself wrote:

> Whoever surveys the work in this field will come to the conviction that among the numerous experiments, not one has been carried out to an extent or in a manner that would make it possible to determine the number of different forms in which hybrid progeny appear, permit classification of these forms in each generation with certainty, and ascertain their numerical interrelationships. . . . [H]owever, this seems to be the one correct way of finally reaching the solution to a question whose significance for the evolutionary history of organic forms must not be underestimated.

Why did Mendel succeed where others before him had failed? First, Mendel's choice of the edible pea as an experimental subject was a good one. The pea flower is built so that pollen normally cannot enter from outside the flower, that is, from another plant (Fig. 9-3). Instead, pea flowers usually self-pollinate. As a result, the egg cells in each flower are fertilized by sperm from the pollen of the same flower (**self-fertilization**). Even in Mendel's time commercial seed dealers sold many different types of peas that were **true-breeding;** that is, all offspring produced through self-fertilization were essentially identical to the parent plant. Even though peas normally self-fertilize, different plants can be mated manually. By carefully picking the flowers apart, pollen can be collected from one plant and transferred to a flower on another plant. The sperm from the "foreign" pollen will then fertilize the egg cells of the recipient flower (**cross-fertilization).** In this way, one can mate two different true-breeding plants and see what types of offspring are produced.

Although peas were an excellent subject for studies of inheritance, they were no guarantee of success. Others had experimented with peas long before Mendel, without discovering the principles of genetics. The key reasons for Mendel's success were his methods of experimentation and analysis.

Mendel began by working with only one trait at a time,

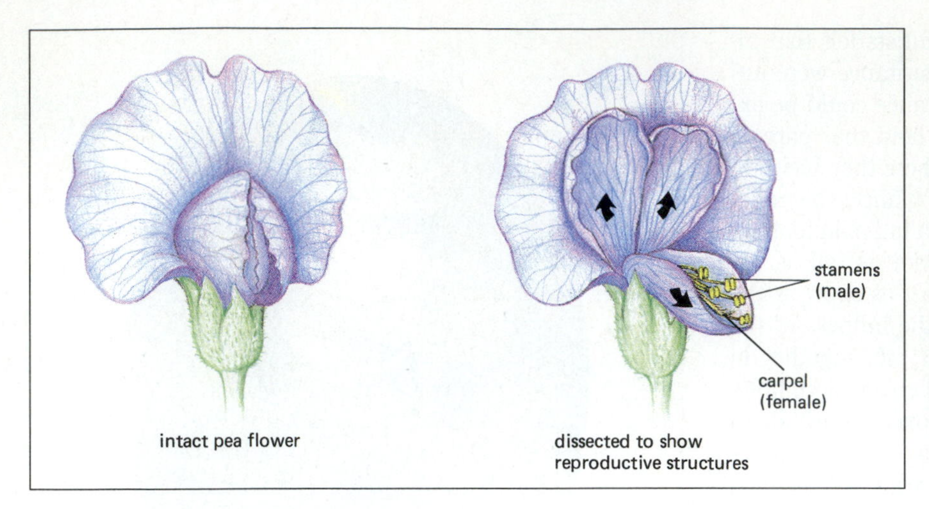

Figure 9-3 *The flower of the edible pea. In the intact flower (left), the lower petals form a container enclosing the reproductive structures, the stamens (male) and carpel (female). Pollen normally cannot enter the flower from outside, and consequently peas usually self-fertilize. Plant breeders, however, can pull apart the petals (right) and remove the stamens so that self-fertilization cannot occur. Dusting the carpels with pollen of their choice results in controlled cross-fertilization.*

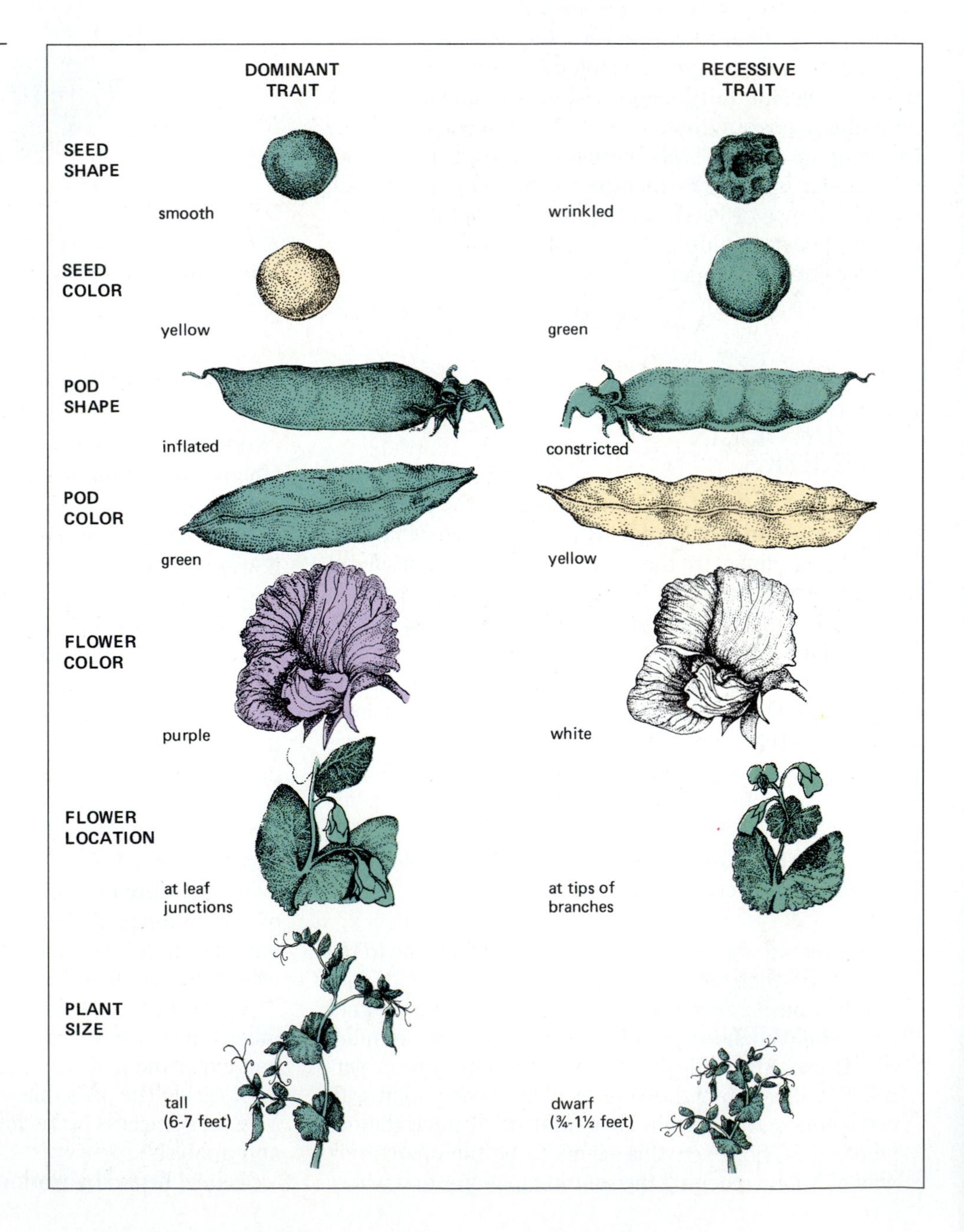

Figure 9-4 *Traits of pea plants used by Mendel in his studies of plant inheritance.*

and he chose traits that had unmistakably different forms of expression (Fig. 9-4). Previous workers had often crossed plants that differed in several traits, or that had different forms of a trait that might be confused with one another. The results were difficult to interpret. Perhaps the most important feature of Mendel's experiments was that he counted the numbers of offspring bearing the various traits and critically analyzed the numbers. The use of numbers as a tool for finding underlying principles has since become an extremely important practice in biology (see the essay, "On the Importance of Statistics in Biology").

Single-Trait Experiments: The Law of Segregation

Mendel started as simply as possible. He raised varieties of pea plants that were true-breeding for different forms of a single trait and cross-fertilized them. Such an experiment, involving organisms that differ in only one trait, is called a **monohybrid cross.** (In this usage, "hybrid" refers not to the offspring of parents of two different species, such as a mule, but to offspring of parents of the same species that differ in one or more inherited traits, such as hybrid corn.) Mendel saved the resulting hybrid seeds and grew them the next year.

In one of these experiments, Mendel cross-fertilized a white-flowered pea with a purple-flowered one. (In modern nomenclature, this was the **parental generation,** denoted by the letter P.) When he grew the resulting seeds, he found that all the first generation offspring ("first filial," or **F_1 generation**) produced purple flowers:

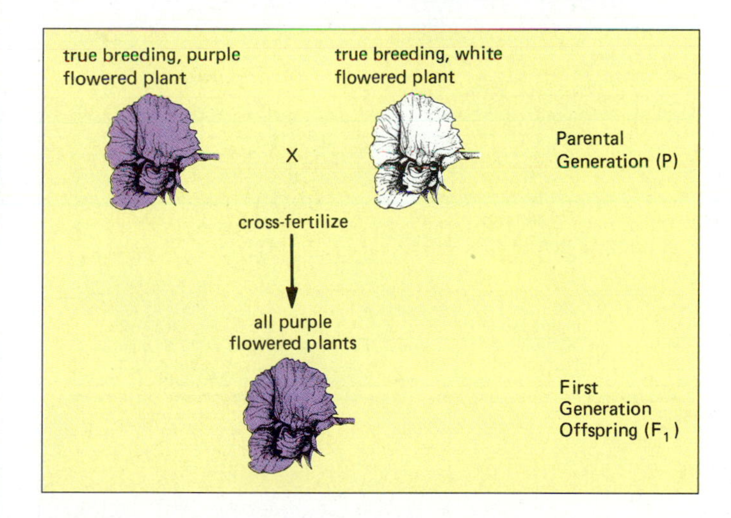

What happened to the white color? It hadn't blended in; the flowers of the hybrids were every bit as purple as the flowers of their purple parent. White seemed to have disappeared completely in the F_1 offspring.

Mendel allowed the F_1 flowers to self-fertilize, collected the seeds, and planted them the next spring. The second generation (F_2) had some plants with flowers of each color:

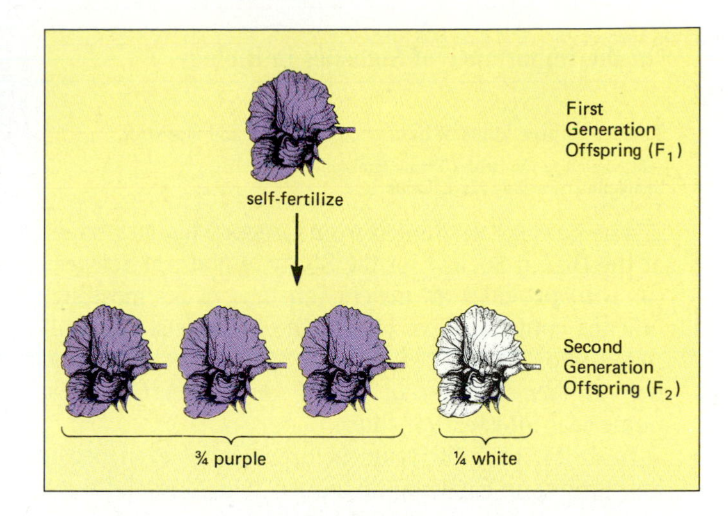

Overall, about three-fourths of the plants had purple flowers and one-fourth bore white flowers. The exact numbers were 705 purple and 224 white, or a ratio of 3.15 purple to 1 white.

Mendel allowed the F_2 plants to self-fertilize and produce yet a third (F_3) generation. He found that all the white-flowered F_2 plants produced white-flowered offspring; that is, they bred true. This was true for as many generations as he had time and patience to raise: white-flowered parents always gave rise to white-flowered offspring. About one-third of the purple-flowered F_2 plants were also true-breeding. The remaining two-thirds were hybrids and produced both purple and white-flowered offspring, again in the ratio 3:1. Therefore, the F_2 generation included $\frac{1}{4}$ true-breeding purple plants, $\frac{1}{2}$ hybrid purple, and $\frac{1}{4}$ true-breeding white.

Note Mendel's careful attention to detail. He kept track of each individual seed, recording which plants were its "parents" and "grandparents" and which plants were its "children" and "grandchildren." Otherwise, he would never have realized that some of the purple F_2 plants bred true, while others were hybrids.

MENDEL'S EXPLANATION. Mendel formed a five-part hypothesis to explain these results.

1. *Each trait is determined by pairs of discrete physical units.* Modern geneticists call these units **genes.** Each individual organism has two genes (e.g., two genes for flower color) that together control the expression of a given trait.
2. *Pairs of genes separate from each other during gamete formation.* This is Mendel's **law of segregation:** each gamete receives only one of an organism's pair of genes. When a sperm fertilizes an egg, the resulting offspring receives one gene from the father and one from the mother.
3. *Which member of a pair of genes becomes included in a gamete is determined by chance.* If we assign letters to each member of a pair of genes, say G1 and G2, each gamete is just as likely to receive a G1 as a G2. The

On the Importance of Statistics in Biology

There are three kinds of lies: lies, damn lies, and statistics.

—attributed to the British Prime Minister Benjamin Disraeli by Mark Twain

The passage we quoted from Gregor Mendel's paper at the Brunn Society for the Study of Natural Sciences (p. 155) provides an insight into one of his most significant contributions. Unlike the biologists who preceded him, Mendel realized that he had to "determine the number of different forms" and "ascertain their numerical interrelationships."

In 1824, over 30 years before Mendel even began his studies, an Englishman named John Goss did some of the same crosses that Mendel later repeated. Goss cross-fertilized true-breeding yellow-seeded peas with true-breeding green-seeded peas, and obtained all yellow seeds in the F_1 generation. Further, he allowed the F_1 to self-fertilize and raised an F_2 generation. He found that the F_2 plants bore both yellow peas and green peas. But if Goss realized the significance of the *numbers* of green and yellow peas, or even counted them at all, he never reported it. Consequently, we have Mendelian genetics, not Gossian genetics.

Since Mendel's time, numbers have become increasingly important, not just in genetics, but in all of biology and medicine. Most modern studies are quantitative: the interpretation of the results depends on the numbers obtained in the experiment.

As an example, let's suppose that a biologist suspects that a toxic substance in the wastewater from an industrial plant stunts the growth of fish. To test his hypothesis, he might set up two aquaria, each stocked with 10 small fish of the same size and age. All environmental conditions, such as light, temperature, and food, would be kept the same for both tanks, except that the suspected toxin would be added to the water in one tank. At the end of a predetermined time, say six months, he would measure the fish in each tank.

What the biologist would probably find is that some fish in the toxic tank would be larger than some fish in the control tank, and vice versa (see the table). How can he reach any conclusions? He would calculate the average size of fish in each tank, and the standard deviation, which is a measure of the variability of sizes. He would then perform statistical tests to see if the difference in the average size of the fish in the two tanks reflects chance variation, or might be caused by the toxic substance.

Statistical analyses are extremely important in biology and medicine, especially when a given effect might have several causes. For example, it has been known for many years that people who smoke cigarettes are more likely to suffer from lung cancer and heart attacks than are nonsmokers. This is what might be called a statistical fact: obviously, some nonsmokers die from heart attacks, and some smokers do not contract lung cancer. Nevertheless, on the average, nonsmokers run a much lower risk of these diseases.

Mark Twain expressed a common feeling when he implied that statistics is worse than a "damn lie." Statistics can be confusing, and statistical facts often seem to be contradicted by personal experience, such as knowing someone who smoked three packs of cigarettes a day for 60 years and never developed lung cancer. However, in a world in which effects may have many causes, statistics is an essential tool in our efforts to understand the workings of natural phenomena. It was only through statistical analysis that Mendel could understand pea genetics or the U.S. Surgeon General could draw conclusions about smoking and health. Without the marriage of biology and mathematics begun by Gregor Mendel, we could not evaluate the results of new drug therapies, crop varieties, or thousands of other advances in biology and medicine.

	Length of Fish (cm)	
	Control	**Experimental**
	6.0	5.0
	6.0	5.0
	7.0	5.0
	7.0	6.0
	7.0	6.0
	8.0	7.0
	8.0	7.0
	9.0	7.0
	10.0	8.0
	10.0	8.0
Average	7.8	6.4
Standard deviation	1.5	1.2

As determined by a statistical procedure called Student's *t*-test, the probability of obtaining this difference in average lengths of fish in the two tanks purely by chance, and not due to the effect of the pollutant, is less than 1 in 20. This is the generally accepted level of "statistical significance," that is, that the toxin really did stunt the growth of the fish.

gamete will not receive both, and there will not be a preference for one member of the pair.

4. *There may be two or more alternative forms of a gene.* White-flowered peas, for example, have a different form for the gene controlling flower color than purple-flowered peas do. Alternative forms of a single gene are called **alleles.** In many instances, one allele, called the **dominant** allele, can completely mask the expression of the other, **recessive,** allele. However, although the dominant allele of a gene masks the *expression* of the recessive allele, it does not alter the *physical nature* of the recessive allele, which can be passed unaltered into the individual's gametes.
5. *True-breeding organisms have two of the same alleles for the trait under study; hybrids have two different alleles.* Since a true-breeding or **homozygous** ("same pair") organism has only one type of allele, it can produce only one type of gamete:

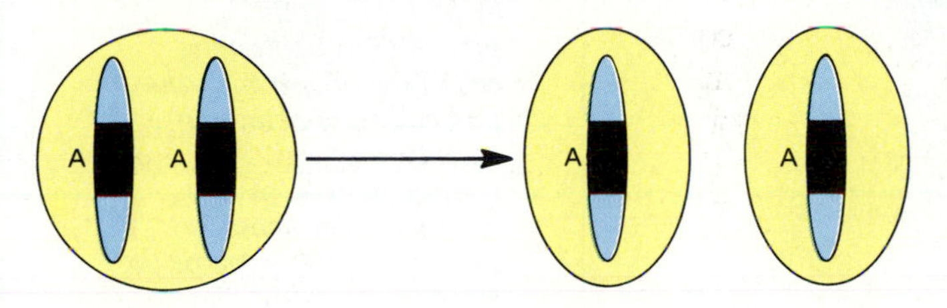

A hybrid or **heterozygous** ("different pair") individual, with two different alleles, produces equal numbers of gametes with each of the two alleles:

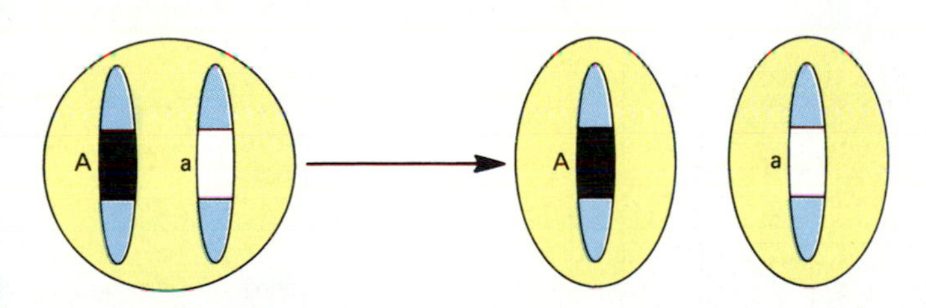

Let's see how Mendel's hypothesis explains the results of his experiments with flower color. Using letters to represent the different alleles, we will assign the uppercase letter *P* to the allele for purple (the dominant allele is usually represented by an uppercase letter), and the lowercase letter *p* to the allele for white (recessive). A homozygous purple-flowered plant has two alleles for purple (*PP*), while a white-flowered plant has two alleles for white (*pp*). A *PP* plant produces all *P* sperm and eggs, while a *pp* plant produces all *p* sperm and eggs:

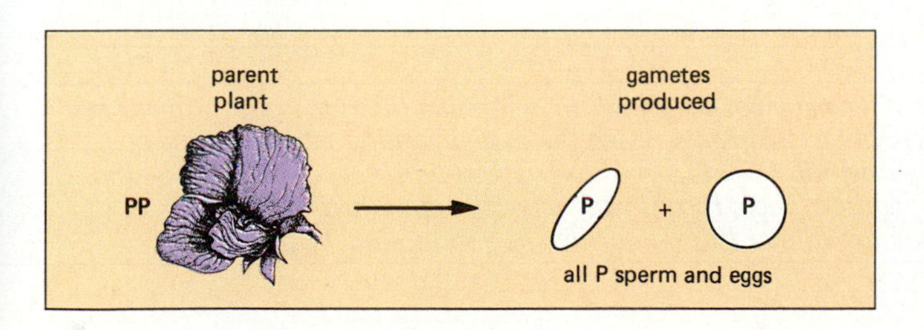

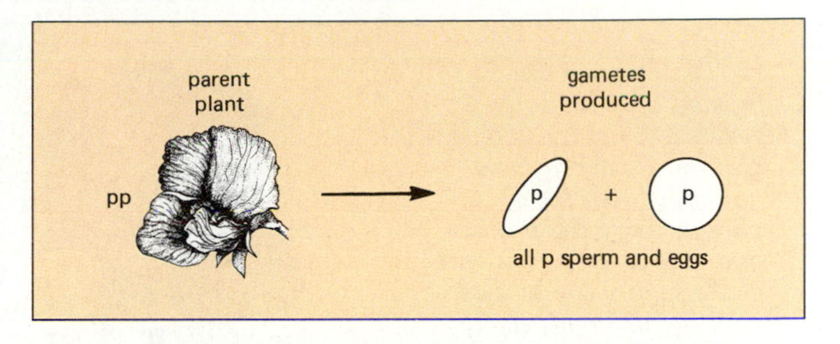

The F_1 hybrid offspring are produced when *P* sperm fertilize *p* eggs, or when *p* sperm fertilize *P* eggs. In either case, the F_1 offspring are *Pp* (purple):

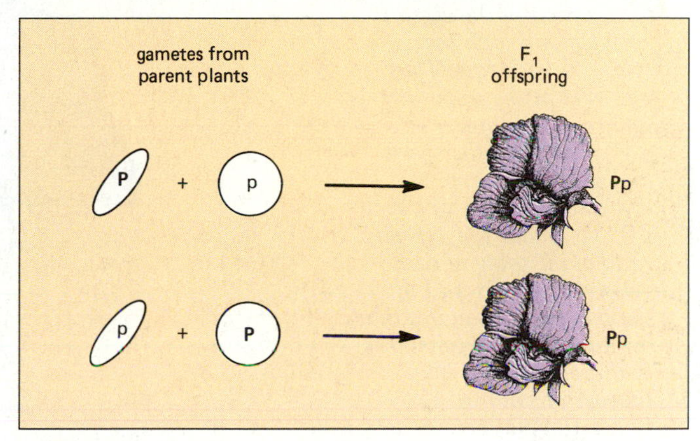

Each gamete produced by a *Pp* plant has an equal chance of receiving either the *P* allele or the *p* allele (i.e., the plant produces equal numbers of *P* and *p* sperm and equal numbers of *P* and *p* eggs). When a *Pp* plant self-fertilizes, both types of sperm have an equal chance of fertilizing both types of eggs:

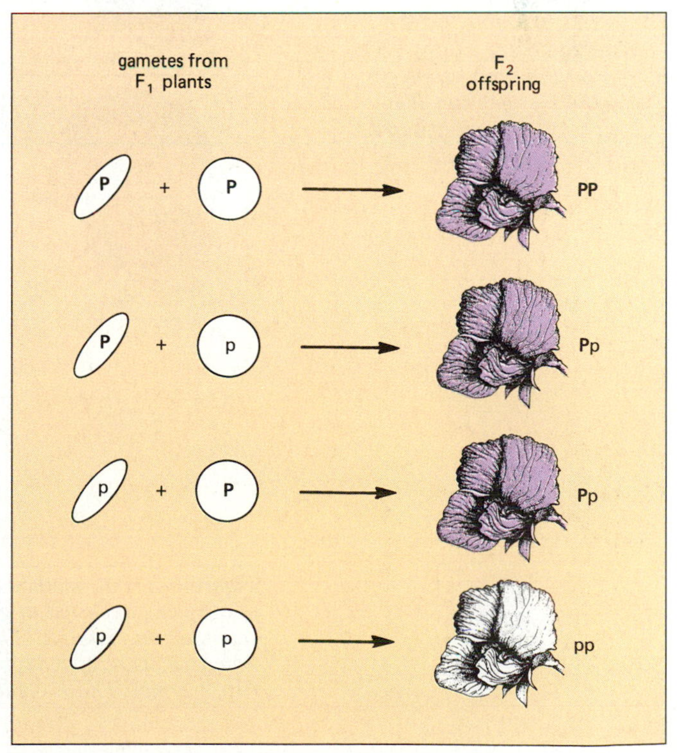

(a) *The Punnett square (named after an early geneticist):* On the average *half the gametes (sperm and eggs) from the* F_1 *parents carry the* P *allele and half carry* p. *The rows of the square are labeled with the different, equally likely, types of sperm; the columns are labeled with the types of eggs. The offspring within the squares are the result of the fertilization of the eggs in each column with the sperm in each row. Thus offspring in the upper left square, for instance, receive a* P *allele from the sperm and another* P *from the egg, and have the genotype* PP. *To obtain the expected ratio of offspring of each genotype, simply add up the numbers of squares that contain identical genotypes, and divide by the total number of squares.* Note that a Punnett square does not mean that every mating results in exactly four offspring, nor that the different types of offspring always occur in exactly the ratio shown in the square. *Rather, the four offspring squares represent genotypes of equally likely occurrence. In this example,* on the average *there will be* $\frac{1}{4}$ PP, $\frac{1}{2}$ Pp, *and* $\frac{1}{4}$ pp *genotypes in the* F_2 *offspring. The expected phenotypes are* $\frac{3}{4}$ *purple and* $\frac{1}{4}$ *white.*

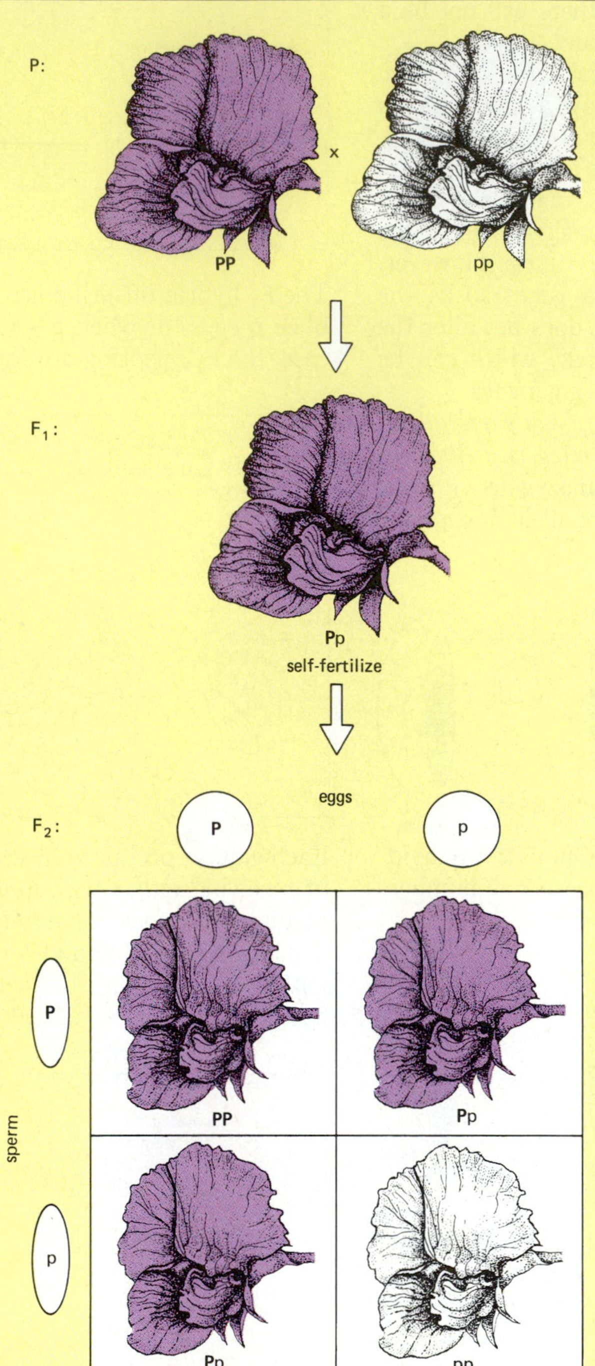

(b) *Probability theory:* The probability of two independent events occurring together is the product (multiplication) of their separate probabilities. *Each type of gamete (*P *or* p*) is equally likely to occur; that is, each has a probability of* $\frac{1}{2}$. *Furthermore, each type of sperm is equally likely to fertilize each type of egg. This means that the first event (which allele ends up in a particular sperm) is independent of the second event (which allele ends up in a particular egg). We can therefore obtain the probable ratio of offspring by multiplying the probability of obtaining each type of sperm by the probability of obtaining each type of egg: Since* Pp *is equivalent to* pP *(i.e., whether a chromosome came from the sperm or the egg does not affect the genotype of the offspring), we can add up the proportions of* Pp *and* pP *offspring. Probability theory thus predicts that the* F_2 *offspring will be* $\frac{1}{4}$PP, $\frac{1}{2}$Pp, *and* $\frac{1}{4}$pp *(*$\frac{3}{4}$ *purple and* $\frac{1}{4}$ *white), the same result produced by the Punnett square:*

Sperm		Egg		Offspring Genotypes	Offspring Phenotypes
½*P*	×	½*P*	=	¼*PP*	¾ purple
½*P*	×	½*p*	=	¼*Pp*	
½*p*	×	½*P*	=	¼*pP*	
½*p*	×	½*p*	=	¼*pp*	¼ white

Figure 9-5 *Two methods of "genetic bookkeeping" to determine the types and proportions of offspring expected in genetics experiments, using the monohybrid cross of pea flower colors as an example. The original parents were homozygous dominant (*PP*, purple) and homozygous recessive (*pp*, white). Their* F_1 *offspring were therefore heterozygous (*Pp*, purple). The* F_1 *plants are allowed to self-fertilize to produce an* F_2 *generation.*

Therefore, three types of offspring can be produced: *PP*, *Pp*, and *pp*. These occur in the approximate proportions of $\frac{1}{4}PP$, $\frac{1}{2}Pp$, and $\frac{1}{4}pp$.

The *actual combination of alleles* carried by an individual (e.g., *PP* or *Pp*) is its **genotype.** The *morphological appearance* of an organism is its **phenotype.** As we have seen, plants with genotypes *PP* and *Pp* both bear purple flowers. Thus, even though they have different genotypes, they have the same phenotype. Therefore, the F_2 generation consists of three genotypes ($\frac{1}{4}PP$, $\frac{1}{2}Pp$, and $\frac{1}{4}pp$) but only two phenotypes ($\frac{3}{4}$ purple and $\frac{1}{4}$ white).

Figure 9-5 presents two different methods of "genetic bookkeeping" for determining the expected proportions of offspring in a monohybrid cross. The first method uses a diagram called a **Punnett square,** named after a famous geneticist of the early twentieth century. The second method relies on probability theory. Both methods yield the same results, so you can use whichever seems easier for you. Whichever method you choose, there is one absolutely crucial point to keep in mind: *these calculations only give the most probable proportions of offspring of different genotypes and phenotypes.* In a real experiment, one would only expect that the offspring will occur in *approximately* the predicted proportions. In the F_2 generation of Mendel's monohybrid cross, note that he did not obtain *exactly* $\frac{3}{4}$ purple-flowered and $\frac{1}{4}$ white-flowered plants.

THE TEST CROSS. By now you have probably recognized that Mendel used the scientific method discussed in Chapter 1. He observed the results of his experiments and then formed a hypothesis. As you know, scientific method has a third step: to use the hypothesis to predict the results of other experiments and see if those experiments support or refute the hypothesis. Mendel did just that. If the hybrid F_1 flowers have one allele for purple and one

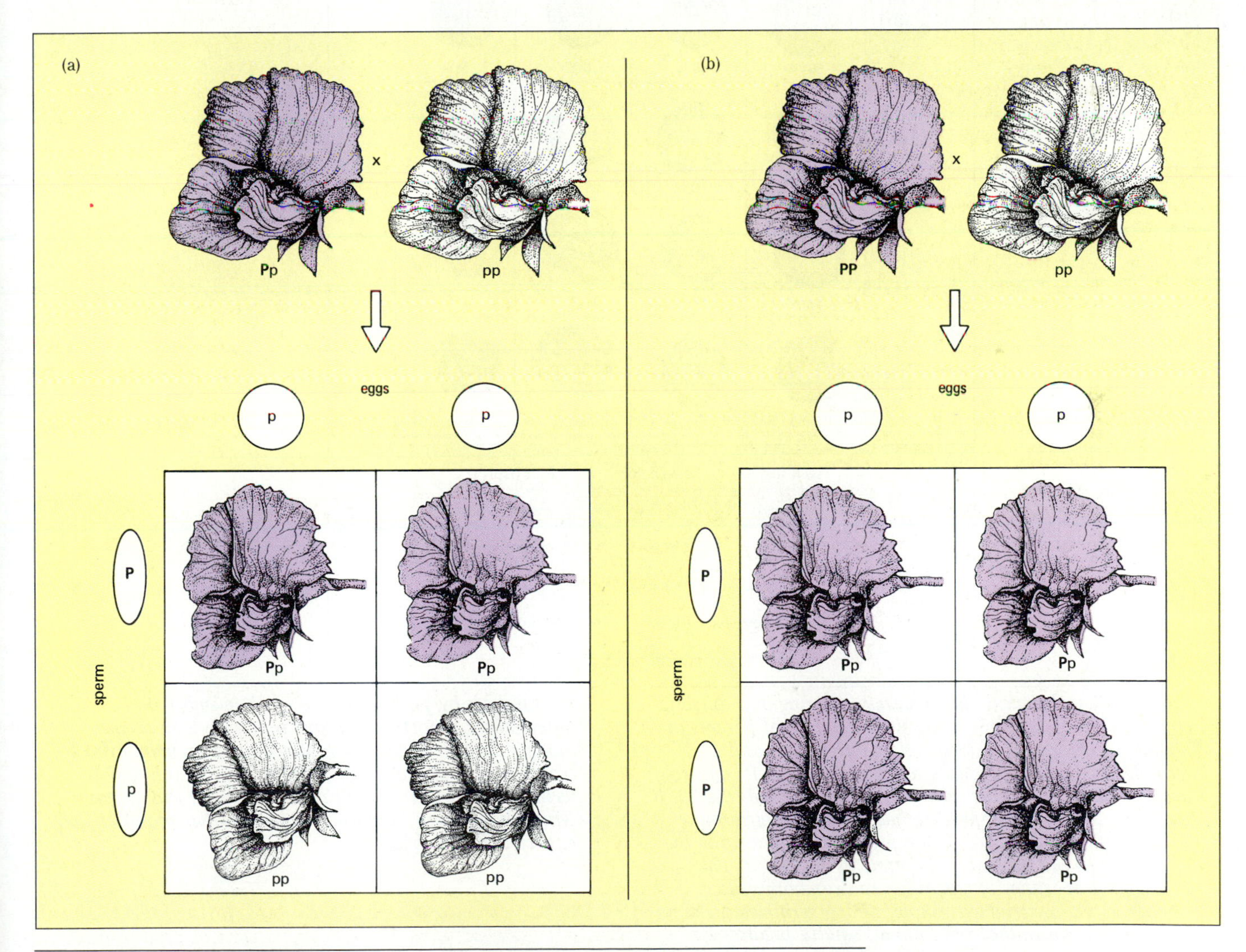

Figure 9-6 Test crosses involve breeding a homozygous recessive organism with a phenotypically dominant organism. The resulting offspring differ, depending on whether the phenotypically dominant parent is heterozygous **(a)** *or homozygous* **(b)**.

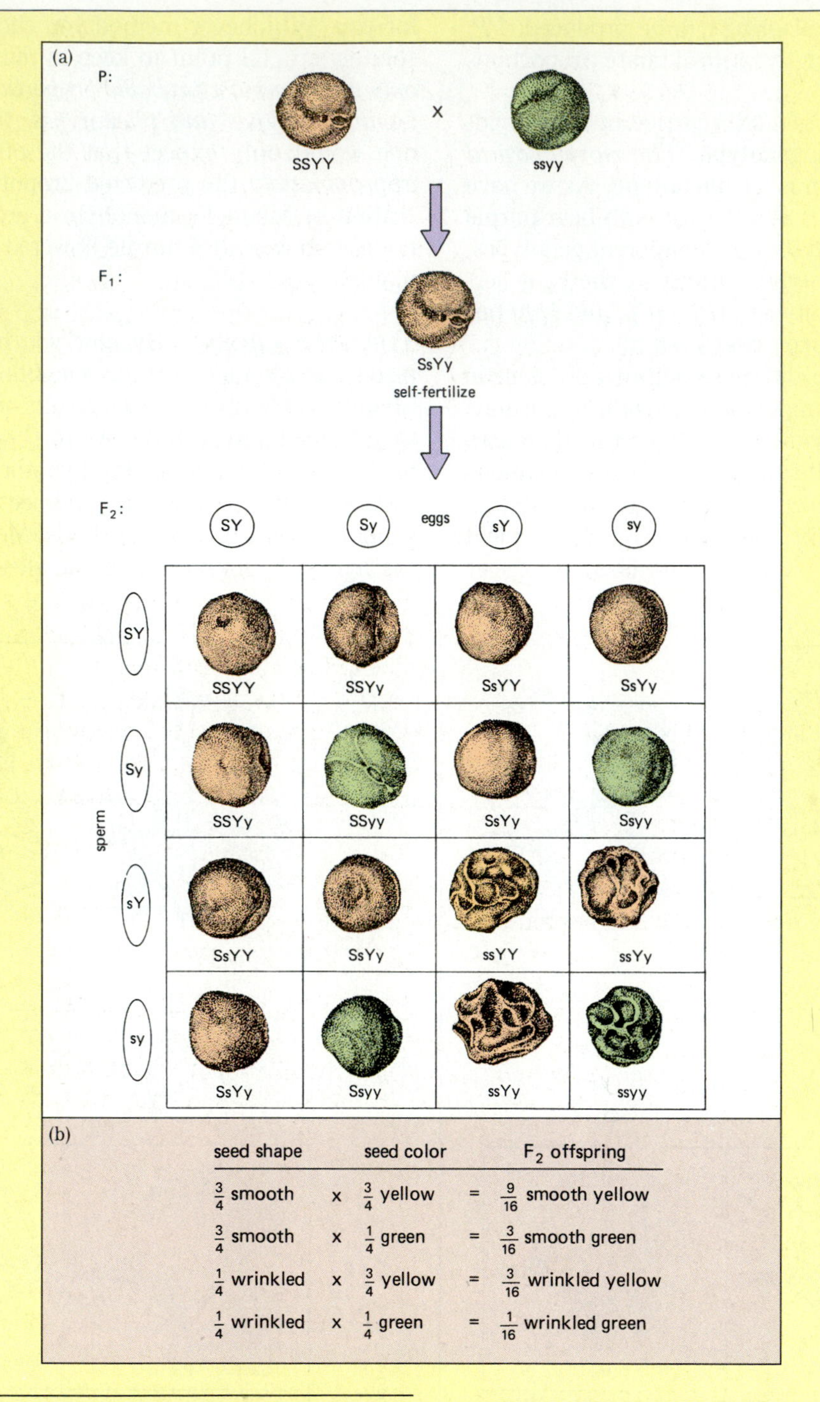

***Figure 9-7** Calculating the expected offspring in a dihybrid cross.* **(a)** *Punnett square analysis. The original parents are homozygous for smooth* (SS) *yellow* (YY) *seeds and wrinkled* (ss) *green* (yy) *seeds, respectively. The F_1 offspring are therefore all hybrids with smooth yellow seeds* (SsYy). *If we allow the F_1 to self-fertilize, we obtain the offspring shown within the square. There are now 16 squares in the analysis, but the method is the same as Fig. 9-5a. The expected proportions of offspring are $\frac{9}{16}$ smooth yellow, $\frac{3}{16}$ smooth green, $\frac{3}{16}$ wrinkled yellow, and $\frac{1}{16}$ wrinkled green, or 9:3:3:1. Note also that the Punnett square predicts 12 smooth:4 wrinkled seeds and 12 yellow:4 green seeds (both 3:1 ratios), just as would be expected from a monohybrid cross if we ignore the second trait.* **(b)** *Analysis by probability theory. Remember that* independent assortment *means that* each trait *assorts independently of the others. We know from monohybrid crosses that $\frac{3}{4}$ of the F_2 offspring will be smooth and $\frac{1}{4}$ wrinkled, and that $\frac{3}{4}$ will be yellow while $\frac{1}{4}$ will be green. Multiplying these independent probabilities produces the expected F_2 offspring.*

for white (Pp), he should be able to predict the outcome of cross-fertilizing these Pp plants with homozygous recessive (pp) white plants. (Can you?) As Fig. 9-6a shows, Mendel's hypothesis predicts that there will be equal numbers of Pp (purple) and pp (white) offspring. This is precisely what happened.

This type of experiment has practical uses, too. Just by looking at an organism that has the dominant phenotype, one usually cannot tell whether it is homozygous or heterozygous. Cross-fertilization of a phenotypically dominant individual with a homozygous recessive individual is called a **test cross**, because it can be used to test whether the dominant parent is homozygous or heterozygous. When crossed with a homozygous recessive, a homozygous dominant produces all phenotypically dominant offspring (Fig. 9-6b), while a heterozygous dominant yields offspring with both dominant and recessive phenotypes, in a 1:1 ratio (Fig. 9-6a).

Multiple-Trait Experiments: Independent Assortment

Having determined the mode of inheritance of single traits, Mendel then turned to the more complex question of multiple traits. He began with a **dihybrid cross;** that is, he cross-bred plants that differed in two traits, say seed color (yellow or green) and seed shape (smooth or wrinkled). If he crossed a plant that was homozygous for smooth yellow seeds with one that was homozygous for wrinkled green seeds, the F_1 offspring all bore smooth yellow seeds. This was no great surprise, because from separate monohybrid crosses of each of these traits he already knew that smooth (S) is dominant to wrinkled (s), and that yellow (Y) is dominant to green (y) (Fig. 9-4). The F_1 offspring, therefore, are all genotypically $SsYy$. Allowing these F_1 plants to self-fertilize, Mendel found that the F_2 generation consisted of 315 smooth yellow seeds, 101 wrinkled yellow seeds, 108 smooth green seeds, and 32 wrinkled green seeds. The ratio of phenotypes was about 9:3:3:1. Other dihybrid crosses also produced offspring in approximately 9:3:3:1 ratios.

Mendel realized that this could be explained if he assumed that the genes for seed color and seed shape are inherited independently of each other and do not influence each other during gamete formation. (Imagine flipping two coins, a dime and a nickel. Whether the dime comes up heads or tails has no effect on which side of the nickel comes up.) Thus the outcome for each trait could be regarded as a simple monohybrid cross, in which a 3:1 ratio of offspring would be expected. The laws of probability state that the independent combination of two 3:1 ratios yields a 9:3:3:1 ratio (Fig. 9-7), and we can see from Mendel's results that this is just what happened. There were 423 smooth seeds (of either color) to 133 wrinkled ones (3.18:1) and 416 yellow seeds (of either shape) to 140 green ones (2.97:1). The Punnett square shows how two 3:1 ratios combine to form an overall 9:3:3:1 ratio (Fig. 9-7).

This independence in the inheritance of two distinct traits is called the **law of independent assortment:** *the distribution of alleles for one trait into the gametes does not affect the distribution of alleles for other traits.*

In Mendel's experiments, independent assortment also seemed to hold true for combinations of three traits, say flower color, seed color and seed shape, although now there were eight different phenotypes in the F_2 generation. Mendel expected that this would hold true for any number of traits, although the numbers of phenotypes of the F_2 offspring would be very large.

In 1865, Gregor Mendel presented his theories of inheritance to the Brunn Society for the Study of Natural Science, and they were published the next year. It did *not* mark the beginning of genetics. In fact, it didn't make any impression at all on the biology of his time. In one of the great ironies of science, Mendel's experiments, which spawned one of the most elegant and important theories in all of science, vanished from the scientific scene. Apparently, very few biologists read his paper, and those who did probably could not understand it. It was not until almost half a century later that biologists rediscovered Gregor Mendel and his principles of genetics.

GENES AND CHROMOSOMES

While Mendel's findings languished in oblivion, other areas of biology, particularly microscopy, flourished. In the 1870s and 1880s, chromosomes were discovered, and their movements during mitosis and meiosis were deduced. Having already studied mitosis and meiosis in Chapter 8, you have probably noticed that Mendel's genes behave much like chromosomes during meiosis:

1. There are two copies of each gene in each organism, and there are two homologous chromosomes in each diploid cell.
2. Only one copy of each gene is passed on to the offspring in each sperm or egg; during meiosis only one of the two homologous chromosomes ends up in each haploid daughter cell.
3. Different genes sort out independently of each other; the different types of chromosomes also appear to assort independently.

In 1900, three biologists, Carl Correns, Hugo de Vries, and Eric von Tschermak, working independently and knowing nothing of Mendel's work, rediscovered the principles of inheritance. No doubt to their intense disappointment, when they searched the scientific literature before publishing their results, they found that Mendel had scooped them over 30 years before.

Figure 9-8 The common fruit fly, Drosophila melanogaster, *has had an uncommon impact on studies of inheritance.*

It wasn't long before someone saw the connection between chromosomes and Mendel's hypotheses. In 1902, William Sutton, a graduate student at Columbia University, was studying sperm formation in grasshoppers. With Mendel's principles of inheritance in mind, he saw the chromosome movements of meiosis in a new light: perhaps Mendel's genes were chromosomes. Of course, there was immediately a great difficulty with this hypothesis, namely that there have to be many more genes than there are chromosomes. So Sutton's hypothesis became: *genes are parts of chromosomes.*

The proof that genes are parts of chromosomes came from the work of Thomas Hunt Morgan of Columbia University and his students, studying the common fruit fly, *Drosophila melanogaster* (Fig. 9-8). You might be surprised that anyone would bother to study the genetics of such a homely little pest. However, fruit flies are nearly ideal organisms for genetics experiments. To study inheritance in any organism, geneticists must cross-breed the organism, raise the F_1 generation, breed these individuals, raise the F_2 generation, and so on, and end up with enough offspring to draw meaningful conclusions about the ratio of phenotypes. As Mendel found, if only one generation a year is possible, and if the organism needs a lot of space or care, it takes a long time and huge amounts of work to carry out complicated crosses for several generations. Fruit flies are small, each female lays hundreds of eggs, and a complete generation from egg to egg takes only two weeks. Moreover, fruit flies have only four pairs of chromosomes, making *Drosophila* nearly perfect for genetics experiments. For these reasons, and because Morgan and his colleagues were careful, persistent, and brilliant scientists, it was the fruit fly that provided many of the advances in genetics in the first half of this century.

The Chromosomal Basis of Sex Determination

Sex determination provided the first direct evidence supporting Sutton's hypothesis that genes are parts of chromosomes. In this case, a trait (sex) is almost invariably associated with a particular distribution of chromosomes.

In the 1890s, microscopists observed that not all chromosomes have exact homologues. In mammals and many insects, males have the same number of chromosomes as females, but one "pair," the **sex chromosomes,** are very different in appearance. Females have two identical sex chromosomes, called **X chromosomes,** while males have one X chromosome and one **Y chromosome.** Parts of the sex chromosomes are homologous to one another, so they pair up during prophase of meiosis I and separate at the end of meiosis I. The remaining chromosomes, which oc-

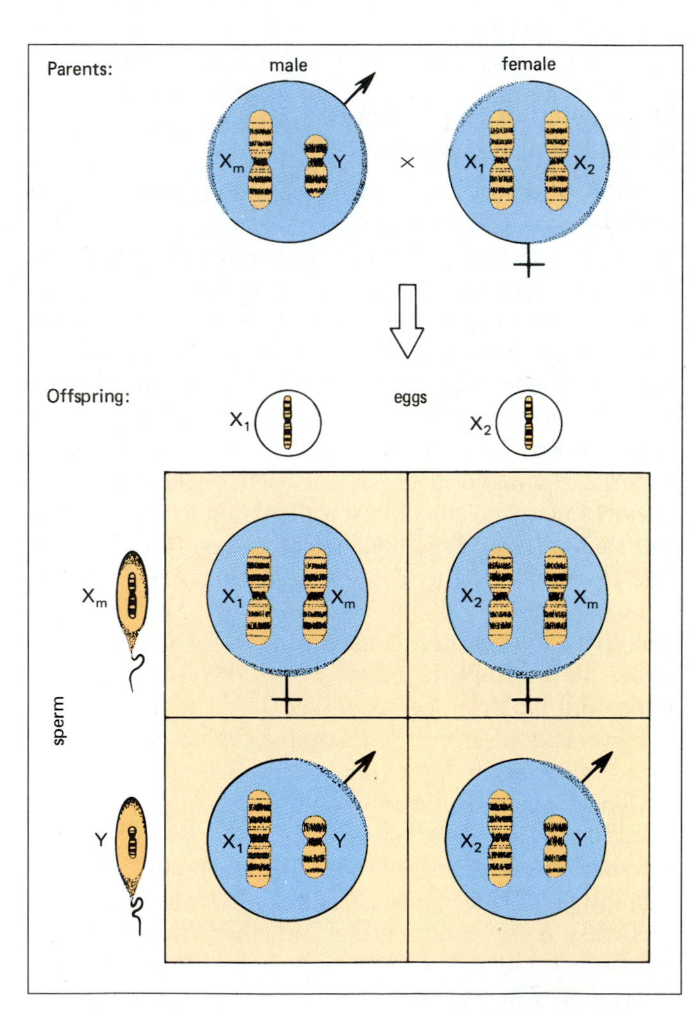

Figure 9-9 Sex determination in species such as humans and fruit flies, in which males carry two dissimilar sex chromosomes (XY) while females carry two similar sex chromosomes (XX). Only the distribution of sex chromosomes is illustrated. Male offspring receive the Y chromosome from the father, while female offspring receive the father's X chromosome (labeled X_m in the drawing). The mother passes one of her X chromosomes (X_1 and X_2 in the drawing) to both male and female offspring.

cur in pairs of identical appearance in both males and females, are called **autosomes.** *Drosophila* has three pairs of autosomes, while humans have 22 pairs.

In organisms in which males are XY and females are XX, the sex of an offspring is determined by which sex chromosome is in the sperm that fertilizes the egg (Fig. 9-9). During sperm formation, the sex chromosomes segregate, and each sperm receives either the X or Y chromosome (plus one of each pair of autosomes). The sex chromosomes also segregate during egg formation, but since females have two X chromosomes, every egg receives one X chromosome. Therefore, an offspring is male if an egg is fertilized by a Y-bearing sperm, or female if an egg is fertilized by an X-bearing sperm.

Sex Linkage

Proof that chromosomes carry more than one gene came early in the course of Morgan's fruit fly studies. A male fly was discovered that had white eyes, instead of the normal red eyes (Fig. 9-10). This white-eyed male was mated to a virgin red-eyed female. The resulting offspring were all red-eyed flies, indicating that white eyes (*w*) is probably recessive to red eyes (*W*). The F_2 generation, however, was a surprise. As expected, the ratio of red to white-eyed flies was about 3 to 1; however, *there were nearly equal numbers of males with red eyes and with white eyes, and no females with white eyes at all* (Fig. 9-10)! A test cross of the F_1 red-eyed females and the original white-eyed male yielded roughly equal numbers of red-eyed and white-eyed males and females.

What was going on? Morgan made the brilliant hypothesis that *the gene for eye color must be located on the X chromosome, while the Y chromosome has no corresponding gene:*

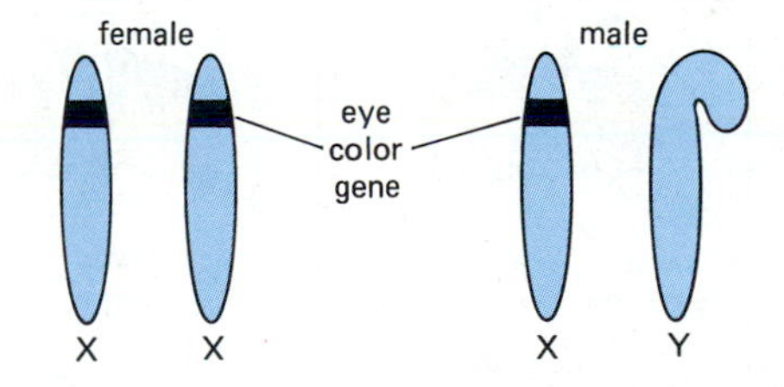

Let's look first at the F_1 generation. Both male and female F_1's receive an X chromosome, with its *W* allele for red eyes, from their mother. The F_1 males receive a Y chromosome from their father, with no allele for eye color at all, so the males have a *WY* genotype. The F_1 females received the father's X chromosome with its *w* allele, so the females have a *Ww* genotype.

Crossing two F_1 flies thus results in an F_2 generation with the chromosome distribution of Fig. 9-10b. All the F_2 females receive one X chromosome from their F_1 male parent with its *W* (red) allele. They therefore have red eyes. All the F_2 males, on the other hand, inherit their single X chromosome from their F_1 female parent. Since the F_1 females are heterozygous for eye color (*Ww*), the F_2 males have a 50–50 chance of receiving either an X chromosome with the *W* allele or one with the *w* allele. *With no corresponding gene on the Y chromosome, the F_2 males must display the phenotype determined by the allele on the X chromosome.* Therefore, half the F_2 males have red eyes and half have white eyes.

Genes on the X chromosome that are not found on the Y chromosome are called **sex linked.** In many animals, the Y chromosome carries relatively few genes other than those determining maleness, whereas the X chromosome bears many genes in addition to those that produce specifically female traits. Therefore, females can be homozygous or heterozygous for X chromosome genes, but males cannot. Whatever genes a male carries on his single X chromosome are the only copies of those genes that he has. They cannot be masked by or interact with their corresponding gene on another X chromosome, as they can in females. Genes on the X chromosome are thus fully expressed in males, whether they are dominant or recessive. As we shall see in Chapter 12, many human traits are sex-linked.

Other Chromosomes, Other Linkages

Like the X chromosome, the autosomes also bear many genes. Sex linkage is merely a special case of the general phenomenon of **linkage,** in which two or more genes tend to be inherited together because they lie on the same chromosome. One of the first pairs of linked autosomal genes to be discovered were those for flower color and pollen grain shape in the sweet pea. Purple flower color (*P*) is dominant over red (*p*), and long pollen shape (*L*) is dominant over round (*l*). When a homozygous purple-flowered, long-pollen pea is crossed with a red-flowered, round-pollen pea, all the F_1 offspring have purple flowers and long pollen. However, a single chromosome carries the genes for flower color and for pollen shape. Since chromosomes assort independently during meiosis, both genes tend to assort together, and therefore to be inherited together. Thus, the phenotypes of the F_2 generation do not occur in a 9:3:3:1 ratio, as Mendel's law of independent assortment would predict. Instead, the F_2 generation has about three purple-flowered, long-pollen plants for every one red-flowered, round-pollen plant (Fig. 9-11).

The discovery of linkage requires that the law of independent assortment be modified, as follows:

1. *Genes located on different chromosomes assort independently during meiosis.* (Mendel was fortunate that the traits he studied were controlled by genes on different chromosomes.)
2. *Genes located on the same chromosome usually will not assort independently.* They tend to be inherited together.

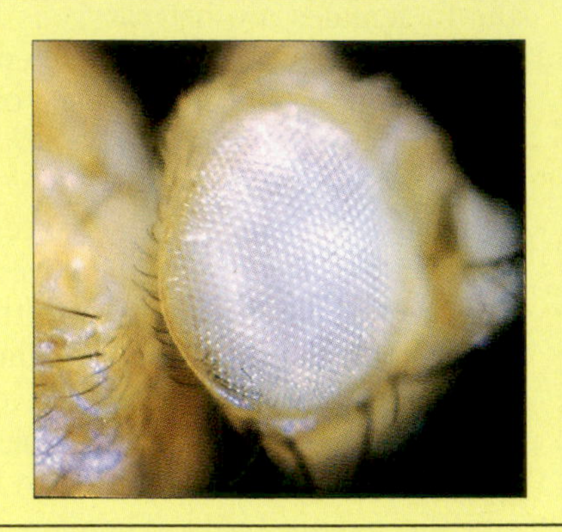

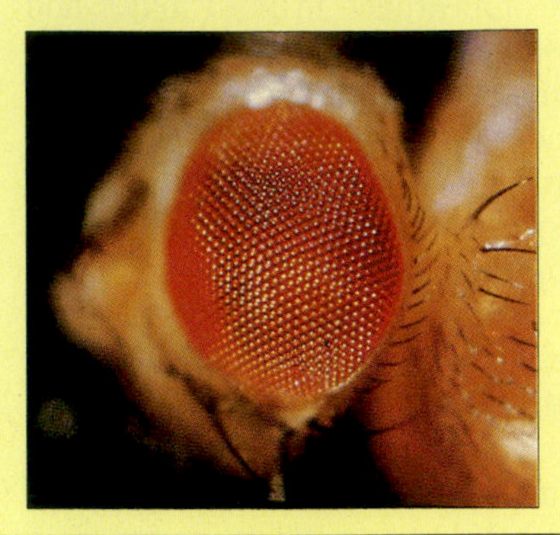

The F_1 generation: *Female offspring receive the* w *allele on their father's X chromosome, but phenotypically this is masked by the dominant* W *allele on their mother's X chromosome. Male offspring receive the* W *allele from their mother, and the Y chromosome with no eye color gene from their father. Therefore, all the females are* Ww *and all the males are* WY. *Both males and females have red eyes.*

The F_2 generation: *All F_1 males carry the* W *allele, so all their F_2 female offspring receive the* W *allele as well. Therefore, the F_2 females are all red eyed. The F_1 females are all heterozygous* Ww. *Consequently, half their F_2 sons receive the* W *allele and half receive the* w *allele. The F_2 males also receive a Y chromosome from their fathers. Therefore, half the F_2 males are* WY *(red eyes) and half are* wY *(white eyes).*

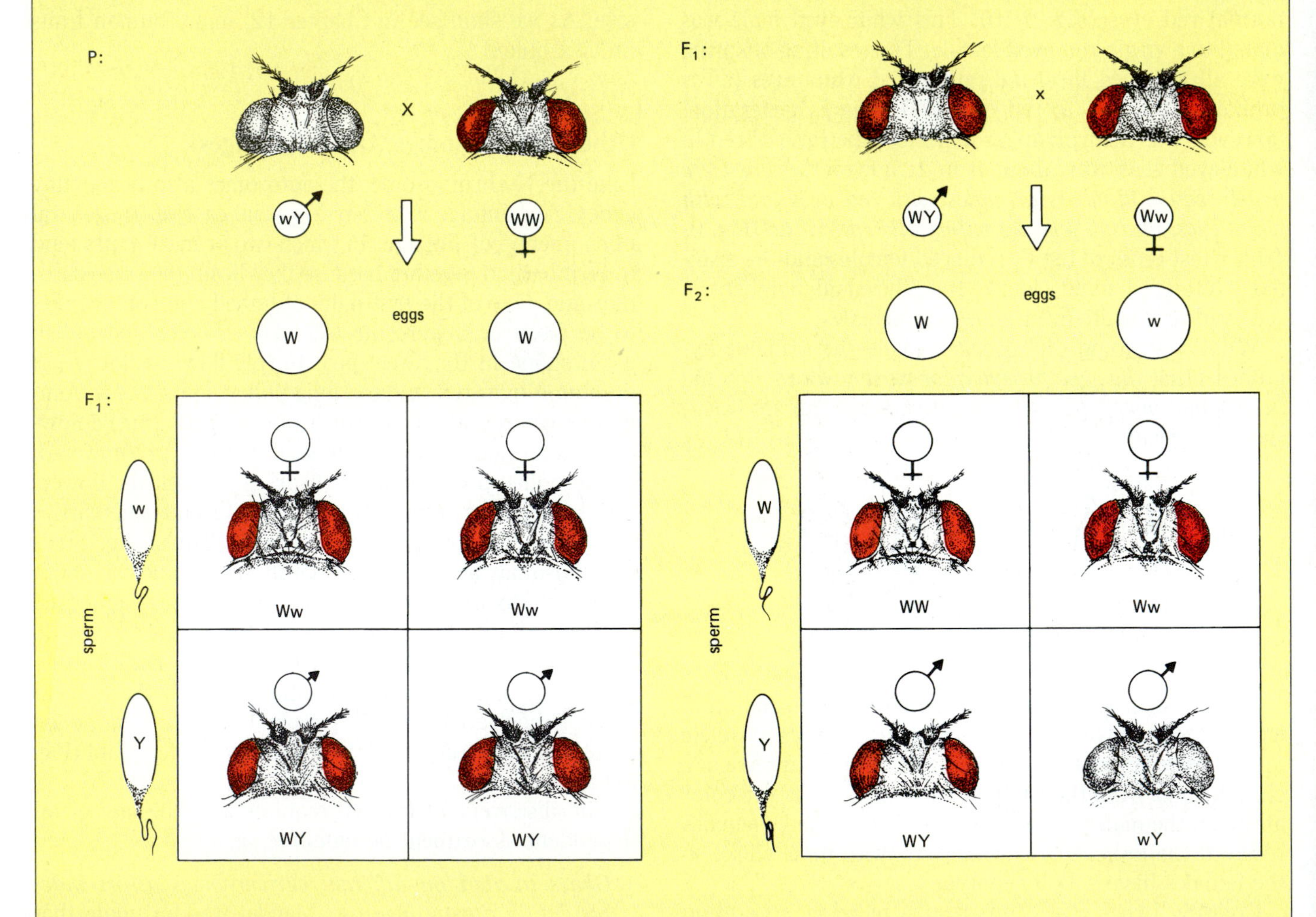

Figure 9-10 *Morgan's interpretation of the results of sex-linked inheritance of white eye color in fruit flies. The gene for eye color is carried on the X chromosome, and has no corresponding gene on the Y chromosome. Normal red eyes (W) is dominant to the mutant allele for white eyes (w).*

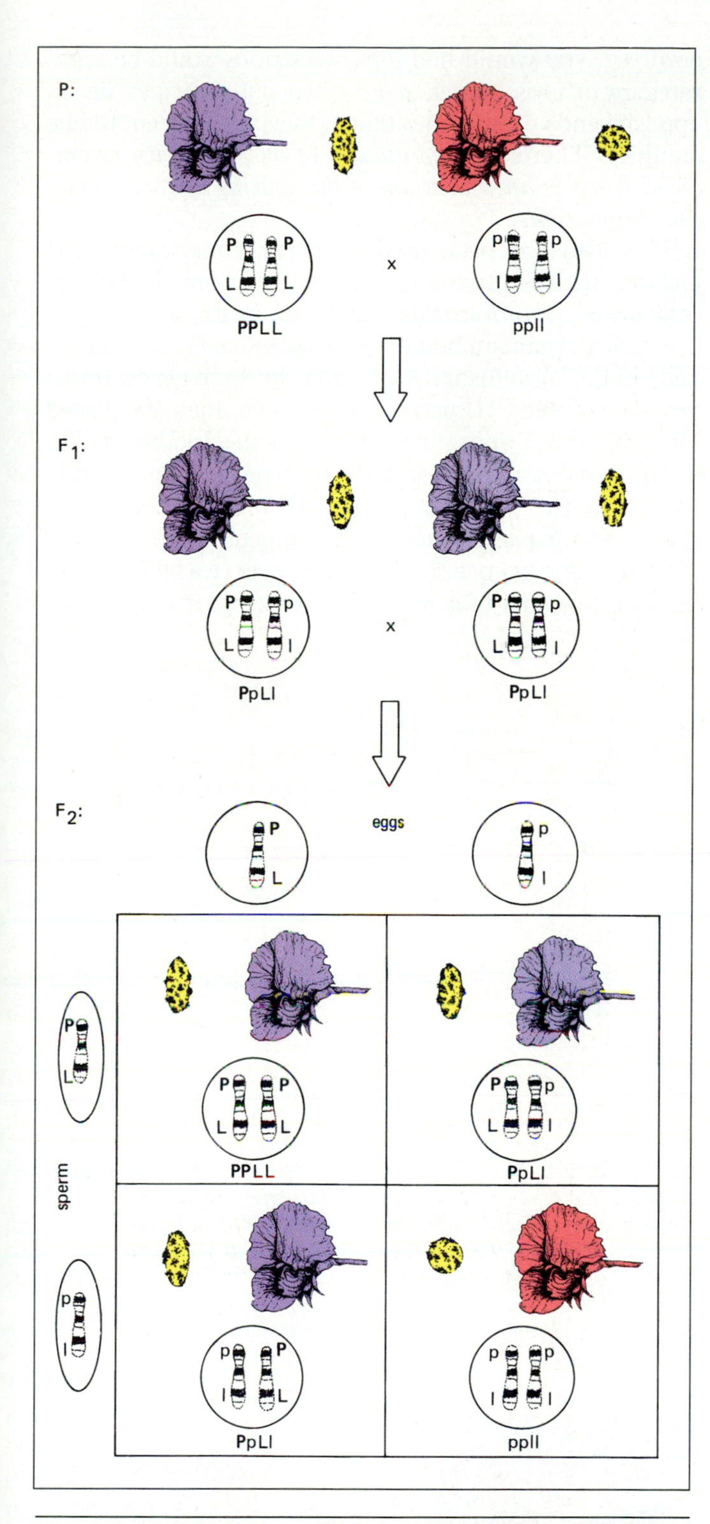

Figure 9-11 Inheritance of linked genes for pollen grain shape and flower color in sweet peas. Since these genes are found on the same chromosome, the configuration PL *and* pl *in the parental generation tends to be preserved in all subsequent generations. The F_2 offspring show a phenotypic ratio of 3 purple, long:1 red, round rather than the 9:3:3:1 ratio one would expect from independently assorted genes.*

Linkage, Crossing Over, and Chromosome Mapping

There is just one thing wrong with this tidy scheme: genes on the same chromosome do not *always* stay together. In the sweet pea cross described above, for example, often the F_2 generation will include a few purple-flowered, round-pollen plants and red-flowered, long-pollen plants. How can this be?

During meiosis, *parts of homologous chromosomes are often exchanged between the chromosomes,* a process called **crossing over** (Fig. 9-12). As you know, the homologous chromosomes pair up together in prophase I of meiosis. The components of a chromosome "recognize" their matching counterparts on the homologous chromosome. Parts of the chromosomes intertwine, producing X-shaped figures called **chiasmata.** At chiasmata, parts of chromosomes may exchange with each other, forming new gene combinations on the chromosomes. When homologous pairs separate at anaphase I, the chromosomes that each haploid daughter cell receives may be different from those of the parental cell.

How does crossing over exchange genes? Chromosomes are long, continuous strands of DNA. Genes are specific segments of a DNA strand that contain the information needed to produce a particular trait. Genes are arranged linearly along each chromosome, like colored stripes painted on a rope, with each stripe corresponding to a gene. When crossing over occurs, the two chromosome "ropes" swap corresponding segments, so that a series of genes may switch over from one chromosome to the other.

Crossing over during meiosis explains why new combinations of traits occurred in the sweet pea cross. At the beginning of prophase I, all the F_1 peas had this pair of homologous chromosomes:

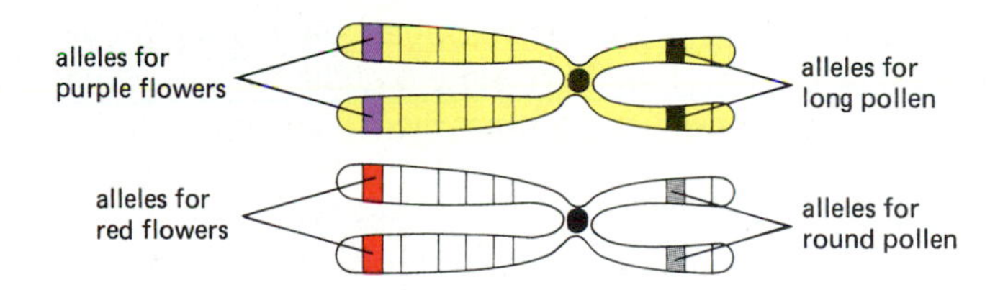

In a few reproductive cells, crossing over occurred between the locations of the genes for flower color and pollen shape:

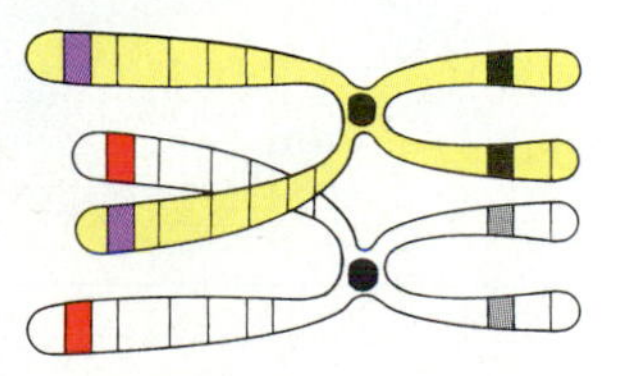

At anaphase I the separated homologous chromosomes had this gene composition:

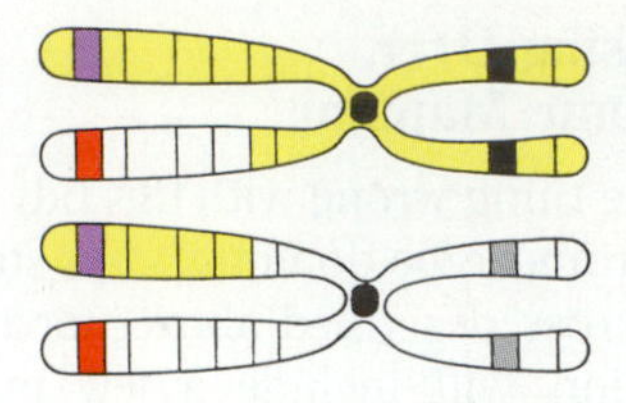

Four different types of chromosomes were distributed to the haploid daughter cells during meiosis II:

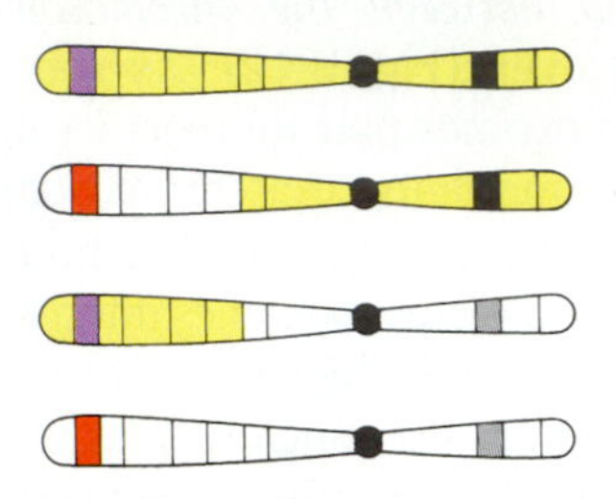

Therefore, some gametes were produced with each of the four chromosome configurations: *PL* and *pl* (the original parental types), and *Pl* and *pL* (recombined chromosomes).

In most of the reproductive cells, crossing over between the flower-color and pollen-shape genes did not occur. Therefore, most of the F_2 offspring received chromosomes with the *PL* and *pl* allele combinations. Fusion of sperm and eggs that formed from cells in which crossing over occurred, however, gave rise to a few purple-flowered, round-pollen and red-flowered, long-pollen plants. This is **genetic recombination:** generating new combinations of genes due to exchange of DNA between homologous chromosomes. Genetic recombination is an important source of genetic variability among organisms.

CHROMOSOME MAPPING. Crossing over occurs more often between genes that are far apart on a chromosome than between genes that are close together. (Once again, think of chromosomes as ropes with "gene stripes." If you throw two ropes together onto the ground, one rope might lie on top so that it crosses the other. If you threw the pair of ropes many times and recorded where crossing occurred, you would find that two stripes would be separated by a cross much more often if the stripes lie on opposite ends of the ropes than if they are adjacent to one another.) Therefore, *the number of offspring with recombined traits is an indication of the spacing of genes along the chromosome.*

This idea has been used to map chromosomes. The distance between two genes on a chromosome is the percentage of recombination observed during a cross between one organism homozygous recessive for both genes and another organism heterozygous for both genes. If two genes recombine 10 percent of the time, they are placed 10 "map units" apart on the chromosome. In *Drosophila,* it has been possible to correlate this type of recombination map with the physical appearance of certain chromosomes, and locate the actual sites of genes (Fig. 9-13).

Using the evidence from crossing over, we can now refine our concepts of genes and alleles (Fig. 9-14). *A gene*

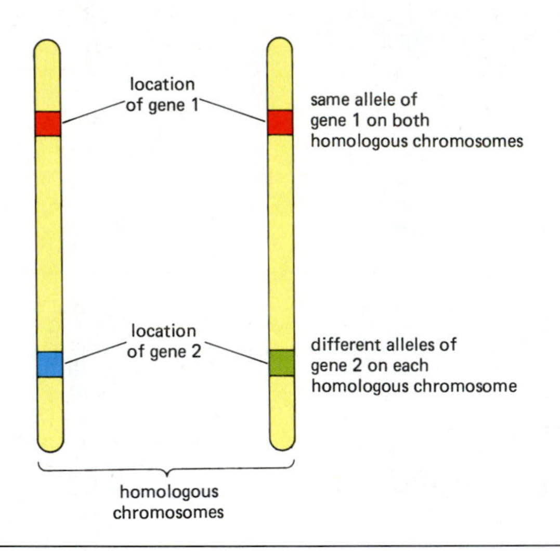

Figure 9-14 *The relationships among genes, alleles, and chromosomes. The drawing shows a pair of homologous chromosomes. A gene is a segment of DNA at the same location on both homologous chromosomes. Differences in DNA composition at the same gene location form different alleles of the gene. A chromosome carries many genes; each gene occupies a different location on the chromosome.*

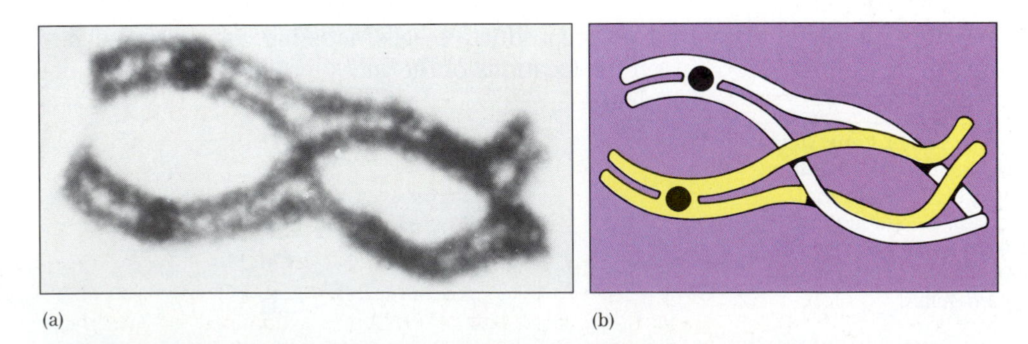

Figure 9-12 *A photomicrograph of crossing over during meiosis* **(a)** *and a drawing* **(b)** *based on the micrograph. The two pairs of chromatids intertwine, forming chiasmata in several places. Chromosomal material is exchanged at the chiasmata.*

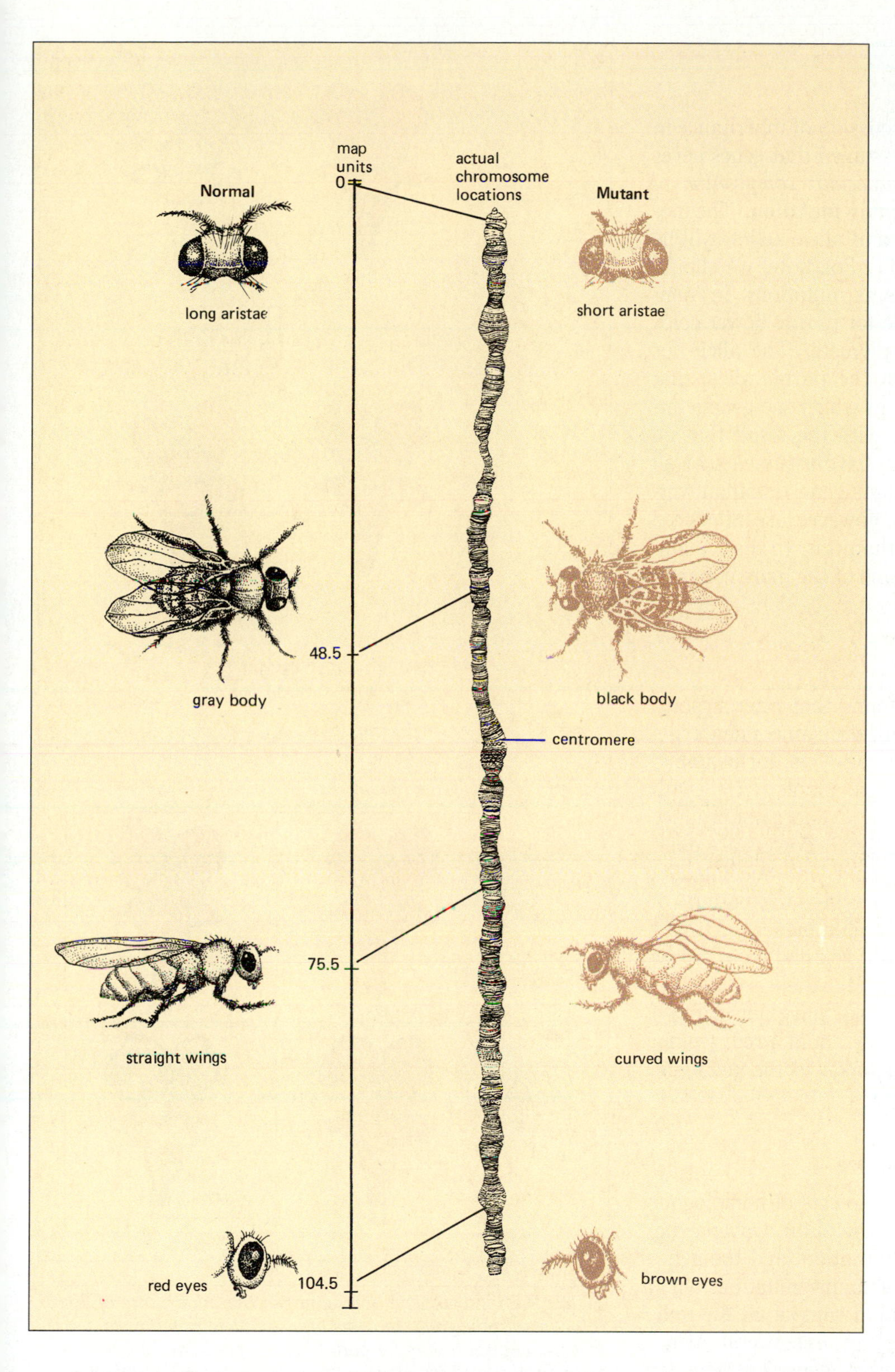

Figure 9-13 *A map of chromosome 2 in the fruit fly, comparing the distances between genes obtained by recombination analysis with the physical locations of genes on the chromosomes. The order of the genes is the same in both maps, but the distances between genes are not. Crossing over is apparently hindered in some regions of the chromosome, especially around the centromere, distorting the recombination map.*

is a segment of DNA on a chromosome that encodes the instructions for producing a specific trait. Homologous chromosomes carry the same genes, located at the same places. However, two homologous chromosomes may have slightly different DNA compositions located at the site of a given gene. Variations in DNA composition may in turn cause variations in the trait, such as purple versus red flowers. *For any gene location, every variation in DNA composition that produces a difference in the form of the trait is a separate allele.* During crossing over, corresponding segments of DNA are exchanged between homologous chromosomes. If these two segments differ in their DNA composition (i.e., if they are different alleles), crossing over can lead to new combinations of alleles.

VARIATIONS ON THE MENDELIAN THEME

So far, we have restricted our discussion of inheritance in two major ways. First, we have assumed that genes never change. This is not true: *the molecular composition of genes can change, a process known as* **mutation.** The molecular mechanisms of mutation and allele action will be described in Chapters 10 and 11. For now, we should merely note that alleles originate as mutations. In Mendel's peas, for example, the allele for purple flower color directs the synthesis of purple pigment. The allele for white is probably a mutation of the purple allele that occurred many generations ago, in which the genetic instructions have been scrambled, with the result that no pigment is produced. Second, we have assumed that all traits are inherited in a simple, single-gene, dominant versus recessive manner. Most traits, however, are influenced in more varied and subtle ways than this. In the remainder of this chapter, we sample some of the more common variations in inheritance.

Incomplete Dominance

In his pea experiments, Mendel found that heterozygotes and homozygous dominants had the same phenotype. This is often not the case. In snapdragons, for example, crossing homozygous red-flowered plants (*RR*) with homozygous white-flowered ones (*R′R′*) does not produce red-flowered F_1 hybrids. Instead, the F_1 flowers are pink. When the heterozygous phenotype is intermediate between the two homozygous phenotypes, the pattern of inheritance is called **incomplete dominance.** At first glance, this looks like blending, as if you had mixed white paint with red paint to make pink. But the F_2 generation shows that the alleles for flower color have not changed (Fig. 9-15). The F_2 offspring include about $\frac{1}{4}$ red, $\frac{1}{2}$ pink, and $\frac{1}{4}$ white flowers. This corresponds to the genotypic ratio of $\frac{1}{4}RR$:$\frac{1}{2}RR'$:$\frac{1}{4}R'R'$.

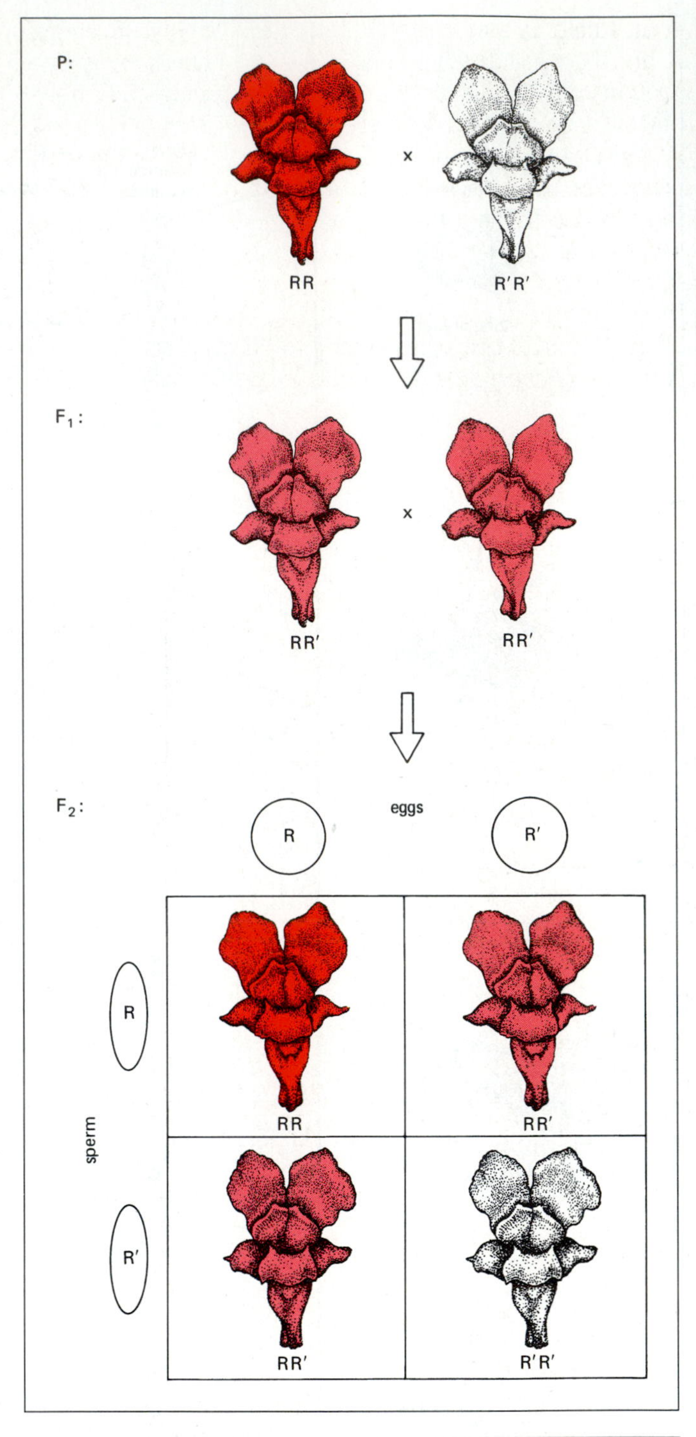

Figure 9-15 Incomplete dominance in the inheritance of flower color in snapdragons. In cases of incomplete dominance, we will use capital letters for both alleles (here R *and* R′*) rather than capital and lowercase letters. Hybrids (*RR′*) have pink flowers, while the homozygotes are red (*RR*) or white (*R′R′*). Since heterozygotes can be distinguished from homozygous dominants, the phenotypic ratio in the F_2 generation (1 red:2 pink:1 white) is the same as the genotypic ratio (1*RR*:2*RR′*:1*R′R′*).*

Multiple Alleles and Codominance

A single individual, having only two sets of homologous chromosomes, can have only two alleles for a given gene. Alleles, however, arise through mutation, and the genes of different organisms may suffer many mutations, each producing a new allele. If we could sample all the individuals of a species, we would often find several, sometimes even dozens, of alleles for every gene. The gene for eye color in fruit flies, for example, has many alleles, each recessive to normal red eyes and producing various shades of yellow or pink when homozygous.

The ABO blood types in humans constitute a familiar system of such **multiple alleles.** The blood types A, B, AB, or O arise as a result of three different alleles (*A, B,* and *O*) of a single gene. This gene directs the synthesis of glycoproteins that protrude from the surfaces of red blood cells. Alleles *A* and *B* direct the synthesis of glycoproteins A and B, respectively, while allele *O* produces no glycoproteins at all (Fig. 9-16). Individual humans may have one of six genotypes: *AA, BB, AB, AO, BO,* or *OO.* Alleles *A* and *B* are dominant to *O.* Therefore, individuals with genotypes *AA* or *AO* have type A glycoproteins on their red blood cells, and have type A blood. Those with genotypes *BB* or *BO* synthesize type B, and have type B blood, while homozygous recessive *OO* individuals lack these glycoproteins, and have type O blood. However, alleles A and B are **codominant** to one another—that is, *both are phenotypically detectable in heterozygotes. AB* individuals have red blood cells with both A and B glycoproteins, and have type AB blood. These glycoproteins react with antibodies in the blood plasma and are responsible for the reactions seen during blood transfusions, explained in Fig. 9-16.

Polygenic Inheritance

Many traits are not controlled by just one gene, but are influenced by the action of many genes. This is called **polygenic inheritance.** The simplest polygenic inheritance occurs when two genes code for the same trait. In wheat,

(a)

Genotype	Blood Type	Red Blood Cells	Plasma Antibodies
AA, AO	A	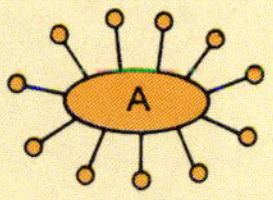 A glycoprotein	anti-B
BB, BO	B	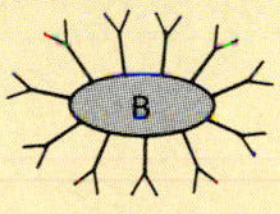 B glycoprotein	anti-A
AB	AB	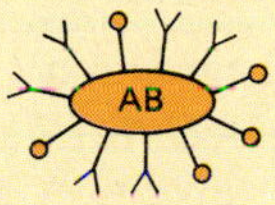 A glycoprotein and B glycoprotein	none
OO	O	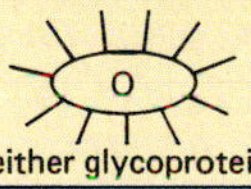 neither glycoprotein	anti-A anti-B

(a) *Red blood cell glycoproteins and plasma antibodies found in individuals with each blood type. Type A individuals have the A glycoprotein on their red blood cell surfaces. Their plasma contains antibodies only against B glycoproteins (anti-B antibodies). Therefore, their plasma antibodies do not bind to their own red cell glycoproteins, and their own cells do not clump. Type B individuals have B glycoproteins and anti-A antibodies. Type AB blood contains red blood cells with both the A and B glycoproteins, but neither antibody. Type O individuals have both anti-A and anti-B antibodies but no reactive glycoproteins on their red blood cells.*

Medical Effects of Blood Transfusions			
Donor Type	Recipient Type	Effect on Recipient	Permissible Blood Donation?
A	A	–	yes
	B	Clumping	no
	AB	–	yes
	O	Clumping	no
B	A	Clumping	no
	B	–	yes
	AB	–	yes
	O	Clumping	no
AB	A	Clumping	no
	B	Clumping	no
	AB	–	yes
	O	Clumping	no
O	A	–	yes
	B	–	yes
	AB	–	yes
	O	–	yes

(b)

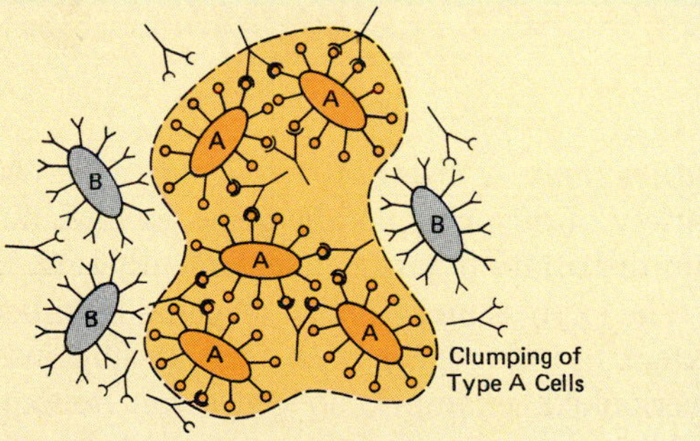

Clumping of Type A Cells

Reaction of Type A Cells in Type B Blood

(b) *The reaction between type A blood cells transfused into type B blood. Type B blood contains many anti-A antibodies. Since each antibody has two sites that can bind the A glycoprotein, the A cells become clumped, held together by the anti-A antibodies. These clumps can become large clots, with serious medical consequences. The permissible donors for each blood type are shown in the table. (Above)*

Figure 9-16 *Human ABO blood group reactions. Glycoproteins on the surfaces of the red blood cells determine blood type. If antibodies in the blood plasma bind to red cell glycoproteins, the red blood cells clump together. Each antibody can bind only to one specific glycoprotein; for example, anti-A antibody binds only to glycoprotein A and not to glycoprotein B.*

for example, there are two genes for kernel color, which we might designate 1 and 2. Each gene has two alleles, R and R'. The R allele directs the synthesis of one "unit" of red pigment in the kernel, while the R' allele causes no pigment synthesis at all. If only gene 1 were active, the inheritance of kernel color would follow simple incomplete dominance: R_1R_1 = red; $R_1R'_1$ = pink; $R'_1R'_1$ = white. Since kernel color is controlled by both genes 1 and 2, the color becomes more finely graded in intensity, depending on the number of R alleles of both genes. Kernels with the genotype $R_1R_1R_2R_2$ synthesize four units of pigment, and therefore are dark red. Kernels with the genotype $R'_1R'_1R'_2R'_2$ produce no pigment, and are white. Intermediate numbers of R alleles yield intermediate intensities of red.

As you can well imagine, the more genes that contribute to a single trait, the greater the number of categories of the trait, with increasingly fine gradation between categories. Continuing our example of color intensity, if three genes are involved, we would have seven phenotypic classes; with four genes, nine classes, and so on. With more than three genes, differences between phenotypes are small, and it is extremely difficult to classify the phenotypes reliably.

Many common human characteristics, such as height, body build, and the color of skin, hair, and eyes, are influenced by many genes. As you know, individual humans show virtually continuous variation in these traits. Such continuous variation is due partly to polygenic inheritance, and as we describe below, partly to environmental influences.

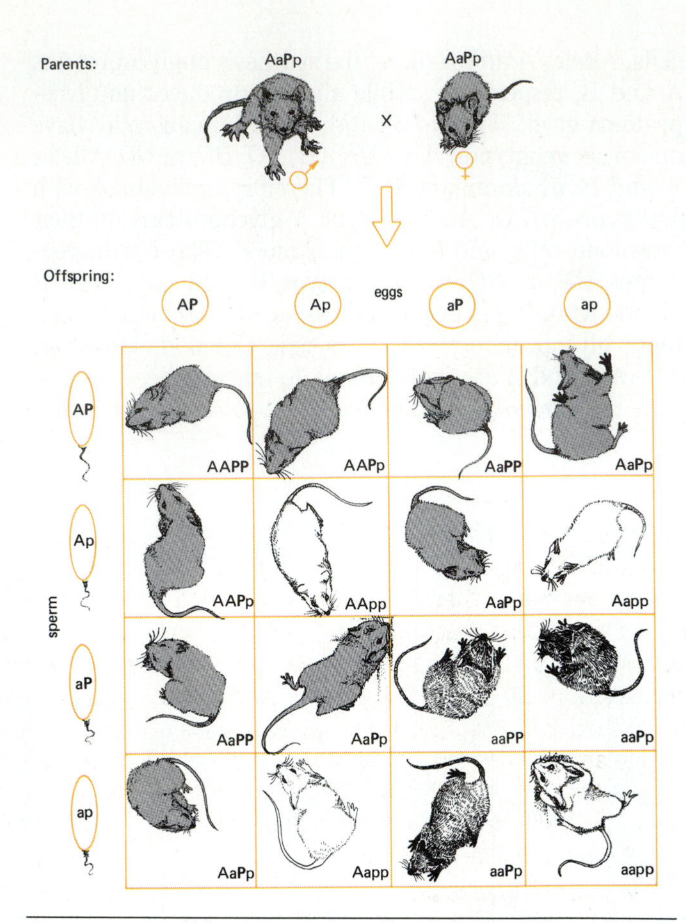

Figure 9-17 Two genes interact in the inheritance of fur color in mice. To produce any melanin pigment at all, a mouse must have at least one dominant allele P *for melanin production. If so, the melanin-distributing gene will produce agouti (*AA *or* Aa *genotypes) or black (*aa *genotype) fur. Homozygous recessive* pp *mice always have white fur. If two mice that are heterozygous for both genes are mated, their offspring do not have a 9:3:3:1 ratio of phenotypes, because of the interaction between the genes. In this cross, the phenotypic ratio of offspring is 9 agouti:3 black:4 albino. Other types of crosses (e.g., a heterozygote mated with a homozygous recessive) yield other modifications of the 9:3:3:1 ratio.*

Gene Interactions

In the inheritance of some traits, there are genes whose actions are required for other genes to be expressed. An example of such gene interaction occurs in the inheritance of hair color in virtually all mammals, including, for example, the mouse. One gene controls the *synthesis* of melanin, the pigment in hair. The dominant allele of this gene (P) allows melanin to be produced, while the recessive allele (p) does not. A second gene controls the *distribution* of pigment in the hair. The normal fur of the wild house mouse, called agouti, has individual hairs that are black with a yellow tip, giving the mouse an overall brownish-gray appearance. Agouti (A) is dominant to plain black fur (a). These two genes control hair color in the following way. If a mouse has the PP or Pp genotype, melanin will be produced. The fur will then be agouti if the mouse is AA or Aa for the melanin-distribution gene, or black if the mouse is aa. If a mouse is pp, it cannot produce melanin. All mice with the pp genotype therefore have white hair and are albino. In an albino, the melanin-distribution gene cannot affect the phenotype, since there is no melanin to distribute. Therefore, the *melanin-synthesis gene* controls the expression of the *melanin-distribution gene.* A cross of two agouti mice can result in a variety of offspring, as the genes for melanin production and distribution assort independently (Fig. 9-17).

The expression of almost all genes is influenced to some extent by other genes. To take an obvious, but usually overlooked, example, no genes can be expressed in an adult organism unless the genes that direct development operate properly. An organism is a cohesive, coordinated whole, and all its genes influence its ultimate anatomy, physiology, and behavior.

Environmental Influences on Gene Actions

An organism is not just the sum of its genes. *Both the genotype and the environment in which an organism lives*

Figure 9-18 A simple case of interaction between genotype and environment in the expression of the gene for black fur in the Himalayan rabbit. Cool areas (nose, ears, feet, or under the cooling coils) allow expression of the gene for black fur.

profoundly affect its phenotype. This holds true even for simple physical traits.

A striking example of environmental effects on gene action occurs in the Himalayan rabbit, which, like the Siamese cat, has pale body fur but black ears, nose, tail, and feet. The Himalayan rabbit actually has the genotype for black fur all over its body. The enzyme that produces the black pigment, however, is temperature sensitive; above about 34°C, the enzyme is inactive. At the temperatures typical of rabbit hutches, extremities such as the ears and feet are cooler than the rest of the body, and black pigment can be produced there. The main body surface is warmer than 34°C, so this fur is pale. By cooling parts of the rabbit's surface artificially, we can produce bizarre patterns of fur color (Fig. 9-18).

Most environmental influences are more complicated and subtle than this. The interactions between complex genetic systems and varied environmental conditions can create a continuum of phenotypes that defies analysis into genetic and environmental components. This is particularly true of human characteristics. Genetic analysis of all but the simplest human traits is notoriously difficult: the human generation time is long, the number of offspring per couple is small, and in any case one cannot very well kidnap a few thousand people and keep them for genetics experiments. Add to these factors the myriad subtle ways in which people respond to their environments, and you can see that a precise determination of the genetic bases of complex traits such as intelligence or musical ability is probably impossible.

REFLECTIONS ON GENETIC DIVERSITY AND HUMAN WELFARE

The science of genetics, which began in a monastery over a century ago, is crucially important to humankind today. Genetics plays a role in many human activities, from possible new sources of energy to inherited human diseases, and from paternity lawsuits to crop improvement. Virtually all the corn grown in the United States today, for example, is hybrid corn, the result of crossing parent strains that are themselves usually not very good as a crop. Faced with high prices, rising populations, and increasing demand for food, agricultural geneticists are constantly trying to develop strains of plants or animals that produce more food per unit input of energy, labor, and money. There are three principal approaches to improving crops and livestock. First, breeders can search for individual plants or animals with superior characteristics and use only these organisms as breeding stock. The superior individuals could arise from mutations or from chance recombinations of preexisting genes. This approach relies on "good" alleles appearing spontaneously and on our ability to recognize them. Second, molecular geneticists can try to create superior genes in the lab or transplant them from one species to another. This is not yet practical, although it may be possible in the near future. The third approach is to look for desirable traits in wild populations of the same or closely related species. Breeders might then crossbreed or otherwise incorporate the desired genetic information into livestock or crop plants. This last approach shows great promise, and in fact is how crops and livestock were developed in the first place, millenia ago. Wheat, for example, was a cross between wild grasses, aided by irregularities in meiosis that resulted in a polyploid plant producing large edible kernels.

Today, our reserves of wild genes are diminishing at a frightening rate, largely due to the pressures of human population and development. An estimated 5 to 20 million species of organisms exist on Earth, and only about a million and a half of these have even been identified. What's worse, many, perhaps most, of the unidentified species may become extinct before science has a chance to study them. Large tracts of wilderness, both in the United States and in other countries, are diminishing rapidly. When these ecosystems go, myriads of species and varieties go with them.

In 1979, biologists Rafael Guzman and Hugh Iltis found a stand of wild corn growing in a mountain forest in a remote area of Mexico. Unlike domesticated corn, which is an annual, this wild variety is a perennial. Imagine if a farmer could plant corn just once every five or ten years and have it bear every year! It may take a lot of work to produce a domesticated perennial corn with high yield and quality. Maybe it will never work out. But just think of the tremendous savings in fuel, labor, and money, and the increase in food supplies if it can be done. If development had destroyed the patch of forest that is the only known habitat of perennial corn (the forest was about to be cleared for a farm), this possibility would have been lost forever.

This is an important, but little recognized, reason to preserve large areas of undisturbed land: to preserve the genes of the plants and animals that live in them. This is also an important goal of efforts to save endangered species, for once a species goes extinct, its genes are forever lost. Genes, with all their various alleles, have evolved over hundreds of millions of years, and represent one of our most valuable and irreplaceable natural resources.

SUMMARY OF KEY CONCEPTS

Gregor Mendel and the Origin of Genetics

Mendel's first experiments dealt with the inheritance of a single trait at a time; these are called monohybrid crosses. From these experiments, Mendel hypothesized that the inheritance of each individual trait is determined by physical entities that we now call genes. Each organism possesses two copies of each gene but includes only one copy in its gametes (the Law of Segregation). An offspring formed by the fusion of two gametes therefore receives two copies of each gene, one from each parent. Each gene may exist in alternative forms, called alleles. Each allele causes a different form of the trait (e.g., purple or white flowers). If an individual possesses two different alleles of the same gene, one allele, called dominant, may completely mask the expression of the other, recessive allele (e.g., a pea with alleles both for purple and white flower color will have purple flowers). In individuals with two different alleles of the same gene, which allele is included in any given gamete is determined by chance. Therefore, we can predict the relative proportions of offspring through the laws of probability.

The physical appearance of an organism (its phenotype), may not always be an infallible indicator of its alleles (the genotype), because of the masking of recessive alleles by dominant alleles. Organisms with two dominant alleles (homozygous dominant) have the same appearance as organisms with one dominant and one recessive allele (heterozygous).

Genes and Chromosomes

Genes are parts of chromosomes. The F_2 generation from a dihybrid cross (parental organisms differing in two traits) may have two fundamentally different outcomes:

1. If the genes for the two traits are on different chromosomes, the F_2 offspring will appear in four different phenotypes, resulting from the independent assortment of the chromosomes (and hence the alleles) during meiosis. This is Mendel's law of independent assortment.
2. If the genes are found on the same chromosome, then (except for crossing over) the F_2 offspring will appear in only the two parental arrangements.

Sex linkage is a special and easily observed case of linkage of traits on the same chromosome. In many animals, females have two X chromosomes, while males have one X and one Y chromosome with many fewer genes. Consequently, males have only one copy of most of the X chromosome genes, and recessive traits are more likely to be phenotypically expressed in males.

In crossing over, corresponding parts of homologous chromosomes are exchanged during meiosis, resulting in different combinations of alleles than existed in the parental chromosomes. The frequency of crossing over can be used to measure the positions of genes on chromosomes.

Variations on the Mendelian Theme

Many inheritance patterns are more complex.

1. In incomplete dominance, heterozygotes have a phenotype intermediate between the two homozygous phenotypes.
2. Codominant alleles, such as those determining blood type, are both phenotypically detectable in heterozygotes. The heterozygotes are not intermediate in phenotype between the two parents but have separately distinguishable features of both parental types.

3. Many traits are determined by several genes, which is called polygenic inheritance. Traits that appear to exist in a continuum of finely graded forms are often determined polygenically.
4. The actions of some genes are required if other genes are to be expressed at all.
5. The environment plays at least some role in the phenotypic expression of all traits.

GLOSSARY

Allele (al-ēl′) one of several alternative forms of a particular gene, usually giving rise to a characteristic form of phenotype (e.g., purple or white flower color).

Autosome (aw′-tō-sōm) a chromosome found in homologous pairs in both males and females, and which does not bear the genes determining sex.

Codominance the relation between two alleles of a gene such that both alleles are phenotypically expressed in heterozygous individuals.

Cross-fertilization union of sperm and egg from two different individuals of the same species.

Crossing over the exchange of corresponding segments of the chromatids of two homologous chromosomes during meiosis.

Dihybrid cross a breeding experiment involving parents that differ in two distinct, genetically determined traits.

Dominant an allele that can determine the phenotype of heterozygotes completely, so that they are indistinguishable from individuals homozygous for the allele. In the heterozygotes, the expression of the other (recessive) allele is completely masked.

Gene a unit of heredity containing the information for a particular characteristic. A gene is a segment of DNA located at a particular place on a chromosome.

Genotype (jēn′-ō-tīp) the genetic composition of an organism; the actual alleles of each gene carried by the organism.

Heterozygote (het-er-ō-zī′-gōt) an organism carrying two different alleles of the gene in question; sometimes called a hybrid.

Homozygote (hō-mō-zī′-gōt) an organism carrying two copies of the same allele of the gene in question; also called a true-breeding organism.

Hybrid an organism that is the offspring of parents differing in at least one genetically determined characteristic; also used to refer to the offspring of parents of different species.

Incomplete dominance a pattern of inheritance in which heterozygotes have a phenotype intermediate between those of the two homozygotes.

Linkage the inheritance of certain genes as a group because they are parts of the same chromosome. Linked genes do not show independent assortment.

Monohybrid cross a breeding experiment in which the parents differ in only one genetically determined trait.

Phenotype (fēn′-ō-tīp) the physical properties of an organism. Phenotype can be defined as outward appearance (e.g., flower color), behavior, or in molecular terms (e.g., ABO glycoproteins on red blood cells).

Polygenic inheritance a pattern of inheritance in which the interactions of two or more genes determine phenotype.

Recessive an allele expressed only in homozygotes and which is completely masked in heterozygotes.

Self-fertilization union of sperm and egg from the same individual.

Sex chromosome one of the pair of chromosomes that differ between the sexes and usually determine the sex of an individual; for example, human females have similar sex chromosomes (XX) while males have dissimilar ones (XY).

Sex linkage a pattern of inheritance characteristic of genes located on one type of sex chromosome (e.g., X) and not found on the other type (e.g., Y).

Test cross a breeding experiment in which an individual showing the dominant phenotype is mated with an individual that is homozygous recessive for the same gene. The ratio of offspring with dominant versus recessive phenotypes can be used to determine the genotype of the phenotypically dominant individual.

True-breeding pertaining to an individual all of whose offspring produced through self-fertilization are identical to the parental type. True-breeding individuals are homozygous for the trait in question.

STUDY QUESTIONS

1. Define gene, allele, dominant, recessive, monohybrid, dihybrid, true-breeding, homozygous, heterozygous, cross-fertilization, and self-fertilization.
2. Explain the meaning of Mendel's law of segregation and law of independent assortment. Under what circumstances does the law of independent assortment apply? When is it violated?
3. Explain why genes located on one chromosome are linked during inheritance. Why do linked genes sometimes separate during meiosis?
4. Explain why human skin color does not occur in just two forms (e.g., black and white).
5. What is sex linkage? In mammals, which sex would be most likely to show recessive sex-linked traits?
6. What is the difference between a phenotype and a genotype? Does knowledge of an organism's phenotype always allow you to determine the genotype? What type of experiment would you perform to determine the genotype of a phenotypically dominant individual?

GENETICS PROBLEMS

(*Note:* An extensive group of genetics problems, with answers, can be found in the *Study Guide.*)

1. In certain cattle, hair color can be red (homozygous *RR*), white (homozygous *R′R′*), or roan (a mixture of red and white hairs; heterozygous *RR′*).
   (a) When a red bull is mated to a white cow, what genotypes and phenotypes of offspring could be obtained?
   (b) If one of these offspring were mated to a white cow, what genotypes and phenotypes of offspring could be produced? In what proportion?
2. The palomino horse is golden in color. Unfortunately for horse fanciers, palominos do not breed true. In a series of matings between palominos, the following offspring were obtained: 65 palominos, 32 cream-colored, 34 chestnut (reddish brown). What is the probable mode of inheritance of palomino coloration?
3. In the edible pea, tall (*T*) is dominant to short (*t*), and green pods (*G*) are dominant to yellow pods (*g*). List the types of gametes and offspring that would be produced in the following crosses:
   (a) *TtGg* × *TtGg*
   (b) *TtGg* × *TTGG*
   (c) *TtGg* × *Ttgg*
4. In tomatoes, round fruit (*R*) is dominant to long fruit (*r*), and smooth skin (*S*) is dominant to fuzzy skin (*s*). A true-breeding round, smooth tomato (*RRSS*) is cross-bred with a true-breeding long, fuzzy tomato (*rrss*). All the F_1 offspring are round and smooth (*RrSs*). When these F_1 plants were bred, the following F_2 generation was obtained:

   Round, smooth: 43
   Long, fuzzy: 13

   Are the genes for skin texture and fruit shape likely to be on the same or on different chromosomes? Explain your answer.
5. In the tomatoes of Question 4, an F_1 offspring (*RrSs*) was mated with a homozygous recessive (*rrss*). The following offspring were obtained:

   Round, smooth: 583
   Long, fuzzy: 602
   Round, fuzzy: 21
   Long, smooth: 16

   What is the most likely explanation for this distribution of phenotypes?
6. In humans, hair color is controlled by two interacting genes. The same pigment, melanin, is present in both brown-haired and blond-haired people, but brown hair has much more of it. Brown hair (*B*) is dominant to blond (*b*). Whether any melanin can be synthesized at all depends on another gene. The dominant (*M*) allows melanin synthesis, while the recessive form (*m*) prevents melanin synthesis. Homozygous recessives *mm* are albino. What will be the expected proportions of phenotypes in the children of the following parents:
   (a) *BBMM* × *BbMm*
   (b) *BbMm* × *BbMm*
   (c) *BbMm* × *bbmm*
7. In humans, one of the genes determining color vision is located on the X chromosome. The dominant form (*C*) produces normal color vision, while red-green color blindness (*c*) is recessive. If a man with normal color vision marries a color-blind woman, what is the probability of their having a color-blind son? A color-blind daughter?
8. In the couple described in Question 7, the woman gives birth to a color-blind daughter. The husband sues for a divorce, on the grounds of adultery. Will his case stand up in court? Explain.

ANSWERS TO GENETICS PROBLEMS

1. (a) A red bull (*RR*) is mated to a white cow (*R′R′*). The bull will produce all *R* sperm, while the cow will produce all *R′* eggs. All the offspring will be *RR′* and have roan hair (codominance).
   (b) A roan bull (*RR′*) is mated to a white cow (*R′R′*). The bull produces $\frac{1}{2}R$ and $\frac{1}{2}R'$ sperm, while the cow produces *R′* eggs. Using the Punnett square method:

sperm \ eggs	R′
R	RR′
R′	R′R′

   Using probabilities:

sperm	egg	offspring
$\frac{1}{2}R$	R'	$\frac{1}{2}RR'$
$\frac{1}{2}R'$	R'	$\frac{1}{2}R'R'$

   The predicted offspring will be $\frac{1}{2}RR'$ (roan) and $\frac{1}{2}R'R'$ (white).
2. The offspring occur in three types, classifiable as dark (chestnut), light (cream), and intermediate (palomino). This suggests incomplete dominance, with the alleles for chestnut (*C*) combining with the allele for cream (*C′*) to produce palomino heterozygotes (*CC′*). We can test this hypothesis by examining the offspring numbers. The ratios are approximately 1 chestnut (*CC*):2 palomino (*CC′*):1 cream (*C′C′*). If palominos are heterozygotes, we would expect the cross *CC′* × *CC′* to yield $\frac{1}{4}CC$, $\frac{1}{2}CC'$, and $\frac{1}{4}C'C'$. Our hypothesis is supported.
3. (a) *TtGg* × *TtGg*: This is a "standard" dihybrid cross. Both parents produce *TG, Tg, tG,* and *tg* gametes. The expected proportions of offspring are $\frac{9}{16}$ tall green:$\frac{3}{16}$ tall yellow:$\frac{3}{16}$ short green:$\frac{1}{16}$ short yellow (see Figure 9-7).
   (b) *TtGg* × *TTGG*: In this cross, the heterozygous parent produces *TG, Tg, tG,* and *tg* gametes. However, the homozygous dominant parent can produce only *TG* gametes. Therefore, all offspring will receive at least one *T* allele for tallness and one *G* allele for green pods, and thus all the offspring will be tall with green pods.
   (c) *TtGg* × *Ttgg*: The second parent will produce two types of gametes, *Tg* and *tg*. Using a Punnett square:

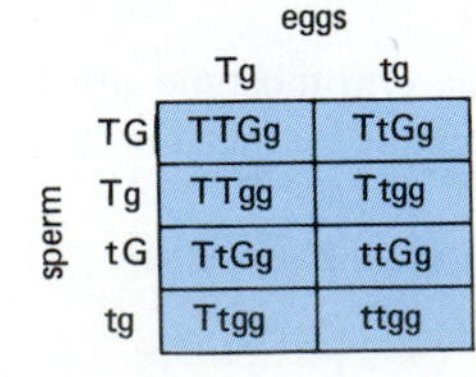

sperm \ eggs	Tg	tg
TG	TTGg	TtGg
Tg	TTgg	Ttgg
tG	TtGg	ttGg
tg	Ttgg	ttgg

The expected proportions of offspring are $\frac{3}{8}$ tall green, $\frac{3}{8}$ tall yellow, $\frac{1}{8}$ short green, $\frac{1}{8}$ short yellow.

4. If the genes are on separate chromosomes, that is, assort independently, this would be a typical dihybrid cross with expected offspring of all four types (approximately $\frac{9}{16}$ round smooth, $\frac{3}{16}$ round fuzzy, $\frac{3}{16}$ long smooth, and $\frac{1}{16}$ long fuzzy). However, only the parental combinations show up in the F_2 offspring, indicating that the genes are on the same chromosome.
5. The genes are on the same chromosome and are quite close together. On rare occasions, crossing over occurs between the two genes, producing recombination of the alleles.
6. (a) *BBMM* (brown) × *BbMm* (brown). The first parent can only produce *BM* gametes, so all offspring will receive at least one dominant for each gene. Therefore, all offspring will be brown haired.

   (b) *BbMm* (brown) × *BbMm* (brown). Both parents can produce four types of gametes: *BM, Bm, bM,* and *bm.* Filling in the Punnett square:

sperm \ eggs	BM	Bm	bM	bm
BM	BBMM	BBMm	BbMM	BbMm
Bm	BBMm	BBmm	BbMm	Bbmm
bM	BbMM	BbMm	bbMM	bbMm
bm	BbMm	Bbmm	bbMm	bbmm

   Remembering that all *mm* offspring are albino, the expected proportions are $\frac{9}{16}$ brown haired, $\frac{3}{16}$ blond haired, and $\frac{4}{16}$ albino.

   (c) *BbMm* (brown) × *bbmm* (albino):

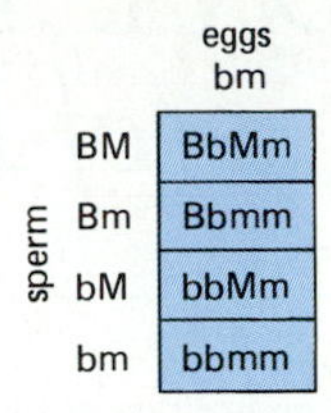

sperm \ eggs	bm
BM	BbMm
Bm	Bbmm
bM	bbMm
bm	bbmm

   The expected proportions of offspring are: $\frac{1}{4}$ brown haired, $\frac{1}{4}$ blond haired, and $\frac{1}{2}$ albino.

7. A man with normal color vision is *CY* (remember, the Y chromosome does not have the gene for color vision). His color-blind wife is *cc*. Their expected offspring will be:

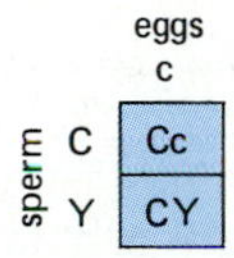

sperm \ eggs	c
C	Cc
Y	CY

   We therefore expect that all the daughters will have normal color vision, while all the sons will be color blind.

8. The husband should win his case. All his daughters must receive one X chromosome, with the *C* allele, from him, and therefore should have normal color vision. If his wife gives birth to a color-blind daughter, her husband cannot be the father (unless there was a new mutation for color blindness in his sperm line, which is very unlikely).

SUGGESTED READINGS

Benzer, S. "Genetic Dissection of Behavior." *Scientific American*, December 1973 (Offprint No. 1285). Not only physical traits, but also behaviors are under the influence of genes. The fruit fly once again proves a useful model system in which to study inheritance.

Crow, J. F. "Genes That Violate Mendel's Rules." *Scientific American*, February 1979 (Offprint No. 1418). A close look at genetic recombination and its role in evolution.

Stern, C., and Sherwood, E. R. *The Origin of Genetics: A Mendel Source Book*, San Francisco: W.H. Freeman and Company, Publishers, 1966. There is no substitute for the real thing, in this case a translation of Mendel's original paper to the Brunn Society.

Strickberger, M. *Genetics*, 3rd ed. New York: Macmillan Publishing Company, 1985. Probably the most complete discussion of all aspects of genetics, including examples of different patterns of inheritance.

10
DNA: The Molecule of Heredity

By the early twentieth century, geneticists knew that chromosomes carry the genetic information that determines the structure and function of organisms. Just how chromosomes exert such profound control remained unknown. Before they could really begin to understand gene function, geneticists had to determine the structure of DNA, the genetic material of life on Earth.

THE COMPOSITION OF CHROMOSOMES

Early microscopists discovered many of the organelles of the cell without, of course, knowing the function of most of them. In the early 1870s, the biochemist Friedrich Miescher analyzed one of these organelles, the nucleus. He found that he could extract a previously unknown chemical substance from nuclei, an acidic material with an unusually high phosphorus content. Due to its location in the nucleus and its acidic properties, this material came to be known as nucleic acid.

Figure 10-1 *"A structure this pretty just had to exist."* *—James D. Watson*

The discovery that chromosomes are the carriers of genetic information focused attention on their chemical composition. Biochemists discovered that a eukaryotic chromosome is composed of protein and a specific kind of nucleic acid, **deoxyribonucleic acid (DNA).** One of these substances, therefore, must carry the cell's hereditary blueprint.

DNA seemed to be a particularly simple molecule, composed of just four kinds of subunits, called **nucleotides,** strung together in a long chain. Each nucleotide consists of three parts: a phosphate group; deoxyribose, a five-carbon sugar; and a nitrogen-containing base:

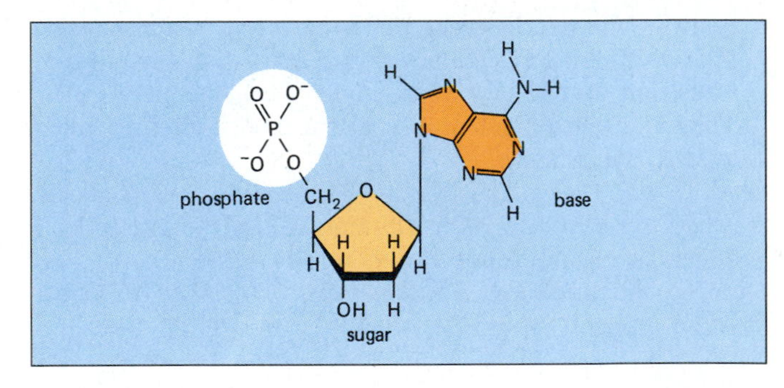

The four DNA nucleotides have the same phosphate and sugar but different bases. The bases come in two types, the single-ringed **pyrimidines,** thymine (abbreviated T) and cytosine (C), and the double-ringed **purines,** adenine (A) and guanine (G):

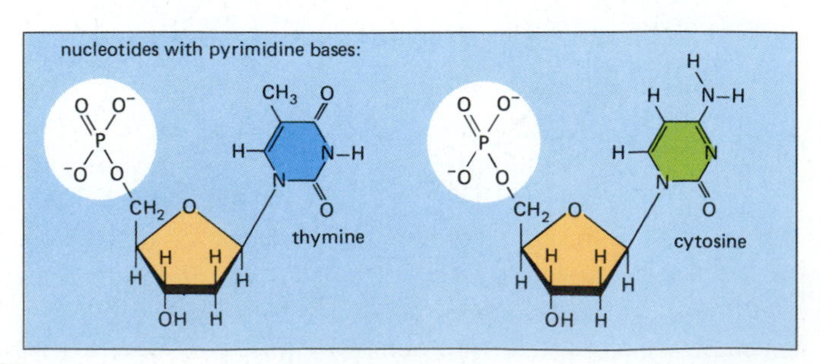

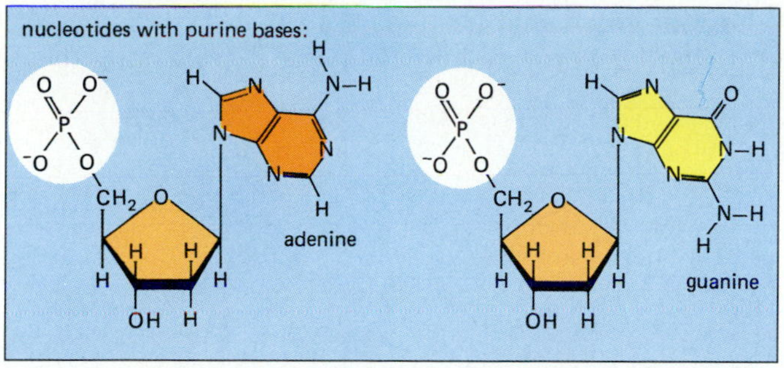

nucleotides with purine bases:

adenine

guanine

In a single strand of DNA, the phosphate of one nucleotide bonds to the sugar of another. This forms a long strand consisting of a "backbone" of sugars and phosphates, with the bases protruding from the backbone:

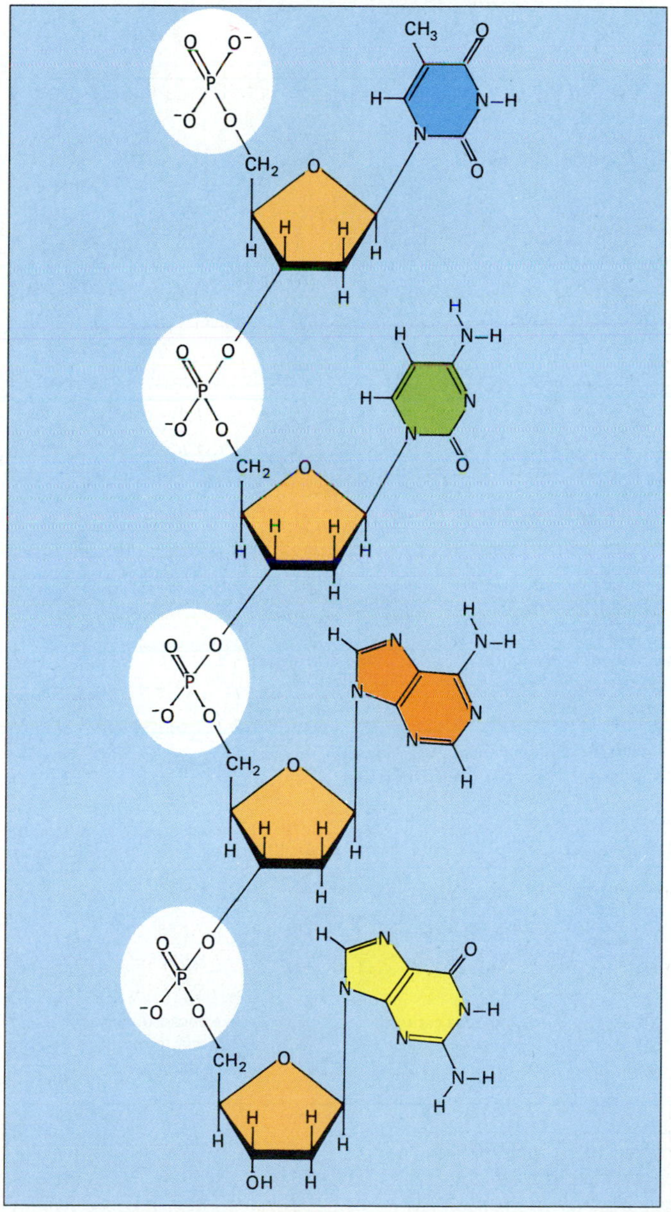

Since DNA has only four subunits, each quite similar to the others, many biologists were skeptical that DNA could contain the information needed to construct all the molecules of complex organisms. Proteins, on the other hand, are constructed of about 20 different amino acids, and intuitively seemed to offer much more opportunity for carrying the enormous amount of genetic information that each chromosome obviously contains. Besides that, most of the substance of living cells, aside from water, is protein or is synthesized through the action of enzymes, which are also proteins. An attractive hypothesis, therefore, was that the proteins of chromosomes are templates for the proteins of the rest of the cell, a kind of mold from which copies could be made.

Gradually, biologists accumulated information that indicated that proteins might not be the genetic material. For example, chromosomes in the sperm of some fish contain only DNA and one protein, called protamine. Like other animals, a fish receives half its genes from its male parent, through the male's sperm. Now protamine is a very simple protein, mostly composed of a single amino acid. If protamine were the genetic material in fish sperm, could it possibly serve as a template for all the other, more complex, proteins of the whole fish? Despite contradictory evidence such as this, many scientists continued to support the protein-as-gene hypothesis. Finally, two experiments with microorganisms showed that DNA must be the hereditary material.

DNA: THE HEREDITARY MOLECULE

Throughout the history of biology, major advances have been made possible by using the "right" organism for a particular experiment. Mendel and Morgan, for instance, discovered the fundamentals of classical genetics through experiments with peas and fruit flies because they realized that peas and flies have advantages for genetics studies. Since all life probably evolved from a common, though distant, ancestor, knowing the mechanisms of inheritance in flies may help us to understand inheritance in other organisms, including people. Often, a particular organism is especially suitable for a certain experiment because it is simple in structure or reproduces very rapidly. Such simple, rapidly reproducing organisms were the keys to proving that DNA, not protein, is the chemical of heredity.

Bacterial Transformation

In 1928, the bacteriologist Frederick Griffith reported an intriguing phenomenon in pneumonia bacteria, *Streptococcus pneumoniae,* which he called **transformation.** Pneumonia bacteria occur in two forms. In one, a complex polysaccharide capsule covers each bacterium; in the other form, the bacteria are naked. When *Streptococcus* invade a mammalian host, for example a mouse, the capsule prevents the mouse's white blood cells from destroying the bacteria. Therefore, the encapsulated form causes pneumonia (Fig. 10-2a), while the naked form cannot

(Fig. 10-2b). The presence or absence of the capsule is genetically determined.

Griffith found that heating encapsulated *Streptococcus* to a high temperature kills the bacteria, rendering them

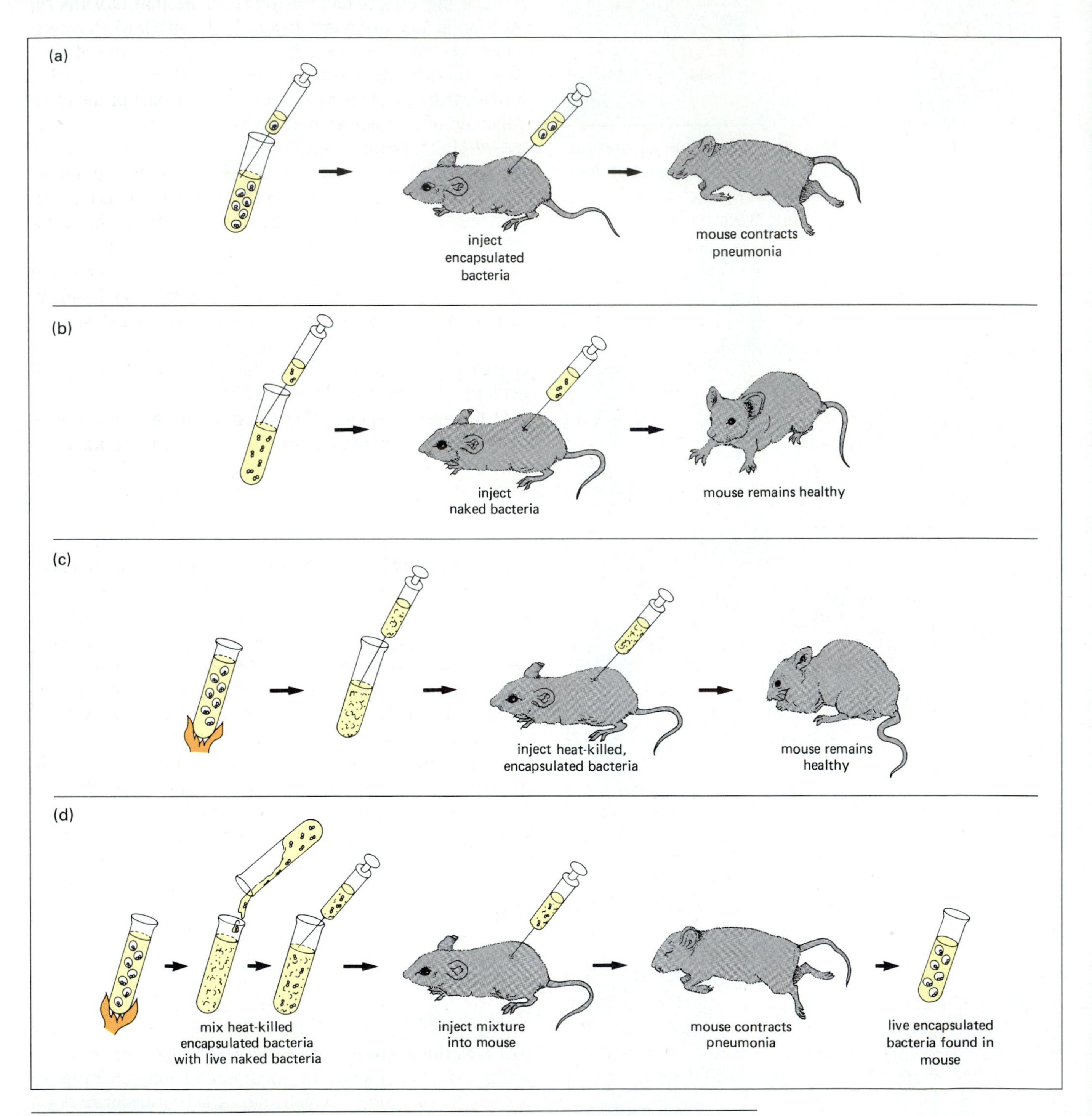

Figure 10-2 *Griffith's discovery of transformation in pneumonia bacteria.* **(a)** *If mice are injected with the encapsulated form of streptococcus, they contract pneumonia and die.* **(b)** *If the mice are injected with the naked form, their white blood cells destroy the invading bacteria and the mice do not develop pneumonia.* **(c)** *Heating kills encapsulated bacteria, so that they cannot cause pneumonia.* **(d)** *If heat-killed encapsulated bacteria are mixed with live naked bacteria, the resulting mixture now can cause pneumonia in mice. Blood samples taken from these mice contain live, encapsulated bacteria.*

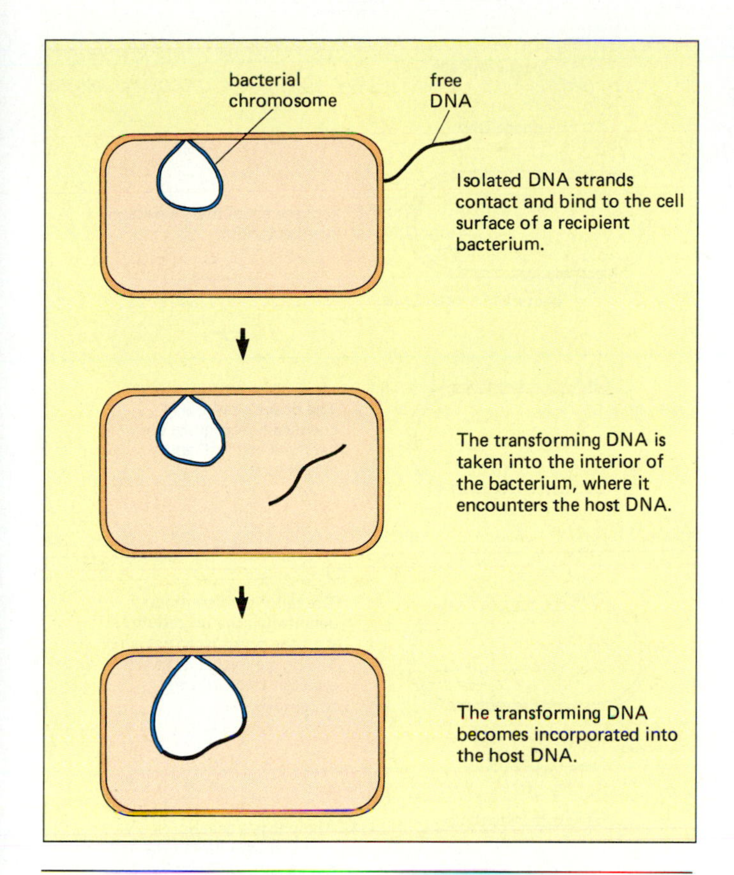

Figure 10-3 *Bacterial transformation by free DNA. Transformation is not the "invasion" of a bacterium by free DNA. Rather, the recipient actively acquires the DNA and incorporates it into its genome. Transformation allows bacteria to acquire new genes and confers a selective advantage on the recipient bacteria, which can now colonize new habitats or hosts with the help of their new genes. For example, transformation aids in the spread of antibiotic resistance from one type of bacteria to another.*

harmless (Fig. 10-2c). However, injecting a mouse with a mixture of heat-killed encapsulated bacteria and live naked bacteria causes pneumonia (Fig. 10-2d). What's more, Griffith found that the infected mouse teemed with live encapsulated bacteria, and these encapsulated bacteria breed true. Finally, the whole process could occur without the mouse at all. Griffith mixed dead encapsulated bacteria with live naked bacteria in a test tube, and produced live, true-breeding, encapsulated bacteria. A likely hypothesis was that the living bacteria had acquired molecules of genetic information from the dead bacteria, which **transformed** the formerly naked bacteria into the encapsulated form.

In 1944, three researchers at Rockefeller University, O. T. Avery, Colin MacLeod, and Maclyn McCarty, discovered that the transforming chemical is DNA. They isolated DNA from encapsulated bacteria, mixed it with live naked bacteria, and produced live encapsulated bacteria. To prove that transformation was caused by DNA, not by small quantities of protein contaminating their extracts, they treated different extracts with enzymes. Protein-destroying enzymes did not affect the transforming ability of the extracts, but DNA-destroying enzymes prevented transformation. They concluded that *DNA is the genetic material of bacteria* and that a live bacterium can take up DNA from its environment and incorporate this DNA into its own chromosome (Fig. 10-3). When DNA carrying the genes for capsules becomes part of the chromosome of another bacterium, the recipient becomes capable of synthesizing capsules.

Despite this evidence, some biologists remained unconvinced that DNA is the universal hereditary material. Perhaps, some thought, DNA induces a mutation in naked bacteria, changing a (protein) gene that does not direct capsule synthesis into one that does. A second experiment convinced most of the skeptics. This experiment used an even simpler organism, a virus that infects bacteria.

The Bacteriophage Experiments

Certain viruses infect only bacteria and are called **bacteriophages** (meaning "bacteria eaters" in Greek). Even though many bacteriophages (phage for short) have elaborate structures (Fig. 10-4), they are chemically very simple, being composed only of DNA and protein. A phage depends on its host bacterium for every aspect of its life cycle (Fig. 10-5). When a phage encounters a bacterium, it attaches to the bacterial cell wall and injects its genetic material into the bacterium. The rest of the phage (head, tail, tail fibers, etc.) remains outside the bacterium. The phage genes subvert the bacterial metabolism into producing more phages. Finally, the genetic material of the phage directs the synthesis of an enzyme that ruptures the bacterium, liberating the newly manufactured phages.

In a brilliant series of experiments published in 1952, Alfred Hershey and Martha Chase utilized the chemical simplicity of bacteriophages to determine whether the genetic material of phages is DNA or protein (Fig. 10-6). Chemically, DNA and proteins both contain carbon, oxygen, hydrogen, and nitrogen. However, DNA also contains phosphorus but not sulfur, while proteins contain sulfur (in the amino acids methionine and cysteine) but not phosphorus. Hershey and Chase forced one population of phages to synthesize DNA using radioactive phosphorus, and another population to synthesize protein using radioactive sulfur. When bacteria were infected by phages containing radioactively labeled protein, radioactivity did not appear inside the bacteria. Offspring phages were not radioactive either (Fig. 10-6a). When bacteria were infected by phages containing radioactive DNA, the radioactivity was subsequently found inside the bacteria. When the bacteria burst, some of the offspring phages also had radioactive DNA (Fig. 10-6b). These experiments showed that the genetic material that phages inject into their hosts is DNA, not protein.

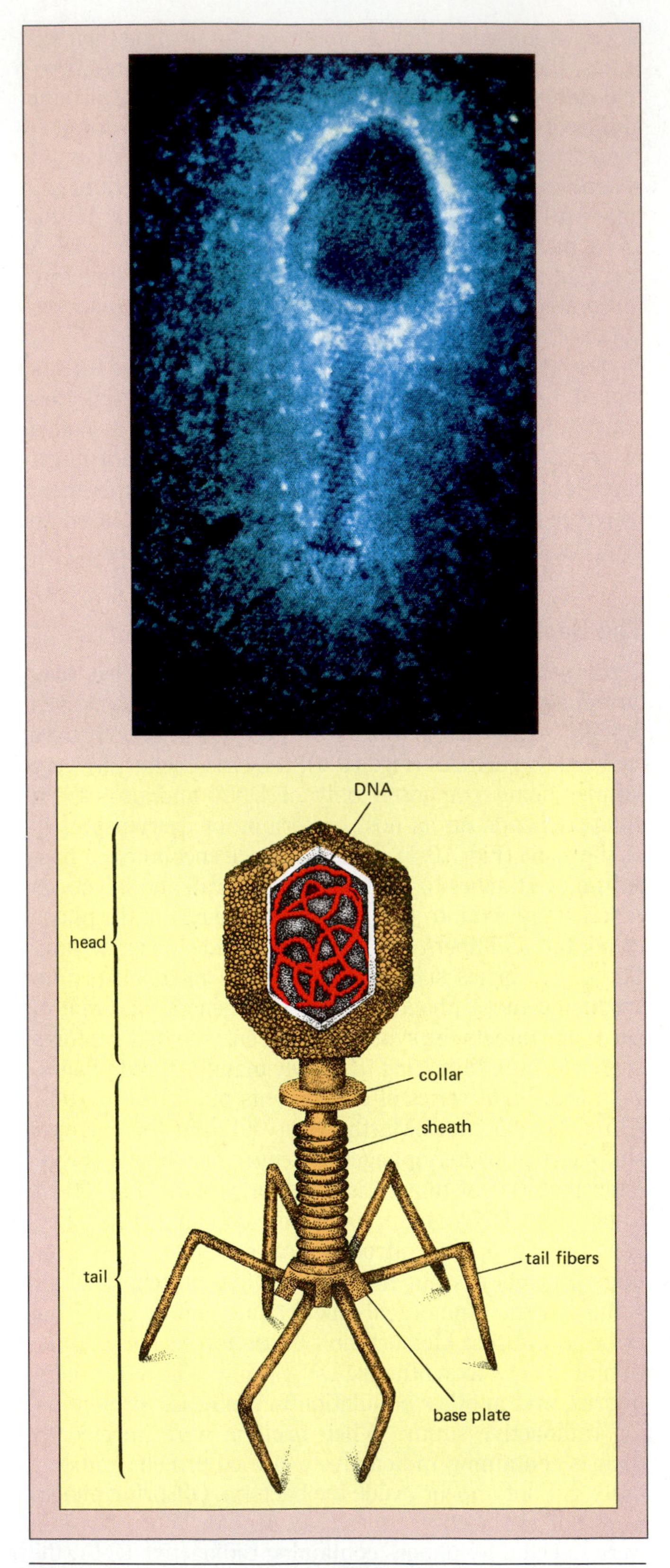

Figure 10-4 *The structure of a T2 bacteriophage. The T2 phage consists of a head region composed of a protein coat containing a single DNA molecule, and a protein tail region (the sheath, base plate, and tail fibers) responsible for attachment to its host bacterium and injection of the genetic material.*

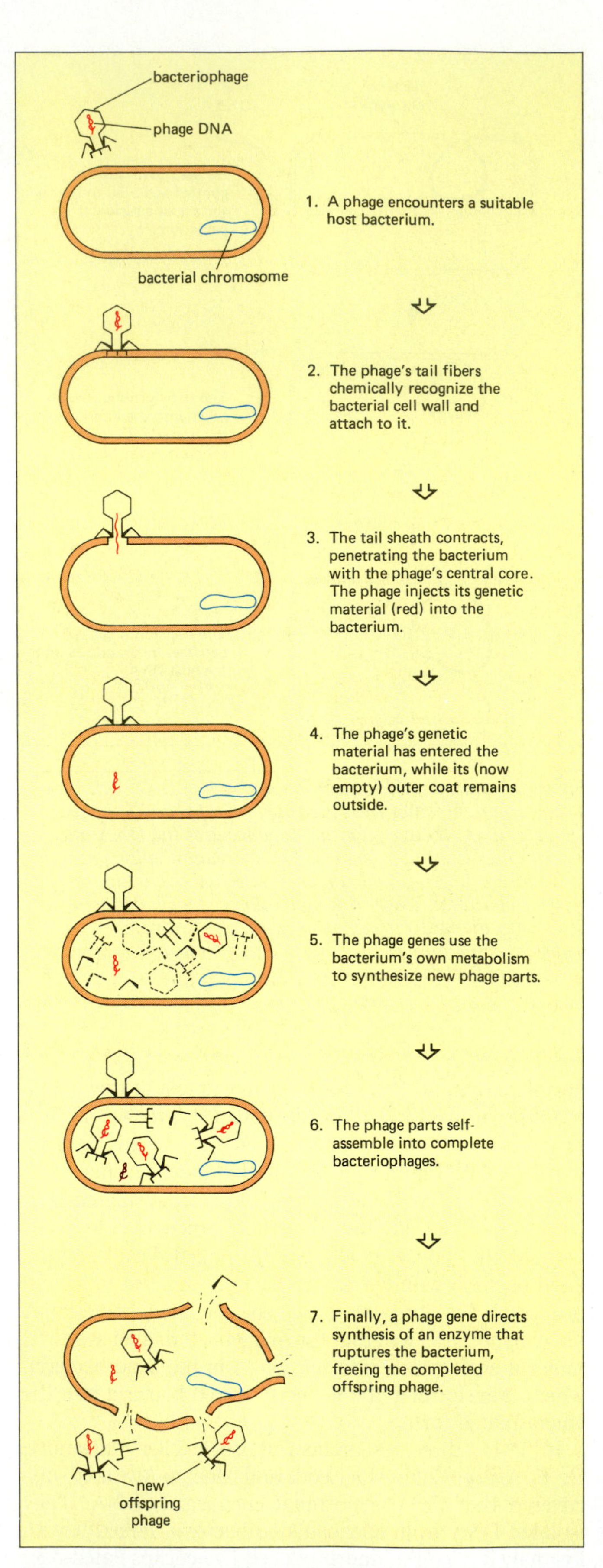

Figure 10-5 *The life cycle of T2 bacteriophages.*

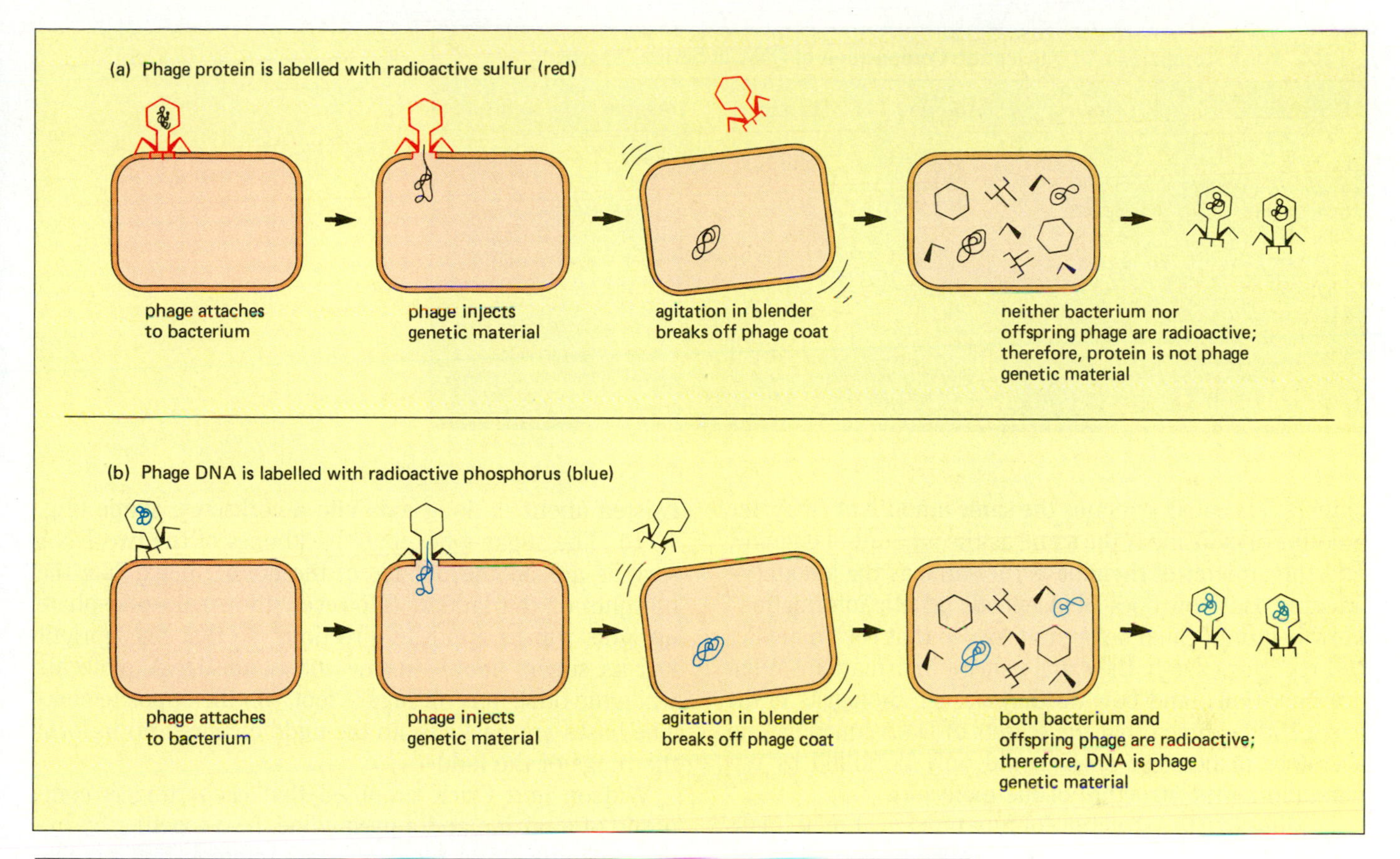

Figure 10-6 *The Hershey–Chase experiment to determine whether DNA or protein is the genetic material of T2 bacteriophages. Hershey and Chase grew one population of phages in a medium containing radioactive sulfur, ^{35}S, which radioactively labels the phage proteins. Another population of phages was grown in radioactive phosphorus, ^{32}P, which labels the phage DNA.* **(a)** *The fate of ^{35}S-labeled protein (red). Phages attach to the cell walls of unlabeled bacteria, and inject their genetic material. Whirling the mixture in a blender breaks off the phage heads and sheaths left on the outside of the bacteria. The ^{35}S radioactivity is found outside the bacteria in the phage coats, while the bacteria, containing the phage genetic material, are not radioactive. New phages resulting from the infection are also not radioactive. The phage genetic material, therefore, is not protein.* **(b)** *The fate of ^{32}P-labeled DNA (blue). When the phage–bacteria combinations are broken apart in a blender, the ^{32}P is found inside the bacteria, while the phage pieces are unlabeled. Some of phages synthesized inside the infected bacteria are ^{32}P-labeled. The genetic material must be DNA, carrying the ^{32}P inside the bacteria and to progeny phage.*

THE STRUCTURE OF DNA

Even before the report of Hershey and Chase in 1952, Avery's demonstration that the bacterial transforming factor was DNA had stimulated a burst of research on the chemical nature of DNA. Alfred Mirsky studied the amount of DNA in the cells of various tissues of several organisms. He found that *the quantity of DNA varies among species, but is constant in each cell of a given species no matter what tissue the cell comes from*. Gametes are a notable exception, in that they have half as much DNA as the other cells of the body. This, of course, is exactly what we would expect if DNA is the hereditary material. (Why?)

What seemed equally significant, if only someone could figure out what it meant, were the results of Erwin Chargaff. Chargaff analyzed the amounts of the four nucleotides of DNA in a variety of species and found a curious consistency (Table 10-1). Although the amounts of each of the four bases vary considerably from species to species, *the DNA of any given species contains equal amounts of adenine and thymine, and equal amounts of cytosine and guanine*, within the limits of experimental measurement.

So: DNA is the genetic material; every cell in the body,

Table 10–1 **Comparison of Nucleotide Composition of DNA in Several Species**[a]

Organism	Tissue	Adenine	Thymine	Guanine	Cytosine
E. coli	—	26.0	23.9	24.9	25.2
S. pneumoniae	—	29.8	31.6	20.5	18.0
Sea urchin	Sperm	32.8	32.1	17.7	17.4
Rat	Marrow	28.6	28.4	21.4	21.5
Human	Sperm	31.0	31.5	19.1	18.4
Human	Thymus	30.9	29.4	19.9	19.8
Human	Liver	30.3	30.3	19.5	19.9

[a]Note that the quantities of the four nucleotides have been determined by biochemical analysis and are subject to errors of measurement. Although we conclude from these data that for each organism the amount of adenine equals the amount of thymine, and the amount of guanine equals the amount of cytosine, the actual measurements are usually not exactly equal.

gametes excepted, contains the same amount of DNA; the amount of cytosine is the same as the amount of guanine; and the amount of thymine is the same as the amount of adenine. But how does DNA encode genetic information? How is it duplicated before mitosis, so that each daughter cell receives exactly the same genetic information? What are mutations, and how do they occur? Biologists in the early 1950s agreed that the secrets of DNA function, and therefore of heredity itself, could only be found by understanding the structure of the molecule.

Finding out the structure of any biological molecule is no simple task. Even the most powerful electron microscopes cannot reveal the structure of molecules in atomic detail. To study the structure of DNA, Maurice Wilkins and Rosalind Franklin turned to X-ray diffraction. They bombarded crystals of purified DNA with X-rays and photographed the resulting diffraction patterns (Fig. 10-7). As you can see, the X-ray pattern of DNA does not provide a direct picture of the structure of the molecule. However, the diffraction pattern suggested to Wilkins and Franklin that a DNA molecule is helical (i.e., twisted like a corkscrew), has a uniform diameter of 2 nanometers, and consists of subunits, each of which is separated from neighboring subunits by 0.34 nanometer. One full turn of the DNA helix occurs every 3.4 nanometers. Finally, the X-ray picture suggests that the sugar–phosphate "backbone" of the molecule (see p. 179) is on the outside of the helix, while the bases are on the inside.

The Double Helix

The chemical and X-ray diffraction data were not nearly enough information with which to work out the structure of DNA. Some good guesses were also needed (see the essay, "Real Science Revisited"). Combining a knowledge of how complex organic molecules bond with one another with an intuition that "important biological objects come in pairs," James Watson and Francis Crick proposed that *the DNA molecule consists of two strands, twisted about each other into a* **"double helix,"** much like a ladder twisted about its long axis into a corkscrew shape (Fig. 10-8). The sugar–phosphate backbones of the two DNA strands are on the outside of the double helix, like the uprights of the ladder. (However, the sugar–phosphate uprights run in opposite directions, so that one upright has its sugar "foot" on one end of the DNA molecule while the other has its sugar "foot" on the opposite end.) The bases are packed into the middle, paired up to form the rungs of the ladder.

Watson and Crick proposed that each rung is composed of a purine and a pyrimidine, held together by hydrogen bonds. If adenine pairs with thymine, and guanine pairs with cytosine, then hydrogen bonds can form between the bases, holding the two halves of the rungs together:

Adenine and thymine are held together by two hydrogen bonds, while cytosine and guanine are held together by three.

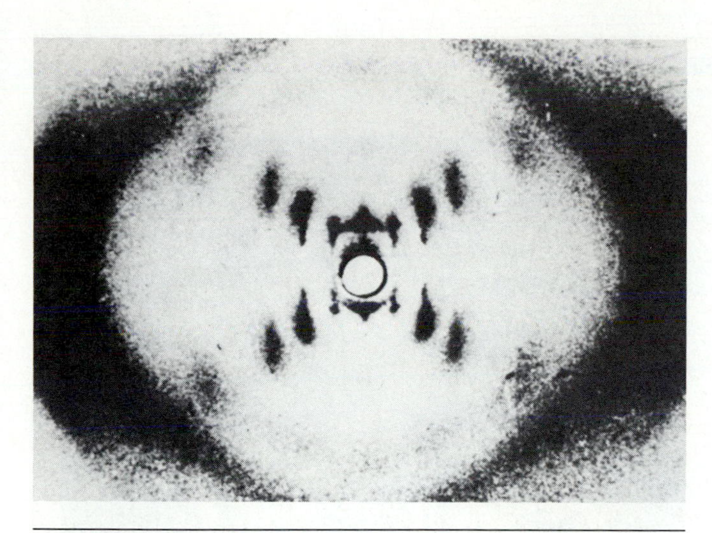

Figure 10-7 *The X-ray diffraction pattern of DNA, taken by Rosalind Franklin. The "cross" formed of dark spots is characteristic of helical molecules such as DNA. Measurements of various aspects of the pattern indicate the dimensions of the DNA helix; for example, the distance between spots in the cross corresponds to the distance between turns of the helix.*

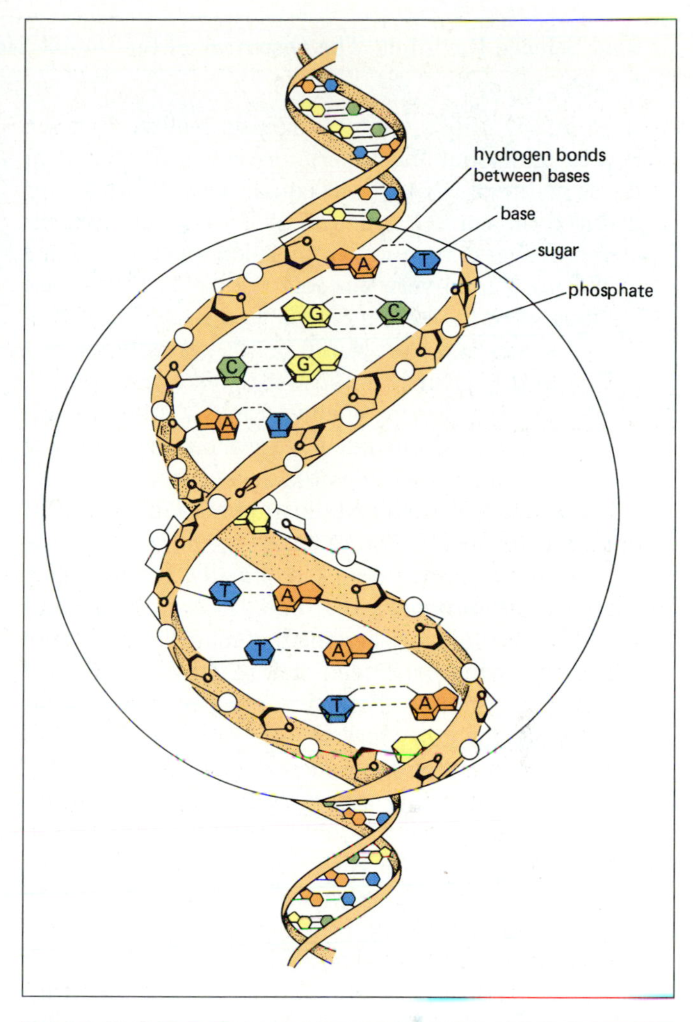

Figure 10-8 *In the Watson–Crick model for DNA, two strands of DNA are wound about each other in a double helix, like a ladder twisted about its long axis. Specific patterns of hydrogen bonding allow complementary bases to pair together in the center of the helix. Three hydrogen bonds hold guanine to cytosine, while two hold adenine to thymine. The two DNA strands run in opposite directions. Note that the "oxygen arrowheads" on the sugars run in opposite directions. As Fig. 10-9 shows, this is important during DNA replication.*

This configuration makes sense of Chargaff's observation that the amount of adenine in DNA equals the amount of thymine, and that the amount of guanine equals the amount of cytosine. In nucleic acids, bases that pair together via hydrogen bonds are called **complementary base pairs.** In DNA, adenine is complementary to thymine, while guanine is complementary to cytosine. This is called the **base-pairing rule.**

The structure of DNA was solved. Although further data would be needed to confirm its details, "a structure this pretty just had to exist," as Watson later put it. Watson and Crick published their double helix model for DNA in 1953, a model that revolutionized all of genetics and practically all of biology in just a few years. In 1962, Watson, Crick, and Maurice Wilkins shared the Nobel Prize in recognition of their brilliant work in deciphering the structure of DNA.

The Watson-Crick model is, of course, not merely pretty; it also helps explain several features of inheritance. First, different types of bases can be arranged in any linear order down one of the DNA strands, matched up with complementary bases on the other strand. Each sequence of bases represents a unique set of genetic instructions, like a biological Morse code. A stretch of DNA just 10 nucleotides long can exist in over 1 million different possible sequences of the four bases. Since an "average" chromosome has millions (in bacteria) to billions (higher plants and animals) of nucleotides, DNA molecules can encode a staggering amount of information.

What remained was to find out how the structure *works:* how the sequence of nucleotides in DNA directs synthesis of cellular components; how identical copies are formed during DNA replication; how mutations occur and how they change the genetic information. In the remainder of this chapter and the following one, we explore how the Watson-Crick model for DNA helps provide answers to these questions.

DNA REPLICATION: THE KEY TO CONSTANCY

The Watson-Crick model immediately suggests a simple, attractive hypothesis for the duplication of chromosomes

Real Science Revisited: The Discovery of the Double Helix

In the early 1950s, many biologists realized that the key to understanding inheritance would be found in the structure of DNA. It was equally clear that whoever deduced the correct structure of DNA would immediately receive recognition from fellow biologists, fame in the press, and very possibly the Nobel Prize. Less obvious were the best methods to discover the structure of DNA, and who would be the person to do it.

The betting favorite in the race to discover the structure of DNA had to be chemist Linus Pauling of Caltech. Pauling probably knew more about the chemistry of large organic molecules than any person alive, he was an expert X-ray crystallographer, and he had hit upon the idea that accurate models could be great aids in deducing molecular structure. Finally, he was almost frighteningly brilliant. In 1950, he amply demonstrated these traits by showing that many proteins were coiled into single-stranded helices (the α-helix; see Chapter 3). Pauling, however, had two main handicaps. First, for years he had concentrated on protein research, and therefore had few data of his own on the structure of DNA. Second, he was active in the peace movement, and therefore, during the reign of Senator Joseph McCarthy in the early 1950s, he was considered to be potentially subversive and possibly dangerous to national security. The latter handicap may have proved decisive, as we shall see.

The second most likely competitors were Maurice Wilkins and Rosalind Franklin, English X-ray crystallographers. They set out to determine the structure of DNA by the most direct method, the careful study of the X-ray diffraction patterns of DNA and its derivatives. They were the only scientists who had really good data indicating what the general shape of the DNA molecule might be. Unfortunately for Wilkins and Franklin, they had a very "proper" notion of the way to do science: the structure of DNA would come along in due time through steady and logical analysis of X-ray and chemical studies.

This slow-but-steady philosophy left the door open for the eventual discoverers of the double helix, James Watson and Francis Crick, two young scientists with neither Pauling's tremendous understanding of chemical bonds nor Franklin and Wilkins' expertise in X-ray analysis. They did have three crucial advantages: the lesson of Pauling's work on proteins, that models could be enormously helpful in studying molecular structure; access to the X-ray data; and a driving ambition to be first.

Watson and Crick did no experiments in the ordinary sense of the word; rather, they spent their time thinking about DNA, trying to construct a molecular model that made sense and fit the available data. Since they were in England and since Wilkins was very open about his and Franklin's data, they were familiar with all the X-ray information relating to DNA. This was just what Pauling lacked. Due to his presumed subversive tendencies, the U.S. State Department refused to issue Pauling a passport to leave the country, and consequently he could not attend meetings at which Wilkins presented the X-ray data, nor visit England to talk with Franklin and Wilkins directly.

Figure E10-1. James Watson and Francis Crick with a model of the structure of DNA.

Watson and Crick knew that Pauling was working on DNA structure and were terrified that he would beat them to it. Watson recounts his belief that if Pauling could have seen the X-ray pictures, "in a week at most, Linus would have the structure."

About this time, you might be thinking, "But wait just a minute! That's not fair at all. If the goal of science is to advance knowledge, everybody should have had access to all the data, and if Pauling was the best, he should have discovered the double helix first." Perhaps so. But science is an activity of scientists, who, after all, are people. Although virtually all scientists want to see the advancement and benefit of humanity, each person also wants to be the one responsible for the advancement and to receive the credit and the glory. We should not overlook the fact that the ambition to be first helps to inspire the intense concen-

Real Science Revisited: The Discovery of the Double Helix

tration, the sleepless nights, and the long days in the laboratory that ultimately produce results.

At any rate, Pauling remained in the dark about the correct X-ray pictures of DNA and was beaten to the correct structure. When Watson and Crick discovered the base pairing rules that were the key to DNA structure, Watson sent off a letter about it to Max Delbrück, a friend and advisor at Caltech. He asked Delbrück not to reveal the contents of the letter to Pauling until their structure was formally published. Delbrück, perhaps more of a model scientist, firmly believed that scientific discoveries belong in the public domain, and promptly told Pauling all about it. With the class of a great scientist and a great person, Pauling graciously congratulated Watson and Crick on their brilliant solution to the structure. The race was over.

prior to mitosis and meiosis. *Since the bases are matched up according to the base pairing rule, the sequence of bases in one strand of the double helix accurately predicts the sequence of bases in the complementary strand.* In brief, DNA duplication occurs in the following way (Fig.10-9). (1) New nucleotides, synthesized in the cytoplasm, are imported into the nucleus. (2) The two DNA strands of the double helix of a parental chromosome unwind and separate. (3) Each parental strand is used as a template upon which new nucleotides are assembled to form a molecule that is complementary to the parental strand. (4) Finally, one parental strand and its newly synthesized, complementary daughter strand rewind into one daughter chromatid, while the other parental strand and its complementary daughter strand rewind into the second daughter chromatid. Since each daughter chromatid consists of one intact parental DNA strand and one newly synthesized complementary strand, the process of DNA duplication is called **semiconservative replication.**

Molecular Events of DNA Replication

How is replication accomplished? Biologists are still discovering the details. However, in eukaryotic cells, the sequence of events seems to be as follows. The hydrogen bonds holding base pairs together are broken, allowing the two strands of DNA to unwind. Enzymes called **DNA polymerase** bind to each unwound strand. DNA polymerase performs a dual function. First, *it bonds together the sugars and phosphates of free nucleotides to form a chain*

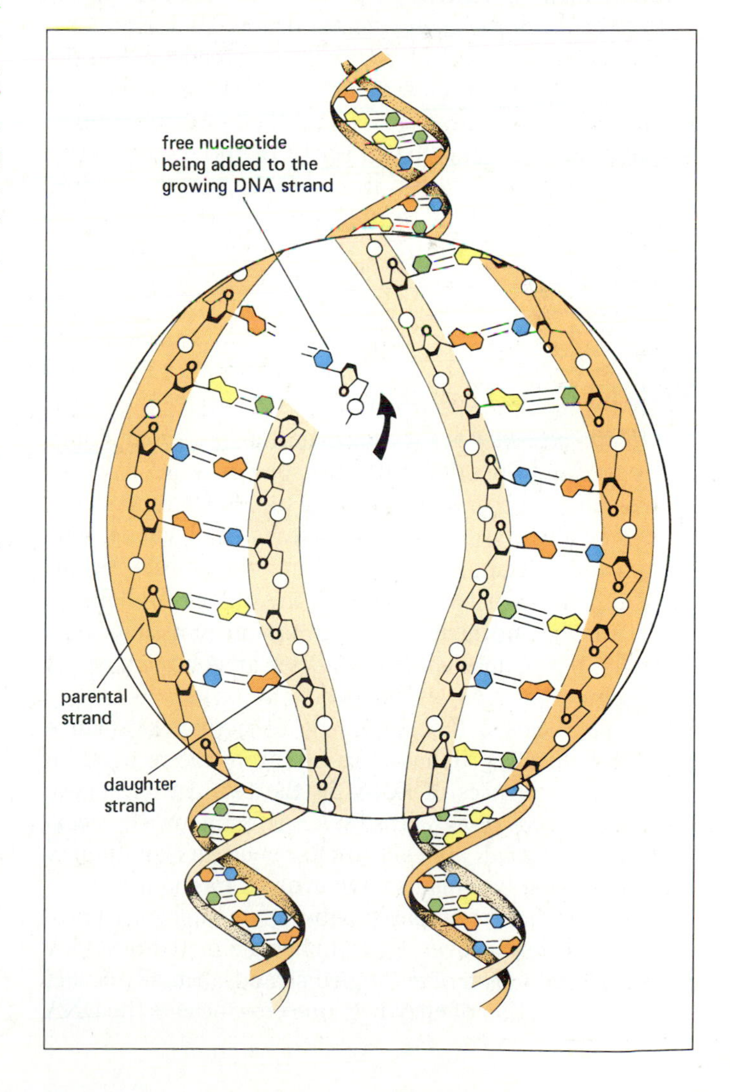

Figure 10-9 Replication of DNA. Individual nucleotides of all four types are synthesized by the cell before replication and are available within the nucleus. The sugar–phosphate backbones of the parental DNA strands are colored dark tan and the backbones of the daughter DNA strands are colored light tan. The central enlargement shows that the parental DNA strands unwind in the middle, and the hydrogen bonds between the complementary bases are broken. DNA polymerase enzymes (not shown) move along the two DNA strands in opposite directions, traveling in the direction of the "oxygen arrowheads." The polymerase enzymes recognize the nucleotides in the parental DNA strands, pair each one with a complementary nucleotide and connect the new nucleotides together into daughter strands. This semiconservative replication process produces two new double helices of DNA, each composed of one parental strand (dark tan backbone) and one daughter strand (light tan backbone) that is an exact copy of the other parental strand. Therefore, each new double helix is an exact copy of the parental DNA.

of DNA. However, it would be useless merely to stitch together nucleotides in an arbitrary order. Therefore, *DNA polymerase must recognize bases exposed in a parental strand, and bond free nucleotides that have bases complementary to those in the parental strand.* In other words, when the DNA polymerase is in contact with an adenine in the parental strand, it adds thymine to the growing daughter DNA strand; when it encounters cytosine in the parental strand, it adds guanine to the daughter strand; and so on.

DNA polymerase is not infallible, and occasionally, it makes mistakes in base pairing. However, other enzymes "proofread" each daughter strand as it is synthesized, to ensure proper base pairing. The directionality of DNA structure allows the proofreading enzymes to recognize the parental strand, running in one direction, as the "right stuff," and to correct any mismatches by changing the daughter strand, which runs in the other direction. As each daughter strand is synthesized, it winds up with its complementary parent strand, forming a new double helix.

During prophase of mitosis, each chromatid consists of a double helix of DNA, composed of one of the original strands of the parent DNA plus one new strand that is an exact copy of the other original strand. When the chromatids separate at anaphase and are parceled out to the daughter cells, each cell receives an exact copy of each of the parental chromosomes. Thus if there are no mistakes in the whole process, the integrity of the genetic information will be maintained from cell division to cell division and from parent to offspring.

MUTATIONS: THE KEY TO VARIABILITY

Complete constancy is not, however, the best evolutionary strategy. As we pointed out in Chapter 8, when environments change, species must change too, or face extinction. We have already discussed two ways of changing gene composition during sexual reproduction: **sexual recombination,** as each parent contributes different alleles to the offspring, and **genetic recombination** (crossing over), in which new combinations of alleles arise when parts of the chromosomes are exchanged.

During meiosis, the alignment of homologous chromosomes during prophase I results from the recognition of similar sequences of DNA on the two chromosomes. Although *which genes* cross over is random, the *rearrangement of genes* appears to offer such an evolutionary advantage that organisms have evolved mechanisms that promote exchange of parts between homologous chromosomes. Crossing over, far from being a matter of sloppy breakage and fusion of entangled strands, actually occurs through the action of enzymes: one enzyme cuts the DNA of the two chromosomes, and another enzyme splices together the pieces.

Sexual recombination and crossing over can generate new combinations of alleles only if there are a number of different alleles in existence. As we pointed out in Chapter 9, new alleles arise as a result of changes in the genetic material, called **mutations.** Since the information in a gene is encoded in the specific sequence of bases, *a mutation must be a change in the sequence of bases.* How can the base sequence change? One way for a mutation to occur is through a *mistake in base pairing* during replication. Base pairing proceeds extremely rapidly, as fast as 200 new base pairs each second. Even with proofreading, replicating 1 billion bases occasions some mistakes; not many, but a few. Certain chemicals (such as the mustard gas used in World War I) and some types of radiation (such as X-rays) increase errors of base pairing during replication or even induce changes in DNA composition between replications.

Random changes in DNA composition are unlikely to code for improvements in the functioning of the gene products (see Chapter 11), much as typing random words in the midst of a script of *Hamlet* will be unlikely to improve on Shakespeare. Therefore, cells have evolved mechanisms to monitor DNA and repair damaged regions (Fig. 10-10a). Nevertheless, mistakes do occur.

Types of Gene Mutations

POINT MUTATIONS. In a **point mutation,** a pair of bases becomes incorrectly matched (Fig. 10-10b). Repair enzymes recognize the mismatch, cut out one of the nucleotides and replace it with a complementary nucleotide. However, the enzymes often replace the wrong nucleotide. Thus, although there is a complementary pair of bases once again, it is not the same pair that the chromosome originally had.

INSERTIONS AND DELETIONS. As their names suggest, an **insertion** occurs when new nucleotide pairs are inserted in the midst of a gene (Fig. 10-10c), while a **deletion** occurs when nucleotide pairs are removed from a gene (Fig. 10-10d).

Effects of Mutations

Although most mutations are harmful or neutral, mutations are the prerequisite for evolution, because ultimately all genetic variation originates as these random changes in DNA sequence. Natural selection tests new alleles in the crucible of competition for survival and reproduction. Occasionally, a mutation proves beneficial in the organism's interactions with its environment. These alleles may spread throughout the population and become common, as their possessors outcompete rivals bearing the old allele.

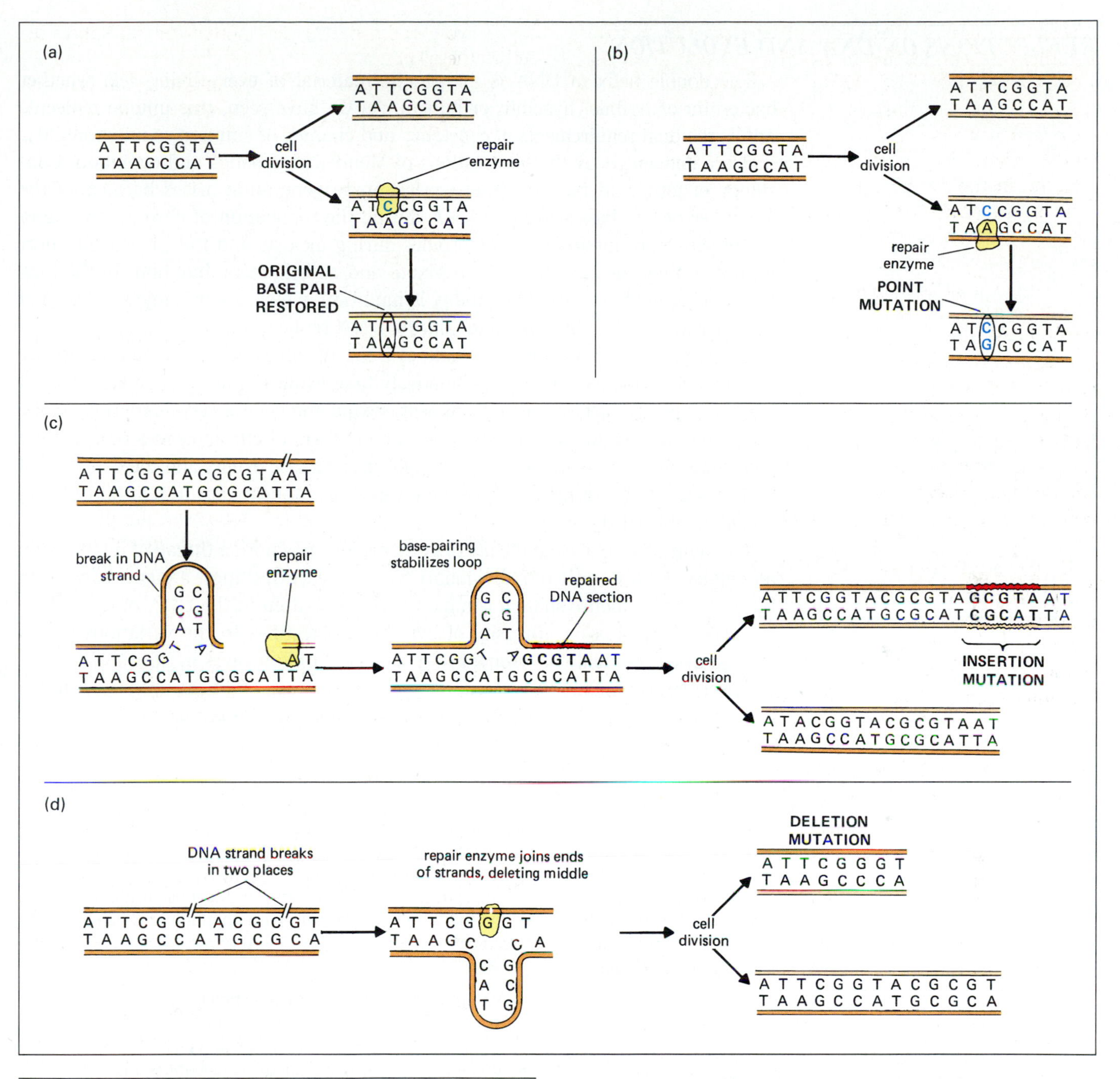

Figure 10-10 *Repair of damaged or mismatched regions of DNA sometimes results in restoration of the original nucleotide sequence and sometimes results in mutations. In this diagram, the sugar–phosphate backbone of DNA is represented by solid lines (dark tan for parental DNA, light tan for daughter DNA), and the bases are represented by the letters A, C, G, and T.*

(a) Correction of mismatched bases: *During DNA replication, a mistake in base pairing occurs, resulting in an incorrect A–C pair instead of the proper A–T pair. Repair enzymes in the daughter cell recognize the mismatch, excise the incorrect base in the daughter DNA, and substitute the correct complementary base (T). No mutation has occurred.*

(b) Point mutation: *With the same replication error as in (a), if the repair enzymes excise the mismatched base on the parental strand (A), the new base pair differs from the original.*

(c) Insertion: *Radiation or chemical mutagens cause a break in one of the DNA strands, and the freed DNA end loops out. If the free loop has the "right" sequences of bases, base pairing may stabilize the loop. Repair enzymes fill in the resulting gap with new nucleotides, making one DNA strand (with the loop) longer than the other (intact) strand. When DNA replication occurs, each strand is used as a template for synthesis of a new complementary strand. The long strand thus carries an insertion mutation to one of the daughter cells.*

(d) Deletion: *If a DNA strand breaks in two places, a repair enzyme may stitch together the remaining pieces without filling in the missing nucleotides. When the DNA is replicated, one daughter cell receives the deletion mutation.*

REFLECTIONS ON DNA AND EVOLUTION

The double helix of DNA is the genetic material of every living cell, whether bacterium or buffalo, hyacinth or human. As we have seen, this unique molecule fulfills the dual requirements of constancy and change, of similarity and variability, that are demanded by the observations of Mendelian genetics and evolution. Constancy of form and function from generation to generation arises because of the almost error-free base pairing of DNA during the replication of chromosomes and the precise transmission of chromosomes during meiosis and mitosis. In the short term, variability is due primarily to sexual and genetic recombination. In the long term, the ultimate source of variability is mutation, which occurs largely because of the occasional errors that are made during DNA replication.

Given what we know about the role of DNA in inheritance, it is not surprising that modern biologists almost unanimously take evolution as an inevitable fact of life. New alleles of genes constantly appear, sexual and genetic recombination shuffle alleles into new combinations, and even the numbers of chromosomes occasionally change through errors in meiosis. Through these chance events and the demands of a changing environment over time, new variations appear, are tested, and are retained or discarded.

The centrality of DNA in the life of every organism is both a thread of relationship among us and a wonder of individuality. At the heart of things, a giant redwood, a butterfly, and a mouse share a common heritage written in the DNA of their chromosomes. The amazing diversity of living forms testifies to the variations that a changing environment has wrought over the last 3 billion years in this most flexible of molecules. If humans remain willing to share the world with the myriad creatures around us, who can say what wonders distant generations may see?

SUMMARY OF KEY CONCEPTS

The Composition of Chromosomes

Eukaryotic chromosomes are composed of DNA and protein. DNA is the molecule that carries genetic information. DNA is composed of subunits called nucleotides, linked together into long strands. Each nucleotide consists of a phosphate group, the five-carbon sugar deoxyribose, and a nitrogen-containing base. Four different bases occur in DNA: adenine, guanine, thymine, and cytosine.

DNA: The Hereditary Molecule

Transformation in bacteria and infection of bacteria by bacteriophages provided the first evidence that DNA is the hereditary molecule. Griffith discovered that living bacteria can acquire genes from dead bacteria, transforming the genotype of the live bacteria. Avery and his colleagues showed that the genes taken up by living bacteria during transformation are composed of DNA.

A bacteriophage consists solely of DNA and protein. When it infects a bacterium, a phage injects its genetic material into the bacterium, where it directs the synthesis of more phage. Hershey and Chase demonstrated that phages inject DNA, but not protein, into bacteria. Therefore, DNA must be the genetic material of phages.

The Structure of DNA

The DNA of chromosomes is composed of two strands, wound about one another in a double helix. The sugars and phosphates that link one nucleotide to the next form the backbone on each side of the double helix, while the bases from each strand pair up in the middle of the helix. Only specific pairs of bases, called complementary base pairs, can link together in the helix, held by hydrogen bonds: adenine with thymine and guanine with cytosine.

DNA Replication: The Key to Constancy

When chromosomes are replicated prior to mitosis or meiosis, the two DNA strands separate. DNA polymerase enzymes move along each strand, linking up free nucleotides into new DNA strands. The sequence of nucleotides in each newly formed strand is complementary to the sequence on a parental strand. As a result, two double helices are synthesized, each consisting of one parental DNA strand plus one newly synthesized, complementary strand that is an exact copy of the other parental strand. The two daughter DNA molecules are therefore duplicates of the parental DNA molecule.

Mutations: The Key to Variability

A gene is a specific sequence of nucleotides in DNA. A mutation is a change in this sequence. Mutations occur from mistakes in base pairing during replication, chemical agents, and certain kinds of radiation. Although mutations are usually harmful, occasionally a mutation will promote better adaptation to the environment and will be favored by natural selection.

GLOSSARY

Bacteriophage (bak-tēr′-ē-ō-fāj) a virus that infects bacteria.

Base in molecular genetics, one of the nitrogen-containing, single- or double-ringed structures that distinguish one nucleotide from another. In DNA, the bases are adenine, guanine, cytosine, and thymine.

Base-pairing rule the rule that only complementary bases can pair during DNA replication (or RNA synthesis from DNA; see Chapter 11).

Complementary referring to a nucleotide that can pair with another nucleotide via hydrogen bonding; in DNA, adenine is complementary to thymine, and guanine is complementary to cytosine.

Deletion a mutation in which one or more nucleotides are removed from a gene.

DNA polymerase an enzyme that covalently bonds DNA nucleotides together into a continuous strand, using a preexisting DNA strand as a template. DNA polymerase catalyzes the duplication of the DNA of chromosomes during interphase prior to mitosis and meiosis.

Genetic recombination assembling a new combination of preexisting genes through crossing over.

Helix (hē′-licks) a corkscrew-shaped object, as if a wire were wrapped around a cylinder.

Insertion a mutation in which one or more nucleotides are inserted within a gene.

Mutation a change in the base sequence of DNA.

Nucleotide an individual subunit of which nucleic acids are composed. A single nucleotide consists of a phosphate group covalently bonded to a sugar (deoxyribose in DNA), which is in turn covalently bonded to a nitrogenous base (adenine, guanine, cytosine, or thymine). Nucleotides are linked together by covalent bonds between the phosphate of one and the sugar of the next to form a strand of nucleic acid.

Purine a nitrogen-containing nucleic acid base consisting of fused six- and five-sided rings. The common purines in nucleic acids are adenine and guanine.

Pyrimidine a nitrogen-containing nucleic acid base consisting of a single six-sided ring. The common pyrimidines in DNA are cytosine and thymine.

Semiconservative replication the process of replication of the DNA double helix; the two DNA strands separate, and each is used as a template for the synthesis of a complementary DNA strand. Each daughter double helix therefore consists of one parental strand and one new strand.

Sexual recombination the formation of new combinations of alleles due to inheritance of chromosomes from two different parental organisms during sexual reproduction.

Transformation a method of genetic recombination whereby DNA from one bacterium (usually released after the death of the bacterium) becomes incorporated into the DNA of another, living, bacterium.

STUDY QUESTIONS

1. Draw the structure of a nucleotide. Which parts are identical in all nucleotides, and which can vary?
2. Name the four types of bases found in DNA.
3. What is transformation? In Griffith's experiments, why were the transformed bacteria able to synthesize capsules?
4. Diagram the life cycle of a bacteriophage.
5. Diagram the Hershey–Chase experiment. Suppose they had found that *both* radioactively labeled DNA and radioactively labeled protein were injected into the bacteria. What would they have been able to conclude from this? (Note that if the bacteriophage chromosome was like the eukaryotic chromosome, that is, if it contained both DNA and protein, this is exactly what would have happened.)
6. Describe the structure of DNA in a chromosome. Where are the bases, sugars, and phosphates in the structure?
7. Which bases are complementary to one another? How are they held together in the double helix of DNA?
8. Describe the process of DNA replication.
9. Define mutation, and give one example of how a mutation might occur. Would you expect most mutations to be beneficial or harmful? Explain your answer.

SUGGESTED READINGS

Felsenfeld, G. "DNA," *Scientific American*. October 1985. The structure of DNA and how its use is regulated in a cell.

Kimura, M. "The Neutral Theory of Molecular Evolution." *Scientific American,* November 1979 (Offprint No. 1451). Kimura persuasively argues that some evolutionary changes can occur without selection.

Mirsky, A. E. "The Discovery of DNA." *Scientific American,* June 1968 (Offprint No. 1109). The early history of DNA.

Olby, R. *The Path to the Double Helix*. Seattle, Wash.: University of Washington Press, 1975. A historical perspective on the development of genetics in this century.

Watson, J. D. *The Double Helix*. New York: Atheneum Publishers, 1968. If you still believe the Hollywood images, that scientists are either maniacs or cold-blooded, logical machines, be sure to read this book. Although scarcely models for the behavior of future scientists, Watson and Crick are certainly human enough!

Wills, C. "Genetic Load." *Scientific American,* March 1970. Although many mutations are harmful, the reservoir of mutations in each species also provides the potential for future evolution.

11
From Genotype to Phenotype: Gene Expression and Regulation

The genotype of an organism, interacting with its environment, generates its actual structure, functioning, and behavior—its phenotype. When Watson and Crick solved the puzzle of the double helix, they provided the framework for understanding the organization of the genotype. But two major questions remained. First, how does a cell use its genotype to create a phenotype? You may recall from previous chapters that most of the organic molecules of a cell are proteins or are synthesized through the actions of protein enzymes. Therefore, to go from genotype to phenotype means synthesizing the appropriate proteins. How does DNA direct protein synthesis and function in a cell? We will see that *the sequence of nucleotides in DNA is a code that is translated into the sequence of amino acids in proteins*. Genes are, in fact, a library containing the information needed to construct all the proteins of a cell.

Second, how does an organism use its genotype in the appropriate manner during its development and subsequent interactions with its environment? For example, if you start out life as a fertilized egg, and if all the cells of your body are derived from that egg through mitosis, it follows that all the cells of your body contain the same genes. How, then, can different cells of the body have different structures and functions? Why do hair follicle cells synthesize hair proteins and not hormones? Further, does the environment influence how the genotype is used? If so, how? The answer is that *cells can regulate which genes are used and which are not, depending on the function of the cell and the environment in which it lives.*

In this chapter we examine how genotypes are transformed into phenotypes. With this knowledge, we can return to the observations of Mendelian genetics and understand what mutations really are, why genes have different alleles, and why some alleles are dominant and some are recessive. Finally, we take a brief glimpse into the world of modern molecular genetics, in which geneticists seek to alter genotypes, even trading genes between organisms as distantly related as bacteria and people, to create phenotypes never before seen on Earth.

CELL FUNCTION, PROTEIN, AND GENES

Long before biologists knew anything about the molecular structure of genes, they tried to find out how genes work. Early in the twentieth century, an English physician, Archibald Garrod, studied the inheritance of human metabolic disorders. Garrod knew that the human body is a chemical cauldron, churning with reactions that convert molecules of food into other molecules the body needs. These biochemical conversions proceed stepwise, with each step catalyzed by a specific enzyme. If one of these enzymes is defective, the affected person lacks the end product, while precursor molecules or the products of other metabolic pathways accumulate. This can often have severe medical consequences. In phenylketonuria (PKU), for example, the enzyme that converts the amino

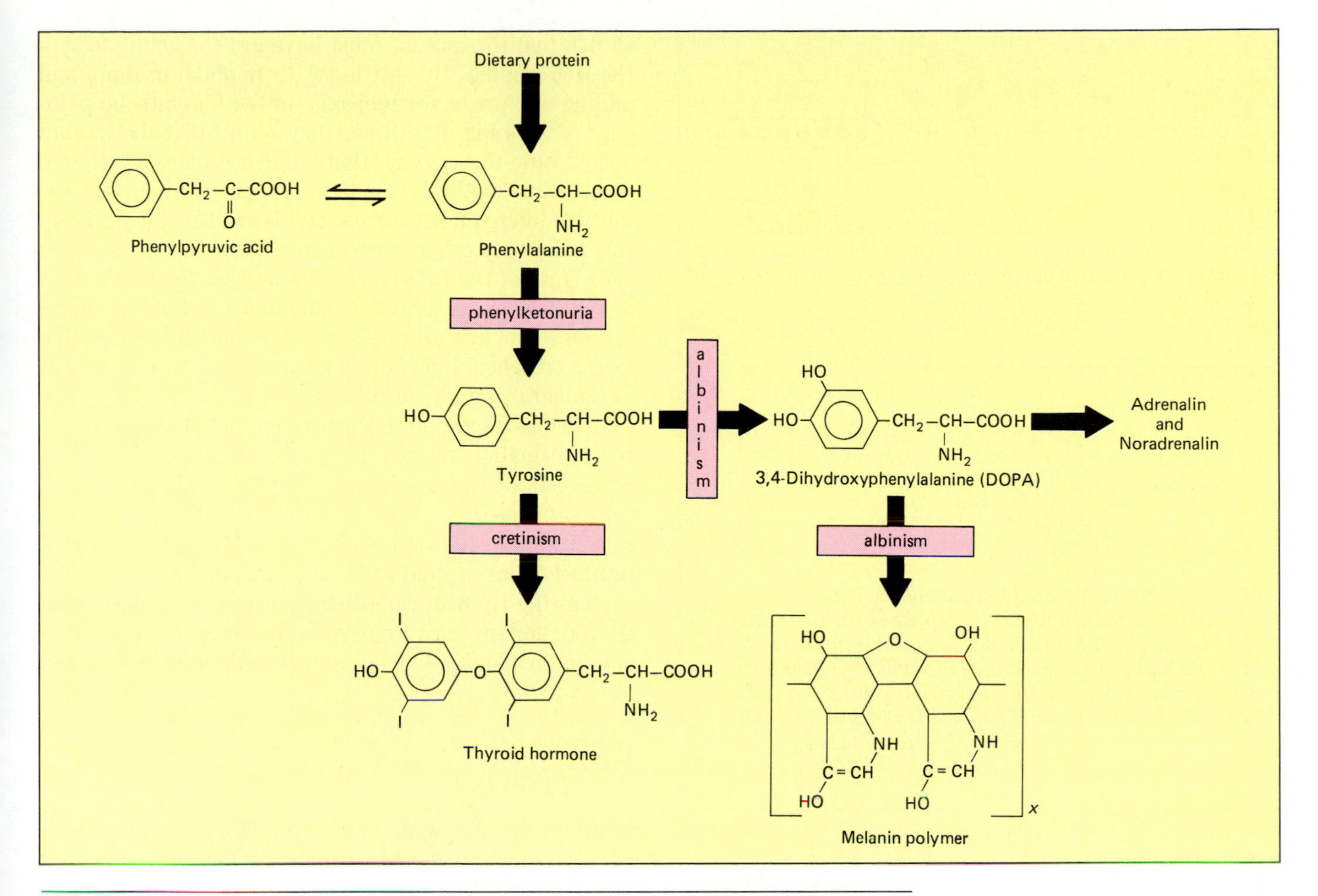

Figure 11-1 Some of the pathways of phenylalanine and tyrosine metabolism in humans. If the enzyme that converts phenylalanine to tyrosine is defective, phenylalanine and phenylpyruvic acid build up, damaging developing brain cells. Tyrosine is a precursor for several other essential compounds, including thyroid hormone, the nervous system chemicals adrenalin and noradrenalin, and the pigment melanin.

acid phenylalanine into tyrosine is defective. Therefore, phenylalanine obtained from dietary proteins accumulates in the body (Fig. 11-1). In infants, high phenylalanine levels injure developing brain cells, causing severe mental retardation. Garrod catalogued a number of these "inborn errors of metabolism," and deduced, from their patterns of inheritance, that each defective enzyme was probably due to a rare allele of a single gene. Of course, the proper genetic experiments to confirm this hypothesis could not be performed with humans.

The One-Gene, One-Protein Hypothesis

The common red bread mold, *Neurospora crassa*, proved to be the perfect organism for studying the relationship between genes and enzymes. Although we commonly see it on stale bread, *Neurospora* is an extremely independent organism that can synthesize almost all the organic compounds it needs. It can grow on a minimal medium containing an energy source such as sucrose, a few minerals to supply essential elements such as nitrogen and phosphorus, and a single vitamin, biotin.

For most of its life cycle, *Neurospora* is haploid, with just one copy of each chromosome. Therefore, mutations in *Neurospora* always show up phenotypically, since they cannot be masked by a normal allele on a homologous chromosome, as they can be in diploid organisms. (A similar phenomenon occurs in male mammals with respect to genes carried on the X chromosome. With no corresponding gene on the Y chromosome, a male must show the phenotype of whatever genes he carries on his X chromosome.)

Geneticists George Beadle and Edward Tatum bombarded *Neurospora* with X-rays, producing mutant molds that lost the ability to synthesize some substances (Fig. 11-2). One particularly interesting mutant could not grow on minimal medium but could grow if the amino acid arginine was added. Beadle and Tatum therefore con-

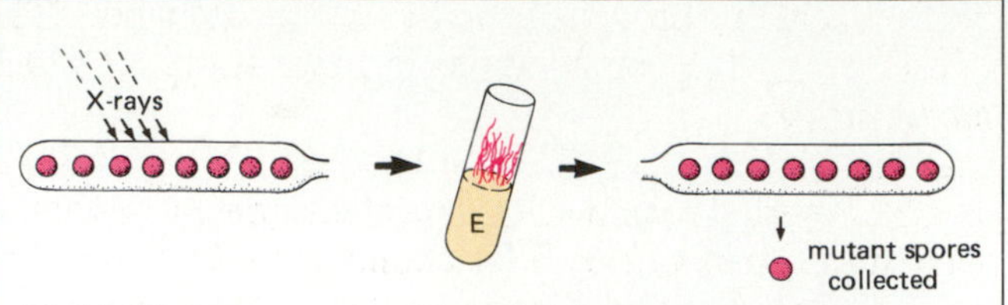

(a) *Mold spores are irradiated with X-rays to induce mutations. They are germinated and grown on an enriched medium. When these molds reproduce, their spores are collected.*

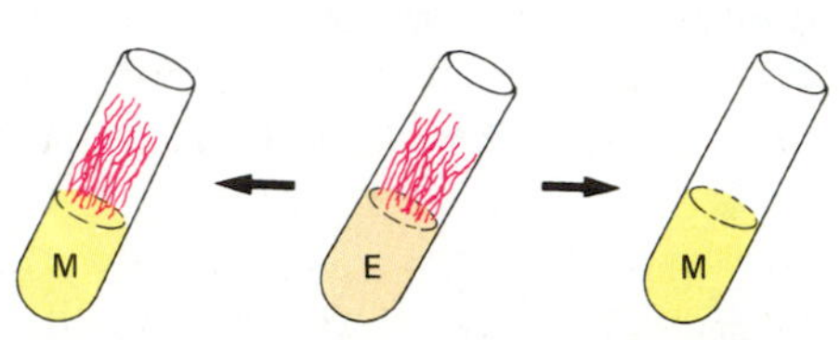

(b) *Spores are germinated individually in enriched medium containing all the amino acids. A piece of each resulting mold is placed on minimal medium lacking amino acids. If the mold still grows on minimal medium (left) then no mutation has occurred, and this mold is discarded. If the mold cannot grow on minimal medium (right), a mutation has occurred and the experiments are continued.*

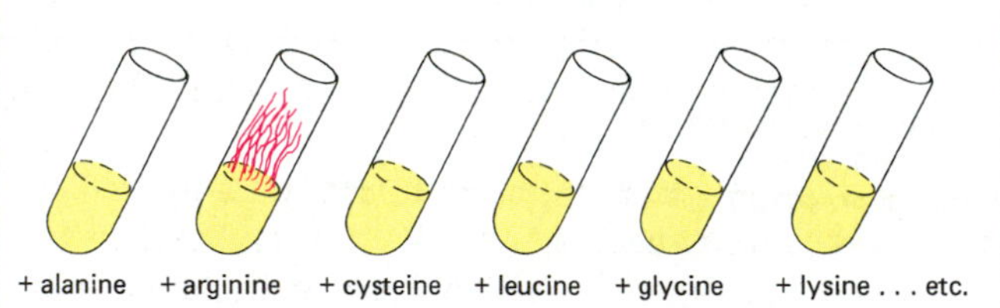

(c) *Pieces of mutated mold are placed in tubes containing minimal medium plus one amino acid. The mutant can grow only if arginine is added.*

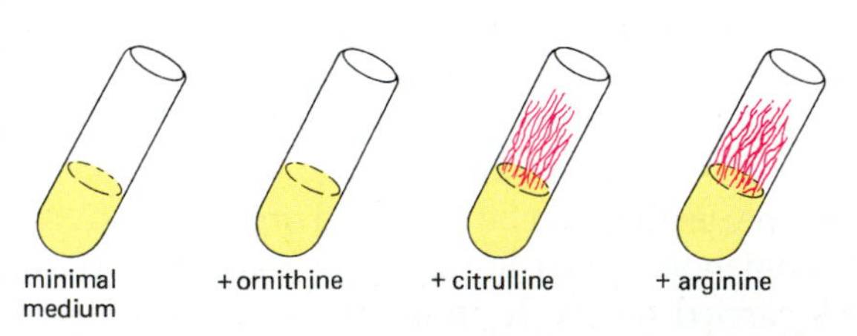

(d) *Mutants are tested in media to which precursors in the biosynthetic pathway of arginine have been added. Since the mold can grow if supplied with citrulline or arginine, but not ornithine, the mutation must have ruined a single enzyme, the one that normally catalyzes the conversion of ornithine to citrulline.*

Figure 11-2 *Beadle and Tatum's experiments with* Neurospora, *showing that a mutation may produce a single defective enzyme.*

cluded that the mutant must have lost the ability to synthesize arginine. By starting with minimal medium and adding one precursor molecule in the biosynthetic pathway of arginine at a time, they found that the mutant lacked only the enzyme that catalyzed one specific step in arginine synthesis. Genetic analysis showed that the mutant differed from normal molds in a single gene. From this and many other experiments, they concluded that *a gene encodes the information needed for the synthesis of a specific enzyme.* This became known as the "one-gene, one-enzyme" hypothesis. This is the same conclusion that Garrod reached, but could not prove, based on his studies of human metabolic disorders.

Geneticists have since learned that not all genes encode the information needed by a cell to produce enzymes. Some genes carry information for the synthesis of structural proteins such as collagen in skin and keratin in hair, or hormones such as insulin. For a few genes, the final product is not protein at all, but ribonucleic acid (RNA), such as the RNA of ribosomes. Nevertheless, most genes do code for proteins or parts of proteins. As a generalization, then, *each gene encodes the information for a single protein.*

FROM DNA TO PROTEIN

Since we now know that genes are DNA, we can reinterpret the experiments of Beadle and Tatum: *a gene is a segment of DNA that contains the information needed to synthesize a protein.* What makes one protein differ from another is its sequence of amino acids. *Therefore, the sequence of nucleotides in DNA must encode the sequence of amino acids in a protein.* How does a cell translate the message of its DNA into proteins?

As you know, the DNA of eukaryotic cells is located in the nucleus. Protein synthesis, on the other hand, occurs on ribosomes in the cytoplasm. Therefore, DNA cannot directly guide protein synthesis. *There must be an intermediate molecule that carries the information from DNA in the nucleus to the ribosomes in the cytoplasm. This molecule is* **RNA.** RNA is similar to DNA, but differs in three respects: (1) RNA is usually single stranded; (2) it has a different type of sugar in its backbone, ribose instead of deoxyribose; and (3) thymine in DNA is replaced by uracil in RNA:

DNA nucleotide

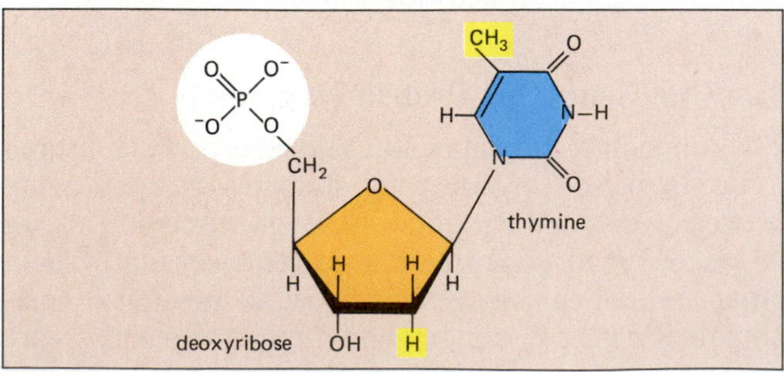

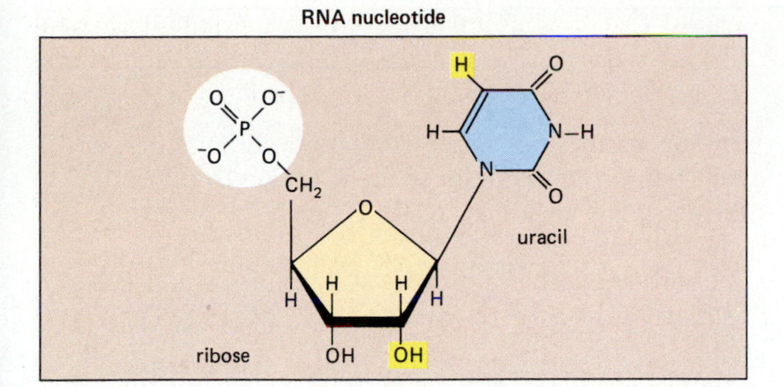

Information flows from DNA to proteins in a two-step process (Fig. 11-3). *In the first step, the information contained in the DNA of a specific gene is copied into RNA, a processed called* **transcription.** There are three types of RNA in a cell, all of which are involved in translating genetic information in DNA to the amino acid sequence of proteins. We will examine these RNAs in more detail shortly; for now, you just need to know that one particular type of RNA, appropriately called **messenger RNA,** is transcribed from DNA in the nucleus and travels to the cytoplasm, carrying information about the sequence of amino acids in the protein to be manufactured. *In the second step, the other two types of RNA,* **transfer RNA** *and* **ribosomal RNA,** *translate the information of messenger RNA into the right amino acids and help to synthesize the protein.*

To understand protein synthesis, geneticists first had to break the language barrier: how does the language of nucleotides in DNA translate into the language of amino acids in proteins?

The Genetic Code

We have used the word "code" several times to refer to the information stored in DNA and ultimately translated into the amino acid sequence of proteins. This **genetic code** is conceptually similar to Morse code: a set of symbols (bases in nucleic acids or dots and dashes in Morse code) that can be translated into another set of symbols (amino acids in proteins or letters of the alphabet). The question is: what combinations of bases stand for which amino acids?

Since there are four different bases in DNA and RNA, and 20 different amino acids in proteins, the bases cannot serve as a one-to-one code for amino acids: there are simply not enough of them. Perhaps, just as Morse code uses a short sequence of dots and dashes to encode the letters of the alphabet, the genetic code might use a short sequence of bases to encode each amino acid. If a sequence of two bases codes for an amino acid, there would be $4^2 = 16$ possible combinations of bases. This isn't enough either. Three bases per amino acid, however, gives $4^3 = 64$ possible combinations, which is more than enough. Under the assumption that nature operates as economically as possible, biologists hypothesized that the genetic code must be **triplet:** *three bases specify one amino acid.* In 1961, Francis Crick and three coworkers demonstrated that this hypothesis is correct (see the essay, "Cracking the Genetic Code").

THE NATURE OF THE CODE. A few requirements must be met for any language to be understood. The users must know what the words mean; where words start and stop; and where sentences start and stop. Shortly after the Crick experiments demonstrated that the "words" of the genetic code are all three nucleotides long, researchers began to decipher the code. They ground up bacteria and isolated the components needed to synthesize proteins. Then, by synthesizing artificial messenger RNA and adding it to their protein-synthesizing mixtures, they could see which amino acids were incorporated into the resulting proteins. For example, an RNA composed entirely of uracil (UUUUUUUUUUU . . .) directed the mixture to synthesize a protein composed solely of phenylalanine. Therefore, the triplet specifying phenylalanine must be UUU. Since the genetic code was deciphered using these artificial RNA's, *the code is usually written as the triplets of messenger RNA that code for each amino acid* (Table 11-1). These messenger RNA triplets are called **codons**.

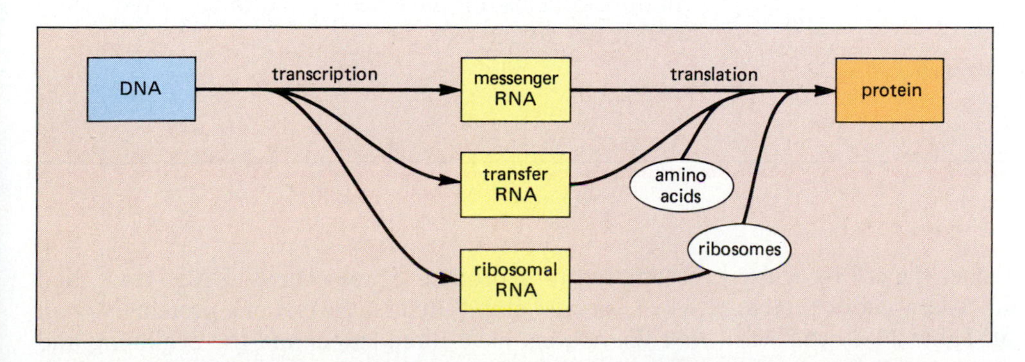

Figure 11-3 A simplified diagram of information flow from DNA to RNA to protein.

Cracking the Genetic Code

The hypothesis that three nucleotides in DNA code for one amino acid in protein is an attractive and logical possibility: fewer than three nucleotides cannot unambiguously code for all 20 amino acids, and more than three are superfluous. But how could you go about proving that nature really uses a triplet code?

As with the Hershey–Chase experiments (see Chapter 10), the simplicity of the bacteriophage proved invaluable in deciphering a fundamental principle of genetics. Francis Crick and his coworkers exposed bacteriophages to a dye called acridine, which causes insertion mutations: one, two, or three nucleotides were inserted into the DNA molecule at random places within a particular gene. They found that inserting one or two nucleotides into the DNA causes the synthesis of defective enzymes that prevent the phages from reproducing in their host bacteria. Inserting three nucleotides, however, sometimes produces phages that synthesize normal or nearly normal enzymes.

Crick concluded that during RNA synthesis the DNA of a gene is "read" in a linear order, starting at the beginning. Each set of three nucleotides comprises a "word"; that is, three nucleotides in DNA encode a single amino acid. The code must specify the beginning and end of a gene, but within a gene there are no spaces or punctuation between words. How did he arrive at these conclusions?

To understand Crick's reasoning, let's suppose that we have a gene with this repetitive sequence of nucleotides:

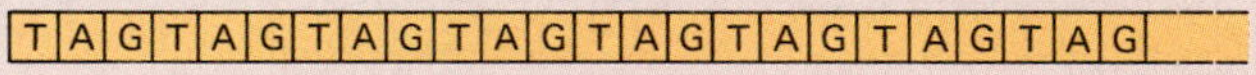

T A G T A G T A G T A G T A G T A G T A G T A G

If DNA always has three-letter words, this nucleotide sequence spells out the English word, "tag," over and over again. You don't need spaces to separate the words if they all have three and only three letters.

Suppose that the acridine dye inserts another thymine somewhere near the beginning of the gene. The gene now reads:

T A G T A T G T A G T A G T A G T A G T A G T A

first insertion

incorrect "reading"

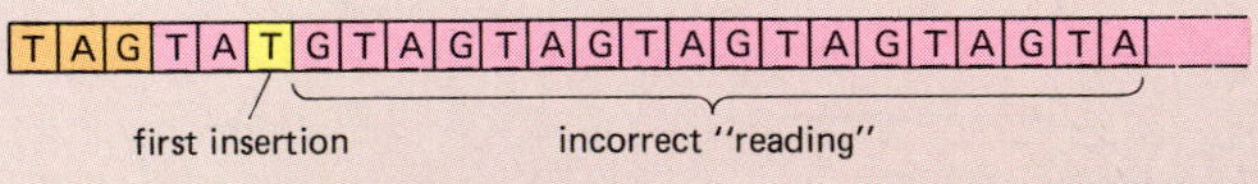

Protein synthesis is like a computer: garbage in, garbage out. From the point of the insertion to the end of the gene, the gene now reads GTA, GTA, GTA . . . , which is nonsense.

Inserting another T nearby still results in nonsense, as most of the gene now reads AGT, AGT, AGT:

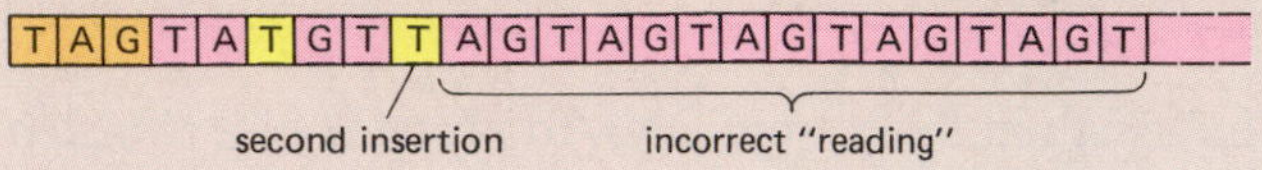

T A G T A T G T T A G T A G T A G T A G T A G T

second insertion

incorrect "reading"

A third insertion, however, results in most of the gene reading TAG, TAG, TAG again:

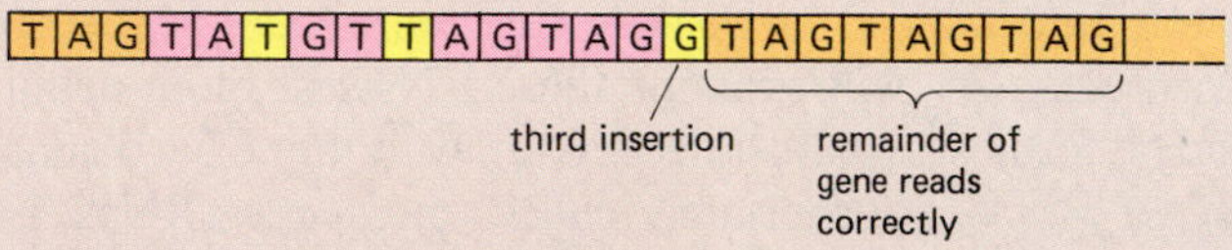

T A G T A T G T T A G T A G G T A G T A G T A G

third insertion

remainder of gene reads correctly

How does this explain Crick's results? One insertion near the beginning of the bacteriophage gene causes the rest of the triplets to encode the wrong amino acids, so naturally the enzyme specified by that gene completely malfunctions. The second insertion still leaves most of the triplets calling for the wrong amino acids. If there are three insertions near the start of the gene, however, then the first few amino acids would be wrong, but all the triplets beyond the third insertion would be correct once again. Most of the enzyme would therefore be synthesized correctly. If the incorrect parts of the enzyme are relatively unimportant to its overall function, it may work well enough to allow the phages to reproduce.

Only a triplet code can account for the fact that three insertions, not one or two or four, restore near-normal enzyme function. Further, only a code without punctuation or spaces between words will be confused at all by any of the insertions. Finally, a code without punctuation between words can work only if the start and stop points of protein synthesis are clearly marked. All three conclusions have been found to be correct, as we describe in the text.

What about punctuation? How does the cell recognize where codons start and stop and where entire protein codes start and stop? Looking at Table 11-1, you will notice that the codon AUG signals "start," that is, the beginning of a protein. Three codons, UAG, UAA, and UGA, signal "stop," that is, the end of a protein. Now, if all the codons have three letters and the beginning and end of a protein are specified, punctuation between co-

Table 11-1 **The Genetic Code (Codons of mRNA)**

First Base	Second Base: G		A		C		U		Third Base
G	GGG	Glycine	GAG	Glutamic Acid	GCG	Alanine	GUG	Valine	G
	GGA	Glycine	GAA	Glutamic Acid	GCA	Alanine	GUA	Valine	A
	GGC	Glycine	GAC	Aspartic Acid	GCC	Alanine	GUC	Valine	C
	GGU	Glycine	GAU	Aspartic Acid	GCU	Alanine	GUU	Valine	U
A	AGG	Arginine	AAG	Lysine	ACG	Threonine	AUG	Start (Methionine)	G
	AGA	Arginine	AAA	Lysine	ACA	Threonine	AUA	Isoleucine	A
	AGC	Serine	AAC	Asparagine	ACC	Threonine	AUC	Isoleucine	C
	AGU	Serine	AAU	Asparagine	ACU	Threonine	AUU	Isoleucine	U
C	CGG	Arginine	CAG	Glutamine	CCG	Proline	CUG	Leucine	G
	CGA	Arginine	CAA	Glutamine	CCA	Proline	CUA	Leucine	A
	CGC	Arginine	CAC	Histidine	CCC	Proline	CUC	Leucine	C
	CGU	Arginine	CAU	Histidine	CCU	Proline	CUU	Leucine	U
U	UGG	Tryptophan	UAG	Stop	UCG	Serine	UUG	Leucine	G
	UGA	Stop	UAA	Stop	UCA	Serine	UUA	Leucine	A
	UGC	Cysteine	UAC	Tyrosine	UCC	Serine	UUC	Phenylalanine	C
	UGU	Cysteine	UAU	Tyrosine	UCU	Serine	UUU	Phenylalanine	U

dons is unnecessary. To see why this is so, consider what would happen if English used only three-letter words: a sentence such as THEMANSAWTHECAT would be perfectly understandable, even without spaces between the words, as long as the reader knew where the sentence started and stopped. The genetic code doesn't need, and doesn't have, punctuation between codons.

Finally, note that there are 60 codons in addition to the start and stop codons, and only 20 amino acids to code for. All 60 codons are used in the genetic code. Therefore, the genetic code is highly redundant or **degenerate**; that is, *several codons may specify the same amino acid*. For example, six different codons all code for arginine (Table 11-1).

RNA: Intermediary in Protein Synthesis

Protein synthesis requires the cooperation of three types of RNA: messenger RNA, transfer RNA, and ribosomal RNA. How are these RNA molecules synthesized, how do they differ, and what are their functions in protein synthesis?

RNA SYNTHESIS. RNA uses nearly the same "nucleotide language" as DNA. RNA is synthesized using molecules of DNA as a template, in a process called **transcription**, meaning "to copy over" (Fig. 11-4). In many respects, RNA synthesis is similar to DNA synthesis. First, an enzyme called **RNA polymerase** binds to DNA and the double helix of DNA unwinds. RNA polymerase travels along one strand of the unwound DNA, and, using ribose nucleotides present within the nucleus, synthesizes an RNA strand that is complementary to the DNA. The same base-pairing rules are used for RNA synthesis as for DNA replication, except that thymine in DNA is replaced by uracil in RNA. The DNA-to-RNA base-pairing rules thus become:

Base in DNA	Complementary Base in RNA
adenine	uracil
cytosine	guanine
guanine	cytosine
thymine	adenine

In other respects, RNA synthesis is quite different from DNA replication. First, the RNA strand does not remain paired with the DNA template; as it is synthesized, it comes free from the DNA. The DNA then either rewinds into its double helix, or a new RNA polymerase molecule may bind to the DNA and synthesize another RNA strand from the same gene. Second, RNA polymerase does not copy all the DNA of an entire chromosome. A short sequence of DNA nucleotides called the **promoter** marks the origin of a gene. RNA polymerase recognizes the promoter sequence as the starting place and transcribes the DNA until it reaches a different DNA sequence at the end of the gene that signals it to stop. Third, RNA polymerase copies only one strand of the double helix within the gene. Finally, a cell only uses a minute fraction of its genes at any given moment, so RNA polymerase must somehow find the right genes to transcribe. In the second part of

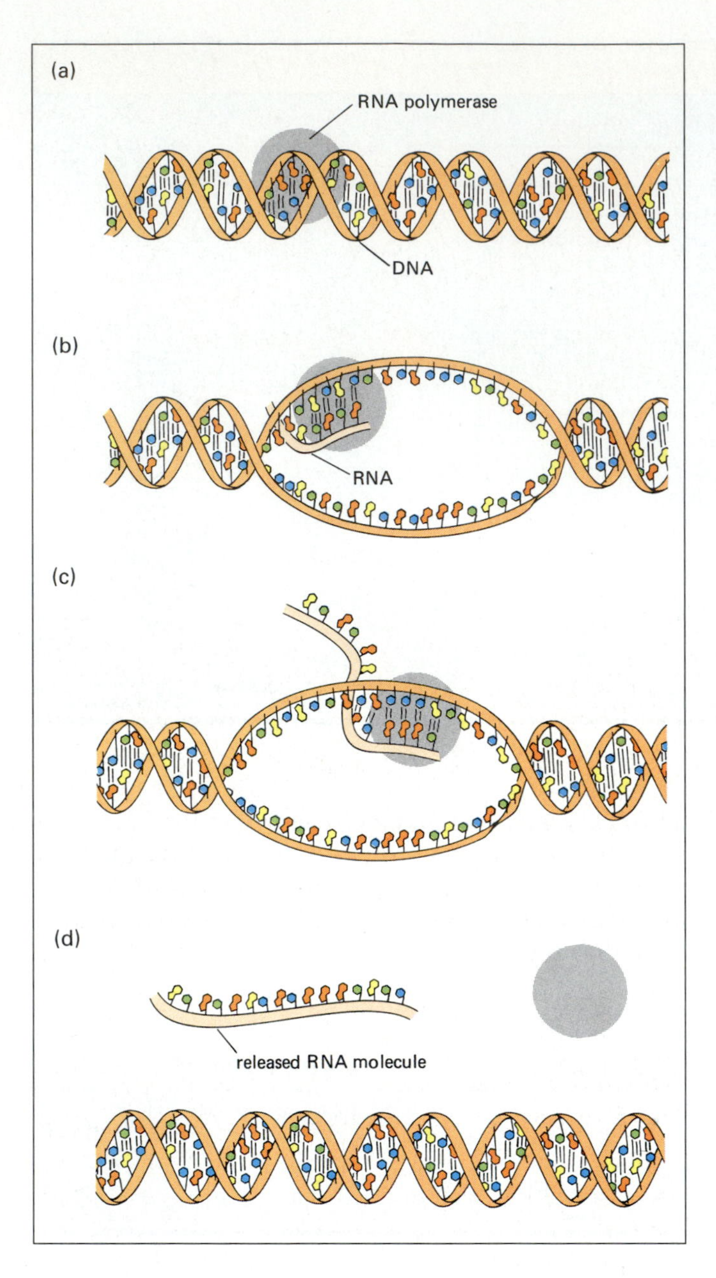

Figure 11-4 RNA transcription.
(a) *The enzyme RNA polymerase binds to the promoter region of DNA near the beginning of a gene.*
(b) *The DNA double helix unwinds. The RNA polymerase travels along one of the DNA strands, catalyzing the formation of a continuous strand of RNA from free RNA nucleotides. The bases incorporated into the growing RNA strand are complementary to the bases in the DNA strand.*
(c) *The RNA polymerase continues to the end of the gene.*
(d) *At the end of the gene, the RNA polymerase leaves the DNA. The DNA rewinds and the RNA molecule is released.*

this chapter we describe some of the mechanisms that regulate which genes are transcribed at any given time.

Types of RNA. **Messenger RNA (mRNA)** is a long, single-stranded molecule that carries the code for the amino acid sequences of proteins. Messenger RNA is transcribed from the DNA of a gene and leaves the nucleus through the pores in the nuclear envelope, carrying its message of nucleotide sequences. In the cytoplasm, mRNA binds to ribosomes, where the codons of mRNA are translated into the language of amino acids in proteins. (You might think of mRNA as a "molecular photocopy" of the gene DNA. The gene itself remains safely stored in the nucleus, like a valuable document in a library, while copies are sent to the cytoplasm to be used in protein synthesis.)

The other two types of RNA do not carry information to be translated into protein. Instead, these RNA molecules are the final products of certain genes, and thus are an exception to the generalization that genes code for proteins. **Ribosomal RNA (rRNA)** combines with specific proteins to form ribosomes (Fig. 11-5). Each ribosome is composed of two subunits of unequal size. The small subunit recognizes and binds messenger RNA. The large subunit contains an enzymatic region that catalyzes the addition of amino acids to the growing protein chain, and

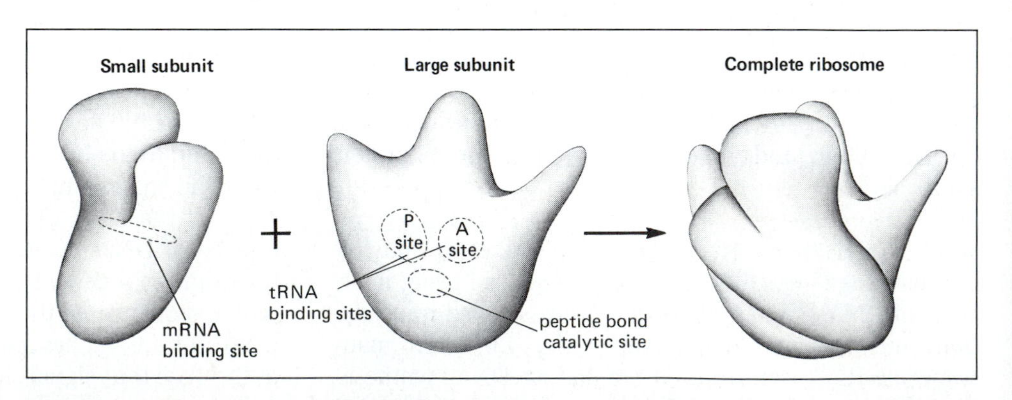

Figure 11-5 A ribosome has two subunits, each composed of protein and rRNA. The small subunit binds messenger RNA. The large subunit has three functional sites. Two, called the P and A sites, bind transfer RNA, and the third catalyzes the formation of the peptide bond between amino acids of the growing protein.

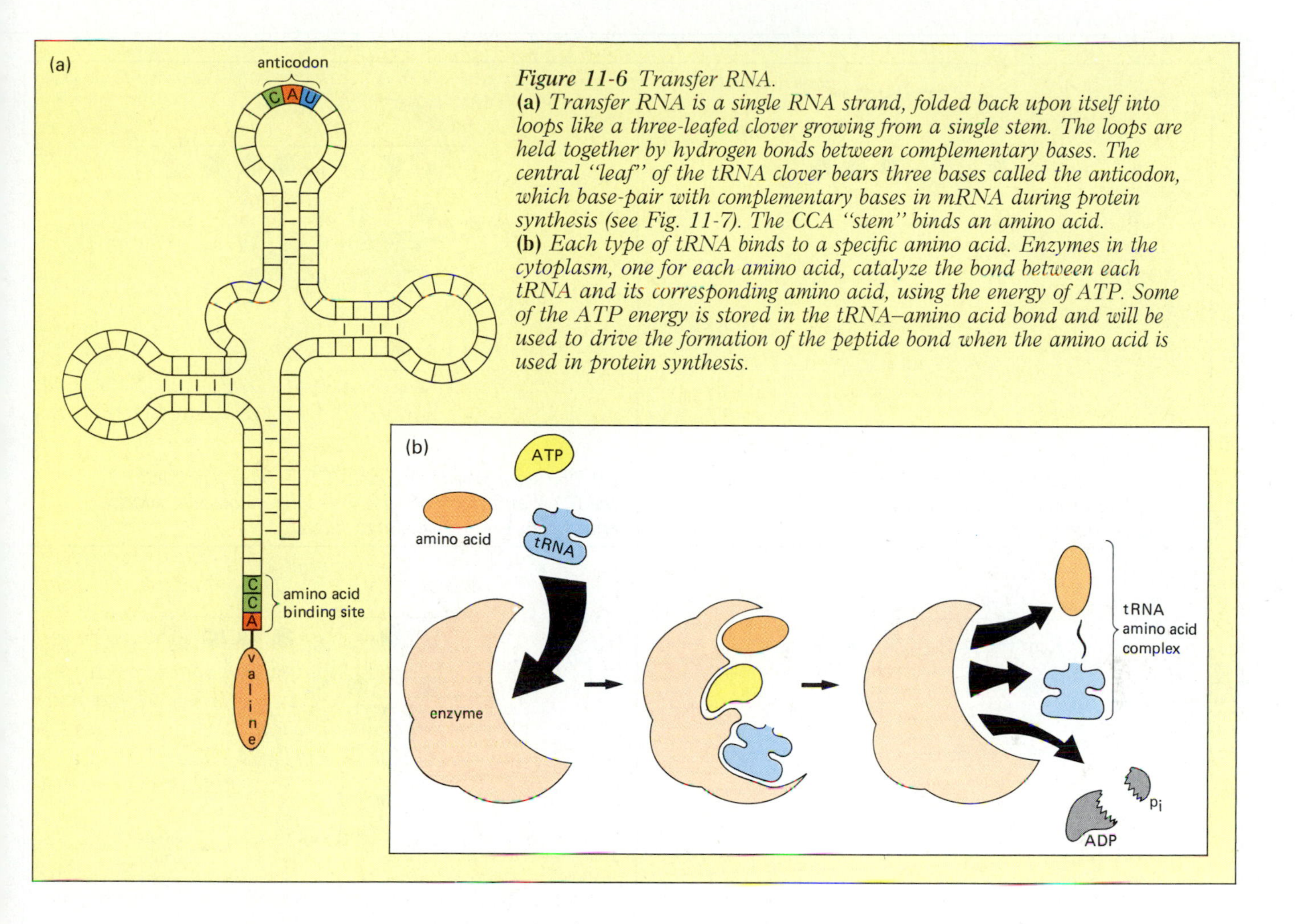

Figure 11-6 Transfer RNA.
(a) *Transfer RNA is a single RNA strand, folded back upon itself into loops like a three-leafed clover growing from a single stem. The loops are held together by hydrogen bonds between complementary bases. The central "leaf" of the tRNA clover bears three bases called the anticodon, which base-pair with complementary bases in mRNA during protein synthesis (see Fig. 11-7). The CCA "stem" binds an amino acid.*
(b) *Each type of tRNA binds to a specific amino acid. Enzymes in the cytoplasm, one for each amino acid, catalyze the bond between each tRNA and its corresponding amino acid, using the energy of ATP. Some of the ATP energy is stored in the tRNA–amino acid bond and will be used to drive the formation of the peptide bond when the amino acid is used in protein synthesis.*

two sites (usually designated P and A) that bind molecules of the third type of RNA, **transfer RNA (tRNA).**

Transfer RNA molecules bind amino acids and deliver them to the ribosome, where they are incorporated into protein chains. There are many different types of transfer RNAs, at least one type for each amino acid (Fig. 11-6). Transfer RNAs are like "code books," the only molecules in the cell that can decipher the codons of messenger RNA and translate them into the amino acids of proteins. Transfer RNAs are complex molecules, twisted about into a shape something like a three-leaf clover with a stem (Fig. 11-6a). For our purposes, the stem and the central leaf are the important parts. Enzymes in the cytoplasm recognize each specific tRNA molecule and attach the correct amino acid to the stem (Fig. 11-6b). The energy of ATP is used to form the tRNA–amino acid bond. Some of the ATP energy is stored in the tRNA–amino acid bond, and this "extra" energy will be used to forge the peptide bond when the amino acid is added to a growing protein molecule. The outside bend of the central tRNA leaf bears three exposed bases, called the **anticodon,** that actually decipher the mRNA code: *the anticodon of each tRNA is complementary to the codon of mRNA that specifies the amino acid attached to that tRNA.*

Protein Synthesis

Now that we have introduced all the actors involved in protein synthesis, let's look at the actual events. Conceptually, protein synthesis occurs in three stages. (1) *Messenger RNA is transcribed from the DNA template of the genes in the nucleus* (Fig. 11-4). The mRNA travels to a ribosome in the cytoplasm. (2) *Meanwhile, amino acids link up with their appropriate transfer RNAs* (Fig. 11-6). (3) *Finally, mRNA and tRNAs bind to a ribosome, where the codons of mRNA are translated into the amino acid sequence of a protein.* We have already discussed the first two stages; here we examine the third stage, **translation** (Fig. 11-7).

Translation has three steps: initiation of protein synthesis, elongation of the protein chain, and termination.

Initiation. The small subunit of a ribosome binds to two codons of mRNA (Fig. 11-7a). The first codon is always the "start codon" AUG. The second codon codes for the next amino acid in the protein. The tRNA bearing the complementary "start anticodon" UAC hydrogen bonds to the start codon (Fig. 11-7b). The large ribosomal subunit then attaches to the small subunit. As it does so, the start tRNA simultaneously binds to the P site on the

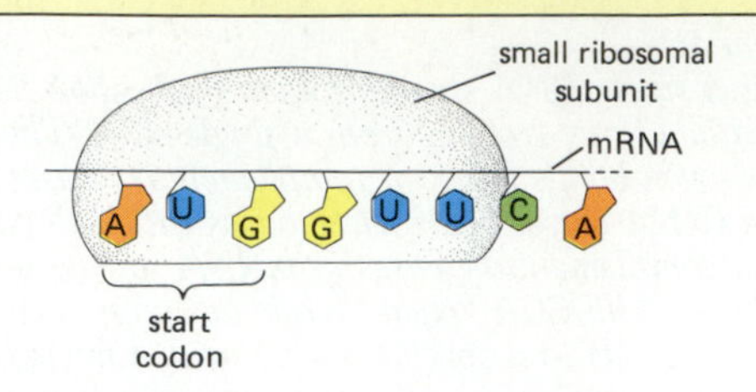

(a) *The small subunit of a ribosome binds to two codons of mRNA. The first codon bound is always the start codon.*

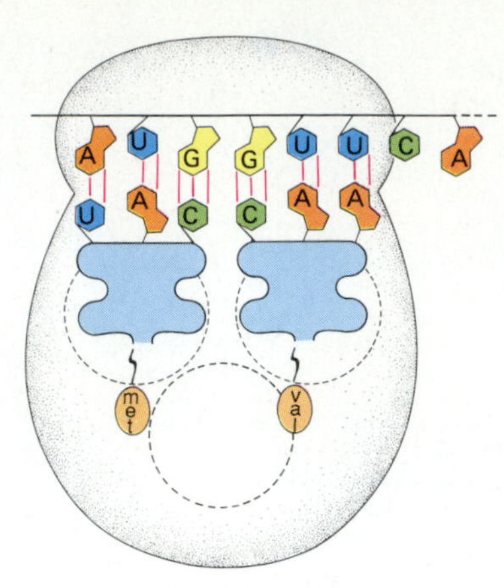

(d) *The second mRNA codon (GUU) base-pairs with the CAA anticodon of a tRNA–valine molecule, which enters the A site of the large subunit.*

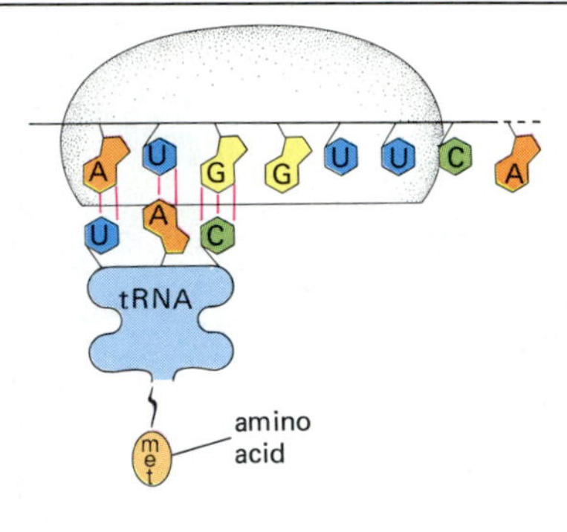

(b) *A tRNA–methionine molecule bearing the anticodon UAC base-pairs with the start codon of the mRNA.*

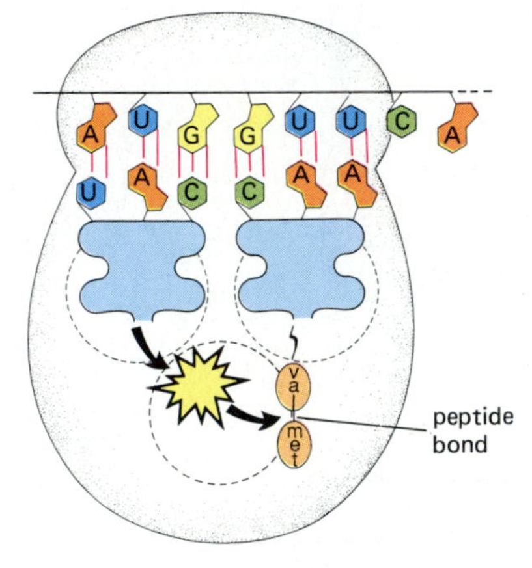

(e) *The catalytic site on the large subunit catalyzes the formation of a peptide bond between the amino acids methionine and valine, using the energy stored in the tRNA–methionine bond. The dipeptide remains attached to the second tRNA.*

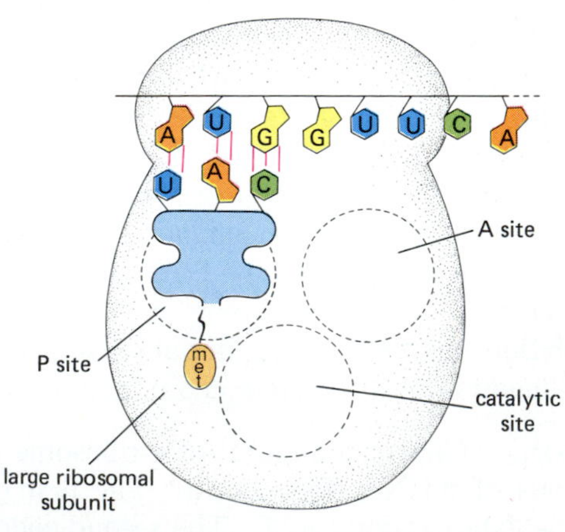

(c) *The large ribosomal subunit joins with the small subunit. The start tRNA–methionine complex binds to the P site of the large subunit.*

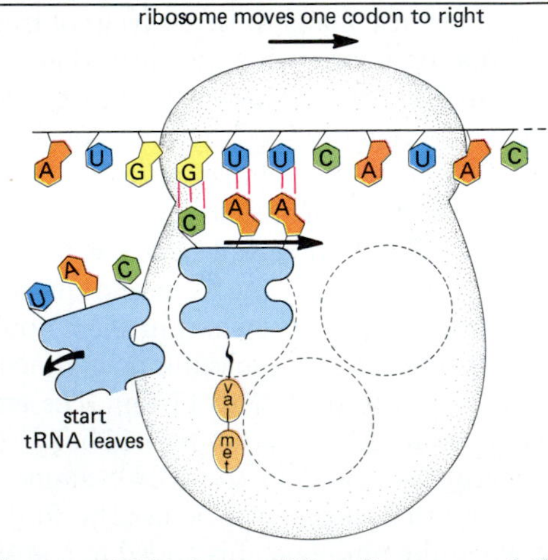

(f) *The start tRNA drops off the ribosome, and the ribosome moves one codon to the right on the mRNA. The tRNA bearing the newly formed dipeptide moves to the P site and the A site is emptied.*

Figure 11-7 *Protein synthesis is the translation of the sequence of nucleotides in mRNA to the sequence of amino acids in the encoded protein.*

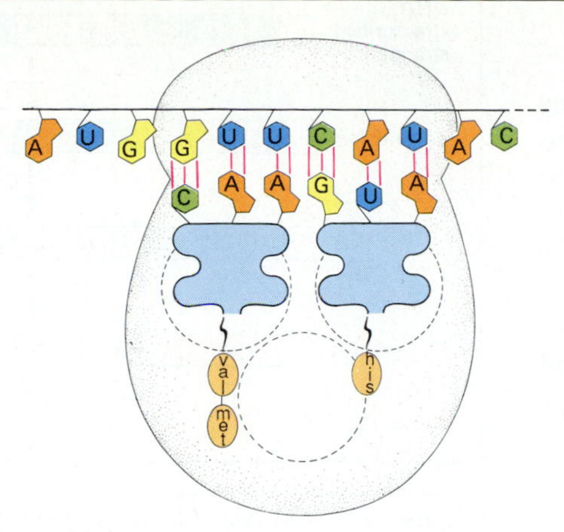

(g) *The next tRNA base-pairs with the third mRNA codon and moves into the A site.*

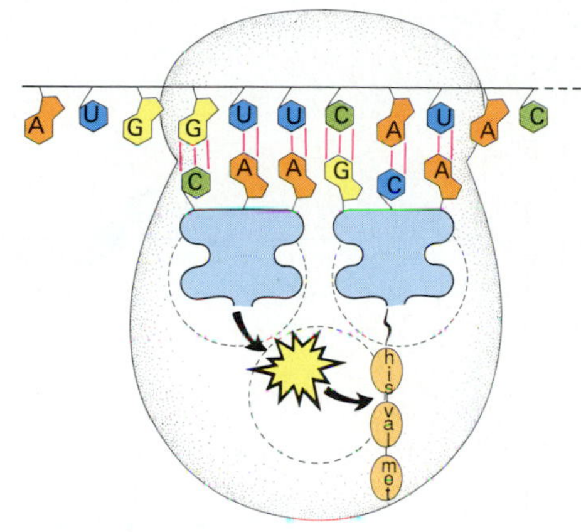

(h) *A peptide bond is forged between the dipeptide and the new amino acid, forming a tripeptide that remains attached to the third tRNA.*

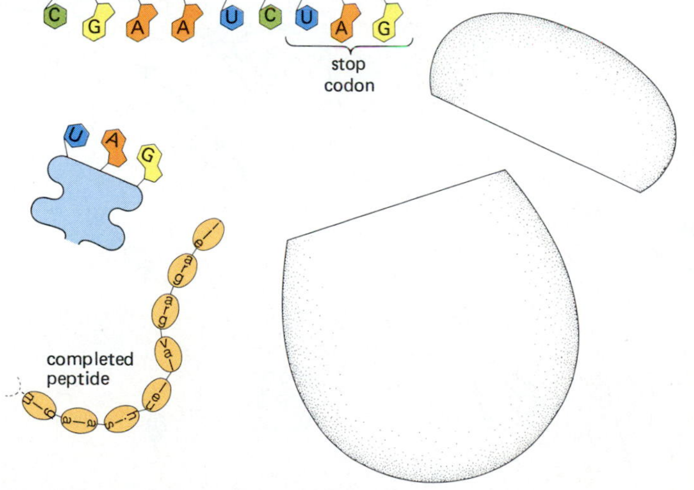

(i) *This process repeats until a "stop" codon (UAG, UAA, UGA) is reached. The finished peptide is released from the ribosome. The ribosomal subunits separate.*

large subunit (Fig. 11-7c). The ribosome is now fully assembled and ready to begin translation.

Protein elongation. Proteins are synthesized one amino acid at a time. The anticodon of a tRNA–amino acid complex recognizes the second mRNA codon, and moves into the A site on the large subunit (Fig. 11-7d). The two amino acids borne by the two tRNAs now lie adjacent to one another. The catalytic site breaks the bond holding the "start" amino acid (methionine) to its tRNA and uses the released energy to form a peptide bond between the methionine and the amino acid borne by the second tRNA. At the end of this step, the start tRNA is now "empty," while the second tRNA bears a short, two-amino acid protein chain (Fig. 11-7e).

At this point, the depleted start tRNA drops off the ribosome, and the ribosome shifts to the next codon on the mRNA molecule (Fig. 11-7f). A new tRNA-amino acid complex binds to the momentarily emptied A site (Fig. 11-7g). The catalytic site on the large subunit breaks

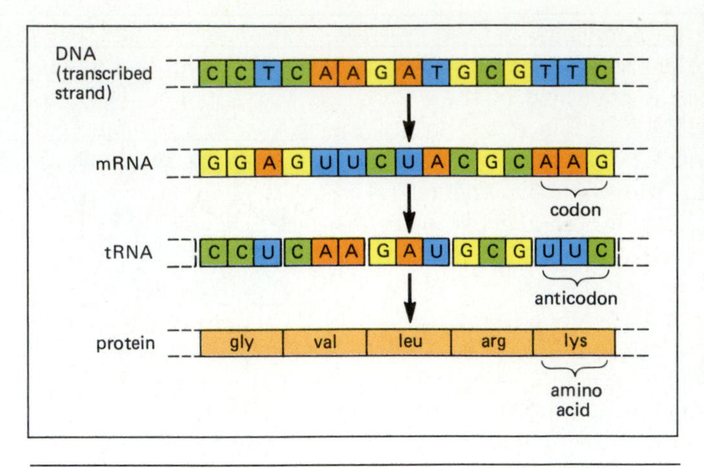

Figure 11-8 The decoding chain from DNA to protein.

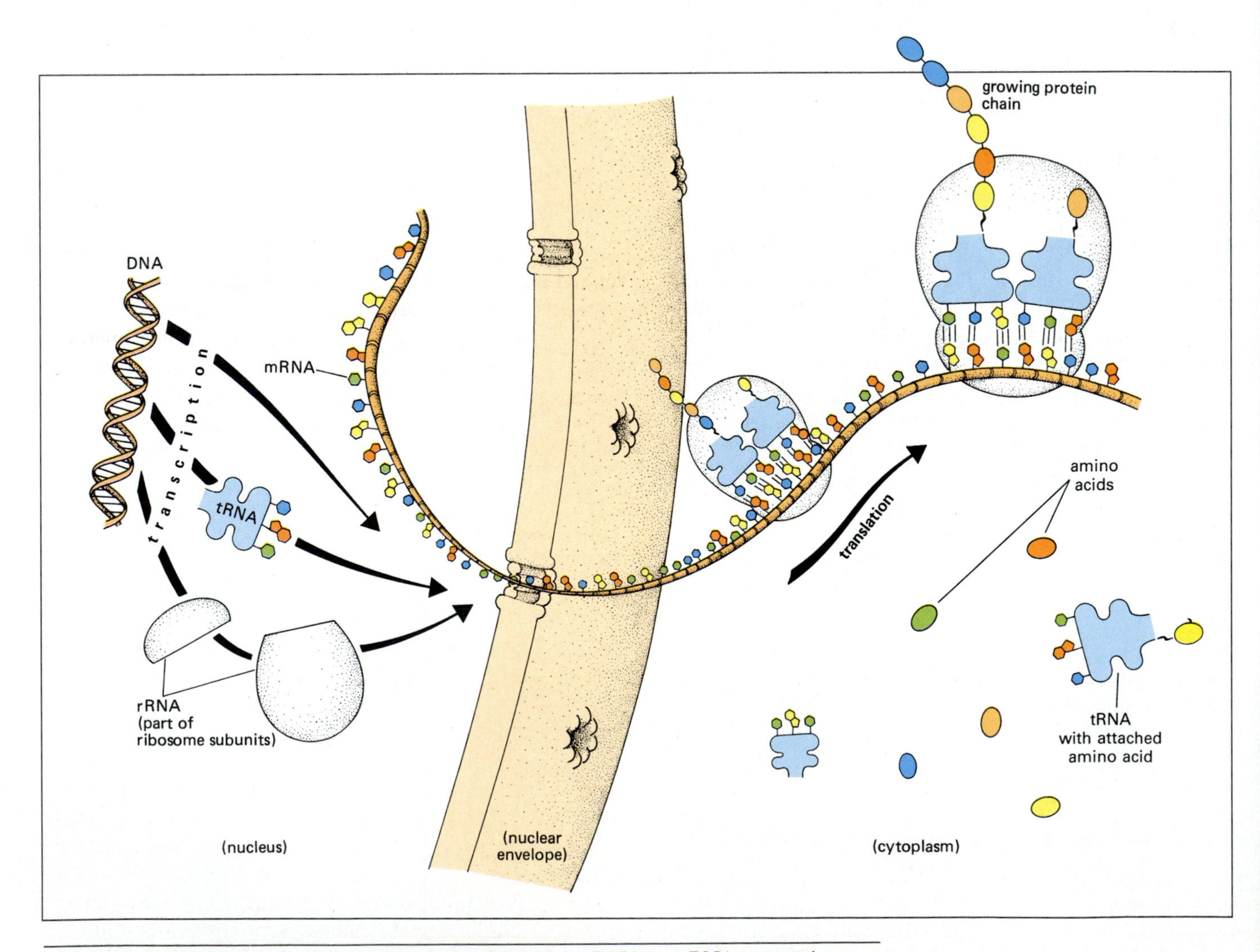

Figure 11-9 A summary diagram of protein synthesis, from DNA to mRNA to protein.

the bond between the dipeptide and its tRNA and uses the energy released to link the dipeptide with the amino acid now held in the A site (Fig. 11-7h). The depleted tRNA in the P site leaves the ribosome, the ribosome shifts over another codon, and the process repeats.

We can now understand how the "decoding" from DNA to amino acids works (Fig. 11-8). First, a codon of mRNA consists of three nucleotides with bases complementary to a set of three bases in the DNA of a gene. An anticodon of tRNA is complementary to a specific codon of mRNA. The tRNA will in turn be carrying a specific amino acid. This decoding chain, from bases in DNA to codon of mRNA to anticodon of tRNA to amino acid, results in the incorporation of the correct amino acid in the growing protein.

Termination. Finally, usually near the end of the mRNA, a "stop" codon is reached. No tRNA recognizes a stop codon. Instead, other molecules cut the finished protein chain off the last tRNA, releasing it from the ribosome (Fig. 11-7i).

Figure 11-9 provides an overall summary of the entire process of protein synthesis, from DNA to RNA to protein.

GENE REGULATION

Knowing how proteins are synthesized does not provide a complete understanding of how an organism's genotype produces its phenotype. For example, most of the cells of your body have the same DNA, but do not use all the DNA all the time. Individual cells express (produce proteins encoded by) only a small fraction of their genes, those that are appropriate to the function of that particular cell type. Muscle cells, for example, synthesize actin and myosin but not insulin or hair proteins. Gene expression also changes over time, depending on the needs of the body from moment to moment. Therefore, understanding gene function requires an understanding of how genes are regulated.

The use of genetic information by a cell is a multistep process, beginning with the transcription of DNA and often ending with an enzyme catalyzing a needed reaction (Fig. 11-10). Regulation can occur at any of these steps. For example, enzyme activity is often controlled by competitive or allosteric inhibition, as we discussed in Chapter 6. The quantity of each enzyme can also be regulated by changes in the rate of mRNA synthesis, degradation, or translation. Here we focus on just one of these steps, the transcription of DNA to mRNA. Because of its importance in understanding gene function, transcriptional regulation is a field of extremely active research. However, in only a few instances do geneticists really understand how cells turn genes on and off. In the following sections, we highlight a few of the better-known examples.

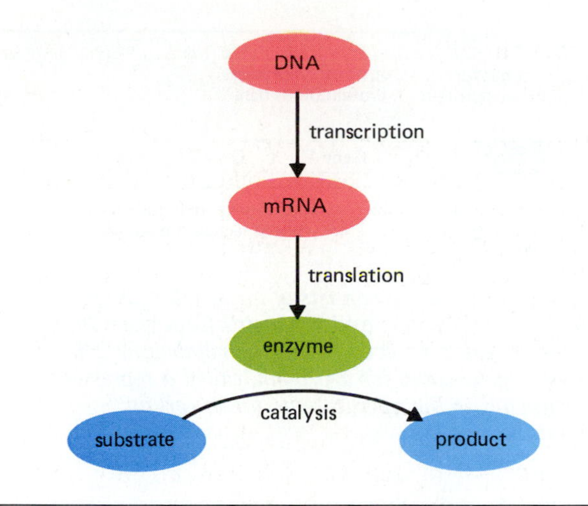

Figure 11-10 A simplified diagram of information flow in a cell, from DNA to the final reaction catalyzed by an enzyme. Information flow may be regulated at any step.

Gene Regulation in Prokaryotes

Prokaryotic DNA is often organized in coherent packages called **operons,** in which the genes for related functions lie next to one another (Fig. 11-11). An operon consists of four parts: (1) a **regulatory gene,** (2) a **promoter** that RNA polymerase recognizes as the place to start transcribing, (3) an **operator** that governs access of RNA polymerase to the promoter, and (4) the **structural genes** that encode the protein sequences of enzymes (Fig. 11-11a). Whole operons are regulated as units, so that related enzymes are synthesized simultaneously when the need arises. Prokaryotic operons are regulated differently, depending on the functions they control. Some operons synthesize enzymes that are needed by the cell just about all the time, such as the enzymes that synthesize amino acids. These operons are usually transcribed continuously, except under unusual circumstances when the bacterium encounters a vast surplus of a particular amino acid. Other operons synthesize enzymes that are only needed occasionally, for example to digest a relatively rare food substance.

As an example of the latter type of operon, consider the common intestinal bacteria *Escherichia coli.* These bacteria have to live on whatever types of nutrients their host eats, and they can synthesize a variety of enzymes to metabolize a potentially wide variety of foods. The genes that code for most of these enzymes are only transcribed when the enzymes are needed. The enzymes that metabolize lactose, the principal sugar in milk, are a case in point. The **lactose operon** contains three structural genes, each coding for an enzyme that aids in lactose metabolism (Fig. 11-11a).

The lactose operon is shut off, or **repressed,** unless specifically activated by lactose. The regulatory gene of the lactose operon directs synthesis of a protein, called a

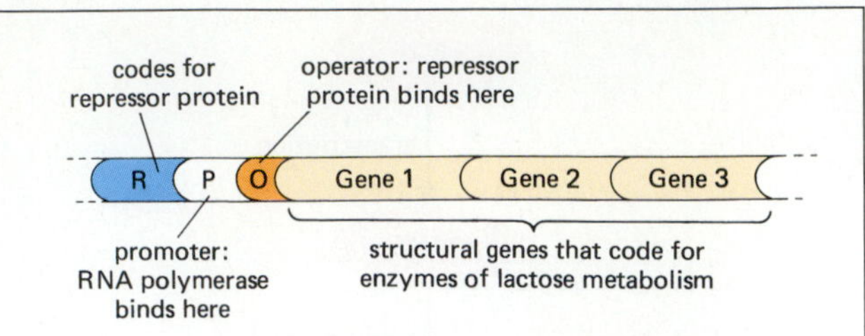

(a) *The lactose operon consists of a regulatory gene, a promoter, an operator, and three structural genes that code for enzymes involved in lactose metabolism. The regulatory gene codes for a protein, called a repressor, that can bind to the operator site under certain circumstances.*

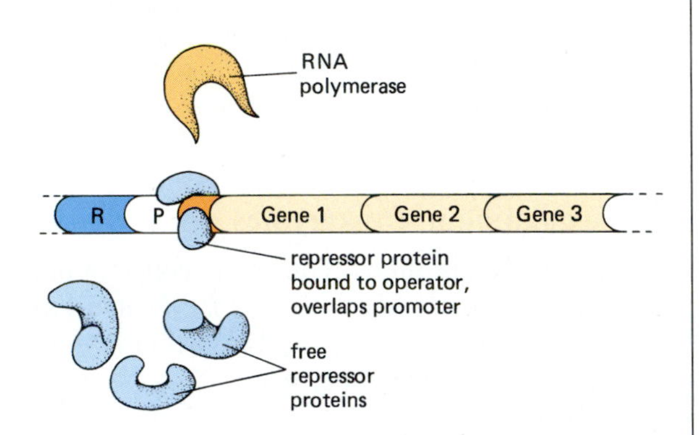

(b) *When lactose is not present, repressor proteins bind to the operator of the lactose operon. The operator region is only a small segment of DNA, but the repressor protein is quite large. When the repressor binds to the operator, it overlaps the promoter site as well. Therefore, RNA polymerase cannot bind to the promoter and the structural genes cannot be transcribed.*

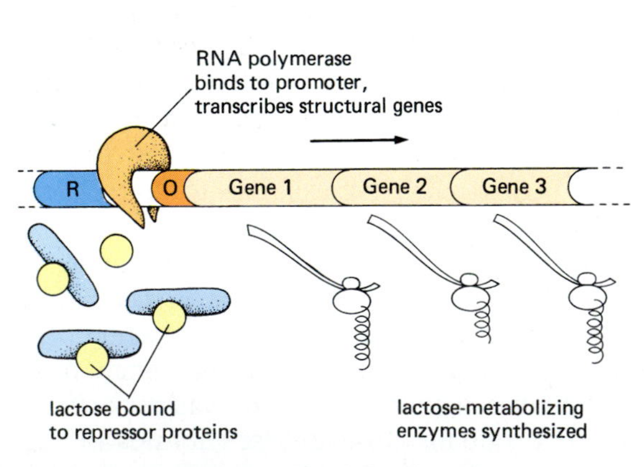

(c) *When lactose is present, it binds to the repressor protein. The lactose-repressor complex cannot bind to the operator, so RNA polymerase has free access to the promoter. The RNA polymerase transcribes the three structural genes coding for the lactose-metabolizing enzymes.*

Figure 11-11 *Structure and regulation of the lactose operon of* E. coli.

repressor protein, that binds to the operator site. Since the operator site on the DNA is small and the repressor protein is large, the repressor overlaps onto the promoter site (Fig. 11-11b). RNA polymerase is physically prevented from binding to the promotor and starting transcription. Consequently, the lactose-metabolizing enzymes are not synthesized.

When *E. coli* colonize the intestines of a newborn mammal, however, they find themselves bathed in a sea of lactose whenever the host nurses from its mother. Lactose molecules enter the bacteria and bind to the repressor proteins, changing their shape (Fig. 11-11c). The lactose–repressor combination cannot attach to the operator site. Therefore, RNA polymerase can bind to the promoter of the lactose operon and transcribe the structural genes. Lactose-metabolizing enzymes are synthesized, allowing the bacterium to use lactose as an energy source.

When the young mammal is weaned, it usually never consumes milk again. The intestinal bacteria no longer encounter lactose, the repressor proteins are free to bind to the operator, and the genes for lactose metabolism are shut down.

Gene Regulation in Eukaryotes

Gene regulation is quite different in eukaryotes. Not only are genes for related functions sometimes found on entirely different chromosomes, but even the individual genes are split up on the chromosome. As a result, transcription and its regulation are more complex.

Eukaryotic gene structure. In the 1970s, molecular geneticists discovered that eukaryotic structural genes have much more DNA than is needed to encode the amino acids of proteins. Each gene consists of two or more DNA segments that encode the protein, interrupted by other DNA segments that apparently code for nothing at all. The coding segments are called **exons** because they are *ex*pressed in protein, while the noncoding segments are called **introns** because they *in*tervene between the exons (Fig. 11-12a).

Each eukaryotic gene has its own promoter. A nearby region of the chromosome, called the **enhancer,** regulates binding of RNA polymerase to the promoter. When specific regulatory proteins bind to the enhancer, they facilitate the binding of RNA polymerase to the promoter, thus enhancing transcription.

When a eukaryotic gene is transcribed, a very long molecule of RNA is synthesized (Fig. 11-12b). This initial RNA transcript contains many more nucleotides than just the codons specifying the amino acids of the encoded protein. Two major steps convert this RNA molecule into messenger RNA. First, RNA nucleotides are added at the beginning (the "cap") and the end (the "tail") of the molecule. Second, cellular enzymes precisely cut the mole-

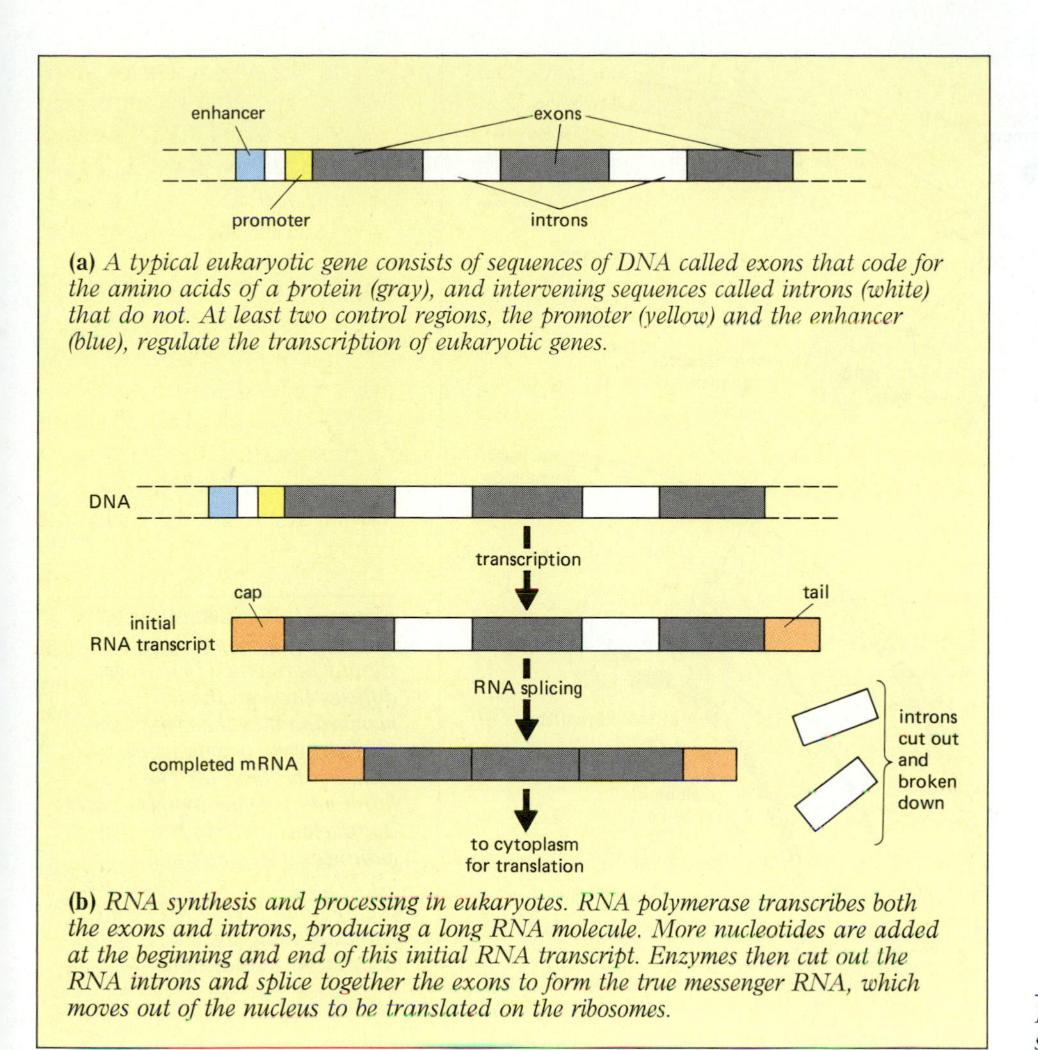

(a) *A typical eukaryotic gene consists of sequences of DNA called exons that code for the amino acids of a protein (gray), and intervening sequences called introns (white) that do not. At least two control regions, the promoter (yellow) and the enhancer (blue), regulate the transcription of eukaryotic genes.*

(b) *RNA synthesis and processing in eukaryotes. RNA polymerase transcribes both the exons and introns, producing a long RNA molecule. More nucleotides are added at the beginning and end of this initial RNA transcript. Enzymes then cut out the RNA introns and splice together the exons to form the true messenger RNA, which moves out of the nucleus to be translated on the ribosomes.*

***Figure 11-12** Eukaryotic gene structure and function.*

cule apart, splice together the sections that code for the protein, and discard the rest.

Why are eukaryotic genes fragmented like this, and are the intervening introns really just "genetic junk" to be snipped out and thrown away? No one knows for sure, but fragmented genes may provide a quick and efficient way for eukaryotes to evolve new proteins with new functions. This possibility is explored in the essay, "Rube Goldberg Genetics."

Regulation of transcription. As you can see, geneticists have learned a great deal about the *parts of the eukaryotic chromosome* that regulate transcription of genes. However, in most cases they do not yet understand *how* regulation occurs or what cellular events control the *timing* of gene transcription. A notable exception is the influence of steroid hormones on transcription, for example the stimulation of albumin (egg white) synthesis in female birds by estrogen (Fig. 11-13). Being lipid soluble, steroid hormones readily penetrate cell membranes and enter the interiors of cells. During the breeding season, estrogen is secreted into the bloodstream by the birds' ovaries and enters the cells of the oviduct. The estrogen binds to receptor proteins in the cytoplasm. The estrogen–protein complex enters the nucleus, where it binds to DNA, probably near the enhancer for the albumin gene. This apparently makes it easier for RNA polymerase to contact the promoter of the albumin gene. Rapid transcription occurs and albumin is synthesized. Similar activation of genes by steroid hormones occurs in other animals, including humans.

Large chunks of DNA, or even entire chromosomes, may also be regulated. Certain parts of chromosomes are in a highly condensed, compact state, in which the DNA seems to be inaccessible to RNA polymerase. Some of these regions are structural parts of chromosomes that do not contain genes. For example, condensed DNA is usually found at the centromeres that hold sister chromatids together during cell division. In other cases, DNA can change from the condensed state to a looser configuration that allows genes to be transcribed, depending on the stage in the life cycle of the animal or the type of cell (Fig. 11-14).

A particularly interesting example of the genetic inactivity of condensed DNA is found in female mammals. Females have two homologous X chromosomes. How-

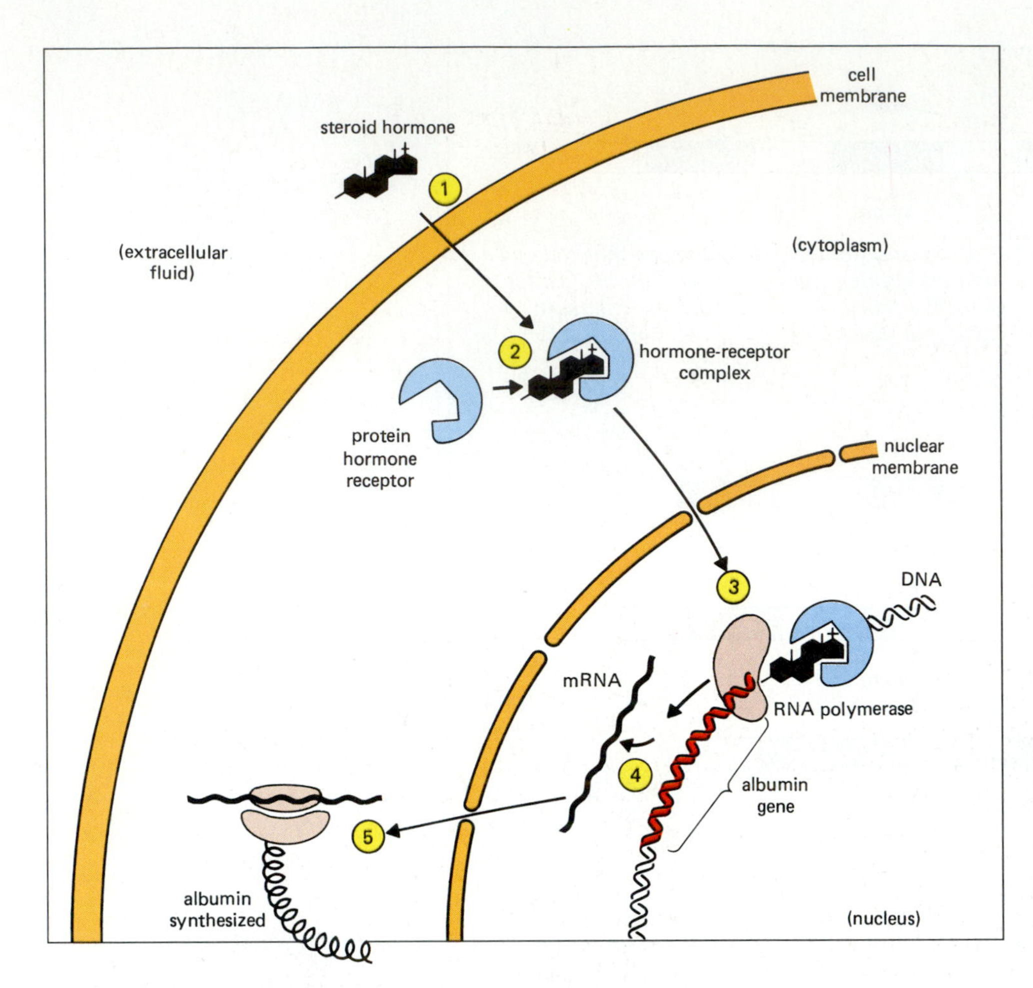

Figure 11-13 Stimulation of transcription by estrogen, a steroid hormone. (1) Estrogen diffuses through the cell membrane into the cytoplasm. (2) Estrogen combines with a receptor protein. (3) The hormone-receptor complex enters the nucleus and binds to the enhancer of the albumin gene. (4) RNA polymerase transcribes the albumin gene. (5) The mRNA leaves the nucleus and is translated into albumin.

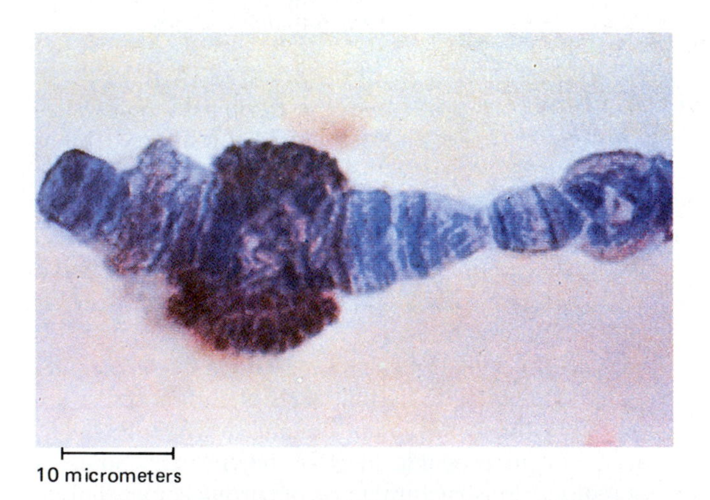

Figure 11-14 A giant chromosome of a midge (a small fly). Most of the chromosome is in a tight, condensed state, while a few regions have puffed out in a much looser configuration. The purple stain binds to RNA, which is mostly synthesized in the loose, puffed spots. Changing patterns of puffing and condensation of the chromosomes reflect different genes turning on and off during development.

ever, only one X chromosome is available for transcription in any given cell. The other entire X chromosome is condensed into a tight mass. In the light microscope, the inactivated X chromosome shows up as a dark spot in the nucleus called a **Barr body,** after its discoverer, Murray Barr (Fig. 11-15). Apparently, both X chromosomes are in the "loose" state in fertilized eggs. After a few cell divisions, one or the other condenses and forms a Barr body. (The germ cells in the ovaries are an exception; here both X chromosomes are in the loose, uncondensed state.) Which X chromosome is inactivated in any given cell is random, but all its daughter cells will then have the same condensed chromosome. As a result, female mammals (including women) are mosaics: patches of cells with one X chromosome active are interspersed with patches in which the other is active. This is strikingly evident in calico and tortoiseshell cats (Fig. 11-16), which are almost invariably female. The X chromosome contains a gene for fur color, with orange and black being the common alleles. Males, with only one X chromosome, are usually either orange *or* black. Females, with two X chromosomes, can be orange (both X chromosomes carry the orange allele), black (both carry the black allele), or orange *and* black in patches (one X chromosome has the orange allele and one has the black allele).

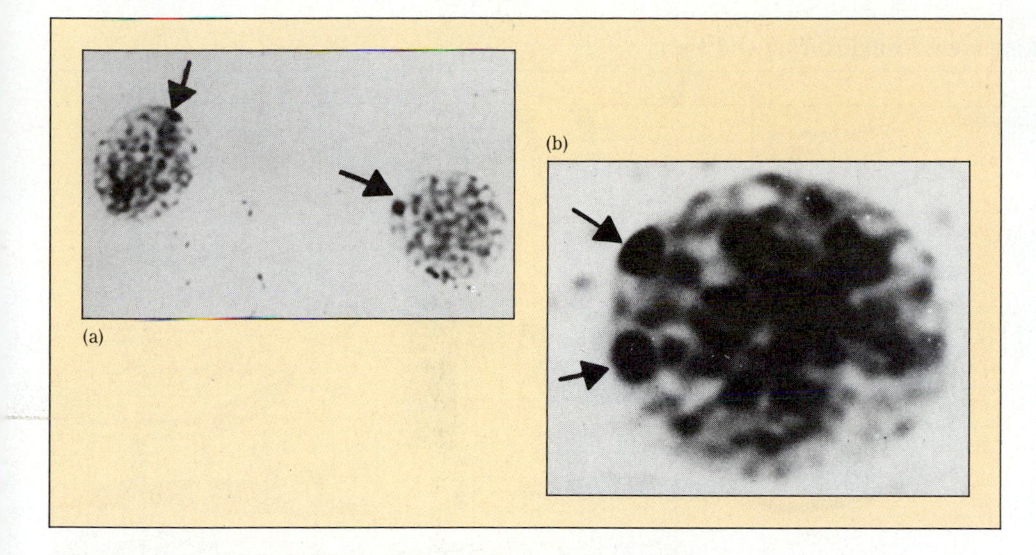

Figure 11-15 *Nuclei of human cells, stained to show Barr bodies (arrows).* **(a)** *These nuclei, from normal females, have one Barr body. Nuclei from males lack Barr bodies.* **(b)** *People with extra X chromosomes have two or more Barr bodies. Female athletes in the international sports are sometimes required to submit to examination for Barr bodies, to ensure that they are normal XX females, not males posing as females.*

Figure 11-16 *Male and female cats from the same litter. The female is a calico, with both orange and black patches of fur. (The predominant white color is due to an entirely different gene, which prevents color formation.) The male is simply orange and white, while another male of the same litter was black and white. An occasional calico or tortoiseshell cat appears to be phenotypically male. These cats are almost always sterile, and chromosome analysis usually finds an XXY genotype.*

MENDELIAN GENETICS REVISITED

You have now learned enough about the mechanisms of heredity to come full circle and reexamine Mendelian genetics, mutations, and evolution from a new, molecular perspective. As you will see, the various types of inheritance and the mechanisms of evolution are outgrowths of the replication and transcription of DNA.

Mutations and the Genetic Code

A **gene mutation** is a change in the sequence of nucleotides in DNA. If a mutation occurs in cells whose progeny become gametes, it may be passed on to future generations. But how does a change in nucleotide sequence affect the organism that inherits the mutated DNA? As the essay "Cracking the Genetic Code" pointed out, deletions and insertions can have catastrophic effects on a gene, since all the codons following the deletion or insertion will be misread. The enzyme synthesized from such misread directions is almost certain to be nonfunctional.

Point mutations, in which one nucleotide is replaced by another, may lead to more subtle effects. Four different categories of effects may result from point mutations (Table 11-2). As a concrete example, let's consider possible mutations of the DNA sequence CTC, which codes for glutamic acid. (1) *A mutation may not change the amino acid sequence of the encoded protein.* Remember that the genetic code is degenerate, so that one amino acid may be encoded by several different codons. If a mutation changes CTC to CTT, the new triplet still codes for glutamic acid. Therefore, the protein synthesized from the mutated gene does not change. (2) *A mutation may code for an amino acid that is functionally equivalent to the original amino acid.* Many proteins have large "background" regions whose exact structure is relatively unimportant. For example, in hemoglobin, the amino acids on the outside of the protein must be hydrophilic to keep the protein dissolved in the cytoplasm of red blood cells. Exactly *which* hydrophilic amino acids are on the outside does not matter too much. A mutation from CTC to CTA, replacing glutamic acid (hydrophilic) with aspartic acid (also hydrophilic), probably would not affect the solubility of hemoglobin. Mutations that do not detectably change the function of the encoded protein are called **neutral mutations.** (3) *A mutation may encode for a functionally different amino acid.* A mutation from CTC to

Rube Goldberg Genetics: Making New Proteins from Old Parts

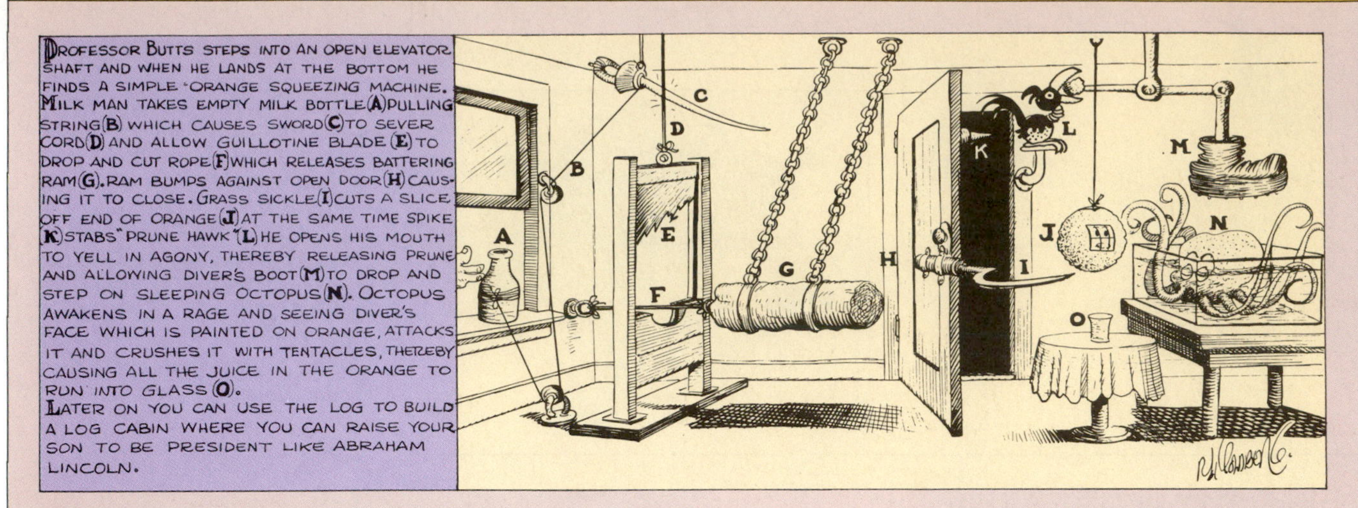

Figure E11-1

The cartoonist Rube Goldberg created marvelous fictional contraptions to perform simple functions, like "squeezing the juice from an orange into a glass" (Fig. E11-1). The beauty of a Goldberg "invention" was that you could have really built it, if you wanted to, from ordinary household items and scrap lumber. In many respects, evolution works a lot like Rube Goldberg, modifying ordinary, preexisting structures to perform new functions. Consider, if you will, an elephant's ears: huge flaps of skin laced with blood vessels. The selective advantage of "ordinary" external ears, like those on a wolf or a deer, is that they funnel sound into the ear canal, helping animals locate the source of sounds. The vast majority of animals bear such hearing-aid ears. From these humble beginnings, the enormous ear flaps on an elephant have been adapted for quite a different role: dissipating heat in the African savanna. They also have the added benefit of looking most impressive when an elephant threatens a rival.

In evolution, it is always difficult to devise anything from scratch. Inventing a new protein, say of 200 amino acids, means putting together a string of 600 nucleotides in DNA in just the right sequence. Mutations cannot assemble a new, useful string of 600 nucleotides in one fell swoop. Instead, evolution has hit upon a very different, much quicker, and quite effective strategy: making new proteins from old parts.

Many proteins consist of several subunits, each with a completely different function. A protein that actively transports potassium across the cell membrane, for example, might have three subunits: one to anchor the protein in the membrane, one to bind potassium ions, and one to bind ATP to power the transport (Fig. E11-2a). The membrane anchor and ATP-binding subunits are modules that might be useful for other transport proteins as well. Exchange the potassium-binding subunit for a calcium-binding subunit, and voilà! the cell has a calcium transport protein (Fig. E11-2b).

How can an organism exchange subunits among various proteins? It turns out that chromosomes are not quite the stable, nearly perfectly replicating structures that geneticists pictured not very long ago. Chromosomes occasionally break in two, and one end may become attached to a completely different chromosome. Sometimes segments pop out of one chromosome and insert themselves in another. These DNA rearrangements provide a mechanism for rearranging protein subunits as well.

Since genes are segments of DNA on chromosomes, scrambling parts of a chromosome will often scramble parts of genes. You might think that this would completely ruin the genetic instructions in the rearranged genes. However, remember that the structure of a eukaryotic gene includes both expressed regions (exons) and intervening regions (introns). Exons often code for individual protein subunits (Fig. E11-2c). What if chromosomes preferentially break within introns, so that intact, functional exons can be moved from one chromosome to another (Fig. E11-2d)? By interchanging exons among genes, a eukaryotic organism can create new genes and thereby adapt more quickly to changing environmental conditions. Molecular biologists have recently found evidence that this is probably just what happened in the case of ATP-powered transport molecules and enzymes, several of which have essentially the same ATP-binding subunit.

Rube Goldberg would have been proud.

Rube Goldberg Genetics: Making New Proteins from Old Parts

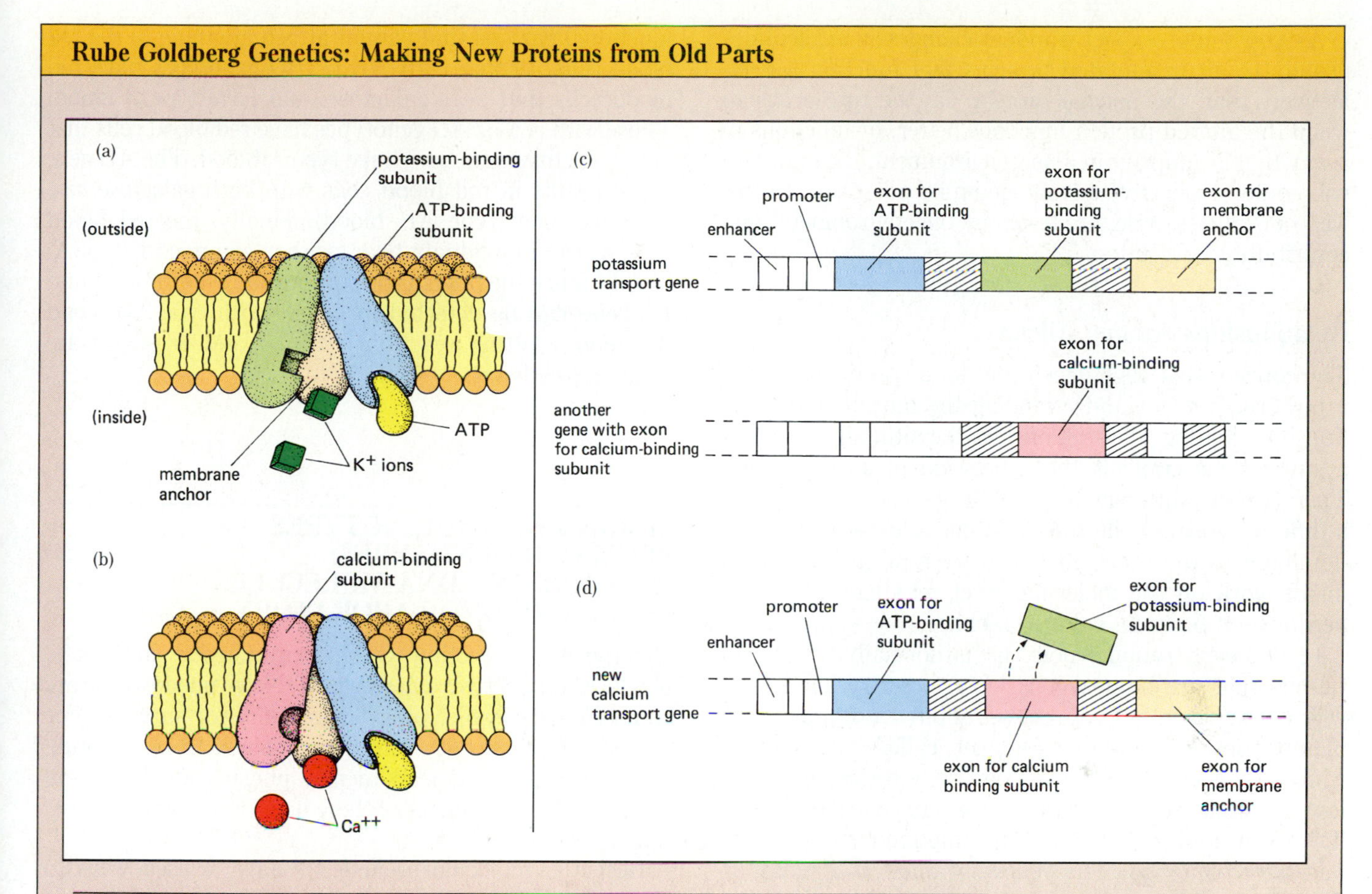

Figure E11-2 *Shuffling exons to make new functional proteins.*
(a) *Hypothetical protein subunits that might make up an active-transport protein for potassium ions.*
(b) *Substituting a calcium-binding subunit for the potassium-binding subunit creates a calcium-transporting protein.*
(c) *The potassium-transport gene might have three exons, one coding for each subunit of the transport protein. Another gene, even on another chromosome, has an exon coding for a calcium-binding protein.*
(d) *If the potassium-binding exon is cut out and replaced with the calcium-binding exon, a new gene is formed, now coding for a calcium-transporting protein.*

CAC replaces glutamic acid (hydrophilic) with valine (hydrophobic). This substitution, which is the genetic defect in sickle-cell anemia (see Chapter 12), causes hemoglobin molecules to stick to each other, clumping up and distorting the shape of the red blood cells. This is a potentially fatal mutation. (4) *A mutation may produce a stop codon.* An inappropriate stop codon will cut short the translation of mRNA before the protein is finished. This is almost certainly catastrophic to protein functioning; if the protein is essential to life, as hemoglobin is, the mutation will be lethal.

Table 11-2 **Examples of Functional Outcomes of Single Substitutions in the Glutamic Acid Codon of DNA**

	DNA	mRNA	Amino Acid	Properties	Effect
Original sequence:	CTC	GAG	Glutamic acid	Hydrophilic, acidic	—
Mutation 1	CT*T*	GAA	Glutamic acid	Hydrophilic, acidic	—
Mutation 2	CT*A*	GAU	Aspartic acid	Hydrophilic, acidic	Neutral
Mutation 3	C*A*C	GUG	Valine	Hydrophobic, neutral	Lose water solubility; possibly catastrophic
Mutation 4	*A*TC	UAG	Stop codon	Ends translation	Only synthesize part of protein; catastrophic

As you might expect, random changes in nucleotides usually result in mutated proteins that function less effectively than the normal protein. On the rare occasion when the altered protein functions better, or functions in a way that is adaptive in a new environment, the mutation will confer a selective advantage on its possessor. In this way new alleles arise, are tested by the environment, and contribute to evolution.

Relationships Among Alleles

The concept that each gene codes for a specific protein is a powerful tool in understanding the multitude of relationships among alleles. Consider the situation in which a single gene controls the expression of a single trait. There are two different alleles of the gene, each producing a different form of the trait, and one allele is completely dominant to the other allele. To see how such a system might work on the molecular level, let's look at the inheritance of body fat color in rabbits.

In domestic rabbits, body fat is normally white, but rabbits that are homozygous for a recessive allele have yellow fat. Why? As you know, rabbits eat plants. Most plants contain a yellow pigment called xanthophyll, which is fat soluble and can potentially color the fat yellow. Normal rabbits synthesize an enzyme that breaks down xanthophyll to a colorless compound, so these rabbits have white fat. The yellow-fat allele is a mutation that renders the enzyme nonfunctional. If a rabbit has one normal allele and one mutant allele, the normal allele directs the synthesis of enough normal enzyme to degrade the xanthophyll in the rabbit's diet completely. The body fat of heterozygotes will still be white, and thus the normal allele is dominant to the mutant allele. If a rabbit is homozygous recessive for the mutant allele, it produces no functional enzymes, so xanthophyll from its diet is not metabolized. The xanthophyll dissolves in the fat, coloring it yellow.

In general, *dominant alleles direct the synthesis of functioning enzymes.* An organism with one dominant allele synthesizes enough enzyme to produce a phenotype that is indistinguishable from the phenotype of organisms with two dominant alleles. In contrast, *a recessive allele usually directs synthesis of a nonfunctional enzyme.*

Mutations may occur anywhere in a gene. Therefore, different organisms may suffer quite different mutations in the same gene. These different mutations may all produce slightly different proteins, giving rise to multiple alleles of the same gene. In some circumstances, each of these multiple alleles may be detectable phenotypically, as is the case with the human ABO blood groups (see Fig. 9-16). The *A* and *B* alleles both code for slightly different, functional enzymes. The *B* enzyme attaches galactose (a monosaccharide) to red blood cell membranes, whereas the *A* enzyme attaches a slightly different compound, galactosamine. The *O* allele is a mutation producing a nonfunctional enzyme that cannot attach anything to the red blood cells. People with *BB* or *BO* genotypes have red blood cells that bear galactose, and have type B blood. Those with *AA* or *AO* genotypes have red blood cells that bear galactosamine, and have type A blood. The *AB* genotype results in red blood cells with both galactose *and* galactosamine (type AB blood). Finally, the red blood cells of *OO* individuals bear neither compound (type O blood). Since the phenotypes produced by both the *A* and *B* alleles can be detected in people with type AB blood (by blood-clotting reactions), the *A* and *B* alleles are called *codominant.*

MAKING NEW PHENOTYPES FROM OLD GENOTYPES: RECOMBINING DNA MOLECULES IN NATURE AND LABORATORY

Genotypes, interacting with environmental influences, produce distinctive phenotypes. New phenotypes are sometimes desirable, or even essential for survival. For billions of years, nature has recombined DNA, creating new genotypes that lead to new phenotypes. Evolution then tests the usefulness of these new DNA combinations: only a few flourish.

Recently, molecular geneticists have devised ways to recombine DNA in the laboratory. The results of this **recombinant DNA** technology are widely touted in the media: easy, inexpensive production of hormones and drugs; cures for genetic diseases; corn or wheat that can produce its own nitrogen fertilizer; bacteria tailored to eat up oil spills. What *is* recombinant DNA? How does it work? Will it really allow us to do all these things?

Naturally Occurring DNA Recombinations

The words "recombinant DNA" seem as if they should mean mixing together genes from two different sources, and so they do. Combining genes from different organisms is nothing new—after all, that's what sexual reproduction is all about. We tend to think of natural recombinations as occurring between members of the same species, and that is certainly the usual case with multicellular organisms. Bacteria, however, employ several exotic methods of recombination that allow gene transfer between unrelated species. In transformation, for example, bacteria pick up free DNA from the environment, including DNA from other species of bacteria. Transformation plays a role in adapting bacteria for life in hospital environments. The frequent use of antibiotics in hospitals allows the survival only of antibiotic-resistant bacteria. Genes for antibiotic resistance can be passed from species to species via transformation. Thus, using antibiotics against one species of bacteria can result in resistant populations of other species that are responsible for entirely different diseases.

Recent evidence suggests that viruses may sometimes transfer genes among eukaryotic organisms. The DNA of certain viruses can insert itself into a chromosome of its eukaryotic host, and exist there quietly for days, months, or even years. Then, perhaps in response to a stressful stimulus to the host, the viral DNA leaves the chromosome, occasionally taking a bit of the eukaryotic DNA along with it. The viral DNA then takes over the host cell metabolism, replicates itself, and directs the synthesis of new viruses. The offspring viruses may thus include some host DNA. If one of these offspring infects a new host of a different species, it may insert itself, along with the piece of DNA from the former host, into a chromosome of its new host. In this way, the new host may acquire some genes that originally belonged to an unrelated species.

Recombinant DNA in the Laboratory: Making Prokaryotic Protein Factories

The laboratory technique of recombinant DNA refers to planned and directed interspecies combinations, often employing bacteria, viruses, or minute circular pieces of DNA, called plasmids, that are found in many bacteria. To understand how recombinant DNA technology came about, we must first consider the evolutionary warfare between bacteria and bacteriophages. The destruction of bacteria by phages has led to evolutionary developments by bacteria that prevent infection by phages. The phages, in return, are under selective pressure to circumvent the bacterial defenses. As a result, each species of bacterium has defenses against phage attack, and only those phages that can overcome those defenses can infect that particular type of bacterium. Among the most potent bacterial defenses are the **restriction enzymes.**

As you know, phage infection begins when a phage injects its DNA into a bacterium (see Fig. 10-5). Many bacteria produce restriction enzymes that cut up phage DNA, and (usually) only phage DNA. To avoid destroying the bacterium's own DNA, the restriction enzymes of each bacterium are very specific, and will only cut apart DNA with certain base sequences. The bacterial DNA either does not have these sequences or chemically protects them in some way. As shown in Figure 11-17a, many restriction enzymes sever palindromic DNA, which reads the same in one direction on one strand as it reads in the reverse direction on the other strand (a palindrome in English is a word that reads the same forward and backward, such as "madam"). Furthermore, the DNA is cut between the same two bases on the two strands, in this case between guanine and adenine. This results in two pieces of DNA, one with a single-stranded end reading TTAA and one with a single-stranded end reading AATT. Complementary DNA regions like these can pair up, held together by hydrogen bonds between the bases. If the appropriate DNA repair enzymes are present, the two pieces can be rejoined.

In the laboratory, recombinant DNA technology uses restriction enzymes to cut open DNA from two different sources, say a human chromosome and a bacterial plasmid (Fig. 11-17b). The single-stranded "sticky ends" join the human and plasmid DNA, and a DNA repair enzyme bonds the sugar-phosphate backbones together, inserting the human gene into the plasmid. (One wag has named such plasmid–human DNA combinations "designer genes": expensive and custom tailored.) The plasmid–human DNA is mixed with bacteria treated to make them permeable to DNA. If all goes well, some of the bacteria take up the plasmid–human DNA. This technique has been used to insert the genes for human insulin, growth hormone, and interferon into bacteria. With some further manipulations, the bacteria produce large quantities of these substances.

Why is this useful? Take the case of human growth hormone. Growth hormone, secreted by the pituitary, governs bodily growth, protein metabolism, and perhaps aging. A few thousand people do not produce enough growth hormone, remain very short, and age prematurely. Formerly, the only source of human growth hormone was human pituitaries. Tens of thousands of pituitaries were removed from corpses and processed to purify the growth hormone, which would then be injected into growth-hormone-deficient children. This, of course, is very expensive. To make matters worse, in 1985 a few recipients died from a rare viral infection, Creutzfeld–Jacob disease, apparently caused by viruses contaminating the hormone preparation. Enter recombinant DNA. A biotechnology company, appropriately called Genentech, inserted the gene for human growth hormone into bacteria, and in late 1985 began to sell bacterially produced hormone. Since no human body parts are involved, no human diseases can be transmitted by the Genentech hormone.

Other successful applications of recombinant DNA have followed similar strategies, incorporating foreign DNA into bacteria, so that the bacteria produce useful products such as insulin or metabolize toxic materials such as oil or pesticides.

Recombinant DNA in the Laboratory: Modifying Eukaryotic DNA

Variations of the recombinant DNA technique have the potential to insert or exchange genes in eukaryotic organisms. One of the most valuable genetic modifications would be the insertion of genes for nitrogen fixation into crop plants. With the exception of soybeans, none of our major crops can use atmospheric nitrogen as a nitrogen source. Consequently, they require nitrogen fertilizers, which are synthesized and spread using the energy of fossil fuels. Rain washes off some of the fertilizer applied to farms into nearby streams and lakes, causing pollution.

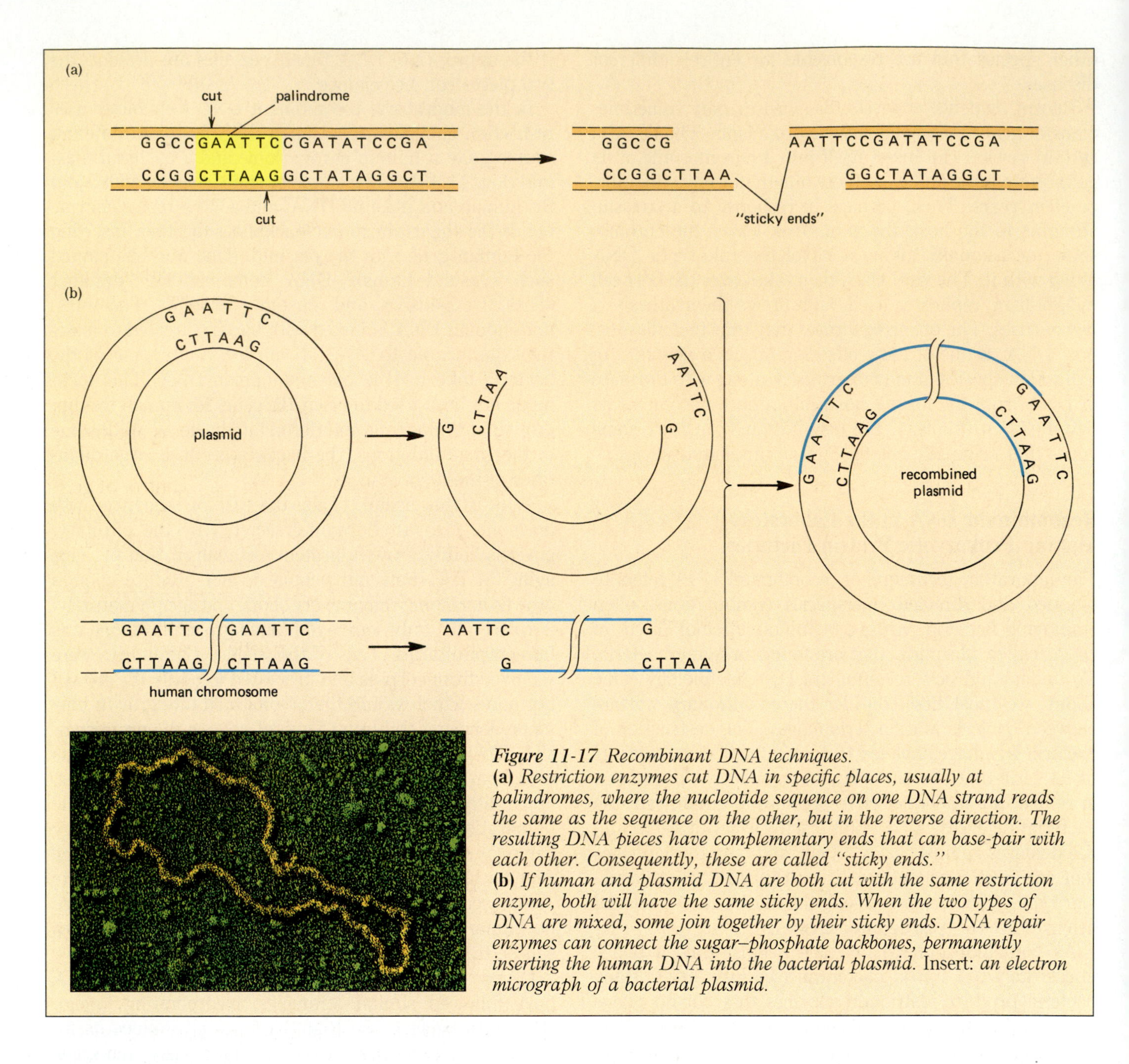

Figure 11-17 Recombinant DNA techniques.
(a) *Restriction enzymes cut DNA in specific places, usually at palindromes, where the nucleotide sequence on one DNA strand reads the same as the sequence on the other, but in the reverse direction. The resulting DNA pieces have complementary ends that can base-pair with each other. Consequently, these are called "sticky ends."*
(b) *If human and plasmid DNA are both cut with the same restriction enzyme, both will have the same sticky ends. When the two types of DNA are mixed, some join together by their sticky ends. DNA repair enzymes can connect the sugar–phosphate backbones, permanently inserting the human DNA into the bacterial plasmid.* Insert: *an electron micrograph of a bacterial plasmid.*

As the price of energy continues to rise, both Third World and American farmers find themselves unable to afford the amounts of fertilizer that would maximize yields, and of course, it is precisely these large amounts that cause the most severe pollution problems.

Several types of bacteria can extract nitrogen from the atmosphere and capture it in amino acids, a process called nitrogen fixation. If the genes for nitrogen fixation could be inserted into wheat and corn, farmers could grow much more food at much lower cost. Especially in poor regions, such as much of Africa, an improvement in food supplies would have a tremendous impact on human health and well-being.

One hitch in this scheme is the proper regulation of the new genes. It is one thing to insert a gene into the familiar *E. coli* and force the bacteria to produce huge quantities of new proteins, and quite another to insert a series of genes into corn and have the correct cells transcribe them all in the right amounts at the right times in the plant's life cycle. Geneticists do not know enough about the regulation of eukaryotic genes to accomplish this yet.

Proper regulation is equally a problem, or perhaps even more so, when we turn to human applications. Imagine the possibility of *curing* pituitary dwarfism or congenital diabetes by gene transfer rather than merely treating the symptoms! But imagine the tragedy if the genes were regulated incorrectly, causing giantism or insulin shock. In any case, how could all the proper cells acquire the new gene? Often, this would have to be done by inserting the gene into an egg. Except in rare cases, how would we know that this particular egg *has* a defective gene and *needs* a new one? Inserting genes into thousands of plant egg nuclei, followed by selection of the few successful individuals and destruction of the rest, would be acceptable agricultural practice. Similar procedures cannot even be considered in medicine.

The first human applications of genetic engineering will probably involve insertion of genes into discrete structures, such as glands or bone marrow, that are normally the only active sites of transcription of those genes. For example, defective insulin-secreting cells might be removed from a patient's pancreas, cultured in the laboratory, and the proper genes inserted into them. These cells would then be reimplanted into the patient.

What about new genes that humans never had before? Can genetic engineering endow humans with the visual acuity of an eagle, the sense of smell of a bloodhound, the intelligence of Einstein? Or, more mundane but more immediately practical, what about giving growth hormone to prospective basketball players to make them taller? Are slam-dunks worth tampering with a child's physiology? Are we ready to direct our own evolution?

REFLECTIONS ON GENE CONTROL

Molecular genetics is one of the most rapidly moving fields of biology. Although it may seem to you that the inner workings of DNA, RNA, and proteins are known in great detail, in fact we have barely begun to scratch the surface. Perhaps the most intensive research effort in biology today is devoted to studying the control of gene expression in eukaryotes. Geneticists really know very little about the regulation of eukaryotic genes, how and why some are activated and transcribed, while others are turned off. These questions are of enormous importance: for example, cancer is probably caused by changes in DNA structure or sequence that permanently turn on growth-inducing genes (see Chapter 27). Understanding gene control is also crucial to understanding development, aging, and inherited genetic diseases. The techniques of recombinant DNA offer the promise of new insights into these problems in the near future. Probably by the time you read this book, there will be new discoveries about gene structure and function that may offer us both the promise of medical advances and unforeseen biological and ethical dilemmas.

SUMMARY OF KEY CONCEPTS

Cell Function, Protein, and Genes

Genes are segments of DNA on chromosomes. The ultimate cellular product encoded by a gene is usually a protein. Therefore, with a few exceptions, the specific nucleotide sequence of a gene encodes the amino acid sequence of a protein or a part of a protein.

From DNA to Protein

Synthesizing proteins from the information in DNA requires RNA molecules as intermediates. RNA is transcribed from one DNA strand by the enzyme RNA polymerase. RNA polymerase recognizes a region of DNA called the promoter as the beginning of a gene. Starting there, RNA polymerase uses free ribose nucleotides to synthesize an RNA strand that is complementary to the DNA of the gene.

There are three types of RNA. The sequence of bases in messenger RNA (mRNA) carries the genetic code for the amino acid sequence in a protein. Three nucleotides in mRNA, called a codon, are used to specify each amino acid of the protein. There are also start and stop codons that signal the beginning and end of protein synthesis. Ribosomal RNA (rRNA) and proteins form ribosomes. Ribosomes consist of large and small subunits. The small subunit has binding sites for two codons of mRNA. The large subunit bears a catalytic site that forges the peptide bond between amino acids as a protein is synthesized, and two binding sites for transfer RNA (tRNA). There are at least 20 different tRNAs. Each binds a specific amino acid and transports it to a ribosome. A set of three bases in tRNA, called the anticodon, is complementary to the codon in mRNA that specifies the amino acid borne by that tRNA.

Protein synthesis occurs in the following sequence:

1. Messenger RNA is transcribed from a gene. The mRNA leaves the nucleus and travels to a ribosome.
2. Two codons of mRNA bind to the small subunit of the ribosome. The first codon is the "start" codon, which signals where protein synthesis is to begin.
3. Transfer RNAs, carrying their amino acids, move to the mRNA. The anticodons of two tRNA molecules base pair with the two codons of mRNA, and the tRNAs bind to the large ribosomal subunit.
4. The large subunit catalyzes the formation of a peptide bond between the amino acids carried by the two tRNA molecules. The "first" amino acid detaches from its tRNA. The chain of two amino acids remains attached to the "second" tRNA.
5. The "first" tRNA leaves the ribosome. The ribosome moves one codon over on the mRNA. A "third" tRNA, with its attached amino acid, base pairs with the third codon on mRNA. A new peptide bond is formed between the amino acid of the "third" tRNA and the dipeptide still attached to the "second" tRNA.

6. This process continues until a "stop" codon is reached, whereupon the mRNA and the newly formed protein leave the ribosome.

Gene Regulation

Which genes are transcribed in a cell at any given time is regulated by the function of the cell, the developmental stage of the organism, and the environment. Access of RNA polymerase to the promoter of a gene may be either prevented or enhanced by other molecules in the cell, including nutrients and hormones. Large parts of chromosomes may also be rendered inaccessible to RNA polymerase by changes in DNA structure.

Mendelian Genetics Revisited

Molecular genetics explains many aspects of Mendelian genetics and evolution. A mutation, for example, is a change in the sequence of nucleotides in DNA. The different alleles of a gene arise by mutation. Mendelian relationships among alleles are a consequence of the molecular nature of the gene and its resulting protein product. For example, a dominant allele codes for an amino acid sequence that results in a functioning enzyme. A recessive allele is usually a mutation that codes for a nonfunctioning enzyme.

Making New Phenotypes from Old Genotypes: Recombining DNA Molecules in Nature and Laboratory

Modern techniques in molecular biology, often called recombinant DNA, have made it possible to move genes from one organism to another. Thus far, most recombinant DNA techniques have involved transplanting eukaryotic genes into bacteria for synthesis of eukaryotic proteins on a commercial scale. In the future, geneticists may be able to insert genes into eukaryotic cells, thus adding new genes to crop plants or domestic animals, or replacing defective human genes.

GLOSSARY

Anticodon a sequence of three nucleotides in transfer RNA that is complementary to the three nucleotides of a codon of messenger RNA.

Barr body an inactive X chromosome found in somatic cells of mammals that have at least two X chromosomes (usually females). The Barr body usually appears as a dark spot in the nucleus.

Codon a sequence of three nucleotides of messenger RNA that specifies a particular amino acid to be incorporated into a protein. Certain codons also signal the beginning and end of protein synthesis.

Degeneracy the property of the genetic code whereby several codons may specify the same amino acid.

Enhancer in eukaryotes, a stretch of DNA that influences the access of RNA polymerase to the promoter region of a structural gene.

Exon a segment of DNA in a eukaryotic gene that codes for amino acids in a protein.

Genetic code the collection of codons of mRNA, each of which directs the incorporation of a particular amino acid into a protein during protein synthesis.

Inactivation a process whereby certain chromosomes or parts of chromosomes are converted into a dense mass, preventing transcription.

Intron a segment of DNA in a eukaryotic gene that does not code for amino acids in a protein.

Messenger RNA (mRNA) a strand of RNA, complementary to DNA, that conveys the genetic information in DNA to the ribosomes to be used during protein synthesis. Sequences of three nucleotides (codons) in mRNA specify particular amino acids to be incorporated into a protein.

Neutral mutation a mutation (change in DNA sequence) that has little or no phenotypic effect.

One-gene, one-protein hypothesis the proposition that each gene encodes the information for the synthesis of a specific protein.

Operator in prokaryotes, a segment of DNA that controls access of RNA polymerase to the promoter.

Operon a unit of organization of prokaryotic chromosomes, in which several genes that specify related functions (e.g., enzymes in the same biosynthetic pathway) are grouped together on the chromosome, are transcribed at the same time, and are regulated together.

Promoter a specific sequence of DNA to which RNA polymerase binds, initiating gene transcription.

Recombinant DNA a laboratory technique of inserting foreign DNA into an organism, producing new gene combinations not previously existing in nature.

Regulatory gene a gene that controls the timing or rate of transcription of other genes.

Ribonucleic acid (RNA) a single-stranded nucleic acid molecule composed of nucleotides, each of which consists of a phosphate group, the sugar ribose, and one of the bases adenine, cytosine, guanine, or uracil.

Ribosomal RNA (rRNA) a type of RNA that combines with proteins to form ribosomes.

Ribosome an organelle consisting of two subunits, each composed of ribosomal RNA and protein. Ribosomes are the site of protein synthesis, in which the sequence of nucleotides of messenger RNA is translated into the sequence of amino acids in a protein.

RNA polymerase an enzyme that catalyzes the covalent bonding of free RNA nucleotides into a continuous strand, using RNA nucleotides that are complementary to those of a strand of DNA.

Start codon a codon in messenger RNA that signals the beginning of protein synthesis on a ribosome.

Stop codon a codon in messenger RNA that stops protein synthesis and causes the completed protein chain to be released from the ribosome.

Structural gene a gene that codes for a protein used by the cell for purposes other than gene regulation. Structural genes code for enzymes or for proteins that are structural parts of a cell.

Transcription the synthesis of an RNA molecule from a DNA template.

Transfer RNA (tRNA) a type of RNA that (1) binds to a specific amino acid and (2) bears a set of three nucleotides (the anticodon) complementary to the mRNA codon for that amino acid. Transfer RNA carries its amino acid to a ribosome during protein synthesis, recognizes a codon of messenger RNA, and positions its amino acid for incorporation into the growing protein chain.

Translation the process whereby the sequence of nucleotides of messenger RNA is converted into the sequence of amino acids of a protein.

STUDY QUESTIONS

1. Draw an RNA nucleotide. How does RNA differ from DNA?
2. Describe RNA synthesis. Where does it occur?
3. What are the three types of RNA? What are their functions?
4. Define genetic code, codon, and anticodon. What is the relationship between the nucleotides in DNA, the codons of mRNA, and the anticodons of tRNA? What does it mean to say that the genetic code is degenerate?
5. Diagram and describe protein synthesis.
6. What is an operon? Are operons found in prokaryotes, eukaryotes, or both?
7. Describe the process of gene regulation in prokaryotes, using the lactose operon as an example.
8. Diagram the structure of a eukaryotic gene, including both the internal structure of the gene and the nearby control regions of the chromosome.
9. How is mRNA formed from a eukaryotic gene?
10. How do steroid hormones regulate eukaryotic genes?
11. Describe the molecular relationships that are involved in genetic dominance and recessiveness. What is the nature of the enzymes synthesized from the directions encoded in many recessive alleles?
12. Describe four functional consequences of substitution mutations.
13. In what ways are laboratory and natural methods of recombining DNA similar? How are they different? What benefits and dangers might you see in recombinant DNA technology?

SUGGESTED READINGS

Anderson, W. F., and Diacumakos, E. G. "Genetic Engineering in Mammalian Cells." *Scientific American,* July 1981. Recombinant DNA techniques have advanced rapidly to the point of inserting genes into eukaryotic cells, with the ultimate goal of curing hereditary diseases.

Chambron, P. "Split Genes." *Scientific American,* May 1981 (Offprint No. 1496). The segmented nature of eukaryotic genes is described.

Crick, F. H. C. "The Genetic Code." *Scientific American,* October 1962 (Offprint No. 123). The determination of the triplet nature of the genetic code.

Crick, F. H. C. "The Genetic Code: III." *Scientific American,* October 1966. The genetic code is completely solved.

Gilbert, W., and Villa-Komaroff, L. "Useful Proteins from Recombinant Bacteria." *Scientific American,* April 1980 (Offprint No. 1466). A description of recombinant DNA techniques and uses, by one of the pioneers in the field.

Murray, A. W. and Szostak, J. W. "Artificial Chromosomes." *Scientific American,* November 1987. Geneticists can now synthesize not only individual genes, but even whole chromosomes.

Tompkins, J. S. "Capitalizing on Life." *Science Digest,* June 1986. The genetic and commercial sides to genetic engineering are explained.

12
Human Genetics

In principle, human genetics is similar to the genetics of peas or fruit flies, but of greater interest to human beings. In practice, however, the study of human genetics is vastly more complex. First, humans have long life spans and few children per couple. Second, people choose their own mates, so crosses that might be genetically informative may not occur very often. Finally, humans interact extensively with their environment, an environment that is incredibly diverse from culture to culture and even within a single culture. These interactions often obscure underlying genetic patterns. This is particularly true for personality traits, in which learning and experience play such a major role.

Nevertheless, human genetics is a flourishing field of study, and a great deal is known about the inheritance of human traits. From the viewpoint of genetics, two factors partially compensate for the difficulties humans present as experimental subjects: the enormous number of human beings on the Earth and the extensive documentation of human families, ranging from family Bibles to the pedigrees of royalty. Given enough time, ingenuity, and patience, geneticists can often find records of matings that help them to determine how particular human traits are inherited, especially physical features and genetic diseases.

METHODS IN HUMAN GENETICS

PEDIGREE ANALYSIS. Since experimental crosses are out of the question, human geneticists must search medical, historical, and family records to find crosses that have already been made voluntarily. Records extending across several generations can be arranged in the form of **family pedigrees,** such as those shown in Fig. 12-1. Careful analysis of pedigrees shows that certain traits, such as an unattached earlobe, are inherited as simple dominants, while other traits, such as albinism, are inherited as recessives.

Pedigree analysis requires that two important criteria be met. First, the trait must be clearly defined. This might seem trivial, yet often proves frustratingly difficult. For instance, "everyone knows" that brown eyes are dominant to blue, but eye color is really much more complex than that. Eyes are not simply blue or brown: they may be light blue, violet-blue, green, golden brown, medium brown, dark brown, or almost black. Carefully defining human eye colors presents quite a different picture than a simple dominant versus recessive pattern of inheritance.

The second criterion is closely related to the first: the people who draw the family tree must be certain who possessed which traits. When traits are difficult to define, or are commonplace, accurate pedigrees are rare. Who remembers, or would think to record, whether their great-grandmother's earlobes were attached or unattached? Thus, much of what we know about human genetics relates to striking physical features such as Kirk Douglas' cleft chin, or to diseases such as phenylketonuria.

MOLECULAR GENETICS. In the past 30 years, great strides have been made in understanding gene function on the molecular level. For instance, geneticists now know the molecular basis of several inherited diseases, such as sickle-cell anemia (see below). Genetic engineering promises to increase our ability to predict genetic diseases and perhaps even to cure them (see the essay, "The Uses of Recombinant DNA in Medical Genetics" on p. 222).

SINGLE-GENE INHERITANCE

Many "normal" human traits, such as freckles, the ability to roll one's tongue, and the length of one's eyelashes, are inherited in a simple Mendelian fashion. Rather than merely cataloging inherited human traits, however, we will concentrate on medically important genetic diseases and defects, and explore their consequences for individuals and society.

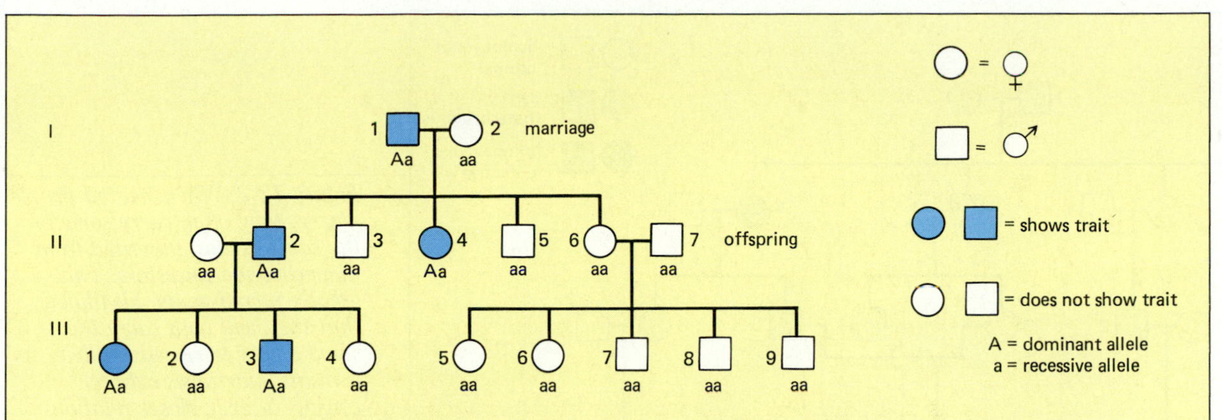

(a) *A pedigree for a dominant trait. Remember that all individuals bearing even one dominant allele will express the trait. Therefore, the trait will appear in every generation, in both males and females, and no child will possess the trait who does not have at least one parent with the trait. On the other hand, since heterozygotes show dominant traits, not all the children of phenotypically dominant parents will necessarily show the trait (e.g., generation III, individuals 2 and 4). In this hypothetical pedigree, the genotypes of the individuals are given; in real pedigrees, genotypes must be deduced from inheritance patterns and cannot always be determined with certainty.*

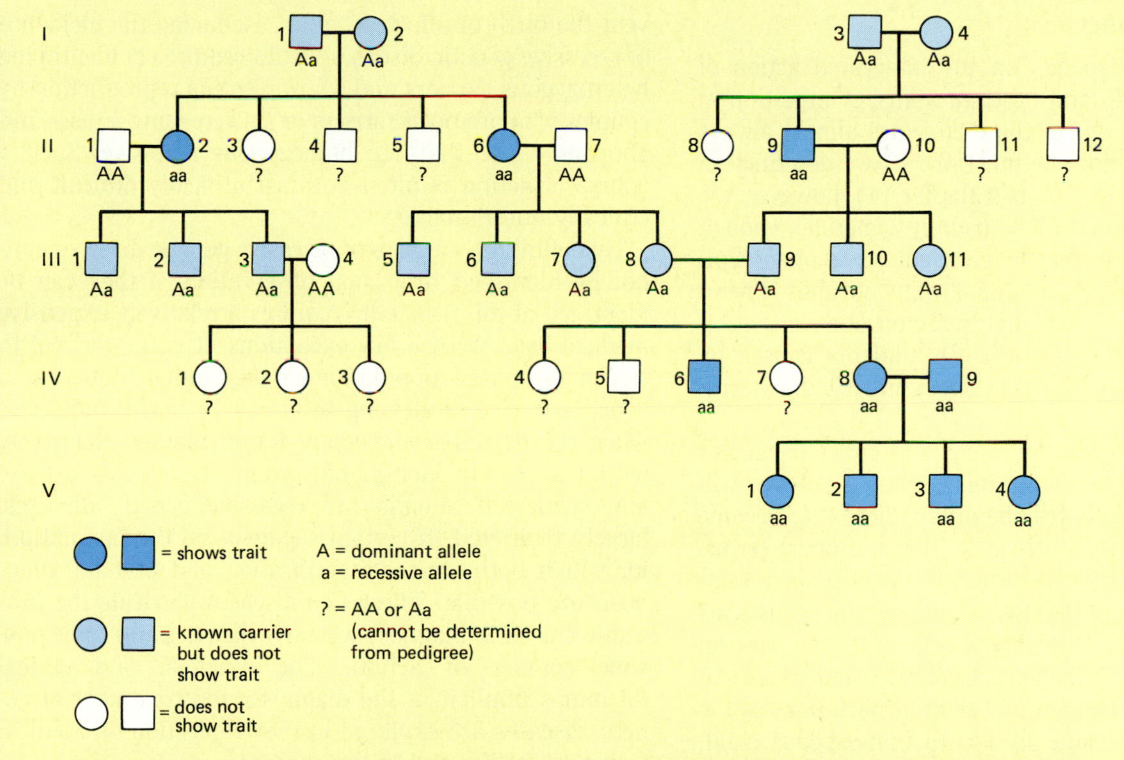

(b) *A pedigree for a recessive trait. Only homozygous recessives will show the trait. People with only one recessive allele will have the dominant phenotype. Parents with the dominant phenotype but who have a child who shows the recessive trait must themselves be heterozygotes. Therefore, the recessive phenotype may not appear in every generation (e.g., generation III, families composed of individuals 5 through 8 and 9 through 11); the recessive allele may remain "hidden" in heterozygotes for several generations until two heterozygotes happen to marry (generation III, individuals 8 and 9) and have a homozygous child (generation IV, individuals 6 and 8). Two parents who both show the recessive phenotype (IV 8 and 9) will have all homozygous recessive children (V 1–4). In a pedigree for a rare trait, it is usually safe to assume that all individuals are homozygous normal (II 1, 7, and 10) unless there is positive evidence of possession of the recessive allele (e.g., an affected child).*

Figure 12-1 *Representative family pedigrees.*

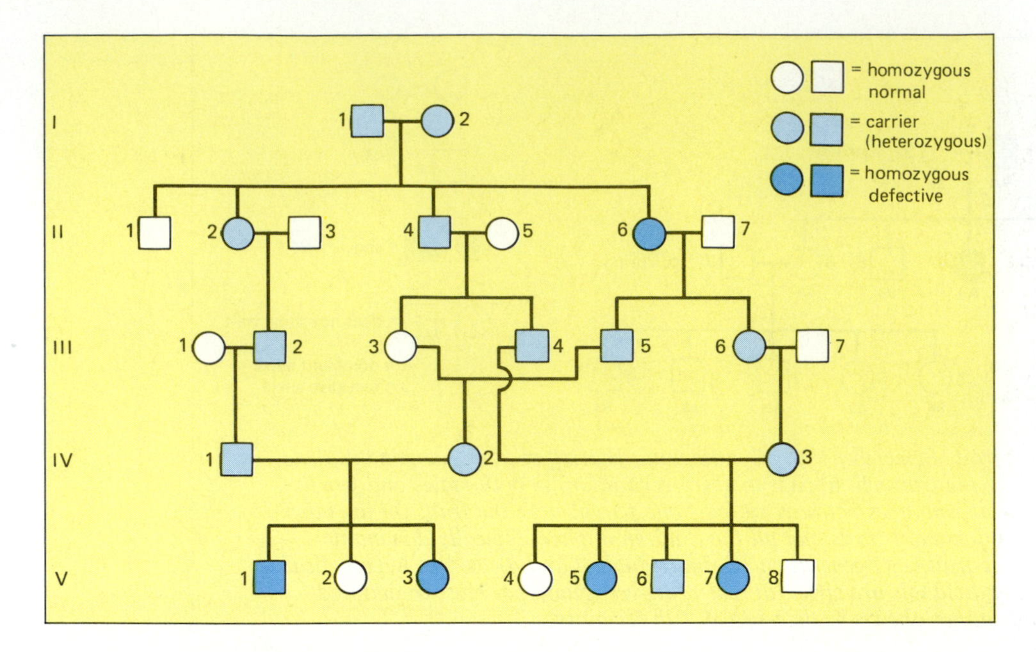

Figure 12-2 When related people marry, they often carry some of the same alleles, inherited from their common ancestors. This greatly increases the likelihood that they will both carry the same defective recessive allele. As a result, marriages between cousins or even closer relations are the cause of a disproportionate number of recessive diseases. In this family, marriages between cousins were common, including those between III 3 and 5, III 4 and IV 3, and IV 1 and 2.

Recessive Inheritance

The human body depends on the integrated action of hundreds of enzymes and other proteins. A mutation in the gene coding for one of these enzymes almost always impairs or destroys enzyme function. As we described in the case of fat color in rabbits (Chapter 11), however, the presence of one normal allele usually generates enough functional enzymes so that heterozygotes are phenotypically indistinguishable from homozygous normals. Therefore, normal alleles are usually dominant and mutant alleles are recessive. Genetic diseases that are caused by lack of an essential enzyme are mostly inherited as recessives.

If the metabolic pathway controlled by a particular gene is essential to survival, homozygous recessives will die. This was especially true before medical care became widely available in recent decades. In evolutionary terms, this means that defective alleles are selected against. Consequently, for serious diseases, even heterozygous **carriers,** who are phenotypically normal but can pass on their defective allele to their children, are usually rare. An unrelated man and woman will seldom both possess the same defective allele and produce a homozygous child. Related couples, however, especially first cousins or closer, have inherited some of their genes from recent common ancestors. Therefore, they are much more likely to carry the same allele and have an affected child (Fig. 12-2).

By their very nature, genetic diseases are an integral part of an affected person. They can neither be prevented with vaccines nor cured with antibiotics. Some, such as diabetes, can be treated, with varying degrees of success. The genetic defect remains, however, and the affected person may pass on the disease to his or her children. The only way to prevent genetic diseases, therefore, is to prevent the birth of affected babies. Reducing the incidence of recessive genetic diseases would require (1) identifying heterozygous carriers and (2) preventing reproduction by couples who are both carriers or (3) screening fetuses and aborting those that are homozygous recessive. Such a course of action is often both scientifically difficult and ethically questionable.

By definition, carriers of recessive genetic diseases cannot be identified by casual observation; if they can be identified at all, it usually requires a relatively expensive medical test. With a few exceptions, it is impractical to screen the entire population of the United States for a genetic disease. Screening tests can be useful, however, when the defective alleles are found almost exclusively within a readily identifiable group. Tay–Sachs disease and sickle-cell anemia are recessive genetic diseases, largely restricted to certain segments of the population, for which both carrier identification and prenatal diagnosis are possible. These two diseases illustrate the procedures used to diagnose recessive diseases and some possible courses of action. The practical and ethical dilemmas implicit in the diagnosis and treatment of genetic diseases are explored in the "Reflection on Medical Genetics" at the end of the chapter.

TAY–SACHS DISEASE. Tay–Sachs disease occurs when brain cells fail to synthesize an enzyme involved in lipid metabolism. Lipid accumulates in the brain of a child who is homozygous for the Tay–Sachs allele, causing progressive mental retardation, blindness, and failure of motor control (Fig. 12-3a). There is no cure, and death occurs in early childhood.

The Tay–Sachs allele is very rare in the U.S. population as a whole; only about 1 in 400 Americans is a carrier.

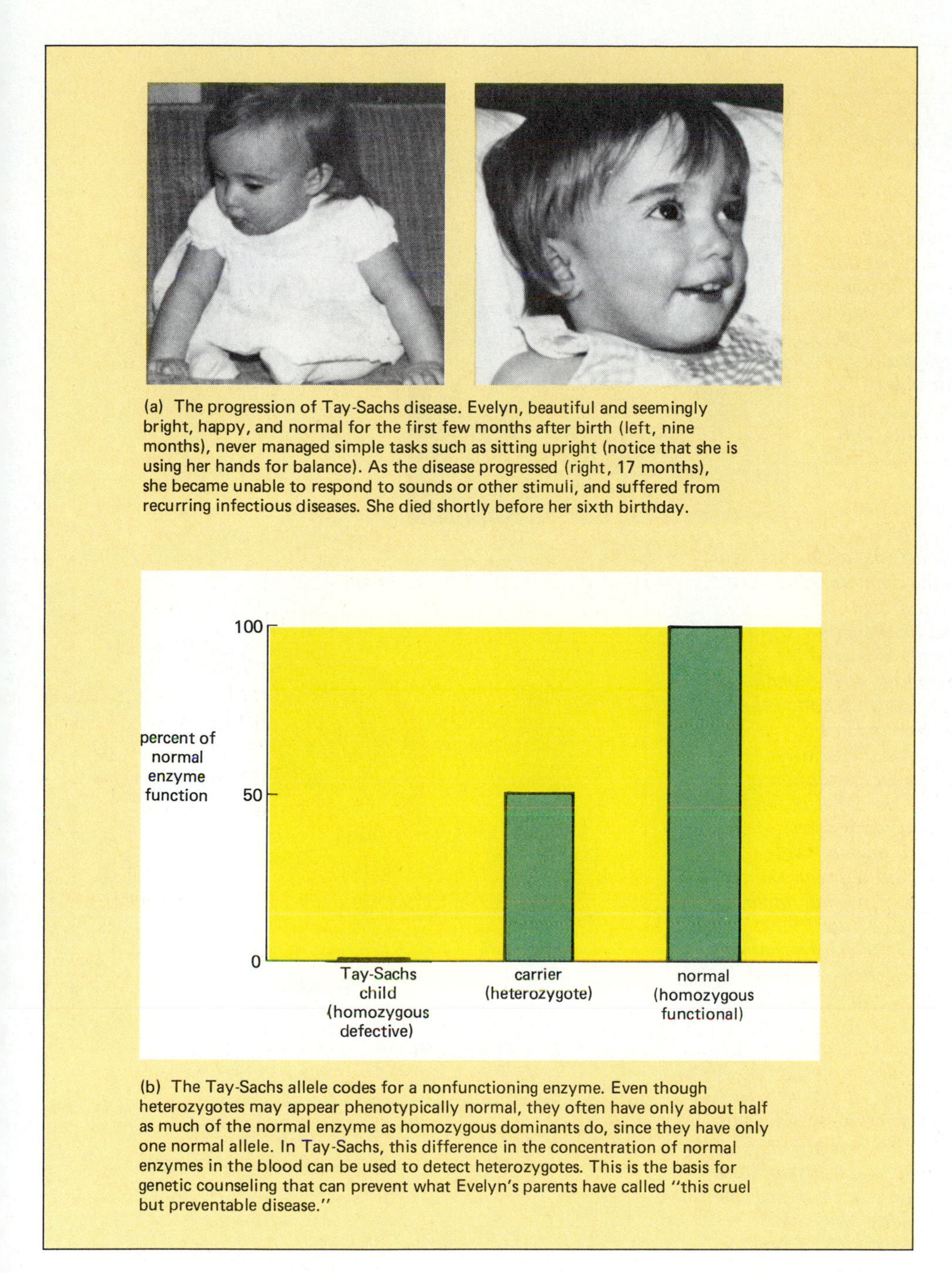

(a) The progression of Tay-Sachs disease. Evelyn, beautiful and seemingly bright, happy, and normal for the first few months after birth (left, nine months), never managed simple tasks such as sitting upright (notice that she is using her hands for balance). As the disease progressed (right, 17 months), she became unable to respond to sounds or other stimuli, and suffered from recurring infectious diseases. She died shortly before her sixth birthday.

(b) The Tay-Sachs allele codes for a nonfunctioning enzyme. Even though heterozygotes may appear phenotypically normal, they often have only about half as much of the normal enzyme as homozygous dominants do, since they have only one normal allele. In Tay-Sachs, this difference in the concentration of normal enzymes in the blood can be used to detect heterozygotes. This is the basis for genetic counseling that can prevent what Evelyn's parents have called "this cruel but preventable disease."

Figure 12-3 *Tay–Sachs disease.*

However, for unknown reasons, about 1 in 30 American Jews is a carrier, which would result in about 1 Tay–Sachs child in 3600 Jewish births. Heterozygotes for Tay–Sachs disease can easily be detected by a blood test (Fig. 12-3b), and many prospective Jewish parents avail themselves of this test. If both husband and wife are identified as carriers, they have three choices: forgo reproduction entirely; take a 25 percent chance of having an affected child who will face suffering and a certain early death; or through **amniocentesis** (Fig. 12-4) determine if each fetus is affected and abort homozygous recessives.

SICKLE-CELL ANEMIA. As with Tay–Sachs disease, almost all the carriers of sickle-cell anemia belong to one ethnic group, in this case blacks. In sickle-cell anemia, a single nucleotide substitution in DNA (adenine for thymine) causes valine to be substituted for glutamic acid at one position on the outside of the hemoglobin molecule. Glutamic acid is highly charged, while valine is neutral and hydrophobic. Glutamic acid is crucially important in keeping hemoglobin molecules dissolved in the cytoplasm of the red blood cells. Substituting valine at this position causes the hemoglobin molecules to clump together, forcing the red blood cells into the sickle-like shape that gives

Figure 12-4. Amniocentesis

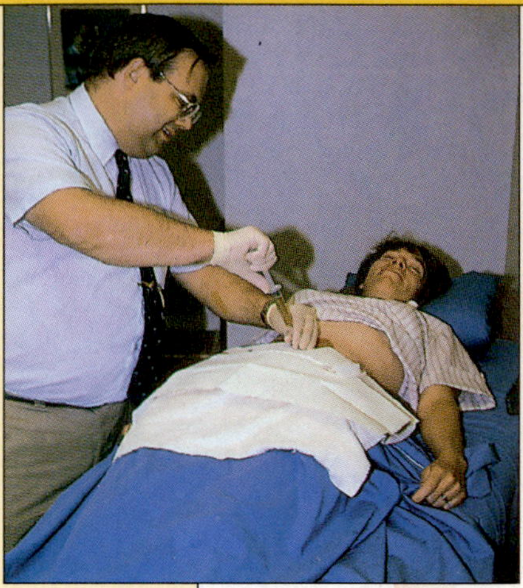

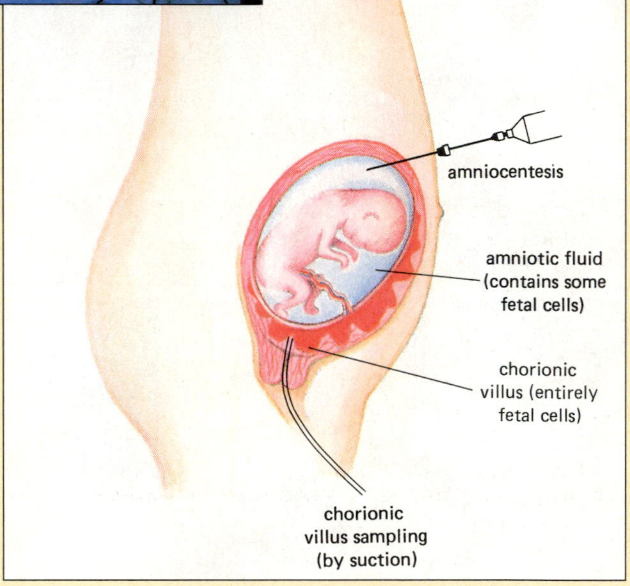

The human fetus, like all animal embryos, develops in a watery environment. A waterproof membrane called the amnion (see Chapter 32) surrounds the fetus and contains the amniotic fluid. This fluid contains fatty acids, steroids, free amino acids, enzymes and other proteins, and cells. The chemicals come from both the mother and the fetus, but the cells are all from the fetus, having been shed by the skin, digestive tract, and respiratory tract. When a fetus is 16 weeks or older, enough amniotic fluid has accumulated, and there is enough space between the fetus and the amniotic membrane, so that amniotic fluid can be collected by a procedure called amniocentesis. A physician determines the position of the fetus by ultrasound scanning and inserts a sterilized needle through the abdominal wall of the pregnant woman and into the amniotic fluid. Ten to twenty milliliters of fluid are withdrawn.

Several analyses can be performed on the fetal cells or on the amniotic fluid. Biochemical analysis is used to determine the concentration of chemicals in the amniotic fluid. For example, Tay–Sachs disease and many other metabolic disorders can be detected by the low concentration of the enzymes that normally catalyze specific metabolic pathways, or by the accumulation of precursors or byproducts. Analysis of the DNA of fetal cells with recombinant DNA techniques can detect some defective alleles, such as the sickle-cell anemia allele. Analysis of the chromosomes of the fetal cells can show if all the chromosomes are present in their normal number, if there are too many or too few of some, and if there are gross structural abnormalities of any of the chromosomes.

Recently, a new procedure has been developed, called chorionic villus sampling. The chorion is a membrane, produced by the fetus, that becomes part of the placenta. Small projections of the chorion, called villi, can be pulled off for analysis. Chorionic villus sampling promises two great advances over amniocentesis. First, it can be done much earlier in pregnancy, perhaps as early as the eighth week. Second, the sample contains a much higher concentration of fetal cells than amniocentesis can obtain, so analyses can be done without first having to allow the cells to multiply in the laboratory. Many researchers predict that chorionic villus sampling will largely replace amniocentesis within a few years.

the disease its name (Fig. 12-5). During exercise or stress, the sickled cells break and clog the capillaries, cutting off circulation. In some instances, this may cause fatal strokes or heart attacks. Heterozygotes have about half normal and half abnormal hemoglobin, which you might expect would lead to partial sickle-cell symptoms. However, heterozygotes usually have few sickled cells and show no symptoms whatever (in fact, many world-class black athletes are heterozygotes).

Sickle-cell anemia is surprisingly common among blacks. In the United States, about 8 percent of the black population is heterozygous; in parts of Africa, two or three times this many may be heterozygous. Given how severe the disease can be to homozygotes, why hasn't natural selection eliminated the sickle-cell allele? The prevalence of the sickle-cell allele is probably an evolutionary compromise: people who are heterozygous for the sickle-cell allele enjoy some protection against malaria. In Africa, where malaria is still all too common, this protection may make the difference between life and death. In Chapter 14 we explore how the dual selective effects of diminished hemoglobin function and protection against

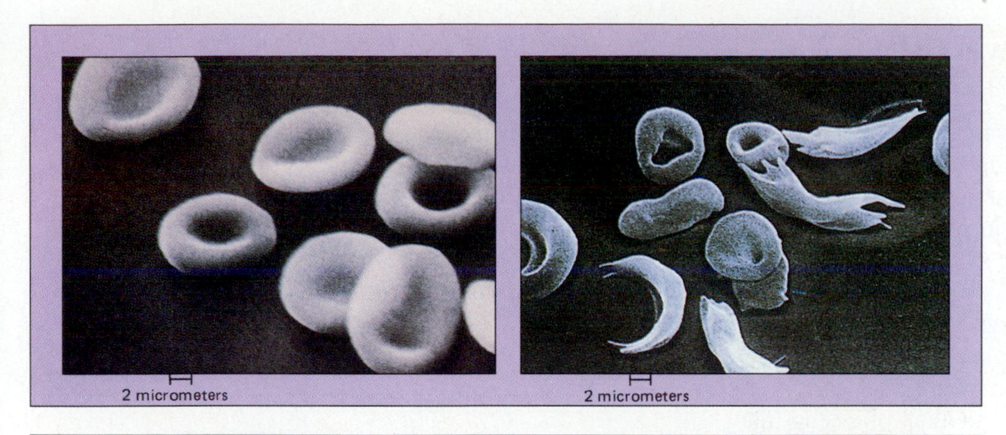

Figure 12-5 Sickle-cell anemia is caused by abnormal hemoglobin molecules that clump together, especially in low oxygen concentrations such as occur in capillaries during exercise. The clumps force the red blood cell out of its normal disk shape **(left)** *into a longer, sickle shape* **(right)**. *The sickled cells are more fragile than normal red blood cells, rendering them likely to break and/or clog in capillaries. Tissues "downstream" from such a clogged capillary do not receive oxygen or have their wastes removed. This can cause pain and, if it occurs in the brain or heart, serious injury.*

malaria may have favored the preservation of both the normal and sickle-cell alleles in African populations.

Heterozygous carriers of the sickle-cell allele can be detected by a blood test, but until recently there was no way to find out if a fetus is homozygous or heterozygous. A few years ago, however, recombinant DNA techniques were devised that could distinguish chromosomes with the normal hemoglobin allele from those with the sickle-cell allele (see the essay, "Uses of Recombinant DNA in Medical Genetics"). Analyzing fetal cells collected by amniocentesis now allows medical geneticists to diagnose sickle-cell anemia in fetuses.

Dominant Inheritance

Many physical traits are inherited as dominants, including cleft chin and freckles. However, for two reasons, few people have serious genetic diseases caused by dominant alleles. First, as we have already explained, when a mutation strikes a normal allele, it usually produces a nonfunctioning, recessive allele. Second, everyone bearing a dominant defective allele will develop the disease: there can be no phenotypically normal carriers. Before the advent of modern medicine, if the disease was serious, these people died without reproducing and did not pass on their defective allele to future generations.

An exception to this rule is Huntington's disease (Fig. 12-6). An incurable disease, Huntington's causes a slow, progressive deterioration of parts of the brain, resulting in loss of motor coordination, personality disturbances, and eventual death. Huntington's is particularly insidious because symptoms usually do not occur until 30 to 50 years of age. Therefore, a person usually has already had children before he or she suffers the first symptoms.

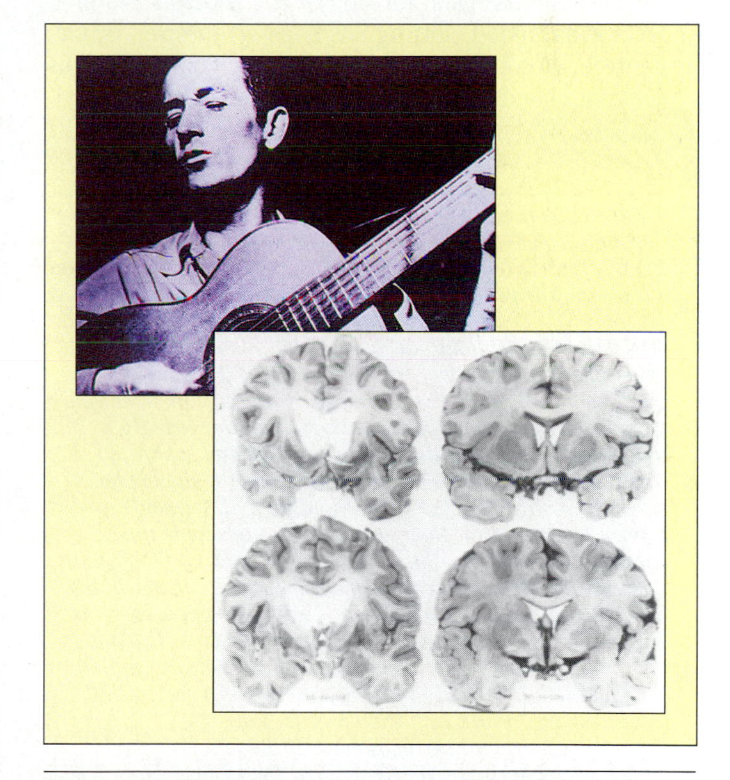

Figure 12-6 The public became aware of Huntington's disease when the folksinger Woody Guthrie **(top)** *developed symptoms in the 1950s, and again in 1981 when* 60 Minutes *interviewed three afflicted families. The reason for the behavioral and motor deficits that Huntington's patients develop is appallingly clear when we compare a cross section of the brain of a person who died from Huntington's* **(bottom**, *left***)** *to a normal adult brain* **(bottom**, *right***)**. *Since symptoms usually do not appear until middle age, children of Huntington's patients, such as Woody Guthrie's son Arlo, live for decades knowing that they too may develop this incurable disease.*

The Uses of Recombinant DNA in Medical Genetics

Although recombinant DNA technology was pioneered just a few years ago, it is proving to be an increasingly valuable tool in human medical genetics. Three applications of genetic engineering in medicine include the synthesis of therapeutic drugs or other agents, early diagnosis of genetic diseases, and possibly even the cure of genetic diseases by implanting genes. As we pointed out in Chapter 11, implanting genes into people is presently impractical, until geneticists learn more about how genes are regulated in eukaryotic cells. The first two applications, however, are already here.

Synthesis of Therapeutic Agents

In Chapter 11 we briefly explored one medical use of recombinant DNA, implanting the genes for human hormones into bacteria to provide an essentially unlimited supply of genuine human hormones for the treatment of diabetes and certain types of dwarfism.

Recently, the gene for one of the proteins required for normal blood clotting has been synthesized by Genentech, a commercial biotechnology company. This protein, called factor VIII, is defective in many hemophiliacs, who must take regular injections of factor VIII so that their blood will clot normally. Factor VIII can be extracted from donated blood at great cost (a year's supply can cost up to $10,000) and some danger (certain diseases, such as hepatitis B and acquired immune deficiency syndrome [AIDS], have sometimes been transmitted to hemophiliacs in Factor VIII extracts).

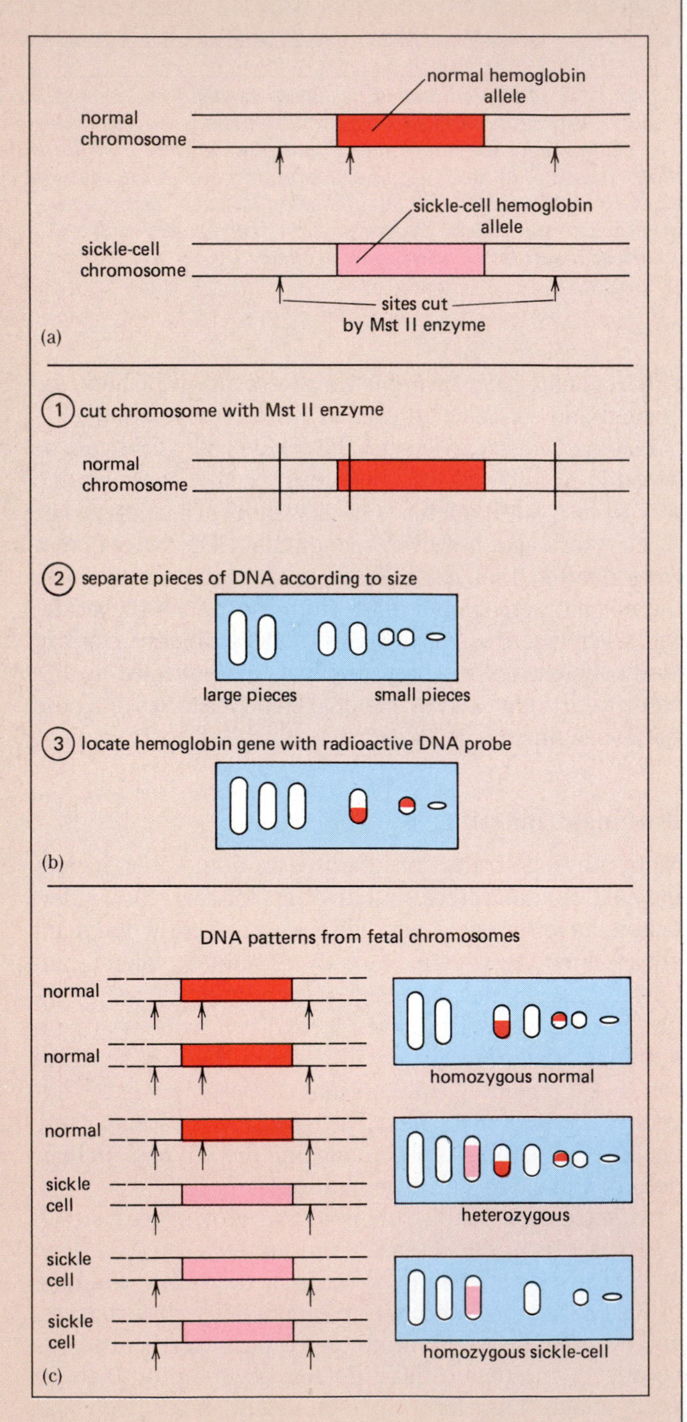

Figure E12-1 Diagnosing sickle-cell anemia with restriction enzymes.
(a) *Representations of chromosomes bearing the normal and sickle-cell alleles of the hemoglobin gene. The restriction enzyme Mst II cuts apart chromosomes in several places. Mst II can cut the normal hemoglobin allele but cannot cut the sickle-cell allele.*
(b) *The use of restriction enyzmes to localize the hemoglobin gene. (1) The chromosome is cut up with the Mst II enzyme. (2) The resulting pieces of DNA are separated according to size. This process also splits the double helix into single-stranded pieces. Although DNA segments are drawn in the figure, they are actually not visible unless labeled in some way. (3) A radioactive piece of DNA, with a nucleotide sequence complementary to one strand of the hemoglobin gene, is synthesized. This radioactive DNA is applied to the separated chromosome segments. It binds to the pieces of the hemoglobin gene by base pairing, so that its radioactivity "lights up" only those DNA pieces that contain parts of the hemoglobin gene (red).*
(c) *Analysis of fetal chromosomes. If the fetus is homozygous normal, Mst II will cut the hemoglobin alleles on both homologous chromosomes into two short pieces. The radioactive DNA probe will light up only short pieces of DNA (red). If the fetus is heterozygous, Mst II will cut the normal hemoglobin allele into two short pieces of DNA but will be unable to cut the sickle-cell allele. The DNA probe will therefore light up both short pieces of DNA (from the normal allele; red) and long pieces (from the sickle-cell allele; pink). If the fetus is homozygous recessive for the sickle-cell allele, the radioactive DNA probe will light up only long DNA segments (pink).*

The Uses of Recombinant DNA in Medical Genetics

The cloned factor VIII gene has been inserted into cultured mammalian cells, which produce vast quantities of the protein as a result. Clinical trials of mass-produced factor VIII began in late 1985; Genentech may be able to market its factor VIII product in 1988.

Early Diagnosis of Genetic Disorders

For some diseases, recombinant DNA technology can assist in diagnosing genetic diseases, even before the first symptoms appear. Probably the most straightforward application is in the prenatal diagnosis of sickle-cell anemia. Sickle-cell anemia is caused by a point mutation in the hemoglobin gene, in which adenine is substituted for the normal thymine. Geneticists have discovered a restriction enzyme called Mst II that can cut apart the DNA of the normal hemoglobin gene, but that cannot cut apart DNA with the mutated base (Fig. E12-1a). How can this enzyme be used to diagnose sickle-cell anemia?

Fetal cells are collected by amniocentesis. Chromosomes are extracted from the cells and exposed to the Mst II enzyme. The chromosome bearing the hemoglobin gene has several places where the Mst II enzyme can cut apart the DNA (Fig. E12-1a). The resulting pieces of DNA are sorted according to size and pieces of hemoglobin gene are identified (Fig. E12-1b). Mst II cuts the normal hemoglobin allele apart but cannot cut apart the sickle-cell allele. Therefore, the normal hemoglobin allele occurs as two short pieces, whereas the sickle-cell allele occurs as one long piece. By examining the pattern of long and short pieces of the hemoglobin genes, geneticists can determine if a fetus is homozygous normal, heterozygous, or homozygous for the sickle-cell allele. If all the pieces are short, the chromosomes contain only the normal allele; if half are long and half are short, the fetus is heterozygous for the sickle-cell allele; if only long pieces are found, the fetus is homozygous recessive and will develop sickle-cell anemia (Fig. E12-1c).

More complicated methods of chromosome analysis can be used to identify people who carry the alleles for other genetic diseases. In 1984, restriction enzymes were used to localize the gene for Huntington's disease on chromosome 4. Although the Huntington's gene itself has not yet been found, a combination of restriction enzyme analysis of chromosome 4 and classical pedigree analysis now allows geneticists to identify most people who carry the Huntington's allele. Equally important, researchers hope that recombinant DNA technology may soon allow them to home in on the Huntington's gene (either the gene itself, or a place on chromosome 4 extremely close to the gene, was located in late 1987). No one knows what the protein encoded by the *normal* Huntington's gene is, or what it does in the brain, much less what is wrong with the protein encoded by the defective allele. By finding the gene and studying the proteins produced by the normal and defective alleles, perhaps it will be possible to devise replacement drug therapy to supply Huntington's sufferers with the missing protein or the substance synthesized by the normal enzyme. We cannot do this yet, but the time may be near.

Since everyone who has even one defective dominant allele shows the defective trait, dominant diseases could be virtually eliminated in a single generation if all the affected people chose not to reproduce. In the case of Huntington's, however, few people know for sure if they have the disease until after they have already had children. Therefore, to eliminate Huntington's, all people with a parent suffering from the disease would themselves have to forgo having children. Since half of these people would be homozygous normal and could not pass on the Huntington's gene, the decision not to reproduce would have to be based on statistics. For some, the choice is clear; for others, the dilemma is excruciating.

In 1984, using pedigree analysis, recombinant DNA techniques, and a lot of luck, geneticists found that the Huntington's gene is on chromosome 4. Although the procedure is not infallible, geneticists can now sample a person's DNA and predict whether he or she has inherited the Huntington's allele *before any symptoms appear.* At first glance, this is a wonderful medical advance, which could result in the elimination of this devastating disease. For the individuals involved, however, the knowledge will be gained at enormous personal cost. What would *you* do if you were told that in 20 or 30 years you would develop Huntington's? Many people do not want to know.

Sex-Linked and Sex-Influenced Inheritance

As we described in Chapter 9, the X chromosome bears many genes that have no counterpart on the Y chromosome. With one X and one Y chromosome, males are effectively haploid for X-chromosome genes. As a result, recessive alleles of X-chromosome genes are always expressed in men, a phenomenon called **sex-linked inheritance.**

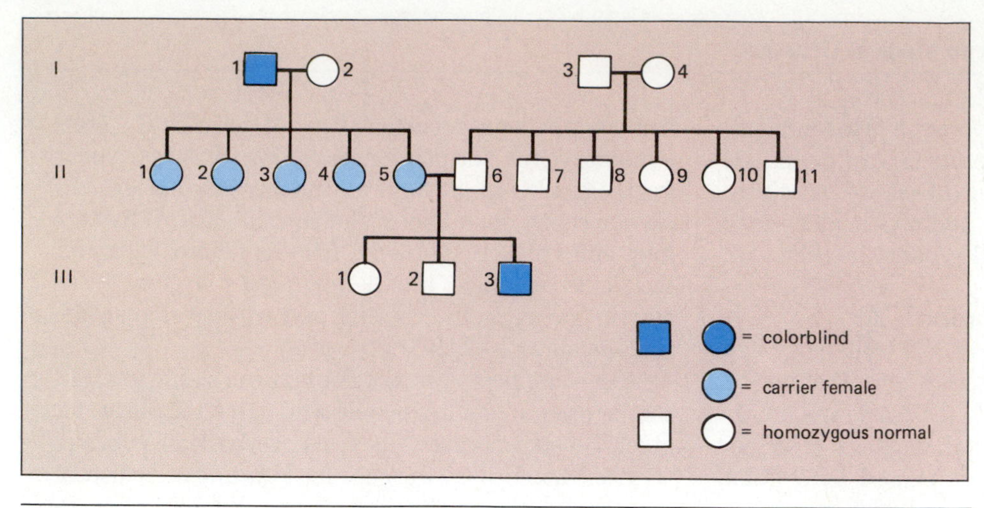

Figure 12-7 *Part of the family tree of one of the authors (GJA; III 3), showing sex-linked inheritance of red-green color vision deficiency. The author and his maternal grandfather are both color deficient, but neither his mother and her sisters, nor any relatives of his father, show the defect. This pattern of skipping generations, a more common occurrence in males, and transmission from affected male to carrier female to affected male is typical of X-linked recessive traits.*

A son receives his X chromosome from his mother and his Y chromosome from his father. A man must therefore inherit X-chromosome genes from his mother, and can only pass them on to his daughters. For rare, recessive alleles of X-chromosome genes, there is usually a striking pattern of inheritance (Fig. 12-7). Recessive traits appear most frequently in males, and skip generations, with an affected male passing the trait on to a phenotypically normal, carrier daughter, who in turn bears affected sons. The most familiar genetic defects due to recessive alleles of X chromosome genes are red-green color blindness (Fig. 12-7) and hemophilia (Fig. 12-8).

Each sex also has its own set of **sex-influenced traits** (in addition to the obvious ones of breast and genitalia development) that occur more commonly or more strongly in that sex but are not coded by genes on the sex chromosomes. The majority of sex-influenced traits affect males more often than females and seem to be enhanced by male sex hormones. A familiar example is baldness, which appears as if it were dominant in men (heterozygotes become bald) but recessive in women (heterozygotes retain their hair). The hormonal connection in baldness is readily apparent in people with abnormal levels of sex hormones. Castrated men almost never become bald, but if given testosterone treatments, castrated men with the allele for baldness usually lose their hair. Some women with tumors of the adrenal cortex produce large amounts of testosterone, and may become bald. Other sex-influenced traits that are more common in men than in women include gout, allergies, harelip, and cleft palate.

Is the Y chromosome totally devoid of genes other than those involved in determining maleness? Probably not, but it turns out to be very difficult to prove Y-linked inheritance. You might think that Y-linked traits should be easy to demonstrate, since *all* the males and *only* the males descended from a given affected male ancestor should show the trait. In practice, the situation is complicated by the influence of male sex hormones. Most traits that once were thought to be due to Y-chromosome genes are actually due to autosomal genes whose expression is strongly influenced by testosterone levels.

COMPLEX INHERITANCE

Many human characteristics are not inherited in a simple either/or fashion. Even traits that are controlled by single genes, such as Huntington's disease, are often influenced by other genes and by environmental factors. For example, although everyone who has at least one allele for Huntington's contracts the disease, the age at which the first symptoms appear is extremely variable: some people are affected in their teens, whereas others do not develop symptoms until their 50s or even 60s. Many, perhaps most, human traits are determined by the interaction of two or more genes. Examples of such polygenic inheritance in humans are the colors of eyes and skin.

The color of the iris in human eyes varies from very pale blue through green to almost black. Nevertheless, there are no blue, green, or black pigments in the human iris. Eye colors are actually caused by the distribution of a single yellowish-brown pigment, melanin, the same pigment that colors skin and hair. The iris contains two lay-

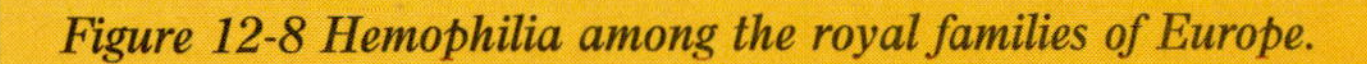

Figure 12-8 Hemophilia among the royal families of Europe.

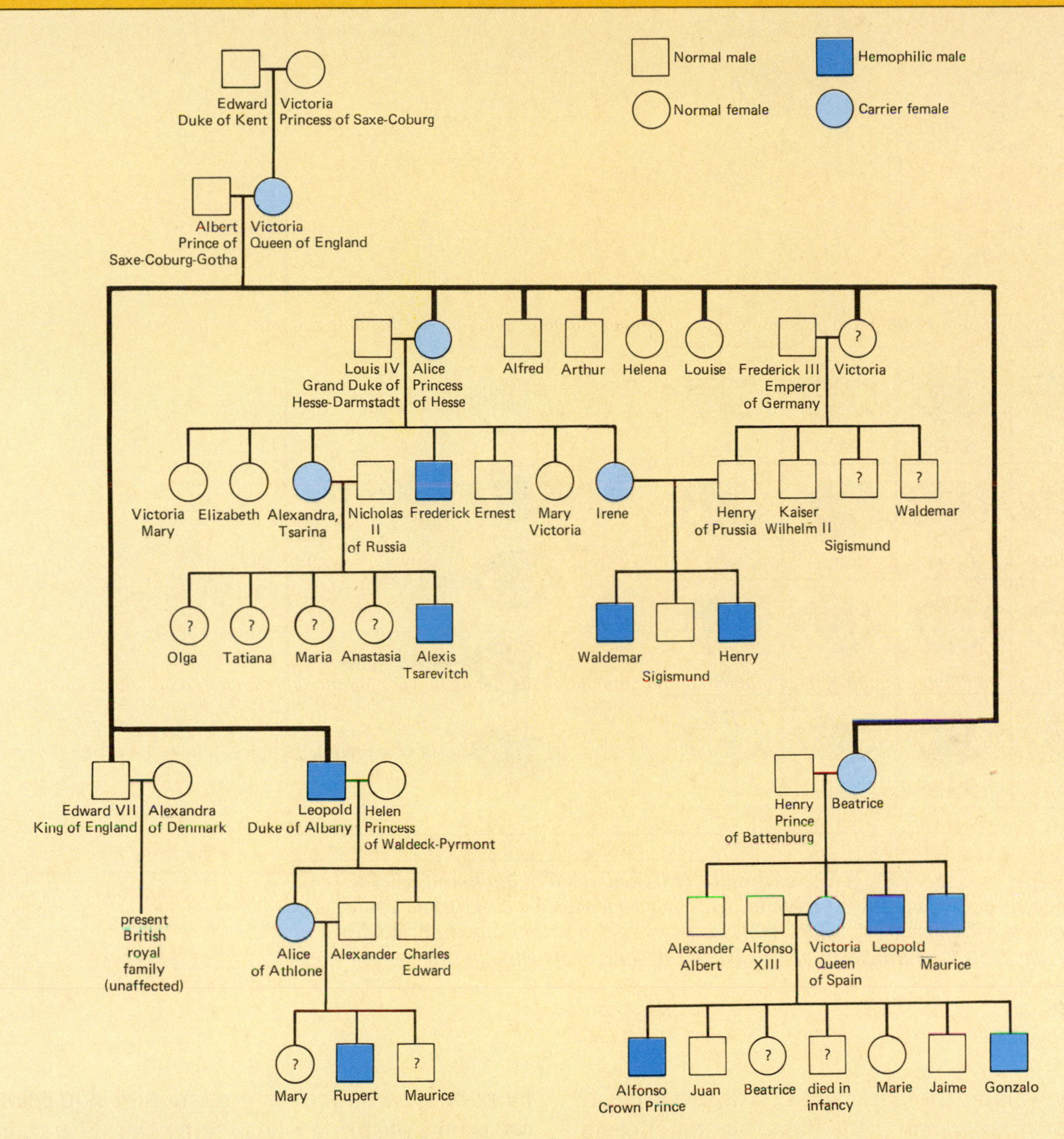

Hemophilia results from a deficiency of one of the factors causing blood clotting, a deficiency inherited as a recessive allele on the X chromosome. Affected males do not necessarily bleed to death from the first scratch or bruise; most hemophiliacs can clot off a minor wound. As a result, males often survive to pass on their allele to their daughters. Homozygous hemophiliac women are rare, but some are known. Surprisingly enough, they do not usually die from menstruation, or even from bleeding following childbirth. In both of these instances, blood flow is stopped not by clotting, but by muscular contraction, shutting off circulation to the uterine wall.

The most famous genetic pedigree in history involves the transmission of sex-linked hemophilia from Queen Victoria of England to her offspring, and eventually to virtually every royal house in Europe. Since all of Victoria's ancestors were free of hemophilia, the hemophilia allele must have arisen as a mutation either in Victoria herself when she was an embryo, or in one of her parents.

Extensive intermarriage among royalty, who, after all, are not supposed to marry commoners, spread Victoria's hemophilia allele throughout Europe. Her most famous hemophiliac descendant was great-grandson Alexis, Tsarevitch (crown prince) of Russia. Bleeding episodes in the only son of the tsar naturally distressed his parents greatly. The Tsarina Alexandra (Victoria's granddaughter) believed that the monk Gregory Rasputin, and no one else, could control Alexis' bleeding. Rasputin may actually have been able to do this through hypnosis, perhaps causing Alexis to cut off circulation to bleeding areas by muscular contraction. Although there were many causes underlying the Russian Revolution, the influence that Rasputin had over the imperial family may have contributed to the downfall of the Tsar. In any event, hemophilia was not to be the cause of Alexis's death; along with the rest of his family, he was killed by the Bolsheviks in 1918.

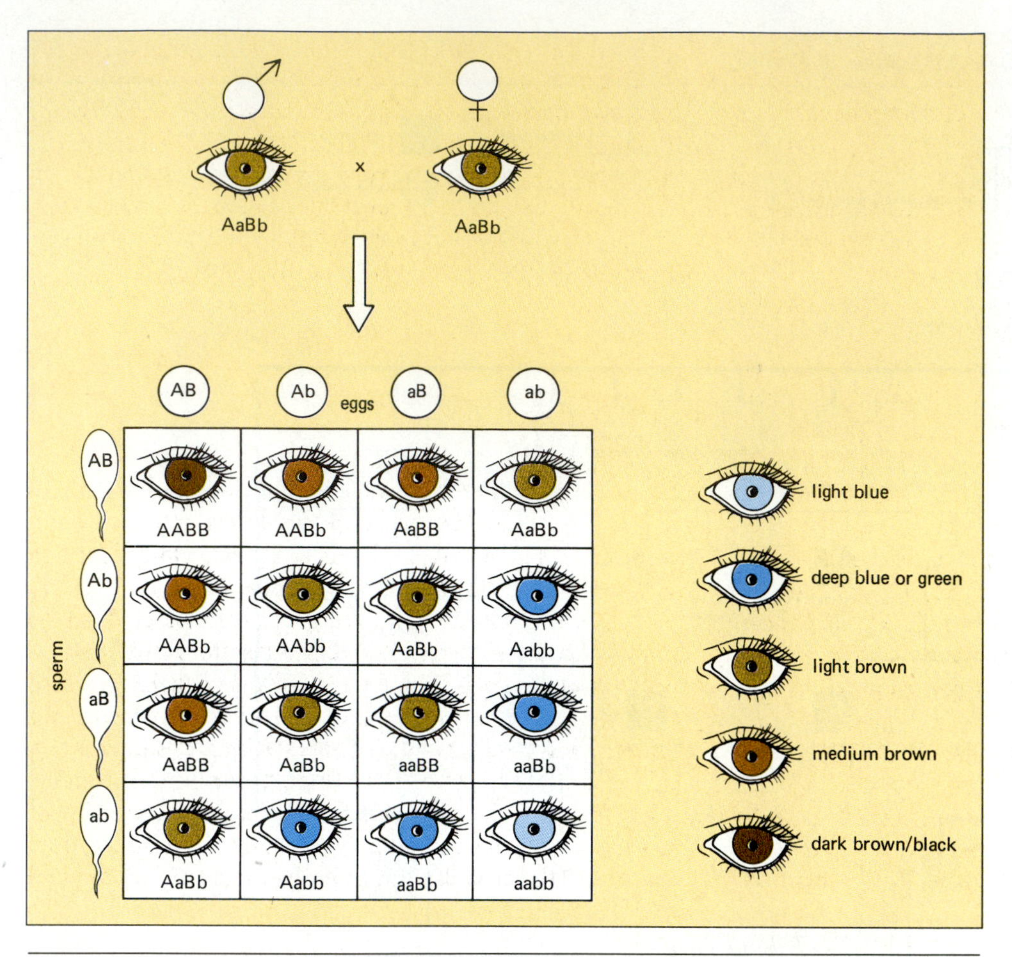

Figure 12-9 At least two separate genes, each with two incompletely dominant alleles, govern human eye color. A man and a woman, each heterozygous for both genes, could have children with five different eye colors, ranging from light blue (no dominant alleles) through light brown (two dominants) to almost black (all four alleles dominant).

ers of pigment, one at the back and one at the front. If there is little or no pigment at the front, the iris appears blue. The blue color arises from the scattering of light in the front layers, viewed against the dark background of melanin in the rear, just as the sky appears blue because of light scattered by the air, seen against the black background of space. In people with greater amounts of melanin in the front layers, the iris color may be green (blue plus yellowish brown), brown, or almost black. At least two, and probably more, genes direct synthesis of melanin in the front of the iris, with each gene having two alleles showing incomplete dominance. The simplest scheme of two genes can create five shades of eye colors (Fig. 12-9).

Skin color is another case of different amounts of melanin. People don't really have white, yellow, red, or black skins: all are various shades of brown, with a pinkish tint from surface blood vessels showing through the paler tones. Skin color is inherited through the action of at least three genes with incompletely dominant pairs of alleles. As with eye color, polygenic inheritance explains both a more-or-less continuous gradation of skin colors and the occasional offspring whose skin color differs considerably from that of either parent.

Environmental Effects

Human physical traits are also affected by the environment. This is obvious to any Caucasian who spends much time in the summer sun: skin color can be darkened by exposure to sunlight. On the other hand, a person's genotype limits the range of environmental effects on phenotype. How fast we run, how high we jump, or how well we sing can all be greatly improved by training, but no amount of practice can convert the average shower singer into another Luciano Pavarotti.

Environmental factors influence the expression of every trait in all organisms, including behavioral traits. Humans, however, with our uniquely developed capacity for personality development, social interactions, and learning, are behaviorally much more flexible than any other animal. So great is the role of learning in human personality development that there are those who maintain that the human mind is almost infinitely changeable and that there are no genetic differences among people in personality or intelligence. However, increasing evidence indicates that at least some mental traits are influenced by both environmental factors and inherited tendencies. Some people are hereditarily predisposed to the development of certain mental diseases, such as schizophrenia, although environmental events may determine whether or not a clinical syndrome develops. Some of these disorders have been traced to imbalances in important chemicals in the brain, and it is likely that affected people bear alleles coding for defective enzymes regulating the synthesis, breakdown, and use of these chemicals.

Intelligence, too, has both genetic and environmental components. Dozens of studies have compared IQ levels in people of varying degrees of relatedness (Fig. 12-10). Even when they have been separated at birth and raised in different environments, identical twins make similar scores on IQ tests, although not as similar as twins who have been raised together. Brothers and sisters who are not twins differ more than twins, but are still fairly similar, whereas parents and children are less so. Thus the more genetically related two people are, the more similar their IQ scores, indicating very strongly that intelligence, or at least whatever it is that IQ tests measure, is partially genetic. At the same time, unrelated people who have been raised together as children (e.g., adoptees) show more similarity on IQ tests than unrelated people reared apart. Therefore, it is fair to say that *both heredity and environment* play major roles in the development of intelligence, and probably other personality traits as well.

The roles of environment and inheritance in human personality are explored further in Chapter 34.

CHROMOSOMAL INHERITANCE

Like other diploid organisms, humans possess two copies of most genes, one on each of two homologous chromosomes. This is not true, of course, for sex-linked genes in men, whose Y chromosome lacks most of the genes borne on the X chromosome. It is also not true for people who inherit abnormal numbers of chromosomes. In Chapter 8 we examined the intricate mechanisms of meiosis, which act to ensure that each sperm and each egg receive one homologous chromosome of each pair. Not surprisingly, this elaborate dance of the chromosomes occasionally misses a step, resulting in gametes that have too many or too few chromosomes. Such errors, called **nondisjunction,** can affect the distribution of both sex chromosomes and

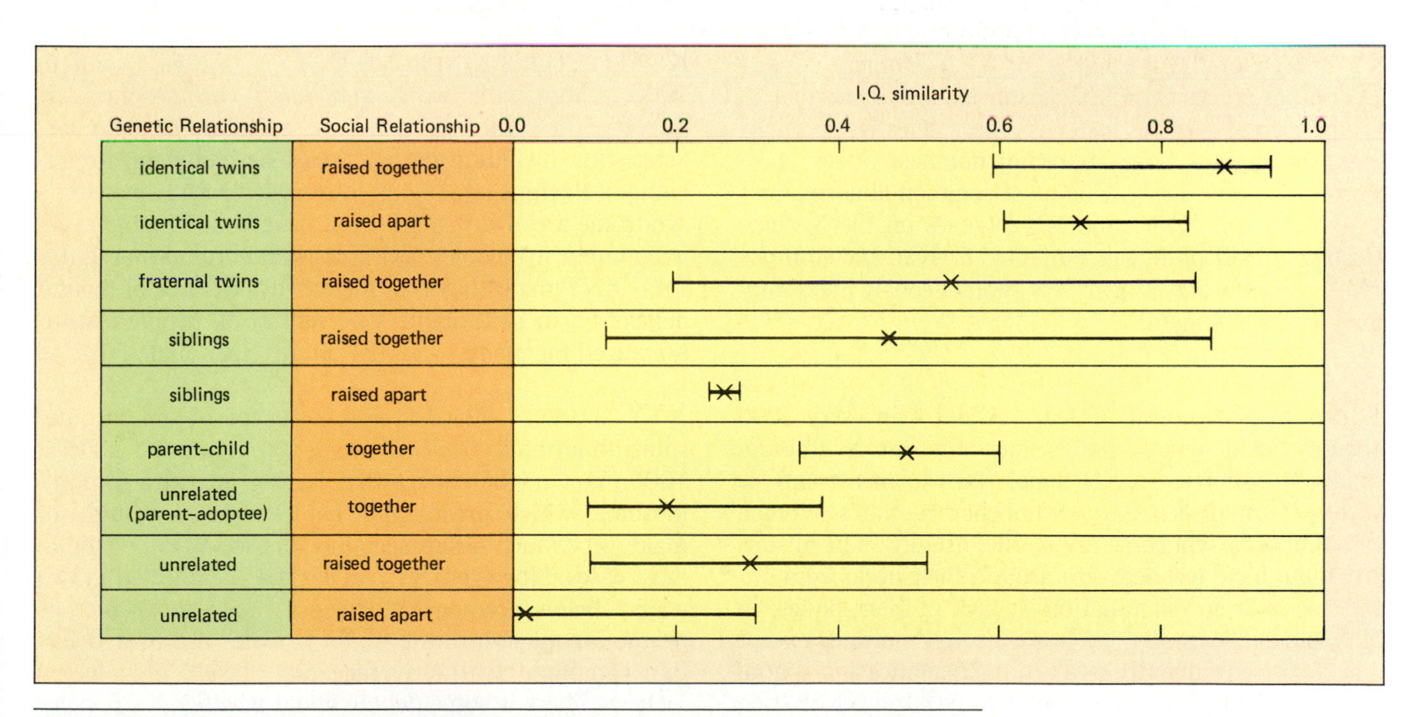

Figure 12-10 *The more closely two people are related, either genetically or socially, the more similar their IQ level is. In this graph, similarity in genetic background and familial upbringing is compared with similarity in IQ, scored on a scale of 0 (no relationship between IQ scores) to 1 (exactly the same score). The bars represent the range of results from different studies, while the crosses mark the average similarity for all the studies.* [*Modified from Bouchard and McGue,* Science *212: 1055–1058 (1981).*]

Table 12-1 **Effects of Nondisjunction of the Sex Chromosomes During Meiosis**

Nondisjunction in Father			
Sex Chromosomes of Defective Sperm	**Sex Chromosomes of Normal Egg**	**Sex Chromosomes of Offspring**	**Phenotype**
O	X	XO	Female
XX	X	XXX	Female
YY	X	XYY	Male
XY	X	XXY	Male
Nondisjunction in Mother			
Sex Chromosomes of Normal Sperm	**Sex Chromosomes of Defective Egg**	**Sex Chromosomes of Offspring**	**Phenotype**
X	O	XO	Female
Y	O	YO	Dies as embryo
X	XX	XXX	Female
Y	XX	XXY	Male

autosomes. Most of the embryos that arise from the fusion of gametes with abnormal chromosome complements spontaneously abort, accounting for 20 to 50 percent of all miscarriages, but some survive.

Abnormal Numbers of Sex Chromosomes

Nondisjunction of the sex chromosomes occurs in both men and women. Nondisjunction in men produces sperm that are O (lacking any sex chromosome), XX, YY, or XY instead of the normal X or Y. In women, nondisjunction produces O or XX eggs instead of the normal X. When normal gametes fuse with these defective sperm or eggs, the zygotes have abnormal numbers of sex chromosomes (Table 12-1). The most common abnormalities are XO, XXX, XXY, and XYY (genes on the X chromosome are absolutely essential to survival, and embryos with no X chromosome always spontaneously abort very early in development).

TURNER'S SYNDROME (XO). About 1 in every 5000 phenotypically female babies has only one X chromosome. At puberty, an XO female fails to menstruate or develop normal secondary sexual characteristics. Women with Turner's syndrome are sterile, usually short in stature, often have webbed skin around their necks, and under microscopic examination, a lack of Barr bodies in their nuclei. Mentally, they are usually normal, except that they are frequently weak in mathematics and spatial perception. The differences between XO and XX women suggest that the theory of X chromosome inactivation in females, presented in Chapter 11, is oversimplified. Some genes on the "inactivated' X chromosome must be functional in XX females, preventing the Turner's syndrome traits.

TRISOMY X (XXX). About 1 in 1000 women have three X chromosomes. These women usually have no detectable defects at all, except for a higher incidence of subnormal intelligence. Unlike women with Turner's syndrome, XXX women are fertile, and, interestingly enough, almost always bear normal XX and XY children. Some force, presently unknown, must operate during meiosis to prevent the extra X chromosome from being included in the egg.

KLINEFELTER'S SYNDROME (XXY). About 1 male in 1000 is born with two X and one Y chromosomes. At puberty, these men show mixed secondary sexual characteristics, including partial breast development, broadening of the hips, and small testes. Men with Klinefelter's syndrome are always sterile, but usually not impotent. As is common in people with abnormal chromosome numbers, XXY males have an increased incidence of mental deficiency; in fact, about 1 percent of all people institutionalized for mental retardation are XXY males.

XYY MALES. The last common type of sex chromosome abnormality is XYY, occurring in about 1 male in 1000. You might expect that having an extra Y chromosome, which presumably has few genes, would not make very much difference, and this seems to be true in most cases. However, XYY males may be affected in two ways: below-average intelligence and above-average height (about two-thirds of XYY males are over 6 feet tall, compared to the average male height of 5 feet 9 inches). There is some debate about whether XYY males are genetically predisposed to violence. For instance, several studies have shown that a higher than expected percentage of men in prison are XYY. In several countries men accused of murder have attempted to use their XYY constitution as a defense, like the insanity plea. They were not acquitted. The juries were probably right: only a minuscule percentage of XYY males ever commit any sort of crime, so an extra Y chromosome certainly does not force anyone into a life of violence.

SEX CHROMOSOMES AND SEX DETERMINATION. Studies of men and women with abnormal numbers of sex chromosomes lead to the inescapable conclusion that the Y chromosome determines maleness in humans. Having only one X chromosome, as both XY males and XO females do, does not automatically lead to maleness. In most respects, including external genitalia, XO individuals are clearly female. On the other hand, having a Y chromosome produces the male phenotype, no matter how many X chromosomes are present. Even the rare XXXY or XXXXY person is male. In computer terminology, femaleness is the "default" condition for sex; explicit instructions encoded on the Y chromosome are required to produce a male.

Abnormal Numbers of Autosomes

Autosomal nondisjunction can also occur, producing eggs or sperm with a missing autosome or two copies of an autosome. Fusion with a normal gamete (one copy of each autosome) leads to an embryo with either one or three copies of the affected autosome. With only one copy of any of the autosomes, the embryo aborts so early in development that the woman never knows that she was pregnant. Three copies of an autosome (trisomy) usually also causes a spontaneous abortion, but often later in pregnancy. Sometimes, however, trisomic babies are born, especially those with three copies of chromosomes 13, 18, or 21. Of these, trisomy 21 is the most common.

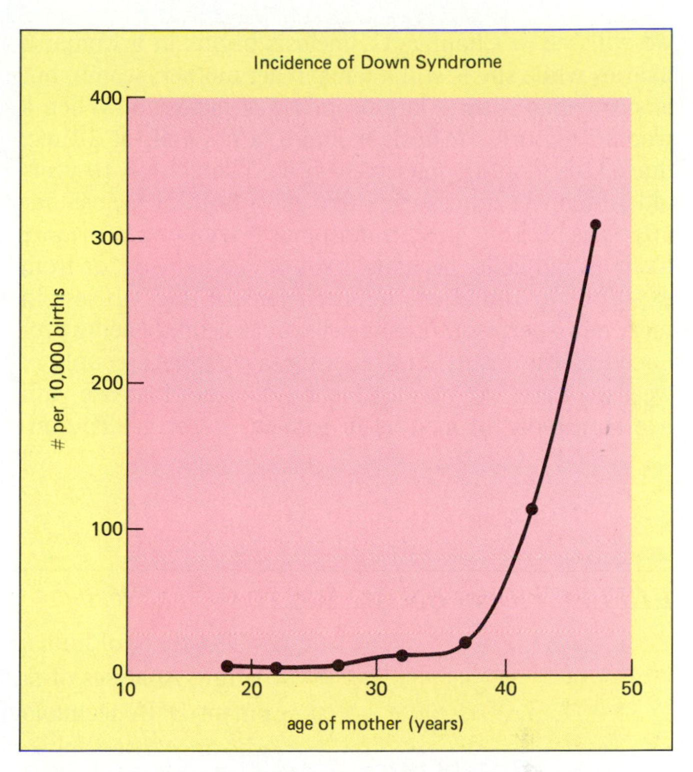

Figure 12-12 The frequency of nondisjunction is influenced by the age of the mother. This graph shows that nondisjunction of chromosome 21, producing children with Down syndrome, greatly increases after the mother reaches age 35. Geneticists have found that this is true even for women whose husbands are much younger. Nondisjunctions of other chromosomes undoubtedly also occur more frequently as women age. All embryos lacking a chromosome, and most with an extra chromosome, abort spontaneously very early during pregnancy, usually before the woman is aware that she is pregnant. This partially accounts for difficulty that many older women experience in becoming pregnant. Data from the U.S. Center for Disease Control, for births during the period 1968 to 1984.

Figure 12-11 Children with abnormal numbers of chromosomes are almost always both physically and mentally defective, in ways that are characteristic of the particular chromosome abnormality. This child has the typical relaxed mouth, "Oriental" eyes, and congenital heart defects usually seen in cases of Down syndrome.

TRISOMY 21 (DOWN SYNDROME). In about 1 of every 900 births, the child has an extra copy of the twenty-first chromosome. These children have several distinctive physical characteristics, including lack of muscle tone, a small mouth held partially open because it cannot accommodate the tongue, and distinctively shaped eyelids (Fig. 12-11). Much more serious defects include low resistance to infectious diseases, heart malformations, and mental retardation so severe that only about 1 in 25 ever learns to read and only 1 in 50 learns to write.

The frequency of nondisjunction is influenced by the age of the parents, especially the mother (Fig. 12-12). As we will see in Chapter 31, meiosis begins in a woman's ovaries while she is still a fetus in her mothers womb, but is suspended during late prophase of meiosis I. When a woman matures, meiosis resumes, a few cells at a time, during the monthly menstrual cycle. Therefore, a 40-year-old woman produces eggs that have been in meiosis for 40 years. Errors in chromosome distribution are more likely in old cells, perhaps just from aging itself, or from exposure to radiation, toxic chemicals, and viruses. In men, new sperm-producing cells are constantly being produced in the testes, and meiosis takes only a couple of weeks. Therefore, nondisjunction does not increase with age as rapidly in men as in women. Nevertheless, nondisjunction in sperm accounts for about 25 percent of the cases of Down syndrome, and there is a small age effect.

In the last two decades, it has become increasingly common for couples to delay having children so that both husband and wife can establish careers. As a result, many women bear children when they are in their late 30s or 40s, and in general they have husbands of similar age. This inevitably leads to more trisomic fetuses than if the same couples had reproduced earlier. Fetal trisomy can be diagnosed by amniocentesis (Fig. 12-4). Fetal cells are collected from the amniotic fluid and examined for abnormal chromosome numbers. A therapeutic abortion may then be used to prevent the birth of a trisomic child.

REFLECTIONS ON MEDICAL GENETICS

The study of human genetics has opened up new vistas of understanding human nature. Analyses of family pedigrees, biochemical tests, amniocentesis, and recombinant DNA technologies have enabled geneticists to identify prospective parents who are carriers for some diseases, such as sickle-cell anemia and Tay-Sachs disease, and to detect fetuses who are homozygous for these and other diseases. These new powers demand a new set of decisions, both ethical and economic, by individuals, physicians, and society. Let us examine these new choices in two specific cases, phenylketonuria and sickle-cell anemia.

Phenylketonuria. As you learned in Chapter 11, people with phenylketonuria are homozygous for a recessive allele coding for a defective version of the enzyme that converts the phenylalanine to tyrosine. As is usual with recessive diseases, most homozygous children result from the mating of two heterozygous (and therefore phenotypically normal) parents. The level of phenylalanine in the developing fetus is regulated by its pregnant mother, so that homozygous recessive infants are born with normal brain development. Once on their own, however, the infants' phenylalanine levels increase rapidly, damaging the brain. Fortunately, homozygous infants can be detected as early as 4 or 5 days after birth by a simple blood test that costs just a few dollars. Treatment consists of maintaining these children on a diet low in phenylalanine for the first five or six years, when most brain development occurs.

This is not the end of the story, however. Women with PKU who were treated in infancy have normal mental development, but they are still deficient in phenylalanine metabolism, and therefore have an elevated concentration of phenylalanine in their blood. Children of such women, fathered by homozygous normal men, will be heterozygous, and therefore should be phenotypically normal. The problem is that, just as the phenylalanine level in a homozygous recessive fetus is kept low by its normal mother, so too the phenylalanine level of a heterozygous fetus is elevated by the high levels of its phenylketonuric mother. By the time these infants are born, it is too late to treat them: they are mentally retarded for life. Only extremely careful regulation of the mother's diet can protect the developing fetus. Thus, treating phenylketonuria in female infants brings with it new responsibilities, including accurate record keeping, so that she will know her condition when she grows up. And then what? Should she refrain from reproduction unless she can pay for expensive prenatal care? Should society pay for it if she cannot? And what about those women who never learned about their condition, or who forgot, or who did not really understand what it meant? To these questions, genetics has no answer.

SICKLE-CELL ANEMIA. Like phenylketonuria, sickle-cell anemia is a recessive genetic disorder. Sickle-cell anemia, however, is almost entirely restricted to blacks. Heterozygotes, who have half normal and half abnormal hemoglobin, are readily detectable by blood analysis, but the red blood cells of almost all heterozygotes function normally.

Some years ago, several states and the federal government set up screening programs for sickle-cell anemia, in which blacks were encouraged to have free blood tests to determine if they were carriers for this disease. Inner-city schools often offered the test to their black students. Although started with the best of intentions, this program was largely a disaster. The reason for the lack of success was simple: no one had a clear idea of what to do when a person was found to be a carrier.

At the time, amniocentesis could not detect fetuses who were homozygous for sickle-cell anemia. This meant that a couple, both of whom were carriers, could only be sure of preventing the birth of a sickle-cell child by not having children at all. This is a decision that is extremely difficult for many people, especially since there is a 75 percent chance of having normal or heterozygous children. Today, recombinant DNA techniques do permit detection of homozygous fetuses. This advance eliminates one dilemma, but creates another: whether or not a couple should abort an affected fetus.

THE CHOICE. Insofar as we cannot cure inherited diseases, or foresee a cure in the near future, what are the responsibilities and options open to individuals and societies? A hundred years ago, when no one understood the nature of inherited diseases, the choices were limited and the responsibilities were few. A prospective parent could not know if he or she carried an inherited defect, could not predict the likelihood of having an affected child, and could not tell if a child was affected until the child was born or perhaps even several years old.

The situation has changed dramatically. Today, many people *do* know that they carry a seriously defective gene, and genetic counselors can predict their chances of having an affected child. Should such an individual refrain from reproduction? For many defects, such as Tay–Sachs disease and the chromosomal abnormalities, amniocentesis can detect the inherited defect in a fetus. Should it be aborted? What are the rights and responsibilities of society in these decisions, a society that, in many Western countries, pays most or all of the costs of medical care for affected children, often running to hundreds of thousands of dollars apiece?

Some people argue that the human species is becoming genetically "loaded" with defective alleles, as modern medicine allows people to survive and reproduce who have the genes for diabetes, sickle-cell anemia, or many other genetically influenced diseases. As a result, some believe that carriers of alleles for serious, incurable diseases should not reproduce, to eliminate not only homozygous recessives who are afflicted with the disease but also future carriers. That is usually the opinion of someone who is not known to be a carrier for anything. Eventually, when we become knowledgeable enough, we may find that almost *all* of us carry recessive alleles for some inherited disease or other, yet few would argue that the majority of the population is morally bound to remain childless. In a free society, probably the best solution is to give people the best information possible about their genetic constitution and that of their future children. The choices, whatever their consequences for society as a whole, almost certainly must remain with the individual.

SUMMARY OF KEY CONCEPTS

Methods in Human Genetics

The genetics of humans is similar to the genetics of other animals, except that experimental crosses are not feasible. Therefore, analysis of family pedigrees and, more recently, molecular genetic techniques must be used to determine the mode of inheritance of human traits.

Single-Gene Inheritance

Many genetic disorders are inherited as recessives; therefore, only homozygous recessive persons show symptoms of the disease, whereas heterozygotes are phenotypically normal and usually cannot be detected by casual observation. Heterozygotes are called carriers, because the offspring of two carriers may be homozygous recessive and show the trait. Recessive genetic disorders include Tay–Sachs disease, sickle-cell anemia, and phenylketonuria.

Many normal traits, and a few diseases, are inherited as simple dominants. Both heterozygous and homozygous dominant individuals show the trait.

The sex chromosomes in humans are paired in women (XX) but unpaired in men (XY). The Y chromosome bears few genes other than those determining maleness. Therefore, women show normal dominant–recessive relationships among alleles of X chromosome genes, while men phenotypically display whatever allele they carry on their single X chromosome, a phenomenon called sex-linked inheritance. Sex-linked conditions include red-green color discrimination and hemophilia.

Sex hormones secreted by the gonads help to determine the expression of sex-influenced traits. The majority of sex-influenced defects are more common in males than in females, and include baldness, harelip, cleft palate, and gout.

Complex Inheritance

Most human traits are polygenic; that is, they are influenced by the action of many genes. Examples include the color of skin and eyes, and probably many personality traits. The environment affects the phenotypic expression of all genes in all organisms. This is particularly apparent in human beings, especially in intellectual and personality traits.

Chromosomal Inheritance

Sex in humans is determined by the presence of a Y chromosome: individuals with one or more Y chromosomes are phenotypically male; individuals with no Y chromosome are phenotypically female. People with abnormal numbers of sex chromosomes often have mental and physical deficiencies. The most common defect is below-normal intelligence.

Abnormal numbers of autosomes usually lead to spontaneous abortion early in pregnancy. In rare instances, the fetus may survive to birth, but severe mental and physical deficiencies are always found. The likelihood of abnormal numbers of chromosomes increases with increasing age of the mother, and to a lesser extent, the father.

GLOSSARY

Amniocentesis (am-nē-ō-sen-tē′-sis) a procedure for sampling the amniotic fluid surrounding a fetus. Various tests may be performed on the fluid and the fetal cells suspended in it to provide information on the developmental and genetic state of the fetus.

Carrier an individual who is heterozygous for a recessive condition. Carriers display the dominant phenotype but can pass on their recessive allele to their offspring.

Down syndrome a genetic disorder caused by the presence of three copies of chromosome 21. Common characteristics include mental retardation, abnormally shaped eyelids, a small mouth with protruding tongue, short fingers, heart defects, and unusual susceptibility to infectious diseases.

Hemophilia a recessive, sex-linked disease in which the blood fails to clot normally.

Klinefelter's syndrome a set of characteristics typically found in individuals who have two X chromosomes and one Y chromosome. These individuals are phenotypically males, but sterile, and have several female-like traits, including narrow shoulders, broad hips, and partial breast development.

Nondisjunction an error in meiosis in which chromosomes fail to segregate properly into the daughter cells.

Pedigree a diagram showing genetic relationships among a set of individuals, usually with respect to a specific genetic trait.

Phenylketonuria (fe-nul-kē-tō-nū′-rē-a) a recessive disease in which the enzyme that catalyzes the conversion of the amino acid phenylalanine to tyrosine is faulty.

Sex-influenced inheritance a mode of inheritance in which traits of a nonsexual nature are more common in one sex than in the other, often due to differing levels of sex hormones.

Sex-linked inheritance inheritance of traits controlled by genes carried on the X chromosome. Females show the dominant trait unless they are homozygous recessive, whereas males will express whatever allele is found on their single X chromosome.

Sickle-cell anemia a recessive disease caused by a single amino acid substitution in the hemoglobin molecule. Sickle-cell hemoglobin molecules tend to cluster together in long chains, distorting the shape of red blood cells and causing them to break and clog the capillaries.

Tay–Sachs disease a recessive disease caused by a deficiency in enzymes regulating lipid metabolism in the brain.

Trisomy 21 *see* Down syndrome.

Trisomy X a condition of females who have three X chromosomes instead of the normal two. Most of these women are phenotypically normal, and are fertile.

Turner's syndrome a set of characteristics typical of a woman with only one X chromosome. These women are sterile, failing to develop normal ovaries. They also tend to be very short, and to lack normal female secondary sexual characteristics.

STUDY QUESTIONS

(The answers to Questions 1 and 2 are given at the end of the question list.)

1. If the frequency of heterozygous carriers for Tay–Sachs disease is one in 30 American Jews, why is the frequency of homozygous recessive babies of Jewish parents 1 in 3600?
2. If one parent of a couple has Huntington's disease (assume that this parent is heterozygous), calculate the fraction of their children that could be expected to develop the disease. What if both parents were heterozygous?
3. Why are most genetic diseases inherited as recessives rather than dominants?
4. How is sex determined in humans? What is the evidence for this?
5. Define polygenic inheritance. Why could polygenic inheritance allow parents to produce offspring that are notably different in eye or skin color than either parent?
6. What is sex-influenced inheritance? What is the evidence that hormonal levels control the expression of sex-influenced traits?
7. Define nondisjunction, and describe the common syndromes caused by nondisjunction of sex chromosomes and autosomes.

ANSWERS TO QUESTIONS 1 AND 2

1. The probability of two or more independent events occurring simultaneously is the product of their individual probabilities. The probability of the husband being heterozygous is $\frac{1}{30}$; the probability that the wife is heterozygous is also $\frac{1}{30}$; the probability that two heterozygotes will produce a homozygous recessive child is $\frac{1}{4}$ (see Chapter 9). Multiplying $\frac{1}{30} \times \frac{1}{30} \times \frac{1}{4}$ gives $\frac{1}{3600}$.
2. Let H = the Huntington's allele (dominant) and h = the normal allele (recessive). Then the first set of parents would be Hh and hh. Half of their offspring would be expected to inherit Huntington's disease. If both parents are heterozygous, then the cross is Hh = Hh. We would expect the offspring to be $\frac{1}{4}$HH, $\frac{1}{2}$Hh, and $\frac{1}{4}$hh; therefore, $\frac{3}{4}$ would develop the disease.

SUGGESTED READINGS

Baskin, Y. "Doctoring Genes." *Science 84,* December 1984. Potential techniques and ethical considerations involved in curing human diseases by gene transplants.

Friedmann, T. "Prenatal Diagnosis of Genetic Disease." *Scientific American,* November 1971. The techniques of amniocentesis are explained, and some of its biological and social consequences are explored.

Mange, A. P., and Mange, E. J. *Genetics: Human Aspects,* 2nd ed. Philadelphia: W. B. Saunders Company, 1986. Clearly and eloquently written descriptions of the biological and social aspects of human heredity.

White, R., and Lalovel, J.-M. "Chromosome Mapping with DNA Markers." *Scientific American,* February 1988 (Offprint No. 1590). Excellent descriptions and diagrams of human chromosome mapping, with practical applications in the diagnosis of genetic disorders.

Pines, M. "In the Shadow of Huntington's," *Science 84,* May 1984, and Grady, D. "The Ticking of a Time Bomb in the Genes," *Discover,* June 1987. The story of the scientific detective work involved in the diagnosis of Huntington's using recombinant DNA techniques, and the social dilemma it created. Particularly poignant because the lead investigator may herself be a victim.

Patterson, D. "The Causes of Down Syndrome." *Scientific American,* August 1987. Specific genes causing the Down defects are being mapped onto chromosome 21.

The top of the "Tower of Time," on display at the Smithsonian Institution's Museum of Natural History, summarizes the last few million years of human evolution.

Unit III
Evolution

13
Evolution: Origin of Species

When on board H.M.S. 'Beagle,' as naturalist, I was much struck with . . . the distribution of the inhabitants of South America, and . . . the geological relations of the present to the past inhabitants of that continent. These facts seemed to me to throw some light on the origin of species—that mystery of mysteries, as it has been called by one of our greatest philosophers.

from *On the Origin of Species by Means of Natural Selection* by Charles Darwin

The lines quoted above introduce perhaps the single most important work in biology. In *Origin of Species,* Charles Darwin proposed that over eons of time, species arise from other, preexisting species through the process of "descent with modification," or evolution.

Before Darwin, how species originated remained the "mystery of mysteries" for a very simple reason. Over the time span of recorded human history, let alone the life of a single human being, no new species had been recognized (although undoubtedly many new species had appeared, especially of plants). It is quite difficult to decide how something happens if there are no witnesses.

Nevertheless, one of the most striking features about our world is the remarkable variety of organisms inhabiting it. Why are there dozens of species of pine trees and scores of species of warblers? With no evidence to go on, nearly all peoples of the world historically turned to hypotheses of **creationism**. The most common of these hypotheses is that a supernatural being created each type of organism separately at the beginning of the world, and that all modern organisms are essentially unchanged descendants of these ancestors (Fig. 13-1).

Figure 13-1 The Garden of Eden, *by Brueghel.*

As we pointed out in Chapter 1, one of the fundamental principles of science is that Earthly phenomena are produced by natural, Earthly causes. A nineteenth-century English essayist wrote: ". . . with regard to the material world, we can at least go so far as this—we can perceive that events are brought about not by insulated interpositions of Divine power exerted in each particular case, but by the establishment of general laws [of nature]." Science cannot say whether or not divine power originally established those "general laws," but science firmly adheres to the principle that natural events have causes that arise from the operation of natural laws.

Therefore, throughout history, scientists have sought natural causes for the origin of species. However, it was only in the nineteenth century that a truly coherent theory—evolution by descent with modification, driven by natural selection—was developed. This theory was published by two British naturalists, Charles Darwin and Alfred Russel Wallace, in 1858, and still forms the foundation of our understanding of evolution. Darwin and Wallace did not work in a vacuum. Centuries of thought and observation preceded them and influenced their ideas. Let us begin, then, with a brief survey of evolutionary thought.

THE HISTORY OF EVOLUTIONARY THOUGHT

Before we can study the origin of species, we must first decide just what a species is. Throughout most of history, "species" was a poorly defined concept. When using the word "species," most people meant one of the originally created "kinds" referred to in the Bible. How could a naturalist tell if two organisms belonged to two different species? Since no one was present at the Creation to record the criteria of the Creator, one had to distinguish among species by visible differences in structure; in fact, the very word "species" is Latin for "appearance." Clearly, pines and warblers are different species, and warblers are different from eagles and ducks. But how do biologists distinguish among species of warblers? Today, biologists define a **species** as *all the populations of organisms that are capable of interbreeding under natural conditions, and that are reproductively isolated from other populations.* In other words, the members of a species can interbreed among themselves, but usually not with members of other species. If interbreeding with another species does occur, the hybrid offspring are usually infertile or handicapped in some way (see Chapter 14).

Figure 13-2 The myrtle warbler (top) and Audubon's warbler (bottom) were formerly considered to be two separate species, but are now considered to be merely local varieties of one widespread species.

Biologists have found that differences in appearance do not always mean that two populations belong to different species. For example, field guides published in the 1970s listed the myrtle warbler and Audubon's warbler (Fig. 13-2) as distinct species; more recently, the American Ornithological Union decided that they are, after all, merely local varieties of the same species. The main reasons for the initial splitting were differences in range and in the color of the throat feathers. Ornithologists now consider them to be a single species because, where their ranges overlap, interbreeding occurs, and the offspring are just as vigorous as the parents.

Early European naturalists didn't have to worry too much about the definition of species. Europe has a rather scanty assortment of flora and fauna, having lost much of its diversity to widespread extinctions during the last Ice Age. Most European species, at least among the more prominent land plants, mammals, and birds, differ quite a bit from their nearest relatives. Therefore, for the first 2000 years of scientific thought about the origin of species, only a limited number of distinct species had to be considered. As we shall see, this comfortable situation changed with the exploration of new lands in southeast Asia, Africa, and the Americas.

The Greek Philosophers

The ancient Greeks, who seem to have thought deeply about nearly everything, pondered the origin of species. Two main lines of thought emerged, one from Plato and one from Aristotle, that influenced later Western ideas. The philosophy of Plato (427–347 B.C.) rested on the foundation of the "ideal Form": each object on Earth, whether animate or inanimate, is a mere temporary reflection of its nonmaterial Form. Thus every dog and every human being is an imperfect version of the ideal Dog or Human Being that exists somewhere beyond the Earth in the world of Forms. The Forms are perfect and unchangeable, having come into existence in some unknown way but persisting without alteration forever into the future. Plato's concept of unchanging Forms greatly influenced early Christian thought and came to be embodied in the idea that every species of living thing was created by God at the beginning of time. Although minor variations may occur among individual members of a given species, each species as a whole remains unchanged, very like a Platonic Form.

Plato's student Aristotle (384–322 B.C.), one of the first great naturalists, categorized all the living things that he encountered. Aristotle thought that all organisms fit into an orderly scheme, later called the *Scala Naturae*, or ladder of nature (Fig. 13-3). The ladder stood, so to speak, on nonliving matter and ascended rung by rung from fungi and mosses to higher plants, through primitive animals such as molluscs and insects, and culminated in human beings. Aristotle's ideas, even more than Plato's, were incorporated into Christian thought. The *Scala Naturae* was considered to be permanent and immutable: each organism has its place on the ladder, ordained by God during creation.

Evolutionary Thought Before Darwin

Creationism, the idea that each species was created individually by God and never changed thereafter, remained unchallenged for nearly 2000 years. The *Scala Naturae* was completely compatible with medieval thought: the Earth was the center of the universe, and man stood atop creation. (In those days, "man" meant white, male man; in an early Church council, the delegates conceded that women were human, but only by a margin of one vote!) With the limited fauna and flora of Europe to consider,

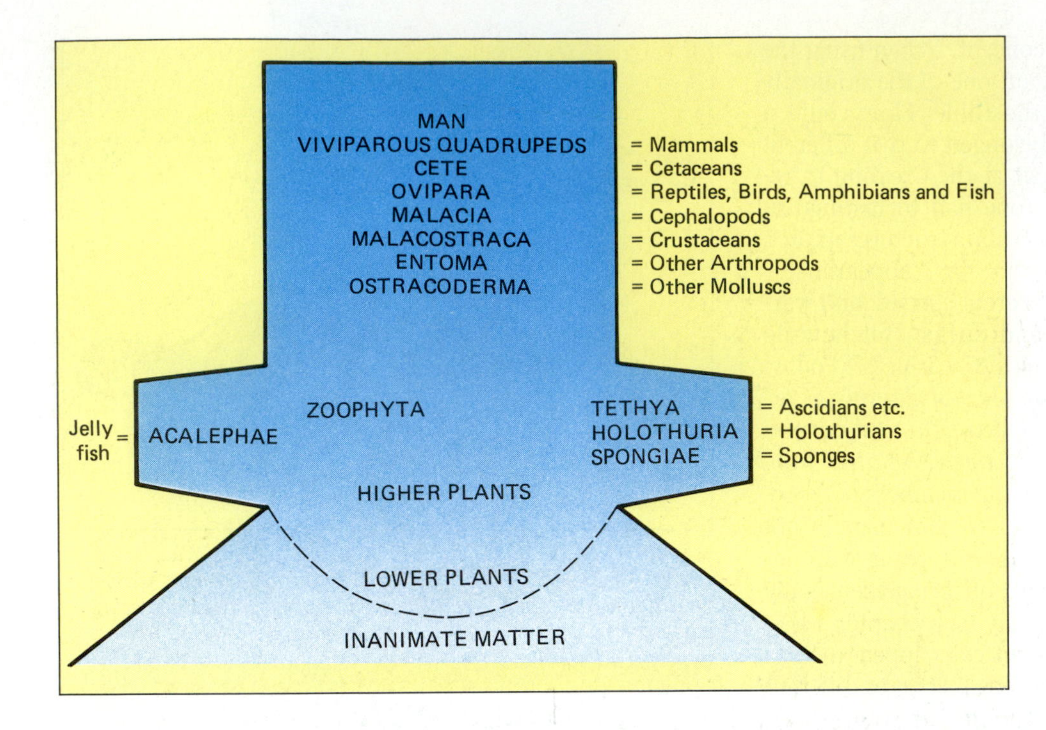

Figure 13-3 *The* Scala Naturae. *It is not clear exactly how Aristotle thought the* Scala *came into being, but he often sounded almost like an evolutionist: "Thus Nature passes from lifeless objects to animals in such unbroken sequence . . . that scarcely any difference seems to exist between two neighboring groups owing to their close proximity." (Drawing and translation from Singer, 1959.)*

a fixed ladder of nature seemed obvious. Naturalists felt it to be their task to catalog the diversity of organisms, describing the glory of God's creation. By the eighteenth century, however, evidence began to accumulate that this view of creation might be incorrect.

As naturalists explored the newly discovered lands of Africa, Asia, and America, they found that the diversity of living things is much greater than anyone had suspected. Further, some of these exotic species closely resemble one another. This embarrassment of riches led some naturalists to consider that perhaps species could change after all, and that some of the similar species might have developed from a common ancestor.

At the same time, excavations for roads, mines, and canals revealed that rocks often occur in layers (Fig. 13-4). Sometimes, a few strangely shaped rocks were found embedded within one of these layers. These rocks, called **fossils** (from the Latin, meaning "dug up"), often resembled parts of living organisms. At first, fossils were thought to be ordinary rocks that wind, water, or people had worked into lifelike forms. As more and more fossils were discovered, however, it became obvious that they were, in fact, plants and animals that had died long ago and been changed into rock (Fig. 13-5). Upon careful study, William Smith (1769–1839) realized that certain fossils were always found in the same layers of rock. Further, the organization of fossils and rock layers was consistent: fossil type A could always be found in a rock layer resting atop an older layer with fossil B, which in turn rested atop a still older layer containing fossil C, and so on.

Fossil remains also showed a remarkable progression in form. Fossils found in the lowest (and therefore oldest) rock layers were invariably primitive looking, with a gradual advancement to greater complexity and greater resemblance to modern species in younger rocks, as if there were a *Scala Naturae* stretching back in time. Many of these fossils were the remains of plants and animals that no longer lived on Earth (Fig. 13-6). Putting these facts

Figure 13-4 *The Grand Canyon of the Colorado River. Layer upon layer of sedimentary rocks form the walls of the canyon, exposed in cliffs and mesas. The Grand Canyon strata cover over a billion years of evolutionary history.*

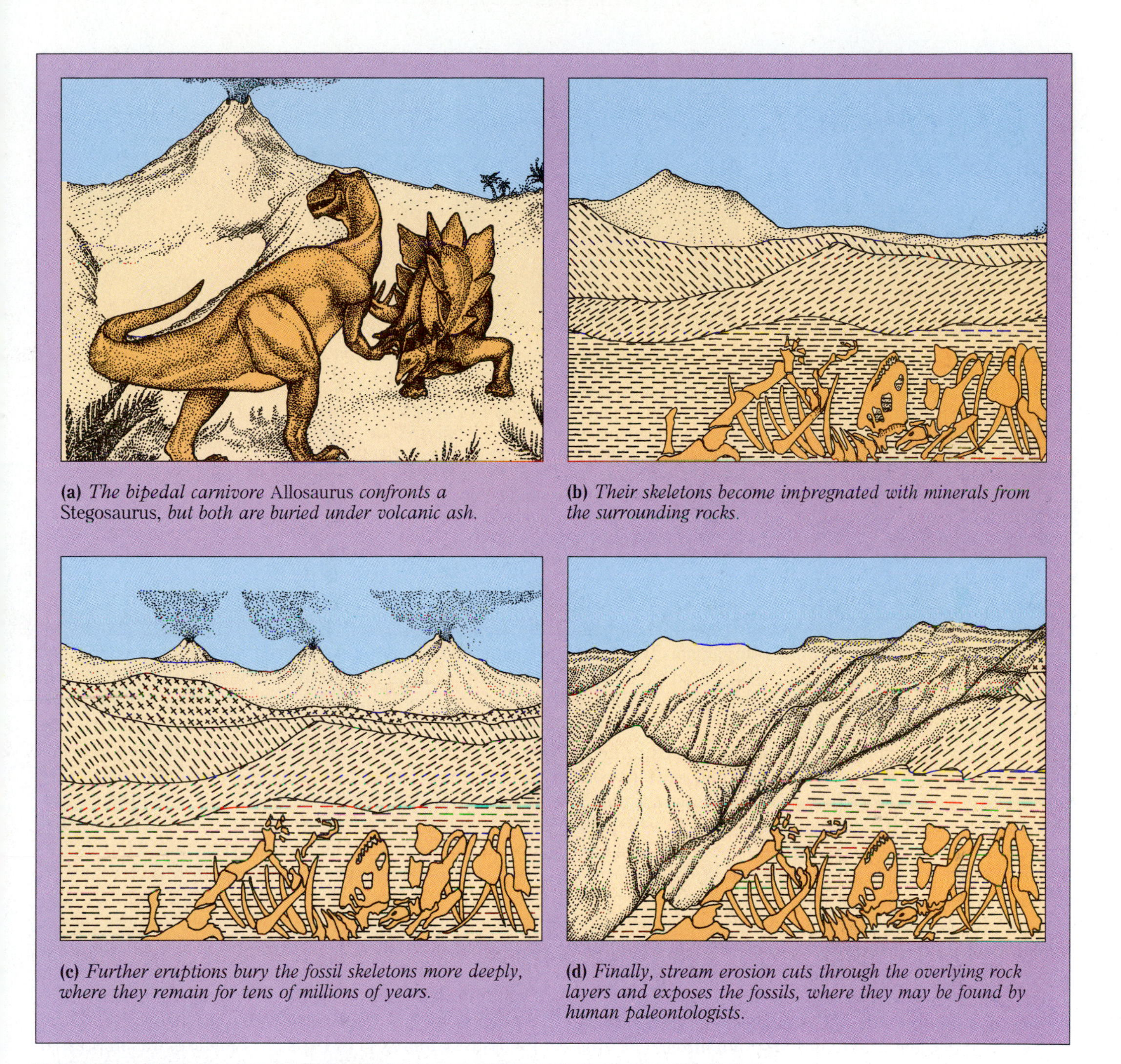

(a) *The bipedal carnivore* Allosaurus *confronts a* Stegosaurus, *but both are buried under volcanic ash.*

(b) *Their skeletons become impregnated with minerals from the surrounding rocks.*

(c) *Further eruptions bury the fossil skeletons more deeply, where they remain for tens of millions of years.*

(d) *Finally, stream erosion cuts through the overlying rock layers and exposes the fossils, where they may be found by human paleontologists.*

Figure 13-5 *Fossilization occurs when living organisms are buried beneath mud, silt, sand, or volcanic ash. This example might have occurred in Montana or Colorado.*

together, the conclusion became inescapable that different types of organisms had lived at various times in the past.

But what did this newfound richness of organisms, both living and extinct, mean? Was each of these organisms produced by a separate act of creation? If so, why? And why bother to create so many, letting thousands become extinct? George-Louis Buffon (1707–1788) suggested that perhaps the original creation provided a relatively small number of founding species, and that some of the modern species had been "conceived by Nature and produced by Time"; that is, they had evolved through natural processes. Most people were not convinced. First, Buffon could not provide any mechanism whereby nature could "conceive" new species. Second, no one thought that there was time enough for their "production."

In the early eighteenth century, few scientists suspected that the Earth could be more than a few thousand years old. Counting generations in the Old Testament, for example, yields a maximum age of 4000 to 6000 years. By reading the descriptions of plants and animals from an-

(a)

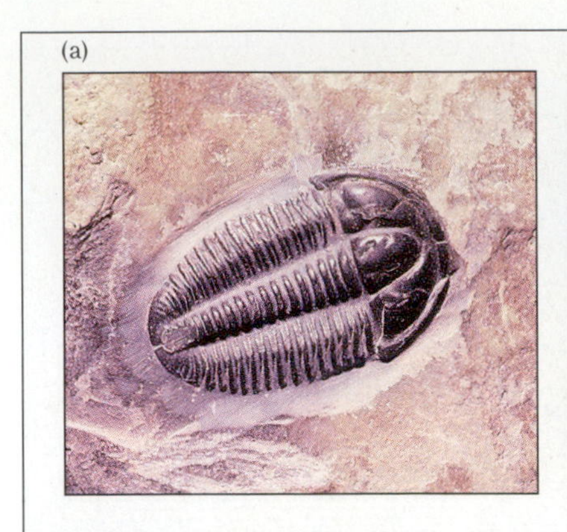

(b)

(c)

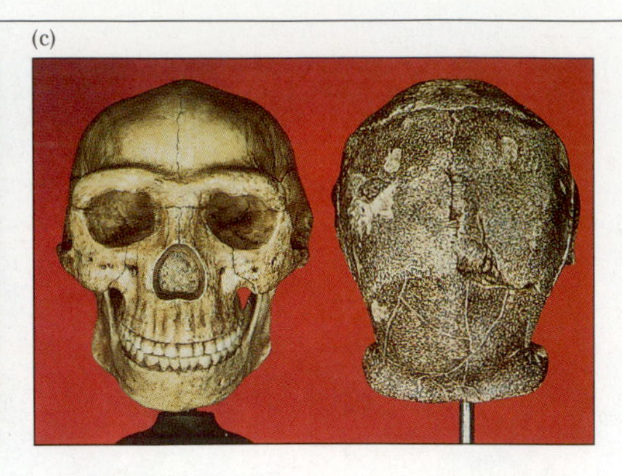

Figure 13-6 *Fossils of extinct animals.* **(a)** *Trilobites were early arthropods (relatives of spiders and crabs) that flourished for scores of millions of years. Dozens of species of fossil trilobites are found in many parts of the world, but there have been no living trilobites for hundreds of millions of years.* **(b)** Triceratops, *a grazing dinosaur. Heavily armored around the head, with three formidable horns, a herd of* Triceratops *must have been extremely dangerous prey, even for the powerful* Tyrannosaurus rex. **(c)** *The skull of* Sinanthropus pekinensis, *an extinct relative of human beings. Features that set* Sinanthropus pekinensis *apart from modern humans include the ridge at the top of the skull, the massive bony brows, and the large jaw.*

cient writers such as Aristotle, it was clear that wolves, deer, lions, and other European organisms had not changed in over 2000 years. How, then, could whole new species arise if the Earth was created only a couple of thousand years before Aristotle?

To account for a multitude of species, both extinct and modern, while preserving creationism, the French paleontologist Georges Cuvier (1769–1832) proposed the theory of **catastrophism**. Cuvier hypothesized that a vast supply of species was created in the beginning. Successive catastrophes (akin to Noah's flood) produced the layers of rock and destroyed many species, fossilizing some of their remains in the process. The reduced flora and fauna of the modern world, he theorized, are the species that survived the catastrophes. However, if modern *species* have survived from an original creation, many *individuals* of those species should have died in the ancient catastrophes. Surely some of them would have been fossilized, and even the lowest and oldest rock layers should contain at least a few fossils of present-day species. Unfortunately for Cuvier's hypothesis, they do not. A rescue attempt by Louis Agassiz (1807–1873) proposed that there was a new creation after each catastrophe, and that modern species result from the most recent creation. The fossil record forced Agassiz to postulate at least 50 separate catastrophes and creations!

Alternatively, perhaps the Earth *is* old enough to allow for the production of new species. Geologists James Hutton (1726–1797) and Charles Lyell (1797–1875) contemplated the forces of wind, water, earthquakes, and volcanism. They concluded that there was no need to invoke catastrophes to explain the findings of geology. Do not rivers in flood lay down layers of sediment? Do not lava flows produce layers of basalt? Why, then, should we assume that layers of rock are evidence of anything but ordinary natural processes, occurring repeatedly over long periods of time? This concept, called **uniformitarianism,** satisfies a scientific axiom often called Occam's razor: the simplest explanation that fits the facts is probably correct. The implications of uniformitarianism were profound. If slow natural processes suffice to produce layers of rock thousands of feet thick, then the Earth must be old indeed, many millions of years old. Hutton and Lyell, in fact, concluded that the Earth was eternal: "no Vestige of a Beginning, no Prospect of an End," in Hutton's words. [Modern geologists estimate that the correct age for the Earth is between 4.5 and 4.6 billion years (Fig. 13-7).] Thus Hutton and Lyell provided the time for evolution, but there was still no convincing mechanism.

One of the first to propose such a mechanism was Jean Baptiste Lamarck (1744–1829). Lamarck was impressed by the progression of forms in the fossil record. Older fossils tend to be simpler, while younger fossils are more complex and more like existing organisms. In 1801, Lamarck hypothesized that organisms evolved through the **inheritance of acquired characteristics:** living organisms can modify their bodies through use or disuse of parts (which is correct to some extent), and these modifications can be inherited by their offspring (which is not correct). Why would organisms modify their bodies? Lamarck proposed that all organisms possess an innate drive for perfection, an urge to climb the ladder of nature. In his most celebrated example, Lamarck hypothesized that ancestral giraffes stretched their necks to feed on leaves growing high up in trees, and as a result their necks became slightly longer. Their offspring inherited these longer

Figure 13-7 Dating Rocks by Radioactivity

Early geologists could only date rock strata and their accompanying fossils in a relative *way: Fossil A would be older than fossil B if the rock layer containing A was beneath that containing B, but neither fossil could be assigned an actual age. With the discovery of radioactivity, it became possible to determine* absolute *dates. The nuclei of radioactive elements spontaneously break down, or decay, into other elements. For example, carbon-14 (usually written* ^{14}C*) emits an electron to become nitrogen-14 (*^{14}N*). Each radioactive element decays at a rate that is independent of temperature, pressure, or the chemical compound of which the element is a part. The rate is geometric, so that half the radioactive nuclei decay in a characteristic time, called the half-life. The half-life of* ^{14}C*, for example, is 5730 years.*

Elements Used in Radioactive Dating

Original Element	Final Element	Half Life (years)	Useful Dating Ages (years)
Carbon-14	Nitrogen-14	5730	100–50,000
Uranium-235	Lead-207	0.70 billion	over 500,000
Potassium-40	Argon-40	1.25 billion	over 500,000
Uranium-238	Lead-206	4.47 billion	over 100 million
Rubidium-87	Strontium-87	48.8 billion	over 100 million
Samarium-147	Neodymium-143	106 billion	over 1 billion

How are radioactive elements used in determining the age of rocks? A particularly straightforward dating technique uses the decay of potassium-40 (^{40}K*) into argon-40 (*^{40}Ar*), which has a half-life of about 1.25 billion years. Potassium is a very reactive element and is a common constituent of igneous rocks such as granite and basalt. Argon, on the other hand, is essentially inert, being unable to enter into chemical bonds with other atoms. What's more, argon is a gas. Let us suppose that a volcano such as Hawaii's Kilauea erupts with a massive lava flow, covering the countryside. All the* ^{40}Ar*, being a gas, will bubble out of the molten lava, so that when the lava solidifies into rock, it will start out with no* ^{40}Ar*. Potassium-40 present in the hardened lava will decay to* ^{40}Ar*, half the* ^{40}K *decaying every 1.25 billion years. The* ^{40}Ar *gas will be trapped in the rock. A geologist could take a sample of the rock and determine the proportion of* ^{40}K *to* ^{40}Ar*. If the analysis finds equal amounts of the two elements, the geologist would conclude that the lava hardened 1.25 billion years ago.*

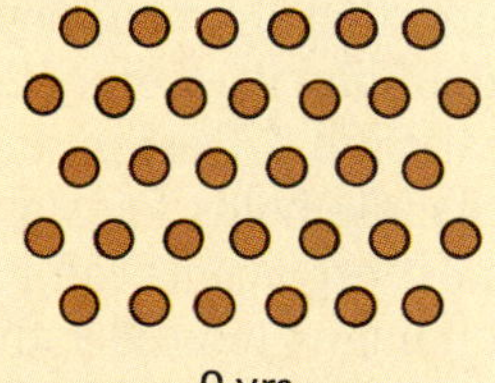

0 yrs

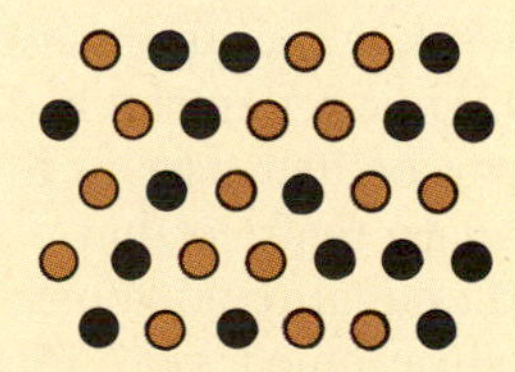

1.25 billion yrs

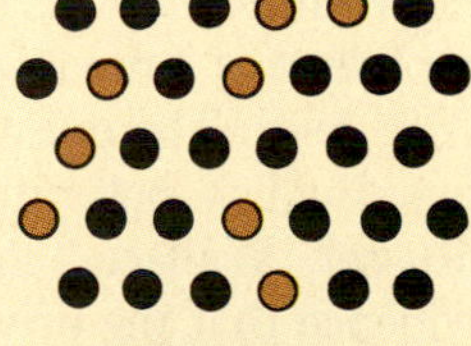

2.50 billion yrs

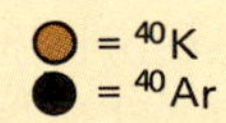

With appropriate care, such age estimates are quite reliable. In the case of $^{40}K/^{40}Ar$ *analysis, the age estimate will be, if anything, too low, because some of the* ^{40}Ar *may have escaped even though the rock was solid, thus increasing the ratio of* ^{40}K *to* ^{40}Ar*. If a fossil is found beneath a lava flow dated at, say, 500 million years, we know that the fossil is at least that old.*

Some radioactive decay pairs, especially uranium-238 to lead-206, can even give an estimation of the age of the solar system. Analysis of Earthly rocks, meteorites, and rocks collected from the moon by the Apollo *astronauts all agree that the solar system, and hence the Earth, is about 4.54 billion years old.*

Figure 13-8 Lamarck's hypothesis for the evolution of the giraffe. Time flows from left to right. Short-necked protogiraffes stripped the leaves from the lower branches of trees, and stretched and strained to reach leaves higher up, making their necks slightly longer. Their offspring inherited these longer necks, did more stretching in their turn, and passed on still longer necks to their offspring. The modern giraffe on the right would be the outcome of this continual striving to feed on tree leaves high off the ground.

necks, and in their turn stretched even farther, to reach still higher leaves. Eventually, this process might produce modern giraffes with very long necks indeed (Fig. 13-8).

Today, Lamarck's theory seems silly: the fact that a prospective father pumps iron doesn't mean that his children will look like Arnold Schwartzenegger. Remember, though, that in Lamarck's day no one had the foggiest idea how inheritance worked. Gregor Mendel would not even be born for another 20 years, and the incorporation of his principles of inheritance into mainstream biology did not happen until the early twentieth century.

Although Lamarck's theory fell by the wayside, by the mid-nineteenth century many biologists realized that the fossil record and the similarities between fossil forms and modern species could best be explained if present-day species had evolved from preexisting ones. The question remained: *but how?* In 1858, two English naturalists, Charles Darwin and Alfred Russel Wallace, independently provided convincing evidence that the driving force behind evolutionary change was natural selection.

EVOLUTION BY NATURAL SELECTION

Although their social and educational backgrounds were very different, Darwin and Wallace were quite similar in some respects. Both had traveled extensively in the tropics (see the essay, "The Voyage of the *Beagle*"), and had studied the staggering variety of plants and animals living there. Both found that some species differed only in some fairly subtle but ecologically important features (Fig. 13-9). Darwin and Wallace were familiar with the fossil record, which showed a trend of increasing complexity through time. Finally, both were aware of the theories of Hutton and Lyell, which proposed that the Earth is extremely ancient. These facts suggested to both men that species change over time (i.e., they evolve). Both sought a mechanism that might direct change over many generations, causing new species to arise.

Part of the answer came to both men from an unlikely source: the writings of an English clergyman, Thomas Malthus. In his *Essay on Population*, Malthus wrote, "It may safely be pronounced, therefore, that [human] population, when unchecked, goes on doubling itself every 25 years, or increases in a geometrical ratio." Darwin and Wallace realized that a similar principle holds for plant and animal populations. In fact, most organisms can reproduce much more rapidly than humans (consider the dandelion and the housefly), and consequently could produce overwhelming populations in short order. Nonetheless, the world is *not* chest-deep in dandelions or flies: natural populations do not grow "unchecked" but tend to remain approximately constant in size. Clearly, vast numbers of organisms must die in each generation, with most never living long enough to reproduce. Population growth is restrained by countless environmental factors, including food, predation, diseases, and weather (see Chapter 35).

From their experience as naturalists, Darwin and Wallace realized that the members of a species often differ from one another in form and function. Further, *which organisms die in each generation is not arbitrary, but depends to some extent on the structures and abilities of the organisms.* As Wallace put it, "... those which, year by year, survived this terrible destruction must be, on the whole, those which have some little superiority enabling them to escape each special form of death to which the great majority succumbed. ..." That "little superiority" might be better resistance to cold, more efficient digestion, or any of hundreds of other advantages, some very subtle. Everything now fell into place. Darwin wrote: "... it at once struck me that under these circumstances favorable variations would tend to be preserved, and unfavorable ones to be destroyed." If the favorable variations were inheritable, the entire species would eventually consist of individuals possessing the favorable trait. With the continual appearance of new variations (due, as we now know, to mutations), which in turn are subject to further selection, "the result of this would be the formation of new species. Here, then, I had at last got a theory by which to work."

In 1858, remarkably similar papers from Darwin and Wallace were presented to the Linnaean Society in London. As with Gregor Mendel's manuscript on the principles of genetics, their papers made little impact. The secretary of the society, in fact, wrote in his annual report

Figure 13-9 Darwin's finches, residents of the Galápagos Islands. The Galápagos are a group of volcanic islands about 600 miles off the coast of Ecuador. On these inhospitable-looking islands, Darwin found many species of finches, all similar to a species found on the South American mainland. Ordinarily, finches are seed-eating birds, with large bills adapted to crushing hard seeds. Apparently, thousands of years ago a finch or small flock of finches became lost during migration or were blown off course by a storm, and arrived at the Galápagos. There they found few other birds. With few competitors, the finches found rich pickings both of seeds and insects. Over time, different groups of finches evolved adaptations to exploit different food sources. Modern biologists agree with Darwin that the variety of finch species on the Galápagos evolved from a common ancestor **(a)**. *Particularly significant are the differences in beaks, which vary from a slender, warbler-like bill in the insect-catching* Certhidia olivacea **(b)** *to the massive crushing bill of* Geospiza magnirostris **(c)**. *Bill differences are also important in mate recognition. From behind, many Galápagos finches are very similar, and a male may approach a female of the wrong species. As soon as he catches sight of her beak, however, he immediately recognizes his mistake and seeks elsewhere for a mate.*

The Voyage of the *Beagle*

Figure E13-1 A painting of Charles Darwin as a young man.

Like many modern students, Charles Darwin excelled only in subjects that intrigued him. Although his father was a physician, Darwin was uninterested in medicine and unable to stand the sight of surgery. He eventually obtained a degree in theology from Cambridge, although this too was of minor interest. What he really liked to do was to tramp over the hills, observing plants and animals, collecting new specimens, scrutinizing their structures, and categorizing them. As he himself later put it, "I was a born naturalist." Fortunately for Darwin (and for the development of biology), some of his professors at Cambridge had similar interests, notably the botanist John Henslow. So constant was their companionship in field studies that Darwin was sometimes called "the man who walks with Henslow."

In 1831, when Darwin was only 22 years old (Fig. E13-1), the British government sent His Majesty's Ship *Beagle* on a surveying expedition along the coast of South America. As was common on such expeditions, the *Beagle* would carry along a naturalist, to observe and collect geological and biological specimens encountered along the route. Thanks to Henslow's recommendation to the captain, Robert FitzRoy, Darwin was offered the position of naturalist aboard the *Beagle*. (When they first met, FitzRoy almost rejected Darwin for the post because of the shape of Darwin's nose! Apparently, FitzRoy felt that a person's personality could be predicted by the shape of the facial features, and Darwin's nose failed to measure up, as it were. Later, Darwin expressed the opinion that FitzRoy was "afterwards well satisfied that my nose had spoken falsely.")

The *Beagle* sailed to South America, making many stops along the coast and visiting the now-famous Galápagos Islands (Fig. E13-2). Along the way, Darwin observed the fauna and flora of the tropics, and was stunned by the diversity of species compared to Europe. Although he boarded the *Beagle* convinced of the permanence of species, his experiences soon led him to doubt this. He discovered a snake with rudimentary hind limbs, calling it "the passage by which Nature joins the lizards to the snakes." Another snake vibrates its tail like a rattlesnake, but has no rattles and therefore makes no noise. Penguins have wings that look like paddles, almost flying through the water. What could be the purpose behind these makeshift arrangements, if the Creator had individually created each animal in its present form, to suit its present environment?

Perhaps the most significant stopover of the voyage was the month spent on the Galápagos Islands. Here Darwin found huge tortoises (*galápagos* in Spanish); different islands bore distinctively different types. On islands without tortoises, prickly pear cactus grew in the common style, with juicy though spiny pads spread out over the ground. On islands where tortoises lived, the prickly pears grew substantial trunks, bearing the succulent pads high above the reach of the voracious and tough-mouthed tortoises (Fig. E13-3). Several varieties of mockingbirds and finches occurred, and as with the tortoises, different islands had subtly different forms. Unfortunately for Darwin, he "was not aware of these facts [about the finches] till my collection was nearly completed," and so did not bring a systematically labeled collection back to England with him. "It is the fate of every voyager, when he has just discovered what object in any place is worth his attention, to be hurried from it." The diversity of tortoises and birds "haunted" him for years afterward, as he pondered how such diversity might have occurred.

In 1836, when Darwin returned to England after five years on the *Beagle*, he was already somewhat famous, based on letters and journals forwarded home on other ships. The specimens and further journals that he brought back with him enhanced his fame, and to all outward appearances he settled down to become one of the foremost naturalists of his day. However, constantly gnawing on his mind was the problem of the origin of species. For over 20 years, Darwin pondered the problem, collecting evidence of any sort that might provide insight into its solution. When he finally published *On the Origin of Species* in 1859, his evidence had become truly overwhelming. Although its full impact would not be realized for decades, Darwin's theory of evolution by natural selection has become a unifying concept for virtually all of biology.

Events that change the world sometimes hinge on minute details, even the shape of a nose!

The Voyage of the *Beagle*

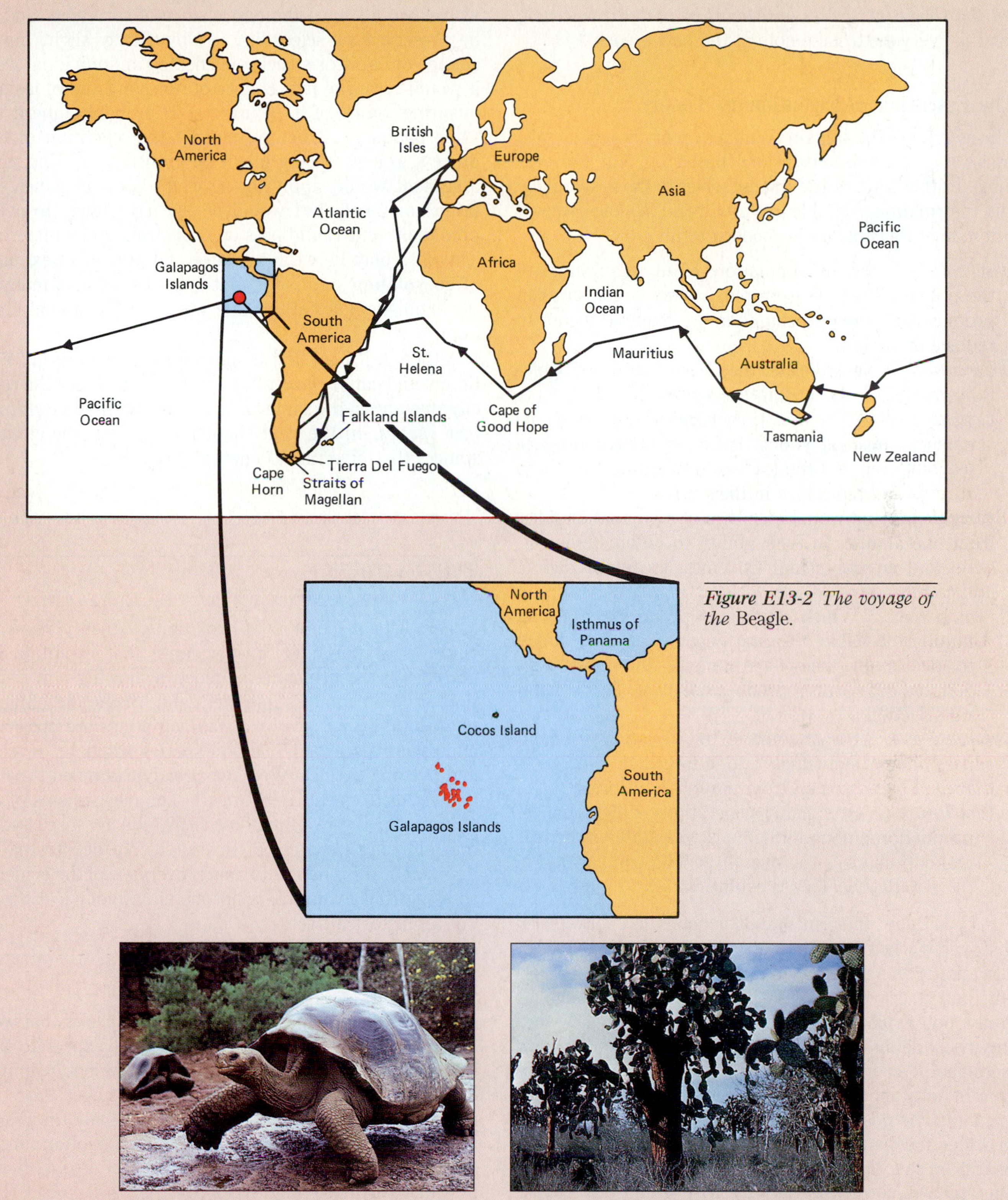

Figure E13-2 The voyage of the Beagle.

Figure E13-3 Galápagos tortoises (left) feed on prickly pear cactuses. On islands with tortoises, a young cactus quickly grows a tall trunk (right), which lifts the succulent pads beyond the reach of the tortoises.

that nothing very interesting happened that year. Fortunately, the next year Darwin published his monumental *On the Origin of Species by Means of Natural Selection*, forcing everyone to take note of the new theory.

The Essentials of Evolutionary Theory

The essence of the Darwin–Wallace theory is very simple, consisting of three conclusions based on four observations. Rather than restricting ourselves to the terminology and information available to Darwin and Wallace, we will summarize their theory in modern terms.

Observation 1: Natural populations of all organisms have the potential to increase rapidly, since organisms can produce far more offspring than are required merely to replace the parents.

Observation 2: Nevertheless, the sizes of natural populations are relatively constant over time.

Conclusion 1: Therefore, in each generation, many organisms must die young, fail to reproduce, produce few offspring, or produce less fit offspring that fail to survive and reproduce in their turn.

Observation 3: Individual members of a population differ from one another in their ability to obtain resources, withstand environmental extremes, escape predation, and so on.

Conclusion 2: Which organisms produce the largest number of viable offspring depends on their ability to deal with their environment; the most well-adapted organisms reproduce the most. This is **natural selection.**

Observation 4: At least some of the variation in adaptedness among individuals is due to genetic differences that may be passed on from parent to offspring.

Conclusion 3: Over many generations, differential reproduction among individuals with different genetic makeup changes the overall genetic composition of the population. This is **evolution.**

As you know, the principles of genetics had not yet been discovered when Darwin wrote *Origin of Species*. Our observation 4 was therefore an untested assumption for Darwin and a grave weakness in his theory. Although he could not explain how inheritance operated, Darwin's theory made an important prediction that we now know is correct. According to Darwin, the variations that appear in natural populations arise purely by chance. Unlike Lamarck, he postulated no internal drives for perfection or other mechanisms that would ensure that variations would be favorable. Molecular genetics has shown that Darwin was correct: variations arise because of chance mutations in DNA (see Chapters 10 and 11).

How could natural selection among chance variations change the makeup of a species? In *Origin of Species*, Darwin proposed the following example. "Let us take the case of a wolf, which preys on various animals, securing [them] by . . . fleetness. I can under such circumstances see no reason to doubt that the swiftest and slimmest wolves would have the best chance of surviving, and so be preserved or selected. . . . Now if any slight innate change of habit or structure benefited an individual wolf, it would have the best chance of surviving and of leaving offspring. Some of its young would probably inherit the same habits or structure, and by the repetition of this process, a new variety might be formed. . . ." The same argument would apply to the wolf's prey, in which the fastest or most alert would be the most likely to avoid predation, and would pass on these traits to its offspring. This interplay between predator and prey is an example of **coevolution** and is thought to be the cause of many of the exquisite adaptations that we see in the animal kingdom.

Although it is easiest to understand how natural selection would cause *changes within a species*, under the right circumstances, the same principles might produce *entirely new species*. In the next chapter we discuss the circumstances that give rise to new species.

THE EVIDENCE FOR EVOLUTION

A complete discussion of evolutionary theory must include three parts: (1) the evidence that evolution occurred; (2) the principles of genetics showing how variation originates and is inherited; and (3) the mechanisms of evolution, that is, why certain variations are preserved and others discarded, and how species originate, develop, and become extinct. We have already examined the fundamentals of genetics in Unit II. The mechanisms of evolution and the resulting diversity of life are discussed in the following chapters of this Unit. We devote the remainder of this chapter to a brief overview of the evidence showing that evolution is not merely a biological theory but an historical fact.

Fossils

If fossils are the remains of members of species that were ancestral to modern species, one would expect to find graded series of fossils leading from an ancient, primitive organism, through several intermediate stages, and culminating in the modern form. (Just how fine the gradations should be is currently a bone of contention among evolutionary biologists; see Chapter 15.) Probably the best-known progressive series are the fossil horses (Fig. 13-10), but giraffes, elephants, and several molluscs all show a gradual evolution of body form over time, suggesting that one species evolved from and replaced previous species. Certain sequences of fossil snails have such slight gradations in form between successive fossils that

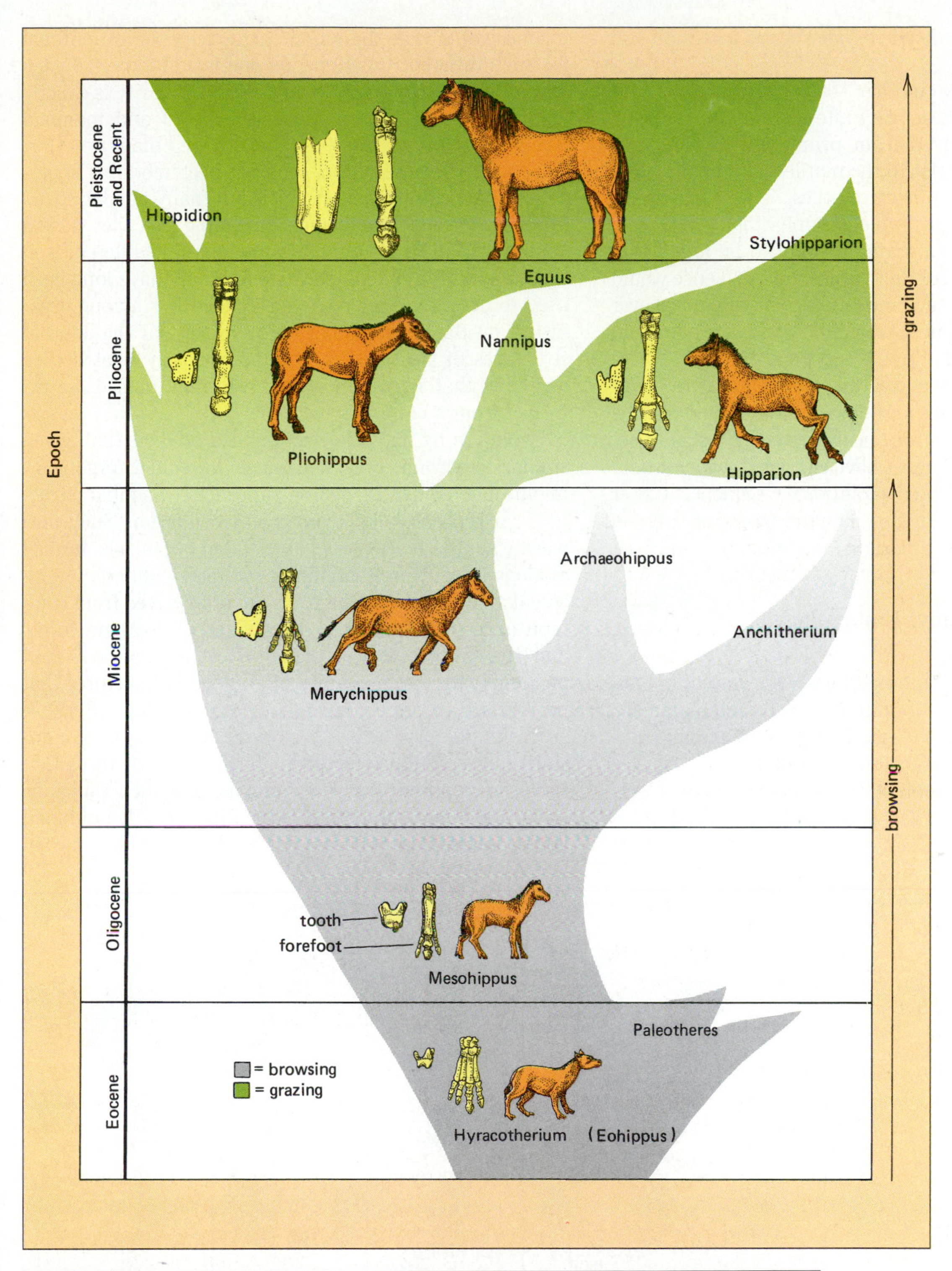

Figure 13-10 Over the last 50 million years, horses evolved from small woodland browsers to large plains-dwelling grazers. Three major changes include size, leg anatomy, and tooth anatomy. No one is certain why horses became larger, but it may have been an antipredator adaptation. In any case, a large body running over hard plains favored the evolution of large, hard hooves, attached by spring-like, shock-absorbing joints to legs with stout bones. Finally, the teeth became larger, with more enamel. This reflects a change in diet from the relatively soft leaves and buds of bushes to the silicon-containing, abrasive blades of grasses. If a modern horse had teeth like Hyracotherium, *they would be ground away while it was still very young, leaving it to starve.*

one cannot easily decide where one species leaves off and the next one begins.

Fossils also provide another sort of evidence for evolution. Modern biologists are quite certain, for example, that amphibians evolved from primitive fish. The amphibians in turn gave rise to the reptiles, and finally birds and mammals evolved from reptiles. This sequence is tested every time a paleontologist unearths a vertebrate fossil. If even one fossil mammal were to be found that predates the evolution of the reptiles, this sequence would be proven wrong. No such out-of-place mammal has ever been found. The evidence is overwhelming that plant and animal species appeared in the sequence predicted by evolution. (Several years ago it was claimed that the footprints of both humans and dinosaurs were found side by side in the same rock strata near a river in Texas. If so, humans and dinosaurs must have coexisted, which would clearly violate the accepted evolutionary sequence. Closer examination proved that the "human" footprints were in fact those of a bipedal dinosaur.)

Comparative Anatomy, Embryology, and Biochemistry

Modern organisms are adapted for a wide variety of habitats and life-styles. The forelimbs of birds and mammals, for example, are variously used for flying, swimming, running over several types of terrain, and grasping objects such as branches and tools. Despite this enormous diversity of function, the internal anatomy of all bird and mammal forelimbs is remarkably similar (Fig. 13-11). It is inconceivable that these bone arrangements are optimal for such different functions, as one would expect if each animal were created separately. Such similarity is exactly what we would expect, however, if bird and mammal forelimbs evolved from a common reptilian ancestor. Through natural selection, each has been modified to perform a particular function. Such internally similar structures are termed **homologous**, meaning that they have a similar evolutionary origin, despite possible differences in function. Studies of comparative anatomy have long been used to determine the degree of relatedness among organisms, on the grounds that the more similar the internal structures of two species, the more closely related they must be, that is, the more recently they must have diverged from a common ancestor.

Evolution by natural selection also predicts that, given similar environmental demands, unrelated organisms might independently evolve superficially similar structures, a process called **convergent evolution**. Such outwardly similar body parts in unrelated organisms, termed **analogous structures**, are often completely different in internal anatomy, since the parts are not derived from common ancestral structures. The wings of flies and birds, and the fat-insulated, streamlined shapes of seals and penguins, are two examples of analogous structures that have arisen through convergent evolution (Fig. 13-12).

Evolution also helps to explain the curious circumstance of **vestigial structures**. These are structures that serve no apparent purpose, and include such things as molar teeth in vampire bats (which live on a diet of blood and therefore do not chew their food) and pelvic bones in

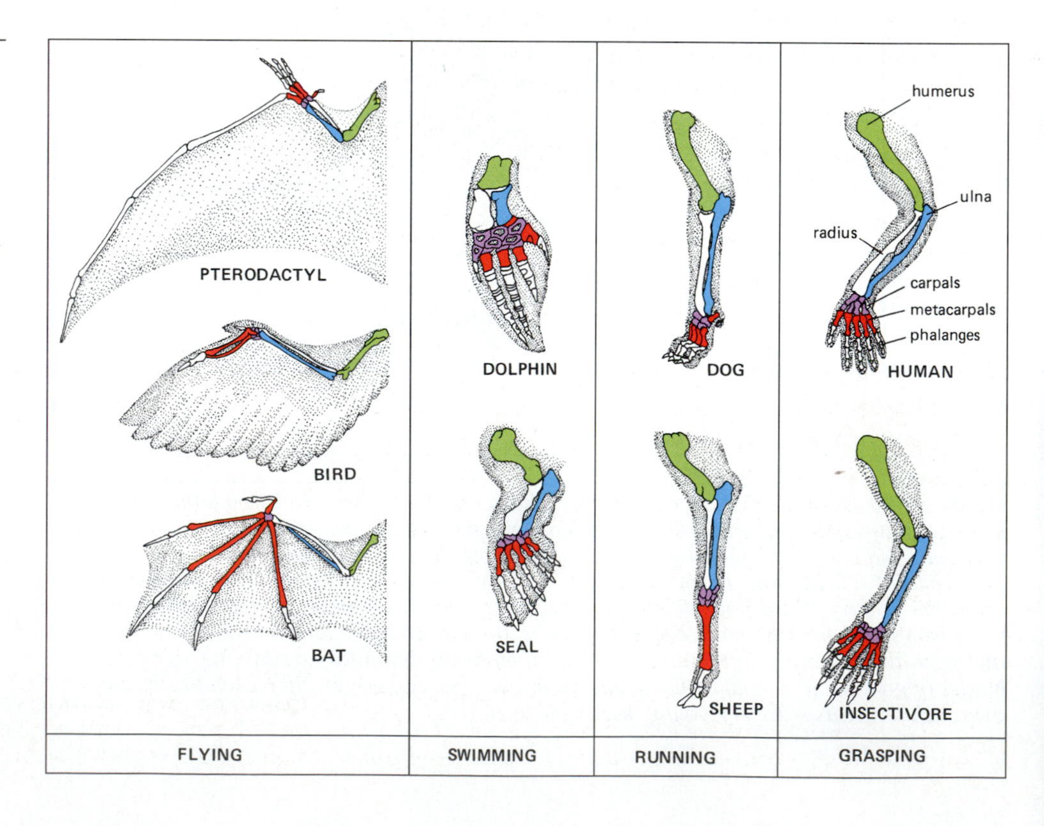

Figure 13-11 *Homologous structures. The bones in the forelimbs of amphibians, reptiles, birds, and mammals are all similar to one another, despite wide differences in function. The bones have been tinted different colors to point out the similarities among the various species.*

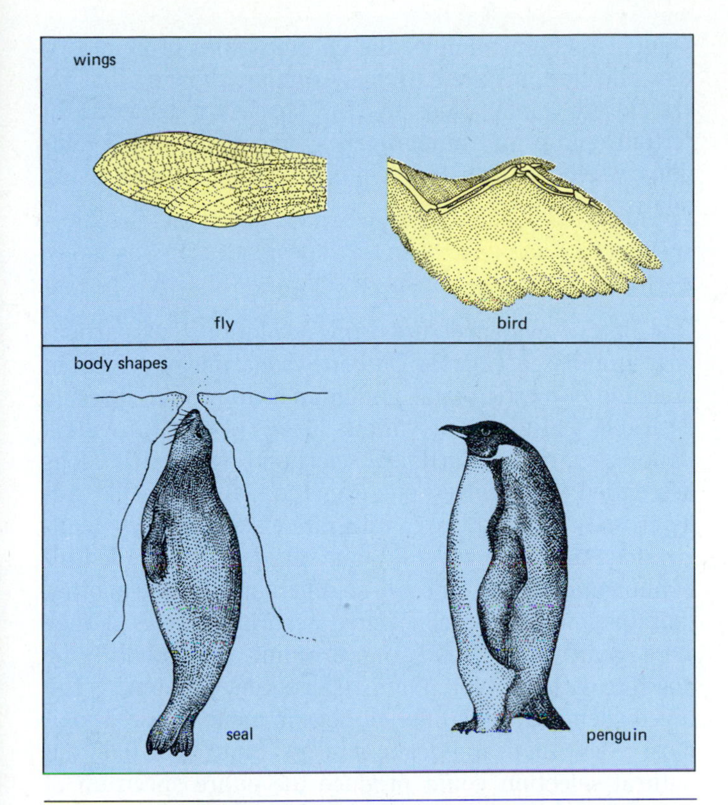

Figure 13-12 *Similar selective pressures acting on unrelated animals may result in the evolution of outwardly similar structures. The wings of insects and birds and the sleek, streamlined shapes of seals and penguins are examples of such analogous structures.*

whales and certain snakes (Fig. 13-13). Both of these vestigial structures are clearly homologous to important structures found in other vertebrates. Their continued existence in animals that have no use for them is best explained as a sort of "evolutionary baggage." For example, the ancestral mammals from which whales evolved had four legs and a well-developed set of pelvic bones. Whales do not have hind legs, yet have small pelvic and leg bones embedded in their sides. During whale evolution, there was a selective advantage to the loss of the hind legs, the better to streamline the body for movement through water. Once mutation and selection reduced the pelvic bones so that they no longer interrupted the smooth line of the body, that selective pressure diminished. The result is the modern whale with small, useless pelvic bones.

Embryological studies also support the theory of evolution. All vertebrate embryos, for example, pass through a stage in development in which they look quite similar to one another (Fig. 13-14). Fish, turtles, chickens, mice, and humans all develop tails and gill arches early in development. Only fish go on to develop gills, and only fish, turtles, and mice retain substantial tails. Why do such diverse vertebrates have similar developmental stages? The only plausible explanation is that ancestral vertebrates had the genes that direct the development of gills and tails. All of their descendants still retain those genes. In fish, these genes are active throughout development, resulting in gill- and tail-bearing adults. In humans and chickens, these genes are active only during early stages, and the structures are lost or inconspicuous in the adults.

An exciting recent development in evolutionary biology is the ability to measure genetic and biochemical similarities. For example, all living organisms use DNA as the carrier of genetic information; all have very similar mechanisms for reading their genes and translating their information into proteins; and all use ATP as an intracellular energy carrier. Analysis of the amino acid sequences of proteins has confirmed that all organisms are related to some degree, and the relationships suggested by the amino acid sequences correspond closely to the evolutionary trees worked out by comparative anatomists. Relatedness among organisms can also be evaluated by examining their chromosomes or, more recently, their DNA

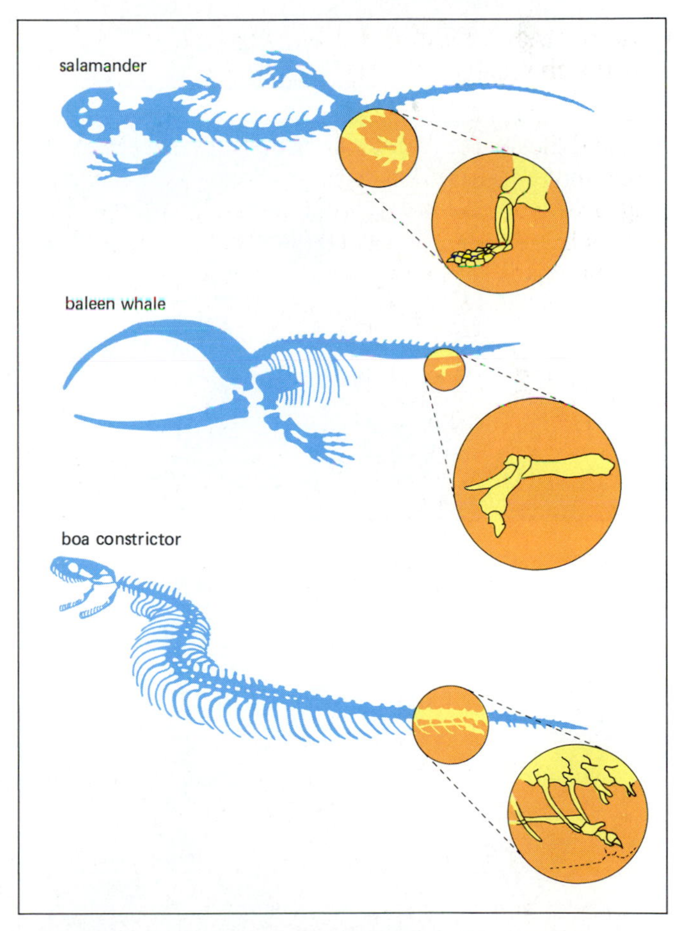

Figure 13-13 *The functional hindlimb of a salamander (top) is probably similar to the limb of the common ancestor of all amphibians, reptiles, birds, and mammals. Baleen whales (center) and boa constrictors (bottom) have no functional legs but still develop vestigial pelvic girdles and even miniature leg bones buried in their sleek sides.*

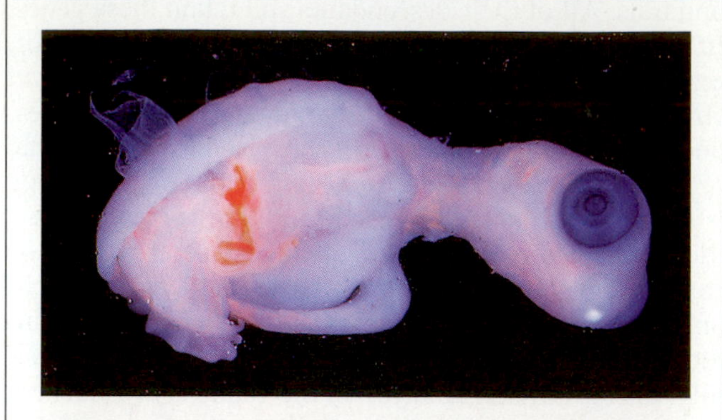

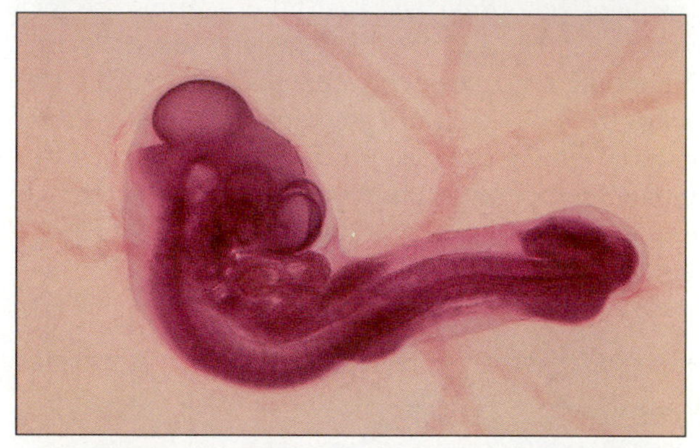

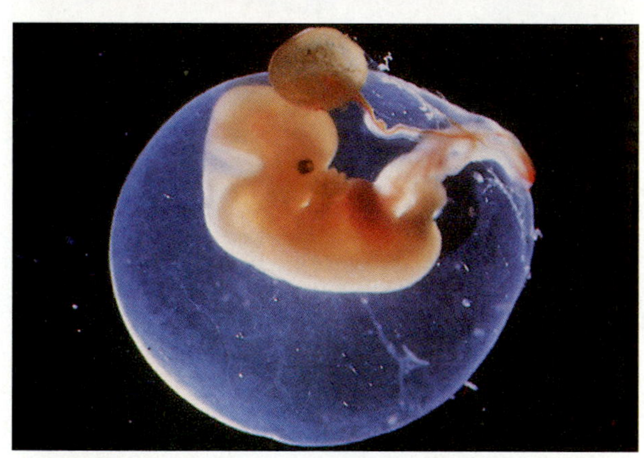

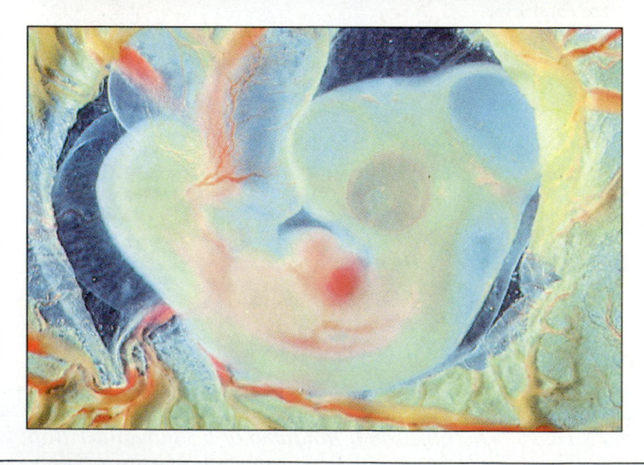

Figure 13-14 *Early embryonic stages of (top to bottom) a turtle, mouse, human, and chicken, showing strikingly similar anatomical features.*

sequences. For example, the chromosomes of chimpanzees and humans are extremely similar, showing that we are closely related (Fig. 13-15). The DNA sequences of certain chimp and human genes suggest to some evolutionary biochemists that chimps are our closest living relatives, and vice versa.

Artificial Selection

One line of evidence supporting evolution that particularly impressed Charles Darwin was **artificial selection**: breeding domestic plants and animals to produce specific desirable features. The various breeds of dogs provide a striking example of artificial selection (Fig. 13-16). Dogs descended from wolves, and even today the two will readily cross-breed. However, with rare exceptions, few modern dogs resemble wolves. Some breeds, such as the Chihuahua and Great Dane, are so different from one another that they would be considered separate species if they were found in the wild. Interbreeding would hardly be possible without a lot of human assistance. If humans can breed such radically different dogs in a few hundred to at most a few thousand years, it seems quite plausible that natural selection could produce the entire spectrum of living organisms in hundreds of millions of years.

Present-Day Evolution

One problem that many people have with the theory of evolution is that they think it all happened in the past. Evolutionary biologists, however, maintain that evolution is *not* merely a phenomenon of the past but that it continues today. Strong evidence for present-day evolution has come as a fortuitous side effect of an otherwise undesirable circumstance: the rise of industrial pollution in the nineteenth and twentieth centuries.

Great Britain is famous for its cool, damp climate. As a result, tree trunks in British woodlands usually support a lush growth of mottled gray lichens (a symbiotic association of an alga and a fungus; see Chapter 19). Before the industrial revolution, most peppered moths, *Biston betularia*, were white with scattered specks of black pigment. This coloration matched the color and pattern of the lichens growing on the trees. Since the moths sat quietly on the lichens during the day, predatory birds could not easily see them (Fig. 13-17a). Occasionally, mutant black individuals appeared. These black moths were extremely conspicuous against the pale lichens, were easily spotted by birds, and did not live long.

The industrial revolution changed everything. The growing industries of nineteenth-century Britain burned coal for fuel. With no pollution control technology, soot from the smokestacks soon blanketed the countryside around the mills and factories (Fig. 13-18). The lichens on the trees became covered with pollutants, died off, and the trunks became sooty black. The pale form of the pep-

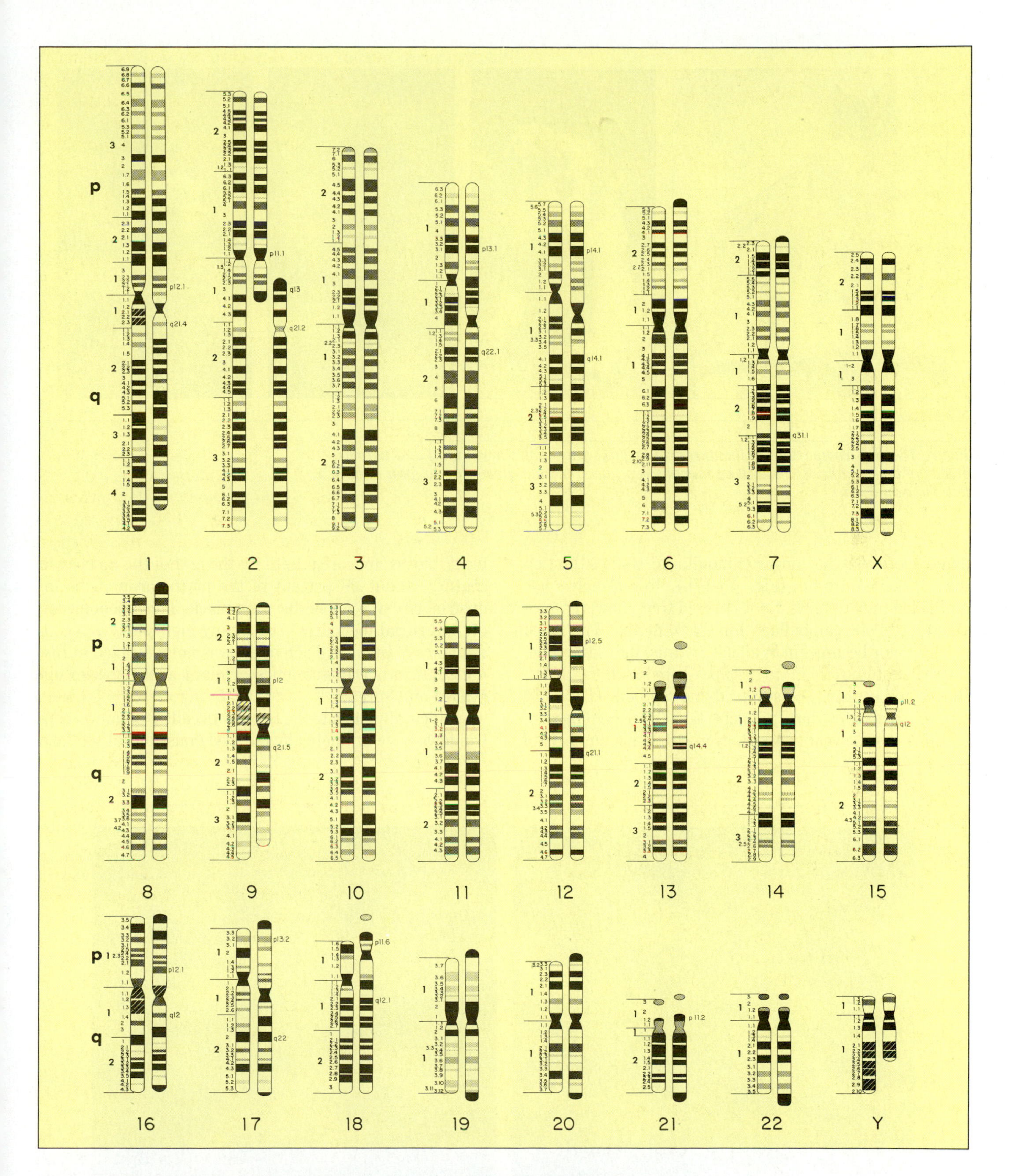

Figure 13-15 A comparison of the chromosomes of humans (left) and chimpanzees (right), stained to reveal detailed banding patterns. The numbers are those assigned to human chromosomes; note that chimps have two chromosomes that, combined, are virtually identical to human chromosome number 2. The close evolutionary relationship between humans and chimps is clearly revealed by the striking similarity of the chromosomes.

Figure 13-16 A comparison of the ancestral dog (the gray wolf, Canis lupus*) and various breeds of dog. Artificial selection by humans has caused a great divergence in form in only a few thousand years.*

pered moth was no longer camouflaged while sitting on the blackened trunks (Fig. 13-17b). Increasingly, pale moths fell prey to birds. Soot-covered trees, however, provided excellent camouflage for black moths. The black moths' color became, in Wallace's words, the "little superiority enabling them to escape" predation by birds. Black moths, once rare, survived and reproduced, passing on their genes for dark pigmentation to succeeding generations. As the years passed, ever-increasing numbers of black moths appeared, until by the end of the nineteenth century, about 98 percent of the moths around the industrial city of Manchester were black. This phenomenon of "industrial melanism" (becoming pigmented as a consequence of industrial activities) was not restricted to *Biston betularia*. Dark varieties of many moth species appeared and spread.

Incidentally, during the last few decades, pollution control laws have dramatically reduced emissions of soot and

Figure 13-17 The two color forms of the peppered moth, resting on a lichen-covered tree trunk **(a)** *and a soot-blackened trunk* **(b)**. *The pale form is well camouflaged on the lichen, but is conspicuous on the sooty trunk, while the reverse is true of the black form.*

Figure 13-18 *The Peak District near Manchester, England, is crisscrossed by black stone fences. If you break one of the stones, you can see that they are actually a bright golden sandstone. The black outer crust is the modern remnant of pollution from the industrial revolution over a half century ago.*

other pollutants, and lichens grow once again on British trees. As predicted by the principles of evolution, the pale moths are making a comeback, and the black form is becoming increasingly scarce.

Although we will come back to these points in Chapter 14, there are a few additional comments that we should make here. First, the black coloration was not *produced* by industrial activities. The mutation for black coloration arises spontaneously now and then in both clean and soot-covered areas. *The variations upon which natural selection works are produced by chance mutations.* This is further shown by other moth species in both England and the eastern United States. Some species had occasional dark mutants, and the dark forms spread as the environment became dark. Other species lacked black mutants and became extinct in industrial areas. A second point to remember, then, is that *selection does not necessarily produce well-adapted species.* If the right raw material isn't there, in this case black mutant moths, selection may drive a species to extinction. Finally, some people tend to regard evolution as a modern mechanism for producing the *Scala Naturae,* with ever-greater degrees of perfection appearing over time. This is not correct. *The processes of evolution select for organisms that are best adapted to a particular environment.* Neither black nor pale colors are "best" in any objective sense, but only with respect to their environment, which may change at any time in an unpredictable fashion.

Industrial melanism is but one of many instances of modern-day evolution. Other examples include pesticide resistance in insects (Chapter 1), antibiotic resistance in bacteria (Chapters 10 and 17), and the appearance of the AIDS virus (Chapter 27). Evolution is easiest to observe in rapidly reproducing, short-lived organisms such as insects or bacteria, but occurs constantly in all the species of life on Earth.

REFLECTIONS ON THE ORIGIN OF SPECIES

> It is interesting to contemplate an entangled bank, clothed with many plants of many kinds, with birds singing on the bushes, with various insects flitting about, and with worms crawling through the damp earth, and to reflect that these elaborately constructed forms . . . have all been produced by laws acting around us. These laws, taken in the highest sense, being Growth with Reproduction; Inheritance [and] Variability. . . ; a Ratio of Increase so high as to lead to a Struggle for Life, and as a consequence to Natural Selection, entailing Divergence of Character and Extinction of less-improved forms. . . . There is grandeur in this view of life, with its several powers, having been originally breathed into a few forms or into one; and that, whilst this planet has gone cycling on according to the fixed law of gravity, from so simple a beginning endless forms most beautiful and most wonderful have been, and are being, evolved.
>
> **—the concluding sentences of *On the Origin of the Species* by Charles Darwin**

SUMMARY OF KEY CONCEPTS

The History of Evolutionary Thought

A species consists of all the populations of organisms that can potentially interbreed and that are reproductively isolated from other populations. Historically, the most common explanation for the origin of species has been creationism, that a divine being created each species in its present form, and that species have not significantly changed since the creation. Since the middle of the nineteenth century, however, scientists have concluded that species originate by the operation of natural laws, as a result of changes in the genetic makeup of the populations of organisms. This process is called evolution.

Evolution by Natural Selection
Charles Darwin and Alfred Russel Wallace independently proposed the theory of evolution by natural selection. Their theory can be expressed concisely as three conclusions based on four observations. In modern biological terms, these are:

Observation 1: Each organism can produce far more offspring than are required merely to replace that organism.

Observation 2: Nevertheless, the sizes of natural populations are usually constant.

Conclusion 1: Therefore, in each generation, many organisms must die young, fail to reproduce, produce few offspring, or produce less fit offspring that fail to reproduce in their turn.

Observation 3: The individuals in a population vary in their ability to obtain resources, withstand environmental extremes, escape predation, and so on.

Conclusion 2: Which organisms produce the largest number of viable offspring depends on their ability to deal with their environment; the best adapted organisms reproduce the most. This is natural selection.

Observation 4: At least some of the variation in adaptedness is due to genetic differences that may be passed on from parent to offspring.

Conclusion 3: Over many generations, differential reproduction among individuals with different genetic makeup changes the overall genetic composition of the population. This is evolution.

The Evidence for Evolution
Many lines of evidence indicate that evolution has occurred, and that natural selection is the chief mechanism driving changes in the characteristics of species over time. The evidence includes:

1. Fossils of ancient organisms are simpler in form than modern organisms. Sequences of fossils have been discovered that show a graded series of changes in form. Both of these facts would be expected if modern forms evolved from older forms.
2. Organisms thought to be related through evolution from a common ancestor show many similar anatomical structures. Examples include the limbs of amphibians, reptiles, birds, and mammals. Similarly, stages in embryological development, similarities in chromosome structure, sequences of amino acids in proteins, and similarities in DNA composition all support the notion of descent of related species through evolution from common ancestors.
3. Rapid, heritable changes have been produced in domestic animals and plants by selectively breeding organisms with desired features (artificial selection). If differences as vast as those between Chihuahuas, bulldogs, and Great Danes can be produced in a few thousand years of artificial selection by humans, it seems likely that much larger changes could be wrought by hundreds of millions of years of natural selection.
4. Both natural and human activities may drastically change the environment over short periods. Significant changes in the characteristics of species have been observed in response to these environmental changes. A well-studied example is the evolution of black coloration among moths in response to the darkening of their environment by industrial pollutants.

GLOSSARY

Analogous structures structures that have similar functions and superficial appearance but very different anatomy, such as the wings of insects and bats. The similarities are due to similar selective pressures.

Catastrophism the hypothesis that the Earth has experienced a series of geological catastrophes, much like Noah's flood, probably imposed by a supernatural being.

Coevolution the evolution of adaptations in two species due to their extensive interactions with one another, so that each species acts as a major force of natural selection upon the other.

Convergent evolution the independent evolution of similar structures among unrelated organisms, due to similar selective pressures.

Creationism the hypothesis that all species of organisms on Earth were created in essentially their present form by a supernatural Being, and that significant modification of those species, specifically their transformation into new species, cannot occur through natural processes.

Evolution a change in the relative proportions of different genotypes in a population.

Fossil the remains of an organism, usually preserved in rock. Fossils include petrified bones or wood; shells; impressions of body forms such as feathers, skin, or leaves; and markings made by organisms such as footprints.

Homologous structures structures that may differ in function but that have similar anatomy, presumably because of descent from common ancestors.

Inheritance of acquired characteristics the hypothesis that organisms' bodies change during their lifetimes by use and disuse, and that these changes are inherited by their offspring.

Natural selection the unequal survival and reproduction of organisms due to environmental forces (e.g., physical factors such as climate and living organisms such as predators or prey) that act differently upon genetically different members of a population.

Species (spē′-sēs) the sum of all the populations of organisms that are potentially capable of interbreeding under natural conditions and that are reproductively isolated from other populations.

Uniformitarianism the hypothesis that the Earth developed gradually through natural forces similar to those at work today.

Vestigial structures (ves-tij′-ē-ul) structures with no known function, but which are homologous to functional structures in related organisms.

STUDY QUESTIONS

1. Distinguish between catastrophism and uniformitarianism. How did these hypotheses contribute to the development of evolutionary theory?
2. What is a fossil? What sorts of evidence do fossils provide about past life on Earth and how it arose?
3. Describe Lamarck's theory of inheritance of acquired characteristics. Why is it invalid?
4. Define the term "species."
5. What is natural selection? Describe how natural selection might have caused differential reproduction among the ancestors of a fast-swimming predatory fish, such as a barracuda.
6. Describe how evolution occurs through the interactions among the reproductive potential of a species, the normally constant size of natural populations, variation among individuals of a species, natural selection, and inheritance.
7. Distinguish between homologous and analogous structures, and give examples of each.
8. What is convergent evolution? Give an example.
9. What is a vestigial structure? Give two examples of vestigial structures.
10. Does evolution through natural selection produce "better" organisms in an absolute sense? Are we climbing the *Scala Naturae?* Defend your answer.

SUGGESTED READINGS

Bishop, J. A., and Cook, L. M. "Moths, Melanism and Clean Air," *Scientific American,* January 1975. This follow-up to Kettlewell's work shows that cleaning up the air reverses industrial melanism in moths, providing graphic evidence that evolution only adapts organisms to existing environments.

Darwin, C., *On the Origin of Species by Means of Natural Selection* (1859). New York: Doubleday & Company, Inc. An impressive array of evidence amassed to convince a skeptical world.

Eiseley, L. C. "Charles Darwin," *Scientific American,* February 1956. (Offprint No. 108). An essay on the life of Darwin, by one of his foremost American biographers. Even if you need no introduction to Darwin, read this anyway, as an introduction to Eiseley, author of many marvelous essays.

Gould, S. J., *Ever Since Darwin,* 1977; *The Panda's Thumb,* 1980; and *The Flamingo's Smile,* 1985. New York: W. W. Norton & Company, Inc. A series of witty, imaginative, and informative essays, mostly from *Natural History* magazine. Many deal with various aspects of evolution.

Kettlewell, H. B. D. "Darwin's Missing Evidence," *Scientific American,* March 1959. Industrial melanism as an example of modern-day evolution.

Lewontin, R. C. "Adaptation," *Scientific American,* September 1978 (Offprint No. 1408). Lewontin shows that perfect adaptation is not always achieved.

14
The Processes and Results of Evolution

In Chapter 13 we discussed the history of the theory of evolution and presented some of the evidence that evolution actually happens. But what processes drive evolutionary change? Is natural selection the only cause of evolution? Does evolution always occur all the time in all populations of organisms? In this chapter we examine evolutionary processes in more detail. As we do, you will see that *evolution is an inevitable consequence of the nature of living things.* It occurs as a direct result of the chemical structure of genes and the interactions between organisms and their environment.

Figure 14-1
"What but the wolf's tooth whittled so fine
The fleet limbs of the antelope?
What but fear winged the birds, and hunger
Jewelled with such eyes the great goshawk's head?"
—from "The Bloody Sire" by Robinson Jeffers

EVOLUTION AND THE GENETICS OF POPULATIONS

Individual organisms live, reproduce, and die. Individuals, however, do not evolve: **evolution** *is genetic change occurring in a population of organisms over many generations.* Inheritance, therefore, is the link between the lives of individual organisms and the evolution of populations. We will begin our discussion of the processes of evolution by reviewing the principles of genetics as they apply to individuals and then extend these principles to the genetics of populations. You may want to refer back to Unit II to refresh your memory on specific points.

Gene Function in Individual Organisms

Each cell of every organism contains a repository of genetic information encoded in the DNA of its chromosomes. A gene is a segment of DNA located at a particular place on a chromosome. Its sequence of nucleotides encodes the sequence of amino acids of a protein, usually an enzyme that catalyzes one particular reaction in the cell. Slightly different sequences of nucleotides at a given gene's location, called alleles, generate different forms of the same enzyme. The specific alleles borne on an organism's chromosomes, interacting with the environment, determine its physical and behavioral traits.

Let's illustrate these principles with an example that should be familiar to you from Unit II. A pea flower is colored purple because a chemical reaction in its petals converts a colorless molecule to a purple pigment. When we say that a pea plant has the allele for purple flowers, we mean that a particular stretch of DNA on one of its chromosomes contains a sequence of nucleotides that codes for the enzyme catalyzing this reaction. A pea with the allele for white flowers has a different sequence of nucleotides at the corresponding place on one of its chromosomes. The resulting enzyme cannot produce purple pigment. If a pea is homozygous for the white allele, its flowers produce no pigment, and are white.

Genes in Populations

In population genetics, the **gene pool** for a particular gene is defined as the total of all the alleles of that gene that occur in a population. (The **total gene pool** for the population is the total complement of alleles for all the genes.) For example, in a population of 100 pea plants, the gene pool for flower color would consist of 200 alleles (peas are diploid, so there are two color alleles per plant, times 100 plants). If we could analyze the genetic composition of every plant in the field, we might find that some have alleles for white flowers, some have alleles for purple flowers, and some have both alleles. If we added up the color alleles of each plant in the population, we could determine the relative proportions of the different alleles, a number called the **allele frequency.** Let's say that the gene pool for flower color consisted of 140 alleles for purple and 60 alleles for white. The allele frequencies would then be: purple, 0.7 (70 percent), and white, 0.3 (30 percent).

Population Genetics and Evolution

What does all this have to do with evolution? Quite a bit. Suppose that a flower-eating cow comes along and, being enamored of purple flowers, eats all the purple flowers before they set seed. As you know from Chapter 9, the allele for purple flowers is dominant to the allele for white. Therefore, all the purple alleles in the entire population are in the purple-flowered plants. If none of these plants reproduce, while the white-flowered plants do reproduce, the next generation will consist entirely of white-flowered peas. The allele frequency for purple will drop to 0, while the allele frequency for white will rise to 1.0. Because of the selective eating habits of the cow, *evolution will have occurred in that field.* The gene pool of the pea population will have changed, and natural selection, in the form of foraging by the cow, will have caused the change.

This simple example illustrates four important points about evolution. (1) *Natural selection does not cause genetic changes in individuals.* The alleles for purple or white flower color arose spontaneously, long before the cow ever found the pea field. The white peas did not possess some sort of "foresight" and acquire the alleles for white flowers in anticipation of the cow. The cow, in turn, did not cause white alleles to appear. It merely favored the differential survival of white alleles compared to purple alleles. (2) *Natural selection befalls individuals, but evolution occurs in populations.* Individual pea plants either reproduced or not, but it was the population as a whole that evolved. (3) *Evolution is a change in the allele frequencies of a population, due to differential reproduction.* In preserving the "favorable variations" that Darwin spoke of, selection increased the frequency of favorable (white) alleles at the expense of unfavorable (purple) alleles. (4) *Evolutionary changes are not "good" or "progressive" in any absolute sense.* The white alleles were favorable only because of the dietary preferences of this particular cow; in another environment, with other predators, the white allele may well be selected against.

EVOLUTIONARY MECHANISMS

To understand the forces that cause populations to evolve, it is helpful first to consider the characteristics of a population that would *not* evolve. In 1908, G. H. Hardy and W. Weinberg defined an **equilibrium population** as one in which neither the allele frequencies nor the distribution of genotypes changes with succeeding generations (Fig. 14-2). Since allele frequencies do not change, evolution does not occur. A population can remain in equilibrium only under several restrictive conditions:

1. There must be no mutation.
2. There must be no differential migration of alleles into the population (immigration) or out of the population (emigration).
3. The population must be large (theoretically infinite).
4. Mating must be completely at random, with no tendency for certain genotypes to mate with specific other genotypes.
5. All alleles must be equally adaptive; that is, all genotypes must reproduce equally well (no natural selection).

If these conditions are met, allele frequencies within a population will remain the same indefinitely. If one or more of these conditions are violated, allele frequencies will change, i.e., evolution will occur. As you might expect, few, if any, natural populations are truly in equilibrium. If so, what is the importance of the Hardy–Weinberg principle? Biologists have found that the Hardy–Weinberg conditions are useful starting points in studying the mechanisms of evolution. In the following sections we study each condition carefully, show why it

*The Hardy–Weinberg equilibrium predicts that if a large population undergoes no mutation, migration, or natural selection, and if all members of the population mate randomly, the frequencies of alleles will not change from generation to generation. To see how this can be so, consider our familiar pea plants. As you know, pea seeds can be round (*R*: dominant) or wrinkled (*r*: recessive). To determine the genotypes of the offspring of two individuals, for example two heterozygotes, we would draw a Punnett square:*

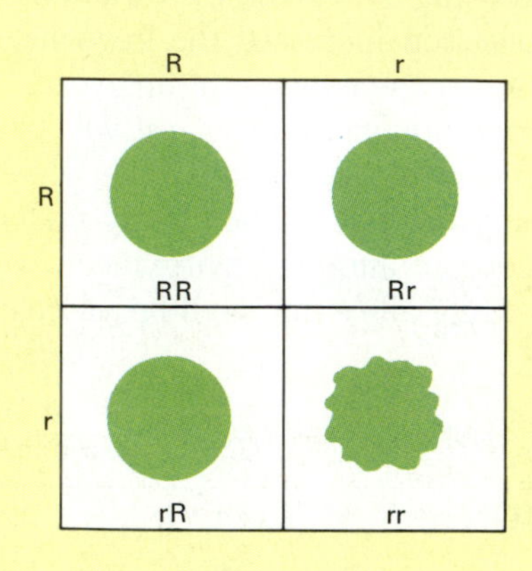

Each parent produces both R *and* r *gametes. The expected offspring are* ¼ RR, ½ Rr, ¼ rr. *There are two ways to arrive at these frequencies. The first is to add up the offspring in each box of the square. Another way of doing it is by probabilities. Each gamete has an equal probability of containing either allele. Therefore, we can assign probabilities to the gametes in the Punnett square:* R = *0.5,* r = *0.5. From the laws of probability,* the probability of two independent events occurring simultaneously is the product of their individual probabilities. *If you flip a coin, the probability of a head is* ½. *If you flip two coins simultaneously, the probability of two heads is* ½ × ½ = ¼. *Similarly, we can obtain the probability of obtaining each type of offspring by multiplying the relative proportions of each allele:*

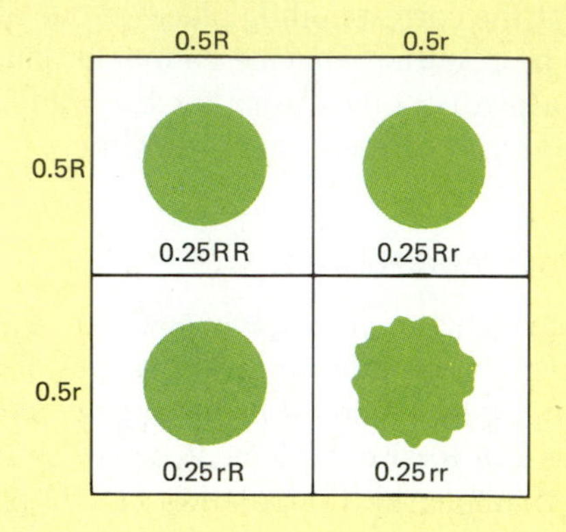

We obtain 0.25 RR, *0.25* Rr, *0.25* rR, *and 0.25* rr.

Let us suppose, now, that we have a population of 100 peas and that we collect sperm and egg cells from all of them and determine their genotypes. We may find, for example, that there are 60 percent R *alleles and 40 percent* r *alleles in the gametes. The relative proportions of the two alleles* R *and* r *are identical to the probability that*

Figure 14-2 *The genetics of equilibrium populations.*

is often violated by natural populations, and illustrate the consequences of its violation. In this way, you can better understand both the inevitability of evolution and the forces that drive evolutionary change.

Mutations: The Raw Material of Evolution

Cells have efficient mechanisms that protect the integrity of their genes. Enzymes constantly scan the DNA, repairing flaws caused by radiation, chemical damage, or mistakes in copying. Nevertheless, changes in nucleotide sequence can happen. These are **mutations,** and they vary tremendously in their impact. As we explained in Chapter 11, some changes in DNA have virtually no effect on the organism; many, perhaps most, are harmful; and a few may be beneficial, or may aid the organism in coping with new or changed environments.

How significant is mutation in altering the gene pool of a population? Mutations are rare, occurring once in 10,000 to 1,000,000 genes per generation per individual. Therefore, mutation is not a major force in evolution by itself. However, *mutations are the source of new alleles,* new heritable variations upon which other evolutionary processes can work. As such, they are the foundation of evolutionary change.

As we mentioned earlier, *mutations are not goal directed.* A mutation does not arise as a result of, or in anticipation of, environmental necessities (Fig. 14-3). A mutation simply happens, and may in turn produce a change in the structure or function of the organism.

any given offspring will receive either R *or* r. *We can thus draw a "population Punnett square":*

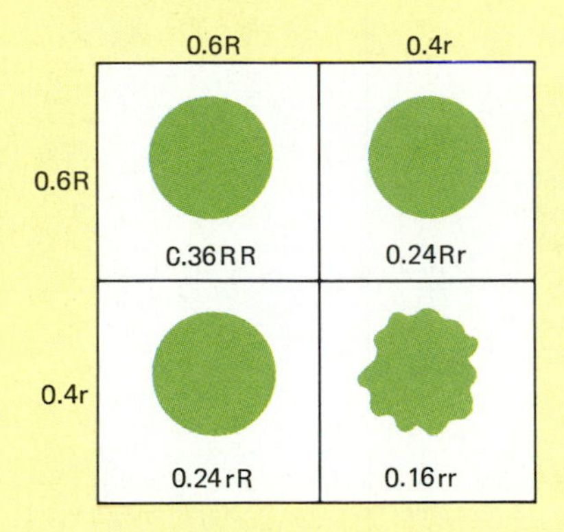

In the "population F_1 generation" we expect the following proportions of genotypes: 0.36 RR, *0.48* Rr, *and 0.16* rr. *If the population remains the same size, 100 peas, we would have 36* RR, *48* Rr, *and 16* rr *peas. What gametes would this F_1 generation in its turn produce? Under the Hardy–Weinberg conditions, each plant produces equal numbers of gametes, and by the principles of Mendelian genetics, each plant produces equal numbers of gametes with each of its two alleles for seed shape. To keep things simple, let's assume that each plant contributes two gametes, one with each of its two alleles. We therefore collect 72* R *alleles from the homozygous dominants, 48* R *alleles and 48* r *alleles from the heterozygotes, and 32* r *alleles from the homozygous recessives, for a total of 120* R *and 80* r *alleles. The allele frequencies of the gametes from the "population F_1 generation," then, are 0.6* R *and 0.4* r, *just as we started out with. Therefore, the F_2 generation has the same distribution of genotypes as the F_1. If there are no disturbances, this process will go on indefinitely: the population remains in equilibrium.*

Rather than going through Punnett squares, there is an easier way of calculating allele and genotype frequencies. The sum of all allele frequencies must equal 1. Let the frequency of the R *allele be represented by* p, *and the frequency of the* r *allele by* q. *Then the sum of the frequencies* $p + q = 1$. *Just as we generated the genotype frequencies in the population Punnett square by multiplying allele frequencies, we can do the same with this equation:*

$$(p + q) \times (p + q) = p^2 + pq + qp + q^2 = p^2 + 2pq + q^2 = 1$$

For our particular example, $p = 0.6$ *and* $q = 0.4$, *so the genotypes of the population F_1 generation will be*

$$(0.6)^2 \text{ RR} + 2 \times (0.6) \times (0.4) \text{ Rr} + (0.4)^2 \text{ rr} = 0.36 \text{ RR}, 0.48 \text{ Rr, and } 0.16 \text{ rr}.$$

This is the same set of frequencies that we calculated with the population Punnett square.

As these calculations show, in an equilibrium population allele frequencies and the distribution of genotypes remain constant, generation after generation. In actual experiments, if measurements of allele frequencies in a population show significant changes over time, evolution is occurring in that population.

Whether that change is helpful or harmful, now or in the future, depends on environmental conditions over which the organism has little or no control. The mutation provides *potential;* it is natural selection, acting on that potential, that may favor the spread of a mutation through the population.

Migration: Redistributing Genes

In biology the word "migration" has two distinct meanings. In the most familiar context, migration refers to the seasonal movement of many species, especially birds, between summer breeding grounds and distant winter refuges. In evolutionary biology, however, **migration** *is the flow of genes between populations.* Baboons, for example, live in social groupings called troops. Within each troop, all the females mate with a handful of dominant males. Juvenile males, however, often leave the troop. If they are lucky, they join and perhaps even become dominant in another troop. Thus the male offspring of one troop carry genes to the gene pool of other troops.

Migration has three significant effects. First, *gene flow helps to spread advantageous alleles throughout the species,* as individuals possessing these alleles survive better and leave more offspring in troop after troop. Second, *gene flow helps to maintain all the organisms over a large area as one species,* since it prevents large differences in allele frequencies from developing between troops. Isolation of populations, with no gene flow to or from other populations of the same species, is an important factor in

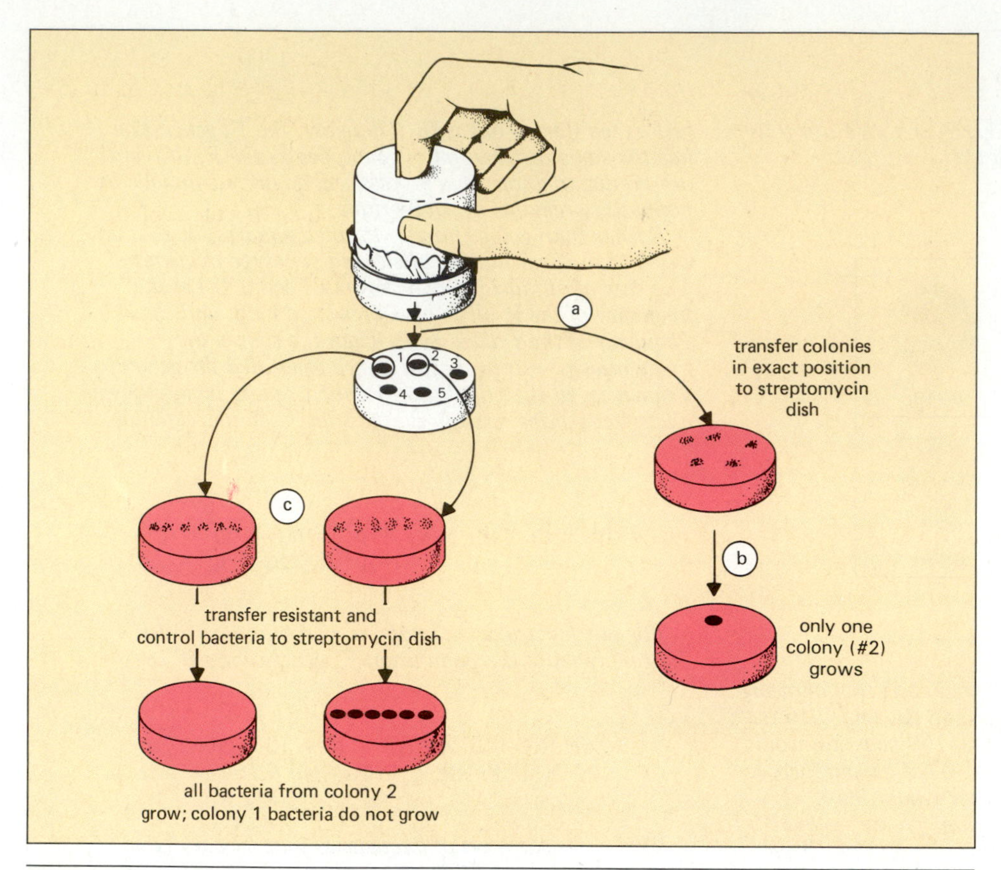

Figure 14-3 *Proof that mutations occur spontaneously, not in response to specific selective pressures.*
(a) *Clumps of bacteria (colonies) are grown on a solid nutrient medium in a dish. These bacteria have never been exposed to antibiotics. A piece of velvet the exact size of the dish is lightly pressed onto the bacterial colonies and then touched to the surface of nutrient medium containing the antibiotic streptomycin in a second dish. A few bacteria from each original colony adhere to the velvet and then come off the velvet onto the second dish. Thus the exact positions of the "parent" colonies are duplicated in the second dish.*
(b) *Only one daughter colony, in position 2, grows on the streptomycin-containing medium in the second dish. There are two possible explanations for this result: First, the bacteria of parent colony 2 may have been already resistant, due to a spontaneous mutation, while the bacteria of the other parent colonies were not. In this case we would expect that the other bacteria of parent colony 2 will also be resistant to streptomycin. Second, perhaps none of the parent bacteria were resistant, and one mutated in the second dish in response to streptomycin. If this happened, the other bacteria from parent colony 2 will not, in general, be resistant to streptomycin.*
(c) *Samples of the original colonies are transferred to streptomycin-containing medium. All bacteria from colony 2 grow, while none from colony 1 grow, proving that the bacteria of colony 2 already possessed the mutation for streptomycin resistance prior to exposure, and that the presence of streptomycin in the medium does not induce an adaptive mutation for streptomycin resistance.*

the origin of new species. Third, as we will discuss shortly, *migration and the subsequent isolation of small numbers of individuals sometimes lead to significant evolutionary change.*

Population Size

To remain in equilibrium, a population must be large. To see why, let's return to the thoughts of Darwin and Wallace in their formulation of the theory of evolution. In general, all populations have tremendous potential for growth but are limited to a relatively constant size by the available resources. This means that most organisms die without reproducing. Put another way, *only a small sample of a population actually serves as parents for the next generation.* Which individuals reproduce depends both on fitness and on chance. Obviously, those individuals that are better adapted to their environment are more likely to survive and leave offspring to carry on their genes. Nevertheless, chance is also important, because disaster may befall even the fittest organism. The maple seed that falls into a pond will never sprout; the deer and elk blasted away by Mt. St. Helens left no descendants.

Population size greatly influences the potential for chance events to change allele frequencies. Consider, for example, two hypothetical populations of ladybugs, in which the carapace is either spotted or solid-colored, controlled by alternate alleles of a single gene. In each population, half the ladybugs are spotted and half are solid-colored (i.e., the frequencies of both alleles are 0.5), but one population has only four bugs while the other has 1000. Let us assume that each individual that survives to maturity produces two offspring identical to itself. Therefore, if population sizes remain constant, exactly half the individuals will reproduce in each generation. Let us further assume that whether an individual reproduces or not is determined entirely by chance. In the larger population, 500 bugs will be parents to the next generation. It would be astronomically unlikely that all 500 parents would be spotted; in fact, it would be extremely unlikely for even 300 parents to be spotted. Thus, because of the large population, we would not expect a major change in allele frequencies to occur from generation to generation (Fig. 14-4). In the small population, on the other hand, only two individuals will reproduce. There is a 25 percent chance that both parents will be spotted (this is the same likelihood as flipping two coins and having both come up heads). If this happens, the next generation will consist entirely of spotted ladybugs. Such a change in allele frequency in small populations purely by chance is termed **genetic drift.**

How much does genetic drift contribute to evolution?

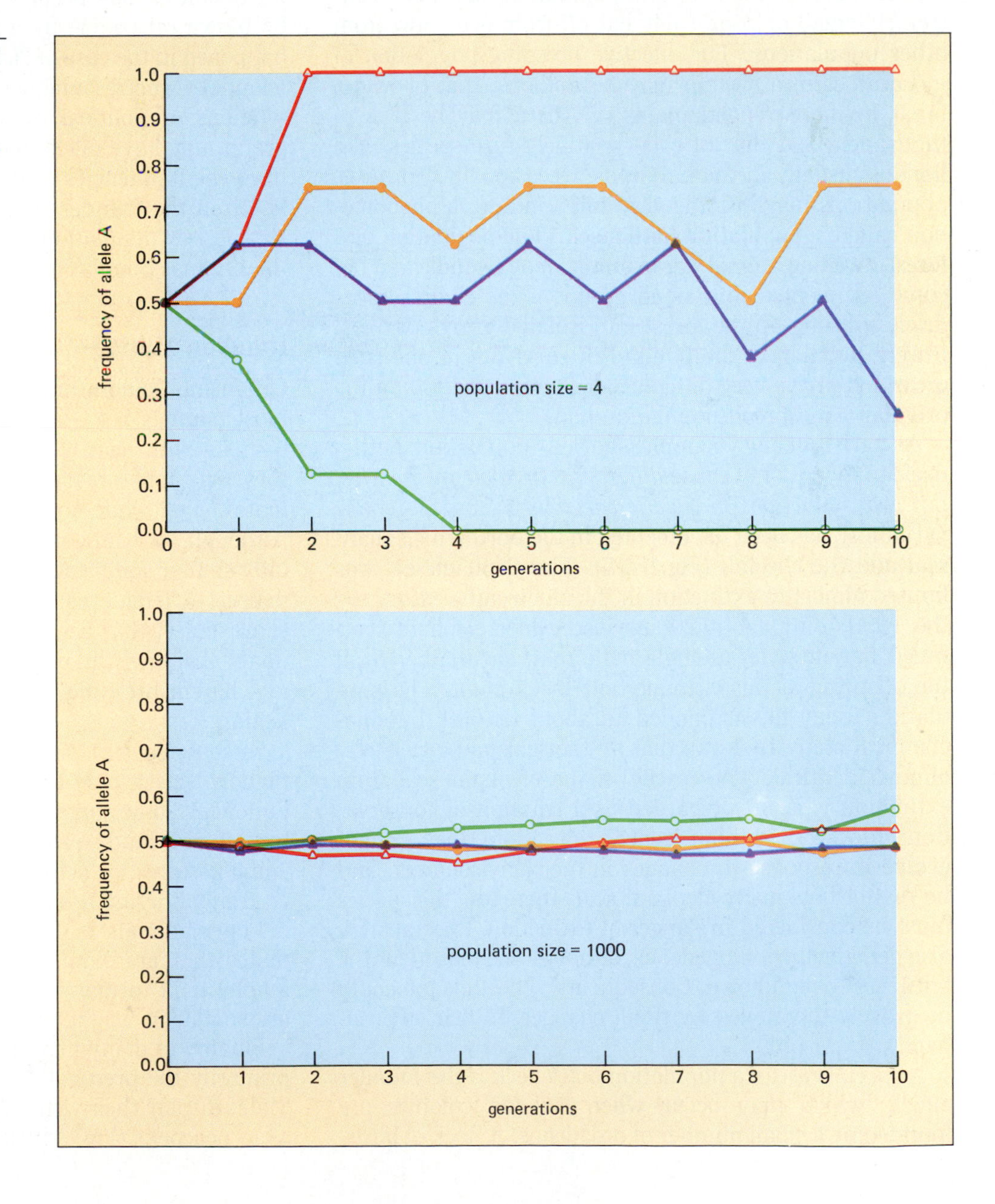

Figure 14-4 Computer-generated graphs illustrating the effect of population size on genetic drift. In both graphs, the initial population was composed of half A *and half* a *alleles, and 10 generations were simulated, with individuals chosen at random to contribute alleles to the next generation. Four simulations were run for each population size, producing the four lines on each graph. With a population size of 4, one allele sometimes became "extinct" due to chance. With a population size of 1000, allele frequencies remained relatively constant.*

Figure 14-5 Both the northern elephant seal **(a)** *and the cheetah* **(b)** *passed through a population bottleneck in the recent past, resulting in an almost total loss of genetic diversity.*

No one really knows. Natural populations are seldom extremely small or completely cut off from gene flow from other populations. The effective breeding population of mosquitoes in a swamp may be millions, that of wildebeest in the Serengeti plains of Africa may be tens of thousands, and that of even relatively rare species, like lions, is usually in the hundreds. Occasionally, however, populations may become very small indeed, a phenomenon called a **population bottleneck.** The population of a forest-dwelling animal, for example, may dwindle and become fragmented into small groups if the climate turns drier, splitting up the forest into isolated groves of trees. Purely by chance, the handful of survivors in one such grove may have very different allele frequencies than the original, widespread population had.

As our ladybug example showed, population bottlenecks may not only cause *differences in allele frequencies,* but also *reductions in genetic variability*. Loss of genetic variability has been documented in the northern elephant seal and the cheetah (Fig. 14-5). The elephant seal was hunted almost to extinction in the nineteenth century; by the 1890s only about 20 survived. Since elephant seals breed harem style, a single male may have fathered all the offspring at this extreme bottleneck point. The population today has expanded to about 30,000, but biochemical analysis shows that all the seals are genetically almost identical. The rescue of the elephant seal from extinction is rightly regarded as a triumph of conservation; however, without genetic variation, they cannot evolve in response to changes in their environment, and no matter how many elephant seals there are, the species must be considered in danger of extinction. Cheetahs are also genetically homogeneous, although the reason for the bottleneck is unknown. Consequently, cheetahs too could be gravely threatened by small changes in their environment.

A special case of a population bottleneck is the **founder effect,** which may occur when isolated colonies are founded by a small number of organisms. A flock of birds, for instance, may become lost during migration, or may be blown off course by a storm (this is thought to have happened in the case of Darwin's finches in the Galápagos Islands). Among humans, small groups may migrate for religious or political reasons (Fig. 14-6). Such a small group may have allele frequencies that are very different from the frequencies of the parent population. If the isolation of the founders is maintained for a long period, a sizable new population may arise that differs greatly from the original population.

Random Mating

Organisms seldom mate strictly randomly. For example, most animals have limited mobility and are most likely to mate with nearby members of their species. Further, they may make behavioral distinctions among potential mates. The white-crowned sparrow is a case in point. Although all white-crowned sparrows sing a fundamentally similar song, each local population has its own song dialect. A female usually chooses a mate that sings the same dialect that her father sang (Fig. 14-7). Among animals, there are three common forms of nonrandom mating: harem breeding, assortative mating, and sexual selection.

In some species, such as elephant seals, baboons, and bighorn sheep, only a few males fertilize all the females. Following some sort of contest, which may involve showing off with loud sounds or flashy colors, making threatening gestures, or actual combat, only certain males succeed in gathering a harem and mating (Fig. 14-8).

Many animals mate assortatively; that is, they select mates that are similar to themselves. Humans, for example, tend to marry members of the opposite sex who are similar in height, race, IQ, and social status.

Finally, in many mammals and birds, mate selection is primarily the prerogative of one sex, usually the female. Males display their virtues, which may be bright plumage, as in peacocks (Fig. 14-9), or rich territories, as in many

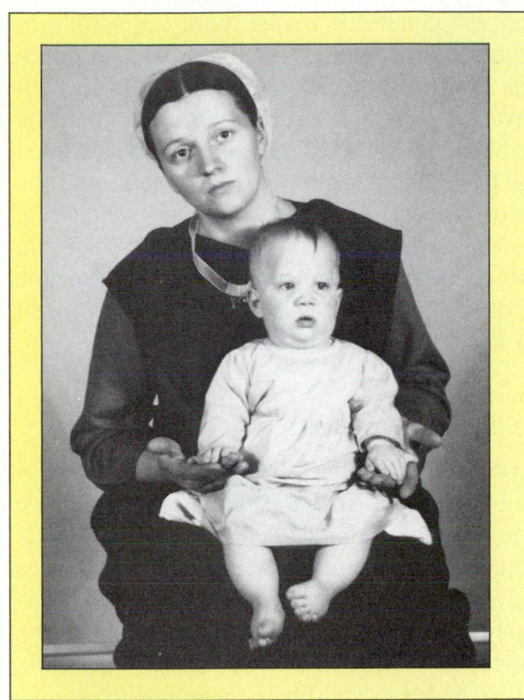

***Figure 14-6.** An Amish woman with her child, suffering from a set of genetic defects known as the Ellis–van Creveld syndrome (short arms and legs, extra fingers, occasionally heart defects). Fleeing from religious persecution, about 200 members of the Amish religion migrated from Switzerland to Pennsylvania between 1720 and 1770. Since that time, virtually all the Pennsylvania Amish moved to Lancaster County and have remained reproductively isolated from non-Amish Americans. The population increased to about 8000 by 1964. In that year, geneticist Victor McKusick surveyed the Lancaster County Amish and discovered that they had an unusually high frequency of Ellis–van Creveld syndrome. McKusick found that the Amish had an allele frequency for Ellis–van Creveld of about 0.07, compared to a frequency of less than 0.001 in the general population. Why? One couple who immigrated in 1744 carried the allele. Inbreeding among the Amish passed the gene along to their descendants, a clear example of a founder effect. In addition, by chance the Ellis–van Creveld carriers had more children than the Amish average, increasing the allele frequency by genetic drift. The combination of an initially high frequency in the immigrants (1 or 2 out of 200) plus genetic drift resulted in the modern situation, in which more cases of Ellis–van Creveld syndrome are known from Lancaster County than from the rest of the world combined.*

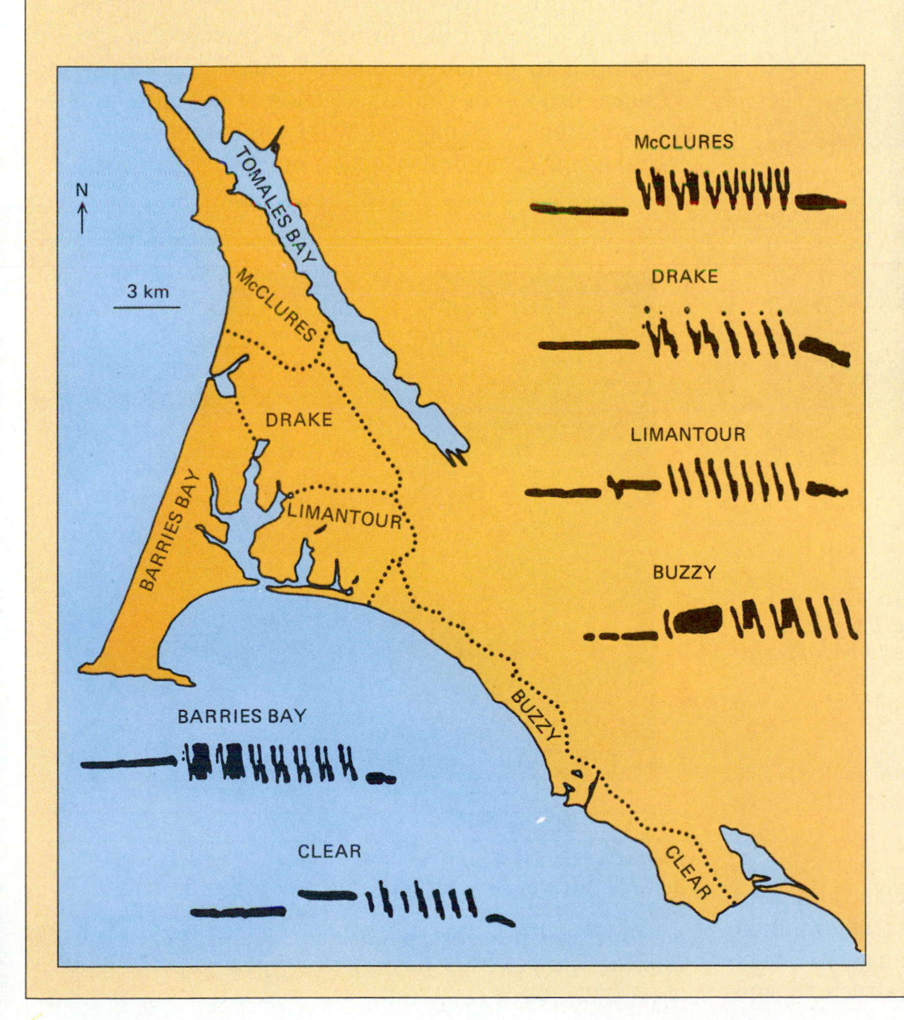

***Figure 14-7** Song dialects among populations of white-crowned sparrows at Point Reyes National Seashore north of San Francisco. As the sonograms show, the songs are fairly similar, but both birds and human listeners can recognize the different dialects. Male birds of each population learn their local dialect while in the nest, and sing it when they mature. Females preferentially mate with males that sing the dialect sung by the females' own fathers (i.e., the females' own local dialect).*

Figure 14-8 Sparring contests between males result in extremely nonrandom mating among many animals, including deer, elk, seals, and many monkeys. Here two male bighorn sheep square off against each other during the fall rutting season. Although the horns are potentially lethal weapons, they are used in ritualized ways that minimize the danger of injury to either contestant.

Figure 14-9 Many male birds, including peacocks, attract mates by displaying their wares. The features evolved for female attraction are often irrelevant, or even harmful, to the day-to-day survival of the males. Sometimes, as this photo suggests, they do not even prove very attractive to the opposite sex!

songbirds. A female evaluates the males and chooses her mate. It is thought that many of the elaborate structures and behaviors found only in males have evolved through the selective pressure of female mate choice. Darwin was so impressed with these structures that he coined the term **sexual selection** to designate the process of evolution through mate choice, and considered it a category distinct from natural selection. Since conspicuous structures and bizarre behaviors render the males more vulnerable to predators, sexual selection often seems to work in opposition to other forms of natural selection. However, nonsexual selective forces may also oppose one another; the height of a giraffe, for example, is a compromise between the advantage of reaching higher leaves for food and the disadvantage of vulnerability while drinking water (Fig. 14-10). In both sexual and nonsexual selection, then, some aspect of the environment (in sexual selection the "opinion" of the opposite sex, which is part of the social environment) influences reproductive success.

Equivalence of Genotypes

Finally, genetic equilibrium requires that all genotypes must be equally adaptive—that is, none has any selective advantage over the others. Some alleles may be adaptively neutral, so that organisms possessing any of several alleles will be equally likely to survive and reproduce. This is clearly not true of all alleles in all environments. Any time an allele confers, in Wallace's words, "some little superiority," natural selection will favor the enhanced reproduction of the individuals possessing it. Although mutation provides the initial variability, and the chance effects of genetic drift drive some changes in allele frequency, it is natural selection that prunes the growth of a species, molding it to fit its environment. Since, as Darwin and Wallace proposed, natural selection is the guiding force for evolutionary change, we will examine the modes and mechanisms of natural selection more closely.

Figure 14-10 The long neck and legs of a giraffe are a decided advantage in feeding on acacia leaves high up in trees. However, a giraffe has to get into an extremely awkward and vulnerable position to drink. Feeding and drinking thus place opposing selective pressures on the length of neck and legs.

NATURAL SELECTION

Natural selection acts on phenotypes: the actual structures and behaviors that the organisms in a population display. If you were to measure the phenotypes of a specific trait in all the individuals in a population, you would find a range of values (Fig. 14-11). This range of phenotypes arises from differences both in the genotypes of the organisms and in the environments in which they live. However, just as bad calls by umpires tend to average out over a long baseball season, so that the best team wins the most games, environmental differences influencing phenotypes average out in a large population. On the average, genotype predicts phenotype: most large plants will have genes promoting large size, while most small plants will have genes promoting small size. In our discussion of selection, therefore, we will ignore environmental causes of variability.

Types of Selection

Biologists recognize three major categories of natural selection: stabilizing selection, disruptive selection, and directional selection (Fig. 14-12).

Stabilizing selection. Natural selection does not necessarily mean continuous change and "improvement" in a species. If a species is already well adapted to a particular environment and the environment does not change, most variations that appear through new mutations or recombination of old alleles will be harmful. Selection will therefore favor the survival and reproduction of "average" individuals, a situation termed **stabilizing selection** (Fig. 14-12a). As an example, consider the wildflowers of the deciduous forests in New England (Fig. 14-13). In early spring, the trees are bare and sunlight penetrates to the forest floor. Once the trees have leafed

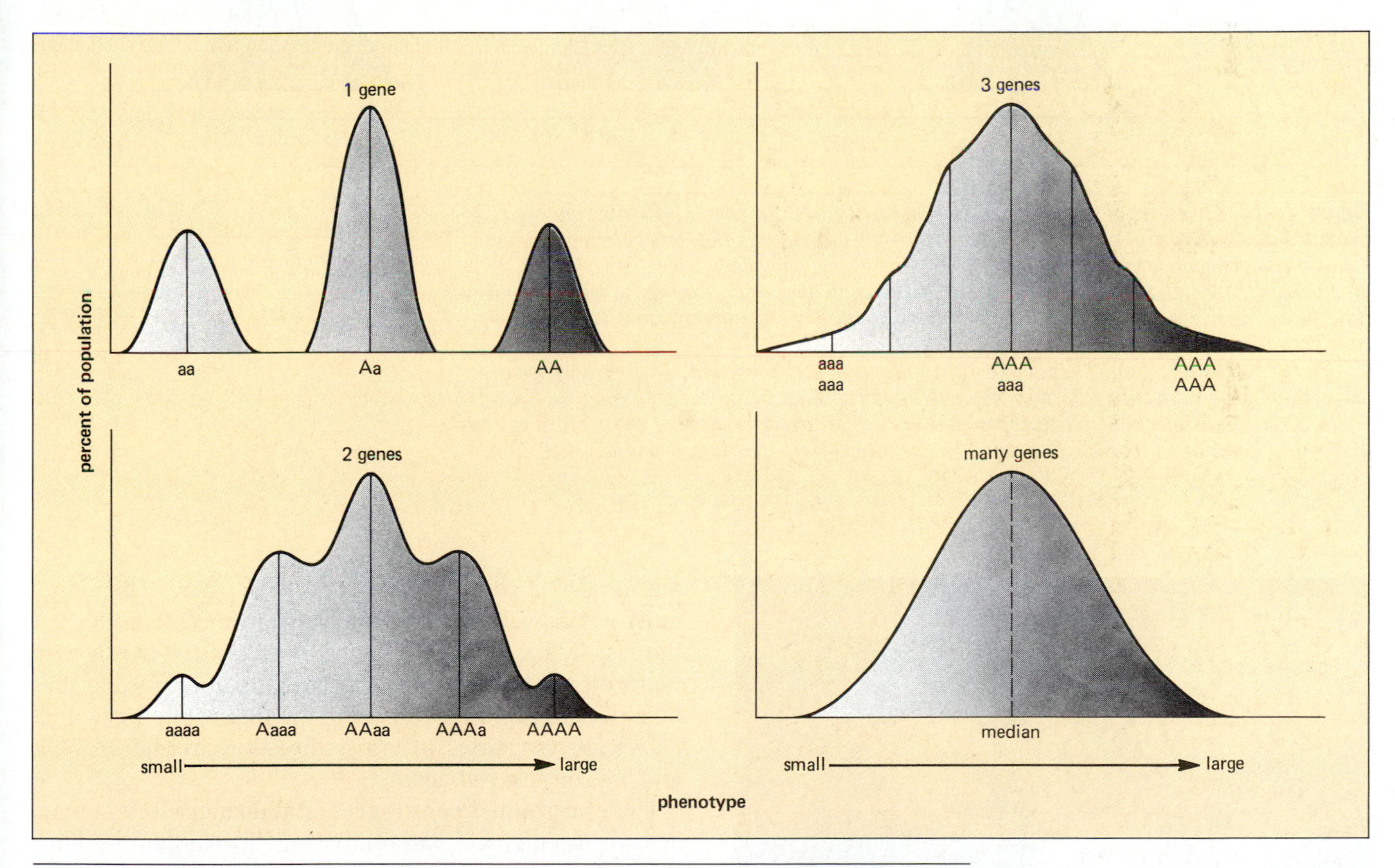

Figure 14-11 *Both genes and environment contribute to the phenotype of an organism. This series of graphs illustrates the distribution of phenotypes which would be expected if one, two, three, or many genes, each with two incompletely dominant alleles (see Chapter 9), contributed to a particular body characteristic (e.g., size). The vertical black lines represent the precise size expected due to genotype alone. In each case, environmental conditions (e.g., amount of available food) create some variation in size, represented by the colored curves. As the number of genes contributing to the characteristic becomes large, the distribution of phenotypes approximates a smooth curve called a normal distribution. The most common value for the phenotype is the middle value, also called the median.*

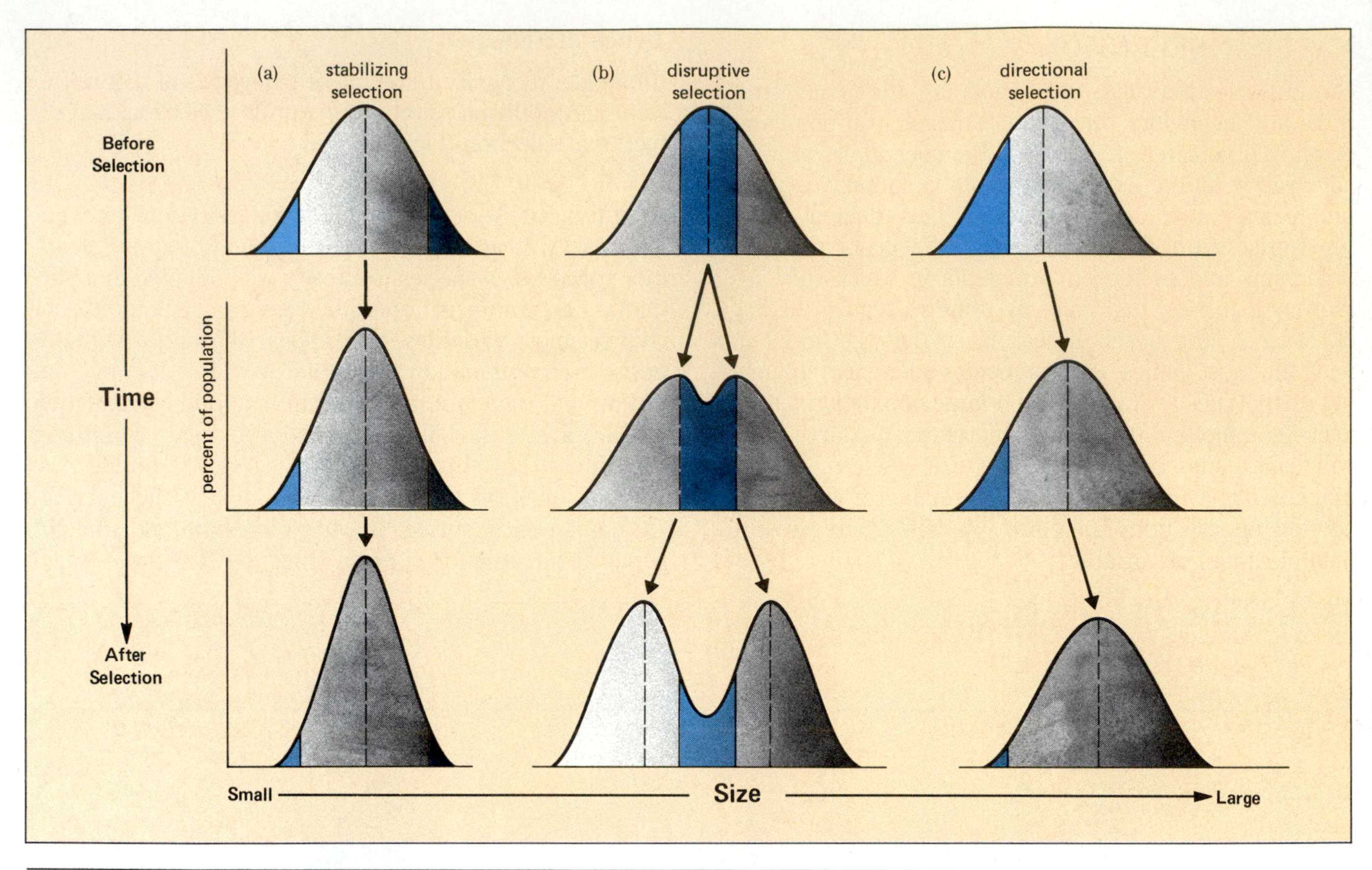

Figure 14-12 *Three types of natural selection, acting upon a normal distribution of phenotypes. In all graphs, the blue area represents individuals that are selected against (i.e., do not reproduce as frequently).*
(a) *In stabilizing selection, the organisms most likely to reproduce are those with phenotypes close to the average for the population. The variability of phenotypes close to the average value remains the same.*
(b) *In disruptive selection, phenotypes that are either larger or smaller than average are favored. The population splits into two phenotypic groups.*
(c) *In directional selection, phenotypes that are either larger or smaller than average (larger illustrated here) are favored. The average phenotype shifts position over the generations.*

Figure 14-13 *Spring beauty, an early spring wildflower common in deciduous woodlands of eastern North America. The flowers are in the sun because the trees have not yet leafed out.*

out, a deep shade settles over the understory. Therefore, most wildflowers must sprout, flower, and set seed after the last killing frost of winter but before the tree leaves block off the sun. Wildflowers are thus under strong stabilizing selection with respect to the timing of growth and flowering: variations in either direction are detrimental and will be selected against.

Under certain circumstances, stabilizing selection may act not to eliminate variability, but to maintain it. This seems to have occurred with the hemoglobin alleles in blacks (see Chapter 12). In people who are homozygous for sickle-cell anemia, hemoglobin molecules clump up into long chains, distorting and weakening their red blood cells. This causes severe anemia, and people homozygous for sickle-cell anemia are strongly selected against. Heterozygotes have only mild anemia, but may still suffer ill effects during strenuous exercise. Under these circumstances, you might wonder why natural selection has not eliminated the sickle-cell allele. Far from being eliminated, however, the sickle-cell allele is carried by nearly half the people in some areas of Africa. This is apparently due to the counterbalancing effects of anemia and malaria, which was formerly very common in equatorial Africa. Malaria parasites multiply rapidly within the red blood cells of homozygous normal individuals. Before effective medical treatments were discovered, homozygous normals consequently often died of malaria. Heterozygotes, on the other hand, enjoy some protection against malaria. Oxygen consumption by the parasites induces sickling in their red blood cells. Shortly after a cell becomes infected, it sickles and is destroyed by the spleen before the parasites can complete their development. Malaria-infected heterozygotes, therefore, have mild anemia but do not succumb to the disease. During the evolution of African populations, heterozygotes survived the best and reproduced the most. As a result, both the normal hemoglobin allele and the sickle-cell allele have been preserved (Fig. 14-14).

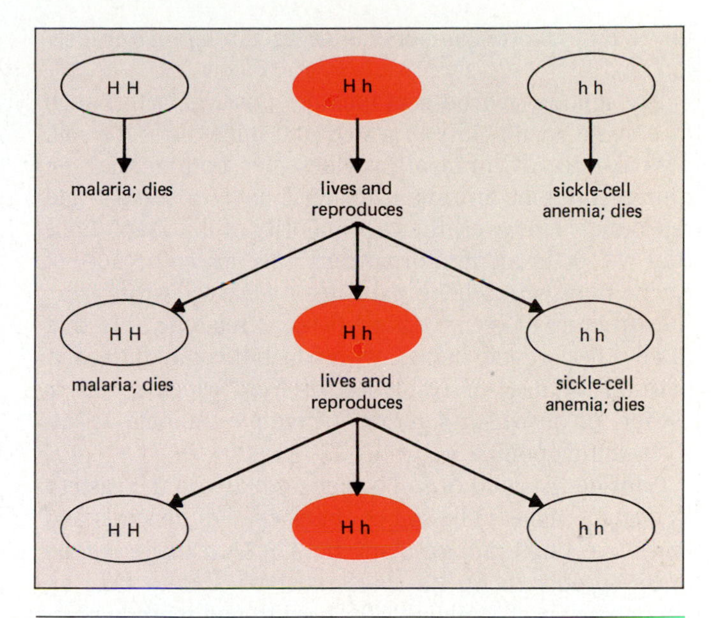

Figure 14-14 Sometimes two or more alleles, each producing a different phenotype, can be maintained in a population by stabilizing selection. The alleles for normal (H) *and sickle-cell* (h) *hemoglobin are maintained by selection against both homozygotes. Heterozygotes* (Hh) *reproduce the most, thereby keeping both alleles present in the population.*

Disruptive Selection. **Disruptive selection** favors *both extremes* at the expense of average individuals (Fig. 14-12b). This may occur when there are different microhabitats available to a species, and different characteristics best adapt individuals to each microhabitat. For example, an island, such as one of the Galápagos, may have several species of plants, some producing large, hard seeds and others small, soft seeds. Large seeds provide the most food per seed, but can be cracked and eaten only by birds with large bills. Although large birds can easily eat small seeds, they would probably spend too much energy lugging their large bodies about looking for tiny seeds. If a single species of bird colonizes such an island, what will happen? We would expect that larger-bodied, larger-beaked birds will specialize on large seeds, while small-bodied, small-beaked birds will specialize on small seeds. Medium-sized birds might not be able to crack open the large seeds, and might not get enough energy from small seeds, and so would be selected against. Disruptive selection would favor the survival and reproduction of both large and small, but not medium-sized, birds.

Directional selection. If environmental conditions change rapidly, perhaps because the climate changes or because new predators or prey appear, a species may evolve rapidly in response. Darwin's illustration of evolutionary trends in wolves, presented in Chapter 13, is an example of **directional selection:** an extreme form, in this case swifter wolves, is more adaptive than the average (Fig. 14-12c). The evolution of long necks in giraffes was almost certainly due to directional selection: pre-giraffes with longer necks obtained more food and therefore reproduced more prolifically than their shorter-necked contemporaries did.

How fast can directional selection change genotypes? That depends on both the genetic nature of the variability in the population and the strength of selection. The increased frequency of the black form of the peppered moth in Britain earlier this century was an extremely rapid case of directional selection. In this instance, the color did not vary in a finely graded manner, but was either black or pale, controlled by a single gene. Predation by birds was also a very strong selective force. Together, these two factors produced a dramatic change in the population in just a few years. If little variability exists in the population, or if the different alleles produce only slightly different phenotypes, directional selection will drive much slower changes. In some instances, a population may not be able to respond fast enough to the selective forces and may become extinct.

THE RESULTS OF NATURAL SELECTION

Natural selection acts by eliminating individuals that do not have the characteristics needed for survival and reproduction in their environment. It follows, then, that *the end result of natural selection is adaptation to the environment.* An organism's environment can be divided into two components: the nonliving *abiotic* part and the *biotic* part, consisting of other organisms. Adaptations to both biotic and abiotic components occur through natural selection.

The abiotic environment includes physical factors such as climate, availability of water, and minerals in the soil. The abiotic environment provides the "bottom line" requirements that an organism must have to survive and reproduce. However, the vast majority of the adaptations that we see in modern organisms have arisen because of interactions with other organisms. As Darwin wrote: ". . . the structure of every organic being is related . . . to that of all other organic beings, with which it comes into competition for food or residence, or from which it has to escape, or on which it preys." A simple example will illustrate this point.

Consider a patch of soil a meter square on the eastern Wyoming plains. The soil contains enough minerals and receives enough rain for plant life. Let's say that a buffalo grass sprouts there. Its roots must be able to take up enough water and minerals for growth and reproduction, and to that extent it must be adapted to its abiotic environment. Even in the dry prairies of Wyoming, this is a rather trivial requirement *provided that the plant is alone and protected in its square meter of soil.* In reality, many plants, including other grasses, sagebrush bushes, and annual wildflowers, also sprout in that same patch of soil. If our buffalo grass is to survive, it must compete for resources with the other plants. Its long, deep roots and efficient uptake mechanisms have evolved not so much because the plains are dry but because it must share the dry prairies with other plants. Further, cattle (formerly bison) graze the prairies. Buffalo grass is extremely tough, with silica compounds reinforcing the blades, an adaptation that discourages grazing. Over millenia, since tougher plants were harder to eat, they survived better and reproduced more—another adaptation to the biotic environment.

When two species or two populations of a single species interact extensively, each exerts strong selective pressures on the other. When one evolves a new feature, or modifies an old one, the other often evolves new adaptations in response. The constant, mutual feedback between two species is called **coevolution** (Table 14-1).

Table 14-1 **Interactions Among Organisms**

Type of Interaction	Effect on Organism A[a]	Effect on Organism B
Competition between A and B	−	−
Predation by A on B	+	−
Symbiosis		
Parasitism by A on B	+	−
Commensalism of A with B	+	0
Mutualism between A and B	+	+
Altruism by A on behalf of B	−	+

[a] +, benefits; −, harms; 0, neutral or no effect.

Competition. One of the major selective forces in the biotic environment is **competition** *with other members of the same species.* Darwin recognized this and wrote in the *Origin of Species* ". . . the struggle almost invariably will be most severe between the individuals of the same species, for they frequent the same districts, require the same food, and are exposed to the same dangers." In other words, no competing organism has such similar requirements for survival as another member of the same species. American robins and cardinals, for example, are both brightly colored small songbirds with widely overlapping ranges, and both build nests in trees in the spring. However, they do not compete very much with each other, because they eat different foods: robins eat worms and insects, while cardinals eat seeds. Each worm and grasshopper that a robin eats make little difference to cardinals, but makes it harder for other robins to find enough to eat.

Different species may also compete for the same resources, although generally to a lesser extent. As we discuss more fully in Chapter 38, whether a particular plot of prairie is covered with grass, sagebrush, or trees is at least partly determined by competition among these plants for scarce soil moisture.

Predation. Although we commonly think of predation as one animal preying upon another animal, **predation** actually includes any situation in which one organism eats another. In some instances, coevolution between predators and prey is a sort of "biological arms race," with each side evolving new adaptations in response to escalations by the other. Using Darwin's example of wolves and deer, wolf predation selects against slow or incautious deer, thus leaving faster, more alert deer to propagate the species. In their turn, alert, swift deer select against slow, clumsy wolves, since such predators cannot acquire enough food.

Symbiosis. **Symbiosis** is any relationship in which individuals of different species closely interact with one another for much of their lives. One species usually benefits from the relationship, but the other may suffer injury, enjoy benefits, or not be affected at all (Table 14-1). The different types of symbiosis are described in Chapter 36. From an evolutionary perspective, symbiosis leads to the most intricate coevolutionary adaptations. Although predators usually utilize several species of prey, and may only interact with a particular species sporadically, partners in symbiosis live together virtually their entire lives (Fig. 14-15). At least one of the partners, and usually both, must continually adjust to any evolutionary changes developed by the other.

Altruism. Evolution is often portrayed in the popular press as being "red in tooth and claw." Although it is true that competitive and predatory interactions influence

Figure 14-15 *Several species of clownfish live in a symbiotic relationship with anemones, each species of fish frequenting its own species of anemone. The fish nestle within the stinging tentacles of the anemone, thus protected from predation by other fish. The clownfish evolved specialized skin secretions and behaviors, protecting it from being eaten by the anemone. The fish may accidentally drop food onto the anemone once in a while, but the benefits are probably pretty one-sided.*

the evolution of most species, cooperation and even self-sacrifice can also be important selective forces. **Altruism** is a behavior that endangers an animal or reduces its reproductive success but that benefits other members of its species. People helping other people immediately springs to mind, but altruistic behaviors are common in the animal kingdom. A mother killdeer flutters just out of reach of a predator, feigning an injured wing and luring the predator away from her nest (Fig. 14-16); female

Figure 14-16 *Altruism between mother and offspring: a female killdeer lures a predator away from its nest by feigning injury. The mother places herself in some small danger by this behavior, but saves her offspring from much greater danger.*

worker bees forego reproduction and devote their lives to raising the offspring of the hive queen (see Chapter 34); and young male baboons scout around the edges of the troop, even though this increases their danger from leopards. You might think that altruism runs counter to natural selection: if altruism is encoded in an organism's genes, those genes are placed at risk every time the altruist performs one of its gallant behaviors. However, natural selection can indeed select for altruistic genes if the altruist benefits relatives who possess the same genes. This special case of natural selection is an example of **kin selection,** and is explored in the essay on p. 270.

SPECIATION

A **species** is the total of all the populations of organisms that can potentially interbreed under natural circumstances and that cannot successfully interbreed with members of other populations, that is, either they do not mate with members of other species or, if they do, the offspring are sterile or less viable. Although we have illustrated a number of different evolutionary forces leading to changes *within* species, we have not yet outlined a mechanism whereby *new species* may be formed. To produce a new species, evolution must generate large enough genetic changes between populations so that mating cannot occur or hybrid offspring are unfit. The two most common causes of speciation are geographical isolation and polyploidy.

Geographical isolation. As we described earlier, migration causes gene flow, which reduces genetic differences between populations and probably eliminates any possibility that two populations could ever become reproductively isolated from one another. **Geographical isolation** occurs when a single population is split up into two separate populations, usually by a physical barrier, such as a river or mountain range (Fig. 14-17). This prevents gene flow. If environmental pressures differ in the two locations, or if the populations are small enough for genetic drift to occur, the two populations may accumulate large genetic differences and become separate species. Evolutionary biologists believe that geographical isolation is involved in most cases of speciation, especially in animals.

Sometimes a species gives rise to many new species in a relatively short time. This process, called **adaptive radiation,** occurs when several populations of a single species invade different habitats and evolve in response to the differing selective pressures in those habitats. Adaptive radiation has occurred many times during evolution, and usually results from one of two causes. First, a species may encounter a wide variety of unoccupied habitats, for example when the ancestors of Darwin's finches colonized

Kin Selection and the Evolution of Altruism

Altruism is any behavior that is potentially harmful to the future reproduction of an individual but that enhances the reproductive potential of other individuals. Altruism includes the defense of offspring by parent birds and mammals, worker bees rearing the offspring of queen bees, and nest assistance by offspring of the previous year in Florida scrub jays (Fig. E14-1). Note that altruism does not imply conscious, voluntary decisions to engage in selfless behavior. Indeed, no one would argue that worker bees consider the welfare of the queen and her offspring and deliberately sacrifice their own reproduction in favor of the queen's. Rather, most altruistic behaviors have a strong instinctive component; in other words, a wide variety of animals have altruism programmed in their genes.

From an evolutionary viewpoint, how can this be? Surely, if a mutation arose that caused altruistic behavior, and the bearers of that mutation lost their lives or failed to reproduce because of their self-sacrificing behaviors, their "altruistic alleles" would disappear from the population. This would certainly be true if *all* the animals with the altruistic allele died. But what if only *some* of them died? *If an altruist benefits related members of its own species that bear the same altruistic allele, the altruistic allele may be favored by natural selection.*

Figure E14-1 In Florida scrub jays, young birds usually do not go off and mate in their first year, even though they are sexually mature. Instead, they remain at their parents' nest and assist in feeding the next year's brood. Genetically, this makes sense. First, an animal donates 50% of its genes to its offspring, the other 50% coming from the other parent. On the average, siblings also share 50% of their genes, so two brothers are usually just as related to each other as either of them would be to his own offspring. Second, a young bird may not be experienced enough to obtain a good territory and successfully rear its own young. By helping to rear their siblings, the yearling birds actually promote the survival of their genes better than they would if they attempted to reproduce on their own.

To see how this might work, let's consider a population of deer that live at most three years, one as a fawn and two as reproductive adults, producing two fawns each adult year (Fig. E14-2). One particular doe, White-tail, undergoes a mutation that causes her to defend her offspring against danger. Trying to fight off a mountain lion, she dies during her first reproductive year, but saves her two fawns, who escape to grow up and reproduce in their turn the next year. Other does, without the altruism allele, leave their fawns when the lion approaches, and as a result most of their fawns are eaten. White-tail has a reproductive output of two, while the other does average less than two. White-tail's altruism allele has therefore increased her reproductive success, even though White-tail herself has died because of it. If White-tail's offspring inherited her altruistic allele, the proportion of deer in the next generation that have that allele will increase. This phenomenon, whereby the death or decreased fitness of an individual benefits its relatives, is called **kin selection.**

At this point, you should remember that evolution is a change in allele frequencies. In our example, the frequency of the altruism allele rises relative to other, "selfish" alleles. Eventually, if the benefit to the relatives of altruistic animals exceeds the harm to the altruists themselves, the altruistic allele may spread throughout the population.

As this example suggests, *kin selection can favor the evolution of altruism only if the altruistic behavior benefits relatives that bear the same allele.* If White-tail dies defending another doe's fawns while her own deserted fawns are eaten, her altruistic allele will vanish with her. Identification of relatives isn't too hard to imagine in the case of mothers and their offspring. Many biologists objected to other proposed instances of altruistic behaviors, however, arguing that animals cannot evaluate degrees of relatedness. This concern has been alleviated by two findings. First, many groups of animals, including wolf packs and baboon troops, are actually family groups. In this situation, an animal would not have to identify relatives in order for its altruistic behaviors to benefit them the most. The second finding is quite astounding. Many animals, including birds, monkeys, tadpoles, bees, and even tunicate

Kin Selection and the Evolution of Altruism

Figure E14-2 Kin selection for altruistic alleles. The overall reproductive success of the altruistic deer is higher than that of its selfish compatriots.

larvae, can indeed identify relatives. Given the choice between relatives and strangers, these animals preferentially associate with their relatives, *even if they were separated at birth and have never seen those relatives before.* If animals selectively form related groups, once again altruistic behaviors will most likely benefit relatives.

Humans display a wide range of altruistic behaviors. In modern society, human altruism most often benefits unrelated individuals. Evolutionarily speaking, this is a recent development, brought about by population growth. For most of our evolutionary history, humans lived in small family groups, and altruistic behavior would have benefited relatives. Like altruism in other social animals, our altruistic tendencies, although greatly enhanced by social forces, have their biological bases in kin selection.

the Galápagos Islands. With no competitors except other members of their own species, all the available ecological roles were rapidly filled with new species of finch. Adaptive radiation also occurs if a species develops a fundamentally new and superior adaptation, enabling it to displace less well adapted species from a variety of habitats. This happened several times during evolutionary history, for example when warmblooded mammals diversified extensively at the expense of the reptiles.

INSTANT SPECIATION THROUGH POLYPLOIDY. In some instances, new species arise nearly instantaneously, without geographic isolation. This occurs through **polyploidy,** the acquisition of multiple copies of each chro-

Figure 14-17. Geographic isolation that prevents gene flow is almost always necessary for animal populations to split into new species. The tassel-eared squirrels on the rims of the Grand Canyon are a classic example. In the distant past, a single population of squirrels became separated as the Canyon was carved into the plateau of northwestern Arizona. The two populations are split by the treeless desert of the canyon depths. Today, the Kaibab squirrel **(a)** *is confined to the north rim, while the Abert squirrel* **(b)** *is found on the south rim and in forests throughout northern Arizona, Utah, New Mexico, and Colorado. It is not known if the two are genetically distinct enough to constitute separate species.*

mosome. As you know, most plants and animals have paired chromosomes and are called diploid. Occasionally, especially in plants, a fertilized egg duplicates its chromosomes but does not divide into two daughter cells. The resulting cell thus becomes tetraploid, with four copies of each chromosome. Tetraploid plants are usually vigorous and healthy, and successfully complete meiosis to form viable gametes. However, if sperm from a tetraploid plant fertilizes an egg cell of the diploid "parental" species, the resulting offspring is triploid. When this triploid offspring begins meiosis, it will be unable to pair up all its chromosomes, since there are an odd number (three) of each. Meiosis fails, gametes are not formed, and the triploid hybrids are sterile. The tetraploid plant is therefore reproductively isolated from its diploid parent. If it can self-fertilize or produce offspring asexually, it may perpetuate itself and form a new species. Speciation by polyploidy is extremely common in plants: in fact, about half of all species of flowering plants are polyploid, many of them tetraploid.

MAINTAINING REPRODUCTIVE ISOLATION BETWEEN SPECIES

Once a species has formed, it may remain reproductively isolated from other species in two ways. First, members of the species may never breed with members of other species. This usually has a clear adaptive value. Separate species are usually genetically different in ways that adapt them to different environments. Any individual that mates with a member of another species will probably produce unfit or sterile offspring, thereby "wasting" its genes and contributing nothing to future generations. Thus there is strong selective pressure to avoid mating between different species. Incompatibilities between species that prevent mating are called **premating isolating mechanisms.** Sometimes premating isolation fails, and members of different species do mate. If the resulting hybrid offspring are less fit or infertile, the two species may remain separate, with no gene flow between them. Incompatibilities between species that prevent the formation of vigorous, fertile hybrids are called **postmating isolating mechanisms.**

Premating Isolating Mechanisms

Mechanisms that prevent mating between different species include geographical isolation, ecological isolation, temporal isolation, behavioral isolation, and mechanical incompatibility.

GEOGRAPHICAL ISOLATION. Members of different species obviously cannot mate if they never get near one another. As we have already seen, geographical isolation often creates new species in the first place. However, we cannot tell if geographically separated populations constitute distinct species. Should the barrier separating the two populations disappear (an intervening river changes course, for example), it may well turn out that the reunited populations will interbreed freely and not be separate species at all. If they cannot interbreed, other mechanisms, such as different courtship rituals, must have developed during their isolation. Geographical isolation is thus usually considered to be a mechanism that creates new species rather than a mechanism that maintains reproductive isolation between species.

Ecological isolation. If two populations have different resource requirements, they may use different local habitats within the same general area. White-crowned and white-throated sparrows, for example, have extensively overlapping ranges. The white-throated sparrow, however, frequents dense thickets, whereas the white-crowned sparrow inhabits fields and meadows, seldom penetrating far into dense growth. The two species may coexist within a few hundred yards of one another and yet seldom meet during breeding season. Although ecological isolation may slow down interbreeding, it seems unlikely that it could prevent gene flow entirely. Other mechanisms usually also contribute to interspecific isolation.

Temporal isolation. Even if two species occupy similar habitats, they cannot mate if they have different breeding seasons. Bishop pines and Monterey pines coexist near Monterey on the California coast. Viable hybrids have been produced between these two species in the laboratory. However, in the wild, they pollinate at different times: the Monterey pine releases pollen in early spring, the bishop pine in summer. Therefore, the two species never crossbreed under natural conditions.

Behavioral isolation. Among animals, the elaborate courtship colors and behaviors that so enthrall human observers have evolved not only as recognition and evaluation signals between male and female, but also to distinguish among species. The striking colors and calls of male songbirds may be siren songs for females of their own species, but are treated with the utmost indifference by other females. Among frogs, males are often impressively indiscriminate, jumping on every female in sight, regardless of the species, when the spirit moves them. Females, however, only approach male frogs croaking the correct "ribbet!" If they do find themselves in an unwanted embrace, they utter the "release call," which causes the male to let go. As a result, few hybrids are produced.

Mechanical incompatibility In rare instances, these isolating mechanisms fail, and male and female of different species attempt to breed. Among animals with internal fertilization, in some cases the male and female genitalia simply won't fit together.

Postmating Isolating Mechanisms

Sometimes premating isolation fails, and mating occurs between members of different species. However, if vigorous, fertile hybrids are not produced, there will be no gene flow between the species.

Gametic incompatibility. Even though a male inseminates a female, his sperm may not fertilize her eggs. For example, the fluids of the female reproductive tract may weaken or kill sperm of other species.

Hybrid inviability. If fertilization does occur, the resulting hybrid may be weak or even unable to survive. The genetic programs directing development of the two species may be so different that hybrids abort early in development. Even if the hybrid survives, it may display behaviors that are mixtures of the two parental types. In attempting to do some things the way that species A does them, and other things the way that species B does, it may be hopelessly uncoordinated. Hybrids between certain species of lovebirds, for example, have great difficulty learning to carry nest materials during flight and probably could not reproduce in the wild (see Chapter 33).

Hybrid infertility. Animal hybrids, such as the mule, are usually sterile. A common reason is the failure of chromosomes to pair properly during meiosis, so that eggs and sperm never develop.

EXTINCTION

Natural selection does not always lead to adaptation. It may also lead to extinction. Trilobites, dinosaurs, saber-toothed cats—all are extinct, known only from fossils. Paleontologists estimate that *at least* 99.9 percent of all the species that ever existed are now extinct. Why? Two characteristics seem to predispose a species to extinction: localized distribution and overspecialization. However, environmental events are usually the immediate cause of extinction.

Susceptibility to Extinction

Species vary widely in their range, and hence in their susceptibility to extinction. Some species, such as herring gulls, white-tailed deer, and humans, inhabit entire continents or even the whole Earth, while others, such as the Devil's Hole pupfish (Fig. 14-18), have extremely limited ranges. Obviously, if a species occurs only in a very small area, any disturbance of that area could easily result in extinction. If Devil's Hole dries up from climatic change

Figure 14-18 The Devil's Hole pupfish is found in only one spring-fed waterhole in the Nevada desert. During the last glacial period, the southwestern deserts received much more rainfall, forming numerous lakes and rivers. As the rainfall decreased, pupfish populations were isolated in shrinking small springs and streams. Isolated small populations and differing environmental conditions caused the ancestral pupfish species to split up into several very restricted modern species, all of which swim on the brink of extinction.

or well drilling nearby, its pupfish will immediately vanish. Wide-ranging species, on the other hand, usually do not succumb to local environmental catastrophes.

Another factor that makes a species vulnerable to extinction is extreme specialization. Each species evolves a set of genetic adaptations in response to pressures from its particular environment. Sometimes these adaptations imprison the organism in a very narrow ecological niche. The Everglades kite, for example, feeds only on a certain freshwater snail (Fig. 14-19). As the swamps of the American southeast are drained for farms and developments, the snail population shrinks. Should the snail become extinct, the kite will surely go extinct along with it. In the fossil record, such behavioral specialization is hard to recognize. Structural specializations, however, may be just as restrictive. A case in point is giantism. For poorly understood reasons, many animals have evolved huge size: certain amphibians, dinosaurs, and giant mammals (including mammoths, ground sloths, and titanotheres). To support their bulk, these animals must have consumed enormous amounts of food. If environmental conditions deteriorated, these giants would be unable to find enough food, and would die. Smaller animals that ate the same food but needed less of it might survive.

Extinction and the Environment

The actual cause of extinction is probably always environmental change, either in the living or the nonliving parts of the environment. Three major changes that drive species to extinction are *competition among species, novel predators or parasites,* and *habitat destruction.*

Competition for limited resources occurs in all environments. If a species' competitors evolve superior adaptations, and it doesn't evolve fast enough to keep up, it may become extinct. A particularly striking example of extinction through competition occurred in South America 2 to 3 million years ago. For millions of years North and South America were isolated from one another, and each developed a distinctive fauna. When the Panamanian land bridge arose, connecting the two continents, massive migrations took place. In general, North American animals displaced their South American counterparts.

When isolated populations encounter one another, not only competitors migrate between the areas: predators and parasites do, too. With the exception of humans, who have exterminated hundreds of species, predators probably cause few extinctions. Parasites, on the other hand, can be devastating. In North America, Dutch elm disease and chestnut blight are well-known instances of introduced parasites that almost completely destroyed widespread native species. We cannot tell much about prehistoric parasite invasions, but the destruction of South American animals, mentioned above, might have been at least partly due to diseases carried south by resistant North American migrants.

Habitat destruction may be the leading cause of extinction, both contemporary and prehistoric. Presently, habitat destruction due to human activities is proceeding at a frightening pace. Perhaps the most rapid extinction in the history of life will occur over the next 50 years, as tropical forests are cut for timber and to clear land for

Figure 14-19 The Everglades kite feeds only on the apple snail, found in swamps of the southeastern United States. Such behavioral specialization renders the kite extremely vulnerable to any environmental change that may exterminate its single prey species.

cattle and crops. As many as half the species presently on Earth may be lost because of tropical deforestation.

Prehistoric habitat destruction usually occurred over a longer time span, but nevertheless had serious consequences. Climate changes, in particular, caused many extinctions. Several times, moist, warm climates gave way to drier, colder climates with more variable temperatures. Many plants and animals failed to adapt to the new rigors and became extinct. One cause of climate change is continental drift (Fig. 14-20). As the continents flow about over the surface of the Earth, they change latitudes. Much of North America was tropical many millions of years ago, but drift carried the continent up into temperate, and even arctic, regions.

An extreme and very sudden type of habitat destruction might be caused by catastrophic geological events, such as massive volcanic eruptions. Several prehistoric eruptions, which would make the Mt. St. Helens explosion look like a firecracker by comparison, wiped out every living thing for scores of miles around, and probably caused global climatic changes as well.

The fossil record reveals several episodes of extensive worldwide extinctions, especially among marine life. A recent speculation suggests that enormous meteorites, several miles in diameter, may have hit the Earth at these times. If a huge meteorite struck land, it would kick up enormous amounts of dust. The dust might be thick enough, and spread widely enough, to block out most of the sun's rays. Many plants would die because they could not photosynthesize. Many animals, all of which ultimately depend on plants for food, would also die. Smaller amounts of dust might still block out enough sunlight to cause global cooling, perhaps even triggering the onset of an ice age. In either case, widespread extinctions would result.

Did such massive meteorite strikes really occur, and if so, would they cause extinctions? No one knows for sure, but considerable evidence points to meteorites as the causes of at least some major extinctions. Gigantic volcanic eruptions may have triggered other episodes of extinction. Modern humans needn't worry very much about volcanic or meteoritic extinction, because the odds of either are very small. Another event, unfortunately much more likely, could have a similar effect, however. In 1983, a group of geologists, climatologists, and biologists warned that a nuclear war might result in the extinction of many forms of life. Their calculations suggest that exploding only a fraction of the nuclear weapons now in existence might produce enough dust and soot to cause worldwide darkness and cold, exterminating many organisms, including humans. This "nuclear winter" hypothesis is hotly debated among climatologists, and as of early 1988 the consensus seems to be that nuclear war would trigger only a less catastrophic "nuclear autumn." Many biologists fear that even a "nuclear autumn" might drive many plant and animal species to extinction.

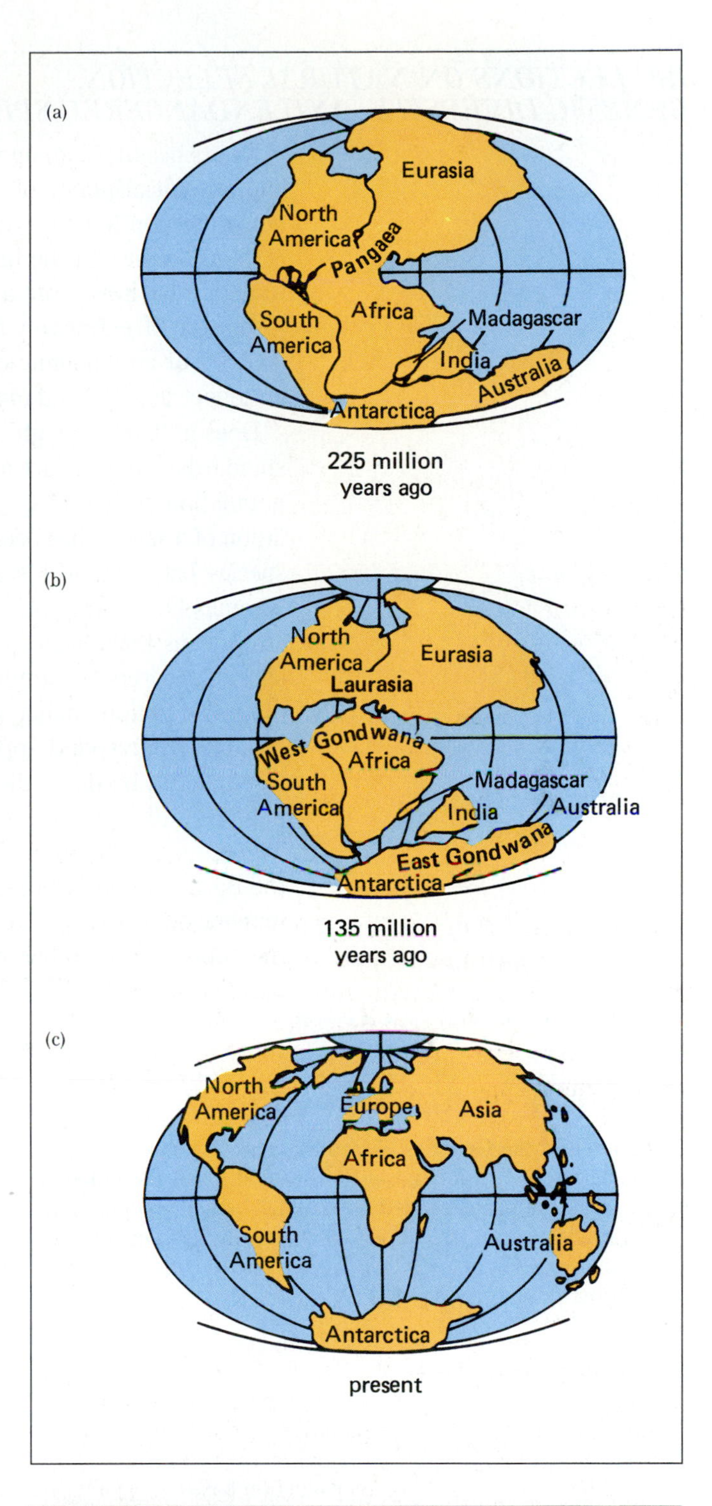

Figure 14-20 Although slow, continental drift can cause tremendous environmental changes, as landmasses are moved about on the surface of the Earth. The solid surfaces of the continents slide about over the viscous, but fluid, subterranean mantle. About 225 million years ago **(a)**, *all the continents were fused together into one gigantic landmass which geologists call Pangaea. Gradually, Pangaea broke up into Laurasia and Gondwana* **(b)**. *Further drift eventually resulted in the modern positions of the continents* **(c)**. *Continental drift continues today: the Atlantic Ocean, for example, widens by a few centimeters each year.*

REFLECTIONS ON NATURAL SELECTION, GENETIC DIVERSITY, AND ENDANGERED SPECIES

Ever since the Endangered Species Act was passed in 1976, the United States has had an official policy of protecting rare species. In fact, the real goal of the act is not protection but recovery; as one U.S. Fish and Wildlife official put it, "the goal is to get species *off* the list." What this usually means is that wildlife biologists try to determine how large a population a species needs to have before it is no longer in danger of extinction from unpredictable events, such as a couple of years of drought or an epidemic of parasites. If a species reaches this population size, it is no longer legally "endangered" with extinction.

Does a "large enough" population (which usually is still very small by historical standards) really ensure a species' survival? From our discussion of genetic drift and population bottlenecks, you probably realize that the answer is "no." If the population of a species has been reduced to the point where it is placed on the endangered species list, it almost surely has lost much of its genetic diversity. As ecologist Thomas Foose aptly put it, loss of habitat and consequent reduction in population size means that "gene pools are being converted into gene puddles." Even if the species recovers in numbers, it cannot recover its gene pool. When the forces of natural selection change at some future time, the species may not have the genetic capability to respond appropriately, and may become extinct.

What can be done about this? The best solution, of course, is to leave enough habitat of diverse types so that species never become endangered in the first place. The human population, however, has grown so large and appropriated so much of the Earth's resources that this is not feasible in many places. If we are to be, as we imagine ourselves, the stewards of the planet and not merely its ultimate consumers, the protection of other life forms will be a continuing responsibility as long as humankind exists.

SUMMARY OF KEY CONCEPTS

Evolution and the Genetics of Populations

The gene pool of a population is the total of all the different alleles of all the genes carried by the members of a population. In its broadest sense, evolution is a change in the frequencies of alleles in the gene pool of a population, due to enhanced reproduction by individuals bearing certain alleles.

Evolutionary Mechanisms

Allele frequencies in a population will remain constant over generations only if the following conditions are met: (1) no mutation; (2) no migration; (3) large population; (4) random mating; (5) no differential survival or reproduction based on genotype.

These conditions are rarely, if ever, met in nature. Understanding why leads to an understanding of the mechanisms of evolution.

1. Mutations are random, undirected changes in DNA composition. Although most mutations are neutral or harmful to the organism, some prove advantageous in certain environments. Mutations are usually rare and do not change allele frequencies very much. However, they provide the raw material for evolution.
2. Migration is the flow of genes between populations. If migrants carry different alleles than the populations from which they come or to which they migrate, migration will cause changes in allele frequencies.
3. In any population, chance events kill or prevent reproduction by some of the individuals. If the population is small, chance events may eliminate a disproportionate number of individuals bearing a particular allele, thereby greatly changing the allele frequency in the population. This is termed genetic drift.
4. Many organisms do not mate randomly. If only certain members of a population can mate, the next generation of organisms in the population will all be offspring of this select group, whose allele frequencies may differ from those of the population as a whole.
5. The survival and reproduction of organisms is influenced by their phenotype. Since phenotype depends at least partly on genotype, natural selection will tend to favor the reproduction of certain alleles at the expense of others.

Natural Selection

Three types of natural selection are:

Stabilizing selection: organisms of the "average value" for a characteristic are favored over organisms of extreme values.

Disruptive selection: organisms with extreme characteristics are favored over organisms with average values.

Directional selection: organisms with characteristics that are different from average in one direction (e.g., smaller) are favored both over average organisms and over those that differ from average in the opposite direction.

The Results of Natural Selection
Natural selection occurs as a result of the interactions of organisms with both the living and nonliving parts of their environments. The living parts, however, usually exert the stronger selective pressures. When two or more species interact extensively so as to exert mutual selective pressures on each other for long periods, they both evolve in response. Coevolution can occur as a result of any type of relationship between organisms, including competition, predation, and symbiosis.

Speciation
A species is all the populations of organisms that can potentially interbreed with one another under natural circumstances and that are reproductively isolated from other populations. Two populations will form different species if there are such large genetic differences that they cannot or will not interbreed, or if their hybrid offspring are inviable or infertile. Most speciation occurs through geographical isolation or polyploidy.

Maintaining Reproductive Isolation Between Species
Reproductive isolation between species may be maintained by one or more of several mechanisms, collectively called premating isolating mechanisms and postmating isolating mechanisms. Premating isolating mechanisms include geographical isolation, ecological isolation, temporal isolation, behavioral isolation, and mechanical incompatibility. Postmating isolating mechanisms include gametic incompatibility, hybrid inviability, and hybrid sterility.

Extinction
Over time, many species become extinct. Two factors that contribute to the likelihood of extinction of a species are localized distribution and overspecialization. Factors that actually cause extinctions include competition among species, novel predators or parasites, and habitat destruction.

GLOSSARY

Adaptation a characteristic of an organism that helps it to survive and reproduce in a particular environment; also the process of acquiring such characteristics.

Adaptive radiation extensive speciation occurring among related populations as a result of adaptation to a wide variety of habitats.

Allele frequency for any given gene, the relative proportion of each allele of that gene found in a population.

Altruism a behavior that benefits another organism, usually at some risk to the altruistic organism.

Behavioral isolation lack of mating between species of animals that differ substantially in courtship and mating rituals.

Coevolution the evolution of adaptations in two different species due to their extensive interactions with one another, so that each acts as a major force of natural selection upon the other.

Competition a relationship between individuals or species in which both require the same resource, which is not available in sufficient quantity to satisfy the needs of all users.

Directional selection a type of natural selection in which one extreme phenotype is favored over all others.

Disruptive selection a type of natural selection in which both extreme phenotypes are favored over the average phenotype.

Ecological isolation lack of mating between organisms belonging to different populations that occupy distinct habitats within the same general area.

Equilibrium population a population in which allele frequencies do not change from generation to generation.

Extinction the death of all members of a species.

Founder effect a type of genetic drift in which an isolated population founded by a small number of individuals may develop allele frequencies that are very different from those of the parent population, because of chance inclusion of disproportionate numbers of certain alleles in the founders.

Gametic incompatibility the inability of sperm from one species to fertilize eggs of another species.

Gene flow the movement of alleles from one population to another due to migration of individual organisms.

Gene pool for a single gene, the total of all the alleles of that gene that occur in a population; the total gene pool is the total of all alleles of all genes in the population.

Genetic drift a change in the allele frequencies of a small population purely by chance.

Geographical isolation the separation of two populations by a physical barrier.

Hybrid infertility reduced fertility (often complete sterility) in hybrid offspring of two different species.

Hybrid inviability the failure of a hybrid offspring of two different species to survive to maturity.

Kin selection selection favoring a certain allele because of benefits accruing to relatives bearing the same allele.

Mechanical incompatibility the inability of male and female animals to mate because their genitalia do not fit together properly.

Migration in population genetics, the flow of genes between populations.

Natural selection the differential survival or reproduction of organisms due to environmental forces that act differently upon genetically different members of a population.

Population a group of individuals of the same species, found in the same time and place, and actually or potentially interbreeding.

Population bottleneck a form of genetic drift in which a population becomes extremely small, which may lead to differences in allele frequencies as compared to other populations of the species, and to a loss in genetic variability.

Postmating isolating mechanism any mechanism that prevents organisms of two different populations, once mating has occurred, from producing vigorous, fertile offspring.

Premating isolating mechanism any mechanism that prevents organisms of two different populations from mating.

Reproductive isolation the failure of organisms of one population to breed successfully with members of another population; may be due to premating or postmating isolating mechanisms.

Sexual selection a type of natural selection in which the choice of mates by one sex is the selective agent.

Stabilizing selection a type of natural selection in which those organisms displaying extreme phenotypes of any sort are selected against.

Symbiosis a sustained relationship between two organisms of different species.

Temporal isolation the inability of organisms to mate if they have significantly different breeding seasons.

STUDY QUESTIONS

1. What is a gene pool? How would you determine the allele frequencies in a gene pool?
2. Define an equilibrium population, and outline the conditions that must be met for a population to remain in equilibrium.
3. How does population size affect the likelihood of changes in allele frequencies by chance alone? Can significant changes in allele frequencies (i.e., evolution) occur due to genetic drift?
4. Describe the three types of natural selection. Which type(s) are most likely to occur in stable environments and which type(s) in rapidly changing environments?
5. Why is geographical isolation so important to the formation of new species?
6. Describe how polyploidy can give rise to "instant species."
7. List and describe the different types of premating and postmating isolating mechanisms.
8. What factors predispose species to extinction? What types of environmental changes actually drive species to extinction?
9. What is adaptation? Is adaptation always the end result of natural selection?

SUGGESTED READINGS

Allison A. C. "Sickle Cells and Evolution." *Scientific American,* August 1956. The story of the interaction between sickle-cell anemia and malaria in Africa.

Ayala, F. "The Mechanisms of Evolution." *Scientific American,* September 1979. Our increasing understanding of the molecular mechanisms of heredity and mutation reveals that organisms are extremely variable, providing tremendous amounts of material on which selection can operate.

Futuyma, D. *Evolutionary Biology,* 2nd ed. Sunderland, Mass.: Sinauer Associates, Inc., 1986. A rather advanced text, but probably the best descriptions of the mechanisms of evolution.

Kimura, M. "The Neutral Theory of Molecular Evolution." *Scientific American,* November 1979. Kimura presents his theory that much evolutionary change is not driven by natural selection but by genetic drift, even in large populations, as chance increases the frequency of alleles that are no more adaptive than the alleles they replace.

May, R. M. "The Evolution of Ecological Systems." *Scientific American,* September 1979 (Offprint No. 1404). Coevolution accounts for much of the structure of natural communities of plants and animals.

O'Brien, S. J., Wildt, D. E., and Bush, M. "The Cheetah in Peril." *Scientific American,* May 1986. According to molecular and immunological techniques, a population bottleneck has reduced the genetic variability of the world's cheetahs almost to zero.

Smith, J. M. "The Evolution of Behavior." *Scientific American,* September 1979 (Offprint No. 1405). A leading mathematical evolutionist explains how complex, apparently unlikely behaviors might be selected for.

Stebbins, G. L., and Ayala, F. "The Evolution of Darwinism." *Scientific American,* July 1985. A synthesis of molecular and classical evolutionary methodologies.

15
The History of Life on Earth

. . . thence life was born,
Its nitrogen from ammonia, carbon from methane,
Water from the cloud and salts from the young seas . . .
. . . the cells of life
Bound themselves together into clans, a multitude of cells
To make one being—as the molecules before
Had made of many one cell. Meanwhile they had invented
Chlorophyll and ate sunlight, cradled in peace
On the warm waves; but certain assassins among them
Discovered that it was easier to eat flesh
Than feed on lean air and sunlight: thence the animals,
Greedy mouths and guts, life robbing life,
Grew from the plants; and as the ocean ebbed and flowed many plants and animals
Were stranded in the great marshes along the shore,
Where many died and some lived. From these grew all land-life,
Plants, beasts, and men; the mountain forest and the mind of Aeschylus
And the mouse in the wall.

from "The Beginning and the End"
by Robinson Jeffers

Fifteen to twenty billion years ago, before the Big Bang, there was no universe as we know it: only a dense, minuscule mass in which energy and matter were melded into a single incomprehensibly ferocious state. Then the mass erupted. Particles of matter and antimatter formed, annihilating each other in great bursts of energy when they collided. So great was the energy released that astrophysicists can still detect its faint radiation.

Gradually, local accumulations of matter formed as gravity drew particle to particle. Many accumulations grew so large that their centers became dense and hot, triggering the thermonuclear reactions that fuel the light of the stars (Fig. 15-1). Small, simple atoms of hydrogen fused to form helium, and these fused to give rise to still larger atoms. The nuclear forces of some stars grew so intense that they exploded, spewing their matter out into space.

Figure 15-1 The Great Nebula in Orion, a cloud of gases in which stars may be forming, as they did in the beginning of the universe.

About 5 billion years ago, far out in one of the arms of a spiral galaxy, a small cloud of matter began to condense, enriched with heavy elements provided by the self-destruction of ancient stars. The center of the cloud collapsed into the yellow dwarf star we call the sun. Farther out, other local aggregations appeared, forming the planets. The first four planets were small and contained a high proportion of heavy elements. The innermost two, being close to the sun, became very hot, while the fourth cooled into an eternal winter of temperatures more than 100 degrees Celsius below zero. The third settled into an orbit that would receive just the right intensity of sunlight to permit water to exist as a liquid. This became the Earth, although perhaps it is more appropriate to name it, in Jacques Cousteau's phrase, the Water Planet.

On this third planet, and as far as we know upon no other in our solar system, life evolved (Table 15-1). Where did life come from? How did it develop into the myriad forms we see today? What is its future?

ORIGINS

How and when did life first appear on Earth? Just a few centuries ago, this question would have been considered trivial. Although no one knew how life *first* arose, people thought that new living things appeared all the time,

Table 15-1 **The History of Life on Earth**

Era	Period	Epoch	Years Ago (millions)	Major Events
Precambrian			4600–3500 3500–600	Origin of solar system and Earth. Origin of first living cells; dominance of bacteria; origin of photosynthesis and evolution of oxygen atmosphere; origin of algae and soft-bodied marine invertebrates.
Paleozoic	Cambrian		600–500	Primitive marine algae flourish; origin of most marine invertebrate types.
	Ordovician		500–430	Invertebrates, especially arthropods and molluscs, dominant in sea; first fish, fungi; invasion of land by plants.
	Silurian		430–400	Many fish, trilobites, molluscs in sea; first vascular plants; invasion of land by arthropods.
	Devonian		400–345	Fishes and trilobites flourish in sea; origin of amphibians and insects.
	Carboniferous		345–280	Swamp forests of tree ferns and club mosses; dominance of amphibians; numerous insects; origin of reptiles.
	Permian		280–230	Origin of gymnosperms; massive marine extinctions, including last of trilobites; flourishing of reptiles and decline of amphibians; continents aggregated into one land mass, Pangaea.
Mesozoic	Triassic		230–180	Origin of mammals and dinosaurs; forests of gymnosperms and tree ferns; breakup of Pangaea begins.
	Jurassic		180–135	Dominance of dinosaurs and conifers; origin of birds and flowering plants; continents partially separated.
	Cretaceous		135–65	Dominance of flowering plants; mass extinctions of marine life and some terrestrial life, including last dinosaurs; modern continents well separated.
Cenozoic	Tertiary	Paleocene Eocene Oligocene Miocene Pliocene	65–58 58–36 36–25 25–12 12–2	Widespread flourishing of birds, mammals, insects, and flowering plants; drift brings continents into modern positions; mild climate at beginning of period, with extensive mountain building and cooling toward end.
	Quaternary	Pleistocene Recent	2–0.01 0.01–present	Evolution of *Homo;* repeated glaciations in northern hemisphere; extinction of many giant mammals.

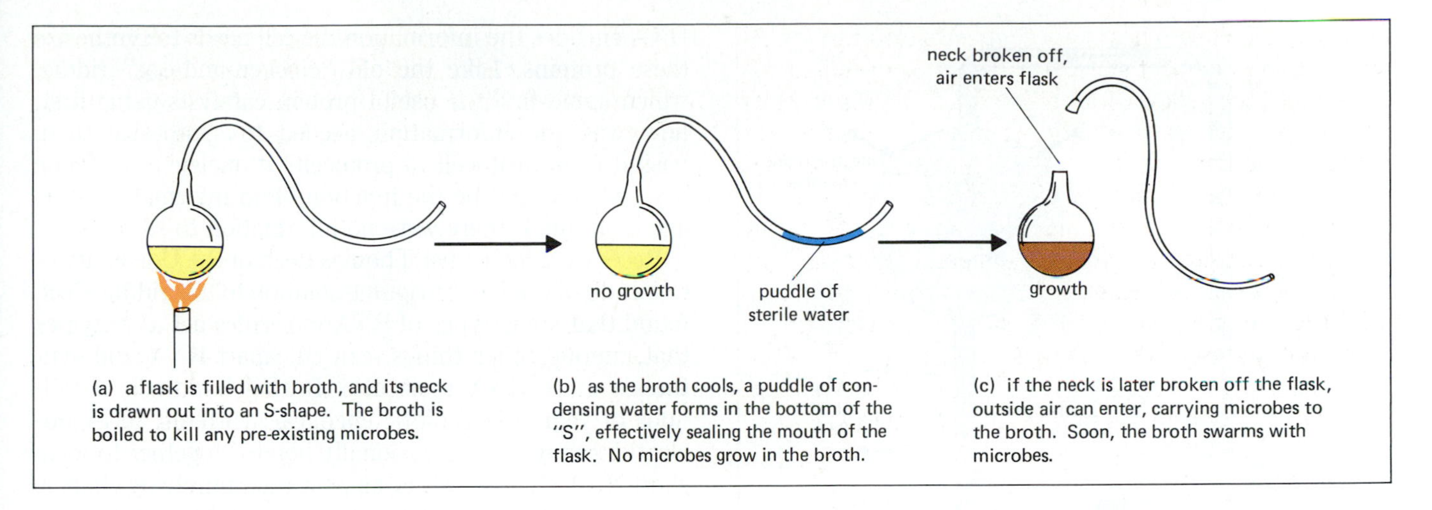

Figure 15-2 *Louis Pasteur's experiment disproving the spontaneous generation of microbes in broth.*

through **spontaneous generation** from both nonliving matter and other, unrelated forms of life. In 1609, a French botanist wrote: "There is a tree—not, it is true, common in France, but frequently observed in Scotland. From this tree leaves are falling; upon one side they strike the water and slowly turn into fishes, upon the other they strike the land and turn into birds." Medieval writings abound with similar observations and delightful prescriptions for creating life—even human beings. Microorganisms were thought to arise spontaneously from broth, maggots from meat, mice from mixtures of sweaty shirts and wheat, and people from sperm injected into a cucumber!

In 1668, the Italian physician Francesco Redi disproved the maggots-from-meat hypothesis simply by keeping flies (whose eggs hatch into maggots) away from uncontaminated meat (see Chapter 1). Then, in the mid-nineteenth century, Louis Pasteur in France and John Tyndall in England disproved the broth-to-microbe idea as well (Fig. 15-2), effectively demolishing the notion of spontaneous generation. After this, speculation about life originating from nonliving matter, now or on the primeval Earth, was not likely to further one's scientific career. For almost half a century the subject lay dormant.

Eventually, biologists returned to the question of the origin of life. In the 1920s and 1930s, Alexander Oparin in the Soviet Union and J. B. S. Haldane in England speculated that given the correct conditions, perhaps life could have arisen from nonliving matter through ordinary chemical reactions. This process is called chemical or **prebiotic evolution,** that is, evolution before life existed.

Prebiotic Evolution

The primordial Earth differed greatly from the equable planet we now enjoy. As rock after rock smashed into the forming planet, their energies of motion were converted into heat. Radioactive atoms decayed, releasing still more heat. Soon the rock melted, and heavier elements such as iron and nickel sank to the center of the mass, where they remain molten still. Gradually the Earth cooled, and elements combined to form compounds of many sorts. Virtually all the oxygen combined with hydrogen to form water, carbon to form carbon dioxide, or heavier elements to form minerals. After millions of years, the Earth cooled enough to allow water to exist as a liquid, and for millenia it must have rained, as water vapor condensed out of the cooling atmosphere. As the water struck the surface, it dissolved many minerals, forming a weakly salty ocean. Lightning from storms, heat from volcanoes, and intense ultraviolet light from the sun all poured energy into the young seas.

Judging from the chemical composition of the rocks formed at this time, geochemists have deduced that the primitive atmosphere probably contained carbon dioxide, methane, ammonia, hydrogen, nitrogen, and water vapor. Since the oxygen atoms were bound up in water, carbon dioxide, and minerals, there was virtually no free oxygen in the early atmosphere. This is an important factor in all hypotheses and experiments dealing with prebiotic evolution.

PREBIOTIC SYNTHESIS OF ORGANIC MOLECULES. In 1953, Stanley Miller, a graduate student at the University of Chicago, mixed water, ammonia, hydrogen, and methane in a flask, and provided energy with heat and electric discharge (to simulate lightning). He found that simple organic molecules appeared after just a few days (Fig. 15-3). In these and similar experiments, Miller and others have produced amino acids, short proteins, nucleotides, ATP, and other molecules characteristic of living things. Interestingly, the composition of the "atmosphere" used

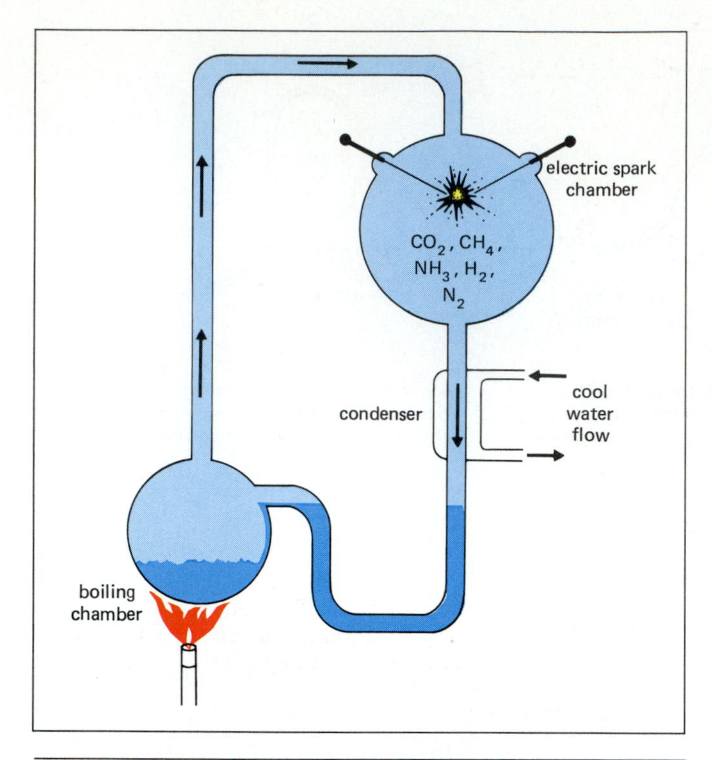

Figure 15-3 *The experimental apparatus of Stanley Miller. Energy from heat and electrical discharge cause amino acids and other organic molecules to form from carbon dioxide, methane, ammonia, hydrogen, nitrogen, and water, all of which are thought to have been present in the atmosphere of the early Earth.*

in these experiments is unimportant, provided that there are sources of hydrogen, carbon, and nitrogen, and that free oxygen is excluded. Similarly, energy sources, including ultraviolet light, electric discharge, and heat, all seem about equally effective. Even though geochemists may never know exactly what the primordial atmosphere was like, it is certain that organic molecules were synthesized on the ancient Earth.

If organic molecules formed today, they would break down very rapidly by reacting with oxygen in the atmosphere. Since the primordial atmosphere lacked free oxygen, however, organic molecules were quite stable. Therefore, they accumulated over time, as they were steadily synthesized but not destroyed. The organic compounds dissolved in the waters of the early Earth, making a dilute nutrient soup. At least in some places the soup probably became quite concentrated, such as in pools of water slowly evaporating in the sunlight. Such concentrations of organic molecules may have been crucial in the evolution of life. First, they may have provided the molecules that would form the first living organisms. Second, the chemical energy stored in these molecules would be food for the first cells.

RNA: THE FIRST LIVING MOLECULES? In modern cells, proteins carry out most of the cellular functions, while DNA encodes the information the cell needs to synthesize these proteins. Like the old "chicken-and-egg" riddle, which came first? If useful protein catalysts came first, how was the information needed to synthesize them passed from protocell to protocell? If nucleic acids came first, what would be the function of an information-storage molecule if there were no information to store?

Recently, biochemist Thomas Cech of the University of Colorado offered an intriguing solution to this riddle. Cech found that some types of RNA molecules act as enzymes that, among other things, can cut apart RNA and synthesize more RNA molecules. During hundreds of millions of years of prebiotic chemical synthesis, RNA nucleotides may have occasionally bonded together to form short RNA chains. Let us suppose that, purely by chance, one of these RNA chains was a catalyst—dubbed a **ribozyme** —that could synthesize copies of itself from the free ribonucleotides in the surrounding waters. This first ribozyme probably wasn't very good at its job and made lots of mistakes. These mistakes, of course, were the first mutations. Like modern mutations, most undoubtedly ruined the catalytic abilities of the "daughter molecules," but a few may have been improvements. Molecular evolution could begin, as ribozymes with increased speed and accuracy of replication reproduced faster, making more and more copies of themselves.

Just as RNA-containing ribosomes are crucial to protein synthesis today, perhaps some early ribozymes began to bind amino acids and catalyze the synthesis of short proteins. Further ribozyme mutations might lead to the formation of the first protein enzymes. At the same time, these protein-synthesizing ribozymes were vulnerable to being cut up by other ribozymes, destroying the information so slowly accumulated over millions of years of random nucleotide substitutions. Further mutations might have allowed certain ribozymes to copy themselves over into DNA molecules that would be safe from the ravages of their fellow ribozymes. In this hypothesis, then, RNA occupies center stage as the first living molecule, with both DNA and proteins evolving later (Fig. 15-4). RNA gradually receded into its present role as an intermediary between DNA and the protein enzymes that carry out most of the work of modern cells.

THE FIRST LIVING CELLS. If proteins and lipids are agitated in water, simulating waves beating against ancient shores, hollow structures called **microspheres** are formed (Fig. 15-5). These hollow balls resemble living cells in several respects. They have a well-defined outer boundary, separating internal contents from the external solution. If the composition of the microsphere is right, a "membrane" is formed that is remarkably similar in appearance to a real cell membrane. Under certain conditions microspheres can absorb more material from the solution ("feed"), grow, and even divide (Fig. 15-5).

If a microsphere happened to surround the right ribo-

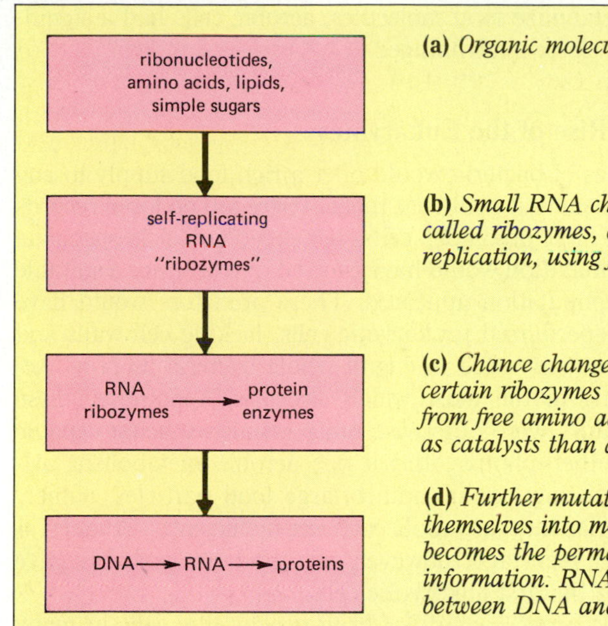

Figure 15-4 A hypothesis for the origin of life, with RNA as the first "living molecule."

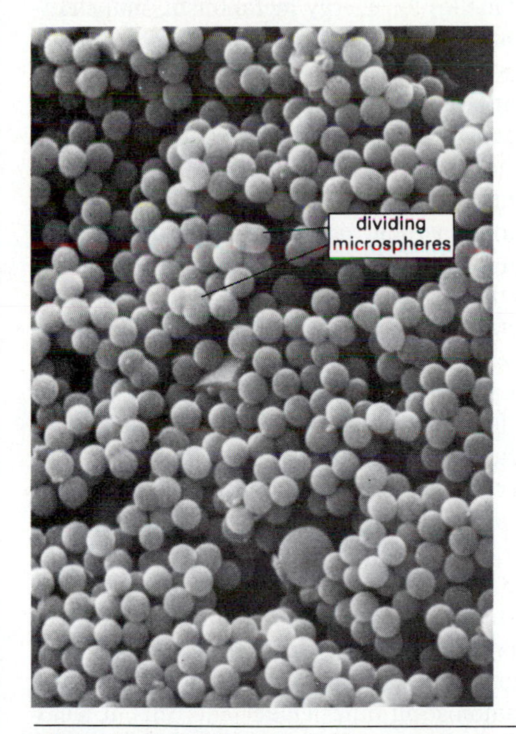

Figure 15-5 Cell-like microspheres can be formed by agitating proteins and lipids in a liquid medium. Such microspheres can take in material from the surrounding solution, grow, and even "reproduce," as seen here.

zymes, something very much like a living cell would have been formed. The ribozymes and their protein products would have been protected from free-roaming ribozymes in the primordial soup. Nucleotides and amino acids might have diffused across the membrane and have been used to synthesize new RNA and protein molecules. After sufficient growth, the microsphere may have divided, with a few copies of both ribozymes and proteins becoming incorporated into each daughter microsphere. The first cells would have evolved.

But did all this happen? What are the odds that RNA nucleotides will spontaneously form a functional ribozyme? What are the odds that a ribozyme will be able to catalyze the synthesis of a functional protein? No one really knows the answers to these questions, but we can offer a few useful observations.

First, the experiments of Miller and others show that amino acids, nucleotides, and other organic molecules would have been formed in abundance on the primordial Earth. Second, no one imagines that prebiotic evolution would have formed any molecules as large or sophisticated as, say, hemoglobin. The first ribozymes, if they existed, may have been only a dozen nucleotides long. Third, ribozymes need not have been very efficient enzymes to have accelerated the synthesis of important molecules. The stone tools of early humans weren't as handy as a Swiss army knife, either, but they were better than fingernails. Finally, several hundred million years elapsed between the appearance of organic molecules and the first cells, and spontaneous synthesis must have proceeded over thousands, if not millions, of square miles of surface waters. The spontaneous synthesis of modestly functional ribozymes is probably quite likely.

Even crudely useful ribozymes and protein enzymes would provide a tremendous advantage in acquiring and processing molecules from the nutrient soup. Microspheres containing such enzymes would grow faster and divide more rapidly than other microspheres: evolution by natural selection could begin. Most biologists conclude that the origin of life is probably an inevitable consequence of the working of natural laws. We should emphasize, however, *that this proposition is not proven* and may never be. Biologists investigating the origin of life have neither millions of years nor trillions of liters of reaction solutions with which to work!

THE AGE OF MICROBES

The fossil record indicates that the earliest living cells arose about 3.5 billion years ago (Table 15-1). What were these early cells like? The first cells were prokaryotic; that is, their genetic material was not sequestered from the rest of the cell within a membrane-limited nucleus. These cells probably obtained nutrients and energy by absorbing organic molecules from the primordial soup. Since there was no free oxygen in the atmosphere, the cells must have metabolized the organic molecules anaerobically. You will recall from Chapter 7 that anaerobic metabolism yields only small amounts of energy.

As you probably recognized already, the earliest cells were primitive anaerobic bacteria. As these protobacteria multiplied, however, they probably used up the organic molecules of the nutrient soup faster than prebiotic synthesis could create more. Simpler molecules, such as carbon dioxide and water, were still very abundant, as was energy, in the "dilute" form of sunlight. What was lacking, then, was not *materials* or *energy itself,* but *energetic molecules.* Eventually, some cells evolved the ability to use the energy of sunlight to drive the synthesis of their own complex, high-energy molecules from simpler molecules: photosynthesis appeared.

Several kinds of photosynthetic bacteria evolved, but the ones that proved most important in the evolution of life were the cyanobacteria. Their photosynthetic reactions converted water and carbon dioxide to organic compounds, releasing oxygen as a by-product. At first, the oxygen reacted with iron atoms in the Earth's crust, forming huge deposits of iron oxide. After all the iron turned to rust, the concentration of free oxygen in the atmosphere rose. Chemical analysis of rocks suggests that appreciable amounts of free oxygen appeared in the atmosphere about 2.2 billion years ago.

Now oxygen is potentially very dangerous to life because it reacts with organic molecules, destroying them and releasing their stored energy. The accumulation of oxygen in the atmosphere provided the selective pressure for the next great advance in the Age of Microbes: the ability to use oxygen in metabolism, channeling its destructive power through aerobic respiration to generate useful energy for the cell. Since the amount of energy available to a cell is vastly increased when oxygen is used to metabolize food molecules, aerobic cells had a significant selective advantage.

The Rise of the Eukaryotes

Hordes of bacteria would offer a rich food supply to any organism that could eat them. There are no fossil records of the first predatory cells, but paleobiologists speculate that predation would have evolved quickly once a suitable prey population appeared. These predators would have been specialized prokaryotic cells, lacking cell walls and consequently able to engulf whole bacteria as prey. According to the most widely accepted hypothesis, these predators were otherwise quite primitive, being capable of neither photosynthesis nor aerobic metabolism. Although they could capture large food particles, namely bacteria, they metabolized them inefficiently. About 1.4 billion years ago, however, one predator probably gave rise to the first eukaryotic cell.

Eukaryotic cells differ from prokaryotic cells in many ways, but perhaps most fundamental are the membrane-bound nucleus containing the genetic material and the inclusion of organelles for energy metabolism, mitochondria and (in plants) chloroplasts. How did these organelles evolve?

The **endosymbiotic hypothesis,** championed most forcefully by Lynn Margulis, proposes that a predatory cell acquired the precursors of chloroplasts and mitochondria by engulfing certain types of bacteria. Let us suppose that an anaerobic predatory cell captured an aerobic bacterium for food, as it often did, but for some reason failed to digest this particular prey (Fig. 15-6a; see also the essay in Chapter 5). The aerobic bacterium remained alive and well. In fact, it was better off than ever, because the cytoplasm of its predator–host was chock full of half-digested food molecules, the remnants of anaerobic metabolism. The aerobe absorbed these molecules and used oxygen to complete their metabolism, gaining enormous amounts of energy as it did. So abundant were its food resources, and so bountiful its energy production, that the aerobe must have leaked energy, probably as ATP or similar molecules, back into its host's cytoplasm. The mitochondrion had been born. Interestingly, the amoeba *Pelomyxa palustris* is almost a living fossil in this respect. Unlike almost all other eukaryotic cells, it lacks mitochondria; however, a permanent population of aerobic bacteria that resides in its cytoplasm carries out much the same role.

The predator cell with its symbiotic bacteria could metabolize food aerobically, gaining a great selective advantage over its anaerobic compatriots. Soon its progeny filled the seas. One of these daughter cells carried out a second coup: it captured a photosynthetic cyanobacterium and similarly failed to digest its prey (Fig. 15-6a). The cyanobacterium flourished in its new host and gradually evolved into the first chloroplast. Some modern or-

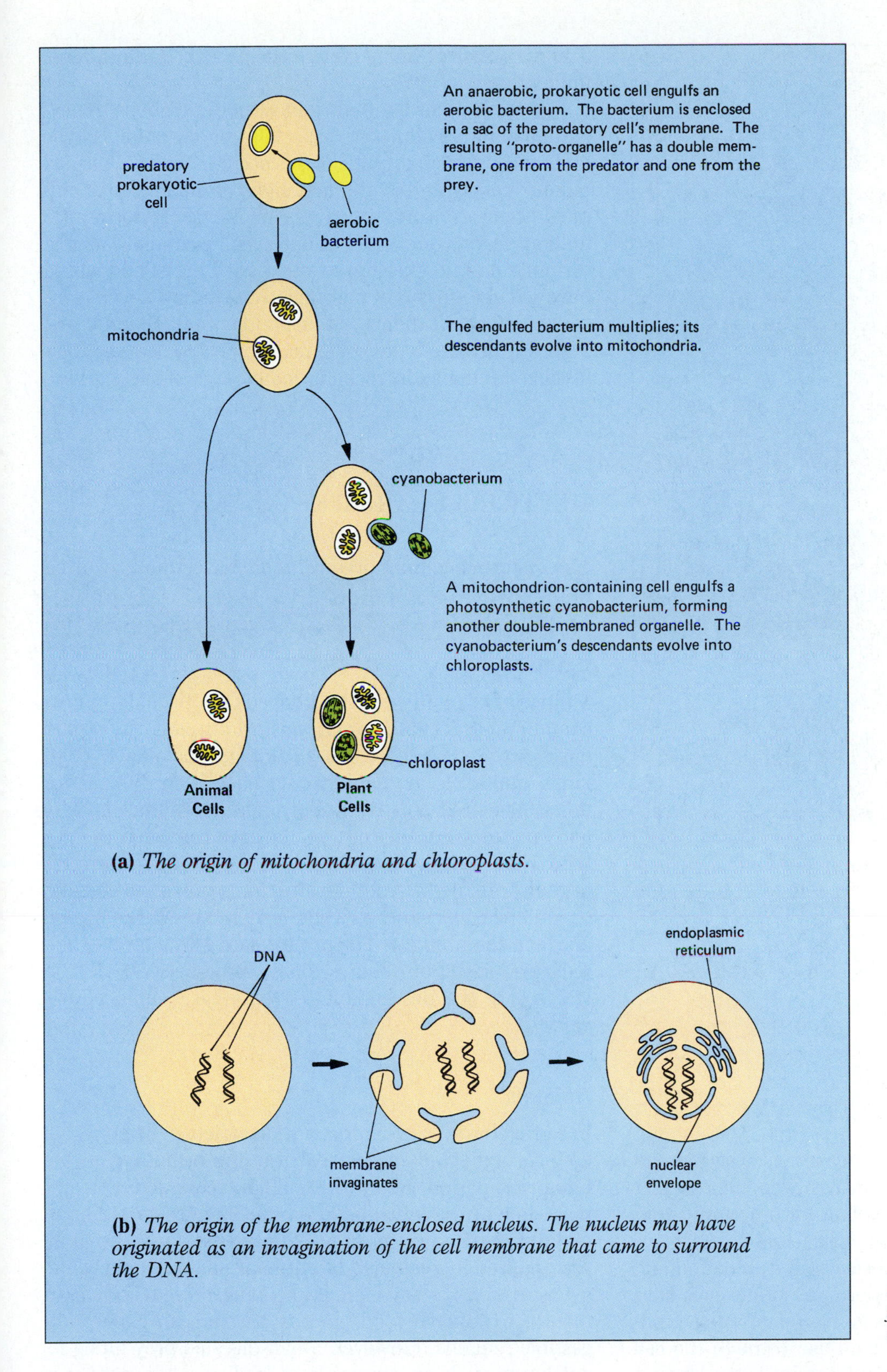

(a) *The origin of mitochondria and chloroplasts.*

(b) *The origin of the membrane-enclosed nucleus. The nucleus may have originated as an invagination of the cell membrane that came to surround the DNA.*

Figure 15-6 *Possible mechanisms for the appearance of organelles in eukaryotic cells.*

ganisms resemble this hypothetical ancestral condition. A variety of corals, some clams, a few snails, and at least one species of *Paramecium* harbor a permanent collection of algae in their cells (Fig. 15-7). These algae share some of their photosynthetically produced food molecules with the host cells.

The origin of the nucleus is more obscure. One possibility is that the cell membrane folded inward, surrounding the DNA (Fig. 15-6b). However the nucleus originated, having the DNA sequestered within the nucleus seems to have conferred great advantages in regulating the use of the genetic material.

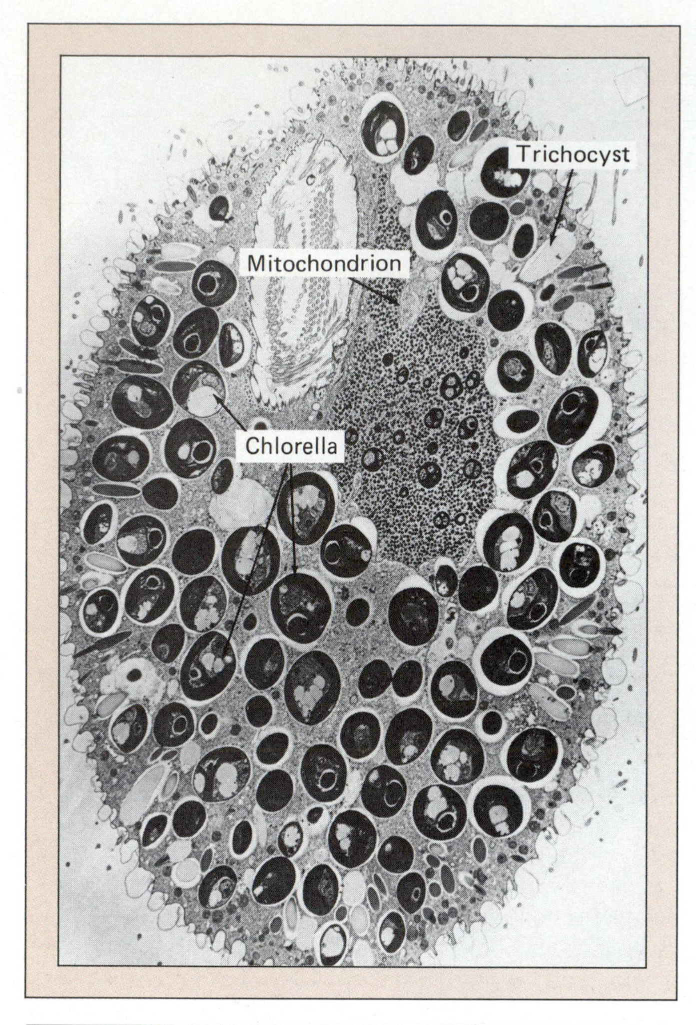

Figure 15-7 Green unicellular algae of the genus Chlorella *live within the cytoplasm of a* Paramecium. *A similar symbiotic relationship may have given rise to the ancestors of modern chloroplasts.*

MULTICELLULARITY

Once predation had evolved, increased size became an advantage. A larger cell could more easily engulf a smaller cell, while in turn being more difficult for other predatory cells to ingest. Larger organisms can usually also move faster than small ones, making successful predation and escape more likely. However, enormous single cells have problems. Oxygen and nutrients going in and waste products going out of the cell must diffuse through the cell membrane. As we pointed out in Chapter 4, the larger a cell becomes, the less surface membrane is available per unit volume of cytoplasm. There are only two ways that an organism larger than a millimeter or so in diameter can survive. First, it can be metabolically sluggish. This seems to work for certain very large unicellular algae. Alternatively, an organism may be multicellular; that is, it may consist of many small cells packaged into a large unified body.

The fossil record reveals almost nothing about the evolution of multicellularity, especially among animals. The first unicellular eukaryotic fossils occur in rocks about 1.4 billion years old, and the first signs of multicellular animals occur as worm burrows and tracks in rocks 400 million years younger. The intervening animals almost certainly had no skeleton or other hard parts, and would have left few fossils. Consequently, we may never learn very much about them. Within another 500 million years, however, diverse types of animals appeared, leaving their fossilized remains in rocks in many parts of the world.

MULTICELLULAR LIFE IN THE SEA

The first multicellular organisms almost certainly evolved in the sea.

Plants

Multicellular plants evolved from eukaryotic, chloroplast-containing unicellular organisms. Multicellularity would have provided at least two advantages for plants. First, large, multicellular plants would have been difficult for unicellular predators to swallow. Second, multicellularity and specialization of cells would have conferred the potential for staying in one place in the brightly lit waters of the shoreline, as rootlike structures burrowed in sand or clutched onto rocks, while leaflike structures floated above in the sunlight. The profusion of green, brown, and red algae lining our shores today, some over 200 feet in length, are the descendants of these early multicellular algae.

Animals

In a great burst of evolution, a wide variety of invertebrate animals appeared in the sea near the beginning of the Cambrian period, about 600 million years ago. For animals, one of the advantages of multicellularity is the potential for eating larger prey. This potential was probably first realized in a jellyfish-like animal, shaped vaguely like a vase (Fig. 15-8). A single opening served both as a mouth, to take in food, and as an anus, to expel indigestible remains. However, a half-digested prey filling its gut keeps such an animal from feeding again until it is finished with its first meal. Soon more efficient means of feeding evolved, employing a separate mouth and anus, found today in almost all animals (Fig. 15-9). With this design, the animal can feed more or less continuously, as earthworms and sea cucumbers do today.

The coevolution of predator and prey rapidly led to

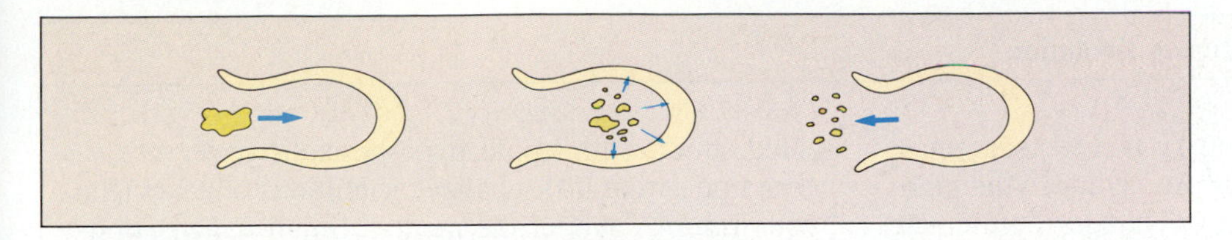

Figure 15-8 The earliest animals may have possessed a digestive system with only one opening to the outside world. Consequently, feeding would have had to be periodic. If one piece of prey was already in the digestive tract, a second piece would have to be stuffed on top. When the indigestible remnants of the first were expelled, they would force out the not-yet-digested second piece as well.

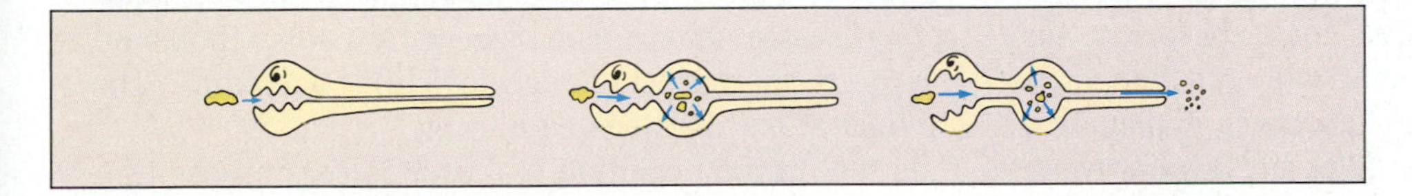

Figure 15-9 A major advance in feeding is to have separate openings for ingesting food and for ejecting the indigestible residue. In principle, this allows more-or-less continuous feeding.

increased sophistication in many kinds of animals, ranging from the mud-skimming, armored trilobites, to their predators, the ammonites and the chambered nautilus, which still survives almost unchanged in deep Pacific waters (Fig. 15-10). A major trend at this time was toward greater mobility. Predators often need to travel over wide areas in search of suitable prey, while speedy escape is also an advantage for prey. Locomotion is usually accomplished by contraction of muscles that move body parts through their attachments to some sort of skeleton. The dominant invertebrates at this time possessed either an internal hydrostatic skeleton, much like a water-filled tube (worms, jellyfish), or an external skeleton covering the body (arthropods such as trilobites). Rapid locomotion, of course, is of little use if an animal cannot tell where it is going or what to do once it gets there. Greater sensory capabilities and more sophisticated nervous systems evolved along with locomotor abilities. Senses for

Figure 15-10 **(a)** *Characteristic life of the oceans during the Silurian period. Among the most common fossils from that time are the trilobite* **(b)** *and its predators the ammonites* **(c)** *and the nautiloids. Although* **(d)** *illustrates a living* Nautilus, *the Silurian nautiloids were very similar in structure, showing that a successful body plan may exist virtually unchanged for hundreds of millions of years.*

How Fast Do Species Arise During Evolution?

For nearly a century, evolutionary biologists assumed that evolution is a slow process. According to the **gradualism** model, many miniscule changes accumulate over millions of years to produce a new species (Fig. E15-1a). For example, consider the evolution of the horse (see Fig. 13-10). The fossil record shows a progression from *Hyracotherium,* the "dawn horse," through many intermediate steps to modern *Equus.* Gradualism proposes that fossil horses are representative samples of changes that occurred gradually throughout the evolution of the horse. However, the gradation is not very fine: significant differences are found between each stage. A reasonable explanation might be that fossilization is, after all, a rare event, and that thousands of other intermediate forms just did not become fossilized.

More recently, a group of evolutionary biologists, led by Niles Eldredge and Stephen Jay Gould, proposed a different explanation for the gaps, called the **punctuated equilibrium** model. They argue that evolution proceeds by a series of jumps (Fig. E15-1b). The gaps in the fossil record are produced by such rapid evolutionary change that fossil remains of intermediate forms would be extremely rare. For the horse, punctuated equilibrium would hypothesize that *Hyracotherium* underwent little change for millions of years (the "equilibrium" part of the model). Small fringe populations of *Hyracotherium* split off now and then. Some evolved very rapidly, perhaps mostly by genetic drift, to become entirely new species. At some point, perhaps because of a major change in the environment, one of these new species, *Mesohippus,* quickly replaced *Hyracotherium* as the dominant form of horse. The speciation of *Mesohippus* and its replacement of *Hyracotherium* may have taken place within a few thousand years; a mere instant, geologically speaking (the "punctuation" between equilibria).

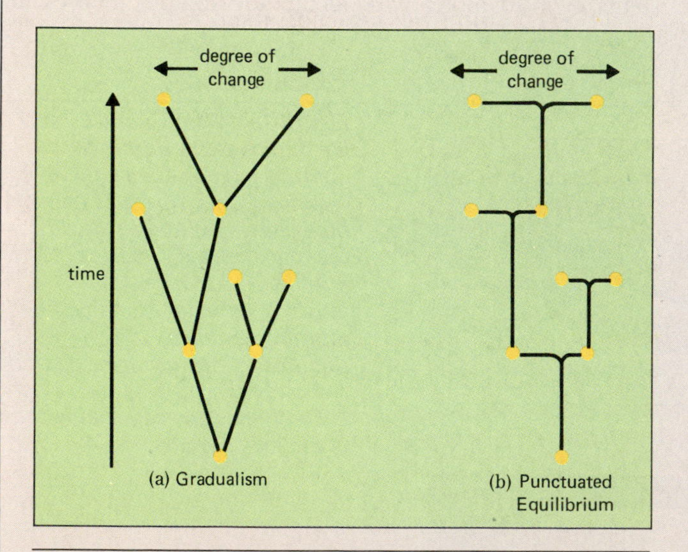

Figure E15-1 Hypothetical evolutionary trees, according to the theories of gradualism **(a)** *and punctuated equilibrium* **(b)**. *Each dot represents a fossil find; lines connecting the dots represent the time course of evolutionary change between fossils. Gradualism maintains that more-or-less steady change occurred between fossil forms. Punctuated equilibrium holds that long periods of constancy were followed by bursts of change leading to the next fossil form.*

Is such rapid evolution realistic? Let's suppose that a 60-centimeter-tall *Hyracotherium* evolves into a 100-centimeter-tall *Mesohippus* in 10,000 years, and that the time between generations is 5 years. This allows 2000 generations to produce a height increase of 40 centimeters, or 0.2 millimeter per generation. This would be far too small a difference to detect with the naked eye and is more than 500 times less than the variability in size among modern horses, even of the same breed. To a geneticist, this rate of change doesn't seem very fast at all. In the fruit fly *Drosophila,* for example, mild selection pressure can cause evolutionary changes that are more than 100 times faster than those required for *Hyracotherium* to evolve into *Mesohippus* in 10,000 years.

Which hypothesis, gradualism or punctuated equilibrum, is correct? There are two aspects to this question: which better explains the observed fossil record, and which provides the more plausible mechanism of evolution? First, paleobiologists are finding that some fossil sequences show the punctuated equilibrium pattern, while other sequences show fairly steady changes over long periods. Therefore, each theory may correctly explain the evolutionary history of certain species but not others. Second, some proponents of punctuated equilibrium say that this hypothesis is a radical departure from past ideas about evolution, not just in the speed of speciation, but in the mechanisms of evolutionary change. Their opponents, however, say that a punctuated fossil record can be produced by nothing more than the "ordinary" mechanisms of gradualism: namely, stabilizing selection during the "equilibrium" periods and strong directional selection during the "punctuational" changes. Given the strong personalities involved, this argument is not likely to be settled anytime soon.

detecting touch, chemicals, and light became highly developed. The senses were usually concentrated in the head end of the animal, along with a nervous system capable of handling the sensory information and directing appropriate behaviors.

About 500 million years ago, an entirely new type of animal appeared, possessing an internal skeleton. For a hundred million years these were inconspicuous members of the ocean community, but by 400 million years ago some had evolved into fishes. By and large, the fishes proved to be faster than their invertebrate compatriots, with more acute senses and larger brains. Eventually, they became the dominant predators of the open seas.

THE INVASION OF THE LAND

Between 600 and 400 million years ago, both plants and animals evolved greatly, but they remained cushioned by the equanimity of the sea. Life in the ocean provides buoyant support against gravity and ready access to life-sustaining water. Reproduction is also simple in the sea. At some point in their life cycles, both plants and animals produce sex cells that must fuse to form the beginnings of a new generation. Sea-dwelling organisms usually have mobile sperm and/or eggs that swim to each other through the water.

In contrast, on land an organism must bear up its weight against the crushing force of gravity; it must find adequate water; and to reproduce, it must ensure that its gametes, particularly the sperm, are protected from drying out. Nevertheless, the land offered great potential, particularly for plants. Water strongly absorbs light; even in the clearest water, photosynthesis is limited to the upper couple of hundred meters, and usually much less. Out of the water, the sun is dazzlingly bright, permitting rapid photosynthesis. Sea water also tends to be low in certain nutrients, particularly nitrogen and phosphorus. By comparison, terrestrial soils are rich storehouses of nutrients. Finally, by this time the sea swarmed with plant-eating animals. Since the land was devoid of animal life, plants that colonized the land would have no predators.

Land Plants

In moist soils near the shore, a few small green algae began to grow, taking advantage of the sunlight and nutrients. They did not have large bodies to support against the constant pressure of gravity, and living right in the film of water on the soil, they could easily obtain water. About 400 million years ago, some of these algae gave rise to the first multicellular land plants. Initially simple, low-growing forms, land plants rapidly evolved solutions to two of the main difficulties of plant life on land: obtaining and conserving water, and staying upright despite gravity and winds. Rootlike structures delved into the soil, mining water and minerals, while waterproof coatings on the aboveground parts reduced water loss by evaporation. Specialized cells formed tubes called vascular tissues to conduct water from roots to leaves. Extra thick walls surrounding certain cells enabled stems to stand erect.

Reproduction out of water seems to have been harder to solve. We examine plant reproduction in more detail in Chapters 19 and 22; the important point to note here is that, like animals, plants produce sperm and eggs. Primitive marine plants have swimming sperm and sometimes swimming eggs as well. The first land plants retained swimming sperm, which restricted them to swamps and marshes where the sperm and eggs could be released into the water, or to areas with abundant rainfall where the ground would occasionally be covered with water.

This strategy sufficed for millions of years. During the Carboniferous period, 345 to 280 million years ago, the climate was warm and moist, and great stretches of the land were covered with forests of giant tree ferns and club mosses that produced swimming sperm (Fig. 15-11). The coal we mine today is the fossilized remains of these forests. Meanwhile, some plants inhabiting drier regions evolved reproductive strategies that no longer depended on films of water. In these protoconifers, the eggs were retained in the parent plant, while the sperm were encased in drought-resistant pollen grains that blew on the wind from plant to plant. Landing on a female cone near the egg, the pollen released sperm cells directly into living tissue, eliminating the need for a surface film of water. About 250 million years ago, mountains rose, swamps drained, and the moist climate dried up. The swimming sperm of tree ferns and giant club mosses doomed most of them to extinction, while the conifers flourished and spread.

About 130 million years ago, the flowering plants appeared, having evolved from a group of conifer-like plants. The initial advantage of the flowering plants seems to have been pollination by insects. The conifers are wind pollinated, which demands an enormous pollen production, since the vast majority of pollen grains fail to reach their target. Flower pollination by insects wastes far less pollen. Flowering plants also evolved other advantages, including more rapid reproduction and, in some cases, much more rapid growth. Today, flowering plants dominate the land, except in cold boreal regions, where conifers still prevail.

Land Animals

Soon after land plants evolved, providing potential food sources for animals, arthropods (probably early relatives of scorpions) emerged from the sea. Why arthropods? The answer seems to be that they were *preadapted* for land life: that is, they already possessed structures,

Figure 15-11 A reconstruction of a swamp forest of the Carboniferous period. The treelike plants are tree ferns and giant club mosses, both now mostly extinct. Note the dragonfly; some Carboniferous dragonflies had wing spans in excess of half a meter!

evolved under totally different selective pressures, that suited life on land. Foremost among these was the exoskeleton. Exoskeletons are both waterproof and strong enough to bear up a small animal under the stresses of gravity.

Land animals encounter another difficulty, also relatively easily solved by arthropods: breathing. Respiratory surfaces must be kept moist, and this is difficult in the dry air. Some arthropods, such as land crabs and spiders, evolved what amounts to an internal gill, kept moist within a waterproof sac (see Chapter 18). The insects developed tracheae, small branching tubes directly penetrating the body, with adjustable openings in the exoskeleton leading to the outside air.

For millions of years, arthropods had the land and its plants to themselves, and for tens of millions of years more, they were the dominant animals. Dragonflies with a wingspan of 70 centimeters flew among the Carboniferous tree ferns (Fig. 15-11), while 2-meter-long millipedes munched their way across the swampy forest floor. Eventually, however, the arthropods' splendid isolation came to an end.

About 400 million years ago, a group of fishes called the lobe-fins appeared, probably in fresh water. Lobe-fins had two important preadaptations to life on land: stout, fleshy fins, with which they crawled about on the bottoms of shallow, quiet waters, and an outpouching of the digestive tract that could be filled with air, like a primitive lung (Fig. 15-12). In one group of lobe-fins, the lung evolved into a swimbladder, with which they could regulate their buoyancy and remain suspended in the water without active exertion. Many of these migrated back to the sea, where they evolved into the modern bony fishes. Another group of lobe-fins colonized very shallow ponds and

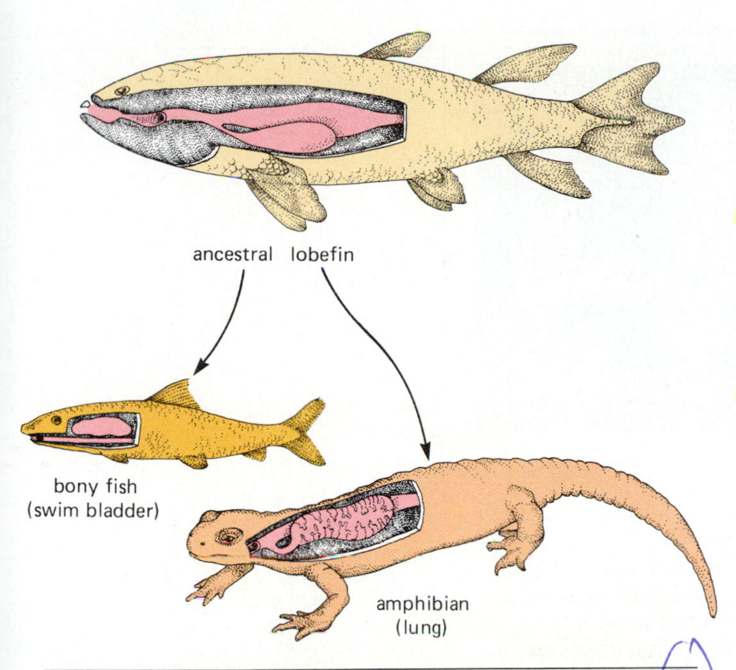

Figure 15-12 A group of primitive fish, called the lobe-fins, gave rise both to the bony fish and the amphibians. Lobe-fins had a pair of lungs, which arise as outpouchings of the digestive tract. Lobe-fins probably used their lungs for breathing air when the pools in which they lived became stagnant and foul. In one group of lobe-fins, one lung evolved into an air bladder, used in bony fishes to regulate buoyancy. The other lung disappeared. Another group of lobe-fins became further adapted for air breathing, evolving more complex, efficient lungs and sturdier fins for walking on land. These were the ancestors of the amphibians.

streams, which shrank during droughts and often became oxygen-poor. By taking air into their lungs, the lobe-fins could obtain oxygen anyway. Some of their descendants began to use their fins to crawl from pond to pond in search of prey or water, as some modern fish can do today (Fig. 15-13).

As the arthropods discovered previously, the land is a rich source of food. Feeding on land and moving from pool to pool favored the evolution of fish that could stay out of water for longer periods and that could move about more effectively on land. With improvements in lungs and legs, the amphibians evolved from lobe-fins, first appearing in the fossil record about 350 million years ago. If an amphibian could have thought about such things, it would have thought that the Carboniferous swamp forests were heaven itself: no predators to speak of, abundant prey, and a warm, moist climate. As with the insects and millipedes, some amphibians evolved gigantic size, including salamanders over 3 meters long.

Despite their success, the early amphibians still lacked complete adaptation to life on land. Their lungs were simple sacs without very much surface area, so they had to obtain some of their oxygen through their skins. Therefore, the skin had to be kept moist, restricting amphibians to swampy habitats where they would not dry out. Further, amphibians shed their sperm and eggs in water. As with the tree ferns and club mosses, when the Carboniferous climate turned dry at the beginning of the Permian, amphibians were in trouble.

Meanwhile, just as the conifers had been evolving on the fringes of the swamp forests, a group of amphibians were evolving adaptations to drier conditions. These became the reptiles, which achieved four great advances over the amphibians. First, they evolved internal fertilization: the reptilian female provides the watery environment for the sperm inside her reproductive tract. Thus sperm transfer could occur on land, without venturing back to the dangerous swamps full of fish and amphibian predators. Second, they developed waterproof eggs enclosing their own supply of water for the developing embryo (see Chapter 18), again providing freedom from the swamps. Third, the protoreptiles developed scaly, waterproof skin. Finally, accompanying the evolution of waterproof skin came improved lungs that could provide the entire oxygen supply for an active animal. As the climate dried in the Permian, reptiles became the dominant vertebrate land fauna, relegating the amphibians to swampy backwaters, where most remain today.

A few tens of millions of years later, the climate returned to more moist and equable conditions, providing for lush plant growth. Once again, gigantism became the rage, as certain families of reptiles evolved into the dinosaurs (Fig. 15-14). These were among the most successful animals ever, if we consider length of dominance as a measure of success. They flourished for over 100 million years, until about 65 million years ago the last

Figure 15-13 Some modern fish resemble their lobe-fin ancestors in their ability to crawl about on land. The walking catfish was imported into Florida a few years ago, escaped into local waters, and has spread across much of southern Florida by walking from pond to pond.

Figure 15-14 A reconstruction of a Jurassic swamp. In the foreground we see, standing in the swamp, the gigantic Brontosaurus, *20 meters long and weighing over 30 tons. On the right, a carnivorous* Allosaurus *uses its meter-long jaws to tear flesh from its prey. In the background grazes the 10-ton* Stegosaurus, *an astoundingly dim-witted herbivore with a brain the size of a walnut.* Stegosaurus *had a swelling of its spinal cord in the hip region which formed a second "brain" of sorts, 20 times larger than the one inside its skull. The plates on its back probably functioned in temperature regulation. Depending on air temperature, the animal's orientation to the sun, and the amount of blood flow across the surface of the plates, they could be used as radiators to cool the blood or as solar panels to pick up warmth.*

dinosaurs became extinct. No one is certain why they died, but a climate change, perhaps initiated by a gigantic meteorite impact, seems to have been the final blow (see Chapter 14). With less luxuriant plant growth, the herbivorous dinosaurs could not find enough food. Without large herbivores as prey, the carnivores were doomed too.

Even during the age of dinosaurs, many reptiles remained quite small. One major difficulty faced by small reptiles is keeping a high body temperature. Being active on land seems to require a rather warm body, maximizing the efficiency of the nervous system and muscles. However, a warm body loses heat to the environment unless the air is also warm. Small reptiles have a relatively large surface area through which heat is lost and a relatively small internal volume in which metabolic heat is generated. If the body is warmed metabolically when the air is cool, an enormous amount of food must be consumed to provide sufficient energy. Apparently, the food requirement is too high, for the naked-skinned small reptiles have a fairly low metabolism, not enough to keep their bodies warm in cool air, especially at night when there is no sun for radiant warmth. Two groups of small reptiles independently evolved insulation that minimizes heat loss: one group evolved feathers, while another group evolved hair.

In the protobirds, insulating feathers retained body heat. Consequently, these animals could be active in cooler habitats and during the night, when their scaly relatives became sluggish. Later, some protobirds developed longer, stronger feathers on their forelimbs, perhaps allowing them to glide from trees or to assist in jumping after insect prey. From this point, the evolution of flight became possible.

The hair evolved by the protomammals also provided insulation. Unlike the birds, which retained the reptilian habit of egg laying, mammals evolved live birth and the ability to feed their young with secretions of the mammary glands. Since these structures do not fossilize, we may never know when the uterus, mammary glands, and hair first appeared, or what their intermediate forms looked like.

The earliest mammals were small creatures, probably living in trees and being active mostly at night. When the dinosaurs became extinct, the mammals radiated out into the vast array of modern forms. While some stayed small and nocturnal, eating mostly seeds and insects, others evolved larger size and different habits, colonizing the habitats left empty by the extinction of the dinosaurs. One group remained in the trees, giving rise to the primates.

HUMAN EVOLUTION

Primate fossils are relatively rare compared to those of many other animals. There are at least three reasons for this. First, most primates did not live in habitats that readily preserve fossils, such as swamps and shallow lagoons. Second, until recently primates were fairly small. Therefore, predators and scavengers would be more likely to break up their bones into unrecognizable fragments than they would the bones of a bison or dinosaur. Third, ancestral primates may have had small populations, thus providing less material for fossilization in the first place.

Humans are intensely interested in their own evolution, especially in trying to determine what conditions led to the evolution of the gigantic human brain. This obsession has led to sweeping speculations despite an often skimpy fossil record. Therefore, although the outline of human evolution that we will present is a synthesis of current thought on the subject, it is by no means as well understood as, say, the genetic code. Paleontologists disagree about the interpretation of the fossil evidence, and many ideas may have to be revised as new fossils are found.

The first protoprimates were the insectivorous tree shrews, whose fossils are found in rocks about 80 million years old. Nimble, probably nocturnal animals, tree shrews were smaller than all but the tiniest modern primates. Over the next 50 million years, the descendants of the tree shrews evolved forms similar to the modern tarsiers, lemurs, and monkeys (Fig. 15-15). These primates stayed in the trees. Some remained nocturnal, but many became active during the day.

Primates evolved several adaptations for life in the trees, feeding on fruits and leaves. The tree shrews already possessed handlike paws for grasping branches, and the primates further refined these grasping appendages, which made life in the trees a little safer. One of the earliest new adaptations seems to have been large, forward-facing eyes (Fig. 15-15). Jumping from branch to branch is risky business unless an animal can accurately judge where the next branch is located. Accurate depth perception was made possible by binocular vision provided by forward-facing eyes with overlapping fields of view. Another adaptation was color vision. We cannot, of course, tell if a fossil animal had color vision, but modern primates have excellent color vision, and it seems reasonable to assume that earlier primates did, too. Many primates feed on fruit, and color vision helps in detecting fruit among a welter of green leaves. Hands, binocular vision, and color vision seem to have served as preadaptations in human evolution—features that, although not evolved for tool use, were indispensable in the evolution of our tool-using ancestors.

Fossil Hominids

Between 20 and 30 million years ago, in the moist tropical forests of Africa, a group of primates called the Dryopithecines diverged from the monkey line. The Dryopithecines appear to be ancestral to the hominids (humans and their fossil relatives) and pongids (the great apes). Around 15 million years ago, global climatic cooling began to shrink the vast expanses of forest, splitting up the woodlands into isolated islands dotted upon a sea of grassland. Diversification of habitat and isolation of small populations led to the diversification of Dryopithecines.

About 4 million years ago, the first true hominids appeared, called the Australopithecines ("southern ape," from their original discovery site in southern Africa). The Australopithecines (Fig. 15-16) seem to have been creatures of the grasslands. Footprints almost 4 million years old, discovered in Tanzania by Mary Leakey, show that the Australopithecines could walk upright. Their brains

Figure 15-15 *Representative primates include the tarsier* **(a)**, *lemur* **(b)**, *and liontail macaque* **(c)**. *Note that all have relatively flat faces, with forward-looking eyes providing binocular vision. All also have color vision and grasping hands. These features served as preadaptations for tool and weapon use by early humans.*

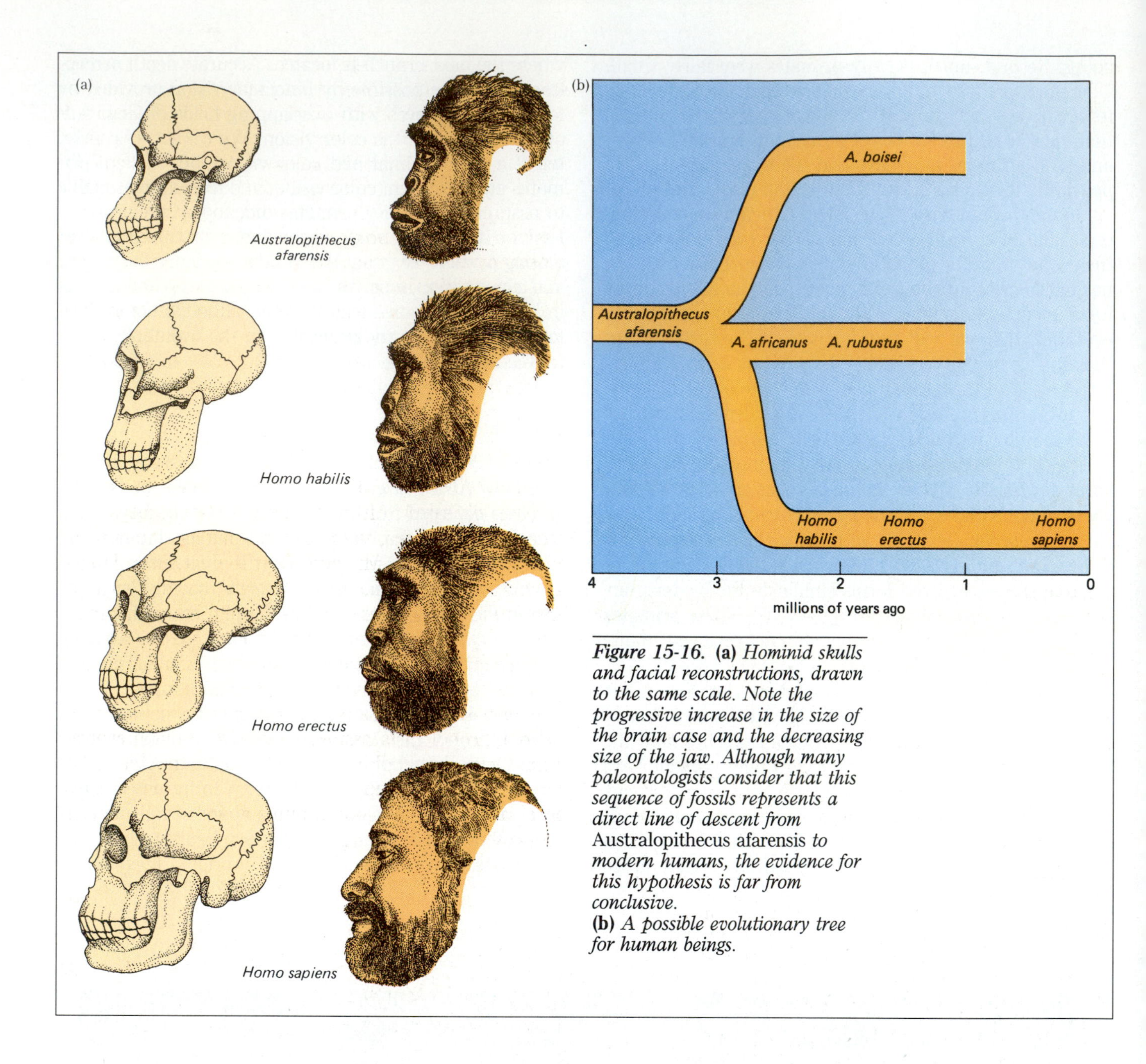

Figure 15-16. **(a)** *Hominid skulls and facial reconstructions, drawn to the same scale. Note the progressive increase in the size of the brain case and the decreasing size of the jaw. Although many paleontologists consider that this sequence of fossils represents a direct line of descent from* Australopithecus afarensis *to modern humans, the evidence for this hypothesis is far from conclusive.*
(b) *A possible evolutionary tree for human beings.*

were fairly large, although still much smaller than those of modern humans. The Australopithecines split into at least two distinct forms: the small, omnivorous *Australopithecus afarensis* and *A. africanus,* and the large, herbivorous *A. robustus* and *A. boisei.* By about 2 million years ago, the smaller Australopithecines (probably *A. afarensis*) seem to have given rise to a new form, *Homo habilis* (Fig. 15-16). *Homo habilis* had a considerably larger brain than the Australopithecines, and was probably the first hominid that used stone and bone tools, mostly crude weapons of various sorts (Fig. 15-17). *Homo habilis* ranged throughout Europe, Asia, and Africa.

About 1.5 million years ago, *Homo erectus* appeared, probably a descendant of *H. habilis.* The brain of *Homo erectus* was as large as the smallest modern adult human brains. The face was notably different from that of modern humans (Fig. 15-16), featuring large brow ridges, a slightly protruding face, and no chin (the protruding tip of the lower jaw). Nevertheless, a suitably dressed *H. erectus* could probably walk the streets of New York without exciting much comment.

Behaviorally, *H. erectus* seems to have been more advanced than *H. habilis. Homo erectus* fashioned sophisticated stone tools, ranging from hand axes used for cutting and chopping to points probably used on spears (Fig. 15-17). These weapons suggest that *H. erectus* ate animal food, probably both hunting on their own and scavenging from the remains of prey killed by lions and other predators. *Homo erectus,* like *H. habilis,* spread throughout much of the Old World—Peking Man and Java Man were

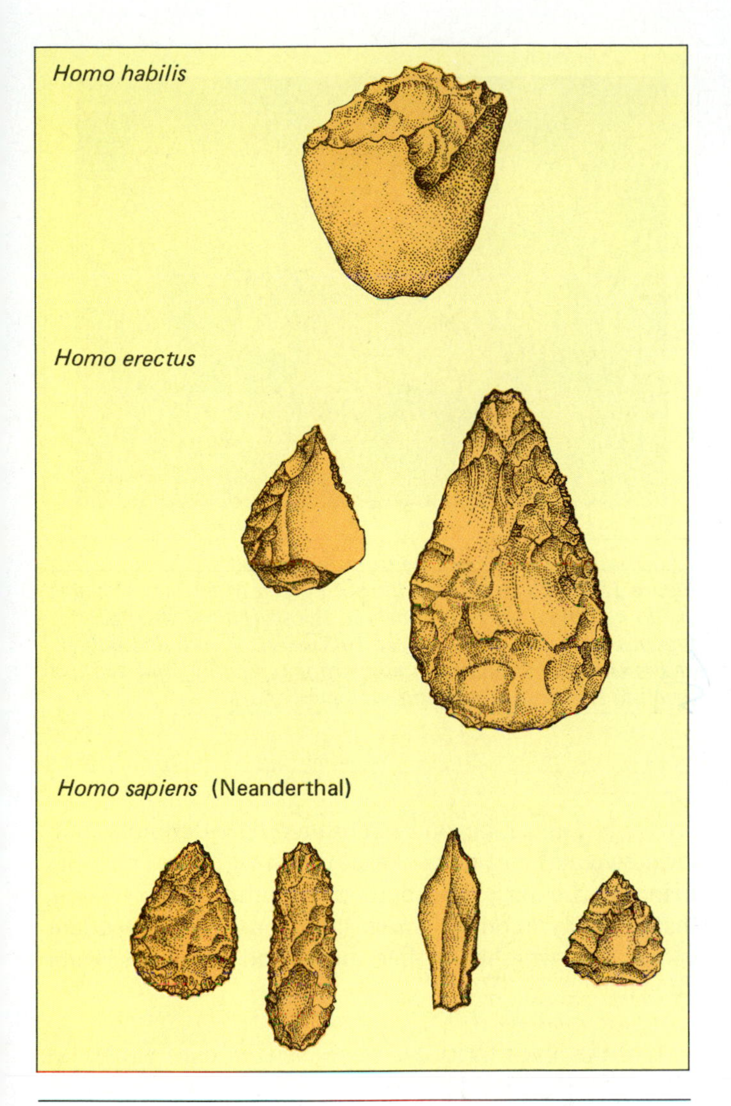

Figure 15-17 Representative hominid tools. Homo habilis *produced only fairly crude chopping tools called hand axes, usually unchipped on one side to hold in the hand.* Homo erectus *manufactured much finer tools. Since the tools were often sharp all the way around the stone, at least some of these blades were probably tied to spears rather than held in the hand. Neanderthal tools were works of art, with extremely sharp edges made by flaking off tiny bits of stone. In comparing these* H. habilis, H. erectus, *and* H. sapiens *weapons, note the progressive increase in the number of flakes taken off the blades, and the corresponding decrease in flake size. Smaller, more numerous flakes produce a sharper blade and suggest more insight into toolmaking (perhaps passed down culturally from experienced toolmakers to apprentices), more patience, finer control of hand movements, or perhaps all three.*

specimens of *Homo erectus.* In China, evidence from 500,000 years ago suggests that *Homo erectus* used fire, although for what purpose (to cook food, keep warm, or ward off predators) no one knows.

The first fossils of our own species, *Homo sapiens,* date from about 200,000 years ago. The earliest fossils are fragmentary, but beginning about 100,000 years ago the Neanderthal variety left many complete fossils, found throughout the Old World. Contrary to the popular image of the hulking, stoop-shouldered "cave man," Neanderthals were actually quite advanced. It is true that they were very heavily muscled, but Neanderthals walked fully erect, and had brains, on the average, slightly larger than those of modern humans. Although many of the European fossils show heavy brow ridges and a broad, flat skull, others, particularly from Africa and the Near East, were remarkably like ourselves.

Neanderthal remains show evidence of modern behaviors, too, particularly ritualistic burial ceremonies (Fig. 15-18). Neanderthal skeletons have been discovered in clearly marked burial sites surrounded with stones, and often include offerings of flowers, bear skulls, and food. Altars found at other Neanderthal sites were probably used in "religious" rites associated with a bear cult.

Finally, about 90,000 to 100,000 years ago, modern humans appeared. Although the oldest specimens have been found in Africa, these anatomically modern humans are often called Cro-Magnon, after the district in France in which their remains were first discovered. Cro-Magnons had domed heads, smooth brows, and prominent chins. Their tools were precision instruments not very different from the stone tools used until recently in many parts of the world. Where did the Cro-Magnons come from and where did the Neanderthals go? Formerly, most paleoanthropologists assumed that Cro-Magnon evolved from the modern-looking Neanderthals of the Near East, and spread into Europe from there. More recently, it has been proposed that modern humans are not the descendants of Neanderthals, but evolved from a distinct ances-

Figure 15-18 A Neanderthal bear ceremony. Power and ferocity, and perhaps their ability to rise up on their hind legs like humans, made cave bears cult objects to Neanderthals. Some Neanderthals were buried with elaborate ceremonies, with bear skulls placed on the grave of the deceased.

tral population. Modern humans coexisted with Neanderthals for thousands of years, but whether the Neanderthals were eliminated by interbreeding, competition, or warfare is unknown.

Behaviorally, Cro-Magnon seems to have been similar to, if more sophisticated than, Neanderthal. Perhaps the most remarkable accomplishment of Cro-Magnon is the magnificent art left in caves such as Altamira in Spain and Lascaux in France (Fig. 15-19). No one knows exactly why these paintings and sculptures were made, but they attest to minds fully as human as our own.

Figure 15-19 *Clay bison fashioned by Cro-Magnons. Although we do not know the reason behind the sculptures, their very presence bespeaks an advanced culture which included skilled artists with the intelligence, motivation, empathy, and "leisure time" to create such magnificent works of art.*

The Evolution of Human Behavior

Perhaps the most contentious subject in human evolution is the development of human behavior. Except in rare instances, such as the Neanderthal burials or the Cro-Magnon sculptures, there is no direct evidence of the behavior of prehistoric hominids. A few hypotheses have been offered by biologists and anthropologists, based on the fossil record and the behavior of modern humans and animals.

Brain development. The fossil record shows that the development of truly huge brains occurred within the last couple of million years. What were the selective advantages of large brains? This subject continues to stir controversy among paleoanthropologists. However, the enlarging brain may have been selected both because it provided for improved hand–eye coordination and because it facilitated complex social interactions.

As the early hominids descended from the trees into the savanna, they began to walk upright. Bipedal locomotion allowed them to carry things in their hands as they walked. Hominid fossils show shoulder joints capable of powerful throwing motions and an opposable thumb that would enhance the ability to manipulate objects with great dexterity. These early hominids probably could see well and judge depth accurately with binocular vision. The brain expansion that occurred in the Australopithecines may have been at least partly related to integration of visual input and control of hand/arm movements.

Homo erectus, and perhaps *H. habilis* before it, was social. So, of course, are many monkeys and apes. The later hominids, however, seem to have engaged in a new type of social activity: cooperative scavenging and hunting. *H. erectus,* in particular, seems to have hunted extremely large game, which obviously calls for cooperation in all phases of the hunt. Some paleoanthropologists believe that selection favored larger and more powerful brains as an adaptation for success in cooperative hunting. (Lest you become too impressed with such speculation, remember that lions and wolves also hunt cooperatively, yet are not noticeably more intelligent than their relatives, such as leopards and coyotes, which do not.)

How, why, and when the use of language appeared is probably forever beyond our knowledge. Certainly, language would be of great assistance in all phases of social living, particularly in cooperative hunting. Cro-Magnon, that is to say, modern humans, obviously developed language, but whether earlier forms had language is not known.

Cultural evolution. Isaac Newton once wrote, "If I have seen further, it is because I stand on the shoulders of giants." All humanity stands on the shoulders of giants, not only pre-Newtonian physicists, but Cro-Magnons, Neanderthals, the hairy reptiles of 100 million years ago, the first lobefinned fishes to crawl out of the swamps, and the first cell. We are the product of $3\frac{1}{2}$ billion years of biological evolution, a fact that can be read in our genes, which are similar to those of bacteria, plants, and fungi.

In recent millenia, however, biological evolution has been outstripped by **cultural evolution:** learned behaviors passed down from previous generations. Although our bodily forms have not substantially changed in thousands of years, our behavior has changed radically. Most human behaviors are strongly influenced by learning. The giants that loom tallest in our heritage are not necessarily our biological ancestors, but the people, living and dead, who shaped our behavior. Indeed, history influences entire societies; the religious and cultural roots of Western civilization can be traced back at least to the ancient Greeks, almost 2500 years ago.

What is the relationship between biological and cultural evolution? How much of our behavior is cultural and how much is genetically determined? Some people believe that human nature is almost infinitely malleable;

that biological evolution endowed us with a mind that can be molded into whatever form society wishes. Others, particularly the adherents of the new field of sociobiology, maintain that millions of years of natural selection must have left us with some genetic tendencies to perform certain types of behaviors.

Which position, if either, is correct? It is likely that the truth lies somewhere in the middle ground, that most human behavior is the product of both genetic influences and extensive learning. First, no reasonable person doubts that human behavior is tremendously influenced by culture. We learn at least part of every behavior we perform. Second, humans do display some obviously instinctive behaviors, including suckling by infants, smiling, and flashing the eyebrows in greeting (see Fig. 34-24). These behaviors, of course, are not what people really argue about, although not so very long ago some would have disavowed any genetic contribution even in these cases. Third, in recent years biologists have found that many forms of aberrant behavior are strongly influenced by innate biological factors. For example, early in 1987, molecular biologists localized the gene that predisposes a person to develop manic depression. Schizophrenia is also at least partially hereditary. If abnormal behavior can be influenced by defective genes, it seems likely that normal behavior is influenced by the functioning of normal, nondefective genes. However, few if any behaviors, normal or abnormal, are *completely controlled* by heredity. Most sociobiologists readily agree that virtually all human behaviors can be enhanced, suppressed, or modified by culture. The roles of culture and heredity in human behavior are explored further in Chapter 34.

SUMMARY OF KEY CONCEPTS

Origins

Before life arose, lightning, ultraviolet light, and heat formed organic molecules from water and the components of the primordial Earth's atmosphere. These molecules probably included nucleic acids, amino acids, short proteins, and lipids. By chance, some molecules of RNA may have had enzymatic properties, catalyzing their own assembly from nucleotides in the Earth's waters. These may have been the precursors of life. Protein–lipid microspheres enclosing these RNA molecules may have formed the first cell-like organisms.

The Age of Microbes

The first fossil cells are found in rocks about 3.5 billion years old. These cells were prokaryotes that fed by absorbing organic molecules that had been synthesized abiotically. Since there was no free oxygen in the atmosphere at this time, energy metabolism must have been anaerobic. As the cells multiplied, they depleted the organic molecules in the waters of the Earth. Some cells evolved the ability to synthesize their own food molecules using simple inorganic molecules and the energy of sunlight. These earliest photosynthetic cells were probably ancestors of today's cyanobacteria.

Photosynthesis releases oxygen as a by-product, and by about 2.2 billion years ago significant amounts of free oxygen had accumulated in the atmosphere. Aerobic metabolism, which generates more cellular energy than anaerobic metabolism, probably arose about this time.

Eukaryotic cells evolved about 1.5 billion years ago. The first eukaryotic cells probably arose as symbiotic associations between predatory prokaryotic cells and bacteria. Mitochondria may have evolved from aerobic bacteria engulfed by predatory cells. Similarly, chloroplasts may have evolved from photosynthetic cyanobacteria.

Multicellularity

Multicellular organisms evolved from eukaryotic cells, first appearing about 1 billion years ago. Multicellularity offers several advantages, including increased speed of locomotion and increased size.

Multicellular Life in the Sea

The first multicellular organisms arose in the sea. In plants, increased size due to multicellularity offered some protection from predation. Specialization of cells allowed plants to anchor themselves in the nutrient-rich, well-lit waters of the shore. For animals, multicellularity allowed more efficient predation and more effective escape from predators.

The Invasion of the Land

The first land organisms were probably plants, appearing 500 to 600 million years ago. Although the land required special adaptations for support of the body, reproduction, and the acquisition, distribution, and retention of water, the land also offered abundant sunlight and protection from aquatic herbivores. Around 400 million years ago, arthropods invaded the land. Absence of predators and abundant land plants for food were probably the selective pressures favoring the invasion of the land by animals.

The earliest land vertebrates evolved from lobe-finned fishes, which had leglike fins and a primitive lung. A group of lobefins evolved into the amphibians about 350 million years ago. Reptiles evolved from amphibians, with several further adaptations for land life: internal fertilization, waterproof eggs that could be laid on land, waterproof skin, and better lungs. Around 150 million years ago, birds and mammals evolved independently from separate groups of reptiles. Major advances included a high, constant body temperature and insulation over the body surface.

Human Evolution

One group of mammals evolved into the tree-dwelling primates. Primates show several preadaptations for human evolution: forward-facing eyes for binocular vision, color vision, and grasping hands. Between 20 and 30 million years ago some primates descended from the trees; these were the ancestors of apes and humans. The Australopithecines arose in Africa about 4 million years ago. These hominids walked erect, had much larger brains than those of their forebears, and made primitive tools. One group of Australopithecines evolved into true humans.

GLOSSARY

Cultural evolution: changes in the behavior of a population of animals, especially humans, by learning behaviors acquired by members of previous generations.

Endosymbiosis: the hypothesis that certain organelles, especially chloroplasts and mitochondria, evolved from bacteria captured by ancient predatory prokaryotic cells.

Gradualism: the hypothesis that species arise through slow, gradual changes over hundreds of thousands, even millions, of years.

Microsphere: a small, hollow sphere formed from proteins or proteins complexed with other compounds.

Preadaptation: a feature evolved under one set of environmental conditions that, purely by chance, helps an organism to adapt to new environmental conditions.

Prebiotic evolution: evolution before life existed; especially abiotic synthesis of organic molecules.

Punctuated equilibrium: the hypothesis that the characteristics of species are constant for long periods of time (e.g., millions of years) and that new species arise rapidly over a few thousand years.

Ribozyme: an RNA molecule that can catalyze certain chemical reactions, especially those involved in synthesis and processing of RNA itself.

STUDY QUESTIONS

1. What is the evidence that life might have originated from nonliving matter on the primordial Earth? What kind of evidence would you like to see before you would consider this hypothesis proven?
2. Explain the endosymbiotic hypothesis for the origin of chloroplasts and mitochondria.
3. Name two advantages of multicellularity in plants and animals.
4. What advantages and disadvantages would terrestrial existence have had for the first plants to invade the land? For the first land animals?
5. Outline the general trends in the evolution of vertebrates, from fish to amphibians to reptiles to birds and mammals. Explain how these adaptations increased the fitness of the various groups for life on land.
6. Outline the evolution of humans from early primates. Include in your discussion such features as binocular vision, grasping hands, bipedal locomotion, social living, toolmaking, and brain expansion.
7. What is cultural evolution? Is cultural evolution more or less rapid than biological evolution? Why?
8. Define and distinguish between gradualism and punctuated equilibrium.

SUGGESTED READINGS

Brownlee, S. "The Great Dyings." *Discover,* May 1984. Some of the evidence that extraterrestrial causes may have triggered some massive extinctions during evolutionary history.

Hay, R. L., and Leakey, M.D. "The Fossil Footprints of Laetoli." *Scientific American,* February 1982. The actual footprints of a hominid family were discovered by Hay and Leakey in volcanic ash 3.5 million years old.

Laporte, L. F. *The Fossil Record and Evolution.* San Francisco: W. H. Freeman and Company, Publishers, 1982. A collection of readings from *Scientific American,* covering the entire sweep of evolution from the first cells to the evolution of humanity.

Morell, V. "Announcing the Birth of a Heresy." *Discover,* March 1987. Paleontologists Robert Bakker and Jack Horner speculate that dinosaurs were not the plodding beasts of monster flicks, but warm-blooded, advanced animals that may have even cared for their young.

Pilbeam, D. "The Descent of Hominoids and Hominids." *Scientific American,* March 1984. A refreshingly unbiased account. Pilbeam, as he puts it, "tiptoes past" the often fruitless controversies in prehuman evolution, focussing instead on the actual fossil finds.

Rensberger, B. "Bones of Our Ancestors." *Science 84,* April 1984. Rensberger and photographer Margo Crabtree traveled around the world to photograph prehuman fossils. A magnificent view of our ancestors.

Stebbins, G. L. *Darwin to DNA, Molecules to Humanity.* San Francisco: W. H. Freeman and Company, Publishers, 1982. An engagingly written account of evolution, this book describes both the events in the history of life and the mechanisms of evolution.

Washburn, S. L. "The Evolution of Man." *Scientific American,* September 1978. (Offprint # 1406). Compare this account to Pilbeam's article six years later, and see how fast science changes.

Weaver, K. F. "The Search for Our Ancestors." *National Geographic,* November 1985. The findings, the interpretations, and the personalities behind prehuman paleontology.

16
Taxonomy: Imposing Order on Diversity

The Ancient Origins of Taxonomy
Modern Criteria for Classification

Taxonomy (from the Greek *taxis*, meaning "arrangement") is the science by which organisms are classified and placed into categories based on their evolutionary relationships. These categories form a hierarchy, that is, a series of levels each more inclusive than the last, like a set of nesting boxes. There are seven major categories: *kingdom, phylum* (or *division* for plants and fungi), *class, order, family, genus*, and *species*. Each category from species to kingdom is increasingly more general and includes organisms whose common ancestor was increasingly remote. Examples of classifications of specific organisms may be found in Table 16-1. The scientific name of an organism is actually formed from the two smallest of these categories, the genus and species. The first word of each scientific name is its **genus,** a category that includes a number of very closely related organisms that do not generally interbreed. The second name assigns the organism to a **species,** a category limited to naturally interbreeding living things. Thus the genus *Canis* (Fig. 16-1) includes coyotes, wolves, and domestic dogs, very similar animals that do not normally interbreed. Each is assigned to a different species: the coyote, *Canis latrans*, the wolf, *Canis lupus*, and the dog, *Canis familiaris*. Scientific names are always underlined or italicized, and the genus is always capitalized. These names are recognized by biologists worldwide, transcending language barriers and allowing very precise communication.

Table 16-1 **Classification Reflects the Degree of Relatedness of Organisms**

	Human	Chimpanzee	Wolf	Fruit Fly
Kingdom	**Animalia**	**Animalia**	**Animalia**	**Animalia**
Phylum	**Chordata**	**Chordata**	**Chordata**	Arthropoda
Class	**Mammalia**	**Mammalia**	**Mammalia**	Insecta
Order	**Primates**	**Primates**	Carnivora	Diptera
Family	Hominidae	Pongidae	Canidae	Drosophilidae
Genus	*Homo*	*Pan*	*Canis*	*Drosophila*
Species	*sapiens*	*troglodites*	*lupus*	*melanogaster*

Figure 16-1 Representatives of the genus Canis. *Similarities among* **(a)** *the dog* Canis domesticus, **(b)** *the coyote* Canis latrans, *and* **(c)** *the wolf* Canis lupus.

THE ANCIENT ORIGINS OF TAXONOMY

Aristotle (384–322 B.C.) was among the first to attempt to formulate a logical, standardized language for naming living things. Using characteristics such as structural complexity, behavior, and degree of development at birth, he classified about 500 different organisms into 11 categories. Aristotle placed organisms into a hierarchy of categories each more inclusive than the one before it, a concept that survives today. Building on this foundation over 2000 years later, the Swedish naturalist Carolus Linnaeus (1707–1778) laid the groundwork for the classification system in use today. He placed each organism into a series of hierarchically arranged categories based on its resemblance to other life forms, and also introduced the scientific name based on genus and species. The publication of Charles Darwin's *Origin of Species* nearly 100 years later added a new significance to these categories—they were seen to reflect evolutionary relatedness of organisms. In Table 16-1, the boldface categories are common to more than one of the organisms classified. The more categories two organisms share, the closer their evolutionary relationship.

MODERN CRITERIA FOR CLASSIFICATION

Four major criteria are used to place living things in taxonomic groupings. These are: *anatomy, developmental stages, biochemical similarities, and behavior*. The most important and useful of these is anatomy. In addition to obvious similarities in external body structure such as those seen in members of the genus *Canis*, taxonomists look carefully at details such as skeletons and tooth structure. The presence of homologous structures (such as the finger bones of dolphins, bats, seals, and humans) provides evidence of a distant common ancestor. To distinguish between closely related species, taxonomists may use microscopes to discern finer details—the number and shape of the "teeth" on the tongue-like radula of a mollusc, the spines on a marine worm, or the projections on a pollen grain.

Clues to common ancestry may also be provided by the developmental stages that animals undergo on their way to adulthood. For example, the tunicate or "sea squirt" (see Fig. E18.1) spends its adult life permanently attached to rocks on the ocean floor, resembling an undersea vase. But its active, free-swimming larva has a nerve cord, tail, and gills, placing it with the vertebrates in the phylum Chordata.

Modern taxonomists also have at their disposal sophisticated biochemical techniques that may be used to determine the amino acid sequences of protein molecules such as hemoglobin, certain hormones, and various enzymes. Since proteins are coded by DNA, the greater the similarity in protein composition between two different species, the closer their evolutionary relationship is likely to be.

A fourth criterion is behavior. Some behaviors can prevent mating between very closely related animals. For example, dogs and coyotes are interfertile but rarely interbreed for behavioral reasons, and are therefore considered separate species. Similarly, some closely related tree frogs remain separate because the males produce distinctive mating calls that only attract females of their own species.

Taxonomy is a challenging and frequently frustrating endeavor. For example, most unicellular organisms reproduce asexually, making their classification difficult, since the criterion of interbreeding cannot be applied to distinguish species. In fact, because the taxonomy of microorganisms is so controversial, this book presents only general descriptive groupings. Not surprisingly, categories are debated and occasionally revised as taxonomists learn more about evolutionary relationships. For example, not long ago all organisms were placed in one of two kingdoms: Plantae or Animalia, even though many fit neither very well. Today, a five-kingdom classification scheme is more widely used and will be followed in this book. The addition of kingdoms Protista, Fungi, and Monera to the traditional Plantae and Animalia create categories for organisms whose cellular structures are fundamentally distinct (Fig. 16-2; Table 16-2). Although the precise evolutionary relationships of many organisms continue to elude us, taxonomy is enormously helpful in ordering our thoughts and investigations into the diversity of life on Earth.

Table 16-2 **Some Characteristics of the Five Kingdoms**

Kingdom	Cell Type	Cell Number	Major Mode of Nutrition
Monera	Prokaryotic	Unicellular	Absorb or photosynthesize
Protista	Eukaryotic	Unicellular	Absorb, ingest, or photosynthesize
Plantae	Eukaryotic	Multicellular	Photosynthesize
Fungi	Eukaryotic	Most multicellular	Absorb
Animalia	Eukaryotic	Multicellular	Ingest

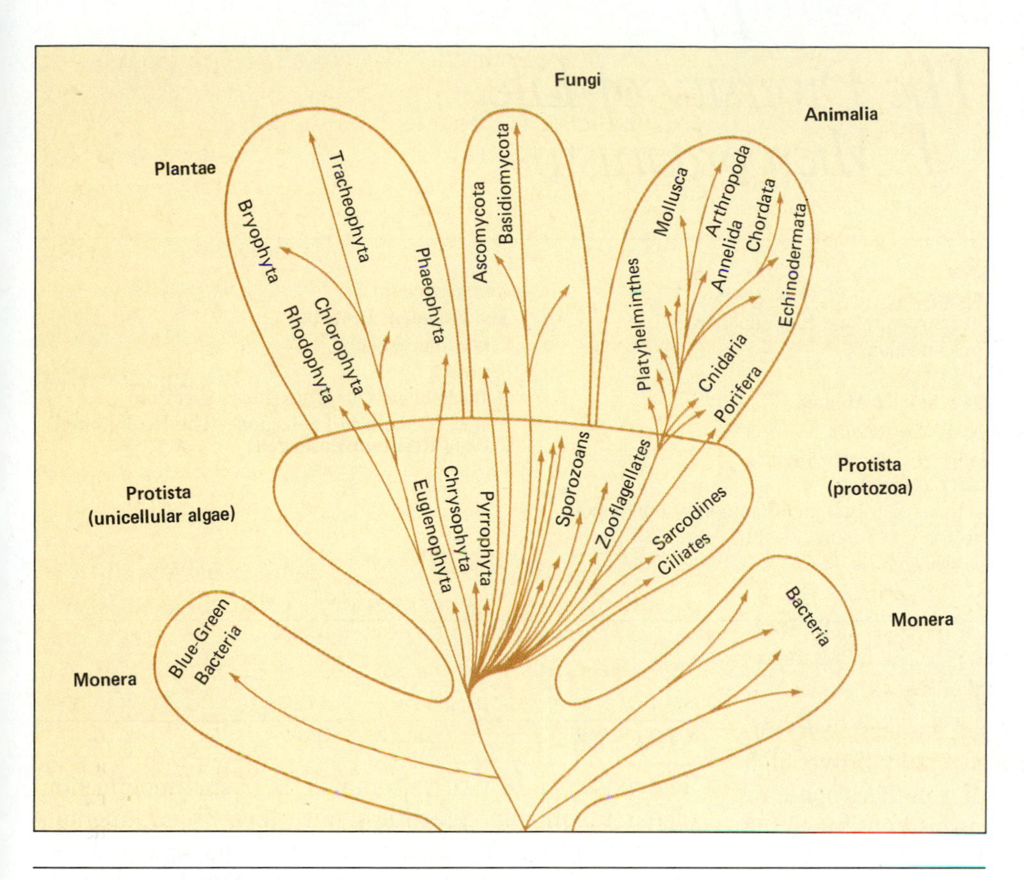

Figure 16-2 The five-kingdom classification of organisms, showing probable evolutionary relationships among the kingdoms and some of their major divisions and phyla.

SUMMARY OF KEY CONCEPTS

Taxonomy is the science by which organisms are classified and placed into hierarchical categories that reflect their evolutionary relationships. The seven major categories in order of increasing inclusiveness are: species, genus, family, order, class, phylum (or division for plants and fungi), and kingdom. The scientific name of an organism is its genus and species.

The hierarchical concept was first used by Aristotle, but Linnaeus in the mid-eighteenth century laid the foundation for modern taxonomy. Evolutionary theory then provided an explanation for the observed similarities and differences between organisms, and taxonomy now is seen to reflect relatedness. Today, taxonomists use features such as anatomy, developmental stages, biochemical similarities, and behavior to categorize organisms. Most biologists recognize five distinct kingdoms. Taxonomy is difficult and controversial where evolutionary relationships are uncertain, but it is essential for precise communication and contributes immensely to our understanding of species diversity and its origins.

GLOSSARY

Genus (jē′-nis) a taxonomic category consisting of very closely related species.

Species (spē′-cēs) a group of organisms that interbreed under natural conditions.

Taxonomy (tax-on′-uh-mē) the science by which organisms are classified into hierarchically arranged categories that reflect their evolutionary relationships.

STUDY QUESTIONS

1. You are in a boat with a companion after a major flood, and you encounter the topmost branch tips of a submerged tree. Your companion challenges you to sketch the branches below the surface based on the location of the exposed tips. How is this situation analogous to the dilemma facing taxonomists? What relative advantages do the taxonomists have?
2. What contributions did Aristotle, Linnaeus, and Darwin each make to modern taxonomy?
3. What features would you investigate to determine whether a dolphin is more closely related to a fish or a bear?

SUGGESTED READING

Margulis, Lynn and Schwartz, Karlene *Five Kingdoms.* New York: W. H. Freeman and Co., 1987.

17
The Diversity of Life. I. Microorganisms

If humans were microbes, the entire world population could thrive in a spadeful of garden soil. The water of a puddle, pond, or sea also teems with unicellular organisms invisible to the naked eye. We are surrounded, coated, and inhabited by life forms that we become aware of only if they make us ill. Although minuscule, microorganisms are of immense importance, and not only as agents of disease. Protists are responsible for the majority of photosynthetic activity on Earth, replenishing oxygen and capturing the sun's energy in food. Of all living things, only certain bacteria can capture atmospheric nitrogen and convert it into a nutrient used by plants. Were microorganisms to disappear, cows, sheep, and other ruminants might starve, unable to break down the energy-rich cellulose in their diet. Without bacteria to decompose them, bodies of animals and plants would accumulate, locking up their valuable nutrients and disrupting the recycling process on which life relies. Life as we know it would cease.

In this chapter we provide an introduction to the unseen world of microorganisms: bacteria, protists, and viruses. Members of the kingdom Monera (the bacteria) are single, **prokaryotic** cells that have changed very little from their fossil ancestors discovered in rocks roughly $3\frac{1}{2}$ billion years old. As described in Chapter 5, they lack organelles such as the nucleus, chloroplasts, and mitochondria. In contrast, the unicellular members of the kingdom Protista are masterpieces of condensed complexity, with organelles serving the functions played by entire organs in animals and plants. Protists (and all more complex organisms) are composed of **eukaryotic** cells (Fig. 5-1), the first of which appeared roughly $2\frac{1}{2}$ billion years after the Monera. This chapter also introduces the viruses, puzzling parasitic particles that hover outside the normal boundaries of life and defy classification.

VIRUSES

The existence of viruses reminds us of the imperfection of our taxonomic categories, since there is no kingdom into which they fit comfortably. Viruses possess no membranes of their own, no ribosomes on which to make proteins, no cytoplasm, and no source of energy. *They cannot move or grow, and they can reproduce only inside a host cell.* The utter simplicity of viruses makes it impossible to call them cells, and indeed, seems to place them outside the realm of living things. Of course, the viruses themselves are unaffected by our confusion regarding them, and continue their successful existence as the ultimate intracellular parasite.

Viral Structure and Reproduction

A virus particle is so small (about 0.1 micrometer) that visualizing it requires the enormous magnification of the electron microscope. Viruses consist of two major parts, a coat of protein surrounding a molecule of hereditary material, either DNA or RNA. An envelope formed from the membrane of the host cell may surround the protein coat (Fig. 17-1a). Even if placed in a rich broth of nutrients at optimal temperature, viruses remain inert, unable to grow or divide, since they lack the complex cellular organization that these activities require. The protein coat, however, is specialized to allow viruses to penetrate the cells of a specific host, where they assume a rather insidious semblance of life. After entering the cells, the viral genetic material takes command. The host cells are forced to read the viral genes, and to use these instructions to produce the components of new viruses. The pieces are rapidly assembled (Fig. 17-1b), and an army of new viruses bursts forth to invade and conquer

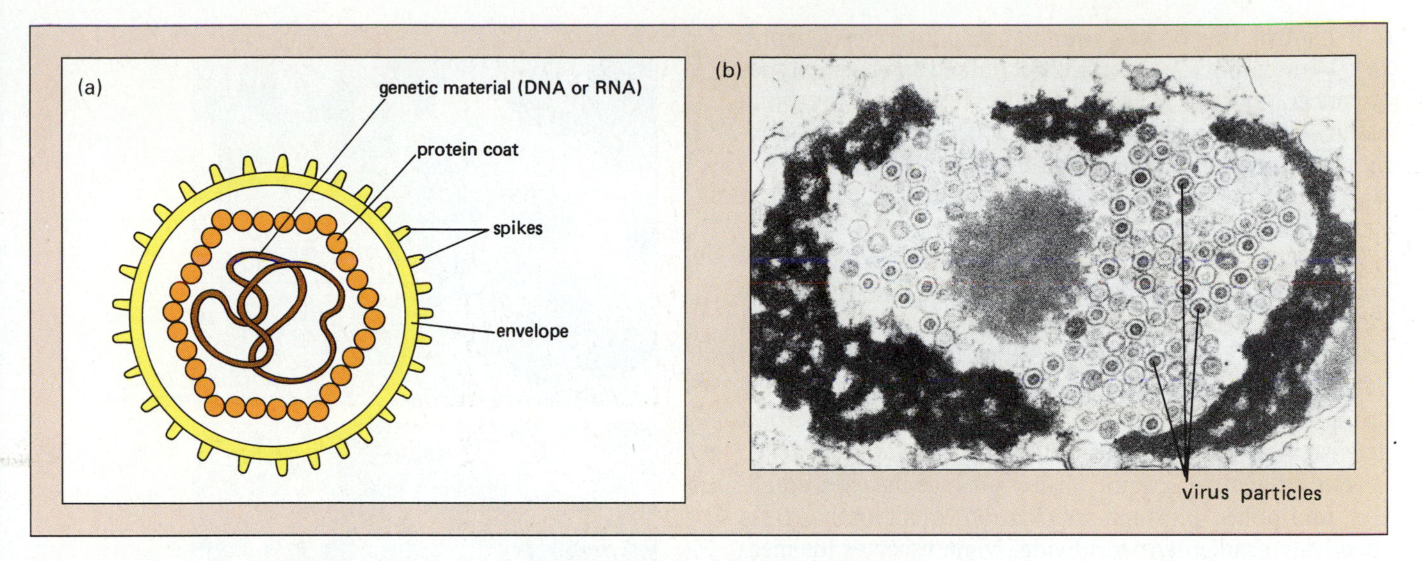

Figure 17-1 **(a)** *Cross section of a virus. Inside is genetic material, surrounded by a protein coat. Some viruses, including those causing herpes, rabies, and AIDS, have an outer envelope that may be formed from the membrane of the host cell. Spikes made of protein and carbohydrate may project from the envelope, and some viruses use these to attach to their host cell.* **(b)** *In this electron micrograph, herpes viruses are seen packed into an infected cell.*

neighboring cells (this cycle is diagrammed for the bacterial virus in Fig. 10-5).

Viral Infections

Each type of virus is specialized to attack a specific host cell (Fig. 17-2a), and probably no organism is immune to all viruses. Even bacteria fall victim to viral invaders called **bacteriophages** (Fig. 17-2b). Within a particular organism, viruses specialize on particular cell types. Those responsible for the common cold attack the membranes of the respiratory tract, those causing measles infect the skin, and the rabies virus attacks nerve cells. One type of herpes virus specializes in the mucous membranes of the mouth and lips, causing cold sores, while a second type (transmitted through sexual contact) produces similar sores on or near the genitals. Herpes viruses, unfortunately, take up permanent residence in the body, erupting periodically (often during times of stress) as infectious sores. The devastating disease AIDS (acquired immune deficiency syndrome), which cripples the body's immune system, is caused by a virus that attacks white blood cells

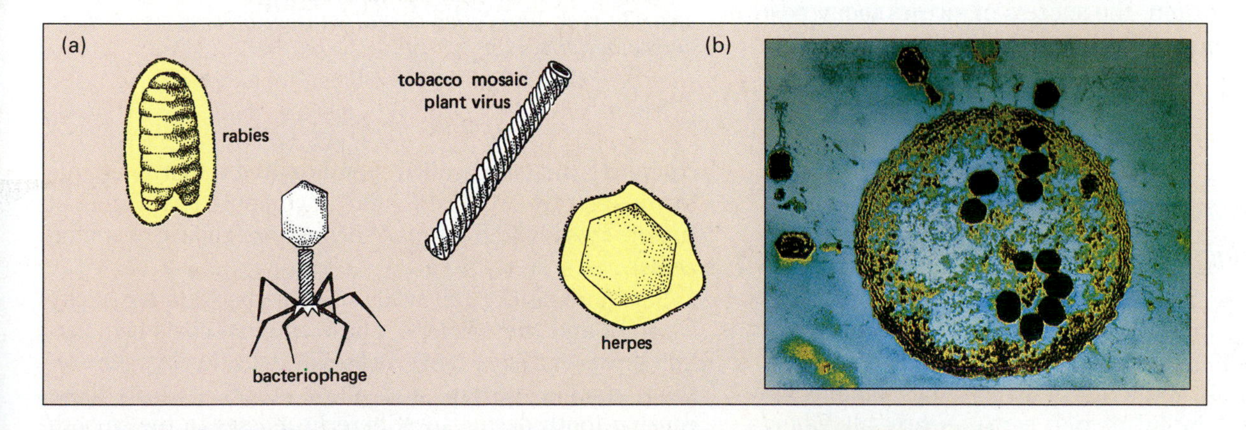

Figure 17-2 **(a)** *Viruses come in a variety of shapes, determined by their protein coats. The rabies and herpes viruses are surrounded by an envelope derived from membranes of the host cell.* **(b)** *In this transmission electron micrograph, bacteriophage viruses are seen attacking a bacterium. They have injected their genetic material inside, leaving their protein coats clinging to the bacterial cell wall. Black objects inside the bacterium are newly forming viruses.*

that control the body's immune response (see Chapter 27). Viruses have been definitely linked to specific types of cancer, such as T-cell leukemia, a cancer of the white blood cells. The papilloma virus, long known for its ability to cause genital warts, has recently been identified in 90 percent of cervical cancers sampled, suggesting a causal relationship.

Since viruses are intracellular parasites utilizing the cellular machinery of their host, the illnesses they cause are difficult to treat because antiviral agents may destroy host cells as well. The antibiotics so effective against bacterial infections are useless against viruses, although some promising antiviral drugs are being developed. The best defense is prevention, and to this end vaccines have been developed against specific viruses such as measles, smallpox, and polio. Through vaccination, smallpox has been effectively eradicated worldwide, while vaccines for measles and mumps have made these childhood diseases preventable. The common cold and flu have proven far more intractable. The viruses causing these infections mutate so frequently that new forms are produced as fast as we develop immunity to the old ones. Thus each cold or flu that you contract during your life is a slightly different illness, caused by a slightly different virus. Rapid mutation of the AIDS virus is also hindering efforts to produce a vaccine.

Viral Origins

The origin of viruses is obscure. It is unlikely that viruses are the forerunners of life, since they cannot reproduce without infecting more complex cells. Some scientists believe that viruses originated from simple parasitic cells which evolved such complete dependence on their hosts that they lost the ability to perform the basic processes of life. Or perhaps viruses originated as loose fragments of genetic material that took up an independent existence. Whatever their origin, the success of viruses today poses a continuing challenge to living things.

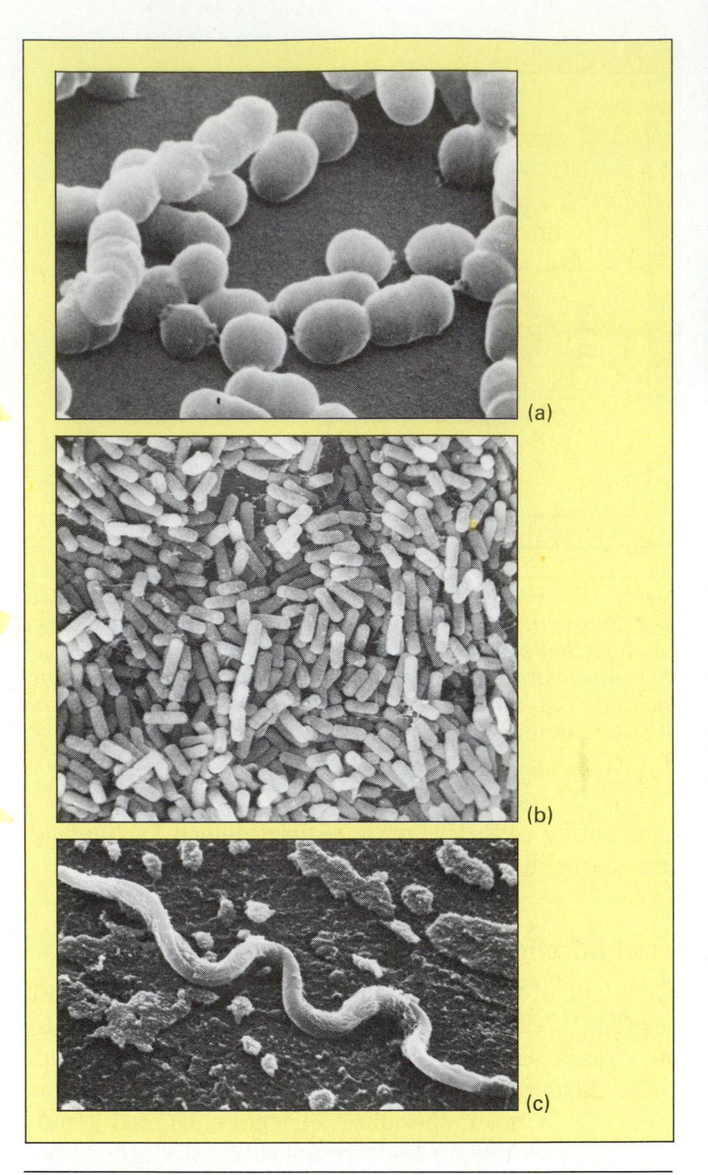

Figure 17-3 Three common bacterial forms as seen under the scanning electron microscope: **(a)** *spherical bacteria, also called cocci;* **(b)** *rods, also called bacilli; and* **(c)** *corkscrew or spirilla-shaped bacteria.*

THE KINGDOM MONERA

Bacterial Structure

The prokaryotic bacteria, often only a few micrometers in diameter, are much larger than viruses but smaller than eukaryotic cells. Although invisible to the naked eye, bacteria are the most abundant of living things; billions may be found in a handful of rich earth. Nearly all Monera are encased in a porous but rigid cell wall that protects them from osmotic rupture in watery environments and gives different types of bacteria characteristic shapes. The most common shapes are rodlike **bacilli,** spheres called **cocci,** and corkscrew-like spirals (Fig. 17-3). The cell wall contains a material called **peptidoglycan** that is unique to prokaryotic organisms and is composed of chains of sugars cross-linked by peptides (short chains of amino acids).

Surrounding the cell walls of some bacteria are sticky capsules and slime layers. These external coats help certain disease-causing bacteria escape detection by their victim's immune system, and allow others, such as those causing tooth decay, to adhere in masses to the smooth surface of a tooth. Some bacteria are equipped with flagella (simpler in structure than eukaryotic flagella). These structures allow bacteria to travel through liquids, thus dispersing into new habitats, moving toward nutrients, or escaping unfavorable conditions.

Bacterial Reproduction

Bacterial reproduction is by simple cell division, which produces genetically identical copies of the original cell (Fig. 17-4a). Under ideal conditions, a bacterium may divide about once every 20 minutes, potentially giving rise to sextillions (1×10^{21}) of offspring in a single day. This rapid reproduction allows bacteria to exploit temporary habitats such as a mud puddle or warm potato salad at a summer picnic. Since mutations, the source of genetic variability, can occur as a result of mistakes in DNA replication during cell division (see Chapter 10), the rapid reproductive rate of prokaryotes provides ample opportunity for new forms to arise and allows mutations that enhance survival to spread quickly (see the essay, "Unnatural Selection—The Evolution of Drug-Resistant Pathogens"). Additional variability can result from **bacterial conjugation,** the bacterial version of sex, in which genetic material is exchanged between cells (Fig. 17-4b).

When conditions become inhospitable, many bacteria form protective resting structures called **spores** (Fig. 17-5). The spore packages a chromosome and the minimum material to sustain life within a thick protective coating of cell-wall material and protein. Metabolic activity ceases. Spores can survive extremely unfavorable conditions. Some can withstand boiling for an hour or more, while others, still living, have been found in the intestines of mummies 2000 years old. Spores may be carried for long distances in air or water, and germinate rapidly when they encounter favorable conditions.

Bacterial Habitats

As striking as their sheer numbers is the diversity of bacterial habitats. Bacteria have been isolated from snowy mountaintops and ocean depths of 4400 meters. Some revel in hot springs found in places such as Yellowstone

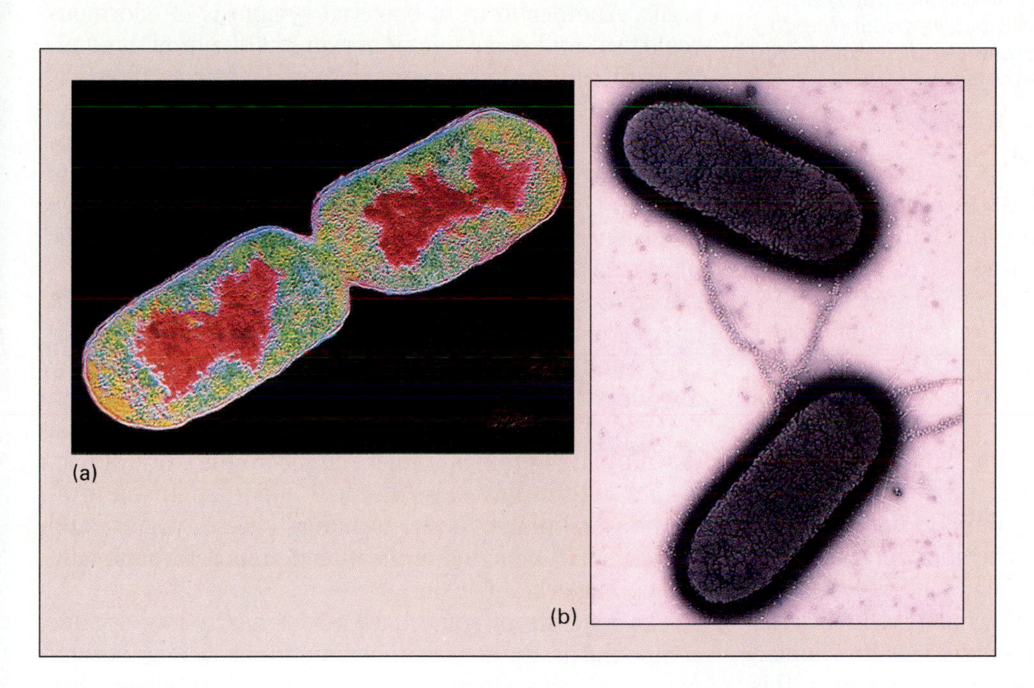

Figure 17-4 **(a)** *Reproduction in prokaryotes is by cell division, as illustrated by this color-enhanced electron micrograph of a dividing* Escherichia coli, *found abundantly in the human intestine. Red areas are genetic material.* **(b)** *Conjugation, the exchange of genetic material, occurs across a thin strand of cytoplasm which is formed temporarily between mating bacterial cells.*

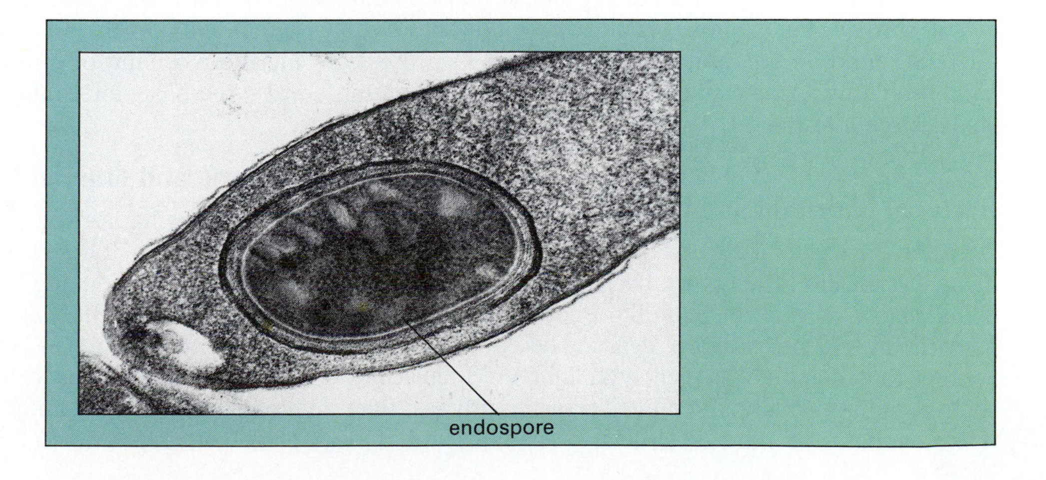

Figure 17-5 A resistant spore, also called an endospore that has formed inside the bacterium Clostridium.

Figure 17-6 Hot springs harbor heat- and mineral-tolerant bacteria. Some cyanobacteria can tolerate temperatures up to 165°C. Several species of cyanobacteria paint these hot springs in Peru with vivid colors. The bacterial pigments aid in photosynthesis.

National Park and Peru (Fig. 17-6). But the near-boiling temperatures of hot springs would be chilly for some of their relatives, who thrive near deep ocean vents where cracks in the Earth's crust spew water heated to 350°C (662°F). Only the tremendous pressure at the 2650-meter (8745-foot) depth keeps the water from exploding into superheated steam. Placed in a broth in the equivalent of a pressure cooker and heated to 250°C (482°F), these bacteria will increase 1000-fold in about 6 hours. If "chilled" to temperatures just below boiling, they are completely unable to grow. Bacteria are also found in the Dead Sea, where the tremendous salt concentration precludes all other life. They are found floating high in the atmosphere and in the fuel tanks of jetliners. Of course, rich bacterial communities are also found in and on the healthy human body. However, no single bacterium is as versatile as this may imply. Indeed, bacteria are specialists; those found in hot springs, for example, could thrive nowhere else. Bacteria found on the human body are often specialized, inhabiting only a single site, such as the skin, the mouth, or the large intestine.

Bacterial Nutrition and Community Interactions

Bacterial invasion of diverse habitats is aided by their dietary versatility. Blue-green bacteria, discussed in more detail below, engage in plantlike photosynthesis. Other bacteria are **chemosynthetic,** deriving energy through reactions that combine oxygen with inorganic molecules such as sulfur, ammonia, or nitrite. In the process, they release sulfates or nitrates, crucial plant nutrients, into the soil. Many bacteria, called **anaerobes,** are not dependent on oxygen to extract energy. Some, such as the bacterium causing tetanus, are poisoned by oxygen. Others are opportunists, engaging in fermentation when oxygen is lacking and switching to cellular respiration (a more efficient process) when oxygen becomes available. Anaerobes such as the sulfur bacteria obtain energy from a unique type of bacterial photosynthesis. They use hydrogen sulfide (H_2S) instead of water (H_2O) in photosynthesis, releasing sulfur instead of oxygen.

Certain bacteria have the unusual ability to break down cellulose, the principal component of plant cell walls. Some of these have entered into a **symbiotic** (literally, "living together") relationship with cows, sheep, and goats, living in their digestive tracts and helping extract otherwise unavailable nutrients from plant fodder. You also host symbiotic bacteria that inhabit your intestines. These feed on undigested food and synthesize nutrients such as vitamin K and vitamin B_{12} which your body absorbs. Another form of bacterial symbiosis of enormous ecological and economic importance is the growth of **nitrogen-fixing** bacteria in specialized nodules on the roots of certain plants (**legumes,** which include alfalfa, soybeans, lupines, and clover; Fig. 17-7). These bacteria capture nitrogen gas (N_2, which the plant cannot use directly) from air trapped in the soil and combine it with hydrogen to produce ammonium (NH_4^+), a form usable by the plant.

Most bacteria obtain energy by breaking down complex organic (carbon-containing) molecules, and the range of compounds attacked by bacteria is staggering. Nearly anything that humans can synthesize, some bacteria can destroy. The term "biodegradable" (meaning "broken down by living things") refers largely to the work of bacteria. Bacteria have also become important in the production of human foods, including cheese, yogurt, and sauerkraut. The aging of meat tenderizes it through controlled bacterial digestion.

The appetite of some bacteria for nearly any organic compound is the key to their important role as decomposers in ecosystems. While feeding themselves, they break down the waste products and dead bodies of more complex life forms, freeing nutrients for reuse and allowing the recycling of nutrients (see Chapter 37) that provides the basis for continued life on Earth.

Bacteria and Human Health

The feeding habits of certain bacteria threaten our health and well-being. These bacteria, called **pathogens** (meaning "disease-producing"), synthesize toxic substances in the human body that cause disease symptoms. An allergic reaction to substances released by the bacterium *Streptococcus pneumoniae* results in the symptoms of pneumonia, in which the lungs become clogged with fluid. The plague or "black death," which killed 100 million people

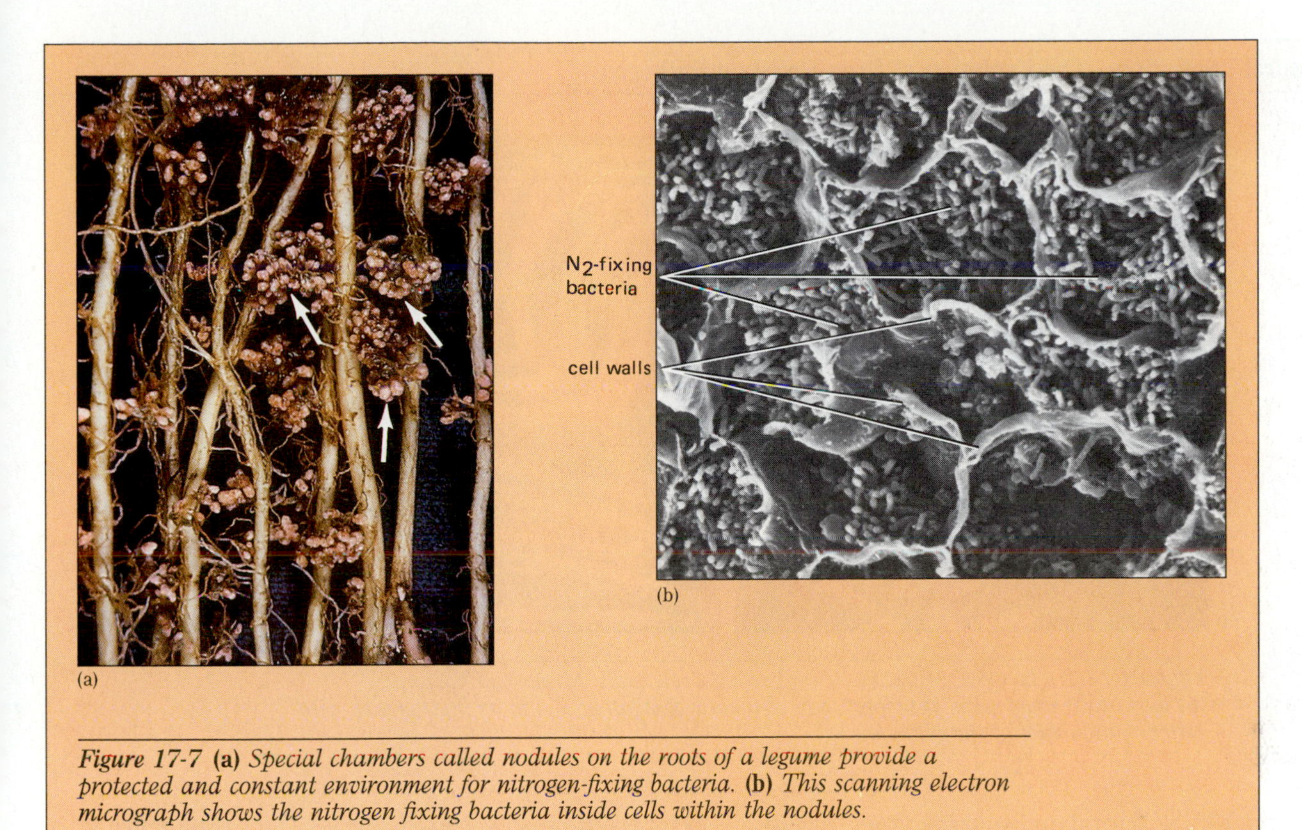

Figure 17-7 **(a)** *Special chambers called nodules on the roots of a legume provide a protected and constant environment for nitrogen-fixing bacteria.* **(b)** *This scanning electron micrograph shows the nitrogen fixing bacteria inside cells within the nodules.*

during the fourteenth century, is caused by highly infectious bacteria spread by fleas carried by infected rats. Tuberculosis and leprosy are also bacterial diseases.

Some bacteria produce deadly toxins that enter the bloodstream and attack the nervous system. One of these causes tetanus, and another botulism, a lethal form of food poisoning. These related bacteria are anaerobes that survive as spores until introduced into a favorable environment. A deep puncture wound protects tetanus bacteria from contact with oxygen, allowing them to multiply and release their paralyzing poison into the bloodstream. A sealed container of canned food that has been improperly sterilized provides a haven for botulism bacteria. These produce a toxin so deadly that a single gram could kill 15 million people.

Two bacterial diseases, gonorrhea and syphilis, have reached epidemic proportions in modern society. Both are transmitted through direct sexual contact, since the bacteria responsible cannot survive for any appreciable time outside the human body. Gonorrhea is currently among the most common infectious diseases in the United States. It enters the sexual organs and progressively spreads through the reproductive and urinary systems. Gonorrhea can scar the tubes that transport sperm in the male (the epididymis and vas deferens) and eggs in the female (the oviduct), resulting in sterility. Penicillin treatment has been very successful against gonorrhea, except for the recent and ominous emergence of penicillin-resistant strains (see the essay, " 'Unnatural' Selection—The Evolution of Drug-Resistant Pathogens"). Syphilis is a less common but far more insidious disease. The spiral-shaped syphilis bacteria enter through the sexual organs, then spread through the circulatory system to all parts of the body. The initial symptoms are sores at the site of entry, usually on or near the genitals. Untreated, these heal within 6 weeks, but the bacteria continue to reside in the body without symptoms for many years, finally emerging and attacking the blood vessels and nervous system. This can lead to heart disease, insanity, and death. Fortunately, syphilis still responds well to penicillin treatment during its early stages.

These descriptions of bacterial assaults on the human body should not lead to an irrational hatred of all bacteria. As Lewis Thomas so aptly put it: "Pathogenicity is, in a sense, a highly skilled trade, and only a tiny minority of all the numberless tons of microbes on the earth has ever been involved in it; most bacteria are busy with their own business, browsing and recycling the rest of life."

Cyanobacteria

Like other bacteria, blue-green or cyanobacteria (*cyan* is Greek for "dark blue") are widespread, making their

"Unnatural" Selection—The Evolution of Drug-Resistant Pathogens

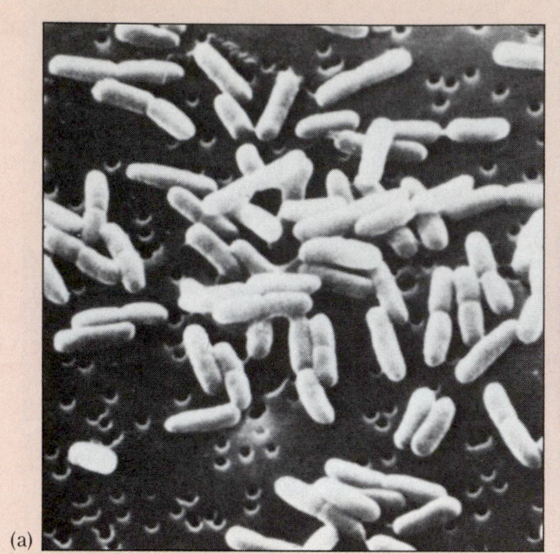
(a)

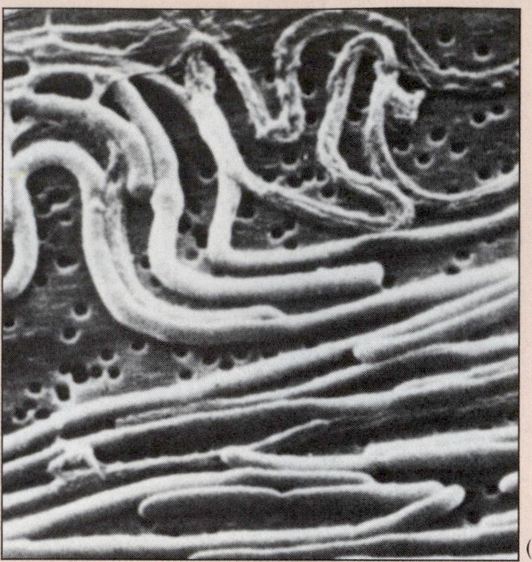
(b)

Figure E17-1 The effect of antibiotic on bacterial cell walls: **(a)** *normal bacteria;* **(b)** *bacteria treated with antibiotic show dramatic elongation due to its interference with proper production of the cell wall.*

In the early 1950s, several antibacterial drugs, including penicillin, streptomycin, and tetracycline, became available to treat bacterial infections. Penicillin and some other antibiotics interfere with synthesis of the bacterial cell wall, causing the bacterium to swell and disintegrate (Fig. E17-1). The discovery of these "wonder drugs" appeared to be an important step toward a disease-free existence. Penicillin was added to toothpaste, mouthwash, and chewing gum, and was also used indiscriminately to treat mild infections of all types. However, it soon became apparent that there were strings attached to the use of antibiotics. Physicians found that certain bacteria, such as *Staphylococcus*, which can cause food poisoning, blood poisoning, and toxic shock syndrome, became increasingly difficult to kill with these drugs—the pathogens had developed antibiotic resistance.

Humans had unwittingly introduced a strong agent of natural selection into the microbial world. As a result of overuse of antibiotics, a variety of bacteria such as *Staphylococcus*, which exist normally and harmlessly on human skin and in the nose and throat, were bathed frequently in weak antibiotic solutions. Occasional mutant bacteria arise that are resistant to the drugs; these survived and flourished, passing on resistance to their descendants. When these bacteria invade the body and cause disease, antibiotics are useless. In the late 1950s, the dangers of drug-resistant bacteria became more evident when researchers discovered that the resistant bacteria could transfer their genes for drug resistance to other bacteria, even to members of other bacterial species. This dramatically shortens the time necessary for drug resistance to spread within and between bacterial populations.

Unfortunately, recognition of the problem has spread more slowly than the resistant bacterial strains. Antibiotics continue to be prescribed inappropriately; for example, for colds and sore throats, even though these infections are usually caused by viruses which are unaffected by antibiotics. Tetracycline, another antibiotic, may be prescribed for teenage acne. Excessive use of antibiotics unnecessarily increases the exposure of bacteria to these powerful selective agents and encourages the spread of resistant strains.

In some countries, prescriptions for such drugs are not even required. One ominous outcome is the emergence of a penicillin-resistant strain of gonorrhea as a result of regular use of the drug as a preventive measure by prostitutes in southeast Asia. This resistant bacteria has now reached the United States and is spreading. The United States itself is far from blameless in hastening the selection of "super-germs." Farmers regularly add both penicillin and tetracycline to animal feeds as growth promoters, despite years of protest by scientists and attempts by the FDA to ban this practice. A strain of *Salmonella*, a bacterium responsible for a virulent form of food poisoning, has now developed resistance to the antibiotics fed to meat animals. These bacteria contaminate raw meat and are believed to have caused an outbreak of severe *Salmonella* poisoning in the midwest in 1983.

Stuart Levy, an investigator from Tufts University

"Unnatural" Selection—The Evolution of Drug-Resistant Pathogens

in Boston, has been investigating the spread of antibiotic resistance. He has found that a typical salad contains about 100 million bacterial cells, including a few strains that can grow in the intestine. The percentage of these that are antibiotic resistant ranges from 40 to 100 pecent, and some have developed resistance to as many as eight different antibiotics. Researchers are working to develop new antibiotics and other drugs to substitute for those whose effectiveness has been diminished, but drug development and testing take many years, whereas resistant bacterial strains spread rapidly. Moreover, not all the drugs are as free of side effects as those they replace. Clearly, we must restrict our use of antibiotics to situations where they are urgently required, rather than relying on a steady flow of new ones. Awareness and restraint may yet enable our children to benefit from the same "miracle drugs" that protected our parents.

home on snowfields, in hot springs that may reach 85°C (186°F), in oceans, lakes, ponds, and moist soils. Since most are aerobic and all are photosynthetic, they can exist only where light and oxygen are available. Like green plants, cyanobacteria possess chlorophyll and produce oxygen as a by-product of photosynthesis. Since the Monera lack organelles, including chloroplasts, chlorophyll is located on special membranes inside the cell (Fig. 17-8a). In addition to trapping solar energy, most cyanobacteria can acquire nitrogen from the atmosphere, making them extremely self-sufficient nutritionally. Some cyanobacteria form chains of cells that are unique among prokaryotes because they have a rudimentary division of labor. In these filamentous colonies (Fig. 17-8b), a few cells capture atmospheric nitrogen while the rest photosynthesize. When new islands are formed from volcanic eruptions, cyanobacteria are among the first colonizers of the bare rock, their activities preparing the way for more complex and less independent life forms.

Archaebacteria

Recently, a unique group of bacteria, the archaebacteria, have come under close scrutiny. Archaebacteria include

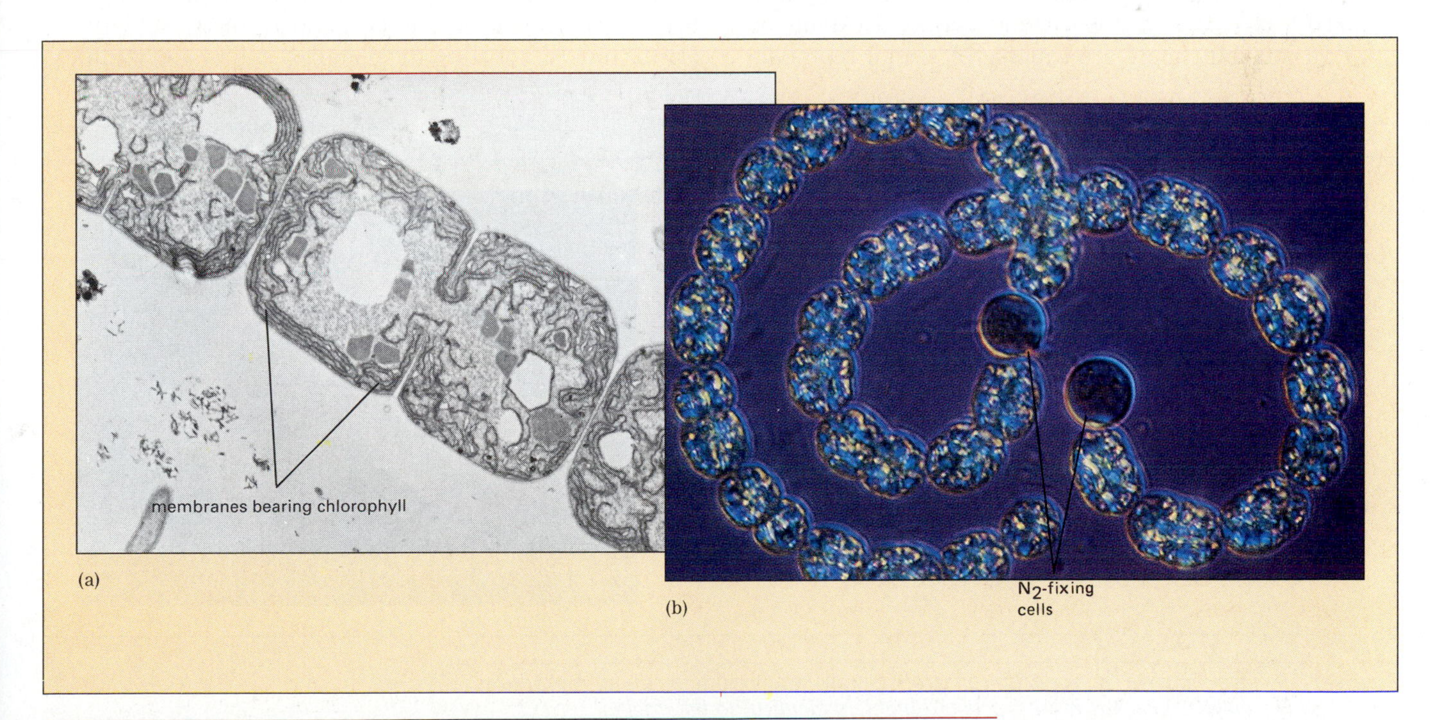

Figure 17-8 **(a)** *Electron micrograph of a section through a cyanobacterial filament (genus* Anabaena*). Chlorophyll is located on the membranes visible within the cells.* **(b)** *Simple division of labor, rare among prokaryotic cells, is seen in this filamentous cyanobacterium (genus* Nostoc*). The larger cells are specialized for nitrogen fixation.*

methanogens, anaerobic bacteria that convert carbon dioxide to methane, sometimes called "swamp gas." These are found in such diverse habitats as swamps, sewage-treatment plants, hot springs, and the stomachs of cows. Others include extreme **halophiles,** bacteria that thrive in concentrated salt solutions such as the Dead Sea, and **thermacidophiles,** which, as their name implies, thrive in hot, acidic environments such as hot sulfur springs.

Experimental findings reveal such dramatic differences between the archaebacteria and other Monera that some scientists believe they deserve the status of a separate kingdom. Although this proposal is hotly disputed, there is no question that the archaebacteria are different. The lipids of their cell membranes differ considerably from those of both eukaryotic and other prokaryotic cells, as does the composition of their cell walls, and the sequence of subunits in their ribosomal RNA. The intriguing suggestion has been raised that this life form is a modern descendant of the type of organism that, billions of years ago, gave rise to the ancestor of both eukaryotic and prokaryotic cells.

THE KINGDOM PROTISTA

Since Anton van Leeuwenhoek first observed protists through his simple homemade microscope in 1674, at least 30,000 different species have been described. Protists have one thing in common: *each consists of a single eukaryotic cell.* Most protists can reproduce asexually by cell division, but many are capable of sexual reproduction, or conjugation, as well (Fig. 17-9). All three major modes of nutrition are represented in this group. The unicellular algae trap solar energy through photosynthesis, predatory protists ingest their food, while parasitic forms, some flagellates, and the versatile euglenoids (see below) can absorb nutrients from their surroundings.

Among the protists, both plant- and animal-like forms are richly represented, and indeed, were once classified as either plants or animals. To the confusion of taxonomists, many protists (such as the euglenoids) fit equally well into either category. *Euglena* (see Fig. 17-12), for example, has a photoreceptor and can swim toward a stimulus, features commonly associated with animals, but it uses these abilities to seek light levels appropriate for photosynthesis. Although placement of the protists in their own kingdom has simplified their classification, confusion persists. Some unicellular forms are often classified with multicellular algae, while other taxonomic schemes place multicellular algae among the protists. This book places some unicellular green algae, which might well be considered protists, with the green algae in the plant kingdom (see Chapter 19). Here we discuss members of the kingdom Protista in two separate categories: the plantlike unicellular algae and the animal-like protozoa.

Unicellular Algae

Often called **phytoplankton** (literally, "floating plants"), these photosynthetic protists are widely distributed in oceans and lakes. Although microscopic in size, their importance is immense. Marine phytoplankton account for nearly 70 percent of all the photosynthetic activity on Earth, thus supporting the complex web of aquatic life.

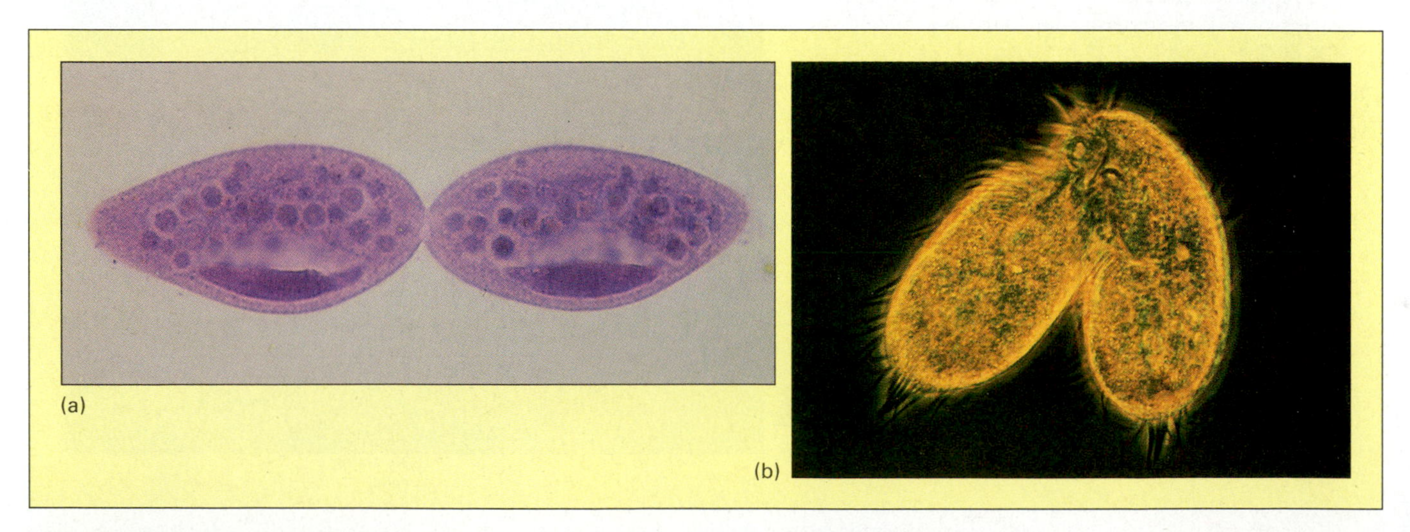

Figure 17-9 Two modes of reproduction in protists. **(a)** Paramecium, *a ciliate, reproduces asexually by cell division that results in two daughters identical to the original parent.* **(b)** *Mating in a ciliate,* Euplotes. *Genetic material is exchanged across a cytoplasmic bridge. After the exchange occurs, new individuals formed by cell division will have gene combinations different from those of either parent cell.*

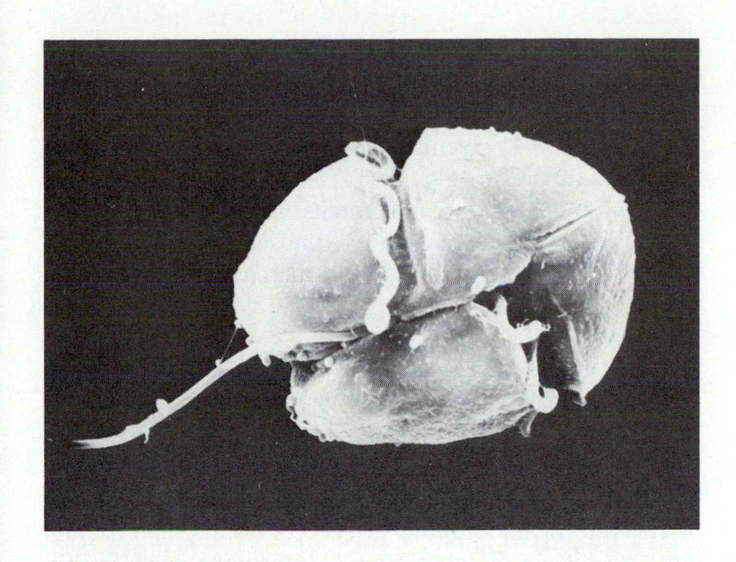

Figure 17-10 A dinoflagellate, covered with protective cellulose armor. Two flagella lie within the grooves encircling the body.

Dinoflagellates. These photosynthetic protists are so named because of their two whiplike flagella, which in many species project through cellulose walls that resemble armor plates (Fig. 17-10). Although some are found in fresh water, dinoflagellates are especially abundant in the ocean, where they provide an important food source for larger organisms. Many are bioluminescent, producing a brilliant blue-green light when disturbed. Clear waters inhabited by these protists take on a magical quality after sunset as the bodies of fish or swimmers are silhouetted in shimmering radiance.

The green chlorophyll in dinoflagellates is often masked by red pigments that help trap light energy. Under certain conditions, when the water is warm and rich in nutrients, a dinoflagellate population explosion occurs. These microorganisms can become so numerous that the waters are dyed red by the color of their bodies, causing a "red tide." Fish die by the thousands, suffocated by clogged gills or by oxygen depletion resulting from the decay of the bodies of billions of dinoflagellates. But oysters, mussels, and clams have a feast, filtering millions from the water for food. In the process, however, they concentrate a nerve poison produced by the dinoflagellates. During red tides, people or other animals feeding on these molluscs may be stricken with potentially lethal paralytic shellfish poisoning.

Diatoms. The photosynthetic diatoms, found in both fresh and salt water, are so important to marine food webs that they have been called the "pastures of the sea." They produce glassy protective coverings, some of exceptional beauty (Fig. 17-11). These consist of top and bottom halves that fit together like a pillbox or Petri dish. Accumulations of the glassy walls of diatoms over thousands of years have produced fossil deposits of "diatomaceous earth" that may be hundreds of meters thick. This slightly abrasive substance is widely used in products such as toothpaste and metal polish. Diatoms store reserve food as oil, whose buoyancy in water helps their bodies to float near the surface, where light is abundant for photosynthesis. Prehistoric accumulations of diatoms and their stored oil may contribute to today's petroleum reserves.

Euglenoids. This group of protists is named after its best-known representative, *Euglena* (Fig. 17-12), a complex single cell that locomotes by whipping its flagellum through the water. Its simple light-sensing organelles consist of a photoreceptor at the base of the flagellum, and an adjacent patch of pigment. The pigment shades the photoreceptor only when light impinges from certain directions, allowing *Euglena* to determine the direction of the light source. Using information from the photoreceptor, the flagellum propels the protist toward light levels appropriate for photosynthesis. All euglenoids live in fresh water, and in contrast to other unicellular algae, they lack a rigid outer covering. This allows some to locomote by wriggling as well as by whipping the flagellum. If *Euglena* is maintained in darkness, it loses its chloroplasts, but can still absorb nutrients from its surroundings. In this state it closely resembles the animal-like zooflagellates described on page 312.

Protozoa

The protozoa (literally "first animals") include four major groups of organisms: zooflagellates, sporozoans, sarcodines, and ciliates. All are unicellular, eukaryotic, and heterotrophic (i.e., they obtain their energy from the bodies of other organisms).

Figure 17-11 Some representative diatoms, illustrating the intricate, microscopic beauty of their glassy walls.

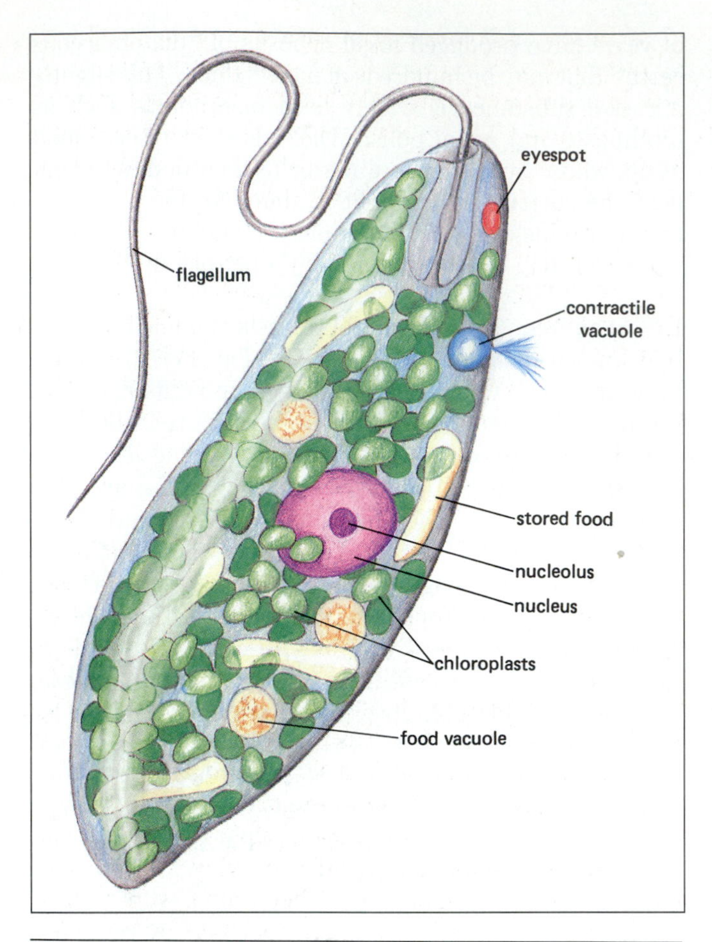

Figure 17-12 Euglena, *a representative euglenoid, showing its elaborate, single-celled structure. The cell is packed with green chloroplasts which will disappear if the protist is kept in darkness.*

ZOOFLAGELLATES. These protists all possess at least one flagellum. This versatile organelle may propel the organism, sense the environment, or ensnare food. The zooflagellates are a diverse group believed to be ancestral to the other protists. Many are free-living, inhabiting soil and water, while others are symbiotic, living inside other organisms in a relationship that may be either mutually beneficial or parasitic. One symbiotic form can digest cellulose and lives in the gut of termites, where it helps them extract energy from wood. A zooflagellate of the genus *Trypanosoma* is responsible for African sleeping sickness, a potentially fatal disease (Fig. 17-13). Like many parasites, this organism has a complex life cycle, part of which is spent in the tsetse fly, which transmits it to mammals while feeding on their blood. The parasite then develops in the host (which may be a person), entering the bloodstream. It may then be ingested by another tsetse fly that bites the host, thus beginning a new cycle of infection.

Another parasitic zooflagellate, *Giardia,* is an increasing problem in the United States, particularly to backpackers who drink from apparently pure mountain streams. Cysts of this flagellate are released in the feces of infected humans or other animals (a single gram of feces may contain 300 million cysts) and enter freshwater streams and even community reservoirs. Cysts develop into the adult form (Fig. 17-14) in the small intestine of their mammalian host. In humans, infections may cause severe diahrrea, dehydration, nausea, vomiting, and cramps. Fortunately, deaths are rare and infections are treatable with drugs.

SPOROZOANS. These specialized protozoa are all parasites, living inside the bodies and sometimes inside the individual cells of their hosts. They are named after their ability to form infectious **spores,** resistant structures transmitted from one host to another through food, water, or the bite of an infected insect. As adults, sporozoans have no means of locomotion. Many have complex life cycles, a common feature of parasites. A well-known example is the malarial parasite *Plasmodium.* Parts of its cycle are spent in the stomach, and later the salivary glands, of the *Anopheles* mosquito. When the mosquito bites a person, it passes the *Plasmodium* to the unfortunate victim. The sporozoan develops in the liver, then enters the blood, where it reproduces rapidly in human red blood cells. The synchronized release of spores through rupture of the blood cells causes the recurrent fever of malaria. Uninfected mosquitos may acquire the parasite by feeding on the blood of a malaria victim, thus continuing the infectious cycle.

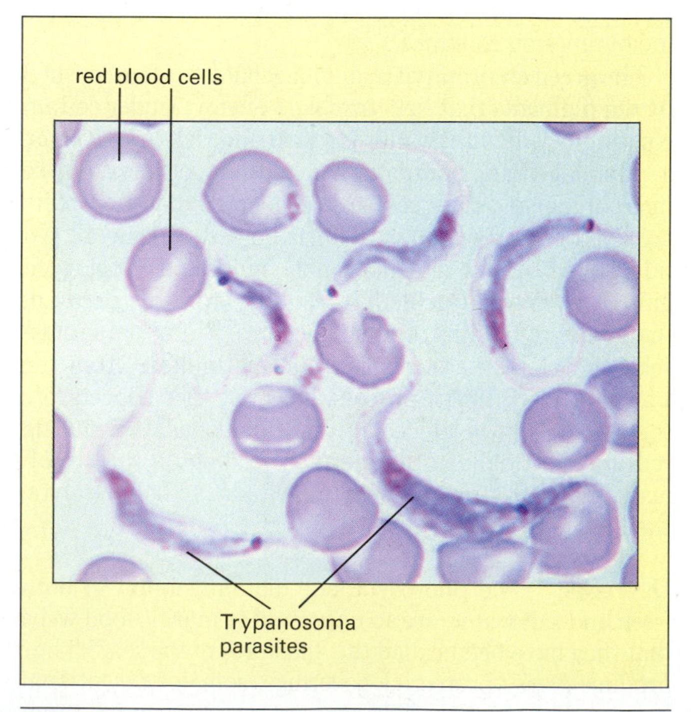

Figure 17-13 A photomicrograph showing human blood that is heavily infested with the parasitic zooflagellate Trypanosoma, *causing African sleeping sickness.*

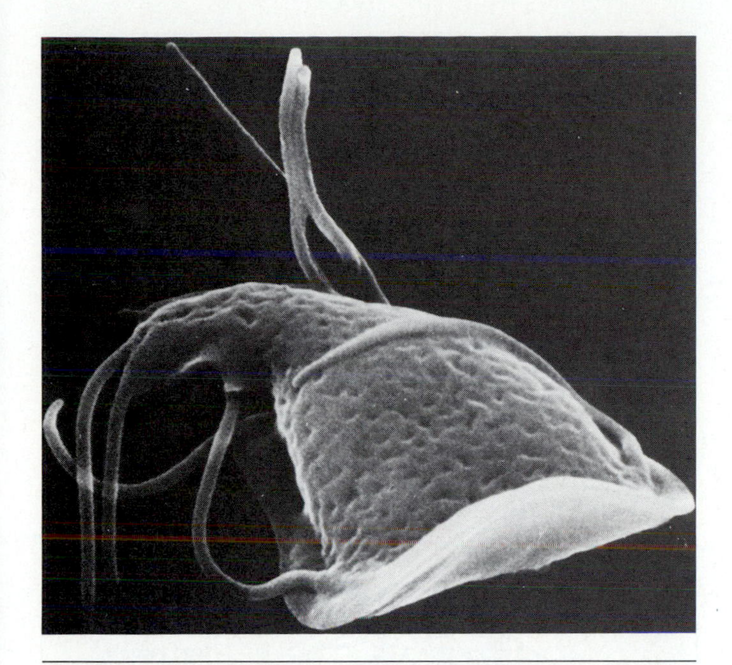

Figure 17-14 A zooflagellate (genus Giardia*) that may infect drinking water, causing gastrointestinal disorders.*

SARCODINES. Also called amoebae, these protists possess flexible cell membranes they can extend in any direction to form pseudopodia (literally "false feet") used for locomotion and for engulfing food (Fig. 17-15a). Amoebae lack many of the specialized organelles found in flagellates and ciliates, but their reputation as "blobs of jelly" is contradicted by their complex internal structure and their sophisticated ability to sense and capture prey. An important parasitic form, particularly common in warm climates, causes amoebic dysentery. Multiplying in the intestinal wall, this parasite causes severe diarrhea, and may perforate the intestine, occasionally causing fatal infections. Not all amoebae are formless; the foraminiferans and radiolarians produce beautiful and elaborate shells of glass or calcium carbonate (Fig. 17-15b,c). These elaborate shells are pierced by myriad openings through which tiny pseudopods extend. The chalky shells of these marine-dwelling sarcodines, accumulating over millions of years, have resulted in immense deposits of limestone such as form the famous white cliffs of Dover, England.

CILIATES. These inhabitants of fresh or salt water represent the peak of unicellular complexity. They possess many specialized organelles, including the cilia after which they are named. Their cilia may cover the cell, or they may be localized, as in *Didinium* (see Fig. 17-17). In the well-known freshwater genus *Paramecium* (Fig. 17-16), rows of cilia cover the entire body surface. Their coordinated beating propels the cell through the water at a protistan speed record of a millimeter per second. Although only a single cell, *Paramecium* responds to its environment as if it had a well-developed nervous system. Confronted with some noxious chemical or with a physical barrier, the cell immediately backs up by reversing

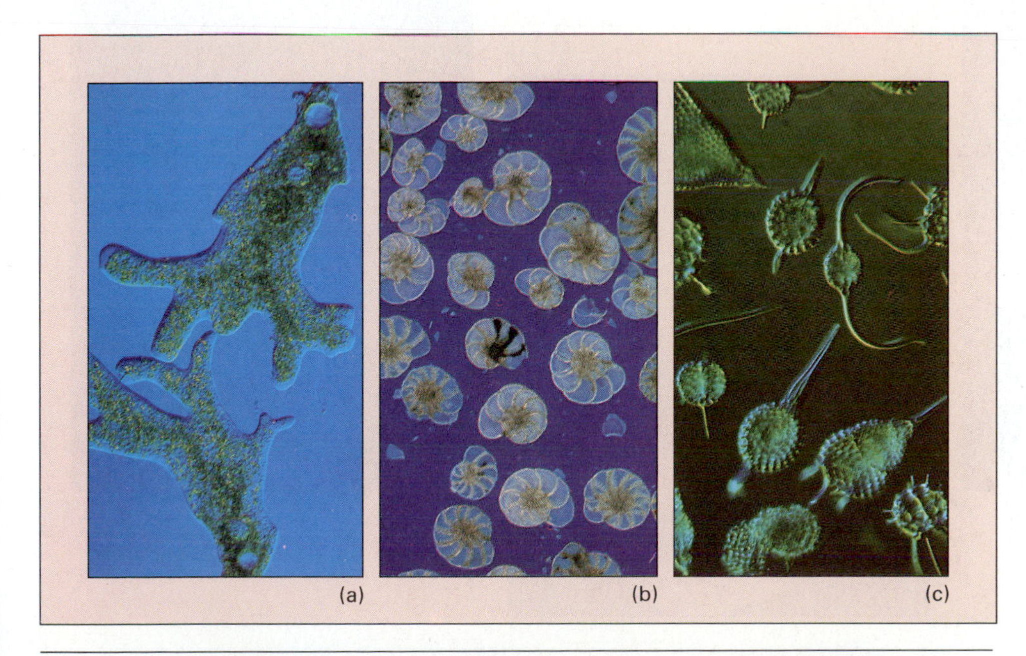

Figure 17-15 Examples of sarcodines. **(a)** Amoeba proteus *sends out cytoplasmic projections called pseudopodia, which serve in locomotion and prey capture.* **(b)** *The chalky shells of foraminiferans show numerous interior compartments.* **(c)** *The delicate, glassy shells of radiolarians. Pseudopodia are extended out through the openings to sense the environment and capture food.*

the beating of its cilia, then proceeds in a new direction. Ciliates are accomplished predators (Fig. 17-17). Some, including *Paramecium* and *Didinium,* immobilize their prey with explosive darts called **trichocysts** embedded in the outer covering of the cell. Prey is escorted to a mouthlike opening, the oral groove. It is digested in a food vacuole, which forms a temporary "stomach," and excreted by exocytosis. Excess water is accumulated in a contractile vacuole, which periodically contracts, emptying the fluid through a pore to the outside.

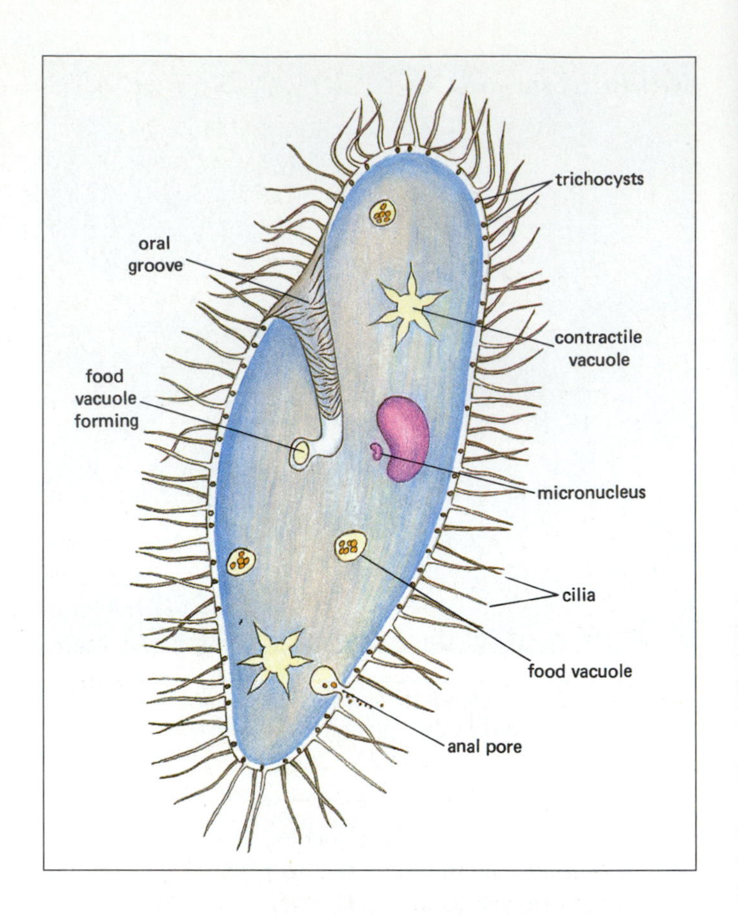

Figure 17-16 Paramecium, *illustrating some important ciliate organelles. The oral groove acts as a mouth, and food vacuoles, miniature digestive systems, form at its apex. The contractile vacuoles regulate water balance. Trichocysts help this predator stun its prey, while cilia propel it rapidly through the water.*

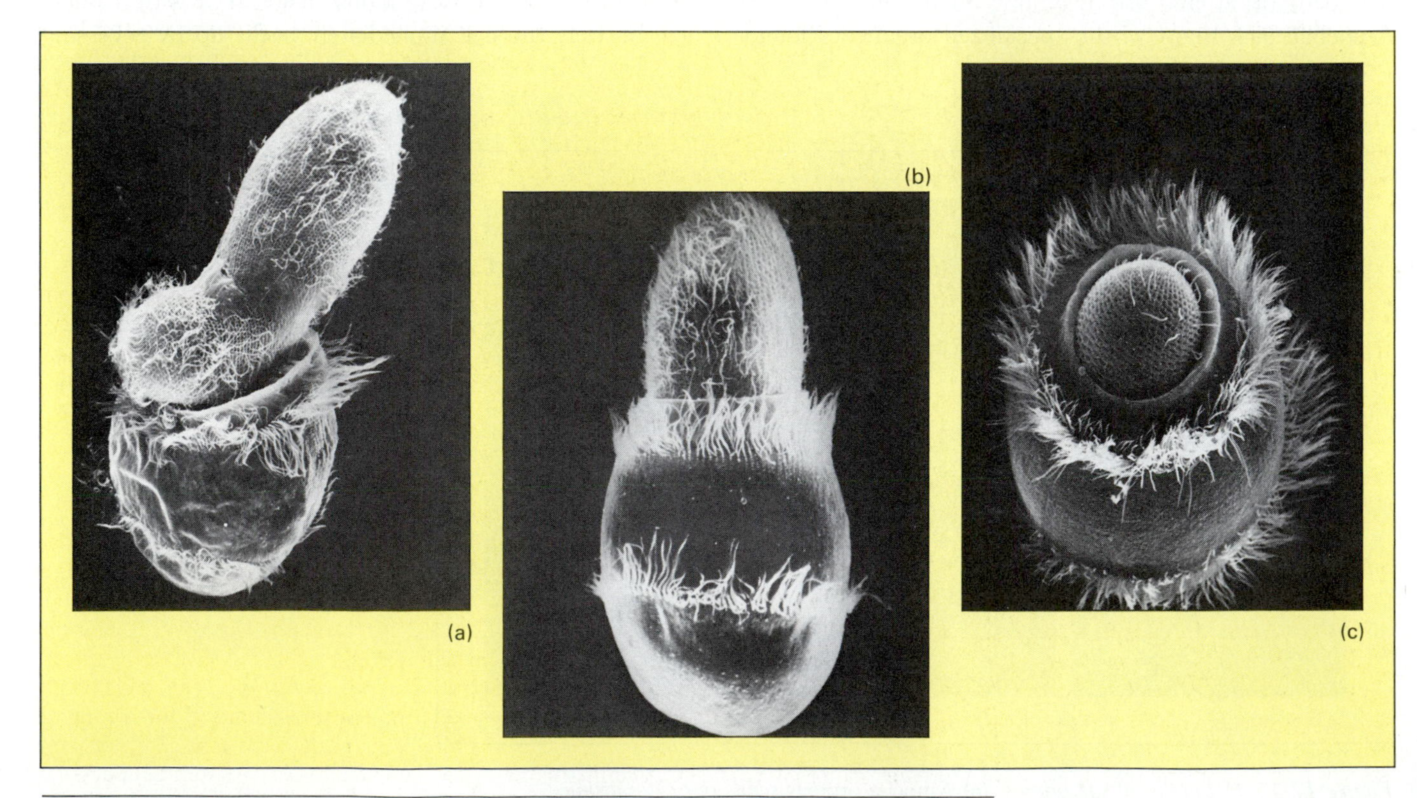

Figure 17-17 In this series of scanning electron micrographs, Paramecium *is stung* **(a)** *and gradually engulfed* **(b, c)** *by another predatory ciliate,* Didinium, *whose cilia are confined to two bands encircling the egg-shaped body. This microscopic drama could occur on a pinpoint with room to spare.*

REFLECTIONS ON OUR UNICELLULAR ANCESTORS

In the kingdoms Monera and Protista, we find persisting today the types of cellular organization that gave rise to the more complex organisms that now dominate the Earth. Modern monerans have changed little from their ancestors, whose fossilized remains date back $3\frac{1}{2}$ billion years. Life might still consist of primitive single cells if the protists, with their radical eukaryotic design, had not appeared on the scene nearly $1\frac{1}{2}$ billion years ago. As you learned in Chapter 15, eukaryotic cells may have originated when one moneran, perhaps a bacterium capable of cellular respiration, took up residence inside a partner, forming the first "mitochondrion." A separate but equally crucial merger may have occurred when a photosynthetic moneran (probably resembling a blue-green bacterium) took up residence within a nonphotosynthetic partner and became the first "chloroplast." The foundations of multicellularity were laid with the eukaryotic cell, whose intricacy allowed specialization of entire cells for specific functions within a multicellular aggregation. Thus primitive protists, some consuming their food in chunks, some photosynthesizing and others absorbing nutrients from the environment, almost certainly followed divergent evolutionary paths that led to the three multicellular kingdoms predominating today: the animals, plants, and fungi.

SUMMARY OF KEY CONCEPTS

Microorganisms are placed in two kingdoms, Monera and Protista, which can be distinguished by their fundamentally different cell types. Monera consist of tiny prokaryotic cells lacking organelles, such as nuclei, mitochondria, and chloroplasts. Protists are eukaryotic cells that possess the full range of organelles and resemble the cells of multicellular organisms. Without the photosynthetic, nitrogen trapping, and decomposing abilities of the bacteria, and the photosynthetic activities of protists, life as we know it would grind to a halt.

Viruses

Viruses are parasites consisting of a protein coat surrounding genetic material. They are noncellular and unable to move, grow, or reproduce outside a living cell. They invade cells of a specific host and use the host cell's energy, enzymes, and ribosomes to produce more virus particles, which are liberated when the cell ruptures. Many viruses are pathogenic to humans, including those causing colds and flu, herpes, AIDS, and certain forms of cancer.

The Kingdom Monera

Members of the kingdom Monera, the bacteria, are unicellular and prokaryotic. A cell wall of peptidoglycan determines their characteristic shape: coccus (round), bacillus (rodlike), or spiral. Many form spores that disperse widely and withstand inhospitable conditions. Bacteria obtain energy in a variety of ways. Some, including the cyanobacteria, rely on photosynthesis. Others are chemosynthetic, breaking down inorganic molecules to obtain energy. Heterotrophic forms are capable of utilizing a wide variety of organic compounds. Many are anaerobic, able to obtain energy from fermentation when oxygen is not available.

Some bacteria are pathogenic, causing disorders including pneumonia, tetanus, botulism, and the venereal diseases gonorrhea and syphilis. Most, however, are harmless to humans and play important roles in natural ecosystems. Bacteria have colonized nearly every habitat on Earth. Some live in the digestive tracts of larger organisms such as cows and sheep, where they break down cellulose. Nitrogen-fixing bacteria enrich the soil and aid in plant growth, while many others live off the dead bodies and wastes of other organisms, liberating nutrients for reuse.

Cyanobacteria engage in plant-like photosynthesis and can also fix nitrogen, making them nutritionally very self-reliant.

The archaebacteria comprise a unique and diverse group that flourish under extreme conditions, including hot, acidic, very salty, and anaerobic environments. They differ in several ways from all other bacteria, including cell wall composition, ribosomal RNA sequence, and cell membrane lipid structure.

The Kingdom Protista

The kingdom Protista consists of organisms composed of single, highly complex eukaryotic cells. They may be divided into two major groups: the plant-like unicellular algae and the animal-like protozoa.

The unicellular algae are important photosynthetic organisms in marine and freshwater ecosystems. They include dinoflagellates, diatoms, and the exclusively freshwater euglenoids.

Protozoa are nonphotosynthetic protists that absorb or ingest their food. They are widely distributed in soil and water, and some are parasitic. They include the zooflagellates, the parasitic sporozoans, the amoeboid sarcodines, and the predatory ciliates.

GLOSSARY

Anaerobic (an-er-ō′-bik): capable of living and obtaining energy in the absence of oxygen.

Bacteriophage (bak-tir′-ē-ō-fāj): a virus specialized to parasitize bacteria.

Bacterial conjugation: the exchange of genetic material between bacteria.

Bacterium (bak-tir′-ē-um; pl. bacteria): an organism consisting of a single prokaryotic cell surrounded by a complex polysaccharide coat.

Bacillus (buh-sil′-us; pl. bacilli): a rod-shaped bacterium.

Chemosynthetic (kēm′-ō-sin-the-tic): capable of oxidizing inorganic molecules to obtain energy.

Ciliate (sil′-ē-et): a category of protozoan characterized by cilia and a complex unicellular structure, including harpoon-like organelles called trichocysts. Members of the genus *Paramecium* are well-known ciliates.

Cilium (sil′-ē-um; pl. cilia): A short, hairlike organelle projecting through the cell membrane, usually numerous and engaged in coordinated beating which moves a cell through a fluid environment or moves the fluid over the surface of the cell.

Coccus (ka′-kus; pl. cocci): a spherical bacterium.

Cyanobacteria: photosynthetic prokaryotic cells, utilizing chlorophyll and releasing oxygen as a photosynthetic by-product, sometimes called "blue-green algae."

Diatom (dī′-e-tom): A category of protist that includes photosynthetic forms with two-part glassy outer coverings which separate when the cell divides. Diatoms are important primary producers in fresh and salt water.

Dinoflagellate (dī-nō-fla′-gel-et): A category of protist that includes photosynthetic forms in which two flagella project through armorlike plates. Abundant in oceans, these sometimes reproduce rapidly, causing "red tides."

Euglenoid (yū′-gle-noid): a category of protist characterized by one or more whiplike flagella used for locomotion and a photoreceptor for detecting light. Euglenoids are photosynthetic, but some are capable of heterotrophic nutrition if deprived of chlorophyll.

Eukaryotic (yū′-kār-ē-ot′-ik): A type of cellular organization characterized by the presence of a membrane-bound nucleus, mitochondria, and other complex membranous organelles.

Flagellum (fla-gel′-um): motile hairlike organelle that propels a cell through a fluid.

Gonorrhea (gon-a-rē′-uh): a sexually transmitted bacterial infection of the reproductive organs. Untreated gonorrhea may result in sterility.

Halophile (hā′-lō-fīl): a salt-loving organism.

Legume (leg′-oom): a family of dicotyledonous plants characterized by root swellings in which nitrogen-fixing bacteria are housed. Includes soybeans, lupines, alfalfa, and clover.

Methanogen (me-than′-ō-gen): a type of anaerobic archaebacterium capable of converting carbon dioxide to methane.

Monera (mō′-ne-ra): a taxonomic kingdom consisting of unicellular prokaryotic organisms, including bacteria, archaebacteria, and cyanobacteria.

Nitrogen-fixing: possessing the ability to remove nitrogen from the atmosphere and combine it with hydrogen to produce ammonia.

Pathogen (path′-ō-gen): an organism capable of producing disease.

Peptidoglycan (pep-tid-ō-glī′-can): Material found in prokaryotic cell walls consisting of chains of sugars cross-linked by short chains of amino acids called peptides.

Phytoplankton (fī′-to-plank-ten): a general term describing photosynthetic protists that are abundant in marine and freshwater environments.

Prokaryotic (prō′-kar-ē-ot′-ik): a type of cellular organization characterized by the lack of a membrane-bound nucleus and other membranous organelles, such as mitochondria, Golgi, ER, and chloroplasts. Prokaryotic cells are restricted to the kingdom Monera.

Protista (prō-tis′-tuh): a taxonomic kingdom including unicellular, eukaryotic organisms.

Protozoan (prō-te-zō′-an; pl. protozoa): a nonphotosynthetic or animal-like protist.

Pseudopod (sūd′-ō-pod): extension of the cell membrane by which certain cells, such as amoebae, locomote and engulf prey.

Sarcodine (sar-kō′-dīn) a category of nonphotosynthetic protist (protozoa) characterized by the ability to form pseudopodia. Some, such as amoebae, are naked, while others have elaborate shells.

Spore (spōr) a resistant or resting structure that disperses readily and withstands unfavorable environmental conditions.

Sporozoan (spōr-ō-zō′-en) a category of parasitic protist. Sporozoans have complex life cycles often involving more than one host, and are named for their ability to form infectious spores. A well-known member (genus *Plasmodium*) causes malaria.

Syphilis (si′-ful-is) a sexually transmitted bacterial infection of the reproductive organs which, if untreated, can damage the nervous and circulatory systems.

Thermoacidophile (ther-mō-a-sid′-eh-fīl) a form of archaebacterium that thrives in hot, acidic environments.

Trichocyst (trik′-eh-sist) a stinging organelle of protists.

Virus (vī′-rus) a noncellular parasitic particle consisting of a protein coat surrounding a strand of genetic material. Viruses can multiply only within the cells of living organisms.

Zooflagellate (zō-ō-fla′-gel-et) a category of nonphotosynthetic protist that move using flagella.

STUDY QUESTIONS

1. List the five kingdoms and differentiate between them on the basis of cell type, cell number, and mode of nutrition.
2. List the major differences between monerans and protistans.
3. Describe some of the ways in which bacteria obtain energy and nutrients.
4. What are nitrogen-fixing bacteria, and what role do they play in ecosystems?
5. What is a spore?
6. Why do bacteria readily become resistant to antibiotics, and what steps can people take to prevent this?
7. Describe some examples of bacterial symbiosis.
8. Argue for and against the statement: "Viruses are not alive."
9. Describe the structure of a typical virus. How do viruses reproduce?
10. What is the importance of dinoflagellates in marine ecosystems? What happens when they reproduce rapidly?
11. What is the major ecological role played by unicellular algae?
12. What protozoan group consists entirely of parasitic forms?
13. Describe the life cycle and mode of transmission of the malarial parasite.
14. What protozoan group is responsible for forming the white limestone cliffs of Dover, England?
15. How do amoebae feed? What disease is caused by a parasitic amoeba?
16. The most complex single cells are found in which protozoan group? How do they feed? From what feature is their name derived?

SUGGESTED READINGS

Adler, J. "The Sensing of Chemicals by Bacteria." *Scientific American,* April 1976.

Atlas, R. M. *Basic and Practical Microbiology.* New York: Macmillan Publishing Company, 1986.

Barghoorn, E. S. "The Oldest Fossils." *Scientific American,* May 1971.

Brill, W. J. "Biological Nitrogen Fixation." *Scientific American,* March 1977.

Butler, P. J. G., and Klug, A. "The Assembly of a Virus." *Scientific American,* November 1978. (Offprint No. 1412).

Dixon, Bernard. "Overdosing on Wonder Drugs." *Science 86,* May, 1986.

Gallo, R. C. "The AIDS Virus." *Scientific American,* January 1987. (Offprint No. 1577).

Jannasch, H. W., and Wilson, C. O. "Microbial Life in the Deep Sea." *Scientific American,* June 1977.

Margulis, L. "Symbiosis and Evolution." *Scientific American,* August 1971.

Marples, M. J. "Life on the Human Skin." *Scientific American,* January 1969. (Offprint No. 1132).

Simons K., Garoff, H., and Helenius, A. "How an Animal Virus Gets into and out of Its Host Cell." *Scientific American,* February 1982. (Offprint No. 1511).

Woese, Carl. "Archaebacteria." *Scientific American,* June 1981.

18
The Diversity of Life. II. Animals

The kingdom Animalia is composed of multicellular heterotrophic organisms. Most animals have nerves and muscles that allow them to react to stimuli, whether to escape danger, seek food, or communicate with others.

For convenience, biologists often place animals in one of two major categories: **vertebrates,** those with a vertebral column or backbone, and **invertebrates,** those lacking a backbone. You are probably more familiar with the vertebrates: fish, amphibians, reptiles, birds, and mammals. The invertebrates include everything else, from sponges to worms to snails to insects. Our human bias is clearly reflected in these categories, since the invertebrates include at least 99 percent of all the animal species on Earth, including 27 different phyla. In contrast, vertebrates constitute only part of a single phylum, the chordates.

Several features are used to classify animals. These include the degree of organization of cells within the body (whether they are relatively independent, or form functional groups such as tissues or organs), body symmetry (indeterminate shape, radial, or bilateral), the presence of an internal space or **coelom,** and other anatomical similarities, some of which are apparent only during particular developmental stages. One very general classification scheme is shown in Figure 18-1.

The first animals to evolve probably originated from colonies of protozoa, whose members had become specialized to perform distinct roles within the body. In our survey of the animals, we begin with the sponges, whose body plan most closely resembles these ancestral protozoan colonies. Our discussion of the animal kingdom will attempt to follow the order in which the various groups appeared on Earth. As we progress from sponges through the Cnidaria (jellyfish and their relatives) through the three phyla of worms, you will see a clear trend toward increasing complexity. This trend culminates in the arthropods, molluscs, echinoderms, and chordates (see Table 18-1). The striking structural differences among the latter four phyla do not reflect further increases in complexity as much as adaptations to different habitats and life-styles. Within the chordates, however, there is another clear trend in complexity: in the size and sophistication of the brain.

PHYLUM PORIFERA: THE SPONGES

In 1907, the embryologist H. V. Wilson mashed a sponge through a piece of silk, dissociating it into single cells and cell clusters. After sitting in seawater for three weeks, the cells had reaggregated into a functional sponge. Because sponge cells are relatively independent, sponges resemble colonies in which single-celled organisms live together for mutual benefit, but lack specialized tissues. Sponges lack a nervous system and each cell engages in intracellular digestion. Sponges are, therefore, the simplest multicellular animals. Although some have a definite size and shape, others grow freely and indeterminately over rocks in their aquatic habitats (Fig. 18-2a).

All sponges have a similar general body plan (Fig. 18-2b). The body is perforated by numerous tiny pores through which water enters, and by fewer, large openings,

Figure 18-2 **(a)** *Sponges come in a wide variety of sizes, shapes and colors, some over a meter tall, others growing in free-form pattern over undersea rocks. The large exit pores are visible in the upper specimen.*
(b) *Sponges all have a similar body plan. Currents created by collar cells draw water in through numerous tiny pores. Microscopic food particles are filtered out by collar cells and shared among the various cell types. Water exits through larger pores, the oscula. Spicules form a supportive internal skeleton.*

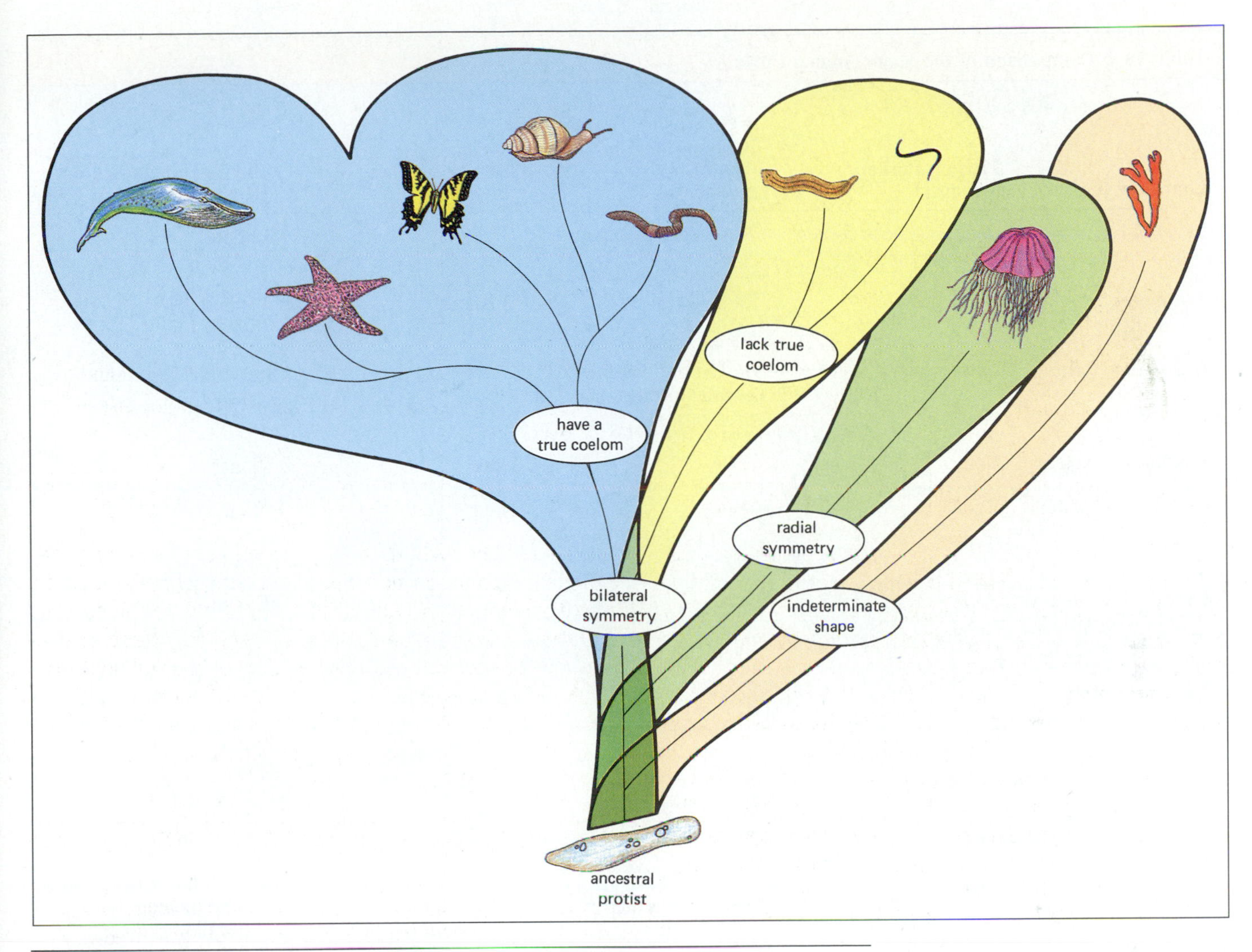

Figure 18-1 *A simple classification scheme for animals based on anatomical features.*

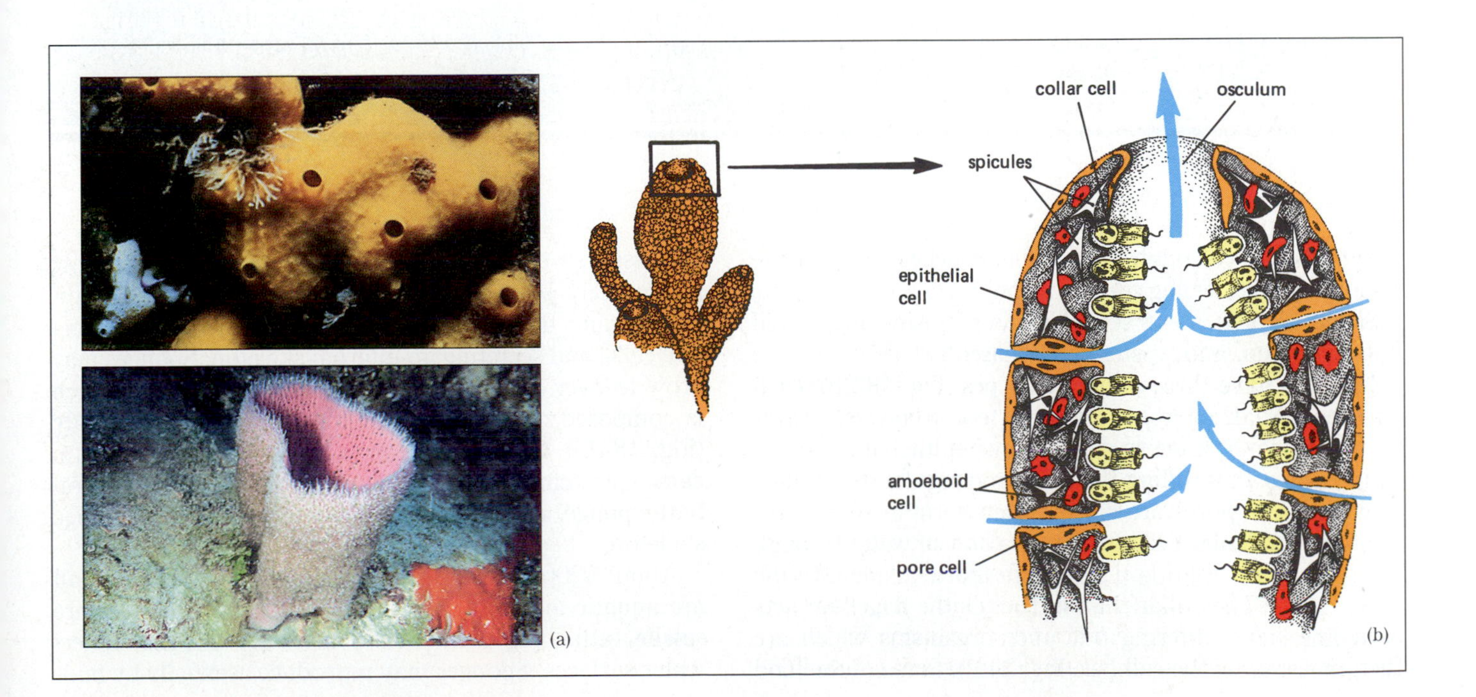

Table 18-1 **Comparison of the Major Animal Phyla**

Characteristic	Phylum Porifera (sponges)	Phylum Cnidaria (Hydra, Jellyfish, Anemones)	Phylum Platyhelminthes (Flatworms)	Phylum Nematoda (Roundworms)
Level of organization	Cellular—lacks tissues and organs	Tissue—lacks organs	Organ	Organ
Symmetry	None	Radial	Bilateral	Bilateral
Digestive system	Intracellular	Gastrovascular cavity; some intracellular	Gastrovascular cavity	Separate mouth and anus
Circulatory system	None	None	None	None
Respiratory System	None	None	None	None
Excretory system (fluid regulation)	None	None	Canals with flame cells	Excretory gland cells
Nervous system	None	Nerve net	Head ganglia with longitudinal nerve cords	Head ganglia with dorsal and ventral nerve cords
Reproduction	Sexual; asexual (budding)	Sexual; asexual (budding)	Sexual (some hermaphroditic); asexual (body splits)	Sexual (some hermaphroditic)
Support	Endoskeleton of spicules	Hydrostatic skeleton	None	Hydrostatic skeleton

called **oscula,** through which it is expelled. Within the hollow sponge, water travels through canals. During its passage, oxygen is extracted, microorganisms are filtered out and eaten, and wastes are released.

Sponges have three major cell types (Fig. 18-2b), each with a specialized role. Flattened epithelial cells cover their inner and outer surfaces. Some epithelial cells surround pores, controlling their size and regulating the flow of water. The pores are closed when harmful substances are present. **Collar cells** maintain a flow of water through the sponge, by beating a flagellum that extends into the inner canal. The collar that surrounds the flagellum acts as a fine sieve, filtering out microorganisms which are then ingested by the cell (see Fig. 24.2). Some of the food is passed to the third cell type, **amoeboid cells.** These roam freely between the epithelial and collar cells, digesting and distributing nutrients, producing reproductive cells, and secreting an internal skeleton. Sponges may grow to over a meter in height, and the skeleton, which is composed of **spicules,** provides support for the body (Fig. 18-2b). The spicules may be formed from calcium carbonate (chalk), silica (glass), or protein. The natural bath sponge, now rarely used, is a proteinaceous sponge skeleton.

About 5000 species of sponges have been identified; all are aquatic and most are marine. All adult sponges are **sessile,** attaching permanently to rocks or other underwater surfaces. Sponges may reproduce asexually by bud-

Table 18-1 **Comparison of the Major Animal Phyla**

Characteristic	Phylum Annelida (Segmented Worms)	Phylum Arthropoda (Insects, Arachnids, Crustaceans)	Phylum Mollusca (Snails, Clams, Squids)	Phylum Echinodermata (Sea Stars, Sea Urchins, Sea Cucumbers)	Phylum Chordata (Tunicates, Vertebrates)
Level of organization	Organ	Organ	Organ	Organ	Organ
Symmetry	Bilateral	Bilateral	Bilateral	Larvae bilateral, adults radial	Bilateral
Digestive system	Separate mouth and anus	Separate mouth and anus	Separate mouth and anus	Separate mouth and anus (usually)	Separate mouth and anus
Circulatory system	Closed	Open	Open	None	Closed
Respiratory system	None	Tracheae, gills, or book lungs	Gills, lungs	Tube feet, skin gills, respiratory tree	Gills, lungs
Excretory system (fluid regulation)	Nephridia	Excretory glands resembling nephridia	Nephridia	None	Kidneys
Nervous system	Head ganglia with paired ventral cords; ganglia in each segment	Head ganglia with paired ventral nerve cords; ganglia in segments, some fused	Well-developed brain in some cephalopods; several paired ganglia, most in the head; nerve network in body wall	Head ganglia absent; nerve ring and radial nerves; nerve network in skin	Well-developed brain; dorsal nerve cord
Reproduction	Sexual (some hermaphroditic)	Usually sexual	Sexual (some hermaphroditic)	Sexual (some hermaphroditic); asexual by regeneration (rare)	Sexual
Support	Hydrostatic skeleton	Exoskeleton	Hydrostatic skeleton	Endoskeleton of plates beneath outer skin	Endoskeleton of cartilage or bone

ding, or sexually through the fusion of sperm and eggs. Fertilized eggs develop inside the adult into active larvae that escape through the oscula. Water currents disperse the larvae to new areas where they settle permanently and develop into adult sponges.

PHYLUM CNIDARIA: THE HYDRA, ANEMONES, AND JELLYFISH

The Cnidaria are clearly a step above the Porifera in complexity. Their cells are organized into distinct **tissues,** including nerves, and contractile tissue that acts like muscle. Nerve cells are organized into a **nerve net** that branches through the body and controls the contractile tissue. Cnidarians lack true organs, however, and have no brain.

Members of the phylum Cnidaria come in a bewildering and beautiful variety of forms (Fig. 18-3), all of which are actually variations on two basic body plans: the **polyp** and the **medusa,** illustrated in Figure 18-4. The polyp, with its foot attached and its tentacles reaching upward, is adapted to a life spent quietly attached to rocks, awaiting prey like a predatory flower. The medusa ("jellyfish") swims weakly by contracting its bell-shaped body, but primarily is carried by ocean currents, trailing its tentacles like multiple fishing lines. Both forms are **radially**

Figure 18-3 *Cnidarian diversity.* **(a)** *A pink anemone from the Gulf of California, about 3″ in diameter.* **(b)** *A medusa found off southern California.* **(c)** *A reef-forming coral.* **(d)** *The Portuguese Man-of-War, a colonial cnidarian dangerous to humans. Notice the stunned fish trapped in the tentacles.*

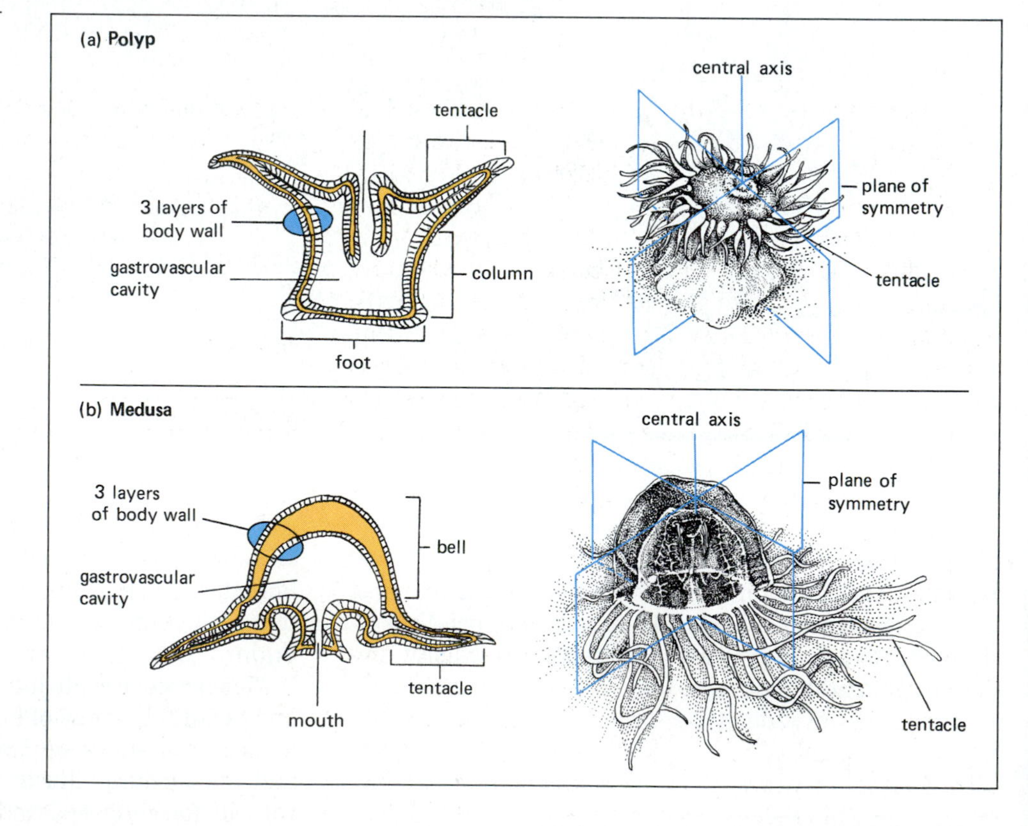

Figure 18-4 *The two basic body forms of cnidarians are actually variations on a single, simple theme. The polyp, shown in detail in* **(a),** *is exemplified by hydra and the sea anemones, while jellyfish, which resemble an inverted polyp, illustrate the medusa form* **(b).** *Both forms exhibit radial symmetry, with body parts arranged in a circle around a central axis. Any plane passing through this central axis will divide the cnidarian body into approximately equal parts.*

symmetrical, with body parts arranged in a circle around an axis drawn through the mouth and digestive cavity (Fig. 18-4). This arrangement is particularly well suited to animals that are sessile, or carried every which way by water currents, since they are prepared to capture prey or defend themselves from any direction.

Although all cnidarians are predatory, none actively hunt. Instead, they rely on their victims blundering by

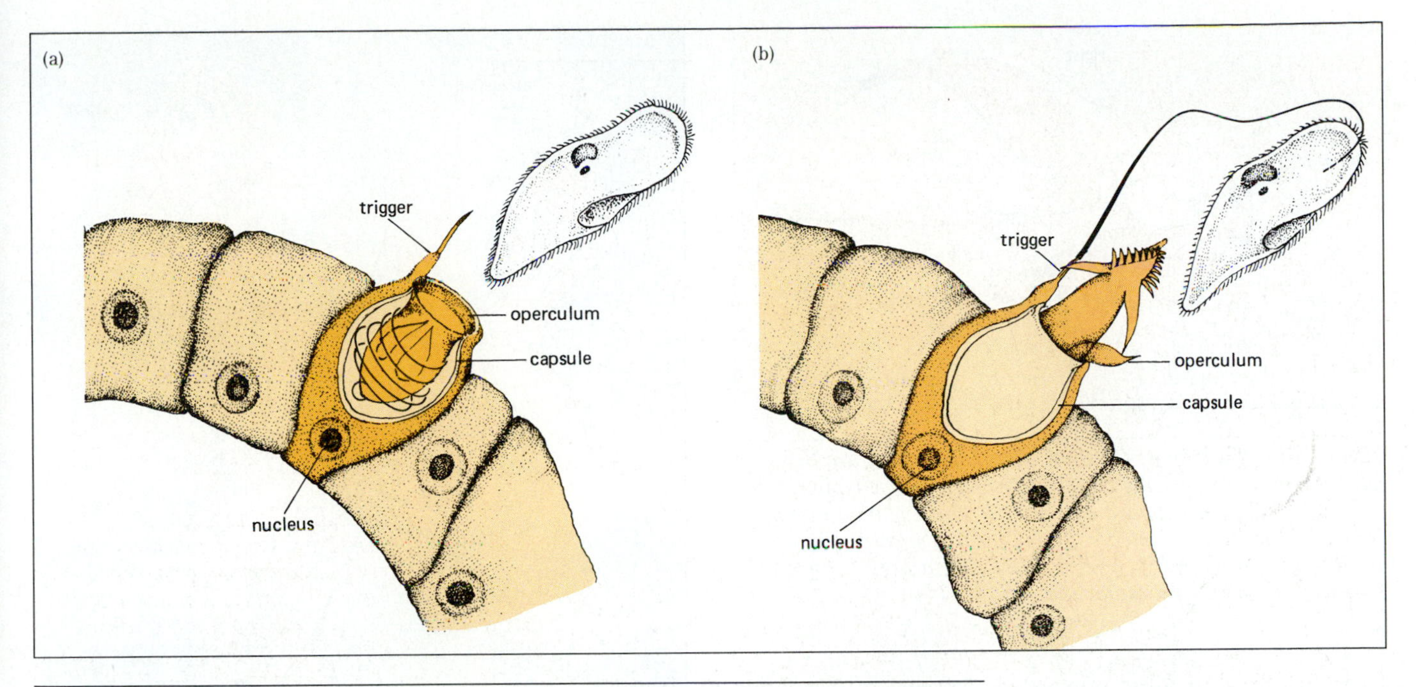

Figure 18-5 **(a)** *At the slightest touch to the trigger, the nematocyst of cnidarians violently expels the poisoned or sticky dart that lies coiled and inverted inside. During the process, the barbs and hollow filament actually turn inside out, impaling the prey and injecting a paralyzing venom* **(b).** *Nematocysts are microscopic, and only a few species inject venom in sufficient quantity to harm a person.*

chance into the grasp of their enveloping tentacles. Cnidarian tentacles are armed with **nematocysts,** cells containing poisonous or sticky darts that are injected explosively into prey upon contact (Fig. 18-5). Stung and firmly grasped, the prey is forced through an expansible mouth into a digestive sac, the **gastrovascular cavity.** Digestive enzymes secreted into this cavity break down some of the food, and further digestion occurs intracellularly, within the cells lining the cavity. Digestion completed, the versatile mouth becomes an anus through which undigested material is expelled, since the gastrovascular cavity has only a single opening. Although this two-way traffic is inefficient because it prevents continuous feeding, it is adequate to support the low energy demands of these animals.

Cnidarians can reproduce both asexually and sexually. Some medusae and some polyps, such as hydra and sea anemones, bud off miniature replicas of themselves. An anemone, crawling slowly over a rock, may also leave behind pieces of foot, which grow into new individuals. Sexual reproduction involves the fusion of sperm and eggs released into the water or retained within the parent. The fertilized egg often develops into a free-swimming ciliated larval stage that settles and becomes a tiny polyp.

Of the 9000 or more species of Cnidaria, all are aquatic and most are marine. One group, the corals, are of particular ecological importance (see Fig. 18-3c). These polyps secrete a hard protective "house" of limestone that persists long after their death, serving as a base for others. The cycle continues until, after thousands of years, massive coral reefs are formed. Corals are restricted to the warm, clear waters of the tropics, where their reefs form undersea habitats, the basis of an ecosystem of stunning diversity and unparalleled beauty (see Chapter 38).

PHYLUM PLATYHELMINTHES: THE FLATWORMS

Although flatworms do not look anything like cnidarians, the two have certain features in common which have led biologists to speculate that they have evolved from a common ancestor. Both have a gastrovascular cavity with a single opening. Certain flatworms also show striking similarities to the larval stage of cnidarians.

Flatworms are clearly more complex than cnidarians. For one thing, they have **bilateral symmetry** (Fig. 18-6), a feature we can recognize in ourselves. Bilaterally symmetrical animals have roughly mirror-image right and left halves, an **anterior** head end, a **posterior** tail end, an upper or **dorsal** surface, and a lower **ventral** surface. This body plan, found in all the more complex animals, is an adaptation to active movement. The anterior end first encounters the environment ahead. The sense organs are concentrated here, where they can inform the orga-

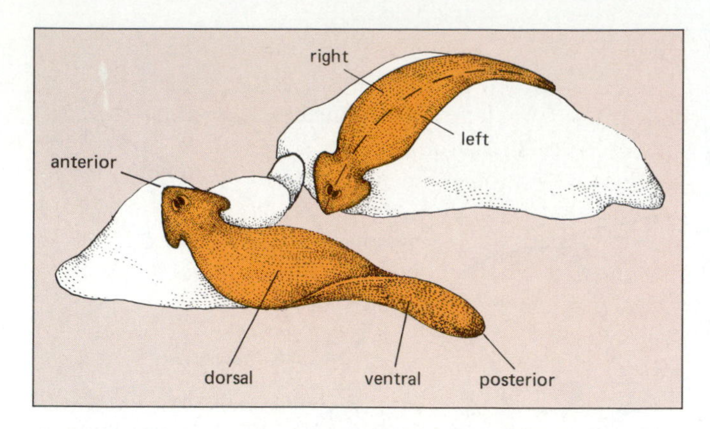

Figure 18-6 Flatworms (phylum Platyhelminthes) are the simplest animals showing bilateral symmetry. The body can be split into two mirror-image halves by a single plane running down the midline. Animals with bilateral symmetry also have an anterior head end, a posterior tail end, a dorsal upper surface, and a ventral underside.

nism whether to forge onward or retreat. In free-living (nonparasitic) flatworms such as the freshwater planarians (Fig. 18-7), sense organs consist of eyespots for detecting light and dark, and cells responsive to chemical and tactile stimuli. To process this information, flatworms have aggregations of nerve cells called **ganglia** in the head that form a simple brain. A pair of nerve cords conducts nervous signals to and from the head ganglia.

Flatworms are the simplest organisms with true **organs,** in which tissues are grouped into functional units. When a free-living flatworm encounters food, usually smaller animals, it sucks up its prey using a muscular **pharynx** located in the middle of the ventral side of its body. The food is digested in an intricately branched gastrovascular cavity that distributes nutrients to all parts of the body (Fig. 18-7). Free-living flatworms have a simple system for excreting and regulating body fluids. It consists of a network of canals ending in bulbs containing beating

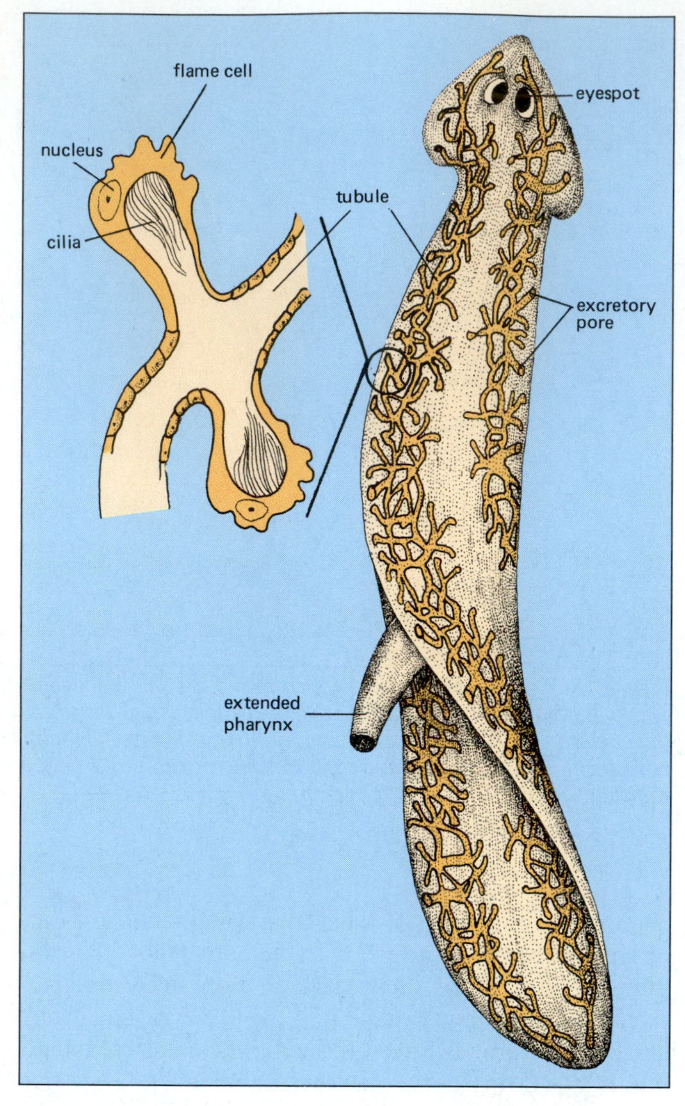

Figure 18-8 The excretory system of Planaria, *with an enlargement of* flame cells, *showing the cilia whose flickering movement gives them their name.*

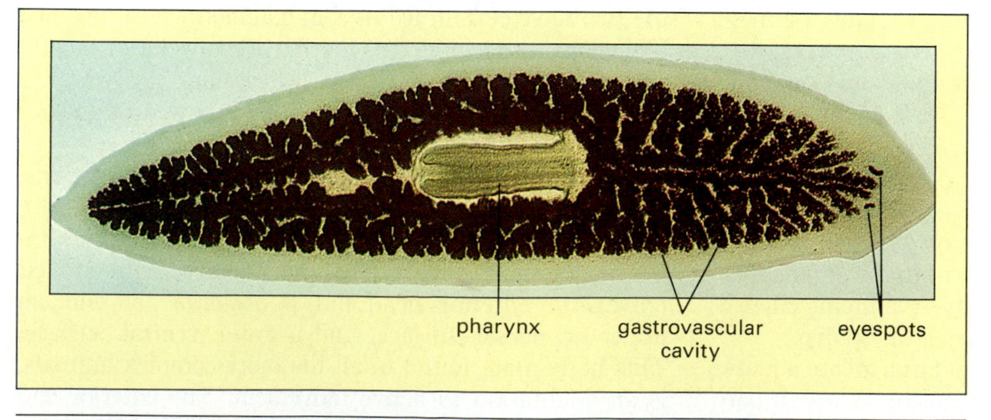

Figure 18-7 In this stained Planaria, *the eyespots and the elaborately branched gastrovascular cavity are clearly visible. The central, ventral pharynx can also be seen.*

cilia. The flickering motion of the cilia has led to the descriptive name: **flame cells** (Fig. 18-8). The beating cilia drive liquids through the system, emptying excess fluids to the outside through numerous tiny pores.

Flatworms lack both respiratory and circulatory systems. Nutrients are distributed by a branching digestive tract, from which they readily diffuse into nearby cells. Gas exchange between the cells and the environment by diffusion is aided by the flattened body, which ensures that all the cells are relatively close to the outside.

Although many flatworms, such as the planarians, are free living, those of major importance to humans are parasites. These include the **tapeworms,** several of which can infect humans. In most cases, infection occurs by eating improperly cooked beef, pork, or fish infected by the worms, whose larvae form **cysts** in the muscles of these animals. The cysts hatch in the human digestive tract, where they attach to the intestine and mature. Here they may grow to a length of 7 meters, absorbing digested nutrients directly through their outer surface, and releasing packets of eggs that are shed in the host's feces. If pigs eat grass contaminated with infected human feces, the eggs hatch in the pig's digestive tract, releasing larvae that burrow into its muscles and form cysts, continuing the infective cycle (Fig. 18-9).

Another group of parasitic flatworms are the flukes. Of these, the most devastating are liver flukes (common in the Orient) and blood flukes, such as those of the genus *Schistosoma,* which cause schistosomiasis. Prevalent in Africa and parts of South America, this disease affects an estimated 200 million people; its symptoms include dysentery, anemia, and possible brain damage. Like most

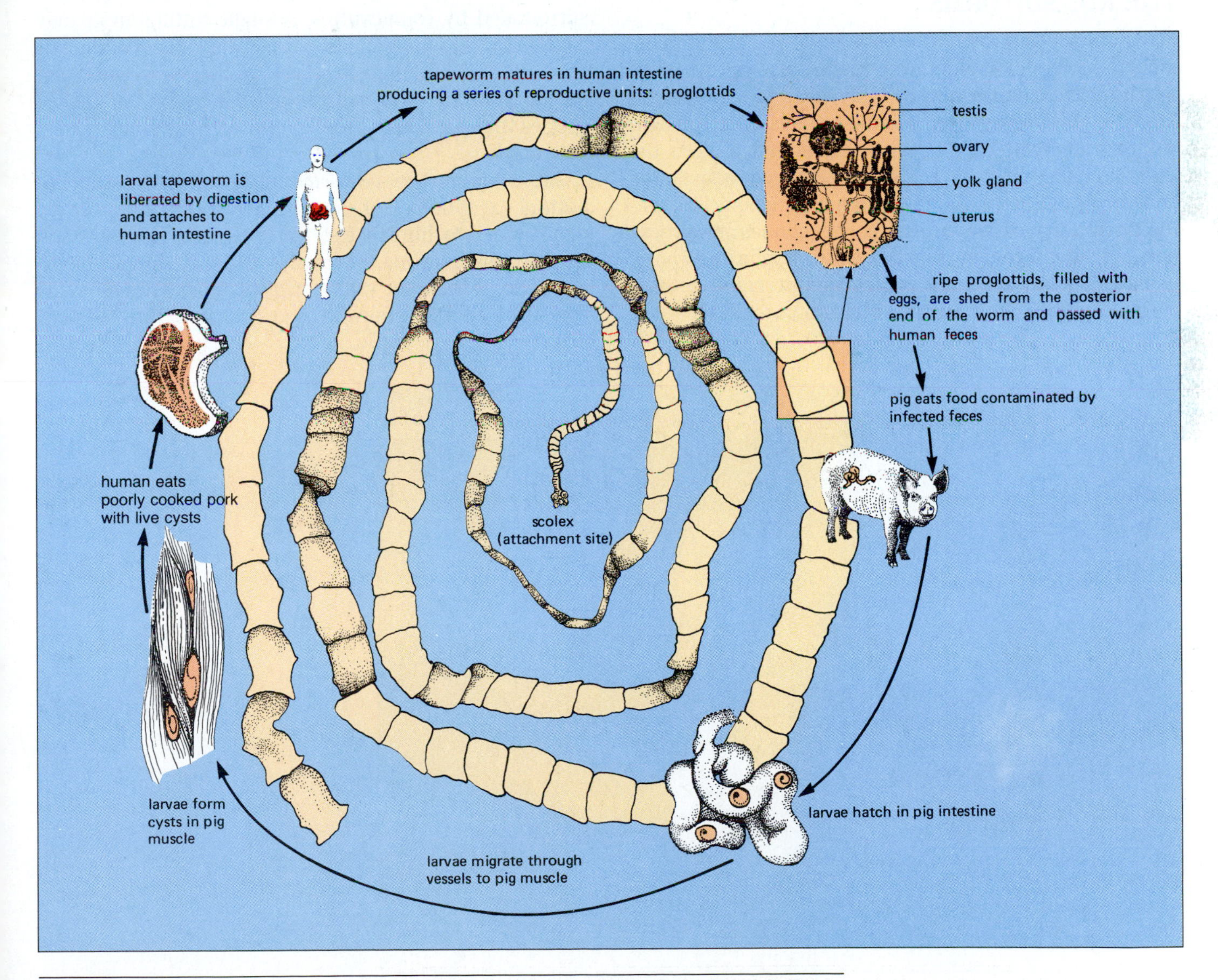

Figure 18-9 The life cycle of the human tapeworm. Each reproductive unit, or proglottid, is a self-contained reproductive factory including both male and female sex organs.

parasites, these flukes have a complex life cycle that includes an intermediate host, in this case a snail. The irrigation ditches filled by the Aswan dam in Egypt have contributed to the spread of schistosomiasis by creating extensive new habitat for the snail host.

Flatworms can reproduce both sexually and asexually. Free-living forms may reproduce by cinching themselves around the middle until they separate into two halves, each of which regenerates its missing parts. All can reproduce sexually, and many are **hermaphroditic** (possessing both male and female sexual organs). This is a great advantage to parasitic forms because it allows an individual living alone in a host's intestine to self-fertilize.

PHYLUM NEMATODA: THE ROUNDWORMS

Few habitats on Earth lack representatives of this enormously successful phylum. Nematodes, also called roundworms, have a rather simple body plan. It consists of a tubular gut running from mouth to anus, and a tough, flexible, nonliving cuticle that encloses and protects the thin, elongated body (Fig. 18-10). Sensory organs in the head transmit information to a simple ganglionic "brain" resembling that of flatworms. Reproduction is always sexual, and the sexes are separate, with the male (who is usually smaller) fertilizing the female internally.

Roundworms show a major evolutionary advance over the flatworms: they possess a separate mouth and anus. This arrangement, which characterizes nearly all complex animals, allows efficient, one-way passage of food through the digestive system. With the mouth at the head end of the animal, food may be ingested as soon as it is detected. One-way movement of food also permits specialization of the digestive tract into compartments that process the food in stages.

Although only about 10,000 species of nematodes have been named, there may be as many as 500,000 species. Although some parasitic forms reach a meter in length, most are microscopic, as shown in Figure 18-10. Nematodes lack both circulatory and respiratory systems. Since most are extremely thin and all have low energy requirements, diffusion suffices for gas exchange and the distribution of nutrients. Free-living forms are important decomposers in terrestrial and aquatic environments. Although blissfully ignorant of their presence, you are surrounded by roundworms. A single rotting apple may contain 90,000 individuals. Nearly all plants and animals are host to several parasitic species. Chances are good that during your life you will be invaded by one of the 50 species that parasitize people, most without doing noticeable damage.

Although most roundworms are harmless, there are important exceptions. For example, hookworm larvae in soil may bore into human feet, enter the bloodstream, and travel to the intestine, where they cause continuous bleeding. The *Trichinella* worm (causing trichinosis) may be ingested by eating improperly cooked pork. Infected pork

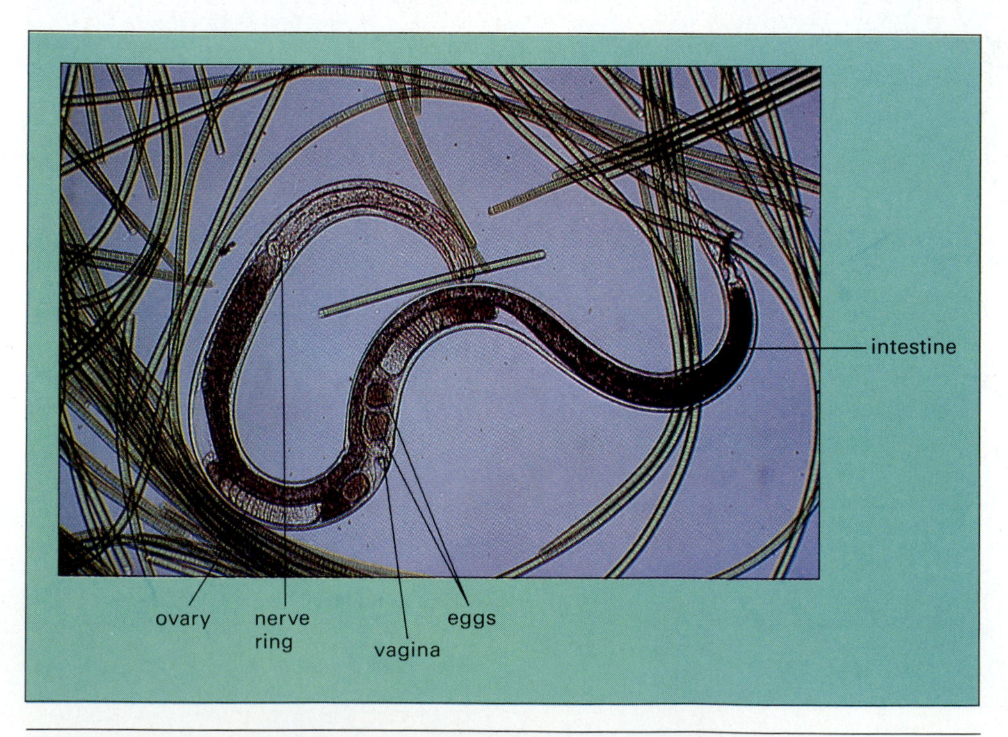

Figure 18-10 A freshwater nematode that feeds on algae. Eggs are visible in this female specimen.

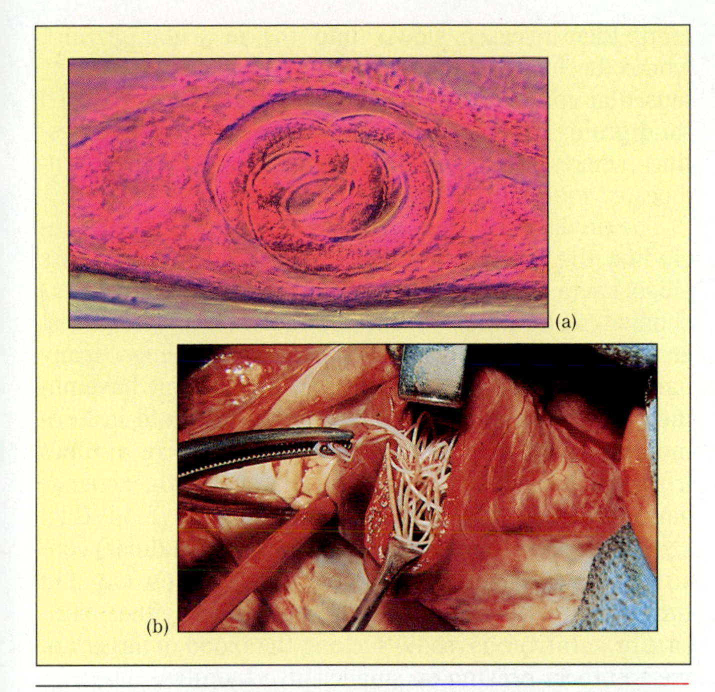

Figure 18-11 Some parasitic nematodes. **(a)** *Encysted larva of the* Trichinella *worm in muscle tissue, where it may live for up to 20 years.* **(b)** *Adult heartworms in the heart of a dog. The juveniles are released into the bloodstream where they may be ingested by mosquitos and passed to another dog by the bite of the infected mosquito.*

may contain up to 15,000 larval cysts per gram (Fig. 18-11a). The cysts hatch in the human digestive tract and invade blood vessels and muscles, causing bleeding and muscle damage. Another dangerous nematode parasite, the heartworm of dogs, is transmitted by mosquitoes (Fig. 18-11b). In the south, and increasingly in other parts of the country, it poses a severe threat to the health of unprotected pets.

PHYLUM ANNELIDA: THE SEGMENTED WORMS

As the name *annelid* (meaning "little ring" in Latin) suggests, a prominent feature of the phylum Annelida is the division of the body into a series of repeating segments. Externally, these are visible as ringlike depressions on the surface. Internally, many of the segments contain identical copies of nerve ganglia, excretory structures, and muscles (Fig. 18-12). Segmentation appears to be an evolutionary device for increasing body size with a minimum of new genetic information, since the "blueprint" for each segment is similar. In addition, the presence of numerous similar segments has allowed some to specialize for specific functions, increasing the complexity of the organism.

Figure 18-12 The earthworm, an annelid, showing an enlargement of segments, many of which are repeating similar units. The digestive system, which has two openings, is divided into a series of compartments specialized to process food in an orderly sequence.

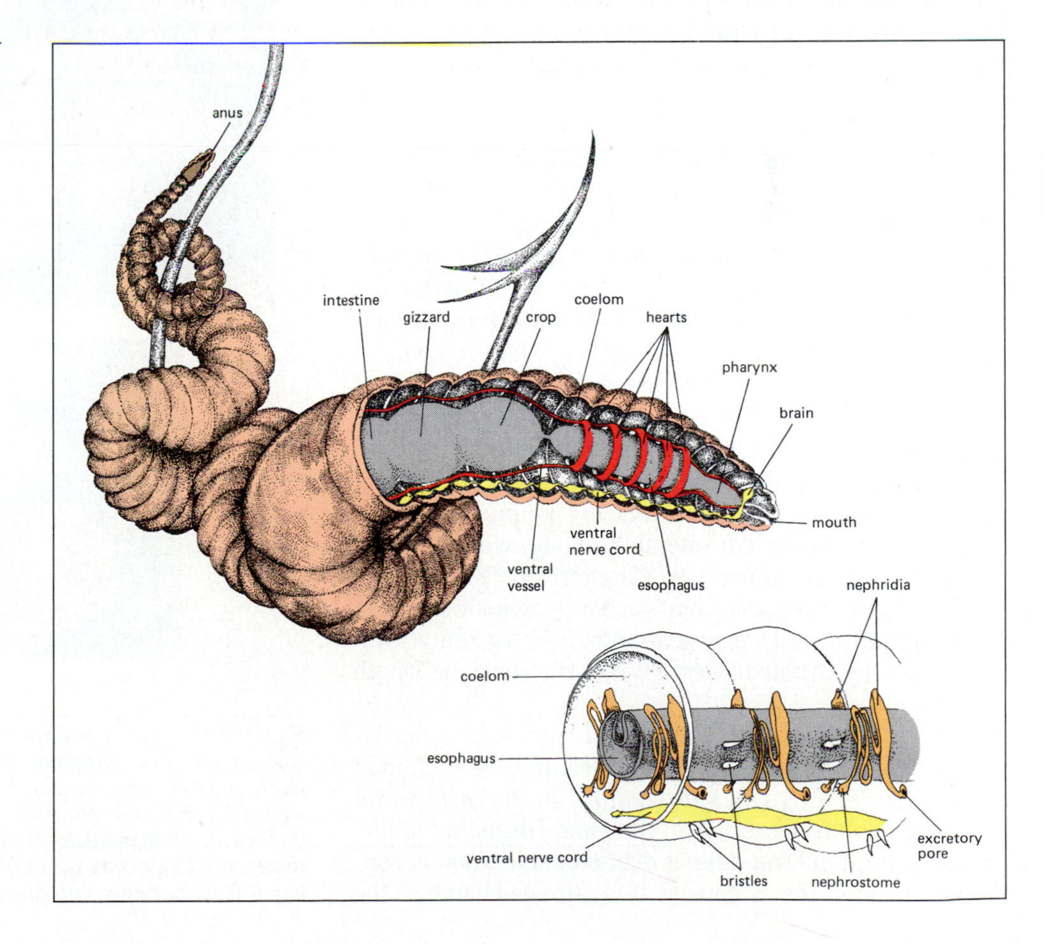

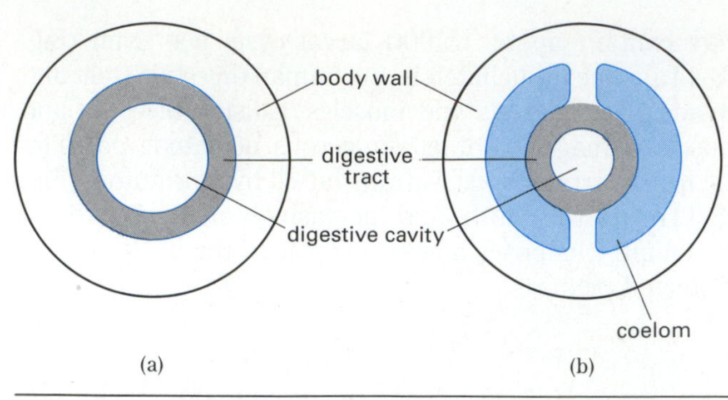

Figure 18-13 *Diagrammatic cross section of animal bodies* **(a)** *without and* **(b)** *with a coelom. Cnidarians and flatworms lack this internal space, having no separation between the digestive tract and body wall. Annelids and most other complex animals have their digestive tracts and other internal organs suspended within the coelom.*

Segmentation is also advantageous for locomotion since the body compartments, each controlled by separate muscles, collectively are capable of far greater complexity of movement than is seen in the nonsegmented worms.

A second evolutionary advance first seen in annelids is a fluid-filled space, called a **coelom,** that separates the body wall from the digestive tract (Fig. 18-13). The coelom serves as a protective buffer between delicate internal structures and the outside world. Its presence allows the internal organs to evolve greater complexity and to move independently of the external body wall. Thus the digestive tract can grind and churn, while the external worm remains calm. The incompressible fluid in the coelom in many annelids is confined by the partitions separating the segments and serves as a **hydrostatic skeleton,** a framework against which muscles can act, allowing feats such as burrowing through soil.

Annelids, in contrast to nematodes, have a well-developed circulatory system that distributes gases and nutrients throughout the body. In the earthworm, for example, blood with oxygen-carrying hemoglobin is pumped through well-developed vessels by five pairs of "hearts." These hearts are actually short, expanded segments of specialized blood vessel that contract rhythmically. The blood is filtered and wastes removed by excretory organs called nephridia that are found in many of the segments. Nephridia resemble the individual tubules of the vertebrate kidney (see Chapter 25). The annelid nervous system consists of a simple ganglionic brain in the head and a series of repeating paired segmental ganglia joined by a pair of nerve cords traveling the length of the body.

Digestion in annelids occurs in a series of compartments, each specialized for a different phase of food processing (Fig. 18-12). For example, in the earthworm, a muscular pharynx draws in the food, consisting of bits of decaying plant and animal debris in soil. Food is conducted through the esophagus to a storage chamber, the crop, then released slowly into the muscular gizzard, where the food is ground into tiny particles assisted by muscular contractions of the gizzard and the sharp-edged sand grains it contains. Food then passes into the intestine, where it is digested and nutrients absorbed. Undigested food and soil exit through the anus.

The phylum Annelida includes about 9000 species, including the familiar earthworm and its relatives (class Oligochaeta, meaning "few hairs"). In general, these exchange gas by diffusion through moist skin. The largest group of annelids (class Polychaeta, meaning "many hairs") are found primarily in the ocean. Some have numerous bristles and paired fleshy paddles, used in locomotion, on most of their segments. Others live in tubes from which they project feathery gills that both exchange gases and sift the water for microscopic food (Fig. 18-14a). A third group of annelids (class Hirudinea) consists of the leeches (Fig. 18-14b). These worms, found in freshwater or moist terrestrial habitats, are either parasitic or carnivorous, some sucking the blood of larger animals, others preying on smaller invertebrates.

Figure 18-14 **(a)** *A polychete annelid projects its brightly-colored spiraling gills from a tube in the ocean bed.* **(b)** *This leech, a freshwater annelid (class Hirudinea) found in a Georgia pond, shows numerous segments. The sucker encircles its mouth, allowing it to attach to its prey. Medicinal leeches were used by doctors up until the 1800's to suck the "tainted" blood from patients suffering from a variety of disorders.*

PHYLUM ARTHROPODA: THE INSECTS, ARACHNIDS, AND CRUSTACEANS

Spread your picnic tablecloth beneath a shading oak beside a stream-fed pond, and prepare to discover the diversity of arthropods. As you shoo flies from the potato salad, a yellowjacket may industriously attack your hamburger, flying off with a small piece of meat in its grasp. While a woolly caterpillar undulating up the tree trunk distracts you, ants will be discovering the cookie crumbs and a spider may lower itself into your midst, suspended by a gossamer thread. Watch carefully as you move a large sheltering stone in the stream for the sudden backward flipping of a crayfish. Dragonflies hover near the water's edge, their wings iridescent in the sunlight. Later, as dusk falls, you'll be glad you brought the mosquito repellent!

In numbers, both of individuals and species, arthropods are the dominant animals on Earth. About 1 million species have been discovered, and scientists estimate that up to 9 million remain undescribed. The phylum Arthropoda includes many classes, three of which are particularly large and important: class Insecta, class Arachnida (spiders and their relatives), and class Crustacea (crabs, shrimp, and their relatives). The success of this group can be attributed to several important adaptations that have allowed them to exploit nearly every possible habitat. Chief among these is the **exoskeleton** (Greek, "outside skeleton").

The exoskeleton, as its name suggests, is an external skeleton that encloses the arthropod body like a suit of armor. In places, it is thin and flexible, to allow movement of the paired, jointed appendages from which the phylum Arthropoda (Greek, "jointed foot") derives its name. The exoskeleton is secreted by the epidermis and composed chiefly of protein and a polysaccharide called **chitin.** It provides an important defense against small predators and is responsible for the greatly increased agility of arthropods over their ancestors, the annelid worms. By providing rigid attachment sites for muscles together with stiff but flexible appendages, the exoskeleton makes possible the flight of the bumblebee and the intricate, delicate manipulations of the spider as it weaves its web (Fig. 18-15). The exoskeleton also contributed enormously to the arthropod invasion of dry terrestrial habitats by providing a watertight covering for delicate, moist tissues, such as those used for gas exchange.

Although it offers some major advantages, the exoskeleton also poses some unique problems, sharing many of the undesirable traits of the suit of armor. First, because it cannot expand as the animal grows, the exoskeleton must be shed or **molted** periodically and replaced with a larger size (Fig. 18-16). This uses energy and leaves the animal temporarily vulnerable before the new skeleton hardens ("soft-shelled" crabs are eaten during this delicate period). The exoskeleton is also heavy; its weight increases exponentially as the animal grows. It is no coincidence that the largest arthropods are found among the crustaceans (crabs and lobsters), whose watery habitat supports much of their weight.

Figure 18-15 *A garden spider, having immobilized its prey with a paralyzing venom, rapidly encases it in web. Such dextrous manipulations are made possible by the exoskeleton and jointed appendages characteristic of arthropods.*

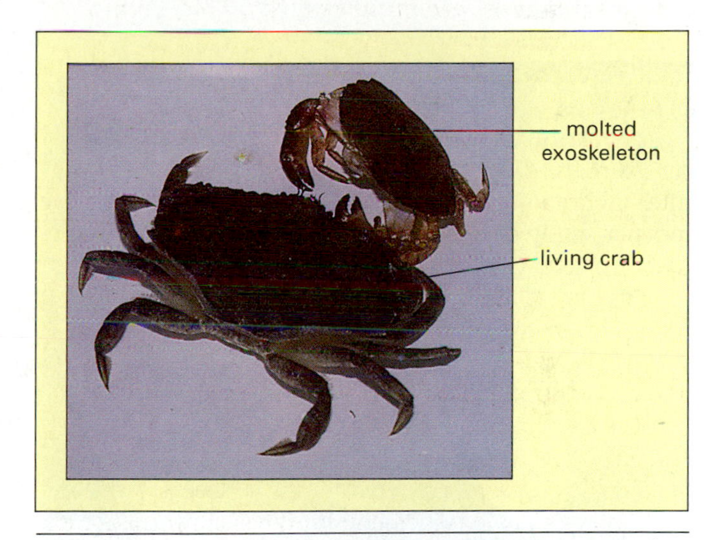

Figure 18-16 *A crab sitting beside its outgrown armor, shed the previous day. The newly emerged crab rapidly expands its new exoskeleton, which is temporarily soft, by taking up water into internal spaces. The exoskeleton then hardens in a size large enough to permit additional growth.*

Segmentation in arthropods is evidence of annelid ancestry. Arthropod segments, however, tend to be reduced in number, fused, and specialized for distinct functions such as locomotion, feeding, and sensing the environment (Fig. 18-17).

Most arthropods possess a well-developed sensory system, including complex compound eyes (Fig. 18-18) and acute chemical and tactile senses. Sensory information is

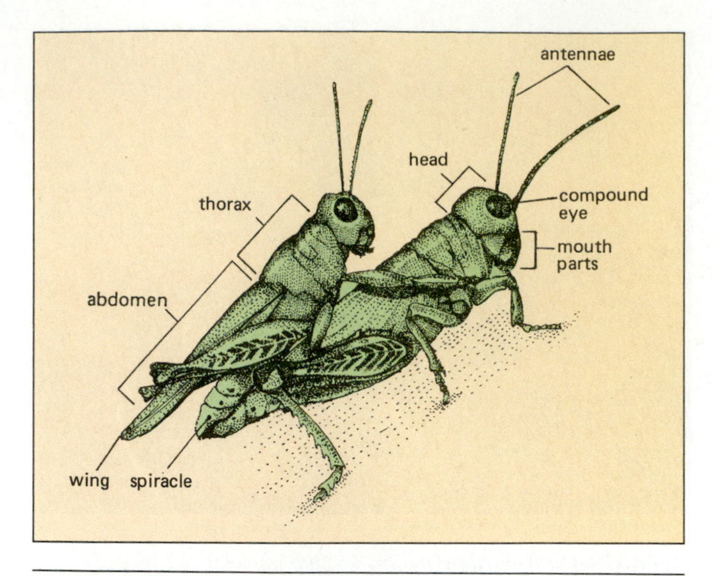

Figure 18-17 The grasshopper, an insect, shows fusion and specialization of body segments into a distinct head, thorax, and abdomen. Segments are visible beneath the wings on the abdomen.

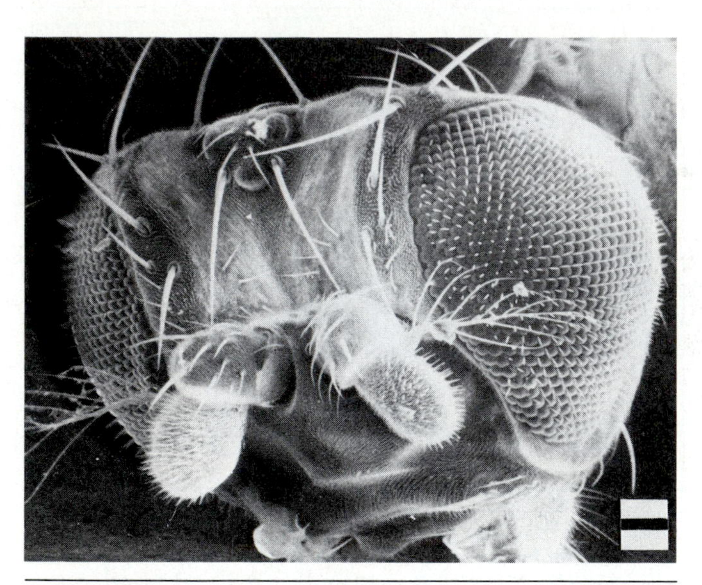

Figure 18-18 This scanning electron micrograph shows the compound eye of a fruitfly. Compound eyes consist of an array of similar light-gathering and sensing elements whose orientation gives the arthropod a panoramic view of the world. Insects have reasonable image-forming ability and good color discrimination.

processed by a nervous system similar to that of annelids, but more complex. The capacity for finely coordinated movement combined with sophisticated sensory abilities and a well-developed nervous system has allowed complex behavior to evolve. In fact, the interactions among certain social insects such as the honeybee are more complex than those of most vertebrate societies. Here, communication and genetically programmed learning play important roles (see Chapter 34).

Efficient gas exchange is necessary for the rapid movement of arthropods. This is accomplished by gills in aquatic forms such as the crustacea, and either **tracheae** or **book lungs** in terrestrial forms (Fig. 18-19). Arthropods have a well-developed circulatory system with a feature not seen in annelids: the **hemocoel,** or blood cavity. Blood not only travels through vessels, but also empties into the hemocoel, where it bathes internal organs directly. This arrangement, known as an **open circulatory system,** is also found in molluscs. Arthropod sexes are separate, and fertilization is internal.

Class Insecta

Insects are by far the most diverse and abundant arthropod class; the number of species is estimated at 800,000 (roughly the same as the total number of species in all other classes of animals combined, Fig. 18-20). Insects have three pairs of legs, usually supplemented by two pairs of wings. The capacity for flight distinguishes them from all other invertebrates and has contributed to their enormous success (Fig. 18-20c). As anyone who has unsuccessfully pursued a fly can testify, flight aids in escape from predators. It also allows the insect to find widely dispersed food. Locusts swarms have been traced from Saskatchewan, Canada, all the way into Texas on the trail of food. Flight requires rapid and efficient gas exchange. Insects use a network of narrow branching tubes called **tracheae,** which conduct air to all parts of the body (Fig. 18-19c).

During their development, insects undergo **metamorphosis,** which frequently involves a radical change in body form from juvenile to adult. The immature form is called a **larva,** which is wormlike in shape (e.g., the maggot of a housefly or the caterpillar of a moth). Metamorphosis may include a change in diet as well as shape, eliminating competition for food between adults and juveniles, and in some cases allowing the insect to exploit different foods when they are most available. For example, the caterpillar feeding on new green shoots in spring metamorphoses into the butterfly drinking nectar from summer flowers.

Class Arachnida

The arachnids comprise about 50,000 species of terrestrial arthropods, including spiders, mites, ticks, and scorpions (Fig. 18-21). All have eight walking legs, and most are carnivorous, many subsisting on a liquid diet consisting of blood or predigested prey. Spiders, the most numerous arachnids, first immobilize their prey with a paralyzing venom. They then inject digestive enzymes into the helpless victim (often an insect), and suck in the resulting soup. Arachnids breathe using tracheae or a unique arachnid "invention," **book lungs,** or both (Fig. 18-19b). Arachnids have simple eyes, each with a single lens, in contrast to the compound eyes of insects and crus-

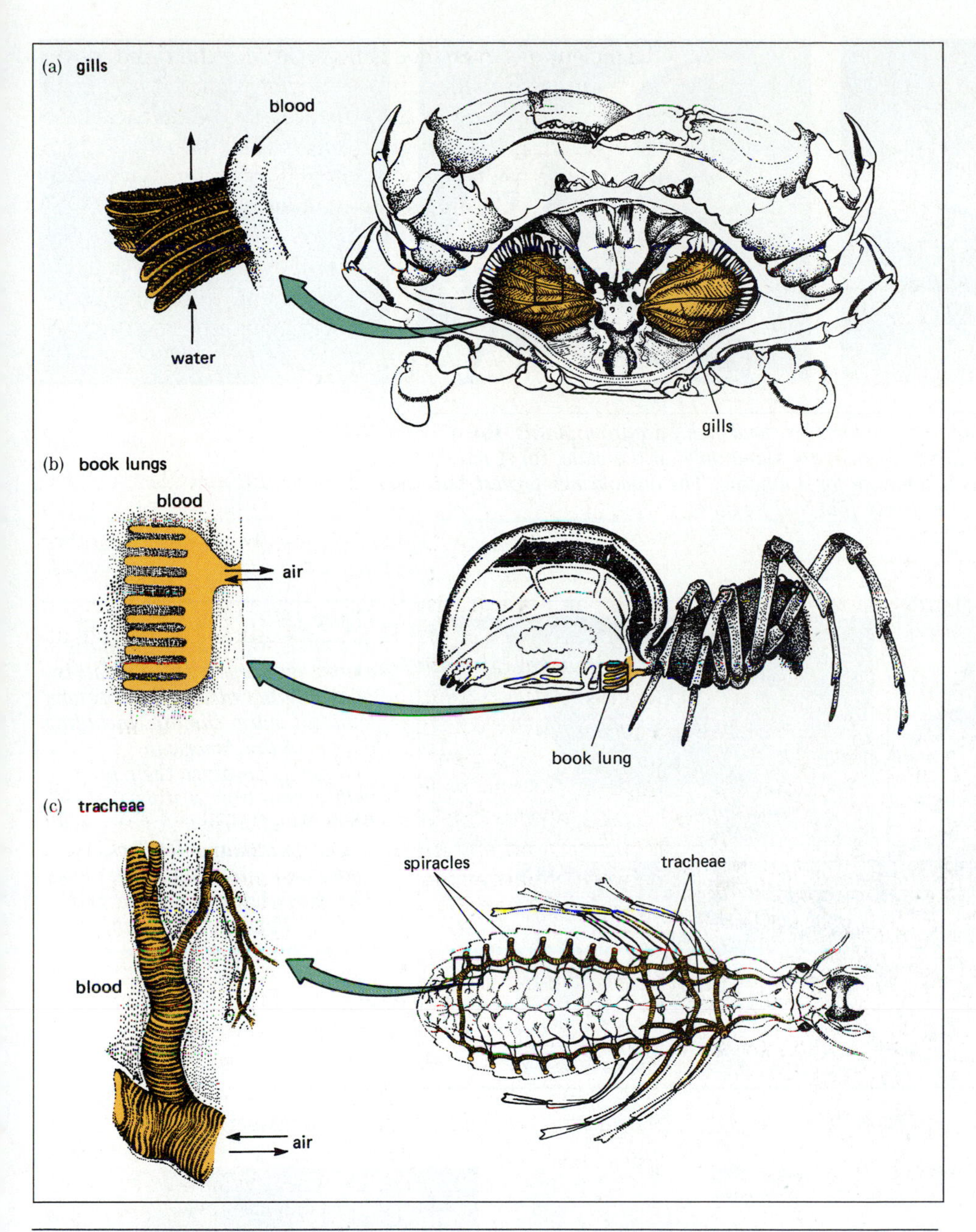

Figure 18-19 Arthropod respiratory structures. **(a)** *Gills, adapted for life in water, expose a large surface area of tissue rich in blood vessels to the water for gas exchange. Life on land demands protection of delicate, moist respiratory surfaces which are placed inside the body, with air entering through a small opening to minimize evaporation.* **(b)** *The book lungs of spiders resemble internal gills.* **(c)** *The internal tracheae of insects branch elaborately, carrying air close to each cell.*

tacea. The eyes are particularly sensitive to movement, but in some species they probably can form images. Most spiders have eight eyes, whose placement gives them a panoramic view of predators and prey.

Class Crustacea

The roughly 30,000 species of crustaceans, including crabs, crayfish, lobster, shrimp, and barnacles, comprise the only class of arthropods that is primarily aquatic (Fig. 18-22). Crustaceans range in size from the microscopic "water flea" (found in ponds) to the largest of all arthropods, the Japanese crab, with legs spanning up to 3.7 meters. Crustaceans have two pairs of antennae, but the rest of their appendages are highly variable in form and number, depending on the habitat and life-style of the species. Most have compound eyes similar to those of insects, and nearly all respire using gills (Fig. 18-19a).

Figure 18-20 *Insect diversity.* **(a)** *The rose aphid sucks sugar-rich juice from plants.* **(b)** *A mating pair of Hercules beetles. The large "horns" are found only on the male.* **(c)** *A may bug displays its two pairs of wings as it comes in for a landing. The outer wings protect the abdomen and inner wings, which are relatively thin and fragile.*

Figure 18-21 *The diversity of arachnids.* **(a)** *The tarantula is among the largest spiders, but is relatively harmless.* **(b)** *Scorpions, found in warm climates including deserts of the American southwest, paralyze their prey with venom from a stinger at the tip of the abdomen. A few species can harm humans.* **(c)** *Ticks before and after feeding on blood. The exoskeleton is flexible and folded, allowing the animal to become grotesquely bloated.*

Figure 18-22 *The diversity of crustacea.* **(a)** *The microscopic water flea* Daphnia, *common in freshwater ponds.* **(b)** *The sow bug, found in dark moist places such as under rocks, leaves, and decaying logs, is one of the few crustaceans to successfully invade the land.* **(c)** *The hermit crab protects its soft abdomen by inhabiting an abandoned snail shell.* **(d)** *The barnacle, an unusual sessile crustacean, anchors itself to rocks (and boats) with a protective shell. Barnacles were believed to be molluscs until the jointed legs were observed.*

PHYLUM MOLLUSCA: THE SNAILS, CLAMS, AND SQUID

Molluscs, like arthropods, are believed to have arisen from an annelid-like ancestor. Their numbers and variety (about 100,000 species have been described) are second only to the arthropods. Molluscs (whose name comes from the Latin *mollis*, meaning "soft") have a moist, muscular body without a skeleton. Some protect their body with a shell of calcium carbonate, while others escape predation by tasting terrible or moving swiftly. The molluscan circulatory system is open, with blood directly bathing the organs in a hemocoel. The nervous system resembles that of annelids and arthropods, but many more of the ganglia are concentrated in the brain. Reproduction is always sexual; both separate sexes and hermaphrodites are represented.

Among the many classes of molluscs, three of outstanding importance will be discussed in more detail: the class Gastropoda (snails and their relatives), the class Pelecypoda (clams and their relatives), and the class Cephalopoda (octopuses and their relatives).

Figure 18-23 The diversity of gastropods. **(a)** *A giant African land snail. Each tentacle bears a simple eye at the tip.* **(b)** *Spanish shawl sea slugs prepare to mate. The brilliant colors of many nudibranchs warn potential predators that they are distasteful.*

Class Gastropoda

This group of about 35,000 known species crawl on a muscular foot (Gastropoda is Greek for "stomach foot"), and many have shells that vary widely in form and color. Some of the most beautiful gastropods, the "sea slugs," are shell-less; their brilliant colors warn predators that they are poisonous or at least bad-tasting (Fig. 18-23). Gastropods feed with a **radula,** a flexible ribbon of tissue studded with spines that is used to scrape algae from rocks or grasp larger plants or prey. Most gastropods respire using gills in addition to their moist skin, through which dissolved gases readily diffuse. The gills may be enclosed in a cavity beneath the shell, or exposed, as in the sea slugs. A few gastropods (including the destructive garden snails and slugs) live in moist terrestrial habitats. These terrestrial gastropods (and some freshwater forms that evolved from them) breathe using a simple lung.

Figure 18-24 The diversity of pelecypods. **(a)** *This swimming scallop from Vancouver parts its hinged shells, revealing an array of blue eyes. The upper shell is covered with an encrusting sponge.* **(b)** *Mussels are sessile and attach to rocks in dense aggregations exposed at low tide.*

Class Pelecypoda

Included in this class are the scallops, oysters, mussels, and clams. Not only do pelecypods lend exotic variety to the human diet, they are extremely important members of the marine intertidal community (Fig. 18-24). Pelecypods possess two shells connected by a flexible hinge. A strong muscle can clamp the shells closed in response to danger (this muscle is what you are served when you order scallops in a restaurant). Most pelecypods are sessile. Since the head is an adaptation to moving about in a directional manner, members of this group have lost their heads during evolution. Pelecypods are filter feeders, drawing water over gills covered with a thin layer of mu-

Figure 18-25 The diversity of cephalopods. **(a)** *An octopus crawls using its eight suckered tentacles. It can alter its color and skin texture to blend with its surroundings. In emergencies this mollusc can jet backwards by vigorously contracting its mantle. Octopuses and squid can emit clouds of dark purple ink to confuse pursuing predators.* **(b)** *The squid moves entirely by jet propulsion by contracting its mantle, pushing the animal backwards through the water. The giant squid is the largest invertebrate, reaching a length of 15 meters, including tentacles.* **(c)** *The chambered nautilus secretes a shell with internal, gas-filled chambers providing buoyancy in the water. Note the well-developed eyes, and tentacles used to capture prey.*

cus that traps microscopic food particles. Food is conveyed by beating cilia on the gills to the mouth. A muscular foot is used by clams for burrowing in sand or mud. In mussels, the foot is reduced in size and used to help secrete a set of threads that anchor the animal to rocks. Scallops lack a foot, moving by a sort of whimsical jet propulsion achieved by flapping their shells together.

Class Cephalopoda

This fascinating group, including octopuses, squid, nautiluses, and cuttlefish (Fig. 18-25), includes the largest, swiftest, and smartest of all invertebrates. All cephalopods are predatory carnivores, and all are marine. The foot has evolved into tentacles with well-developed chemosensory abilities and suction disks for detecting and grasping prey. Prey grasped by tentacles may be immobilized by a paralyzing venom in the saliva before being torn apart by beaklike jaws. The cephalopod eye resembles our own in complexity and exceeds it in efficiency of design (see Fig. 30-7). Cephalopods move rapidly by jet propulsion accomplished by forceful expulsion of water from the mantle cavity. The octopus may also travel along the seafloor using its tentacles like multiple, undulating legs. The cephalopod brain, especially that of the octopus, is exceptional, and in many ways resembles the brain of a vertebrate. It is enclosed in a skull-like case of cartilage and endows the octopus with highly developed capabilities to learn and remember.

PHYLUM ECHINODERMATA: THE SEA STARS, SEA URCHINS, AND SEA CUCUMBERS

Although echinoderms have evolved a bewildering diversity of forms, they have never left their ancestral home on the ocean floor. Their descriptive common names reflect this: sand dollar, sea urchin, sea star (or starfish), sea cucumber, and sea lily (Fig. 18-26). Although their free-swimming embryos are bilaterally symmetrical, echinoderm adults have radial symmetry, an adaptation to a sluggish, in some forms a sessile, existence. Most echinoderms lack a head and move very slowly and in any direction, feeding on algae or small particles sifted from sand or water. The sea star (or starfish) is a predator. It can slowly pursue prey (including pelecypod molluscs) from any direction (Fig. 18-27). Echinoderms move on numerous tiny **tube feet,** delicate cylindrical projections that extend from the ventral surface of the body, terminating in a suction cup. Tube feet are part of a unique echinoderm feature, the **water-vascular system,** which functions in locomotion, respiration, and food capture (Fig. 18-28). Seawater enters through an opening (the sieve plate) on the animal's dorsal surface and is conducted through a ring canal that encircles the esophagus, from which branch a number of radial canals. These conduct water to the tube feet, each of which is controlled by a muscular squeeze bulb (ampulla). Contraction of the bulb forces water into the tube foot, causing it to extend. The suction cup may be pressed against the substrate or a food object, to which it adheres tightly until pressure is released.

Figure 18-26 *The diversity of echinoderms.* **(a)** *A sea cucumber off southern California feeds on debris in the sand.* **(b)** *The sea urchin's spines are actually projections of the internal skeleton.* **(c)** *The sea star has reduced spines, and often has five legs. This specimen is seen amidst colorful cnidarians called cup corals.*

Figure 18-27 *The sea star often feeds on pelecypod molluscs such as this mussel. Numerous tube feet are attached to the shells, exerting a relentless pull. The sea star everts the delicate tissue of its stomach through the centrally located ventral mouth. A gape in the bivalve shells of less than 1 mm. is sufficient for the stomach tissue to insinuate between the shells, secreting digestive enzymes which weaken the mollusc, causing it to gape further. Partially digested food is transported to the upper portion of the stomach where digestion is completed.*

Figure 18-28 *The water-vascular system of echinoderms. Sea-water enters through the sieve plate and is transported into the ring canal, from which it is distributed to each of the arms through radial canals. The water inflates squeeze-bulb-like ampullae which expand and contract to extend or retract the tube feet. The plates of the endoskeleton can be seen embedded in the body wall.*

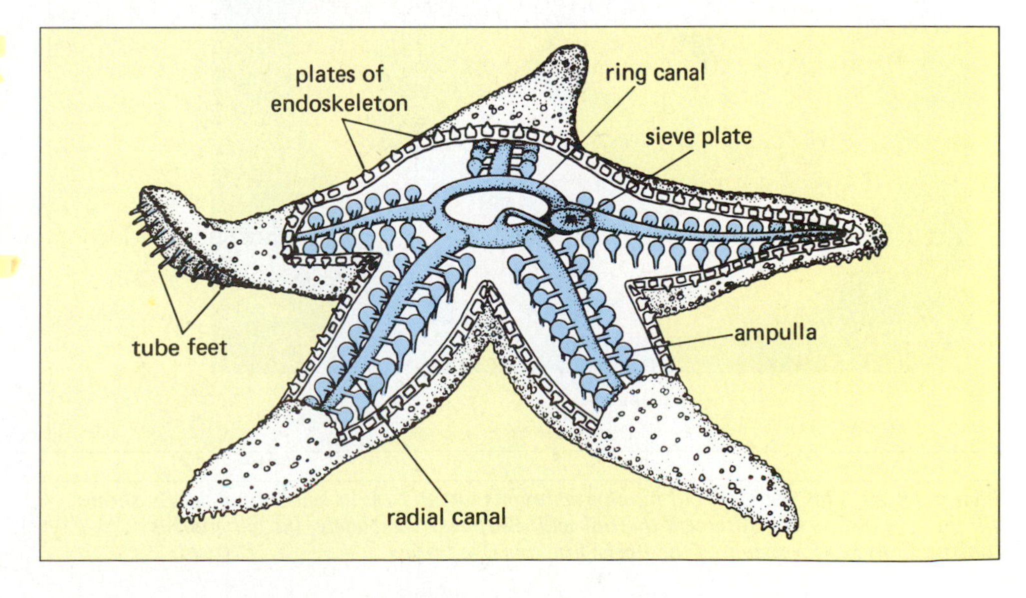

Echinoderms have a relatively simple nervous system with no distinct brain. Movements are loosely coordinated using a nerve ring encircling the esophagus, radial nerves to the rest of the body, and a nerve network through the epidermis. In sea stars, simple receptors for light and chemicals are concentrated on the arm tips, and sensory cells are also scattered over the skin. The echinoderms lack a circulatory system, although movement of the fluid in their well-developed coelom serves this function. Gas exchange occurs through the tube feet, and in some forms, numerous tiny "skin gills" project through the epidermis. Sea cucumbers possess an internal system of canals called a respiratory tree. Most species reproduce by shedding sperm and eggs into the water, where fertilization occurs, and a free-swimming larva develops. The sexes are usually separate. Sea stars have the ability to regenerate lost parts; new individuals may form from a single arm, provided that part of the central body is attached. When mussel fisherman tried to rid their mussel beds of predatory sea stars by hacking them into pieces and throwing them back, needless to say, the stratagem backfired!

Echinoderms possess an **endoskeleton** composed of plates of calcium carbonate formed beneath the outer skin (Fig. 18-28). The name *echinoderm* (meaning "hedgehog skin" in Greek) comes from projections of the endoskeleton that extend as bumps or spines through the epidermis. These are especially pronounced in the sea urchins, and much reduced in the sea stars and sea cucumbers.

PHYLUM CHORDATA: THE TUNICATES, LANCELETS, AND VERTEBRATES

The chordates are an extremely diverse group united by four features that all possess at some stage of their lives:

A notochord: A stiff but flexible rod that extends the length of the body and provides an attachment site for muscles.

A dorsal nerve cord: Lying dorsal to the digestive tract, this hollow, nervous structure develops a thickening at its anterior end which becomes a brain.

Pharyngeal gill grooves: Located in the pharynx (the cavity behind the mouth), these may form functional openings or may appear only as grooves during an early stage of development.

A tail: An extension of the body past the anus.

This list must seem particularly puzzling since humans are chordates, and at first glance we seem to lack every feature except the second. But evolutionary relationships are sometimes seen most clearly during early stages of development, and it is then that we develop, and lose, our **notochord,** our gill grooves, and finally, our tail (Fig. 18-29). We share these chordate features with other vertebrates and with two invertebrate chordate groups: the lancelets and the tunicates.

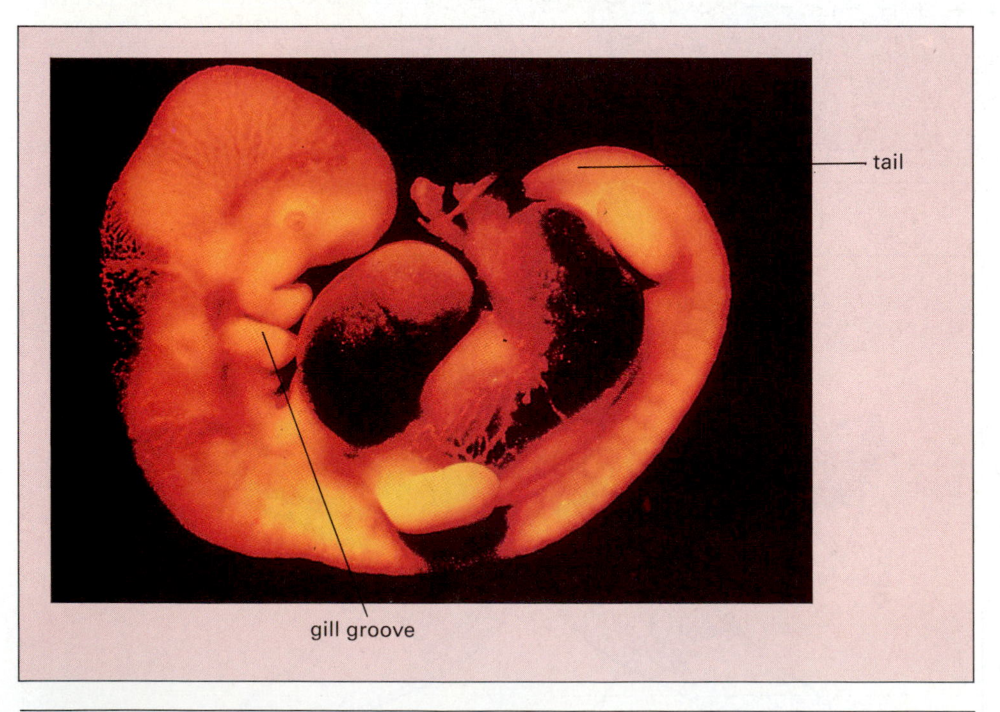

Figure 18-29 *The five-week-old human embryo is about 1 cm in length and clearly shows gill grooves and a tail. Although the tail will disappear completely, the gill grooves contribute to the formation of the lower jaw and the larynx.*

The Invertebrate Chordates

The invertebrate chordates lack a head, and of course, lack the backbone that distinguishes the vertebrates. The small (5 centimeters) fishlike lancelet (also called *Amphioxus*) is an invertebrate chordate that spends most of its time half-buried in the sandy sea bottom, filtering tiny food particles from the water. As seen in Figure 18-30a, all the typical chordate features are present in the adult organism.

The tunicates form a larger group of marine invertebrate chordates. It is difficult to imagine a less likely relative than this sessile, filter-feeding vase (Fig. 18-30b,c). Its ability to move is limited to a forceful contraction of the saclike body (sending a jet of seawater into the face of anyone who plucks it from its undersea home, hence the common name "sea squirt"). However, tunicates produce actively swimming tadpole-like larvae that possess all the proper chordate features (Fig. 18-30d). Since fossils of intermediate forms have never been discovered, the exact sequence of evolutionary events that led from invertebrate chordates to the first, fishlike vertebrates remains shrouded in mystery.

Subphylum Vertebrata

In the vertebrates, the embryonic notochord is replaced during development by a backbone or **vertebral column** composed of **cartilage** or bone. This structure provides support for the body, an attachment site for muscles, and protection for the delicate nerve cord and brain. The backbone is part of a living endoskeleton, capable of growth and self-repair. Because the internal skeleton provides support without the armorlike weight of the arthropod exoskeleton, it has allowed vertebrates to achieve great size and mobility and has contributed to their invasion of the land and the air. Today, vertebrates are represented by six major classes: cartilaginous fishes (class Chondrichthyes), bony fishes (class Osteichthyes), amphibians (class Amphibia), reptiles (class Reptilia), birds (class Aves), and mammals (class Mammalia).

Vertebrates arose in the sea; the earliest vertebrate fossils are those of strange jawless fishes (ancestors of today's lamprey eels), protected by bony armor plates. These gave rise to a group that possessed an important new structure found in all more advanced vertebrates: jaws. Although these first jawed fishes have been extinct for

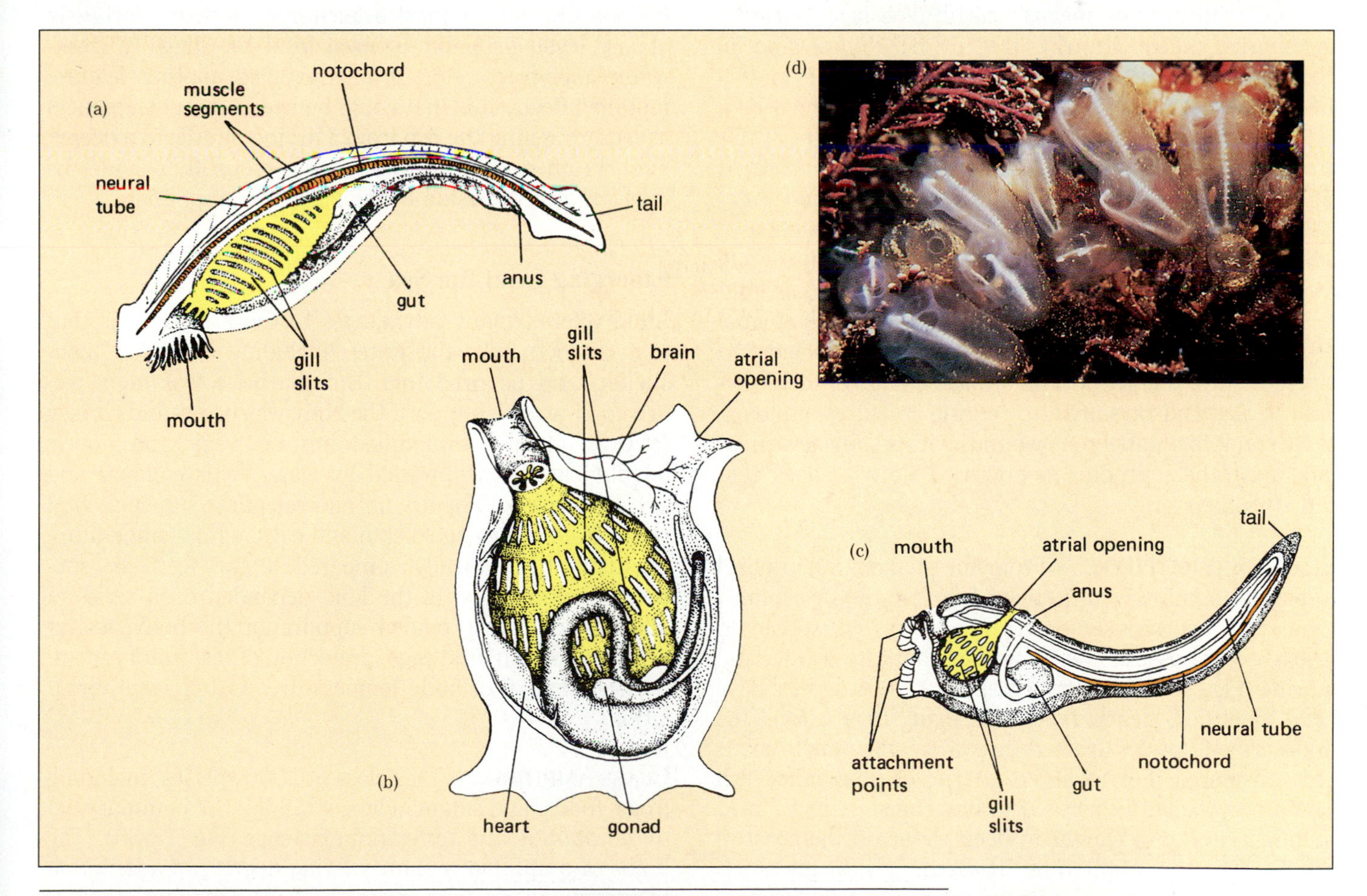

Figure 18-30 **(a)** Amphioxus, *a fishlike invertebrate chordate.* **(b)** *An adult tunicate, showing internal anatomy.* **(c)** *The larval tunicate has all the chordate features.* **(d)** *A living tunicate off southern California.*

Figure 18-31 *Two members of the class Chondrichthyes.* **(a)** *The tropical blue-spotted sting ray swims by graceful undulations of lateral extensions of the body.* **(b)** *A sand tiger shark displaying several rows of teeth. As outer teeth are lost, they are replaced by new ones formed behind them. Both sharks and rays lack a swim bladder, and tend to sink toward the bottom when they stop swimming.*

230 million years, they are the ancestors of the two major classes of fishes that survive today.

CLASS CHONDRICHTHYES. This marine group of 625 species, whose name means "cartilage fishes," includes the sharks, skates, and rays (Fig. 18-31). These graceful predators lack any bone in their skeleton, which is formed entirely of flexible cartilage. The body is protected by a leathery skin roughened by tiny scales. Members of this group respire using gills. Although some must swim to circulate water through the gills, most can pump water across their gills. These and all fish have a two-chambered heart (see Chapter 26). Sharks may have several rows of razor-sharp teeth, the back rows moving forward as front teeth are lost. Although a few species consider us potential prey, most sharks are shy of humans. Sharks include the largest fishes; the gentle whale shark can grow to 15 meters. Skates and rays are also retiring creatures, although some can inflict dangerous wounds with a spine near their tail, and others produce a powerful electric shock that can stun their prey.

CLASS OSTEICHTHYES. From the snakelike moray eel to bizarre, luminescent deep-sea forms to the streamlined tuna, this enormously successful group has spread to nearly every possible watery habitat, both freshwater and marine (Fig. 18-32). Although about 17,000 species have been identified, nearly twice this many may exist if the undescribed species from deep water and remote areas are considered. For example, a type of lobe-finned fish called a coelacanth, believed to have been extinct for 75 million years, was caught in deep water off the coast of South Africa in 1939 (Fig. 18-32d). The name osteichthyes, literally "bony fishes," refers to their skeleton, which is composed of bone rather than cartilage. Another feature found in early representatives of this group and retained in a few modern species is the presence of lungs that supplement the gills. Lungs are adaptations allowing life in fresh water, which could become foul and stagnant or dry up entirely. The swimbladder, a sort of internal balloon that allows most osteichthyes to float effortlessly at any level, probably evolved from the lungs of freshwater ancestors. Some groups evolved another feature, modified fleshy fins that could be used (in an emergency) as legs, dragging the fish from a drying puddle to a deeper pool. From such ancestors arose a group that made the first tentative invasion of the land: the amphibians.

Emerging from the Sea

Land offered many advantages to those organisms that first crawled from the water, including abundant food, shelter, and no predators. But the price was high. Deprived of water's support, the body was heavy and clumsy to drag along on modified fins or weak and poorly adapted legs. Unsupported by water, gills collapse and become useless. The dry air and relentless sun suck vital water from unprotected skin and eggs, while temperature fluctuates dramatically compared to the sea. The successful colonization of the land depended on a series of adaptations that provided support for the body, waterproofing for skin and eggs, protection of respiratory membranes, control of body temperature, and efficient circulation.

CLASS AMPHIBIA. This class of 2500 species, including frogs, toads, and salamanders, straddles the boundary between aquatic and terrestrial existence (Fig. 18-33). The limbs of amphibians show varying degrees of adaptation to movement on land, from the belly-dragging crawl of salamanders to the efficient leaping of frogs. Lungs replace gills in most adult forms, and a three-chambered

Figure 18-32 The diversity of the bony fishes attests to their successful invasion of nearly every aquatic habitat. **(a)** *This female deep sea angler fish attracts prey with a living lure projecting just above her mouth. In the 2000m depth where anglers live, no light penetrates. Thus colors are superfluous, and the fish is ghostly white. Male deep sea anglers are extremely small, and attach to the female early in life. Here they remain as permanent parasites, always available to fertilize her eggs. Two parasitic males can be seen attached to this female.* **(b)** *The tropical green moray is rid of parasites by another fish, the banded cleaner goby.* **(c)** *The tropical sea horse may anchor itself with its prehensile tail while feeding on small crustaceans.* **(d)** *A rare photo of a coelacanth, a "living fossil," in its natural habitat.*

Figure 18-33 The "double life" of amphibians is illustrated by the transition from tadpole **(a)** *to bullfrog* **(b)**. **(c)** *The red salamander is restricted to moist habitats in the eastern U.S.*

heart (in contrast to the two-chambered heart of fishes) circulates blood more efficiently (see Chapter 26). However, the skin of frogs and salamanders must remain moist, since it serves as an additional respiratory organ to supplement poorly developed lungs. This greatly restricts their habitats on land, and frogs are rarely far from water. Fertilization of amphibian eggs is external and must occur in water so that the sperm can swim to the eggs. The eggs are particularly vulnerable to water loss, being surrounded only with a jellylike coating. Thus they are laid in water and develop into aquatic larvae—the tadpoles of frogs and toads, for example. The dramatic

Figure 18-34 The diversity of reptiles. **(a)** *The mountain king snake has evolved a color pattern very similar to the poisonous coral snake, which potential predators avoid. This mimicry helps the harmless king snake avoid predation.* **(b)** *The American alligator, found in swampy areas of the south, has survived with little change for 150 million years.*

transition from completely aquatic larva to semiterrestrial adult gives the class Amphibia its name, meaning "double life."

CLASS REPTILIA. The approximately 7000 species of reptile include the lizards and snakes (by far the most successful groups), and the turtles, alligators, and crocodiles, which have survived virtually unchanged from prehistoric times (Fig. 18-34). Reptiles evolved from an amphibian ancestor about 250 million years ago. Their descendants, the dinosaurs, ruled the land for nearly 150 million years. Some reptiles, particularly desert dwellers such as tortoises and lizards, have achieved complete independence from their aquatic origins. This was achieved through a series of adaptations of which three are outstanding. First, reptiles evolved a dry, tough, scaly skin that resists water loss while protecting the body. Second, reptiles evolved internal fertilization, in which the male deposits sperm within the female's body. Third, the shelled **amniotic egg** of reptiles can be buried in sand or dirt, far from water with its hungry predators. The shell prevents the egg from drying, while an internal membrane, the **amnion,** encloses the embryo in the watery environment that all developing animals require (Fig. 18-35). To supplement these features, reptiles evolved more efficient lungs, dispensing with the skin as a respiratory organ. The three-chambered heart improved to allow better separation of oxygenated and deoxygenated blood, and the limbs and skeleton were modified to provide better support and more efficient movement on land.

CLASS AVES. Having conquered the sea and the land, vertebrates took to the air, a source of abundant insect food and a haven from predators. The 8600 species of birds attest to the success of this strategy (Fig. 18-36). The first birdlike creatures, reptiles modified for flight, appeared roughly 150 million years ago (Fig. 18-37). Body scales were dramatically modified to form feathers, while those on the legs remained, testimony to their reptilian origin.

Many aspects of bird anatomy and physiology support the rigorous demands of flight. In contrast to reptiles, birds maintain an elevated body temperature. "Warm-bloodedness" allows both muscles and metabolic processes to operate at peak efficiency, supplying the power and the energy necessary to fly. The high metabolic rate demands efficient oxygenation of tissues. To meet this need, birds have a four-chambered heart that completely separates oxygenated and deoxygenated blood. The respiratory system is supplemented by air sacs that supply oxygenated air to the lungs even as the bird exhales. Feathers protect and insulate the body; they also form lightweight extensions to the wings and tail for the lift and control demanded by flight. Hollow bones reduce the weight of the skeleton to a fraction of that in other vertebrates. Reproductive organs are considerably reduced in size during nonbreeding periods, and female birds possess only a single ovary, minimizing weight. The shelled egg that contributed to the reptiles' success on land frees the mother bird from carrying her developing offspring. The nervous system of birds accommodates the special demands of flight with extraordinary coordination and balance combined with acute eyesight.

CLASS MAMMALIA. As one line of reptiles was developing feathers, a different group, the mammals, was evolving hair, also a modification of scales. The mammals came into prominence after the extinction of the dinosaurs roughly 70 million years ago, and today are represented by some 4500 species. Like birds, mammals are warm-blooded, with high metabolic rates. In most mammals, fur protects and insulates the warm body. Like birds, mammals have four-chambered hearts that increase the amount of oxygen delivered to the tissues. Legs designed for running rather than crawling, and a high metabolic rate, make many mammals fast and agile. In contrast to birds, whose bodies are almost uniformly molded to the requirement of flight, mammals have

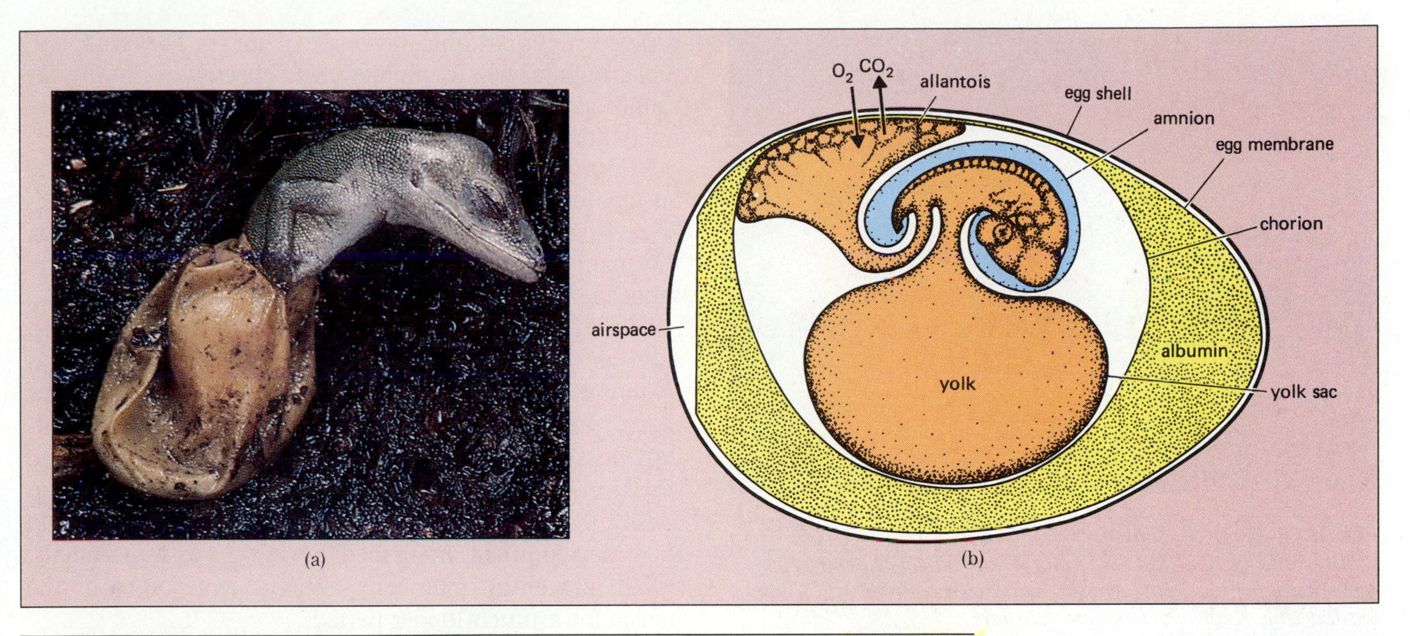

Figure 18-35 **(a)** *An anole lizard from Carolina struggles free of its egg.* **(b)** *The amniotic egg of reptiles and birds is shown diagramatically. In addition to the shell which helps prevent dehydration, the egg contains several membranes. Enclosing the embryo in a watery "pond" is the amnion. The allantois stores the urinary wastes of the embryo, and exchanges gases which can diffuse in and out through the shell. A yolk sac surrounds the fatty yolk, a high-energy food source. The chorion encloses the embryo with its membranes. Outside the chorion is the albumin, a protein food source surrounded by the egg membrane.*

Figure 18-36 *The diversity of birds.* **(a)** *The delicate hummingbird beats its wings about 60 times/sec and weighs about 4 grams.* **(b)** *This young frigate bird, a fish-eater from the Galapagos islands, has nearly outgrown its nest.* **(c)** *The ostrich is the largest of all birds, weighing 317 lbs (144 kilograms) and producing eggs weighing over 3 lbs (1500 grams).*

Figure 18-37 Archaeopteryx, *the "missing link" between reptiles and birds. This 150 million year old fossil shows a remarkable animal possessing a beak with sharp teeth, a long jointed tail, clawed wings, and feathers.*

evolved a remarkable diversity of form. The seal, bat, mole, impala, whale, monkey, and cheetah exemplify the radiation of mammals into nearly all habitats, with bodies finely adapted to their varied life-styles (Fig. 18-38).

This group is named for the **mammary glands** used by all members of this class to suckle their young (Fig. 18-38c). In addition to these unique milk-producing glands, the mammalian body is arrayed with sweat, scent, and sebaceous (oil-producing) glands, not found in reptiles.

With the exception of the egg-laying **monotremes,** such as the platypus and spiny anteater (Fig. 18-39a), mammals give birth to live young that develop in the uterus. In one specialized group, the **marsupials** (including opossums, koalas, and kangaroos), the young develop briefly in the uterus, then crawl to a protective pouch (Fig. 18-39b). Here they firmly grasp a nipple and complete their development nourished by milk. Most mammals, called **placental** mammals, retain their young in the uterus for a much longer period.

Figure 18-38 The diversity of mammals. **(a)** *A humpback whale gives its offspring a boost.* **(b)** *A bat, the only mammal capable of true flight, navigates at night using a kind of sonar. Large ears aid in detecting echoes as its high-pitched cries bounce off nearby objects.* **(c)** *Mammals are named after the mammary glands with which females nurse their young, as illustrated by this mother cheetah.*

Figure 18-39 Non-placental mammals. **(a)** *Monotremes, such as this platypus from Australia, lay leathery eggs resembling those of reptiles. The newly hatched young obtain milk from slitlike openings in the mother's abdomen.* **(b)** *Marsupials, such as the kangaroo, give birth to extremely immature young who immediately grasp a nipple and develop within the mother's protective pouch.*

The mammalian nervous system has contributed significantly to the success of this group by allowing behavioral adaptation to changing and varied environments. The cerebral cortex is more highly developed than in any other class, endowing mammals with unparalleled curiosity and learning ability. This allows them to alter their behavior based on experience, and helps them survive in a changing world. Recently, much of that change is due to the activities of a single mammalian species, *Homo sapiens,* whose intellectual development has led to domination of the environment.

REFLECTIONS: ARE HUMANS A BIOLOGICAL SUCCESS?

"... for now we have taken
The primal powers, creation and annihilation; we make
new elements such as God never saw.
We can explode atoms and annul the fragments, nothing left
but pure energy, we shall use it
In peace and in war ...
We have minds like the tusks of those forgotten tigers
hypertrophied and terrible ...

From "Passenger Pigeons" by Robinson Jeffers

Physically, humans are fairly unimpressive biological specimens. We are not very strong for such large animals, nor very fast, and we lack the natural weapons of fang and claw. It is the human mind, with its tremendously developed cerebral cortex, that truly sets us apart from other animals. Our minds, in single bursts of brilliance, and in collective pursuit of common goals, have created wonders. No other animal could even appreciate the Parthenon, much less sculpt its graceful columns. We alone can eradicate smallpox and polio, domesticate other life forms, penetrate space with our rockets, and fly to the stars in our imaginations.

And yet, are we, as it appears at first glance, the most successful of all living things? The few-hundred-thousand-year duration of human existence is a mere instant in the 3-billion-year span of life on Earth. But during the last 300 years, our population has increased from 500 million to 5 billion, and may double within the next 35 years. Such continued growth is unprecedented among natural populations. Is this a measure of our success? As we have expanded our range over the globe, we have driven at least 300 major species to extinction. Rapid destruction of tropical rain forests and other diverse habitats may wipe out an additional 500,000 species of plants, invertebrates, and vertebrates, including many we will never know, within your lifetime.

Many of our activities have altered the environment in ways inimical to life, including our own. Acid pollution from power plants and autos rains down on the land, threatening our forests and lakes—and eroding the Parthenon. Deserts spread as land is stripped of its cover by overgrazing and the demand for firewood. Parts of the land and the ocean have been rendered devoid of life by the poisonous wastes of modern civilization. Our aggressive tendencies, spurred by pressures of expanding wants and needs, their scope magnified by our technological prowess, have given us the capacity to destroy ourselves and most other life forms as well.

The human mind is the source of our power over the environment. Our brains, "hypertrophied and terrible," are simultaneously the source of our most pressing problems—and our greatest hope for solving them. Are we a phenomenal biological success, or are we a brilliant, but short-lived, "flash in the pan"? Perhaps the next few centuries will tell.

SUMMARY OF KEY CONCEPTS

Animals are multicellular heterotrophic organisms, most of which can perceive and react rapidly to environmental stimuli. For convenience, animals are often placed into one of two major groups. Vertebrates, animals with backbones, comprise a single subphylum within the phylum Chordata. All other animals lack a backbone and are called invertebrates. Invertebrates comprise over 99 percent of the animal species on Earth.

Phylum Porifera: The Sponges

Sponge bodies are often free-form in shape and are always sessile. Sponges have relatively few types of cells, and although division of labor among the cell types is present, there is little coordination of activity. Sponges lack the muscles and nerves required for coordinated movement, and digestion occurs exclusively within the individual cells.

Phylum Cnidaria: The Hydra, Anemones, and Jellyfish

Far more coordination is evident among cells of cnidarians than occurs in sponges. A simple network of neurons is present, directing loosely coordinated activity of contractile cells, allowing coordinated movements. Digestion is extracellular, occurring in a central gastrovascular cavity with a single opening serving as mouth and anus. Cnidarians exhibit radial symmetry, an adaptation to the free-floating life-style of the medusa or the sedentary existence of the polyp.

The Worms: Phylum Platyhelminthes, Phylum Nematoda, and Phylum Annelida

Several advances in complexity are evident in the worms. Flatworms (phylum Platyhelminthes) are the simplest to show a distinct head with sensory organs and a simple brain. A system of canals forming a network through the body aids in excretion. Phylum Nematoda (the roundworms) are the first to possess a separate mouth and anus. The segmented worms (phylum Annelida) are the most complex, with a well-developed, closed circulatory system and excretory organs resembling the basic unit of the vertebrate kidney. The segmented worms have a compartmentalized digestive system, like that of vertebrates, which processes food in a sequence. A final evolutionary advance found in annelids is the coelom, a fluid-filled space between the body wall and the internal organs, found in most complex animals, including vertebrates.

Phylum Arthropoda: The Insects, Arachnids, and Crustaceans

Arthropods are the most diverse and abundant organisms on Earth. They have invaded nearly every available terrestrial and aquatic habitat. Jointed appendages and well-developed nervous systems allow complex, finely coordinated behavior. The exoskeleton (which conserves water and provides support) and specialized respiratory structures (which remain moist and protected) allow the insects and arachnids to inhabit dry land. The diversification of insects has been enhanced by their ability to fly. Crustaceans are restricted to moist, usually aquatic habitats, respire using gills, and include the largest arthropods.

Phylum Mollusca: The Snails, Clams, and Squids

This highly successful group lacks a skeleton, sometimes protecting the soft, moist, muscular body with a single shell (as in many gastropods and a few cephalopods) or a pair of hinged shells (as in the pelecypods). The lack of a waterproof external covering limits this phylum to aquatic and moist terrestrial habitats. Although the body plan of gastropods and pelecypods limits the complexity of their behavior, the cephalopod's tentacles are capable of precisely controlled movements. The octopus has the most complex brain and the best-developed learning capacity of any invertebrate.

Phylum Echinodermata: The Sea Stars, Sea Urchins, and Sea Cucumbers

This is an exclusively marine group. Although like other complex invertebrates and chordates, echinoderm larvae are bilaterally symmetrical, the adults show radial symmetry. This, in addition to their primitive nervous system lacking any definite brain, adapts them to a relatively sedentary existence. Echinoderm bodies are supported by a nonliving internal skeleton that sends projections through the skin. The water-vascular system, which functions in locomotion, feeding, and respiration, is a unique echinoderm feature.

Phylum Chordata: The Tunicates, Lancelets, and Vertebrates

The phylum Chordata includes two invertebrate groups, the lancelets and tunicates, as well as the familiar vertebrates. All possess a notochord, a dorsal nerve cord, gill grooves, and a tail at some stage in their development. Vertebrates are a subphylum of chordates that have a backbone which is part of a living endoskeleton. Vertebrate evolution is believed to have proceeded from the fishes to amphibians to reptiles, which gave rise to both birds and mammals. The heart increases in complexity from the two-chambered heart of fishes, to three in amphibians and most reptiles, to four in the warm-blooded birds and mammals. During the progression from fishes to amphibians to reptiles, a series of adaptations evolved that helped vertebrates colonize dry land. Amphibians have legs and most have simple lungs for air breathing, but most are confined to relatively damp terrestrial habitats by their moist skin, use of external fertilization, and the requirement that their eggs and larvae develop in water. Reptiles, with well-developed lungs, dry skin covered with relatively waterproof scales, internal fertilization, and the amniotic egg with its own water supply, are well adapted to the driest terrestrial habitats. Birds and mammals are also fully terrestrial and have additional adaptations, such as an elevated body temperature that allows the muscles to respond rapidly regardless of the temperature of the environment. The bird body is molded for flight, with feathers, hollow bones, efficient circulatory and respiratory systems, and well-developed eyes. Mammals have insulating hair and give birth to live young that are nourished with milk. The mammalian nervous system is the most complex in the animal kingdom, providing mammals with a unique learning ability that helps them adapt to changing environments.

GLOSSARY

Amniotic egg (am-nē-ot'-ik): the egg of reptiles and birds. It contains an amnion that encloses the embryo in a watery environment; this allows the egg to be laid on dry land.

Anterior (an-tēr'-ē-ur): the front, forward, or head end of an animal.

Asexual reproduction: reproduction that does not involve the fusion of haploid sex cells. The parent body may divide and new parts regenerate, or a new, smaller individual may be formed attached to the parent, to drop off when complete.

Bilateral symmetry: body plan in which only a single plane drawn through the central axis will divide the body into mirror-image halves.

Book lungs: thin layers of tissue resembling pages in a book, enclosed in a chamber and used as a respiratory organ by certain types of arachnids.

Budding: a form of asexual reproduction in which the adult produces miniature versions of itself that drop off and assume independent existence.

Carnivorous (kar-niv'-e-rus): feeding on the bodies of other living animals.

Cartilage (kart'-lij): flexible, translucent tissue that serves as the forerunner of bone during embryonic development in most vertebrates. In the class Chondrichthyes, cartilage is retained and forms the entire skeleton.

Chemoreceptors (kē'-mō-rē-sep-ters): sensory cells specialized for detecting chemicals, such as those in food.

Chitin (kī'-tin): a tough, flexible polysaccharide; an important constituent of the arthropod exoskeleton.

Closed circulatory system: a type of circulatory system in which the blood is always enclosed in the heart and vessels.

Coelom (sē'-lōm): a space or cavity within the body separating the body wall from the inner organs.

Collar cells: specialized cells lining the inside channels of sponges. Flagella extend from a sievelike collar, creating a water current that draws microscopic organisms through the collar to be trapped.

Compound eye: an image-forming eye consisting of numerous similar light-gathering and light-detecting elements.

Cyst (sist): an encapsulated resting stage in the life cycle of certain invertebrates, such as parasitic flatworms and roundworms.

Dorsal (dōr'-sul): the top, back, or uppermost surface of an animal oriented with its head forward.

Endoskeleton: a supportive structure within the body; an internal skeleton. It may be nonliving, as in echinoderms and sponges, or living, as in vertebrates.

Exoskeleton: an external, nonliving supporting structure; an external skeleton.

Extracellular (ex-tra-sel'-ya-ler): occurring outside cells, usually in a specialized chamber surrounded by cells.

Flame cells: cells in flatworms specialized for excretion and fluid regulation. They enclose a small chamber full of beating cilia, whose flickering appearance gives them their name.

Free-living: not parasitic.

Ganglion (gan'-glē-un): an aggregation of neurons.

Gastrovascular cavity (gas'-trō-vas'-kū-lar): a saclike chamber in the bodies of some invertebrates (such as cnidarians and flatworms) with a single opening serving as both mouth and anus.

Hemocoel (hē'-mō-sēl): a blood cavity within the bodies of certain invertebrates in which blood bathes tissues directly. A hemocoel is part of an open circulatory system.

Hermaphroditic (her-maf'-ruh-dit'-ik): (Hermes and Aphrodite were male and female Greek gods) possessing both male and female sexual organs. Some hermaphroditic animals can fertilize themselves; others must exchange sex cells with a mate.

Hydrostatic skeleton (hī-drō-stat'-ik): the use of fluid contained in body compartments to provide support for the body and mass against which muscles can contract.

Intracellular (in-tra-sel'-ya-ler): occurring within individual cells.

Invertebrate (in-vert'-uh-bret): a category of animals that never possess a vertebral column.

Larva (lar'-vuh) an immature form of an organism prior to metamorphosis into its adult form. The caterpillars of moths and butterflies, and the maggots of flies, are larvae.

Mammary glands (mam'-uh-rē) milk-producing organs used by female mammals to nourish their young.

Mantle (man'-tul) an extension of the body wall in certain invertebrates, such as molluscs. It may secrete a shell, protect the gills, and, as in cephalopods, aid in locomotion.

Marsupial (mar-sū'-pē-ul) a type of mammal whose young are born at an extremely immature stage and undergo further development in a pouch while they remain attached to a mammary gland. Includes kangaroos, opossums, and koalas.

Medusa (meh-dū'-suh) a bell-shaped, often free-swimming stage in the life cycle of many cnidarians. Jellyfish are one example.

Metamorphosis (met-uh-mor'-fuh-ses) a dramatic change in body form during development, as seen in amphibians (tadpole to frog) and insects (caterpillar to butterfly).

Molt to shed an external body covering, such as an exoskeleton, skin, feathers, or fur.

Nematocyst (nēm-āt'-ō-sist) a specialized cell found in cnidarians which, when disturbed, ejects a sticky or poisoned thread. Used by cnidarians to trap and sting their prey.

Nerve cord also called the spinal cord of vertebrates, a hollow nervous structure lying along the dorsal side of the body of chordates.

Nerve net a loosely coordinated network of neurons.

Notochord (nōt'-ō-kōrd) a stiff but somewhat flexible, supportive rod found in all members of the phylum Chordata at some stage of development.

Open circulatory system a type of circulatory system in arthropods and molluscs in which the blood is pumped through an open space (the hemocoel), where it bathes the internal organs directly.

Osculum (os'-kya-lum) relatively large opening in the sponge body through which water is expelled.

Parasitic (par-uh-sit'-ik) living in or on the body of another organism, and causing it harm as a result.

Pharynx (fār'-inx) a portion of the digestive system between the mouth and the esophagus. In flatworms, it is developed as an extensible, muscular organ.

Photoreceptors (fō'-tō-rē-sep-ters) sensory cells specialized for detecting light and differences in light intensity.

Placenta (pluh-sen'-ta) a tissue rich in blood vessels which develops in the mammalian uterus during pregnancy. Here nutrients and oxygen from maternal blood are exchanged for wastes from the developing embryo.

Polyp (pol′-ip) the sedentary, vase-shaped stage in the life cycle of many cnidarians. Hydra and sea anemones are examples.

Posterior (pos-tēr′-ē-ur) the tail, hindmost, or rear end of an animal.

Radial symmetry a body plan in which any plane drawn along a central axis will divide the body into approximately mirror-image halves. Cnidarians and many adult echinoderms show radial symmetry.

Radula (ra′-dū-luh) a ribbon of tissue in the mouth of gastropod molluscs that bears numerous teeth on its outer surface and is used to scrape and drag food into the mouth.

Segmentation (seg-men-tā′-shun) division of the body into repeated, often similar units.

Sessile (ses′-ul) not free to move about, usually permanently attached to a surface.

Tentacle (ten′-te-kul) an elongate, extensible projection of the body of cnidarians and cephalopod molluscs that may be used for grasping, stinging, and immobilizing prey, and locomotion.

Tube feet cylindrical extensions of the water-vascular system of echinoderms, used for locomotion, grasping food, and respiration.

Ventral (ven′-trul) the lower, or underside of an animal whose head is oriented forward.

Water-vascular system a system in echinoderms consisting of a series of canals through which seawater is conducted and used to inflate tube feet for locomotion, grasping food, and respiration.

STUDY QUESTIONS

1. After each characteristic listed, state the phylum or phyla to which it applies.
   a. radial symmetry
   b. a gastrovascular cavity
   c. an open circulatory system
   d. some members are sessile
   e. lacks nerves and muscles
   f. a water-vascular system
   g. an internal skeleton
   h. an external skeleton
   i. some members have a radula
   j. some members have tentacles
   k. a notochord
   l. more species than all others combined
   m. segmentation
   n. includes both vertebrate and invertebrate members
   o. includes members that can fly
2. Of the phyla described in the text, list the first (simplest) to possess each of the following features:
   a. a gastrovascular cavity
   b. bilateral symmetry
   c. a coelom
   d. nerves
   e. a one-way digestive tract
   f. a closed circulatory system
   g. radial symmetry
3. Describe and compare respiratory systems in the three major arthropod classes.
4. Describe the advantages and disadvantages of the arthropod exoskeleton.
5. List three ways in which the ability to fly has contributed to the success and diversity of insects.
6. State in which of the three major mollusc classes each of the following characteristics is found.
   a. two hinged shells
   b. a radula
   c. tentacles
   d. some sessile members
   e. the best-developed brains
   f. numerous eyes
7. Give three functions of the water-vascular system of echinoderms.
8. Explain the origin of echinoderm spines.
9. To what life-style is radial symmetry an adaptation? bilateral symmetry?
10. List the vertebrate class (or classes) in which we find each of the following.
    a. a skeleton of cartilage
    b. a two-chambered heart
    c. a four-chambered heart
    d. lungs supplemented by air sacs
    e. the amniotic egg
    f. warm-bloodedness
    g. a placenta
11. Distinguish between vertebrates and invertebrates. List the major phyla found in each broad grouping.
12. List four distinguishing features of chordates.
13. Describe the ways in which amphibians are adapted to life on land, and in what ways they are still restricted to a watery or moist environment.
14. List the adaptations that distinguish reptiles from amphibians and help them adapt to life in dry terrestrial environments.
15. List the adaptations of birds that contribute to their ability to fly.
16. How do mammals differ from birds, and what adaptations do they share?
17. How has the mammalian nervous system contributed to the success of this group?

SUGGESTED READINGS

Barnes, R. D. *Invertebrate Zoology*. 5th ed. Philadelphia: Saunders Publishing Co., 1987.

Goreau, T. F., Goreau, N. I., and Goreau, T. J. "Corals and Coral Reefs." *Scientific American*, August 1979.

Hickman, C. P., Roberts, L. S., and Hickman, F. M. *Biology of Animals*. St. Louis: The C. V. Mosby Company, 1982.

Horridge, G. A., "The Compound Eye of Insects." *Scientific American*, July 1977.

Pough, F. H., Heiser, J. B., and McFarland, W. N. *Vertebrate Life*. 3rd ed. New York: Macmillan Publishing Company, 1989.

Rahn, H., Ar, A., and Paganelli, C. V. "How Bird Eggs Breathe." *Scientific American*, February 1979. (Offprint No. 1420).

Russell-Hunter, W. D. *A Life of Invertebrates*. New York: Macmillan Publishing Company, 1979.

19
The Diversity of Life. III. Fungi and Plants

Grouping plants and fungi into a single chapter has a certain logic to it—the same logic that originally caused fungi to be placed in the plant kingdom. Faced with the pale, sedentary mushroom growing out of the ground and asked to classify it as animal or plant, the human animal knew instinctively that it wasn't one of *us* and must therefore be one of *them:* a plant. In reality, fungi are as different from plants as they are from animals. The assumption that they are merely a "lower" form of plant has hampered human understanding and appreciation of their complexity, diversity, and uniqueness. The roles of fungi and plants in living communities differ significantly. Plants are the "builders," using energy captured from sunlight to synthesize complex molecules from simple raw materials in the air and soil. Fungi are the "decomposers," insinuating the microscopic filaments of their bodies throughout the dead bodies and wastes of plants and animals. Here they break down complex molecules, obtaining energy and releasing simpler molecules back into the air, water, and soil. Plants and fungi thus participate in complementary ways in recycling the raw materials of life. In this chapter we explore the variety of each of these two kingdoms.

THE KINGDOM FUNGI

Several features distinguish fungi from animals and plants. The fungal body, with rare exceptions, is filamentous. It consists of microscopically thin, threadlike structures called **hyphae** (singular: hypha), that grow together in an interwoven mass called a **mycelium** (Fig. 19-1). The hyphae of some fungi consist of single elongated cells with numerous nuclei, while in others the hyphae may be subdivided by partitions into many cells, each containing one or a few nuclei. Pores in the partitions allow the cytoplasm to stream between cells, distributing nutrients.

In contrast to the nuclei found in the cells of animals and the dominant forms of most plants, those of the fungal body are usually haploid (possessing only a single set of chromosomes). Most fungal cells are surrounded by an outer wall composed not of cellulose as in plants, but of chitin, the same substance as that found in the exoskeletons of arthropods.

The fungal body spreads thinly beneath the soil, under the bark of decaying logs, or inside aging bread or cheese. Periodically, however, the hyphae aggregate and differentiate into reproductive structures that project above the surface (see Figs. 19-2a; 19-3). These structures, seen as mushrooms, puffballs, or the powdery molds on food, are often the only visible part of the fungus.

Fungi obtain their nutrients by breaking down the bodies of dead and sometimes living organisms. Like animals, they are heterotrophic, but their means of obtaining nutrients resembles that of bacteria. As in bacteria, the cell walls of fungi prevent them from ingesting food. Fungal cells thus must release enzymes outside their bodies to break down surrounding food. They can then absorb the simple nutrients directly into the cells. The fungal body, composed of long filaments only a single cell thick, presents an enormous surface area for nutrient absorption. Rapidly growing filaments penetrate deeply into a source of nutrients—a dead animal, decaying log, or the soil itself—hastening decomposition through digestion of the material. External digestion makes both fungi and bacteria important decomposers in ecosystems.

Not all fungi are restricted to nonliving food sources; fungi are among the most devastating plant parasites. Animals, including humans, are also attacked by several parasitic species, including "ringworm," "athlete's foot," and yeast infections of the vagina. Although over 50,000 species of fungi have been described, we have only begun to comprehend fungal diversity; at least 1000 additional species are described yearly. Estimates of the total, including undiscovered species, range up to 250,000, equal to the number of species of flowering plants.

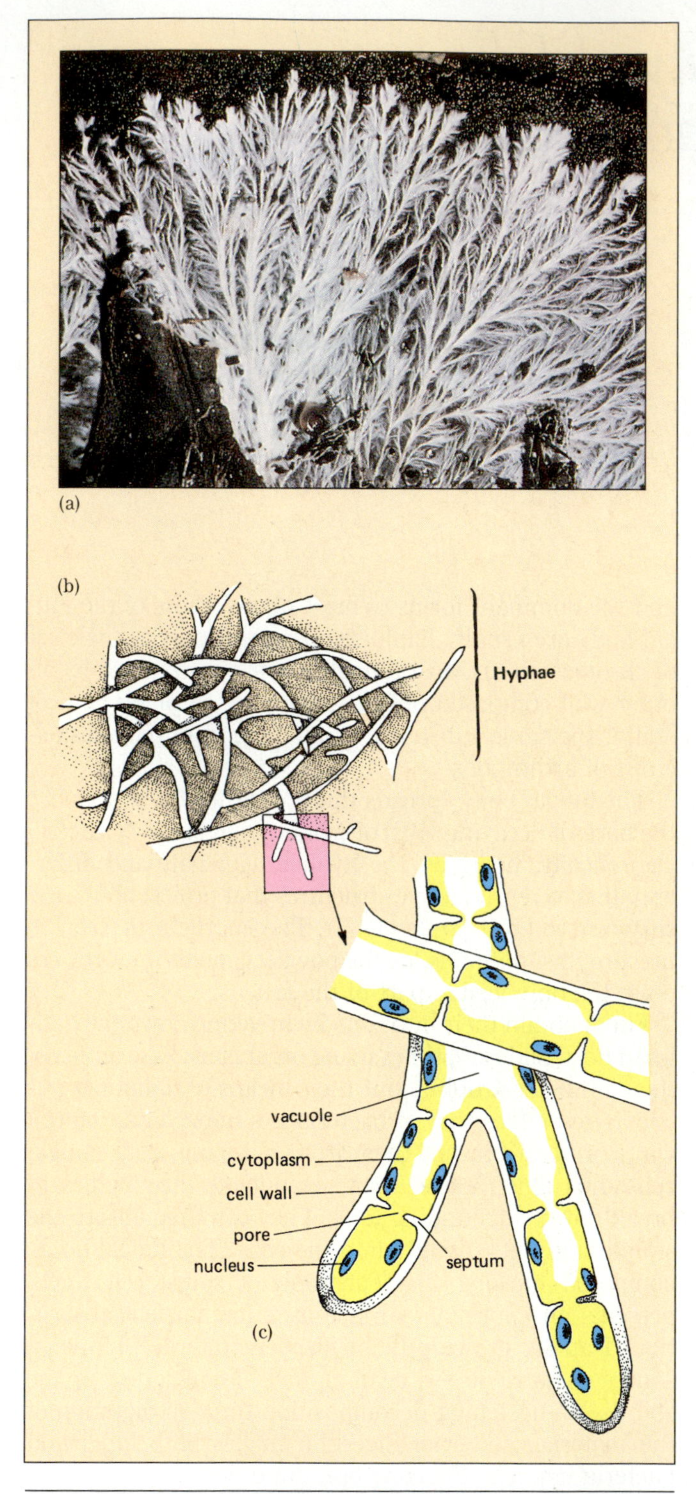

Figure 19-1 **(a)** *A fungal mycelium, about the size of your hand, spreads over a decaying log. The mycelium is composed of a tangle of microscopic hyphae, drawn in* **(b)** *and enlarged in* **(c)** *to show their internal organization.*

Fungal reproduction is varied and often complex. In its simplest asexual form, a mycelium breaks into pieces, each of which grows into a new individual. Fungi frequently reproduce by spores formed either sexually or asexually inside special spore cases. These microscopic spores are dispersed by wind or water. The reproductive capacity of fungi is prodigious: giant puffballs may contain 5 *trillion* spores.

Fungal taxonomy, like that of microorganisms, is the subject of considerable controversy. For convenience, we will divide the kingdom Fungi into the true fungi, which have cell walls, filamentous bodies and feed by absorption, and the protist-like slime molds, which resemble amoebae and engulf their food. Major divisions of true fungi include the zygomycetes, ascomycetes, fungi imperfecti, basidiomycetes, and oomycetes. There are two major types of slime mold: cellular and acellular.

True Fungi

ZYGOMYCETES OR "ZYGOTE FUNGI." This group of about 600 species is named for the ability of its haploid hyphae to "mate," fusing their nuclei to produce a diploid **zygospore** (Fig. 19-2a). This resistant structure is dispersed through the air and can remain dormant until conditions are favorable for growth. Then it undergoes meiosis, producing haploid cells that grow into new hyphae. These hyphae may either reproduce asexually by forming haploid spores in black spore cases (Fig. 19-2b), or sexually by fusing to produce more zygospores. The zygomycetes include the dung fungus *Pilobolus* (Greek for

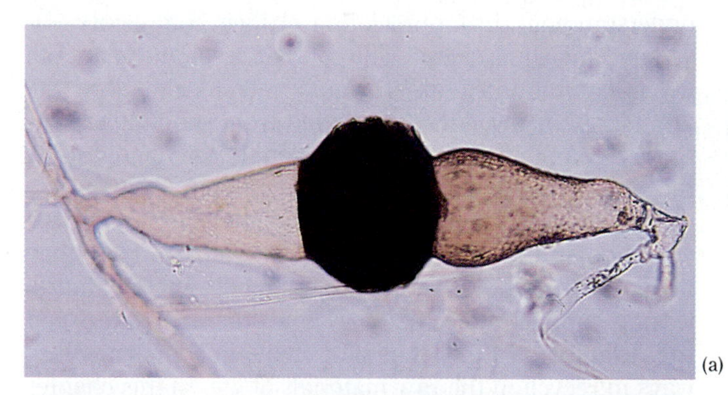
(a)

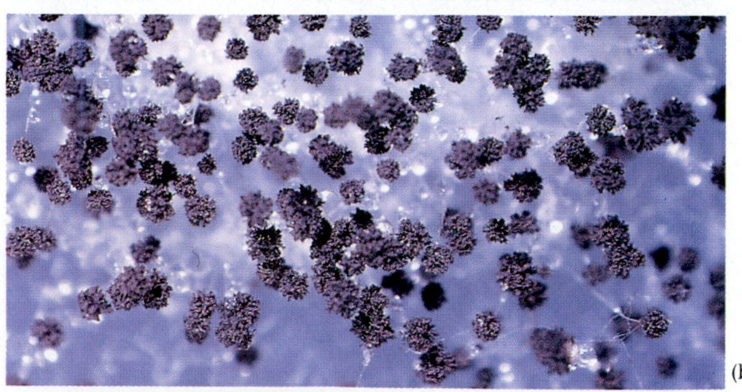
(b)

Figure 19-2 Reproductive structures of the black bread mold Rhizopus, *a zygomycete.* **(a)** *Sexual reproduction involves fusion of hyphae of two different mating types. Where the two meet, the black, diploid zygospore is formed.* **(b)** *Close-up of the asexual reproductive sporangia.*

Figure 19-3 Diverse ascomycetes. **(a)** *The cup-shaped fruiting body of the scarlet cup fungus.* **(b)** *The morel, an edible delicacy. (But consult an expert before sampling any wild fungus—some are deadly!)*

"cap thrower"; see Fig. E19-2). A more familiar and annoying zygomycete is the black bread mold of the genus *Rhizopus* (Fig. 19-2).

Ascomycetes or "sac fungi." The 30,000 species of ascomycetes are named after the saclike case in which spores form during sexual reproduction. This group includes many of the colorful molds that decorate decaying food or attack and destroy fruit crops. The fungus causing Dutch elm disease, which has destroyed nearly all the elms in the United States, is an ascomycete. Some ascomycetes live in decaying forest vegetation and form beautiful cup-shaped reproductive structures or corrugated, mushroom-like fruiting bodies (Fig. 19-3). In this group we also find the yeasts, one of the few unicellular fungi, without which we would lack not only our loaf of bread but our jug of wine as well! Another gastronomic delicacy, the truffle, is also a member of this diverse division (see the essay, "Evolutionary Ingenuity in Fungi").

Fungi imperfecti or "imperfect fungi." The "imperfect fungi" were originally given their name because none had been observed engaging in sexual reproduction. This large division includes 25,000 described species of great diversity and considerable importance to humans. It was a member of this group that Alexander Fleming discovered contaminating and killing his bacterial cultures. His keen observations led to the isolation of penicillin from the fungus *Penicillium* (Fig. 19-4). To these fungi we also owe the indescribable flavor and aroma of Roquefort and Camembert cheese. Other fungi imperfecti are human parasites, such as ringworm and athlete's foot. Some are not content to live on dead organisms or even to parasitize live ones; they act as predators, laying deadly traps for unsuspecting roundworms (see the essay).

Basidiomycetes or "club fungi." This division contains about 25,000 species, including the familiar mushrooms, puffballs, and shelf fungus ("monkey-stools"), as well as some devastating plant pests descriptively called "rusts" and "smuts" that cause billions of dollars worth of damage to grain crops yearly (Fig. 19-5).

Mushrooms and puffballs are actually reproductive structures, dense aggregations of mycelia that emerge under proper conditions from a massive underground network to release billions of spores for dispersal by the wind

Figure 19-4 Penicillium *growing on an orange. Reproductive structures coat the fruit's surface, while hyphae draw nourishment from inside.*

Evolutionary Ingenuity in Fungi

Natural selection, operating over millennia on the diverse forms of fungi, has produced some remarkable adaptations by which fungi solve the problems of dispersing their spores and obtaining nutrients. A few of these are highlighted in this essay.

The Rare, Sexy Truffle

Although many fungi are prized as food, none are as avidly sought as the truffle. A single specimen resembling a small, shriveled, blackened apple (Fig. E19-1) may sell for $100. Truffles are the reproductive structure of an ascomycete that forms a mycorrhizal association with the roots of oak trees. Formed underground, the truffle is faced with the problem of dispersing its spores. Its solution is to entice animals to dig it up. Although humans cannot smell underground truffles, their odor serves as an irresistible attractant to certain mammals, especially wild pigs. Why? It was recently discovered that the truffle releases a chemical that closely resembles the pig's sex attractant. As aroused pigs dig up and devour the truffle, millions of spores are scattered to the winds.

Figure E19-1 *The truffle. This rare ascomycete is a gastronomic delicacy.*

Human truffle hunters use muzzled pigs to hunt their quarry; a good truffle-pig can smell an underground truffle 50 meters away! Although a truffle growth can be encouraged by sowing spores throughout oak groves, they have eluded commercial cultivation. Thus they remain so rare and costly that few of us will ever taste one.

Figure E19-2 *The delicate translucent reproductive structures of the zygomycete* Pilobolus *will literally blow their tops when ripe, dispersing the black caps with their payload of spores.*

The Shotgun Approach to Spore Dispersal

The delicate structures in Fig. E19-2 are actually fungal shotguns, the reproductive structures of the zygomycete *Pilobolus.* Only by closely scrutinizing piles of horse manure are you likely to observe this miniature beauty. Hyphae penetrating the dung send up clear bulbs capped with sticky black spore cases. As they mature, the sugar concentration in the bulbs increases, drawing in water by osmosis. Meanwhile, the bulb begins to weaken just below its cap. Suddenly, like an overinflated balloon, it bursts, blowing its spore-carrying top up to a meter away. *Pilobolus* bends toward the light. This may increase the probability that its spores will land on open pasture. Here they adhere to grass blades until consumed by a grazing herbivore, perhaps a horse. Later (some distance away) the spores are deposited unharmed in a pile of their favorite food: horse manure. Growing hyphae penetrate this rich source of nutrients, sending up new projectiles to continue this ingenious cycle.

The Nematode Nemesis

You may recall from Chapter 18 the billions of microscopic nematode roundworms that thrive in rich soil. Several types of fungi have evolved deadly snares that exploit this rich source of protein. Natural selection has resulted in a variety of ingenious modifications of nematode-nabbing hyphae, studied by George Barron of Guelph, Ontario. Some produce sticky pods that adhere to passing nematodes. These will germinate into hyphae that penetrate the nematode body, digest-

Evolutionary Ingenuity in Fungi

Figure E19-3 Arthrobotrys, *the nematode strangler, traps its nematode (roundworm) prey in a modified hypha that swells when the inside of the loop is contacted.*

ing it from within. Others ensnare the worms in sticky tangled hyphae. In 1978, Barron discovered a new species that shoots a harpoonlike projectile 1/10,000 of a centimeter in diameter into passing nematodes. The projectile develops into a new fungus inside the worm, using it as food. The fungal strangler, *Arthrobotrys,* a member of the fungi imperfecti, produces nooses formed from three hyphal cells. When a nematode blunders into the noose, its contact with the inner parts of the noose stimulates a sudden increase in permeability to water that causes the noose cells to swell (Fig. E19-3). The resulting constriction of the hole takes only $\frac{1}{10}$ of a second and anchors the worm firmly. Fungal hyphae then penetrate and feast on the hapless prey.

Figure 19-5 *Diverse basidiomycetes.* **(a)** *Shelf fungi, the size of dessert plates, are conspicuous on trees.* **(b)** *A poisonous mushroom of the genus* Amanita. **(c)** *Puffballs, when ripe, liberate clouds of spores when touched.* **(d)** *Corn smut causes major losses to this crop each year.*

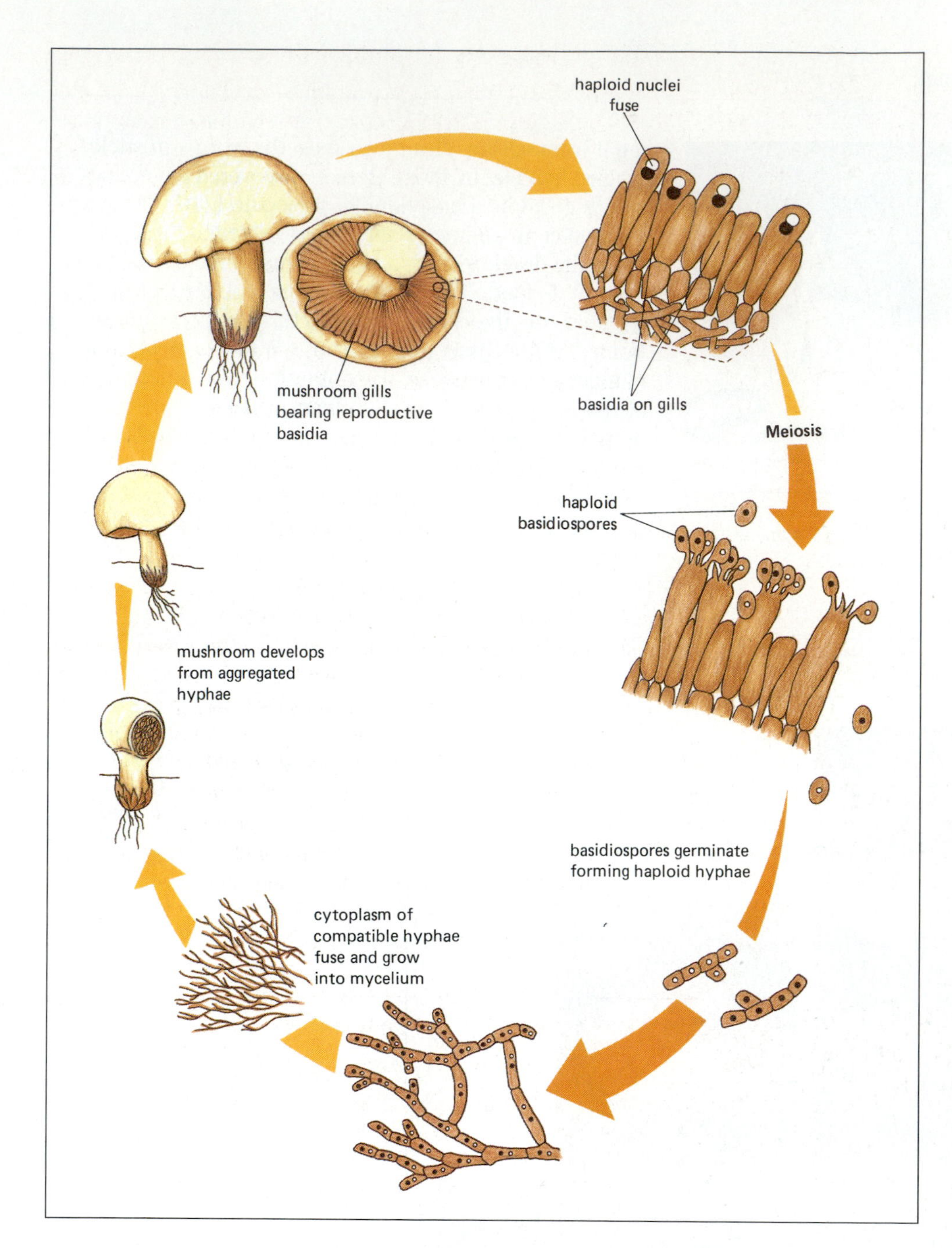

Figure 19-6 The life cycle of the mushroom, a basidiomycete. The mushroom is a reproductive structure, bearing cells called basidia on gills underneath the cap. The basidia form haploid basidiospores by meiosis. These disperse by wind and germinate into haploid hyphae. Hyphae of different but compatible mating types fuse, then grow into a mycelium whose cells are diploid (although the two haploid nuclei remain separate). In the presence of adequate water and nutrients, portions of the mycelium differentiate and swell, emerging through the soil as mushrooms to complete the cycle.

(Fig. 19-6). Falling on fertile ground, a spore may germinate and form its own underground network of mycelia. These grow outward from the original spore in a roughly circular pattern as the older mycelia in the center die. The circle diameter increases yearly by an average of 15 to 30 centimeters as new mycelia advance outward seeking nutrients. The subterranean body sends up numerous mushrooms, which emerge in a pattern reflecting the underground ring. Thus a "fairy ring" is formed, and its diameter can reveal the approximate age of the fungus—some are estimated to be 700 years old (Fig. 19-7).

OOMYCETES OR "EGG FUNGI." This group differs significantly from other true fungi. For example, their cell walls often contain cellulose, the main ingredient in plant cell walls, instead of (or in addition to) chitin. Sexual reproduction involves the fertilization of a large egg cell, hence their name "egg fungi." This rather small division of 475 species includes some of profound economic importance. One oomycete causes a disease called downy mildew of grapes. Its accidental introduction into France from the United States in the late 1870s nearly destroyed the French wine industry. Another member of this group has destroyed millions of avocado trees in California, while still another is responsible for "late blight," a devastating disease of potatoes. When accidentally introduced into Ireland, this fungus destroyed nearly the entire potato crop, on which most of the Irish lived. The result-

Figure 19-7 Mushrooms emerge in a "fairy ring" from an underground fungal mycelium growing outward from a central point where a spore germinated, perhaps decades ago.

ing famines of 1845–1847 reduced the Irish population by half, from 8 million to 4 million. Over 1 million starved outright, while the rest emigrated (many to the United States) or fell victim to diseases spreading through the weakened populace. The oomycetes also include inoffensive water molds that live in water and damp soil.

Fungal Partners in Symbiosis

LICHENS. Lichens are a unique **symbiotic** (literally "living together") association between fungi, usually ascomycetes, and unicellular algae or cyanobacteria. Together, these organisms form a living unit so tough and undemanding of nutrients that it is among the first to colonize newly formed volcanic islands. The algal partner provides food formed during photosynthesis, while the fungus provides support and protection in this mutually beneficial relationship. Lichens in a variety of bright colors can be found growing on bare rock, and have invaded habitats from deserts to the arctic. Based on their size and slow rate of growth, some lichens in the arctic are believed to be 4000 years old. Twenty-five thousand types of lichens have been identified, each with a different combination of fungus and alga or cyanobacterium and a unique color and growth pattern (Fig. 19-8).

MYCORRHIZAE. These are fungi (over 5000 species, mostly basidiomycetes or ascomycetes) that are found growing in intimate symbiotic association with the roots of about 80 percent of all vascular plants. These associations benefit both the plant and its fungal partner. The hyphae of mycorrhizae surround the root and frequently invade the root cells. Plants that participate in this unique relationship tend to grow larger and more vigorously, especially in poor soils, than those deprived of the fungus. The fungus digests organic compounds in the soil, some of which are passed directly from fungal cells into the root

Figure 19-8 Diverse lichens. **(a)** *Goat's beard, a lichen found hanging from tree limbs.* **(b)** *A colorful encrusting lichen, growing on dry rock, illustrates the tough independence of this symbiotic combination of fungus and alga.* **(c)** *A leafy lichen grows from a dead tree branch.*

(a)

(b)

Figure 19-9 **(a)** *The acellular slime mold* Physarum *oozes over a stone on the damp forest floor.* **(b)** *When food becomes scarce, the mass differentiates into black fruiting bodies in which spores are formed.*

cells they invade. Water is also absorbed by the fungus and passed to the plant, an advantage in dry, sandy soils. In return, food produced photosynthetically by the plant is passed from the root to the fungus.

The Slime Molds

These unique organisms are not considered "true fungi" and actually resemble amoebae in their means of feeding and locomotion. Slime molds come in two major groups, cellular and acellular. The **cellular slime molds** live in soil as independent amoeboid cells, extending pseudopods to engulf food such as bacteria. Unlike true fungi, they lack cell walls and feed by ingestion. When food becomes scarce, individual cells release a chemical signal that causes nearby cells to swarm together in dense aggregations. These aggregates then behave like a multicellular organism. The cells in the aggregation take on specific roles and form a reproductive structure called a fruiting body. Spores formed within the fruiting body are dispersed by the wind and grow into new amoeba-like individuals.

The **acellular slime molds** consist of a mass of cytoplasm that may spread thinly over an area of several square meters. Although the mass contains thousands of nuclei, the nuclei are not confined in discrete cells surrounded by cell membranes, as in most multicellular organisms: hence the term "acellular." These "supercells" ooze through decaying leaves and rotting logs, engulfing food such as bacteria and protists. The mass may be bright yellow or orange, and a large mass can be rather startling (Fig. 19-9a). The acellular slime molds reproduce by spores formed in fruiting bodies that differentiate from the formless mass (Fig. 19-9b).

THE KINGDOM PLANTAE

Plants are multicellular photosynthetic organisms. Like the unicellular algae of the kingdom Protista and the cyanobacteria of the kingdom Monera, plants capture sunlight and synthesize organic molecules from water and carbon dioxide. Collectively, the energy these organisms harvest from sunlight powers all life on Earth.

Plant life cycles all exhibit **alternation of generations.** A haploid plant, the **gametophyte,** produces by mitosis sex cells that unite and develop into a diploid plant, the **sporophyte.** The sporophyte, in turn, produces haploid spores by meiosis, each spore developing into a haploid gametophyte plant, completing the cycle (see Chapter 22). Tracing plants from the evolutionarily ancient algae to the more recently evolved seed plants, we see a general trend toward decreased size of the gametophyte generation (Table 19-1). In some algae, the gametophyte is similar in size and appearance to the sporophyte (Fig. 19-10). In seed plants, however, the gametophyte is microscopic, barely recognizable as an alternate generation.

Watery Origins: The Algae

Plant life arose in the sea, a womblike environment that supports the plant body, provides rather constant temperature, and bathes the entire plant in nutrients. Early plants were relatively simple, partly because the sea provided so many of their needs. Today, we still find the simplest plants, and the plantlike protists in watery environments; collectively, they are called **algae.**

Land plants show clear structural differences within their roots, stems, and leaves, discussed in more detail in Chapter 21. Algae, in contrast, lack this internal differentiation, although many algae have structures that appear rootlike or leaflike. Algae also lack complex reproductive structures such as flowers and cones. Their gametes are usually shed directly into the water, where they unite and develop.

Table 19-1 **Features of the Major Plant Groups**

	Relationship of Sporophyte and Gametophyte	Transfer of Gametes	Early Embryonic Development	Dispersal	Water and Nutrient Transport Structures	Typical Habitat
Algae (e.g., kelp)	Independent organisms; may be of equal size, or gametophyte reduced	Gametes released into water	Occurs independently after zygote settles to substrate	Water currents carry variety of reproductive cells	None (usually)	Aquatic
Bryophytes (e.g., mosses)	Gametophyte dominant—sporophyte develops from zygote retained on gametophyte	Motile sperm swims to stationary egg retained on gametophyte	Occurs within archegonium of gametophyte	Haploid spores carried by wind	None	Moist terrestrial
Simple vascular plants (e.g., ferns)	Sporophyte dominant—develops from zygote retained on gametophyte	Motile sperm swims to stationary egg retained on gametophyte	Occurs within archegonium of gametophyte	Haploid spores carried by wind	Vessels	Moist terrestrial
Gymnosperms—naked seed plants (e.g., conifers)	Sporophyte dominant—microscopic gametophyte develops within sporophyte	Wind-dispersed pollen carries sperm to stationary egg in cone	Occurs within a protective seed containing a food supply	Seeds containing diploid sporophyte embryo dispersed by wind or animals	Vessels	Varied terrestrial habitats—dominate in dry, cold climates
Angiosperms—flowering plants	Sporophyte dominant—microscopic gametophyte develops within sporophyte	Pollen, dispersed by wind or animals, carries sperm to stationary egg within flower	Occurs within a protective seed containing a food supply; seed encased in fruit	Fruit, carrying seeds, dispersed by animals, wind, or water	Vessels	Varied terrestrial habitats—dominant terrestrial plant

Algal life cycles are complex, and vary considerably between, and even within, the various algal groups. The sporophyte and gametophyte may appear nearly identical, as in the green alga *Ulva* (Fig. 19-10) or they may be very different in size and appearance, as in the giant kelp, whose sporophyte is illustrated (Fig. 19-12b,c) and whose gametophyte is very small.

Algae may be classified into three divisions, each named for its characteristic color. Algae are colored by pigments that help them capture light energy for photosynthesis. These pigments are often red or brown, absorbing the green, violet, and blue light that most readily penetrates deep water. The combination of these pigments and green chlorophyll lend algae their distinctive colors, and their names—rhodophyta: the red algae; phaeophyta: the brown algae; and chlorophyta: the green algae.

DIVISION RHODOPHYTA: THE RED ALGAE. The 4000 species of red algae, ranging from bright red to nearly black, derive their color from red pigments that mask their green chlorophyll (Fig. 19-11). Red algae are mostly marine. They dominate in deep, clear tropical waters, where their red pigments absorb the deeply penetrating blue-green light.

Some red algae contribute to the formation of reefs by depositing calcium carbonate (limestone) in their tissues. Others are harvested as food in the Orient, and from some the gelatinous substance carrageenan is extracted and used as an emulsifier in such products as paints, cosmetics, and ice cream. However, the major importance of these and all other algae is their photosynthetic ability: they form the food-producing foundation of marine ecosystems.

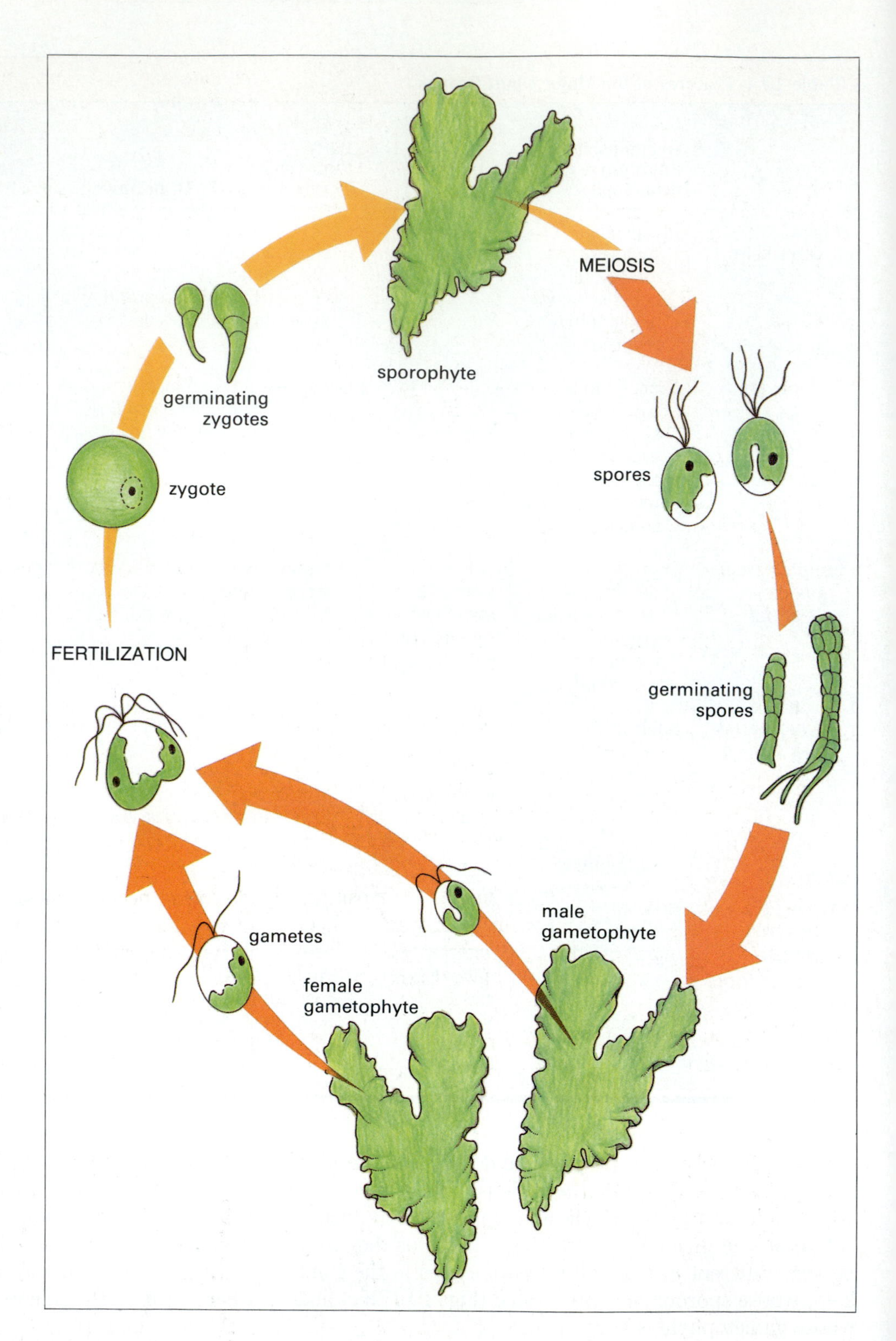

Figure 19-10 Life cycle of Ulva, *a large marine alga. In this case the sporophyte and gametophyte plants are equal in size and nearly indistinguishable. However, the sporophyte is diploid and produces haploid spores through the process of meiosis, while the male and female gametophytes are haploid and produce gametes by mitosis. The fusion of these gametes results in a diploid zygote that develops into the sporophyte plant.*

DIVISION PHAEOPHYTA: THE BROWN ALGAE. These plants increase their light-gathering ability by using brownish-yellow pigments which (in combination with green chlorophyll) produce their brown to olive-green color. Like the red algae, the 1500 species of brown algae are almost entirely marine. This group forms the dominant "seaweed" along rocky shores in the temperate (cooler) oceans of the world, including the east and west coast of the United States. Brown algae are found from the intertidal zone, where they cling to rocks exposed at low tide, to far offshore. Several types use gas-filled floats to support their bodies (Fig. 19-12). The giant kelp plants found along the Pacific coast occasionally reach heights of 100 meters, and may grow over 15 centimeters daily. With their dense growth and towering height, kelp forms undersea forests that provide food, shelter, and breeding

Figure 19-11 The red alga Plocamium, *whose flattened body is delicately branched and strikingly red.*

areas for a variety of marine animals (Fig. 19-12b).

Division Chlorophyta: The green algae. This large (7000 species) and extremely diverse group is of special interest because green algal ancestors are believed to have given rise to the terrestrial plants. Three lines of evidence support this hypothesis. *First*, green algae use the same type of chlorophyll and accessory pigments as do land plants. *Second*, they store food as starch (as do land plants) and have cell walls of similar composition. *Third*, nearly all green algae live in fresh water, where they form the dominant plant life. In contrast to the nearly constant conditions of the ocean, freshwater habitats (ponds, swamps, streams, and lakes) are highly variable. Dramatic fluctuations in temperature and rainfall over evolutionary time have exerted intense selection pressure on ancient freshwater algae to withstand extremes of temperature and periods of dryness. The adaptations that resulted served their descendents well in their invasion of the land.

Green algae are mainly multicellular, but include a few unicellular forms and **colonies** —clusters of cells that are somewhat interdependent. These colonies range from a few cells to a few thousand cells, as in members of the genus *Volvox*, where rudimentary division of labor is seen (Fig. 19-13a). Some green algae, such as members of the genus *Spirogyra*, form long chains of cells (Fig. 19-13b). Most are quite small, but some exceptions are found in the sea. The green alga *Ulva* or "sea lettuce" (see Fig. 19-10) is similar in size to its namesake.

Land: The New Frontier

For the descendants of the ancient green algae there were considerable advantages to moving onto the land. The raw materials for photosynthesis, carbon dioxide and sunlight, are present in far higher concentration in the air than dissolved in water. The ponds and seas had grown crowded, teeming with hungry animals and with other plants competing for light and nutrients. The land, in contrast, offered abundant space and resources to the new colonists. But as the pioneers of the American West discovered, the riches of land are not won easily. A large plant, adapted to the supporting aquatic environment, would have sprawled like beached seaweed, unable to support its weight. New structures to give the body rigidity were needed. These land pioneers were also limited in size since (like algae) they were forced to rely on diffusion to carry substances to and from each cell. Special vessels were needed to transport water and minerals upward to the leaves, and food produced by photosynthesis down to the roots. Further, a truly terrestrial plant could not rely on water to carry its sex cells or to disperse the

Figure 19-12 Diverse brown algae. **(a)** Fucus, *an intertidal genus, is shown here exposed at low tide. Notice the gas-filled bladders, which confer bouyancy in water.* **(b)** *The sporophyte generation of the giant kelp* Macrocystis *forms underwater forests off southern California. In* **(c)**, *the growing tip of the plant is shown. Gas-filled bladders are clearly visible at the base of leaflike blades.*

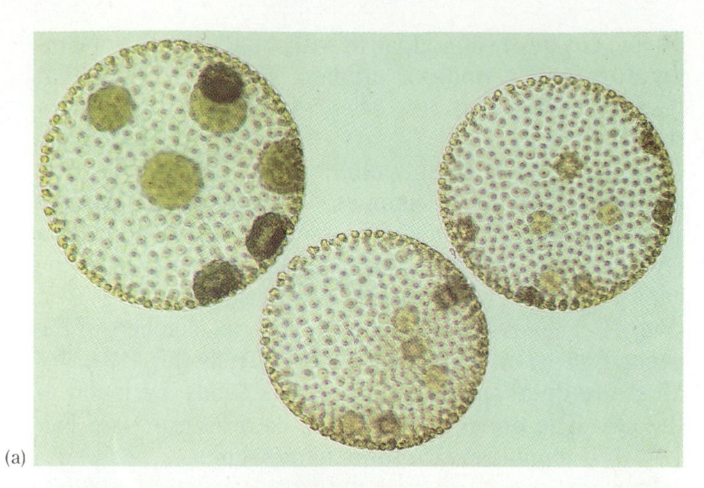
(a)

(b)

Figure 19-13 Diverse green algae. **(a)** Volvox, *a colonial green alga composed of a sphere of cells embedded in a gelatinous matrix. New daughter colonies can be seen developing inside the sphere.* **(b)** Spirogyra, *a filamentous green alga composed of strands only one cell thick. Inside these microscopic transparent cells, spiral chloroplasts are clearly visible.*

fertilized eggs; special adaptations were needed to reproduce. Another important hurdle faced by the new colonizers of land was the constant threat of drying—the plant body needed waterproofing.

Two major groups of land plant arose from the ancient algal ancestor. One group, the **vascular plants,** are completely adapted to land. A second group, the **bryophytes,** straddle the boundary between aquatic and terrestrial life.

Division Bryophyta: The Mosses and Liverworts

The 16,000 species of bryophytes show some, but not all, of the adaptations necessary for a completely terrestrial existence. Bryophytes lack well-developed vessels for conducting water and nutrients through the plant body. Since the presence of conducting vessels defines true roots, stems, and leaves, the bryophytes, like algae, lack these as well (although they have structures serving similar functions). The absence of an efficient means of moving substances through the body limits the size of bryophytes; most are about 2 centimeters tall. Bryophytes vary in their ability to withstand desiccation. Some possess a waterproof **cuticle** that retains moisture. Although the liverworts (Fig. 19-14) and most mosses (Fig. 19-15) are confined to moist areas, a few mosses can survive in deserts, on bare rock, and in far northern and southern latitudes where water is scarce.

Bryophytes have adapted to terrestrial existence by developing enclosed reproductive structures that protect the gametes from desiccation. These are the **archegonia,** in which eggs develop, and **antheridia,** where sperm are formed. They may be located on the same plant, or the entire plant may be either male or female, depending on the species. In all bryophytes, the sperm must swim to the egg (which emits a chemical attractant) through a film of water. Those living in dry areas must therefore time their reproduction to coincide with the limited rainfall. As in many algae, the larger "leafy" plant body is the haploid gametophyte, which forms sperm and eggs by mitosis (Fig. 19-15). Bryophytes retain the fertilized egg in the archegonium. Here the embryo grows and matures into a small diploid sporophyte attached to the parent plant. At maturity, the sporophyte produces haploid spores by meiosis. These are released explosively when the capsule in which they are formed bursts open. Landing on fertile soil, each can develop into a new, haploid plant.

The Vascular Plants

Although the bryophytes were quite successful, they left vast areas of the land unoccupied. When the moister regions were covered with the short green fuzz of bryophytes, any plant that could stand taller would benefit by basking in sunlight while shading its short competitors. To grow tall, two new features were needed: support for the body, and specialized cells to conduct materials between the food-making leaves and the roots. The **vascular** (vessel-bearing) plants solved the two problems si-

Figure 19-14 The liverwort grows inconspicuously in moist, shaded areas. This is the female gametophyte plant, bearing umbrella-like archegonia, which hold the eggs. Sperm must swim up the stalks through a film of water to fertilize the eggs.

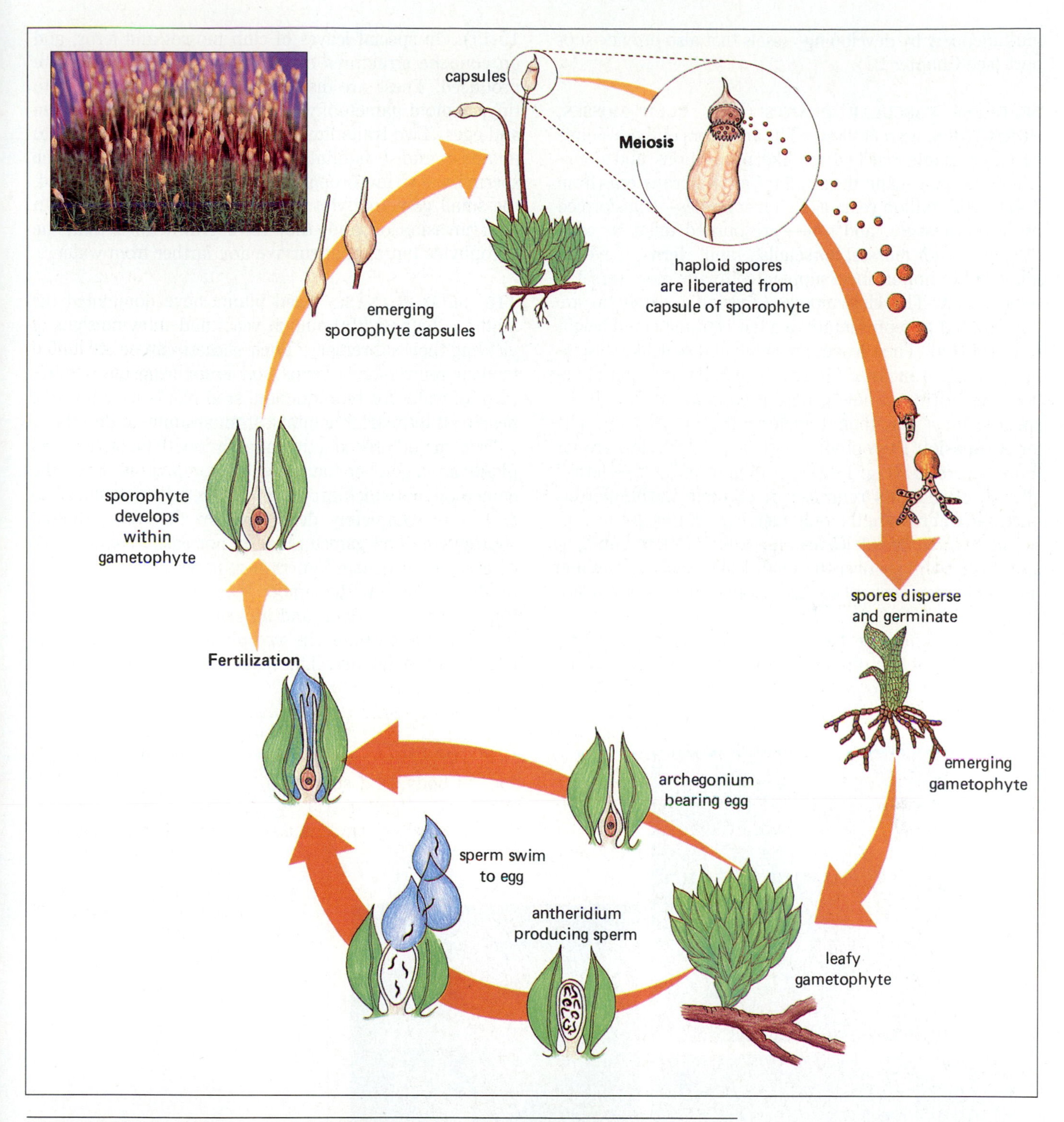

Figure 19-15 Life cycle of a moss, showing alternation of diploid and haploid generations. The leafy green cushion (lower right) is actually the haploid gametophyte generation that produces sperm and eggs. The sperm develop in the antheridium and must swim to the egg (which remains in the archegonium where it is formed) through a film of water. The zygote develops into a stalked, diploid sporophyte that emerges from the gametophyte plant. The sporophyte is topped by a brown capsule in which haploid spores are produced by meiosis. These are dispersed and germinate, producing another green gametophyte generation. (Inset) Moss plants showing both stages in the life cycle. The short, leafy green plants are the haploid gametophytes, while the reddish-brown stalks are the diploid sporophyte generation. The stalks are about 1 centimeter in height.

multaneously by developing vessels that also provide support (see Chapter 21).

Seedless vascular plants: The club mosses, horsetails, and ferns. The ancestors of these simple vascular plants reached treelike proportions and dominated the landscape during the Carboniferous era (from 355 to 265 million years ago). Their bodies—transformed by heat, pressure, and time—are burned today as coal. Modern **club mosses, horsetails,** and **ferns** have diminished in importance, supplanted by the more versatile seed plants. The club mosses (division Lycophyta) are now limited to representatives a few centimeters in height (Fig. 19-16a). Their leaves are small and scalelike, resembling those of mosses. Modern horsetails (division Sphenophyta) form a single genus, *Equisetum*, with only 15 species, most less than 1 meter tall (Fig. 19-16b). The ferns (division Pterophyta), with 12,000 species, are far more successful (Fig. 19-16c). In the tropics, "tree ferns" still reach heights reminiscent of their Carboniferous ancestors. Ferns are the only members of this group with broad leaves. Broad leaves can capture more sunlight, and this advantage over the small-leaved club mosses and horsetails may account for the relative success of modern ferns.

A major difference between vascular plants and bryophytes is that the diploid sporophyte is dominant (Fig. 19-17). On special leaves of club mosses and ferns, and on conelike structures of horsetails, haploid spores are produced. These are dispersed by wind and give rise to tiny, haploid gametophyte plants, which produce sperm and eggs. Two traits limit the seedless vascular plants to relatively moist habitats. First, as in bryophytes, the sperm must swim through water to reach the egg. Second, the small gametophytes also lack conducting vessels. In their invasion of land, they have grown taller than the bryophytes but cannot survive any farther from water.

The seed plants. Seed plants have dominated the land for the past 350 million years and show no signs of yielding their supremacy. Their success can be attributed to their reproductive versatility. Freed from the requirement of water for reproduction, seed plants have invaded nearly all terrestrial habitats, from swamps to deserts.

Two major reproductive adaptations have given seed plants an edge over their seedless competitors. First, the gametophyte (which produces the sex cells) is reduced so that it is completely dependent on the large, diploid sporophyte. The gametophyte is not easily recognizable as a separate generation; rather, it resembles a tiny reproductive organ. Remember that in mosses the gametophyte was dominant, and in ferns it existed as a small but independent plant. In seed plants the gametophyte is reduced even further. The female is a small tissue that

Figure 19-16 Some simple vascular plants. All are found in moist woodland habitats. **(a)** *The club mosses (sometimes called "ground pines") grow in temperate forests. This specimen is liberating spores.* **(b)** *The horsetail* (Equisetum) *has long, needlelike leaves.* **(c)** *The sword fern, named for the swordlike shape of its leaves.*

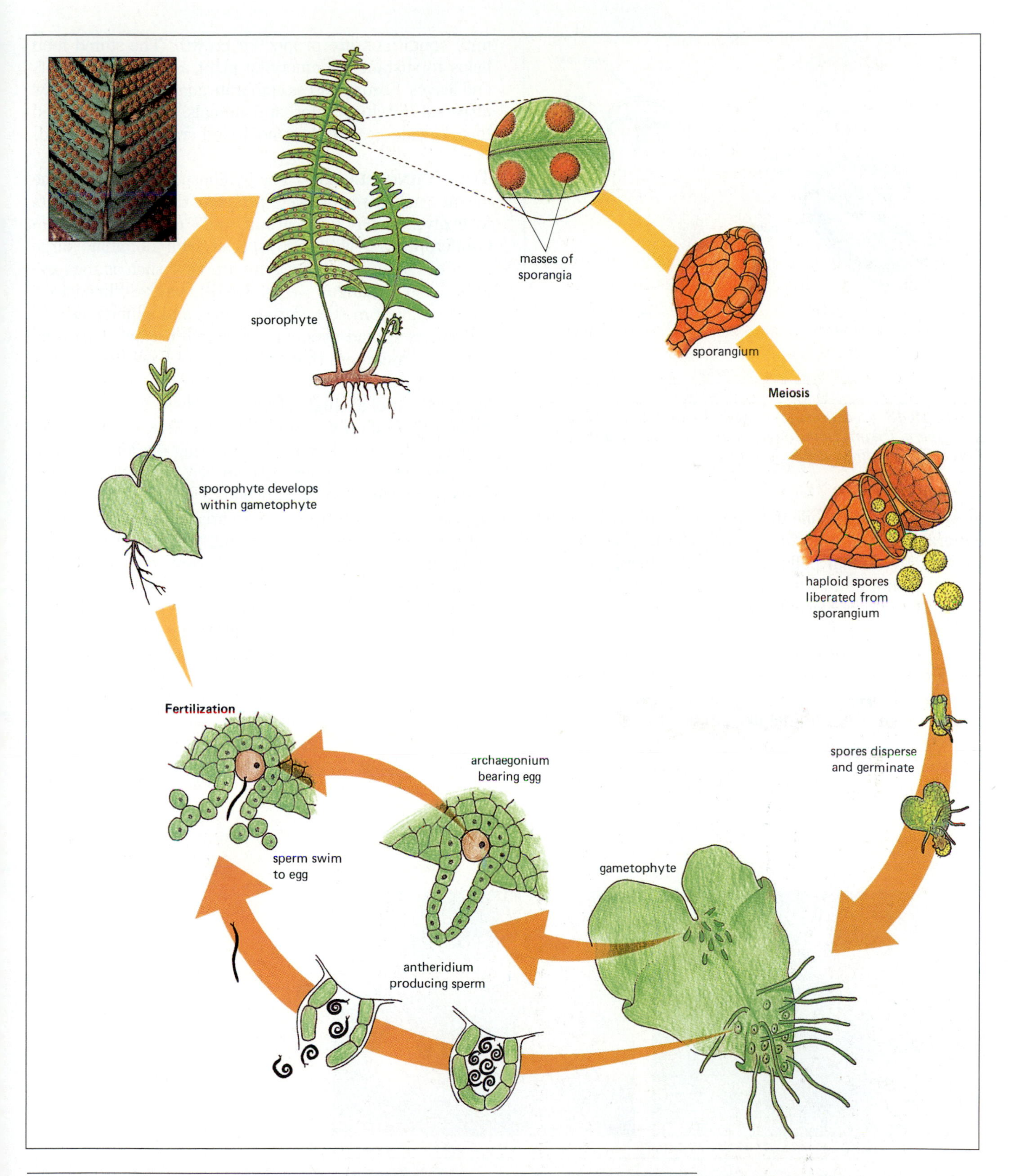

Figure 19-17 The life cycle of a fern, showing alternation of generations. The dominant plant body (upper left) is the diploid sporophyte. Haploid spores, formed in clusters of sporangia located on the underside of certain leaves, are dispersed by wind to germinate on the moist forest floor into inconspicuous haploid gametophyte plants. On the lower surface of these small, sheetlike gametophytes, male antheridia and female archegonia produce sperm and eggs. The sperm must swim to the egg, which remains in the archegonium. The zygote develops into the large sporophyte plant. (Inset) Underside of a fern leaf, showing clusters of sporangia.

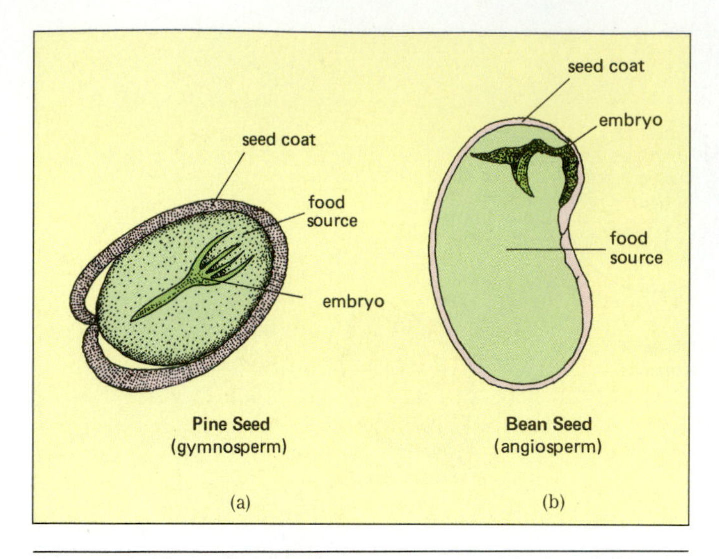

Figure 19-18 *Seeds from* **(a)** *a gymnosperm (pine) and* **(b)** *an angiosperm (bean). Both consist of an embryonic plant and stored food confined within a seed coat.*

produces the egg, while the male gametophyte is the pollen grain. Within the pollen grain, sperm is dispersed by wind or by animal pollinators, such as bees. Thus seed plants are not limited in their distribution by the need for water that allows sperm to swim to the egg; they are fully adapted to dry land.

The second reproductive adaptation is the seed itself. Seeds are remarkable structures—somewhat analogous to the eggs of birds and reptiles. The seed (whose structure is covered in detail in Chapter 22) consists of an embryonic plant, a supply of food for the embryo, and a protective coat (Fig. 19-18). The seed coat maintains the embryo in a state of suspended animation or dormancy until conditions are proper for growth. The stored food helps to sustain the emerging plant as it develops roots and leaves. Seeds possess elaborate adaptations that allow dispersal by wind, water, and animals. These have helped them invade nearly every nook and cranny of the world.

THE GYMNOSPERMS. The gymnosperms (whose name means "naked seed") were probably the first seed plants to evolve. Today, one group, the **conifers** (division Coniferophyta), with 500 species, still dominate large areas of the globe. Other gymnosperms, such as the **cycads** and **ginkgos** (Fig. 19-19), have declined to a small remnant of their former range and abundance.

Conifers spread widely as the Earth became drier during the Permian era that followed the Carboniferous. Today they are most abundant in the cold latitudes of the far north and at high elevations where conditions are rather dry. Not only is rainfall limited here, but soil water remains frozen and unavailable during the long winters. Conifers, including pines, firs, spruce, hemlocks, and cypresses, are adapted to withstand dry, cold conditions in several ways. *First*, their leaves are thin needles covered with a thick cuticle, whose small, waterproofed surface allows little evaporation. *Second*, conifers (often called "evergreens") retain their leaves year-round, extending the relatively short northern growing season. *Third*, they produce a fragrant, resinous "antifreeze" in their sap that allows it to continue transporting nutrients in subfreezing temperatures.

Reproduction is similar in all conifers, with pines serving as a good illustration (Fig. 19-20). The tree itself is the diploid sporophyte. It develops male and female cones. The male cones are often found in clusters at the ends of lower branches, while the larger woody female

Figure 19-19 *Two uncommon gymnosperms.* **(a)** *A cycad. Common in the age of dinosaurs, these are now limited to about 100 species living in warm, moist climates.* **(b)** *The ginkgo, or "maidenhair tree," has been kept alive by cultivation in China and Japan. Extremely resistant to pollution, these have become popular in American cities. Both ginkgos and cycads have separate sexes. The ginkgo shown here is female and bears fleshy seeds the size of large cherries, which are noted for their foul smell when ripe. For obvious reasons, most ginkgos grown for decorative purposes are male.*

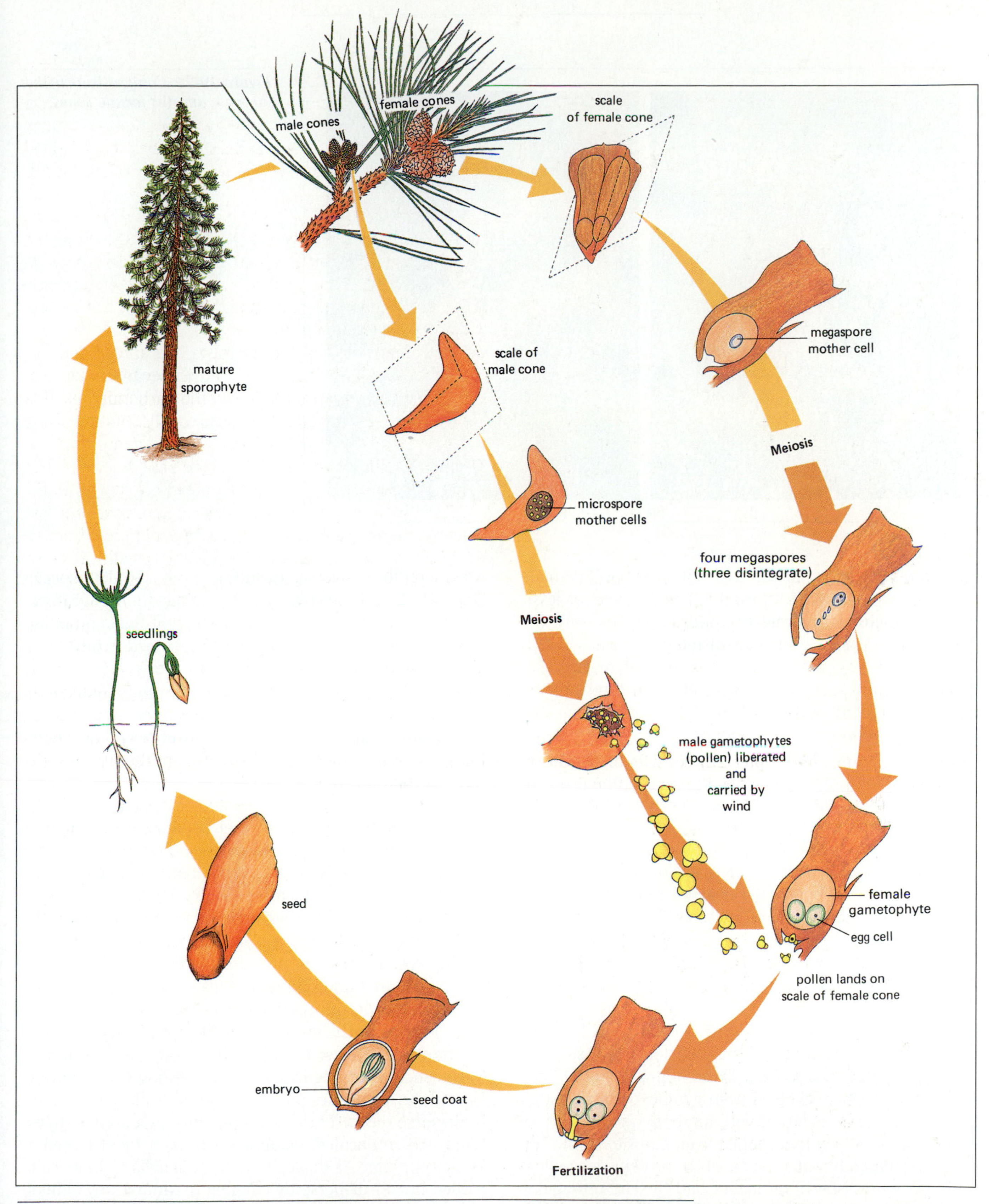

Figure 19-20 *Life cycle of the pine, a seed plant of the division Coniferophyta. The tree is the dominant sporophyte generation, bearing both male and female cones. Megaspore mother cells within the forming seeds on the scales of the female cones undergo meiosis to form megaspores, one of which develops into the female gametophyte. The female gametophyte in turn produces egg cells. Meanwhile, in the male cones, microspore mother cells undergo meiosis to produce the male gametophytes: the pollen. Pollen, carrying sperm nuclei, are dispersed by wind and land on the scales of the female cone. The pollen produces a pollen tube that penetrates the female gametophyte and conducts the sperm to the egg. The fertilized egg develops into an embryonic plant enclosed in a seed which is eventually released from the cone. The seed germinates and grows into the sporophyte tree.*

Figure 19-21 Conifers bear both **(a)** *male and* **(b)** *female cones.*

cones may be higher in the tree (Fig. 19-21). The male cones are relatively small (usually 2 centimeters or less), delicate structures that release clouds of pollen during the reproductive season and then disintegrate. Each pollen grain is a male gametophyte, consisting of several specialized haploid cells, some of which form tiny winglike structures that allow the pollen to be carried long distances by the wind. Immense clouds of pollen are released by the male cones, so inevitably some lands by chance on the female cone. Each female cone consists of a series of woody scales arranged spirally around a central axis. At the base of the scale are two haploid female gametophytes, each producing an egg cell. A pollen grain landing nearby sends out a pollen tube that slowly burrows into the female gametophyte. After nearly 14 months, the tube finally reaches the egg cell and releases sperm that fertilize it. The fertilized egg becomes enclosed in a seed as it develops into a tiny embryonic plant. The seed is liberated when the cone matures and the scales separate.

THE ANGIOSPERMS—FLOWERING PLANTS. The earliest fossils of flowering plants (division Angiospermophyta) are estimated to be 127 million years old. They are believed to have evolved from gymnosperm ancestors that formed an association with animals (most likely insects), which carried their pollen from plant to plant. The insects benefited by eating some of the protein-rich pollen, while the gymnosperm no longer had to send prodigious quantities flying to the winds to ensure fertilization. The relationship was so beneficial that through natural selection, plants evolved structures specifically to attract insects and other animals—these are called flowers. One hundred million years ago, flowering plants already dominated the Earth, as they do today. Modern angiosperms are incredibly diverse, including over 250,000 species (Fig. 19-22). They range in size from the diminutive duckweed (a few millimeters in diameter) that floats on ponds to the mighty eucalyptus tree over 100 meters tall. From desert cactus to tropical orchids to grasses to parasitic mistletoe, angiosperms dominate the plant kingdom. Their success can be attributed to several adaptations, the most important of which is the flower. The angiosperm life cycle and flower structure are discussed in detail in Chapter 22.

Like their gymnosperm ancestors, angiosperms have a dominant sporophyte plant that produces and nurtures tiny male and female gametophytes. These in turn produce the sex cells. In angiosperms, both male and female gametophytes are formed within the flower (see Chapter 22). Fertilization of the egg occurs within the ovary of the flower, which surrounds the female gametophyte. The resulting zygote develops into an embryo enclosed in a seed. The ovary surrounding the seed matures into a **fruit.** The word "angiosperm" (derived from the Greek for "vessel seed") refers to the enclosure of the seed within a fruit.

The fruit is a second adaptation that has contributed to the success of angiosperms. Just as flowers encourage animals to transport pollen, so many fruits entice them to disperse seeds. These seeds pass through animal digestive tracts unharmed; examples can easily be observed in bird droppings. As dog owners are well aware, some fruits (called burrs) disperse by clinging to animal fur. Others, like those of maples, form wings that carry the seed through the air. The variety of dispersal mechanisms made possible by the fruit has helped the angiosperms invade nearly all possible habitats.

A third feature that gives angiosperms an adaptive advantage in warmer, wetter climates is broad leaves that

Figure 19-22 Diverse angiosperms. Both grasses **(a)**, *and many trees such as this birch* **(b)**, *have inconspicuous flowers and rely on wind for pollination.* **(c)** *Flowers such as found on the hedgehog cactus entice insects to carry pollen between individuals.*

are shed under adverse conditions, such as periods of cold and drought. When water is plentiful, as during the long, warm growing season of temperate and tropical climates, broad leaves give trees an advantage by collecting more sunlight for photosynthesis. The extra energy gained during the spring and summer allows the trees to drop their leaves and enter a dormant period during the fall and winter. In the north, the growing season is considerably shorter, so angiosperm trees remain leafless and dormant for a much greater part of the year, avoiding excess water loss through their broad leaves. The water-conserving conifers dominate here because they can continue photosynthesizing and growing slowly during the long winters (Fig. 19-23).

The flowering plants are grouped into two broad classes based on their internal and external structure, discussed in more detail in Chapter 21. The **monocots** (class Monocotyledoneae,) are a group of about 65,000 species, including grasses, corn and other grains, irises, lilies, and palms. The **dicots** (class Dicotyledoneae) are a considerably larger group with about 170,000 species, including most of the angiosperm trees, shrubs, and herbs.

Figure 19-23 Two ways of coping with the dryness of winter. The evergreen Engelmann spruce (a conifer) retains its needles throughout the year. The small surface area and heavy cuticle of the needles retard water loss. In contrast, the aspen (an angiosperm) sheds its leaves each fall. The dying leaves turn brilliant shades of gold as accessory pigments are exposed when the chlorophyll disintegrates.

REFLECTIONS ON CONTINUITY

You walk softly on a cushion of springy leaf litter, shaded by tall trees. Mosses cling to the damp bases of the old tree trunks, and ferns uncoil, spreading feathery leaves to catch filtered sunlight. Mushrooms poke through the leaves, while decaying logs crumble to reveal a tangle of pale fibers and, occasionally, a bright yellow gelatinous mass. Leathery lichens emerge from the flaking bark. Images of decay and rebirth merge inextricably in the complexity of the forest ecosystem, where fungi and plants exist in mutual dependence. Here we find the starting point for life as we know it: the ability of plants to harness the energy of sunlight to run the machinery of life, and also the link between death and new life: the fungi. These, in feeding themselves, release the raw materials from the discards of life for reuse by the emerging green shoots.

SUMMARY OF KEY CONCEPTS

The Kingdom Fungi

True fungi are heterotrophic, generally with multicellular, filamentous bodies. Most secrete digestive enzymes outside their bodies and absorb the liberated nutrients. Most have cell walls containing chitin. The fungal body, or mycelium, is composed of a mass of filamentous hyphae. Fungal reproduction is varied and complex, with the dominant generation usually haploid. There are five major divisions of fungi. The zygomycetes include the dung fungus *Pilobolus* and the black bread mold *Rhizopus*. The ascomycetes include morels, truffles, molds, Dutch elm disease, and several crop parasites. The fungi imperfecti are diverse; many appear to lack a sexual stage. Included are penicillin, ringworm and athlete's foot, and some predators of nematodes. The basidiomycetes include mushrooms, puffballs, and food crop parasites called rusts and smuts. Oomycetes include members causing diseases of grape and potato crops, as well as water molds.

Slime molds more closely resemble the amoeboid protists. Cellular slime molds spend part of their life as independent amoeboid cells that lack cell walls and engulf their food but aggregate to form a multicellular reproductive structure. Acellular slime molds consist of a mass of cytoplasm with millions of nuclei but no internal membranes separating it into cells.

Fungi are one of the principal decomposers in ecosystems. In addition, some enter mutually beneficial relationships with plants. These include lichens, a symbiotic association between a fungus and an alga, and mycorrhizae, fungi that surround and invade the root cells of many vascular plants, sharing nutrients.

The Kingdom Plantae

The kingdom Plantae is comprised of photosynthetic, usually multicellular organisms. The ability of plants to capture sunlight and convert it to high-energy food is the basis of nearly all life on Earth. Plants have complex life cycles, exhibiting alternation of generations in which haploid and diploid individuals alternately give rise to one another.

Watery Origins: The Algae The simplest plants are the aquatic algae. There are three major divisions of algae. The red algae (rhodophyta) dominate in clear tropical waters, while the brown algae (phaeophyta) rule temperate oceans. The green algae (chlorophyta) are primarily small, freshwater forms, and are believed to be ancestral to modern land plants. Algal divisions are named for their predominant colors, resulting from a combination of green chlorophyll and accessory light-trapping pigments. Algae lack true roots, stems, and leaves, and rely on the water to carry sex cells.

Land: The New Frontier Plants that invaded land found abundant space, nutrients, and light, but four major adaptations were required before they could fully exploit terrestrial habitats: (1) support for the body, (2) conducting vessels, (3) independence from water for reproduction, and (4) a water-retaining coating over parts exposed to air.

Division Bryophyta: The Mosses and Liverworts Bryophytes, which include the mosses and liverworts, are small, simple land plants that lack conducting vessels. They are generally confined to moist areas since their reproduction requires the presence of liquid water through which the sperm must swim to the egg.

The Vascular Plants Vascular plants have evolved a system of vessels that also supports the body. Seedless vascular plants such as mosses, horsetails, and ferns are confined to moist areas since they, like bryophytes, require a film of water for reproduction.

Gymnosperms include ginkgos, cycads, and the highly successful conifers. These were the first fully terrestrial plants to evolve. Their success on dry land is partially due to the evolution of pollen. Pollen protects and transports the male gamete, eliminating the need for the sperm to swim to the egg. The seed (a protective resting structure containing an embryo and a supply of food) is a second important adaptation contributing to their success.

Angiosperms, the flowering plants, dominate much of the land today. The flower has allowed them to utilize animals as pollinators. In contrast to wind dispersal, animals can carry pollen longer distances with greater accuracy and less waste.

GLOSSARY

Accessory pigments: colored molecules other than chlorophyll that absorb light energy and pass it to chlorophyll.

Algae (al′-gē; sing. alga): a general term for simple aquatic plants lacking vascular tissue.

Alternation of generations: a life cycle typical of plants in which a diploid sporophyte (spore-producing) generation alternates with a haploid gametophyte (gamete-producing) generation.

Angiosperm (an′-gē-ō-sperm): a flowering vascular plant.

Antheridium (an-ther-id′-ē-um): a structure in which male sex cells are produced, found in the bryophytes and certain seedless vascular plants.

Archegonium (ar-ke-gō′-nē-um): structure in which female sex cells are produced, found in the bryophytes and certain seedless vascular plants.

Bryophyte (brī′-ō-fīt): a division of simple nonvascular plants including mosses and liverworts.

Conifer (kon′-eh-fer): a class of tracheophyte that reproduces using cones and retains its leaves throughout the year.

Cuticle (kū'-ti-kul): a waxy or fatty coating on the exposed epidermal cells of many land plants, which aids in the retention of water.

Dicotyledon (dī'-kot-ul-ēd'-un): a class of angiosperm whose embryo has two cotyledons, or seed leaves.

Division: a taxonomic category in botany; the equivalent of an animal phylum.

Endosperm (en'-dō-sperm): the stored food within a seed used to nourish the developing plant embryo.

Flower: the reproductive structure of an angiosperm plant.

Fruit: the mature ripened ovary of an angiosperm plant. This structure contains the seeds.

Fruiting body: a reproductive structure of fungi in which spores are formed.

Gametophyte (ga-mēt'-eh-fīt): a multicellular haploid plant that produces haploid sex cells by mitosis.

Hypha (hī'-pha; pl. hyphae): threadlike structure consisting of elongated cells, often with many nucleii. The body of a fungus is composed of numerous hyphae.

Lichen (lī'-ken): a symbiotic association between an alga or cyanobacterium and a fungus, resulting in a composite organism.

Monocotyledon (mahn'-eh-kot-ul-ēd'-un): a class of angiosperm plant in which the embryo has one cotyledon, or seed leaf.

Mycelium (mī-sēl'-ē-um): the body of a fungus, consisting of a mass of hyphae.

Mycorrhizae (mī-ke-rī'-ze): a fungus, often a basidiomycete or an ascomycete, that grows intimately associated with the roots of vascular plants in a mutually beneficial association.

Pollen: the male gametophyte of gymnosperms and angiosperms.

Seed: the reproductive structure of a seed plant. The seed is protected by a seed coat and contains an embryonic plant and a supply of food for it.

Sporophyte (spōr'-ō-fīt): the diploid form of a plant that produces haploid, asexual spores throgh meiosis.

Symbiosis (sim-bī-ō'-sis): a close relationship between two types of organisms. Either or both may benefit from the association, and in some cases, such as parasitism, one of the participants is harmed.

Vascular (vas'-ku-ler): possessing or composed of conducting tissue or vessels.

Zygospore (zī'-gō-spōr): a fungal spore surrounded by a thick, resistant wall, which forms from a diploid zygote.

Zygote (zī'-gōt): a diploid cell resulting from the fusion of male and female sex cells.

STUDY QUESTIONS

1. Compare the energy-obtaining methods of fungi and plants. Describe how each is specialized for its particular method of obtaining energy.
2. Describe the structure of the fungal body. What portion is represented by mushrooms, puffballs, and similar structures? Why are these elevated above the ground?
3. List the major categories of fungi and give one example of each.
4. Describe how a "fairy ring" of mushrooms is produced. Why is the diameter related to its age?
5. What is meant by "alternation of generations"? What two generations are involved? How does each reproduce?
6. Assuming that green algae have fewer accessory pigments than red or brown algae, would you expect to find them in shallow or deep water? Where would you find the red algae, and why?
7. From which algal division did green plants probably arise? Explain the evidence supporting this hypothesis.
8. List the various adaptations necessary for the invasion of dry habitats on land by plants. Which of these are possessed by bryophytes? by ferns? By gymnosperms and angiosperms?
9. What single feature is probably most responsible for the enormous success of angiosperms? Explain why.
10. List the adaptations of gymnosperms that have helped them become the dominant tree in dry, cold climates.
11. What is a pollen grain? What role has it played in helping plants colonize dry land?
12. Compare the life cycle of a moss and a pine. Use information from Table 19-1 to help identify differences.
13. Describe two different symbiotic associations between plants and fungi. In each case, explain how each partner in these associations is affected.

SUGGESTED READINGS

Brodie, H. J. *Fungi: Delight of Curiosity*. Toronto: Univ. of Toronto Press, 1978. Highlights amazing fungal adaptations in a lively text written for the interested layperson.

Cooke, R. C. *Fungi, Man and His Environment.* New York. Longman, 1977. A discussion of fungi with emphasis on their impact on human activities.

Emerson, R. "Molds and men." *Scientific American,* January 1952. Traces the impact of fungi on human civilization.

Muller, W. H. *Botany. A Functional Approach.* New York. Macmillan, 1979. A general text for both science and non-science majors; includes both plants and fungi.

Raven, P., Evart, R. F. and Eichhorn, S. E. *Biology of Plants.* 4th edition. New York: Worth, 1986. A beautifully illustrated and comprehensive botany text, stressing evolution and environmental adaptations.

"The Society of Amoebas." *Science 84.* December, 1984. Beautiful photography and clear descriptions highlight this article for the interested layperson.

Successful reproduction—from an evolutionary perspective, the *raison d'etre* for anatomical and physiological adaptions that allow an organism to flourish in its environment.

Unit IV
Anatomy and Physiology of Plants and Animals

20
Unifying Concepts in Physiology

All organisms face similar challenges in life: obtaining food and water, defending themselves against predators and parasites, and reproducing, to name a few. How organisms cope with these problems is the province of **physiology,** which Webster's dictionary defines as "a branch of biology dealing with the processes, activities, and phenomena characteristic of life." In this chapter we provide an overview of the basic physiological processes common to all living organisms. In the remaining chapters of this unit, we will examine the physiology of plants and animals more closely.

LIFE PROCESSES

Picture a "living organism" in the abstract (Fig. 20-1): an entity composed of one or more cells, isolated in many respects from its environment, yet intimately dependent on that environment for energy and materials. As it goes through life, the organism requires certain things from its environment, and the environment places stresses on the organism that must be overcome if the organism is to survive and perpetuate its kind. Whether unicellular or multicellular; bacterium, protist, plant, animal, or fungus; marine, freshwater, terrestrial, or parasitic, all organisms perform a common set of activities that sustain life:

1. Obtain materials to construct the body.
2. Obtain energy for construction, maintenance, and reproduction.
3. Exchange gases, principally carbon dioxide and oxygen, with the environment.
4. Regulate body composition by eliminating excess or undesirable materials and the waste products of body metabolism.
5. Distribute materials throughout the body.
6. Coordinate the activities of the various parts of the body.
7. Defend the body against predators, parasites, and disease.
8. Reproduce.
9. Regulate the growth and development of body parts.

Terrestrial organisms face additional challenges peculiar to life on land. They must also:

10. Obtain water.
11. Reduce water loss.
12. Protect the body against the temperature extremes encountered on land.
13. Support the body against the force of gravity, unaided by the buoyancy provided by water.
14. Reproduce without a watery environment through which the sperm can swim to meet the egg, and in which the embryo can develop.

Although we discuss each activity separately, many of these processes are interrelated. We will emphasize the adaptations of the most familiar organisms, plants and animals.

Processes Common to All Organisms

Obtaining materials. Every organism must procure the materials required to construct its body. Animals typically obtain materials by ingesting food, usually plants, protists, or other animals. The complex molecules of protein, fat, and carbohydrate that make up their prey are broken down into simpler molecules of amino acids, fatty acids, and sugars in the digestive tract. The various parts of the body then use these simple molecules to synthesize new, larger molecules that are specific to the animal.

Plants obtain their materials as simple inorganic molecules of carbon dioxide, water, and minerals. Plants synthesize their amino acids, fatty acids, and sugars starting from scratch, so to speak, since they cannot acquire the ready-made "convenience foods" available to animals.

Obtaining energy. Plants are the entry point for energy in ecosystems. Green plants capture the energy of sunlight and use it to drive the synthesis of organic molecules from minerals, carbon dioxide, and water. Energy

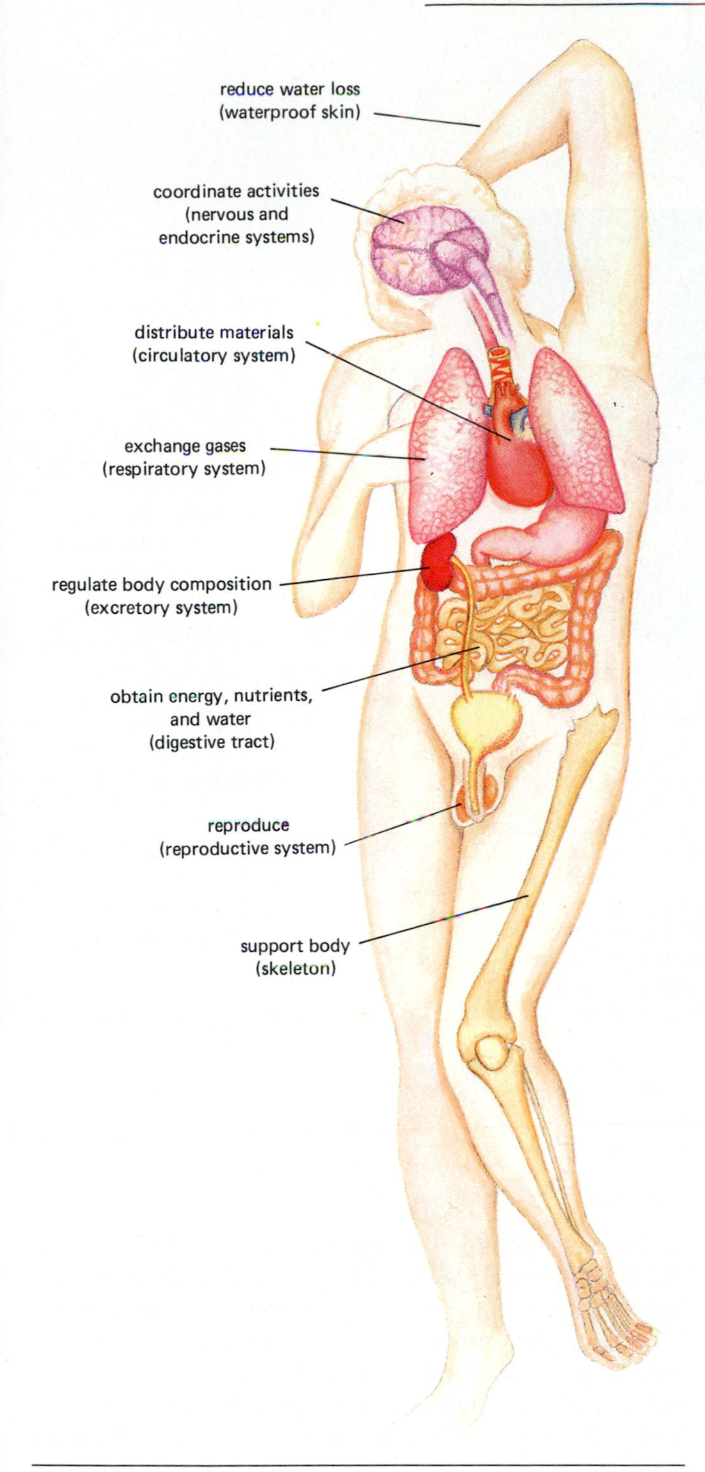

Figure 20-1 *Multicellular organisms, including human beings, have evolved a variety of physiological systems that carry out their necessary life processes. Each is coordinated with the others to ensure proper overall functioning of the entire organism.*

acquisition and material acquisition, although intimately connected, are therefore separate processes.

Animals, on the other hand, acquire both materials and energy in the food they eat. By guiding the natural tendency for complex molecules of fat, protein, and sugar to break down into simpler molecules, animals capture the released energy and direct it into their own bodily activities.

EXCHANGING GASES. All animals and plants exchange oxygen and carbon dioxide with the air or water. Plants absorb carbon dioxide and release oxygen during photosynthesis. During cellular respiration, both plants and animals use up oxygen and produce carbon dioxide as a waste product (see Chapter 7).

For very small organisms, all parts of the body are close enough to the surface so that diffusion suffices to move gases to and from the cells. Active, large organisms, such as many animals, must have respiratory systems: elaborations of the body surface that provide a large area for gas exchange. Most also have some means of pumping air or water over the respiratory surface, ensuring maximum availability of oxygen. Plants typically have a much slower metabolism. Further, the deep internal tissues of large roots and stems are usually composed of dead tissues that do not use oxygen. Therefore, most plants do not have specialized respiratory systems.

REGULATING BODY COMPOSITION. In the process of obtaining enough of all materials, most organisms acquire too much of at least some of them. Since the cells of the body function properly only within a narrow range of composition of both intracellular and extracellular fluids, the body must rid itself of surplus materials while retaining desirable ones. Most organisms also create toxic wastes as the result of normal metabolic processes, and must excrete these as well. Different organisms have evolved a variety of excretory organs, from the contractile vacuoles of *Paramecium,* for whom water is a surplus material, to the elaborate kidneys of terrestrial vertebrates, for whom water is a scarce resource. Plants sequester many surplus or toxic products in the central vacuoles of their cells. Some plants also excrete wastes in droplets of water that often appear at the tips of leaves in the morning.

DISTRIBUTING MATERIALS. In a large body, most of the cells are located far from the sources of food and oxygen. Consequently, these materials must be distributed throughout the body. Even small and sluggish animals such as snails have circulatory systems in which a fluid is pumped around the body by contractions of the heart. Plants, too, have circulatory systems, to move sugars from leaves to roots and fruits, and to move minerals and water from roots to leaves and fruits. The fluids in plant circulatory systems are not pumped, however. Instead, plants have evolved mechanisms based on osmosis and the evaporation of water that transport fluids within rigid tubes.

COORDINATING THE PARTS OF THE BODY. Early in the history of life, cells evolved the ability to influence other

Figure 20-2 Defensive adaptations in plants and animals.
(a) *A monarch butterfly larva feeding on a milkweed leaf. The milkweed synthesizes toxic compounds that interfere with the functioning of the nervous systems and hearts of potential animal predators. Consequently, few animals eat milkweeds. One of the few that can is the caterpillar of the monarch butterfly. It incorporates the poisonous compounds in its tissues and advertises this fact with its distinctive coloration. As a result, birds that prey on other butterflies leave the monarch alone.*
(b) *The flesh of the cactus is often one of the few sources of moisture available in a desert. The leaves of cacti are reduced to thin spines, which both reduce evaporation and discourage predation.*
(c) *The porcupine displays an antipredator adaptation similar to that of the cactus.*

cells by releasing chemicals that trigger responses in their recipients. Both plants and animals still use chemicals, called hormones, to coordinate the growth, development, and functioning of the body. Animals also possess nervous systems, which in essence are rapid transmission lines that send information from one part of the body to another. At the receiving end, the nerve cells release a chemical called a neurotransmitter that evokes appropriate responses in target cells.

DEFENDING THE BODY. "Big fleas have little fleas upon their backs to bite 'em," wrote Jonathan Swift, expressing the principle that every organism is food for another. Some plants and animals have evolved similar adaptations that discourage predation, including bad taste or poisons (Fig. 20-2a) and defensive armament (Fig. 20-2b,c). Many animals, being more mobile than plants, have also evolved more active techniques, such as running away, hiding, or fighting back. On the other hand, most plants can survive even if parts of them are eaten.

Equally serious is the threat of disease, that is, parasitism by microorganisms. Impervious outer coverings and walling off infected regions from the rest of the body are common defenses of both animals and plants. Animals also have roaming defensive cells that devour invaders, and circulating chemical warfare molecules, the antibodies, that target microbes for destruction or neutralize their toxins.

REPRODUCTION. Most plants and animals reproduce sexually. Animals often seek out mates, defend mates and/or nest sites, perform elaborate courtship displays, and provide parental care for their offspring. Plants, being mostly stationary, use wind, water, and animals to carry sperm or pollen (containing sperm) to the egg. Although plants have their own forms of "courtship" (pollen- and nectar-laden flowers that attract animal pollinators), territorial defense (e.g., shading competitors from the sunlight needed for photosynthesis), and even passive parental care (packaging food together with the embryo in a seed), the adaptations involved are quite unlike those of animals.

Regulating growth and development. In sexually reproducing organisms, all the structures of the adult develop from a fertilized egg. During embryonic development, the descendants of this single cell differentiate into specialized cell types and form organs that carry out specific bodily functions. Both plants and animals use chemical signals, either on the surfaces of cells or in the circulation, to govern cell differentiation and body development.

Processes Required for Life on Land

The first organisms probably lived in the sea, and for over 2 billion years, organisms adapted to continuous immersion in water. Besides ample water for cellular metabolism, the sea also provides a relatively constant temperature and buoyant support for bodies against the pull of gravity. When plants and animals invaded the land, they had to evolve adaptations to compensate for the loss of their watery habitat.

Obtaining water. The first essential for land organisms is to obtain sufficient water. Plants usually mine water from the soil with their roots, using the power of water evaporating from their leaves to pull water into the plant from the soil. In a few favorable locations, water may also be available as fog or dew. Animals acquire water in three different ways. Most animals drink liquid water from pools, lakes, or streams. Since the foods they eat contain water, some water is also obtained as a by-product of acquiring materials and energy. Finally, as you learned in Chapter 7, the end products of cellular respiration are carbon dioxide and water. This "metabolic water" actually provides some desert animals with a significant percentage of their required daily supply.

Preventing water loss. Terrestrial organisms must also minimize the loss of water by evaporation into the air. Virtually all land plants and animals are covered with a waterproof coating that reduces evaporation (Fig. 20-3a). Animals and plants also lose water during gas exchange, and desert dwellers usually have adaptations that minimize this loss. Many desert animals remain inactive during the heat of the day, which reduces the need for gas exchange when the air is hottest, driest, and most effective at evaporating water. Animals also lose significant amounts of water during excretion of urine. As you might expect, desert animals such as the kangaroo rat (Fig. 20-3b) have specialized kidneys that excrete wastes dissolved in the smallest possible volume of water.

Protecting against temperature extremes. Both high and low temperatures can be fatal to cells. High temperatures denature proteins (like cooked egg whites), whereas freezing temperatures cause ice crystals to form within the cytoplasm, piercing through membranes and destroying the cell. Life does not exist in regions of ex-

Figure 20-3 Since water is so scarce, desert organisms must reduce water loss to a minimum. **(a)** *Succulents store water in fleshy leaves, covered with a thick waxy coating that reduces evaporation.* **(b)** *Kangaroo rats stay underground during the heat of the desert day, emerging out of their burrows to forage at night. The kidneys of kangaroo rats are remarkably efficient, producing extremely concentrated urine and thereby eliminating wastes without losing much water.*

Figure 20-4 High temperatures can be lethal to animals. If the temperature is very high, excess metabolic heat cannot be given off directly to the air. Surplus heat can be disposed of, however, by using its energy to evaporate water, as dogs do when they pant.

tremely high temperatures, such as the sources of volcanic hot springs, or low temperatures, such as in the perpetually frozen hearts of glaciers. In between, including the outflows of hot springs and the surfaces of glaciers, living organisms have adapted to a remarkable range of temperatures. Plants and animals have independently evolved such features as heat-resistant proteins and antifreeze molecules. Many produce heat- or cold-resistant reproductive stages (eggs, larvae, or seeds) which can survive temperature extremes that kill off the adults. Many animals display one of two other adaptations to cold. By eating enough either before winter (bears) or during winter (snowshoe hares, arctic foxes, wolves), some animals can maintain their body temperatures well above freezing even in the coldest weather. Others, such as most North American songbirds, migrate to warmer climes. High temperatures also evoke adaptations such as burrowing into the cool ground during the hottest part of the day, seeking out shade, and evaporating water from the skin or respiratory system to carry away excess heat (Fig. 20-4).

SUPPORTING THE BODY AGAINST GRAVITY. The bodies of aquatic organisms are supported by the water all around them, and some very large organisms, such as octopus and giant squid, can get along with little or no skeleton for support. The air, however, is too thin to support any but the flimsiest of bodies, so terrestrial plants and animals must have some type of skeleton. Plant cells are surrounded by strong cell walls made of cellulose. These cellulose "skeletons," together with the pressure of intracellular water pushing up against the inside of the wall, hold up the bodies of many small plants, such as herbs and grasses. Larger plants have tracts of specialized cells with extra-thick walls, forming the wood that makes up and supports the massive trunks of trees. Animal skeletons are extremely diverse in form and materials, and include internal skeletons of bone (vertebrates), external skeletons of chitin (insects and other arthropods), and hydrostatic skeletons of water confined within a muscular tube (earthworms).

REPRODUCING ON LAND. For perhaps a billion years, plants and animals reproduced in water: sperm and eggs were released, the sperm swam to meet the egg, and the offspring developed in water. The first organisms invading the land retained these characteristics. Even today, ferns reproduce only when a film of water covers the small sexually reproducing plant, allowing the sperm to swim to the egg (see Chapter 19).

For animals, true independence of water for reproduction requires two adaptations: internal fertilization and waterproof eggs. During internal fertilization, the male deposits sperm directly into the reproductive tract of the female, which provides the watery environment for sperm movement. Waterproof eggs in insects, reptiles, and birds allow the offspring to develop in its own private pond until it can face the rigors of land existence. Mammals have taken the process one step further: the embryo develops within the mother's body, enclosed in the waters of the amniotic fluid.

Land plants have evolved a different solution. The truly terrestrial plants, including grasses, wildflowers, and trees, have done away with swimming sperm entirely. Sperm cells are packaged in pollen grains that do not need water for survival. Pollen can remain viable for weeks, months, or even centuries. Most plants enclose the embryo in a seed that in some instances may be almost as enduring as pollen. In some desert plants, seeds can remain dormant for years, through drought and frost, sprouting only when liquid water finally becomes available following the infrequent rains.

THE STUDY OF PLANT AND ANIMAL PHYSIOLOGY

As you can see from our brief survey of the essential processes of living organisms, plants and animals face similar challenges in similar environments. The solutions that each has evolved, however, are extremely different. To take but one example, the heart, arteries, and veins of an animal circulatory system provide some of the same functions as the xylem and phloem tubes of plants, namely to move materials from one part of the body to another. However, they are extremely different in almost every other respect, including the embryonic development of the systems, the forces that drive fluid movement through the vessels, the directionality of flow, the composition of the fluids, and the presence of cells within the fluids in animals.

Plant and animal physiology may thus be considered together, unified by the similarity of function of the various systems, or individually, separated by the enormously different mechanisms that each has developed over millenia of independent evolution. For the sake of clarity, we have chosen to discuss plant and animal physiology separately. As you progress through these chapters, however, keep in mind that the different physiological solutions evolved by plants and animals are responses to virtually identical challenges posed by their environments.

SUMMARY OF KEY CONCEPTS

Life Processes
All organisms face many of the same challenges, including obtaining food, water, and energy; exchanging gases with the environment; regulating body composition; distributing materials throughout the body; coordinating bodily activities; defending themselves against predators, parasites, and disease; regulating the development of the body; and reproducing. Terrestrial organisms face further problems of water acquisition and loss; gravity stresses; temperature extremes; and protection of gametes and embryos from desiccation.

The Study of Plant and Animal Physiology
In response to these challenges, plants and animals have evolved adaptations that, while they may be fundamentally similar in principle, are often strikingly different in detail. The nature of these adaptations and how they function are the subject matter of the study of physiology.

SUGGESTED READINGS

Readings in particular areas of plant and animal physiology are listed at the ends of the appropriate chapters. Here we offer a few interesting articles that explore adaptations to unusual body forms or rigorous environments.

Degabriele, R. "The Physiology of the Koala." *Scientific American,* June 1980 (Offprint No. 1476). Koalas eat poisonous eucalyptus leaves and almost never drink water. How can they survive on such a bizarre diet?

Roger, C. F. E., and Boss, K. J. "The Giant Squid," *Scientific American,* April 1982 (Offprint No. 1515). Yes, there really are giant squid in the ocean depths, doing battle with sperm whales. This article explores new findings on the anatomy, physiology, and ecology of this almost-mythical beast.

Schmidt-Nielson, K. "The Physiology of the Camel." *Scientific American,* December 1959. Do camels really store water in their humps? How long can one go without drinking?

Schmidt-Nielson, K., and Schmidt-Nielson, B. "The Desert Rat." *Scientific American,* July 1953 (Offprint No. 1050). The desert dweller *par excellence,* the kangaroo rat never needs to drink water.

Went, F. W. "The Ecology of Desert Plants." *Scientific American,* April 1955. Like the kangaroo rat, desert plants must also acquire and conserve water.

21
Plant Structure and Function

As you learned in Chapter 15, life probably arose in the shallows of the sea, and coastal waters still teem with plant life (Fig. 21-1). However, about 400 million years ago, the first plants invaded the land, and today more species of plants inhabit the land than live in the oceans. Why? What advantages are offered by the land that the oceans lack?

To survive and prosper, a living organism requires a suitable range of four environmental conditions: *temperature, water, energy,* and *nutrients*. Generally, the maximum density of organisms that a habitat can support is determined by the least favorable of these four conditions. In the sea, temperature extremes are rare and water, of course, is everywhere. However, almost all plants obtain energy from sunlight and absorb nutrients dissolved in water, whether in the ocean, streams, or water trapped within the soil. These two factors, light and nutrients, are usually the limiting factors for marine plants. If you have ever gone scuba diving in the ocean, you know that it's pretty dark down there, and the deeper you dive, the darker it gets. The light needed for photosynthesis quickly fades with depth, restricting most marine plants to the upper 10 or 20 meters at the surface. Further, although nutrients in the ocean are conveniently in solution, they are relatively scarce.

Figure 21-1 The cradle of life, coastal waters abound with large algae, including forests of giant kelp that provide a habitat for many fish and invertebrates.

The land, on the other hand, fairly explodes with light, and even poor soils are veritable storehouses of minerals compared with the sea. The plant that can inhabit the land reaps a rich harvest of energy and nutrients. However, leaving the sea means abandoning an endless supply of water, braving extremes of heat and cold, and acquiring nutrients that no longer bathe the entire plant but are localized in the soil.

The first three chapters of this unit examine the adaptations of plants that have colonized the land. The most important of these adaptations are roots that anchor the plant and acquire water and nutrients from the soil; stems that lift leaves up above competitors to reach the sunlight; tubelike vascular tissues that transport water, nutrients, and sugars among roots, stems, and leaves; reproductive structures, especially pollen, that allow land plants to reproduce independently of water; and seeds that protect and nourish the developing embryo.

Not all land plants have evolved this full complement of adaptations; ferns, for example, can reproduce only when a film of water covers the ground, so their sperm can swim from one plant to the next. By far the most successful terrestrial plants are the seed plants, which *can* reproduce independently of liquid water. These are the **gymnosperms** (conifers, cycads, and ginkgos) and the **angiosperms** (flowering plants; see Chapter 19 for descriptions of the diversity of seed plants).

In this chapter we examine the structure and function of roots, stems, and leaves. In Chapter 22 we discuss reproductive strategies, and in Chapter 23 we describe how land plants control their growth and development. Because of their evolutionary success, abundance on land, and importance to humans, we will emphasize the flowering plants.

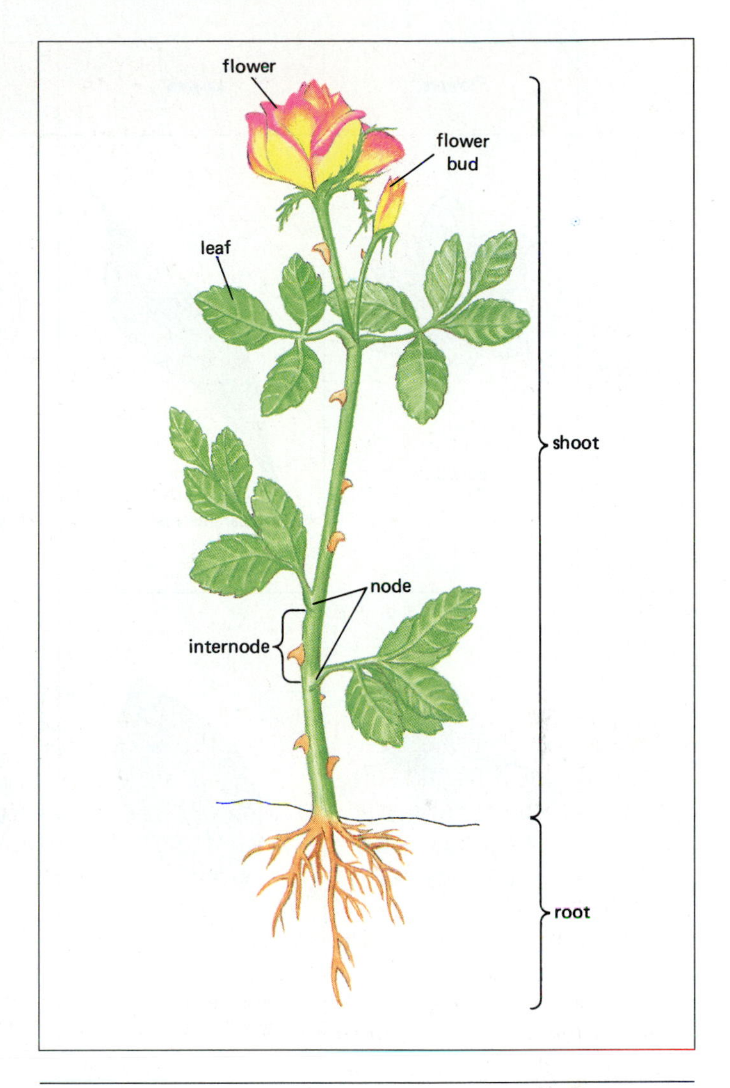

Figure 21-2 The external structure of a flowering plant.

THE GROWTH AND STRUCTURE OF LAND PLANTS

The bodies of angiosperms and gymnosperms consist of two major regions, the root and the shoot (Fig. 21-2). The **roots** are usually below ground and serve five functions. Roots (1) anchor the plant in the ground; (2) absorb water and minerals; (3) store surplus sugars manufactured during photosynthesis; (4) transport water, minerals, sugars, and hormones to and from the shoot; and (5) produce some hormones.

The rest of the plant is the **shoot,** usually found above ground. The shoot consists of stems, leaves, and reproductive parts of the plant (flowers and fruit in angiosperms, cones in gymnosperms). The functions of the shoots include (1) photosynthesis, mainly in leaves and young green stems; (2) transport of materials among leaves, flowers, fruits, and roots; (3) reproduction; and (4) hormone synthesis.

Figure 21-3 illustrates the two types of flowering plants, monocots (class Monocotyledonae: grasses, lilies, orchids) and dicots (class Dicotyledonae: deciduous trees and bushes, most garden flowers). Formally, the monocots and dicots are distinguished by the number of "seed leaves," or cotyledons, found in their seeds (see Chapter 22), but they differ in a variety of other ways as well. Don't worry about terms that are not yet familiar to you; just look over the figure for now, and refer back to it as we examine the parts of flowering plants in more detail.

Plant Cell Types

From the moment of sprouting, plants are composites of two fundamentally different types of cells: embryonic, undifferentiated **meristem cells** that remain capable of cell division, and mature, **differentiated cells** that are specialized in structure and function, and that usually do not divide. Normally, *each time a meristem cell divides, one daughter cell develops specialized structures and becomes a differentiated cell, while the other daughter remains meristematic* (Fig. 21-4). Continued divisions of meristem cells, then, keep a plant growing throughout its life, while their differentiated daughter cells form relatively permanent parts of the plant, such as mature leaves or the trunks of trees. Rather than cataloging all the types of differentiated cells here, we will wait until we encounter them in roots, stems, and leaves.

Plant Growth

Plant growth occurs by division and differentiation of meristem cells located in two regions of the plant: **apical meristems** at the tips of roots and shoots (including main stems and branches; Figs. 21-6 and 21-9) and **lateral meristems** or **cambia** (singular **cambium**) forming cylinders running parallel to the long axis of roots and stems (Figs. 21-9 and 21-15).

Division in apical meristems followed by differentiation of the resulting daughter cells is called **primary growth.** This occurs in young plants and the growing tips of roots and shoots in older plants, as they simultaneously grow longer and develop the structures essential for life.

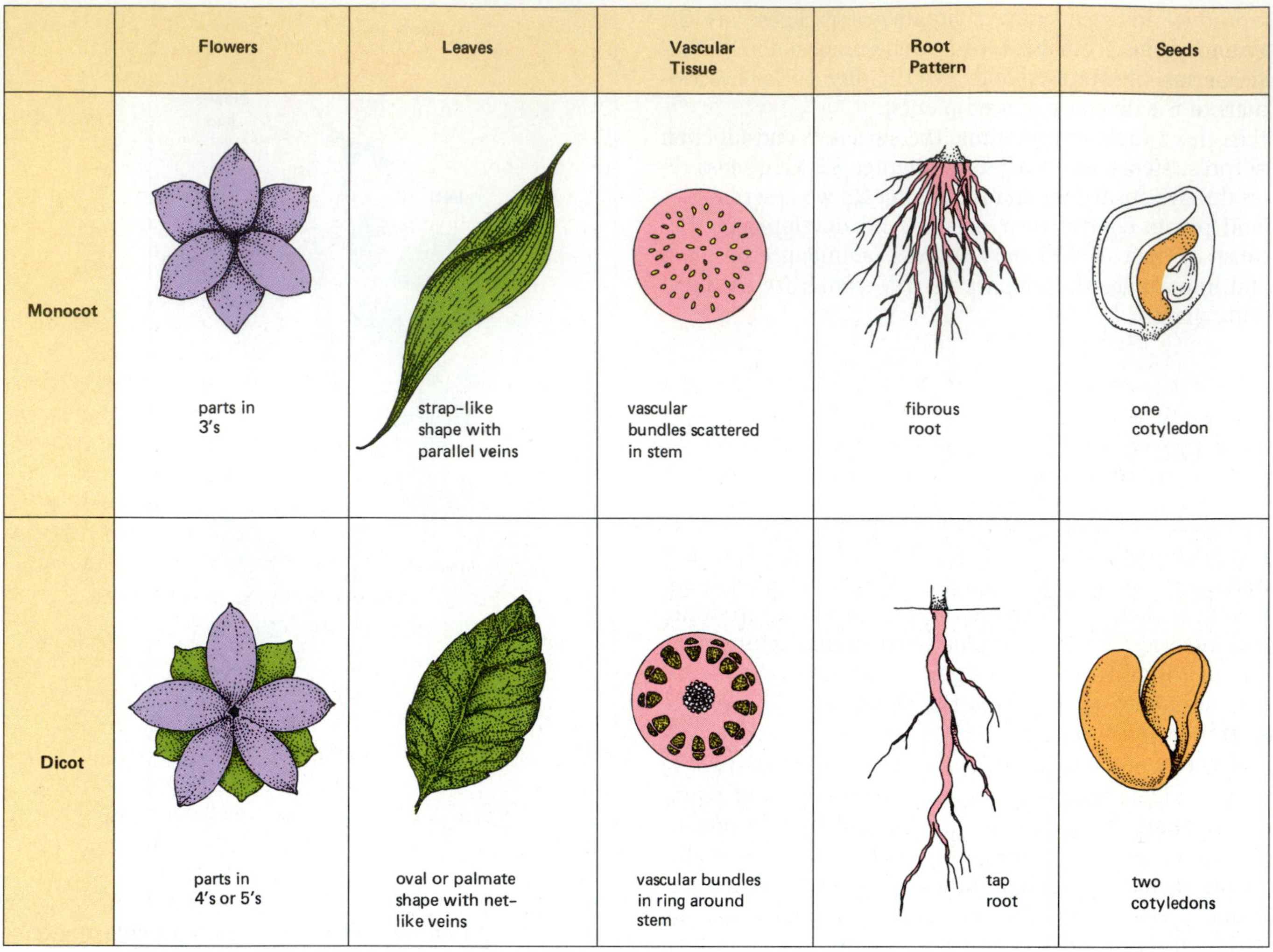

Flowers
Leaves
Vascular Tissue
Root Pattern
Seeds
Monocot
parts in 3's
strap-like shape with parallel veins
vascular bundles scattered in stem
fibrous root
one cotyledon
Dicot
parts in 4's or 5's
oval or palmate shape with net-like veins
vascular bundles in ring around stem
tap root
two cotyledons

Figure 21-3 *Distinguishing traits of the two major classes of flowering plants, the monocots and the dicots.*

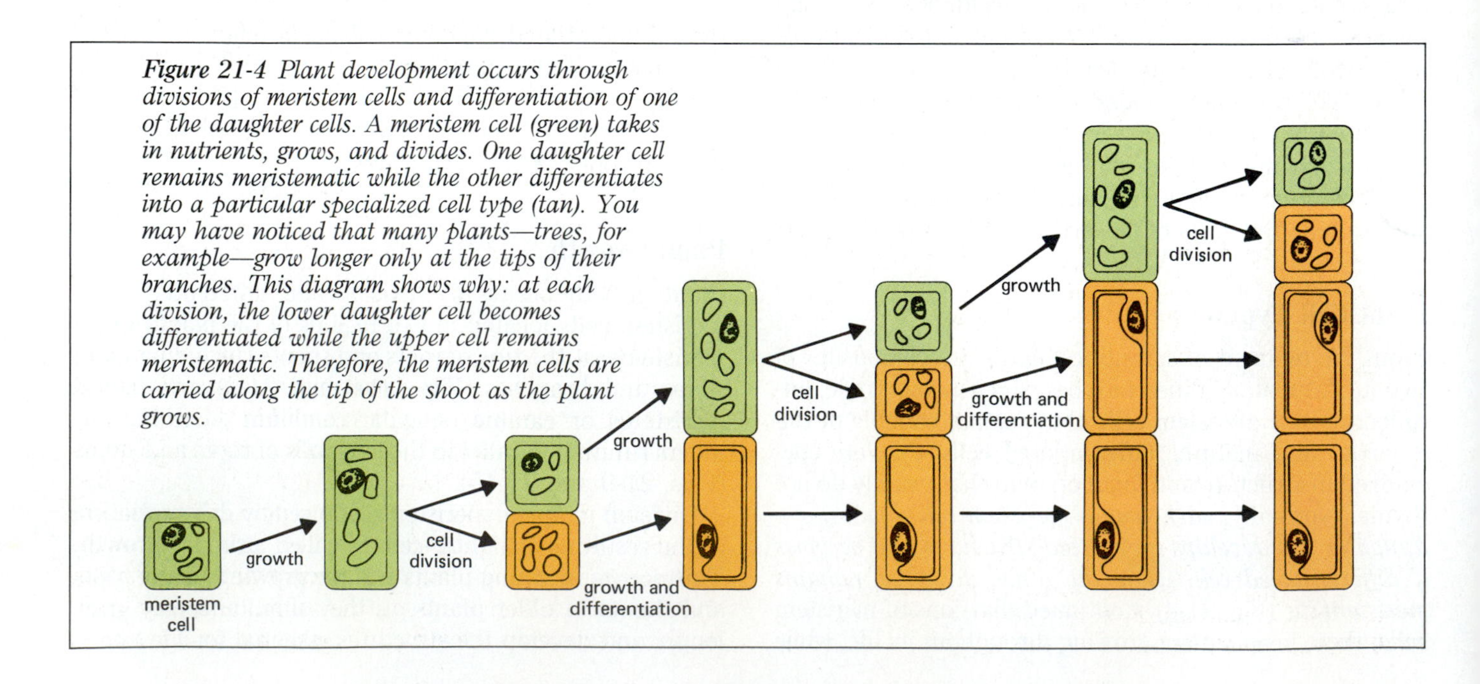

Figure 21-4 *Plant development occurs through divisions of meristem cells and differentiation of one of the daughter cells. A meristem cell (green) takes in nutrients, grows, and divides. One daughter cell remains meristematic while the other differentiates into a particular specialized cell type (tan). You may have noticed that many plants—trees, for example—grow longer only at the tips of their branches. This diagram shows why: at each division, the lower daughter cell becomes differentiated while the upper cell remains meristematic. Therefore, the meristem cells are carried along the tip of the shoot as the plant grows.*

meristem cell
growth
cell division
growth
growth and differentiation
cell division
growth
growth and differentiation
cell division

The stems and roots of most gymnosperms and dicots become thicker and woody as they age. This is called **secondary growth,** and occurs through division of lateral meristem cells and maturation of their daughter cells. Although both stems and roots undergo secondary growth, we will discuss secondary growth only in stems.

ROOTS: ANCHORAGE, ABSORPTION, AND STORAGE

As a seed sprouts, the **primary root** grows down into the soil. In dicots such as carrots and dandelions, this primary root usually becomes longer and stouter with time, forming a **taproot,** which gives rise to smaller branches that make up the rest of the root system (Fig. 21-5a). In monocots such as grasses, on the other hand, the primary root soon dies off, replaced by many new roots that emerge from the base of the stem. These secondary roots are nearly equal in size, forming a **fibrous root system** (Fig. 21-5b).

Primary Growth in Roots

In young roots of both tap and fibrous systems, divisions of the apical meristem give rise to four tissues (Fig. 21-6). At the very tip of the root, daughter cells produced on the "soil side" of the apical meristem differentiate into the **root cap.** The root cap protects the apical meristem from being scraped off as the root pushes down through the rocky soil particles. Root cap cells have thick cell walls and secrete a slimy lubricant that helps to ease the way between soil particles. Nevertheless, root cap cells wear away, and must be replaced continuously by new cells from the meristem.

(a)

(b)

Figure 21-5 Typical root systems in dicots and monocots. **(a)** *Dicots often have a taproot system, consisting of a long central root with many smaller, secondary roots branching from it.* **(b)** *Monocots usually have a fibrous system, with many roots of equal size.*

Daughter cells produced on the "shoot side" of the apical meristem differentiate into an outer envelope of **epidermis,** a **vascular cylinder** containing conducting tissues at the core of the root, and between the two, the **cortex** (Fig. 21-6).

EPIDERMIS. The outermost covering of cells is the epidermis, which is in contact with the soil and any air or water trapped among soil particles. The cell walls of the epidermal cells are highly water permeable. Water can also penetrate into the root interior by passing between cells of the epidermis. Many epidermal cells grow long projections, called **root hairs,** into the surrounding soil (Fig. 21-6). By increasing the surface area, root hairs increase the root's ability to absorb water and minerals. Root hairs may add dozens of square meters of surface area to the roots of even small plants.

CORTEX. Cortex occupies most of the inside of a young root. The cortex consists of two very different types of cells: an outer mass of large, loosely packed cells just beneath the epidermis, and an inner layer of smaller, close-fitting cells forming a ring around the vascular cylinder, called the **endodermis** (Fig. 21-6). Sugars produced by photosynthesis in the shoot are transported down to the large cortex cells, where they are converted to starch and stored. These cells are particularly abundant in roots specialized for carbohydrate storage, such as the thick roots of carrots and dandelions.

The endodermis is a layer of cells with highly specialized cell walls (Fig. 21-7). Where endodermal cells contact each other, their cell walls are impregnated with a waxy material, forming the **Casparian strip.** In three dimensions, the Casparian strip resembles the mortar in a brick wall: the waxy waterproofing covers the top, bottom, and sides of the endodermal cells, but not the outer surfaces (facing the rest of the cortex) or inner surfaces (facing the vascular cylinder; Fig. 21-7). As we shall see later in this chapter, the Casparian strip plays an important role in water and mineral absorption in a root.

VASCULAR CYLINDER. The vascular cylinder contains the conducting tissues of xylem and phloem that transport water and dissolved materials within the plant (Fig. 21-6). These two tissues, which will be described more fully when we discuss the structure of stems, serve distinctly different functions. **Xylem** conducts water and dissolved minerals from the roots up to the aerial parts of the plant. **Phloem** carries a concentrated sugar solution from parts of the plant that produce sugars, such as photosynthesizing leaves, to parts that use up sugars,

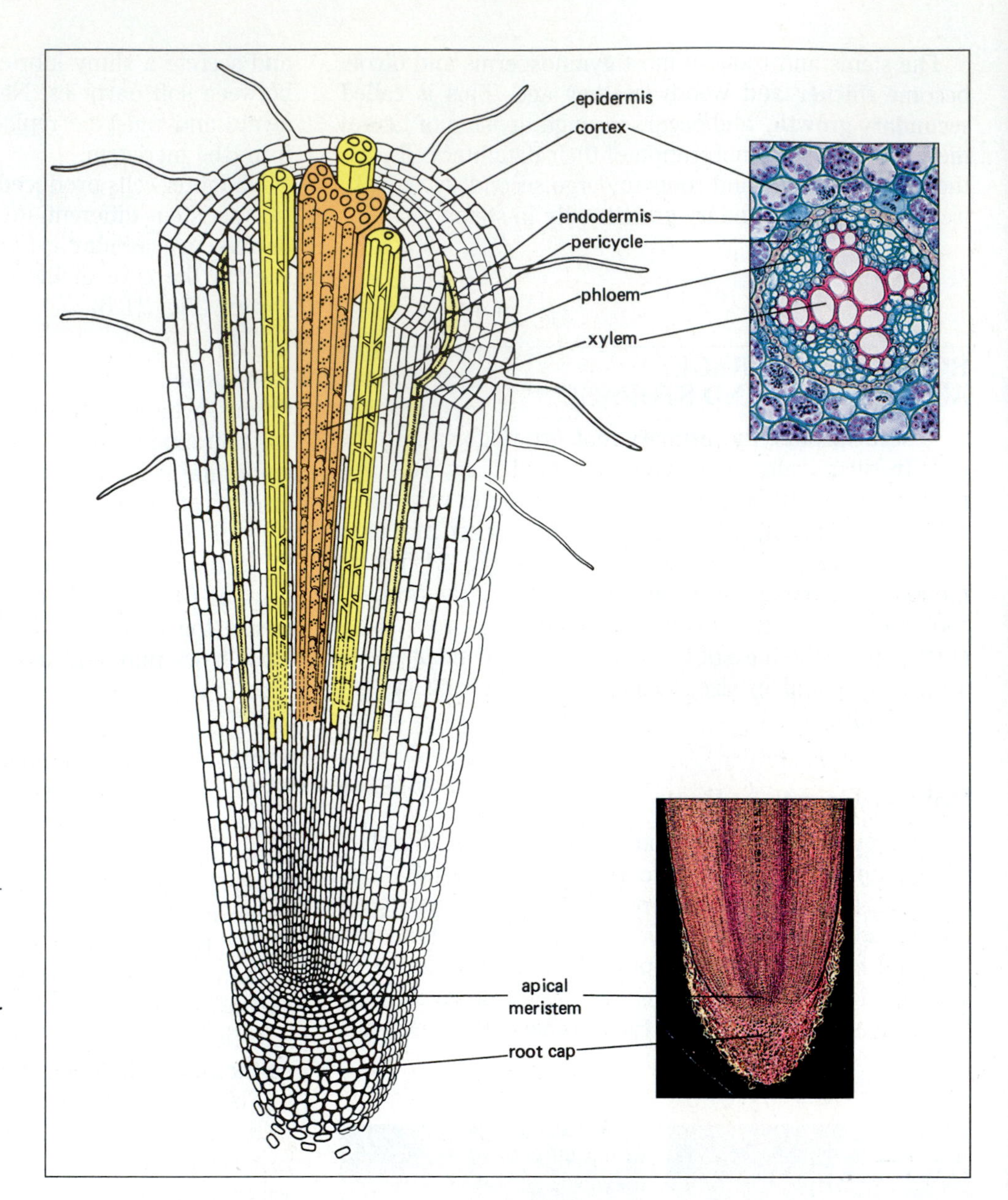

Figure 21-6 *Primary growth in roots results from cell divisions in the apical meristem, located near the tip of the root, and subsequent elongation of daughter cells as they differentiate into epidermis, cortex, and vascular cylinder. The root cap forms at the very tip of the root, also from divisions of cells in the apical meristem.*

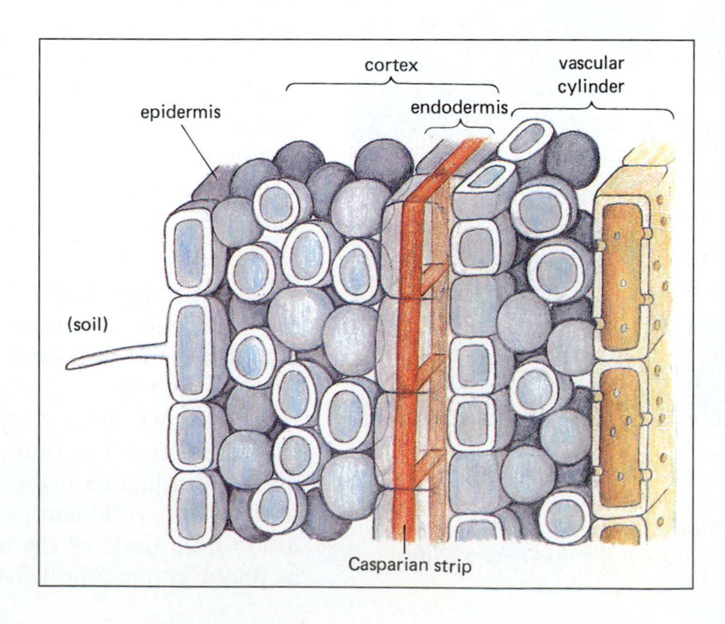

Figure 21-7 *The Casparian strip is a band of waterproof material in the walls between cells of the endodermis. In three dimensions, the Casparian strip is arranged like the mortar in a brick wall, sealing off the top, bottom, and sides of the endodermal cells, thereby preventing water movement between the cells.*

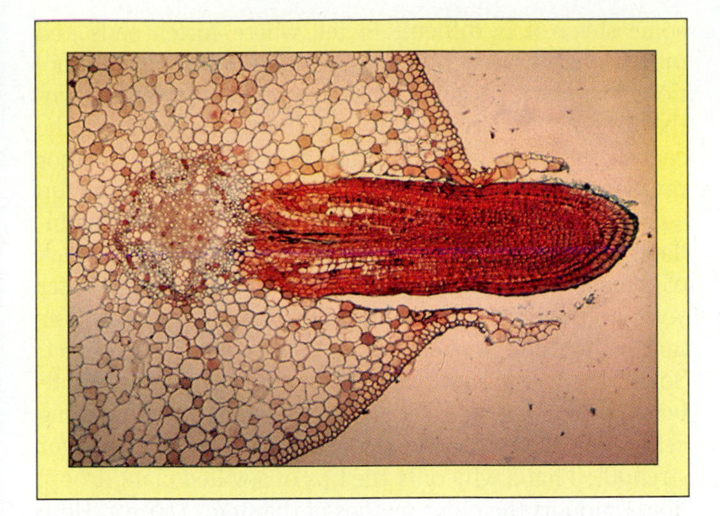

Figure 21-8 *Branch roots emerge from the pericycle of a root. Notice that the central axis of the branch is already differentiating into vascular tissue.*

such as growing root and shoot tips, and to storage regions, such as root cortex cells.

The outermost layer of the vascular cylinder, the **pericycle,** is a remnant of meristem that retains the capacity for cell division. Under the influence of plant hormones, pericycle cells divide and form the apical meristem of a **branch root** (Fig. 21-8). Branch root development is similar to primary root development, except that the branch must first break out through the cortex and epidermis of the primary root. It does this partly by crushing the cells that lie in its path, and partly by secreting enzymes that digest them away.

STEMS: REACHING FOR THE LIGHT

One of the major benefits to plants that invaded the land was access to light. The first land plants had no competition for their place in the sun, but soon some plant species evolved the ability to raise themselves above the rest. Tall plants enjoyed a great selective advantage, being exposed to full sunlight while shading out competitors below.

Erect stems, however, have their costs. First, a tall stem separates the leaves from the roots. The roots supply the entire plant with nutrients and water, while the leaves supply the roots with the sugar they must have to survive and function. The most successful land plants evolved specialized structures, xylem and phloem, that move materials up and down the stem. Second, an upright stem must be strong enough to withstand the forces of gravity and wind without collapsing. The stems of most land plants are reinforced with tracts of cells that develop extra-strong, rigid cell walls. These tracts become thicker as the stem grows taller, culminating in the massive columns of redwoods and eucalyptus trees that tower 100 meters above the forest floor.

Primary Growth and the Structure of Stems

Like roots, stems develop from a small group of actively dividing cells, the **apical meristem,** that lie at the tip of the young shoot. The daughter cells of the apical meristem differentiate into the specialized cell types of stem, buds, leaves, and flowers (Fig. 21-9).

SURFACE STRUCTURES OF THE STEM. As the shoot grows, small clusters of meristem cells are "left behind" at the surface of the stem. These meristem cells form the

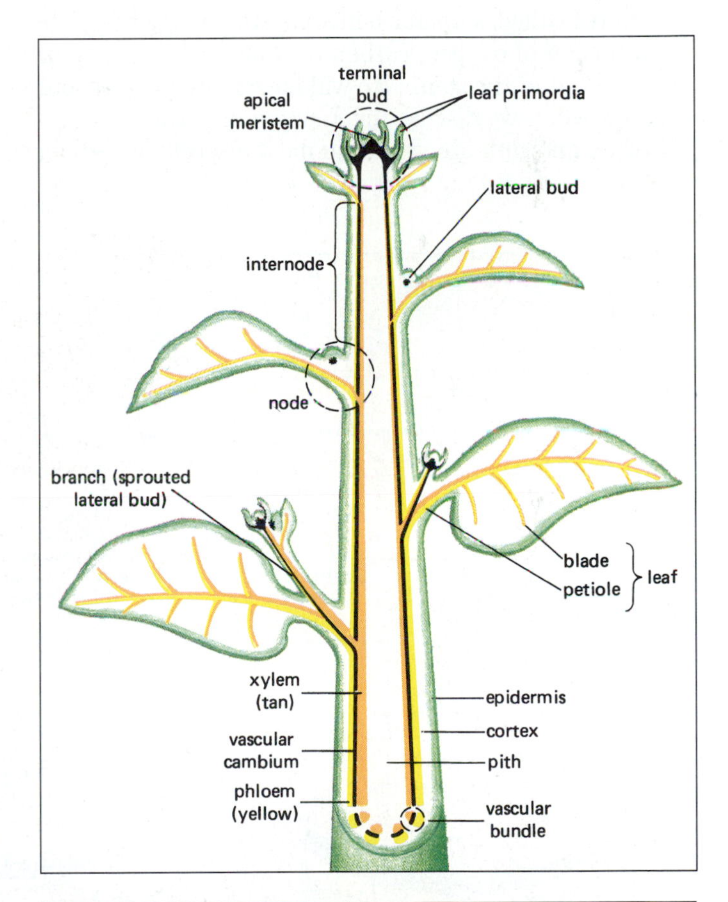

Figure 21-9 *The structure of a young dicot shoot. At the tip of the stem, the terminal bud includes the apical meristem and several leaf primordia, produced by the meristem. Other daughter cells of the apical meristem differentiate into epidermis, cortex, pith, and vascular tissues. As the young stem grows, the leaf primordia develop into mature leaves. Meanwhile, epidermis, cortex, pith, and vascular cells elongate between the points of attachment of leaf to stem, effectively pushing the leaves apart. A remnant of meristem tissue, called a lateral bud, remains in the crotch between each leaf and the stem. Points on the shoot where leaves and lateral buds are located are called nodes; the naked stem between nodes is an internode.*

leaf primordia and **lateral buds** that appear at characteristic locations, called **nodes,** on the stem; regions of stem between these nodes are called **internodes** (Fig. 21-9). Leaf primordia develop into the mature leaves typical of the species of plant. Under appropriate conditions, lateral buds grow into branches. We will discuss the growth of branches shortly.

INTERNAL ORGANIZATION OF THE STEM. Most young stems are composed of four tissues: **epidermis, cortex, vascular tissues,** and **pith.** As Figure 21-10 illustrates, monocots and dicots differ somewhat in the arrangement of vascular tissues. We discuss only dicot stems here.

The stem epidermis, unlike the root epidermis, secretes a waxy covering, the **cuticle,** that reduces evaporation. The epidermis, however, is often perforated with adjustable pores called **stomata** (singular **stoma**) that regulate the diffusion of oxygen, carbon dioxide, and water vapor into and out of the stem. We will have more to say about stomata when we discuss the function of leaves.

Cortex and pith are similar in most respects; in fact, in some stems it is difficult to tell where cortex ends and pith begins. Cortex and pith perform three major functions: support, storage, and in some cases, photosynthesis. *Support:* In very young stems, water filling the central vacuoles of cortex and pith cells causes turgor pressure (see Chapter 4) that pushes the cytoplasm up against the cell wall, stiffening the cells much as air inflates a tire. Just as an underinflated tire goes flat, lack of water causes the cells to go limp; if you forget to water your houseplants, their drooping tips show the importance of turgor pressure in keeping young stems erect. Somewhat older stems also have support cells called **fibers,** with thickened cell walls. Because of their strong cell walls, fibers do not depend on turgor pressure for strength. That's why only the tips of a wilted plant droop: fibers support the older regions of the stem. *Storage:* Both cortex and pith convert sugar into starch and store the starch as a food reserve. *Photosynthesis:* In many stems, the outer layers of cortex cells contain chloroplasts and actively photosynthesize. In some desert plants, such as cacti, the leaves are reduced or absent, and the stem cortex is the only green, photosynthetic part of the plant.

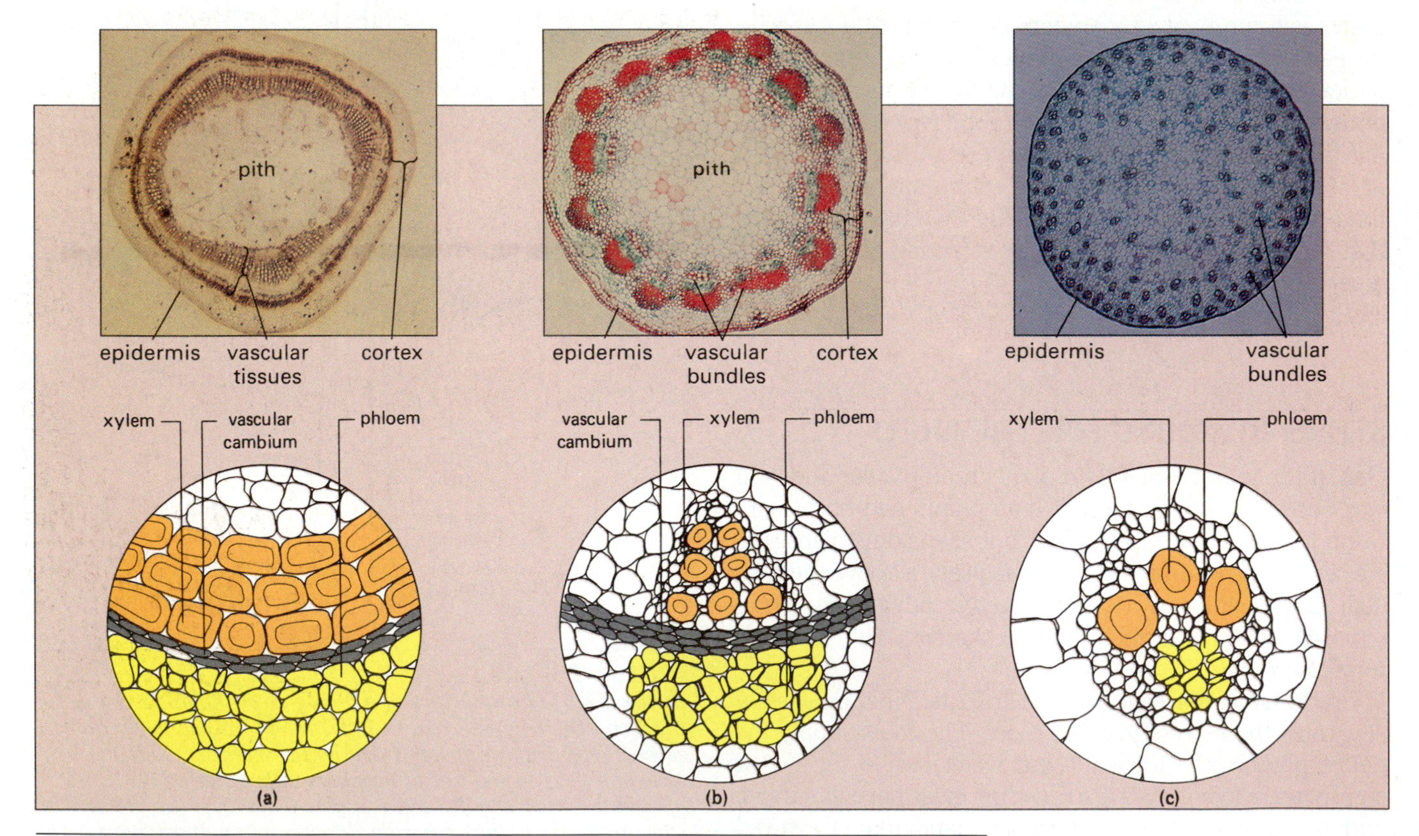

Figure 21-10 Cross sections of stems (top row) of dicots **(a, b)** *and monocots* **(c)** *and the detailed anatomy of their vascular tissues (bottom row). Dicots have vascular tissues arranged in a cylinder surrounding a central "filler" of pith. The linden stem* **(a)** *has a continuous band of xylem on the inside lined with phloem on the outside, while the sunflower* **(b)** *has a beaded "bracelet" of bundles. Within each bundle of the sunflower bracelet, xylem lies to the inside and phloem to the outside. Monocots such as corn* **(c)** *have vascular bundles scattered throughout the stem.*

As in roots, the vascular tissues transport water, minerals, sugars, and hormones. These tissues are continuous in root, stem, and leaf, interconnecting all the parts of the plant. The **primary xylem** and **primary phloem** found in young stems arise from the apical meristem. In young dicot stems, the xylem and phloem are usually arranged as concentric cylinders (Fig. 21-10a) or as a ring of bundles running up the stem, with each bundle containing both phloem and xylem (Fig. 21-10b). Secondary growth in dicot stems, as we will discuss later, always results in concentric cylinders of xylem and phloem (see Fig. 21-14).

Xylem conducts water and minerals in tubes made from one of two types of cells: **tracheids** *and* **vessel elements** (Fig. 21-11). Both cell types are dead at maturity: the cytoplasm and cell membrane disintegrate, leaving behind a skeleton of cell wall. Most gymnosperms only have tracheids, while angiosperms usually have both tracheids and vessel elements.

Tracheids are thin cells with slanted ends like the tips

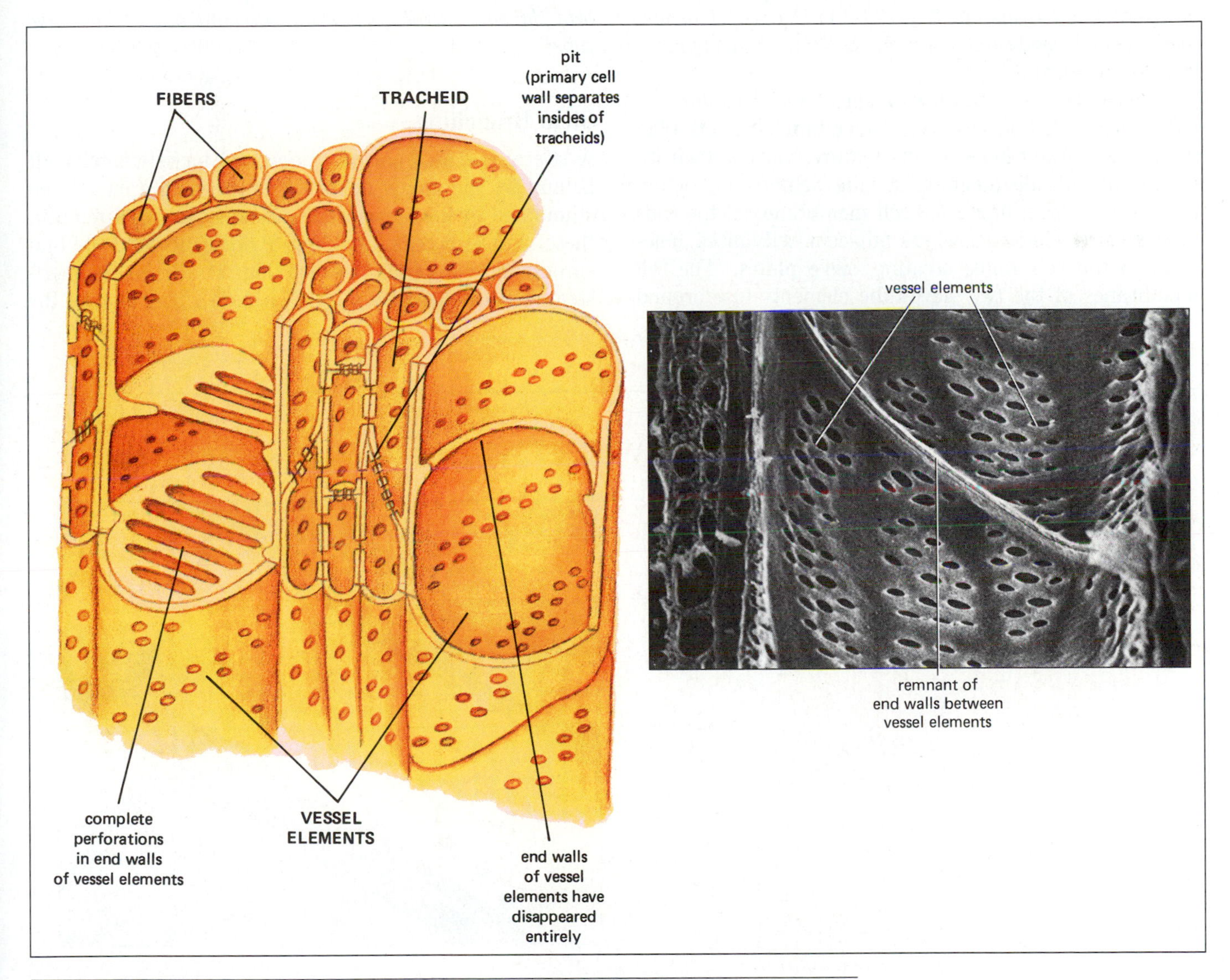

Figure 21-11 Xylem is a mixture of cell types, including fibers and conducting cells. There are two types of conducting cells, tracheids and vessel elements. Tracheids are thin, with overlapping ends punctuated by pits. Notice that the pits are not holes all the way through from one cell to the next, but still have a primary cell wall separating the interiors of the two cells. The primary cell wall is very water permeable, and does not seriously impede water flow between tracheids. Vessels consist of vessel elements stacked atop one another. In many vessel elements, the end walls virtually disappear. Both tracheids and vessel elements have pits in their sidewalls, allowing movement of water and dissolved minerals sideways between adjacent conducting systems.

of hypodermic needles. Tracheids are stacked atop one another with the slanted ends overlapping (Fig. 21-11). The overlapping walls contain **pits** where secondary cell walls failed to form, so water and minerals can pass from one tracheid to the next by crossing only the thin and water-permeable primary cell wall.

Just as tracheids resemble double-ended needles, vessel elements are like soup cans: larger in diameter, with blunt ends. Further, complete perforations form in the ends of abutting vessel elements, with both the primary and secondary cell walls disappearing. In some cases, the ends almost completely disintegrate, making an open pipe the same way that plumbers do, by sticking together a series of open-ended cylinders (Fig. 21-11). Vessel elements, then, form large-bore, relatively unobstructed pipelines from root to leaf.

Phloem carries concentrated sugar solutions through tubes constructed of cells called **sieve-tube elements** (Fig. 21-12). As sieve-tube elements mature, most of their internal contents disintegrate, leaving behind only a thin rind of cytoplasm lining the cell membrane. At the ends of sieve-tube elements, where adjacent cells meet, holes form in the cell walls, creating **sieve plates.** The cell membranes of the two sieve-tube elements fuse around the lips of the sieve plate pores, forming membrane-lined channels connecting the interiors of the two cells. A continuous pipe is forged by many sieve-tube elements linking up end to end in this way.

Since they still have cell membranes and shreds of cytoplasm, sieve-tube elements are considered to be alive, although they usually have no ribosomes, mitochondria, or nuclei. How, then, can they remain alive? *Each sieve-tube element is nourished by a smaller, adjacent* **companion cell.** These companion cells maintain the integrity of the sieve-tube elements by donating high-energy compounds and perhaps even by repairing the sieve-tube cell membrane. As we shall see below, *companion cells also regulate the movements of sugars into and out of the sieve tubes.*

Stem Branching

A lateral bud is a cluster of dormant meristem cells left behind by the apical meristem as the stem grew. When stimulated by the appropriate hormones (see Chapter 23), these meristem cells break out of dormancy and the bud sprouts, growing into a branch (Fig. 21-13). As the meristem cells divide, they release hormones that change the

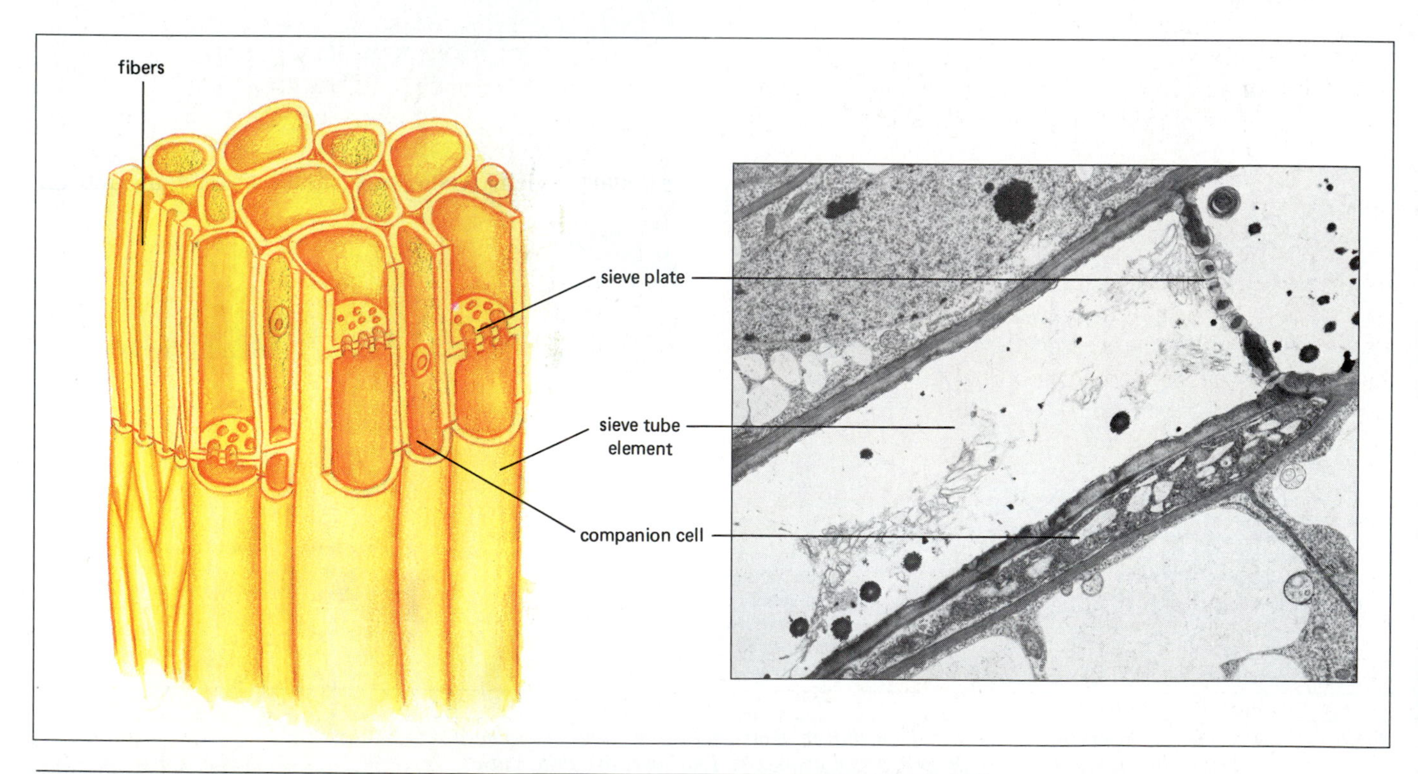

Figure 21-12 Phloem is a mixture of cell types, including sieve-tube elements and companion cells. Note in the micrograph that the sieve-tube elements are almost empty, with only a rind of cytoplasm lining the cell membrane. Sieve-tube elements, stacked end to end, form the conducting system of phloem. Where they join, sieve-tube elements form sieve plates, where membrane-lined pores allow fluids to pass from cell to cell. Each sieve-tube element has a companion cell (the "filled" cell in the micrograph) that nourishes it and regulates its function.

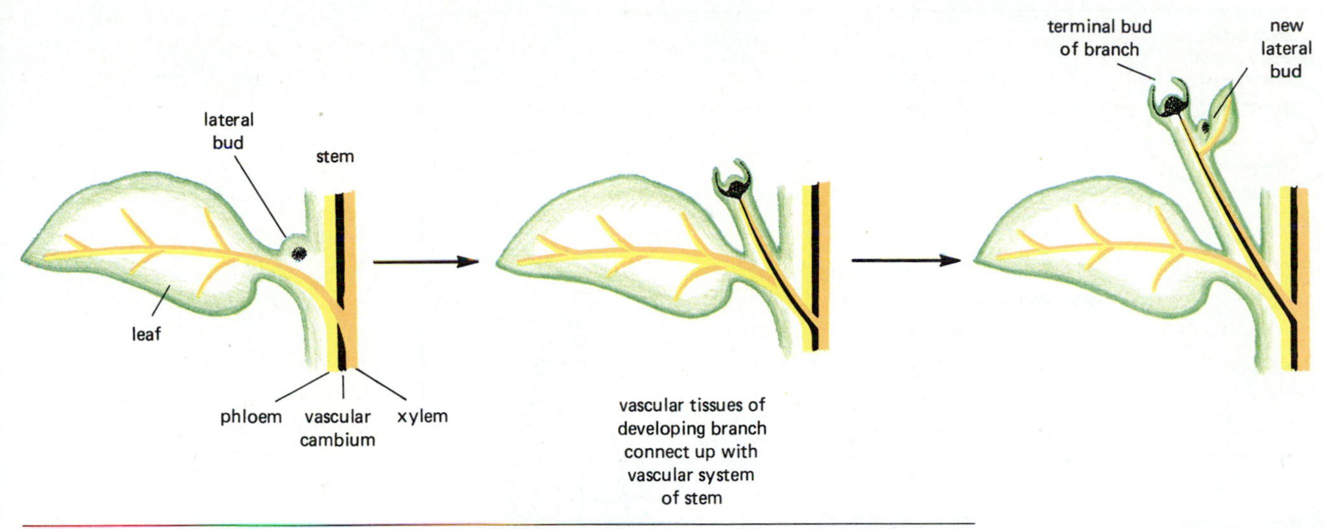

Figure 21-13 *Stem branches grow from lateral buds located at the outer surface of a stem. The bud apical meristem generates an outward-growing branch, replicating the pattern of nodes and internodes characteristic of the type of plant. Meanwhile, cortex cells beneath the sprouting bud differentiate into vascular tissues and connect up with the vascular system of the stem.*

developmental fate of the cells between the bud and the vascular tissues of the stem. Former cortex cells differentiate into xylem and phloem, ultimately connecting up with the main vascular systems in the stem. As the branch grows, it duplicates the development of the stem, complete with an apical meristem at its growing tip and new nodes of leaf primordia and lateral buds left behind at intervals.

Secondary Growth in Stems

In gymnosperms and perennial dicots, stems last for several years, becoming thicker and stronger each year. This **secondary growth** in stem thickness results from cell division in the lateral meristems of the **vascular cambium** and **cork cambium** (Fig. 21-14).

VASCULAR TISSUES. The **vascular cambium** is a cylinder of meristem cells located between the primary xylem and primary phloem. Daughter cells of the vascular cambium produced toward the inside of the stem differentiate into **secondary xylem,** while those produced toward the outside of the stem differentiate into **secondary phloem** (Fig. 21-15). Since the center of the stem is already filled with pith and primary xylem, newly formed secondary xylem pushes the vascular cambium and all outer tissues farther out, increasing the diameter of the stem. The secondary xylem, with its thick cell walls, forms the wood that makes up most of the trunk of a tree. Young xylem (the sapwood just inside the vascular cambium) transports water and minerals; older xylem (the heartwood nearest the pith) only contributes to the strength of the trunk.

Figure 21-14 *Secondary growth in a dicot stem. A vascular cambium forms between the primary xylem and primary phloem. Cell divisions in the vascular cambium produce new secondary xylem and secondary phloem, increasing the diameter of the stem (see Fig. 21-15). The cork cambium produces cork cells that cover the outside of the stem.*

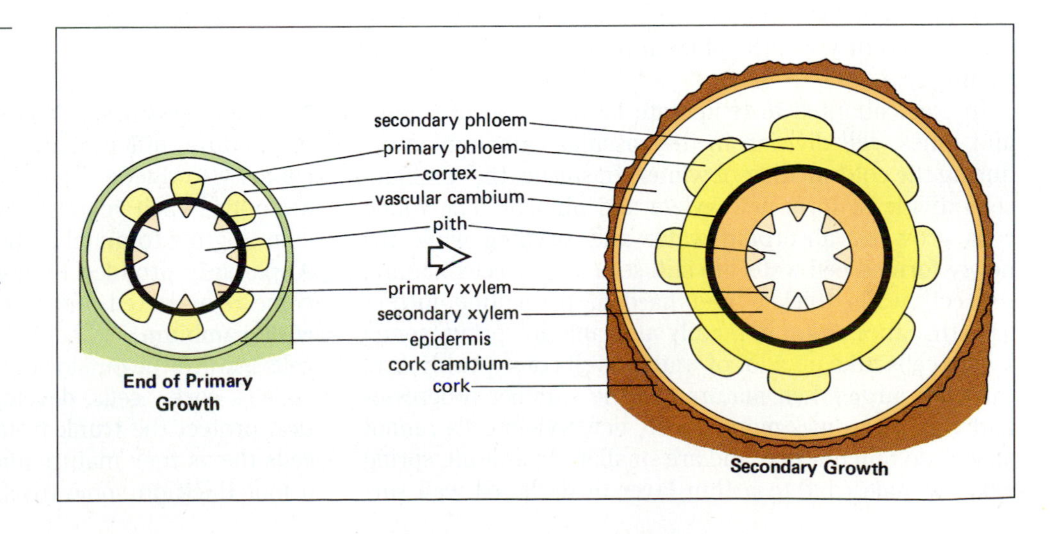

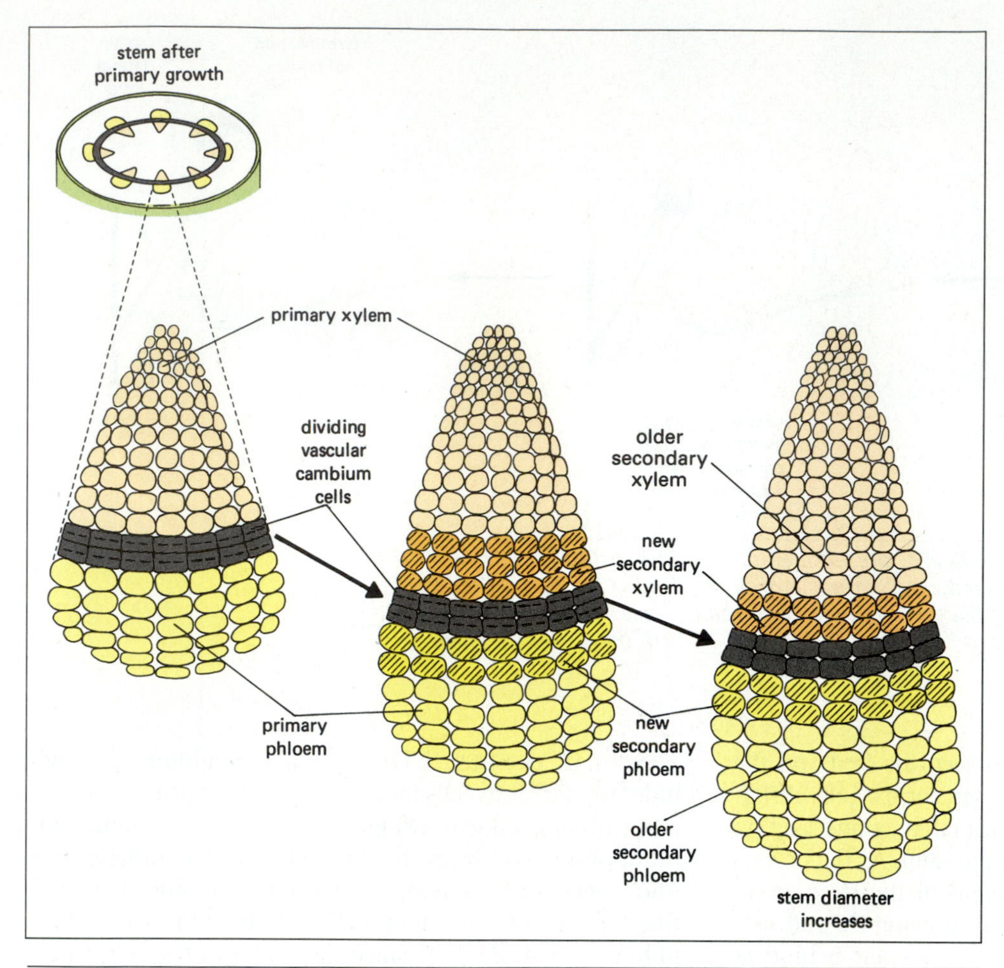

Figure 21-15 *The formation of secondary xylem and phloem from vascular cambium. Daughter cells formed on the inside of the cambium differentiate into xylem, while cells formed on the outside of the cambium differentiate into phloem. Note that since xylem and pith already fill the inside of the stem, newly formed secondary xylem forces the cambium, phloem, and all outer tissues farther out, increasing the diameter of the stem.*

Phloem cells are much weaker than xylem. As they die with age, the sieve-tube elements and companion cells are crushed between the hard xylem on the inside of the trunk and the tough cork on the outside (see below). Only a thin strip of recently formed phloem remains alive and functioning.

In trees adapted to temperate latitudes, such as oaks and pines, cell division in the vascular cambium ceases during the cold weather of winter. In spring, the cambium cells divide to form new xylem and phloem. The young cells grow by absorbing water and swelling while the newly formed cell walls are still soft. As the cells mature, the cell walls thicken and harden, preventing further growth. Since water is readily available in spring, young xylem cells take up a lot of water, swell considerably, and thus are large when mature. As the summer progresses and water becomes more scarce, new xylem cells cannot absorb so much water, and are smaller. As a result, spring wood is pale (due to a thin layer of dark cell wall surrounding a large pale interior) and summer wood is dark (with a thick cell wall and very little interior space; Fig. 21-16). This pattern of alternating light and dark xylem forms the familiar **annual rings** of growth in temperate trees.

SURFACE TISSUES. You will recall that epidermal cells are mature, differentiated cells that no longer retain the capacity to divide. Therefore, as new xylem and phloem are added each year, expanding the stem, the epidermis cannot grow to match. The epidermis splits off and dies. Apparently prodded by hormones, some cortex cells become rejuvenated and form a new lateral meristem, the **cork cambium.** These cells divide, forming daughter cells toward the outside of the stem. These daughter cells, called **cork** cells, develop tough, waterproof cell walls that protect the trunk from desiccation and abuse. Cork cells die as they mature and may form a protective layer a foot thick on some trees, such as the fire-resistant se-

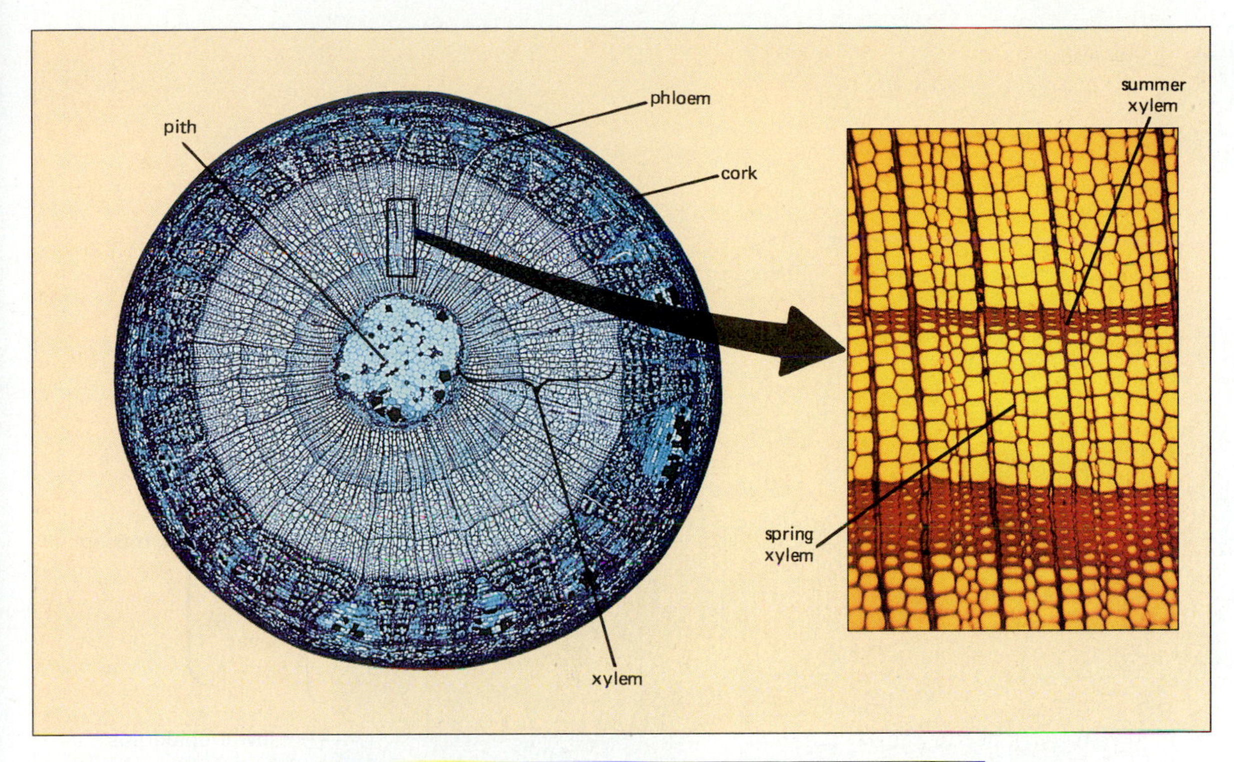

Figure 21-16 *Many trees, such as this pine, form annual rings of xylem. As this micrograph shows, xylem cells formed during the wet spring are large, while xylem cells formed the dry summer are small. The ratio of cell wall to "hole" determines the color of the wood: spring wood, with lots of "hole," is pale, while summer wood, with lots of wall, is dark.*

Figure 21-17 *An ancient sequoia in the Sierra Nevada of California. The cork cambium of a sequoia produces new layers of cork each year, eventually producing a protective, fire-resistant outer covering a foot or two thick. This massive cork layer contributes to a sequoia's great longevity; forest fires that kill lesser trees merely burn off a few inches of sequoia cork, leaving the living parts of the tree inside unharmed.*

quoia (Fig. 21-17). As the trunk expands from year to year, the outermost layers of the cork split apart or peel off, accommodating the growth. (The "cork" in the neck of a wine bottle is part of the outermost layer of cork from a certain type of oak, carefully peeled off by harvesters. Since the cork cambium is left behind, the trees continue to produce new cork throughout their lives.)

The common term **bark** includes all the tissues outside the vascular cambium, namely phloem, cork cambium, and cork. Complete removal of a strip of bark all the way around a tree, called girdling, is invariably fatal to a tree because it severs the phloem. With the phloem gone, sugars synthesized in the leaves cannot reach the roots. Hence the roots die and no longer take up water and minerals, resulting in the death of the entire tree.

LEAVES: NATURE'S SOLAR COLLECTORS

Leaves are the major photosynthetic structures of most plants. As you may remember from Chapter 7, photosynthesis uses the energy of sunlight to convert water and carbon dioxide to sugars, releasing oxygen as a by-product:

$$6\,H_2O + 6\,CO_2 + \text{energy} \rightarrow C_6H_{12}O_6\ \text{(glucose)} + 6\,O_2$$

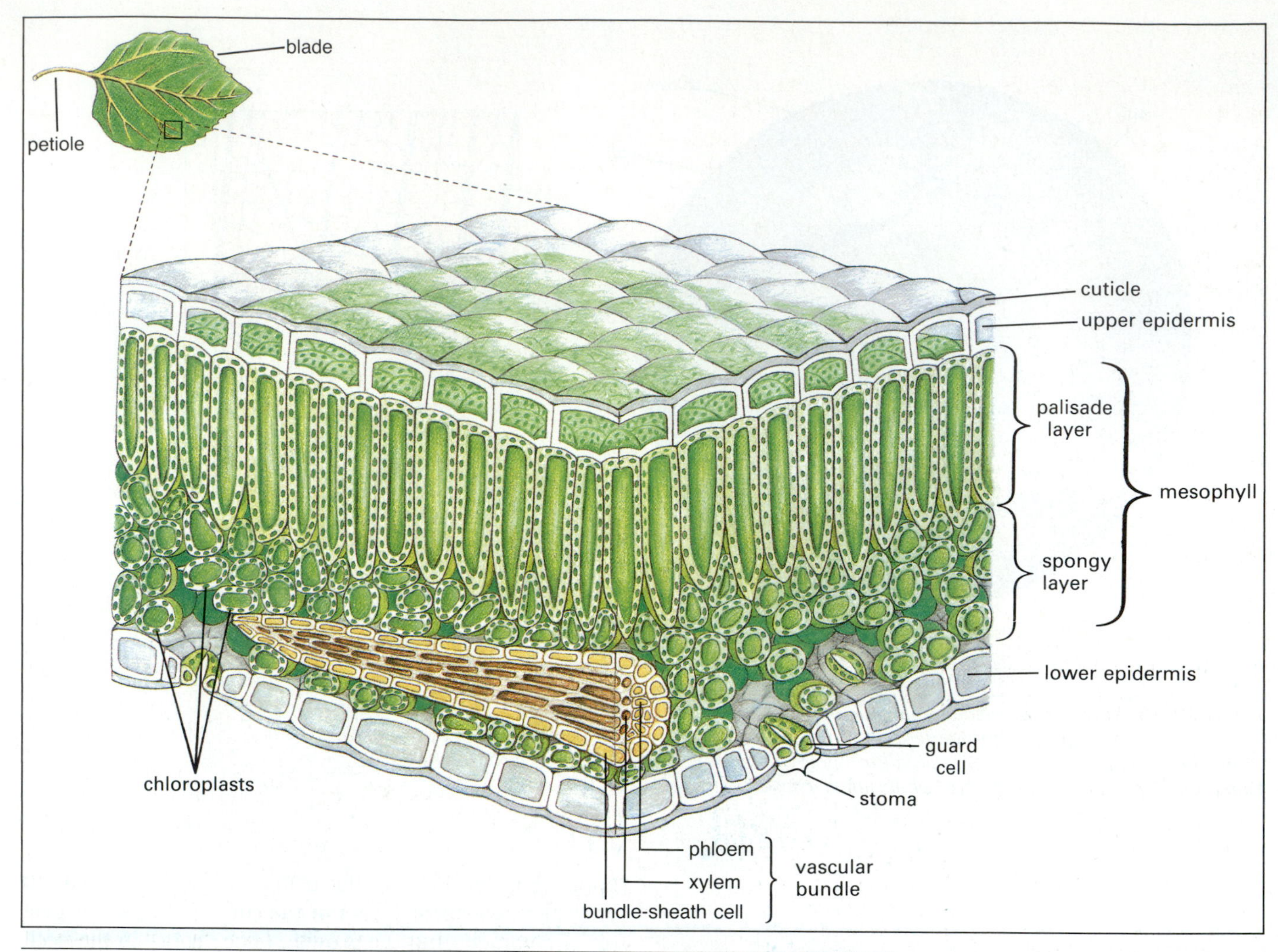

Figure 21-18 The structure of a typical dicot leaf. Note that the cells of the epidermis lack chloroplasts and are transparent, allowing sunlight to penetrate to the chloroplast-containing mesophyll cells beneath. The stomata that pierce the epidermis and the loose, open arrangement of the mesophyll cells ensure that CO_2 can diffuse into the leaf from the air and reach all the photosynthetic cells.

Therefore, the cells of a leaf must be provided with water and CO_2. Water is obtained from the soil and transported to the leaf through the xylem, but CO_2 must diffuse into the leaf from the air. Thus an ideal leaf should have a large surface area for gathering light and should be porous to permit CO_2 to enter from the air for photosynthesis. However, a large, porous leaf would also lose large amounts of water through evaporation, while waterproofing the entire surface would reduce the diffusion of CO_2 into the leaf. The structure of an angiosperm leaf represents an elegant compromise among these conflicting demands.

Leaf Structure

A typical angiosperm leaf consists of a flat **blade** connected to the stem by a stalk called the **petiole** (Fig. 21-18). The petiole positions the blade in space, usually orienting the leaf for maximum exposure to the sun. Inside the petiole are vessels of xylem and phloem that are continuous with those in the stem, root, and blade. Within the blade, the vascular tissues branch into **vascular bundles** or **veins.**

The leaf epidermis consists of a layer of nonphotosynthetic, transparent cells that secrete a waxy waterproof cuticle on their outer surfaces. The epidermis and its cuticle are pierced by adjustable pores called **stomata** (singular **stoma**) that regulate the diffusion of water and CO_2 into and out of the leaf. A stoma consists of two sausage-shaped **guard cells** that surround and regulate the size of a central pore opening into the interior of the leaf (Fig. 21-18). Unlike the surrounding epidermal cells, guard cells contain chloroplasts and can carry out photosynthesis. As we shall see later in this chapter, their photosynthetic ability is important for adjusting the size of the pore.

Beneath the epidermis lies the loosely packed **mesophyll** ("middle of the leaf"). In many leaves, mesophyll cells are of two types, a layer of columnar **palisade cells** just beneath the upper epidermis, and a layer of irregularly shaped **spongy cells** above the lower epidermis. Both palisade and spongy cells contain chloroplasts and usually perform most of the photosynthesis of the leaf. The openness of the leaf interior allows CO_2 to diffuse easily to all the mesophyll cells. Vascular bundles are embedded within the mesophyll, with fine veins reaching very close to each photosynthetic cell. Thus each mesophyll cell receives energy from sunlight transmitted through the clear epidermis; carbon dioxide from the air, diffusing through the stomata; and water from the xylem. The sugars it produces are carried away to the rest of the plant by the phloem.

PLANT NUTRITION

Like all living organisms, a plant takes in materials from its environment, thus acquiring the atoms needed for the synthesis of its own body. Plants acquire nutrients as simple, inorganic compounds. A few elements are needed in relatively large quantities by the plant, usually as constituents of common organic molecules or as ions dissolved in the plant cytoplasm; these are called **macronutrients.** Most plants also need trace amounts of **micronutrients,** mostly elements that aid enzyme functioning. Table 21-1 briefly describes the sources and functions of macro- and micronutrients.

Carbon dioxide and oxygen normally enter a plant by diffusion from the air into leaves, stem, and roots. Water and all other nutrients (collectively called **minerals**) are extracted from the soil by the roots.

Table 21-1 **Essential Plant Nutrients**

Element	Molecular Form Absorbed	Source	Major Functions
MACRONUTRIENTS (usually 0.1% or more of plant dry weight)			
Carbon	CO_2	Air	Major element in organic molecules
Oxygen	O_2	Air	Major element in organic molecules; cellular respiration
Hydrogen	H_2O	Soil	Major element in organic molecules
Nitrogen	NO_3^- or NH_4^+	Soil	Major element in proteins, nucleic acids, and chlorophyll
Potassium	K^+	Soil	Principal positive ion inside cells; control of stomatal opening and closing; enzyme activation
Phosphorus	$H_2PO_4^-$ or HPO_4^{2-}	Soil	Major element in nucleic acids, phospholipids, and electron carriers in chloroplasts and mitochondria
Calcium	Ca^{2+}	Soil	Component of adhesive compounds in cell walls; important in control of membrane permeability; enzyme activation
Magnesium	Mg^{2+}	Soil	Component of chlorophyll; enzyme activation; ribosome stability
Sulfur	SO_4^{2-}	Soil	Component of proteins and many coenzymes
MICRONUTRIENTS (usually less than 0.01% of dry weight of plant)			
Iron	Fe^{2+} or Fe^{3+}	Soil	Needed for synthesis of chlorophyll; component of many electron carriers
Chlorine	Cl^-	Soil	Required for photosynthesis
Manganese	Mn^{2+}	Soil	Required for photosynthesis; enzyme activation
Molybdenum	MoO_4^{2-}	Soil	Required for nitrogen metabolism
Copper	Cu^{2+}	Soil	Enzyme activation; component of electron carriers in chloroplasts
Boron	BO_3^{3-} or $B_4O_7^{2-}$	Soil	Involved in sugar transport
Zinc	Zn^{2+}	Soil	Enzyme activation; protein synthesis; hormone synthesis

Mineral and Water Uptake

MINERAL ABSORPTION. Soil consists of bits of pulverized rock, air, water, and organic matter (Fig. 21-19). Although both the rock particles and the organic matter contain many essential nutrients, only minerals dissolved in the soil water are accessible to the roots. The concentration of minerals in the soil water is very low, usually much lower than the concentration within plant cells and fluids. For example, potassium is the most abundant positive ion in the cytoplasm of most living cells. Its concentration within root cells is at least tenfold greater than in soil water. As a result, diffusion cannot move potassium into the root. *Most minerals, like potassium, are moved into a root against their concentration gradients by active transport.* Since roots are below ground, they cannot photosynthesize. Sugar synthesized in the leaves is transported in the phloem to the roots, where mitochondria in the root cells produce ATP by cellular respiration. Some of this ATP is used to drive the active transport of minerals.

Figure 21-20 shows the process of mineral absorption by roots. Root hairs projecting from the epidermal cells have most of the surface area of the root and are in intimate contact with the soil water. The cell membranes of the root hairs use the energy of ATP to transport minerals from the soil water, concentrating the minerals in the root hair cytoplasm (step 1 in Fig. 21-20). You may remember from Chapter 4 that the cytoplasm of adjacent living plant cells is interconnected by pores called **plasmodesmata.** Therefore, *minerals can diffuse through plasmodesmata from the epidermal cells into the cortex, endodermis, and pericycle cells* (step 2).

At the center of the vascular cylinder lies the xylem, into which the minerals must ultimately be transported. The tracheids and vessel elements of xylem are dead, without cytoplasm or cell membrane—merely an outer skeleton of cell wall shot full of holes (see Fig. 21-11). Therefore, plasmodesmata do not connect pericycle cells with the inside of the xylem. On the other hand, any minerals that enter the extracellular space surrounding

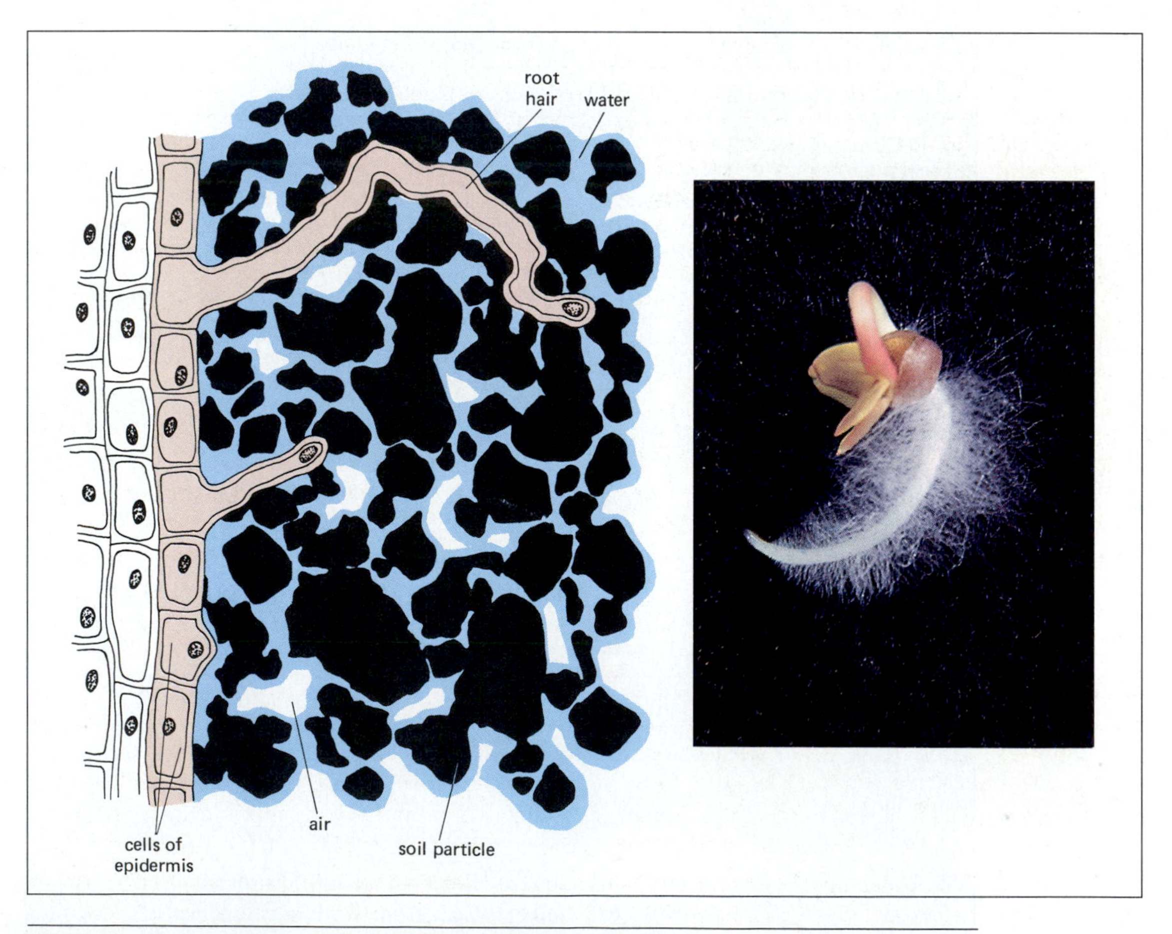

Figure 21-19 *Soil has a complex structure, consisting mostly of rock particles, water, and air, along with some organic matter and living organisms. Only substances dissolved in the soil water are immediately accessible for uptake by roots. Root hairs penetrate between rock particles and increase the surface area of the root in contact with the soil water. Photo: Root hairs cover the root of a sprouting radish seedling.*

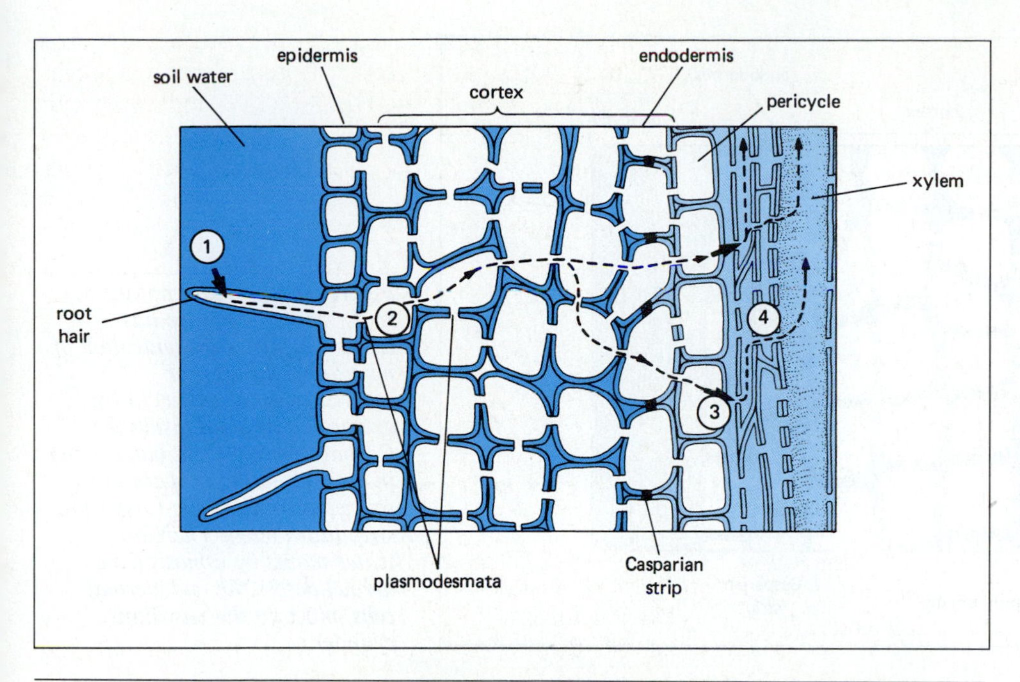

Figure 21-20 Mineral uptake by roots. The Casparian strip separates the extracellular space in the root into two compartments: an outer compartment (dark blue) that is continuous with the soil water and an inner compartment (light blue) that is continuous with the inside of the conducting cells of the xylem. Mineral uptake occurs in a four-step process. (1) Active transport proteins in root hair membranes pump minerals into the root hair cytoplasm. (2) Minerals diffuse inward from cell to cell through plasmodesmata that interconnect the cytoplasm of all living cells in the root. (3) The last living cells at the root interior, the pericycle cells, actively transport minerals out of their cytoplasm into the extracellular space surrounding the xylem. (4) This raises the concentration of minerals in the extracellular space, so the minerals diffuse into the xylem cells through the pits in their walls.

the xylem can easily diffuse into the xylem cells through the holes in their walls. *Pericycle cells actively transport minerals out of their cytoplasm into the extracellular space surrounding the xylem* (step 3). This raises the concentration of minerals in the extracellular space, creating a gradient that allows the minerals to diffuse into the xylem (step 4).

You can now appreciate one of the functions of the Casparian strip in the root endodermis. If water and minerals could pass between endodermal cells, then minerals would leak back out of the extracellular space of the vascular cylinder as fast as they were pumped in. *The Casparian strip, then, retains the concentrated mineral solution within the extracellular space of the vascular cylinder.*

Water absorption. Once a high concentration of minerals builds up in the vascular cylinder, water absorption becomes very straightforward. Water moves by osmosis from regions of high water concentration to regions of low water concentration. Dissolved minerals tie up water molecules, lowering the effective water concentration. Therefore, *the low-mineral solution in the soil water has a high water concentration, while the high-mineral solution in the vascular cylinder has a low water concentration* (Fig. 21-21). The epidermal and cortex cells of a young root are loosely packed, so that the soil water is continuous with the extracellular space within the outer layers of the root *until the water reaches the Casparian strip,* where the waterproofing blocks further passage between cells. The faces of the endodermal cells, however, are not waterproofed. Therefore, *the cell membranes of the endodermis form a pair of semipermeable membranes separating an outer solution of high water concentration from an inner solution of low water concentration.* Water moves across the membranes by osmosis from high to low concentration, that is, from the soil water outside the Casparian strip into the vascular cylinder and xylem inside the Casparian strip.

As you will see when we discuss transport in xylem, water is also pulled up the xylem, powered by the force of water evaporating from the leaves. This further lowers the water concentration within the vascular cylinder and promotes water entry across the endodermis.

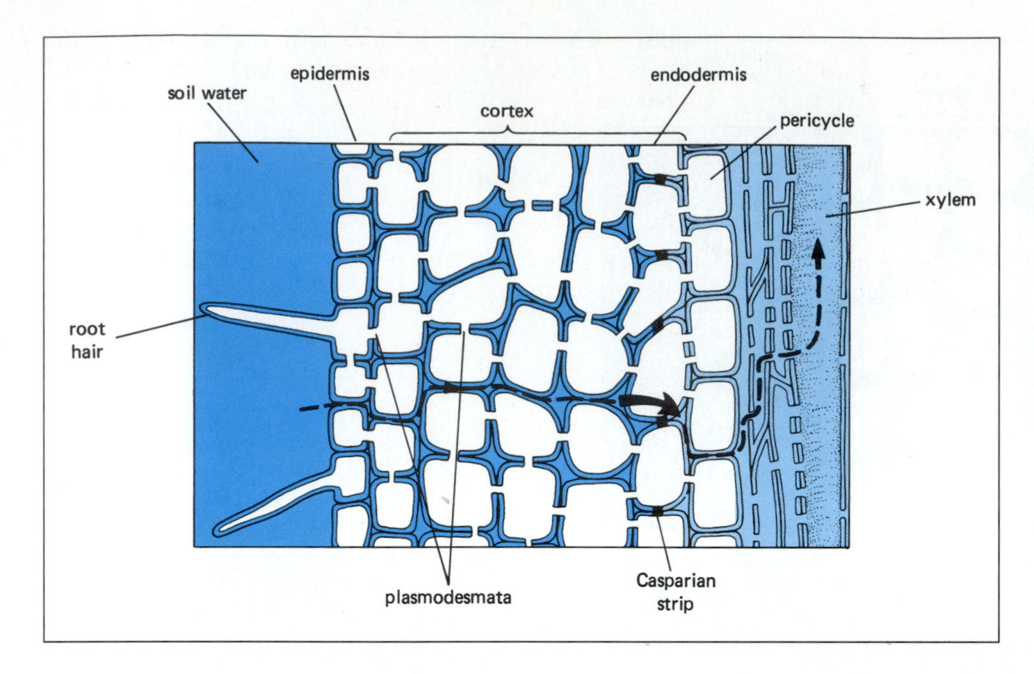

Figure 21-21 Water uptake by roots. Mineral uptake has created a higher concentration of minerals (and a lower concentration of water) in the extracellular space inside the Casparian strip (light blue) than in the extracellular space and soil water outside the Casparian strip (dark blue). Therefore, water moves by osmosis (bold arrow) across the endodermal cells and into the vascular cylinder.

Symbiotic Relationships in Plant Nutrition

Many minerals are too scarce in soil water to support plant growth, although there may be plenty bound up in the rock particles all around. Further, one nutrient, nitrogen, is almost always in short supply in both rock particles and the soil water. Many plants enter into mutually beneficial relationships with other organisms that help them to acquire these scarce nutrients.

MINERAL ACCESSIBILITY AND MYCORRHIZAE. Under normal conditions, the spontaneous rate of release of water-soluble minerals from rock particles is very slow. Further, the chemical forms of the minerals may not be suitable for uptake by plant roots. Most plants form symbiotic relationships with fungi to form root–fungus complexes called **mycorrhizae** that facilitate mineral extraction and absorption. Fungal strands intertwine between the root cells and extend out into the soil (Fig. 21-22), sometimes even contacting nearby roots of other plants. In some way that is not yet understood, the fungus renders nutrients accessible for uptake by the roots, perhaps by converting rock-bound minerals into simple soluble compounds that root cell membranes can transport. The fungus, in return, receives sugars and amino acids from the plant. In this way, both the fungus and the plant can grow in places where neither could survive alone, including deserts and high-altitude, rocky soils.

NITROGEN ACQUISITION. Amino acids, nucleic acids, and chlorophyll all contain nitrogen, so plants need prodigious amounts of this element. Unfortunately, although nitrogen is abundant in the biosphere, most of it is not readily available to plants. About 80 percent of the atmosphere is molecular nitrogen, N_2, but plants can only

(a)

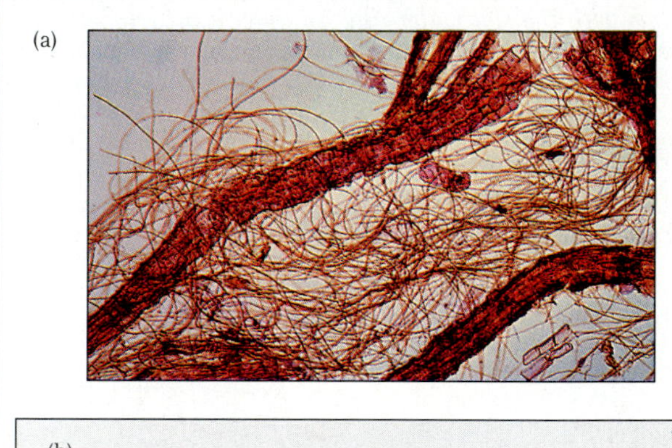

(b)

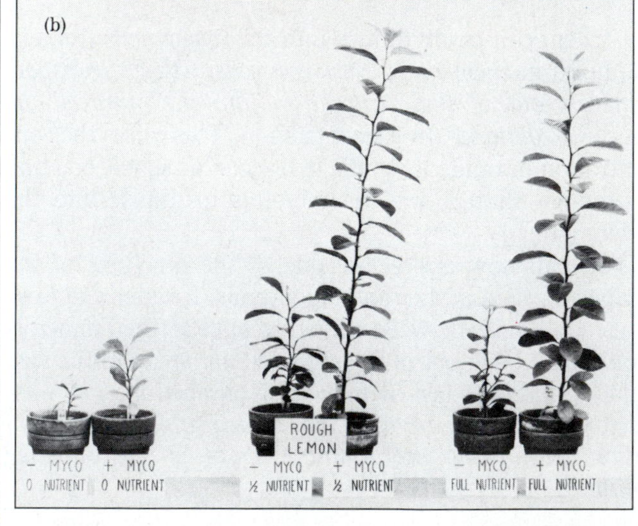

Figure 21-22 Mycorrhizae, a root-fungus symbiosis. **(a)** A tangled meshwork of fungal strands surrounds and penetrates into the infected root. **(b)** Seedlings growing with and without mycorrhizal infections show the importance of mycorrhizae in plant nutrition.

take up nitrogen via their roots, usually in the form of ammonium ion (NH_4^+) or nitrate ion (NO_3^-). Although N_2 diffuses from the atmosphere into the air spaces in the soil, it cannot be used by plants because they do not have the enzymes needed to convert N_2 into ammonium or nitrate ions.

A variety of *nitrogen-fixing bacteria* do have these enzymes. Some of these bacteria are free-living in the soil. However, nitrogen fixation is very costly, energetically speaking, using at least 12 ATPs per ammonium ion synthesized. Bacteria do not routinely manufacture a lot of extra ammonium and liberate it into the soil. Some plants, the **legumes,** including peas, clover, and soybeans, enter into a mutually beneficial relationship with specific species of nitrogen-fixing bacteria. By secreting chemicals into the soil, legumes attract nitrogen-fixing bacteria to their roots (Fig. 21-23). Once there, the bacteria enter into the root hairs. The bacteria then digest

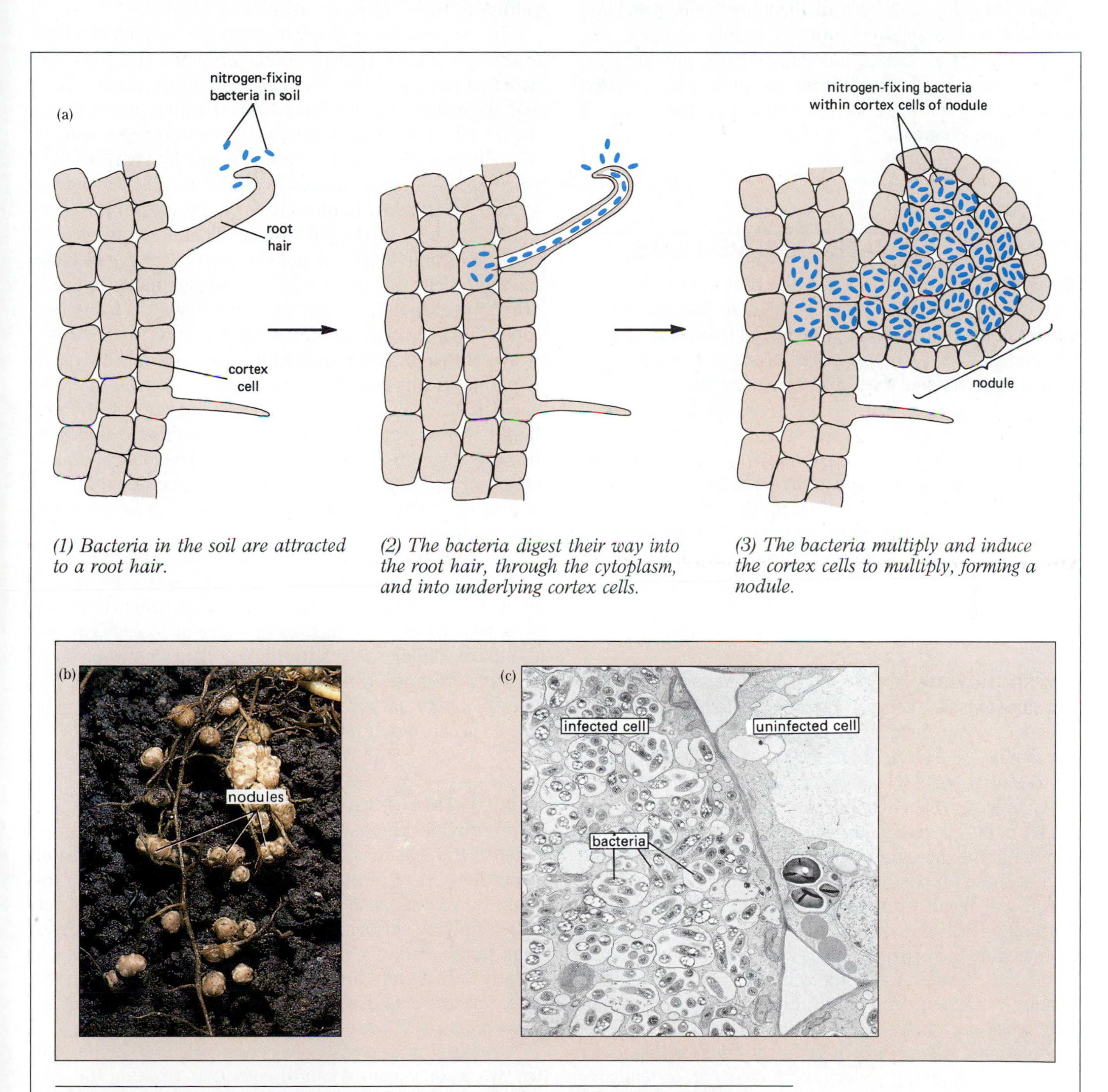

Figure 21-23 **(a)** *Nitrogen-fixing bacteria penetrate a legume root through the root hairs.* **(b)** *and* **(c)** *Nodules in legume roots consist of cortex cells filled with nitrogen-fixing bacteria.*

channels through the epidermal cells and into the cortex. Both bacteria and their host cortex cells multiply to form a **nodule.** A cooperative relationship ensues. The plant transports sugars from its leaves down to the cortex, just as it normally would for storage. The bacteria within the cortex cells take up the sugar and use its energy for all of their metabolic processes, including nitrogen fixation. The bacteria obtain so much energy that they produce more ammonium than they need. The surplus ammonium diffuses into the cytoplasm of their host cells, providing the plant with a steady supply of usable nitrogen. The complex interplay among bacteria, plants, and animals, whereby nitrogen is fixed from the atmosphere, cycled through living matter, and returned to the atmosphere, is described in Chapter 37.

TRANSPORT OF WATER AND MINERALS

Once water and minerals enter the root xylem, there remains the problem of moving them to the uppermost reaches of the plant, which, in redwood trees, may be over 100 meters away. The processes of active transport, diffusion, and osmosis that suffice for the few millimeters from soil to root xylem, however, would be hopelessly slow for getting water and minerals to the top of a tree. Land plants accordingly move fluids up the xylem from root to stem and leaf *en masse* by bulk flow.

Both minerals and water move up the xylem. However, the minerals are dissolved in the water, and are passively carried along as the water is transported. Therefore, we only have to concern ourselves with the mechanisms of water transport.

WATER MOVEMENT IN XYLEM. According to the **cohesion–tension theory,** *water is pulled up the xylem, powered by the force of evaporation of water from the leaves* (Fig. 21-24). As its name suggests, this theory has two essential parts: (1) *cohesion*: water within the xylem holds together like a solid rope; and (2) *tension*: this "water rope" is pulled up the xylem, with evaporation providing the necessary energy. Let's briefly examine both of these propositions.

You will recall from Chapter 2 that water is a polar molecule, with the oxygen carrying a slight negative charge while the hydrogens carry a slight positive charge. As a result, nearby water molecules attract one another, forming weak *hydrogen bonds*. However, just as individually weak cotton threads together make a strong seam in your jeans, *the network of hydrogen bonds within water is quite strong*, giving water a high *cohesion*, or tendency to resist being separated. This is important, because if the water is pulled up the xylem from the top, the column of water has to be strong enough to bear its own weight without breaking—all the water molecules within the xylem are "hanging on" to the uppermost molecules by the hydrogen bonds connecting this handful of molecules with the rest of the column. Actual experiments have found that the column of water within the xylem is at least as strong as a steel wire of the same diameter, and therefore does not break (see the essay, "How Do We Know What Goes on Inside Xylem?"). This is the "cohesion" part of the theory: cohesion between water molecules holds together a "rope" of water within the xylem.

The evaporation of water through the stomata of a leaf, a process called **transpiration,** provides the force for water movement—the "tension" part of the theory. As a leaf transpires, the concentration of water in the mesophyll cells falls. This lower water concentration causes osmosis of water from xylem in the nearby veins into the dehydrating mesophyll cells. Water molecules leaving the xylem are attached to other water molecules in the same xylem tube by hydrogen bonds. Therefore, when one water molecule leaves, it pulls adjacent water molecules up the xylem. As these water molecules move upward, other water molecules farther down move up to replace them. This process continues all the way to the roots, where water in the extracellular space around the xylem is pulled in through the holes in vessel element and tracheid walls. This upward and inward movement of water finally causes water to move into the vascular cylinder by osmosis through the endodermal cells. The force of evaporation of water from the leaves transmitted down the xylem to the roots is so strong that water can be absorbed from quite dry soils.

In summary, transpiration from the leaves removes water from the top of a xylem tube. This water is replaced by water lower down in the tube, so that water moves by bulk flow up the xylem. Loss of water in root xylem and the extracellular space surrounding it promotes osmosis of water from the soil water into the root. Note that *the flow of water in xylem is unidirectional*, from root to shoot, because only the shoot can transpire.

THE CONTROL OF TRANSPIRATION. Although transpiration provides the motive force that gets water and minerals to the leaves, it is also by far the largest source of water loss. Most of the water transpires through the stomata, so you might think that a plant could prevent water loss simply by closing its stomata. However, do not forget that photosynthesis requires CO_2 from the air, which mostly diffuses into the leaf through open stomata. Therefore, *opening and closing the stomata must achieve a balance between CO_2 acquisition and water loss.*

A single stoma consists of a central opening surrounded by two kidney-shaped guard cells that regulate the size of the opening (Fig. 21-25). With some exceptions, stomata open during the day and close at night, but they will also close if the leaf begins to dehydrate. The selective

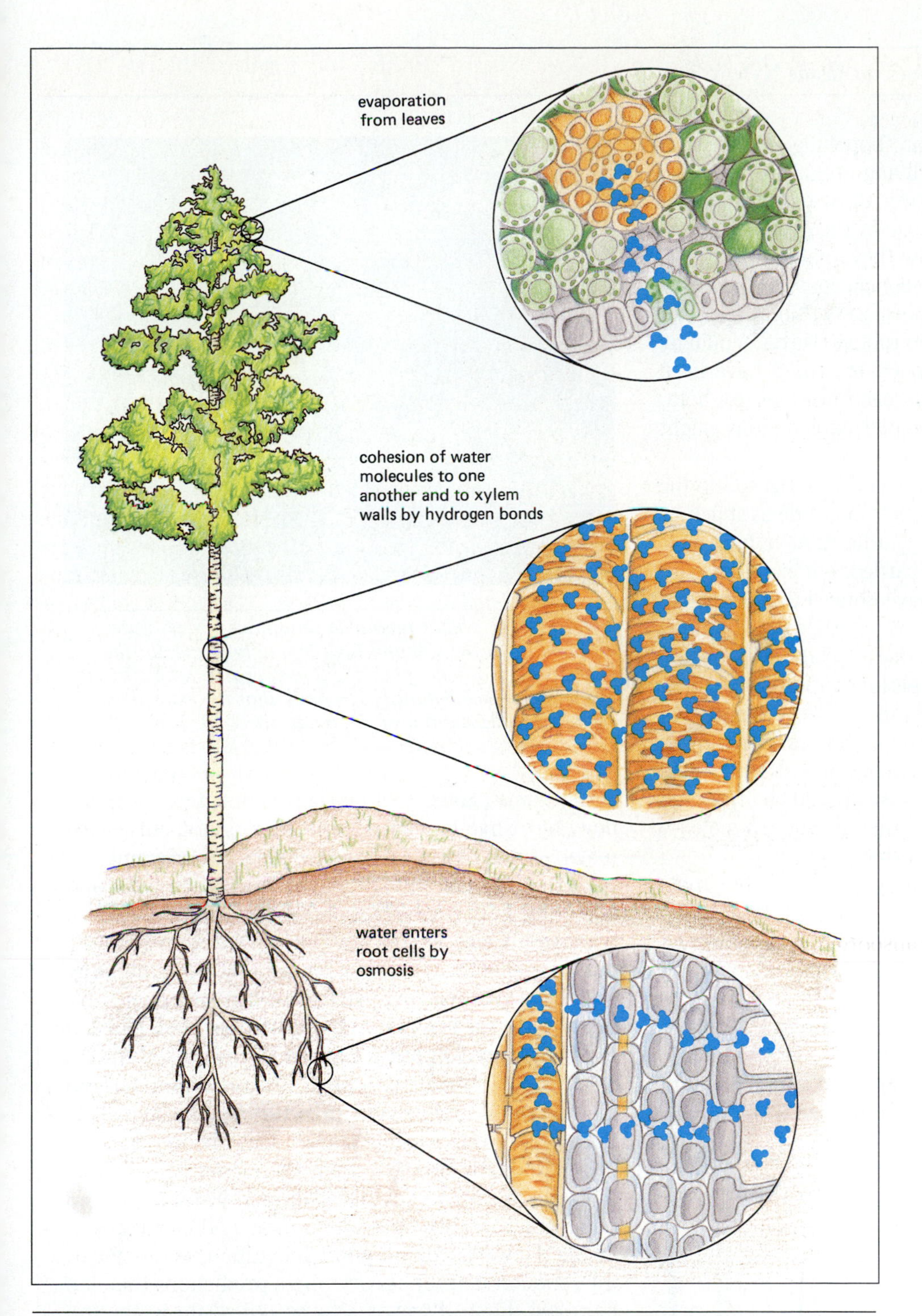

Figure 21-24 *The cohesion–tension theory of water flow from root to leaf in xylem. Water evaporates out of the leaves, and other water molecules replace them from the xylem of the leaf veins. Within the xylem, hydrogen bonding holds nearby water molecules together so firmly that the column of water behaves very much like a rope. The top of the "water rope" is pulled up by evaporation, and the rest of the rope comes along as well, all the way down to the roots. As the last molecules of the rope retreat up the xylem away from the root cells, the decreased water concentration within the root xylem and the surrounding extracellular space causes water to enter from the soil water by osmosis, thus steadily replenishing the bottom of the rope.*

How Do We Know What Goes on Inside Xylem?

As you study biology, you are hopefully not only absorbing a lot of facts about living organisms, but also learning how biologists discover those facts in the first place: what tools biologists use in their research, how they design experiments, how they analyze the resulting data, and how they draw conclusions. In this chapter we have presented the cohesion–tension theory for water transport in xylem from roots to leaves. You may have wondered how such a scheme could have been thought up at all, why it has been accepted by botanists, and whether other, simpler explanations might not suffice.

The problem is how to get water from the soil to the topmost leaves of a plant, which might be as much as 100 meters away. Various hypotheses have been proposed, usually soon to be discarded. For example, consider **capillarity.** You may have noticed that if one end of a thin tube is immersed in water while the other end sticks out in the air, water rises a short way up the tube (Fig. E21-1). A little experimentation reveals that water ascends farther in thinner tubes. Perhaps in tubes as thin as those of xylem, water simply creeps up by capillarity. However, xylem cells with diameters of 30 to 50 micrometers (about 1/500 inch) would allow capillarity to raise water up only about a meter—not even close to the top of a tree.

A second, more serious proposal is that the roots might pump water up the xylem, a phenomenon called **root pressure.** If you cut the top off a tomato plant and seal a pressure-monitoring device onto the remaining stump, you would find that the roots do indeed push water up: hard enough to reach 30 or 40 meters. In some plants—strawberries, for example—root pressure occasionally forces water droplets out of the leaf xylem (Fig. E21-2). Under some conditions,

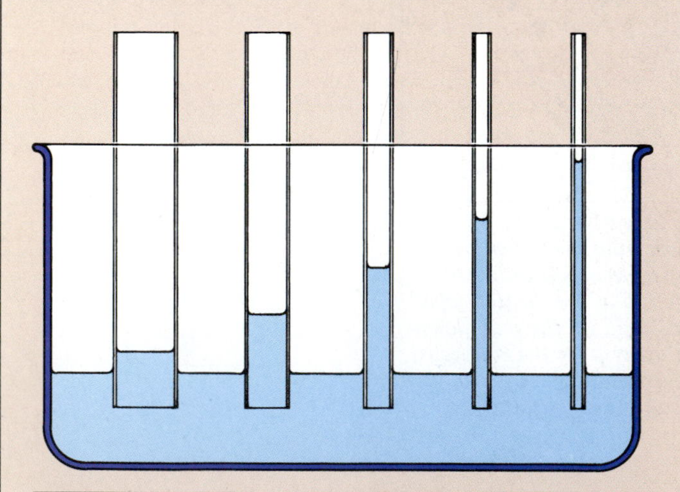

Figure E21-1

Figure E21-2 *Under favorable conditions of high water availability in the soil and high humidity in the air, the root pressure of some plants will force water out the tips of the leaves. Some botanists speculate that the exuded water might carry wastes out of the plant.*

and in some plants, root pressures contribute to xylem flow, but three crucial experiments rule it out as the major factor. First, it is too slow to account for measured xylem flow rates. Second, redwoods and other tall conifers usually have no measurable root pressures. Third, if you saw down a tree and stick the bottom end of the trunk in a bucket of water containing a suitable dye, you will find that the water and dye travel up to the leaves without any roots at all (this, of course, is why you put a Christmas tree in water).

We have already described the **cohesion–tension theory** for the movement of water up the xylem. Is there any positive evidence for this theory, or is it merely the only one left when the others have failed? Needless to say, no one has ever seen water molecules pulling one another up a xylem tube by their hydrogen bonds. However, as all good scientific theories must, the cohesion–tension theory offers predictions that can be tested. First, if water is to be pulled up to the top of a redwood, a 100-meter-tall tube of water must be strong enough to hold up its own weight. Actual experiments show that the cohesion of water is more than strong enough to hold water molecules together on their way up a tree (Fig. E21-3).

Second, since water is being *pulled up*, the water inside the xylem must be under tension. As anyone who has ever cut down a tree knows, when xylem (wood) is cut with an axe, water does *not* flow out of

How Do We Know What Goes on Inside Xylem?

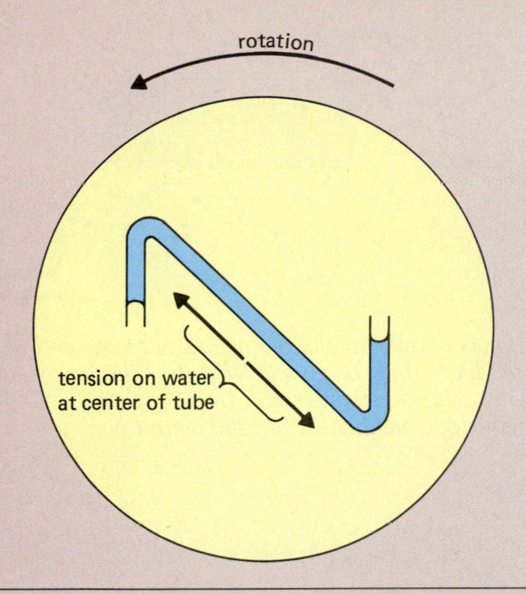

Figure E21-3 An experiment to test if water cohesion is strong enough to support the weight of a column of water in the xylem of a tall tree. A capillary tube is bent into a Z shape, filled with water, and spun at high speed in a centrifuge. Centrifugal forces push the water toward the ends of the Z. At high enough speeds, the column of water breaks apart in the middle. However, the forces required to break the water column in the Z tube are equivalent to the weight of water in a xylem tube more than 500 meters tall, far taller than any tree that ever lived.

the stump (as the root pressure hypothesis would predict). Water under tension should really *pull back into the cut* like a snapped rubber band. This is exactly what happens: if a drop of water is placed in a fresh axe cut, tension in the xylem will draw the water into the wood. Another prediction based on tension within the xylem is that the diameter of a tree trunk should decrease during periods of high flow rates, just as a soda straw collapses if you suck on it too hard. Although it is difficult to imagine a tree trunk shrinking, measurements do show that trunks are thinner during the day (when water is evaporating rapidly from the leaves and therefore is being rapidly pulled up the xylem) than at night (when evaporation and flow rates are low). Finally, botanists can measure the tension within a stem, and have found tensions strong enough to pull water up *200 meters*!

Clearly, the cohesion-tension theory has passed several tests with flying colors, while competing hypotheses have failed. Nevertheless, as with all scientific theories, future experiments may someday come up with data that the cohesion-tension theory cannot explain. Such is the nature of science: a theory stands only as long as experimental data allow it to. So far, the cohesion-tension theory stands as tall as the redwood trees whose water transport it explains.

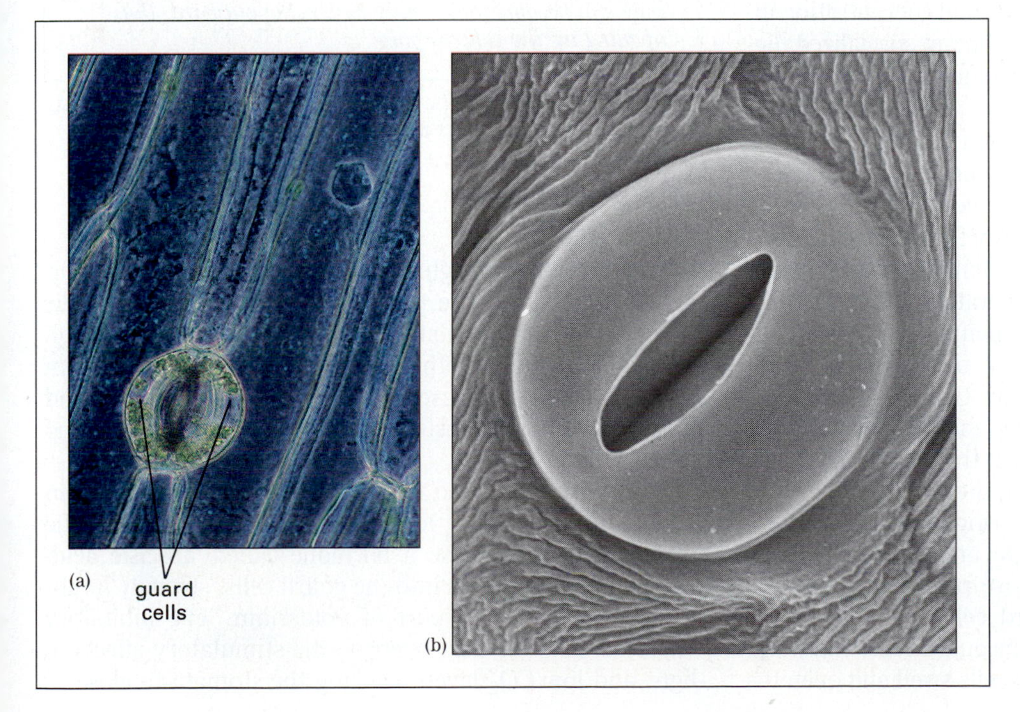

Figure 21-25 Stomata seen through the light microscope **(a)** *and scanning electron microscope* **(b).** *In the light micrograph, note that the guard cells contain chloroplasts (the green ovals within the cells), but the other epidermal cells do not.*

value of this arrangement is obvious. First, the stomata only open during the day, when sunlight allows photosynthesis. Even then, the stomata close if water loss becomes too great.

How do the guard cells regulate the size of the opening between them? There are two levels to this question: first, how does changing the shape of the guard cells open and close the opening; and second, what physiological causes lead to the shape change?

Stomata open when the guard cells take up water and swell, and close when guard cells lose water and shrink. This might seem paradoxical, since swollen guard cells must take up a larger volume than shrunken ones, and therefore you might expect that the potential central hole would be shut more tightly than ever. The key lies in the construction of the cell wall of guard cells (Fig. 21-26a). Cellulose fibers in the wall encircle the guard cells like a series of inelastic belts. Thus, when water enters the guard cells and their volume increases, they cannot become fatter but must become longer. Each pair of guard cells is attached at both ends, so the only way the cells can become longer is by bowing outward like a cooked sausage, opening a hole between them (Fig. 21-26b).

According to the principles of osmosis, water will enter a guard cell if its cytoplasm has a lower water concentration than the cytoplasm of surrounding cells, and will leave the guard cell if it has a higher concentration of water. Large changes in potassium concentration within the guard cells cause correspondingly large changes in water concentration, driving the osmotic fluxes that open and close a stoma: when potassium enters the guard cells, water follows by osmosis, opening the stoma; when potassium leaves, water leaves again by osmosis, and the stoma closes (Fig. 21-27).

Several factors regulate the potassium concentration inside guard cells. First, guard cells contain specialized pigments that absorb light. When light strikes these pigments, they trigger a series of reactions that cause potassium to be actively transported from the extracellular fluid into the guard cells. Second, potassium transport is also influenced by the CO_2 concentration within the guard cells, which is regulated by the balance between photosynthesis and respiration. Remember that plant cells have mitochondria that generate ATP by aerobic respiration, using up O_2 and producing CO_2. Guard cells also have chloroplasts and carry out photosynthesis, which consumes CO_2 and produces O_2. During the day, photosynthesis uses up CO_2 faster than respiration produces it, so the CO_2 concentration in the guard cells drops. Low CO_2 concentrations stimulate the active transport of potassium into the guard cells. During the day, therefore, both the direct perception of light and the reduction of CO_2 levels by photosynthesis cause potassium to be actively transported into the guard cells. This raises the potassium concentration within the guard cells, and water follows in by osmosis. The guard cells swell and open the stoma. During the night, photosynthesis stops but respiration continues. The CO_2 level within the guard cells rises. No longer stimulated by light, and inhibited by high CO_2 concentrations, the active transport of potassium stops. Potassium diffuses back out of the guard cells, and water leaves by osmosis. The guard cells shrink, closing the stoma.

Water loss can also cause the stomata to close, even during the day. If a leaf is losing water too fast, the mesophyll cells release a hormone called **abscisic acid.** The hormone diffuses into the guard cells, strongly inhibiting the active transport of potassium. The inhibitory effects of abscisic acid override the stimulatory effects of light and low CO_2 levels, causing the stomata to close.

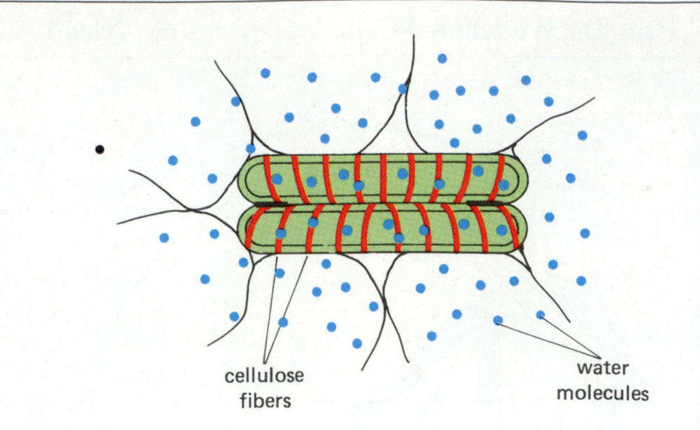

(a) *Bands of cellulose fibers in guard cell walls enclose the cell like strong, inexpandable belts. When the guard cells are relatively dehydrated, they lie straight alongside one another, closing the central pore.*

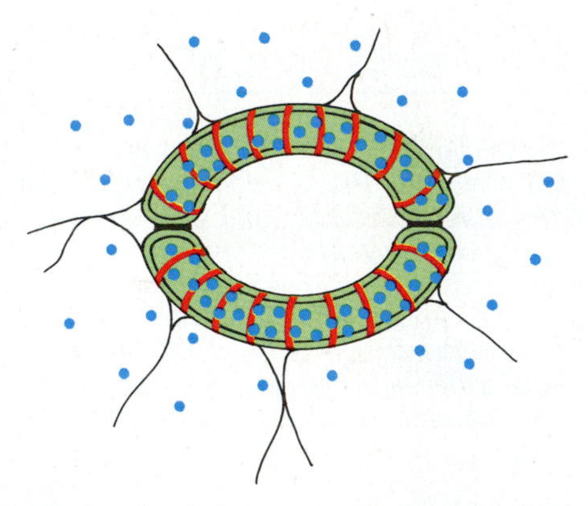

(b) *When the guard cells take up water, they swell. However, the cellulose belts do not allow the cells to become fatter as they swell, so they must become longer. Since the cells are attached to each other at the ends, they can become longer only by bowing outward, thus opening up the central pore.*

Figure 21-26 *Mechanics of stomatal opening and closing.*

Autumn in Colorado

We began this chapter with the observation that plant growth depends on the environmental factors of light, nutrients, water, and temperature. Those of us who live in temperate climates, including most of North America, are familiar with the yearly cycle of deciduous trees: the bare limbs of winter put forth new green leaves in spring, which turn flaming red and yellow in autumn, only to drop off, once again leaving the trees naked for winter. This cycle is an evolutionary response to temperature and water availability.

The leaves of deciduous trees present a large surface area to the sunlight, maximizing their photosynthetic abilities. These large leaves also transpire a lot of water, which must be replaced by water absorbed from the soil. What happens to water acquisition and loss when winter arrives? In temperate climates, the ground freezes for at least part of the winter, and roots cannot absorb water out of frozen soil. Simultaneously, *low temperatures reduce but do not eliminate evaporation of water from leaves*. By dropping their leaves, deciduous plants avoid the double bind of continued water loss without water replacement.

How are leaves shed? As a leaf grows, a layer of thin-walled cells develops at the base of the petiole, near its attachment to the stem (Fig. E21-4). In the fall, these cells, called the **abscission layer,** produce an enzyme that digests the cell walls holding them together. Only a few strands of xylem and phloem hold the leaf to the stem. These eventually break, allowing the leaf to fall.

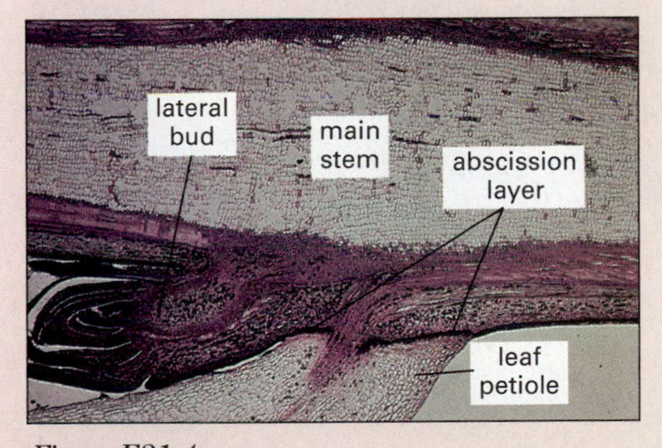

Figure E21-4

Leaf fall is not, however, a direct response to freezing temperatures, nor the result of leaf death. Rather, leaf fall is an active event directed by the plant, evolutionarily programmed into its life cycle. This is shown by two observations. First, if a branch breaks from a tree in late summer, its leaves die and turn brown, but they do not drop off the branch. In fact, the branch may hold its dry, withered leaves all winter long and on into spring, until they disintegrate in the natural course of decay. Second, many deciduous trees have been transplanted to lawns and gardens far out of their native habitats. Maples in southern California are watered year round by their owners and never experience freezing weather, yet they drop their leaves in autumn anyway. Why? The trees seem to respond to daylength: abscission is triggered by the shorter days of autumn, which would be the sure sign of approaching winter in colder climes.

Figure E21-5 Aspens, Gore Range, Colorado

This leaves us with one unanswered question: why the changing colors? In summer, leaves are green because of the chlorophyll in their chloroplasts, which reflects or transmits green light while absorbing red and blue (see Chapter 7). Chloroplasts also contain other pigments, notably red and yellow carotenoids, but these are normally almost completely masked by the chlorophyll. As the plant prepares for winter, hormonal signals trigger the degradation of chlorophyll while the leaf is still alive. Many of the atoms that formerly composed the chlorophyll molecules are reclaimed and stored in other, permanent parts of the tree. As the chlorophyll disappears, the yellows and reds of carotenoids are unveiled. Meanwhile, blue and red anthocyanins are synthesized and stored in the central vacuoles of the leaf cells. Depending on the types and amounts of carotenoids and anthocyanins, autumn displays the clear golds of birches, tuliptrees, and aspens (Fig. E21-5), the oranges and reds of maples, or the bronzes and purples of oaks. After the abscission layer breaks and the leaf falls off, it truly dies. The gaudy pigments are degraded, and the fallen leaves fade to the quiet brown of November.

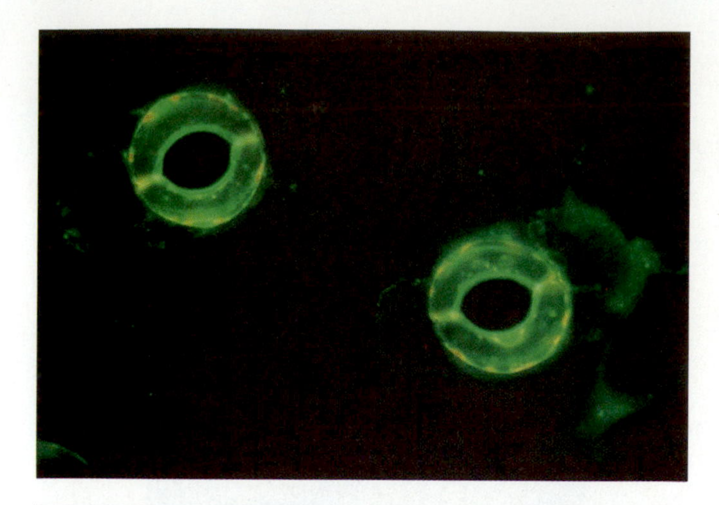

Figure 21-27 The concentration of potassium ions drives osmosis of water into and out of guard cells. In this micrograph, the potassium concentration is indicated by the brightness of the yellow dye, clearly showing that the guard cells surrounding open stomata have a higher potassium concentration than the other epidermal cells.

THE TRANSPORT OF SUGARS

Water and minerals transported into the leaves allow them to carry out photosynthesis, producing sugars from water and carbon dioxide. These sugars must be moved to other parts of the plant, to nourish nonphotosynthetic structures such as roots or flowers, and to be stored in the cortex cells of root and stem. Sugar transport is the function of phloem.

Botanists employ a most unlikely lab assistant in studying phloem function: the aphid. Aphids are insects that specialize in feeding on the fluid contained in phloem sieve tubes. An aphid inserts a pointed, hollow tube, the stylet, through the epidermis and cortex of a young stem into a sieve tube (Fig. 21-28). The aphid can then relax and let the plant do the work. The fluid in the phloem is under pressure, and actively pushes up through the stylet into the digestive tract of the aphid (sometimes with enough pressure to force its way out the other end!). By cutting off the aphid but leaving its stylet in place, botanists have collected phloem fluid and found that it consists mostly of sucrose and water, as much as 25% sucrose by weight. How are these high sugar concentrations moved about the plant, and in which directions?

The most widely accepted mechanism for transport of sugars in phloem is the **pressure flow** hypothesis (Fig. 21-29). Let's suppose that a leaf is photosynthesizing rapidly, manufacturing lots of sugar. Much of this sugar is actively transported into the leaf sieve-tube elements by companion cells, raising the concentration of sugar in the leaf sieve tube. The added sugar lowers the water concentration, causing water to enter the sieve tube by os-

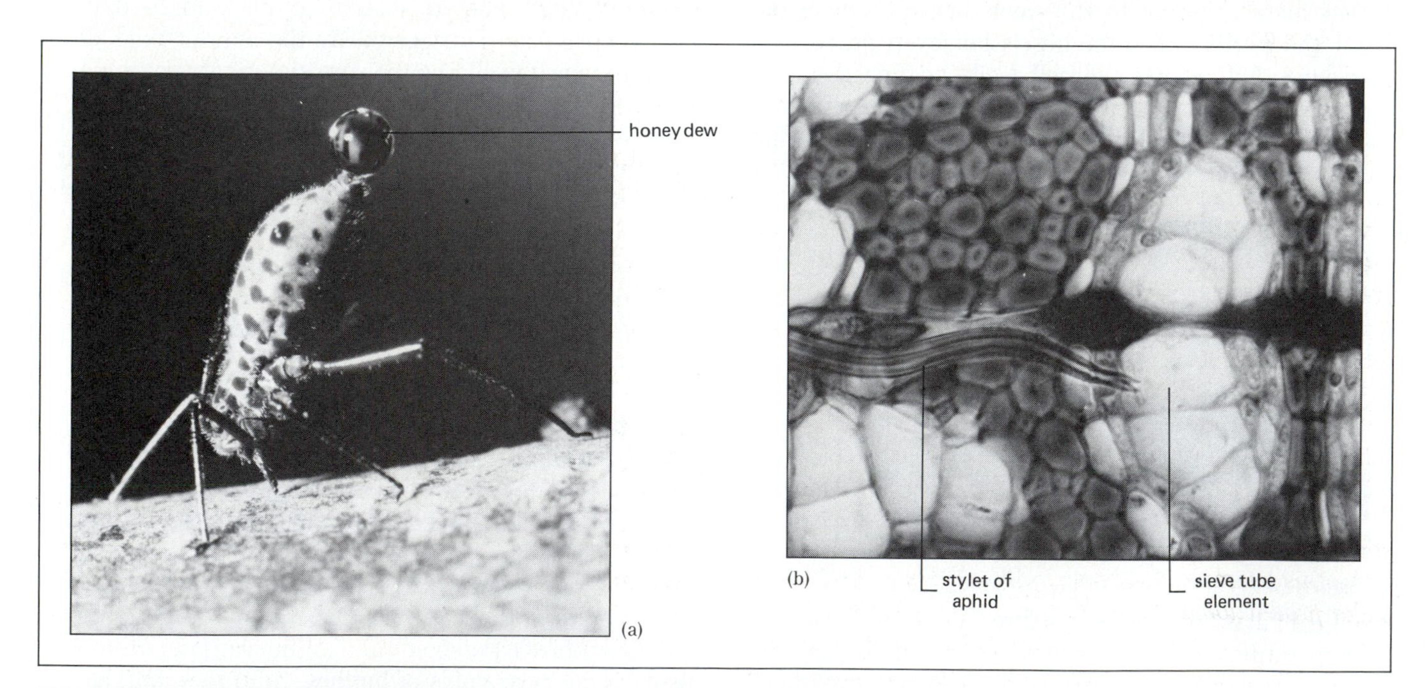

Figure 21-28 Aphids feed on the sugary fluid in phloem sieve tubes.
(a) *When an aphid pierces a sieve tube, pressure in the tube forces the fluid out of the plant and into the digestive tract of the aphid. Sometimes the pressure is so great that fluid is forced completely through the aphid and out its anus, as "honeydew." This exuded fluid is collected by certain species of ants that act as "shepherds" to the aphids, defending them from predators in return for a diet of sweet honeydew.*
(b) *The flexible stylet of an aphid, passing through many layers of cells to penetrate a sieve tube cell.*

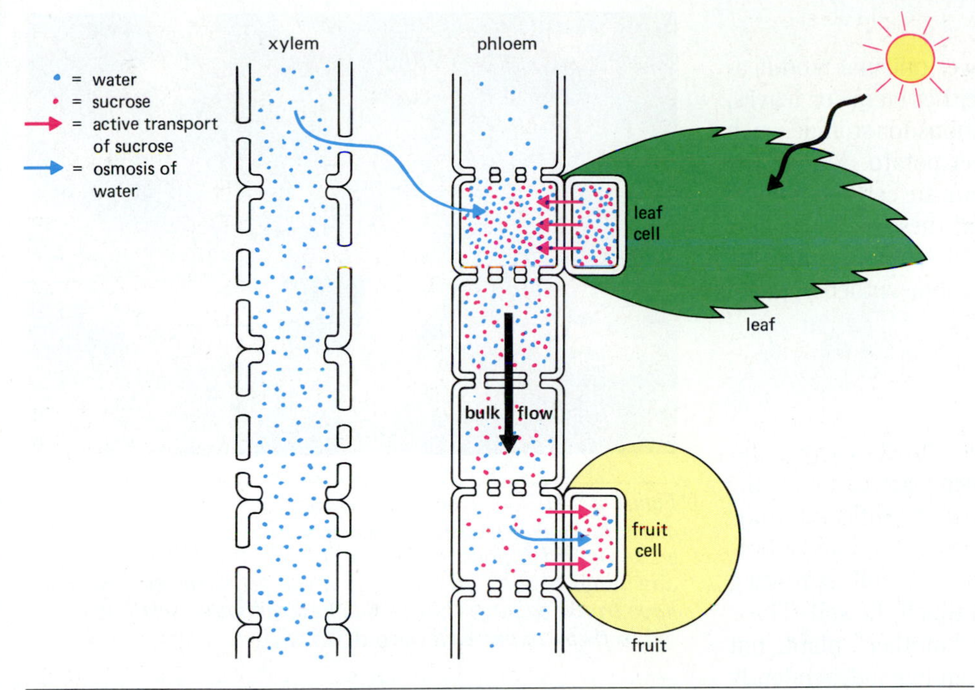

Figure 21-29 *The pressure flow theory relies on differences in hydrostatic pressure to move fluid through phloem sieve tubes. Sucrose (red) is actively transported from a photosynthesizing leaf into a nearby sieve tube. Water (blue) follows into the tube by osmosis, raising the hydrostatic pressure as increasing numbers of water molecules enter the fixed volume of the tube. The same sieve tube connects to a developing fruit, where sugar is pumped out of the tube and into the fruit cells. Water moves out of the tube by osmosis, lowering the hydrostatic pressure. High pressure in the leaf end of the phloem and low pressure in the fruit end causes water, together with any dissolved solutes, to flow in bulk from leaf to fruit (thick black arrow).*

mosis from nearby xylem. Meanwhile, some distance away but connected by the same sieve tube, a fruit is developing. Sugar is actively transported out of the nearby sieve-tube elements into the fruit cells, thus raising the concentration of sugar in the fruit and lowering the sugar concentration in that end of the sieve tube. Water therefore leaves the sieve tube by osmosis and follows the sugar into the fruit. If water enters the leaf end of the sieve tube and leaves the fruit end of the same tube, water will flow *in bulk* from the leaf to the fruit, driven by the difference in hydrostatic pressure between the two ends of the sieve tube. The bulk flow of water carries the dissolved sugar along with it.

As this example illustrates, *phloem flow is directed by sugar production and use. Any structure that actively synthesizes sugar will be a source of phloem flow, and any structure that uses up sugar or converts sugar to starch will be a sink toward which phloem fluids will flow.* A newly forming leaf will be a sink as it develops, with phloem flow up into it from more mature leaves farther down on the plant. When the leaf matures, it will photosynthesize and produce sugar, becoming a source for phloem flow to other newly developing leaves above it, to flowers or fruits, or to the roots below it. Therefore, fluid can move in phloem either up or down the plant, depending on the metabolic demands of the various parts of the plant at any given time.

SPECIAL ADAPTATIONS OF ROOTS, STEMS, AND LEAVES

Not all roots are sinuous fibers, not all stems are smooth and upright, and not all leaves are flat and fanlike. Just as evolution has changed the basic shape of the vertebrate forelimb to suit the demands of running, swimming, and flying, plant parts have become modified in response to environmental demands. You may be surprised to learn that many familiar structures are derived from unlikely parts of a plant.

Although we will highlight unusual adaptations, don't forget that *all* plants are adapted to their environments. The "typical" leaf of an oak or maple is just as much a "special adaptation" as a cactus spine or daffodil bulb.

Root Adaptations

Roots have probably undergone fewer unusual modifications of their basic structure than either stems or leaves. Some roots have extreme specializations for storage, such as the familiar beet, carrot, or sweet potato. Among the most bizarre root adaptations occur in certain orchids that grow perched on trees. A few of these aerial orchids have green, photosynthetic roots; in fact, for some orchids, the green roots are the only photosynthetic part of the plant.

Stem Adaptations

Many plants have stems modified for functions very different from the original one of raising leaves up to the light. Strawberries, for example, have two different kinds of stems (Fig. 21-30): the regular upright kind and a horizontal **runner** that snakes out over the soil, sprouting new strawberry plants where nodes touch the soil. These new plants are connected with the "mother" plant, but once the plantlets form roots, they can live independently if the runner is severed.

Some plants, such as the saguaro cactus and the baobab tree (Fig. 21-31), store water in aboveground stems. Many other plants store carbohydrates in underground stems. The common white potato is actually a storage stem; each eye is a lateral bud, ready to send up a branch next year, using the energy stored as starch in the potato to power the growth of the branch. Irises have underground stems called **rhizomes** that store carbohydrates produced during the summer. Most rhizomes have ridges formed by the closely spaced nodes and internodes of the fat stem. Irises can be propagated by cutting up the rhizome; if it contains enough stored food, each piece with a node can generate a complete plant.

Figure 21-30 *The beach strawberry can reproduce with horizontal stems called runners that course out over the surface of the sand. If a node of a runner touches the soil, it will sprout roots and develop into a complete plant.*

Figure 21-31 *The baobab tree is usually not very tall, but develops an enormously fat, water-storing trunk. The baobab grows in arid regions, and when it rains, it is to the tree's advantage to store all the water it can get. Some of these trees have trunks so large that people have hollowed small houses out in them, in one case even a jail cell!*

Many aboveground stems grow modified branches with special functions. One common branch adaptation is the **thorn,** usually growing from the normal branch location of the leaf axil (Fig. 21-32a). Thorns, of course, discourage animals from dining on the branches. Grapes and Boston ivy have some of their branches modified into grasping **tendrils** that hold the otherwise prostrate plant up on supporting objects such as trees, trellises, or buildings, providing better access to the sunlight (Fig. 21-32b).

Leaf Adaptations

The most important environmental factors that affect the evolution of leaves are light, temperature, and water availability. For example, plants growing on the floor of a tropical rain forest have plenty of water year round, but very little light, due to the deep shade cast by several layers of trees above them. Consequently, their leaves tend to be extremely large, an adaptation demanded by the low light level and permitted by the abundant water (Fig. 21-33a).

At the other extreme, deserts receive bright sunlight virtually every day of the year, but have limited water and scorching temperatures. Desert plants have evolved two strikingly different adaptations to this situation. One group of plants, the succulents, have very thick leaves, with large cells that store water from the infrequent rains against the inevitable long droughts (Fig. 21-33b). Succulent leaves are covered with a thick cuticle that greatly reduces the evaporation of water. The cacti display the opposite strategy, reducing the leaves to thin spines that protect the plant from herbivores and reduce water loss

Figure 21-32 *The branches growing from a single stem may differ in structure or function.* **(a)** *The hawthorn has both "regular" branches forming the crown of the tree, and many smaller branches modified into sharp thorns. That these thorns are actually branches in disguise is clearly shown by their position in the normal branch location, the crotch between a leaf and its stem, and by the fact that the thorns themselves often branch.* **(b)** *Grape tendrils are long, soft, leafless branches. When a tendril contacts an object in the environment, it curls in the direction of the touching surface, wrapping itself firmly around the object.*

(a) *In tropical rain forests, light levels near the ground are very low. Consequently, understory plants often bear huge, dark green leaves for maximal light absorption. Many jungle plants, such as this philodendron, have found new ecological niches in modern houses, where they are grown because of their deep green foliage and tolerance of dim light.*
(b) *Desert plants receive plenty of light but are chronically short on water. Succulents have evolved fleshy leaves that store water from the occasional rains, just as the baobab tree does in its trunk.*
(c) *The cactus takes the other extreme in desert adaptations: reduce the leaves to vanishingly thin lines, which cannot store water but do not have much surface area for evaporation either. These spines also have very sharp points that discourage animals who might be tempted to munch on the tasty stem beneath.*
(d) *Pea tendrils are modified leaflets that respond to contact with objects by coiling, much like stem tendrils do.*
(e) *Daffodil bulbs are enormous buds, with short central stems surrounded by thick water- and food-storing leaves. Like other monocots, these bulbs form roots as outgrowths of the base of the stem.*

Figure 21-33 *Differing environments have produced a myriad of leaf types.*

(Fig. 21-33c). Photosynthesis in cacti occurs in cortex cells of the green, water-storing stems.

Modified leaves in other plants function in ways having nothing to do with photosynthesis or water conservation. The common pea climbs by grasping fences, mailbox posts, or other plants with clinging tendrils (Fig. 21-33d). Unlike grape tendrils, which are derived from branches, pea tendrils are slender, supple leaflets. Some plants, such as onions, daffodils, and tulips, use thick, fleshy leaves as storage organs (Fig. 21-33e). A daffodil bulb consists of a short stem bearing thick, overlapping leaves that store nutrients and energy over the winter.

Finally, a few plants have turned the table on the animals and have become predators themselves. Venus flytraps (Fig. 21-34a) and sundews (Fig. 21-34b) both have leaves modified into snares for unwary insects. These plants live in nitrogen-poor swamps and derive most of their nitrogen supply from the bodies of their prey which they trap with their leaves.

As varied and sometimes bizarre as these leaf specializations are, the most extreme and most important leaf modification is the flower. As we shall see in Chapter 22, these "reproductive leaves" enabled the angiosperms to become the dominant plants on land.

Figure 21-34 Predatory leaves adorn the Venus flytrap **(a)** *and the sundew* **(b)**. *When an insect blunders into the fringed, hinged leaves of the flytrap, the leaves close up rapidly, trapping their victim. Digestive enzymes reduce the insect to nourishing molecules that are absorbed by the plant. The sundew leaf bears glistening droplets that attract hungry insects. However, the droplets are incredibly sticky, and the unsuspecting insect becomes, not the eater, but the eaten. As in the flytrap, enzymes secreted by the leaf digest the insect and the leaf absorbs the resulting nutrients.*

SUMMARY OF KEY CONCEPTS

The Growth and Structure of Land Plants

The body of a land plant consists of root and shoot. Roots, which are usually underground, have five functions. They (1) anchor the plant in the soil; (2) absorb water and minerals from the soil; (3) store surplus photosynthetic products; (4) transport water, minerals, photosynthetic products, and hormones; and (5) produce hormones. Shoots are usually above ground, and consist of stem, leaves, and reproductive structures (including flowers and fruit in angiosperms). Shoot functions include (1) photosynthesis; (2) transport of materials; (3) reproduction; and (4) hormone synthesis.

Plant bodies are composed of two main classes of cells. Meristem cells are embryonic cells that are capable of cell division throughout the life of the plant. Differentiated cells arise from divisions of meristem cells, become specialized for particular functions, and normally do not divide.

Most meristem cells are located in apical meristems at the tips of roots and shoots, and lateral meristems in the shafts of roots and shoots. Primary growth (growth in length and differentiation of parts) results from division and differentiation of cells from apical meristems, while secondary growth (growth in diameter) results from division and differentiation of cells from lateral meristems.

Roots: Anchorage, Absorption, and Storage

Primary growth in roots results in a structure consisting of an outer epidermis, an inner vascular cylinder of conducting tissues, and cortex between the two. The apical meristem near the tip of the root is protected by the root cap. Cells of the root epidermis absorb water and minerals from the soil. Root hairs are projections of epidermal cells that increase the surface area for absorption. Most cortex cells store surplus sugars produced through photosynthesis, usually as starch. The innermost layer of cortex cells is the endodermis, which forms a waterproof barrier preventing movement of water and minerals to and from the soil and the vascular cylinder via the extracellular space.

The vascular cylinder contains the conducting tissues of xylem and phloem. The cells of the outermost layer of the vascular cylinder of roots, the pericycle, may divide and produce branch roots.

Stems: Reaching for the Light

Primary growth in dicot stems results in a structure consisting of an outer, waterproof epidermis, supporting and photosynthetic cells of cortex beneath the epidermis, vascular tissues of xylem and phloem, and supporting and storage cells of pith at the center of the stem. Xylem transports water and minerals from the roots up the stem to leaves, flowers, and fruits. Xylem contains two types of transport cells, tracheids and vessel elements. These cells are dead at maturity, forming hollow tubes composed of the cell walls that remain after the cell contents disintegrate. Phloem transports sugar solutions up and down the plant. Phloem contains two cell types involved in transport: sieve-tube elements that form conducting tubes, and companion cells that nourish the sieve-tube elements and regulate sugar movement in and out of the tubes.

Secondary growth in stems results from cell divisions in the vascular cambium and cork cambium. Vascular cambium produces secondary xylem and secondary phloem, increasing the diameter of the stem. Cork cambium produces waterproof cork cells that cover the outside of the stem.

Leaves: Nature's Solar Collectors

Most shoots produce leaves, the main photosynthetic organs of plants. The blade of a leaf consists of a waterproof outer epidermis surrounding mesophyll cells that have chloroplasts and carry out photosynthesis, and vascular bundles of xylem and phloem that carry water, minerals, and photosynthetic products to and from the leaf. The epidermis is punctuated by adjustable pores called stomata. A pair of guard cells forms the pore and, by changes in their water content, regulates the size of the opening.

Plant Nutrition

Plants acquire carbon dioxide and oxygen from the air, and water and all other essential nutrients from the soil. Most minerals are taken up from the soil water by active transport. Water is absorbed by osmosis, following the uptake of minerals. Mineral accessibility is enhanced by fungi associated with the roots of many plants. Nitrogen, a necessary nutrient for plants, can only be absorbed in the form of ammonium or nitrate, and these ions are scarce in most soils. Some plants, the legumes, have evolved a cooperative relationship with nitrogen-fixing bacteria that invade legume roots. The plant provides the bacteria with sugars and the bacteria use some of the energy to convert atmospheric nitrogen to ammonium, which is then absorbed by the plant.

Transport of Water and Minerals

The cohesion–tension theory explains xylem function: cohesion of water molecules to one another by hydrogen bonds holds together the water within xylem tubes as if it were a solid. As water molecules evaporate from the leaves, the hydrogen bonds pull other water molecules up the xylem to replace them. This movement is transmitted down the xylem to the root, where water loss from the vascular cylinder promotes water movement across the endodermis from the soil water by osmosis.

Transport of Sugars

The pressure flow theory explains sugar transport in phloem: parts of the plant that synthesize sugar (e.g., leaves) export sugar into the sieve tube. Increasing sugar concentrations attract water entry by osmosis, causing high hydrostatic pressure in that part of the phloem. Parts of the plant that consume sugar (e.g., fruits) remove sugar from the sieve tube. Loss of sugar causes loss of water by osmosis, resulting in low hydrostatic pressure. Water and dissolved sugar flow in the sieve tube from high to low pressure.

Special Adaptations of Roots, Stems, and Leaves

Through evolution, the roots, stems, and leaves of many plants have been modified into diverse structures such as spines, tendrils, thorns, and bulbs. These unusual structures are often involved in water or energy storage, support, or protection.

GLOSSARY

Abscisic acid (ab-sis′-ik) a plant hormone that is generally inhibitory, enforcing dormancy in seeds and buds, and closing stomata.

Abscission (ab-si′-shun) separation of leaves, flowers, or fruits from a stem, due to formation of a weakened layer of cells at the site of attachment to the stem.

Abscission layer a layer of thin-walled cells at the base of the petiole of a leaf, flower, or fruit, the usual site of separation from the stem.

Angiosperm (an′-jē-ō-sperm) a flowering plant (division Angiospermophyta); produces seeds within a ripened ovary.

Apical meristem (āp′-i-kul mer′-i-stem) the cluster of meristematic cells found at the tip of a shoot or root (or one of their branches).

Bark the outer layer of a woody stem, consisting of cork cells, cork cambium, and phloem.

Blade the flat part of a leaf.

Bud an embryonic shoot, usually very short and consisting of an apical meristem with several leaf primordia.

Cambium (kam′-bē-um) a lateral meristem that causes secondary growth of woody plant stems and roots. *See also* Cork cambium; Vascular cambium.

Casparian strip (kas-par′-ē-an) a waxy, waterproof band in the cell walls between endodermal cells in a root, which prevents the movement of water and minerals in and out of the vascular cylinder via extracellular space.

Companion cell a cell adjacent to a sieve-tube element in phloem, involved in control and nutrition of the sieve-tube element.

Cork cambium a lateral meristem in woody roots and stems that gives rise to cork cells.

Cork cell a protective cell of the bark of woody stems and roots; at maturity, cork cells are dead, with thick, waterproofed cell walls.

Cortex the part of a primary root or stem located between the epidermis and the vascular cylinder.
Cuticle a waxy layer secreted by epidermal cells of shoots onto their outside surfaces.
Dicot short for dicotyledon; a type of flowering plant characterized by embryos with two food-storage organs, called cotyledons.
Differentiated cell a mature cell specialized for a specific function; in plants, differentiated cells usually do not divide.
Endodermis (en-dō-der′-mis) the innermost layer of cells of the cortex of a root.
Epidermis (ep-i-der′-mis) the outermost layer of cells of a leaf, young root, or young stem.
Fibrous root system a root system characterized by many roots of approximately the same diameter arising from the base of the stem.
Guard cell one of a pair of specialized epidermal cells surrounding the central opening of a stoma of a leaf, which regulates the size of the opening.
Gymnosperm (jim′-nō-sperm) a vascular plant that produces seeds not enclosed in an ovary.
Internode the part of a stem between two nodes.
Lateral bud a bud located at a node of a stem, usually in the crotch between the stem and the petiole of the leaf found at the same node.
Lateral meristem also called cambium; a meristematic tissue in dicot stems and roots, usually found between the xylem and phloem (vascular cambium) and just outside the phloem (cork cambium).
Leaf an outgrowth of a stem, flattened and photosynthetic.
Leaf primordium (prī-mor′-dē-um) the outgrowth of a shoot that develops into a leaf.
Macronutrient a nutrient needed in relatively large quantities (often defined as composing more than 0.1 percent of an organism's body).
Meristem cell (mer′-i-stem) an undifferentiated cell that remains capable of cell division throughout the life of a plant.
Mesophyll (mez′-ō-fil) cells located between the epidermal layers of a leaf.
Micronutrient a nutrient needed in relatively small quantities (often defined as composing less than 0.01 percent of an organism's body).
Mineral an inorganic substance, found in rocks or soil.
Monocot short for monocotyledon; a type of flowering plant characterized by embryos with one food-storage organ, called a cotyledon.
Mycorrhiza (mī-kō-rī′za; pl. mycorrhizae) a symbiotic relationship between a fungus and the roots of a land plant.
Nitrogen fixation the process of converting atmospheric nitrogen (N_2) to ammonium (NH_4^+).
Node a region of a stem at which leaves and lateral buds are located.
Nutrient a substance acquired from the environment and needed for survival, growth, and development of an organism.
Palisade cells mesophyll cells just beneath the upper epidermis of a leaf; usually elongated perpendicularly to the epidermis.
Pericycle (per′-i-sī-kul) the outermost layer of cells of the vascular cylinder of roots.
Petiole (pet′-ē-ōl) the stalk that connects the blade of a leaf to the stem.
Phloem (flō′-um) the vascular tissue that conducts a concentrated sugar solution.
Pith cells at the center of a root or stem.
Primary growth growth in length and development of initial structures of plant roots and shoots, due to cell division of apical meristems and differentiation of the daughter cells.
Primary phloem phloem produced from an apical meristem.
Primary root the first root that develops from a seed.
Primary xylem xylem produced from an apical meristem.
Rhizome (rī′-zōm) an underground stem, usually horizontal and functioning in food storage.
Root the part of a plant, usually below ground, that anchors the plant in the soil, absorbs and transports water and minerals, stores food, and produces certain hormones.
Root cap a cluster of cells at the tip of a growing root, derived from the apical meristem. The root cap protects the growing tip from damage as it burrows through the soil.
Root hair a fine projection from the epidermal cells of a young root.
Runner a horizontally growing stem that may develop new plants at nodes that touch the soil.
Secondary growth growth in diameter of a stem or root due to cell divisions in lateral meristems.
Secondary phloem phloem produced from cells arising at the outside of the vascular cambium.
Secondary xylem xylem produced from cells arising at the inside of the vascular cambium.
Shoot all the parts of a vascular plant exclusive of the root. Usually above ground, consisting of stem, leaves, and reproductive structures.
Sieve plate a part of the cell wall between two sieve-tube elements in phloem, with large pores interconnecting the cytoplasm of the elements.
Sieve tube in phloem, a tube made from sieve-tube elements, which transports sugar solutions.
Sieve-tube element one of the cells of a sieve tube.
Spongy layer irregularly shaped cells located just above the lower epidermis of a leaf.
Stem the normally vertical, aboveground part of a plant body that bears leaves.
Stoma (stō′-ma; pl. stomata) an opening in the epidermis of a leaf, surrounded by a pair of guard cells.
Taproot a relatively thick, vertically oriented root that develops from the primary root of dicots.
Tendril a slender outgrowth of a stem that coils about external objects and supports the stem; usually a modified leaf or branch.
Terminal bud the bud at the extreme end of a stem or branch.
Thorn a hard, pointed outgrowth of a stem; usually a modified branch.
Tracheid (trā′-kē-id) an elongated xylem cell with tapering ends containing pits in the cell walls; forms tubes that transport water.
Transpiration evaporation of water from a leaf.
Vascular bundle (vas′-kū-lar) a strand of xylem and phloem found in a leaf; commonly called a vein.
Vascular cambium a lateral meristem located between the xylem and phloem of a woody root or stem.
Vascular cylinder the centrally located conducting tissue of a young root, consisting of primary xylem and phloem.

Vessel a tube of xylem composed of vertically stacked vessel elements, with perforated or missing end walls, leaving a continuous, uninterrupted hollow cylinder.

Vessel element one of the cells of a xylem vessel; elongated, dead at maturity, with thick lateral cell walls for support but with end walls either lacking entirely or heavily perforated.

Xylem (zī′-lum) a conducting tissue of vascular plants that transports water and minerals from root to shoot.

STUDY QUESTIONS

1. List five functions of roots and four functions of shoots.
2. Distinguish between meristem cells and differentiated cells.
3. Distinguish between primary growth and secondary growth, and describe the tissues involved in each.
4. Diagram the internal structure of a root after primary growth, labeling and describing the function of epidermis, cortex, endodermis, pericycle, xylem, and phloem. What tissues are located in the vascular cylinder?
5. What types of cells form root hairs? What is the function of root hairs?
6. Diagram the internal structure of a shoot after primary growth. What is the function of each part of the shoot?
7. What cell types make up the conducting tubes of xylem? Are these cells alive or dead at maturity? What cell types make up the conducting tubes of phloem? Are these cells alive or dead at maturity?
8. Diagram the internal structure of leaves.
9. How are minerals and water taken up by roots? Diagram the structures involved, the pathways for water and minerals from soil water to xylem, and the transport processes at each step.
10. What is nitrogen fixation? Describe the formation of a root nodule in a legume, and the nature of the relationship between legume and bacteria in the nodule.
11. Describe the cohesion–tension theory of water movement in xylem.
12. Describe the pressure flow theory of movement in phloem.
13. Describe the daily cycle of guard cells, and explain how the movement of water into and out of guard cells causes them to open and close the central pore between them.

SUGGESTED READINGS

Brill, W. J. "Biological Nitrogen Fixation." *Scientific American,* March 1977. A description of the bacteria and algae that are the sole natural source of usable nitrogen for plant life.

Epstein, E. "Roots." *Scientific American,* May 1973. An in-depth look at root structure and function.

Frits, H. C. "Tree Rings and Climate," *Scientific American,* May 1972. The width of the annual rings of trees reflects the length of the growing season and the amount of rainfall, and can be used to determine prehistoric climate.

Raven, P., Evert, R. F., and Eichhorn, S. *Biology of Plants,* 4th ed. New York: Worth Publishers, Inc., 1986. A beautifully illustrated botany textbook, particularly strong on the evolution of flowering plants and the ecological adaptations of plants to their environments.

Zimmerman, M. H. "How Sap Moves in Trees." *Scientific American,* March 1963. A delightful description of the use of aphids as research tools in botany.

22
Plant Reproduction and Development

As you walk through a wildflower-strewn meadow (Fig. 22-1), you may be tempted to think that the floral display was created just for your enjoyment. Unfortunately for the human ego, plants do not develop flowers for us, but for the birds and the bees—and beetles, moths, and even bats. The flower, you see, is a sexual display that enhances the reproductive output of a plant. By enticing animals to transfer pollen from one plant to another, flowers enable stationary plants to "court" distant members of their own species. This critical selective advantage has allowed the angiosperms to become the dominant plants on land.

As you know, evolution commonly produces new structures by modifying old ones, and flowers are no exception. Flowers are not wholly new adaptations; rather, they are the most recent and most sophisticated elaboration of the reproductive strategies common to all plants. You will be able to understand the structure and function of flowers more easily if we begin, therefore, by briefly reviewing the essentials of the plant life cycle.

Figure 22-1 In early spring, wildflowers carpet the California hills. The colors, scents, and shapes attract insects that pollinate the flowers, ensuring a new display next year.

REPRODUCTION AND THE PLANT LIFE CYCLE

Sexual Versus Asexual Reproduction

Many plants can reproduce either sexually or asexually. Asexual reproduction in plants usually involves part of a single plant, say a stem, giving rise to a new plant. The cells that form an asexually produced offspring arise by mitosis from cells of the parent plant. Therefore, these offspring are genetically identical to the parent (see Chapter 8). In Chapter 21 we encountered several methods of asexual reproduction, including runners in strawberries and rhizome sprouting in irises. Asexual reproduction is often a highly effective reproductive strategy. For example, an offspring strawberry connected to its parent by a runner (see Fig. 21-30) draws nourishment from the parent until it grows large enough to fend for itself.

However, genetic identity between parent and asexually produced offspring means that the offspring is only as well adapted to the environment as its parent was. What if the environment changes? Sexually produced offspring usually combine genes from two different parents, and therefore they may be endowed with traits that differ from those of either parent. This new combination of traits may help the offspring to adapt to the environment better than either parent could. As a result, most organisms reproduce sexually, at least some of the time.

Plant Life Cycles

The sexual life cycle of plants is more complex than the familiar animal life cycle. In animals, cells in the gonads of a diploid adult undergo meiosis to produce haploid gametes, either sperm or eggs. Sperm and egg fuse to create a diploid fertilized egg, the **zygote.** Through repeated mitosis and differentiation of the daughter cells, the zygote develops into another diploid adult. *Plants, on the other hand, have two distinct, multicellular "adult" forms, one diploid and one haploid.* For this reason, the plant life cycle is named **alternation of generations,** as diploid adults alternate with haploid adults (Fig. 22-2).

Let's examine the life cycle of a fern (Fig. 22-2a), starting with the diploid adult form. This stage of the life cycle, the **sporophyte** ("spore plant" in Greek), bears reproductive cells that undergo meiosis to produce haploid cells that are **spores,** not gametes. The difference between spores and gametes is that spores do not fuse together to re-form a diploid cell. Instead, fern spores are blown off the parent frond by the wind and land on the soil. There the spore sprouts, dividing repeatedly by mitosis to form a multicellular, haploid organism. This organism produces gametes, and hence is called the **gametophyte** (Greek for "gamete plant"). Since its cells are haploid already, the "gonads" of the gametophyte can produce sperm and eggs without further meiosis. Usually, a single gametophyte produces both sperm and eggs, although often at different times, thereby preventing self-fertilization. Sperm and egg fuse to form a zygote that develops into a new diploid sporophyte plant.

Alternation of generations occurs in all plants. In primitive land plants, including mosses and ferns, the gametophyte is an independent, although usually small, plant. It liberates mobile sperm cells that reach an egg by swimming through thin films of water covering adjacent gametophytes. Therefore, ferns and mosses can live only in moist habitats where continuous sheets of water exist at some time during the year. Most terrestrial habitats, however, are not so liberally supplied with water. To reproduce in these drier places, a plant must surround its sperm in a desiccation-proof package, transport that package to another plant, and liberate the sperm directly into the egg-bearing structures of the second plant.

The seed plants (angiosperms and gymnosperms) do just that. In the angiosperms, separate male and female spores are formed by meiosis within the flowers borne by the sporophyte generation (Fig. 22-2b). The spores de-

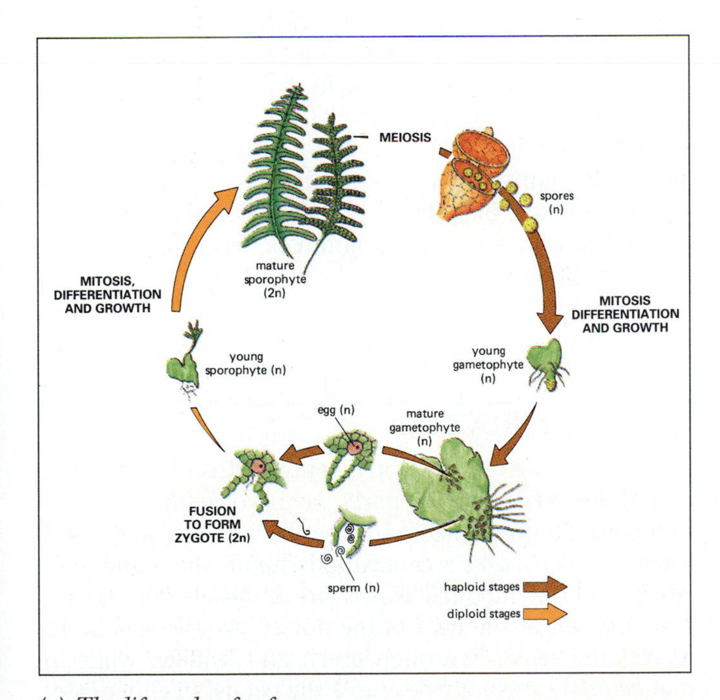

(a) *The life cycle of a fern.*

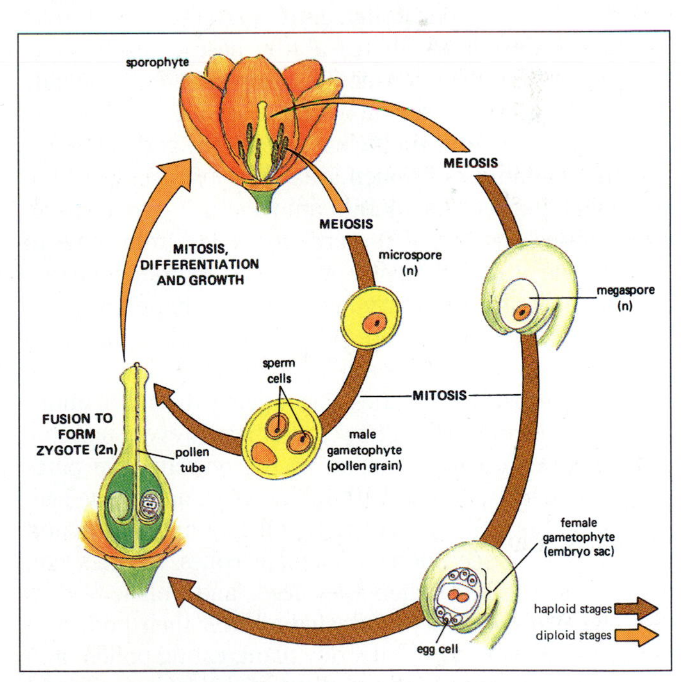

(b) *The life cycle of a flowering plant.*

Figure 22-2 The life cycles of plants, called alternation of generations, include separate multicellular haploid and diploid adult stages. The events occurring at various stages of the life cycle are described in the text.

velop into gametophytes not in the soil, but within the flower. The female gametophyte is a small cluster of cells permanently retained within the flower, while the male gametophyte becomes a sperm-transport package: a tough, watertight **pollen grain** that drifts on the wind or is carried by an animal from one flower to another. The pollen grain germinates on the recipient flower and burrows through the flower tissues to the female gametophyte within. This miniature male gametophyte liberates its sperm inside the female gametophyte, where fertilization occurs. The zygote becomes enclosed in a drought-resistant **seed** that may lie dormant for months, years, or even centuries waiting for favorable conditions for growth. In this chapter we examine sexual reproduction in flowering plants, from the evolution of the flower through the formation of the seed and the development of the new seedling.

Figure 22-3 Conifers are wind-pollinated. Even slight breezes blow thick clouds of pollen from ripe male cones.

THE EVOLUTION OF FLOWERS

The earliest seed plants were the gymnosperms, represented today mainly by pines, firs, and other conifers. As we described in Chaper 19, conifers bear male and female gametophytes on separate cones. During early spring, the small male cones release hordes of pollen grains that waft about on the breezes (Fig. 22-3). Most blow uselessly away, but with so many grains floating around, some enter the pollen chambers located on the scales of the female cones, where they are captured by sticky coatings of sugars and resins. The pollen grains germinate and tunnel to the female gametophytes at the base of each scale. Sperm are liberated, fertilize the eggs within the female gametophyte, and a new generation begins.

Clearly, this is an inefficient operation, and the overwhelming majority of pollen grains are lost. In a world of stationary plants and mobile animals, if a gymnosperm could entice an animal to carry its pollen from male to female cone, it would greatly enhance its reproductive rate and hence its evolutionary success. As it happens, gymnosperms and insects were poised to establish just such a relationship about 150 million years ago.

Insects, especially beetles, are among the most abundant animals on Earth. They exploit nearly every possible food resource on land, including the reproductive parts of gymnosperms. About 150 million years ago, some beetles fed on both the protein-rich pollen of male cones and the sugar-rich secretions of female cones. Beetles can make quite a mess when they feed, and pollen feeders often wind up with pollen dusted all over their bodies. If the same beetle were to visit one plant, eating pollen, and then wander over to another plant of the same species to dine on the sugary secretions of a female cone, some of the loose pollen would probably rub off on the female cone.

The stage was set for the evolution of angiosperms. Efficient pollination by insects requires that a given insect visit several plants of the same species, pollinating them on the way. For the plants, two key adaptations were necessary. First, enough pollen and/or sugary secretions (nectar) must be produced within the reproductive structures (the future flowers) so that insects would regularly visit them to feed. Second, the location and richness of these storehouses of pollen and nectar must be advertised to the insects, both to show them where to go and to entice them to specialize on that specific plant species. Any mutation that contributed to these adaptations would enhance the reproductive potential of the plant carrying the mutation, and would be favored by natural selection. By about 130 million years ago, flowers had evolved with exactly these adaptations. The advantages of flowers are so great that in today's temperate and tropical zones, flowering plants are overwhelmingly dominant, and a host of animals, including bees, moths, butterflies, hummingbirds, and even some mammals, feed almost exclusively at flowers.

FLOWER STRUCTURE

Flowers, like leaves, develop as outgrowths of stems, and in fact the various flower parts have evolved from leaves. **Complete flowers,** such as those of crocuses, lilies, and tomatoes, consist of a central axis upon which four successive sets of modifed leaves are attached (Fig. 22-4). The first set, at the base of the flower, are the **sepals.** In dicots, the sepals are often green and leaflike, while in monocots they usually resemble the petals. In either case, sepals surround and protect the flower bud as the remaining three structures develop. Just above the sepals are the **petals,** which are usually brightly colored and fragrant, advertising the location of the flower.

The male reproductive structures, the **stamens,** are at-

Figure 22-4 **(a)** *A complete flower has four parts: sepals, petals, stamens (the male reproductive structures) and at least one carpel (the female reproductive structure).* **(b)** *The lily is a complete monocot flower, with three sepals (virtually identical to the petals), three petals, six stamens, and three fused carpels. Each stamen consists of a filament bearing an anther at its tip. The carpels consist of an ovary hidden in the base of the flower, with a long style protruding out, ending in a sticky stigma. Note that the anthers are considerably below the stigma. This is probably an adaptation preventing self-pollination: pollen cannot simply fall from the anther onto the stigma.*

tached just above the petals. Stamens usually consist of a long slender **filament** bearing at its tip an **anther** that produces pollen. The female reproductive structures, the **carpels,** occupy the uppermost position in the flower. An idealized carpel is somewhat vase-shaped, with a sticky **stigma** for catching pollen mounted atop an elongated **style** connecting it with the bulbous **ovary.** Inside the ovary are one or more **ovules,** in which the female gametophytes develop. When mature, the ovule will become the seed, while the ovary will develop into a protective, adhesive, or edible covering, the fruit.

As you may know from your own gardening experience, not all flowers are complete. **Incomplete flowers** lack one or more of the four floral parts. For example, many plants have separate male and female flowers, which may be borne on the same plant, as in cucumbers and squashes (Fig. 22-5), or on different plants, as in the American holly. Male flowers lack carpels, while female flowers lack stamens. Incomplete flowers may also lack sepals or petals.

Figure 22-5 Plants of the squash family, such as these zucchinis, bear separate female (left) and male (right) flowers. Obviously, individual flowers cannot be self-pollinated, but the plant could still be self-pollinated if an insect carried pollen from a male flower to a female flower of the same plant. However, since each plant initially produces only male flowers, some cross-pollination between plants maturing at slightly different times is virtually assured.

COEVOLUTION OF FLOWERS AND POLLINATORS

Wind-pollinated flowers, such as those of grasses and oaks, are usually inconspicuous and unscented, often scarcely more than naked stamens that liberate pollen to the wind (Fig. 22-6). Animal-pollinated flowers, on the other hand, are more elaborate, shaped by millions of years of natural selection to attract useful animal pollinators and frustrate undesirable visitors who might eat nectar or pollen without fertilizing the flower in return. The distinctive shapes, colors, and odors of flowers complement the sensory capabilities and life-styles of their animal pollinators.

Animal-pollinated flowers can be loosely grouped into three categories, depending on the benefits (real or imagined) that they offer to potential pollinators: food, sex, or a nursery.

***Figure 22-7** The carrion flower is pollinated by flies, which are attracted by its aroma of rotting meat.*

Food

Many flowers provide food for foraging animals such as bees or hummingbirds. In return, the animals unwittingly distribute pollen from flower to flower.

BEETLE-POLLINATED FLOWERS. Beetles dine on a wide variety of foods, including both living and dead plants and animals, and animal dung. Since animal tissue usually has more protein, fat, and calories than plant tissue, many beetles preferentially feed on animal material. Beetle-pollinated flowers often smell like rotting carrion or dung, which attracts scavenging beetles. Since most beetles are quite clumsy and unspecialized for feeding at flowers, these flowers are usually simple, open affairs, with everything except the ovaries in plain sight and within easy reach. A beetle may eat nectar, pollen, and even petals, strewing everything about in the process. Some pollen sticks to its body and may rub off on the next flower it visits. Flies have much the same taste in foods (or lack thereof!) as beetles, and most fly-pollinated flowers, such as the carrion flower (Fig. 22-7), also emit a powerful stench.

***Figure 22-6** The flowers of grasses and many deciduous trees are wind pollinated, with anthers (yellow structures hanging beneath flowers) exposed to the wind. Petals are usually reduced or absent.*

BEE-POLLINATED FLOWERS. Bees are culinary specialists, often feeding only on nectar and pollen. Bees scout out prospective dinners from the air, using both scent and sight to locate and identify flowers. We can thank the bees for most of the sweet-smelling flowers, since sweet "flowery" odors attract these pollinators. Bees also have good color vision, but do not see exactly the same range of colors that humans do (Fig. 22-8). Although unable to distinguish red from gray or black, their color vision extends into the ultraviolet. To attract a bee from afar, bee-pollinated flowers must look brightly colored *to a bee.* Typically, these flowers are white, yellow, or blue, and often have other markings, such as central spots or lines pointing toward the center, which reflect ultraviolet light (Fig. 22-8).

Bee-pollinated flowers have several structural adaptations that help to ensure pollen transfer. Many bee-pollinated flowers, such as nasturtiums and floxgloves, produce nectar at the bottom of a tube (Fig. 22-9). Either pollen-laden stamens (usually in newly opened flowers) or the sticky stigma of the carpel (in older flowers) protrude out of the top of the tube. When a bee visits a young flower, she lands on the lip of the flower and thrusts her

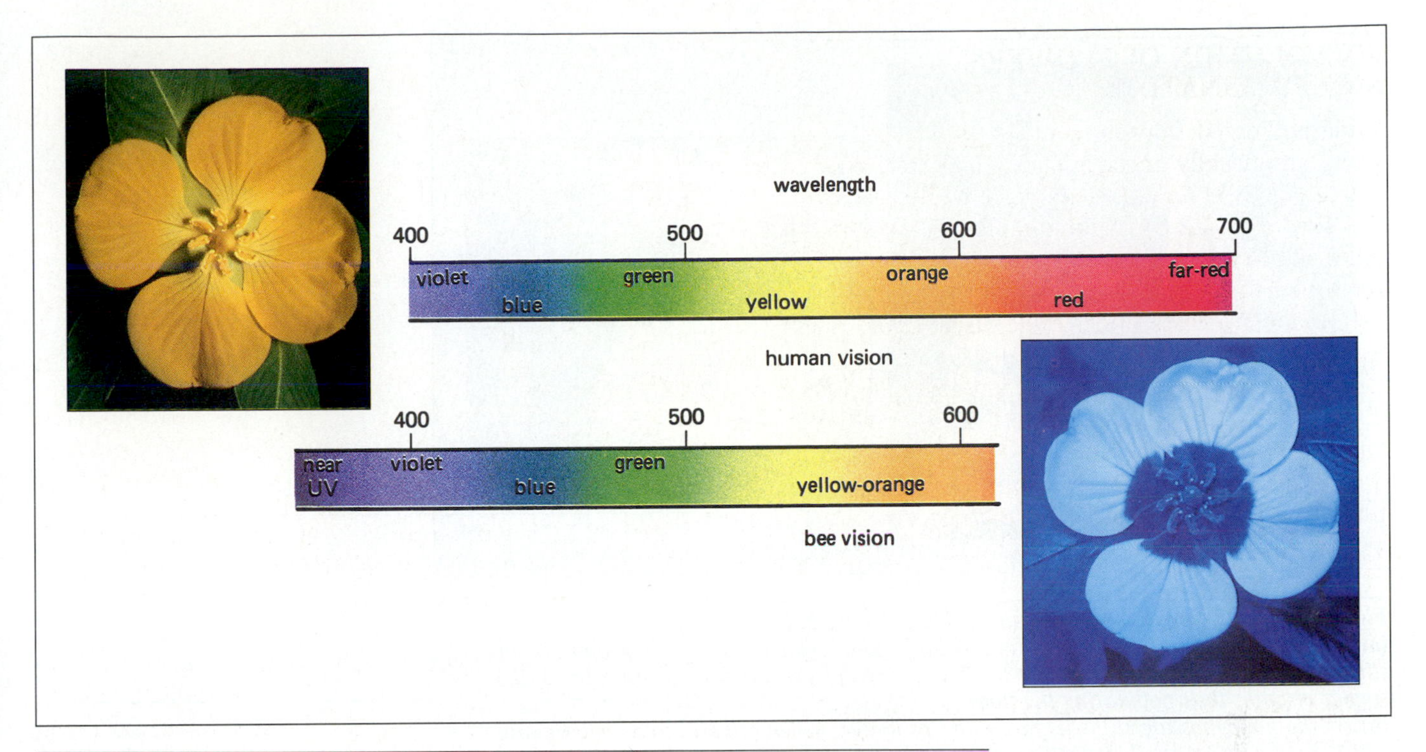

Figure 22-8 *The spectra of color vision for humans and bees overlap considerably in the blue, green, and yellow ranges, but differ on the edges. Humans are sensitive to orange and red, which bees do not perceive, whereas bees can see ultraviolet, which is invisible to the human eye. Flowers photographed under ordinary daylight (upper left) and under ultraviolet light (lower right) show striking differences in color patterns. Bees can see the ultraviolet patterns that presumably lead them to the nectar- and pollen-containing centers of the flowers.*

head into the tube to reach the nectar. Simultaneously, the stamens brush pollen onto her back. She may then visit an older flower and repeat her foraging behavior. This time, she leaves pollen behind on the stigma.

Many moths and butterflies also feed on nectar. Flowers adapted for these pollinators are often superficially similar to bee-pollinated flowers, except that the nectar tubes are deeper, which suits the long (up to 25 centimeters) tongues of moths and butterflies, and prevents access by shorter-tongued bees. Day-flying butterflies are attracted by white, yellow, blue, and orange flowers with mild, sweet fragrances. Flowers pollinated by night-flying moths open only in the evening, are usually white, and exude strong, musky odors that help the moth to locate the flower in the dark.

HUMMINGBIRD-POLLINATED FLOWERS. Hummingbirds are one of the few vertebrates that are important pollinators, although several mammals also visit flowers (Fig. 22-10). Birds have notoriously poor senses of smell, and hummingbird-pollinated flowers seldom synthesize fragrant chemicals. However, hummingbirds need lots of energy, and these flowers always produce large amounts of nectar. If they did not, the hummers would go elsewhere and the flower would remain unpollinated. On the

Figure 22-9 *Many bee-pollinated flowers, such as these foxgloves, are tubular. Nectar is produced at the base of the tube, while the anthers and stigmas protrude out the open end. Often a lower "lip" on the tube serves as a landing platform for bees, since bees are not very good at hovering in midair.*

Figure 22-10 Among the more unusual pollinators are bats and honeypossums.
(a) *A tropical bat feeds at a cluster of tubular flowers. Note the protruding stamens and stigma. As the bat hovers before the flower, the top of its head touches either the anthers or stigma or both, thus pollinating the flower.*
(b) *As the honeypossum stuffs its face into this flower, pollen adheres to its muzzle and whiskers. A visit to another flower may result in transfer of pollen.*

other hand, a large supply of nectar would also attract insects, and a bee might return again and again to the same flower, never transferring pollen to another flower. Not surprisingly, hummingbird-pollinated flowers have evolved several adaptations that keep insects from drinking their nectar. These flowers are usually tubular, matching the long bills and tongues of hummers. The tube is much too deep for bees to reach the nectar at its base. In addition, most are red or orange, which is brightly attractive to a bird but drab to a bee (Fig. 22-11).

Figure 22-11 A hummingbird hovers before a flower. Note that the flower is red and lacks the landing platform commonly found in bee-pollinated flowers.

Sex

A few plants, most notably the orchids, take advantage of the insatiable libido and stereotyped behaviors of male wasps and flies to pollinate their flowers. Some orchid flowers mimic female wasps both in scent and shape (Fig. 22-12). The males land atop these "females" and attempt to copulate, but only get a packet of pollen for their efforts. Further rendezvous with other orchids result in pollination for the flowers.

Nursery Flowers

Perhaps the most elaborate relationships between plants and pollinators occur in a few cases in which insects fertilize a flower and then lay their eggs in the flower's ovary. This arrangement is found between milkweeds and milkweed bugs, figs and certain wasps, and yuccas and yucca moths (Fig. 22-13). The yucca moth performs a remarkable series of behaviors, resulting in pollination of yuccas and a well-stocked pantry for its own offspring. A female moth visits a yucca flower, collects pollen, and rolls it into a compact ball. The moth flies off with the pollen ball to another yucca flower, drills a hole in the ovary wall, and lays its eggs inside the ovary. Then it takes its pollen ball and smears pollen all over the stigma of the flower! By pollinating the yucca, the moth ensures that the plant will provide a supply of developing seeds for its offspring caterpillars to eat. It is most unlikely that the moth "knows" that it must fertilize the yucca so that its young

Figure 22-12 *This male wasp is actually trying to copulate with an orchid flower. The result is successful reproduction, not for the wasp, but for the orchid.*

have enough to eat, but nonetheless it performs the genetically programmed behavior flawlessly. Since the caterpillars eat only a small fraction of the seeds, the yucca also reproduces successfully. The mutual adaptation of yucca and moth is so complete that neither can reproduce without the other.

GAMETOPHYTE DEVELOPMENT IN FLOWERING PLANTS

As Fig. 22-2b illustrates, in the life cycle of flowering plants, the familiar plant of meadow, garden, and farm is the diploid sporophyte. The **pollen grain** (male) and the **embryo sac** (female) are the haploid gametophytes that develop within the flowers borne by the sporophytes. Both are much smaller than the gametophyte stages of ferns and mosses, and cannot live independently of the sporophyte.

POLLEN. An anther consists of four chambers called pollen sacs (Fig. 22-14). Within each sac, hundreds to thousands of diploid **microspore mother cells** develop. Each microspore mother cell undergoes meiosis (see Chapter 8) to produce four haploid **microspores** (just as animal sperm are smaller than eggs, the male spores of flowering plants are smaller than the female spores; hence the name "microspore"). Each microspore divides once, by mitosis, to produce a male gametophyte, or pollen grain, consisting of only two cells: a large **tube cell** and a smaller **generative cell** residing *within the cytoplasm* of the tube cell (Fig. 22-14). A tough surface coat develops around the pollen grain, protecting the cells within during their journey to the carpel (Fig. 22-15).

When ripe, the pollen sacs split open. In wind-pollinated flowers such as those of grasses and oaks, the pollen spills out, a fortunate few to be carried by wind currents to other flowers of the same species. In animal-pollinated flowers, the pollen adheres weakly to the anther case until the pollinator comes along and brushes or picks it off.

Figure 22-13 **(a)** *Yuccas bloom on the dry plains of eastern Colorado in early summer.* **(b)** *Within many of the yucca flowers, yucca moths carry out their part in one of nature's most unusual and most effective cooperative relationships between plant and animal.*

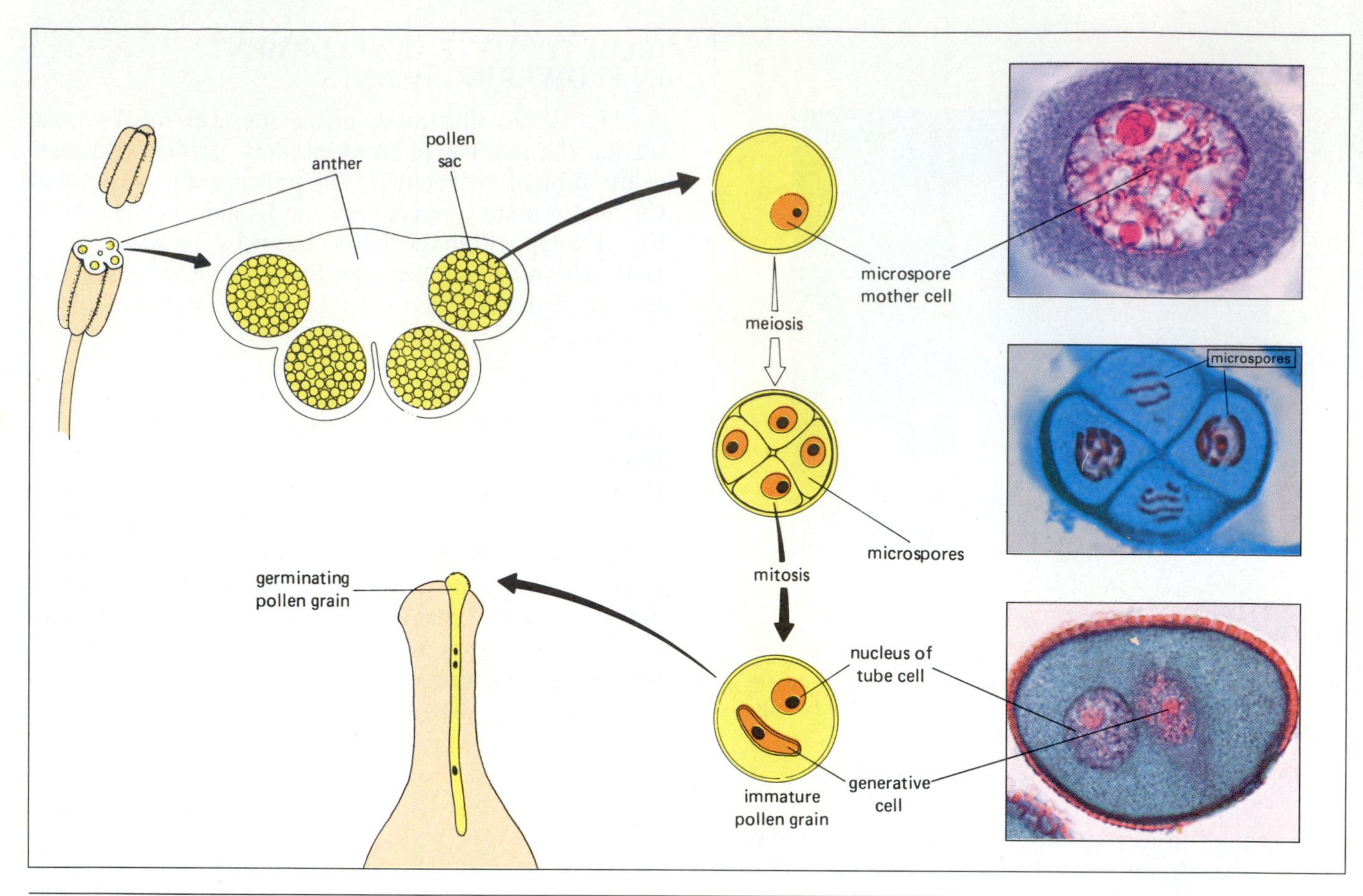

Figure 22-14 Pollen development in angiosperms. Within the anthers, diploid microspore mother cells form. These cells undergo meiosis to produce four haploid microspores. Through mitosis, each microspore divides into two (still haploid) cells that form the juvenile pollen grain, the male gametophyte stage of flowering plants.

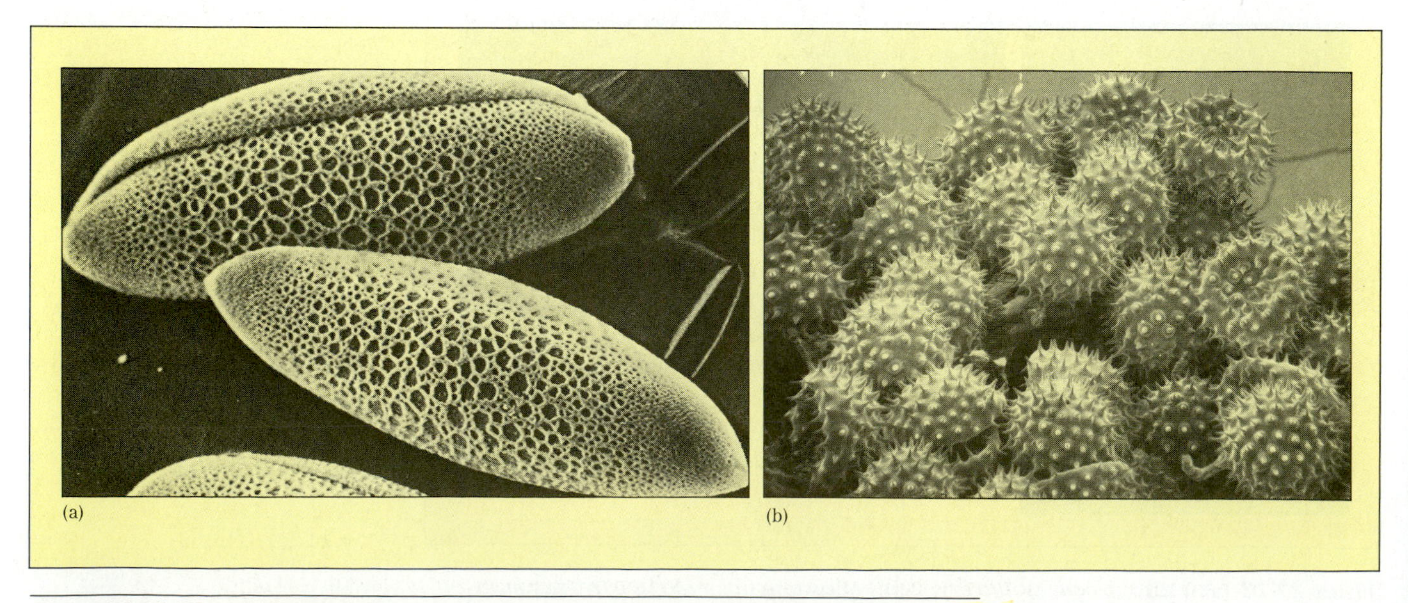

Figure 22-15 The tough outer coverings of pollen grains are often elaborately sculptured in species-specific shapes and patterns. Shown here are pollen from amaryllis **(a)** *and cotton* **(b)**.

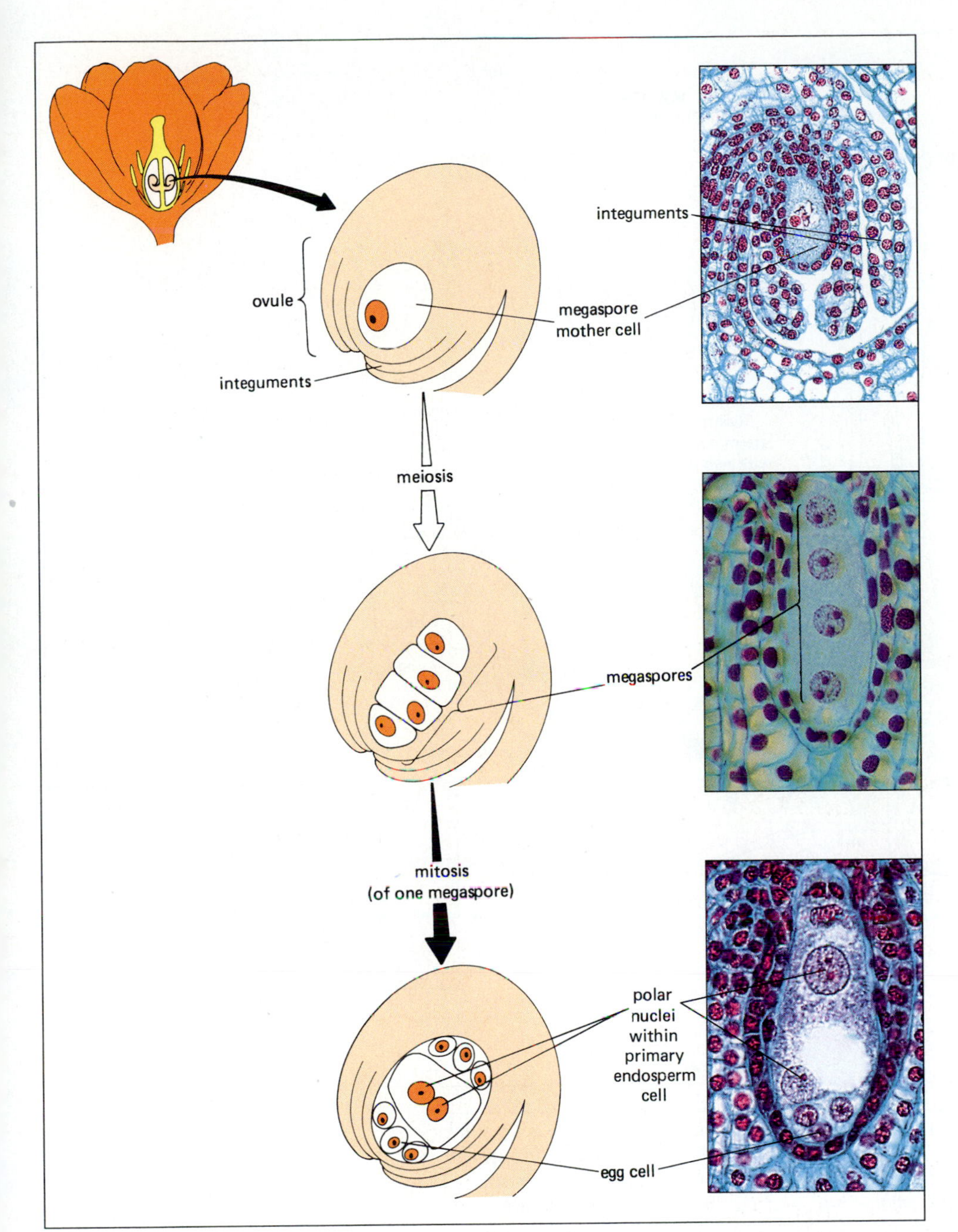

Figure 22-16 *Development of the female gametophyte. A single diploid megaspore mother cell matures within the integuments of an ovule. Through meiosis, it gives rise to four haploid megaspores. Three of these degenerate, while the fourth passes through three mitotic divisions to produce eight haploid nuclei. Two of these, the polar nuclei, become enclosed within the large primary endosperm cell, one becomes the nucleus of the egg cell, and the other five pass into peripheral cells that degenerate soon after fertilization.*

EMBRYO SAC. Within the ovary of a carpel, one or more dome-shaped masses of cells differentiate into **ovules.** Each ovule consists of outer layers of cells called **integuments** that surround a single, diploid **megaspore mother cell** (Fig. 22-16). The megaspore mother cell divides by meiosis to produce four large haploid **megaspores.** Three megaspores degenerate, and only one survives. This remaining megaspore undergoes an unusual set of mitotic divisions. Three nuclear divisions produce a total of eight haploid nuclei. Cell membranes then divide up the cytoplasm into *seven,* not eight, cells: three small cells at each end, with one nucleus apiece, and one remaining large cell in the middle with two nuclei called polar nuclei. This seven-celled organism, called the **embryo sac,** is the haploid female gametophyte. The central, binucleate cell is the **primary endosperm cell.** The **egg** is one of the cells at the bottom of the embryo sac, near a pore in the integuments.

POLLINATION AND FERTILIZATON

When a pollen grain lands on the stigma of a carpel of the same species of plant, a remarkable chain of events occurs (Fig. 22-17). The pollen grain absorbs water,

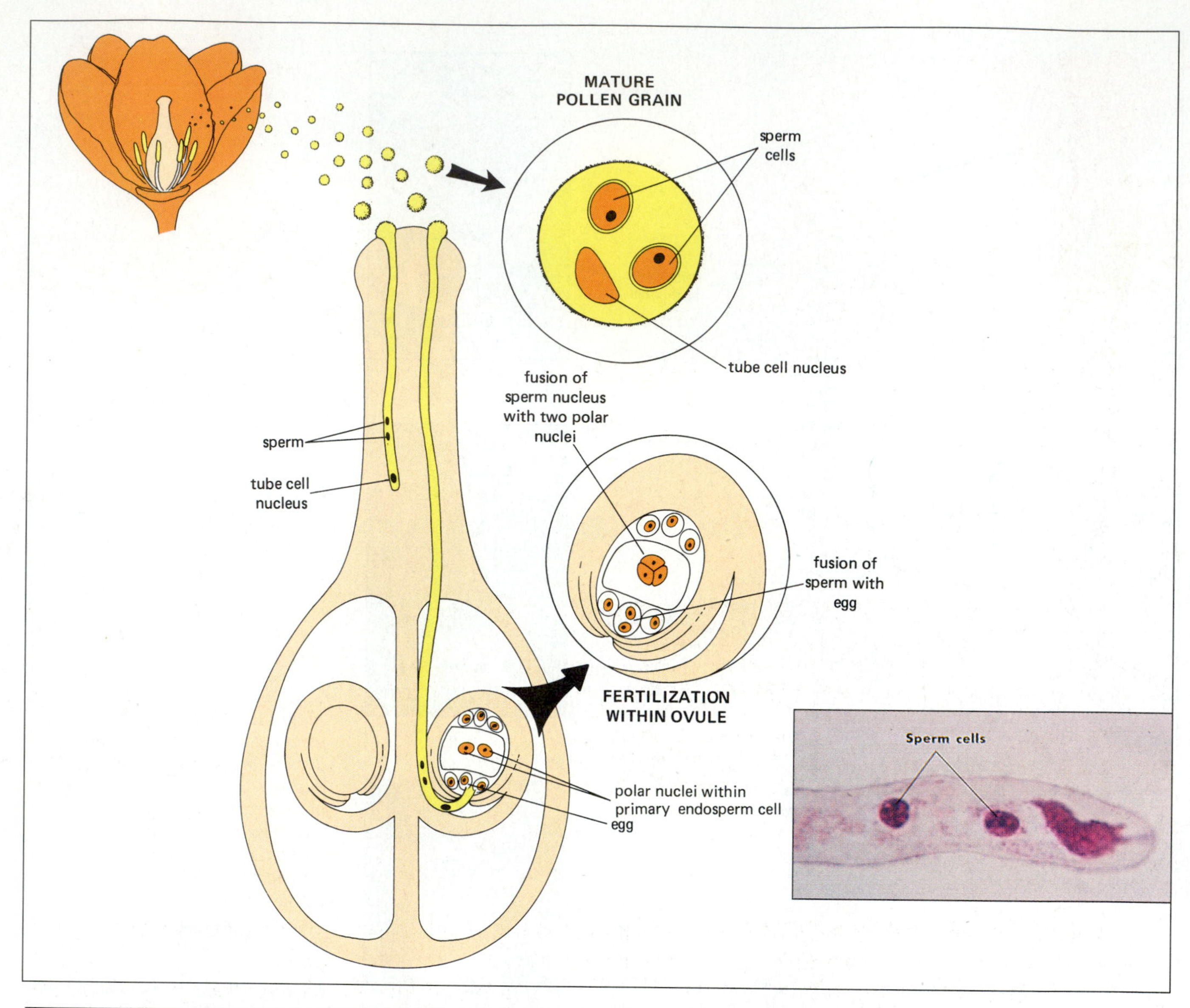

Figure 22-17 Pollination and fertilization of a flower. The pollen grain lands on the sticky surface of the stigma and germinates. The tube cell elongates, burrowing down through the style. The sperm nuclei follow, inside the cytoplasm of the tube cell. Upon reaching an ovule, the tube breaks through into the embryo sac, and the tube nucleus degenerates. One sperm enters the primary endosperm cell; its nucleus fuses with both polar nuclei to form a triploid cell. The other sperm fertilizes the egg cell to form the diploid zygote that will develop into the new embryonic plant. The other cells of the embryo sac degenerate.

swells, and splits its outer coat. The tube cell grows, digesting a tunnel within the style leading down toward an ovule in the ovary. Meanwhile, the generative cell achieves puberty, so to speak, and divides mitotically to form two **sperm cells.** This three-celled organism (one tube cell containing two sperm cells) is the mature male gametophyte. If all goes well, the pollen tube reaches the pore in the integument of an ovule and breaks into the embryo sac. Its tip ruptures, releasing the two sperm. One sperm fertilizes the egg cell to form the diploid zygote that will develop into a new sporophyte. The second sperm enters the primary endosperm cell. Its nucleus fuses with *both polar nuclei* to form a triploid cell that will develop into the **endosperm,** a food storage organ within the seed. The fusion of the egg with one sperm and the primary endosperm cell with the second sperm is often called **double fertilization** and is unique to flowering plants. The other five cells of the embryo sac degenerate.

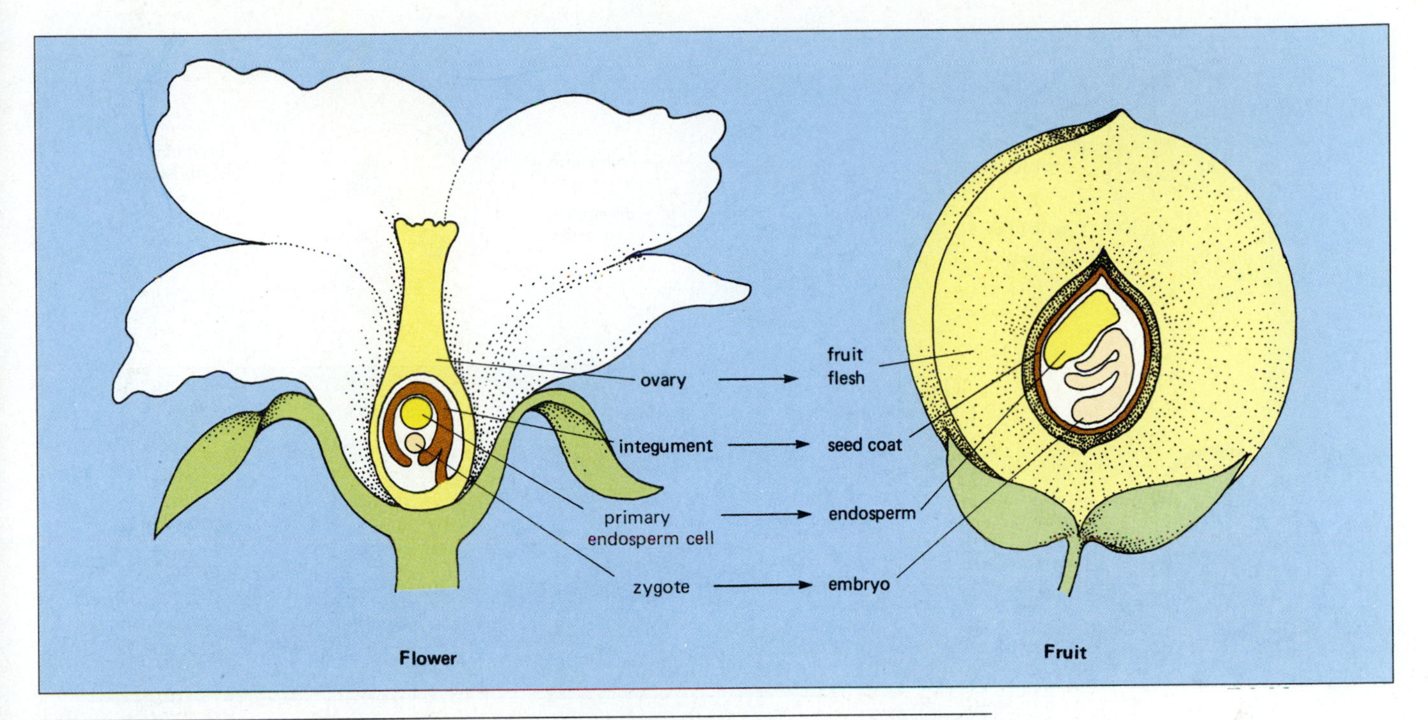

Figure 22-18 The fertilized flower develops into a fruit. Starting at the outside, the ovary wall ripens into the fruit flesh, which may be soft and tasty like a peach, or variously hard, hooked, or tufted in other fruits. The integuments of the individual ovule, which surround the embryo sac, harden and become waterproof, forming the seed coat. These two fruit parts are derived from tissues of the parent sporophyte plant. Within the seed, two structures develop from cells of the fertilized female gametophyte. The triploid endosperm cell divides repeatedly, absorbs nutrients from the parent plant, and becomes the endosperm, a food storage structure within the seed. Finally the zygote develops into the embryo.

Incidentally, you should note the distinction between pollination and fertilization. **Pollination** occurs when a pollen grain lands on a stigma, while **fertilization** is the fusion of sperm and egg. Although pollination is obviously a prerequisite for fertilization, they are two separate events. For example, pollination will not lead to fertilization if the tube cell fails to grow properly, if the embryo sac is sterile, or if the sperm from another pollen grain have already reached the egg.

THE DEVELOPMENT OF SEEDS AND FRUITS

Drawing on the resources of the parent plant, the embryo sac develops into a seed, surrounded by the accessory tissues of a fruit (Fig. 22-18).

Seed Development

The integuments of the ovule develop into the **seed coat,** a thin, tough, waterproof outer covering. As we shall see, the characteristics of the seed coat play a role in regulating when the seed will sprout.

Meanwhile, within the integuments, two distinct developmental processes occur (Fig. 22-19). First, the triploid endosperm cell divides rapidly. Its daughter cells absorb nutrients from the parent plant, forming a large, food-filled endosperm. Second, the zygote develops into the embryo. Both dicot and monocot embryos consist of three parts: the shoot, the root, and the **cotyledons,** or seed leaves. The cotyledons absorb food molecules from the endosperm and transfer them to the embryo. In dicots ("two cotyledons"), the cotyledons usually absorb most of the endosperm during seed development, so that the mature seed is virtually filled with embryo (Fig. 22-19b). In monocots ("one cotyledon"), the cotyledon absorbs some of the endosperm during seed development, but most of the endosperm remains in the mature seed (Fig. 22-19c).

The future primary root develops at one end of the main embryonic axis. The future shoot, at the other end, is usually divided into two regions by the site of attachment of the cotyledons. Below the cotyledons, but above the root, is the **hypocotyl** (*hypo* in Greek means "be-

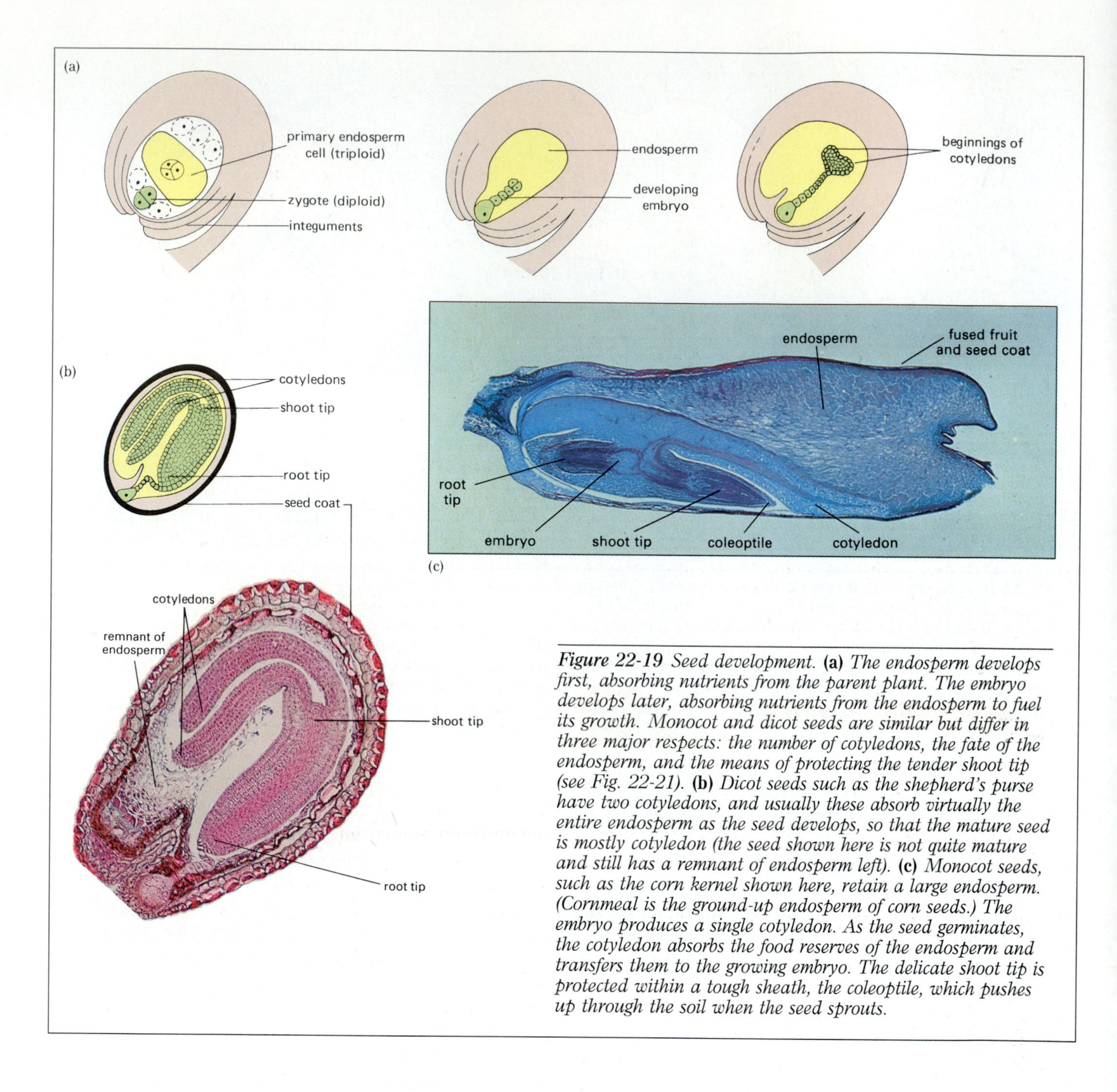

Figure 22-19 Seed development. **(a)** *The endosperm develops first, absorbing nutrients from the parent plant. The embryo develops later, absorbing nutrients from the endosperm to fuel its growth. Monocot and dicot seeds are similar but differ in three major respects: the number of cotyledons, the fate of the endosperm, and the means of protecting the tender shoot tip (see Fig. 22-21).* **(b)** *Dicot seeds such as the shepherd's purse have two cotyledons, and usually these absorb virtually the entire endosperm as the seed develops, so that the mature seed is mostly cotyledon (the seed shown here is not quite mature and still has a remnant of endosperm left).* **(c)** *Monocot seeds, such as the corn kernel shown here, retain a large endosperm. (Cornmeal is the ground-up endosperm of corn seeds.) The embryo produces a single cotyledon. As the seed germinates, the cotyledon absorbs the food reserves of the endosperm and transfers them to the growing embryo. The delicate shoot tip is protected within a tough sheath, the coleoptile, which pushes up through the soil when the seed sprouts.*

neath" or "lower"), while above the cotyledons the shoot is called the **epicotyl** (*epi* means "above"). At the tip of the epicotyl lies the apical meristem of the shoot, often already with one or two developing leaves.

Fruit Development

The wall of the ovary develops into a **fruit** (Fig. 22-18). There are a bewildering variety of fruits, with outer layers that are variously fleshy, hard, winged, or even spiked like a medieval mace. The selective forces favoring the evolution of all fruits, however, are similar: *fruits help to disperse the seeds to distant locations away from the parent plant* (see the essay, "Leaving Home: Adaptations for Seed Dispersal").

Seed Dormancy

All seeds need warmth and moisture to germinate. However, most newly matured seeds will not germinate immediately, even under ideal conditions. Instead, they enter a period of **dormancy** during which they will not sprout. Seed dormancy solves two problems, one intrinsic to the plant itself and one related to environmental factors. First, many seeds develop inside juicy fruits, such as apples, grapes, or oranges. If a seed germinates while still

Leaving Home: Adaptations for Seed Dispersal

Successful plant reproduction requires more than fertilizing a flower and developing a seed. It also requires a suitable site for the seed to germinate and the young plant to grow. Gardeners usually think of "suitable sites" only in terms of fertile soil, equable weather, and moisture. From an evolutionary perspective, however, suitability is also a function of distance from the parent plant. The growth of a seedling that sprouts right next to its parent will be inhibited by the larger plant's shade. If the offspring should manage to survive, it will compete with its parent for water and nutrients, to the detriment of both plants. Finally, a plant species will become more widespread if its members send out some seeds to distant habitats. In flowering plants, seed dispersal is the function of fruits. A wide variety of fruits have evolved, each dispersing seeds in a different way.

Shotgun Dispersal

A few plants disperse their seeds in a very straightforward way: they develop explosive fruits that eject their seed meters away from the parent plant. Mistletoes, for example, are common parasites of trees. Mistletoe fruits shoot out sticky seeds. If one strikes a nearby tree, it adheres to the bark and germinates, sending rootlike fibers into the vascular tissues of its host, from which it draws its nourishment. Since the proper germination site for a mistletoe seed is not the ground, which it can reach simply by falling, but a tree limb many meters above, it is clearly useful to shoot the seeds up, up, and away.

Wind Dispersal

Dandelions, milkweeds, and maples (Fig. E22-1) produce lightweight fruits with surfaces that catch the wind. (Yes, each individual hairy tuft on a dandelion ball is a separate fruit!) Each fruit typically contains a single small seed, which reduces weight and lets the fruit remain aloft longer. These featherweight fruits aid the seed in traveling away from the parent plant, from a few meters for maples to miles for a milkweed or dandelion on a windy day.

Water Dispersal

Many fruits can float on water for a time, and may be dispersed by streams and rivers. The coconut fruit, however, is a floater *par excellence.* Round, buoyant, and watertight, the coconut drops off its parent palm, rolls to the sea, and floats for weeks or months until it washes ashore on some distant isle (Fig. E22-2). There it germinates, perhaps establishing a new coconut colony on a formerly barren island.

Animal Dispersal

Perhaps the majority of fruits use animals as agents of seed dispersal. Two quite distinct strategies have evolved for dispersal by animals: grab an animal as it passes by, or entice it to eat the fruit but not digest the seeds.

Anyone who takes a long-haired dog on a walk through an abandoned field knows about fruits that hitchhike on animal fur. Burdocks, burr clover, fox-

Figure E22-1 Wind-dispersed fruits usually contain only one or two lightweight seeds. Some, such as dandelions **(a)** *and milkweeds* **(b)**, *have filamentous tufts that catch the breezes. Others, such as maple fruits* **(c)**, *are actually miniature glider-helicopters, silently whirling away from the tree as they fall. To see how the wings aid in seed dispersal, take two maple fruits and pluck the wing off one. Hold both fruits over your head and drop them. The wingless fruit will fall at your feet, while the winged one will glide some distance away.*

Leaving Home: Adaptations for Seed Dispersal

Figure E22-2 *After a long journey at sea, this coconut was washed high onto a beach by a storm. The large size and massive food reserves of coconuts are probably adaptations required for successful germination and seedling growth on barren, sandy beaches.*

tails, and sticktights all develop fruits with prongs, hooks, spines, or adhesive hairs (Fig. E22-3). The parent plants hold these fruits very loosely, so that even slight contact with fur pulls the fruit free of the plant and leaves it stuck on the animal. Some of these fruits don't hold on to the fur very tightly either, and may fall off the next time the animal brushes against a tree or rock, or may come out when the animal grooms its fur. Others, such as burdocks, embed themselves very firmly, and it takes some determined effort to dislodge the fruit.

Figure E22-3 *The burdock fruit hitches a ride on animal fur. This mouse is trying to rid itself of its troublesome passengers.*

Unlike these hitchhiker fruits, edible fruits benefit both animal and plant. The plant stores sugars and tasty flavors in a fleshy fruit surrounding the seeds, luring hungry animals into eating the fruit (Fig. E22-4). Some fruits, such as peaches and plums, contain large, hard seeds that animals usually do not eat. After eating the flesh of the fruit, the animals discard the seeds. Other fruits, including blackberries, raspberries, strawberries, and tomatoes, have small seeds that are swallowed along with the fruit flesh. The seeds then pass through the animal's digestive tract without harm. In some cases, passing through an animal's gut may even be essential to seed germination, by abrading or digesting away part of the seed coat. Besides transport away from its parent, a seed that is swallowed and excreted benefits in another way: it ends up with its own supply of fertilizer!

Figure E22-4 *These bright red fruits have attracted a cedar waxwing. Only ripe fruits with mature seeds inside have bright coloration, sweet tastes, and exotic flavors. Unripe fruits are usually green, hard, and bitter, which makes them unpalatable to birds and mammals. This too is an evolutionary adaptation. The immature seeds within unripe fruit may not survive passage through an animal's gut; if the fruits ripen before the seeds, the plant is doomed to reproductive failure.*

enclosed in a fruit and hanging from the tree or vine, it may exhaust its food reserves before it ever touches the ground. When the fruit finally falls, the seedling would then lack the energy to burrow its roots into the soil, and would die. Second, environmental conditions of temperature and precipitation suitable for seedling growth may not coincide with seed maturation. Seeds that mature in late summer in temperate climates, for example, face the harsh winter to come. Spending the winter as a dormant seed is clearly preferable to death by freezing as a tender young sprout.

Plants have evolved many different mechanisms for seed dormancy, and the seeds of each species have their own set of requirements that must be met before germination can occur. Perhaps the three most common requirements are *an initial desiccation, exposure to cold,* and *disruption of the seed coat.* By requiring a period of dryness, a seed prevents germination while still within the fruit. Such seeds are often dispersed by animals that eat fruit but cannot digest the seeds. Seeds of many temperate and arctic plants will not germinate unless exposed to prolonged subfreezing temperatures, followed by warmth and moisture. This ensures that they stay dormant during the balmy days of autumn, and sprout only after winter yields to spring. Finally, the seed coat itself is often a barrier to seed germination. It may be impermeable to water and oxygen, it may bind the developing embryo so tightly that growth simply cannot occur, or it may contain chemicals that inhibit germination (see Chapter 23). In deserts, for example, rainfall is spotty and scarce. Years may go by without enough water for plants to germinate, grow, flower, and set more seed. Therefore, the seed must not sprout unless a given rainfall is heavy enough to allow the plant to complete its life cycle, since another rain may not fall in time. The seeds of most desert plants have water-soluble chemicals in the seed coat that inhibit germination. Only a hard rainfall can wash away enough of the inhibitors to allow sprouting.

GERMINATION AND GROWTH OF SEEDLINGS

When a seed germinates, it absorbs water, swells, and bursts its seed coat. The root is usually the first structure to emerge from the seed coat, growing rapidly and absorbing water and minerals from the soil. Much of the water is transported to cells in the shoot. As its cells elongate, the stem lengthens, pushing up through the soil.

The growing shoot faces a serious difficulty: it must push through the soil without scraping away the apical meristem and tender leaflets at its tip. A root, of course, must always contend with tip abrasion, and its apical meristem is protected by a root cap (see Fig. 21-6). Shoots spend most of their time in the air and do not develop permanent protective caps. Instead, germinating shoots have other mechanisms that cope with the abrasion of sprouting (Fig. 22-20). In monocots, a tough sheath, the

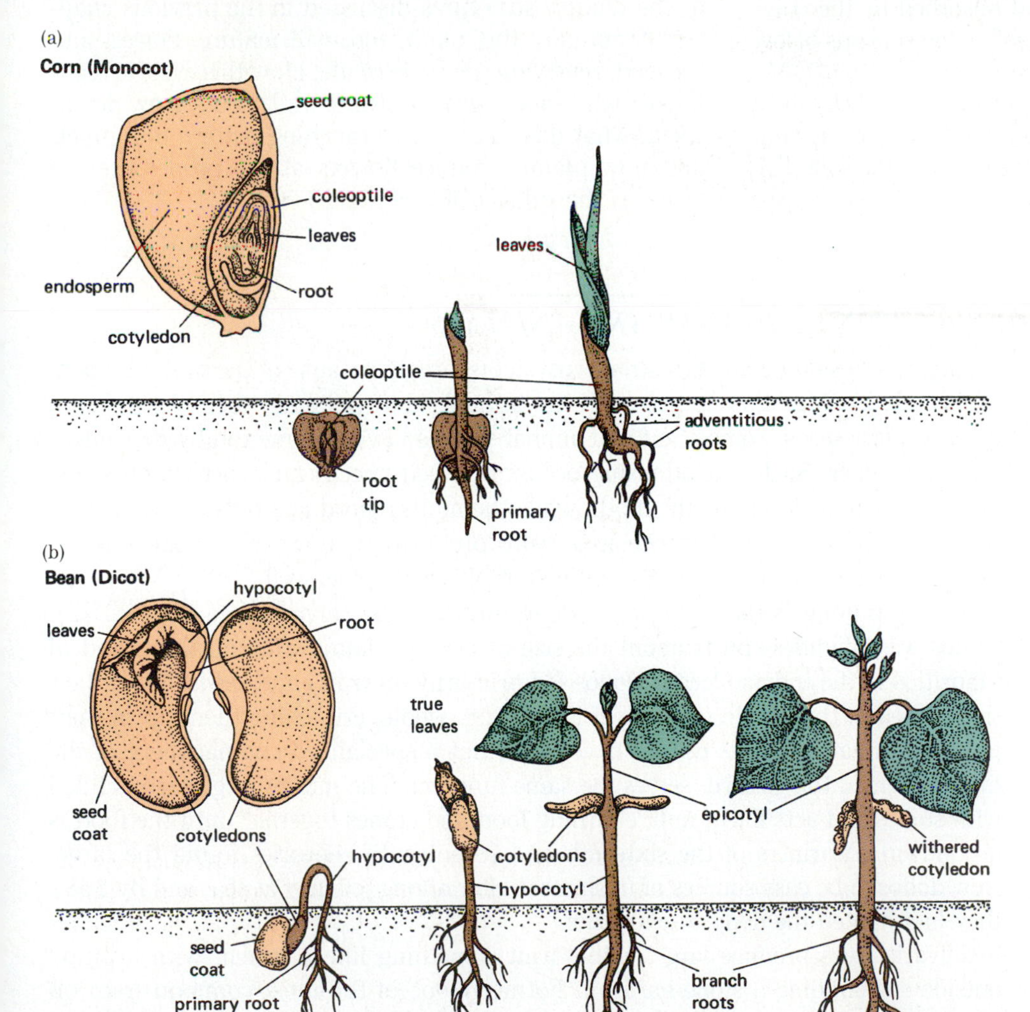

Figure 22-20 Seed germination in corn, a monocot **(a)**, *and the common bean, a dicot* **(b)**, *differ in detail, but involve the same three principles: (1) use of stored food to provide energy for seedling growth until photosynthesis can take over; (2) rapid growth and branching of the root to absorb water and nutrients; and (3) protection of the delicate shoot tip as it moves upward through the soil.*

coleoptile, encloses the shoot tip as a glove encloses a gardener's fingers (Fig. 22-20a). The coleoptile "glove" pushes aside the soil particles as it grows. Once out into the air, the coleoptile tip degenerates, allowing the tender "finger" of shoot to emerge. Dicots do not have coleoptiles. Instead, the dicot shoot forms a hook in either the hypocotyl or epicotyl (Fig. 22-20b). The bend of the hook, encased in epidermal cells with tough cell walls, leads the way through the soil, clearing the way for the downward-pointing apical meristem with its delicate new leaves.

Food stored in the seed provides the energy for sprouting. You will recall that the cotyledons of dicots have already absorbed the endosperm while the seed was developing, and are now fat and full of food. In dicots with hypocotyl hooks, the elongating shoot carries the cotyledons out of the soil into the air. These aboveground cotyledons often become green and photosynthetic (Fig. 22-21) and transfer both previously stored food and newly synthesized sugars to the shoot. In dicots with epicotyl hooks, the cotyledons stay below ground, shriveling up as the embryo absorbs their stored food. Monocots retain most of their food reserves in the endosperm until germination, when it is digested and absorbed by the cotyledon as the embryo grows. The cotyledon remains below ground in the remnants of the seed.

Figure 22-21 *In the squash family, a hypocotyl hook carries the cotyledons out of the soil. The cotyledons expand into photosynthetic leaves (the pair of smooth oval leaves). The first true leaf (crinkled single leaf) develops a little later. Eventually, the cotyledons shrivel up and die.*

Once out in the air, the hook in the stem straightens out, and the shoot rapidly spreads its leaves to the sun. Simultaneously, the root system delves into the soil. The apical meristem cells of shoot and root divide, giving rise to the mature structures discussed in the previous chapter. Eventually this plant, too, will mature, flower, and set seed, renewing the cycle of life. How this cycle is regulated—why shoots grow upward while roots grow downward, what determines the branching pattern of a plant, and how plants produce flowers at the proper time of year—is the subject of Chapter 23.

REFLECTIONS ON THE COEVOLUTION OF PLANTS AND ANIMALS

Angiosperms dominate terrestrial ecosystems largely because of the mutually beneficial relationships they forged with the animals that pollinate their flowers and disperse their seeds. Some plant and animal species have become totally dependent on one another. Such a relationship occurs between yuccas and yucca moths: the yucca is pollinated only by the moth, while the moth reproduces only in the fruit of the yucca. Absolute dependence is a risky proposition, however; if one partner becomes extinct, it may mean the extinction of the other as well.

A case in point is the story of the dodo bird and the calvaria tree (Fig. 22-22). Dodos were flightless birds about the size of turkeys, found only on the island of Mauritius in the Indian Ocean. Dodos fed primarily on fruits and seeds. With their sizable beaks, they could gulp down large fruits whole, crunching them up in their powerful gizzards. (Birds do not have teeth, but a specialized compartment of the digestive tract, the gizzard, serves the same function. The muscular gizzard is filled with stones and acts like a mill, churning food and stones together until the food is ground up.) Mariners of the sixteenth and seventeenth centuries found the large, slow dodos to be easy sources of fresh meat after a long journey at sea, and by 1681 they had hunted the dodo to extinction.

Calvaria trees produce large, edible fruit something like a peach, with a pulpy outside surrounding a stone-hard pit. Formerly one of the most common trees on Mauritius, only a few ancient specimens survive today. Judging by the ages of these patriarchs, no calvaria tree has successfully reproduced in about 300 years, even though the remaining trees still set normal fruit each year. Ecologist Stanley Temple has found out why.

Temple believes that the calvaria tree is endangered because it depends exclusively on the dodo for seed dispersal and germination. Eons ago, ancestral dodos fed on the fruits of ancestral calvaria trees. Only fruits with strong pits had any chance of surviving the trial-by-gizzard within a dodo. As dodos became larger with ever-stronger gizzards, calvarias were selected for harder and harder pits. Supremely strong pit walls, however, also inhibited germination. In fact, the calvaria pit came to *require* processing through the dodo before it could germinate: the pit walls must be partially ground away before the seed can sprout. By eliminating the dodo, humans inadvertently doomed the calvaria tree as well. Now that we know the problem, a few calvaria trees can be propagated in nurseries, by filing away the pit wall, or, as Temple discovered, by force-feeding calvaria pits to domestic turkeys. However, the forests of calvarias that formerly clothed Mauritius, providing habitat for wildlife and lumber for people, are gone forever.

Figure 22-22 *The dodo bird and calvaria tree were formerly abundant denizens of the island of Mauritius in the Indian Ocean. Following the extinction of the dodo by European sailors, the calvaria tree also declined, because its large, hard seeds must be processed through the digestive tract of the dodo before they can germinate.*

SUMMARY OF KEY CONCEPTS

Reproduction and the Plant Life Cycle

The sexual life cycle of plants, called alternation of generations, includes both a multicellular diploid form (the sporophyte generation) and a multicellular haploid form (the gametophyte generation). In the seed plants, the gametophyte stage is greatly reduced. The male gametophyte is the pollen grain, a drought-resistant structure that can be carried from plant to plant by wind or animals. The female gametophyte is also reduced, and is retained within the body of the sporophyte stage. In this way, seed plants can reproduce independently of liquid water.

The Evolution of Flowers

Flowering plants evolved from gymnosperms. In gymnosperms, pollen blows on the wind from male cones to female cones. Flowering plants enjoy a selective advantage over gymnosperms because the flowers attract insects that carry pollen from plant to plant.

Flower Structure

A complete flower consists of four parts: sepals, petals, stamens (male reproductive structures), and carpels (female reproductive structures). The sepals form the outer covering of the flower bud. The petals (and sometimes the sepals) are usually brightly colored and attract pollinators to the flower. The stamen consists of a filament that bears at its tip an anther in which pollen (the male gametophyte) develops. The carpel consists of the ovary in which one or more embryo sacs (the female gametophytes) develop, and a style that bears at its end a sticky stigma to which pollen adheres during pollination. Incomplete flowers lack one or more of the four floral parts.

Coevolution of Flowers and Pollinators

Most flowers are pollinated by the wind or by animals, usually insects or birds. Flowers show specific adaptations to pollination by particular pollinators, due to the extensive coevolution between flowering plants and their pollinators.

Gametophyte Development in Flowering Plants

Pollen develops in the anthers. A diploid cell, the microspore mother cell, undergoes meiosis to produce four haploid microspores. Each of these divides mitotically to form pollen grains. An immature pollen grain consists of two cells: the tube cell and the generative cell. The generative cell divides once to produce two sperm cells.

The embryo sac develops within the ovules of the ovary. A diploid megaspore mother cell undergoes meiosis to form four haploid megaspores. Three of these degenerate; the fourth undergoes three sets of mitotic divisions to produce the eight nuclei of the embryo sac. These eight nuclei come to reside in only seven cells. One of these cells, with a single nucleus, is the egg cell; another, with two nuclei, is the primary endosperm cell. These two cells are involved in seed formation; the rest of the cells degenerate.

Pollination and Fertilization

Pollination is the transfer of pollen from anther to stigma. When a pollen grain lands on a stigma, its tube cell tunnels through the style down to the embryo sac. The generative cell divides to form two sperm cells that travel down the style within the tube cell, eventually entering the embryo sac. One sperm fuses with the egg to form a diploid zygote, which will give rise to the embryo. The other sperm fuses with the binucleate primary endosperm cell to produce a triploid cell. This cell will give rise to the endosperm, a food storage organ within the seed.

The Development of Seeds and Fruits

The embryo develops a root, shoot, and cotyledons. Cotyledons digest and absorb food from the endosperm, and transfer it to the growing embryo. Monocot embryos have one cotyledon, while dicot embryos have two. The seed is enclosed within a fruit that develops from the ovary wall. The function of the fruit is to disperse the seeds away from the parent plant.

Seeds often remain dormant for some time after fruit ripening. Environmental conditions involved in breaking dormancy may include an initial desiccation, exposure to cold, or exposure to water.

Germination and Growth of Seedlings

Seed germination requires warmth and moisture. Energy for germination comes from food stored in the endosperm, transferred to the embryo by the cotyledons.

GLOSSARY

Anther (an′-ther) the uppermost part of the stamen in which pollen develops.

Carpel (kar′pel) the female reproductive structure of a flower, composed of stigma, style, and ovary.

Coleoptile (kō-lē-op′tīl) a protective sheath surrounding the shoot in monocot seeds.

Complete flower a flower that has all four floral parts (sepals, petals, stamens, and carpels).

Cotyledon (kot-ul-ē′don) also called a seed leaf; a leaflike structure within a seed that absorbs food from the endosperm and transfers it to the growing embryo.

Dormancy a state in which an organism does not grow or develop; usually marked by lowered metabolic activity and resistance to adverse environmental conditions.

Double fertilization in flowering plants, a phenomenon in which two sperm nuclei fuse with the nuclei of two cells of the female gametophyte. One sperm fuses with the egg to form the zygote, while the second sperm nucleus fuses with the two haploid nuclei of the primary endosperm cell to form a triploid endosperm cell.

Embryo sac the haploid female gametophyte of flowering plants.

Endosperm a triploid food storage organ found in the seeds of flowering plants.

Epicotyl (ep′-ē-kot-ul) the part of the embryonic shoot located between the tip of the shoot and the attachment point of the cotyledons.

Flower the reproductive structure of an angiosperm.

Fruit in angiosperms, the ripened ovary (plus, in some cases, other parts of the flower).

Gametophyte (ga-mēt′-ō-fīt) the multicellular haploid stage in the life cycle of plants.

Generative cell in flowering plants, one of the haploid cells of a pollen grain. The generative cell undergoes mitosis to form two sperm cells.

Germination the growth and development of a seed, spore, or pollen grain.

Hypocotyl (hī′-pō-kot-ul) the part of the embryonic shoot located between the attachment point of the cotyledons and the root.

Incomplete flower a flower that is missing one of the four floral parts (sepals, petals, stamens, or carpels).

Integument (in-teg′-ū-ment) the layers of the ovule immediately surrounding the embryo sac; develops into the seed coat.

Megaspore a haploid cell formed by meiosis from a diploid megaspore mother cell. Through mitosis and differentiation, the megaspore develops into the female gametophyte.

Megaspore mother cell a diploid cell contained within the ovule of a flowering plant, which undergoes meiosis to produce four haploid megaspores.

Microspore a haploid cell formed by meiosis from a microspore mother cell. Through mitosis and differentiation, the microspore develops into the male gametophyte.

Microspore mother cell a diploid cell contained within an anther of a flowering plant, which undergoes meiosis to produce four microspores.

Ovary in flowering plants, a structure at the base of the carpel containing one or more ovules; develops into the fruit.

Ovule a structure within the ovary of a flower, inside which the female gametophyte develops. After fertilization, the ovule develops into the seed.

Petal part of a flower, often flat and brightly colored, serving to attract potential animal pollinators.

Pollen the male gametophyte of a gymnosperm or flowering plant.

Pollination in angiosperms, the deposition of pollen onto the stigma of a flower of the same species; in gymnosperms, the deposition of pollen within the pollen chamber of a female cone of the same species.

Seed the reproductive stage of gymnosperms and angiosperms, usually including an embryonic plant and a food reserve, enclosed within a resistant outer covering.

Seed coat the outermost covering of a seed, formed from the integuments of the ovule.

Sepal (sē′-pul) one of the protective outer coverings of a flower bud, often opening into green, leaflike structures when the flower blooms.

Sporophyte (spor′-ō-fīt) the multicellular diploid stage of the life cycle of plants.

Stamen (stā′-men) the male reproductive structure of a flower, consisting of a filament and an anther in which pollen grains develop.

Stigma (stig′-ma) the pollen-capturing tip of a carpel.

Style a stalk connecting the stigma of a carpel with the ovary at its base.

Tube cell the outermost cell of a pollen grain; the tube cell digests a tube through the tissues of the carpel, ultimately penetrating into the female gametophyte.

STUDY QUESTIONS

1. Diagram the plant life cycle, comparing ferns with flowering plants. Which stages are haploid and which are diploid? At which stage are gametes formed?
2. What are the advantages of the reduced gametophyte stages in flowering plants compared to the more substantial gametophytes of ferns?
3. Diagram a complete flower. Where are the male and female gametophytes formed? What are the male and female gametophytes called?
4. Diagram the development of the pollen grain. At what point does meiosis occur?
5. Diagram the development of the embryo sac. At what point does meiosis occur? How many cells and how many nuclei are in a mature embryo sac? Which cells and nuclei ultimately fuse with sperm?
6. Describe the characteristics you would expect to find in flowers that are pollinated by the wind, beetles, bees, and hummingbirds, respectively.
7. What is the endosperm? From which cell of the embryo sac is it derived? How much endosperm is usually present in a mature seed of a dicot? Of a monocot?
8. Describe the function of cotyledons.
9. Describe three mechanisms whereby seed dormancy is broken in different types of seeds. How do these mechanisms relate to the normal environment of the plant?
10. How do monocot and dicot seedlings protect the delicate shoot tip during seed germination?
11. Describe three types of fruits and the mechanisms whereby these fruit structures help to disperse their seeds.

SUGGESTED READINGS

Barth, F. G. *Insects and Flowers: The Biology of a Partnership.* Princeton, N.J.: Princeton University Press, 1985. A fascinating, well-illustrated book that describes how flowers ensure their pollination through their interactions with insects.

Jordan, W. "The Bee Complex." *Science 84,* May 1984. Although mostly about bumblebees, this article vividly portrays the evolutionary forces behind insect–flower relationships.

Newman, C. "Pollen: Breath of Life and Sneezes." *National Geographic,* October 1984. An interesting look at pollen from the perspectives of plants and allergic humans.

Raven, P. H., Evert, R. F., and Eichhorn, S. *Biology of Plants,* 4th ed. New York: Worth Publishers, Inc., 1986. Lavishly illustrated, with a particularly interesting section on the evolution of flowers and the coevolution between flowers and pollinating animals.

Weier, T. E., Stocking, C. R., Barbour, M. G., and Rost, T. L. *Botany,* 6th ed. New York: John Wiley & Sons, Inc., 1982. Complete coverage of the variations found in flowers and fruits, with excellent photographs and diagrams.

23
Control of the Plant Life Cycle

The time is late fall, the place a woodlot in Ohio. Masses of acorns burden the oaks, and gray squirrels busily gather and bury the nuts, storing them against the coming winter. With so many acorns "squirreled away," some are forgotten or are not needed. They sprout next spring as the sun warms the forest soil.

The squirrels, of course, haven't planted the acorns so that new oaks will grow. Each acorn is oriented randomly, perhaps with the future stem end down and root end up. Buried beneath several inches of soil, how does the germinating seedling "know" which way is up? If it does sprout and send its shoot up into the air, how does it develop into a mature oak? How does it distinguish the seasons, flowering in spring and going dormant in fall?

Plants do not have specialized organs for perceiving the environment, comparable to our eyes or ears. Nevertheless, plants *do* perceive many features of their world: the direction of gravity; the direction, intensity, and duration of sunlight; the strength of the wind; and in some cases, even the touch of a fly upon a leaf. Plants also respond to these stimuli by regulating their growth and development in appropriate ways. Usually, plants control their bodily functions through the action of simple chemicals called hormones or growth regulators. In this chapter we explore how plants detect external stimuli, produce and distribute hormones in response, and how these hormones govern the development of the plant body.

THE DISCOVERY OF PLANT HORMONES

Everyone who keeps houseplants on a windowsill knows that they bend toward the window as they grow, in response to the sunlight streaming in. Over a hundred years ago Charles Darwin and his son Francis studied this phenomenon of growth toward the light, or **phototropism.** The Darwins illuminated grass coleoptiles from various angles (a coleoptile is the protective sheath surrounding a monocot seedling; see Fig. 22-20). They noted that a region of the coleoptile a few millimeters below the tip bent toward the light until the tip faced directly into the light source (Fig. 23-1). If they covered the very tip of the coleoptile with a lightproof cap, the coleoptile did not bend. Covering the bending region, on the other hand, did not keep it from bending. The Darwins concluded that (1) the tip of the coleoptile perceives the direction of light and (2) bending occurs farther down the coleoptile; therefore, (3) the tip must transmit information about the light direction down to the bending region.

How does the coleoptile bend? Although the Darwins didn't know this, growth in the bending region is entirely due to *elongation of preexisting cells.* Therefore, the coleoptile must bend due to *differential elongation of cells* (Fig. 23-2). The cells on the outside of the bend (away from the light) elongate more than the cells on the inside of the bend (toward the light). If one side of the coleoptile becomes longer than the opposite side, the whole shaft must bend away from the longer side. Thus we can reinterpret the Darwins' results and say that *the information transmitted from the tip to the bending region causes greater elongation of cells on the side of the coleoptile shaft away from the light.*

About 30 years later, Peter Boysen-Jensen showed that the information transferred down from the tip is chemical in nature (Fig. 23-3). He cut the tips off coleoptiles and found that the remaining stump neither elongated nor bent toward the light. If he replaced the tip and put the patched-together coleoptile in the dark, it elongated straight up. In the light, it showed normal phototropism. Placing a thin layer of porous gelatin between the severed tip and the stump still allowed elongation and bending, but an impervious barrier of mica eliminated these responses. Boysen-Jensen concluded that a chemical is produced in the tip and moves down the shaft, causing cell

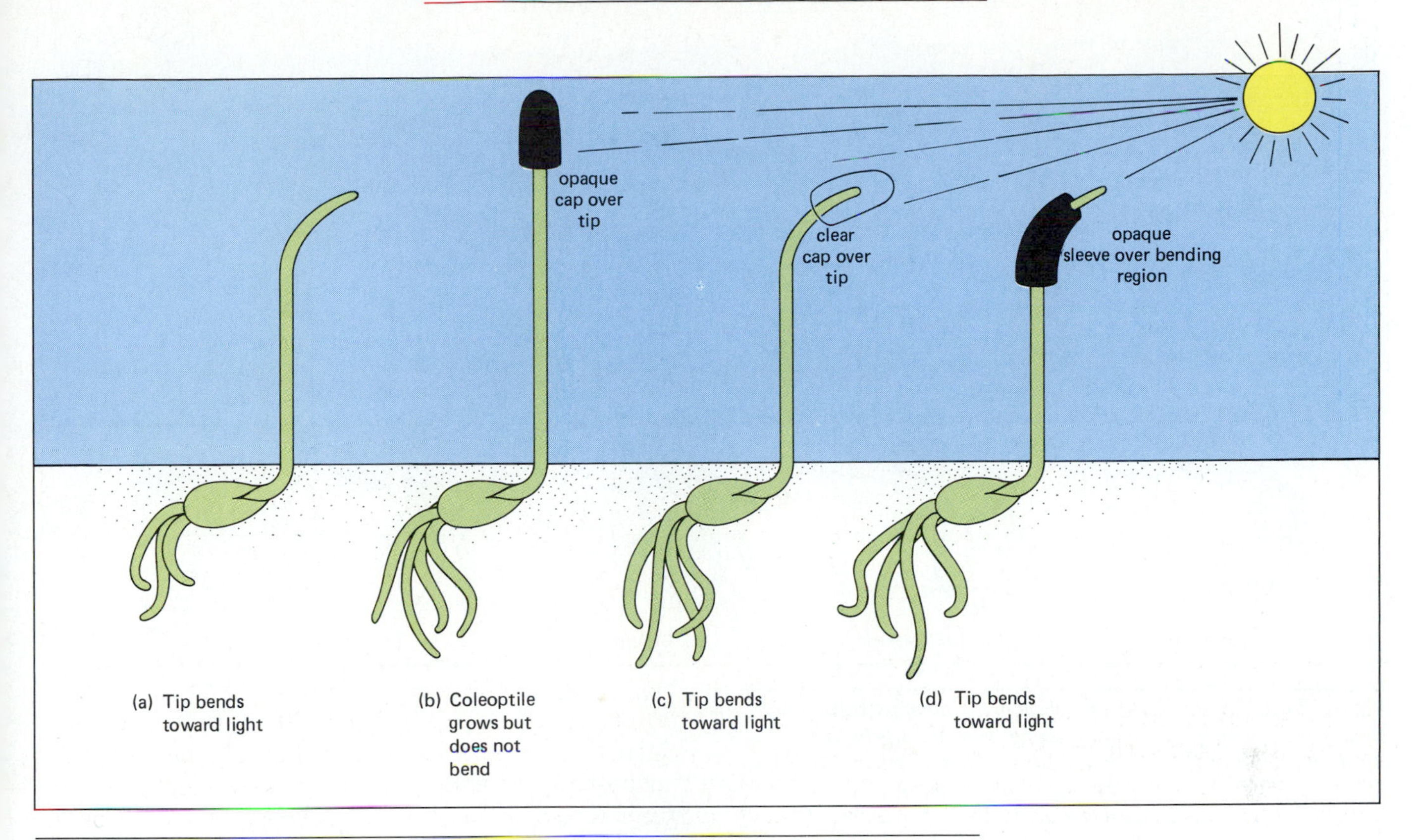

Figure 23-1 The phototropism experiments of Charles and Francis Darwin. **(a)** *If light strikes a coleoptile from the side, a region a few millimeters down from the tip bends as it grows, until the tip points toward the light source.* **(b)** *If the tip is covered with a lightproof cap, the coleoptile does not bend.* **(c)** *If a clear cap is placed over the tip, bending still occurs, proving that the mere presence of a cap does not prevent bending.* **(d)** *If a flexible light-proof collar is placed around the bending region, the coleoptile still bends toward the light.*

elongation. In the dark, the elongation-producing chemical diffuses straight down from the tip and causes the coleoptile to elongate straight up. Presumably, light causes the chemical to become more concentrated on the "shady" side of the shaft, so that cells on the shady side elongate faster than cells on the "sunny" side, causing the shaft to bend toward the light.

The next step was to isolate and identify the chemical. In the 1920s, Frits Went devised a way to collect the elongation-promoting chemical. He cut off the tips of oat coleoptiles and placed them on a block of agar (a porous, gelatinous material) for a few hours (Fig. 23-4a). Went hoped that the chemical would migrate out of the coleoptiles into the agar. Removing the tips, he cut up the agar, now presumably loaded with the chemical, and placed small pieces on the tops of coleoptile stumps. If he put a piece of agar squarely atop a stump, the stump elongated straight up (Fig. 23-4b). If he placed a piece on one side of a cut stump, the stump would invariably bend away from the side with the agar (Fig. 23-4c). Went called the elongation-promoting chemical **auxin,** from a Greek word meaning "to increase." Kenneth Thimann later purified auxin and determined its molecular structure.

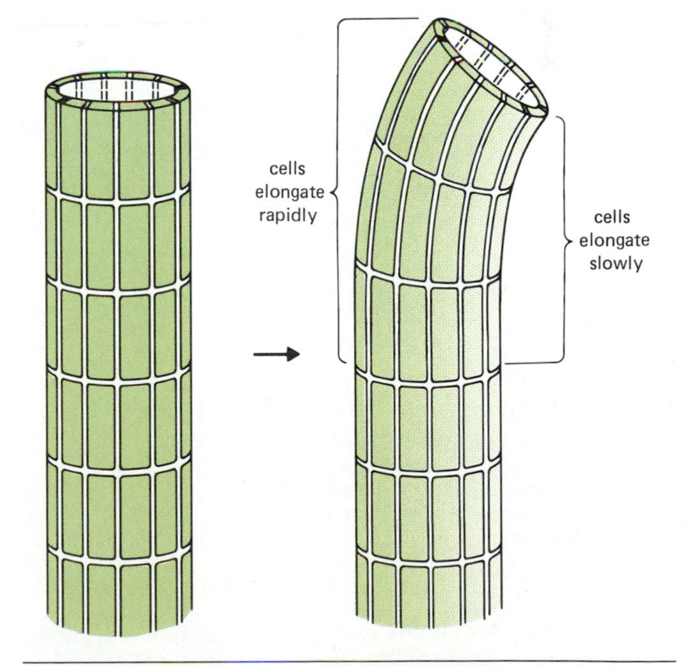

Figure 23-2 Coleoptiles bend by differential elongation of cells on opposite sides of the coleoptile. If cells on one side elongate more rapidly than cells on the other side, the coleoptile bends away from the longer side.

(a) Tip removed
(b) Tip replaced
(c) Gelatin
(d) Mica

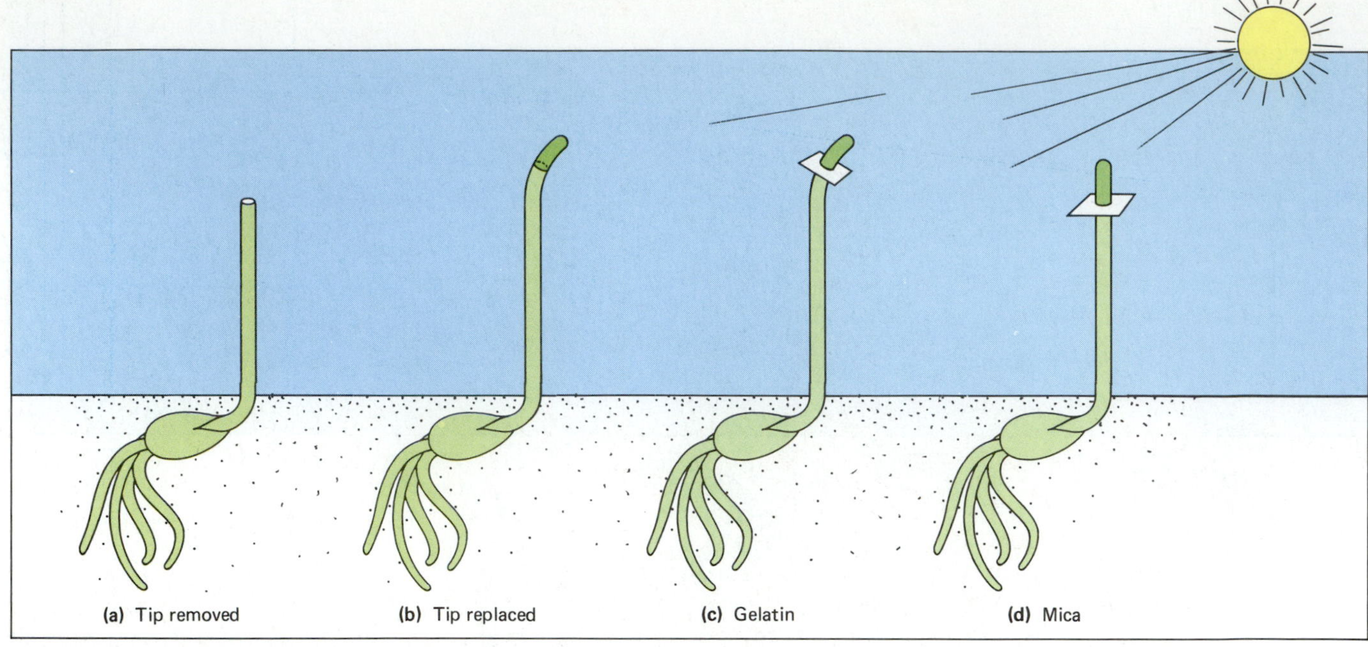

Figure 23-3 Boysen-Jensen's experiments demonstrating the chemical nature of the signal passing from the tip to the bending region of the coleoptile during phototropism. **(a)** *If the tip is cut off a coleoptile, the remaining stump neither elongates nor bends toward the light.* **(b)** *Replacing the tip restores both elongation and phototropism.* **(c)** *A gelatin "filter" between the tip and stump does not prevent phototropism, but* **(d)** *An impenetrable slice of mica eliminates the response.*

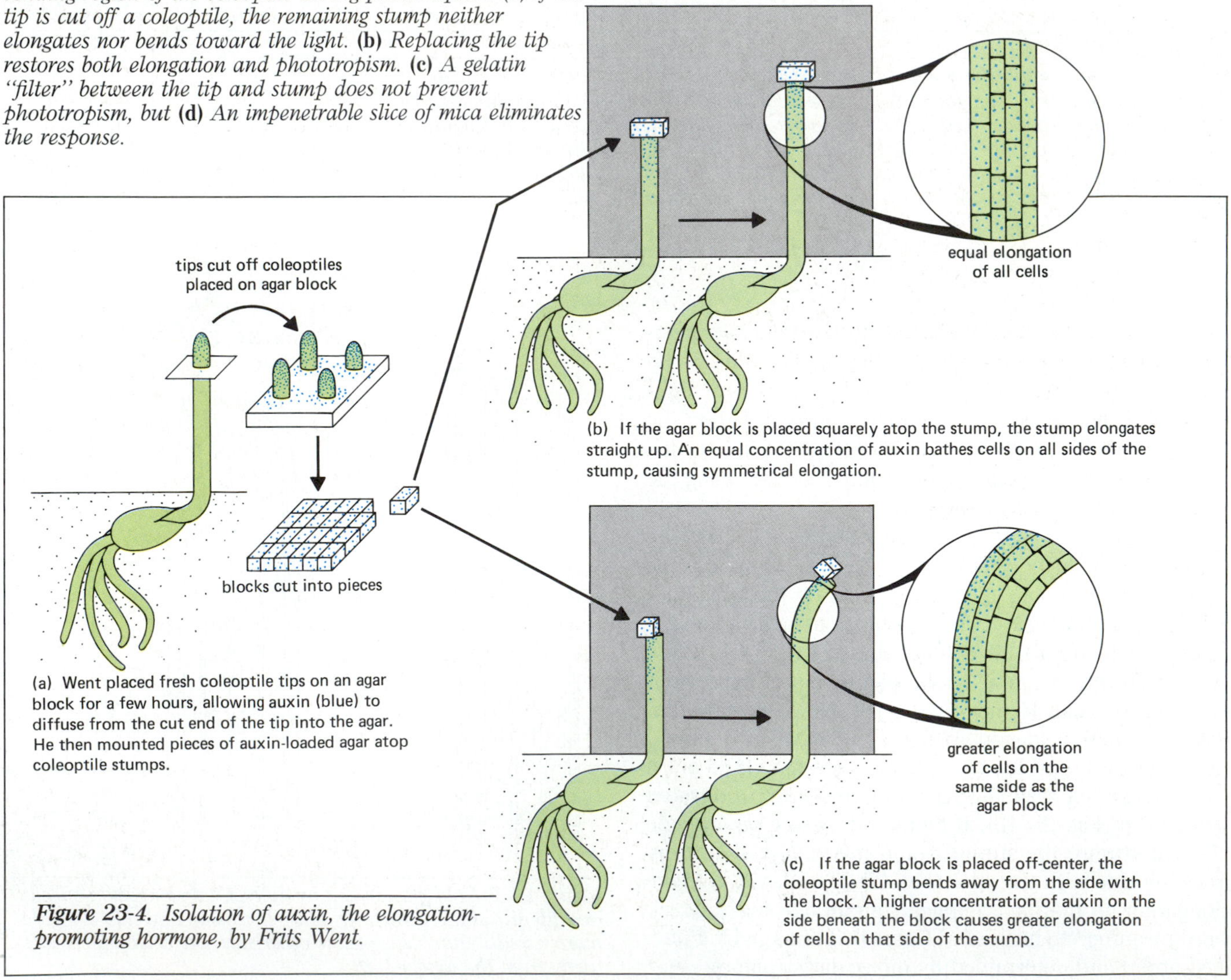

tips cut off coleoptiles placed on agar block
blocks cut into pieces
(a) Went placed fresh coleoptile tips on an agar block for a few hours, allowing auxin (blue) to diffuse from the cut end of the tip into the agar. He then mounted pieces of auxin-loaded agar atop coleoptile stumps.
equal elongation of all cells
(b) If the agar block is placed squarely atop the stump, the stump elongates straight up. An equal concentration of auxin bathes cells on all sides of the stump, causing symmetrical elongation.
greater elongation of cells on the same side as the agar block
(c) If the agar block is placed off-center, the coleoptile stump bends away from the side with the block. A higher concentration of auxin on the side beneath the block causes greater elongation of cells on that side of the stump.

Figure 23-4. Isolation of auxin, the elongation-promoting hormone, by Frits Went.

PLANT HORMONES AND THEIR ACTIONS

Animal physiologists have long recognized that chemicals called **hormones** are produced in one location and transported to other parts of the body, where they exert specific effects. By analogy, auxin and other plant-regulating chemicals are called **plant hormones.** So far, plant physiologists have identified five major classes of plant hormones: auxins, gibberellins, cytokinins, ethylene, and abscisic acid (Table 23-1). Several other types of hormones are suspected but have not yet been isolated and identified. Each hormone can elicit a variety of responses from plant cells, depending on the type of cell, its physiological state, and the presence of other hormones. In this chapter we outline the major actions of each hormone and then describe a year in the life of a plant, illustrating how hormones regulate its growth and development.

As we have seen, **auxin** promotes elongation of cells in coleoptiles and other parts of the shoot. In roots, low concentrations of auxin stimulate elongation, while slightly higher concentrations inhibit elongation. Both light and gravity affect the distribution of auxin in roots and shoots, so that auxin plays a major role in both phototropism and gravitropism (directional growth with respect to gravity). Auxin affects many other aspects of plant development, too. It stimulates root branching, the differentiation of vascular tissues, and the development of fruits. Auxin also prevents sprouting by lateral buds.

Gibberellins are a group of chemically similar molecules that, like auxin, promote elongation of cells in stems. In some plants, gibberellins stimulate flowering, fruit development, seed germination, and bud sprouting.

Cytokinins promote cell division in many plant tissues; consequently, they stimulate fruit, endosperm, and embryonic development, and sprouting of buds. Cytokinins also stimulate plant metabolism, preventing or at least delaying aging of plant parts, especially leaves.

Ethylene is the only known hormone in plants or animals that is a gas at normal environmental temperatures. Ethylene is best known, and most commercially valuable, for its ability to cause fruit to ripen. It also stimulates the breakdown of cell walls in abscission layers, allowing leaves, flowers, and fruit to drop off at the appropriate time.

Abscisic acid is an inhibitory hormone that helps plants to withstand unfavorable environmental conditions. It causes stomata to close when water availability is low. It inhibits the activity of gibberellin, thus helping to maintain dormancy in buds and seeds during times when germination would be dangerous.

THE PLANT LIFE CYCLE: RECEPTION, RESPONSE, AND REGULATION

The life cycle of a plant results from a complex interplay between its genetic information and its environment. Hormones mediate many of the genetic determinants of plant growth and development, as well as nearly all responses to environmental factors. At each stage in its life cycle, a plant produces a distinctive set of hormones that interact with one another in directing the growth of the plant body.

Seed Dormancy and Germination

As we pointed out in Chapter 22, a seed maturing within a juicy fruit on a warm autumn day has ideal conditions for germination, yet remains dormant until the following

Table 23-1 **Hormone Actions in Plants**

Hormone	Functions
Abscisic acid	Closing of stomata; seed dormancy; bud dormancy
Auxin	Elongation of cells in coleoptiles and shoots; phototropism; gravitropism in shoots and roots; root branching; apical dominance; development of vascular tissue; fruit development; retarding senescence in leaves and fruit; ethylene production in fruit
Cytokinin	Promotion of sprouting of lateral buds; prevention of leaf senescence; promotion of cell division; stimulation of fruit, endosperm, and embryo development
Ethylene	Ripening of fruit; abscission of fruits, flowers, and leaves; inhibition of stem elongation; formation of hook in dicot seedlings
Gibberellin	Germination of seeds and sprouting of buds; elongation of stems; stimulation of flowering; development of fruit

spring. In many seeds, *abscisic acid enforces dormancy.* Abscisic acid slows down the metabolism of the embryo within the seed, preventing its growth. The seeds of some desert plants contain high concentrations of abscisic acid; only a really hard rain can wash out the abscisic acid, freeing the embryo from inhibition and allowing the seed to germinate. Seeds of northern plants usually require a prolonged period of cold to break dormancy; in these seeds chilling induces the destruction of abscisic acid.

Germination is stimulated by other hormones, especially gibberellin. The same environmental conditions that cause abscisic acid breakdown also promote gibberellin synthesis in the embryo. Gibberellin causes transcription of genes that code for the enzymes that digest the food reserves of the endosperm and cotyledons, making sugars, lipids, and amino acids available to the growing embryo.

Growth of the Seedling

When the growing embryo breaks out of the seed coat, it immediately faces a crucial problem: in the eternal darkness of the soil, which way is up? The roots must burrow downward, while the shoot must grow upward to emerge into the light. Auxin apparently controls the responses of both roots and shoots to light and gravity.

SHOOT GROWTH. Let's begin by looking at the growth of a shoot as it first emerges from the seed, buried underground. As we described earlier, auxin is synthesized in shoot tips, moves down the shaft of the stem, and stimulates cell elongation. If the stem is not exactly vertical, organelles in the cells of the stem detect the direction of gravity and somehow cause auxin to accumulate on the stem's lower side (Fig. 23-5). Therefore, the lower cells elongate rapidly, forcing the stem to bend upward. When the shoot tip is vertical, the auxin distribution becomes symmetrical. Now the stem grows straight up, emerging from the soil into the light.

Auxin also mediates phototropism. Ordinarily, the auxin distribution caused by light is the same as the distribution caused by gravity, because the direction of brightest light (the sun) is directly opposite that of gravity. For example, if a young shoot still buried underground is close enough to the surface so that some light penetrates down to it, both light and gravity cause auxin to be transported to the lower side of the shoot and promote upward bending. Thus, under normal conditions, gravitropism and phototropism augment each other.

ROOT GROWTH. Gravitropism in roots is less well understood. According to a recently proposed model, auxin controls the direction of root growth (Fig. 23-6). Auxin is transported from the shoot down to the root. If the root is not vertical, the root cap senses the direction of gravity and causes the auxin to be concentrated on the lower side of the root, where it *slows down cell elongation.* The upper side of the root, with less auxin, elongates more rapidly, causing the root to bend downward. When the root tip points directly downward, the auxin distribution becomes equal on all sides, and the root continues to grow straight down. Note that auxin *slows down* but does not *eliminate* root cell elongation: a vertical root continues to grow.

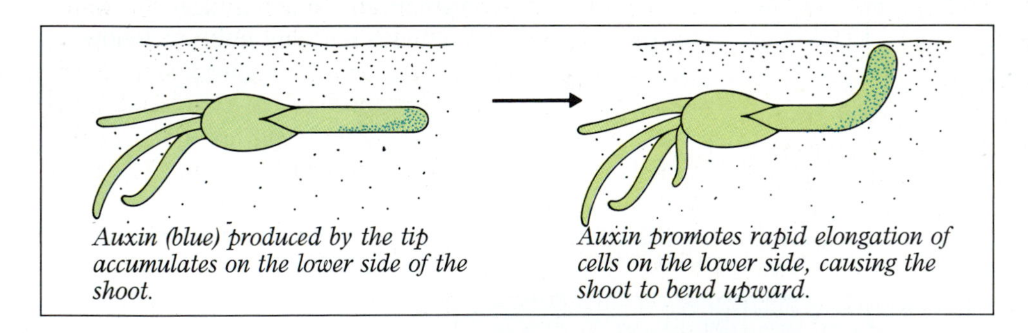

Figure 23-5 *The mechanism of gravitropism in shoots.*

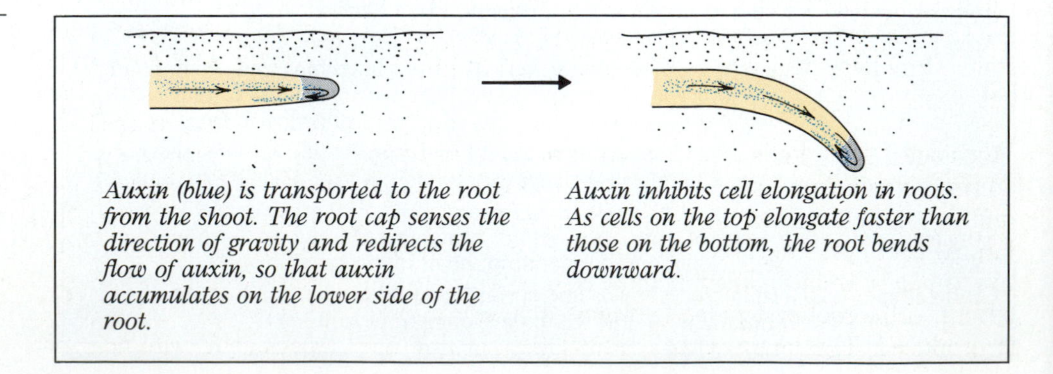

Figure 23-6 *The mechanism of gravitropism in roots.*

Development of the Mature Plant Form

As a plant grows, both its root and shoot develop branching patterns that are largely determined by its genetic heritage. For example, the stems of some plants, such as sunflowers, hardly branch at all; others, such as oaks and cottonwoods, branch profusely in seeming confusion; still others branch in a very regular pattern, producing the conical shapes of firs and spruces.

The amount of growth in shoot and root systems must also be kept in balance. The shoot must be large enough to supply the roots with sugars, while the roots must be large enough to provide the shoot with water and minerals. Interactions between auxin and cytokinin regulate root and stem branching, thereby regulating the relative sizes of root and shoot systems.

STEM BRANCHING. Gardeners know that pinching back the tip of a growing plant makes it become bushier. The botanical explanation for this practice is that the growing tip suppresses the sprouting of lateral buds, a phenomenon known as **apical dominance.** Although it is not known for sure how lateral bud sprouting is controlled, there is some evidence that the proper levels of auxin and cytokinin must be present. Auxin is produced by the shoot tip and transported down the stem. Cytokinin is produced by the roots and is transported up the stem. Therefore, the relative concentrations of these two hormones will vary, depending on where a bud is located on the stem.

Auxin alone appears to inhibit the sprouting of lateral buds, while auxin and cytokinin together stimulate bud sprouting. The lateral buds closest to the shoot tip receive a great deal of auxin, probably enough to inhibit their growth, but receive very little cytokinin because they are so far from the roots. Therefore, they remain dormant. Lower buds receive less auxin while receiving much more cytokinin. They are stimulated by optimal concentrations of both hormones, so they sprout. In many plants, this interaction between auxin and cytokinin produces an orderly progression of bud sprouting, from the bottom to the top of the shoot.

ROOT BRANCHING. Auxin stimulates root branching, even in extremely low concentrations. As we described in Chapter 21, lateral roots arise from the pericycle layer of the vascular cylinder. Auxin, transported down from the stem, stimulates pericycle cells to divide and form a lateral root.

BALANCE BETWEEN ROOT AND SHOOT SYSTEMS. Through the interaction of auxin and cytokinin, the root and shoot systems regulate each other's growth. An enlarging root system synthesizes large amounts of cytokinin, which stimulates lateral buds to break dormancy and sprout. If the root system isn't keeping up, less cytokinin is produced. The lateral buds sprout later, slowing the growth of the shoot system. Simultaneously, as the stem grows and branches, it produces lots of auxin, which stimulates root branching and growth. Neither system can get too far ahead or behind, and the plant is adequately supplied with all its needs.

Differentiation of Vascular Tissues

As a plant grows, its parts must be interconnected by the vascular tissues of xylem and phloem. Differentiation of cells into vascular tissues appears to be yet another function of auxin and perhaps gibberellin as well. Apical meristems release auxin and gibberellin into cells that are maturing just behind them. High hormone levels stimulate these cells to differentiate into xylem and phloem. As leaves grow and lateral buds sprout, they too release auxin and gibberellin. These hormones cause cortex cells just beneath the leaves and buds to differentiate into strands of xylem and phloem that connect up with the main vascular systems of the stem.

Control of Flowering

Ultimately, the plant matures enough to reproduce. The timing of flowering and seed production are finely tuned to the physiology of the plant and the rigors of its environment. In temperate climates, plants must flower early enough so that their seeds can mature before the killing frosts of autumn. Depending on how quickly the seed and fruit develop, flowering may occur in spring, as it does in oaks, in summer, as in lettuce, or even in autumn, as in asters.

What environmental cues do plants use to determine the season? Most cues, such as temperature or water availability, are quite variable: October can be warm, a late snow may fall in May, or the summer might be unusually cool and wet. *The only reliable cue is daylength:* longer days always mean that spring and summer are coming, while shorter days foretell the onset of autumn and winter.

With respect to flowering, plants are classified as day-neutral, long-day, or short-day plants (Fig. 23-7). A **day-neutral plant** is one that flowers as soon as it has grown and developed enough, regardless of the length of the day. Day-neutral plants include tomatoes, corn, and snapdragons. A **long-day plant** flowers when the day length is *greater than some critical value,* while a **short-day plant** flowers when the daylength is *less than some critical value.* Thus spinach is classified as a long-day plant, because it flowers only if the day is *longer than* 13 hours, and cockleburs are short-day plants because they flower only if the day is *less than* 15.5 hours long. Note that both will flower with 14 hours of light and 10 hours of darkness. The spinach, however, will also flower in much longer daylengths (e.g., 16 hours of light), whereas the cocklebur will not, and conversely the cocklebur will

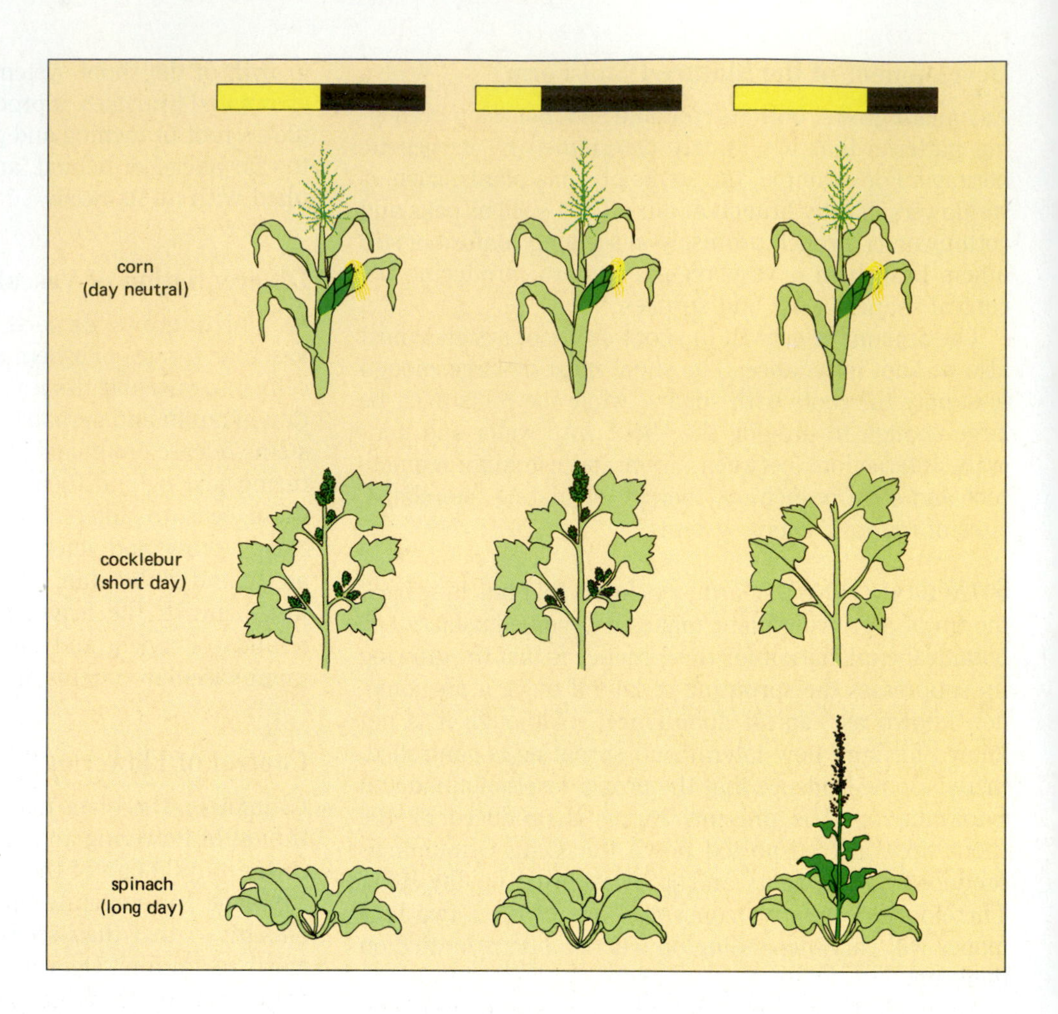

Figure 23-7 The effects of daylength on flowering in plants. Yellow bars indicate day and black bars indicate darkness.

flower in much shorter daylengths (e.g., 12 hours), whereas the spinach will not.

Mechanisms for measuring daylength. To measure daylength, a plant needs a *clock* to measure time (how long has it been light or dark) and a *light-detecting system* to set the clock. Virtually all organisms have an internal **biological clock** that measures time even without environmental cues. For example, the sorrel raises its leaves during the day and lowers them again in the evening (Fig. 23-8). If a sorrel is brought indoors and kept in total darkness, its leaves still rise and fall on a daily cycle. Activities that recur approximately every 24 hours in the absence of external cues are called **circadian rhythms.**

The light-detecting system of plants is a pigment in the leaves called **phytochrome** (meaning simply "plant color"). Phytochrome occurs in two interchangeable forms (Fig. 23-9). One form strongly absorbs red light, and is called P_r; the other form absorbs far-red light (almost infrared) and is accordingly called P_{fr}. In most plants, P_{fr} is the active form of phytochrome; that is, a suitable concentration of P_{fr} stimulates or inhibits physiological processes, such as flowering or setting the biological clock. P_r exerts no influence on these processes.

Phytochrome flips back and forth from one form to the other when it absorbs light of the appropriate color: when P_r absorbs red light, it is converted into P_{fr}, and when P_{fr} absorbs far-red light, it is transformed back into P_r. Daylight consists of all wavelengths of visible light, including both red and far-red. Therefore, during the day a leaf contains both forms of phytochrome. In the dark, P_{fr} rather rapidly breaks down or reverts to P_r.

Daylength and the control of flowering. Plants seem to use the phytochrome system and their internal biological clocks to control flowering. Cockleburs, for example, flower under a lighting regime of 8 hours of light and 16 hours of darkness. However, interrupting the middle of the dark period with just a minute or two of light prevents flowering. Thus, although cockleburs are usually classified as short-day plants, what really matters is not how long the day is, but how long the continuous darkness lasts. The color of the light used for the night flash is also important. A midnight flash of red light inhibits flowering, but a far-red flash allows flowering. This, of course, implicates phytochrome in the control of flowering. Unfortunately, no one knows how the response of phytochrome to light determines whether or not a plant will flower. It seems likely that the biological

Figure 23-8 Under normal conditions, the sorrel *Oxalis* raises its leaves during the day (left) and lowers them at night (right). If the plant is placed in constant darkness, it continues to raise its leaves during the "expected" daytime and lower them during the "expected" night, proving that the timing of leaf movement is intrinsic to the plant.

clock measures the length of the night, and that light reception by phytochrome tells the clock when sunrise and sunset have occurred, but this is not certain.

OTHER PHYTOCHROME-MEDIATED PROCESSES. Phytochrome is involved in many plant responses. For example, P_{fr} inhibits elongation of seedlings, with profound and obviously adaptive results. Since P_{fr} breaks down or reverts to P_r in the dark, seedlings germinating in the darkness of the soil contain no P_{fr}, and consequently elongate very rapidly, emerging out from the soil. Seedlings growing beneath other plants will be exposed largely to far-red light, because the green chlorophyll of the leaves above them will absorb most of the red light but transmit the far-red. Far-red light converts P_{fr} to P_r, so shaded seedlings grow rapidly, which may bring them out of the shade. Once out in the sunlight, P_{fr} forms. P_{fr} slows down elongation, which prevents the seedlings from becoming too spindly.

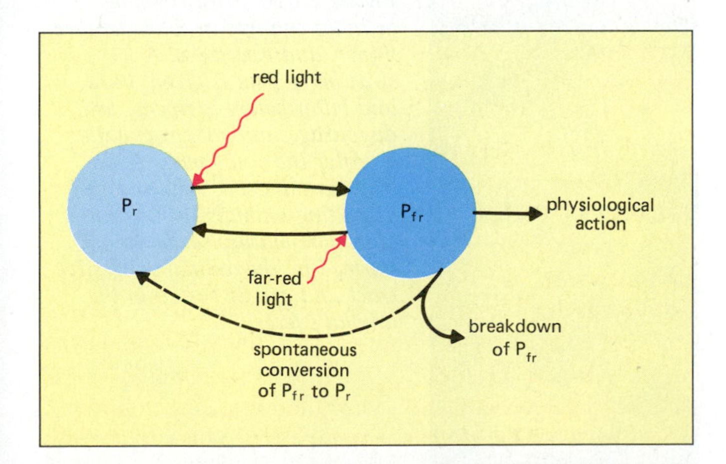

Figure 23-9 The light-sensitive pigment phytochrome exists in two forms, inactive (P_r) and active (P_{fr}). P_r is converted to P_{fr} by red light. P_{fr} may then participate in physiological responses, be converted to P_r by far-red light, revert spontaneously to P_r, or break down to other, inactive compounds.

Other plant responses governed by phytochrome include straightening the epicotyl or hypocotyl hook of dicot seedlings, leaf growth, and chlorophyll synthesis, all of which are stimulated by P_{fr}. As with stem elongation, these responses are adaptations related to burial in the soil or shading by the leaves of other plants. For example, a newly germinating shoot needs to stay in its protective bend while still in the soil (i.e., in the dark), and only straighten out in the open air, where sunlight converts P_r to P_{fr}.

Development of Seeds and Fruit

When a flower is pollinated, auxin or gibberellin released by the pollen stimulates the ovary to begin developing into a fruit. If fertilization also occurs, the developing seeds release still more auxin and/or gibberellin into the surrounding ovary tissues. Cells of the ovary multiply and grow larger, often storing starches and other food materials, forming a mature fruit.

Seeds and fruits acquire nutrients for growth and development from their parent plant. If the seed is separated from the parent too soon, it may not complete its development. Not surprisingly, seed maturation and fruit ripening are closely coordinated. Unripe fruits are often inconspicuously colored (usually green, like the rest of the plant), hard, bitter, and sometimes even poisonous. As a result, animals seldom eat unripe fruit. When the seeds mature, the fruit ripens: it becomes brightly colored, softer, and sweeter, and therefore more noticeable and attractive to animals (Fig. 23-10).

Ripening is stimulated by ethylene, which is synthesized by fruit cells in response to a surge of auxin released by the seeds. Since ethylene is a gas, a ripe fruit continually leaks ethylene into the air. In nature this probably doesn't make much difference. However, when people store fruit in closed containers, ethylene released from one fruit will hasten ripening in the rest, which is why "one rotten apple spoils the barrel." Although this may inconvenience the consumer whose entire carton of fruit ripens

Rapid-Fire Plant Responses

All plants are alive, but some are definitely more lively than others. Poke a sycamore tree, for example, and nothing much seems to happen. But watch a fly brush against the sensory hairs in a Venus flytrap, and you will see a response that is almost animal-like in its purposefulness and speed of movement (Fig. E23-1). Why is it useful for a Venus flytrap to catch a fly, and how does it accomplish this task?

You may recall from Chapter 21 that many soils are nitrogen-poor. This is particularly true in acid bogs, where nitrogen is practically non-existent. As an evolutionary response to chronic nitrogen shortages, several bog plants, including the Venus flytrap, pitcher plant, and sundew, have resorted to carnivory. By snaring a beetle or fly now and then, the plant obtains nitrogen from the chitin, proteins, and nucleic acids of its prey. While the evolutionary advantage seems clear, the *mechanism* of movement is much less obvious. Let's examine the Venus flytrap more closely, and find the answers to two intriguing aspects of its rapid responses: (1) how does the plant perceive the touch of a fly, and (2) how does it move its leaves to catch it?

Sensory Perception

Each lobe of the fringed trapping leaves of a Venus flytrap bears three sensory "hairs" on its inside surface (Fig. E23-1a). These hairs act as triggers, stimulating the leaf to close around its prey. However, a single touch to a single hair will not close the trap. This is clearly adaptive behavior. Dead leaves, twigs, or bits of dirt might fall into the open trap, but being inanimate, would probably only brush against a hair once before settling to the bottom of the leaf. An insect, on the other hand, wanders around inside the trap, sipping the nectar secreted by the leaves. If it touches one hair twice in rapid succession, or touches two different hairs, the hairs initiate an electrical potential analogous to the action potential of animal nerve cells. The electrical potential sets off a rapid chain of events that causes the trap to close (Fig. 23-E1b).

Figure E23-1a

in just a few days, the discovery of the role of ethylene in ripening revolutionized modern fruit and vegetable marketing. Bananas, for instance, are grown in Central America and shipped by boat to North American markets. By picking and shipping the bananas green and then exposing them to ethylene at their destination, grocers can market perfectly ripe fruit. Unfortunately, not all fruits seem to ripen properly when separated from the plant, resulting in such supermarket delicacies as the infamous pink cardboard tomato: ripening tomatoes probably continue to acquire nutrients from the parent plant, so gassing a green tomato with ethylene turns it red but does not really duplicate natural ripening.

Figure 23-10 *Fruit ripening includes changes in color, texture, flavor, and sweetness. A strawberry fruit is green, hard, and bitter before it ripens, which discourages animals from eating it. After the seeds mature, the fruit becomes soft, red, and tasty, attracting animals such as this vole. The mature seeds are not harmed by the animal's digestive tract and are dispersed in the animal's feces.*

Senescence and Dormancy

The season is autumn. If animals have not eaten the fruits yet, the time has come to let them drop to the ground. For perennial broadleafed plants, the leaves must be shed as well, because they will be a liability in winter, unable to photosynthesize but still evaporating water (see the essay "Autumn in Colorado" in Chapter 21). Both leaves and fruits undergo a rapid aging called **senescence.** The culmination of senescence is the formation of the **abscission layer** at the base of the petiole, allowing the leaf or fruit to drop off.

Rapid-Fire Plant Responses

Figure E23-1b

Leaf movement

In a beautiful set of experiments, botanists Stephen Williams and Alan Bennett found that the flytrap leaf closes because of *irreversible, differential growth.* The flytrap leaves can be pictured most simply as two layers of cells, outer and inner (Fig. E23-2). The electrical potential triggered by hair movement stimulates cells of the outer layer to pump hydrogen ions (H^+) extremely rapidly into their cell walls. Enzymes in the cell walls are activated by acid conditions and loosen the cellulose fibers of the walls. As the walls weaken, the high osmotic pressure inside the cells causes them to absorb water from extracellular fluids, swiftly growing by about 25 percent. Since the outer layer expands while the inner layer does not, the leaf is pushed closed. Reopening the trap occurs much more slowly, taking several hours. However, the fundamental mechanism is similar: during opening, the cells on the inside of the leaf expand, pushing apart the lobes of the trap.

So much energy is used up by the hydrogen pumps that *closing the trap consumes nearly a third of all the ATP within the entire leaf.* Since only the outer layer of cells are involved in closure, they must virtually empty themselves of ATP. It is therefore very important that something digestible actually be in the leaf before it closes the trap.

Puzzles to Solve

Although a lot has been learned about the mechanisms producing movement in lively plants such as the Venus flytrap, mysteries remain. How is a touch stimulus transformed into an electrical stimulus by the sensory hairs? What is the nature of the electrical potential change? How does the electrical signal cause the cells to begin pumping hydrogen ions? As so often happens in biology, the answer to one question immediately poses several new, usually tougher, questions.

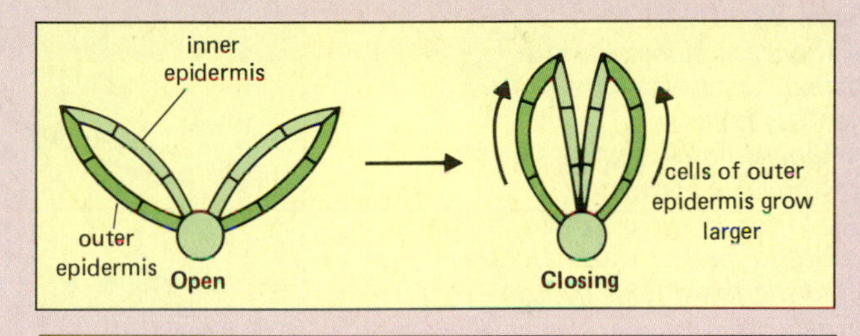

Figure E23-2

Senescence and abscission are complex processes controlled by several different hormones. In most plants, healthy leaves and developing seeds produce auxin, which in turn helps to maintain the health of the leaf or fruit. Simultaneously, the roots synthesize cytokinin, which is transported up the stem and out the branches. Cytokinin also prevents senescence—a leaf plucked from a tree and floated in a dilute solution of cytokinin stays green for weeks. As winter approaches, cytokinin production in the roots slows down, and fruits and leaves produce less auxin. Perhaps driven by these hormonal changes, much of the organic material in leaves is broken down to simple molecules that are transported to the roots for winter storage. Meanwhile, ethylene is released by both aging leaves and ripening fruit, stimulating the production of enzymes that destroy the cell walls holding together the abscission layer at the base of the petiole. When the abscission layers weaken, leaves and fruits fall from the branches.

Other changes also occur that prepare the plant for winter. Buds, which developed into new leaves and branches during spring and summer, now become dormant, waiting out the winter tightly wrapped up. Dormancy in buds, as in seeds, is enforced by abscisic acid. Metabolism slows to a crawl, and the plant enters its long winter sleep, waiting for the signals of warmth and longer days in spring before awakening again.

REFLECTIONS ON PLANT PHYSIOLOGY AND ECOLOGY

Plants have evolved the ability to regulate their growth and development in response to the demands of their environment. The mechanisms by which plants perceive environmental stimuli and respond to them are genetically determined, and

Figure 23-11
(a) *Ragweed blooms in late summer, when days are shorter than 14½ hours.*
(b) *Violets flower during short days in early spring, an adaptation to their habitat in deciduous forests, where the trees shade the forest floor by late spring.*
(c) *The length of the longest day in June progressively increases with latitude, varying from just over 12 hours at the equator to continuous day at the North Pole.*

vary considerably among plant species. These genetic differences in turn permit some species to thrive in a particular environment while other species cannot.

Consider, for example, that bane of hay-fever sufferers, ragweed (Fig. 23-11a). A ragweed seed germinates in spring, but the resulting plant does not mature enough to flower until summer. In addition, no matter how large the plant becomes, ragweed will not flower if the days are longer than about 14½ hours. Now, summer days are longer the farther north you go (Fig. 23-11c). In nearly all of Canada, the longest day of the year (June 21) has more than 16 hours of daylight, and the daylength does not fall below 14½ hours until mid-August. By that time, the killing frosts of autumn are not far away, and few ragweeds can complete seed development before dying. Therefore, most of Canada is free of ragweed. In New Jersey, on the other hand, summer days shorten to 14½ hours by late July, and frosts occur later than they do in Canada. The result, as the authors can sneezily testify, is a prolific ragweed crop.

Other plants, such as the buffalo grass of the plains, can grow and reproduce successfully from Texas to Alberta. These plant species show geographically distinct varieties that differ in the critical daylength for flowering. Northern varieties of buffalo grass flower during the long days of early summer, which leaves plenty of time to complete seed development before frost. Southern varieties flower under much shorter days. A Canadian grass planted in Texas will never flower, while a transplanted Texan will continue blooming into late summer, spending energy on new flowers instead of maturing seeds.

The local environment of a plant also requires adaptations in flowering. Many wildflowers of the eastern deciduous forests, such as the violet (Fig. 23-11b), have an extremely short growing season, sandwiched between the end of winter and the dense growth of leaves on the trees above them in spring. Once the trees have leafed out, not much light reaches the forest floor, and the violets cannot photosynthesize rapidly enough to set seeds. Thus violets flower only when daylengths are short in early spring. By late spring, producing more flowers would waste energy better spent in maturing the seeds already set. Not surprisingly, the longer days of May inhibit flowering.

The daylength requirements for flowering found in various plants allow us to predict where they will be found. For example, we can confidently assert that Alaska will be a haven for ragweed sufferers, because ragweed cannot set seeds there. Do not forget, however, that daylength requirements have not evolved to *prevent* plants from colonizing new habitats. Today, ragweed is restricted to places where summer days are not too long. On an evolutionary time scale, however, there are probably other factors that would normally prevent completion of the ragweed life cycle in the far north. The genetically determined daylength requirements of ragweed and other plants have evolved, not to limit their distribution, but because they ensure that flowering is properly timed for the normal environment of the plant.

SUMMARY OF KEY CONCEPTS

The Discovery of Plant Hormones

Most responses of plants to their environment are produced through the actions of simple chemicals called plant hormones. The first plant hormone, auxin, was discovered as a result of many years of experiments on the mechanism of phototropism, the growth of plants toward the light.

Plant Hormones and Their Actions

Plant hormones are chemicals that are produced in one part of a plant body and exert actions on the same or distant parts of the plant. The five major classes of plant hormones are abscisic acid, auxin, cytokinin, ethylene, and gibberellin. The major functions of these hormones are summarized in Table 23-1.

The Plant Life Cycle: Reception, Response, and Regulation

Dormancy in seeds is enforced by abscisic acid. Falling levels of abscisic acid, and rising levels of gibberellin, trigger germination. As the seedling grows, it shows differential growth with respect to the direction of light (phototropism) and gravity (gravitropism). Auxin mediates phototropism and gravitropism in shoots, and gravitropism in roots, but by different mechanisms. In shoots, auxin stimulates elongation of cells; by accumulating on the lower side of a shoot and the side away from light, auxin causes the shoot to bend away from gravity and toward the light. In roots, auxin inhibits elongation. By accumulating on the lower side of a root, auxin causes the root to bend toward gravity.

Branching in stems results from the interplay of two hormones, auxin and cytokinin. High concentrations of auxin (produced in shoot tips and transported downward) inhibits the growth of lateral buds. An optimum concentration of both auxin and cytokinin (synthesized in roots and transported up the shoot) stimulates growth of lateral buds. Auxin also stimulates the growth of branch roots.

The timing of flowering is usually controlled by daylength. Plants appear to detect light and dark by changes in phytochrome, a pigment in the leaves. Plant processes influenced by phytochrome responses to light include flowering, straightening the epicotyl or hypocotyl hook, seedling elongation, leaf growth, and chlorophyll development.

Developing seeds produce auxin, which diffuses into the surrounding ovary tissues and causes production of a fruit. A surge of auxin as the seed matures stimulates fruit cells to release another hormone, ethylene, which causes the fruit to ripen. Ripening includes the conversion of starches to sugars, softening of the fruit, development of bright colors, and often the formation of an abscission layer at the base of the petiole.

Several changes prepare perennial plants of temperate zones for winter. Leaves and fruits undergo a rapid aging process called senescence, including formation of an abscission layer. Senescence occurs due to a fall in levels of auxin and cytokinin, and perhaps to a rise in ethylene concentrations. Other parts of the plant, including buds, become dormant. Dormancy in buds is enforced by high concentrations of abscisic acid.

GLOSSARY

Abscisic acid (ab-sis'-ik) a plant hormone that generally inhibits the action of other hormones, enforcing dormancy in seeds and buds and causing closing of stomata.

Apical dominance the phenomenon whereby a growing shoot tip inhibits the sprouting of lateral buds.

Auxin (awk'-sin) a plant hormone that influences many plant functions, including phototropism, apical dominance, and root branching. Auxin generally stimulates cell elongation and, in some cases, cell division and differentiation.

Biological clock a metabolic timekeeping mechanism found in most organisms, whereby the organism measures the approximate length of a (24-hour) day even without external environmental cues such as light and dark.

Circadian rhythm (sir-kā'-dē-un) an event that recurs with a period of about 24 hours, even in the absence of environmental cues.

Cytokinin (sī-tō-kī'-nin) a plant hormone that promotes cell division, fruit growth, and sprouting of lateral buds, and prevents leaf aging and leaf drop.

Day-neutral plant a plant in which flowering occurs under a wide range of daylengths.

Ethylene a plant hormone that promotes ripening of fruits, and leaf and fruit drop.

Gibberellin (jib-er-el'-in) a plant hormone that stimulates seed germination, fruit development, and cell division and elongation.

Gravitropism growth with respect to the direction of gravity.

Long-day plant a plant that will flower only if the length of daylight is greater than some species-specific duration.

Phototropism growth with respect to the direction of light.

Phytochrome (fī'-tō-krōm) a light-sensitive plant pigment that mediates many plant responses to light, including flowering, stem elongation, and seed germination.

Senescence in plants, a specific aging process, often including deterioration and dropping of leaves and flowers.

Short-day plant a plant that will flower only if the length of daylight is shorter than some species-specific duration.

STUDY QUESTIONS

1. Describe the experiments of the Darwins, Boysen-Jensen, and Went. Do these experiments truly prove that auxin is the hormone controlling phototropism? What other experiments would you like to see?
2. What two hormones are involved in seed dormancy and germination? What are their roles?
3. Describe the mechanisms of action of auxin in shoot phototropism and gravitropism and root gravitropism.
4. What is apical dominance? How do auxin and cytokinin interact in determining the growth of lateral buds?
5. Define *day-neutral plant, long-day plant,* and *short-day plant.* What pigment is thought to be involved in light perception in plants?
6. What is a biological clock?
7. Describe the role of phytochrome in stem elongation in seedlings growing in the shade of other plants. What is the likely adaptive significance of this response?
8. What hormone(s) cause fruit development? Where does this hormone come from? What hormone causes fruit ripening?
9. What hormone(s) are involved in leaf and fruit drop? In bud dormancy?

SUGGESTED READINGS

Evans, M. L., Moore, R., and Hasenstein, K.-H. "How Roots Respond to Gravity." *Scientific American,* December 1986. Although botanists still dispute the mechanisms of root gravitropism, these authors argue convincingly that it is mediated by auxin.

Heslop-Harrison, Y. "Carnivorous Plants." *Scientific American,* February 1978. Some plants living in nitrogen-poor environments have turned carnivore, evolving surprising adaptations for capturing insect food.

Raven, P. H., Evert, R. F., and Eichhorn, S. *Biology of Plants,* 4th ed. New York: Worth Publishers, Inc., 1986. One of the best botany textbooks available, with a lucid discussion of plant hormones.

Salisbury, R. B., and Ross, C. W. *Plant Physiology,* 3rd ed. Belmont, Calif.: 1985. Wadsworth Publishing Company, Inc., The sections on plant hormones and the control of flowering may at times seem confusing, but accurately reflect the uncertainties botanists face in interpreting conflicting experimental data.

24
Nutrition and Digestion

In one sense, animals and plants have similar nutritional needs: materials to synthesize the components of their bodies and a source of energy to power that synthesis. However, plants and animals satisfy these needs in quite different ways. A plant takes water and minerals from the soil and carbon dioxide from the air, and uses the radiant energy of sunlight to form these materials into complex molecules such as starch and cellulose. Animals, on the other hand, eat plants or each other. The food that animals eat provides both energy and the basic building blocks of complex molecules: amino acids, fatty acids, and simple sugars. These building blocks are combined in unique ways by each organism. The proteins of a mule deer, for example, are slightly different from those of the timber wolf that may consume it. Thus the deer proteins must first be broken down into amino acids before the wolf can use them to make wolf proteins. The digestive systems of animals first grind up and then chemically break down the complex molecules of their food into simpler components that can be reassembled into the unique molecules of the consumer.

Frontispiece: For many animals, the first challenge of digestion is to capture elusive and unwilling prey.

NUTRITION

The nutrient requirements of animals fall into four categories: (1) *sources of energy,* mostly fats and carbohydrates; (2) *amino acids* to construct proteins; (3) *minerals;* and (4) *vitamins.*

Sources of Energy

Each living cell in the animal body relies on a continuous expenditure of energy to maintain its incredible complexity, to transport molecules, and to grow and divide. Three major nutrients provide dietary energy (measured in **Calories**) for animals: lipids (fats and oils), carbohydrates, and proteins (see Chapter 3). These molecules can be broken down during cellular respiration, providing chemical energy (ATP) (see Chapter 7).

Of the three, lipids are the most concentrated energy source. Lipids provide about 9.3 Calories/gram, over twice the energy per unit weight of carbohydrates and protein (approximately 4 Calories/gram). By storing energy as fat, animals minimize their body weight, which in turn allows them greater mobility. Lipids are more than a source of energy; they are also important for the synthesis of certain hormones, the coverings of neurons, and cell membranes. Some animals can synthesize all the specialized lipids they need; others, including humans, require specific types of lipid in their diet. In the average North American diet, fats and oils provide about 45 percent of the caloric intake.

Carbohydrates, consisting of simple sugars linked together in chains of varying lengths, are the principal energy-storage material of plants. They contribute about 45 percent of the energy in our diets, although it is possible to do without them entirely if sufficient calories and other nutrients are available. Eskimos, for example, have traditionally thrived on a diet of seal, fish, and caribou, which is very high in fat and protein, with little carbohydrate. The human body stores carbohydrate as **glycogen** in the liver and muscles. During intense exercise such as running, the body draws on this store of glycogen as a source of quick energy. When the activity is prolonged, as in the case of a marathon runner, the glycogen store may be totally depleted. "Hitting the wall" describes the extreme fatigue a marathon runner may experience when she exhausts her glycogen supply about 18 miles into the race.

Only about 10 percent of our energy requirement is obtained from protein. The breakdown of protein produces the waste product urea, which is filtered from the blood by the kidneys. Specialized diets in which protein is the major energy source place extra stress on the kidneys. The major role of protein is as a source of amino acids, described below.

Amino Acids to Make Protein

In the digestive tract, proteins are broken down into their building blocks, amino acids (see Chapter 3). Then, in the body cells, the amino acids are linked together in specific sequences to form new proteins. Animals are capable of synthesizing some but not all of the 20 different types of amino acids used in proteins. Those that cannot be synthesized, called **essential amino acids,** must be supplied by the diet. There are 11 essential amino acids for humans.

Minerals

A wide variety of minerals are required by animals (see Table 24-1), and all must be obtained in the diet, either from food or dissolved in drinking water. These include calcium, magnesium, and phosphorus, which are major constituents of bones and teeth. Others, including sodium and potassium, are essential for nerve impulse conduction and muscle contraction. Iron is used in the production of hemoglobin, and iodine for hormones produced by the thyroid gland. In addition, trace amounts of several other minerals, including zinc, copper, and selenium, are required, often as parts of enzymes.

Vitamins

Vitamins are a diverse group of organic compounds required in small amounts for normal bodily functioning. Generally, vitamins cannot be synthesized by the body and must be obtained from food. Since the ability of animals to manufacture substances varies considerably, the same compound might be a vitamin for one animal but not for another. For example, ascorbic acid is a vitamin (vitamin C) for guinea pigs and for humans and other primates who cannot manufacture it, but not to most other animals, whose bodies can synthesize it. Although our skin can manufacture some vitamin D when it is exposed to sunlight, because we spend so much time indoors most of us do not synthesize enough. Thus we must supplement it through our diet. Whether or not a substance is a vitamin, then, depends on the animal species and the conditions under which it lives. The vitamins considered essential in human nutrition are listed in Table 24-2.

Human vitamins are often grouped into two categories, water soluble and fat soluble. Water-soluble vitamins include vitamin C and the 11 different compounds comprising the B vitamin complex. These substances dissolve in the water of the blood plasma and are excreted by the kidneys. They are not stored in the body in any appreciable amount. Water-soluble vitamins generally work in conjunction with enzymes to promote chemical reactions in the body that supply energy or synthesize materials. Since each vitamin participates in several metabolic processes, a deficiency of a single vitamin can produce wide-ranging effects (see Table 24-2).

The fat-soluble vitamins, A, D, E, and K, have even more varied roles, from the regulation of blood clotting by vitamin E to the formation of visual pigment by vitamin A. Fat-soluble vitamins can be stored in body fat and may accumulate in the body over time. Excessive intake of vitamins A and D can produce toxic effects (see Table 24-2).

THE CHALLENGE OF DIGESTION

Animals eat the bodies of other organisms, bodies that were not designed to provide food. The plant body, for example, armors each cell with a wall of indigestible cellulose, whereas animals may be covered with equally indigestible fur, scales, or feathers. In addition, the complex lipids, carbohydrates, and proteins of plants and animals are not in a form that can be used directly. They must be broken down before they can be absorbed into the bloodstream of the consumer and distributed to its own cells. This is the challenge of digestion. To meet it, animals have evolved a variety of digestive tracts; each is finely tuned to a unique life-style and diet.

Diverse Digestive Systems

DIGESTION WITHIN SINGLE CELLS. **Intracellular digestion** occurs after microscopic food particles are engulfed by single cells. Once engulfed by a cell, the food is enclosed in a **food vacuole,** a space surrounded by membrane which serves as a temporary stomach. The vacuole

Table 24-1 **Human Mineral Requirements**[a]

Mineral	RDA for Healthy Adult Male (milligrams)	Dietary Source	Major Functions	Deficiency
Water	1.5 liters/day	Solid foods, liquids, drinking water	Transport of nutrients; Temperature regulation; Participates in metabolic reactions	Thirst, dehydration
Calcium	800	Milk, cheese, dark-green vegetables, dried legumes	Bone and tooth formation; Blood clotting; Nerve transmission	Stunted growth; Rickets, osteoporosis; Convulsions
Phosphorus	800	Milk, cheese, meat, poultry, grains	Bone and tooth formation; Acid-base balance	Weakness, demineralization of bone; Loss of calcium
Sulfur	(Provided by sulfur amino acids)	Sulfur amino acids (methionine and cystine) in dietary proteins	Constituent of active tissue compounds, cartilage and tendons	Related to intake and deficiency of sulfur amino acids
Potassium	2,500	Meats, milk, many fruits	Acid-base balance; Body water balance; Nerve function	Muscular weakness; Paralysis
Chlorine	2,000	Table salt	Formation of gastric juice; Acid-base balance	Muscle cramps; Mental apathy; Reduced appetite
Sodium	2,500	Table salt	Acid-base balance; Body-water balance; Nerve function	Muscle cramps; Mental apathy; Reduced appetite
Magnesium	350	Whole grains, green leafy vegetables	Activates enzymes involved in protein synthesis	Growth failure; Behavioral disturbances; Weakness, spasms
Iron	10	Eggs, lean meats, legumes, whole grains, green leafy vegetables	Constituent of hemoglobin and enzymes involved in energy metabolism	Iron-deficiency anemia (weakness, reduced resistance to infection)
Fluorine	2	Drinking water, tea, seafood	May be important in maintenance of bone structure	Higher frequency of tooth decay
Zinc	15	Widely distributed in foods	Constituent of enzymes involved in digestion	Growth failure; Small sex glands
Iodine	0.14	Seafish and shellfish, dairy products, many vegetables, iodized salt	Constituent of thyroid hormones	Goiter
Copper, Silicon, Vanadium, Tin, Nickel, Selenium, Manganese	Not established (trace amounts)	Widely distributed in foods	Some unknown; some work in conjunction with enzymes	Occurs rarely

[a]Modified from "The Requirements of Human Nutrition," by Nevin S. Scrimshaw and Vernon R. Young, *Scientific American*, September 1976.

is fused with small packets of digestive enzymes called lysosomes, and food is broken down within the vacuole into smaller molecules that can be absorbed into the cell cytoplasm. Undigested remnants remain in the vacuole, which eventually dumps its contents outside the cell. This system is found in both the single-celled protists and the simplest animals. Sponges, for example, rely entirely on intracellular digestion (Fig. 24-1). This limits their menu to microscopic food particles, often protists filtered from the surrounding sea.

Digestion in a simple sac. Larger, more complex organisms evolved a chamber within the body where chunks of food are broken down by enzymes acting *outside the cells*. This is called **extracellular digestion.** One of the simplest of these chambers is found in sea ane-

Table 24-2 **Human Vitamin Requirements**[a]

Vitamin	RDA for Health Adult Male (milligrams)	Dietary Sources	Major Functions	Deficiency and Symptoms	Symptoms of Excess
Water Soluble					
Vitamin B-1 (thiamin)	1.5	Pork, organ meats, whole grains, legumes	Coenzyme in reactions involving the removal of carbon	Beriberi (peripheral nerve changes, edema, heart failure)	None reported
Vitamin B-2 (riboflavin)	1.8	Widely distributed in foods	Constituent of two coenzymes involved in energy metabolism	Reddened lips, cracks at corner of mouth, lesions of eye	None reported
Niacin	20	Liver, lean meats, grains, legumes (can be formed from tryptophan)	Constituent of two coenzymes involved in energy metabolism	Pellagra (skin and gastrointestinal lesions, nervous, mental disorders)	Flushing, burning, tingling around neck and hands
Vitamin B-6 (pyridoxine)	2	Meats, vegetables, whole-grain cereals	Coenzyme involved in amino acid metabolism	Irritability, convulsions, muscular twitching, dermatitis near eyes, kidney stones	None reported
Pantothenic acid	5–10	Widely distributed in foods	Constituent of coenzyme A, which plays a central role in energy metabolism	Fatigue, sleep disturbances, impaired coordination, nausea (rare in humans)	None reported
Folacin	0.4	Legumes, green vegetables, whole-wheat products	Coenzyme involved in nucleic acid and amino acid metabolism	Anemia, gastrointestinal disturbances, diarrhea, red tongue	None reported
Vitamin B-12	0.003	Muscle meats, eggs, dairy products, (not present in plant foods)	Coenzyme involved in nucleic acid metabolism	Pernicious anemia, neurological disorders	None reported
Biotin	Not established Usual diet provides 0.15–0.3	Legumes, vegetables, meats	Coenzyme required for fat synthesis amino acids metabolism and glycogen (animal-starch) formation	Fatigue, depression, nausea, dermatitis, muscular pains	None reported
Choline	Not established Usual diet provides 500–900	All foods containing phospholipids (egg yolk, liver, grains, legumes)	Constituent of phospholipids, precursor of the neurotransmitter acetylcholine	None reported in humans	None reported
Vitamin C (ascorbic acid)	45	Citrus fruits, tomatoes, green peppers	Maintains intercellular matrix of cartilage, bone and dentine, important in collagen synthesis	Scurvy (degeneration of skin, teeth, blood vessels, epithelial hemorrhages)	Relatively nontoxic, possibility of kidney stones

Table 24-2 **Human Vitamin Requirements**[a] ***(cont'd.)***

Vitamin	RDA for Health Adult Male (milligrams)	Dietary Sources	Major Functions	Deficiency and Symptoms	Symptoms of Excess
Fat Soluble					
Vitamin A (retinol)	1	Provitamin A (beta-carotene) widely distributed in green vegetables. Retinol present in milk, butter, cheese, fortified margarine	Constituent of rhodopsin (visual pigment). Maintenance of epithelial tissues	Xerophthalmia (keratinization of ocular tissue), night blindness, permanent blindness	Headache, vomiting, peeling of skin, anorexia, swelling of long bones
Vitamin D	0.01	Cod-liver oil, eggs, dairy products, fortified milk and margarine.	Promotes growth and mineralization of bones; increases absorption of calcium	Rickets (bone deformities) in children Osteomalacia in adults	Vomiting, diarrhea, loss of weight, kidney damage
Vitamin E (tocopherol)	15	Seeds, green leafy vegetables, margarines, shortenings	Functions as an antioxidant to prevent cell membrane damage	Possibly anemia	Relatively nontoxic
Vitamin K	0.03	Green leafy vegetables. Small amount in cereals, fruits and meats	Important in blood clotting	Bleeding, internal hemorrhages	Relatively nontoxic

[a]Modified from "The Requirements of Human Nutrients," by Nevin S. Scrimshaw and Vernon R. Young, *Scientific American*, September 1976.

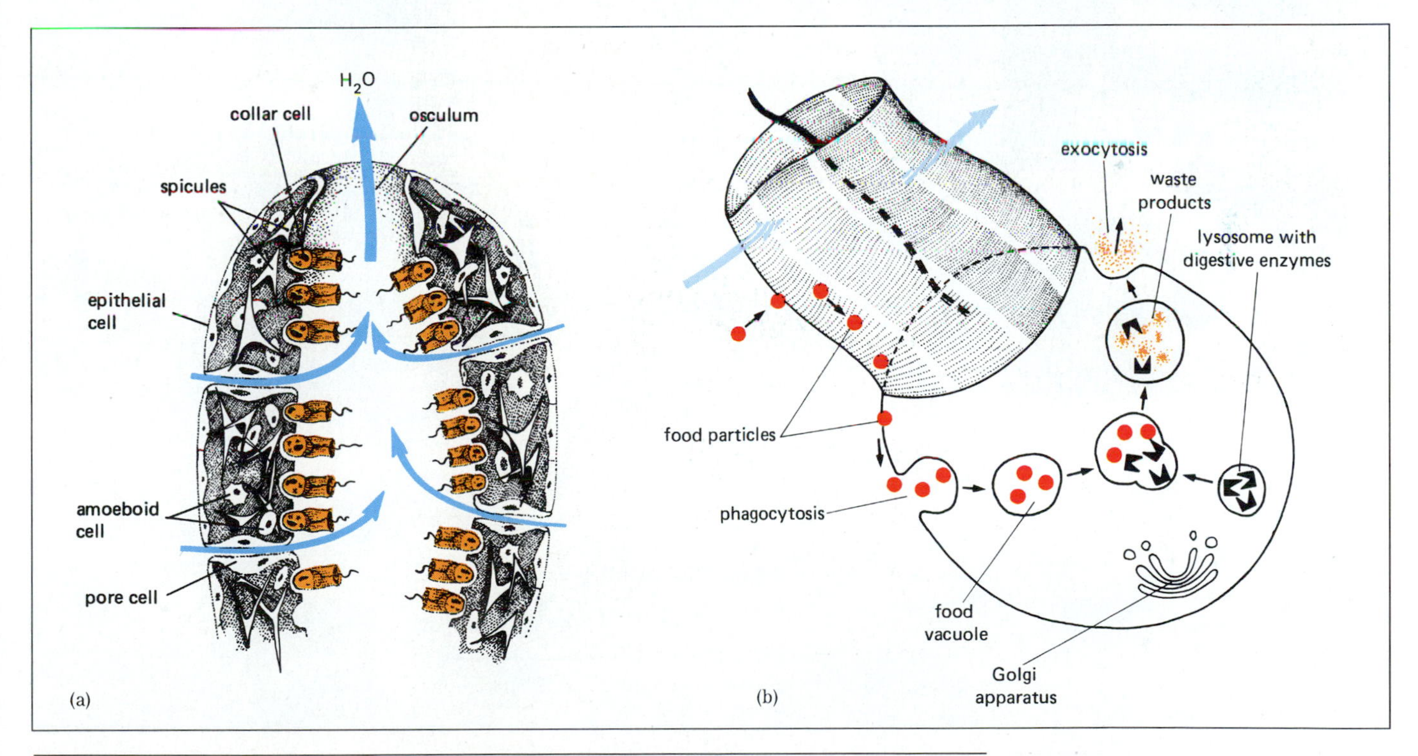

Figure 24-1 Intracellular digestion in a sponge. **(a)** *Internal anatomy of a simple sponge showing the direction of water flow and the location of the collar cells.* **(b)** *Enlargement of a single collar cell. Water is filtered through the collar and food particles (single-celled organisms) are trapped. Within its cell body, food is engulfed, digested, and wastes are expelled.*

mones, hydra, and jellyfish, members of the phylum Cnidaria. These animals possess a digestive sac called a **gastrovascular cavity,** with a single opening through which foods are ingested and wastes are ejected (Fig. 24-2). Although generally referred to as the mouth, this opening is equally an anus. Food captured by stinging tentacles is escorted into the gastrovascular cavity, where enzymes break it down. Cells lining the cavity absorb the nutrients and engulf small food particles. Further digestion occurs using the intracellular processes described above. The undigested remains are eventually voided through the same opening by which they entered. While one meal is being digested, a second cannot be processed efficiently, since the same chamber is used. Thus this type of digestive system is unsuited to active animals requiring frequent meals or to animals whose food supplies so little nutrition that they must feed continuously. For such animals, a one-way tube with two openings is the answer.

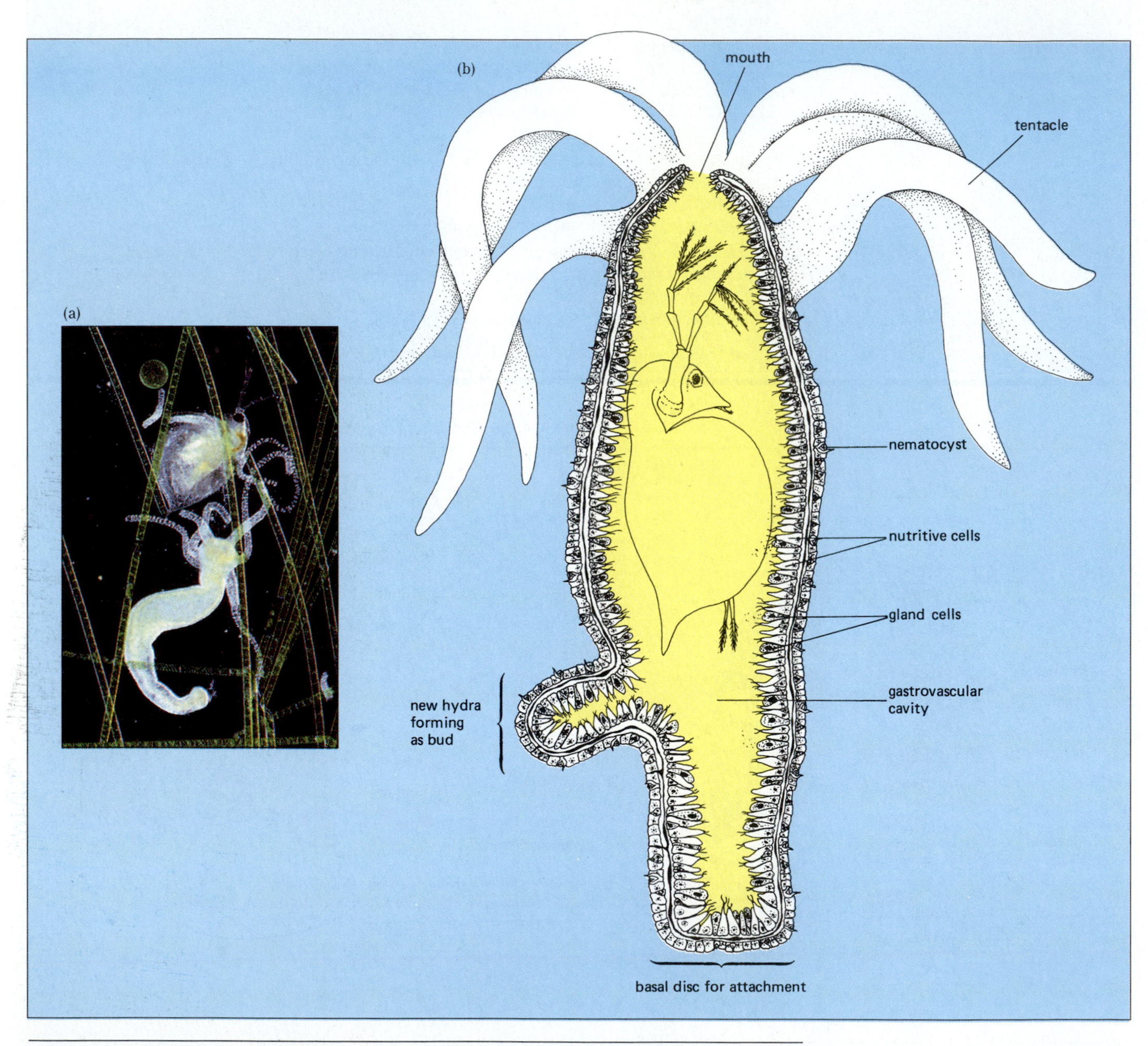

Figure 24-2 *Digestion in a sac.* **(a)** *A* Hydra *has ensnared a small crustacean in its tentacles.* **(b)** *Within the gastrovascular cavity, gland cells secrete enzymes that digest the prey into smaller particles and nutrients. Elongated cells lining the cavity ingest these particles as described for the sponge, and digestion is completed intracellularly. Undigested waste is then expelled through the single opening.*

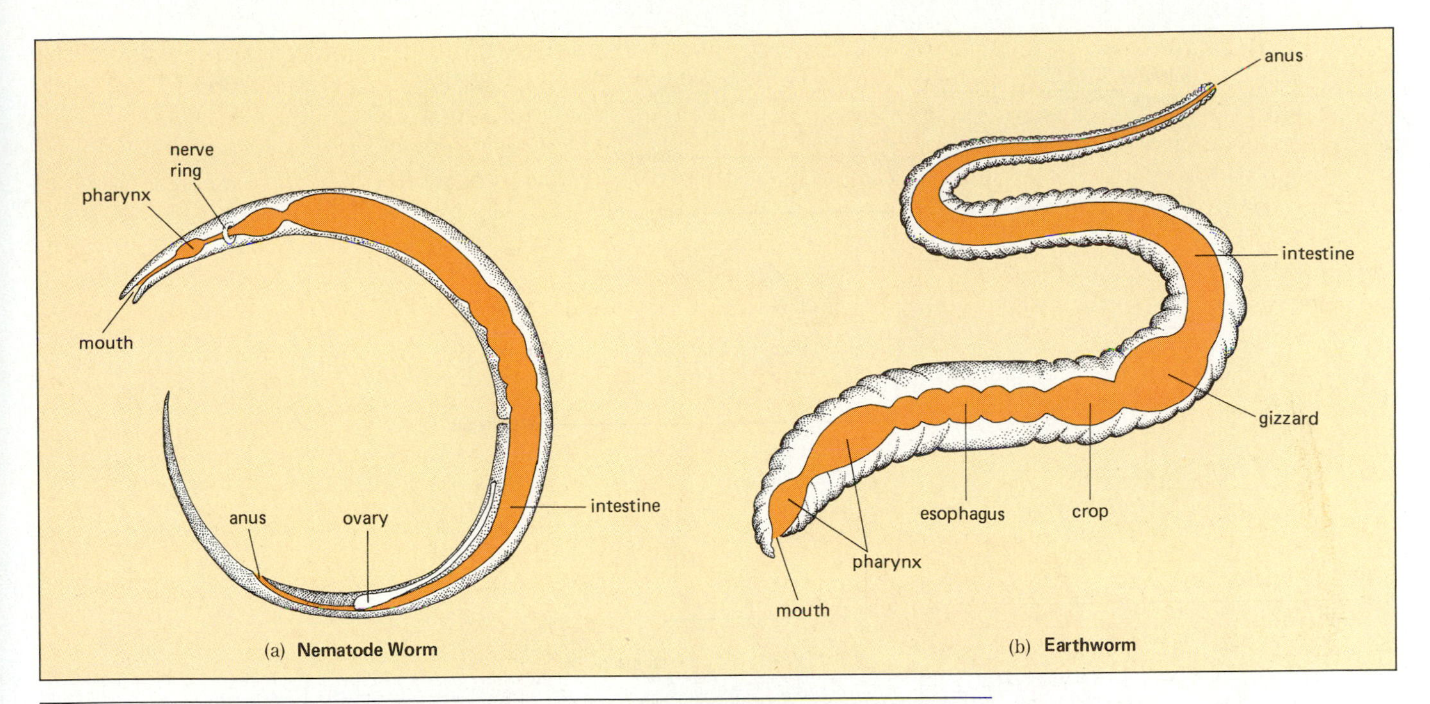

Figure 24-3 Tubular digestive tracts. **(a)** *The nematode worm. This abundant, usually microscopic, animal is one of the simplest animals with a tubular digestive system. Nematodes often live within their food source (several hundred thousand may be found in a rotting apple) and may eat almost continuously.* **(b)** *The earthworm. An advantage of the one-way digestive system is that food can be passed along a series of compartments, each specialized to play a specific role in food breakdown.*

DIGESTION IN A TUBE. Most animals, from nematode worms to earthworms, molluscs, arthropods, echinoderms, and vertebrates, have a digestive system that is basically a tube running through the body. In its simplest form, as seen in the threadlike nematode worm, the tube is relatively unspecialized along its length (Fig. 24-3a). In more complex organisms, such as the earthworm, the tube consists of a series of compartments, each with a specific role in the breakdown of food (Fig. 24-3b). The earthworm extracts nutrients from decaying organic material in the soil, and also feeds on particles of leaves and bits of animal remains or wastes which it gathers during nightly forays to the surface. A tubular digestive system is essential to the earthworm, which continuously ingests soil as it burrows through the earth, passing it out one end while taking it in the other. A muscular **pharynx** draws in soil and bits of vegetation, which are passed through the **esophagus** to a thin-walled storage organ, the **crop.** The crop collects the food and gradually passes it to the **gizzard.** Here bits of sand and the contraction of muscles physically break the food down into smaller particles. Ground-up food from the gizzard then travels to the intestine, where enzymes break it down into simple molecules that can be absorbed by the cells lining the intestine. Animals with tubular digestive systems utilize extracellular digestion to dismantle their food outside the body cells.

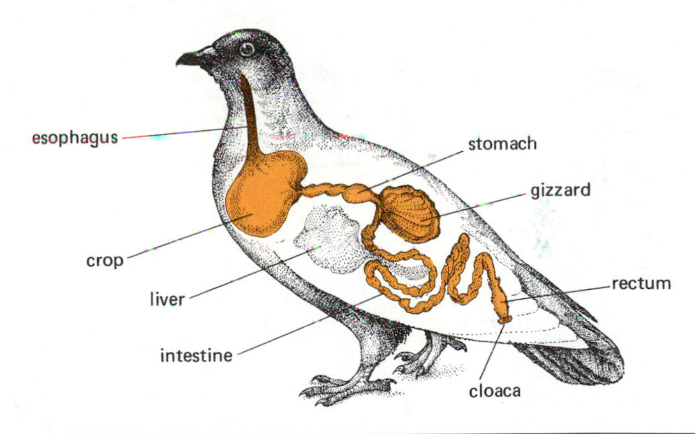

Figure 24-4 The digestive system of birds is adapted to the demands of flight. Teeth, present in primitive birds, have been lost through evolution. This shifts the weight from the head to a point closer to the bird's center of gravity. The expansible crop serves as a storage organ, allowing the bird to store food to meet the enormous caloric demands of flight. The gizzard replaces the teeth, using small stones that are stored in this organ, and muscular action to break down the hard seeds and insect exoskeletons prevalent in the diet of many birds.

As in the earthworm, humans and other vertebrates have tubular digestive tracts with several compartments in which food is first physically, then chemically, broken down prior to absorption by individual cells. Vertebrate

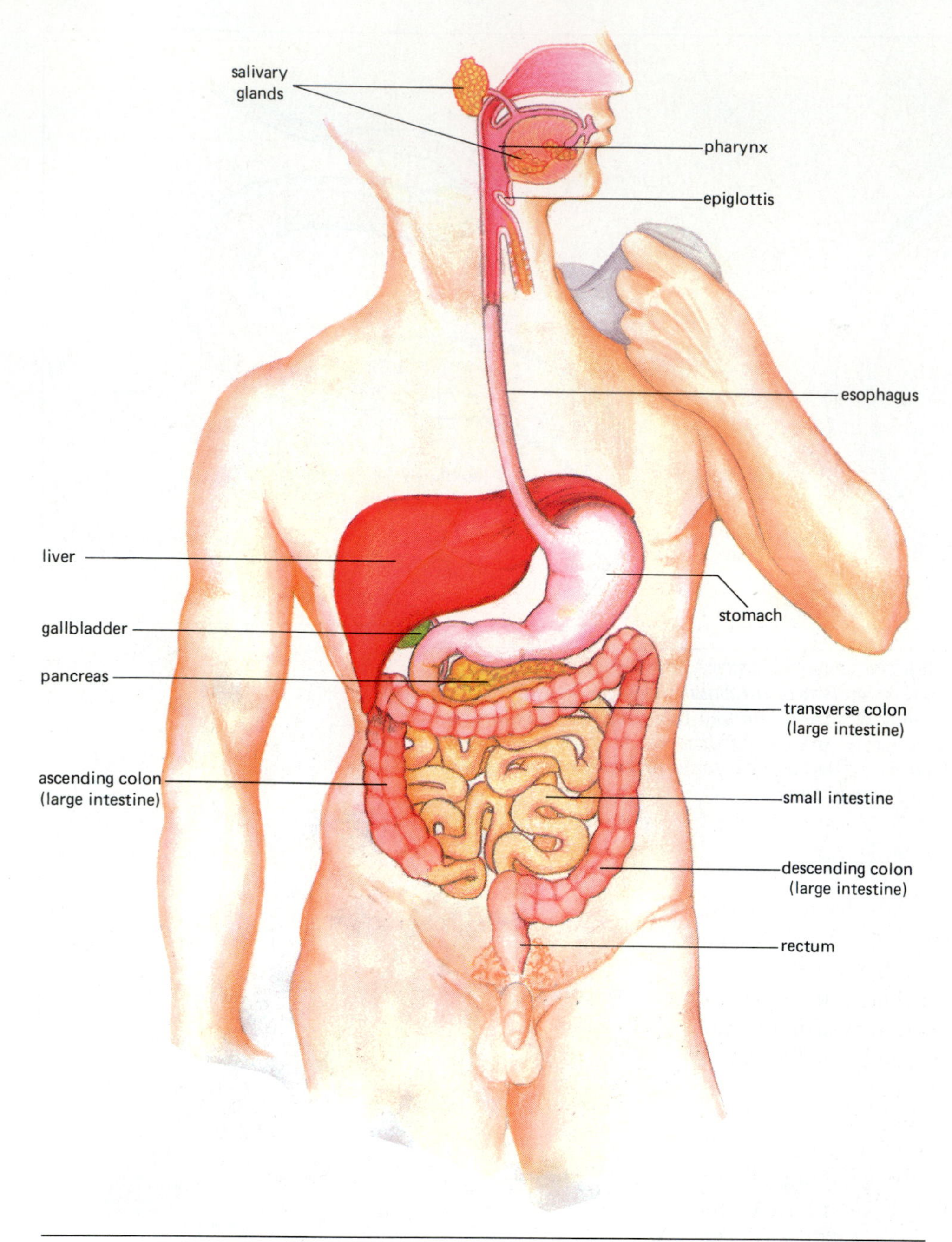

Figure 24-5 The human digestive tract. Some organs that produce and store digestive secretions, such as the salivary glands, liver, gallbladder, and pancreas, are included.

digestive tracts, as illustrated by those of the bird (Fig. 24-4), the cow (Fig. E24-1), and the human (Fig. 24-5), are specialized for the particular diet of the animal.

HUMAN DIGESTION

The Mouth

Both mechanical and chemical breakdown of food begins in the mouth. In the adult human, 32 teeth of varying sizes and shapes cut and grind the food into small pieces. The teeth of humans and other vertebrates are specialized to the diet of the animal (Fig. 24-6). As the food is pulverized by the teeth, the first phase of chemical digestion occurs as three pairs of salivary glands pour out saliva in response to the smell, feel, taste, and (if you're hungry) even the thought of food (Fig. 24-7).

Saliva contains the digestive enzyme **amylase,** which begins the breakdown of starches into sugar (see Table 24-3). Saliva has other functions as well. It contains bacteria-killing enzymes and antibodies that help guard against infection, it lubricates the food to facilitate swal-

Table 24-3 Digestive Secretions

Site	Source	Secretion	Role in Digestion
Mouth	Salivary glands	Amylase	Breaks down starch into disaccharides
	Salivary glands	Mucus, water	Lubricates, dissolves food
Stomach	Cells lining stomach	Hydrochloric acid	Allows pepsin to work, kills bacteria, solubilizes minerals
	Cells lining stomach	Pepsin	Breaks down proteins into large peptides
	Cells lining stomach	Mucus	Protects stomach
Small intestine	Pancreas	Sodium bicarbonate	Neutralizes acidic chyme from stomach
	Pancreas	Amylase	Breaks down starch into disaccharides
	Pancreas	Peptidases	Splits large peptides into small peptides
	Pancreas	Trypsin	Breaks down proteins into large peptides
	Pancreas	Chymotrypsin	Breaks down proteins into large peptides
	Pancreas	Lipase	Breaks down lipids into fatty acids and glycerol
	Liver	Bile	Emulsifies lipids
	Cells lining small intestine	Peptidases	Splits small peptides into amino acids
	Cells lining small intestine	Disaccharidases	Splits disaccharides into monosaccharides

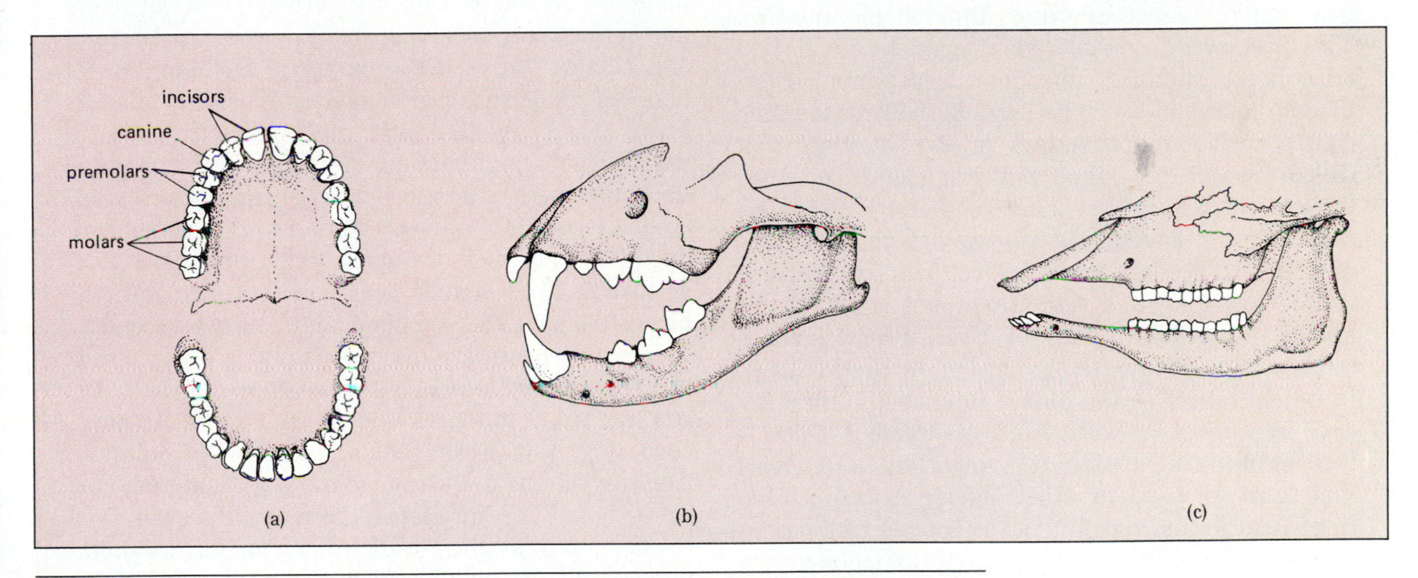

Figure 24-6 **(a)** *The upper and lower jaws of a human adult showing teeth specialized for cutting and grasping (incisors and canines) and grinding (molars and premolars). The shape of human teeth reflects our varied omnivorous diet, which includes both plant and animal material.* **(b)** *The teeth of a tiger, a carnivore (meat eater), are specialized for grasping and tearing flesh and for shearing bone.* **(c)** *The teeth of the herbivorous (plant-eating) cow have large flat surfaces for grinding tough plant material.*

lowing, and it dissolves some food molecules, such as acids and sugars, carrying them to taste buds on the tongue. The taste buds help identify the type and quality of the food.

With the help of the muscular tongue, food is manipulated into a mass and pressed backward into the **pharynx,** a cavity connecting the mouth with the esophagus (Fig 24-8). The pharynx also connects the nose and mouth with the trachea, which conducts air to the lungs. As anyone who has ever choked on a piece of food can attest, this anatomical arrangement occasionally results in problems. Normally, however, the swallowing reflex elevates the larynx so that it meets the **epiglottis,** a flap of tissue that blocks off the respiratory passages. Food is thus directed into the esophagus (Fig. 24-9).

The Esophagus and Stomach

The **esophagus** is a muscular tube that propels food from the mouth to the stomach. Circular muscles surrounding

Coping with Cellulose

Among all the adaptations that evolution has produced, one that is conspicuously absent in most animals is a digestive enzyme to break down cellulose. As explained in Chapter 3, cellulose, like starch, consists of long chains of glucose molecules, but it differs from starch in the way these molecules are linked together. This linkage resists the attack of animal digestive enzymes. Cellulose surrounds each plant cell and thus is potentially one of the most abundant energy sources on Earth. But instead, it passes untouched through most animal digestive systems.

How Humans Benefit

In humans, cellulose, passing unscathed through the intestine, provides important "roughage" or fiber. Dietary fiber increases the volume of feces, increases their water content (making the feces softer), and promotes more rapid passage of wastes through the intestine. This has several possible advantages. In addition to relieving constipation, fiber may help prevent diverticular disease, in which pressure in the large intestine causes saclike outpocketings of the intestinal wall called diverticula. These pouches can become inflamed, infected, or they may even burst, with serious consequences. Diverticular disease has increased dramatically in Western society since the introduction of refined foods such as white flour, but is almost unknown in some rural parts of Africa, where the fiber intake is at least twice that typical of the United States. Intestinal cancer is also almost unknown in rural Africa, in contrast to North America, where it afflicts 3 percent of the population. It is important to recognize that there are so many other differences between the rural African and the American that the relationship between dietary fiber and intestinal disease is a very suggestive correlation but not hard scientific evidence.

However, scientists recently discovered that a potent mutagenic (mutation-causing) substance is actually produced in the large intestine by certain intestinal bacteria (mutagenic substances have the potential to cause cancer because of their ability to alter normal cells). Further, it has been discovered that fiber in the diet decreases the production of this mutagen. By speeding travel of wastes through the intestine, fiber may both limit the time available for the mutagen to form, and reduce the time during which the intestinal cells are exposed to it.

In contrast to most animals, certain bacteria and protists have evolved the enzyme cellulase, which is able to break down cellulose into its component sugar molecules. These single-celled organisms could not by themselves chew up grass, leaves, or wood, but have solved the problem by teaming up with animals. Cows and other ruminants, cockroaches, and termites have all entered into symbiotic partnerships with microbes. The animal provides ground-up cellulose, the microbe produces the necessary enzymes, and both share the abundant energy harvest.

The Remarkable Ruminants

Ruminant animals are the cud chewers: cows, sheep, goats, camels, and hippos, to name a few. Rumination, or cud chewing, is the process of regurgitating food and rechewing it, one of several adaptations of these animals for digesting plant material. The digestive systems of ruminants include three compartments preceding the stomach (Fig. E24-1). The first and largest of these, the rumen, is a massive fermentation vat. Here microorganisms, including many species of bacteria and ciliates, thrive, their numbers reaching several hundred thousand per milliliter. These symbionts break down cellulose and other carbohydrates to sugar. Since conditions in the rumen are anaerobic (lacking oxygen), sugar is broken down further by microbial fermentation. Fermentation in the rumen produces several types of organic acids, methane, CO_2, and water. The organic acids are absorbed into the bloodstream of the ruminant and utilized as energy sources. To neutralize these acids and provide a suitable growth medium for its microscopic partners, the cow each day produces 100 to 200 liters of saliva, consisting of a weak sodium bicarbonate solution. After fermenting in the rumen, the plant material, now called "cud," is regurgitated, chewed, and reswallowed (along with the saliva) to the rumen for further digestion. Gradually, the processed and reprocessed cud is released into the rest of the digestive tract, passing into the reticulum, the omasum, and finally into the abomasum or stomach. In addition to digesting cellulose, the microscopic partners are able to synthesize proteins from urea, normally a waste product. These proteins, as well as those derived from digesting the microorganisms themselves, contribute significantly to the diet of the ruminant. The microorganisms in the digestive tract also produce the ruminant animal's entire supply of vitamin B_{12}.

The Terrible Termites

Termites are among the most remarkable of social insects, a fact easily overlooked as one surveys the wreckage they can make of a wood home. Some species de-

Coping with Cellulose

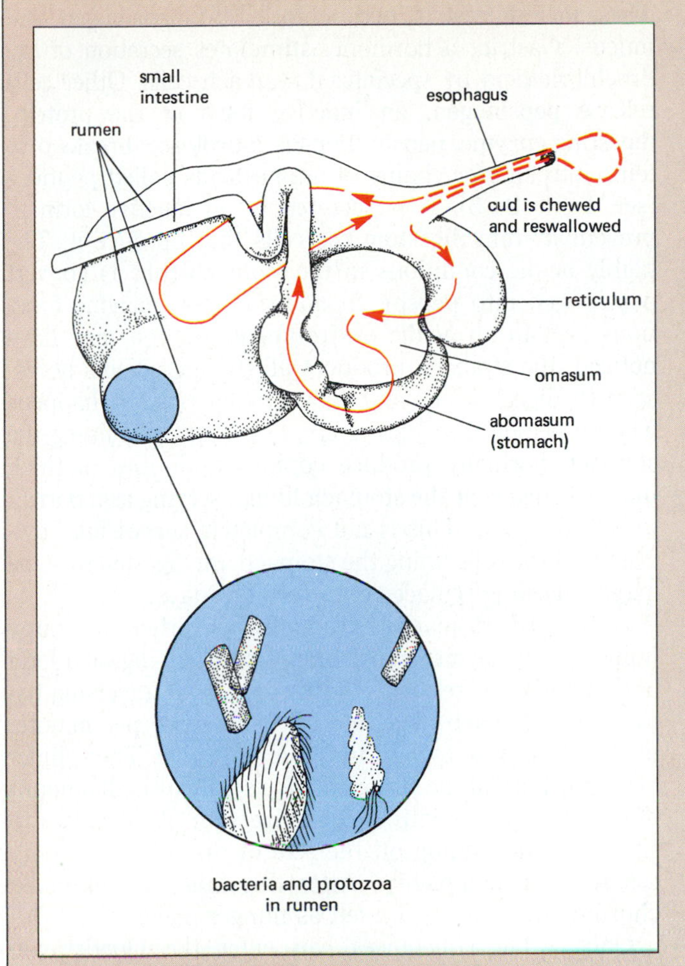

Figure E24-1 The stomach of the cow, the abomasum, is preceded by three other chambers. The largest is the rumen, which houses a flourishing population of microorganisms that digest the cellulose in the cow's vegetarian diet. Arrows trace the path of food through the digestive tract.

rive their ability to digest wood from their partnership with several species of flagellates (protists) living packed inside their intestines (Fig. E24-2). When the protists are removed and cultured separately, they can digest cellulose, but the termite cannot. Only after reinfection with its digestive partners is the termite able to survive on a diet of cellulose. As in ruminants, the microbial partners of termites are digested in large numbers, providing an important source of protein for the termite.

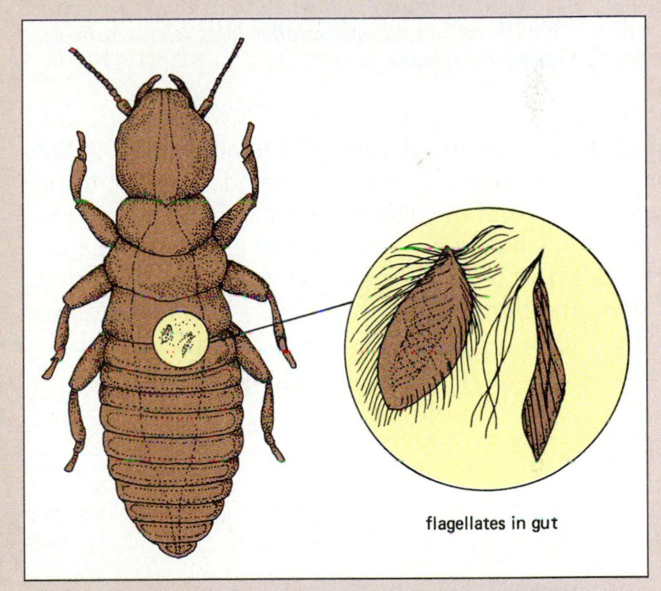

Figure E24-2 The termite derives its dismaying ability to digest wood from the presence of several types of protists, called flagellates, in its intestine.

the esophagus contract in sequence above the swallowed food mass, squeezing it downward toward the stomach (Fig. 24-9). This muscular action, called **peristalsis,** also occurs in the stomach and intestines, where it helps move food along the digestive tract. Peristalsis is so effective that a person can actually swallow when upside down. Mucus secretions by the esophagus help protect it from abrasion and lubricate the food during its passage.

The stomach is an expansible, muscular sac capable of holding from 2 to 4 liters of food and liquids. A ring of muscle, called the **gastroesophageal sphincter,** opens to allow passage of food from the esophagus, then closes to retain it. A second sphincter at the base of the stomach, the **pyloric sphincter,** regulates passage of food into the small intestine. The stomach has three major functions. First, it stores food and releases it gradually into the small intestine, at a rate suitable for proper digestion and absorption. Thus the stomach allows us to eat large, infre-

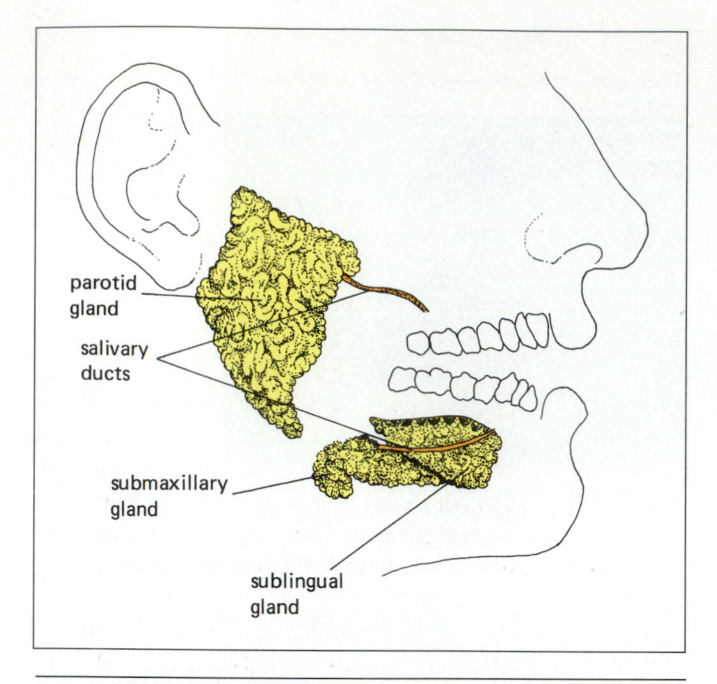

Figure 24-7 Digestion begins in the mouth as three large pairs of salivary glands and numerous smaller ones release saliva in response to the food stimuli.

quent meals. Carnivores carry this to an extreme. A lion, for instance, may consume 40 pounds of meat at one meal, then spend the next few days quietly digesting it.

A second function of the stomach is the mechanical breakdown of food. Its muscular walls exhibit a variety of churning movements which help disrupt large pieces of food.

Third, glands in the lining of the stomach secrete enzymes and other substances that facilitate digestion. These include gastrin, hydrochloric acid, pepsinogen, and mucus. Gastrin, a hormone, stimulates secretion of hydrochloric acid by specialized stomach cells. Other cells release pepsinogen, an inactive form of the protein-digesting enzyme, pepsin. Pepsin, a protease, breaks proteins into shorter chains of amino acids called peptides (see Table 24-3). It is secreted in an inactive form to prevent it from digesting the cells that produce it. The highly acidic conditions in the stomach (pH 1) convert pepsinogen into pepsin. In addition, this enzyme functions best in an acidic environment. As you may have noticed, the stomach produces all the ingredients necessary to digest itself, and indeed, this is what happens when a person develops ulcers. However, cells lining the stomach normally produce copious quantities of thick mucus which coat the stomach lining, serving as a barrier to self-digestion. This is not completely successful, however, and the cells lining the stomach are digested to some extent, needing replacement every few days.

Food in the stomach is gradually converted to a soupy liquid called **chyme,** consisting of partially digested food and digestive secretions. Peristaltic waves, traversing the muscular stomach at a rate of about three per minute, propel the chyme toward the small intestine. The sphincter at the base of the stomach allows only a small amount of chyme to leave with each contraction. It takes two to six hours, depending on the size of the meal, to empty the stomach completely. After this time, its continued churning movements are felt as hunger pangs.

Only a few substances can enter the bloodstream

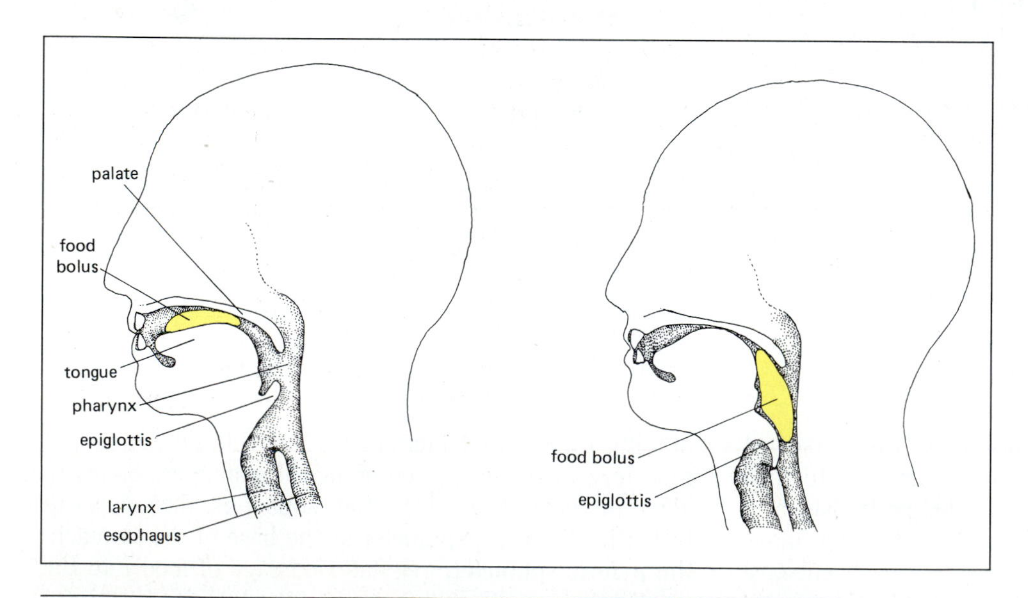

Figure 24-8 Swallowing is complicated by the fact that both the esophagus and the larynx (part of the respiratory system) open into the pharynx. During swallowing, the larynx moves upward beneath a small flap of cartilage, the epiglottis. This seals off the opening to the respiratory system, directing food down the esophagus.

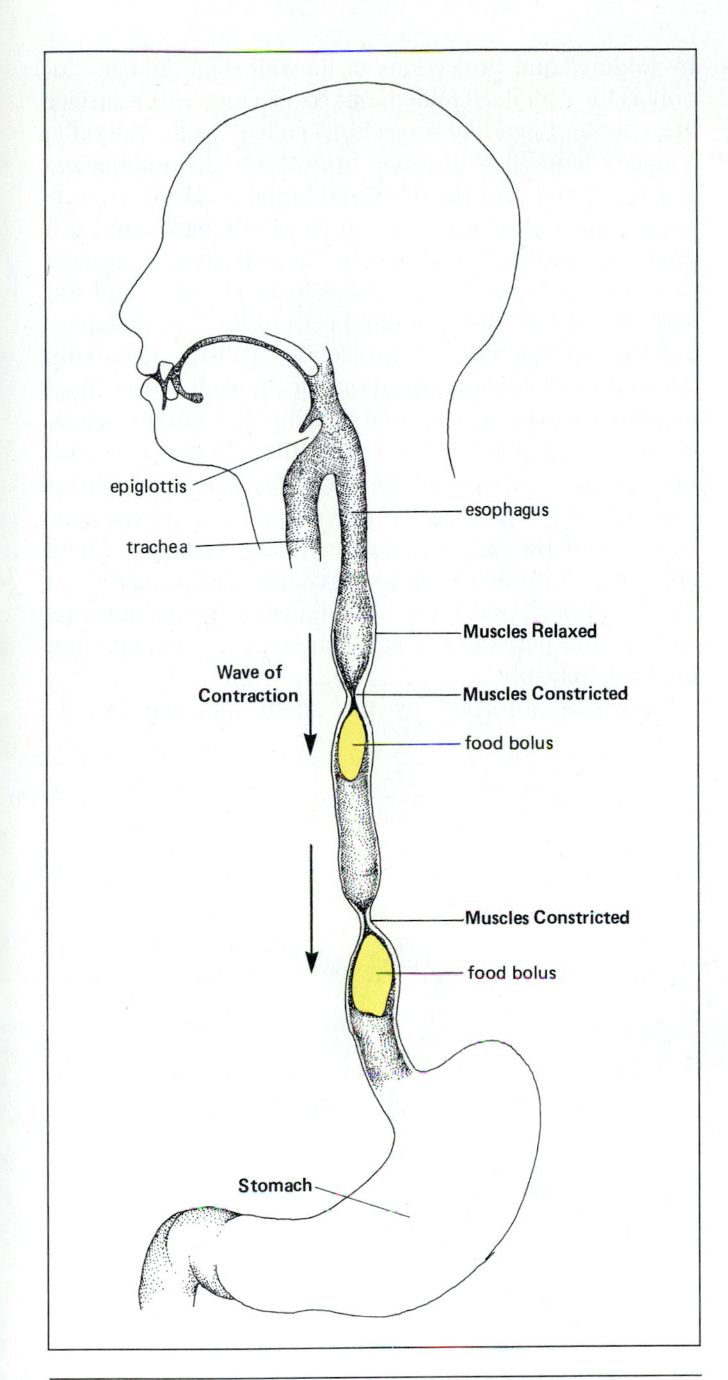

Figure 24-9 Food is propelled through the digestive system by peristaltic contractions of circular muscles which proceed downward, forcing the food along in front of them. These first occur in the esophagus as illustrated, but will also assist the movement of food through the stomach and intestines.

through the stomach wall, including water, some drugs, and alcohol. Alcohol consumed when the stomach is empty is immediately absorbed into the bloodstream, with strong and rapid effects. Since food in the stomach slows alcohol absorption, the advice "never drink on an empty stomach" is based on sound physiological principles.

The Small Intestine

This coiled, narrow (2 to 4 centimeters in diameter) tube is the longest portion of the digestive tract, reaching about 275 centimeters (9 feet) in an adult. The small intestine has two major functions: it digests food into small molecules, and absorbs these molecules, passing them to the bloodstream or lymph vessels. The small intestine receives digestive secretions from three sources, the liver, the pancreas, and the cells of the small intestine itself, each discussed in more detail below.

THE LIVER. This is the largest and perhaps the most versatile organ in the body. Its many functions include the storage of fats and carbohydrates for energy, the regulation of glucose levels in the blood, the synthesis of blood proteins, the storage of iron and certain vitamins, and the detoxification of harmful substances such as nicotine and alcohol. Its role in digestion is to produce **bile,** which is stored in the gallbladder and released into the small intestine through the bile duct (see Fig. 24-5). Bile assists in the breakdown of lipids, but it is not a digestive enzyme. Instead, bile acts as a detergent or emulsifying agent. Since lipids and water do not mix, the oils and fat in chyme tend to accumulate in globs. Bile contains molecules with both hydrophilic (literally, "water-loving") and hydrophobic ("water-fearing" or lipid-soluble) portions. One end binds to the lipid while the other dissolves in the surrounding water. This suspends the lipids in the chyme as microscopic particles. These particles expose a large surface area for attack by **lipases,** lipid-digesting enzymes.

THE PANCREAS. This small organ lies in the loop between the stomach and small intestine (see Fig. 24-5). It consists of two major cell types. One type produces hormones involved in blood sugar regulation (insulin and glucagon) and the other type produces digestive secretions released into the small intestine. The 1200 to 1500 milliliters (1.0 to 1.5 quarts) of pancreatic juice released into the small intestine includes water, sodium bicarbonate, and a number of digestive enzymes (see Table 24-3). Sodium bicarbonate (the active ingredient in baking soda) neutralizes the acidic chyme, producing the slightly alkaline pH required for proper functioning of the pancreatic digestive enzymes. These enzymes are specialized to break down three major types of food: an amylase breaks down carbohydrates, lipases digest lipids, and several proteases disrupt proteins and peptides. The pancreatic proteases include trypsin, chymotrypsin, and carboxypeptidase. Both trypsin and chymotrypsin break proteins and peptides into shorter peptide chains. Carboxypeptidase completes protein digestion by liberating individual amino acids from the ends of the peptides. These proteases are secreted in an inactive form and become activated after reaching the small intestine.

The intestinal wall. The wall of the small intestine is studded with cells specialized to complete the digestive process and absorb the small molecules that result. These cells have on their external membranes several types of enzymes, including proteases, which complete the breakdown of peptides into amino acids, and sucrase, lactase, and maltase, which break down disaccharides into monosaccharides (see Chapter 3), and small amounts of lipase. These enzymes are actually embedded in the membranes of the cells lining the small intestine, so this final phase of digestion occurs as the nutrient is being absorbed into the cell. As in the stomach, the small intestine is protected by copious mucus secretions from specialized cells in its lining.

Absorption in the small intestine. The small intestine is not only the principal site of chemical digestion, it is also the major site of nutrient absorption into the blood or lymph. To facilitate absorption, the small intestine, in addition to its length, has an internal surface area which is increased 600-fold over that of a smooth tube by foldings and projections of its wall (Fig. 24-10). Not only is the wall itself folded, but covering its entire surface are minute, fingerlike projections called **villi** (literally, "shaggy hairs"). Villi range from 0.5 to 1.5 millimeters in length, and give the intestinal lining a velvety appearance to the naked eye. They move gently back and forth amid the digested food within the intestine, increasing their exposure to the molecules to be digested and absorbed. Further, the individual cells of the villi each bear a fringe of microscopic projections called **microvilli.** Altogether, these specializations of the wall of the small intestine give it a surface area of about 250 square meters, about the size of a tennis court. Contact of nutrients with the absorptive surface of the small intestine is further facilitated by **segmentation movements,** rhythmic contractions of the circular muscles of the intestine. These are not synchronized, as in peristalsis, but rather slosh the liquid back and forth. When absorption is complete, coordinated peristaltic waves conduct the remnants into the large intestine.

Nutrients absorbed by the small intestine include

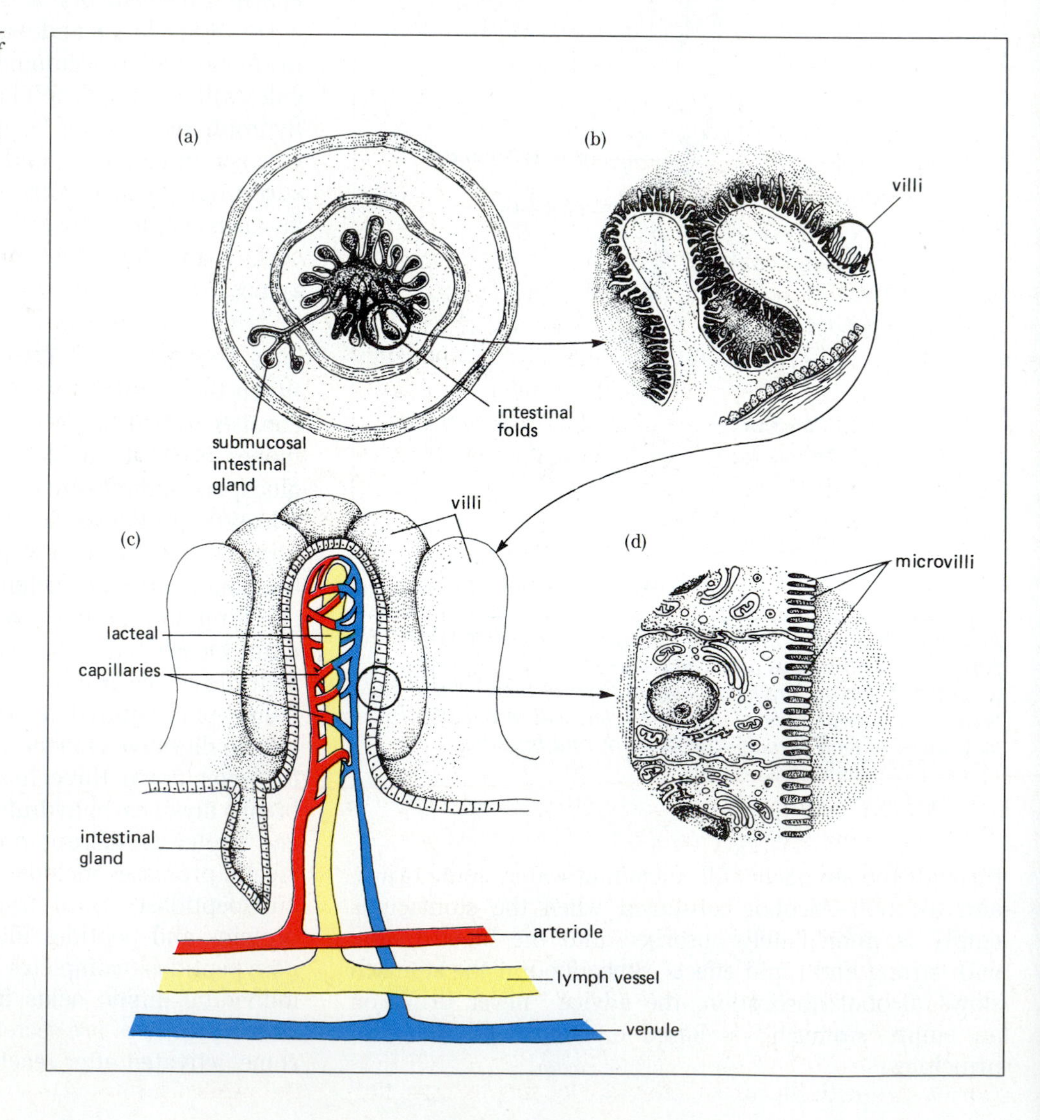

Figure 24-10 The surface area of the small intestine is greatly increased by **(a)** *foldings of the intestinal lining,* **(b, c)** *projections called* villi *(singular* villus*) that extend from the folds, and* **(d)** *microscopic projections of individual cell membranes called* microvilli. **(c)** *The absorbed nutrients enter blood capillaries and lymph vessels that form a network within each villus.*

monosaccharides, amino acids and short peptides, fatty acids produced by lipid digestion, vitamins, and minerals. The mechanisms by which this absorption occurs are varied and complex and are currently the subject of active investigation. In most cases, energy is expended to transport nutrients into the intestinal cells. The nutrients then diffuse out of the intestinal cells into the **interstitial fluid,** from whence they enter the bloodstream. Each villus of the small intestine is provided with a rich supply of blood and lymph capillaries to carry off the absorbed nutrients and distribute them throughout the body (Fig. 24-10c). Most of the nutrients enter the bloodstream via the capillaries, but fatty acids take a different route. After diffusing into the epithelial cells, they are synthesized into larger triglycerides, mixed with cholesterol and phospholipids, and then released as globules into the interstitial fluid by exocytosis. Blood capillaries are relatively impermeable to these globules, so most enter the lymph capillaries, whose walls allow larger particles to enter (see Fig. 26-15). The lymph vessels eventually empty into the veins, delivering the lipids to the bloodstream.

The Large Intestine

The large intestine is not named for its length (about 150 centimeters or 5 feet in the living adult human), but for its diameter (about 6.5 centimeters or 2.5 inches), which is considerably larger than the small intestine. It receives the leftovers of digestion: a mixture of water, undigested fats and proteins, and indigestible fibers such as the cell walls of vegetables and fruits. The large intestine contains a flourishing population of bacteria living on unabsorbed nutrients. These bacteria earn their keep by synthesizing vitamins B_{12}, thiamine, riboflavin, and most important, vitamin K, which otherwise is often deficient in a normal diet. The absorption of residual water and salts occurs in the large intestine. The result is a semisolid fecel mass, consisting not only of indigestible wastes but also the dead bodies of bacteria, which account for about one-third the dry weight of feces. The feces are transported by peristaltic movements until they reach the rectum. Distention of this chamber stimulates the defecation reflex, which, fortunately, can be regulated voluntarily.

The Control of Digestion

Considerable coordination is required to break down a chef's salad into amino acid and peptides, water, sugars, fatty acids, vitamins, minerals, and indigestible cellulose. As the mouth responds to the first bite, the stomach must be warned of the imminent influx of food. In addition, the enzymes of the stomach and small intestine require different environments for proper functioning (highly acidic in the stomach, slightly alkaline in the small intestine), and secretions into various parts of the digestive tract must be coordinated with the arrival of food. Not surprisingly, the secretions and activity of the digestive tract are coordinated by both nerves and hormones (Table 24-4). Here we examine a few of these control mechanisms.

The initial phase of digestion is under the control of the nervous system and involves responses to signals originating in the head. These signals include the sight, smell, taste, and sometimes the thought of food, as well as the muscular activity of chewing. In response to these stimuli, saliva is secreted to the mouth, while nervous signals to the stomach walls initiate secretion of acid and the hormone **gastrin,** which stimulates further acid secretion. The concentration of acid is regulated by a **negative feedback** mechanism. When acid levels reach a certain point, they inhibit gastrin secretion, thus inhibiting further acid production. The control of hormones by negative feedback is discussed in more detail in Chapter 28.

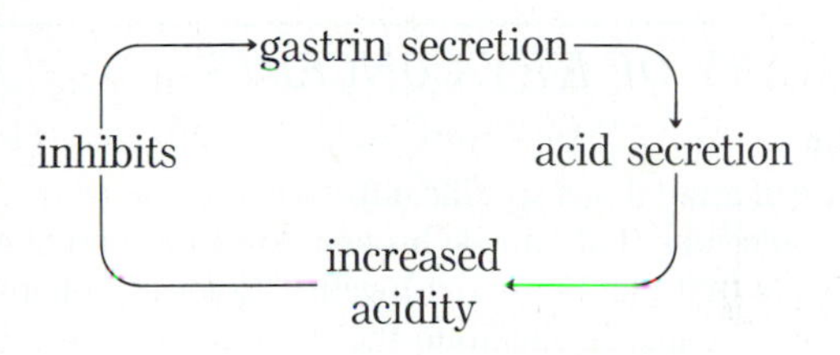

The arrival of food in the stomach triggers the second phase of digestion. Irritation of the stomach wall causes copious production of mucus to protect against self-digestion. The acidity of the stomach converts pepsinogen to its active form, pepsin, which begins protein digestion. Protein in the food tends to buffer or reduce the concentration of stomach acid. The release of gastrin is no longer inhibited, and gastrin release stimulates further acid production. The cells secreting stomach acid are also activated by distension of the stomach and by the presence of peptides produced by protein digestion.

As the liquid chyme is gradually released into the small intestine, its acidity stimulates the release of a second hormone, **secretin,** by cells of the upper small intestine. Secretin causes the pancreas and liver to pour bicarbonate into the small intestine. Bicarbonate neutralizes the acidity of the incoming chyme and creates an environment in which the pancreatic enzymes can function. A third hormone, **cholecystokinin,** is also produced by cells of the upper small intestine in response to the presence of chyme. This hormone stimulates the release of various digestive enzymes by the pancreas into the small intestine. It also stimulates the gallbladder to contract, squeezing bile through the bile duct to the small intestine. Bile assists in fat breakdown, as described earlier. **Gastric inhibitory peptide,** a hormone secreted by the small intestine in response to fatty acids and sugars in chyme, inhibits acid production and peristalsis in the stomach. This slows down the rate at which chyme is dumped into the small intestine, providing additional time for digestion and absorption to occur.

Table 24-4 **Some Important Digestive Hormones**

Hormone	Site of Production	Stimulus for Production	Effect
Gastrin	Stomach	Food in mouth Distention of stomach Peptides in stomach	Stimulates acid secretion by cells in stomach
Secretin	Small intestine	Acid in small intestine	Stimulates bicarbonate production by pancreas and liver, increases bile output by liver
Cholecystokinin	Small intestine	Amino acids, fatty acids in small intestine	Stimulates secretion of pancreatic enzymes and release of bile by gallbladder
Gastric-inhibitory peptide	Small intestine	Fatty acids and sugars in small intestine	Inhibits stomach movements and release of stomach acid

SUMMARY OF KEY CONCEPTS

Nutrition

Each type of animal has specific nutritional requirements. These include molecules that can be broken down to liberate energy, amino acids that can be linked together to form proteins, minerals, and vitamins to facilitate the diverse chemical reactions of metabolism.

The Challenge of Digestion

Diverse Digestive Systems Digestive systems are designed to convert the complex molecules of the bodies of other animals or plants into simpler molecules which can be utilized by the consumer. Animal digestion at its simplest is intracellular, as occurs within the individual cells of a sponge. Extracellular digestion, utilized by all more complex animals, occurs in a body cavity. The simplest form is a "dead end," saclike gastrovascular cavity in organisms such as flatworms and hydra. Still more complex animals utilize a tubular compartment with specialized chambers where food is processed in a well-defined sequence.

Human Digestion In humans, digestion begins in the mouth, where food is physically broken down by chewing and chemical digestion is initiated by saliva. Food is then conducted to the stomach by peristaltic waves of the esophagus. In the acidic environment of the stomach, food is churned into smaller particles, and protein digestion begins. Gradually, the liquefied food, now called chyme, is released to the small intestine. Here it is neutralized by bicarbonate from the pancreas. Secretions from the pancreas, liver, and the cells of the intestine complete the breakdown of proteins, fats, and carbohydrates. The small intestine is also the site where the simple molecular products of digestion are absorbed into the bloodstream for distribution to the body cells. The large intestine absorbs the remaining water and converts indigestible material to feces.

Digestion is regulated by the nervous system and hormones. The smell and taste of food and the actions of chewing causes salivary secretion and production of gastrin by the stomach. Gastrin stimulates stomach acid production. As chyme enters the small intestine, three additional hormones are produced by intestinal cells. These are: secretin, which causes bicarbonate production to neutralize the acid chyme; cholecystokinin, which stimulates bile release and causes the pancreas to secrete digestive enzymes into the small intestine; and gastric inhibitory peptide, which inhibits acid production and peristalsis by the stomach. This slows the movement of food into the intestine.

GLOSSARY

Absorption the movement of nutrients into cells.

Amylase (am′-ē-lās) an enzyme that catalyzes the breakdown of starch, found in saliva and pancreatic secretions.

Bile (bīl′) a liquid secretion of the liver stored in the gallbladder and released into the small intestine during digestion. Its role is to emulsify or disperse fats into small particles on which fat-digesting enzymes may act.

Calorie (kal′-ōr-ē) a measure of the energy derived from food. When capitalized (i.e., Calorie) this unit is the amount of energy required to raise the temperature of 1 liter of water 1 degree Celsius. It represents 1000 calories (with a lowercase "c"). The energy content of foods is measured in Calories.

Carbohydrate (kar-bō-hī′-drāt) a class of nutrient including simple sugars, disaccharides, and polysaccharides (starches, glycogen, and cellulose). Sugars and starches are used by animal cells as a source of energy.

Cholecystokinin (kō′-lē-sis-tō-kī′-nin) a digestive hormone produced by the small intestine that stimulates release of pancreatic enzymes.

Chyme (kīme) an acidic, souplike mixture of partially digested food, water, and digestive secretions that is released from the stomach into the small intestine.

Crop an organ found in both earthworms and birds in which ingested food is stored temporarily before passing to the gizzard, where it is pulverized.

Digestion the process by which food is physically and chemically broken down into molecules that can be absorbed by cells.

Epiglottis (ep-eh-gla′-tis) a flap of cartilage in the lower pharynx that covers the opening to the larynx during swallowing. This directs the food down the esophagus.

Esophagus (eh-sof′-eh-gus) a muscular passageway connecting the pharynx to the next chamber of the digestive tract, the stomach in humans and other mammals.

Extracellular digestion the physical and chemical breakdown of food that occurs in a digestive cavity.

Food vacuole a membranc-bound space within a single cell in which food is enclosed. Digestive enzymes are released into the vacuole and intracellular digestion occurs here.

Gallbladder a small sac adjacent to the liver in which the bile secreted by the liver is stored. Bile is released from the gallbladder via the bile duct to the small intestine.

Gastrovascular cavity a chamber that has both digestive and circulatory functions, found in simple invertebrates. A single opening serves as both mouth and anus, while the chamber provides direct access of nutrients to the cells.

Gizzard a muscular organ found in earthworms and birds in which food is mechanically broken down prior to chemical digestion.

Intracellular digestion the chemical breakdown of food, which occurs within single cells.

Lipase (lī′-pāse) an enzyme that catalyzes the breakdown of lipids, such as fats.

Lymph (limpf) a fluid resembling blood plasma that collects in special lymph vessels and eventually returns to the bloodstream.

Microvilli (mī-krō-vi′-lī) a series of folded projections of the cell membrane which increase its surface area.

Pancreas (pan′-krē-is) an organ lying adjacent to the stomach which secretes enzymes for fat, carbohydrate, and protein digestion into the small intestine. Other cells in the pancreas are responsible for production of the hormones glucagon and insulin.

Pharynx (fār′-inx) a chamber located behind the mouth. In vertebrates, it is common to both the respiratory and digestive systems.

Protease (prō′-tē-ās) an enzyme that digests proteins.

Segmentation movements asynchronous contractions of the small intestine which result in mixing of the partially digested food and digestive enzymes. The movements also bring nutrients into contact with the absorptive intestinal wall.

Villus (vi-lus) projections of the wall of the small intestine which increase its absorptive surface area.

Vitamin any one of a group of diverse chemicals that must be present in trace amounts in the diet to maintain health. Vitamins are used by the body in conjunction with enzymes in a variety of metabolic reactions; a few are involved in growth and differentiation.

STUDY QUESTIONS

1. List four general types of nutrients and describe the role of each in nutrition.
2. What general role do water-soluble vitamins play in human metabolism?
3. Vitamin C is a vitamin for humans but not for dogs. Explain.
4. Trace the pathway of an indigestible tomato seed through the human digestive tract.
5. List and describe the function of the three principal secretions of the stomach.
6. Explain how surface area is increased in the small intestine, and why this is important.
7. List the substances secreted into the small intestine and describe the origin and function of each.
8. Describe the function and origin of four major digestive hormones.
9. Name and describe the muscular movements that usher food through the human digestive tract.

SUGGESTED READINGS

Davenport, H. "Why the Stomach Does Not Digest Itself." *Scientific American,* January 1972. How the stomach protects itself from its own strongly acidic secretions, and what happens when the defenses fail.

Eckert, R., Randall, D. and Augustine, G. *Animal Physiology: Mechanisms and Adaptations,* 3rd ed. New York: W. H. Freeman and Company, Publishers, 1988. An excellent and complete textbook of comparative animal physiology.

Hole, Jr, J. W. *Essentials of Human Anatomy and Physiology.* 2nd ed. Dubuque, Iowa: Wm. C. Brown Company, 1986. Descriptions of both human anatomy and physiology. Beautiful, clear illustrations.

Moog, F. "The Lining of the Small Intestine." *Scientific American,* November 1981. A description of the structure and function of this intricate tissue, which is responsible for absorbing nutrients into the body.

Vander, A., and Sherman, J. *Human Physiology,* 4th ed. New York: McGraw-Hill Book Company, 1985. Complete coverage of human physiology.

25
Excretion

Animal cells can function only under a relatively narrow range of conditions. Thus, while we load our digestive systems with pepperoni pizza and hot fudge sundaes, our cells remain bathed in a precisely regulated solution of salts and nutrients maintained by the body despite our dietary eccentricities. This precise internal regulation is called **homeostasis.**

Homeostasis is the end result of the coordinated activity of the liver, the nervous system, the endocrine system, and the circulatory system. These are aided by organs that exchange materials directly with the external environment, such as lungs (or gills), skin, the digestive organs, and most important, the excretory organs. The cells of the digestive system, through which most substances enter the body, are relatively unselective; any molecule that can move into the body through the intestinal lining does so, including an excess of water, nutrients, salts, and minerals, and nonnutritive substances such as drugs. The burden of restoring and maintaining proper internal balance then falls on the organs of homeostasis, particularly the excretory system.

Excretory systems have two major functions: excretion of cellular waste products such as urea, and maintenance of body fluid composition. They perform both of these functions simultaneously by filtering the blood. Excretory systems first collect the fluid portion of the blood. From this fluid, water and important nutrients are reabsorbed into the blood, while toxic substances, wastes, excess nutrients and hormones, and some water are left behind to be eliminated as urine (Fig. 25-1). Urine is produced by the kidneys of vertebrates and various simpler excretory organs of invertebrates, as described below. The type of urine produced is intimately related to the animals' need to conserve water, as discussed in the essay, "Excretion and Animal Life-Style."

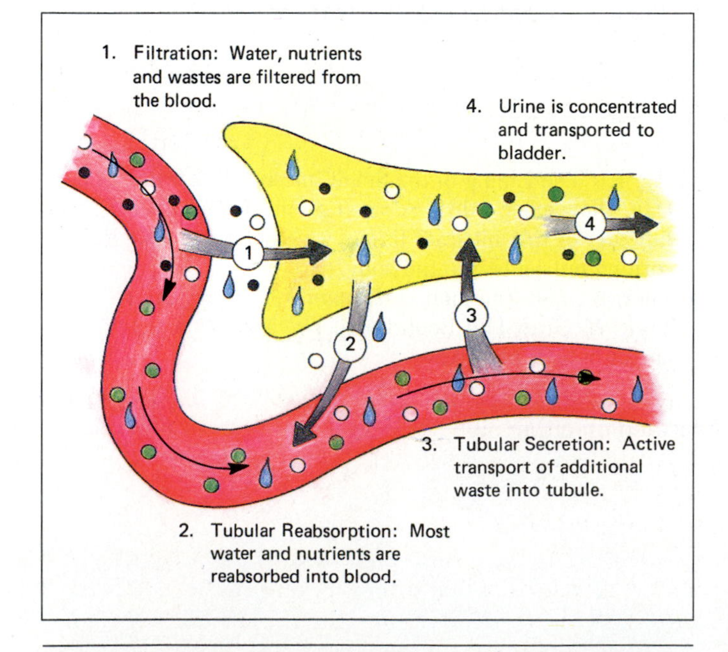

Figure 25-1 Diagrammatic illustration of the major processes of excretion. Red indicates the bloodstream, yellow the excretory system.

SIMPLE EXCRETORY SYSTEMS

Flame Cells in Flatworms

One of the simplest animals with an excretory system is the flatworm; an example is *Planaria,* common under rocks in streams. Its excretory system consists of a network of tubes that branch throughout the body (Fig. 25-2). At intervals the tubes end blindly in single-celled bulbs called "flame cells," named after the tuft of beating cilia extending into the hollow bulb. Under the microscope, the beating of the cilia resembles a flickering flame. Water and some dissolved wastes are filtered into the bulbs, where the beating cilia produce a current that conducts the fluid through the tubular network. Here waste products may be added to the filtrate and nutrients withdrawn. Eventually, the waste liquid reaches one of numerous pores which release it to the outside. Flatworms also rely extensively on their large skin surface, through which wastes leave by diffusion.

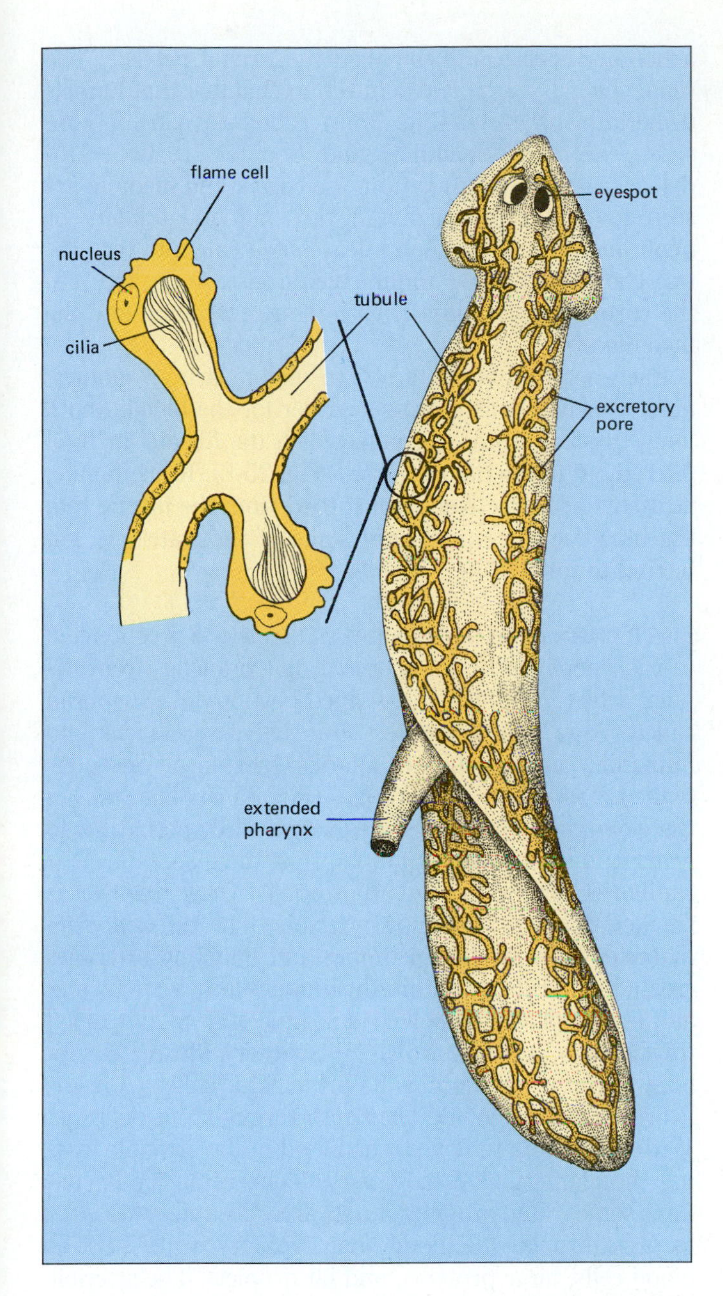

Figure 25-2 The simple excretory system of a flatworm. Hollow flame cells direct excess water and dissolved wastes into a network of tubes. The beating cilia of the flame cells help circulate the fluid to excretory pores.

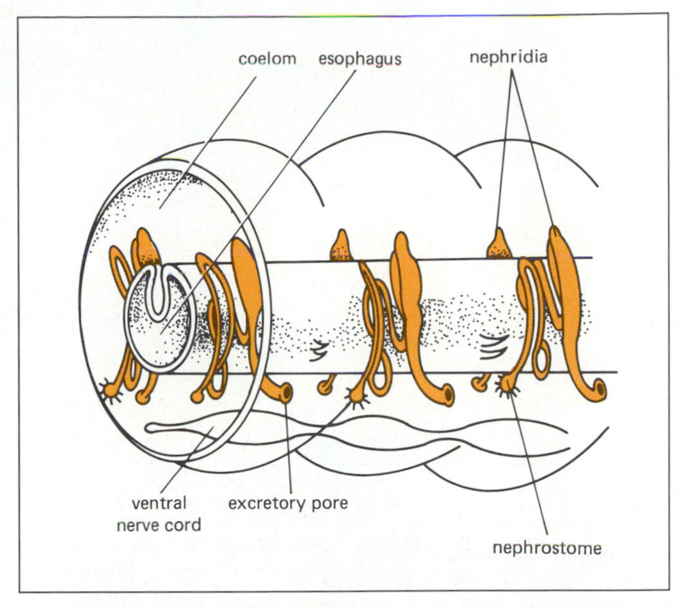

Figure 25-3 The excretory system of the earthworm consists of a series of structures called nephridia, one pair per segment. Coelomic fluid is drawn into the nephrostome and urine is released through the excretory pore. Each nephridium resembles a vertebrate nephron.

Nephridia in Earthworms

Earthworms, molluscs, and several other types of invertebrates have simple kidneys called **nephridia** (singular **nephridium**). In the earthworm, coelomic fluid fills the body cavity surrounding the internal organs, collecting both wastes and nutrients from the blood and tissues. This fluid is conducted into the funnel-shaped opening of the nephridium and swept by cilia along a narrow, twisted tube (Fig. 25-3). Here important salts and other dissolved substances are absorbed back into the blood, leaving water and wastes behind. The resulting urine is stored in an enlarged bladder-like portion of the nephridium, and then excreted through a pore in the body wall. The earthworm body is composed of repeating segments, nearly every one of which contains its own pair of nephridia.

HUMAN EXCRETION

Humans and other vertebrates filter their blood through **kidneys,** complex organs that in some ways resemble dense collections of nephridia. The kidneys are part of a larger group of structures called the **excretory system** (Fig. 25-4). While the kidneys actually produce the urine, the rest of the system serves to transport, store, and eliminate it. We will first examine the major structures of the excretory system, tracing the pathway of waste products, before proceeding to a more detailed account of kidney function.

Human kidneys are paired, kidney bean shaped organs located on either side of the spinal column and extending slightly above the waist. Each is approximately 13 centimeters (5 inches) long, 8 centimeters (3 inches) wide and $2\frac{1}{2}$ centimeters (1 inch) thick.

Blood carrying dissolved cellular wastes enters each kidney through a **renal artery.** After it has been filtered, the blood exits via the **renal vein** (Fig. 25-4). Urine (consisting of waste substances and water filtered from the blood) leaves each kidney through a tube called the **ureter.** The ureters transport urine by peristaltic contrac-

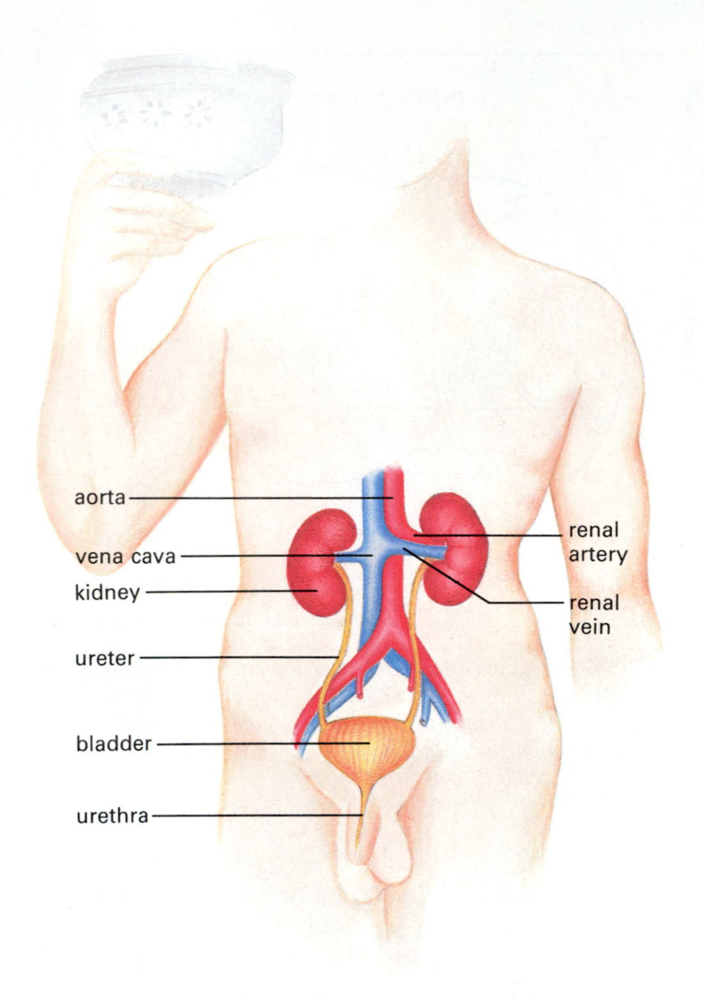

Figure 25-4 The human excretory system and its blood supply.

tions to the **bladder.** This hollow muscular chamber collects and stores the urine. Urine completes its journey to the outside via the **urethra,** a single narrow tube about 3.8 centimeters (1.5 inches) long in the female and about 20 centimeters (8 inches) long in the male. The walls of the bladder, composed of smooth muscle, are capable of considerable expansion. Urine is retained in the bladder by two sphincter muscles located at its base just above the junction with the urethra. When the bladder becomes distended, receptors in the walls signal its condition and trigger reflexive contractions. The sphincter nearest the bladder, the internal sphincter, is opened during this reflex. The lower or external sphincter, however, is under voluntary control, so the reflex can be suppressed by the brain unless distension becomes acute. The average adult bladder will hold 700 to 800 milliliters (about $2\frac{1}{2}$ cups) of urine, but the desire to urinate is triggered by accumulations of 200 to 400 milliliters of urine.

Human Kidney Structure and Function

KIDNEY STRUCTURE. In cross section, the kidney consists of a solid outer layer where the urine is formed, and a hollow inner chamber called the **renal pelvis.** The renal pelvis is a branched collecting chamber that funnels urine into the ureter (Fig. 25-5). The solid portion consists of an inner **medulla** and an outer **cortex.** Under microscopic examination, the cortex and medulla are seen to contain an array of tiny individual filters or **nephrons,** each of which releases its urine into the central chamber. Over 1 million nephrons are packed into the cortex of each kidney, with many extending into the medulla.

Each nephron consists of two major parts: a **glomerulus,** which acts as a pressure filter for the blood; and a long, twisted **tubule** through which the filtrate, or fluid filtered out of the blood, passes (Fig. 25-6). In the tubule, nutrients are selectively reabsorbed from the filtrate into the blood, while wastes and some of the water are left behind to form urine.

FILTRATION BY THE GLOMERULUS. Blood is conducted to each nephron by an arteriole that branches from the renal artery. Within a cup-shaped portion of the nephron called **Bowman's capsule,** this vessel subdivides into numerous microscopic capillaries that form an intertwined mass, the **glomerulus** (Fig. 25-6). The walls of the glomerular capillaries are exceptionally permeable to water and dissolved substances. Past the glomerulus, the capillaries reunite to form an arteriole, whose diameter is smaller than the incoming arteriole. The differences in diameter between the incoming and outgoing arterioles create high pressure within the glomerulus, driving water and many of the dissolved substances from the blood through the capillary walls. This watery filtrate, resembling blood plasma minus its proteins, is collected in the Bowman's capsule for transport through the nephron. With the filtrate removed, the blood in the arteriole leaving the glomerulus is now very "concentrated." It contains some water with its solutes, and substances too large to pass through the glomerular capillary walls, such as blood cells, large proteins, and fat droplets. The arteriole now branches into smaller, highly porous capillaries. These capillaries surround the tubule, forming intimate contacts with it. Here water and nutrients are reabsorbed from the filtrate as it passes through the nephron and are returned to the blood.

URINE FORMATION IN THE NEPHRON. The blood filtrate collected in Bowman's capsule contains a mixture of both wastes and essential nutrients, including most of the blood's vital water. The nephron restores the nutrients and most of the water to the blood, while retaining wastes for elimination. This is accomplished by two processes: tubular reabsorption and tubular secretion.

Tubular reabsorption is the process by which cells of the tubule remove water and nutrients from the filtrate within the tubule and pass them back into the blood. Reabsorption of salts and other nutrients, such as amino

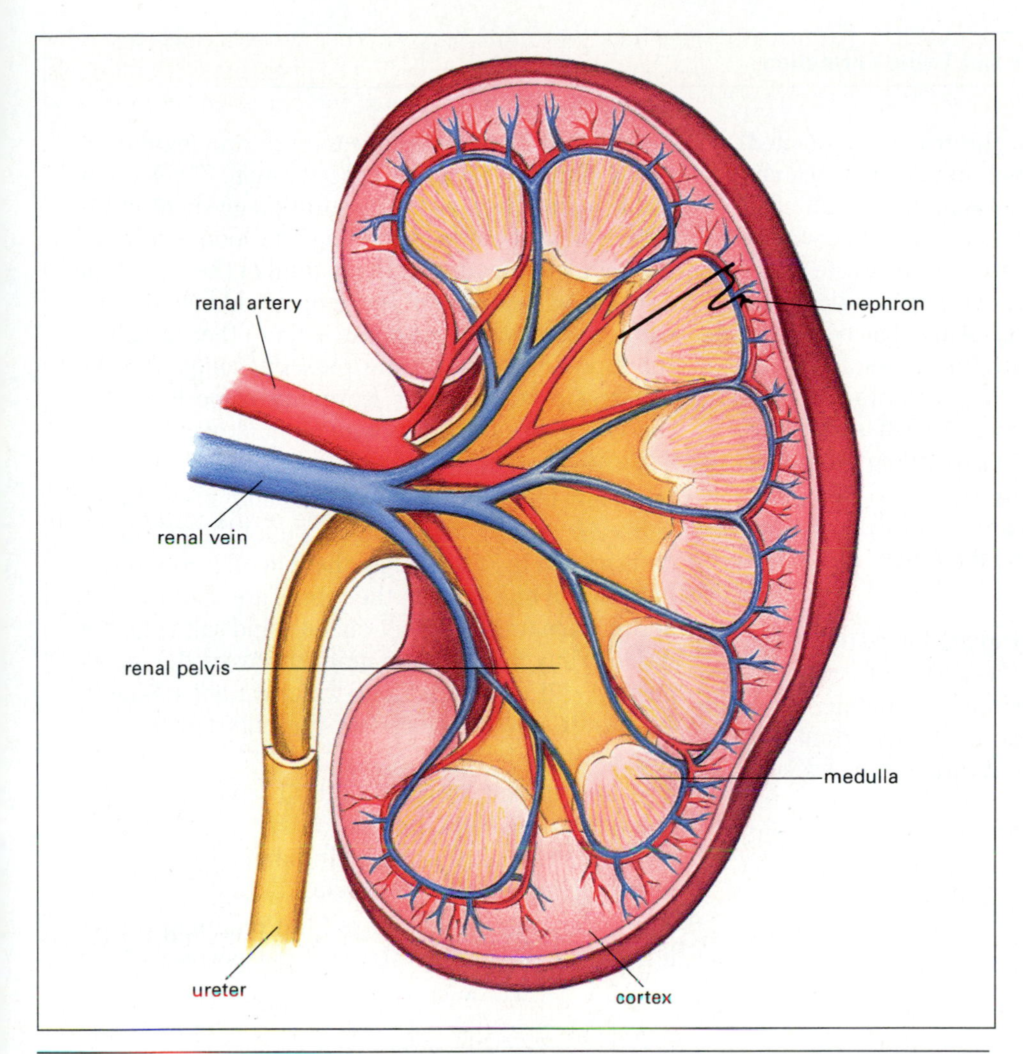

Figure 25-5 *Cross section of a kidney, showing its blood supply and gross internal structure. The renal artery, which brings blood to the kidney, and the renal vein, which carries the filtered blood away, branch extensively within the kidney. The two are joined by a highly permeable capillary network through which substances are exchanged between the blood and the nephrons. A nephron, considerably enlarged, is drawn to show its orientation in the kidney.*

acids and glucose, generally occurs by active transport (cells of the tubule expend energy to transport these substances out of the tubule). These nutrients then enter adjacent capillaries by diffusion. Water is reabsorbed passively, following the nutrients out of the tubule in response to the osmotic gradient created in the fluid surrounding the tubule. Wastes such as urea (a product of protein digestion) remain in the tubule and become concentrated as water leaves.

Tubular secretion is the process by which waste substances remaining in the blood are actively secreted *into* the tubule by tubule cells. These substances include hydrogen and potassium ions, various foreign substances, and drugs such as penicillin. In this way, excess substances which were not initially filtered out into the Bowman's capsule are removed from the blood for excretion.

THE PRODUCTION AND CONCENTRATION OF URINE. The kidneys of mammals and birds are able to produce urine that is more concentrated than the blood. This ability is determined by the structure of both the nephron and the **collecting duct** into which several nephrons empty. The nephron consists of five major parts. In the direction of fluid flow, these are: the glomerulus, Bowman's capsule, the **proximal tubule,** the **loop of Henle,** and the **distal tubule,** which leads to the collecting duct (see Fig. 25-6). Each section is specialized, and the filtrate is processed sequentially as it moves through the nephron. Details are presented in the essay; "Probing Deeper: The Nephron and Urine Formation."

Concentrated urine results from the osmotic concentration gradient of salts and urea in the fluid surrounding the loop of Henle. The most osmotically concentrated

Probing Deeper: The Nephron and Urine Formation

The complex structure of the nephron is finely adapted to its function. However, despite decades of research, the precise mechanism of urine formation is still controversial. Our description reflects a recent understanding which includes a prominent role for the waste product urea. The nephron is presented diagramatically in Fig. E25-1 to illustrate the processing that occurs in each part. The osmotic concentration of the filtrate is shown (the higher the number, the greater the concentration of solutes). The graph to the left shows the concentration of solutes in the surrounding fluid. Notice that the primary solutes are NaCl and urea, and that their concentrations increase toward the bottom of the loop of Henle. Circled numbers refer to the following descriptions.

1. Water and dissolved substances are forced out of the glomerular capillaries into the Bowman's capsule, from which they are funneled into the tubule.
2. In the proximal tubule, most of the important nutrients are actively pumped out through the walls of the tubule and are reabsorbed into the blood (tubular reabsorption). These include about 75 percent of the salts and water, as well as amino acids, sugars, and vitamins. The proximal tubule is highly permeable to water, so water follows the nutrients, moving by osmosis along its concentration gradient.
3. The loop of Henle is unique to birds and mammals, and is essential for urine concentration. The loop of Henle maintains a salt concentration gradient in the extracellular fluid surrounding the loop, with the highest concentration at the bottom of the loop. The descending portion of the loop of Henle is very permeable to water but not to salt or other dissolved substances. As the filtrate passes through the descending portion, water leaves by osmosis as the concentration of the surrounding fluid increases.
4. The thin portion of the ascending loop of Henle is relatively impermeable to water and urea but is permeable to salt, which moves out of the filtrate by diffusion. Why? Although the osmotic concentrations inside and outside the tubule are about equal, at this stage urea is higher outside, and salt is higher inside. Thus the diffusion gradient favors the movement of salt outward. Since water cannot follow it, the filtrate now becomes less concentrated than its surroundings.
5. The loss of salt from the filtrate continues as it moves into the thick portion of the ascending loop of Henle. Here salt is actively pumped out of the filtrate, leaving water and wastes behind.
6. The watery filtrate, low in salt but retaining wastes like urea, now arrives at the distal portion of the

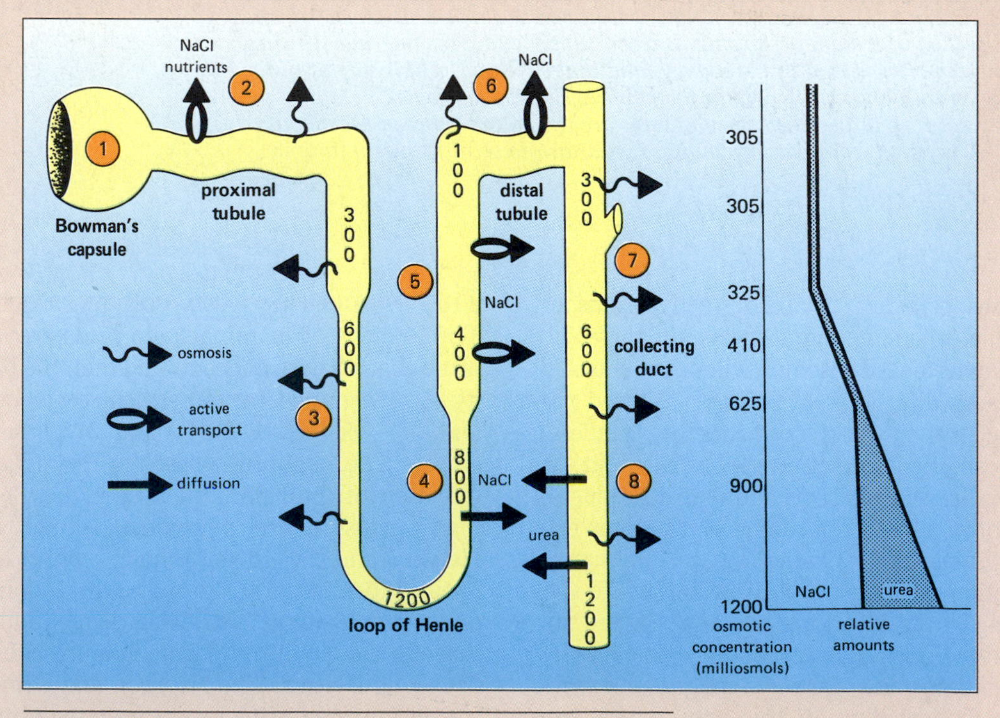

Figure E25-1

Probing Deeper: The Nephron and Urine Formation

tubule. Here more salt is pumped out. Since this portion is permeable to water, water follows by osmosis.

7. By the time the filtrate reaches the collecting duct, very little salt is left and about 99 percent of the water has been reabsorbed into the bloodstream. The collecting duct conducts the urine down through the increasingly concentrated extracellular fluid gradient created by the loop of Henle. The entire collecting duct is permeable to water, but the lower portion of the collecting duct is also permeable to urea. Water moves out in response to the increasing external concentration gradient. As the filtrate moves farther down the collecting duct, some urea diffuses out as well, contributing to the osmotic concentration of the surrounding fluid. As water and urea move out, the concentration of dissolved wastes such as urea in the duct gradually approaches equilibrium with the high osmotic concentration of the external fluid surrounding it.

fluid surrounds the bottom of the loop. The collecting duct passes through this gradient. As the filtrate moves through the nephron, nutrients and water are reabsorbed into the blood, while wastes are left behind. Finally, the filtrate passes through the collecting duct, traversing the osmotic gradient created by the loop of Henle. During this passage, additional water leaves the filtrate by osmosis. As it moves through the collecting duct, the urine can reach osmotic equilibrium with the highly concentrated surrounding fluid. The rest of the excretory system is not permeable to water or urea, so the urine remains concentrated. Antidiuretic hormone (ADH) regulates the permeability of the collecting duct to water. This determines how much water leaves the filtrate and how concentrated the urine becomes, as described below.

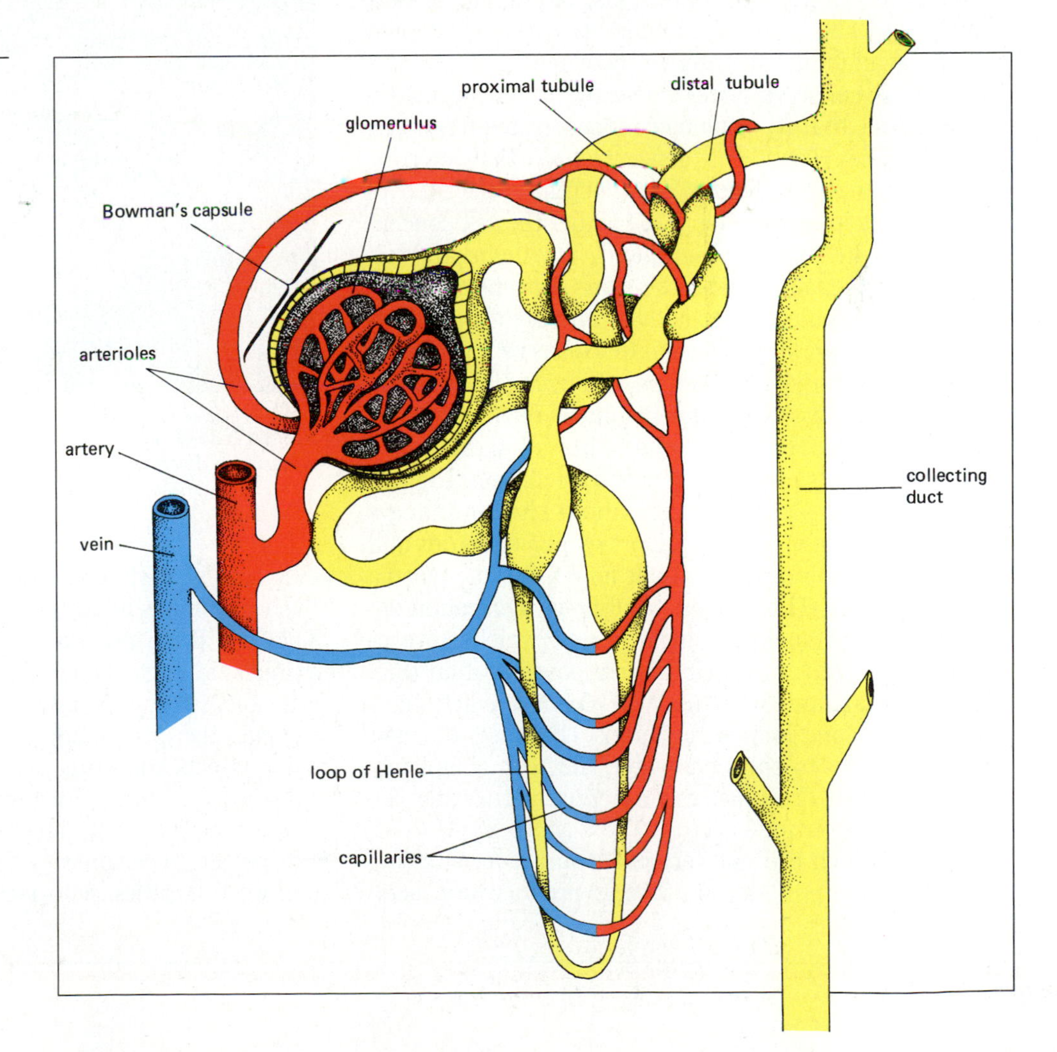

Figure 25-6 *An individual nephron and its blood supply.*

Excretion and Animal Life-Style

Most animals consume far more amino acids than they require for the synthesis of new proteins. Since these excess amino acids cannot be stored, they are broken down to provide energy, or converted to storable fats or carbohydrates. In each of these cases, the amino group (—NH_2) must be removed from the amino acid. This occurs in the liver, and the initial waste product formed from the amino group is a highly poisonous substance: ammonia (NH_3). Because ammonia is so toxic, it must be excreted immediately or converted to a less toxic substance such as urea or uric acid. The solution used by different animals is intimately related to their environment. Aquatic animals—freshwater fish, certain amphibians, and many invertebrates—release ammonia into the water as it is formed. Land dwellers, on the other hand, cannot afford to waste water by producing urine continuously as is necessary for the excretion of ammonia. Instead, they convert the ammonia to uric acid or urea, substances that may be stored and concentrated.

All mammals produce urea, and since urea is highly soluble in water, some water must be excreted simultaneously. The kidneys of mammals are finely adapted to the availability of water in their habitat. Mammals that must conserve water do so by producing hypertonic urine, that is, urine more concentrated than their blood. The degree of concentration that can be achieved is determined by the length of the loop of Henle. The longer the loop, the higher the salt concentration in the fluid surrounding it. The higher the salt concentration, the greater the urine concentration. As you might predict, animals living in very dry climates and having an urgent need to conserve their body water have the longest loops of Henle. In contrast, those in watery environments have relatively short loops. The beaver, for example, has all short-looped nephrons and is unable to concentrate its urine to more than twice its plasma concentration. Human kidneys have a mixture of long and short-looped nephrons and can concentrate urine to about four times the plasma concentration. The masters of urine concentration are desert rodents such as kangaroo rats, which can produce urine 14 times their plasma concentration (Fig. E25-2). Kangaroo rats (as you might predict) have only very long-looped nephrons. Because of their unique ability to conserve water, they can completely dispense with drinking, relying entirely on water derived from their food.

Birds (which are not very efficient at concentrating their urine), reptiles, and insects produce uric acid. Uric acid is a white crystalline substance that is relatively insoluble in water and is excreted as a paste. This has three potential advantages. First, all these animals lay eggs. The embryo, by depositing uric acid as insoluble crystals, avoids stewing in its own wastes. Second, the production of nearly dry urine allows flying insects and birds to avoid carrying the weight of extra water. Third, since much of the uric acid is not in solution, the urine can contain a great deal of it and still not be hypertonic to the blood. This is very important to reptiles, who lack a loop of Henle and are

Figure E25-2 The desert kangaroo rat of the southwestern United States can dispense with drinking partly because its long loops of Henle allow it to produce very concentrated urine.

Figure E25-3 This green sea turtle shows tearlike deposits caused by discharge of its salt glands.

Excretion and Animal Life-Style

unable to produce hypertonic urine. By excreting uric acid in crystalline form, they are both producing a concentrated nitrogenous waste and conserving water. This ability helps reptiles thrive in dry desert environments.

Marine reptiles such as sea turtles, sea snakes, and marine iguanas, and birds such as seagulls, have another problem: they take in excessive salt from their diet and their surroundings. Since they cannot produce concentrated urine, these groups have evolved other excretory organs that efficiently eliminate salt. These organs are located in the head and contain cells that secrete salt by active transport. The highly concentrated salt solution may drain out through the nasal passages or from the corners of the eyes. Sea turtles may be seen "crying crocodile tears" as these glands discharge their contents near the eye (Fig. E25-3). Ironically, the salt glands of crocodiles discharge under the tongue, so the origin of this colorful expression remains obscure.

The Kidneys as Organs of Homeostasis

Each drop of blood in your body passes through a kidney about 350 times daily; thus the kidney is able to "fine tune" the composition of the blood. The importance of this task is illustrated by the fact that kidney failure is rapidly fatal.

WATER BALANCE. One of the most important functions of the kidney is to regulate the water content of the blood. Human kidneys filter about 125 milliliters of fluid from the blood each minute. This means that without reabsorption of water, you would produce 180 liters (over 45 gallons) of urine daily! Water reabsorption occurs passively by osmosis as the filtrate travels through the tubule and the collecting duct. How much water is reabsorbed into the blood is controlled by **antidiuretic hormone** (ADH; also called *vasopressin*) circulating in the blood. This hormone increases the permeability of the distal tubule and the collecting duct to water, allowing more water to be reabsorbed from the urine. ADH is produced by cells in the hypothalamus and is released by the posterior pituitary gland (see Chapter 28). ADH release is regulated by receptor cells in the hypothalamus that monitor the osmotic concentration of the blood, and by receptors in the heart that monitor blood volume. For example, as the lost traveler staggers through the searing desert sun, dehydration occurs. The osmotic concentration of his blood rises and his blood volume falls, triggering the release of more ADH (Fig. 25-7). This increases water reabsorption and produces urine more concentrated than the blood. In contrast, a partygoer overindulging in beer will experience a decrease in blood concentration and an increase in blood volume, and her receptors will cause a decrease in ADH output. Reduced ADH concentration will make the distal tubule and collecting duct less permeable to water. When ADH is very low, little water is reabsorbed after the urine leaves the loop of Henle, and the urine

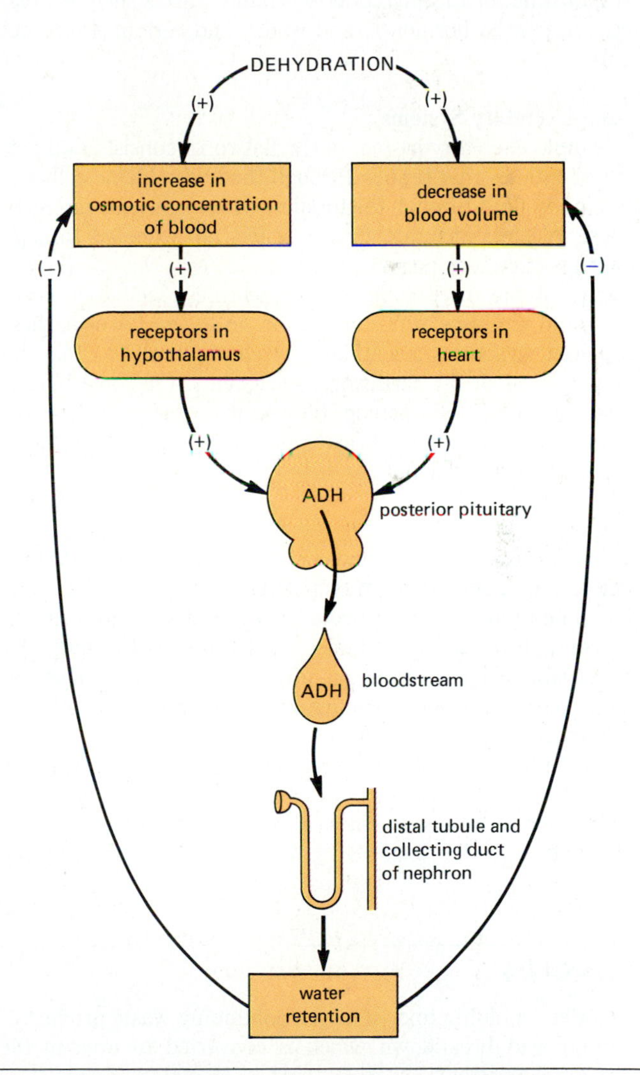

Figure 25-7 Regulation of the water content of the blood is hormonally controlled by ADH through a negative feedback process. Stimulation is indicated by (+), reduction by (–). Dehydration, in addition to ADH release, triggers the sensation of thirst, leading to increased water intake. This is necessary to restore blood volume and diminish ADH secretion.

produced will be more dilute than the blood. In extreme cases, urine flow may exceed 1 liter (over a quart) per hour. As homeostasis is restored, the increasing osmotic concentration of the blood and decreased blood volume will stimulate increased ADH production. This is an example of the negative feedback system which regulates the levels of most hormones (see Chapter 28).

REGULATION OF DISSOLVED SUBSTANCES. As the kidney filters the blood, it monitors and regulates blood composition to maintain a constant internal environment. Substances regulated by the kidney, in addition to water, include nutrients such as glucose, amino acids, vitamins, urea, and a variety of ions, including sodium, potassium, chloride, and sulfate. The kidney maintains a constant blood pH by regulating the content of hydrogen and bicarbonate ions. This remarkable organ also eliminates potentially harmful substances, including some drugs, food additives, pesticides, and some of the toxins produced by cigarette smoke.

SUMMARY OF KEY CONCEPTS

The digestive system plays a crucial role in homeostasis, the precise regulation of the composition of body fluids. The kidney is responsible for the excretion of cellular wastes such as urea, excess nutrients, hormones, and water, and certain drugs and toxins.

Simple Excretory Systems

The simple excretory system of the flatworm consists of a network of tubules that branch through the body. Flame cells circulate body fluid through the tubules, where nutrients are reabsorbed. Wastes, including excess water, are excreted through numerous excretory pores.

Many of the more complex invertebrates, including earthworms and molluscs, utilize nephridia. In the earthworm, these are paired structures resembling vertebrate nephrons that are found in most of the earthworm's segments. Coelomic fluid is drawn into a ciliated opening, the nephrostome, and nutrients and water are reabsorbed. Wastes and excess water are released through the excretory pore.

Human Excretion

The human excretory system consists of elaborate blood filters called kidneys, and other structures that transport and store the urine. The ureters conduct urine from the kidneys to a distensible storage organ, the bladder. Distention of the muscular bladder walls triggers urination, a reflex that is also under voluntary control. From the bladder, urine traverses the urethra, a single narrow tube that opens to the outside.

Each kidney consists of over a million individual nephrons in an outer cortex, with many extending into an inner layer, the medulla. Urine formed in the nephrons enters collecting ducts that empty into the renal pelvis, a central chamber. From the renal pelvis, urine is funneled into the ureter. Each nephron is served by an arteriole that branches from the renal artery. The arteriole further branches into a mass of capillaries called the glomerulus. Here water and dissolved substances are filtered from the blood by pressure, which forces them through the porous capillary walls. The glomerulus is surrounded by a cuplike portion of the nephron called the Bowman's capsule, which collects the filtrate. From the Bowman's capsule, the filtrate is conducted along the tubular portion of the nephron. During its transit, nutrients are actively pumped out of the filtrate through the walls of the tubule. Nutrients then enter capillaries that surround the tubule, and water follows by osmosis. Wastes and excess water remain in the filtrate. The tubule forms the loop of Henle, which creates a salt concentration gradient surrounding it. After completing its passage through the tubule, the filtrate enters the collecting duct, which passes through the concentration gradient. Final passage of the filtrate through this gradient via the collecting duct allows concentration of the urine.

The kidneys are important organs of homeostasis. The water content of the blood is regulated by antidiuretic hormone (ADH), produced in the hypothalamus and released by the posterior pituitary gland. Low blood volume and high osmotic concentration of the blood signal dehydration and stimulate release of ADH into the bloodstream. ADH increases the permeability of the distal tubule and the collecting duct to water, allowing more water to be reabsorbed into the blood. In addition to its role in water balance, the kidney also controls blood pH, removes toxins, and regulates ions such as sodium, chloride, potassium, and sulfate. Excess glucose, vitamins, and amino acids are also excreted by the kidney.

GLOSSARY

Ammonia a highly toxic nitrogen-containing waste product of amino acid breakdown which is converted to urea in the mammalian liver.

Antidiuretic hormone (an-tē-dī-ūr-et'-ik) also called ADH; a hormone produced by the hypothalamus and released by the posterior pituitary gland into the bloodstream. It acts on the nephron of the kidney and causes more water to be reabsorbed into the bloodstream.

Bladder a muscular storage organ for urine.

Bowman's capsule the portion of the nephron in which blood filtrate is collected from the glomerulus.

Flame cell a specialized cell containing beating cilia that conducts water and wastes through the branching tubes that serve as an excretory system in flatworms.

Glomerulus (glō-mer'-ū-lus) a dense network of thin-walled capillaries located within the Bowman's capsule of each nephron. Here blood pressure forces water and dissolved nutrients through capillary walls for filtration by the nephron.

Homeostasis (hōm-ē-ō-stā′sis) the precise regulation of the composition of fluid bathing the body cells. The relatively constant environment required for optimal functioning of cells is maintained by the coordinated activity of numerous regulatory mechanisms, including the respiratory, endocrine, circulatory, and excretory systems.

Loop of Henle (hen′-lē) a specialized portion of the tubule of the nephron in birds and mammals which creates an osmotic concentration gradient in the fluid immediately surrounding it. This in turn allows the production of urine more osmotically concentrated than blood plasma.

Nephridium (nef-rid′-ē-um) a type of excretory organ found in earthworms, molluscs, and certain other invertebrates. A nephridium somewhat resembles a single vertebrate nephron.

Nephron (nef′-ron) the functional unit of the kidney, where blood is filtered and urine formed.

Tubule (tūb′-ūle) the tubular portion of the nephron. It includes a proximal portion, the loop of Henle, and a distal portion. Urine is formed from the blood filtrate as it passes through the tubule.

Urea (ū-rē′-uh) a water-soluble, nitrogen-containing waste product of amino acid breakdown which is one of the principal components of mammalian urine.

Ureter (ū′-re-tur) a tube that conducts urine from each kidney to the bladder.

Urethra (ū-rē′-thruh) a tube that conducts urine from the bladder to the outside of the body.

Uric acid (ūr′-ik acid) a nitrogen-containing waste product of amino acid breakdown which is a relatively insoluble white crystal. Uric acid is excreted by birds, reptiles, and insects.

STUDY QUESTIONS

1. Trace an amino acid from the renal artery to the renal vein.
2. Trace a urea molecule from the bloodstream to the external environment.
3. What is the function of the loop of Henle? The collecting duct? Antidiuretic hormone?
4. List two phyla whose members produce uric acid as a waste product and describe the advantages for each.
5. Why don't humans excrete ammonia?
6. Would you predict that the loop of Henle would be longer in a river otter or in a jackrabbit? Explain your answer.

SUGGESTED READINGS

Eckert, R., Randall, D., and Augustine, G. *Animal Physiology: Mechanisms and Adaptations.* New York: W. H. Freeman and Company, Publishers, 3rd ed., 1988. An excellent and complete textbook of comparative animal physiology.

Hole, Jr, J. W., *Essentials of Human Anatomy and Physiology.* Dubuque, Iowa: Wm. C. Brown Company, 2nd ed. 1986. Excellent illustrations.

26
Circulation and Respiration

Billions of years ago, the first living cells were nurtured by the sea, where they evolved. The waters brought them nutrients that diffused into the cell, and washed away the wastes that diffused out. Diffusion is a slow process, so to satisfy the demands of a living cell, diffusion distances must be kept short. Today, microorganisms and some simple multicellular animals rely almost exclusively on diffusion for exchange of wastes and nutrients with the environment. Sponges, for example, circulate seawater through pores in their bodies, bringing the environment within diffusing distance of each cell (see Fig. 18-2). The threadlike bodies of nematode worms provide an enormous surface area for gas and nutrient exchange with their surroundings.

As larger, more complex animals evolved, individual cells became increasingly distant from the outside world. To avoid starving and stewing in its own wastes, a source of nutrients and a sink for wastes had to be brought within diffusing distance of each cell. With the evolution of the **circulatory system,** an internal sea was created, bringing each cell into close proximity with a source of food and oxygen. The circulatory system also provided a means to carry wastes away from cells. These multicelled animals also needed to exchange gases, acquiring oxygen and releasing carbon dioxide. For relatively small animals, such as the earthworm, the moist skin surface is adequate for gas exchange, supplemented by a circulatory system to transport gases to and from the cells. But for large animals, the surface area of the skin is small relative to the total body volume and is inadequate for gas exchange. In addition, some animals evolved waterproof skin that prevents desiccation in dry air. This prevents gas exchange through the skin as well. A large moist surface for gas exchange, the **respiratory system,** is crucial to support large size and waterproof skin. Most animals possess both circulatory and respiratory systems. These mutually supporting adaptations are the subject of this chapter.

CIRCULATORY SYSTEMS

All circulatory systems have three major parts: (1) a fluid (blood) that serves as a medium of transport, (2) a system of channels or vessels to conduct the fluid throughout the body, and (3) a pump or heart to keep it moving. Two major types of circulatory system are found in animals: **open** and **closed.**

Open circulatory systems include an open space within the body, the **hemocoel,** into which vessels empty and from which they pick up blood (Fig. 26-1). Within this space, tissues are directly bathed in blood. This type of system is found in arthropods (insects, spiders, and crustaceans) and most molluscs (snails, clams).

In closed circulatory systems, blood is confined to the

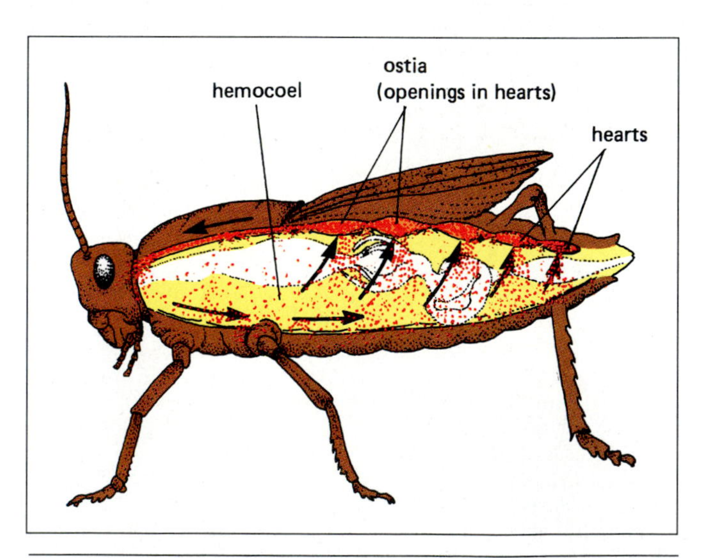

Figure 26-1 In the open circulatory system of insects, a series of hearts pumps blood through vessels into the hemocoel, where blood directly bathes the organs. When the hearts relax, blood is sucked back into them through openings called ostia, guarded by one-way valves. When the hearts contract, the valves are pressed shut, forcing the blood to travel out through the vessels returning to the hemocoel.

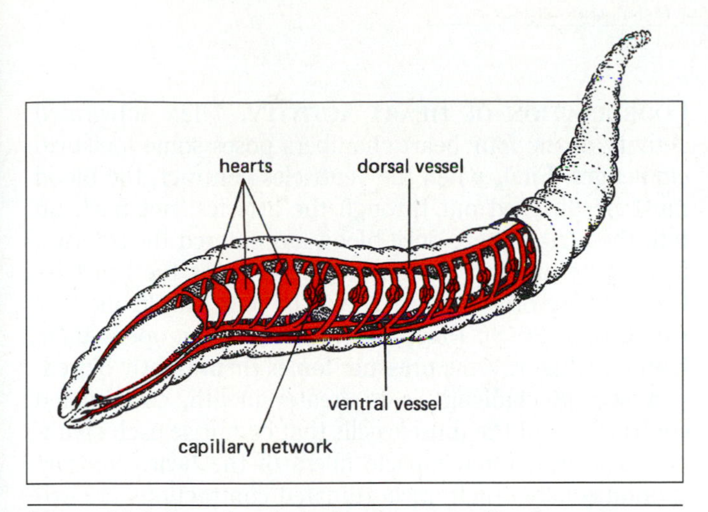

Figure 26-2 In a closed circulatory system, blood remains confined to the heart and the blood vessels. In the earthworm, five contractile vessels serve as hearts.

heart and a continuous series of vessels (Fig. 26-2), allowing more rapid blood flow and more efficient transport than in an open system. Closed systems are found in some invertebrates, such as earthworms and cephalopod molluscs (squids, octopuses), and in all vertebrates.

THE VERTEBRATE CIRCULATORY SYSTEM

The circulatory system has many diverse roles and reaches its greatest development in the vertebrates. Some of the most important functions of the vertebrate circulatory system are listed below.

1. The transport of oxygen from the lungs to the tissues and transport of carbon dioxide from the tissues to the lungs or gills.
2. The distribution of nutrients from the digestive system to all body cells.
3. The transport of waste products and toxic substances to the liver, where many are detoxified, and to the kidneys for excretion.
4. The distribution of hormones from the organs that produce them to the tissues on which they act.
5. The regulation of body temperature, which is achieved partly by adjustments in blood flow. For example, to cool the body, blood flow in the skin and extremities is increased.
6. The defense of the body against blood loss through clotting, and protection against bacteria and viruses by circulating antibodies and phagocytic white blood cells.

In the following sections we examine the three parts of the circulatory system—the heart, vessels, and blood—considering the human as a representative vertebrate.

The Heart

HEART STRUCTURE. The vertebrate heart consists of muscular chambers capable of strong contractions that circulate blood through the body. During the course of vertebrate evolution, the heart has increased in complexity. Starting with the two-chambered heart of fish, it reaches its greatest complexity in the four-chambered hearts of birds and mammals (Fig. 26-3). These warm-

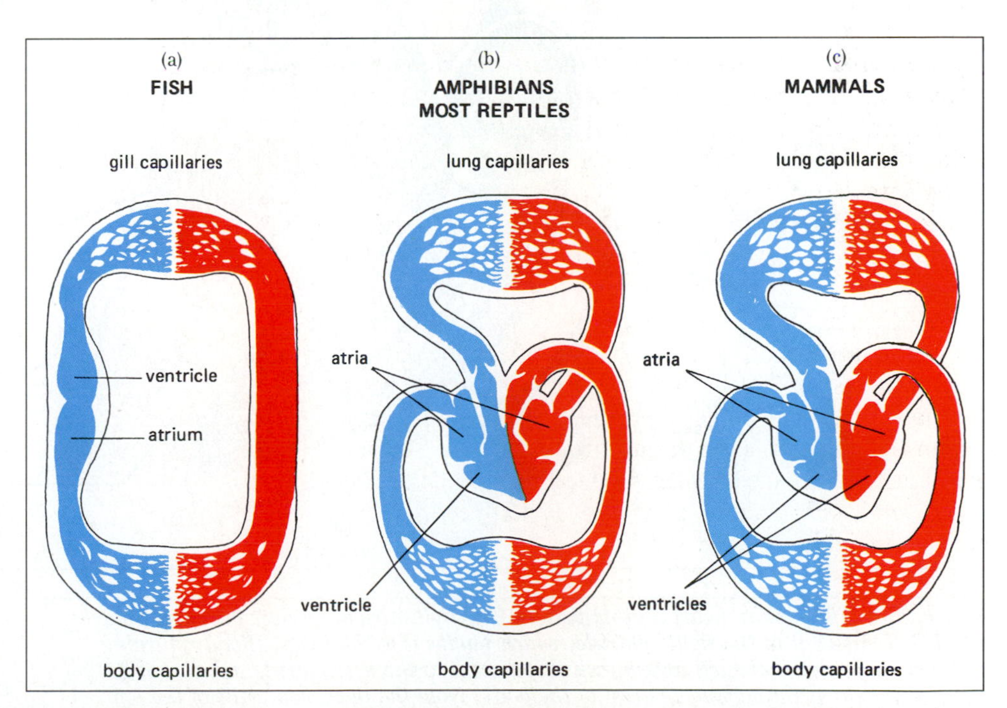

Figure 26-3 **(a)** *The evolution of the vertebrate heart begins with the two-chambered heart of fishes. Blood from the body tissues is collected in the atrium and transferred to the single ventricle. Contraction of the ventricle sends blood through the gill capillaries, where it picks up oxygen and gives off carbon dioxide, and then to the body capillaries, where it delivers oxygen to the tissues and picks up carbon dioxide.* **(b)** *In amphibians and most reptiles, the heart has two atria, the left receiving oxygenated blood from the lungs, the right receiving deoxygenated blood from body tissues. Although both empty into a single ventricle, the deoxygenated blood tends to remain on the right, where it is directed to the lungs, while most of the oxygenated blood stays on the left and is sent to the body tissues. In reptiles, there is often a partial wall down the middle of the ventricle, enhancing this separation.* **(c)** *The hearts of birds and mammals are actually two separate pumps, with no possible mixing of oxygenated and deoxygenated blood.*

blooded animals have high metabolic demands and require the complete separation of oxygenated and deoxygenated blood that the four-chambered heart provides.

The heart of mammals and birds consists of two separate pumps, each with two chambers. In each pump, an **atrium** receives and briefly stores the blood, passing it to a **ventricle** that propels it through the body (Fig. 26-4). One pump is for **pulmonary circulation** and consists of the right atrium and ventricle. Oxygen-depleted blood from the body is collected in the right atrium, transferred to the right ventricle and pumped to the lungs, where it picks up oxygen. The other pump, consisting of the left atrium and ventricle, powers **systemic circulation.** Newly oxygenated blood from the lungs is collected in the left atrium, then passed to the left ventricle, which sends it coursing through the rest of the body.

Coordination of heart activity. The integrated activity of the four heart chambers poses some logistical challenges. First, when the ventricles contract, the blood must be directed out through the arteries, not back up into the atria. Then, once blood has entered the arteries, it must be prevented from flowing back as the heart relaxes. These problems are solved by four simple one-way valves (Fig. 26-5). Pressure in one direction opens them readily, while reverse pressure forces them tightly closed.

A second challenge is to create smooth, coordinated contractions of the muscle cells that comprise each chamber. The individual muscle fibers of the heart contract spontaneously. Such uncoordinated contractions are evident when a heart goes into fibrillation, which can rapidly lead to death because blood is not pumped out of the

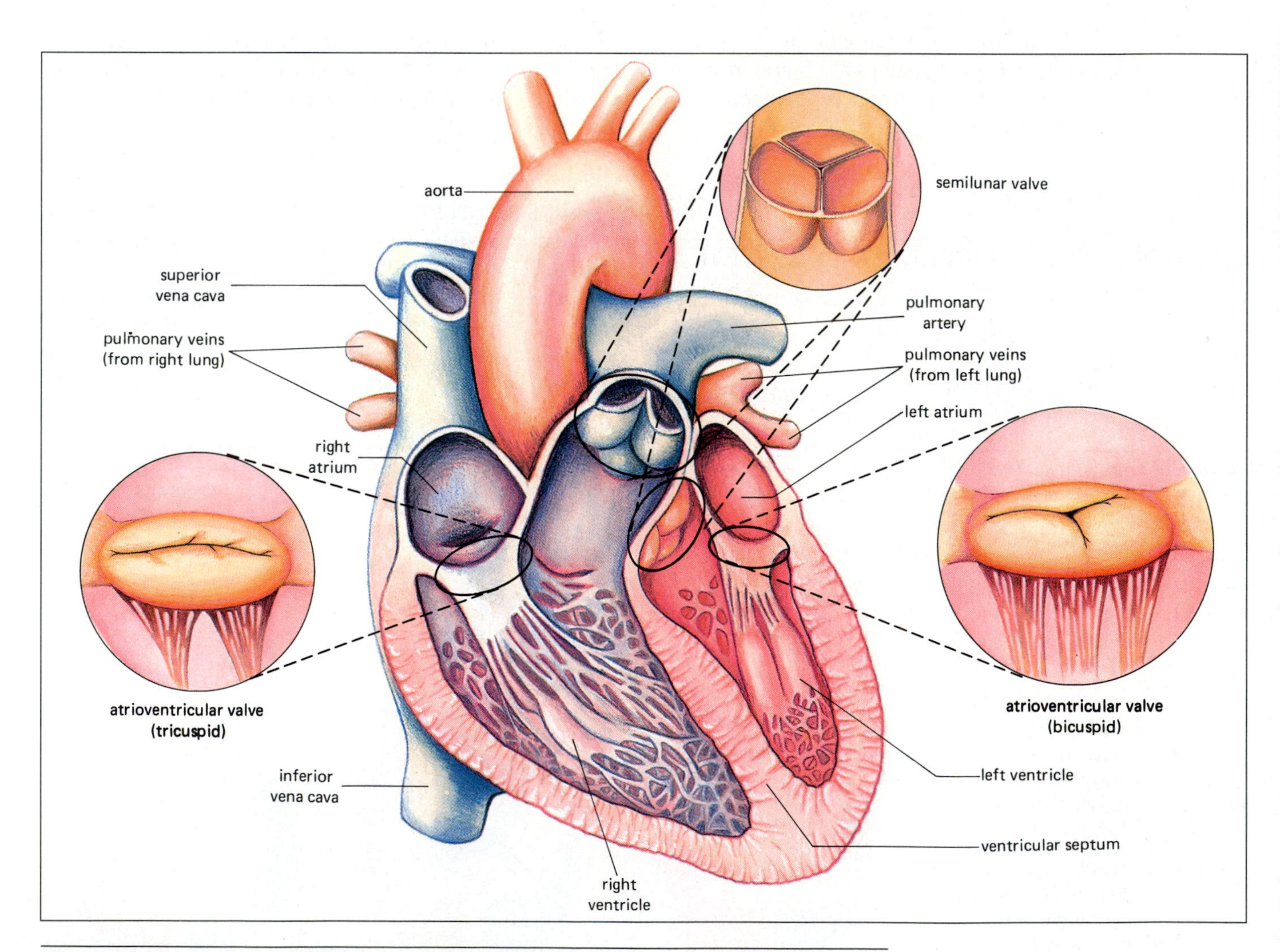

Figure 26-4 *The human heart and its vessels. The right atrium receives deoxygenated blood and passes it to the right ventricle, which pumps it to the lungs. Blood returning from the lungs enters the left atrium, which passes it to the left ventricle, which pumps oxygenated blood throughout the rest of the body. Note the thickened walls of the left ventricle, which must pump blood over a considerably longer distance. One-way valves are located between the aorta and pulmonary artery and the ventricle, and between the atria and ventricles.*

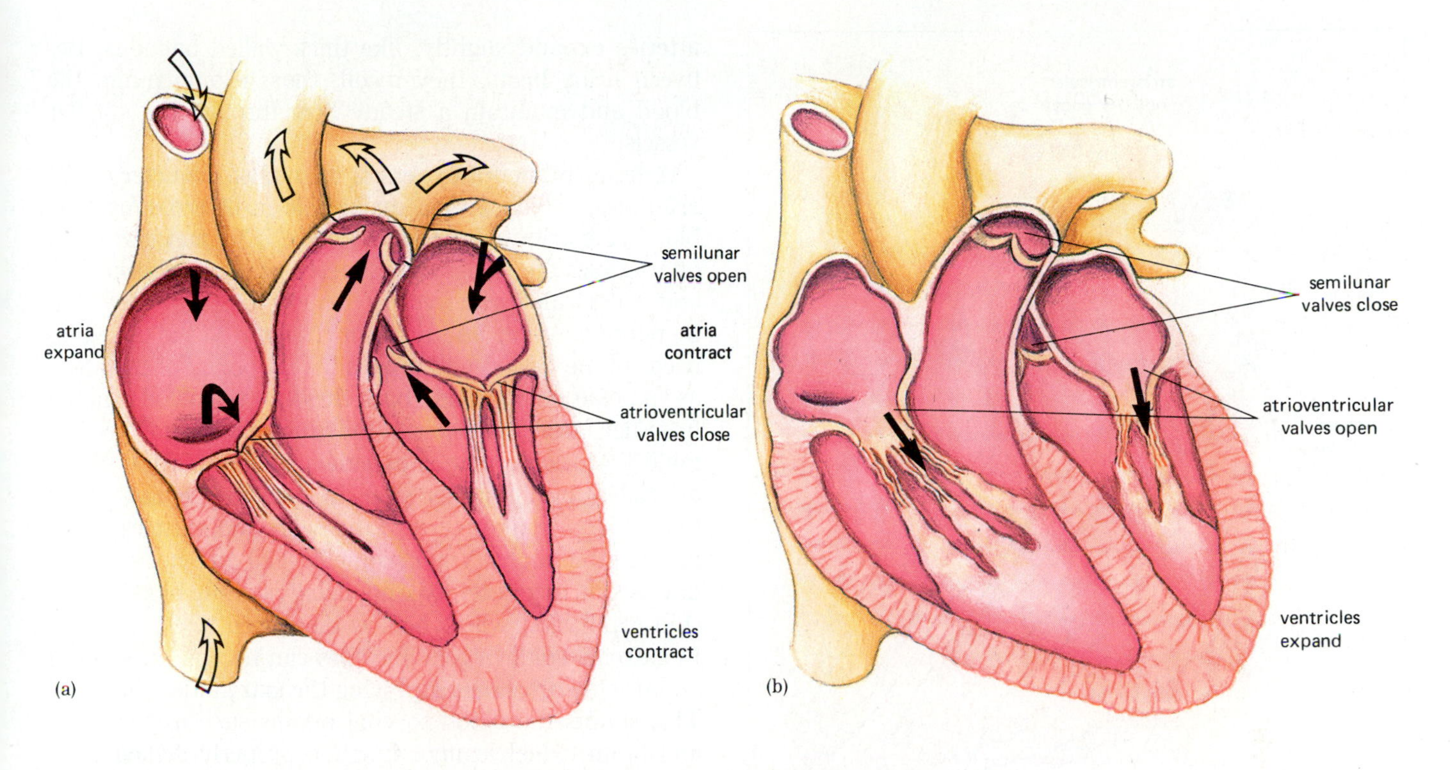

Figure 26-5 The heart valves in action, with arrows indicating direction of blood flow. **(a)** *During ventricular contraction, the pressure within the ventricles forces the semilunar valves open, allowing blood to flow into the aorta and pulmonary arteries. The atrioventricular valves are simultaneously pressed shut, preventing blood flow back into the atria.* **(b)** *As the ventricles reexpand, they would tend to draw blood back from the arteries, but this backpressure forces the semilunar valves closed. Simultaneously, contraction of the atria forces open the atrioventricular valves, allowing blood to flow from the atria into the ventricles.*

heart, but merely sloshed around. Coordinated contraction requires a **pacemaker,** an area of muscle whose rapid contractions set the pace for the other muscle cells. The individual heart muscle cells communicate directly with one another through special pores in their adjacent membranes. These allow electrical signals from the pacemaker to pass freely and rapidly between heart cells, initiating and coordinating their contractions. The heart's primary pacemaker is the **sinoatrial node** (SA node), a small mass of specialized muscle cells located in the wall of the right atrium (Fig. 26-6). The SA node generates electrical impulses at a higher rate than do the individual muscle fibers. Signals from the SA node spread rapidly through both the right and left atria, superseding the spontaneous contractions of individual fibers and causing the atria to contract in smooth synchrony.

A third challenge is to coordinate contractions of the four chambers. The atria must contract first, emptying their contents into the ventricles and then refilling while the ventricles contract. Thus there must be a delay between the contractions of the atria and the ventricles. From the SA node, the wave of contraction sweeps through the atria until it reaches a barrier of inexcitable tissue separating the atria from the ventricles. Here the excitation is channeled through a second small mass of specialized muscle cells, the **atrioventricular node** (AV node), located on the floor of the right atrium (Fig. 26-6). The impulse is delayed at the AV node, postponing the ventricular contraction for about $\frac{1}{10}$ of a second after contraction of the atria. This delay gives the atria time to complete the transfer of blood into the ventricles before ventricular contraction begins. From the AV node, the signal to contract spreads to the base of the two ventricles along tracts of excitable fibers. From these, the impulse travels rapidly through the communicating muscle fibers, causing the ventricles to contract in unison.

OUTSIDE INFLUENCES ON HEART RATE. Left on its own, the SA node pacemaker would maintain a steady rhythm of about 100 beats per minute. However, the heart rate is significantly altered by the influence of nervous impulses and hormones. In the resting person, activity of the parasympathetic nervous system (see Chapter 29) slows it to around 70 beats per minute. When exercise or stress creates a demand for greater blood flow to the muscles, the parasympathetic influence is reduced

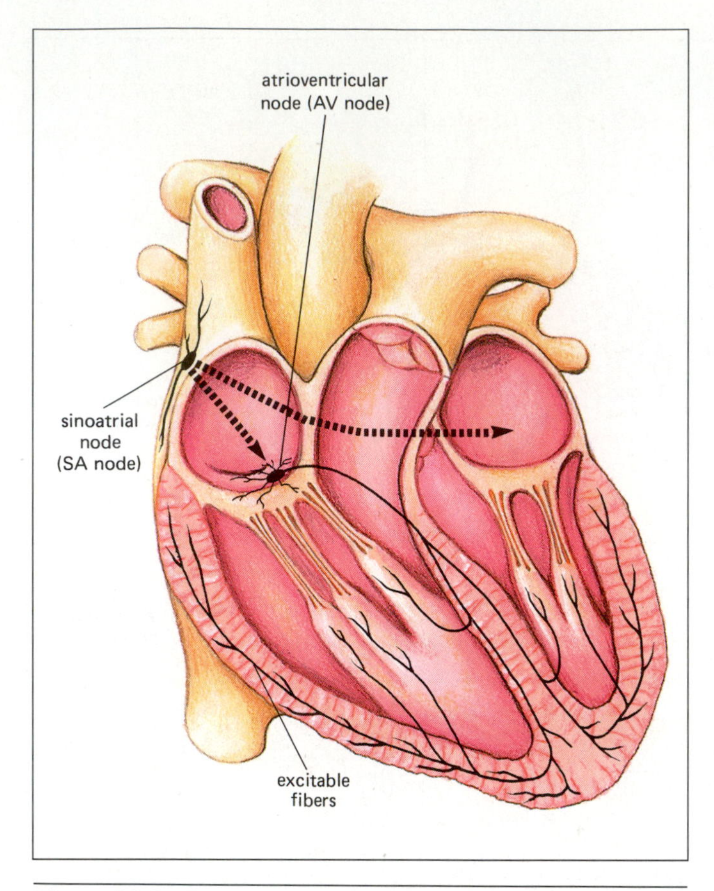

Figure 26-6 The pacemaker of the heart is a spontaneously active mass of modified muscle fibers in the right atrium called the sinoatrial (SA) node. The signal to contract spreads from the SA node through the muscle fibers of both atria (dashed arrows), finally exciting the atrioventricular (AV) node in the right atrium. The AV node then transmits the signal to contract through bundles of excitable fibers which stimulate the ventricular muscle.

and the sympathetic nervous system (which prepares the body for emergency action) accelerates the heart rate. Similarly, the hormone epinephrine increases heart rate as it mobilizes the entire body for response to threatening situations. When astronauts were landing on the moon, their heartrates were over 170 beats per minute, even though they were sitting still!

The Blood Vessels

Blood leaving the heart travels through a series of vessels in the following order: arteries to arterioles to capillaries to venules to veins, which return it to the heart. Let's look at each in more detail.

ARTERIES AND ARTERIOLES. Blood leaving the heart enters large vessels called **arteries.** These have thick walls containing smooth muscle and elastic tissue (Fig. 26-7a). With each surge of blood from the ventricles, the arteries expand slightly, like thick-walled balloons. Between heart beats, they recoil, thus helping pump the blood and maintain a steady flow through the smaller vessels.

Arteries branch into vessels of smaller diameter called **arterioles,** which play a major role in determining how blood is distributed within the body. The muscular walls of arterioles are under the influence of nerves, hormones, and chemicals produced by nearby tissues. They can therefore contract and relax in response to the changing needs of the tissues and organs they supply. For instance, as you read in your paperback thriller, "the blood drained from her face as she beheld the gruesome sight," keep in mind that the heroine is experiencing constriction of the arterioles supplying her skin. In such threatening situations, blood is redirected to the heart and muscles, facilitating rapid flight. You become flushed on a hot summer day as skin arterioles expand, bringing more blood to the skin capillaries, where heat is dissipated to the outside. In extreme cold, fingers and toes can become frostbitten because the arterioles supplying the extremities constrict. This shunts the blood to vital organs such as the heart and brain, which cannot function properly if their temperature drops. By minimizing blood flow to the heat-radiating extremities, the body conserves heat.

CAPILLARIES. One can envision the circulatory system as an elaborate device for getting blood into the **capillaries,** the tiniest of all vessels. Here wastes, nutrients, gases, and hormones are exchanged between blood and the body cells. Capillaries are finely adapted to their role of exchange. Their walls are only a single cell thick (Fig. 26-7a). Dissolved substances readily diffuse through the capillary cell membranes or move through the spaces between adjacent capillary cells. Capillaries are microscopically narrow, so narrow that red blood cells must pass through them single file (Fig. 26-8). This ensures that all the blood passes very close to the capillary walls, where exchange occurs. In addition, capillaries are so numerous that no body cell is more than 10 micrometers from a capillary; this facilitates the exchange of materials by diffusion. It is estimated that the total length of capillaries in a human is over 50,000 miles, enough to encircle the globe twice if placed end to end! The speed of blood flow drops precipitously as blood is forced through this narrow, almost interminable network, and this is important because diffusion is a slow process. The slower the blood flow, the greater the exchange of materials. The flow of blood in capillaries is regulated by tiny rings of muscle surrounding the junctions between arterioles and capillaries. These open and close in response to local changes that signal the needs of nearby tissues, precisely regulating capillary blood flow.

VENULES AND VEINS. Blood from the capillaries drains into larger vessels called **venules** that empty into still

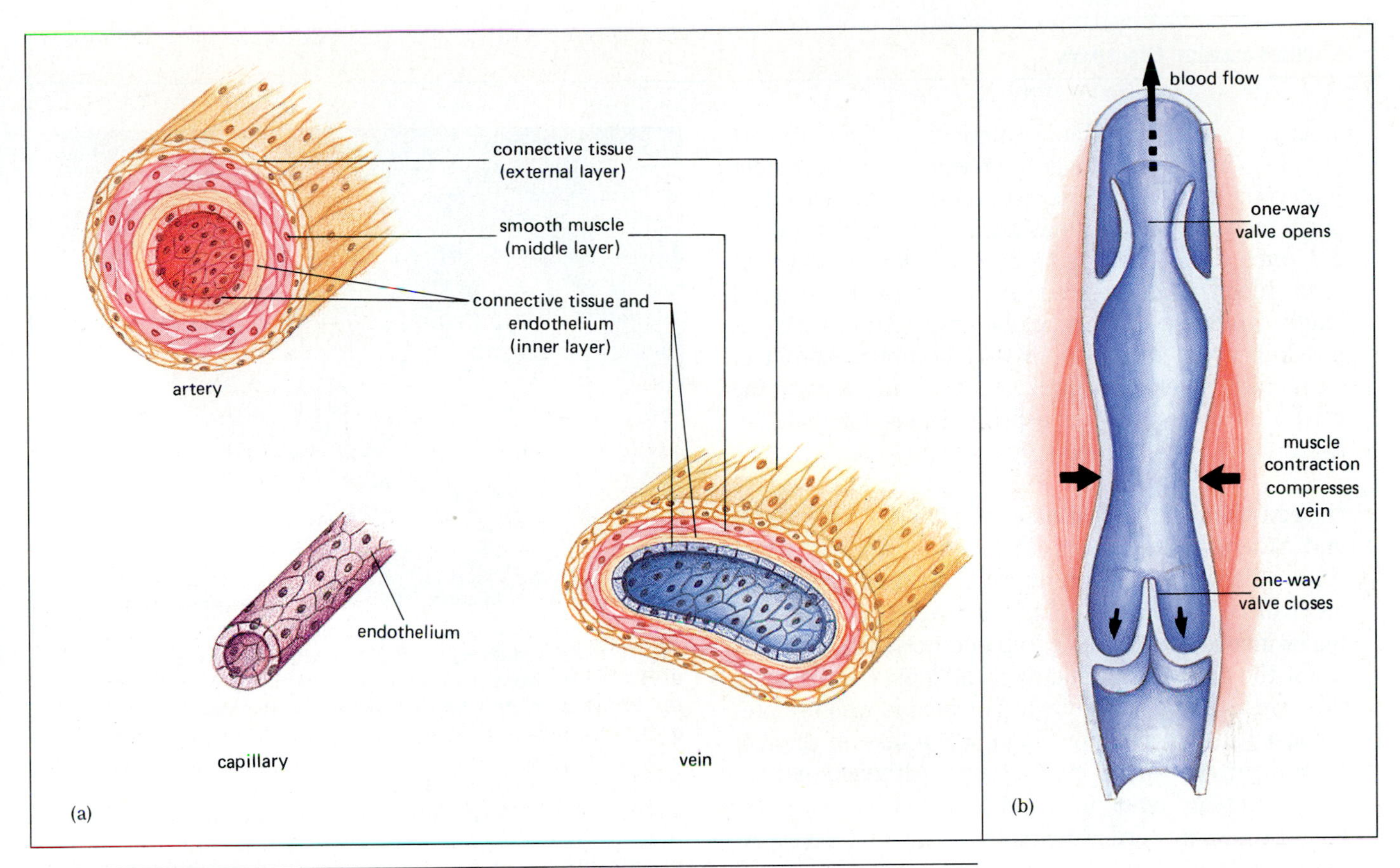

Figure 26-7 **(a)** *Cross sections through arteries, capillaries, and veins reveal structural differences. Arteries and arterioles are more muscular and maintain a greater muscular tension than veins and venules, whose walls are thinner and more distensible. Capillaries have walls only a single cell thick, allowing movement of dissolved substances and white blood cells across the capillary wall.* **(b)** *To maintain flow in the proper direction, veins and venules have one-way valves. When the vein is compressed by nearby muscles, the valves allow blood to flow toward the heart but clamp shut to prevent backflow.*

larger **veins** (Fig. 26-7a). Veins provide a low-resistance pathway for blood to return to the heart.

The walls of veins are much thinner and more distensible than those of arteries, although both contain a layer of smooth muscle. Blood pressure in the veins is low, and the return of blood to the heart is assisted by contractions of skeletal muscles during exercise and breathing. These muscular movements squeeze the veins, forcing blood through them. When veins are compressed, you might predict that blood would be forced *away* from the heart as well as toward it. To prevent this, most veins are equipped with one-way valves that allow blood flow only toward the heart (Fig. 26-7b). When you sit or stand for long periods, the lack of muscular activity allows blood to pool in the veins of the lower legs. This accounts for the swollen feet of airline passengers. It can also contribute to varicose veins, in which the valves become stretched and weakened.

If blood pressure should fall, for instance after extensive bleeding, veins can help restore it. The sympathetic nervous system stimulates contraction of the smooth muscles in the vein walls. This decreases their volume and raises blood pressure, speeding up the return of blood to the heart.

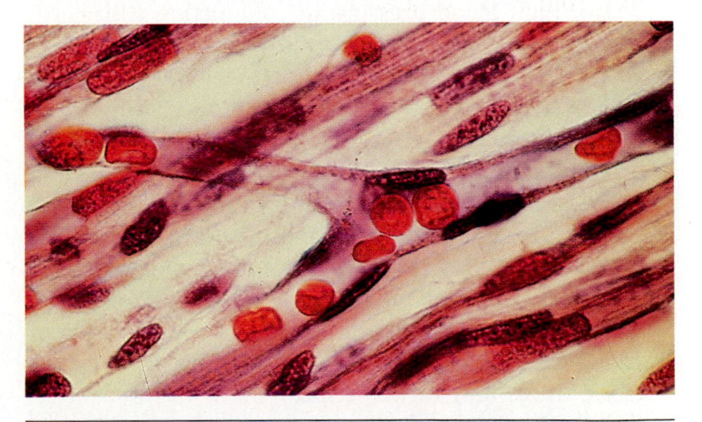

Figure 26-8 Capillaries are so narrow that red blood cells must pass through them single file. This facilitates exchange of gases by diffusion.

Cardiovascular Disorders

This year, nearly a million Americans will die of heart attacks, strokes, and congestive heart failure. Consider that your heart muscle is expected to contract vigorously over 2.5 billion times during your lifetime without once stopping to rest, and that it is expected to force blood through a series of vessels whose total length would encircle the globe twice. Add to this the possibility that these vessels may become constricted, weakened, or clogged, and it is easy to see why the cardiovascular system is a prime target for malfunction.

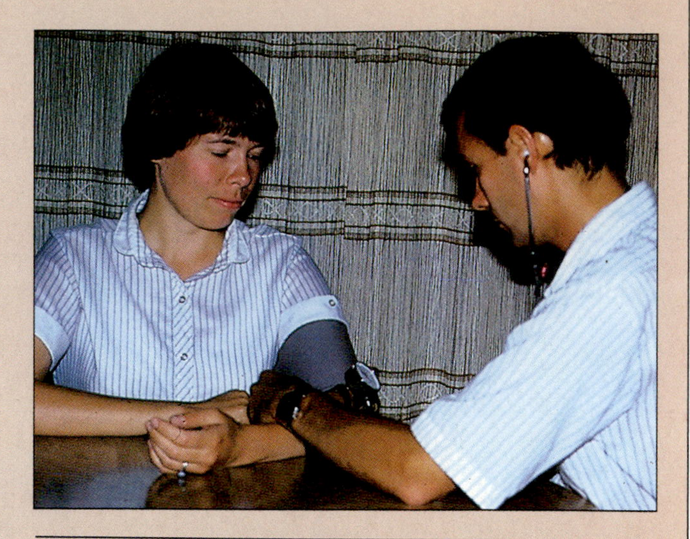

Figure E26-1 *Blood pressure is measured using an inflatable cuff and a stethoscope. The cuff is placed around the upper arm, and the stethoscope is positioned over an artery just below the cuff. The cuff is inflated until no pulse is detected in the artery. The pressure in the cuff is then gradually reduced. When the sound of the pulse is first audible in the artery, this indicates that the pulses of pressure created by the contracting ventricles are just able to overcome the pressure exerted by the cuff and force blood through. The pressure in the cuff at this point is the upper or* systolic *pressure (the pressure created by the contraction of the ventricles). Pressure is then released further until no pulse is audible. At this point, blood is flowing continuously through the artery, indicating that the pressure between heartbeats is sufficient to overcome pressure in the cuff. This pressure is recorded as the lower or* diastolic *pressure (the baseline pressure in the arteries between heartbeats). Like barometric pressure, blood pressure is measured in millimeters of mercury. A fairly typical reading would be 140/70 mmHg (systolic/diastolic).*

Unseen Villains: High Blood Pressure and Atherosclerosis

Also called hypertension, high blood pressure is usually caused by constriction of the arterioles, which produces increased resistance to blood flow. In the majority of the 45 million Americans afflicted, the cause of this constriction is unknown. For people who are predisposed to hypertension, high salt intake in the diet may aggravate it, as can obesity. Although normal blood pressure tends to increase somewhat with age, an approximate borderline reading for high blood pressure is 140/95 mm Hg. A diastolic reading (the lower reading, see Fig. E26-1) consistently over 95 is the best indicator of high blood pressure.

High blood pressure may give few warning signals, but it undermines the cardiovascular system in several ways. First, it causes strain on the heart by increasing resistance to blood flow. Although the heart may enlarge in response to this added demand, its own blood supply may not increase proportionately. The heart muscle is then inadequately supplied with blood, especially during exercise. Lack of sufficient oxygen to the heart results in chest pain called **angina pectoris.** Second, high blood pressure contributes to "hardening of the arteries," or atherosclerosis, described below. Third, high blood pressure in conjunction with hardened arteries can lead to rupture of an artery and internal bleeding. Large areas of the kidneys may be destroyed by rupture of the renal artery. Rupture of vessels supplying the brain causes **stroke,** in which brain function is lost in the area deprived of blood and the vital oxygen and nutrients that blood carries.

Hypertension may be treated in several ways. Mild hypertension may respond to weight reduction, exercise, and reduction of dietary salt. Stress reduction therapy, such as relaxation techniques, meditation, and biofeedback, may also be helpful. For more severe cases, drugs may be prescribed. These include diuretics, which cause increased urination and reduced blood volume, and drugs that dilate the arteries and arterioles. Since 1971, the number of deaths due to hypertension has fallen by over one-third, thanks to increased public awareness, early detection, and successful treatment.

Atherosclerosis. Atherosclerosis kills an estimated 700,000 Americans each year. This disease is characterized by a loss of elasticity in the large arteries and a thickening of the arterial walls. The thickening results from deposits of cholesterol and other substances and an overgrowth of cells engorged with cholesterol. The irregularities caused by these deposits in the artery

Cardiovascular Disorders

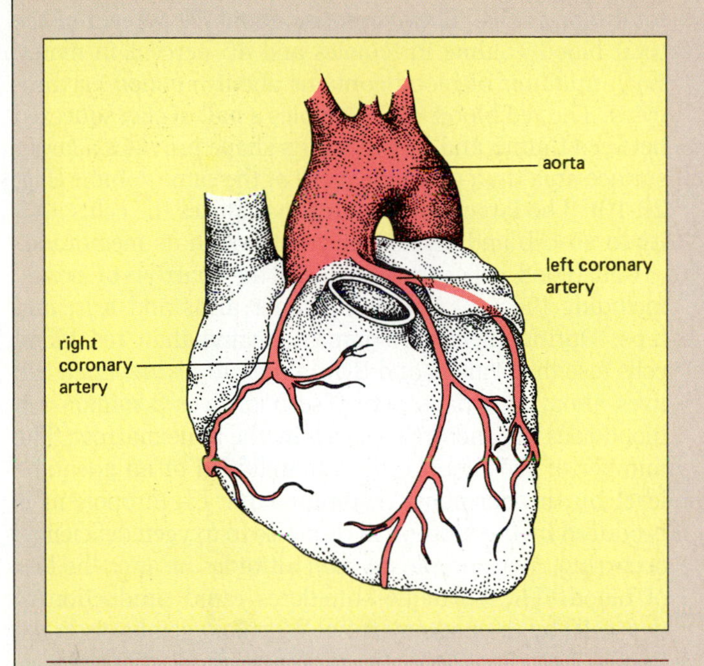

Figure E26-2 An external view of the heart showing the coronary arteries that supply the heart muscle. Heart attacks result when these arteries become obstructed.

wall stimulate the blood platelets (which interpret them as injuries) to initiate blood clots. These clots further obstruct the vessels, setting off a clotting cycle that can completely block the artery. Arterial clots are responsible for the most serious consequence of atherosclerosis: heart attack.

Heart attacks occur when one of the coronary arteries (arteries that supply the heart muscle itself; Fig. E26-2) is blocked. If a blood clot suddenly breaks loose, it may be carried to a narrower part of the artery, obstructing blood flow. Deprived of the nutrients and oxygen it needs to function, the heart muscle once served by the blocked artery rapidly and painfully dies. If the area is small, the heart may be able to continue without it, and the victim may recover. Death of large areas of heart muscle is almost instantly fatal.

Although heart attacks are the major cause of death from atherosclerosis, cholesterol deposits and clots form in arteries throughout the body. A clot that obstructs an artery supplying the brain causes a stroke, with the same results as if the artery has burst.

As with hypertension, the exact cause of atherosclerosis is unclear, but several factors are known to encourage it. These include hypertension, high blood cholesterol, cigarette smoking, genetic predisposition, obesity, diabetes, and a sedentary life-style. By exercising regularly, controlling weight, and not smoking, people can greatly reduce their chances of contracting atherosclerosis. Moderation in the consumption of animal and other saturated fats is also important, since recent evidence links dietary intake of these substances and blood cholesterol levels. There is also some evidence that dietary fiber may help reduce cholesterol.

Traditional treatment of atherosclerosis includes the use of drugs to lower blood pressure and blood cholesterol. In extreme cases, nitroglycerin is used to dilate blood vessels and ease the pain of angina caused by constriction of the coronary arteries. Coronary bypass surgery has been used increasingly in recent years, with 100,000 to 200,000 operations per year being performed in the United States. This procedure consists of bypassing an obstructed coronary artery with a piece of a vein, usually removed from the patient's leg. The new vessel may also clog eventually if the underlying problem, high blood cholesterol, is not corrected. At a cost of about $15,000 to $20,000 per person, continued expansion of this procedure may place overwhelming demands on the country's health-care resources. Recent studies indicate that this radical operation is probably overused; about 30 percent of the patients might benefit equally well from other less drastic treatments.

Although prevention is by far the most cost-effective and successful strategy, a variety of high-technology treatments are under development for this devastating condition. Blood clots can now be dissolved by injecting an enzyme, streptokinase, into the coronary artery. When performed immediately after a heart attack, this procedure can significantly increase the victim's chances of survival. A more promising drug, tissue plasminogen activator (t-PA) has recently been approved for use in the United States. t-PA selectively dissolves clots in vessels without interfering with the body's ability to form clots when needed to prevent bleeding. Another procedure involves squashing the cholesterol deposits flat against the artery walls by inserting a catheter with a tiny balloon into the obstructed artery. The balloon is inflated, crushing the deposits and restoring greater blood flow. Alternatively, a catheter bearing a tiny laser is manipulated up the artery to the clot and the laser is fired, destroying the obstruction. Hopefully, a combination of these new techniques in conjunction with changes in our eating habits and life-style will significantly reduce early deaths from atherosclerosis in the future.

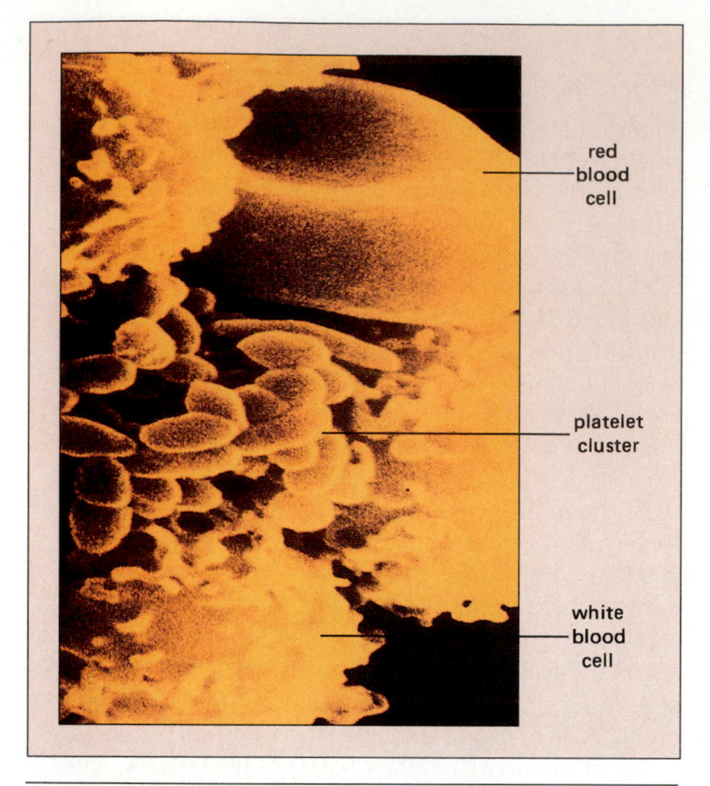

Figure 26-9 In this scanning electron micrograph, the three major blood cell types can all be seen together. Platelets are the cluster of small fragments near the center, white blood cells are large and round with irregular surfaces, and the red cell has a biconcave shape.

The Blood

Blood is the fluid medium in which dissolved nutrients, gases, hormones, and wastes are transported. It has two major components: specialized cells, including red blood cells, white blood cells, and platelets (Fig. 26-9; Table 26.1), and the fluid, called **plasma,** in which they are suspended. On the average, the cellular components of blood comprise 40 to 45 percent of its volume; the other 55 to 60 percent is plasma. The average person has 5 to 6 liters of blood, constituting about 8 percent of his or her total body weight.

PLASMA. The straw-colored fluid called plasma is about 90 percent water, in which a number of substances are dissolved. Dissolved substances include proteins, hormones, nutrients (glucose, vitamins, amino acids, lipids), gases (carbon dioxide, oxygen), ions (sodium, chloride, calcium, potassium, magnesium), and wastes such as urea.

Plasma proteins are the most abundant of the dissolved substances. The three major plasma proteins are (1) albumins, small proteins that help maintain the osmotic pressure of the blood; (2) globulins, which help transport nutrients and also function in immunity; and (3) fibrinogen, a major factor in blood clotting, discussed later in this chapter.

RED BLOOD CELLS. Also called **erythrocytes,** these oxygen-carrying cells make up about 99 percent of the total blood cells. They comprise about 40 percent of the total blood volume in females and 45 percent in males. Each milliliter of blood contains about 5 *billion* erythrocytes. The red blood cell resembles a ball of clay squeezed between thumb and forefinger. Its shape provides a larger surface area than a spherical cell of the same volume (Fig. 26-10). The larger surface area maximizes the cell's ability to absorb and release oxygen through its membrane.

Red blood cells are formed in the marrow of bones, including those of the chest, upper arms and legs, and hips. During their development, mammalian red blood cells lose their nuclei and their ability to divide. Each cell lives about 120 days. Every second, over 2 million red blood cells die and are replaced by the bone marrow. The number of red blood cells is maintained at an adequate level by the hormone **erythropoietin.** Erythropoietin is produced by the kidneys in response to oxygen deficiency, occurring, for example, at high altitudes or after the loss of blood. The hormone stimulates rapid production of new cells by the bone marrow. When oxygen levels in the tissues become adequate, hormone production ceases, and the rate of red cell production returns to normal.

One of the most striking features of erythrocytes is their red color, caused by the pigment **hemoglobin** (Fig. 26-11). This large, iron-containing protein makes up about one-third the weight of the blood cell. About 97 percent

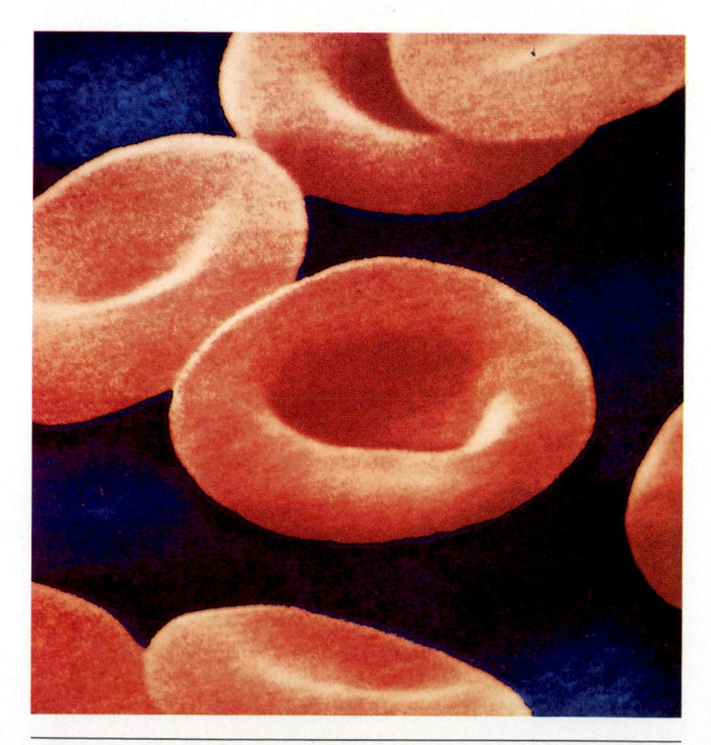

Figure 26-10 Under the scanning electron microscope, the biconcave disk shape of red blood cells is clearly visible. These cells have been artificially colored.

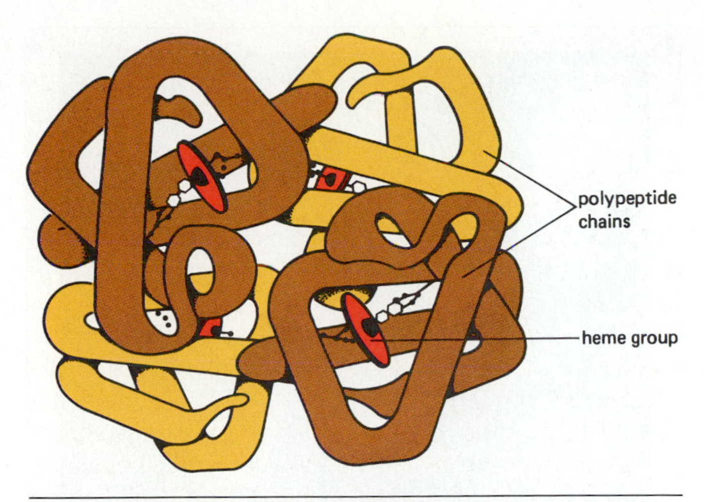

Figure 26-11 A molecule of hemoglobin is composed of four polypeptide chains, each surrounding a heme group. The heme group contains an iron atom and is the site of oxygen binding. When saturated, each hemoglobin molecule can carry four oxygen molecules or eight oxygen atoms.

of the oxygen carried by the blood is bound to hemoglobin. The hemoglobin molecule picks up oxygen where the concentration is high, as in the capillaries of the lungs, and releases it where the concentration is low, in other tissues of the body. After releasing its oxygen, some of the hemoglobin picks up carbon dioxide from the tissues for transport back to the lungs. The role of blood in gas exchange is discussed later in this chapter.

Dead or damaged red blood cells are removed from circulation, primarily in the liver and spleen, and are broken down to release their iron. The salvaged iron is carried in the blood to the bone marrow, where it is used to make more hemoglobin and packaged into new red blood cells. Although the recycling process is efficient, small amounts of iron are excreted daily and must be replenished by the diet. Bleeding from injury or menstruation also tends to deplete iron stores.

WHITE BLOOD CELLS. There are five common types of white blood cells, or **leukocytes,** which together constitute less than 1 percent of the total cellular component of the blood. These cells, described in more detail in Table 26-1, are distinguished from one another by their staining characteristics, size, and the shape of their nuclei. Most function in some way to protect the body against foreign invaders and use the circulatory system to travel to the site of invasion. For example, **monocytes** and **neutrophils** are active white blood cells that travel via capillaries to wounds where bacteria have gained entry. Here they ooze out through the capillary walls like tiny amoebas. In tissues, monocytes differentiate into **macrophages,** amoeba-like cells that engulf foreign particles. Macrophages and neutrophils feed on the bacterial invaders (or other cells recognized as foreign, including cancer cells; Fig. 26-12). They often lose their lives in the process, and their dead bodies accumulate and contribute to the white substance we call pus, seen most abundantly at sites of infection. **Lymphocytes** help provide immunity against disease (see Chapter 27). Least abundant

Table 26.1 **Blood Cells**

Cell Type	Description	Average Number Present	Major Function
Red blood cell (erythrocyte)	Biconcave disk without nucleus, about one-third hemoglobin Approximately 8μm[b] in diameter.	5,000,000 per mm[3a]	Transports oxygen and a small amount of carbon dioxide
White blood cells (leukocytes)		7,500 per mm^3	
1. Neutrophil	About twice the size of red cells, nucleus with two to five lobes	62% of white cells	Destroys relatively small particles by phagocytosis
2. Eosinophil	About twice the size of red cells, nucleus with two lobes	2% of white cells	Inactivates inflammation-producing substances; attacks parasites
3. Basophil	About twice the size of red cells, nucleus with two lobes	Less than 1% of white cells	Releases anticoagulant, to prevent blood clots; and histamine, causing inflammation
4. Monocyte	Two to three times larger than red cells, nuclear shape varies from round to lobed	3% of white cells	Gives rise to macrophage, which destroys relatively large particles by phagocytosis
5. Lymphocyte	Only slightly larger than red cell, nucleus nearly fills cell	32% of white cells	Functions in the immune response
Platelet	Cytoplasmic fragment of cells in bone marrow called megakaryocytes	250,000 per mm^3	Important in blood clotting

[a]mm^3 = cubic millimeter.
[b]μm = micrometer.

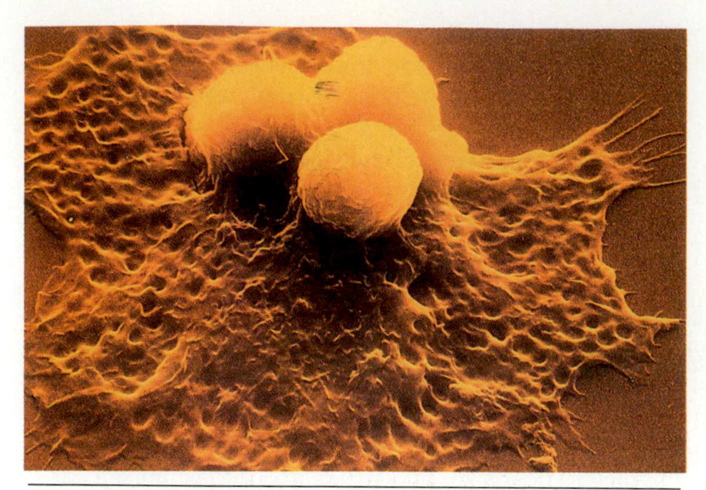

Figure 26-12 An amoeboid white blood cell has left its capillary and engulfed three round cancer cells.

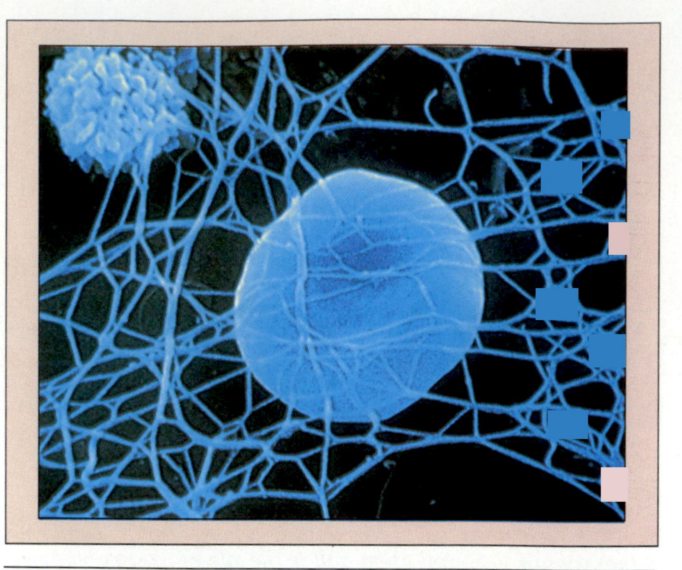

Figure 26-13 In response to damage to a blood vessel, threadlike proteins called fibrin form in the blood. These produce a tangled sticky mass that traps red blood cells and eventually forms a clot.

are the **basophils** and **eosinophils.** Eosinophil production is stimulated by parasitic infections. Eosinophils converge on the parasitic invaders, releasing substances that kill the parasite. Basophils release both substances that inhibit blood clotting and chemicals like histamine, which participate in inflammatory reactions.

PLATELETS AND BLOOD CLOTTING. The third group of blood "cells" are actually fragments of much larger cells. These large cells, called **megakaryocytes,** remain in the bone marrow, pinching off pieces of themselves which enter the circulation. The fragments, called **platelets,** play a central role in blood clotting. Clot formation is a complex process. It starts when platelets and other factors in the plasma contact an irregular surface, such as a damaged blood vessel. Platelets tend to stick to irregular surfaces, and they may build up and plug the damaged vessel if it is narrow enough. This mechanism is supplemented by blood coagulation, or clotting, which is the most important of the body's defenses againt bleeding. The ruptured surface of an injured blood vessel not only causes platelets to adhere, but also initiates a complex sequence of events among circulating plasma proteins. These events culminate in production of the enzyme **thrombin.** Thrombin catalyzes the conversion of the plasma protein fibrinogen into stringlike molecules called **fibrin.** Fibrin molecules adhere to one another, end to end and side to side, forming a fibrous matrix. This protein web immobilizes the fluid portion of the blood, causing it to solidify much like cooling gelatin. The web traps red blood cells, increasing the density of the clot (Fig. 26-13). Platelets then adhere to the fibrous mass and send out sticky projections that attach to one another. Within half an hour, the platelets contract, pulling the mesh tighter and forcing liquid out. This creates a denser, stronger clot and also constricts the wound, pulling the damaged surfaces closer together to promote healing.

The Lymphatic System

Exchange of materials between capillary blood and nearby cells occurs through a liquid medium called **interstitial fluid,** which bathes nearly all the cells of the body. This fluid, derived from the blood plasma, leaks through the permeable walls of the capillaries. Interstitial fluid contains nutrients, hormones, dissolved gases, wastes, and small proteins from the blood. The large plasma proteins, red blood cells, and platelets are unable to leave the capillaries because of their size. In contrast, white blood cells freely enter the interstitial fluid using amoeboid movement. As small proteins, ions, and other dissolved substances leak out of the capillaries and accumulate in the interstitial fluid, water tends to leave the capillaries by osmosis. This causes the volume of interstitial fluid to increase. Although most is eventually reabsorbed by the capillaries, in an average person, about 3 liters more fluid leaves the capillaries than is reabsorbed each day. One role of the **lymphatic system** is to return this excess fluid and dissolved nutrients to the circulation.

The lymphatic system provides a crucial link between the bloodstream and the fluid bathing the cells. In addition, it helps defend the body against foreign invaders such as bacteria and viruses. The lymphatic system includes an elaborate system of vessels resembling capillaries and veins, which empty into the large veins of the circulatory system (Fig. 26-14). In contrast to blood capillaries, lymph capillaries end blindly in the tiny spaces between cells (Fig. 26-15). As shown in Chapter 24 (Fig. 24-10), each villus of the intestine contains a blindly ending lymph vessel. Like blood capillaries, lymph capillaries form a complex network of very thin-walled vessels into

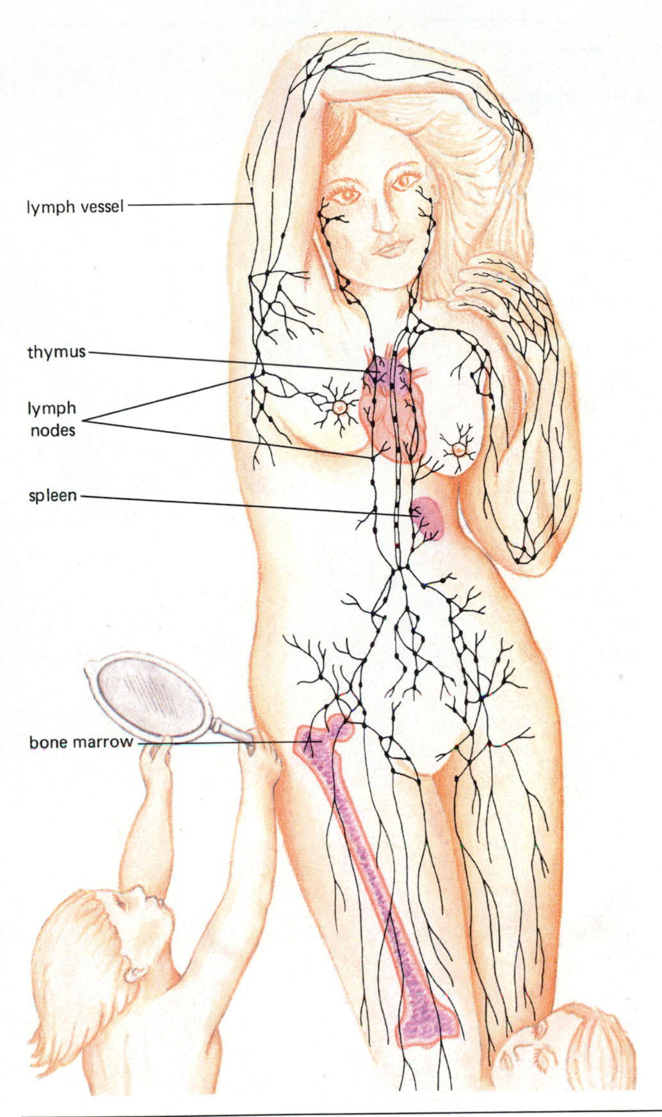

Figure 26-14 *The human lymphatic system, illustrating lymph vessels, lymph nodes, and two auxilliary lymph organs, the thymus and spleen.*

which substances can move readily. From the intestine, fat globules that are too large to enter blood capillaries enter the lymph capillaries. The lymph capillary walls are comprised of cells with openings between them that act as one-way valves. As interstitial fluid accumulates, its pressure forces it through the openings in the lymph capillaries, where it is trapped (Fig. 26-15). Once inside the lymph vessel, the interstitial fluid is called **lymph.** Lymph is forced through its vessels as blood is through veins. Large lymph vessels have somewhat muscular walls, but much of the impetus for lymph flow comes from contraction of nearby muscles, such as those used in breathing and walking. As in blood veins, the direction of flow is regulated by one-way valves (Fig. 26-16.)

The large lymph vessels are interrupted periodically by kidney-bean-shaped structures, about 2.5 centimeters (around 1 inch) long called **lymph nodes** (Fig. 26-14). Lymph is forced through channels within the node. These channels are lined with masses of macrophages. Lymphocytes are also produced in the nodes. Both of these white blood cells recognize and inactivate foreign particles such as bacteria and viruses, and are killed in the process. The painful swelling of lymph nodes in certain diseases such as mumps is largely a result of the accumulation of dead lymphocytes, macrophages, and the virus-infested cells they have engulfed.

Two additional organs, the **thymus** and the **spleen,** are often discussed as part of the lymphatic system, although they have no direct connection to the lymphatic vessels (Fig. 26-14). The thymus, which lies in the chest above the heart, is responsible for the production of lymphocytes that function in the immune response, described in detail in Chapter 27. The thymus is particularly active in infants and young children, regressing in early adulthood. The spleen is located in the left side of the abdominal cavity, between the stomach and diaphragm. Just as the lymph nodes filter lymph, the spleen filters

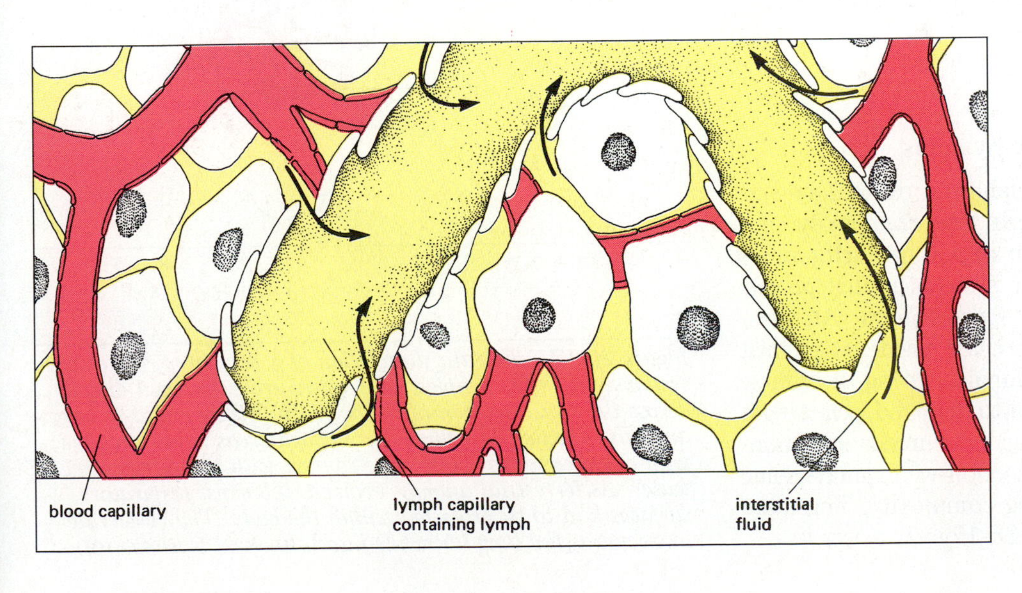

Figure 26-15 *Lymph capillaries end blindly in the body tissues, where pressure from the accumulation of interstitial fluid forces the fluid into the lymph capillaries through valvelike openings between lymph capillary cells.*

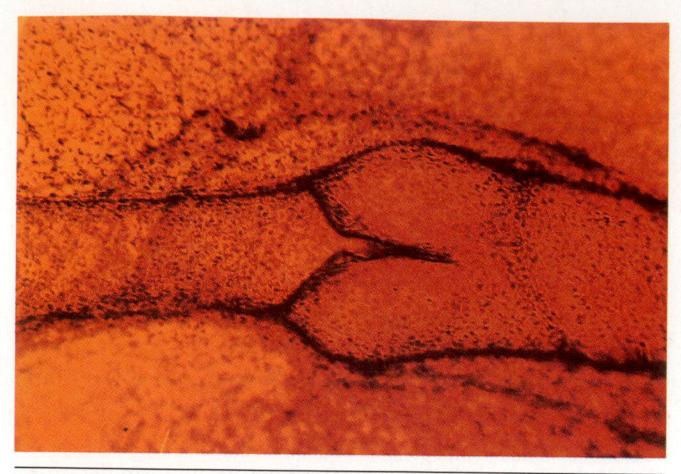

Figure 26-16 Like blood-carrying veins, lymph vessels have internal one-way valves that direct the flow of lymph toward the large veins into which they empty.

blood, exposing it to macrophages and lymphocytes that destroy foreign particles and aged red blood cells.

RESPIRATORY SYSTEMS

Each cell in the animal body is a tiny factory, demanding a continuous influx of energy to maintain itself. The production of energy requires a steady supply of oxygen and generates carbon dioxide as a waste product. These gases move across the cell membrane by diffusion, but as mentioned earlier, diffusion is efficient only over very short distances. Since the cells inside large organisms do not come in contact with the environment, most animals have evolved a pair of systems that effectively bring the environment to the cells: the circulatory and respiratory systems. In the circulatory system, capillaries bring blood within diffusing distance of each cell. Respiratory systems bring a large surface area into intimate contact with both the blood and the external environment so that gases may be exchanged by diffusion between them.

The Evolution of Respiratory Systems

The evolution of increasingly elaborate respiratory systems has paralleled the increase in size, complexity, and activity of the animal body. An important factor influencing the design of respiratory systems is that *all gas exchange occurs across moist membranes*. All gas-exchange systems therefore share the need for a moist respiratory surface. As some animals, during the course of evolution, moved from the sea to dry land, retaining a moist gas-exchange surface provided a major evolutionary challenge. In the following section we examine some respiratory systems of increasing complexity. These are illustrated schematically in Fig. 26-17.

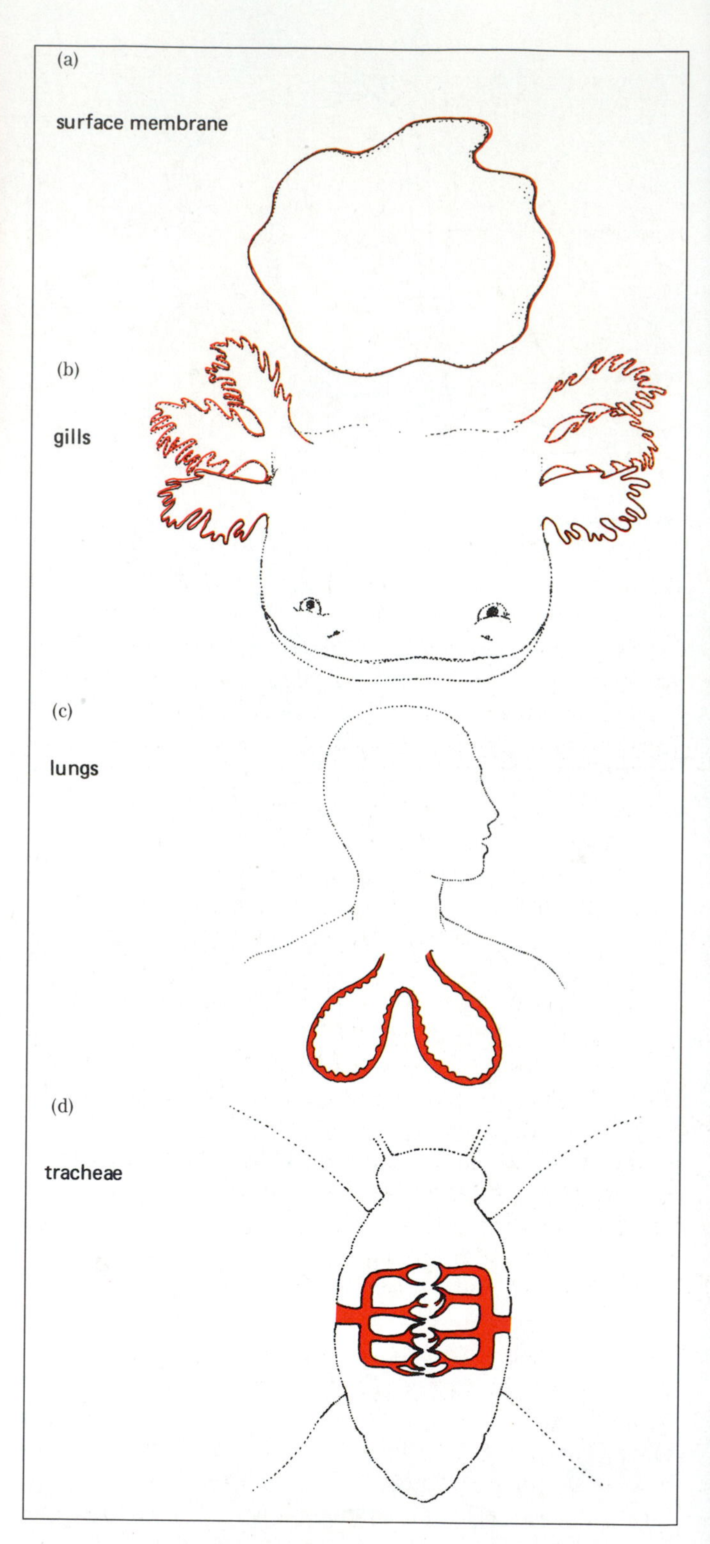

Figure 26-17 Schematic illustration of a variety of respiratory systems. **(a)** *In the simplest case, gases are exchanged directly across the moist exterior of the body.* **(b)** *Gills evolved in larger animals in water, bringing large surface areas richly supplied with blood capillaries into close contact with the surrounding water. As terrestrial animals evolved, the moist respiratory surfaces had to be protected within the body. The vertebrate answer was the lung* **(c)**, *while insects evolved tracheae* **(d)**.

Figure 26-18 A variety of aquatic animals, from the nudibranch (left) with its numerous gill projections, to amphibians such as the axolotl (right), use gills that extend unprotected into the water.

Respiration by diffusion. Diffusion is adequate as a means of gas exchange in moist environments under certain conditions. First, if the animal is extremely small (such as the nematode worm), gases have only a short distance to diffuse to reach all parts of the body. Second, the animal may have a flattened body with a large surface area (such as the flatworm) in which most body cells are close to the outside. Third, the body may consist largely of cells with very low energy demands, such as those of a jellyfish. A fourth mechanism is used by sponges, which bring the environment close to all their cells by circulating seawater throughout their perforated bodies. A fifth condition in which diffusion may be adequate in a relatively large animal is illustrated by the earthworm. These animals have a well-developed closed circulatory system to carry gases throughout the body (Fig. 26-2). As blood rapidly carries off oxygen which has diffused through the skin, a strong concentration gradient is maintained favoring the diffusion of oxygen inward. The elongated shape of the worm assures a relatively large skin surface relative to the internal volume, and the worm's sluggish metabolism has relatively low oxygen demands. The skin must stay moist to remain effective as a gas-exchange organ; a dried-out earthworm will suffocate.

Respiration using gills. In large, active animals, or those living in dry environments, diffusion of gases through the skin is inadequate. In these organisms, specialized respiratory systems have evolved. A wide variety of aquatic animals have evolved **gills.** In its simplest form, as seen in certain molluscs and amphibians (Fig. 26-18), the gill is a projection of the body wall into the surrounding water. In the crab (Fig. 26-20a), the gill is protected inside a rigid exoskeleton. Generally, gills are elaborately branched to maximize surface area, and the size of the gill may reflect environmental conditions. For example, salamanders living in stagnant water have larger gills than those that dwell in well-aerated water. Just beneath the delicate outer membrane, a dense profusion of capillaries brings blood close to the surface, enhancing diffusion tremendously.

The gills of fish (Fig. 26-19) are covered by a protective flap, the **operculum.** This protects the delicate membranes from being nibbled off, and also streamlines the fish body for faster swimming. The fish creates a contin-

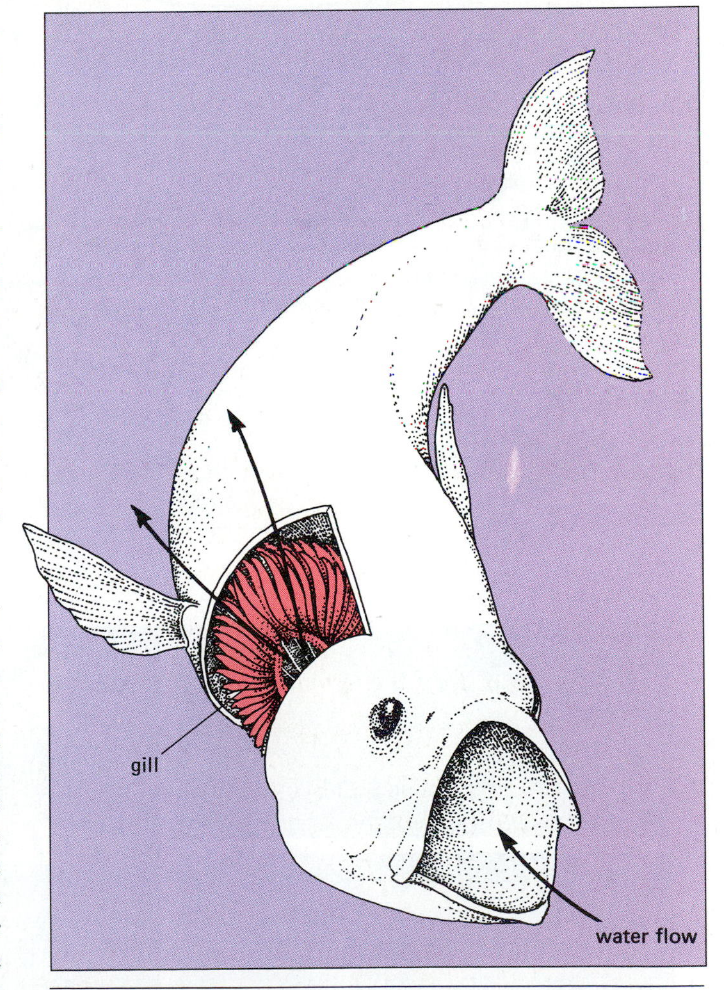

Figure 26-19 The gill reaches its greatest complexity in the fish, where it is protected under a bony flap, the operculum. A one-way flow of water is maintained over the gill by pumping water in through the mouth and out the opercular opening.

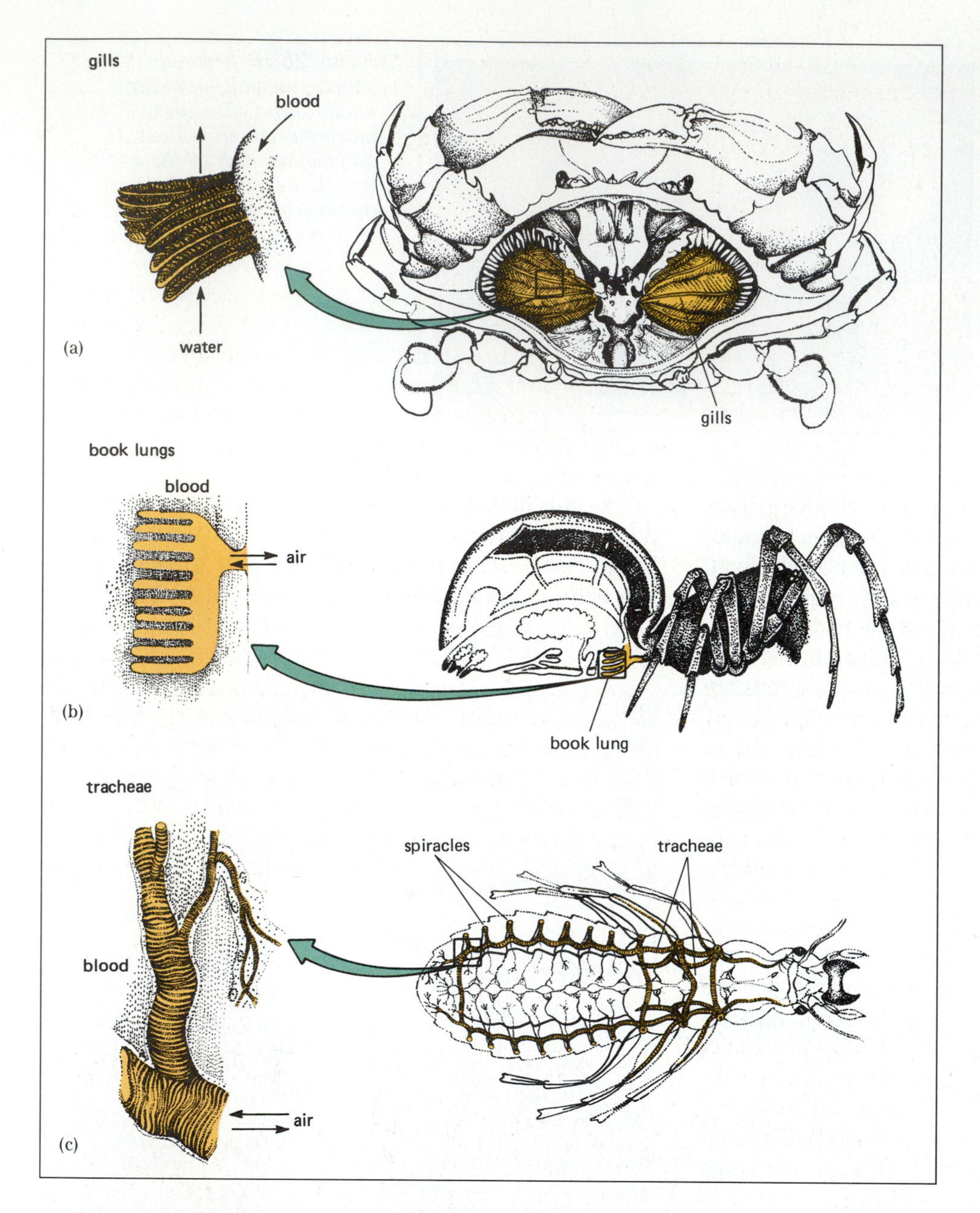

Figure 26-20 Arthropod respiratory structures. **(a)** Crabs protect finely divided gills beneath a rigid exoskeleton. **(b)** The book lungs of spiders are sheetlike layers of thin tissue enclosed in a protective chamber. **(c)** The tracheae of insects open through spiracles in the abdominal wall and branch intricately throughout the body.

Figure 26-21 **(a)** The bullfrog, an amphibian, begins life as a fully aquatic tadpole with feathery external gills **(b)**. During metamorphosis into an air-breathing adult frog, the gills are lost and replaced by simple saclike lungs. In both tadpole and adult, gas exchange also occurs by diffusion through the moist skin. **(c)** The fully terrestrial reptile, illustrated by this mangrove snake, is covered with dry scales that restrict gas exchange through the skin. Reptilian lungs are more efficient than those of amphibians.

uous current over the gills by gulping water into its mouth and ejecting it through the opercular openings. Fish can augment the flow by swimming with their mouths open, and some fast swimmers such as the tuna and certain sharks may rely exclusively on swimming to ventilate their gills.

As described in Chapter 18, the transition to land required dramatic modifications of gas-exchange organs. Gills are useless out of water, collapsing and drying in the air. Terrestrial gas-exchange organs, therefore, need both support and protection from desiccation. Three solutions have evolved: book lungs and tracheae in arthropods, and lungs in vertebrates.

BOOK LUNGS IN ARACHNIDS, TRACHEAE IN INSECTS. Although the external skeleton of terrestrial arthropods helps retain body water, it eliminates the skin as a respiratory surface. Spiders (and some other arachnids, such as scorpions) enclose a series of moist pagelike membranes within a chamber of the exoskeleton, forming a structure called a **book lung** (Fig. 26-20b).

Insects were the first and are still the most successful colonizers of dry land. To convey air to each body cell, insects utilize a system of elaborately branching tubes called **tracheae** (Fig. 26-20c). These subdivide into tiny channels and penetrate the entire body. Each body cell is close to a tracheal tube, minimizing diffusion distances. Tracheae communicate with the outside through openings called **spiracles** located along the side of the abdomen. Muscular pumping movements of the abdomen assist air movement through the tracheae in some large insects, while small insects rely on passive air circulation.

RESPIRATION USING LUNGS. Lungs protect the moist, delicate respiratory surfaces deep within the body, where water loss is minimized and the body wall provides support. The first vertebrate lung probably appeared in a freshwater fish, and consisted of an outpocketing of the digestive tract. Gas exchange in this simple lung helped the fish survive in stagnant water, where oxygen was scarce. Amphibians, straddling the boundary between aquatic and terrestrial life, may use lungs or gills (Figs. 26-18 and 26-21a, b). The aquatic tadpole exchanges its gills for lungs as it develops into a more terresterial frog. Frogs and salamanders utilize their moist skin as an additional respiratory surface.

The scaly, dry skin of reptiles minimizes its use as a gas-exchange organ (Fig. 26-21c). Thus the lungs of reptiles are better developed than those of amphibia. Birds and mammals are exclusively lung breathers. The bird lung has been modified for increased efficiency during strenuous flight (Fig. 26-22). Air sacs inflate during inhalation, storing fresh air. This air passes through the lung during exhalation, providing the bird with a continuous supply of fresh air during both phases of breathing.

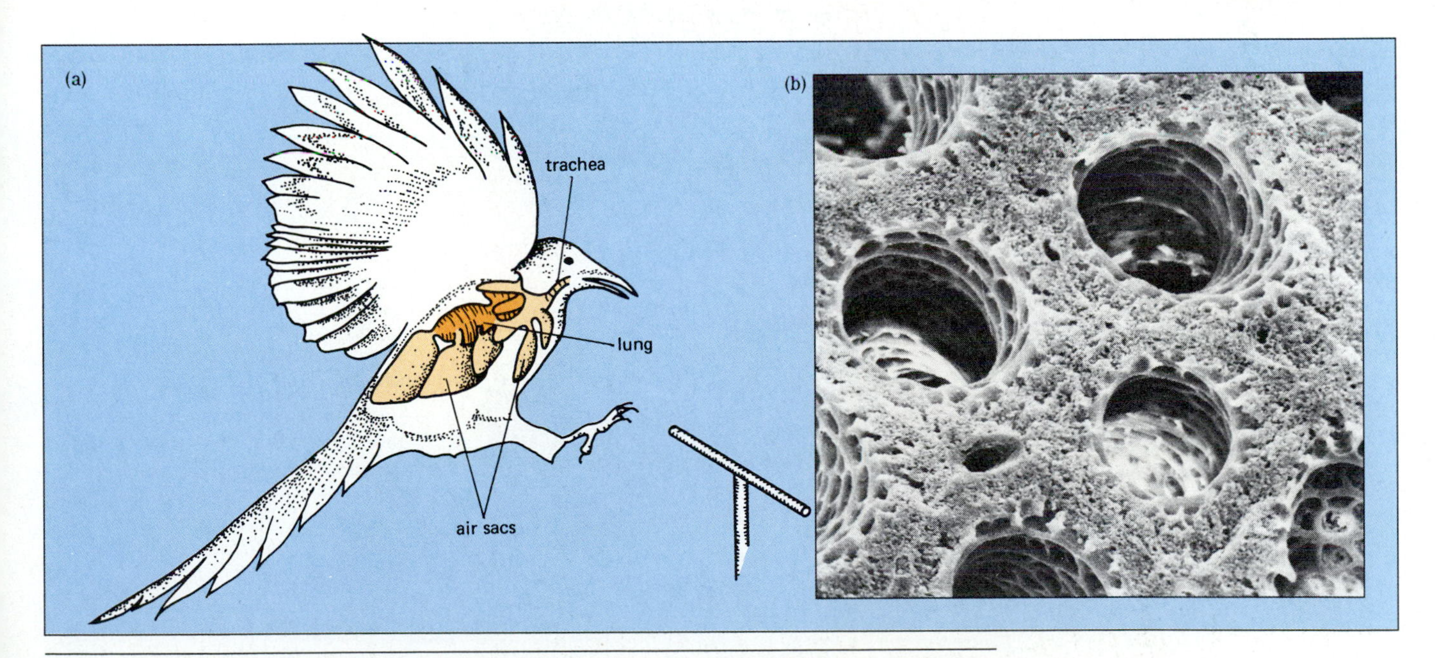

Figure 26-22 **(a)** *The respiratory system of the bird allows extremely efficient gas exchange. When air is inhaled, some fills the lungs while the rest travels past the lungs to fill air sacs. As air is exhaled, the fresh air which has been temporarily stored in the air sacs fills the lungs on its way out.* **(b)** *The bird uses tubular gas-exchange organs called parabronchi rather than saclike alveoli. This allows the air to flow through the lung continuously.*

The Human Respiratory System

The respiratory system in humans and other vertebrates can be divided into two parts: the **conducting portion** and the **gas-exchange portion.** The conducting portion consists of a series of passageways that carry air into the gas-exchange portion. Here gas is exchanged with the blood in tiny sacs called **alveoli.**

THE CONDUCTING PORTION. Air enters through the nose or the mouth, passes through a common chamber, the **pharynx,** and then travels through the **larynx** (Fig. 26-23). In the larynx, the vocal cords, bands of elastic tissue, partially obstruct the opening. Air exhaled past these structures causes them to vibrate, giving rise to the tones of speech or song. The tones are articulated into words by movements of the tongue and lips.

Inhaled air continues past the larynx into the **trachea,** a rigid yet flexible tube whose walls are reinforced with semicircular bands of cartilage. Within the chest, the trachea splits into two large branches called **bronchi** (singular **bronchus**), one leading to each lung. Inside the lung, each bronchus branches repeatedly into ever-smaller tubes called **bronchioles.** These lead finally to the microscopic alveoli, tiny air pockets where gas exchange occurs (Fig. 26-23).

During its passage through the conducting system, the air is warmed and moistened. Much of the dust and bac-

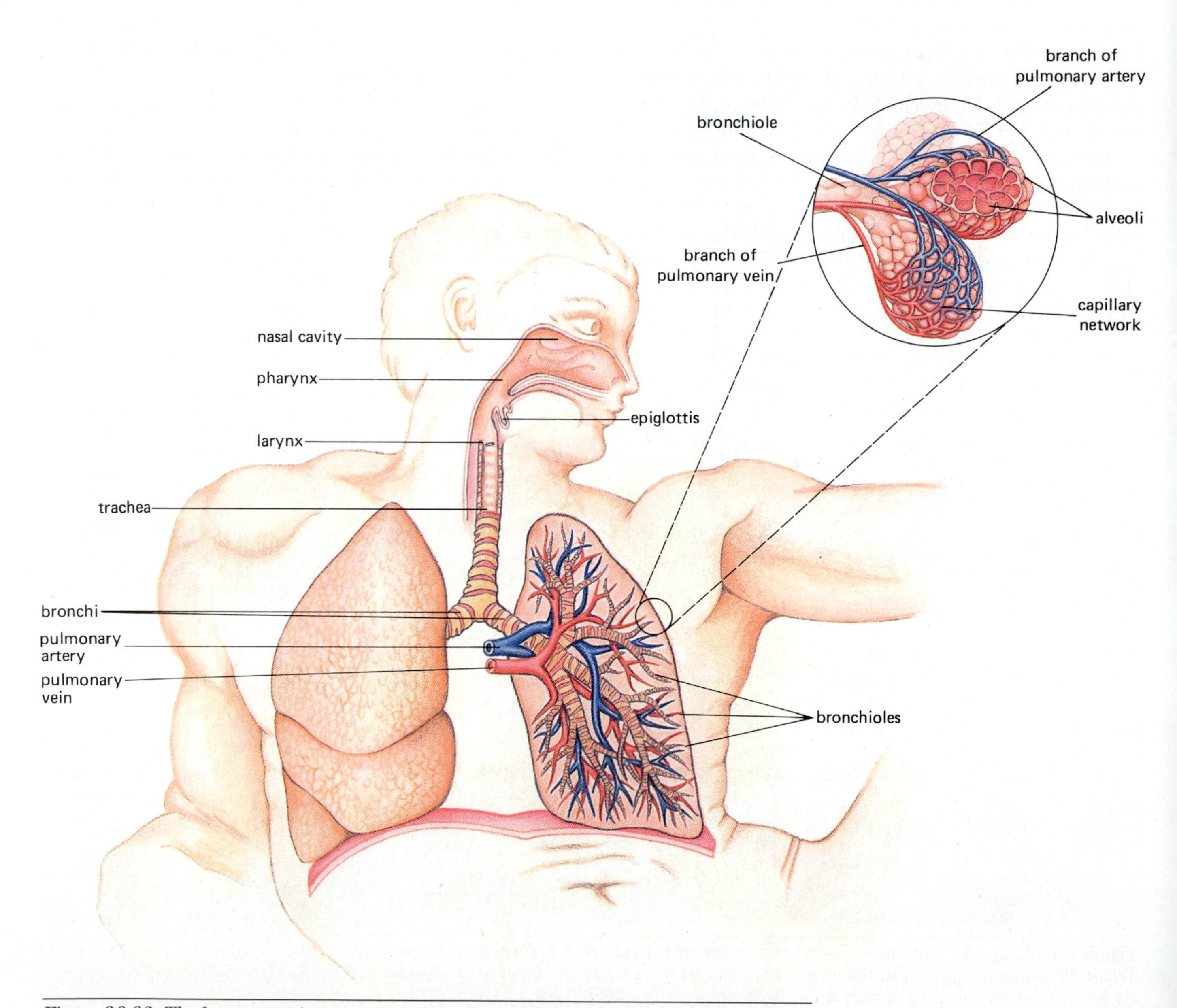

Figure 26-23 The human respiratory system showing a detail of the alveoli surrounded by capillaries. As blood flows through the capillaries, carbon dioxide is exchanged for oxygen.

teria it carries is trapped in mucus secreted by cells lining the conducting passage. The mucus with its trapped debris is continuously swept upward toward the pharynx by cilia that line the bronchioles, bronchi, and trachea. Upon reaching the pharynx, the mucus is coughed up or swallowed. Smoking interferes with this cleansing process by paralyzing the cilia (see the essay, "Smoking and Respiratory Disease").

Gas exchange in the alveoli. The minute (0.2-millimeter) chambers formed by the 150 million alveoli in each lung give the inside of this organ a texture somewhat resembling that of sponge cake. The thin-walled alveoli provide an enormous surface area for diffusion, estimated at 160 square meters in the adult human. The alveoli, which cluster about the end of each bronchiole like a bunch of grapes, are entirely enmeshed in microscopic capillaries (Fig. 26-23). Since the alveolar wall and the capillary wall are each only one cell thick, the air is extremely close to the blood in the capillaries. The lung cells remain moist, coated by a thin layer of water lining each alveolus. Gases dissolve in this water and diffuse through the alveolar and capillary membranes.

As discussed in the preceding section, blood is pumped to the lungs from the right ventricle after having circulated through body tissues. The incoming blood surrounding the alveoli is therefore low in oxygen (since the body cells have used it up) and high in carbon dioxide (released by the cells). Oxygen diffuses from the air in the alveoli, where its concentration is high, into the blood, where its concentration is low. Hemoglobin in the red blood cells greatly increases the diffusion gradient and therefore the tendency of oxygen to move into the blood. By removing the oxygen from solution in the plasma, hemoglobin maintains a high concentration gradient of oxygen from the air to the blood. Conversely, the carbon dioxide content of the blood as it enters the lung capillaries is higher than that of the air, so carbon dioxide diffuses out of the blood into the air in the alveoli.

Blood, oxygenated and purged of carbon dioxide, returns to the left side of the heart, which pumps it to the body tissues. In the tissues, the concentration of oxygen is lower than in the blood, and oxygen diffuses into the cells. Carbon dioxide, which has built up in the cells, diffuses into the blood. While 97 percent of the oxygen in the blood is bound to hemoglobin, only about 23 percent of the carbon dioxide is returned to the lungs bound to hemoglobin. Most of the rest is carried as bicarbonate ion in the blood plasma. The exchange of gases between the cells, blood, and alveoli is illustrated in Fig. 26-24.

The mechanics of breathing. Outside the lungs, the chest cavity is airtight, bounded by neck muscles and connective tissue on top and the dome-shaped muscular **diaphragm** on the bottom. Surrounding and protecting the lungs is the rib cage.

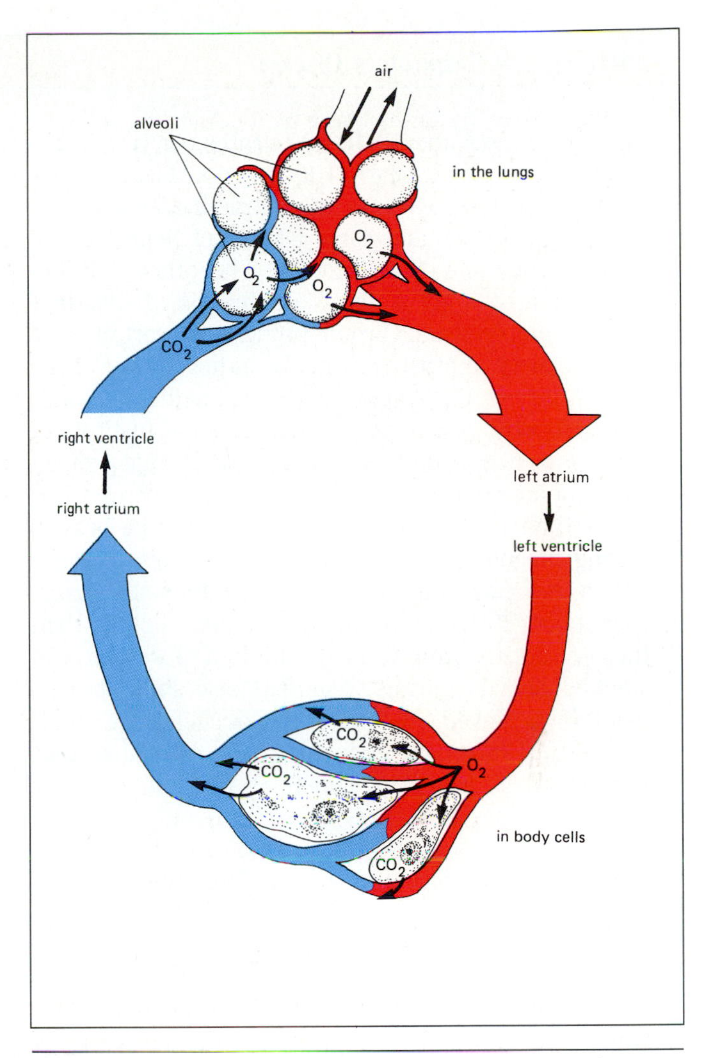

Figure 26-24 *A schematic drawing illustrating gas exchange between the air in the alveoli and the blood, and between the cells of the body and the blood.*

Breathing occurs in two stages: **inspiration,** during which air is actively inhaled, and **expiration,** during which it is passively exhaled. Inspiration is accomplished by making the chest cavity larger. To do this, the diaphragm muscles are contracted, drawing the diaphragm downward, and the rib muscles (intercostals) are contracted, lifting the ribs up and outward (Fig. 26-25). When the chest cavity is expanded, the lungs expand with it, since a vacuum holds them tightly against the inner wall of the chest. (If the chest is punctured and air leaks in, the lung will collapse.) As the lungs expand, air moves into them to fill the additional space. Expiration occurs automatically when the muscles causing inspiration are relaxed, although it can be assisted by contraction of abdominal muscles. The diaphragm is relaxed during expiration and domes upward, and the ribs fall down and inward, decreasing the size of the chest cavity and forcing air out of the lungs.

After expiration, the lungs still contain air. This pre-

Smoking and Respiratory Disease

Smoking causes about 340,000 deaths per year in the United States alone, nearly 1000 per day. Heavy smokers increase their chances of contracting lung cancer by a factor of 20, and some 94,000 of them will die from it this year (Fig. E26-3). The other 226,000 smoking deaths are from a combination of smoking-induced emphysema, chronic bronchitis, heart disease, and a variety of other cancers, including cancer of the mouth, larynx, esophagus, pancreas, bladder, and kidney. Recently, researchers at the National Institute of Environmental Health Sciences reported that women who smoke more than a pack a day are 50 percent less fertile than nonsmokers. Women who smoke heavily during pregnancy have twice the rate of miscarriages. When they carry their babies to term, the babies average about 200 grams (about $\frac{1}{2}$ pound) lighter than those of nonsmoking women. Children of women who smoked heavily during pregnancy also show impairment in achievement test scores during early childhood. A 1985 congressional study estimated that smoking costs American society $65 billion yearly, or $10 million per hour. This figure refers only to dollars spent on health-care services related to smoking and lost job productivity and wages. The human cost is incalculable.

Let us explore briefly the effects of tobacco smoke on the human respiratory tract. As smoke is inhaled through the nose, trachea, and bronchi, toxic substances such as nicotine and sulfur dioxide paralyze the cilia lining the respiratory tract; a single cigarette can inactivate them for a full hour. Since these ciliary sweepers remove inhaled particles, smoking poisons them just when they are most needed. The visible portion of cigarette smoke consists of billions of microscopic carbon particles. Adhering to them are a wide variety of toxic compounds, a dozen or more of which are carcinogenic (cancer causing). With the cilia out of action, the particles stick to the walls of the respiratory tract or enter the lungs. Thus smokers encounter a higher risk of cancer in all areas of the respiratory tract touched by smoke.

A second line of defense is the presence in the respiratory tract of large numbers of amoeba-like white blood cells (macrophages) which engulf foreign particles and bacteria. Cigarette smoke also impairs these cells, allowing still more bacteria, dust, and smoke particles into the lungs. In response to the irritation of cigarette smoke, the respiratory tract increases the production of mucus, a third method of trapping foreign particles. But without the cilia to sweep it along, the mucus builds up and can obstruct the airways; the familiar "smoker's cough" is an attempt to clear the airways. Microscopic smoke particles find a secure lodging place in the tiny, moist alveoli deep within the lungs. There they accumulate over the years until the lungs of a heavy smoker are literally blackened (Figs. E26-3 and E26-4). The longer the delicate tissues of the lungs are exposed to the carcinogens on the trapped particles, the greater the chance of cancer developing.

E26-3 Lung cancer in a smoker. The cancer is seen as a whitish mass, while the tissue around it is blackened by trapped smoke particles.

Smoking also gives rise to chronic bronchitis. This persistent lung infection is characterized by cough, swelling of the lining of the respiratory tract, an increase in mucus production, and a decrease in the number and activity of cilia. The result: a decrease in airflow to the alveoli.

Toxins in cigarette smoke, such as nitrogen oxides and sulfur dioxide, cause the body to produce substances that reduce the elasticity and increase the brittleness of lung tissue. The brittle alveoli eventually rupture, and the lung gradually loses its normal sponge cake appearance and more closely resembles blackened Swiss cheese (Fig. E26-4). This condition is called emphysema. The loss of the alveoli, where gas exchange occurs, leads to oxygen deprivation of all body tissues. The emphysema victim's breathing is labored, and grows increasingly worse until death. Chronic bronchitis and emphysema kill about 50,000 persons yearly in the United States, and smoking is the most important contributing factor to each.

Meanwhile, carbon monoxide, present in high levels in cigarette smoke, is eagerly taken up by the red blood cells in place of oxygen. In fact, hemoglobin binds to

Smoking and Respiratory Disease

carbon monoxide about 210 times as tenaciously as it does to oxygen. This inactivates the hemoglobin and seriously reduces the blood's oxygen-carrying capacity, placing a strain on the heart. Chronic bronchitis and emphysema compound this difficulty by restricting the amount of oxygen that enters the lungs and by destroying the alveoli where oxygen enters the bloodstream. As a result, smokers are 70 percent more likely than nonsmokers to die of heart disease. Although the reasons are not fully understood, reduced oxygen in the blood supplying the heart muscle is undoubtedly a major contributor. The carbon monoxide in cigarette smoke may also contribute to the increased incidence of stillbirths in pregnant women who smoke, the lower birthweight of their babies, and the learning impairment in early childhood of children born to heavy smokers.

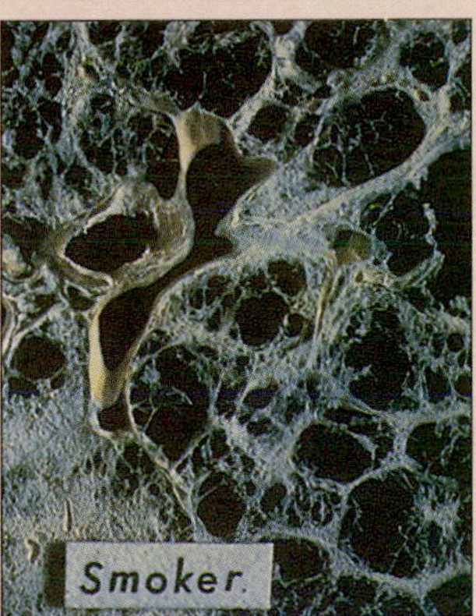

Figure E26-4 *A section through a normal lung appears almost opaque, while the lung of a smoker suffering from emphysema is full of large holes, each caused by the rupture of hundreds of alveoli.*

Evidence is gradually accumulating that nonsmokers exposed to cigarette smoke may suffer more than temporary emotional and physical distress. Several studies have concluded that infants whose mothers smoke are more likely to suffer from bronchitis and pneumonia. A 1983 study from Harvard Medical School showed a 7 percent decrease in the lung capacity of children whose mothers smoke, a condition that may predispose them to lung disease later in life. Abnormally thickened and stiffened heart walls were found in 11- and 12-year-old boys whose parents smoked, in a study conducted at the Medical College of Virginia. People suffering from angina have been found to experience chest pain significantly more readily after sitting in a smoke-filled room. The *New England Journal of Medicine* reported a study of 2100 nonsmokers who were chronically exposed to cigarette smoke. These individuals showed impaired lung function similar to that of people smoking between 1 and 10 cigarettes daily. A 1985 EPA study singled out "passive" tobacco smoke as the major cause of cancer caused by airborne carcinogens. The study estimates that passive smoking is responsible for between 500 and 5000 deaths of nonsmokers each year. A 1986 National Academy of Sciences study concluded that nonsmoking spouses of smokers face a 30 percent higher risk of lung cancer as a result of "passive smoking."

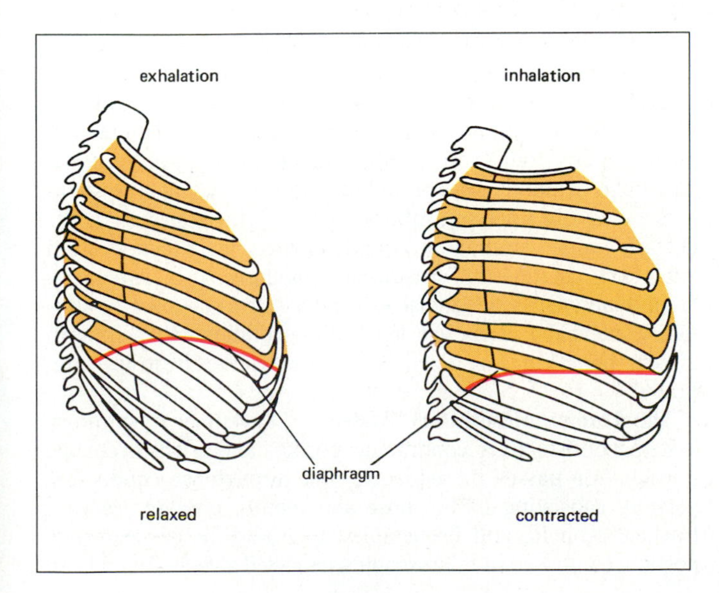

vents the thin alveoli from collapsing, and fills the space within the conducting portion of the respiratory system. A normal breath moves only about 500 milliliters of air into the respiratory system. Of this, only about 350 milliliters reaches the alveoli for gas exchange. Deeper breathing during exercise causes several times this volume to be exchanged.

Figure 26-25 *The mechanics of breathing. Rhythmic nerve impulses from the brain stimulate the diaphragm muscle to contract (pulling it downward) and the muscles surrounding the ribs to contract (moving them up and outward). The result is an increase in the size of the chest cavity, causing air to rush in. Relaxation of these muscles allows the diaphragm to dome upward, and the rib cage to collapse, forcing air out of the lungs.*

CONTROL OF RESPIRATION. Breathing occurs rhythmically and automatically without conscious thought. But the muscles used in breathing are not self-activating; each contraction is stimulated by impulses from nerve cells. These impulses originate in the **respiratory center** located in the brainstem just above the spinal cord. Nerve cells of the respiratory center generate cyclic bursts of impulses, causing alternating contraction and relaxation of the respiratory muscles.

The respiratory center receives input from a variety of sources that adjust breathing rate and volume to meet the body's changing needs. For example, if the lungs become overexpanded, stretch receptors in the lungs inhibit further inspiration. In addition, the brainstem contains receptor neurons that monitor the concentration of carbon dioxide in the blood. Elevated carbon dioxide levels indicate an increase in cellular activity and a corresponding requirement for more oxygen. The receptors therefore stimulate an increase in the rate and depth of breathing. These receptors are extremely sensitive; an increase in carbon dioxide of only 0.3 percent can cause a doubling of the breathing rate. Thus the respiratory rate is regulated to maintain a constant level of carbon dioxide in the blood. The respiratory rate is much less sensitive to changes in oxygen concentration, since normal breathing supplies an overabundance of oxygen. However, should blood oxygen levels fall drastically, receptors in the aorta and carotid arteries stimulate the respiratory center. When a person begins strenuous activity, such as running, an increase in breathing rate *precedes* any changes in blood gas levels. Apparently, when higher brain centers activate muscles during heavy exercise, they simultaneously stimulate the respiratory center to increase breathing rate. Breathing activity is then "fine-tuned" by the receptors monitoring carbon dioxide concentrations.

SUMMARY OF KEY CONCEPTS

Circulatory Systems

Unicellular and some simple multicellular organisms rely exclusively on diffusion to exchange wastes, nutrients, and gases with the environment. As larger, more active multicellular organisms evolved, circulatory and respiratory systems became necessary. These transport a fluid rich in dissolved nutrients and oxygen close to the cells, where nutrients can be released to the cells and wastes absorbed by diffusion. Invertebrates may have either open or closed circulatory systems, while all vertebrates have closed systems. In open systems, blood is pumped into a hemocoel, where it directly bathes internal organs. In closed systems, the blood is confined to the heart and vessels.

Vertebrate Circulatory Systems Vertebrate circulatory systems play many roles, including transport of gases, distribution of nutrients, and transport of wastes. In addition, circulatory systems transport hormones, help regulate body temperature, and defend the body against disease. The circulatory system has three parts: the heart, the vessels, and the blood.

The vertebrate heart evolved from two chambers in fishes, to three in amphibians and most reptiles, to four in birds and mammals. In the four-chambered heart, blood is pumped separately to the lungs and through the body, maintaining complete separation of oxygenated and deoxygenated blood. The heart muscle is spontaneously active, with contractions coordinated by two specialized masses of muscle cells. The SA node of the right atrium serves as the heart's pacemaker. The wave of atrial contraction activates the AV node, also in the right atrium, from which it spreads through the ventricles along specialized fiber tracts. Heart rate can be modified by the nervous system and hormones such as epinephrine.

Blood leaving the heart travels in sequence through arteries, arterioles, capillaries, venules, veins, and back to the heart. Each vessel is specialized for its specific role. Elastic, muscular arteries help pump the blood along. The thin-walled capillaries are the sites of exchange of materials between the body cells and the blood. Veins provide a low-resistance path back to the heart, with one-way valves maintaining the direction of blood flow.

The blood consists of both fluid and cellular portions. Red blood cells are specialized to carry oxygen, platelets aid in blood clotting, and most of the five types of white cells fight infection. The fluid or plasma is primarily water in which proteins, gases, nutrients, and wastes are dissolved.

The human lymphatic system consists of vessels resembling veins and capillaries, lymph nodes, and the thymus and spleen. It transports fats from the intestine to the circulatory system, returns to the blood excess fluids and small proteins which leak through the blood capillary walls, and fights infection by filtering the lymph through lymph nodes, where white blood cells ingest foreign invaders such as bacteria.

Respiratory Systems

Like circulatory systems, respiratory systems evolved in response to the need to facilitate exchange between the individual cells and the external environment. Respiratory and circulatory systems work in intricate harmony, with the blood transporting the gases exchanged across the respiratory membranes.

Evolution of Respiratory Systems Animals have evolved a diverse array of respiratory systems. Animals in moist environments whose bodies are small, flattened, and have very low metabolic demands and/or well-developed circulatory systems sometimes rely solely on diffusion through the skin for gas exchange. Larger animals in aquatic environments have evolved gills, such as those of crustaceans, molluscs, fish, and many amphibians. On land, moist respiratory surfaces must be protected internally. This has led to the evolution of tracheae in insects, book lungs in arachnids, and lungs in terrestrial vertebrates.

The Human Respiratory System The human respiratory system consists of a conducting portion and a gas-exchange portion. Air passes in sequence, first through the conducting portion, consisting of the nose and mouth, pharynx, larynx, trachea, bronchi, and bronchioles, then into the gas-exchange portion, composed of microscopic sacs called alveoli. Blood from a dense capillary network surrounding the alveoli releases carbon dioxide and absorbs oxygen from the air.

Breathing in humans involves actively drawing air into the lungs by contracting the diaphragm and the rib muscles, which expand the chest cavity. Relaxing these muscles causes the air to be expelled passively. Breathing is controlled by nerve impulses originating in the brain, and its rate is modified by a variety of receptors, the most important of which are brain cells that monitor carbon dioxide levels in the blood.

GLOSSARY

Alveolus (al-vē′-o-lus; pl. alveoli) a tiny air sac within the lungs surrounded by capillaries where gas exchange with the blood occurs.

Angina pectoris (an-jī′-na pek-tōr′-is) chest pain associated with reduced blood flow to the heart muscle caused by obstruction of coronary arteries.

Arteriole (ar-tēr′-ē-ōl) a small artery that empties into capillaries. Contraction of the arteriole regulates blood flow to various parts of the body.

Artery (ar′-tur-ē) a vessel with muscular, elastic walls that conducts blood away from the heart.

Atherosclerosis (ath′-er-ō-skler-ō′-sis) a disease characterized by obstruction of arteries by cholesterol deposits and thickening of the arterial walls.

Atrioventricular node (ā′-trē-ō-ven-trik′-ū-lar nōd) a specialized mass of muscle at the base of the right atrium through which the electrical activity initiated in the SA node is transmitted to the ventricles.

Atrium (ā′-trē-um) a chamber of the heart that receives venous blood and passes it to a ventricle.

Bronchiole (bron′-kē-ōl) a narrow tube formed by repeated branching of the bronchi, which conducts air into the alveoli.

Bronchus (bron′-kus) a tube that conducts air from the trachea to each lung.

Closed circulatory system the type of circulatory system found in certain worms and vertebrates in which the blood is always confined within the heart and vessels.

Diaphragm (dī′uh-fram) a dome-shaped muscle forming the floor of the chest cavity. Contraction of this muscle pulls it downward, enlarging the cavity and causing air to be drawn into the lungs.

Emphysema (em-fuh-sē′-muh) a condition in which the alveoli become brittle and rupture, causing decreased area for gas exchange.

Erythrocytes (ē-rith′-rō-sītes) red blood cells active in oxygen transport, which contain the red pigment hemoglobin.

Expiration (ex-per-ā′-shun) the act of exhaling, which results from relaxation of the respiratory muscles.

Fibrin (fī′-brin) a clotting protein formed in the blood in response to a wound. Fibrin binds with other fibrin molecules and provides a matrix around which a blood clot forms.

Gills in aquatic animals, a branched tissue richly supplied with capillaries around which water is circulated for gas exchange.

Hemocoel (hē′-mō-sēl) the blood cavity in an open circulatory system.

Hemoglobin (hē′mō-glō-bin) an iron-containing protein that gives red blood cells their color. Hemoglobin binds to oxygen in the lungs and releases it to the tissues.

Hypertension arterial blood pressure that is chronically elevated above the normal level.

Inspiration the act of inhaling air into the lungs by enlarging the chest cavity.

Interstitial fluid (in-tur-sti′-shul) fluid similar in composition to plasma (except lacking large proteins) which surrounds the body cells. This fluid leaks from the capillaries and acts as a medium of exchange between the body cells and the capillaries.

Larynx (lār′-inx) that portion of the air passage between the pharynx and the trachea. The larynx contains the vocal cords.

Leukocyte (loo′-kō-sīt) any of the white blood cells circulating in the blood.

Lymph (limf) pale fluid within the lymphatic system composed primarily of interstitial fluid and lymphocytes.

Lymphatic system a system consisting of lymph vessels, lymph capillaries, lymph nodes, and the thymus and spleen. The system helps protect the body against infection, absorbs fats, and returns excess fluid and small proteins to the blood circulatory system.

Lymph nodes small structures that act as filters for lymph. These contain both lymphocytes and macrophages, which inactivate foreign bodies such as bacteria.

Lymphocyte (lim′-fō-sīt) white blood cell type important in the immune response.

Macrophage (mak′-rō-faj) a cell derived from white blood cells called monocytes, whose function is to consume foreign particles including bacteria.

Megakaryocyte (meg-a-kār′-ē-ō-sīt) a large cell type that remains in the bone marrow, pinching off pieces of itself. These cytoplasmic fragments enter the circulation as platelets.

Open circulatory system a type of circulatory system found in some invertebrates, such as arthropods and molluscs, which includes an open space in which blood directly bathes body tissues.

Pharynx (fār′-inx) a chamber at the back of the mouth shared by the digestive and respiratory systems.

Plasma the fluid, noncellular portion of the blood.

Platelets (plāt′-lets) cell fragments formed from megakaryocytes in bone marrow. Platelets, which lack nuclei, circulate in the blood and play a role in blood clotting.

Pulmonary circulation the pathway of blood from the right ventricle through the pulmonary artery to the lung capillaries and back to the left atrium through the pulmonary vein.

Respiratory center a location in the brainstem that sends rhythmic bursts of nerve impulses to the respiratory muscles, resulting in breathing.

Sinoatrial node (sī′-nō-āt′-rē-ul nōd) also called the SA node, this is a small mass of specialized muscle in the wall of the right atrium. It generates electrical signals rhythmically and spontaneously and serves as the heart's pacemaker.

Spiracles (spī′-re-kul) openings in the abdominal segments of insects through which air enters the tracheae.

Stroke an interruption of blood flow to part of the brain, caused by the rupture of an artery or the blocking of an artery by a blood clot. Loss of blood supply leads to rapid death of the area of the brain affected.

Systemic circulation the pathway of the blood from the left ventricle through the aorta and other large arteries to the capillaries of the body tissues and back through the large veins to the right atrium.

Trachea (trā′-kē-uh) a rigid but flexible tube supported by rings of cartilage, which conducts air between the larynx and the bronchi.

Tracheae (trā′-kē) elaborately branching tubes that ramify through the bodies of insects and carry air close to each body cell. Air enters the tracheae through openings called spiracles.

Vein a large-diameter, thin-walled vessel that carries blood from venules back to the heart.

Ventricle (ven′-trē-kul) the lower muscular chamber on each side of the heart, which pumps blood out through the arteries. The right ventricle sends blood to the lungs, and the left to the rest of the body.

Venule (ven′-yul) a narrow vessel with thin walls that carries blood from capillaries to veins.

STUDY QUESTIONS

1. Trace the flow of blood through the circulatory system starting and ending with the capillaries surrounding the alveoli.
2. List three types of blood cell and describe their principal functions.
3. What are five functions of the vertebrate circulatory system?
4. In what way do veins and lymph vessels resemble one another? Describe how fluid is transported in each of these vessels.
5. What two important functions are served by the lymphatic system?
6. Distinguish among plasma, interstitial fluid, and lymph.
7. Describe veins, capillaries, and arteries, noting their similarities and differences.
8. Trace the evolution of the vertebrate heart from two to four chambers.
9. What is the role of atherosclerosis in heart attack?
10. Describe three different arthropod respiratory systems and two different vertebrate respiratory systems.
11. Trace the route taken by air in the vertebrate respiratory system, listing the structure through which it flows and the point where gas exchange occurs.
12. What evolutionary changes in the "life-style" of animals led to the evolution of diverse respiratory structures? Explain how each of these adapts the animal to its environment.
13. How are human respiratory movements initiated? Modified?
14. What events occur during human inspiration? Expiration? Which is always an active process?
15. Trace the pathway of an oxygen molecule in the human body starting with the nose and ending with a body cell.
16. Describe the effects of smoking on the human respiratory system.

SUGGESTED READINGS

"Cholesterol; The Villain Revealed." *Discover,* March 1984. A beautifully illustrated three-part special report on cholesterol, its effects on human arteries, and methods of treatment.

Comroe, J. "The Lung." *Scientific American,* February 1966. Describes the intricate relationship between air sacs and blood vessels in the human lung.

Eckert, R., Randall, D., and Augustine, G. New York: W. H. Freeman and Company, Publishers, *Animal Physiology: Mechanisms and Adaptations.* 3rd ed. 1988. An excellent and complete textbook of comparative animal physiology; see Chapters 13 and 14.

Hole, J. W., Jr. *Essentials of Human Anatomy and Physiology.* Dubuque, Iowa: Wm. C. Brown Company, 1986. Descriptions of both human anatomy and physiology. Beautiful, clear illustrations.

Mayerson, H. "The Lymphatic System." *Scientific American,* 1963. Describes how the lymphatic system works in conjunction with the circulatory systems to maintain homeostasis.

Perutz, M. "Hemoglobin Structure and Respiratory Transport." *Scientific American,* December 1978. How hemoglobin plays the dual role of transporting oxygen to the tissues and carbon dioxide back to the lungs.

"Replacing the Heart." *Discover,* February 1983. An excellent four-part report on the artificial heart.

Schmidt-Nielsen, K. "How Birds Breathe." *Scientific American,* December 1971. Specializations of the bird respiratory system include additional air sacs and even hollow bones.

Zucker, M. "The Functioning of Blood Platelets." *Scientific American,* June 1980. Describes the complex role of platelets in blood clotting.

27
The Immune System: Defense Against Disease

If you wish to be well and keep well, take Braggs Vegetable Charcoal and Charcoal Biscuits. Absorbs all impurities in the stomach and bowels, effectually warding off cholera, smallpox, typhoid, and all malignant fevers. Eradicate worms in children. Sweeten the breath.

from an early twentieth century newspaper ad

As incubators for microbial growth, human bodies are nearly ideal. They maintain a constant temperature of 37°C, and their cells contain abundant water and nutrients. Trillions of microbes lurk out there—in the air, in the water, in the foods we eat, and on the objects we touch—ready to colonize our bodies. Nevertheless, day after day we shrug off the assaults of these miniature parasites, usually completely unaware of them. Even when an infection does occur, we almost always fend it off after a few days. How does the human body keep from becoming a warm, moist incubator for microbes? If infected, how does it destroy its invaders?

DEFENSES AGAINST MICROBIAL INVASION

The human body has three lines of defense against microbial attack. The first, and obviously best, defense is to keep microbes out of the body in the first place. The human body has two surfaces exposed to the environment: the **skin** and the **mucous membranes** of the digestive and respiratory tracts. These surfaces are barriers to microbial entry. Two internal defenses are mustered against microbes that penetrate through these surfaces. The first is the nonspecific **inflammatory response** to injury, which, if unsuccessful, is followed by the highly specific **immune response** directed against the particular invading organism.

Barriers to Entry

The skin is both a physical barrier to microbial entry and an inhospitable environment for microbial growth. The outer surface of the skin consists of dry, dead cells filled with horny proteins similar to those in hair and nails. Consequently, most microbes that land on the skin cannot obtain the water and nutrients they need. Secretions from sweat glands and oil-producing sebaceous glands also cover the skin. These secretions contain acids and natural antibiotics, such as lactic acid, that inhibit the growth of bacteria and fungi. These multiple defenses make the unbroken skin an extremely effective barrier against microbial invasion.

The membranes of the digestive and respiratory tracts are also well defended. First, they secrete mucus that contains antibacterial enzymes such as lysozyme, which destroys bacterial cell walls. Second, the mucus physically traps microbes entering through the nose or mouth, and cilia on the membranes sweep up the mucus, microbes and all, until it is either coughed or sneezed out of the body, or swallowed. If microbes are swallowed, they enter the stomach, where they encounter a combination of extreme acidity (about pH 2) and protein-digesting enzymes. Farther along in the digestive tract, the intestine is inhabited by bacteria that are harmless to the human body, but which secrete substances that destroy invading foreign bacteria or fungi. Despite these defenses, the warm, moist mucous membranes are much more vulnerable than the dry, oily skin, and many disease organisms succeed in entering the body through these membranes.

Internal Defenses

A barrier is an effective defense only as long as it remains intact. If you cut your skin or mucous membranes, microbes easily enter the wound. Fortunately, this invasion immediately evokes a second line of defense, the **inflammatory response.** White blood cells detect the presence of bacteria in the wound (see below for the mechanism of recognition). These cells, along with other cells directly damaged by the cut, release the chemical **histamine** into the wounded area. Histamine renders capillary walls leaky, and relaxes the smooth muscle surrounding arterioles, leading to increased blood flow. With extra blood flowing through leaky capillaries, fluid seeps from the capillaries into the tissues around the wound. The wound becomes red, swollen, and warm. Meanwhile, other chemicals released by injured cells initiate blood clotting (see Chapter 26), which "walls off" the wounded area from the rest of the body, thereby preventing microbes from escaping into the bloodstream.

Still other chemicals, some released by wounded cells and others produced by the microbes themselves, attract phagocytic white blood cells to the wound. Like amoebas, these white blood cells move by pseudopodia. They squeeze out through the capillary walls, enter the wound, and engulf bacteria, dirt, and tissue debris (Fig. 27-1). Unfortunately, each white blood cell can eat just so many microbes, and then it dies. The pus that collects around a wound consists largely of bacteria, tissue debris, and living and dead white blood cells.

If the wound is not too large and the microbes reproduce slowly enough, the inflammatory response will keep most of the invaders out of the bloodstream. The few that escape are eaten by white blood cells in the blood vessels and lymph nodes, promptly halting the infection. However, inflammatory responses cannot handle all assaults. The wound may be large, or the microbes may multiply too quickly. Many viruses enter the body through the respiratory tract and do not provoke much, if any, immediate inflammatory response. In these cases, microbes may enter the bloodstream, reproducing rapidly and endangering the entire body. If this happens, the highly specific **immune response** comes into play, directing the destruction of the particular type of microbe that has passed through the first two defenses.

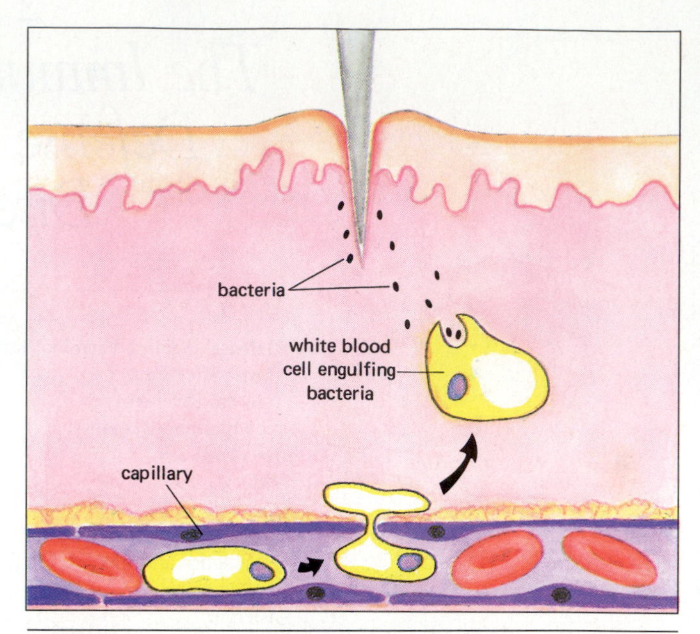

Figure 27-1 During an inflammatory response, the cells forming capillary walls separate slightly. White blood cells squeeze through the capillary wall to join the fray against bacteria that have entered a cut.

THE IMMUNE RESPONSE

Over 2000 years ago, the Greek historian Thucydides recognized the essential features of the immune response. He observed that occasionally someone contracts a disease, recovers, and thereafter is no longer susceptible to that particular disease—the person has become immune. With rare exceptions, however, immunity to one disease confers no protection against other diseases. Thus the immune response attacks one specific type of microbe, overcomes it, and provides future protection against this microbe but no others. How does the immune system accomplish this feat?

The immune system consists of two types of lymphocytes, called **B cells** and **T cells** (Fig. 27-2). As with all white blood cells, B and T cells arise from precursor cells in the bone marrow. Early in embryonic development, the newly forming T cells migrate to the thymus (hence the name T cell) and differentiate into their mature forms. B cells mature in the bone marrow itself. As we shall see shortly, B and T cells play quite different roles in the immune response. However, immune responses produced by both B and T cells consist of the same three fundamental steps: (1) recognizing the invader, (2) launching a successful attack, and (3) retaining the memory of the invader to ward off future infections.

Fever: Friend or Foe?

Mother in TV ad: "Poor Johnny has a fever! Thank heaven we have children's aspirin!"

How many times have we all heard the refrain on TV that we should take aspirin to reduce fevers? Until a few years ago, the medical dogma was that all fevers are undesirable, and extremely high fevers [40°C (104°F) or above] are dangerous; therefore, fevers should be reduced as soon as possible. Recently, however, researchers have begun to raise some important questions: Is fever caused by disease organisms, or is it initiated by the body itself in response to disease? If the body produces its own fevers, might they be useful? Otherwise, why would we have evolved the ability to elevate our body temperatures during disease?

The hypothalamus, a part of the brain, contains temperature-sensing nerve cells that act as a thermostat for the body. Normally, the thermostat is set at 37°C (98.6°F). When disease organisms invade, however, the thermostat is turned up. Certain white blood cells respond to the infection by releasing a hormone called **endogenous pyrogen** ("self-produced fire-maker"). Endogenous pyrogen travels in the bloodstream to the hypothalamus and raises the thermostat's set point. Hypothalamic nerve cells send out signals causing behaviors that increase body temperature: shivering, increased fat metabolism, or feeling cold, so that more clothing is put on. Endogenous pyrogen also causes other cells to reduce the concentration of iron and zinc in the blood.

Fever has both beneficial effects for the body's defenses and detrimental effects on the invading microbes. Many bacteria require more iron to reproduce at temperatures of 38 or 39°C than at 37°C, so fever and reduced iron in the blood combine to slow down their rate of reproduction. Simultaneously, fever increases the activity of immune cells that attack the bacteria, thereby producing a shorter and less serious infection.

Fever also helps to fight viral infections. When certain cells of the body are invaded by viruses, they synthesize and release a protein called **interferon,** which travels to other cells and increases their resistance to viral attack. Fevers increase the production of interferon. In one study, patients with colds were treated with aspirin (to reduce fever) or a placebo (an inactive substance that looks like the drug in question so that the patients won't know whether or not they have been given the drug). Those given aspirin released far more viruses from their noses and throats than did those in the placebo group. This means that the fever-reduced patients were not controlling their own infections very well and that they were significantly more infectious to other people.

Severe fevers are dangerous, or even fatal, but clearly "average" fevers of 38 or 39°C are beneficial. For your own health, and that of the people around you, perhaps you should be slower to reach for the aspirin the next time you get a cold.

Recognition

To understand how the immune system recognizes invading microbes, we must answer three related questions: (1) How do the immune cells "read" the structure of foreign substances? (2) How can they recognize and respond to so many foreign molecules? (3) How do they recognize that a substance is "foreign" and not "self," thereby avoiding destruction of the tissues of their own body?

"READING" FOREIGN MOLECULES. The key to the immune system's ability to attack invading microbes lies in the structure and function of large proteins called **antibodies.** Antibodies are Y-shaped molecules composed of two pairs of peptide chains, one large (heavy) chain and one small (light) chain on each side of the Y (Fig. 27-3). Both heavy and light chains consist of a **constant region** that is the same (or nearly the same) in all antibodies, and a **variable region** that differs among antibodies. The pair of variable regions at the tip of each arm form a binding site for large molecules. These binding sites are a lot like the active sites of enzymes (Chapter 6): each binding site has its own peculiar shape and electrical charge, so only certain molecules can fit in and bind. *Antibody binding sites are so specific, in fact, that each antibody can bind just a few different types of molecules, perhaps only one.*

Generally, only large, complex molecules, such as proteins, polysaccharides, and glycoproteins, can bind to antibodies. Such molecules are called **antigens** (meaning "molecules that generate an antibody response"). Molecules attached to the surfaces of cells constitute an especially important class of antigens. Microbes and the cells of one's own body that have become cancerous or have been infected by viruses often display a variety of antigens on their surfaces. As we will discuss shortly, the binding of antigen to antibody triggers the immune response.

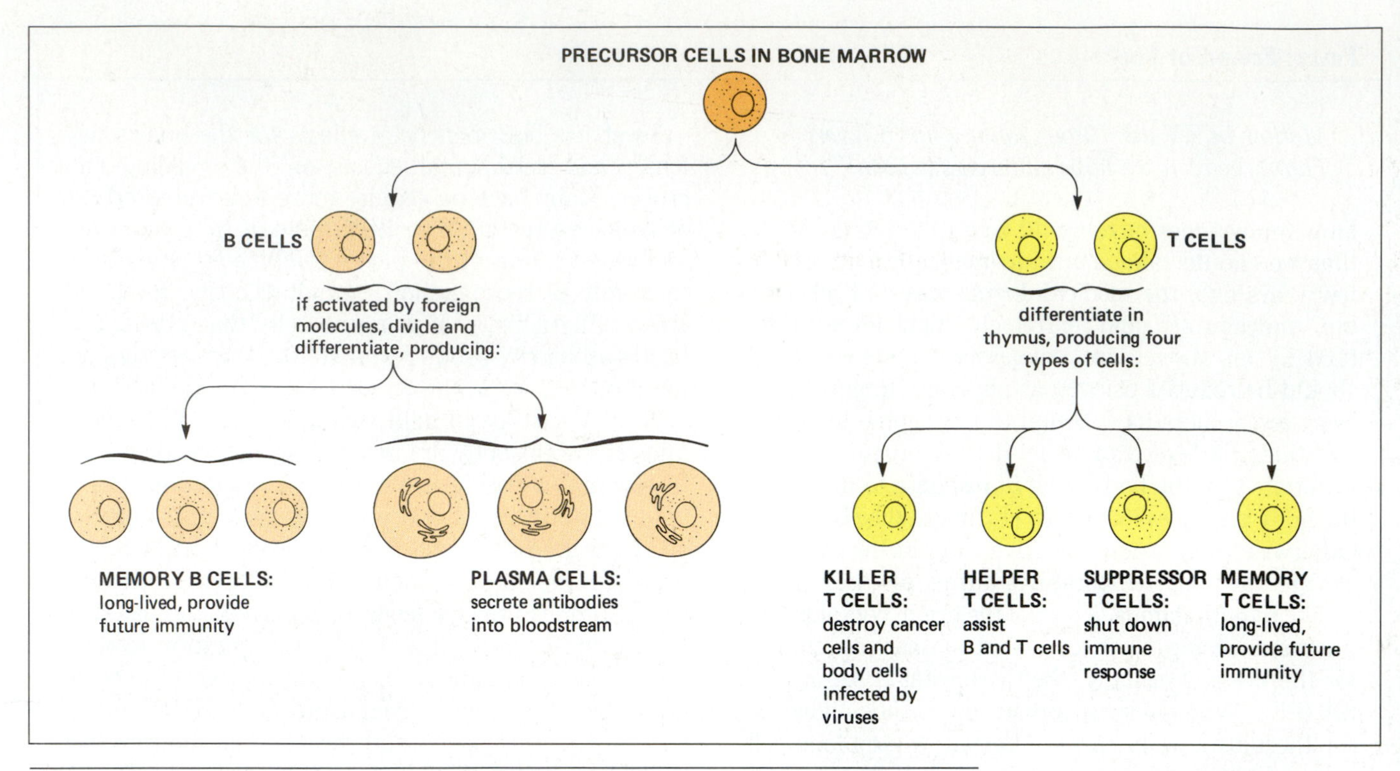

Figure 27-2 *The major cells of the immune system and their roles in the immune response.*

Recognizing a multitude of foreign molecules. A human being may encounter millions of different types of invaders in a lifetime, from pollen grains and mold spores through flu viruses and botulinus toxin. *The immune system recognizes and responds to this multiplicity of antigens because it produces millions of different antibodies, each capable of binding a different antigenic molecule.* This fact presented immunologists with a major problem: since antibodies are proteins, and proteins are encoded by the genes, it would seem that a human being must have millions of antibody genes. However, there are probably fewer than a hundred thousand genes in the entire human genome. How, then, can the genes code for millions of antibodies? Two distinct but complementary mechanisms join forces to produce an enormous diversity of antibodies from a relative handful of antibody genes.

First, *genes encode parts of antibodies, not entire antibodies. These "antibody part" genes join together during immune cell development to form complete genes for the*

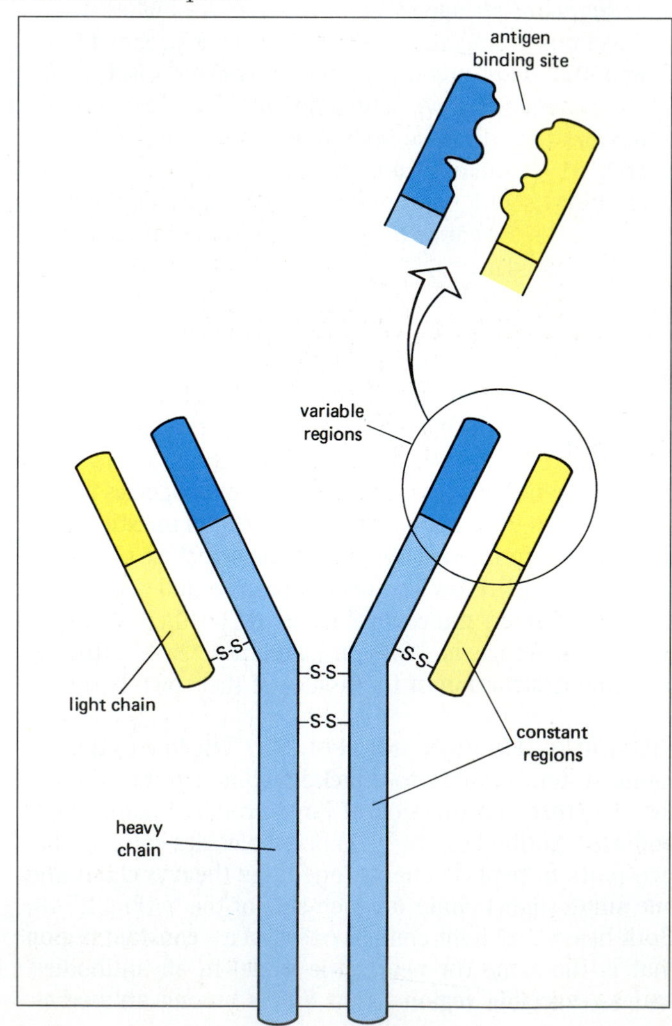

Figure 27-3 *Antibodies are proteins composed of two pairs of peptide chains, called light and heavy chains, arranged something like a double letter Y. The variable regions on the two chains form a specific binding site at the end of each arm of the Y. Different antibodies have different variable regions, forming unique binding sites. The human body synthesizes millions of distinct antibodies, each binding a different antigen.*

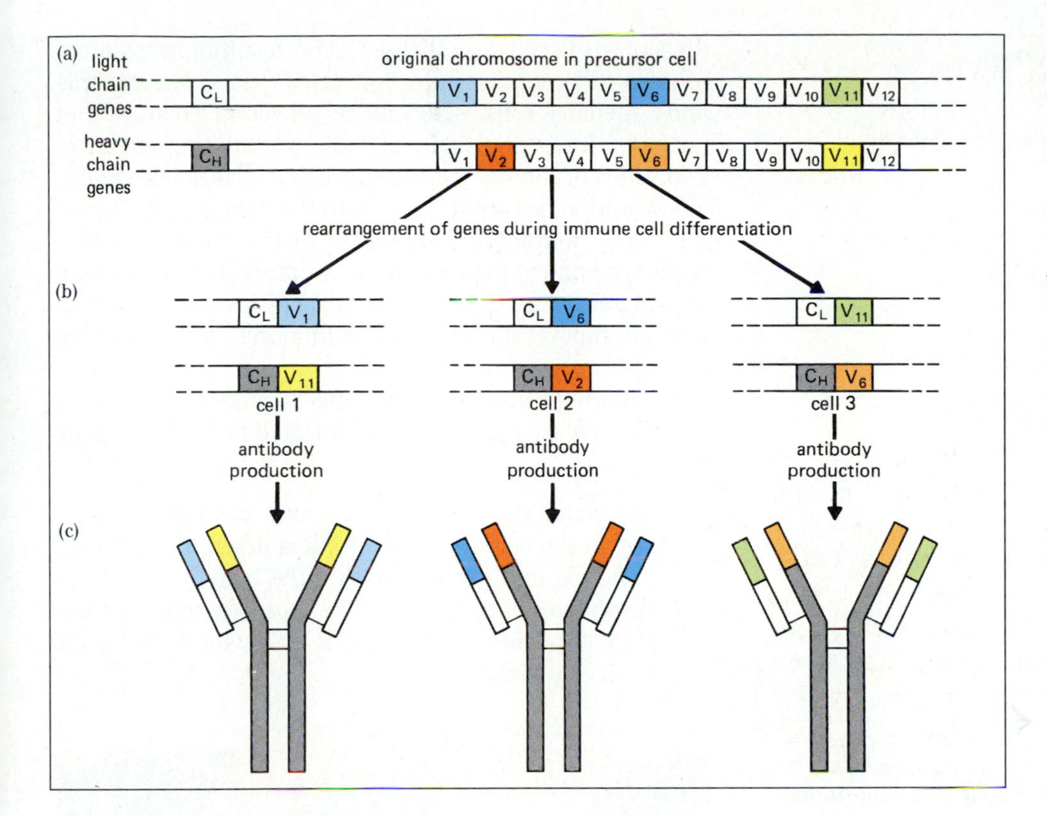

Figure 27-4 Recombination during the construction of antibody genes. **(a)** *The precursor cells of the immune system each contain one or perhaps a few genes for the constant regions of the light and heavy chains of antibodies, and many genes for the variable regions.* **(b)** *During the development of each immune cell, these genes are rearranged, moving one of the variable-region genes next to a constant-region gene. Each cell thus generates a "recombined antibody gene" for each chain, which differs from the recombined antibody genes generated by other immune cells.* **(c)** *Representations of the different antibodies synthesized by each immune cell.*

light and heavy chains of antibodies (Fig. 27-4). Each immature B and T cell contains a few genes for the constant regions of antibodies, and many, perhaps several hundred, genes for the variable regions (Fig. 27-4a). During immune cell development, these genes move around on the chromosomes, so that one variable-region gene ends up adjacent to one constant-region gene. Each cell comes to possess one "constant + variable" light-chain gene, and one "constant + variable" heavy-chain gene (Fig. 27-4b). Which variable-region gene winds up alongside the constant-region gene is completely random. Only the constant and variable genes that are next to one another are transcribed and ultimately provide the genetic information for antibody molecules; the rest remain unused. Therefore, each immune cell produces an antibody, specified by the chance recombination of variable- and constant-region genes, that is different from the antibody produced by most or all other immune cells. (It may help you to think of antibody gene formation in terms of card playing. Each immune cell is dealt a "hand" of two variable-region genes, one for the light chain and one for the heavy chain, randomly chosen from two large "decks" of genes. With each deck containing over a hundred "cards" (genes), virtually every cell will synthesize its own unique antibody.)

Second, *the genes for certain antibody parts are incredibly prone to mutate, constantly generating new antibody genes.* As B and T cells reproduce, some of their daughter cells suffer mutations in their antibody genes. Thus two sister cells may produce different antibodies.

The end result of mutation and gene recombination is that each immune cell has its own particular antibody genes, different from those of most other immune cells (except for its own not-yet-mutated sisters and descendants). Each microbe entering the body may bear several complex proteins or polysaccharides on its surface that can act as antigens. Usually, a few immune cells will have an antibody that binds some of these antigens. Antigen–antibody binding will trigger changes in the immune cells, usually leading to the destruction of the microbe. We will examine these changes in more detail in a moment.

DISTINGUISHING "SELF" FROM "NONSELF." The surfaces of the body's own cells also bear large proteins and polysaccharides, just as microbes do. These molecules can act as antigens in other people's bodies (this is why transplants are usually rejected; the recipient's immune system recognizes molecules on the donor's cells as foreign, and destroys the transplant). Why, then, doesn't a person's immune system respond to these "self" antigens and destroy his own cells? The key seems to be the continuous presence of the body's antigens while the immune cells mature. As an embryo develops, some differentiating immune cells do indeed produce antibodies to the body's own proteins and polysaccharides. However, if *immature* immune cells contact molecules that bind to their surface antibodies, the cells are inactivated or destroyed. In this way, antigens present during immune system development eliminate potentially destructive immune cells. Thus

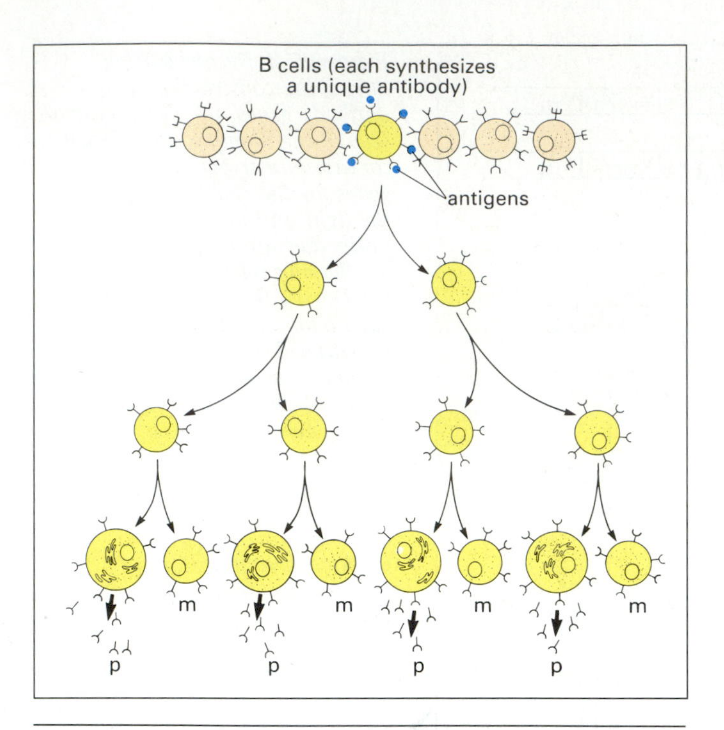

Figure 27-5 *Antigens select a particular type of B cell for replication. Each B cell synthesizes a unique antibody, determined by the "antibody gene hand" it was dealt during its development (see Fig. 27-4). When an antigen enters the bloodstream, it binds to one or a few B cells, triggering them to multiply rapidly. The daughter cells develop either into antibody-producing plasma cells (p) that churn out antibodies directed against the specific activating antigen or into long-lived memory cells (m).*

the immune system distinguishes "self" from "nonself" by retaining only those immune cells that do not respond to the body's own molecules.

Attack

If the body is invaded by microbes, the immune system mounts two types of attack: B cells provide **humoral immunity,** mediated by free antibodies circulating in the bloodstream, while T cells produce **cell-mediated immunity,** as killer T cells destroy infected body cells. Recent research shows that the two types of responses are not as distinct as this implies. However, the humoral and cellular responses are most easily understood when they are considered separately.

Humoral immunity. Each B cell bears a specific antibody on its surface (Fig. 27-5). When a microbial infection occurs, one or a few B cells have an antibody that binds to a microbial antigen. (In reality, the process is considerably more complex than this, but this is the end result.) Antigen–antibody binding causes these B cells to divide rapidly (a type of T cell also helps to stimulate cell division in B cells; see below). The resulting population of cells differentiates into two cell types: **plasma cells** and **memory cells.** Plasma cells become enlarged and packed with endoplasmic reticulum, churning out huge quantities of that cell type's specific antibody (Fig. 27-6). These antibodies are released into the bloodstream (hence the name "humoral" immunity; to the ancient Greeks, blood was one of the four body "humors"). Memory cells do not release antibodies. As we will see, memory cells play an important role in future immunity to this specific microbe.

Circulating antibodies may affect antigenic molecules, microbes bearing antigens, and infected body cells in four ways:

1. *Neutralization.* The antibody may combine with or cover up the active site of a toxic antigen, thereby preventing the toxin from harming the body.
2. *Promotion of phagocytosis.* The antibody may coat the surface of a microbe, making it easier for white blood cells to engulf the microbe.

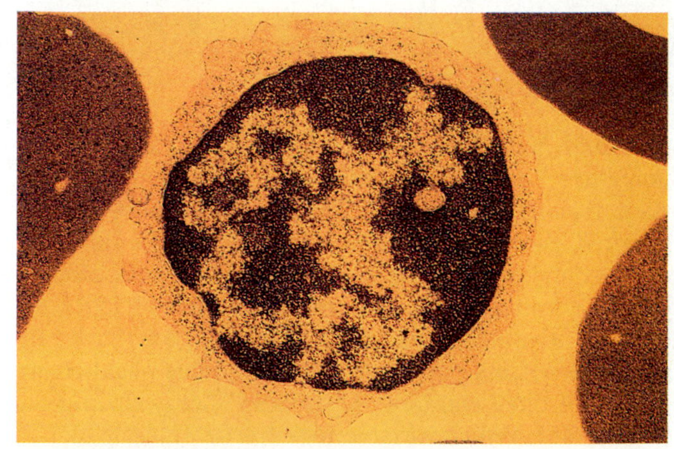

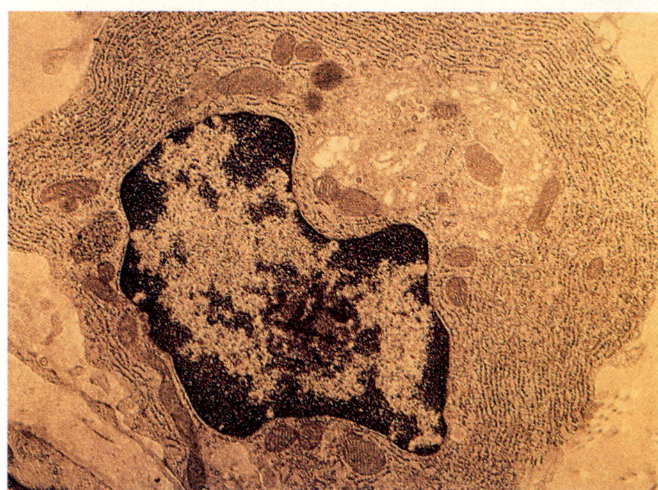

Figure 27-6 *B cells before (top) and after (bottom) conversion to plasma cells. Note that the plasma cell is much larger and is virtually filled with rough endoplasmic reticulum that synthesizes antibodies.*

Figure 27-7 Cell-mediated immunity at work. A T cell contacts a larger tumor cell and causes it to disintegrate. Note the membrane bubbles forming on the tumor cell as it is destroyed.

3. *Agglutination.* Each antibody has two binding sites for antigen, one on each arm (Fig. 27-3). These binding sites may attach to antigens on two different microbes, holding them together. As more and more antibodies link up with antigens on different microbes, the microbes clump together. Agglutination seems to enhance phagocytosis by white blood cells.
4. *Complement reactions.* The antibody–antigen complex on the surface of an invading cell may trigger a series of reactions with other blood proteins called the **complement system.** These proteins attract phagocytic white blood cells to the site, promote phagocytosis of the foreign cells, and in some instances directly destroy the invaders.

CELL-MEDIATED IMMUNITY. The major function of cell-mediated immunity is to destroy the body's own cells when they have become cancerous or have been infected by viruses. Like B cells, the surfaces of T cells bear antibodies. When antigen binds to the antibodies on a T cell, the cell divides rapidly, producing two populations of offspring cells: effector cells and memory cells. There are several types of effector T cells: killer cells, helper cells, and suppressor cells. These all participate in the immune response, but they do not release antibodies into the bloodstream. When antibodies on the surface of a **killer T cell** bind to antigens on the surface of an infected cell, the killer cell releases proteins that disrupt the infected cell's membrane (Fig. 27-7). **Helper T cells** release chemicals that attract phagocytic white blood cells to the area of the infection. These phagocytic cells then engulf the foreign cells. Helper T cells also stimulate cell division in B cells that respond to the same microbial invasion. After the infection has been conquered, **suppressor T cells** appear to shut off the immune response in both B and killer T cells.

The second population of T cell offspring are memory cells. Like memory B cells, these protect the body against future infection.

Memory

As Thucydides observed two millenia ago, a person who overcomes a disease often remains immune to future encounters with that specific disease for many years. Retaining immunity is the function of memory cells. Plasma cells and killer T cells do the immediate job of fighting disease organisms, but they usually live only a few days. B and T memory cells, on the other hand, survive for many years. If foreign cells bearing the same antigens reenter the body, they will be recognized by the appropriate memory cells. These memory cells will multiply rapidly, generating huge populations of plasma cells and killer cells and producing a second immune response.

In the first encounter with a disease microbe, only a few B and T cells respond. Each of these, however, leaves behind hundreds or thousands of memory cells. Further, memory cells respond to antigen much more rapidly than their progenitor B and T cells could. Therefore, the second immune response is very rapid (Fig. 27-8). In most instances, second or subsequent invasions by the same microbe are overcome so quickly that there are no noticeable symptoms of infection at all. Although cold and flu viruses appear to be exceptions to this rule, the immune system does in fact provide lasting protection against these viruses. The problem, as we describe in the essay, "Flu: The Unbeatable Bug," is that next year's flu is not caused by the same virus that you fought off this year.

Flu: The Unbeatable Bug

Every winter, a wave of influenza sweeps across the world. Thousands of the elderly, the newborn, and those already suffering from illness succumb, and hundreds of millions more suffer the fever and muscle aches of milder cases. Occasionally, devastating flu varieties appear. In 1968, the Hong Kong flu infected 50 million Americans, causing 70,000 deaths in six weeks. In the great flu pandemic of 1918, the worldwide toll was 20 million dead in one winter.

Flu is caused by a virus that invades the cells of the respiratory tract, turning each one into a factory for manufacturing new viruses. The outer surface of the virus is studded with proteins, several of which serve as antigens recognized by the immune system. People survive the flu because their immune systems inactivate the viruses or kill off virus-infected body cells before the viruses finish reproducing. This is the same mechanism by which other viruses, such as mumps or measles, are conquered. So why don't people become immune to the flu, as they do to measles?

The answer lies in the flu virus's uncanny ability to change. The viral genes that code for the antigenic proteins mutate rapidly: there are 10 mutations in every million newly synthesized viruses. A single mutation usually does not change the properties of the antigen very much. Four or five, however, may change it enough so that the immune system does not fully recognize it as the same old flu that was beaten off last year. Some of the memory cells don't recognize it at all, and the immune response produced by the rest doesn't work as well as it should. The virus, although slowed down somewhat, gets a foothold in the body, and multiplies until a new set of immune cells recognize the mutated antigen and start up a new immune response. So you get the flu again this year.

Far more deadly are the dramatically new flu viruses that occasionally appear: the epidemic of 1918, the Asian flu of 1957, and the Hong Kong flu of 1968. In these, entirely new antigens seem to show up all at once. These are not just mutations of the old set, but completely novel antigens that the human immune system has never encountered before. Where do they come from? Believe it or not, from birds and pigs. Viruses strikingly like the human flu virus infect the intestinal tracts of birds, especially ducks, without causing any noticeable effect. The avian viruses don't infect people, and human flu viruses don't infect birds. However, both can infect pigs, with the result that both viruses may reproduce simultaneously in the same pig cell. Once in a great while (perhaps only three times this century), offspring viruses end up with a mixture of genes from human and bird viruses (Fig. E27-1). Some combine the worst genes of both types: from the human virus, the genes needed to subvert human cellular metabolism to produce new viruses; from the bird virus, genes for new surface antigens.

If infected by such a hybrid virus, the immune system must start from scratch, selecting out entirely new

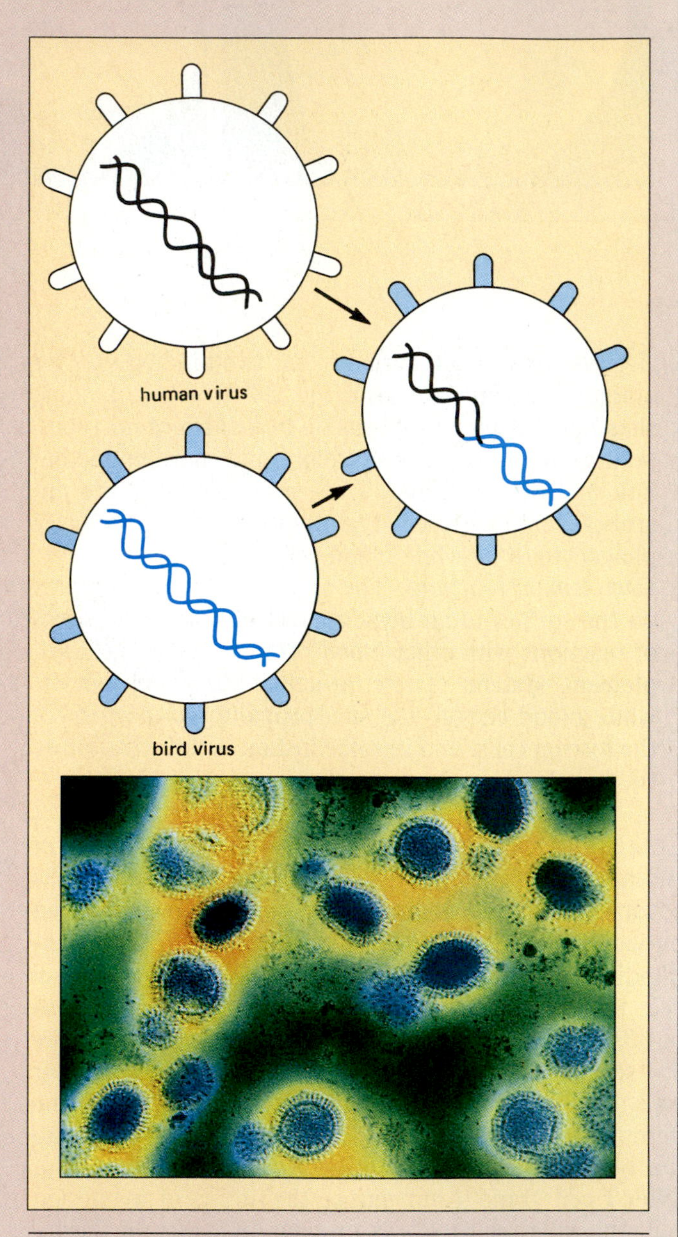

Figure E27-1 Recombination of genes from bird and human viruses. Photo: false color electron micrographs of influenza viruses. Note the protein "spikes" projecting from the virus coats. These attach to human cell membranes, helping the virus to gain entry into the cells.

Flu: The Unbeatable Bug

lines of B and T cells to attack the intruder. But the virus multiplies so rapidly in the meantime that many people die, or become so weakened that they contract some other disease and die of that. The rest of the people recover, with immune systems now primed to resist any further assault from the new virus. Next year a few point mutations allow a slightly altered strain to infect millions of people, but with a partial immune response ready, few fatalities occur. Once again, for most of us, the flu becomes a routine annoyance.

Until next time, somewhere, the improbable happens again. Maybe this year.

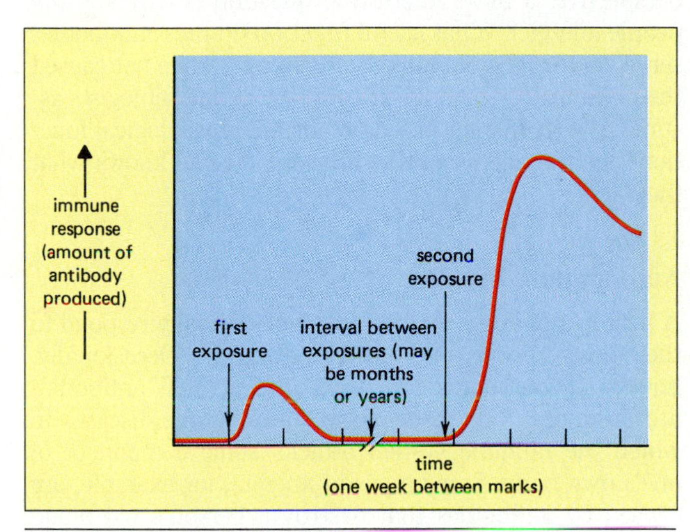

Figure 27-8 The immune response to the first exposure to a disease organism is fairly slow and not very large, as B and T cells are selected and multiply. A second exposure activates memory cells formed during the first response, and consequently the second response is both faster and larger.

MEDICAL IMPLICATIONS OF THE IMMUNE RESPONSE

Antibiotics and Immune Responses

Our description of the body's responses to infection may seem to suggest that nothing can harm us: the immune system conquers all. Unfortunately, this is not the case. If untreated, many diseases kill their victims, usually for a simple reason: the body provides ideal conditions for the growth and reproduction of disease microbes, which can multiply rapidly, sometimes dividing as fast as once an hour. The infection thus becomes a race between the invading microbes and the immune response. If the initial infection is massive, or if the microbes produce particularly toxic products, the full activation of the immune response may come too late.

Antibiotics help to combat infection by slowing down the growth and multiplication of many microbes, including bacteria, fungi, and protists. Although antibiotics do not destroy every single microbe, they give the immune system enough time to finish the job.

Vaccinations

As early as the year 1000, people in India, China, and Africa deliberately exposed themselves to mild cases of smallpox, to acquire immunity to the disease. In 1798, Edward Jenner discovered that infection with cowpox confers immunity to smallpox, thus initiating the modern practice of immunization. In the late nineteenth century, Louis Pasteur extended the use of immunization to several other diseases, by injecting weakened or dead microbes into healthy people. The weakened microbes do not cause disease (or at least not a severe case) but bear antigens that elicit vigorous immune responses. These injections of weakened or killed microbes to confer immunity are called **vaccinations,** from the Latin word for "cow," in honor of Jenner's pioneering efforts. Today, many diseases, including polio, diphtheria, typhoid fever, and measles, can be controlled through vaccination. Smallpox, one of the most deadly diseases of all, has been completely eradicated from the Earth, owing to a vaccination program sponsored by the World Health Organization.

Through genetic engineering (see Chapters 11 and 12), we now enjoy the prospect of manufacturing tailor-made vaccines. One method is to synthesize the antigenic proteins from disease-causing microbes. These antigens can then be used as vaccines without having to raise, isolate, and weaken the disease microbes themselves. A vaccine against anthrax, a severe disease of livestock, has been manufactured with this procedure. A second technique may be to insert the genes for antigens of, say, herpes into the genome of harmless microbes such as the cowpox virus. These designer microbes would produce herpes antigens without being able to cause the disease, and could be used for vaccination.

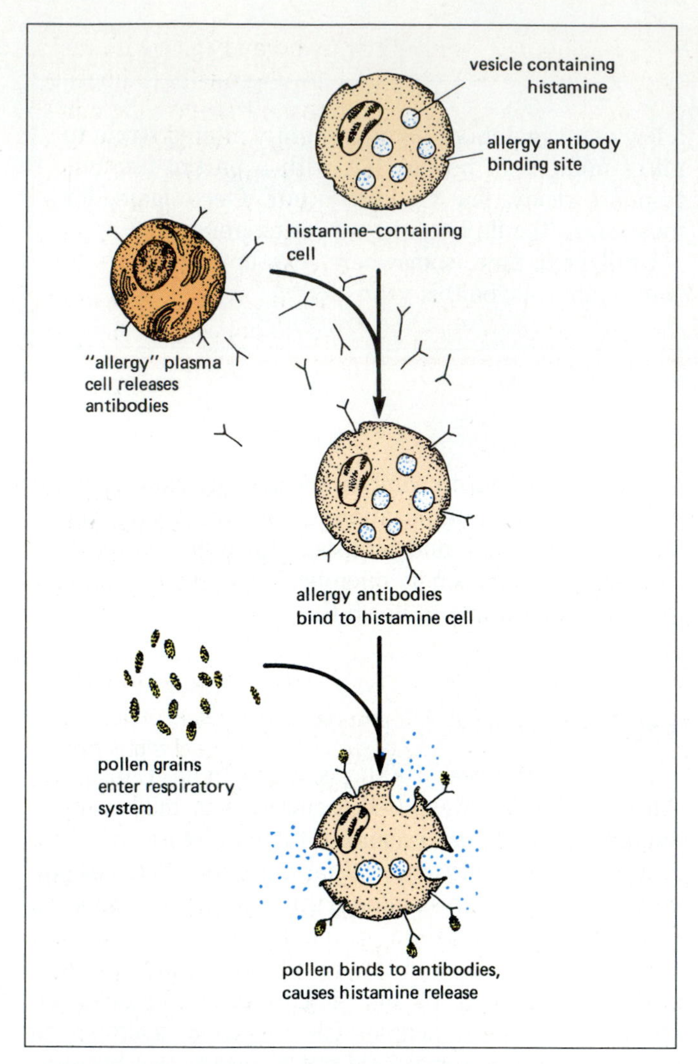

Figure 27-9 Allergic reactions. Upon exposure to certain antigens, for instance pollen grains, some plasma cells synthesize antibodies that bind both to histamine-containing cells (middle) and to the antigen (bottom). When this dual binding occurs, the cells release histamine into their surroundings, causing local inflammation and the symptoms of allergy.

Allergies

Many people suffer from **allergies:** reactions to substances that are not harmful in themselves, and to which other people do not respond. Common allergies include those to pollen, dust, mold spores, and bee stings. Allergies are actually a form of immune response (Fig. 27-9). A foreign substance, say a pollen grain, enters the bloodstream, and is recognized as an antigen by a particular type of B cell. This B cell proliferates, producing plasma cells that pour out antibodies against the pollen antigens. The antibodies responsible for allergic reactions differ from other antibodies in that they attach themselves to histamine-containing cells located in connective tissue throughout the body. When pollen grains encounter these attached antibodies, they trigger the release of histamine, which causes increased mucus secretion, leaky capillaries, and other symptoms of inflammation. Since pollen grains most often enter the nose and throat, the major reactions occur in these locations, resulting in the runny nose, sneezing, and congestion typical of hay fever. Antihistamine drugs block some of the effects of histamine, relieving the symptoms of allergies.

Why are some people allergic and not others? You can probably guess the answer. We are all exposed to the same antigens in pollens, molds, and foods. Therefore, *people without allergies either must lack the genes for the allergy-causing antibodies, or do not produce as much antibody as allergic individuals do.* From an evolutionary perspective, a more interesting question is why are *any* people allergic? What useful function do these antibodies serve, so that the obvious disadvantages have not caused their elimination through natural selection? Biologists assume that there must be some normal physiological functions for the allergy antibodies, but no one knows what they are.

Autoimmune Diseases

A person's immune system does not normally respond to the antigens borne on the body's own cells. Occasionally, however, something goes awry, and "anti-self" antibodies are produced. The result is an **autoimmune disease,** in which the immune system attacks some component of one's own body. Some types of anemia, for example, are caused by antibodies that destroy a person's red blood cells. Many cases of juvenile-onset diabetes occur because the insulin-secreting cells of the pancreas are the victims of a misdirected immune response. Unfortunately, at present there is no way to cure autoimmune diseases. For some types, replacement therapy can alleviate the symptoms, for instance by administering insulin to diabetics or blood transfusions to anemics. Alternatively, the autoimmune response can be suppressed with drugs. Immune suppression, however, also reduces immune responses to the everyday assaults of disease microbes, so this therapy cannot be used except in the most life-threatening cases.

Immune Deficiency Diseases

Rarely, a child is born with a defect in which no immune cells, or very few, are formed. Such a child may survive fetal life and even the first few months of postnatal life, protected by antibodies acquired from the mother during pregnancy or in her milk. Once these antibodies are lost, however, common bacterial infections may prove fatal. Some immune-deficient children have to live in a germ-proof "bubble," isolated from contact with every unsterilized object, including other people. Experimental therapy involves transplanting bone marrow (from which immune cells arise) from a normal donor into the child.

In some children, marrow transplants have resulted in some antibody production, occasionally enough to confer normal immune responses.

Within the last decade, a new type of immune deficiency disease has appeared, not due to a genetic defect in immune cell development, but acquired during the life of the victim: acquired immune deficiency syndrome, or AIDS. AIDS has become a major public health threat (see the essay).

CANCER

Cancer—perhaps the most dreaded word in the English language, and with good reason. Nearly one out of three Americans will contract some form of cancer (Table 27-1). For many, there will be no cure, only a slow wasting away to death. What *is* cancer? If we can prevent smallpox and polio, and cure dozens of other diseases, why can't we cure or prevent cancer?

Unlike most other diseases, cancer is not a straightforward invasion of the body by a foreign organism. Although some cancers may be triggered by viruses, in essence cancer is a malfunctioning of the growth controls of the body's own cells, a disease in which we destroy ourselves. Since it is a case of "self" fighting "self," most treatments designed to combat cancer also damage normal, healthy cells.

Table 27-1 **Cancer Deaths Among Americans in 1987**[a]

Site of Cancer	Men	Women
Lung	92,000	44,000
Breast	300	41,000
Oral cavity	6,350	3,050
Esophagus	6,400	2,400
Stomach	8,300	5,900
Intestine/rectum	31,300	31,300
Pancreas	12,300	12,000
Liver	5,300	5,300
Urinary tract	12,900	7,100
Skin	4,800	3,000
Prostate	27,000	—
Ovary/uterus	—	21,400
Nervous system	5,500	4,700
Leukemia	9,800	8,000
Blood and lymph tissues	12,800	12,100
Other	25,750	22,750
Total	259,000	224,000

[a]Estimated cancer deaths for 1987 (data from the American Cancer Society). Note the large differences between men and women for some cancers. Some differences are biological (e.g., breast, ovary, and prostate cancers). The difference in lung cancer, however, is social. More men than women smoke cigarettes, and since nearly all lung cancer is caused by smoking, men outstrip women in lung cancer deaths (this accounts for most of the difference in total cancer deaths between men and women). In recent years, women are catching up to men in smoking, and lung cancer rates in the year 2000 will be more nearly equal.

The usual development of any organ begins with rapid growth during embryonic life, slower growth as a juvenile, and finally maintenance of a constant size during adulthood. Individual cells may die and be replaced (as happens constantly in the lining of the digestive tract), but most organs remain about the same size throughout adult life. *A cancer is a population of cells that has escaped from normal regulatory processes, and grows without control.* As the cancer grows, it uses increasing amounts of the body's energy and nutrient supplies, and literally squeezes out vital organs nearby.

Causes of Cancer

To learn the causes of cancer, we must answer two related but distinct questions: *what changes occur in a cancerous cell that allow it to escape normal growth controls,* and *what agents (genetic, viral, or environmental) initiate these cellular changes?*

CANCER GENES. In the early 1980s, cancer researchers discovered **oncogenes:** genes that cause cancer. Although research is just beginning to unravel what promises to be a long and complicated plot, there seem to be two principal mechanisms by which oncogenes produce cancer.

First, *a potentially dangerous oncogene may be present in all cells, but causes cancer only when activated by some external trigger.* All cells have genes that can stimulate growth and cell division. These genes may be active during embryonic development, but usually they are turned off or transcribed more slowly in mature organisms. Other genes, such as those for the enzymes needed to metabolize glucose, are actively transcribed in many adult cells. Cancer researchers suspect that some oncogenes may be growth genes that have mistakenly been turned on full speed. For instance, chromosomes may become rearranged so that an embryonic growth gene is transferred to a part of a chromosome that is normally transcribed rapidly (Fig. 27-10). The protein synthesized under the direction of the growth gene then stimulates growth and cell division. The daughter cells inherit the same rearranged chromosome, resulting in explosive, cancerous growth.

Second, *a harmless gene may mutate into an oncogene.* Consider a gene that normally directs the synthesis of a protein that promotes cell reproduction at "maintenance levels," such as those needed to replace cells lost through normal body wear and tear. A mutation in this "pre-oncogene" may change the protein so that it greatly accelerates the rate of cell division. This would probably create a cancer.

What causes activation of oncogenes or mutations in pre-oncogenes? Some types of cancer are caused by viral infections (Fig. 27-11). The viral genes are composed of RNA, which the virus forces the cell to "reverse tran-

Acquired Immune Deficiency Syndrome: AIDS

In January of 1981, a man entered the UCLA Medical Center with a fungal infection in his throat. A few weeks later, he developed a rare form of pneumonia, one almost never seen except in cancer patients and people undergoing immune system suppression to prevent rejection of organ transplants. Following a series of infections, he died that December. Soon, doctors across the country were encountering similar cases: patients who suffered debilitating effects from rare diseases, or from common diseases that are not usually serious in normal adults. Although the particular diseases varied, all the patients had one feature in common: a failure of the immune system to ward off invading microbes. This new disease was named acquired immune deficiency syndrome, or AIDS.

Early in 1984, scientists at the Pasteur Institute in France and the National Institutes of Health in the United States isolated the organism that causes AIDS (Fig. E27-2): a virus that infects helper T cells, preventing them from responding to infections. As a result, rather ordinary diseases can be extremely serious in AIDS patients, and sooner or later, one of them is fatal.

Investigators soon discovered that AIDS victims are not a typical cross section of society. Initially, almost all AIDS victims in the United States were homosexual men or intravenous drug users. There were also a few hemophiliacs. The AIDS virus is apparently transmitted only through direct exchange of body fluids, including blood, saliva, and semen. The disease spreads through the homosexual and drug user populations by sexual encounters and unsterilized hypodermic needles, and to hemophiliacs through contaminated blood transfusions. Although homosexual males and intravenous drug users still comprise the vast majority of AIDS victims, women and heterosexual men who are not drug users have also contracted AIDS. Many health researchers therefore suspect that AIDS can be transmitted through heterosexual contact, although probably not as efficiently as through homosexual contact. In fact, in Africa, where the virus seems to have originated, heterosexual intercourse is probably the most common means of infection.

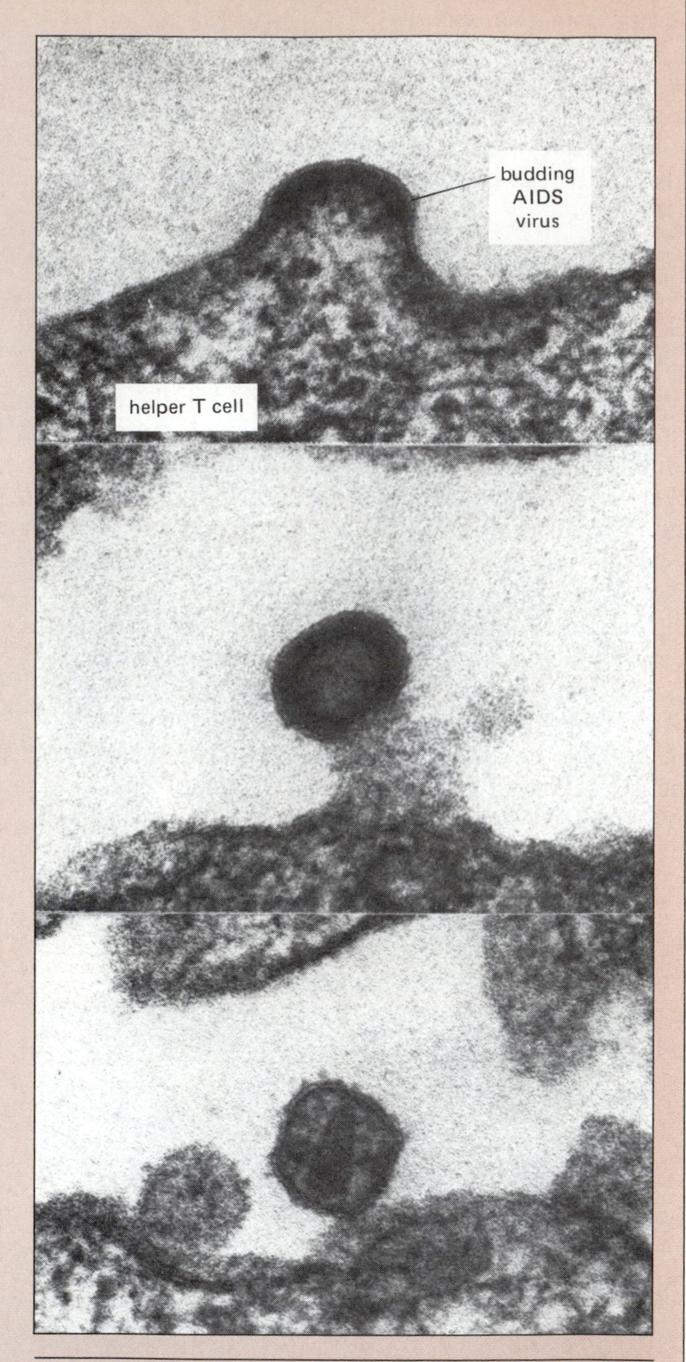

Figure E27-2 *An AIDS virus emerges from an infected helper T cell.*

In some respects, AIDS is not a very powerful virus. For one thing, it cannot survive for very long outside the body. Further, a normally functioning immune system probably destroys most AIDS microbes quickly and efficiently. People at highest risk for AIDS are those whose immune systems are already overloaded. Homosexual AIDS victims, for example, often have had a series of sexually transmitted diseases before they contract AIDS. Therefore, health officials often state, quite accurately, that AIDS is not very infectious to healthy people. Given the deadly nature of the disease, however, such assurances are small comfort to people who do contract AIDS, such as the nurse who picked

Acquired Immune Deficiency Syndrome: AIDS

up the virus through her chapped hands when she contacted blood from a patient undergoing emergency care. *Any* exposure to AIDS carries with it some risk of infection, even to healthy people. The only sure prevention is to avoid contact with the body fluids of infected people. Since the interval between infection and overt AIDS symptoms averages 6 to 8 years, identification of AIDS carriers is difficult.

On a more optimistic note, researchers throughout the world are racing to produce a vaccine against AIDS, although this is turning out to be more difficult than some had assumed just a few years ago. The most formidable problem is the flexibility of the AIDS virus. The AIDS virus mutates rapidly, perhaps a thousand times more rapidly than the flu virus. Different people have different strains of AIDS viruses, perhaps so different that a vaccine against one strain will not work against others. However, a report early in 1987 that part of the outer surface of the AIDS virus is nearly the same in many different strains offers hope that a successful vaccine may be developed. Until then, this deadly disease will continue to take its toll: as of December 1987, 49,793 cases of AIDS have been diagnosed in the United States, causing 27,909 deaths. The Centers for Disease Control estimate that there will be over a quarter of a million cases by 1991, accounting for 180,000 deaths. By that time, AIDS will be the most common cause of death among people 25 to 44 years old.

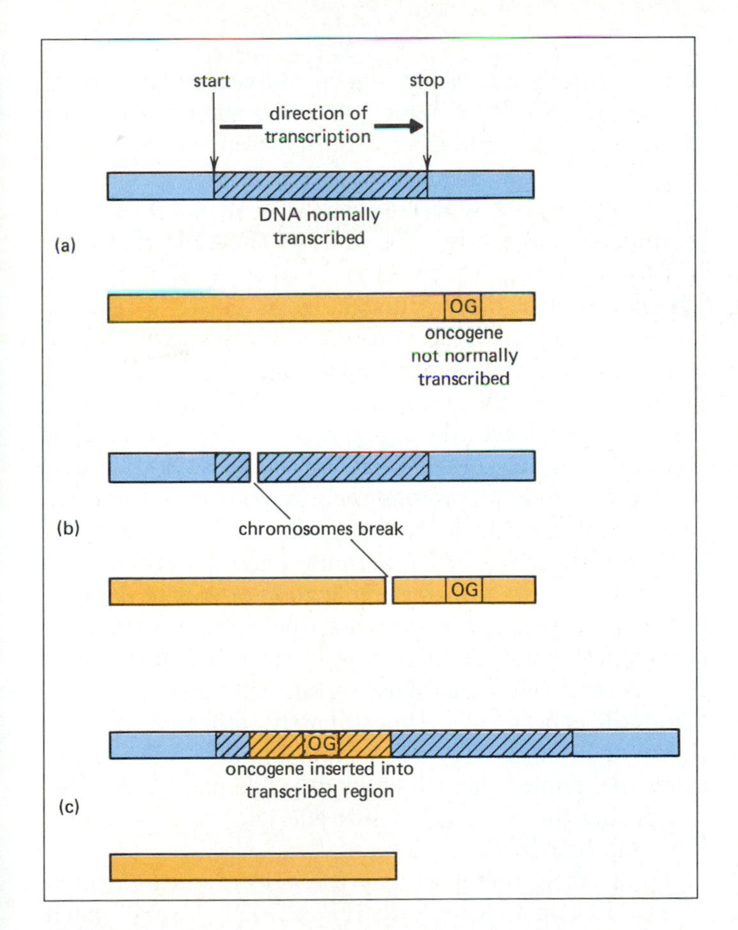

Figure 27-10 Activation of an oncogene (OG) by gene jumping. **(a)** *Only the hatched region of the blue chromosome is normally transcribed.* **(b)** *The tan chromosome bearing the oncogene breaks and* **(c)** *the oncogene is inserted within the transcribed region of the blue chromosome.*

scribe" into DNA. This new DNA is then inserted into chromosomes of the host cell, where it is usually transcribed continuously to synthesize new viral RNA. Nearby DNA on the host chromosome may also be incidentally transcribed. If this host DNA happens to include a previously silent oncogene, the cell becomes cancerous.

Probably the most common causes of cancer are environmental insults, chiefly chemicals and radiation. We are besieged by cancer-causing chemicals (carcinogens), not only the eminently avoidable ones in cigarettes and various industrial processes, but also some in the most innocent foods and even some synthesized in our own digestive tracts. Some chemicals and certain types of radiation induce point mutations in DNA. Others cause chromosomes to break in two and possibly rejoin in new and lethal combinations, by transferring oncogenes into actively transcribed regions of the chromosomes.

If these seemingly simple mechanisms cause cancer, why do cancers take so long to develop? No one knows the answer for certain, but most cancers seem to require two or more distinct steps. Exposure to radiation early in life, perhaps, may mutate a pre-oncogene to a true oncogene. If this oncogene is located in a region of DNA that is not normally transcribed, no cancer will occur. Many years later, a viral infection or exposure to chemicals may move this oncogene to an active region of DNA, and cancer begins.

Defenses Against Cancer

Cancer prevention. Cancer cells form in our bodies every day, and not even the best of preventive measures can eliminate cancer completely. Gamma rays from the

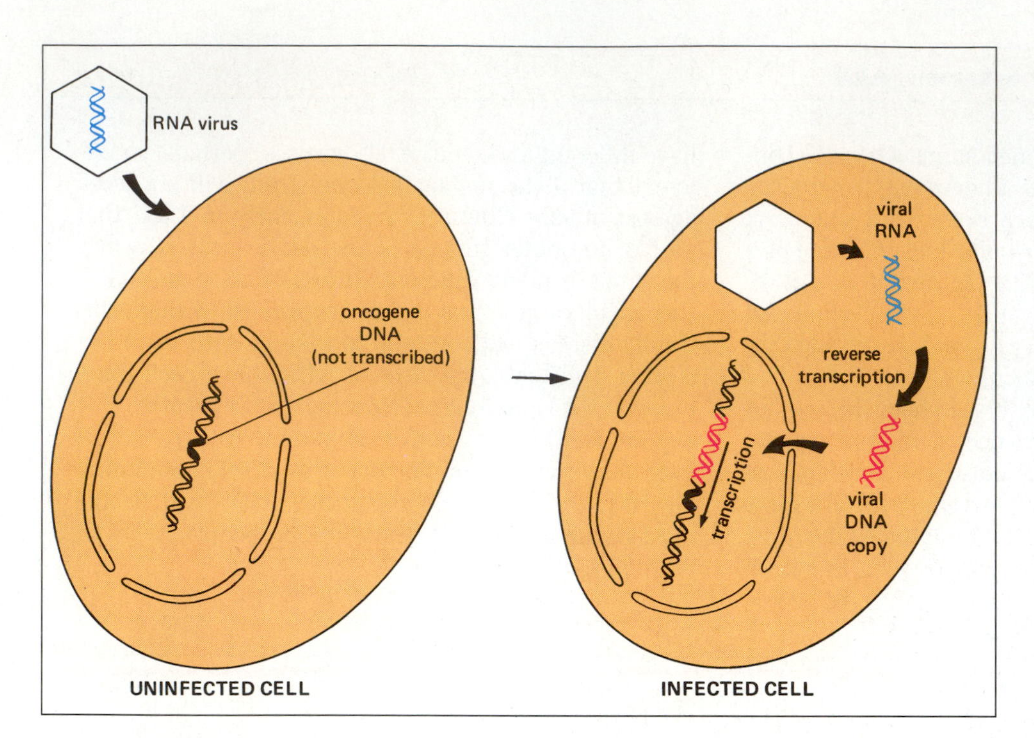

Figure 27-11 Viral activation of an oncogene. Certain RNA viruses invade animal cells and cause "reverse transcription" of their RNA genes into DNA. This DNA then inserts itself into the DNA of the host cell. The viral DNA copy includes nucleotides that promote transcription by host RNA polymerase. Nearby host genes, including oncogenes in some cells, will be transcribed along with the viral DNA copy, and the cell may become cancerous.

sun, radioactivity from the rocks beneath our feet, and naturally produced carcinogens in our food cannot be avoided. However, each of us can reduce his or her own chances of developing cancer. Some chemicals, including carotene and vitamins C and E, appear to offer protection against some forms of cancer. We can also avoid many well-known carcinogens. Cigarette smoking, for example, causes most of the lung cancers in the United States. Other chemicals, including those emitted from oil refineries and those used in certain industrial processes, can cause cancer in exposed workers. The ultraviolet rays from the sun that produce fashionable suntans are also a leading cause of skin cancer. Certain molds produce the most potent carcinogens known and can be avoided by storing food properly. And, of course, preventing nuclear war will prevent an increase in radiation-induced cancer.

Fortunately, killer T cells screen the body for cancer cells and destroy nearly all of them before they have a chance to proliferate and spread. Since cancer cells are "self" cells, and the immune system does not respond to "self," how are cancer cells weeded out? Probably, the very processes that cause cancer also cause new and slightly different proteins to appear on the surfaces of cancer cells. T cells encounter these new proteins, recognize them as "nonself" antigens, and destroy the cancer cells. Without constant surveillance by T cells, it is unlikely that any of us would survive more than a few years.

Cancer treatment. Sometimes, however, the immune system does *not* recognize cancer cells as "nonself." Ignored by the immune system, the cancer grows and spreads. What can medical science do to cure cancer? The rate of cure is increasing, but is still scarcely a third of all cancers. The three main approaches taken are all quite crude: burn the cancer out with radiation, cut it out with surgery, or poison it with drugs.

If a cancer is discovered when it is small enough, radiation or surgery may be able to eliminate it. Breast cancer, for example, can almost always be eliminated by surgery if detected early enough. However, while the body is rid of cancer, surgery and radiation therapies may be traumatic, dangerous, and disfiguring.

In principle, chemotherapy might be able to destroy cancer cells without damaging normal cells, since cancer cells are, after all, quite different in many ways from normal cells. The most common chemotherapies involve drugs that are "artificial nucleotides." These are incorporated into DNA during chromosome replication. They then either prevent further replication or cannot be transcribed correctly. In either case, the cell dies or fails to reproduce. These drugs obviously have their main effect on dividing cells, and since cancer cells divide rapidly, they kill cancer cells. Unfortunately, other cells of the body divide too, such as those in the hair follicles and intestinal lining. Chemotherapy drugs damage those cells, producing the well-known side effects of nausea, vomiting, and hair loss.

Future drug therapies may use *monoclonal antibodies* directed against cancer cells (Fig. 27-12). A small patch of cancer cells may be snipped out of a patient and injected into a mouse. Recognizing the human cancer as "nonself," B cells of the mouse proliferate. Anticancer B cells are extracted from the mouse's spleen and fused with a particular strain of cancerous white blood cells, called

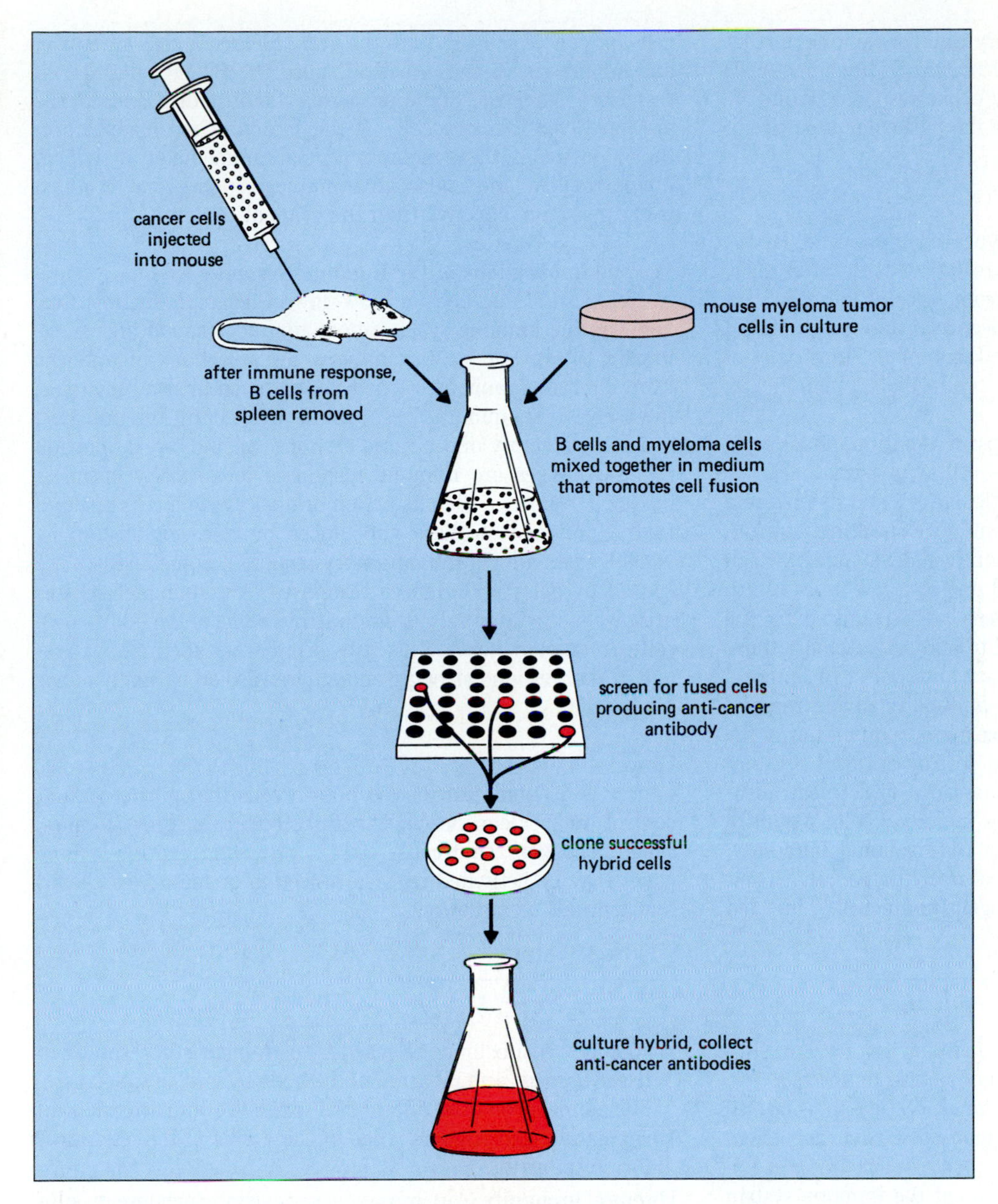

Figure 27-12 The production of monoclonal antibodies against a cancer.

myeloma cells. Some of the fused cells combine the desired properties of both original cell types: from the mouse B-cells, specific antibody production against the patient's cancer, and from the myeloma cells, rapid cell division. All the descendant cells from one such cell (a clone from a single cell) produce the same anti-cancer antibody (hence the term "monoclonal antibody"). Massive amounts of antibodies specific for the cancer cells, and for no other cells of the patient, could be produced in the lab. Particularly lethal drugs may be attached to the antibodies, which are then injected into the patient. The antibodies bind only to the cancer cells, so the drug destroys them without harming normal body cells. Although this technique has not yet been perfected, it offers great promise for future cancer treatments.

SUMMARY OF KEY CONCEPTS

Defenses Against Microbial Invasion

The human body has three lines of defense against invasion by microbes: (1) the barriers of skin and mucous membranes, (2) the inflammatory response, and (3) the immune response. *Barriers:* The skin physically blocks the entry of microbes into the body. It is also covered with secretions from sweat and sebaceous glands that inhibit bacterial and fungal growth. The mucous membranes of the respiratory and digestive tracts secrete antibiotic substances and mucus. Microbes are trapped in the mucus, swept up to the throat by cilia, and expelled or swallowed. *Inflammatory response:* If microbes enter the body, white blood cells travel to the site of entry and engulf the invading cells. Blood clots wall off the injury site, preventing further spread of the microbes. *Immune response:* If invading cells spread to the rest of the body, they evoke responses from the immune system that are specific for the particular type of microbe involved.

The Immune Response

The immune response involves two types of lymphocytes, B cells and T cells. Plasma cells, which are descendants of B cells, secrete antibodies into the bloodstream, causing humoral immunity. T cells destroy microbes, cancer cells, and virus-infected cells on contact, causing cell-mediated immunity. Some types of T cells regulate both humoral and cell-mediated immune responses.

Immune responses have three steps: recognition, attack, and memory. *Recognition:* each immune cell synthesizes only one type of antibody, unique to that particular cell and its progeny. The diversity of antibodies arises from gene shuffling and mutation of antibody genes during immune cell development. Each antibody has specific sites that bind one or a few types of antigen. Normally, only foreign antigens are recognized by the immune cells. *Attack:* antigens bind to and activate only those B and T cells with the complementary antibodies. In humoral immunity, B cells with the proper antibodies divide rapidly, producing plasma cells that synthesize massive quantities of the antibody. The circulating antibodies destroy antigens and antigen-bearing microbes by four mechanisms: direct neutralization, promotion of phagocytosis by white blood cells, agglutination, and complement reactions. In cell-mediated immunity, T cells with the proper antibodies also divide rapidly. Their descendant killer T cells bind to antigens on microbes, infected cells, or cancer cells and kill the cells. Helper T cells stimulate, and suppressor T cells turn off, both the B and killer T cell responses. *Memory:* Some progeny cells of both B and T cells are long-lived memory cells. If the same antigen reappears in the bloodstream, these memory cells are immediately activated, divide rapidly, and cause an immune response that is much faster and more effective than the original response.

Medical Implications of the Immune Response

Antibiotics kill microbes or slow down their reproduction, thus allowing the immune system more time to respond and exterminate the invaders. Vaccinations are injections of antigens from disease organisms, often the weakened or dead microbes themselves. An immune response is evoked by the antigens, providing memory and a rapid response should a real infection occur. Allergies are immune responses to normally harmless foreign substances, such as pollen or dust. Certain cells respond to the presence of these substances by releasing histamine, which causes a local inflammatory response. Some diseases are caused by defective immune responses. Autoimmune diseases arise when the immune system destroys some of the body's own cells. Immune deficiency diseases occur when the immune system cannot respond strongly enough to ward off normally minor diseases.

Cancer

Cancer is a population of the body's cells that grows without control. Cancer can be caused by activation of growth genes, called oncogenes, that cause cells to grow and multiply. These genes may be activated by viral infection or mutations caused by chemicals or radiation.

GLOSSARY

Agglutination (a-glū-tin-ā′-shun) clumping of foreign substances or microbes, caused by binding with antibodies.

Allergy an inflammatory response produced by the body in response to invasion by foreign materials, such as pollen, which are themselves harmless.

Antibody a protein produced by cells of the immune system which combines with a specific antigen and usually facilitates its destruction.

Antigen a complex molecule, usually protein or polysaccharide, that stimulates the production of a specific antibody.

Autoimmune disease a disorder in which the immune system produces antibodies against the body's own cells.

B cell a type of lymphocyte that secretes antibodies into the circulatory system in response to stimulation by antigens.

Cancer a disease in which some of the body's cells grow without control.

Cell-mediated immunity an immune response in which foreign cells or substances are destroyed by contact with T cells.

Complement a group of blood-borne proteins that participate in the destruction of foreign cells to which antibodies have bound.

Complement reactions interactions among foreign cells, antibodies, and complement proteins, resulting in the destruction of the foreign cells.

Helper T cell a type of T cell that aids other immune cells to recognize and act against antigens.

Histamine a substance released by certain cells in response to tissue damage and invasion of the body by foreign substances. Histamine promotes dilation of arterioles and leakiness of capillaries, and triggers some of the events of the inflammatory response.

Humoral immunity an immune response in which foreign substances are inactivated or destroyed by antibodies circulating in the blood.

Immune deficiency disease a disorder in which the immune system is incapable of responding properly to invading disease organisms.

Immune response a specific response by the immune system to invasion of the body by a particular foreign substance or microorganism, characterized by recognition of the foreign material by immune cells and its subsequent destruction by antibodies or cellular attack.

Inflammatory response a nonspecific, local response to injury to the body, characterized by phagocytosis of foreign substances and tissue debris by white blood cells, and "walling off" of the injury site by clotting of fluids escaping from nearby blood vessels.

Killer T cell a type of T cell that directly destroys foreign cells upon contacting them.

Memory cell a long-lived descendant of a B or T cell that has been activated by contact with antigen. Memory cells are a reservoir of cells that rapidly respond to reexposure to the same antigen.

Mucous membrane the lining of the inside of the respiratory and digestive tracts.

Neutralization the process of covering up or inactivating a toxic substance with antibody.

Oncogene a gene that, when transcribed, causes a cell to become cancerous.

Plasma cell an antibody-secreting descendant of a B cell.

Suppressor T cell a type of T cell that depresses the response of other immune cells to foreign antigens.

T cell a type of lymphocyte that recognizes and destroys specific foreign cells or substances, or that regulates other cells of the immune system.

Vaccine a material injected into the body that contains antigens characteristic of a particular disease organism, and that stimulates an immune response.

STUDY QUESTIONS

1. List the three lines of defense of the human body against invading microbes. Which are nonspecific (i.e., act against all types of invaders) and which are specific (i.e., only act against a particular type of invader)? Explain your answer.
2. Name three anti-infection properties of the skin and three of the mucous membranes.
3. Describe the inflammatory response.
4. Describe humoral immunity and cell-mediated immunity. Include in your answer the types of immune cells involved in each, the location of antibodies that attach to foreign antigens, and the mechanisms by which invading cells are destroyed.
5. How does the immune system construct so many different antibodies?
6. How does the body distinguish "self" from "nonself"?
7. Diagram the structure of an antibody. What parts bind to antigens? Why does each antibody bind only to a specific antigen?
8. What are memory cells? How do they contribute to long-lasting immunity to specific diseases?
9. What is a vaccine? How does it confer immunity to a disease?
10. Describe the allergic reaction.
11. Distinguish between autoimmune diseases and immune deficiency diseases, and give one example of each.
12. What is cancer? How do oncogenes cause cancer? How can environmental factors "turn on" oncogenes to cause cancer?

SUGGESTED READINGS

Bishop, J. M. "The Molecular Genetics of Cancer." *Science,* January 1987. A bit technical, but an excellent summary of current knowledge of the causes of cancer.

Buisseret, P. "Allergy." *Scientific American,* August 1982 (Offprint No. 1522). Allergy is now understood as a malfunctioning of the immune system.

Gallo, R. C. "The First Human Retrovirus" and "The AIDS Virus." *Scientific American,* December 1986 (Offprint No. 1576) and January 1987 (Offprint No. 1577). Readable descriptions of viruses of the immune system, by their chief discoverer.

Golub, E. *The Cellular Basis of the Immune Response,* 2nd ed. Sunderland, Mass: Sinauer Associates, Inc., 1981. A relatively advanced, complete treatment of the immune response.

Jaret, P. "The Wars Within." *National Geographic,* June 1986. Lucid diagrams of the complexity of the immune response, accompanied by incredible photographs by Lennart Nilsson.

Leder, P. "The Genetics of Antibody Diversity." *Scientific American,* May 1982. How only a few hundred genes can be used to make millions of antibodies.

Milstein, C. "Monoclonal Antibodies." *Scientific American,* October 1980 (Offprint No. 1479). One of the inventors of the monoclonal antibody technique explains the process and some of its uses.

Young, J. D-E. and Cohn, Z. A. "How Killer Cells Kill." *Scientific American,* January 1988 (Offprint No. 1589). Killer T cells strike by secreting proteins that form large holes in the cell membrane of their targets.

28
Chemical Control of the Animal Body

Large, multicellular organisms have some advantages over small, unicellular organisms: they are more likely to be the eater than the eaten, and they usually have better control over their internal environments. However, a large, multicellular body presents several challenges, not the least of which is the control and coordination of millions to trillions of cells.

Early in the evolution of life on Earth, cells developed ways of communicating with and influencing one another. The simplest method is by direct contact. Molecules protruding from the surface identify cells as members of a particular species, as parts of an individual organism, or even as specific cell types, such as skin or liver. Surface contacts are important in the development of embryos, in which cells migrate around one another to arrive at their proper destination in the adult form. Direct contact also plays a key role in defense against disease organisms, in which immune cells recognize invaders by their surface molecules.

Coordination by cell contact has two obvious limitations. First, communication is restricted to the number of cells that can touch one another simultaneously. Second, if a cell must move to contact new cells, communication over even small distances is difficult or impossible. However, if a cell can release a chemical into its environment, whether that be a pond (external environment) or extracellular fluid (internal environment), and have that chemical influence other cells as it contacts them, then the secreting cell can communicate with vast numbers of other cells over large distances. Such chemical communication has indeed evolved, from "come hither" messages in slime molds (see the essay, "Chemical Communication Creates an Organism" on page 516) to the sophisticated signals released by the endocrine and nervous systems of mammals.

CHEMICAL COMMUNICATION WITHIN THE ANIMAL BODY

Although it is convenient to discuss hormonal control separately from nervous control, the two operate in strikingly similar ways. Both hormone-producing cells and nerve cells synthesize "messenger" chemicals that they release into extracellular spaces (Fig. 28-1). There are three main differences between the two: *distance, number of cells contacted,* and *speed.* First, nerve cells usually release their chemical (called a neurotransmitter) very close to the cells they influence, often from less than a micrometer away. Hormone-producing cells, on the other hand, release their chemicals into the bloodstream, which often carries them many centimeters to their target cells. Second, this difference in distance creates a great difference in the number of cells that the chemicals contact. A nerve cell very precisely squirts particular cells with its chemicals, whereas hormones bathe millions of cells indiscriminately. Finally, a nerve cell speeds information from one part of the body to another via electrical signals traveling within the nerve cell itself, and only then releases its neurotransmitter. Hormones traveling in the bloodstream, of course, move much more slowly.

Even these differences blur as we learn more about the hormonal and nervous systems of animals. Some hormones are in fact produced and released into the bloodstream by nerve cells, and are appropriately called **neurohormones** (Fig. 28-1b). Other chemicals that biologists used to think were strictly hormones have recently been discovered in the brain, where they are synthesized and released by nerve cells.

In this chapter we discuss the "classical" hormones and neurohormones. Although the functioning of the nervous system will follow in Chapter 29, remember that no system in your body works alone. The hormonal and nervous

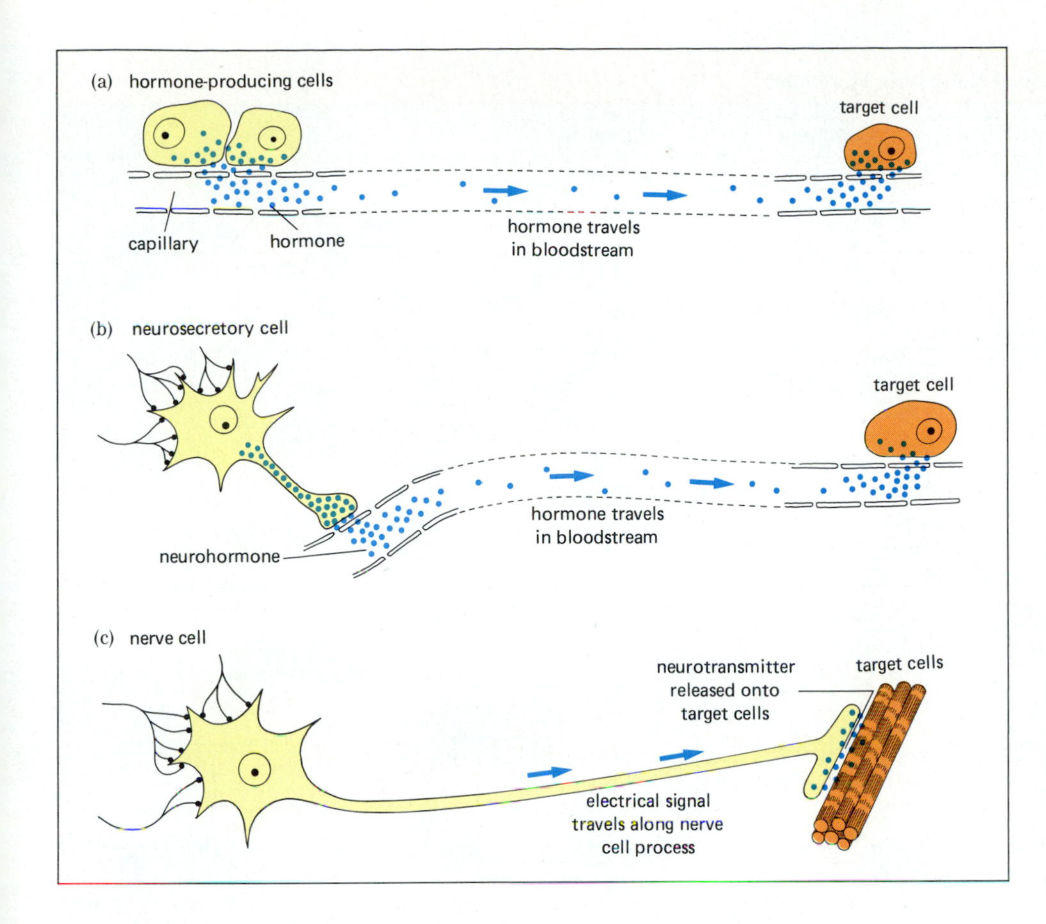

Figure 28-1 Three major types of "control cells" release chemicals that influence the activity of other cells of the body: **(a)** *"classical" hormone-producing cells,* **(b)** *neurosecretory cells, and* **(c)** *ordinary nerve cells. Three common steps occur in each system: (1) The chemical-releasing cell is stimulated. (2) It sends a message to distant cells. (3) Selected target cells respond. Both classical hormone-producing cells and neurosecretory cells release their chemical messages into the bloodstream, which transports the chemicals to distant target cells. Regular nerve cells grow long processes to their target cells and release their chemical messages directly onto the target cell.*

systems are closely coordinated in their control of bodily functions.

HORMONE FUNCTION IN ANIMALS

A hormone is a chemical secreted by cells in one part of the body that is transported in the bloodstream to other parts of the body, where it affects particular target cells. As we shall see, the same hormone may have several different effects, depending on the nature of the target cells it contacts.

Types of Animal Hormones

There are four classes of chemicals used as hormones in the animal kingdom (Table 28-1). A few hormones, and many neurotransmitters, are **modified amino acids,** such as adrenalin, noradrenalin, and the thyroid hormones. Most hormones are **proteins,** ranging from just a few to over a hundred amino acids in length. A few glands synthesize **steroid** hormones, such as estrogen and testosterone, using cholesterol as the starting material. Finally, many tissues of the body, perhaps all, release modified fatty acid hormones known as **prostaglandins.**

Gland Structure and Function

Mammals have two types of glands, **exocrine** or ducted glands and **endocrine** or ductless glands (Fig. 28-2). The exocrine glands, which include the sweat and mammary glands and portions of the pancreas and liver, produce secretions that are released into ducts leading *outside the body* (in Greek, "exo" means "out of") or *into the digestive tract* (which is, in a way, also outside the body, since it is a hollow tube continuous with the outside via mouth and anus). The endocrine glands release hormones *within the body* ("endo" means "inside of"). An endocrine gland consists of clusters of hormone-producing cells embedded within a meshwork of capillaries. The cells secrete their hormones into the extracellular fluid surrounding the capillaries. The hormones enter the capillaries by diffusion, and the circulatory system then distributes the hormones to other parts of the body.

Hormone Actions on Target Cells

Once hormones enter the bloodstream, the glands cannot control where they travel within the body. Since all cells have a blood supply, most hormones contact nearly every cell of the body. Nevertheless, not all cells respond to all hormones. *Whether or not a cell responds to a given hormone depends on the properties of the cell. If the cell has an appropriate receptor for the hormone, it can respond,*

Table 28-1 The Chemical Diversity of Vertebrate Hormones

Chemical Type	Synthesized From	Examples	Structures
Modified amino acids	Amino acid	Noradrenalin	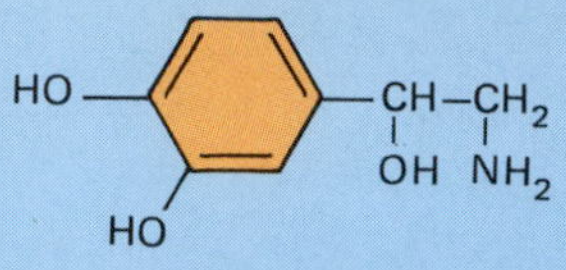 HO, HO, CH–CH$_2$, OH, NH$_2$
		Thyroxine	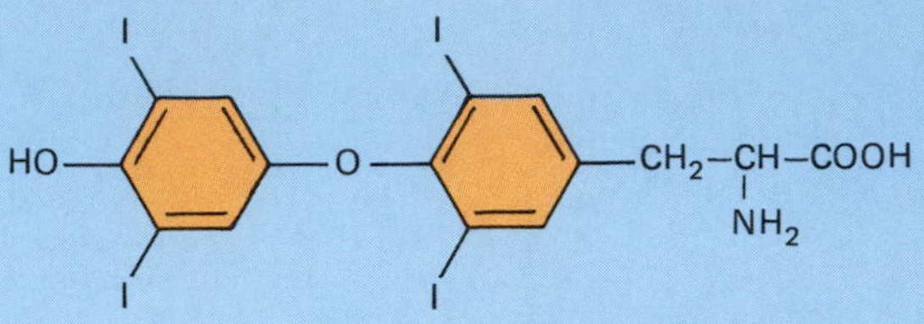 I, I, HO, O, CH$_2$–CH–COOH, NH$_2$, I, I
Proteins and peptides	Amino acids	Oxytocin	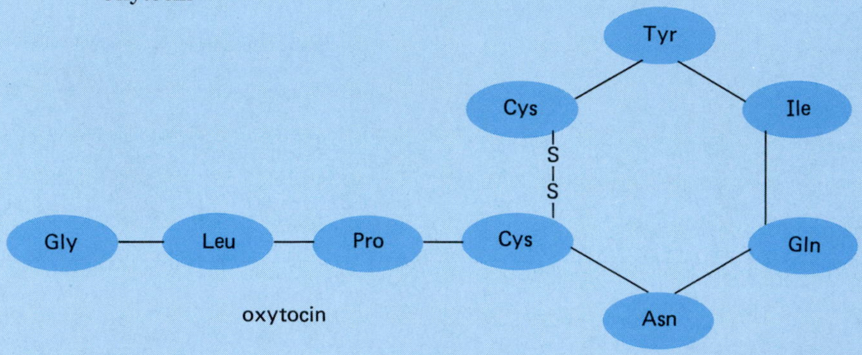 Tyr, Cys, Ile, S, S, Gly, Leu, Pro, Cys, Gln, oxytocin, Asn
Steroids	Cholesterol	Testosterone	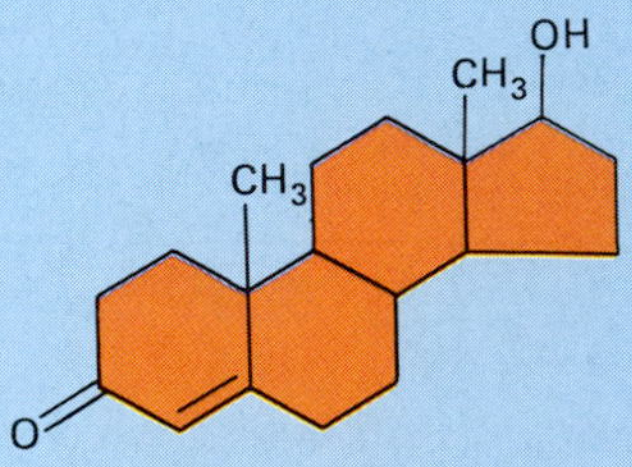 OH, CH$_3$, CH$_3$, O
		Estradiol	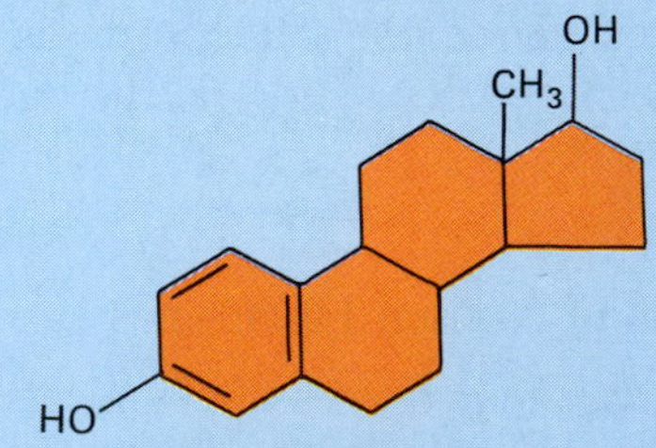 OH, CH$_3$, HO
Prostaglandins	Fatty acids	Prostaglandin E_1	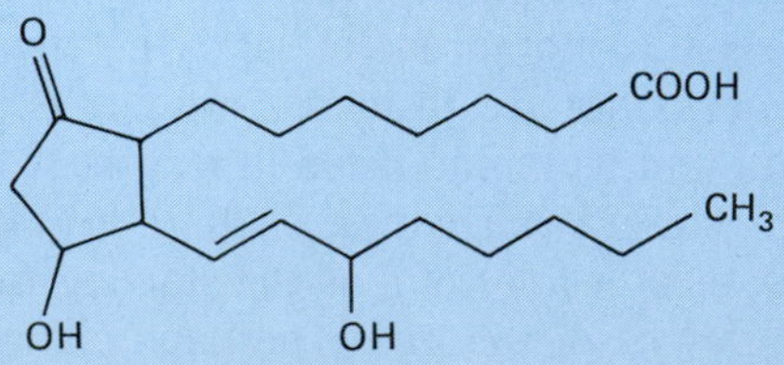 O, COOH, CH$_3$, OH, OH

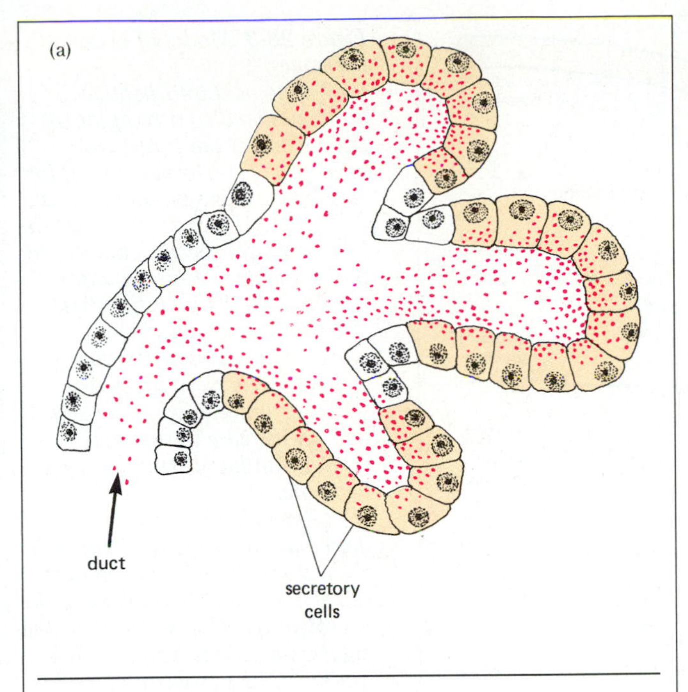

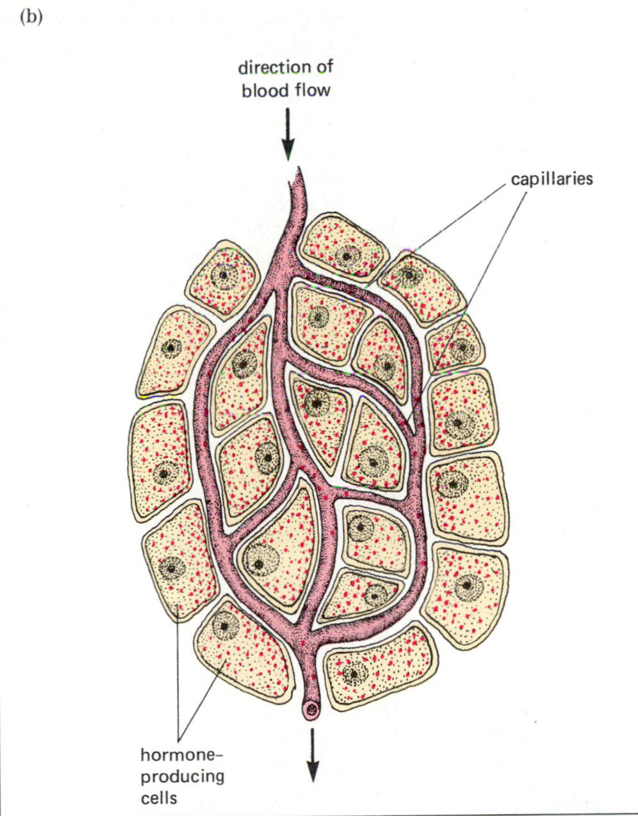

Figure 28-2 The structure of exocrine and endocrine glands. **(a)** *The secretory cells of exocrine glands form the sources of ducts that usually open outside the body (sweat glands, mammary glands) or into the digestive tract (pancreas, salivary glands).* **(b)** *Endocrine glands consist of hormone-producing cells embedded with a network of capillaries. The cells secrete hormones (red dots) into the extracellular space, from which they diffuse into the capillaries.*

and is a target cell for that hormone. If it does not have the right receptor, the hormone will have no effect.

Different types of hormones act on target cells in different ways. Most modified amino acid and peptide hormones are water soluble but not lipid soluble, and hence cannot cross cell membranes. These hormones react with protein receptors protruding from the outside surface of target cell membranes and trigger a series of events that change the activity of the cell (Fig. 28-3a). Many of these hormones stimulate the synthesis of **cyclic AMP** (see Chapter 3), a nucleotide that regulates many cellular activities by activating enzymes. Cyclic AMP is often called a **second messenger,** since it transfers information from the first messenger, the hormone, to target molecules within the cell.

Steroid hormones are lipid soluble and pass easily through cell membranes (Fig. 28-3b). Once in the cytoplasm, they combine with soluble protein receptors and travel to the nucleus. The steroid-receptor complexes bind to specific places on the chromosomes and facilitate the binding of RNA polymerase to appropriate genes (see Chapter 11). These genes are transcribed into messenger RNA, which then directs the synthesis of a new set of proteins.

Hormone Regulation

Hormones, together with the nervous system, comprise the integration and control system of the body. Most hormones exert such powerful effects on the body that it would be harmful to have too much hormone working for too long. For example, let's suppose that you have jogged several miles on a hot, sunny day and have lost a liter of water through perspiration. To prevent further water loss, your pituitary gland releases antidiuretic hormone (ADH), which causes your kidneys to reabsorb water and produce a very concentrated urine (see Chapter 25). However, if you drink a gallon of Gatorade, you will more than replace the water you lost in sweat. Continued retention of water by the kidneys would overload the bloodstream, raising blood pressure and possibly damaging your heart. Therefore, when your body water level returns to normal, ADH secretion stops. As this example illustrates, hormone secretion must be *regulated,* so that just the right amounts are released at the right times.

Animals usually regulate hormone release through **negative feedback,** whereby *hormone secretion causes effects in target cells that inhibit further secretion of the hormone* (Fig. 28-4a). Negative feedback is one way of maintaining **homeostasis,** that is, keeping conditions within the body relatively constant over time. Negative feedback may sound complicated, but in fact you are already familiar with it in a different context, namely temperature control in your house during the winter (Fig. 28-4b). You start out by setting the thermostat to the temperature you want. If the house is too cold, the ther-

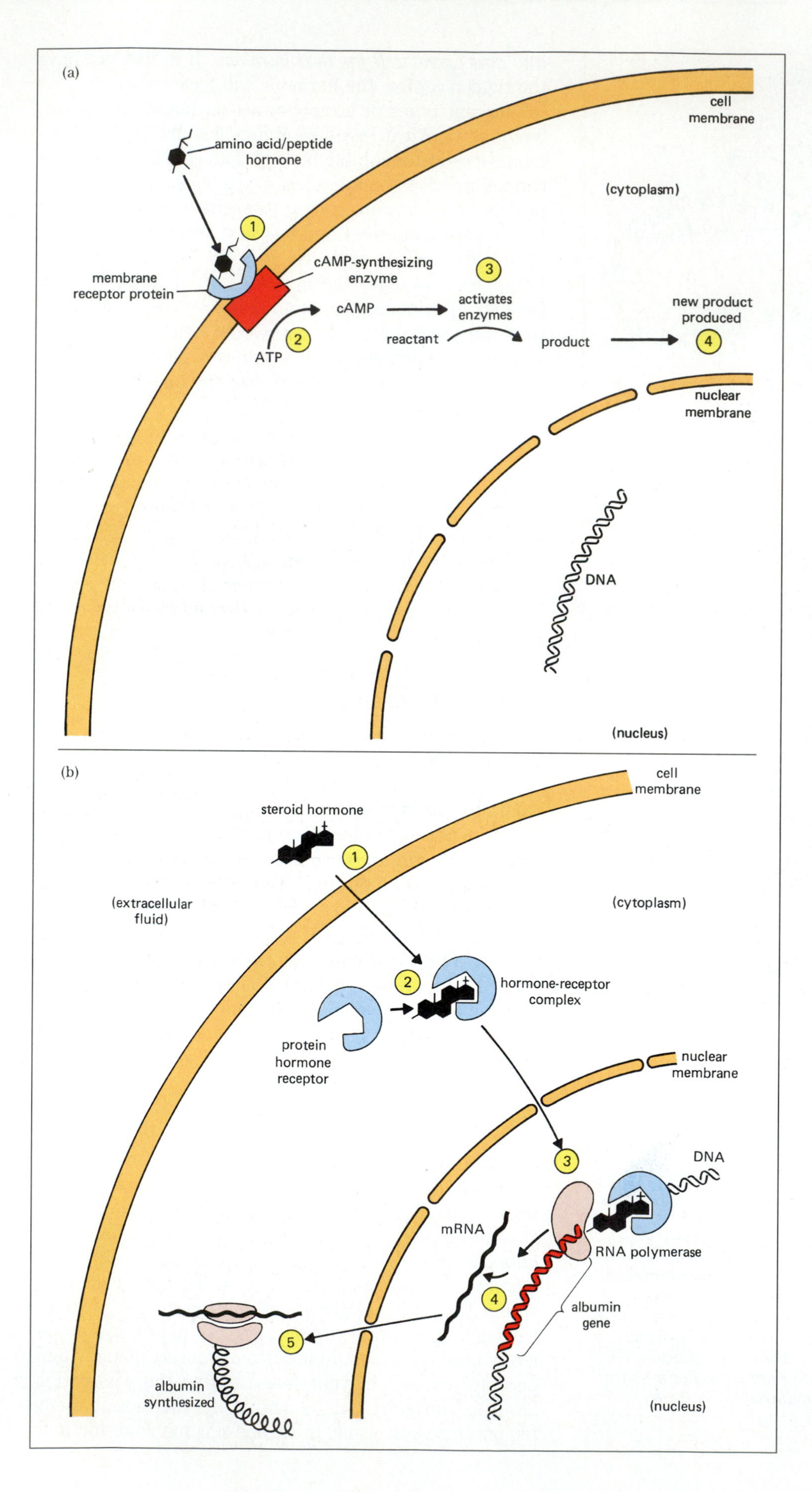

Figure 28-3 Modes of action of hormones.
(a) *Amino acid and peptide hormones bind to a receptor on the outside of the target cell membrane (1). Hormone-receptor binding triggers synthesis of cyclic AMP (cAMP; 2). Cyclic AMP in turn activates specific enzymes (3) that promote specific cellular reactions (4). This cyclic AMP "cascade" may generate a variety of responses. Examples include an increase in glucose synthesis induced by adrenalin and an increase in estrogen synthesis induced by luteinizing hormone.*
(b) *Lipid-soluble steroid hormones diffuse readily through the cell membrane into the target cell (1), where they combine with a protein receptor molecule in the cytoplasm (2) and travel to the nucleus. The steroid-receptor complex facilitates the binding of RNA polymerase to promoter sites on specific genes (3), accelerating transcription (4) of DNA into messenger RNA. The mRNA then directs protein synthesis (5). In hens, for example, estrogen promotes transcription of the albumin gene, causing synthesis of albumin (egg white), which is packaged in the egg as a food supply for the developing chick.*

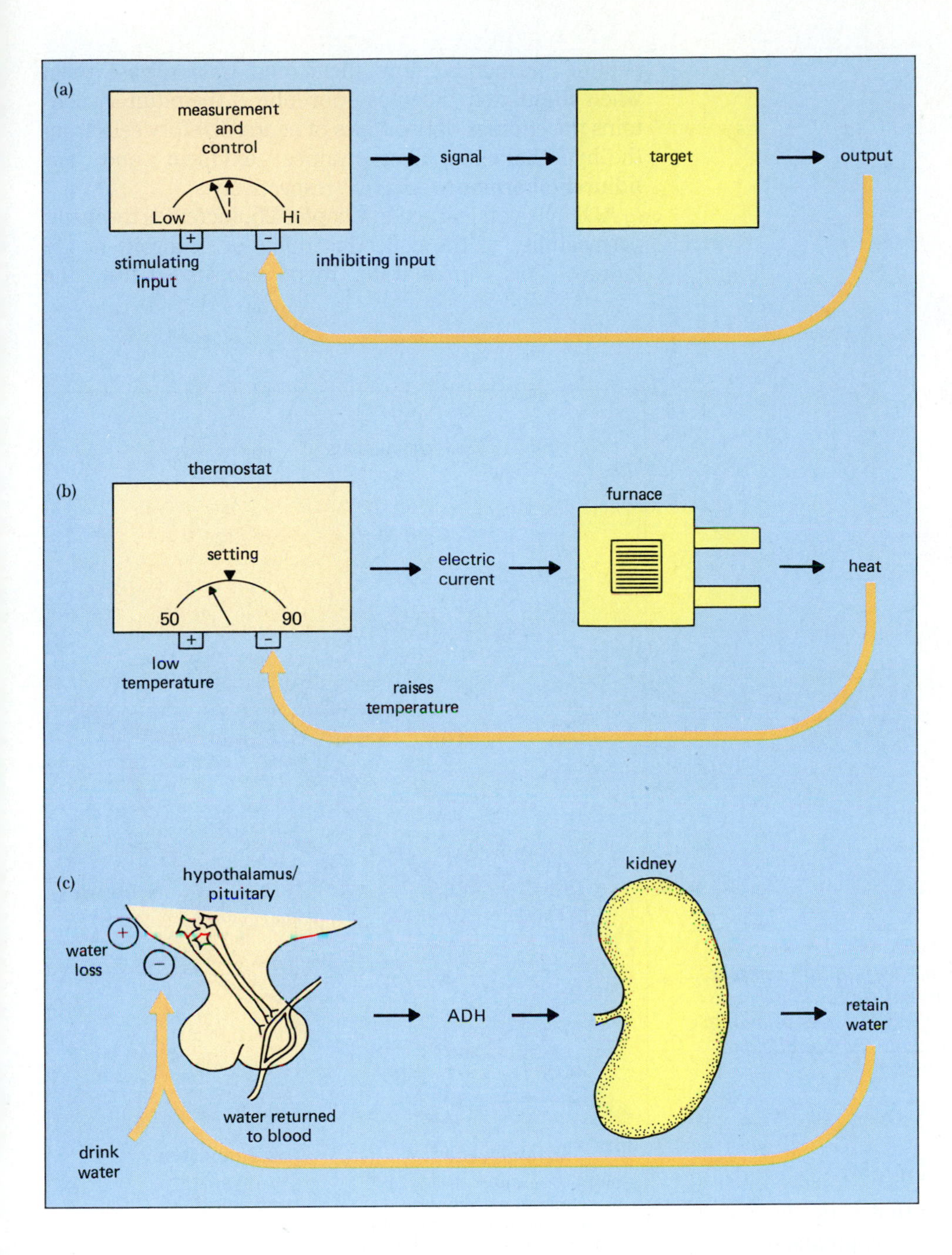

Figure 28-4 Hormone secretion is regulated by negative feedback. In this and subsequent illustrations of hormonal regulation, ⊕ denotes stimulation and ⊖ denotes inhibition. **(a)** *In negative feedback, a control device measures the function to be controlled and compares this level to the "desired" level. An incorrect level stimulates the control device to send a signal to a target device. The target produces an output in response, which brings the function back to the desired level, inhibiting further signal production by the control device.* **(b)** *The control of house temperature in winter is an example of negative feedback. The thermostat is set to the desired temperature. If the temperature drops below the set point, the thermostat turns on, sending a signal to the furnace to produce heat. The heat output of the furnace raises the temperature, which turns off the thermostat, shutting down the furnace.* **(c)** *Most hormones are also regulated by negative feedback. In response to water loss through perspiration, the pituitary gland produces antidiuretic hormone (ADH), which stimulates the kidneys to conserve water. Water conservation and water intake combine to bring the water concentration of the blood back to normal, inhibiting further ADH secretion.*

mostat sends an electric signal to the furnace, turning it on. The furnace generates heat, and the house temperature rises. When the temperature equals the thermostat setting, the thermostat switches off, thus shutting off the furnace. Hormonal control through negative feedback works in a similar fashion (Fig. 28-4c). In our water-balance example, water loss decreases the water concentration in the blood. This stimulates the pituitary to release ADH, which signals the kidney to reabsorb maximal amounts of water from the urine and return it to the blood. Water retention in the kidney, combined with drinking fluids, replenishes the blood's water supply, shutting down the release of ADH by the pituitary.

MAMMALIAN ENDOCRINE SYSTEMS

Endocrinologists are far from attaining a complete understanding of hormonal control in mammals. New hormones, or new functional roles for previously known hormones, are discovered virtually every year. What we might call the seven major endocrine systems, however, have been known for many years: these are the hypothalamus/pituitary complex, thyroid, parathyroid, pancreas, adrenal cortex, adrenal medulla, and gonads (Fig. 28-5). Table 28-2 (pp. 522–523) lists these and other glands, their major hormones, and their principal control functions.

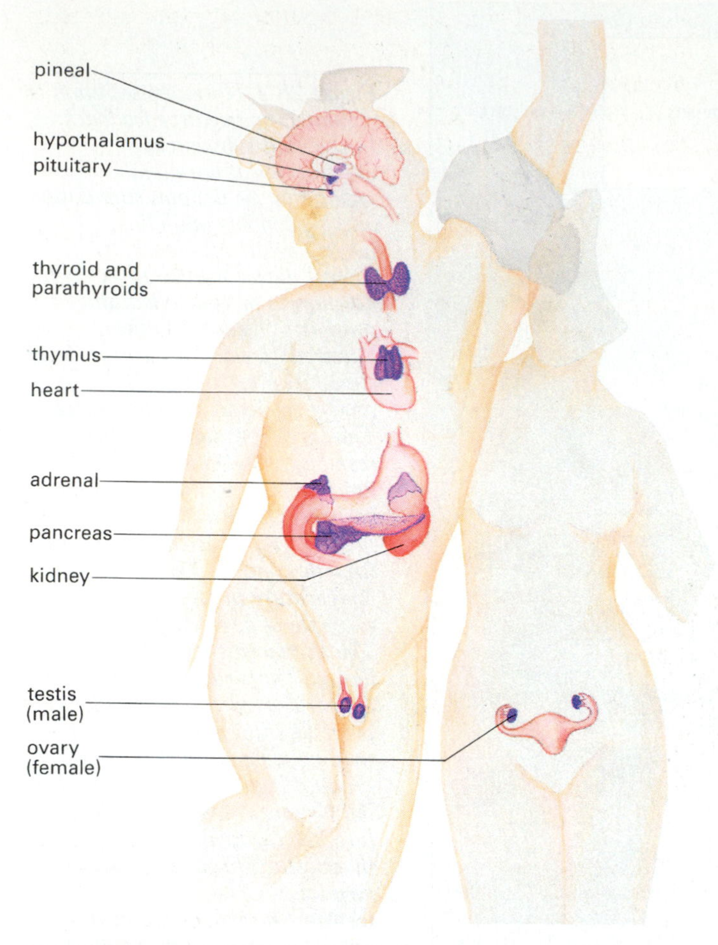

Figure 28-5 The major mammalian endocrine glands discussed in the text are the hypothalamus/pituitary complex, the thyroid and parathyroids, the adrenal glands, the pancreas, and the gonads (ovaries in females, testes in males). Other organs that secrete hormones include the pineal gland, thymus, heart, kidney, and digestive tract.

The Hypothalamus/Pituitary Complex

The **pituitary** is a pea-sized gland that hangs by a stalk from a part of the brain called the **hypothalamus** (Fig. 28-6). Anatomically, the pituitary consists of two parts. The **anterior pituitary** is a true endocrine gland, composed of several types of hormone-secreting cells enmeshed in a network of capillaries. The **posterior pituitary,** on the other hand, should probably be considered part of the brain, because it is derived from nervous tissue during development, and because its hormones are secreted by endings of nerve cells that originate in the hypothalamus. These nerve endings lie in the midst of another capillary bed. The hypothalamus controls the release of hormones from both the anterior and posterior pituitary.

POSTERIOR PITUITARY. The hypothalamus contains clusters of specialized nerve cells, called **neurosecretory cells** (see Fig. 28-1b). Neurosecretory cells synthesize peptide hormones, store them, and then release them when stimulated. The posterior lobe of the pituitary contains the endings of two types of neurosecretory cells from the hypothalamus. They produce **oxytocin** and **antidiuretic hormone (ADH),** respectively.

ADH, as you learned in Chapter 25, increases the water permeability of the collecting ducts of nephrons in the kidney. This causes water to be reabsorbed from the

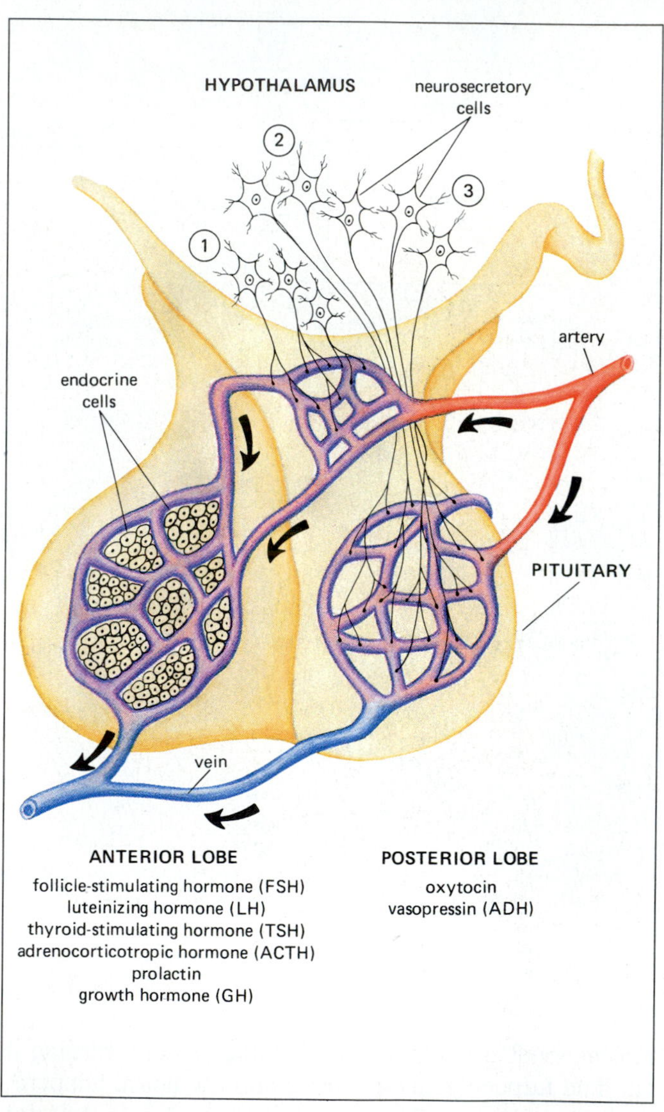

Figure 28-6 Anatomical relationships between the hypothalamus and pituitary. The anterior lobe of the pituitary (left) consists of secretory cells enmeshed in a capillary bed. Release of hormones from these cells is controlled by releasing and inhibiting hormones produced by neurosecretory cells (1) of the hypothalamus. These neurosecretory cells release their hormones into a capillary network directly "upstream" from the anterior pituitary (see Fig. 28-8). The posterior lobe of the pituitary (right) is an extension of the hypothalamus. Two types of neurosecretory cells (2 and 3) send processes from the hypothalamus into the posterior lobe, where they end on a capillary bed into which they release their hormones.

forming urine, so that less water is lost through excretion. This helps the body avoid dehydration ("antidiuretic hormone" literally means "hormone that prevents urination"). Oxytocin causes contraction of the muscles of the uterus during childbirth, helping to expel the infant from the womb. It also causes contraction of muscle cells within the breasts during lactation, squeezing milk out of storage bulbs and into ducts leading to the nipples (Fig. 28-7). Recent evidence in laboratory animals indicates that oxytocin has behavioral effects, too, inducing maternal behaviors even in virgin females. In rats, for example, oxytocin injections cause virgin females to build a nest, lick pups, and retrieve "lost" pups. Oxytocin may also have a role in male reproductive behavior. In several animals, oxytocin stimulates the contraction of muscles surrounding the tubes that conduct sperm from the testes to the penis, thus causing ejaculation.

Anterior pituitary. The anterior lobe of the pituitary produces six different hormones, four of which in turn help to regulate hormone production in other glands. Two of these, **follicle-stimulating hormone (FSH)** and **luteinizing hormone (LH),** stimulate production of sperm and testosterone in males, and eggs, estrogen, and

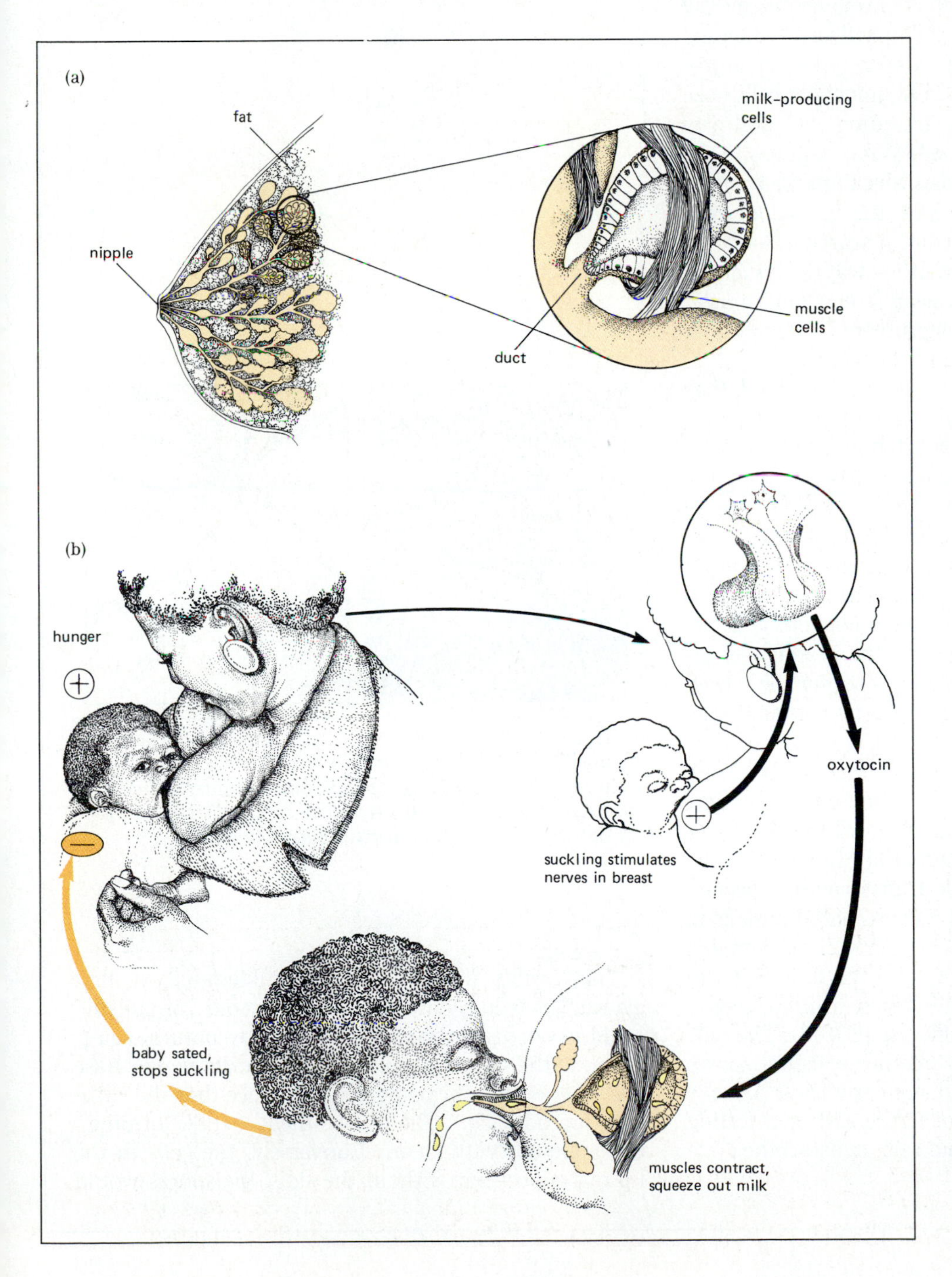

Figure 28-7 Oxytocin controls breast-feeding.
(a) *The structure of the human breast (an exocrine gland). Clusters of milk-producing cells surround hollow bulbs, and ducts lead from the bulbs to the nipple. In lactating women, milk is secreted into the bulbs and stored there until the baby suckles, when muscles surrounding the bulbs contract, squeezing milk out into the ducts.*
(b) *Milk release is controlled by a complex negative feedback loop between baby and mother. When the infant is hungry, it starts to suckle. This stimulates nerve endings in the nipple, sending signals to the mother's brain. Neurosecretory cells in her hypothalamus respond by secreting oxytocin, which travels in her bloodstream to the breast. There, oxytocin stimulates contraction of the muscles surrounding the milk bulbs, forcing milk into the ducts so that the baby can suck it out of the nipple. This continues until the baby is full and stops suckling. With the nipple no longer being stimulated, oxytocin release stops, the muscles relax, and milk flow ceases.*

Chemical Communication Creates an Organism

Most multicellular animals, humans included, produce many chemical secretions that are used for communication among cells. Biologists usually classify these secretions as **hormones** if they influence cells in the same body that produced them, or **pheromones** if they influence cells in another animal's body (see Chapter 33). This dichotomy may be a product of our own biases, since each human is a self-contained organism. But consider the case of the cellular slime mold.

If you go out to your garden and pick up a pinch of soil, you will be holding millions of organisms, mostly bacteria, between your fingers. You will also be holding thousands of single-celled, amoeba-like creatures called cellular slime molds. These amoeboid cells ooze along in the film of water surrounding each soil particle, engulfing bacteria for food. What happens to the slime molds if they eat bacteria faster than the bacteria can reproduce? Needless to say, the amoeboid cells cannot move very fast or very far in search of food. To survive, they must find some other way to travel the relatively vast distance to the nearest new food source. Cellular slime molds solve this problem by temporarily merging to create a multicellular organism.

If slime mold cells go long enough without food, they release a chemical into the soil water. This chemical is called **acrasin,** after Acrasia, a witch in Edmund Spenser's poem, *Faerie Queene.* Acrasia attracted men and turned them into animals, which, as you will see, is very similar to what acrasin does to slime mold cells. The concentration of acrasin is highest right around the cell releasing it, and decreases with distance as it diffuses away (Fig. E28-1). Other amoeboid cells respond to acrasin by slithering up the gradient of increasing concentration until they encounter the releasing cell. The new cells remain with the original one, releasing their own acrasin, until thousands of cells have gathered. (Different species of slime molds release different acrasins. The acrasin of one common slime mold is cyclic AMP, the same chemical that mediates cellular responses to many animal hormones.)

At this point, a remarkable transformation occurs. The mass of cells behaves as a coordinated organism, crawling around for a while (Fig. E28-2). Eventually, the "slime slug" stops moving, and the cells begin to climb atop one another, creating first a lump, then a knob, and eventually a tall stalk (Fig. E28-3). The cells at the tip metamorphose into weather-resistant spores that disperse on the wind to distant environs. A lucky spore may land on a patch of fertile soil or a rotting log, revert back to an amoeboid cell, and start the cycle over again.

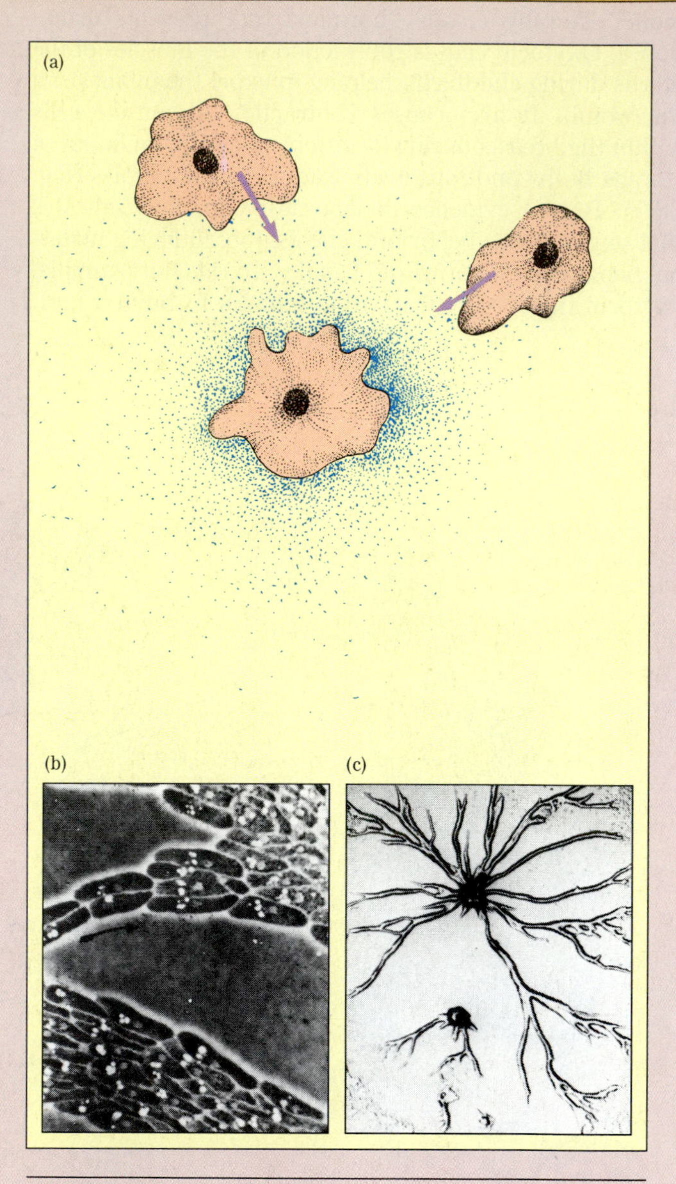

Figure E28-1 **(a)** *A slime mold cell releases acrasin (blue dots) into its surroundings, establishing a gradient that is followed by other cells.* **(b) (c)** *Slime mold cells stream from all directions toward the acrasin-secreting center.*

This brings up an interesting dilemma: at first glance, it would seem that the slime mold "organism" could never arise through evolution by natural selection, for the following reasons. The cells in the base and stalk do not form spores, and hence they die without reproducing. Shouldn't their "stalk-forming" genes die out with them? Conversely, the cells in the tip reproduce, but without the stalk the spores would

Chemical Communication Creates an Organism

not disperse effectively. How can evolution generate some cells that form a stalk but do not reproduce, and other cells that reproduce but do not form stalks? The answer probably lies in kinship among the cells. A patch of soil is colonized by one or at most a few spores, and the resulting amoeboid cells reproduce by simple cell division. The multicellular "slug" therefore contains thousands of genetically identical descendants of these few original founders. Each cell probably has *both* "stalk-forming" genes *and* "stalk-climbing" genes. During stalk formation, the cells that form the base and stalk are really helping to perpetuate their own genes, embodied in their spore-forming sister cells. Vive la mold!

Figure E28-2 The aggregated mold cells form a sluglike mass that often crawls around for a while. If a rich food supply is encountered, the slug may revert to amoeboid cells.

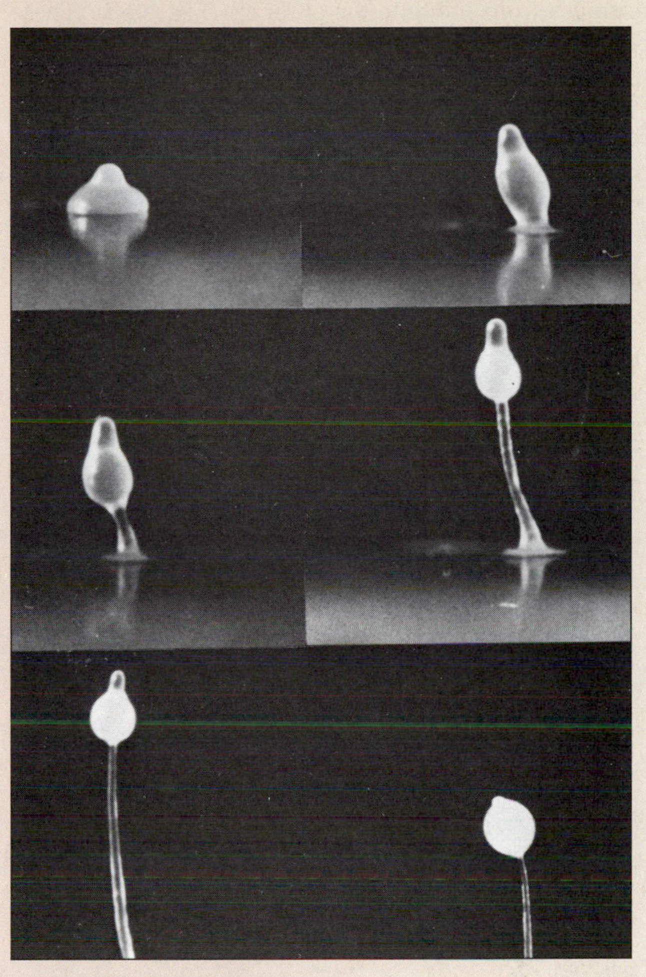

Figure E28-3 The dispersal stage of the cellular slime mold life cycle: out of the ooze rises a tall stalk of cells, upon which sits a capsule containing numerous spores.

progesterone in females. We will defer further examination of the roles of FSH and LH to Chapter 31. **Thyroid-stimulating hormone (TSH)** stimulates the thyroid gland to release its hormones, while **adrenocorticotropic hormone** ("hormone that stimulates the adrenal cortex," or **ACTH**) causes release of hormones from the adrenal cortex. We will discuss the effects of thyroid and adrenal cortical hormones a little later.

The remaining two hormones of the anterior pituitary do not act on other endocrine glands. **Prolactin** stimulates the development of the mammary glands during pregnancy. After the baby is born, suckling by the infant stimulates further release of prolactin, which in turn stimulates milk production. When the infant stops nursing, prolactin secretion is turned off, and milk synthesis ceases.

Growth hormone (also called somatotropin) regulates the growth of the body. Growth hormone acts on all the body's cells, increasing protein synthesis, but it is particularly vital to bone growth. Much of the normal variation in height is due to differences in secretion of growth hormone from the anterior pituitary, as are some cases of both dwarfism (too little) and giantism (too much). Although adults normally do not continue to grow (except perhaps in girth!), growth hormone continues to be produced and secreted, helping to regulate protein, fat, and sugar metabolism.

Although these six hormones are produced in the anterior pituitary, the hypothalamus controls their rate of release. Neurosecretory cells of the hypothalamus produce at least seven peptides that regulate the release of hormones from the anterior pituitary (Fig. 28-8). These are called **releasing hormones** or **inhibiting hormones** depending on whether they stimulate or prevent release of pituitary hormones. Releasing and inhibiting hormones are synthesized in nerve cells in the hypothalamus, secreted into a capillary bed, and travel a short distance in the bloodstream down the pituitary stalk to another capillary bed surrounding the endocrine cells of the anterior pituitary. There, the releasers and inhibitors diffuse out of the bloodstream and influence hormone secretion. Since the releasing and inhibiting hormones are secreted very close to the anterior pituitary and have a very small patch of target tissue, they are only synthesized in minute amounts. Not surprisingly, they were therefore difficult to find and study—Andrew Schally and Roger Guillemin (who shared a Nobel Prize in 1977 for characterizing several of these hormones) needed the brains of millions of sheep and pigs to obtain enough releasing hormone to work with. (No, they did not sacrifice millions of animals for their experiments; brains are almost free for the taking at commercial slaughterhouses.)

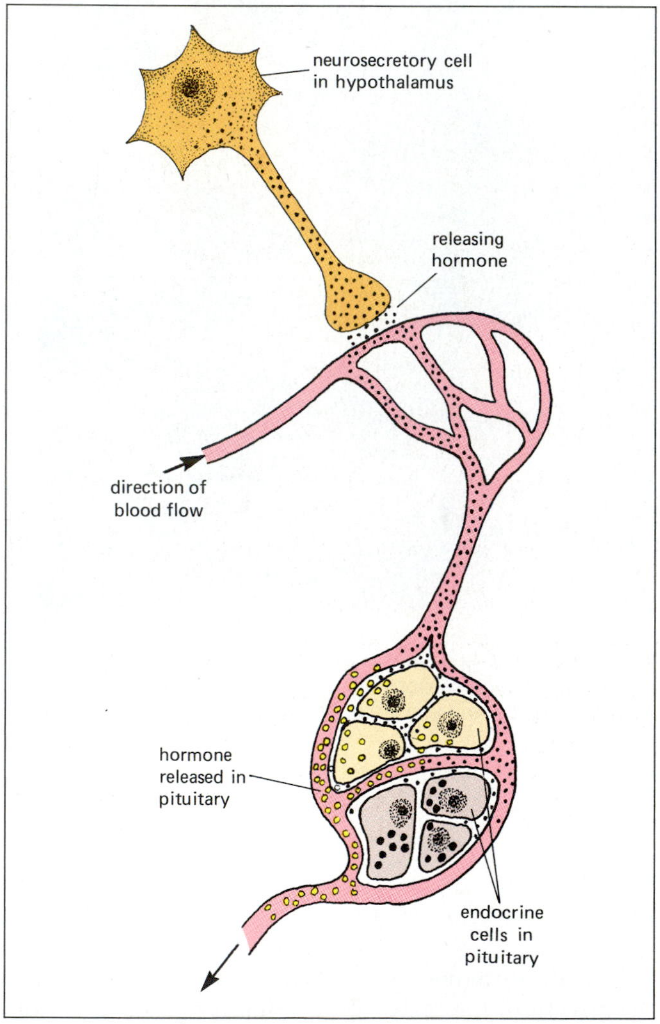

Figure 28-8 Hormone release from the anterior pituitary is under the control of releasing and inhibiting hormones from neurosecretory cells of the hypothalamus (black dots). Releasing hormones enter a capillary bed in the hypothalamus and travel downstream to capillaries in the anterior pituitary. There they contact the various endocrine cells of the pituitary. Only endocrine cells with matching cell membrane receptors respond to a given releasing hormone (see Fig. 28-3a), so each releasing hormone stimulates a particular endocrine cell type (tan) to release its hormone, while leaving other types (gray) unaffected.

Thyroid and Parathyroid

In the front of the neck, nestled around the larynx, lies the **thyroid gland** (Fig. 28-9). The four small disks of the **parathyroid gland** are embedded in the back of the thyroid.

The thyroid produces two hormones, **thyroxine** and **calcitonin.** Thyroxine is an iodine-containing, modified amino acid that raises the metabolic rate of most body cells. In juvenile mammals, thyroxine cooperates with growth hormone in regulating growth. In adults, the elevated metabolic rate seems to be involved in regulating body temperature and stress reactions. Thyroxine release is stimulated by thyroid-stimulating hormone (TSH) from the anterior pituitary, which in turn is stimulated by a releasing hormone from the hypothalamus. The amount of TSH released from the pituitary is regulated by thyroxine levels in the blood (Fig. 28-10): high concentrations of thyroxine inhibit secretion of both the releasing hormone and TSH, thus inhibiting further release of thyroxine from the thyroid. You will recognize this as an example of negative feedback.

Calcitonin, together with **parathormone,** the hormone secreted by the parathyroids, controls the concentration of calcium in the blood and other body fluids. Calcium is essential for many processes, including nerve and muscle function. Therefore, the calcium concentration in body fluids must be kept within narrow limits. Calcitonin and parathormone regulate calcium absorption and release by the bones, which serve both as a skeleton and as a bank into which calcium can be deposited or withdrawn as necessary. In response to low blood calcium, the parathyroids release parathormone, which causes release of calcium from bones. If blood calcium levels become too high, the thyroid releases calcitonin, which inhibits release of calcium from bone.

Pancreas

The pancreas is a double gland. The exocrine part synthesizes digestive enzymes that are released into the pancreatic duct and flow into the small intestine. The endo-

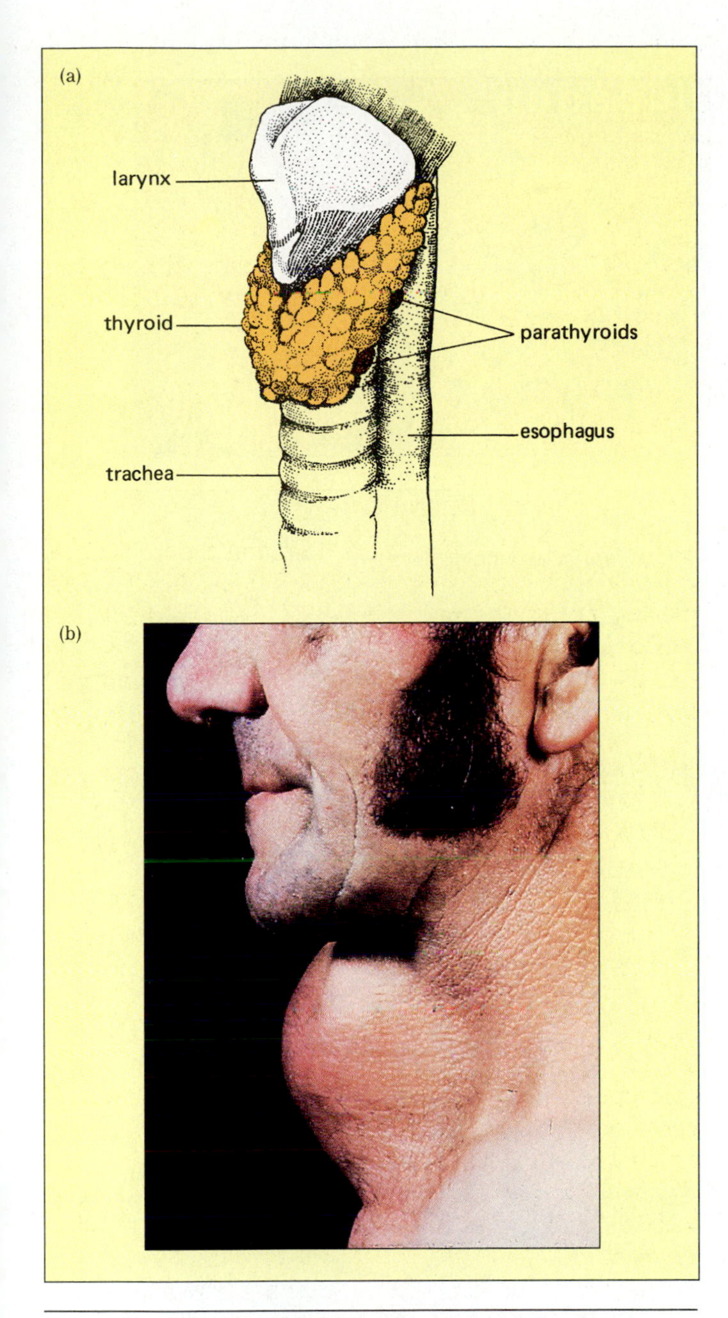

Figure 28-9 **(a)** *The thyroid and parathyroid glands are located around the front of the larynx in the neck. Thyroxine contains four iodine atoms per molecule.* **(b)** *Individuals with iodine-deficient diets may suffer from goiter, a condition in which the thyroids become greatly enlarged.*

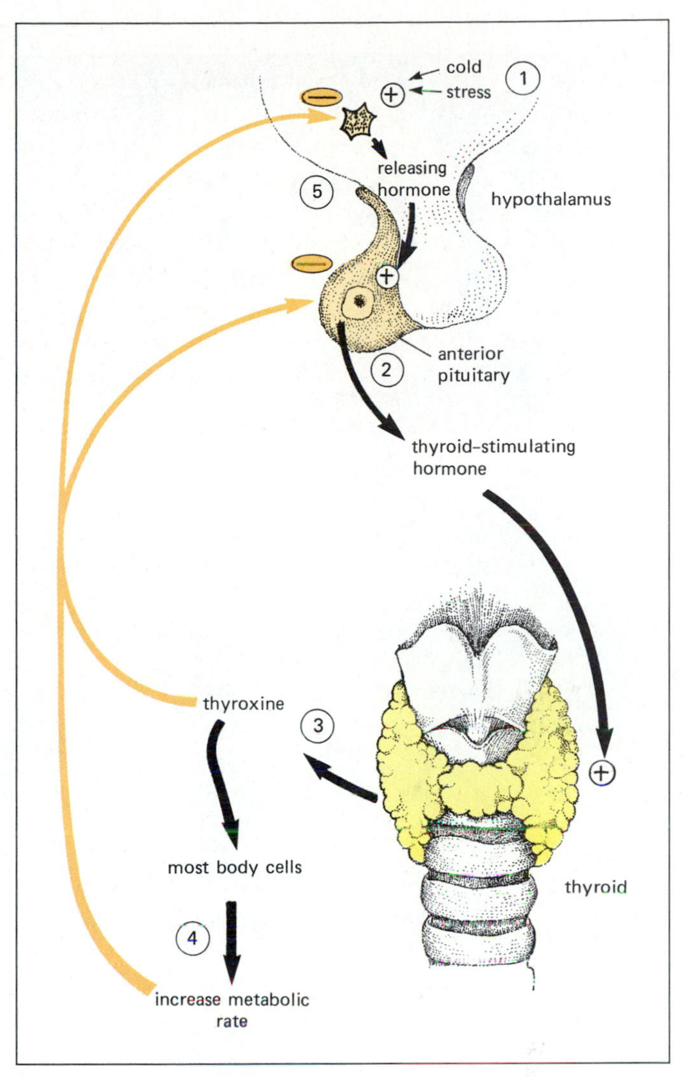

Figure 28-10 Negative feedback in thyroid function. The thyroid releases thyroxine in response to thyroid-stimulating hormone (TSH) from the anterior pituitary. TSH release is in turn controlled by releasing hormones of the hypothalamus. Low body temperature or stress stimulates neurosecretory cells of the hypothalamus (1), whose releasing hormones trigger TSH release in the pituitary (2). TSH then stimulates the thyroid to release thyroxine (3). Thyroxine causes increased metabolic activity in most cells of the body, generating ATP energy and heat (4). Both the raised body temperature and high thyroxine levels in the blood inhibit the releasing-hormone cells and the TSH-producing cells (5).

crine part consists of clusters of cells called **islets.** Each islet contains two types of cells, one type producing **insulin** and another synthesizing **glucagon.** These hormones work antagonistically to regulate carbohydrate and fat metabolism (Fig. 28-11). The islet cells are sensitive to blood glucose levels. When blood glucose rises (perhaps after eating a Twinkie), insulin is released. Insulin causes most of the cells of the body to take up glucose and either metabolize it for energy or convert it to fats or glycogen (a starch-like storage molecule). Conversely, if blood glucose falls (say while running a marathon), glucagon is released, mobilizing fat and glycogen stores. This helps to raise blood glucose in two ways. First, burning fats for energy means that glucose is not being used up. Second, breaking glycogen apart releases glucose into the blood. Insulin, then, reduces blood glucose while glucagon increases it, helping to keep the glucose concentration nearly constant.

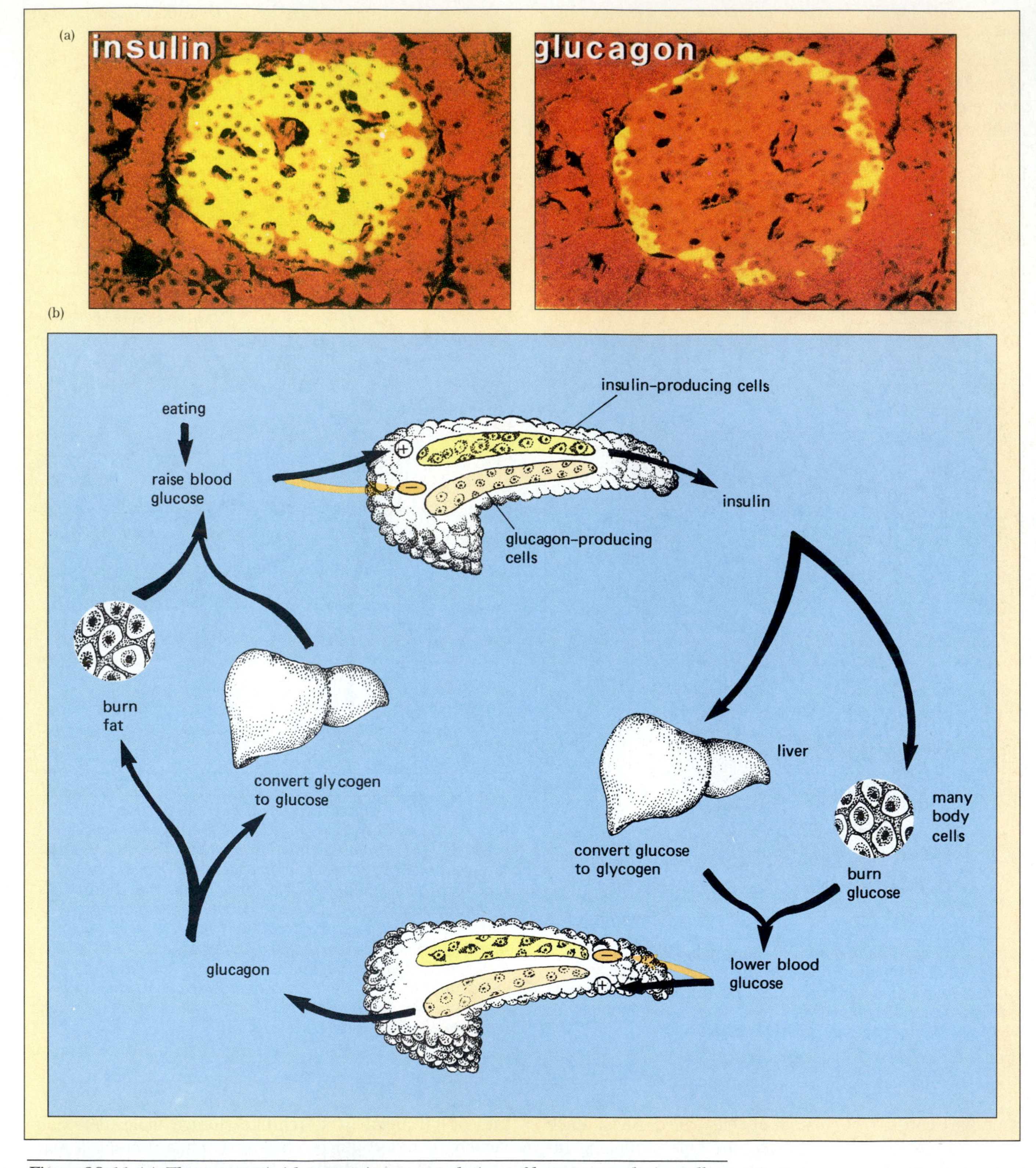

Figure 28-11 **(a)** *The pancreatic islets contain two populations of hormone-producing cells, one producing insulin, the other glucagon.* **(b)** *These two hormones cooperate in a two-part negative feedback loop to control blood glucose concentrations. High blood glucose stimulates the insulin cells (yellow) and inhibits the glucagon cells (tan), while low blood glucose stimulates the glucagon cells and inhibits the insulin cells. This dual control quickly corrects either high or low glucose levels.*

Defects in insulin production, release, or reception by target cells result in *diabetes mellitus,* in which blood glucose levels are high and fluctuate wildly with sugar intake in the diet. Diabetes also interferes with lipid metabolism. Severe diabetes causes fat deposits in the blood vessels, resulting in high blood pressure and heart disease. In fact, diabetes is one of the leading causes of heart attacks in the United States.

Adrenal Glands

Like the pituitary, the adrenals (Latin for "on the kidney") are two glands in one (Fig. 28-12). The center of the gland, the **adrenal medulla** ("medulla" means "marrow" in Latin), is actually an overgrown knot of nerve cells, and its hormone secretion is controlled directly by the nervous system. The medulla produces two hormones, **adrenalin** and **noradrenalin** (also called epinephrine and norepinephrine), in response to stress. These hormones prepare the body for action, raising the rates of heartbeat and breathing, causing blood glucose levels to rise, and directing blood flow away from the digestive tract and toward the brain and muscles. Activation of the adrenal medulla is triggered by, and has effects that are similar to, activation of the sympathetic nervous system, which is discussed in Chapter 29.

The outer **adrenal cortex** ("cortex" is Latin for "bark") is totally unlike the medulla. The cortex secretes three types of hormones, all steroids. The **glucocorticoids** are a group of hormones that help to control glucose metabolism. Glucocorticoid release is stimulated by ACTH from the anterior pituitary; ACTH release, in turn, is stimulated by hypothalamic releasing hormones that are produced in response to stress. In some respects, the glucocorticoids act similarly to glucagon, raising blood glucose concentrations by stimulating glucose synthesis and promoting use of proteins and fat for energy use. (You may be wondering why there are so many different hormones involved in glucose metabolism: thyroxine, insulin, glucagon, adrenalin, and the glucocorticoids. The reason is probably a metabolic quirk of the brain. Although most body cells can produce energy from fats and proteins as well as carbohydrates, the brain cells can only burn glucose. Blood glucose levels cannot be allowed to fall too

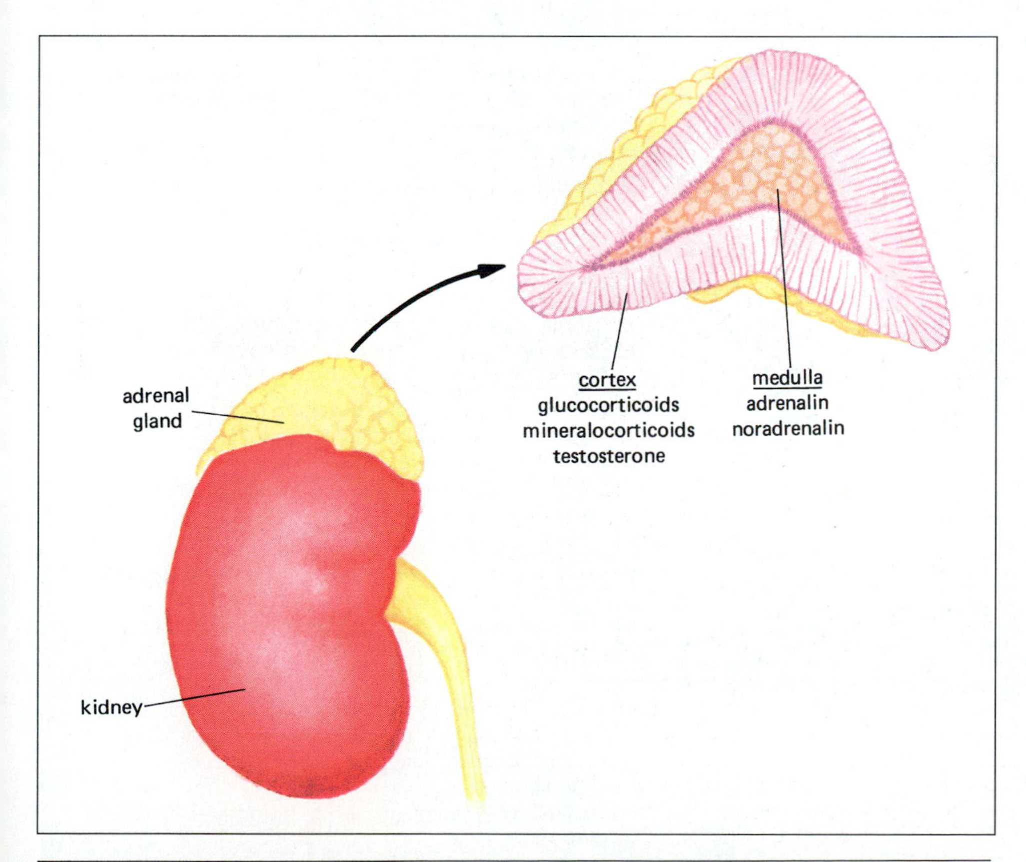

Figure 28-12 *Atop each kidney sits an adrenal gland, which is a two-part gland composed of very dissimilar cells. The outer cortex consists of ordinary endocrine cells that secrete steroid hormones. The inner medulla is derived from nervous tissue during development and secretes the typically nervous products adrenalin and noradrenalin.*

Table 28-2 **Mammalian Endocrine Glands and Hormones**

Source	Hormone	Type of Chemical	Principal Function
Major Endocrine Glands			
Hypothalamus (via posterior pituitary)	Antidiuretic hormone (ADH)	Protein	Promotes reabsorption of water in kidneys and sweat glands; constricts arterioles
	Oxytocin	Protein	In females, stimulates contraction of uterine muscles during childbirth, milk ejection, and maternal behaviors; in males, causes sperm ejection
Hypothalamus (to anterior pituitary)	Releasing and inhibiting hormones	Proteins	At least seven hormones; releasing hormones stimulate release of hormones from anterior pituitary; inhibiting hormones inhibit release of hormones from anterior pituitary
Anterior pituitary	Follicle-stimulating hormone (FSH)	Glycoprotein	In females, stimulates growth of follicle, secretion of estrogen, and perhaps ovulation; in males, stimulates spermatogenesis
	Luteinizing hormone (LH)	Glycoprotein	In females, stimulates ovulation, growth of corpus luteum, and secretion of estrogen and progesterone; in males, stimulates secretion of testosterone
	Thyroid-stimulating hormone (TSH)	Glycoprotein	Stimulates thyroid to release thyroxine
	Growth hormone (somatotropin)	Protein	Stimulates growth, protein synthesis, and fat metabolism; inhibits sugar metabolism
	Adrenocorticotropic hormone (ACTH)	Protein	Stimulates adrenal cortex to release hormones, especially glucocorticoids
	Prolactin	Protein	Stimulates milk synthesis in and secretion from mammary glands
Thyroid	Thyroxine	Modified amino acid	Increases metabolic rate of most body cells; increases body temperature; regulates growth and development
	Calcitonin	Protein	Inhibits release of calcium from bones
Parathyroid	Parathormone	Protein	Stimulates release of calcium from bone; promotes absorption of calcium by intestines; promotes reabsorption of calcium by kidneys
Adrenal medulla	Adrenalin and noradrenalin	Modified amino acids	Increase levels of sugar and fatty acids in blood; increase metabolic rate; increase rate and force of contractions of the heart; constrict some blood vessels
Adrenal cortex	Glucocorticoids	Steroid	Increase blood sugar; regulate sugar, lipid, and fat metabolism; anti-inflammatory effects
	Aldosterone	Steroid	Increases reabsorption of salt in kidney
	Testosterone	Steroid	Causes masculinization of body features, growth
Pancreas	Insulin	Protein	Decreases blood glucose levels by increasing uptake of glucose into cells and converting glucose to glycogen, especially in liver; regulates fat metabolism
	Glucagon	Protein	Converts glycogen to glucose, thereby raising blood glucose levels
Ovaries[a]	Estrogen	Steroid	Causes development of female secondary sexual characteristics and maturation of eggs; promotes growth of uterine lining; has general effects on metabolism
	Progesterone	Steroid	Stimulates development of uterine lining and formation of placenta
Testes[a]	Testosterone	Steroid	Stimulates development of genitalia and male secondary sexual characteristics; stimulates spermatogenesis and growth; has general effects on metabolism

Table 28-2 **Mammalian Endocrine Glands and Hormones**

Source	Hormone	Type of Chemical	Principal Function
Other Sources of Hormones			
Digestive tract[b]	Secretin, gastrin, cholecystokinin, and others	Proteins	Control secretion of mucus, enzymes, and salts in digestive tract; regulate peristalsis
Thymus[c]	Thymosin	Protein	Stimulates maturation of cells of immune system
Pineal	Melatonin	Modified amino acid	Regulates biological clock; may regulate onset of puberty
Kidney	Renin	Protein	Acts on blood proteins to produce hormone (angiotensin) that regulates blood pressure
	Erythropoietin	Protein	Activates a blood protein to stimulate red blood cell synthesis in bone marrow
Heart	Atrial natriuretic hormone	Protein	Increases salt and water excretion by kidney; lowers blood pressure

[a]See Chapter 31. [b]See Chapter 24. [c]See Chapter 27.

low, or brain cells malfunction, bringing on unconsciousness and, all too soon, death.)

The second hormone from the adrenal cortex is **aldosterone.** Aldosterone secretion is regulated by sodium levels in the blood. If blood sodium falls, the cortex releases aldosterone, which causes the kidneys and sweat glands to retain sodium. Even small amounts of salt in the diet, combined with aldosterone-induced salt conservation, raise blood sodium levels again and shut off further aldosterone secretion.

Finally, the adrenal cortex also produces the male sex hormone **testosterone,** although normally in much smaller amounts than the testes produce. Tumors of the adrenal medulla sometimes lead to excessive testosterone release, causing masculinization of women, a malady once cruelly exploited by circus operators displaying "bearded ladies" in their sideshows.

Prostaglandins

In 1930, gynecologists Raphael Kurzok and Charles Lieb discovered that semen causes uterine muscles to contract. A few years later, the Swedish physiologist Ulf von Euler isolated some of the active compounds. Thinking that they were secreted by the prostate gland, he named them prostaglandins. Later research showed that most of the prostaglandins in semen come from the seminal vesicles, not the prostate, but by then the name had stuck.

Prostaglandins are not synthesized only by the male reproductive tract. In fact, prostaglandins are produced by many, perhaps all, cells of animal bodies, both male and female. There are at least 16 distinct types of prostaglandin molecules. Although all are synthesized from a single type of fatty acid, called arachidonic acid, small differences in chemical structure allow the various types of prostaglandins to exert a wide variety of effects. Most prostaglandins cause effects in the cell that produced them or in nearby cells.

As examples of the diversity of prostaglandins, let's consider pain and childbirth. The ancient Greeks knew that salicylic acid (an extract of the bark of the willow, *Salix*) relieves pain. Today we use a derivative, acetylsalicylic acid (aspirin), as a pain reliever. How does aspirin relieve pain? Some years ago, John Vane discovered that aspirin works by inhibiting the enzymes that synthesize prostaglandins. Since some prostaglandins cause pain, probably by making pain receptors more sensitive to painful stimuli, inhibiting prostaglandin synthesis helps to relieve pain.

Some prostaglandins cause smooth muscles to contract. During childbirth, prostaglandins stimulate contraction of the uterine muscles (see Chapter 31). Oxytocin, also released at this time, causes uterine muscles to contract too, and together they squeeze the baby out into the world. During menstruation, a similar series of events occurs, but usually on a smaller scale. Menstrual cramps are caused by excessive contraction of the muscles, which squeezes off the blood supply to the uterus, depriving it of oxygen and resulting in pain. Aspirin can alleviate menstrual cramps to some extent by inhibiting prostaglandin synthesis. For severe cramps, the rather mild inhibitory action of aspirin is insufficient. Other pain relievers, such as ibuprofen, reduce cramping by blocking prostaglandin production more thoroughly.

Hormonal Control of the Insect Life Cycle

Just as hormones control many aspects of vertebrate life cycles, including reproduction, growth, and the development of adult sexual characteristics, they are also important in the life cycles of invertebrates. The most thoroughly studied example of hormonal regulation in invertebrates is the control of molting and maturation in insects.

Many insects, including flies, beetles, moths, and butterflies, have complex life cycles (Fig. E28-4). Females lay eggs that hatch out into wormlike larvae, variously called caterpillars, grubs, or maggots. The larvae grow and eventually form pupae, which, although they look dormant, seethe with activity as the larval body is rearranged and transformed into an adult. This complicated life cycle incurs three major challenges. First, all insects, even caterpillars, are enclosed in more-or-less rigid exoskeletons. Although the exoskeletons of larvae are somewhat expandable, eventually a larva must shed its old exoskeleton, or *molt,* and develop a new, larger one, thereby permitting continued growth. Second, there is a major rearrangement of body parts during the pupal stage. Finally, the adult must break through the pupal shell to emerge into the world. Each of these events is induced by a different hormone, as we can see by following the life cycle of a silk moth, beginning with a young caterpillar (Fig. E28-5).

Larval Molting

The caterpillar brain determines when molting is necessary, using several cues. One stimulus for larval molting is abdominal stretching: to a caterpillar, having its "clothing" get tight doesn't mean that it's time to quit eating, it means that it's time to molt. The brain secretes a protein hormone often called simply **brain hormone.** This hormone circulates in the blood to the prothoracic glands, which, in response, release **molting hormone** (ecdysone). Molting hormone is a steroid that triggers a series of behavioral and biochemical activities: the larva splits the old exoskeleton apart, wriggles out, and produces a new one. As with vertebrate steroid hormones, many of these effects are mediated by the actions of molting hormone on DNA, promoting transcription of selected genes (see Fig. 28-3b).

Figure E28-4 Many insects, such as this convergent ladybug, undergo complete metamorphosis. The egg **(a)** *hatches into a larva* **(b)**. *The larva molts several times, growing at each molt, and finally forms a pupa* **(c)**. *During pupation, the larval body is reorganized into the adult ladybug* **(d)**.

Hormonal Control of the Insect Life Cycle

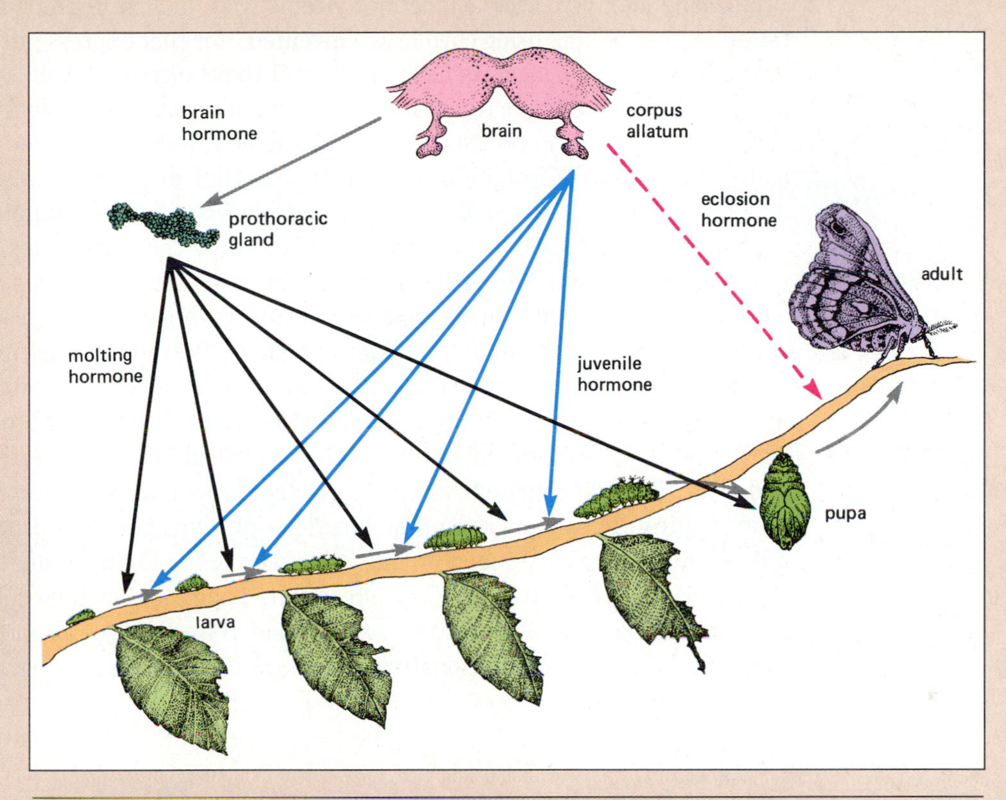

Figure E28-5 Hormonal control of the silk moth life cycle. At each molt, the brain stimulates release of molting hormone (black arrows) by the prothoracic glands. During the caterpillar stage, the brain also stimulates secretion of juvenile hormone (blue arrows), which causes the caterpillar to molt into another caterpillar. Eventually, the brain inhibits further juvenile hormone release, and at the next molt the caterpillar forms a pupa. When the adult moth has completed development inside the pupa, its brain releases eclosion hormone (red arrow), causing emergence.

Pupation

When it molts, a larval silk moth either becomes a new, larger caterpillar or develops into a pupa. Which of these events occurs depends on its level of **juvenile hormone.** A pair of glands, the corpora allata, attached to the brain and under its direct control, synthesize juvenile hormone. Early in the caterpillar stage, these glands release lots of juvenile hormone, so at each molt a new, larger caterpillar is formed. When the caterpillar has matured, however, the brain inhibits further juvenile hormone release. Apparently, juvenile hormone prevents formation of the pupa; when it is no longer secreted, the caterpillar pupates at the next molt.

Emergence

During pupation, the insect undergoes incredible reorganization of its body, digesting away some parts while forming entirely new structures, such as wings. When the rebuilding program has been completed, the silk moth emerges from the pupa. The moth goes through a ritual series of twists and turns, splitting open the pupal case and squirming out. The moth's nervous system is genetically programmed with all the right moves, and this behavioral program is turned on by another peptide hormone from the brain, **eclosion hormone.** The brain secretes eclosion hormone as soon as development is complete, subject to one constraint: silk moths always emerge in the late afternoon.

As you can see, the principles underlying hormonal control in insects and humans are really very similar: most hormones are peptides or steroids; nerve cells in the brain produce a number of hormones, including some that control the release of other, nonnervous hormones; major changes in life, such as maturation and reproduction, are under hormonal control. Although sometimes hidden beneath tremendous diversity, the unity of life on Earth is overwhelming.

REFLECTIONS ON THE EVOLUTION OF HORMONES

Not long ago, vertebrate endocrine systems were looked upon as a special adaptation of our phylum, using chemicals "invented" for that express purpose. In recent years, however, physiologists have discovered that hormones, even our "very own" hormones, are evolutionarily ancient. Insulin, for example, is found not only in vertebrates but also in protists, fungi, and bacteria, although no one has an inkling as to the function of insulin in most of these organisms. Protists also manufacture ACTH, although of course they have no adrenal glands to stimulate. Yeasts have receptors for estrogen, but no ovaries. Even among the vertebrates, identical hormone molecules, secreted by the same glands, may produce dramatically different effects. Let's look briefly at the diversity of thyroid hormone effects.

Some fish undergo radical physiological changes during their lifetimes. A salmon, for example, begins life in fresh water, migrates to the ocean, and finally returns to fresh water to spawn. In the stream where it hatched, fresh water tends to enter the fish's tissues by osmosis; in salt water, the fish tends to lose water, becoming dehydrated. The fish's migrations, therefore, require complete revamping of salt and water control. In salmon, one of the functions of thyroxine is to produce the metabolic changes necessary to go from life in streams to life in the ocean. In amphibians, thyroxine has an even more dramatic effect, triggering metamorphosis from an aquatic tadpole into a terrestrial frog or toad (Fig. 28-13). In most vertebrates, thyroxine also regulates seasonal molting. From salamanders to snakes, and from

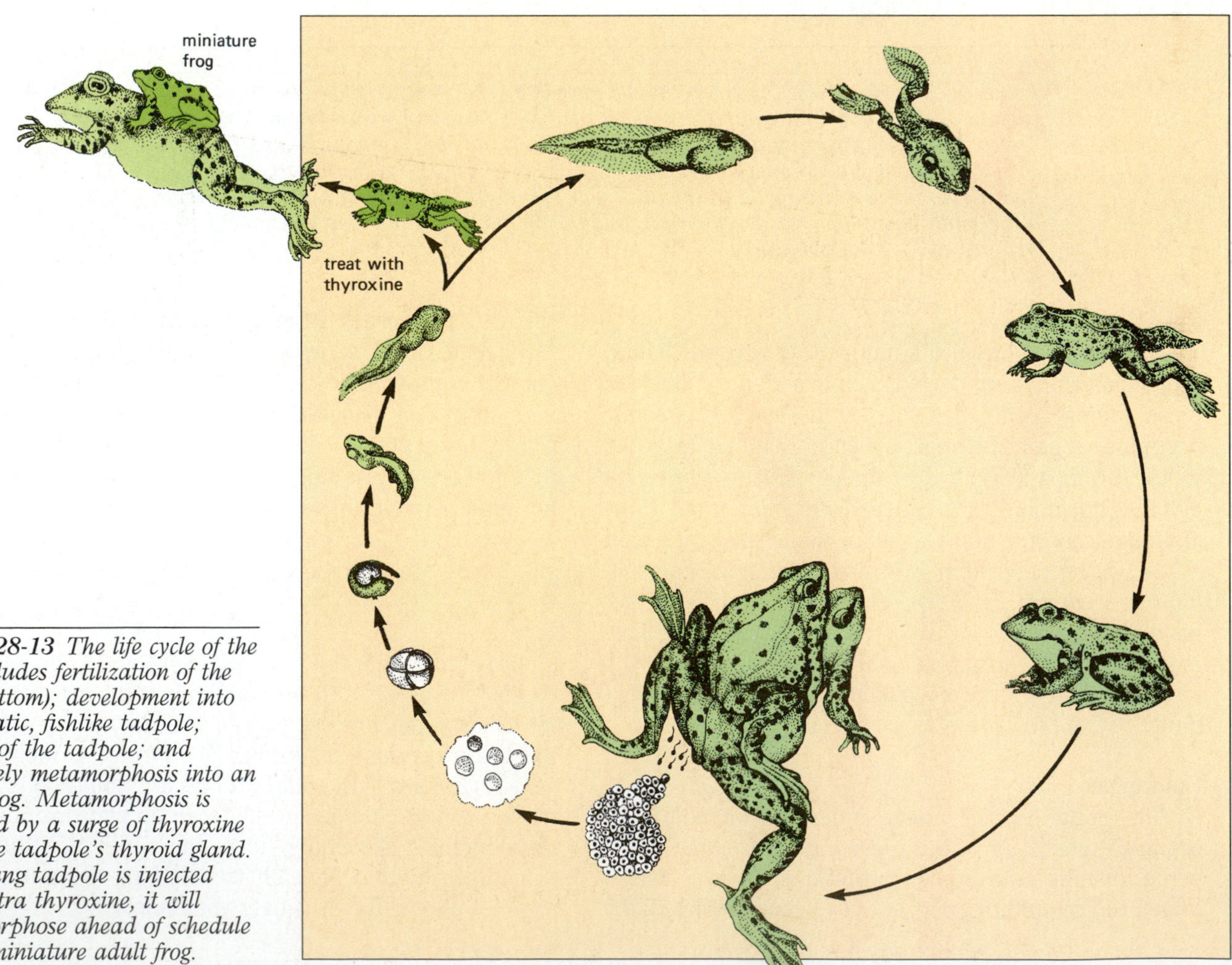

Figure 28-13 The life cycle of the frog includes fertilization of the eggs (bottom); development into an aquatic, fishlike tadpole; growth of the tadpole; and ultimately metamorphosis into an adult frog. Metamorphosis is triggered by a surge of thyroxine from the tadpole's thyroid gland. If a young tadpole is injected with extra thyroxine, it will metamorphose ahead of schedule into a miniature adult frog.

birds to elephant seals, surges of thyroxin stimulate shedding of the skin and/or its associated hair or feathers. Humans, of course, do not undergo metamorphoses and moltings. In people, thyroxine has been stripped of these exotic functions and reduced to the seemingly mundane business of regulating growth and metabolism.

The use of chemicals to regulate cellular activity is extremely ancient. Life on Earth has been built upon a conservative foundation: a relative handful of chemicals coordinate activities within single cells and among groups of cells. Life's diversity originated in part by changing the systems used to deliver the chemicals, and by experimenting with new types of responses. Early in their evolution, animals developed organs for fast, precise delivery of chemicals: the nervous system. As we shall explore in Chapter 29, the development of the nervous system permitted rapid responses to environmental stimuli, flexibility in response options, and ultimately consciousness itself.

SUMMARY OF KEY CONCEPTS

Chemical Communication Within the Animal Body

In multicellular organisms, cellular activity is often coordinated by chemicals that are released by one type of cell and cause effects in other types of cells. In animals, two main types of intercellular communication are found: hormonal and neural.

Hormone Function in Animals

A hormone is a chemical secreted by cells in one part of the body that is transported in the bloodstream to other parts of the body, where it affects the activity of specific target cells.

The vertebrate body has two types of glands, exocrine and endocrine. Exocrine glands release their secretions into ducts leading outside the body or into the digestive tract. Endocrine glands are clusters of hormone-producing cells embedded within a network of capillaries. Hormones are secreted into the extracellular fluid and diffuse into the capillaries. Four types of molecules are known to act as hormones: modified amino acids, proteins, steroids, and prostaglandins.

Most hormones act on their target cells in one of two ways. Modified amino acids and proteins bind to receptors on the surfaces of target cells and cause changes in the concentration of intracellular second messengers such as cyclic AMP. The second messengers then cause changes in the metabolism of the cell. Steroid hormones diffuse through the cell membranes of the target cells, and bind with receptor proteins in the cytoplasm. The hormone/receptor complex travels to the nucleus and promotes transcription of specific genes.

Hormone action is often regulated through negative feedback, a process in which the hormone secreted by a gland causes changes that inhibit further secretion of that hormone.

Mammalian Endocrine Systems

The major endocrine glands of the human body are the hypothalamus/pituitary complex, thyroid, parathyroid, pancreas, adrenal cortex, adrenal medulla, and gonads. The hormones released by these glands and their actions are summarized in Table 28-2. Unlike other hormones, prostaglandins are not secreted by discrete glands, but are synthesized and released by many cells of the body.

GLOSSARY

(*Note:* The major mammalian hormones, their sources, chemical nature, and functions are summarized in Table 28-2.)

Adrenal gland an endocrine gland consisting of an outer cortex and inner medulla. The cortex secretes steroid hormones that regulate metabolism and salt balance. The medulla secretes adrenalin and noradrenalin.

Cyclic AMP a cyclic nucleotide formed within many target cells as a result of the reception of modified amino acid or protein hormones, and which causes metabolic changes in the cell; often called a second messenger.

Endocrine gland a ductless, hormone-producing gland that releases its secretions into the extracellular fluid within the body, from which the secretions diffuse into nearby capillaries.

Exocrine gland a gland that releases its secretions into ducts that lead to the outside of the body or into the digestive tract.

Homeostasis (hō-mē-ō-stā′-sis) the process of maintaining a relatively constant internal environment in the face of variations in the external environment.

Hormone a chemical synthesized by one group of cells and carried in the bloodstream to other cells, whose activity is influenced by reception of the hormone.

Hypothalamus (hī-pō-thal′-a-mus) a region of the brain that controls the secretory activity of the pituitary gland, and also synthesizes oxytocin and antidiuretic hormone.

Inhibiting hormone a hormone secreted by the hypothalamus that inhibits the release of specific hormones from the anterior pituitary gland.

Negative feedback a type of control mechanism in which the output of a system causes actions that suppress further output.

Neurohormone a chemical synthesized by a specialized nerve cell (called a neurosecretory cell) and secreted into the bloodstream as a hormone.

Neurosecretory cell a specialized nerve cell that synthesizes and releases hormones.

Neurotransmitter a chemical released by a nerve cell close to a second nerve cell, a muscle, or a gland cell, and that influences the activity of the second cell.

Pancreas (pan′-krē-as) a combined exocrine and endocrine gland located in the abdominal cavity. Its endocrine parts secrete the hormones insulin and glucagon, which regulate glucose concentrations in the blood.

Parathyroid a set of four small endocrine glands embedded in the surface of the thyroid gland that produce parathormone, which (with calcitonin from the thyroid) regulates calcium ion concentration in the blood.

Pituitary an endocrine gland located at the base of the brain that produces several hormones, many of which influence the activity of other glands.

Prostaglandin (pro-sta-glan′-din) a family of modified fatty acid hormones manufactured by many cells of the body.

Releasing hormone a hormone secreted by the hypothalamus that causes the release of specific hormones by the anterior pituitary gland.

Second messenger a term applied to intracellular chemicals, such as cyclic AMP, that are synthesized or released within a cell in response to the binding of a hormone or neurotransmitter (the first messenger) to receptors on the cell surface. Second messengers bring about specific changes in the metabolism of the cell.

Target cell a cell upon which a particular hormone exerts its effect.

Thyroid an endocrine gland, located in front of the larynx in the neck, that secretes the hormones thyroxine (affecting metabolic rate) and calcitonin (regulating calcium ion concentration in the blood).

STUDY QUESTIONS

1. What are the four types of molecules used as hormones in vertebrates? Give an example of each.
2. What is the difference between an endocrine and an exocrine gland? Which ones release hormones?
3. Describe the mechanisms of action of peptide and steroid hormones. Since hormones bathe many or even all of the cells of the body, why do only specific target cells respond to the hormone?
4. Diagram the process of negative feedback, and give an example of negative feedback in the control of hormone action.
5. What are the major endocrine glands in the human body, and where are they located?
6. Describe the structure of the hypothalamus/pituitary complex. Which pituitary hormones are neurosecretory? What are their functions?
7. Describe how releasing hormones regulate the secretion of hormones by cells of the anterior pituitary. Name the hormones of the anterior pituitary and give one function of each.
8. Describe how the hormones of the pancreas act together to regulate the concentration of glucose in the blood.
9. What are the two parts of the adrenal gland? What hormones does each release?

SUGGESTED READINGS

Berridge, M. J. "The Molecular Basis of Communication Within the Cell." *Scientific American,* October 1985. Both the "classical" cyclic AMP and more recently discovered second messengers convey information from cell surface receptors to DNA and cellular metabolism.

Guillemin, R., and Burgus, R. "The Hormones of the Hypothalamus." *Scientific American,* November 1972. The interaction between hypothalamus and pituitary is explored.

Hole, J. W., Jr. *Essentials of Human Anatomy and Physiology,* 2nd ed. Dubuque, Iowa; Wm. C. Brown Company, 1986. A beautifully illustrated textbook. Besides basic information, Hole also describes the clinical result of malfunctions of each gland.

Pike, J. E., "Prostaglandins." *Scientific American,* November 1971 (Offprint No. 1235). Although somewhat out of date, this article provides a good introduction to this large and poorly understood class of chemicals.

Snyder, S. H. "The Molecular Basis of Communication Between Cells." *Scientific American,* October 1985. Snyder describes the similarities and differences between neural and hormonal control systems in the body.

Vander, A. J., and Sherman, J. H. *Human Physiology,* 4th ed. New York: McGraw-Hill Book Company, 1985. Hormonal actions are considered throughout this text, with clear explanations of how hormones help to regulate virtually every body function.

29
Information Processing: The Nervous System

As you read this, light reflected from this page bombards your eyes and sound waves from the stereo assault your ears. Your senses of smell and taste are stimulated as you drink coffee, touch-sensitive receptors all over your body are activated as your clothes, the chair, and the pages of this book all push at your skin. Meanwhile, your hand dextrously manipulates a pen as you take notes. You breathe, swallow, and shift positions in the chair. Untaxed by these ordinary but marvelous activities, your mind calls upon its reading skills, stores memories of what you read, and still has the capacity to appreciate the music from the stereo.

All this is controlled by your brain, an organ weighing about 3 pounds encased within your skull. Composed of billions of nerve cells, the brain is connected to the rest of your body by millions of slender cables, the nerves. In this chapter we explore how nerve cells create electrical signals, how they use these stereotyped signals to convey information and communicate with one another, and how they activate muscles. In Chapter 30 we will examine the senses through which the brain experiences the outside world, and the muscles that carry out its commands.

"Know thyself."—inscribed above the entrance to the temple of Apollo, home of the Oracle at Delphi. (Statue is "The Thinker" by Rodin.)

NERVE CELL STRUCTURE

Our study of the nervous system begins with the individual nerve cell, or **neuron.** As the fundamental unit of the nervous system, each neuron must perform five functions. It must (1) *receive information* from the internal or external environment or from other neurons; (2) *integrate the information it receives and produce an appropriate output signal;* (3) *conduct the signal* to its output terminal; (4) *transmit the signal* to other nerve cells, glands, or muscles; and (5) *coordinate its metabolic activities,* maintaining the integrity of the cell. Although nerve cells vary enormously in structure, a "typical" vertebrate neuron has four distinct structural regions that carry out these functions (Fig. 29-1).

Information from the outside world or other neurons is

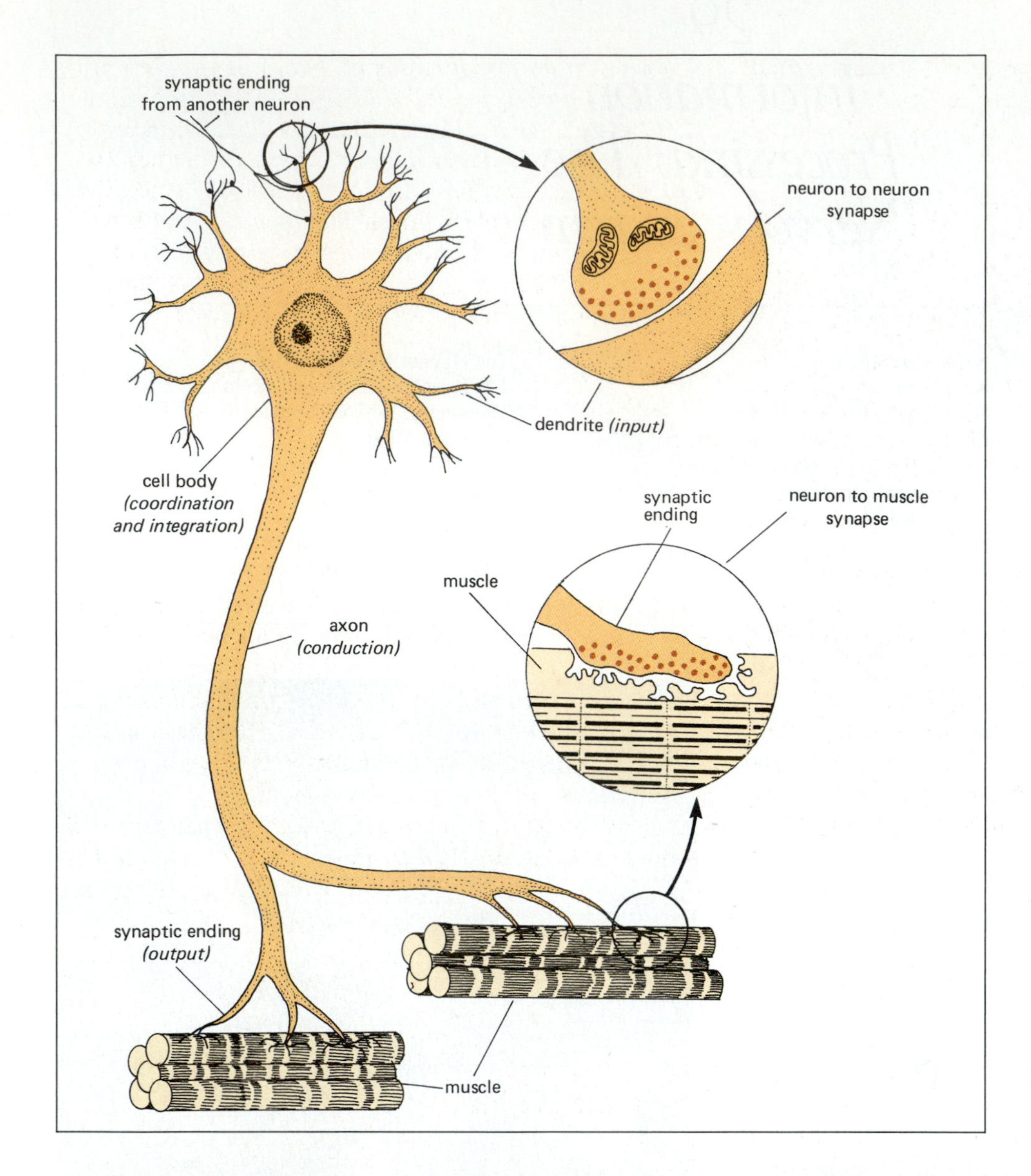

Figure 29-1 *A nerve cell showing its specialized parts and their functions. The neuron to muscle synapse is described in Chapter 30.*

received by the **dendrites,** a tangle of fibers that branch from the cell body. Whether the information is heat, light, pressure, or chemical stimuli, dendrites of neurons specialized to respond to these stimuli *convert them into electrical signals.*

Signals from the dendrites travel to the **cell body** of the neuron, which serves as an integration center. It adds up the various signals from the dendrites and "decides" whether to produce an **action potential,** the electrical output signal of the neuron. The cell body, containing the usual assortment of organelles, also synthesizes proteins, lipids, and carbohydrates, and coordinates the metabolic activities of the cell.

One or more long, thin fibers, called **axons,** extend outward from the cell body. Axons are distribution lines, carrying action potentials from the cell body to the output terminals of the neuron. Like the strands of wire in an electric cable, axons are usually bundled together into **nerves.** In vertebrates, nerves emerge from the brain and spinal cord, extending out to all regions of the body. However, unlike electric power distribution cables, in which energy is lost along the way from power station to customer, the cell membranes of axons are specialized to conduct action potentials *undiminished in size* from the cell body to the synaptic endings.

Synaptic terminals are sites at which signals are transmitted. The synaptic terminals of one neuron may communicate with a gland, a muscle, or the dendrites of a second neuron, so that the output of the first cell becomes the input to the second.

NERVE CELL FUNCTION

About 40 years ago, using the giant axon of a mollusc, the squid, biologists developed ways to record inside individual neurons (Fig. 29-2). They found that unstimulated, inactive neurons maintain a constant electrical potential across their cell membranes, similar to that found across the poles of a battery. This **resting potential** is always negative inside the cell and ranges from −40 to −90 millivolts (thousandths of a volt). If the neuron is stimulated, either naturally or with an electric current, this potential can be altered. Depending on the nature of the stimulus, it can be made either more or less negative. If the resting potential is made sufficiently less negative, it reaches a level called **threshold.** Upon reaching threshold, the neuron's potential suddenly reverses, becoming 20 to 50 millivolts *positive inside.* This is the **action potential.** Action potentials last a few milliseconds (thousandths of a second) before the cell reverts back to its negative resting potential. Let us look more closely at these electrical signals, the language of the nervous system.

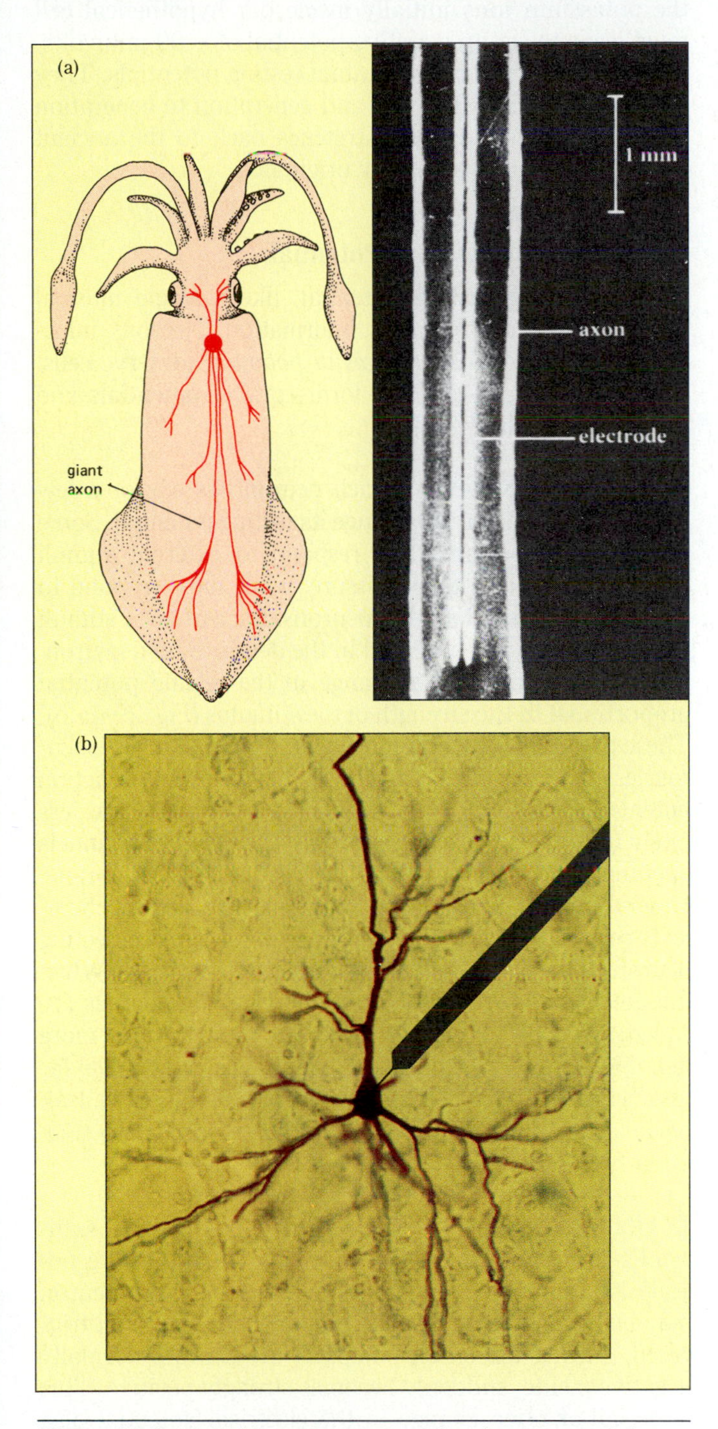

Figure 29-2 **(a)** *Taking advantage of the giant axon of the squid, the British physiologists Bernard Katz, Alan Hodgkin, and Andrew Huxley pushed thin wires or narrow saline-filled tubes down the inside of the axon. These electrodes were connected to voltmeters to record the electrical potential difference between the inside and outside of the axon.* **(b)** *Modern neurobiologists use hollow glass electrodes drawn to a needlelike tip less than 1 micrometer in diameter and filled with a salt solution. The sharp tip penetrates the neuron without damaging it.*

Origin of the Resting Potential

How can a cell behave as a battery, separating electrical charge across its cell membrane? To understand this, recall two physical principles, *diffusion* and *electrical attraction,* and one property of cell membranes, *differential permeability,* all discussed in Chapter 4. These factors interact with concentration differences inside and outside the cell to produce the resting potential.

The cell membrane of a neuron encloses cytoplasm with various ions dissolved in it. The neuron itself is immersed in a salt solution, the extracellular fluid (Fig. 29-3). The

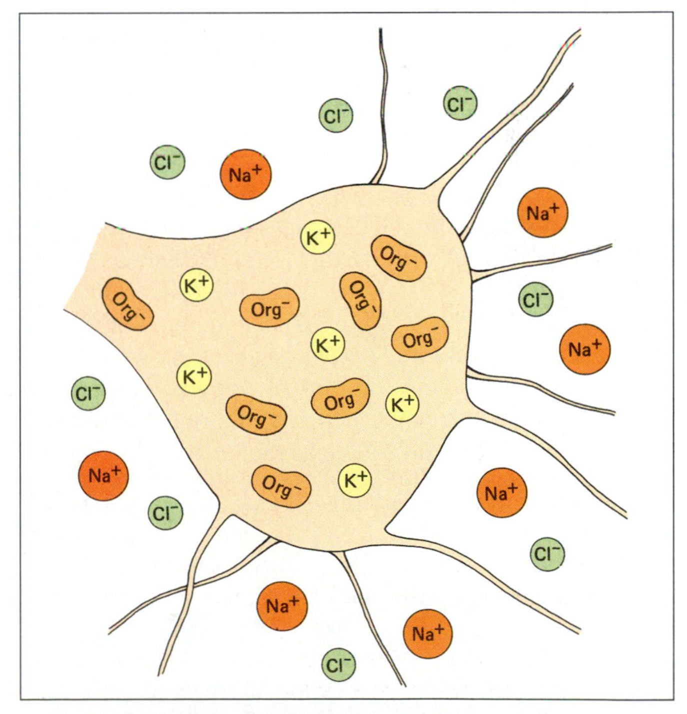

Figure 29-3 The ionic composition of the neuron cytoplasm is significantly different from the extracellular fluid. The nerve cell contains a high concentration of potassium ions (K^+) and large organic anions (Org^-), while the extracelluar fluid is high in sodium chloride (Na^+ and Cl^-).

ions of the cytoplasm consist mainly of positively charged potassium ions (K^+) and large, negatively charged organic molecules, such as proteins and the molecules of the citric acid cycle (Chapter 7). Outside the cell, the extracellular fluid contains mostly positively charged sodium ions (Na^+) and negatively charged chloride ions (Cl^-). In a moment we will see how these concentration differences are maintained.

As you have learned, a cell membrane is a lipid ocean in which protein icebergs are embedded. Since charged particles cannot pass through the lipids, they must travel through tunnel-shaped proteins called **channels** extending through the membrane. In an unstimulated neuron, shown below, only potassium ions can cross the membrane. They travel through specific proteins called **potassium channels,** shown in yellow. Although **sodium channels** (shown in blue) are also present, in unstimulated neurons they remain closed. Since only potassium ions can cross the membrane, and potassium ions are most concentrated inside the cell, potassium ions will diffuse out of the cell, leaving the large, negatively charged organic ions behind.

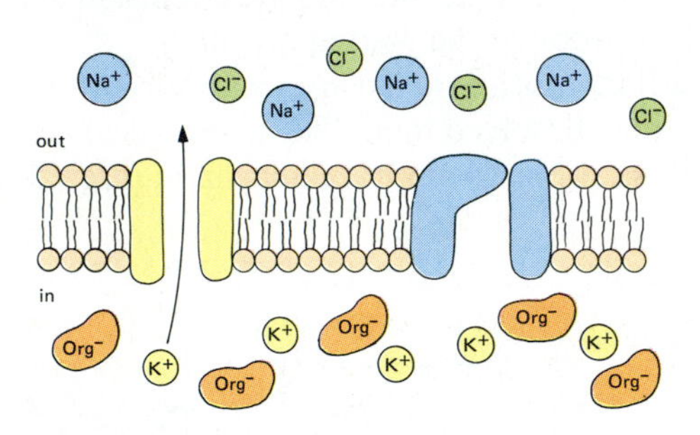

As more and more positively charged potassium ions leave, the inside of the cell becomes increasingly negative. But since unlike charges attract one another, as potassium ions diffuse out, an electrical force develops that tends to pull them back inside. *At some point the diffusion of potassium ions out due to concentration differences will be balanced by the electrical attraction tending to pull them back in.* At this point there is no more net movement of potassium ions, and the cell reaches a stable resting potential, negative inside.

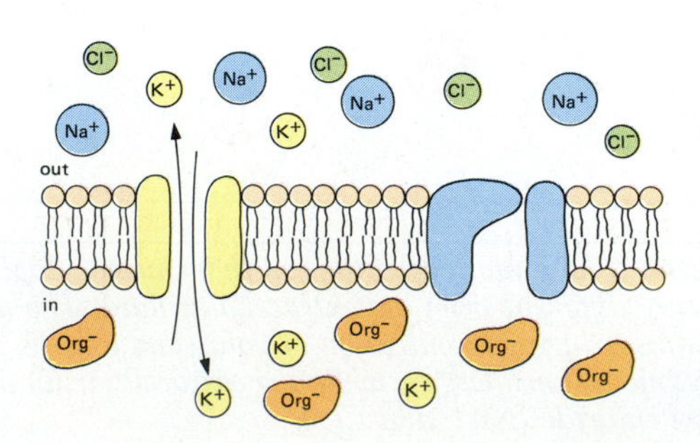

Establishing a resting potential in this way does not require significant changes in the potassium concentration inside and outside the cell. Only about 1/10,000 of the potassium ions initially inside our hypothetical cell must leave to set up a resting potential of −60 millivolts.

Nearly all living cells maintain resting potentials. They are passed from cell to cell and generation to generation in an unbroken line that stretches back to the ancient ancestral cells in which they originated.

Changes in the Resting Potential

An unchanging resting potential, like a single musical note, cannot convey much information. Nervous information is encoded in *changes in potential* in nerve cells. These changes can take two forms, **graded potentials** and **action potentials.**

GRADED POTENTIALS. Each neuron has one or a few types of stimuli that can change its resting potential. Some neurons are specialized to respond to external stimuli such as light, sound, or pressure, while most respond to chemicals released by other neurons. Each of these stimuli can cause a graded potential in the dendrites of a neuron. A graded potential is a change in the resting potential proportional to the strength of the stimulus (Fig. 29-4a,b). These potentials can either be excitatory, making the neuron less negative inside and more likely to fire an action potential, or inhibitory, making it more negative and less likely to fire. Graded potentials are the result of channels opening, allowing ions to flow across the membrane. Graded potentials are most commonly initiated by chemicals released from the synaptic terminal of a connecting neuron. These are called **synaptic potentials.** (When initiated by environmental stimuli, graded potentials are called receptor potentials; these are described in more detail in Chapter 30.) Graded potentials cannot travel far in a neuron; after a few millimeters at most, the ions leak back across the membrane and the signal is lost (Fig. 29-4c).

ACTION POTENTIALS. If a graded potential is sufficiently large and positive, or if many small positive potentials impinge on a neuron simultaneously, the neuron may be brought to threshold and an action potential triggered. During an action potential, the neuron, normally negative inside, suddenly becomes strongly positive. This large, all-or-none change in the electrical potential of a neuron does not diminish with distance. The events during an action potential as recorded by an electrode inside the neuron are shown in Figure 29.5.

A neuron at rest, diagrammed below, is something like a loaded musket, charged and ready to fire if the trigger is pulled. In a neuron, the "charge" is the concentration gradient of sodium ions, which are found mostly outside the cell. The "trigger" is a set of membrane proteins, the

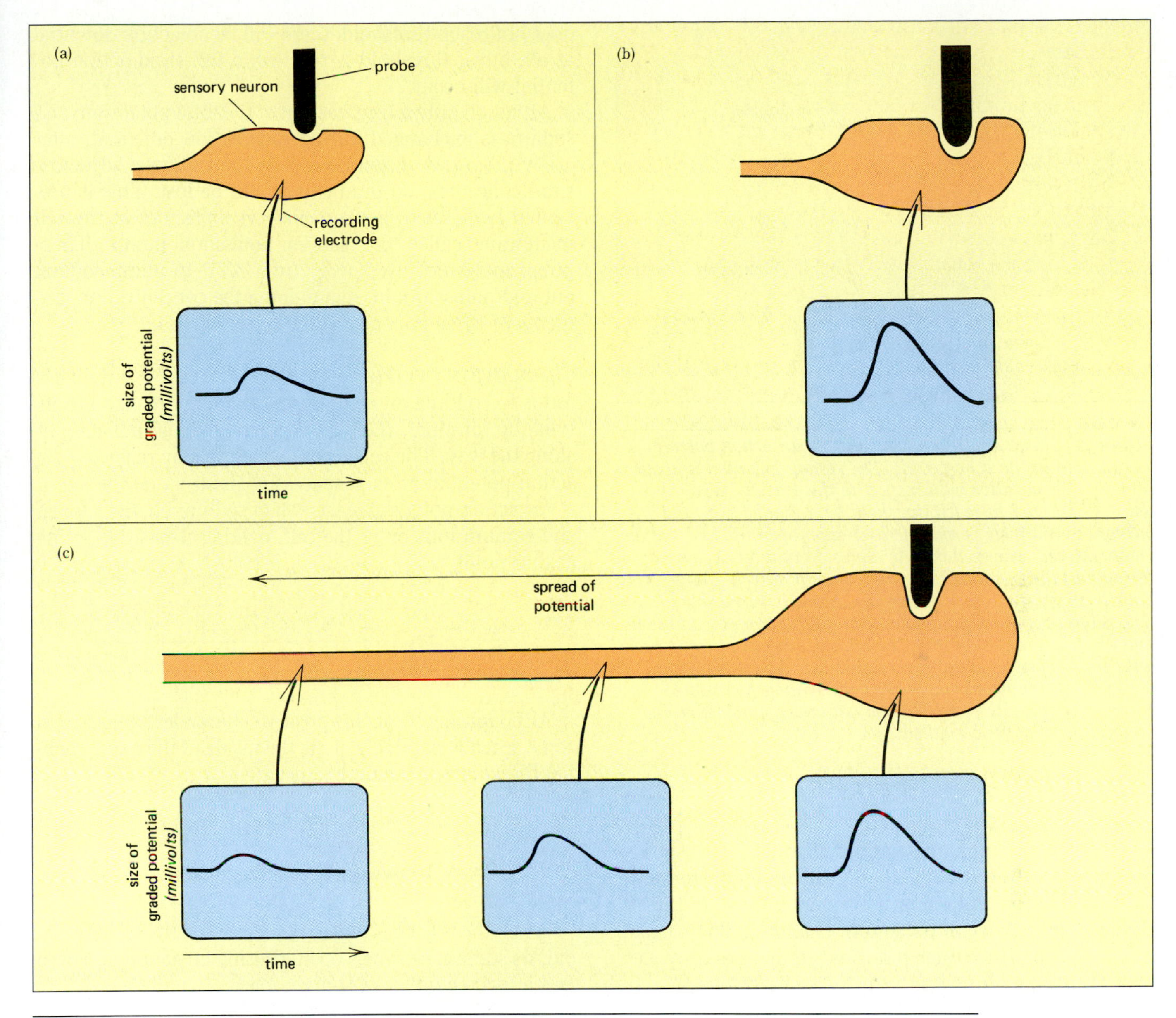

Figure 29-4 Graded potentials in nerve cells. Pressure receptor neurons respond with graded signals when the cell membrane is compressed. **(a)** *Small dents in the cell membrane produce small changes in electrical potential, while larger dents produce larger changes, shown in* **(b)**. **(c)** *Graded potentials diminish as they travel along a neuron. The potential is largest at the site of stimulation and disappears entirely within a few millimeters.*

sodium channels (shown in blue). These proteins are selectively permeable to sodium and are closed in a resting neuron. The energy to pull the trigger is provided by positive graded potentials that bring the neuron to threshold.

At threshold, sodium channels are suddenly opened. Positively charged sodium ions flood into the cell, making the cell's interior momentarily positive.

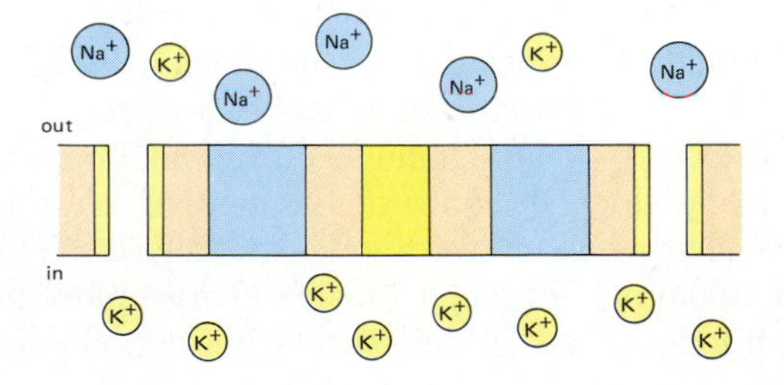

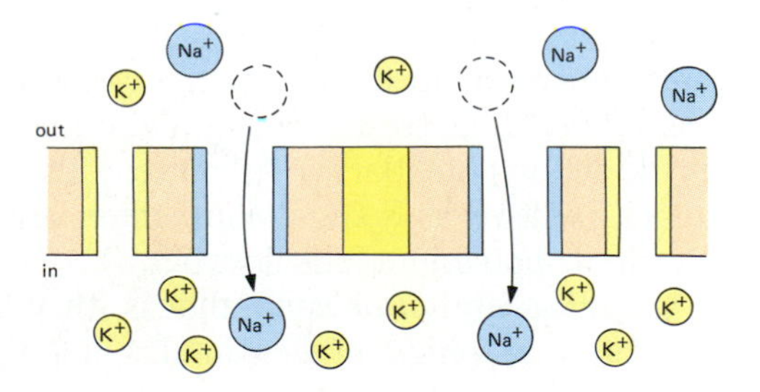

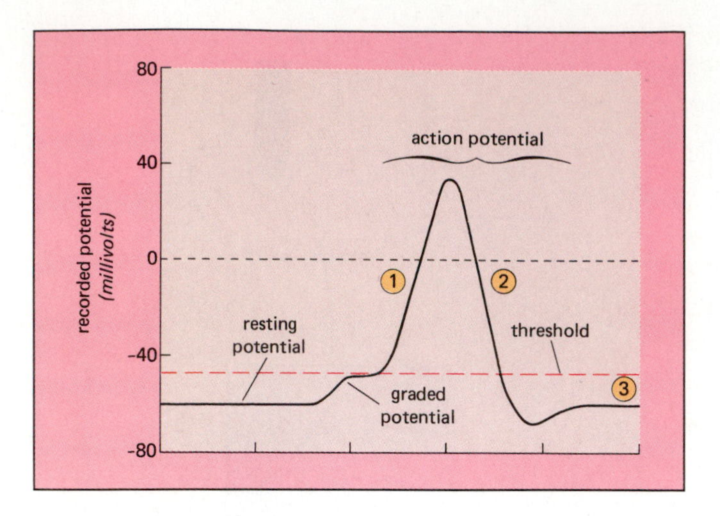

Figure 29-5 *The electrical events during an action potential as recorded inside a nerve cell. The resting potential is about 60 millivolts negative with respect to the outside. When the cell is stimulated to reach threshold by a graded potential, membrane channels permeable to sodium open up, and sodium enters the cell, powered both by diffusion and by electrical attraction; the inside of the cell becomes positively charged (1). Shortly thereafter, other membrane channels permeable to potassium open, and potassium leaves (2), driven by diffusion and electrical repulsion from the now-positive inside of the cell, until the resting potential is reestablished. Active transport molecules in the membrane, called the sodium-potassium pump, continuously pump sodium out and potassium in (3), maintaining the ionic gradient.*

After a short time, the sodium channels close spontaneously and a new set of potassium channels open, shown in yellow below. Potassium ions now flow out of the cell through both types of potassium channels. The ions are driven out both by their diffusion gradient and by electrical repulsion from the positive sodium ions that recently entered.

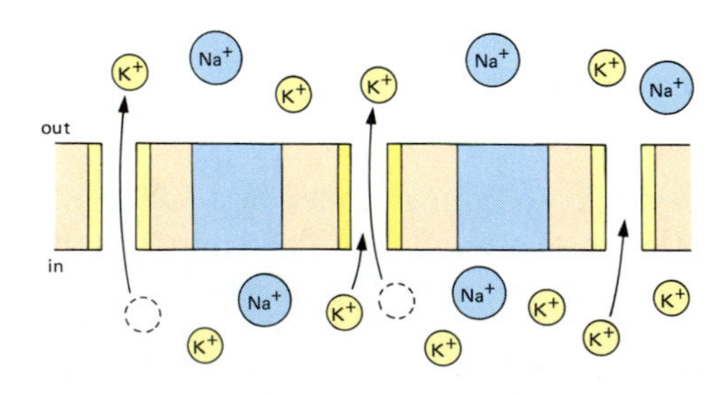

So many potassium ions leave that the inside once again becomes negative, reestablishing the resting potential. Thus the action potential is a brief event; the neuron first becomes positive as sodium ions enter, and then negative again as potassium ions flow out.

Action potentials are *all-or-none,* that is, they do not vary in size with the strength of the stimulus. If a stimulus does not reach threshold, there will be no action potential at all, but if threshold is reached, a full-sized action potential will occur.

Although only a tiny fraction of the total potassium and sodium is exchanged during each action potential, after a few thousand action potentials the sodium and potassium concentration gradients would be lost. This is prevented by a set of active transport molecules in the cell membrane called the **sodium-potassium pump.** These pump molecules use energy from ATP to pump sodium out and potassium in, maintaining the concentration gradients of these ions across the cell membrane.

CONDUCTION OF THE ACTION POTENTIAL. If the neuron is to conduct an action potential along its axon to its synaptic terminal, the action potential must not die out along the way. The cell accomplishes this by renewing the action potential at each successive point along the axon.

An action potential starts when sodium channels open and sodium ions enter the cell, making it positive inside at that point.

Although much of this positive charge leaks back out, some spreads passively along the inside of the axon, making the adjacent region less negative.

When the adjacent region of membrane reaches threshold, its sodium channels open, causing an action potential and a further influx of sodium ions.

The positive charge spreads still further, generating another action potential in the immediately adjacent membrane. This process continues along the entire length of the axon. Meanwhile, the sodium channels at the site of the original action potential close, and the resting potential is reestablished there.

It is important that action potentials travel rapidly (a giraffe couldn't run from a lion if it took 10 seconds for a signal to travel from brain to hoof). However, the opening and closing of ion channels during action potentials is relatively slow. In contrast, the passive movement of positive charges inside the axon is much faster. This creates a dilemma: although passive charge movement is rapid, it fades away with distance, while action potentials are relatively slow but do not grow smaller with distance. For a signal to travel as rapidly as possible, it should combine the speed of passive charge movement with the persistence of the action potential. In vertebrates, this is accomplished by wrapping many of the axons with insulating layers of membrane called **myelin,** interrupted at intervals with naked areas called **nodes of Ranvier** (Fig. 29-6). When an action potential occurs in a myelinated axon, the positive charge that enters the axon cannot leak back out through the myelin, but instead flows rapidly to the next node, setting up a new action potential there. The insulating myelin provides a compromise between the conflicting needs to maintain both the speed and the magnitude of the signal. Myelin allows the rapid, passive spreads of charge to continue as far as possible, but maintains signal size by initiating new action potentials at each node. This is called **saltatory conduction** (literally, "jumping" conduction), since the action potential appears to jump from node to node, speeding down the axon. Myelin is formed from nonnervous **Schwann cells** that wrap themselves around the axon (Fig. 29-6).

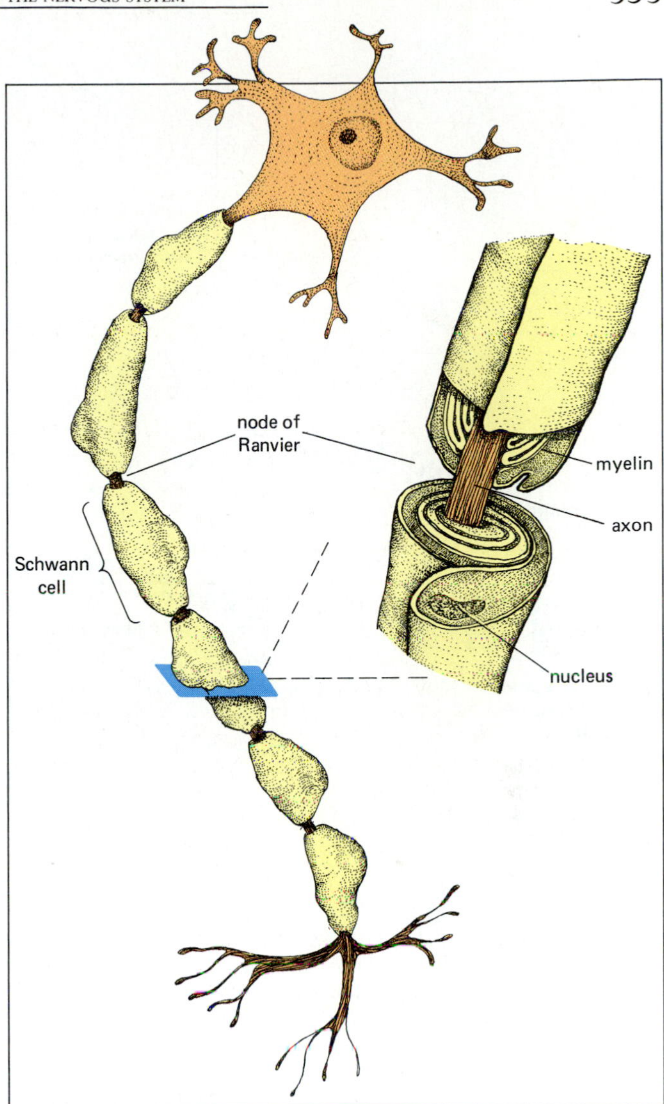

Figure 29-6 Many vertbrate axons are wrapped in a membrane jellyroll of myelin, formed from windings of membrane from specialized nonnervous cells called Schwann cells. At intervals of around a millimeter, the wrapping is interrupted by bare places, called nodes of Ranvier, where action potentials occur.

SUMMARY OF THE ELECTRICAL ACTIVITY IN A NEURON

1. The neuron maintains high concentrations of potassium ions inside and sodium ions outside by active transport. The membrane at rest is permeable only to potassium ions. Some of these positive ions diffuse out, leaving a negative charge inside called the resting potential.
2. Each neuron is specialized so that a particular type of stimulus can elicit graded potentials proportional to the strength of the stimulus. Some graded potentials are inhibitory, making the neuron more negative inside, while others are excitatory, making it less negative.
3. An action potential occurs when a graded potential makes the inside of the neuron substantially less negative, bringing it to threshold. At this point sodium channels open, allowing sodium ions to diffuse in and making the inside of the cell briefly positive.
4. The local positive charge created by an action potential spreads passively to the adjacent membrane, bringing it to threshold. This triggers a new action potential. The process continues down the entire length of the axon.
5. The resting potential is restored when the sodium channels close spontaneously and additional potassium channels open, allowing potassium to diffuse out.

Communication Between Neurons

An action potential can travel undiminished down an axon several meters long (in giraffes and whales, for example). When this electrical signal reaches the synaptic terminal of the axon, it encounters a **synapse:** a place where two neurons come close together but do not touch one another (Fig. 29-7). A gap of 0.05 micrometer separates the first, or **presynaptic,** neuron from the second, or **postsynaptic,** neuron.

When an action potential reaches a synaptic terminal, the inside of the terminal becomes positively charged. This triggers the terminal to release a hormone-like chemical called a **neurotransmitter** into the space separating the presynaptic and postsynaptic neurons (Fig. 29-7). The neurotransmitter molecules rapidly diffuse across the gap and bind to **receptors,** specialized proteins in the membrane of the dendrites of the postsynaptic cell. The receptors are linked to ion channels in the membrane. When the neurotransmitter binds to the receptors, the channels open. This allows ions (such as sodium or potassium) to flow across the postsynaptic membrane. The movement

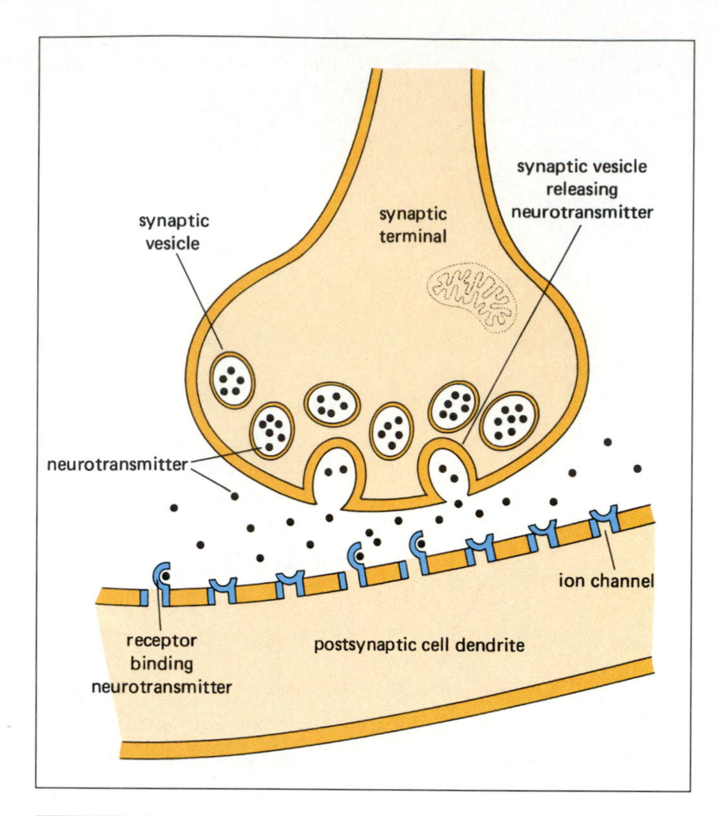

Figure 29-7 The structure and function of the synapse. The ending of the presynaptic cell contains numerous membrane-bound spheres (called synaptic vesicles) that contain neurotransmitter. The postsynaptic cell has membrane receptors for the transmitter. When an action potential enters the synaptic terminal of the presynaptic cell, the vesicles dump their neurotransmitter into the space between the neurons. The neurotransmitter diffuses rapidly across the space, binds to the postsynaptic receptors, and causes ion channels to open. Ions flow through these open channels, causing a graded potential in the postsynaptic cell.

of ions across the membrane creates a graded signal called a **synaptic potential** in the dendrite of the postsynaptic cell. Depending on what type of channels are opened and what type of ions flow, the synaptic potential may make the resting potential of the postsynaptic cell either more negative or less negative. At an **excitatory synapse,** the postsynaptic neuron becomes *less negative,* and may reach threshold, firing an action potential. At an **inhibitory synapse,** the postsynaptic cell becomes *more negative* and less likely to produce action potentials.

The dendrites of a single neuron can receive excitatory and inhibitory synaptic potentials from the synaptic terminals of thousands of presynaptic neurons. The graded synaptic potentials caused by different presynaptic neurons are then added up or **integrated** in the cell body of the postsynaptic neuron. The postsynaptic cell will produce an action potential only if the excitatory and inhibitory potentials, added together, bring the neuron above threshold.

Most neurotransmitters act only briefly on the postsynaptic cell. A few transmitters are *destroyed by enzymes* found in the gap between the two neurons; others are removed from the gap by active transport back into the presynaptic neuron or simply diffuse away into the extracellular fluid. Some transmitters, in contrast, act as **neuromodulators.** These cause long-term changes in the excitability of neurons, and may participate in behavioral states such as hunger, sexual excitement, anger, or depression.

BUILDING AND OPERATING A NERVOUS SYSTEM

The individual neuron uses a simple language of action potentials. Yet somehow this basic language allows even simple animals to perform an impressive variety of complex behaviors. One key to the versatility of the nervous system is the presence of complex networks of neurons. These neural networks range from a few to billions of cells. As in computers, small, simple elements can perform amazing feats when connected properly.

Information Processing in the Nervous System

Before we delve into the construction of nervous systems, we should first examine the operating principles. At a minimum, a nervous system must be able to perform four operations: (1) *signal the intensity of a stimulus,* (2) *determine the type of stimulus,* (3) *integrate information from many sources,* and (4) *initiate and direct the response.*

(1) SIGNAL INTENSITY. Since all action potentials are of the same magnitude and duration, no information about the intensity of a stimulus (e.g., the loudness of a sound or the amount of heat or pressure) can be encoded in a single action potential. Instead, intensity is coded in two other ways. *First,* intensity can be signaled by the *frequency of action potentials in a single neuron.* The more intense the stimulus, the faster the neuron fires. *Second,* a nervous system usually has many neurons that can respond to the same input. Stronger stimuli tend to excite more of these neurons, weaker stimuli excite fewer. Thus *intensity can also be signaled by the number of similar neurons firing at the same time* (Fig. 29-8).

(2) DETERMINE TYPE OF STIMULUS. Besides signaling intensity, the nervous system must have a way of identifying the type of stimulus (e.g., light, touch, sound). Here again, the type of stimulus cannot be coded by the properties of individual action potentials. Instead, the nervous system monitors *which neurons* are firing action potentials. Thus your brain identifies action potentials occurring in the axons of your optic nerves (originating in the eye) as information about light, action potentials in ol-

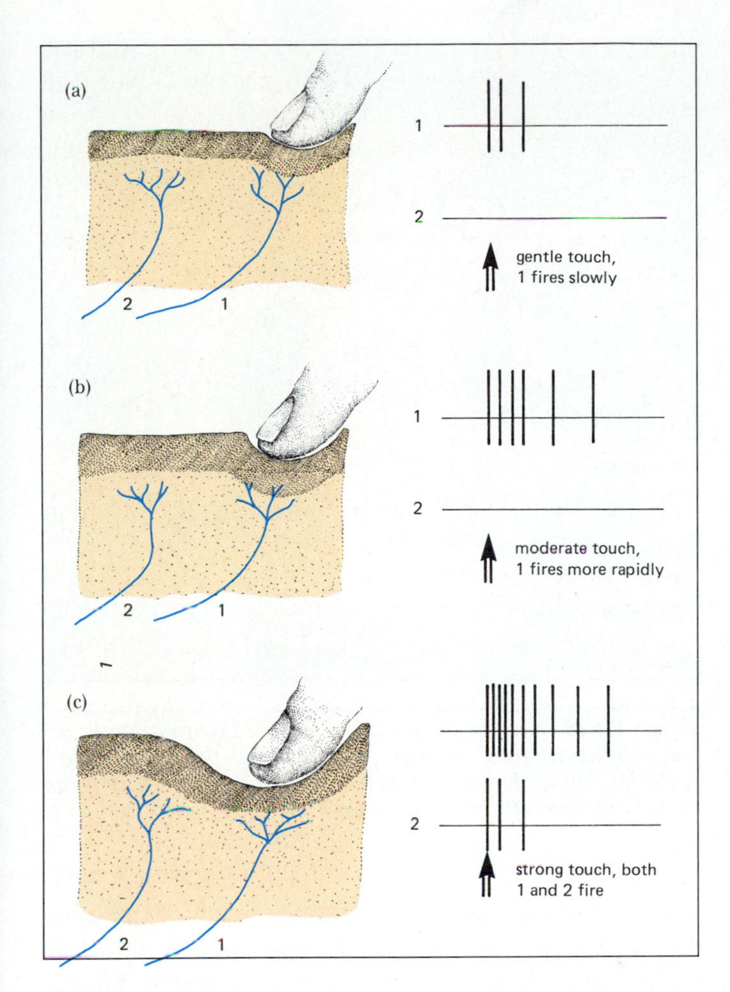

Figure 29-8 The intensity of a stimulus is signaled by the rate at which individual neurons produce action potentials, and by the number of neurons firing. For example, two touch receptors may have endings in adjacent patches of skin. **(a)** *A gentle touch elicits only a few action potentials and from only one of the sensory neurons.* **(b)** *A moderate touch still stimulates only one receptor, but this receptor now fires faster, informing the brain that the touch is more intense than before.* **(c)** *A strong touch activates both receptors, firing one very fast, and the other more slowly, thus signaling to the brain that the touch is very intense.*

factory nerves as odors, and so on. This genetic wiring may occasionally yield false information. Being poked in the eye may cause action potentials in the optic nerve. Even though the stimulus is mechanical, your brain nevertheless interprets all optic nerve activity as light, and you "see stars."

(3) INTEGRATE INFORMATION. Your brain is continually bombarded by a variety of sensory stimuli originating both inside and outside the body. The brain must filter all these inputs, determine which are important, and decide what actions to take in response. Nervous systems integrate information much as do individual neurons, through **convergence.** In this process, many neurons funnel their signals to fewer neurons. For example, many sensory neurons may converge onto a smaller number of brain cells (Fig. 29-9). The brain cells sum up the graded

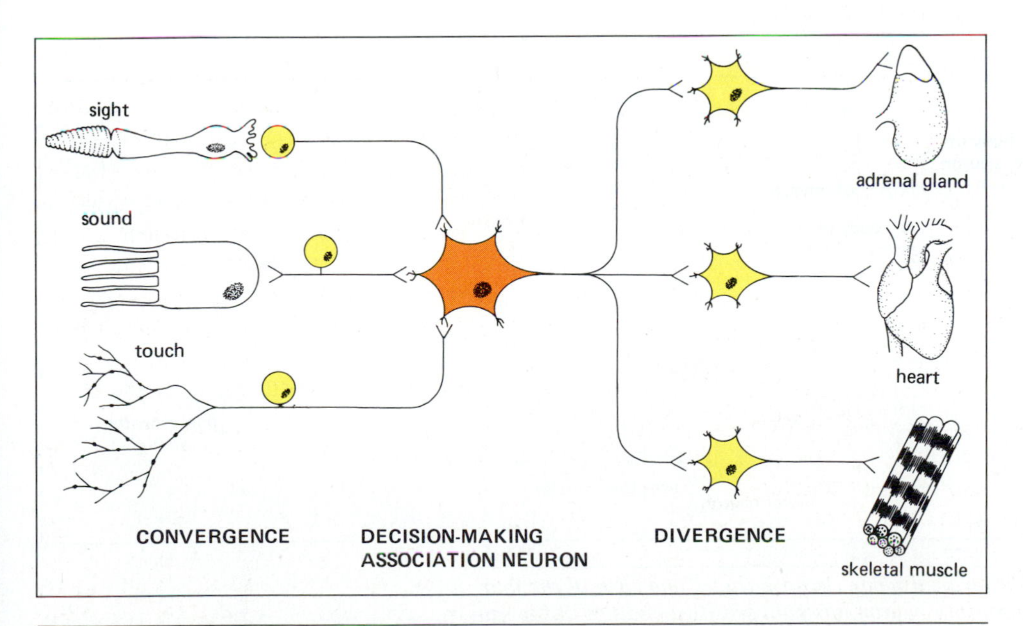

Figure 29-9 Integration of information and initiation of coordinated action involves convergence of inputs to and divergence of outputs from neurons. In this simplified example, inputs of sight, sound, and touch (say, being attacked by a swarm of bees) all converge on a "decision-making" neuron, which is strongly stimulated as a result. Its outputs go to the adrenal glands ("pump out adrenalin"), the heart ("beat faster and stronger"), and skeletal muscles ("move, legs!").

potentials resulting from the synaptic activity of these sensory neurons, and depending on their relative strengths (and other internal factors such as hormones, time of day, or metabolic activity), they produce appropriate outputs.

(4) Initiate and control activity. The output of the integrating cells is responsible for initiating activity. The actions directed by the brain may involve many parts of the body, and require **divergence,** the flow of electrical signals from a relatively small number of decision-making cells onto many different neurons controlling muscle or glandular activity (Fig. 29-9).

Neural Networks

Most behaviors are controlled by nervous–muscular pathways composed of four elements: (1) **receptors** or **sensory neurons,** which respond to a stimulus, either internal or external to the body; (2) **association neurons,** which "decide" what to do, based on input from many sensory neurons, stored memories, hormonal states, and other factors; (3) **motor neurons,** which receive instructions from the association neurons and activate the muscles; and (4) **effectors,** usually muscles or glands, which perform the behavior.

The simplest type of behavior is the **reflex,** a relatively involuntary movement of a body part in response to a stimulus. Examples of human reflexes include the familiar knee jerk and withdrawal reflex. The withdrawal reflex, (which moves a body part away from a painful stimulus) is particularly instructive, since it uses only one neuron of each type (Fig. 29-10). Reflexes of this sort do not require the brain, although, as we know, other pathways do inform the brain of pricked fingers and may in fact trigger other more complex behaviors (cursing, for example!).

Figure 29-11 The diffuse nervous system of Hydra *contains a few concentrations of neurons, particularly at the bases of the tentacles, but no brain. Conduction of neural signals may occur in virtually any direction throughout the body.*

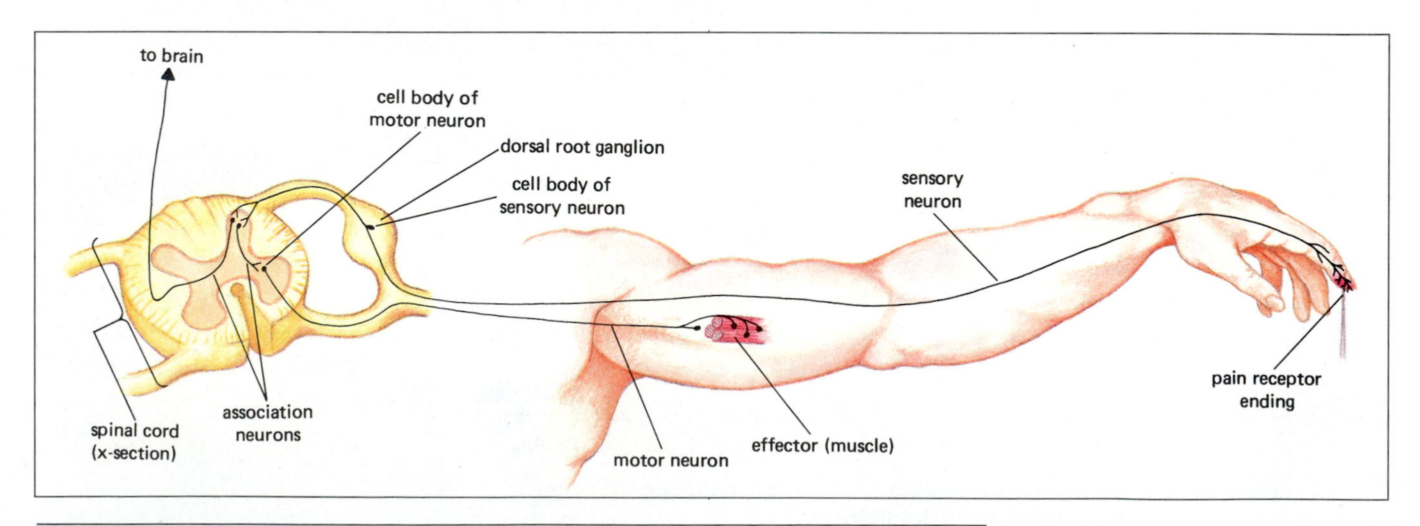

Figure 29-10 The vertebrate pain/withdrawal reflex circuit includes one each of the four elements of a nervous pathway. The sensory neuron has pain-sensitive endings in the skin and a long fiber leading to the spinal cord. The sensory neuron stimulates an association neuron in the spinal cord, which in turn stimulates a motor neuron, also in the cord. The axon of the motor neuron carries action potentials to muscles, causing them to contract and withdraw the body part from the damaging stimulus. Note that the sensory neuron also synapses on other association neurons not directly involved in the reflex, which carry signals to the brain, informing it of the danger below.

Nearly all animals are capable of much more subtle and varied behaviors than can be accounted for by simple reflexes. In principle, these more complex behaviors can be organized by *interconnected nervous pathways,* in which several types of sensory input (along with memories, hormones, etc.) converge on a set of association neurons (Fig. 29-9). By integrating the graded synaptic potentials from several sources, the association neurons can "decide" what to do, and stimulate the motor neurons to direct appropriate activity in muscles and glands.

Nervous System Design

In all the animal kingdom, there are really only two designs for nervous systems: **diffuse nervous systems,** found in the cnidarians (*Hydra,* jellyfish, and their relatives; Fig. 29-11), and **centralized nervous systems,** found to varying degrees in more complex organisms (Fig. 29-12). Not surprisingly, nervous system design is highly correlated with the life-style of the animal. In the radially symmetrical cnidarians, there is no "front end," so there has been no evolutionary pressure to concentrate the senses in one

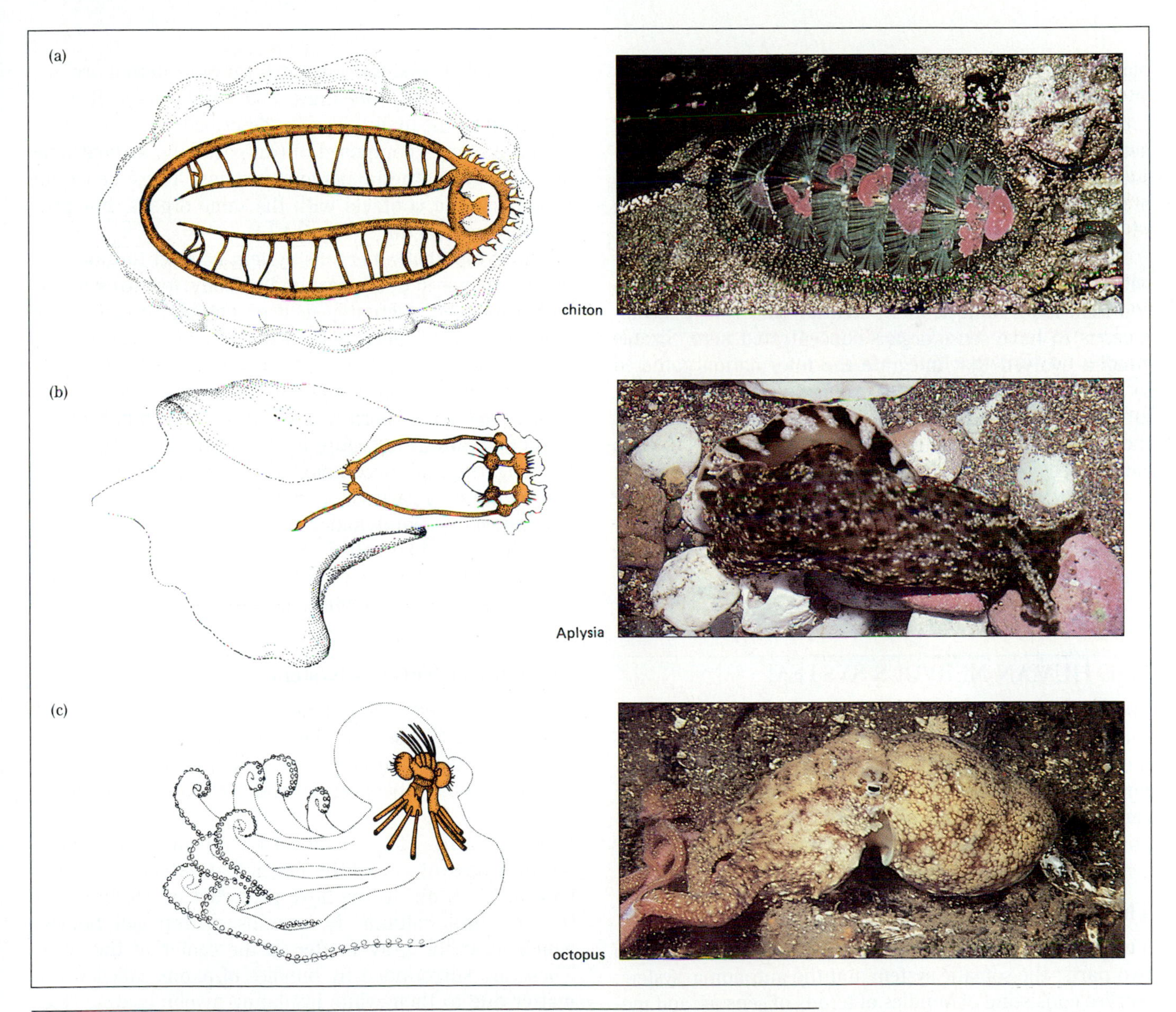

Figure 29-12 Bilaterally symmetrical animals usually have nervous systems concentrated in the head. The trend toward increasing concentration of the nervous system in the head is clearly illustrated by various molluscs. **(a)** *The chiton, although it does have a head end, seldom crawls in any direction and has had little selective pressure to concentrate sense organs and brain in the head.* **(b)** *Some marine snails, such as the shell-less* Aplysia, *can crawl quite rapidly, or even swim. Still more of their neurons are aggregated into a brain.* **(c)** *Mollusc mobility and intelligence culminates in* Octopus, *with its large, complex brain and behavioral capabilities rivaling those of some mammals.*

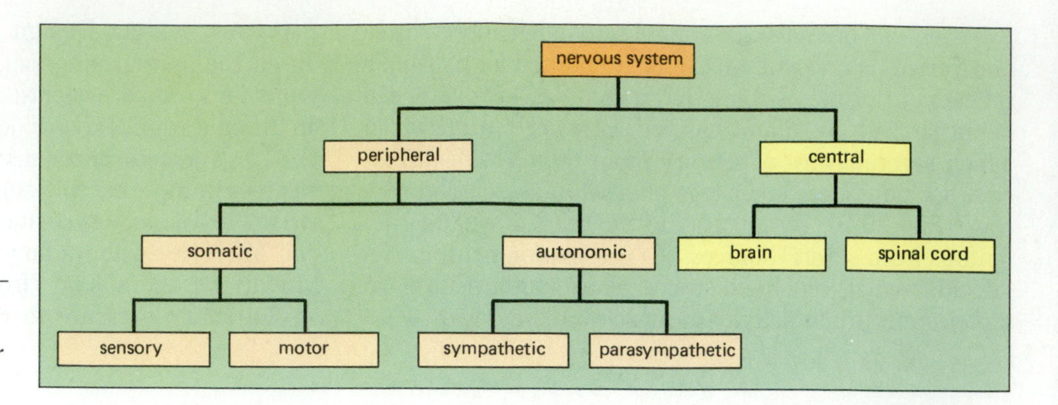

Figure 29-13 The organization of the vertebrate nervous system, showing its major parts and their subdivisions.

place. A *Hydra* sits anchored to the substrate, and prey or danger are equally likely to come from any direction. Cnidarian nervous systems are composed of a network of neurons, often called a **nerve net,** woven through the tissue of the animal. Here and there we can find a cluster of neurons, called a **ganglion** (plural *ganglia*), but nothing resembling a real brain.

Almost all other animals are bilaterally symmetrical, with definite head and tail ends. Since the head first encounters food, danger, and potential mates, it is advantageous to have sense organs concentrated here. Sizable ganglia evolved that integrate the information gathered by the senses and initiate appropriate action. Over evolutionary time, the sense organs gathered in the head and the ganglia became centralized into a brain. This trend is clearly seen in the molluscs (Fig. 29-12). Centralization reaches its peak in the vertebrates, where nearly all the cell bodies of the nervous system are localized in the brain and spinal cord. The organization of the vertebrate nervous system is shown in Figure 29-13.

THE HUMAN NERVOUS SYSTEM

The human nervous system may be divided into two parts. The **central nervous system** consists of a **brain** encased in a bony skull and a similarly protected **spinal cord** extending down the dorsal surface of the torso. The **peripheral nervous system** consists of nerves extending from or leading to the central nervous system.

The Peripheral Nervous System

The peripheral nervous system may be subdivided into two parts: the somatic system and the autonomic system. Nerves consisting of bundles of axons of sensory and motor neurons make up the **somatic nervous system.** These nerves act as transmission lines between the internal and external environment and the central nervous system, conducting messages encoded as all-or-none action potentials to and from the brain and spinal cord. The remaining part of the peripheral nervous system, the **autonomic nervous system,** regulates various internal organs, such as the heart, intestines, and kidneys, and itself is regulated by the hypothalamus, described later. The autonomic system is composed of two parts, the **sympathetic** and **parasympathetic** nervous systems. These generally make synaptic contacts with the same organs, but produce opposite actions (Fig. 29-14).

The sympathetic nervous system acts on the internal organs in ways that prepare the body for stressful or highly energetic activity, such as fighting, escaping, or hunting. During such "fight or flight" activities, the sympathetic nervous system curtails activity of the digestive tract, freeing its blood supply to be used by the muscles of arms and legs. Heart rate speeds up. The pupils of the eyes open, admitting more light, and the air passages in the lungs expand, accommodating more air.

The parasympathetic nervous system, on the other hand, governs maintenance activities that can be carried on at leisure, often called "rest and rumination." Under its control the digestive tract becomes active, heart rate slows, and urine production increases.

The Central Nervous System

The central nervous system consists of the brain and spinal cord. These are the integrating centers of the nervous system, consisting primarily of association neurons (somewhere between 10 and 100 billion of them!).

Spinal cord. The spinal cord is a neural cable about the thickness of your little finger that extends from the base of the brain to the hips, protected by the bones of the vertebral column. It contains neuron cell bodies, which form the **gray matter** in the center of the cord. These are surrounded by bundles of axons called **white matter** due to their white insulating myelin coating (Fig. 29-15). To illustrate some of the functions of the parts of the spinal cord, let's examine the simple spinal reflex illustrated in Figure 29-10. The cell bodies of the sensory neurons from the skin (in this case signaling pain) are found just outside the spinal cord in a row of ganglia. Each of these **dorsal root ganglia** is located on a spinal

Of Squids and Snails and Sea Hares (or, Why Do Biologists Study Such Bizarre Animals?)

If you wander through the biology laboratories at your college, you may find your professors studying some most unlikely organisms, such as insects, snails, bacteria, or algae. Sometimes, the reason is obvious: many insects, for example, are crop pests or carry diseases, and by learning about their life histories scientists may discover ways to control them. You might think, however, that neurobiologists would concentrate on humans, or at least mammals, since their primary interest is surely to find out how the human brain works. Nevertheless, many neurobiologists study snails, lobsters, or even leeches. Aside from the fact that if you work on lobsters, you can eat your experiment when you are through (no small consideration), why would anyone study such creatures?

Figure 29-2 provides a clue. While the largest mammalian axons are about 20 micrometers in diameter, the squid giant axon is about a *millimeter* in diameter! This is extremely convenient for neurobiologists. Tiny electrodes placed with painstaking precision are required to obtain good recordings from mammalian axons. In contrast, relatively crude electrodes can be inserted inside a squid axon without damaging it. Scientists have even taken little rubber rolling pins and squeezed the cytoplasm from a squid axon. This allowed them to analyze the chemical composition of the cytoplasm, and also to reinflate the axon with artificial cytoplasm to test theories about how action potentials are produced. Lobsters, crayfish, and some worms also have giant axons, and are studied for similar reasons.

Another advantage of certain animals is simplicity. Depending on whose guess you believe, the human brain has between 10 billion and 1 trillion tiny nerve cells. In contrast, some leeches and snails have only a few thousand neurons. Many of these neurons are very large and identifiable as individuals (Fig. E29-1). If one wants to find out general principles about how circuits of neurons control behaviors, it is obviously easier to study a nervous system in which an entire behavior may be governed by only a few dozen neurons.

But, you may argue, once you have found out how a squid axon works, or what the circuit for snail feeding looks like, have you learned anything about human nervous systems? As unlikely as it may seem, the answer is unequivocally "yes!". The basis for action potential production in axons was worked using the giant axon of the squid in the early 1950s. Subsequent research has shown that with a few modifications, the same principles operate in mammalian axons. Without the guidance offered by the detailed analyses of squid axons, we still might not know how our own axons work. Similarly, synaptic function is understood largely through studies of the frog nerve-muscle synapse and a particularly large synapse in the squid.

Even complex behaviors such as learning may well have the same neuronal basis in humans and snails. In a simple form of learning called habituation, an animal ceases responding to a harmless, repeated stimulus. Habituation allows you to sleep through your roommate's snoring. Habituation has been studied thoroughly in the sea hare, *Aplysia* (Fig. 29-12b). A few years later, psychologists examined habituation in the frog spinal cord and concluded that very similar mechanisms were at work.

Evolution is basically a conservative process. Organisms that are apparently very different share the same fundamental traits. Whether we consider humans or leeches, snails or crabs, the similarities are compelling. They include the structure of the gene, the mechanism of protein synthesis, the use of ATP for energy transfers, even hormone molecules. While the same mechanisms and molecules are used, they may be assembled in different ways to achieve very different results, such as the wings of the condor and the human hand. We should not be surprised if the human brain turns out to be another instance of an imposing new structure built with the standard issue of materials.

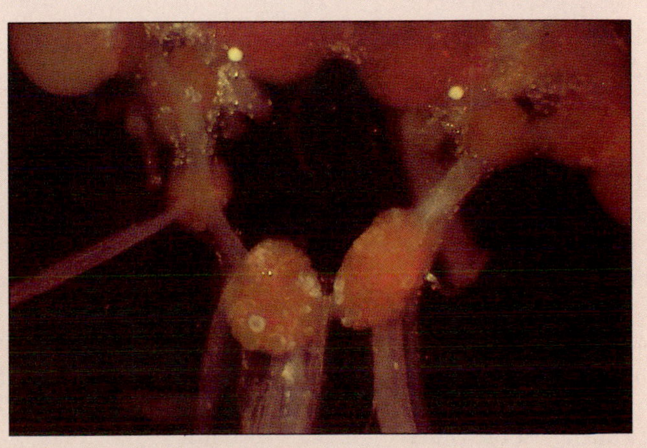

Figure E29-1 The brain of the giant pond snail, Lymnaea stagnalis, *is convenient for the study of neuronal function. Shown here are several ganglia from the brain. The circles within the ganglia are individual nerve cell bodies, often 100 to 200 micrometers in diameter. Some of these neurons can be identified as unique individuals and can be found in every snail brain.*

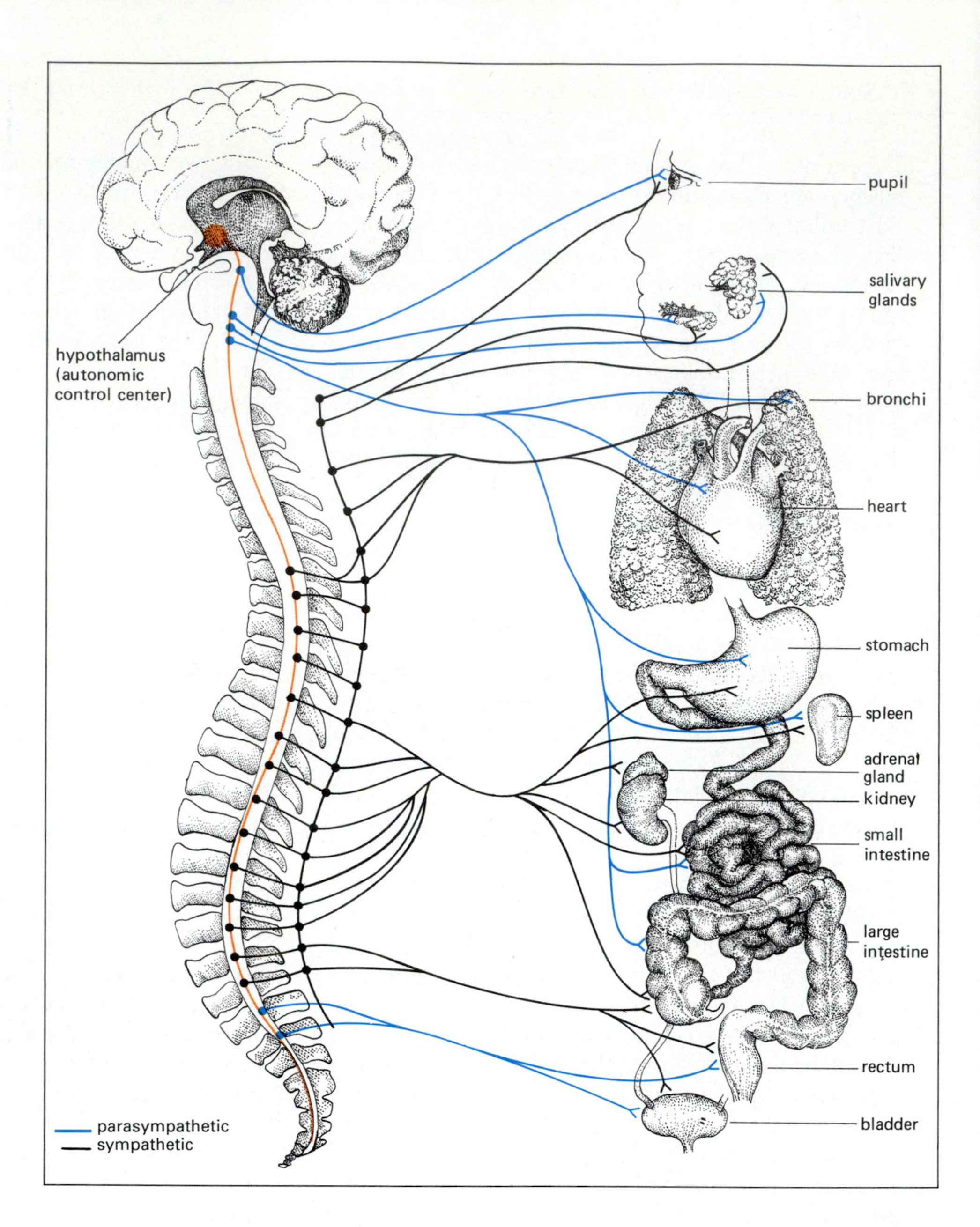

Figure 29-14 The autonomic nervous system has two divisions, the sympathetic and parasympathetic. Both divisions supply nerves to many of the same organs, but produce opposite effects. Activation of the autonomic nervous system is mostly involuntary, produced by nervous outputs from the hypothalamus.

nerve and nestled close to the vertebral column. Both association and motor neuron cell bodies are found in the gray matter in the center of the spinal cord. The axons of the surrounding white matter communicate with the brain. Association neurons for the pain reflex, for example, not only synapse on motor neurons, but have axons extending up to the brain. Signals carried along these axons alert the brain to the painful event. The brain, in turn, sends impulses down axons of the white matter to cells in the grey matter. These signals can modify spinal reflexes. For example, with sufficient motivation (or foolhardiness), you can suppress the pain reflex and hold your hand in a flame.

In addition to simple reflexes, the entire program for operating some fairly complex activities also resides within the spinal cord. All the neurons and interconnections needed to walk and run, for example, are found within the cord. In these cases, the role of the brain is to initiate and guide the spinal activity. The advantage of this semi-independent arrangement is probably an increase in speed and coordination, since messages do not have to travel all the way up the cord to the brain and back down again (in the case of walking) merely to swing one of your legs forward.

The motor neurons of the spinal cord also control the muscles involved in conscious, voluntary activities such as eating, writing, or playing tennis. Axons of the brain cells directing these activities carry signals down the cord and stimulate the appropriate motor cells.

THE BRAIN. All vertebrate brains have the same general structure, with major modifications corresponding to life-

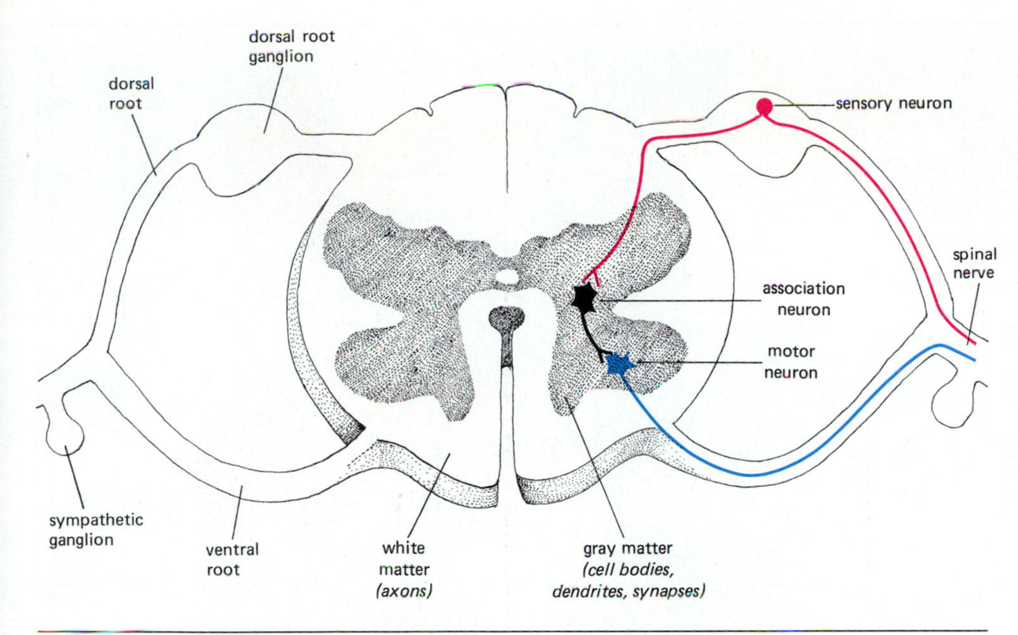

Figure 29-15 The spinal cord runs from the base of the brain to the hips, protected by the vertebrae of the spine. Most of the body below the neck is supplied by paired spinal nerves which emerge from between the vertebrae. The nerves split before entering the spinal cord, with axons of sensory neurons running in the dorsal root, while motor axons comprise the ventral root. A cross section of the spinal cord reveals an outer region of myelinated axons (white matter) traveling to and from the brain, surrounding an inner, butterfly-shaped region of dendrites and the cell bodies of association and motor neurons (gray matter). The cell bodies of the sensory neurons are located outside the cord in the dorsal root ganglion.

style and intelligence. Embryologically, the vertebrate brain begins as a simple tube, which soon develops into three parts: the hindbrain, midbrain, and forebrain (Fig. 29-16). It is believed that in the earliest vertebrates, these three anatomical divisions were also functional divisions: the hindbrain governed automatic behaviors such as breathing and heart rate, the midbrain controlled vision, and the forebrain largely dealt with the sense of smell. In nonmammalian vertebrates, these three divisions remain prominent (Fig. 29-16). However, in mammals, and particularly in humans, the brain regions are significantly modified. Some have been reduced in size and importance, and others, especially the forebrain, greatly enlarged.

In humans, the hindbrain is represented by the **medulla** and the **cerebellum.** In both structure and function, the medulla is very much like an enlarged extension of the spinal cord. Like the cord, the medulla has neuron cell bodies at its center, surrounded by a layer of myelin-covered axons. The medulla controls several automatic functions, such as breathing, heart rate, blood pressure, and swallowing. The cerebellum is crucially important in coordinating movements of the body. It receives information from command centers in the higher, conscious areas of the brain that control movement and from position sensors in muscles and joints. By comparing what the command centers ordered with information from the position sensors, the cerebellum guides smooth, accurate motions and body position. Not surprisingly, the cerebellum is largest in animals whose activities require fine coordination. It is best developed in birds (Fig. 29-16b), which engage in the enormously complex activity of flight.

The midbrain is extremely reduced in humans but contains an important relay center, the **reticular activating formation** (which actually extends all the way from the central core of the medulla, through the midbrain, and on into lower regions of the forebrain). The reticular formation (Fig. 29-17) receives input from virtually every sense and every part of the body, and from many areas of the brain as well. It filters sensory inputs before they reach the conscious regions of the brain, although the selectivity of the filtering seems to be set by higher brain centers. Through a combination of genetically determined wiring and learning, the reticular formation "decides" which stimuli require attention. Important stimuli are forwarded to the conscious centers for processing, while unimportant stimuli are suppressed. The fact that a mother wakens upon hearing the faint cry of her infant, but sleeps through loud traffic noise outside her bedroom window, testifies to the effectiveness of the reticular formation in screening inputs to the brain, and to the powerful role of learning in determining the importance of sensory stimulation.

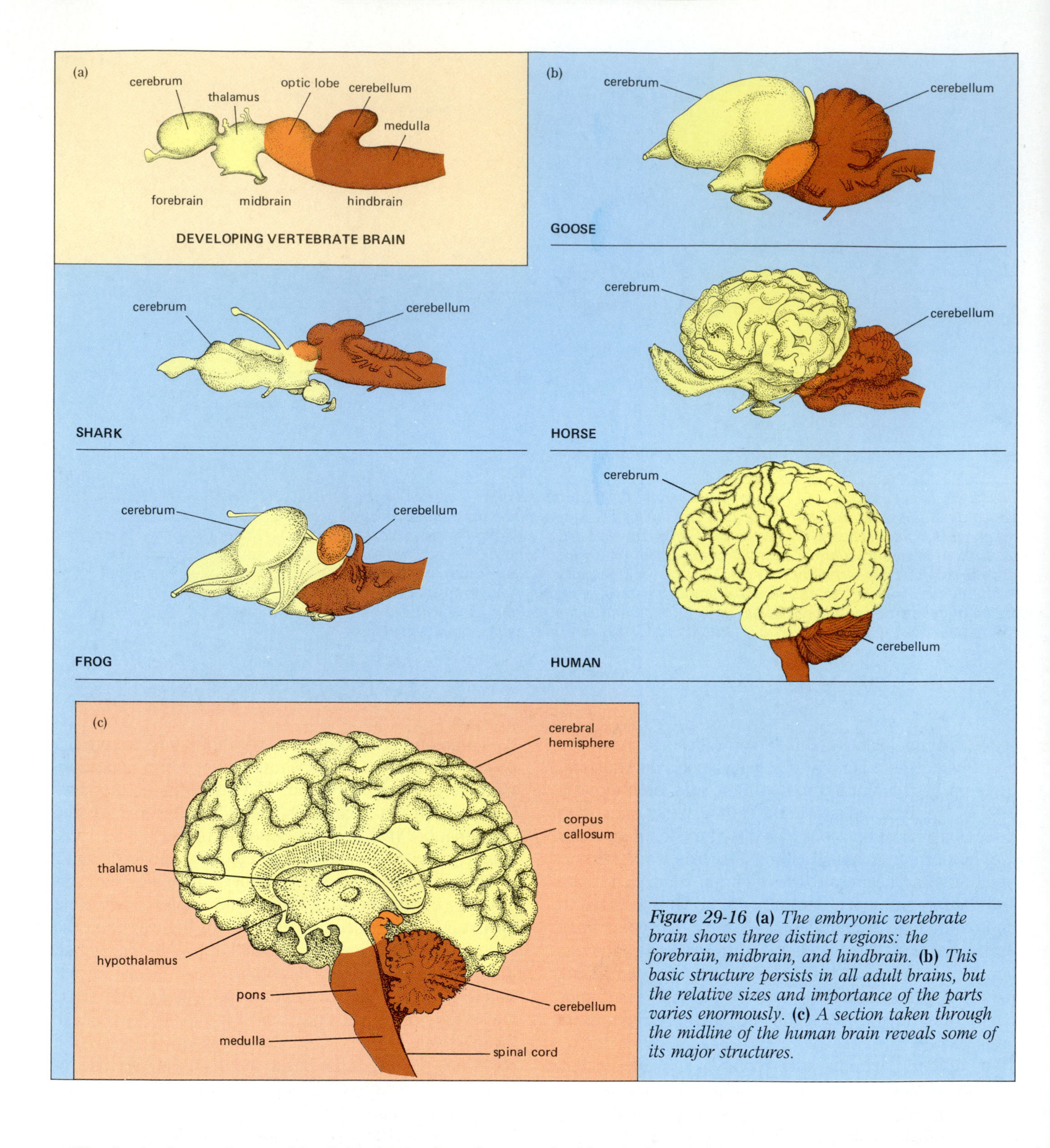

Figure 29-16 **(a)** *The embryonic vertebrate brain shows three distinct regions: the forebrain, midbrain, and hindbrain.* **(b)** *This basic structure persists in all adult brains, but the relative sizes and importance of the parts varies enormously.* **(c)** *A section taken through the midline of the human brain reveals some of its major structures.*

The forebrain can be roughly divided into three functional parts: (1) a relay center, the **thalamus,** which shuttles sensory information to (2) a set of instinctive/emotional centers, **the limbic system,** and (3) an information processing/control center, the **cerebrum.** In fish, amphibians, and reptiles, these three areas are of roughly equal size and importance. In mammals, however, the cerebrum is enlarged, culminating in the enormous cerebral hemispheres of humans, which almost completely envelop the thalamus and limbic system (Fig. 29-16c).

The human **thalamus** and nearby structures channel sensory information to the limbic system and especially up to the cerebrum. Little information processing goes on in the thalamus.

Anatomically, the **limbic system** is a diverse grab bag of structures, including the **hypothalamus, hippocampus,**

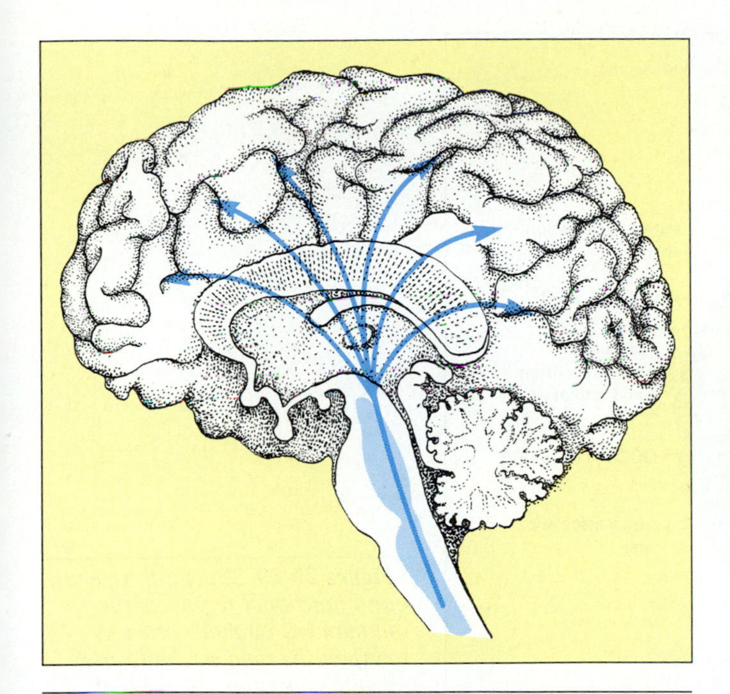

Figure 29-17 The reticular activating system is a diffuse network of neurons running through the lower regions of the brain from the hindbrain, through the midbrain, and up into the thalamus of the forebrain. It receives input from most of the senses and sends outputs to many higher brain centers, filtering the sensory information that reaches the conscious brain.

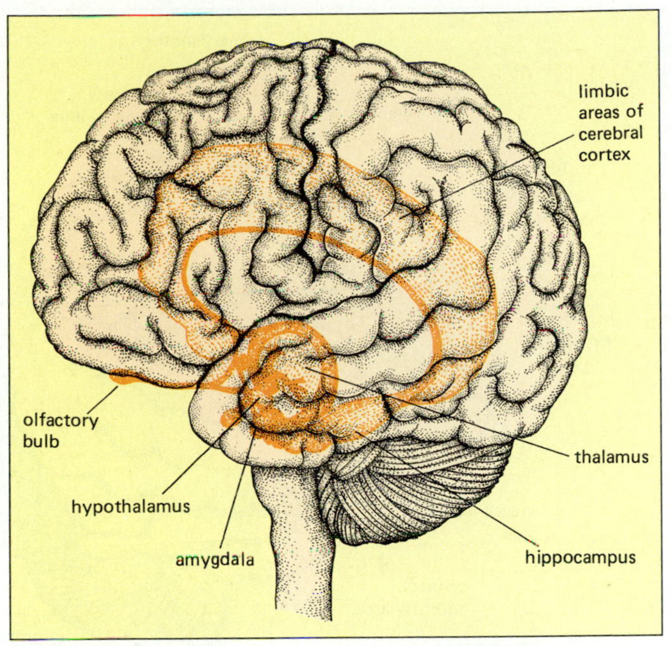

Figure 29-18 The limbic system extends through several brain regions. It seems to be the center of most unconscious, emotional behaviors, such as love, hate, hunger, sex, and fear.

and **amygdala,** located in an arc between the thalamus and cerebrum (Fig. 29-18). Functionally, however, the limbic system forms an interrelated unit. The hypothalamus activates the autonomic nervous system and controls unconscious bodily functions such as body temperature, pituitary hormone secretion, salt and water balance, and the menstrual cycle. In addition, the hypothalamus is the output center for feelings such as rage, fear, or sexual arousal. These emotions actually seem to originate in the amygdala. If the amygdala is electrically stimulated in conscious humans, they report feelings of rage or fear, depending on the exact locations stimulated.

In humans, by far the largest part of the brain is the **cerebrum,** which contains about half the nerve cells in the brain. The cerebrum is split into two halves, called the **cerebral hemispheres,** which communicate with each other via a large band of axons, the **corpus callosum.** The cerebral hemispheres are the most sophisticated information-processing centers known. Roughly *5 billion* neurons are packed into a thin layer at the surface, called the **cortex.** These neurons receive sensory information, process it, store some in memory for future use, and direct voluntary motor output.

The functions of the cerebral hemispheres are localized in discrete regions of the cortex (Fig. 29-19). Brain damage due to trauma, strokes, or tumors results in particular deficits, such as speech, reading, and the ability to sense or move specific parts of the body. Since brain cells cannot reproduce, once a brain region is destroyed, it cannot be repaired or replaced, and such deficits are often permanent. However, in some cases, diligent training can cause undamaged regions of the cortex to take over some of the lost functions.

BRAIN AND MIND

Historically, people have always had difficulty reconciling the physical presence of a few pounds of grayish material in the skull with the range of thoughts, emotions, and memories of the human mind. This "mind–brain problem" has occupied generations of philosophers and, more recently, neurobiologists. Beginning with observations of patients with head injuries and progressing to sophisticated surgical, physiological, and biochemical experiments, the outlines of how the brain creates the mind are beginning to emerge. Here, we will only be able to touch upon a few of the more fascinating features.

Left Brain–Right Brain

The human brain appears bilaterally symmetrical, particularly the cerebrum, which consists of two extremely similar-looking hemispheres. However, it has been known since the early twentieth century that this symmetry does not extend to brain function (Fig. 29-20). Much of what

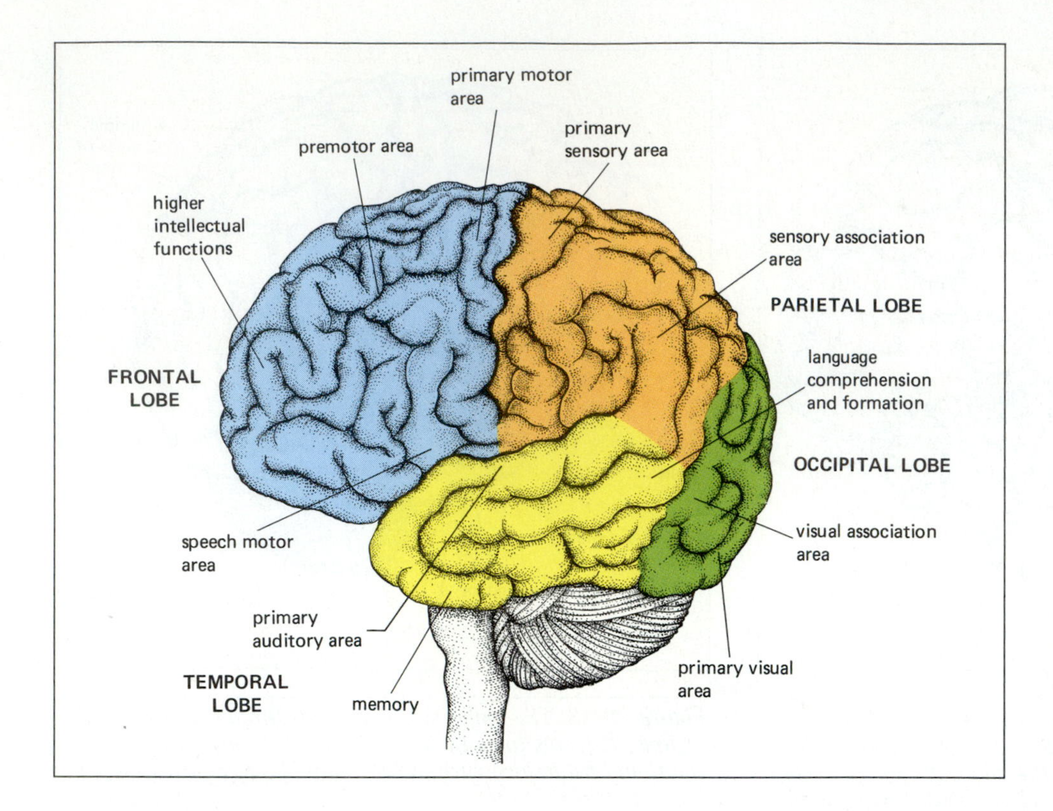

Figure 29-19 Structural (colored) and functional regions of the human left cerebral cortex. A map of the right cerebral cortex would be similar, except that speech and language are less well developed on the right side (see Fig. 29-20).

is known of the differences in hemisphere function comes from two sources: studies of accident victims with localized damage to one hemisphere, and studies of patients who have had the corpus callosum (which connects the two hemispheres) severed. This surgical procedure is performed in rare cases of uncontrollable epilepsy to prevent the spread of seizures through the brain.

People suffering damage to localized areas of the left hemisphere, but not the right, often become unable to speak, read, or understand spoken language. In addition, the left hemisphere for most people is dominant in mathematical ability and in logical problem-solving tasks. Musical skills, artistic ability, recognition of faces, and spatial visualization typically reside in the right hemisphere, as does the ability to recognize and express emotions. Axons from each optic nerve follow a pathway that causes the left half of each visual field to be projected on the right cerebral hemisphere, and vice versa (Fig. 29-20). Roger Sperry, of the California Institute of Technology, worked with epilepsy patients whose hemispheres had been surgically separated. If he projected an image of a nude figure onto the left visual field only, these patients would blush and smile, but would claim to have seen nothing, since the image had only reached the nonverbal right side of the brain! The same figure projected onto the right visual field was readily described verbally.

Recent experiments indicate that the left–right dichotomy is not as rigid as once believed. Patients who have suffered a stroke that disrupted blood supply to the left hemisphere typically show symptoms such as loss of speaking ability. Frequently, however, training can partially overcome speech or reading deficits caused by damage to the left hemisphere, even though the hemisphere itself has not recovered. This suggests that the right hemisphere has some latent language capabilities. Interestingly, female stroke victims recover function more often than males, suggesting a sex difference in hemisphere function.

Learning and Memory

Although theories abound as to the cellular mechanisms of learning and memory, we are a long way from understanding these phenomena. In mammals, and particularly in humans, however, we do know a fair amount about two other aspects of learning and memory: the time course of learning, and some of the brain sites involved in learning, memory storage, and recall.

TIME COURSE OF LEARNING. Experiments show that learning occurs in two phases: an initial **short-term memory** followed by **long-term memory.** For example, if you look up a number in the phone book, you will probably remember the number long enough to dial but forget it promptly thereafter. This is short-term memory. But if you call the number frequently, eventually you will remember the number more or less permanently. This is long-term memory.

Short-term memory seems to be *electrical* in nature, involving the repeated activity of a particular neural circuit in the brain. As long as the circuit is active, the memory stays. If the brain is distracted by other thoughts, or if electrical activity is interrupted, such as by electroconvulsive shock or by a concussion, the memory disappears and cannot be retrieved no matter how hard you try.

Long-term memory, on the other hand, seems to be *structural,* involving, perhaps, the formation of new, permanent synaptic connections between specific neurons, or the strengthening of existing but weak synaptic connections. These new or strengthened synapses last indefinitely, and the long-term memory persists unless certain brain structures are destroyed. How short-term memory is converted to long-term memory remains a mystery, although intensive study, particularly in certain invertebrates, may provide some answers soon.

LEARNING, MEMORY, AND RETRIEVAL SITES IN THE BRAIN. Learning, memory, and retrieval seem to be separate phenomena, mediated by separate areas of the brain.

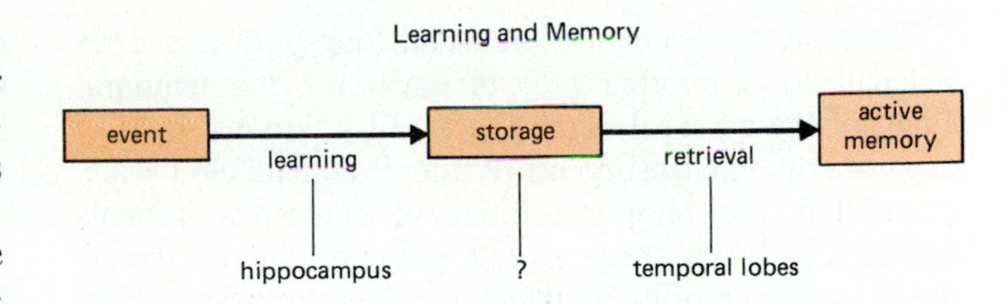

Ample evidence shows that the **hippocampus** (part of the limbic system) is involved in **learning.** For example, intense electrical activity occurs in the hippocampus during learning. Even more striking is the effect of hippocampal damage. Hippocampal destruction does not affect retention of old memories, but results in an inability to learn new things. One patient with hippocampal damage, while retaining old memories, could read the same magazine article day after day for years, never remembering that he had read the article before. Each time the physician came for a visit, the patient would require a new introduction.

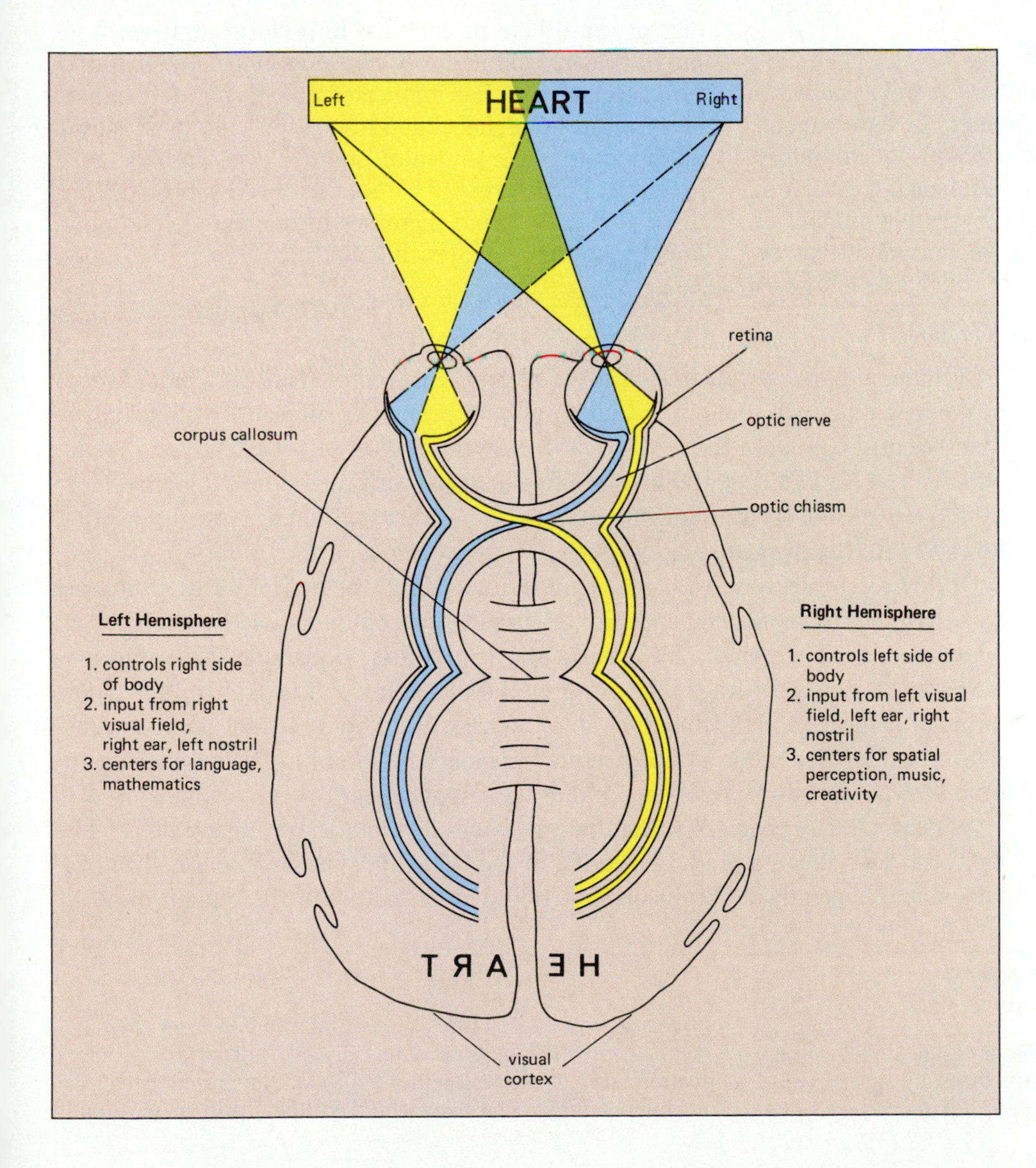

***Figure 29-20** Specializations of the two cerebral hemispheres. Generally, each hemisphere controls sensory and motor functions of the opposite side of the body. Further, the left side seems to predominate in rational and computational activities, while the right side governs creative and spatial abilities.*

Retrieval or recall of established long-term memories is localized in another area of the brain, the **temporal lobes** of the cerebral hemispheres. In a famous series of experiments, neurosurgeon Wilder Penfield electrically stimulated the temporal lobes of conscious patients undergoing brain surgery. The patients did not merely *recall* past memories, but felt that they *experienced* the memories right there in the operating room!

The site of *storage* of complex long-term memories is much less clear. The psychologist Karl Lashley spent many years training rats and subsequently lesioning parts of the brains in an effort to learn the site of the memory trace, but failed. None of Lashley's small lesions could erase a memory completely. Perhaps a given memory is stored in numerous distinct places in the brain. Perhaps memories are stored, like a hologram image, both everywhere and nowhere at the same time: the memory is more precise if the whole brain is intact, but each "bit" (probably several thousands of neurons) of cerebral hemisphere can store an essentially complete memory.

Neurochemicals and Behavior

Twenty years ago biologists thought that the brain functioned with just a few neurotransmitters at its synapses: a couple of excitatory ones and a couple of inhibitory ones. Now we know that the brain is a teeming cauldron, synthesizing and responding to an incredible number of chemicals, including most or all of the hormones that we once thought were unique to the endocrine system. For example, hormones that control digestive tract function are now known to be synthesized in the brain as well, where they control appetite!

Perhaps the most striking findings of the last decade deal with the brain's perception of and response to pain. For centuries, people have known that certain chemicals, such as opium or morphine, reduce pain perception, but did not know why. In the early 1970s, researchers found that these **opiates** bind strongly to certain neurons in the brain, suggesting that these cells have specific receptors for opiates. Since opium and morphine are plant products, it seemed unlikely that the human brain would have evolved receptors specifically for these chemicals. Perhaps, they reasoned, chemically the opiates happen to resemble some naturally occurring brain transmitters that modulate pain perception. Within a few years, several naturally occurring **opioids** (meaning "opiate-like substances") such as **endorphins** were discovered, all peptides of varying length. Endorphins modify our perception of pain. They are believed responsible for allowing a soldier or an athlete to continue to perform even when injured. Their production is also stimulated by acupuncture. Recent work has implicated opioids not only in pain perception and relief, but in appetite, emotional states, attention to environmental stimuli, and perhaps even learning. Future research promises even more startling discoveries and rapid increases in our understanding of how the brain generates the mind.

REFLECTIONS ON MIND AND BRAIN

Humans are intensely interested in the workings of their own minds. Until perhaps a hundred years ago, the mind was more appropriately a subject for philosophers than scientists, because the tools to study the brain did not yet exist. As a result, adherents of disparate viewpoints were free to argue with each other, impelled more by the elegance of their arguments than by the evidence at hand. New discoveries, however, are rapidly changing our views of the workings of the brain.

During recent decades we have begun to understand the neural bases of at least some psychological phenomena. For example, schizophrenics seem to have an overabundance of receptors for a particular transmitter, dopamine. This illness is partially alleviated by drugs that suppress their response to this transmitter. Some forms of clinical depression may be due to excessively low levels of another transmitter, serotonin. Certain antidepressant drugs act by inhibiting reuptake of serotinin at synapses, thus prolonging and enhancing its effect.

As neurobiology progresses, the brain becomes less and less a "black box." The next decades will see a gradual merging of the fields of psychology and neurophysiology, and hopefully, a clearer understanding of the nature of the human mind.

SUMMARY OF KEY CONCEPTS

Nerve Cell Structure

Nervous systems are composed of billions of individual neurons. A neuron is a complex cell whose structures are specialized to perform specific tasks. The cell body coordinates the cell's metabolic activities. Dendrites receive information from the environment or other neurons. The cell body integrates this information and "decides" whether to produce an output signal, the action potential. The axon conducts the signal to its output terminal, the synapse. Synaptic terminals transmit the signal to other nerve cells, glands, or muscles.

Nerve Cell Function

An electrical potential exists between the interiors of neurons and the extracellular fluid surrounding them. In an unstimulated neuron, this is called a resting potential; the inside of the cell is negative with respect to the extracellular fluid. Signals received by the neuron take the form of small, rapidly fading changes in potential called graded potentials. If the graded potential makes the cell substantially less negative, the neuron may reach threshold. This triggers an action potential, during which the inside of the cell becomes momentarily positive. Action potentials are essentially identical, regardless of the strength of the stimulus that evoked them. The action potential is conducted without any change in size from the cell body down the axon to the synaptic terminals.

A synapse consists of the ending of one (presynaptic) neuron and a dendrite of a second (postsynaptic) neuron separated by a small gap. When an action potential reaches the presynaptic ending, it causes the release of neurotransmitter from the presynaptic cell. The neurotransmitter diffuses across the gap and binds to receptor proteins on the surface of the postsynaptic cell. In binding to the receptor, the neurotransmitter causes ion channels to open, allowing ions to flow across the postsynaptic cell membrane. This produces a graded potential which may be excitatory (tending to cause action potentials) or inhibitory (tending to suppress action potentials). The postsynaptic neuron adds up the excitatory and inhibitory graded potentials from many presynaptic cells and produces an action potential if the graded potentials bring it above threshold.

Building and Operating a Nervous System

Information processing in the nervous system requires four operations: (1) detect intensity, (2) detect the type of stimulus, (3) integrate information and (4) initiate action. Nervous systems detect intensity by the frequency of action potentials in single neurons and by the number of similar neurons firing at the same time. The type of stimulus is determined by which neurons fire action potentials. The nervous system collects and processes information from many sources. These may converge on fewer neurons, whose activity in response to these inputs determines action. The "decision" to act may then be transmitted to many more neurons (divergence) which direct the activity.

Neural pathways normally have four elements: a sensory neuron, an association neuron, a motor neuron, and an effector. Overall, nervous systems consist of numerous interconnected nervous pathways, and may be diffuse or centralized.

The Human Nervous System

The nervous system of humans consists of the central nervous system (brain and spinal cord) and the peripheral nervous system. The peripheral nervous system is further subdivided into the somatic nervous system (with sensory and motor components) and the autonomic nervous system (with sympathetic and parasympathetic components). The sympathetic nervous system prepares the body to "fight or take flight," while the parasympathetic nervous system promotes maintenance activities such as digestion.

Within the central nervous system, the spinal cord contains (1) neural pathways for reflexes and certain simple behaviors; (2) motor neurons controlling voluntary muscles; and (3) axons leading to and from the brain. The brain consists of three parts, the hindbrain, midbrain, and forebrain, each further subdivided into distinct structures.

The hindbrain in humans consists mainly of the medulla, which controls involuntary functions (such as breathing), and the cerebellum, which coordinates muscular activities (such as walking). In humans, the small midbrain contains the reticular activating formation, a filter and relay for sensory stimuli. The forebrain includes the thalamus, a relay station that shuttles information to and from higher conscious centers in the forebrain. The limbic system of the forebrain is a collection of diverse structures involved in emotion and the control of instinctive behaviors such as sex, feeding, and aggression. The last part of the forebrain, the cerebrum, is the center for information processing, memory, and initiation of voluntary actions.

Brain and Mind

The cerebral hemispheres are each specialized. The left hemisphere is dominant in speech, reading, writing, langauge comprehension, mathematical ability, and logical problem solving. The right hemisphere specializes in recognizing faces and spatial relationships, artistic and musical abilities, and recognition and expression of emotions.

Memory takes two forms. Short-term memory is electrical, apparently linked to continuous activation of a neural circuit. Long-term memory, in contrast, probably involves structural changes that increase the effectiveness of synapses. The hippocampus is an important site for learning, while the temporal lobes are important for memory retrieval.

The human brain teems with chemicals that act as neurotransmitters or as neuromodulators. Recently, peptides called endorphins have been discovered that bind to the same receptors as do opiates. These influence pain perception, appetite, emotional states, and perhaps learning.

GLOSSARY

Action potential a rapid change from a negative to a positive electrical potential in a nerve cell. This signal travels along an axon without change in size.

Amygdala (am-ig′-da-la) part of the forebrain of vertebrates, involved in control of emotions and instinctive behaviors.

Association neuron in nervous circuits, a nerve cell that is postsynaptic to a sensory neuron and presynaptic to a motor neuron. In actual circuits, there may be many association neurons between individual sensory and motor neurons.

Autonomic nervous system part of the peripheral nervous system of vertebrates that innervates mostly glands and internal organs and produces largely involuntary responses.

Axon a long process of a nerve cell, usually extending from the cell body to synaptic endings on other nerve cells or on muscles.

Brain the part of the central nervous system of vertebrates enclosed within the skull.

Cell body part of a nerve cell in which most of the common cellular organelles are located. Also often a site of integration of inputs to the nerve cell.

Central nervous system in vertebrates, the brain and spinal cord.

Cerebellum (ser-uh-bel′-um) part of the hindbrain of vertebrates, concerned with coordination of motor activities.

Cerebral cortex (ser-ē′-brel kōr′-tex) a thin layer of neurons on the surface of the vertebrate cerebrum, in which most neural processing and coordination of activity occurs.

Cerebral hemisphere one of two nearly symmetrical halves of the cerebrum, connected by a broad band of axons, the corpus callosum.

Cerebrum (ser-ē′-brum) Part of the forebrain of vertebrates concerned with sensory processing, direction of motor output, and coordination of most bodily activities. The cerebrum consists of two nearly symmetrical halves (the hemispheres) connected by a broad band of axons, the corpus callosum.

Convergence a condition in which a large number of nerve cells provide input to a smaller number of cells.

Corpus callosum (kōr′pus kal-ō′-sum) the tract of axons that connect the two cerebral hemispheres of vertebrates.

Dendrite (den′-drīt) the site of signal input to a nerve cell, usually takes the form of branched fibers located close to the cell body.

Divergence a condition in which a small number of nerve cells provide input to a larger number of cells.

Dorsal root ganglion a ganglion located on the dorsal (sensory) branch of each spinal nerve, containing the cell bodies of sensory neurons.

Effector (ē-fek′-tōr) a part of the body (usually a muscle or gland) that carries out responses as directed by the nervous system.

Endorphin (en-dōr′-fin) one of a group of peptides in the vertebrate brain that mimics some of the actions of opiates. Endorphins reduce the sensation of pain.

Excitatory synapse a synapse between two nerve cells in which the resting potential of the postsynaptic cell becomes less negative due to the activity of the presynaptic cell.

Ganglion (gang′-lē-un) a collection of nerve cells.

Graded potential in a nerve cell, an electrical response due to sensory input or synaptic input from another nerve cell. Graded potentials may be positive or negative and vary in amplitude with the strength of stimulation.

Hippocampus (hip-ō-cam′-pus) part of the forebrain of vertebrates, important in motivation, emotion, and especially learning.

Hypothalamus (hī-pō-thal′-uh-mus) part of the forebrain of vertebrates, located just below the thalamus, involved in regulation of hormonal activities (especially of the pituitary gland), and many behaviors such as feeding, drinking, sex, aggression, and fear responses, largely through activation of the autonomic nervous system.

Inhibitory synapse a synapse between two nerve cells in which the resting potential of the postsynaptic cell becomes more negative as a result of the activity of the presynaptic cell.

Integration in nerve cells, the process of adding up electrical signals from sensory inputs or other nerve cells, to determine the overall electrical activity of the nerve cell.

Intensity the strength of stimulation or response.

Limbic system a diverse group of brain structures, mostly in the lower forebrain, including the thalamus, hypothalamus, amygdala, hippocampus, and parts of the cerebrum, involved in emotion, motivation, and learning.

Medulla (med-oo′-la) part of the hindbrain of vertebrates that controls automatic activities such as breathing, swallowing, and heartbeat.

Motor neuron a neuron that carries information from the central nervous system and stimulates effector organs such as muscles or glands.

Myelin (mī′-eh-lin) a wrapping of lipid-rich membranes of Schwann cells around the axon of a vertebrate nerve cell. Myelin increases the speed of conduction of action potentials.

Nerve a bundle of axons of nerve cells, bound together in a sheath.

Neuron (nur′-on) a single nerve cell.

Neurotransmitter a chemical released by a presynaptic cell at a synapse, which binds to receptors on the postsynaptic cell, causing changes in the electrical potential of the second cell.

Node of Ranvier (nōd of ron′-vē-ā) an interruption of the myelin on a vertebrate myelinated axon, at which action potentials are generated.

Opioid (ōp′-ē-ōyd) a group of peptides found in the vertebrate brain that mimic some of the actions of opiates (such as opium, heroin, morphine). Besides analgesia (pain relief), opioids seem to be involved in many behaviors, including emotion, learning, and the control of appetite.

Parasympathetic nervous system the division of the autonomic nervous system that produces largely involuntary responses related to maintenance of normal body functions, such as digestion.

Peripheral nervous system in vertebrates, that part of the nervous system located outside the brain and spinal cord, consisting of the nerves leading to and from the brain and spinal cord, and the ganglia of the autonomic nervous system.

Postsynaptic referring to the nerve cell at a synapse which changes its electrical potential in response to a chemical (the neurotransmitter) released by another (presynaptic) cell.

Potassium channel a hollow-cored protein that spans a nerve cell membrane, forming a pore, and which only allows potassium ions to flow through the pore.

Presynaptic referring to a nerve cell that releases a chemical (the neurotransmitter) at a synapse, which causes changes in the electrical activity of another (postsynaptic) cell.

Receptor (1) a cell that responds to an environmental stimulus (chemicals, sound, light, pH, etc.) by changing its electrical potential; (2) a protein molecule in or on a cell membrane that reacts with another molecule (hormone, neurotransmitter, odorous compound, etc.) to trigger metabolic or electrical changes in the cell.

Receptor potential an electrical change in a cell in response to reception of a stimulus (chemicals, sound, light, pH, heat, cold, etc.) from the internal or external environment.

Reflex a simple, automatic, usually unconscious behavior performed by part of the body in response to a stimulus.

Resting potential an electrical potential found in unstimulated nerve cells; the inside of the cell is negatively charged.

Reticular activating formation (reh-tik′-ū-lar) a diffuse network of neurons extending from the hindbrain, through the midbrain, and into the lower reaches of the forebrain, involved in filtering sensory input and regulating what information is relayed to higher centers in the cerebrum for further attention.

Schwann cell a cell that forms the myelin coating of vertebrate axons by wrapping its cell membrane many times around the axon. *See also* Myelin.

Sensory neuron a nerve cell that carries information about internal or external environmental conditions to the central nervous system.

Sodium channel a protein that spans a nerve cell membrane, forming a pore, and which allows only sodium ions to flow through the pore.

Sodium-potassium pump an enzyme-like protein in nerve cell membranes which actively transports potassium ions into the cell and sodium ions out of the cell.

Spinal cord part of the central nervous system of vertebrates, extending from the base of the brain to the hips, protected by the vertebrae of the spine; contains the cell bodies of motor neurons innervating skeletal muscles, the circuitry for some simple reflex behaviors, and axons communicating with the brain.

Sympathetic nervous system the division of the autonomic nervous system that produces largely involuntary responses that prepare the body for stressful situations.

Synapse (sin′-apz) the site of communication between nerve cells. One cell (presynaptic) usually releases a chemical (the neurotransmitter) which changes the electrical potential of the second (postsynaptic) cell.

Synaptic potential at a synapse between two nerve cells, the electrical change occurring in the postsynaptic cell as a result of reception of neurotransmitter released by the presynaptic cell.

Synaptic terminal the terminal branches of an axon, where the axon forms a synapse, usually with the dendrites of the second nerve cell.

Temporal lobe part of a cerebral hemisphere of the human brain, involved in recall of learned events.

Thalamus part of the forebrain, the thalamus serves as a relay network between other parts of the nervous system and the cerebrum

Threshold the electrical potential (less negative than the resting potential) at which an action potential is initiated.

STUDY QUESTIONS

1. What are five functions of individual neurons? Which structural parts of neurons are specialized for each of these functions?
2. Describe how the resting potential is produced. Include in your answer: which ion(s) are involved; their distributions inside and outside of neurons; and what physical forces cause ion movements across the neural membrane.
3. Describe the events during an action potential. Be sure to include the terms "threshold" and "graded potential."
4. What is the difference between a graded potential and an action potential?
5. Diagram the structure of a synapse. How are signals transmitted from one neuron to another at a synapse?
6. How does the brain perceive the intensity of a stimulus? The type of stimulus?
7. Name three ways in which neurotransmitters are eliminated from a synapse.
8. What are the four elements to a simple nervous pathway? Describe how these elements function in the human pain reflex.
9. Describe and distinguish between convergence and divergence in nervous systems.
10. What is the difference between a diffuse and a concentrated nervous system? What types of animals possess each type?
11. Describe the autonomic nervous system. What are its two subdivisions, and what role does each play in daily life?
12. Draw a cross section of the spinal cord. What types of neurons are located in the spinal cord? What types of behaviors may have their neural circuitry located entirely within the cord?
13. What are the three embryological divisions of the vertebrate brain? In the earliest vertebrates, what were the functions of each part? What structures in the human brain comprise each part?
14. Describe the functions of the following parts of the human brain: medulla, cerebellum, reticular activating formation, thalamus, limbic system, cerebrum.
15. What structure connects the two cerebral hemispheres? Describe the evidence that the two hemispheres are specialized for distinct intellectual functions.
16. Distinguish between long-term and short-term memory.
17. What is an opioid? What types of functions may opioids serve in the human brain?

SUGGESTED READINGS

Thompson, R. F. *The Brain: An Introduction to Neuroscience.* San Francisco: W. H. Freeman and Company, Publishers, 1985. A brief, lucid summary of modern neuroscience.

Eckert, R., Randall, D., and Augustine, G. *Animal Physiology,* 3rd ed. New York: W. H. Freeman and Company, Publishers, 1988. Eckert's descriptions of how nervous systems function are unsurpassed, as are the *Scientific American*–style illustrations.

Lester, H. "The Response of Acetylcholine." *Scientific American,* February 1977. Lester clearly explains the interactions of neurotransmitters with receptors and how ion permeabilities lead to changes in the activity of postsynaptic neurons.

Pearson, K. "The Control of Walking." *Scientific American,* December 1976 (Offprint No. 1346). The neural mechanisms underlying rhythmic movements such as walking seem to be very similar in animals as diverse as cats and cockroaches.

Routtenberg, A. "The Reward System of the Brain." *Scientific American,* November 1978 (Offprint No. 584). This article vividly portrays the crucial roles of limbic system structures in pleasure, learning, and memory.

30
Perception and Action: Senses and Muscles

To interact with its environment and control its body, an animal must perform three operations. First, it must *perceive* its external and internal environment, for example a rabbit behind a nearby bush (external) and an empty stomach (internal). Second, it must *integrate* this information, determine whether conditions deviate from optimum (optimum = full stomach), and decide upon an appropriate response (stalk six steps northeast, then pounce!). Third, it must *act* (stalk, pounce, and if successful, eat). Specialized systems carry out each of these operations: the senses perceive, the central nervous system integrates; and the muscles and skeleton act. In the previous chapter, we examined the central nervous system. Now we turn to the sensory and muscle/skeletal systems responsible for perception and action.

PERCEPTION

Receptor Mechanisms

Despite the diversity of sensory stimuli, almost all sensory systems share similar mechanisms. The dendrites of sensory neurons are specialized to detect a particular stimulus, and only that stimulus. Stimulation of the receptor causes a type of graded potential called a **receptor potential** (Fig. 30-1). The size of the receptor potential is proportional to the size of the stimulus. If the stimulus is sufficiently strong, it will cause a receptor potential that exceeds threshold, initiating action potentials in the sensory neuron. The frequency of action potentials is proportional to the size of the receptor potential, and therefore is also proportional to the intensity of the stimulus (Fig. 30-1). Some receptor cells, such as those specialized for taste, hearing, and light detection, do not have axons or produce action potentials. The receptor potentials produced by these cells cause transmitter release onto a post-

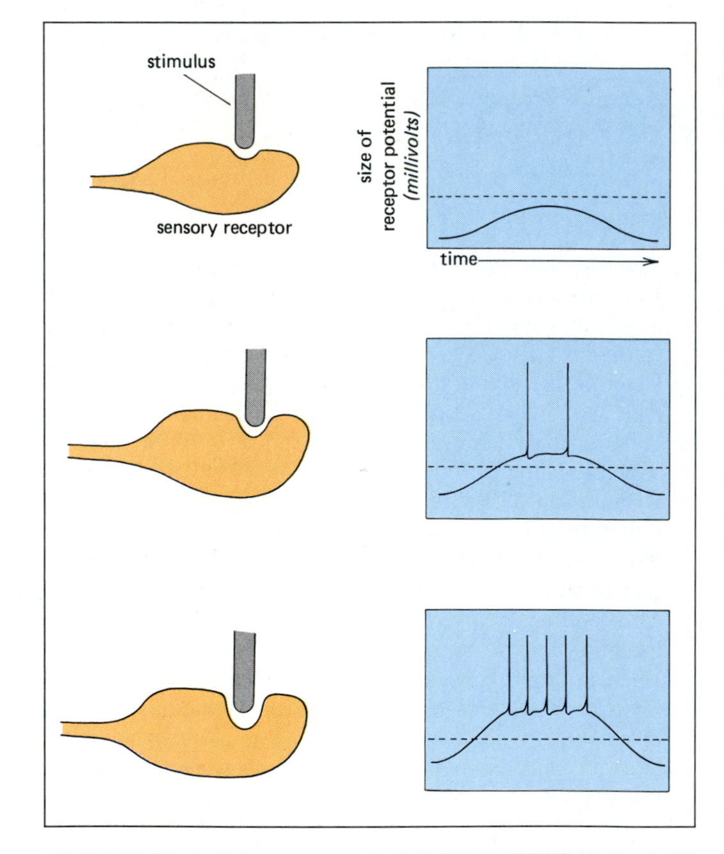

Figure 30-1 Receptor potentials are graded; their size is proportional to the intensity of the stimulus. If the receptor potential exceeds threshold, action potentials are produced whose frequency is proportional to the size of the receptor potential.

synaptic neuron. The amount of transmitter is proportional to the size of the receptor potential. The transmitter in turn causes a graded potential in the postsynaptic neuron that may trigger action potentials which travel to the brain. In this case the frequency of action potentials in

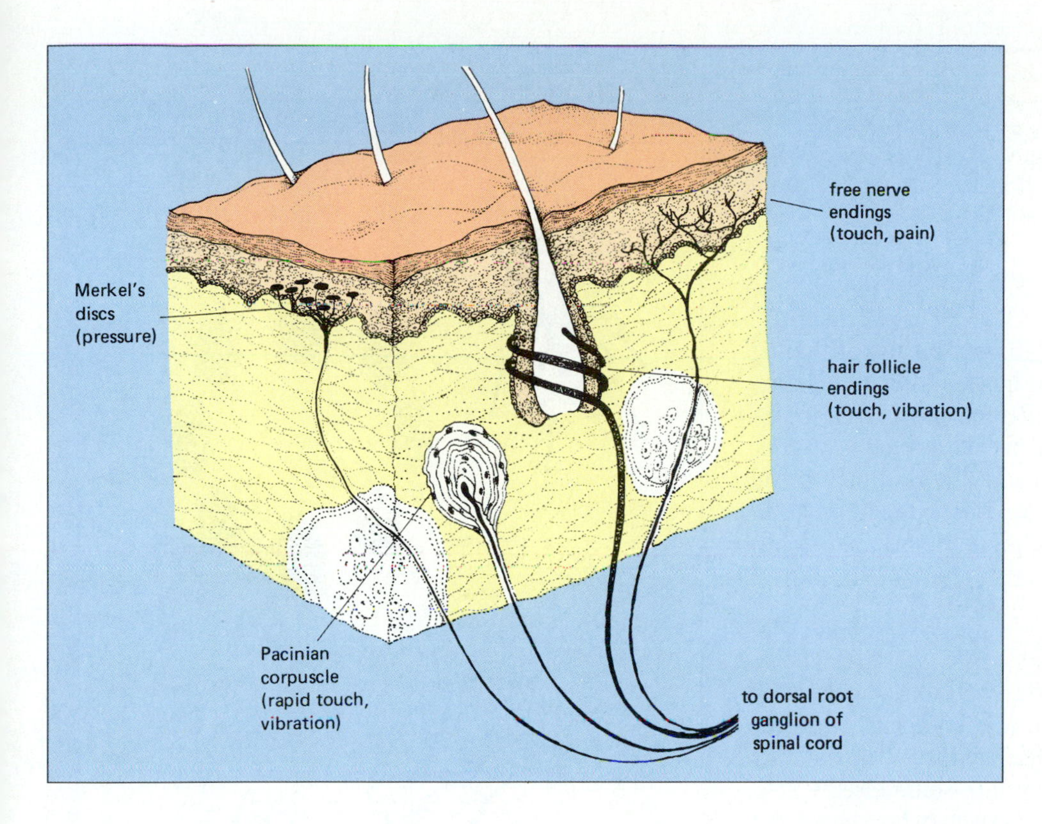

Figure 30-2 *A sampling of receptors in the skin that are sensitive to mechanical stimulation. All are based on nerve endings of sensory neurons that respond to deformation.*

the sensory neuron is once again proportional to the intensity of the stimulus. Most stimuli produce receptor potentials either by mechanical deformation of the receptor cell membrane or by interaction with specialized receptor molecules in the receptor cell membrane.

MECHANICAL DEFORMATION. Mechanical deformation is sensed by cells called **mechanoreceptors.** A receptor potential may be initiated in one of two ways: *stretching the membrane* of the receptor cell or *bending "hairs"* that project from the receptor membrane. The touch receptors in skin, for example, consist of seemingly unspecialized dendrites of neurons (Fig. 30-2). If these dendrites are bent or dented, an area of membrane is stretched. This produces a receptor potential whose size depends on the amount of stretch.

Receptors for sound, motion, and gravity bear "hairs." Currents of fluid or the weight of dense objects bend the hairs, initiating a receptor potential.

MEMBRANE RECEPTOR MOLECULES. Receptors for light, chemicals, and pain possess extensive areas of membrane studded with specialized receptor molecules. When light energy or a chemical stimulus hits a receptor molecule specialized to respond to that stimulus, the receptor molecule changes shape. This alters the permeability of the cell membrane, generating a receptor potential. Keep in mind that the term "receptor" can refer either to an individual sensory cell or to a specialized protein on the cell membrane.

OTHER RECEPTOR MECHANISMS. As usual, not everything fits neatly into our classification scheme, including the receptors responsible for perceiving temperature, electric fields, and magnetic fields. To take one intriguing example, many animals, including insects, birds, and dolphins, have miniature magnets in or near their brains (see the essay, "Road Maps of the Mind"). Although no one knows how these animals "read" their internal magnetic compasses, behavioral experiments show that many animals can navigate using the Earth's magnetic fields.

THE MAJOR SENSES

Mechanical Senses

As the boundary between self and external world, the outer surface of most animals is exquisitely sensitive to touch. Embedded in human skin are several distinct types of sensory structures, all of which contain a sensory ending which produces a receptor potential when its membrane is stretched. These receptors include free nerve endings sensitive to touch or pressure, endings wound around hairs that detect hair movement (such as produced by a crawling insect), and endings enclosed within elaborate capsules, such as the pressure-sensitive Pacinian corpuscles (Fig. 30-2). The density of receptors in the skin varies tremendously over the surface of the body: each square centimeter of fingertip has dozens of touch receptors,

Road Maps of the Mind

The Road goes ever on and on
Down from the door where it began.
Now far ahead the Road has gone,
And I must follow, if I can. . . .
—Bilbo Baggins, in *The Fellowship of the Ring*
by J. R. R. Tolkien

We humans have devised elaborate means of finding our way about: street names, house numbers, odometers to measure distance, and road maps. Without a map, many of us could not drive across town without getting lost. Yet many birds and fish routinely travel hundreds of miles during migrations, often at night, in cloudy weather, or (perhaps most remarkable of all) across the seemingly endless sameness of the ocean. How do they find the Road?

Navigation by Vision

To a human, the most obvious way to navigate is by sight. Close to home, at least, many animals learn the appearance of their surroundings. This is true not only of birds and mammals, but even of some insects.

Many years ago, Niko Tinbergen (who shared the Nobel Prize in 1973 for his pioneering work in animal behavior) studied the homing abilities of digger wasps. A female digger wasp digs a burrow in the soil, lays her eggs, and then goes off to capture insects to stock the nest, providing her offspring with food when they hatch. Tinbergen located such a burrow, and while the female was inside, he surrounded the nest with a ring of pine cones (Fig. E30-1). On her next foraging flight, the female briefly flew about near the entrance before flying out of sight. While she was gone, Tinbergen moved the cones about a foot away. When the wasp returned, she flew to the ring of cones, not to the nest. Even though the burrow was in plain sight nearby, she could not locate it, continuing to search in the center of the cone ring. These and similar experiments showed clearly that the digger wasps memorize certain prominent features of the nest surroundings, which they use to relocate the nests.

Birds can also use landmarks, such as rivers and seashores, but several species fly at night or over large expanses of ocean, using the position of the sun or stars to tell direction. Many species seem to have genetically programmed information about the direction of the sun at various times of day, and also possess a biological clock that measures off a roughly 24-hour day. Other birds have the remarkable ability to "read" the night sky. Indigo buntings, for example, seem to have a built-in star map that enables them to find north during spring migration by looking at the stars.

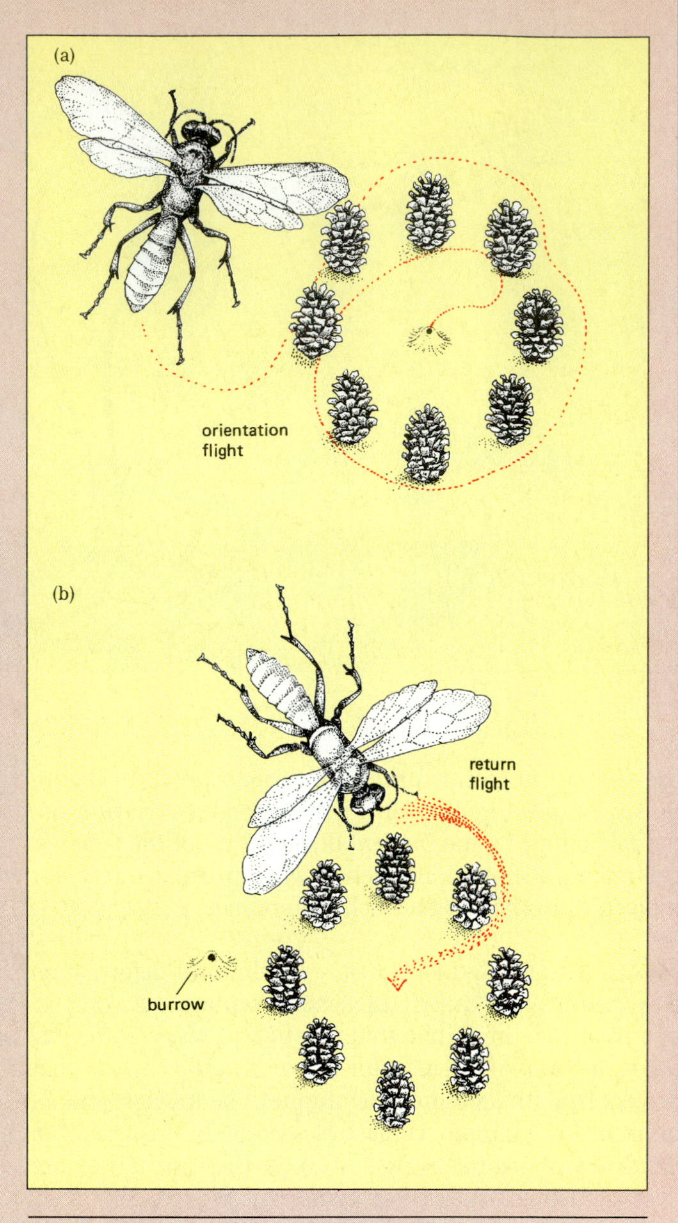

Figure E30-1 **(a)** *When a female digger wasp leaves her burrow, she takes a brief orientation flight, memorizing landmarks such as this ring of pine cones.* **(b)** *The ring of cones is moved. When the wasp returns, she searches for the burrow opening within the ring.*

Navigation by Scent

Since it is difficult to see far under water, aquatic animals tend to rely on cues other than vision to travel long distances. Pacific salmon, for example, migrate hundreds of miles, using scent to locate their final destination. Salmon hatch in freshwater streams, migrate downstream to the ocean to feed and mature, then return to their native stream to spawn (Fig. E30-2). Ex-

Road Maps of the Mind

Figure E30-2 Salmon migrating home to spawn. They will arrive ravaged by the rigors of migration, and die shortly after spawning, their mission complete.

periments by Arthur Hasler have shown conclusively that salmon find their home stream by scent. Hasler raised young salmon in a hatchery with a trace of an odorous chemical, morpholine, added to the water. After a month, the salmon were released into the ocean. When they were scheduled to return, Hasler added morpholine to a stream past which the fish would swim on their migration. The salmon stopped at the treated stream, even though they had never encountered that stream before. Other salmon not raised in morpholine ignored the odor.

Navigation by Magnetic Fields

Migrating with the aid of odors and the position of sun and stars, while remarkable, still cannot explain all the feats of animal navigation. Radar operators, for example, have observed flocks of birds migrating at night under heavy cloud cover, when landmarks, sun, and stars were all unavailable, and at altitudes where scent could not possibly be used (birds smell poorly anyway). Eels of eastern North America and western Europe swim out of streams and rivers into the Atlantic Ocean, and migrate to the Sargasso Sea to spawn. In all that expanse of ocean, it seems unlikely that there could be any consistent chemical cues, and vision certainly cannot be used. How do these birds and eels find their way? The answer seems to lie in responses to the Earth's magnetic field.

Homing pigeons are, of course, famous for their ability to fly home after being released some distance away. They can accurately locate their home roost even under cloudy skies in terrain with few landmarks. If a small magnet is strapped to a pigeon's back, it still homes successfully in sunny weather, but loses its way under overcast skies. Apparently, pigeons can navigate either by the sun or by magnetic fields. In sunny weather, the pigeon can orient by the sun, ignoring the magnetic field. In cloudy weather, the magnets throw off the pigeon's magnetic compass, leaving them no way to find home. How do the pigeons detect magnetic fields? No one knows for sure, but in 1979 it was discovered that pigeons have deposits of magnetite (a magnetic iron compound) located just beneath the skull. These deposits may act as a built-in magnet that the pigeons use to tell direction.

Eels probably also use magnetic fields for navigation, but in a different way. You may recall from high school physics that when an electrical conductor is moved through a magnetic field, an electric current is induced in the conductor (this is how we generate electricity commercially). Seawater, with its high salt concentration, is a fairly good conductor, and the currents of the Gulf Stream provide movement through the Earth's magnetic field. The Gulf Stream generates extremely weak electric fields, roughly equivalent to a potential created by a 1-volt battery with its poles 20 kilometers apart. At first, this was thought to be far too weak to be detected by eels or any other animals. Remarkable behavioral experiments have shown, however, that eels can do much better. Eels were trained to slow their heart rates in response to electrical fields, and the researchers found that eels can detect electric fields as weak as a 1-volt battery with poles *5000 kilometers* apart! For an eel, finding the Gulf Stream must be a piece of cake!

---

while on the back there may be fewer than one receptor per square centimeter. (You can demonstrate this by touching a person with one or two blunt objects simultaneously, varying the separation between the objects. On parts of the back, the subject may perceive only one touch even when the objects are 2 or 3 centimeters apart.)

Similar mechanoreceptive endings in the joints sense the orientation and direction of movement of various body parts. These position sensors, called **proprioceptors,** allow you to walk without watching your feet or eat without watching the fork on its way to your mouth. Still other mechanoreceptors in the walls of the stom-

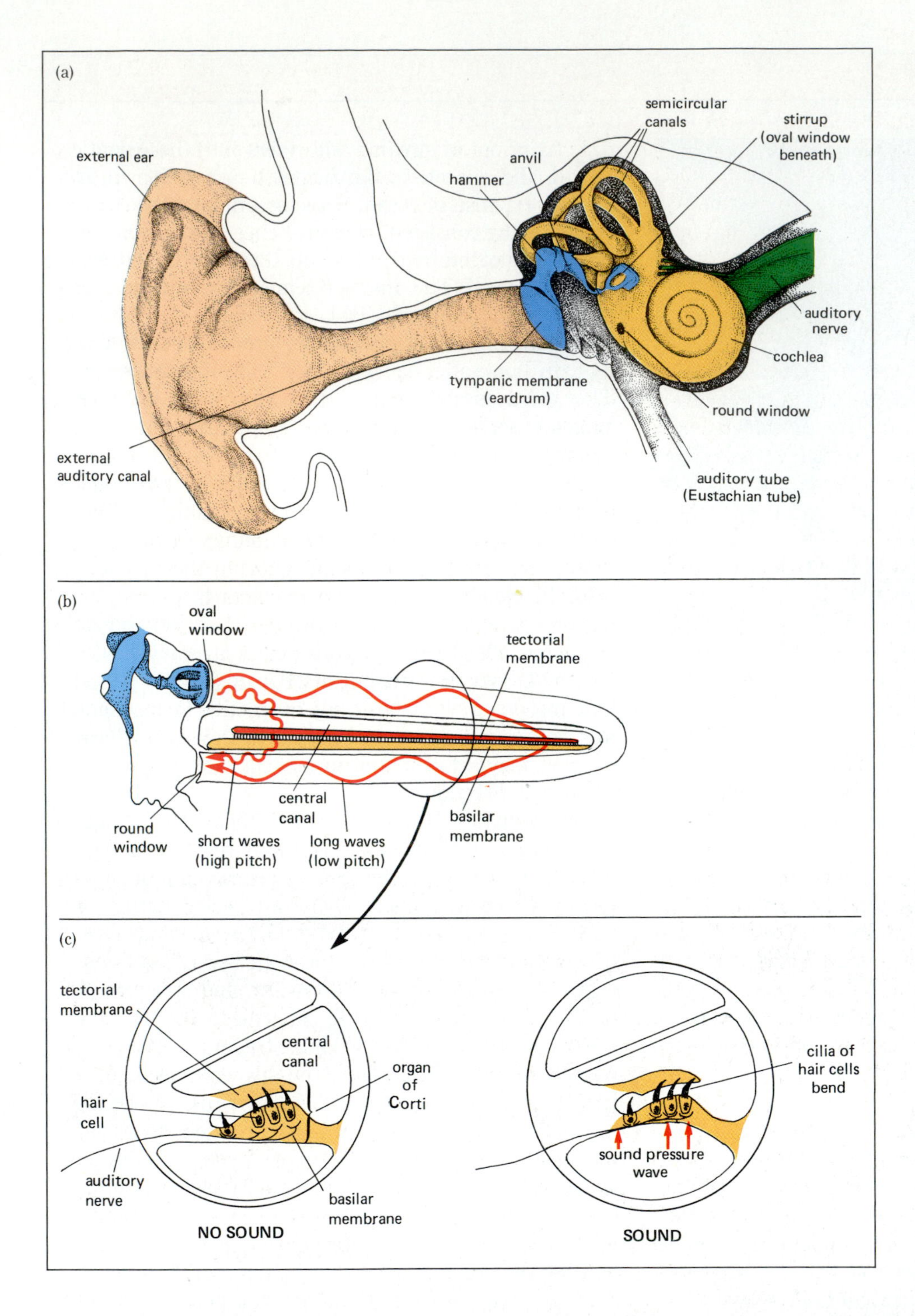

Figure 30-3 **(a)** *The human ear contains elaborate structures for detecting sound, gravity, and movement. Sound waves enter the auditory canal and vibrate the tympanic membrane. These vibrations are transmitted through the bones of the middle ear to the membrane of the oval window, which connects to the fluid-filled cochlea.*
(b) *Uncoiled, the cochlea consists of a bony outer tube surrounding the central canal. Vibrations of the fluid cause the membranes of the central canal to vibrate, stimulating their hair cells. Long wavelengths (low notes) activate receptors toward the tip of the (uncoiled) cochlea, while shorter wavelengths (high notes) activate receptors closer to the oval window.*
(c) *The hairs of hair cells (sound receptors) span the gap between the two membranes of the organ of Corti in the central canal. Sound vibrations move the membranes relative to one another, bending the hairs and producing a receptor potential in the hair cells. The hair cells then cause action potentials in the auditory nerve.*

ach, rectum, and bladder signal fullness by responding to stretch.

Hearing

From both a physical and an evolutionary perspective, hearing is nothing more than a specialized sensitivity to vibration. Virtually all animals have receptors sensitive to vibration: earthworms and insects detect ground vibrations as humans stomp by, while fish detect vibrations of water from approaching predators or prey. Sound, of course, is merely high-frequency vibration of air or water, and many animals use modified vibration receptors for hearing. Both vertebrate and invertebrate sound receptors, although evolutionarily unrelated, utilize the same basic structure consisting of a flexible membrane with receptor cells connected to the membrane. Sound waves cause the membrane to vibrate, bending special "hairs" on the receptor membrane.

The human ear is a remarkable elaboration on this

theme. The ear consists of three parts, the outer, middle, and inner ear (Fig. 30-3). The **outer ear** includes all the structures outside the **tympanic membrane** (eardrum). Sound waves in the air are funneled by the **external ear** into the **auditory canal.** The middle ear is also air-filled, but here vibrations are transmitted through membranes and bones. Sound first vibrates the tympanic membrane, which in turn vibrates three bones, the **hammer, anvil,** and **stirrup.** These bones transmit vibrations from the tympanic membrane to the much smaller membrane of the **oval window.** The oval window covers the opening of the **inner ear,** where the sound vibrations travel through fluid. The fluid-filled, hollow bones of the inner ear form the **cochlea,** where vibrations are translated into neural signals, and the **semicircular canals** (described below). The **Eustachian tube** connects the middle ear to the pharynx, allowing equalization of air pressure between the middle ear and the atmosphere.

The receptor cells for hearing are located within the inner ear, in the spiral-shaped cochlea (Fig. 30-3a). If we mentally straighten out the cochlea (Fig. 30-3b), we can see that it consists of two fluid-filled tubes, an outer U-shaped canal and a straight **central canal.** The central canal contains the **basilar membrane,** which bears receptors called **hair cells.** Protruding into the central canal is another membrane, the **tectorial membrane,** in which the hairs of the hair cells are embedded. The basilar membrane, tectorial membrane, and hair cells together are called the **organ of Corti.**

Let's see how these structures allow the perception of sound. When sound waves enter the ear, they vibrate in turn the tympanic membrane, the bones of the middle ear, the membrane of the oval window, and the fluid in the cochlea. Once in the cochlea, the waves vibrate the basilar membrane, causing it to move relative to the tectorial membrane. This bends the hairs spanning the gap between the membranes (Fig. 30-3c) and causes receptor potentials in the hair cells. The hair cells release transmitter onto neurons of the auditory nerve. Action potentials are triggered in the auditory nerve axons and travel to the brain.

The organ of Corti also allows us to perceive loudness and pitch. A weak sound causes small vibrations, which bend the hairs only slightly. This produces small receptor potentials in the hair cells and a low frequency of action potentials in the auditory nerve axons. A loud sound causes large vibrations, which cause greater bending of the hairs and a larger receptor potential. This leads to a high frequency of action potentials in axons of the auditory nerve. Very loud sounds sustained for a long time can actually damage the hairs (Fig. 30-4), resulting in hearing loss.

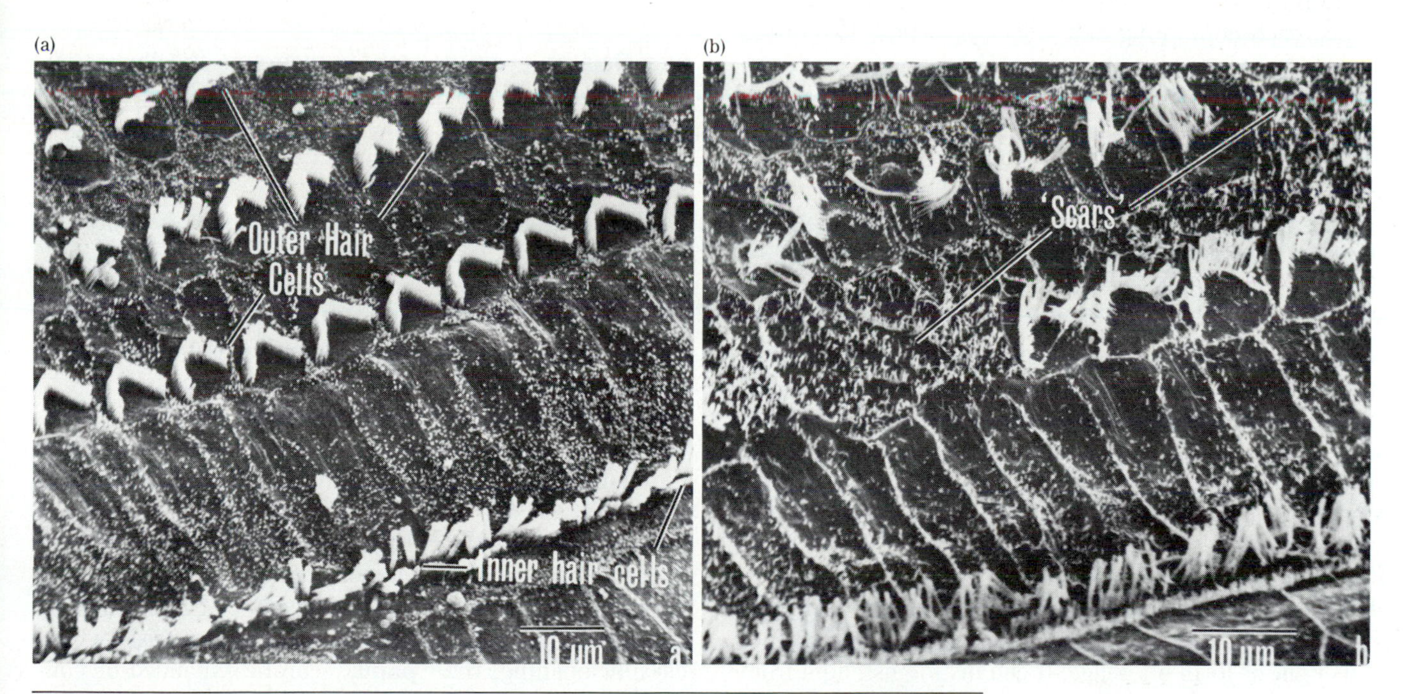

Figure 30-4 Scanning electron micrographs show the effect of intense sound on the hair cells of the inner ear. **(a)** *Organ of Corti of a normal guinea pig, showing three rows of outer hair cells with the hairs of each receptor arranged in a V-shaped pattern.* **(b)** *Organ of Corti of a guinea pig after 24-hour exposure to a sound level approached by loud rock music (2000 hertz at 120 decibels). Note that many of the hairs are damaged or missing entirely. Since hair cells do not normally regenerate, hearing loss is permanent. (Scanning electron micrographs by Robert S. Preston, courtesy of Professor J. E. Hawkins, Kresge Hearing Research Institute, University of Michigan Medical School.)*

The perception of pitch is a little more complex. The basilar membrane is stiff and narrow at the end near the oval window and more flexible and wider near the tip of the cochlea. This progressive change in structure causes different parts of the membrane to resonate best to particular frequencies of sound. High-frequency sound waves (high notes) cause the greatest vibration of the end near the oval window, while lower frequencies (lower notes) cause greatest vibration progressively farther toward the tip (Fig. 30-3b). Thus *where the basilar membrane vibrates most, and consequently which receptors are stimulated most, varies with the frequency of sound.* The brain interprets signals from receptors near the oval window as high-pitched sound, and signals from receptors farther along as lower in pitch.

OTHER FUNCTIONS OF THE INNER EAR. Besides hearing, the mammalian inner ear also contains two other sets of organs that are modified mechanical senses, one for detecting motion and the other for detecting gravity. Motion perception occurs via the **semicircular canals,** fluid-filled channels containing clusters of hair cells similar to those of the inner ear (Fig. 30-3a). Sudden acceleration of the head causes the fluid to lag behind the movement of the canals, bending the hairs in one direction, while cessation of motion causes it to slosh ahead, bending the hairs oppositely. Gravity perception is the function of two chambers at the base of the semicircular canals, each containing hair cells atop whose hairs rests a gelatinous mass containing stones of calcium carbonate. The heavy stones are pulled downward by gravity, bending the hairs according to the orientation of the head.

Vision

Sight is extremely important in obtaining food, avoiding predation, and maneuvering in the environment. Animal vision varies in acuity, and several types of eyes have evolved independently. All forms of vision, however, utilize **photoreceptors.** These sensory cells contain receptor molecules (called photopigments because they are colored) that absorb light and chemically change in the process. This chemical change alters ion channels in the receptor cell membrane, producing a receptor potential.

TYPES OF EYES. The simplest form of light receptor in the animal kingdom is the **eyespot,** found in flatworms (Fig. 30-5). The eyespot usually has no lens and cannot focus light or form an image. It can distinguish light from dark and may also perceive its direction and intensity. This information can be important to the animal: a passing shadow, for example, could be a predator, and often triggers withdrawal or escape responses.

The insects evolved **compound eyes** that consist of a mosaic of numerous individual light-sensitive subunits called **ommatidia** (Fig. 30-6). Although much more complex than an eyespot, each ommatidium functions similarly as an on/off, bright/dim detector. Using a large number of individual units (up to 30,000 per eye in a dragonfly), the insect obtains a reasonably faithful, although grainy image of the world.

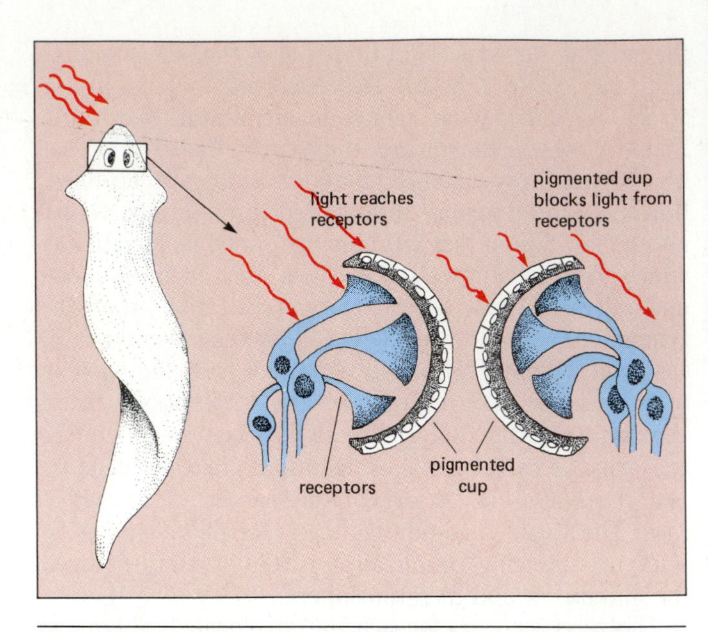

Figure 30-5 The most primitive eyes in the animal kingdom are eyespots, such as those found in flatworms. Eyespots consist of photoreceptors in a pigmented cup. Light entering the open end of the cup stimulates the photoreceptors, but the pigmented back side of the cup prevents light from other directions from reaching them. This allows detection of the direction of light, guiding simple behaviors such as finding a dark place to hide.

The molluscs and vertebrates independently evolved a third type of eye, often called the **camera eye** (Figs. 30-7 and 30-8a). The camera eye consists of three basic parts: a light-sensitive layer, the retina, a lens for focusing light, and a set of muscles for adjusting focus by moving or changing the shape of the lens.

STRUCTURE AND FUNCTION OF THE HUMAN EYE. As illustrated in Fig. 30-8a, incoming light first encounters the **cornea,** a transparent covering over the front of the eyeball. Behind the cornea is a chamber filled with a watery fluid called **aqueous humor,** which provides nourishment for the lens. The amount of light entering the eye is adjusted by a muscular tissue, the **iris,** whose circular opening, the **pupil,** can be expanded or contracted. Light passing through the pupil encounters the **lens,** a structure resembling a flattened sphere and composed of transparent proteinaceous fibers. The lens is suspended behind the pupil by ligaments and muscles that regulate its shape. Behind the lens is another, much larger chamber filled with a clear jellylike substance, the **vitreous humor,** which helps maintain the shape of the

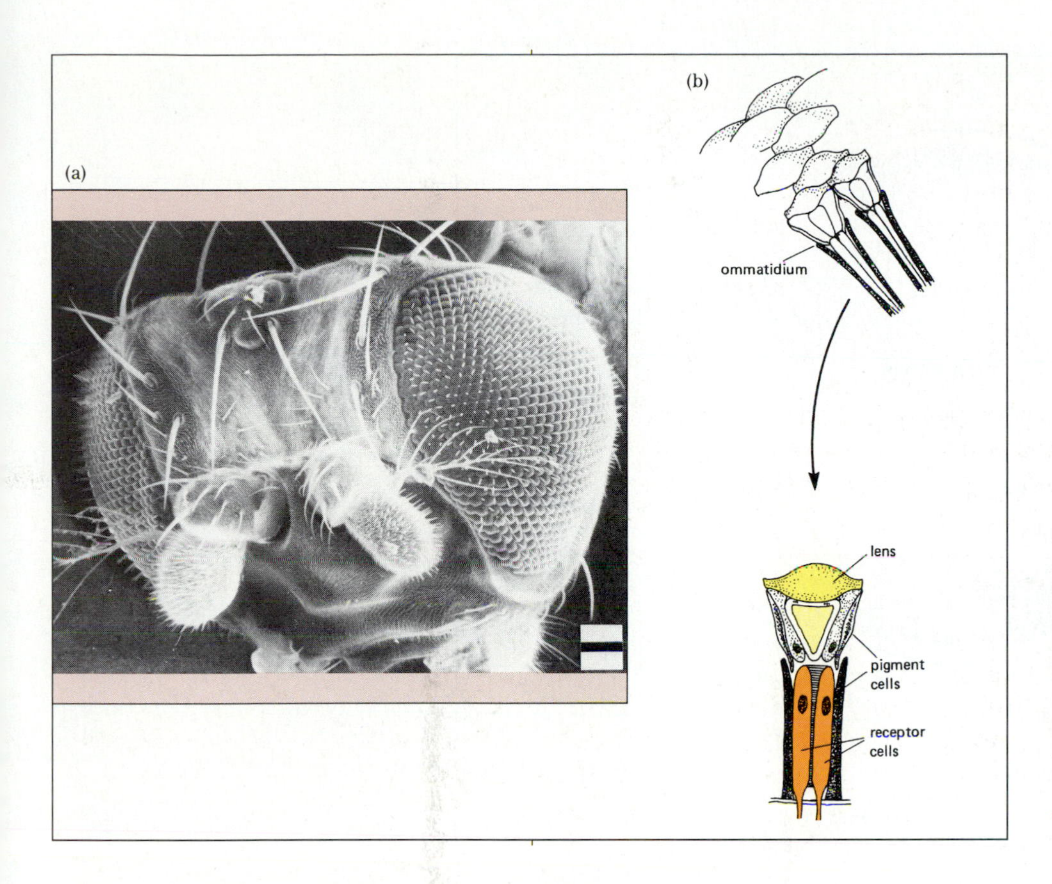

Figure 30-6 **(a)** *Scanning electron micrograph of the compound eye of a fruit fly.* **(b)** *Each eye is made up of numerous individual light-receptive ommatidia. Within each ommatidium are several receptor cells, capped by a lens. Pigmented cells surrounding each ommatidium prevent light from passing through to adjacent receptors.*

Figure 30-7 Molluscs such as this octopus have evolved camera eyes independently of the vertebrates. The octopus eye is very similar in structure to the vertebrate eye, with one improvement: the receptors are at the front of the retina, facing the light, and the optic nerve leaves via the back.

eye. After passing through the vitreous humor, light reaches the retina, a multilayered nervous tissue where the light energy is converted into electrical nerve impulses that are transmitted to the brain (Fig. 30-8b). The retina is richly supplied with blood vessels and contains a layer of pigment that absorbs stray light rays that escape the photoreceptors. Behind the retina is a darkly pigmented tissue, the **choroid.** The choroid's rich blood supply helps nourish the cells of the retina. Its dark pigment absorbs stray light whose reflection inside the eyeball would interfere with clear vision. Surrounding the outer portion of the eyeball is a tough connective tissue layer, the **sclera,** visible as the white of the eye.

In the eye, the image is focused not only by the lens but also by the cornea. The cornea actually does most of the bending of incoming light rays, producing an image of approximately the right size in the general vicinity of the retina. However, the shape of the cornea cannot be adjusted, and the lens is responsible for final, sharp focusing. In the relaxed eye, the ligaments stretch the lens into a relatively flattened shape that allows focusing of distant objects. The lens muscles are arranged so that contracting them releases some of the tension on the ligaments, allowing the lens to become rounder, and to focus on nearby objects (Fig. 30-9a). Nearsighted people cannot focus on distant objects; those who are farsighted cannot focus on nearby objects. These conditions, which are usually caused by abnormally long or short eyeballs, can be corrected by external lenses of the appropriate shape (Fig. 30-9b,c).

The vertebrate eye provides the sharpest vision in the animal kingdom, even though the retina, a complex, multilayered structure, is "built backward" (Fig. 30-8b). The photoreceptors, called **rods** and **cones** after their shapes, have their light-gathering elements farthest away from the

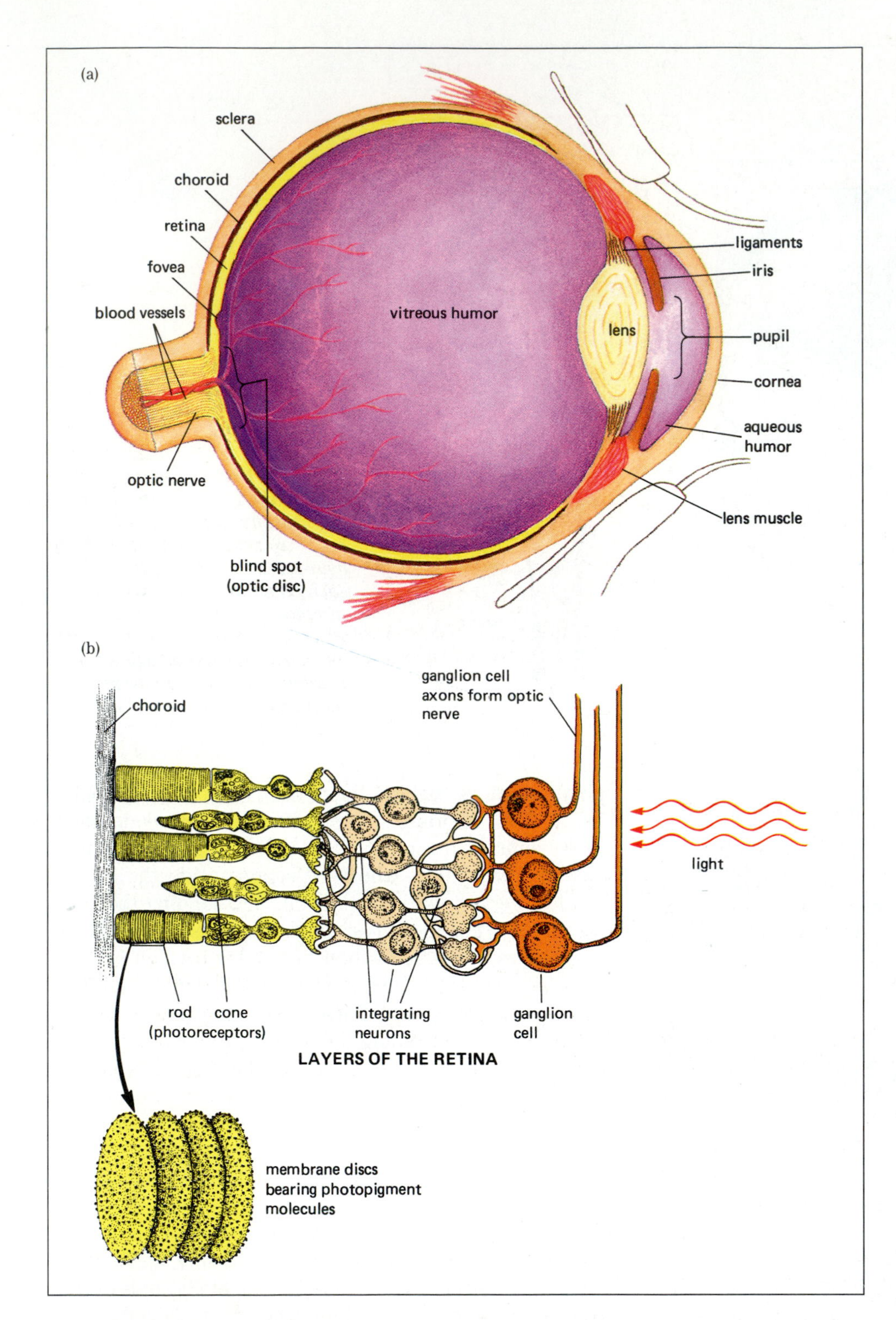

Figure 30-8 **(a)** *The anatomy of the human eye.* **(b)** *The human retina has rods and cones (photoreceptors), integrating cells, and ganglion cells. Each rod and cone bears a long extension packed with membranes in which the light-sensitive molecules are embedded.*

light, at the rear of the retina. Between the receptors and incoming light lie several layers of neurons that process the signals from the photoreceptors. The outermost layer consists of **ganglion cells,** whose axons make up the optic nerve. Action potentials in the ganglion cell axons travel to the brain, providing the sensation of vision. These axons emerge from the front of the retina, then must pass back through the retina to reach the brain. The point where axons pass through the retina is called the optic disk or blind spot (Figs. 30-8a and 30-10). This area lacks receptors, and objects focused here seem to disappear.

Photoreception in both rods and cones begins with absorption of light by a pigment composed of a protein called **opsin** and a vitamin A derivative, **retinal.** When light hits the pigment molecule, it changes the shape of the retinal part, causing it to detach from the opsin. This causes a receptor potential in the photoreceptor cell. The pigment molecules are embedded in the membranes at

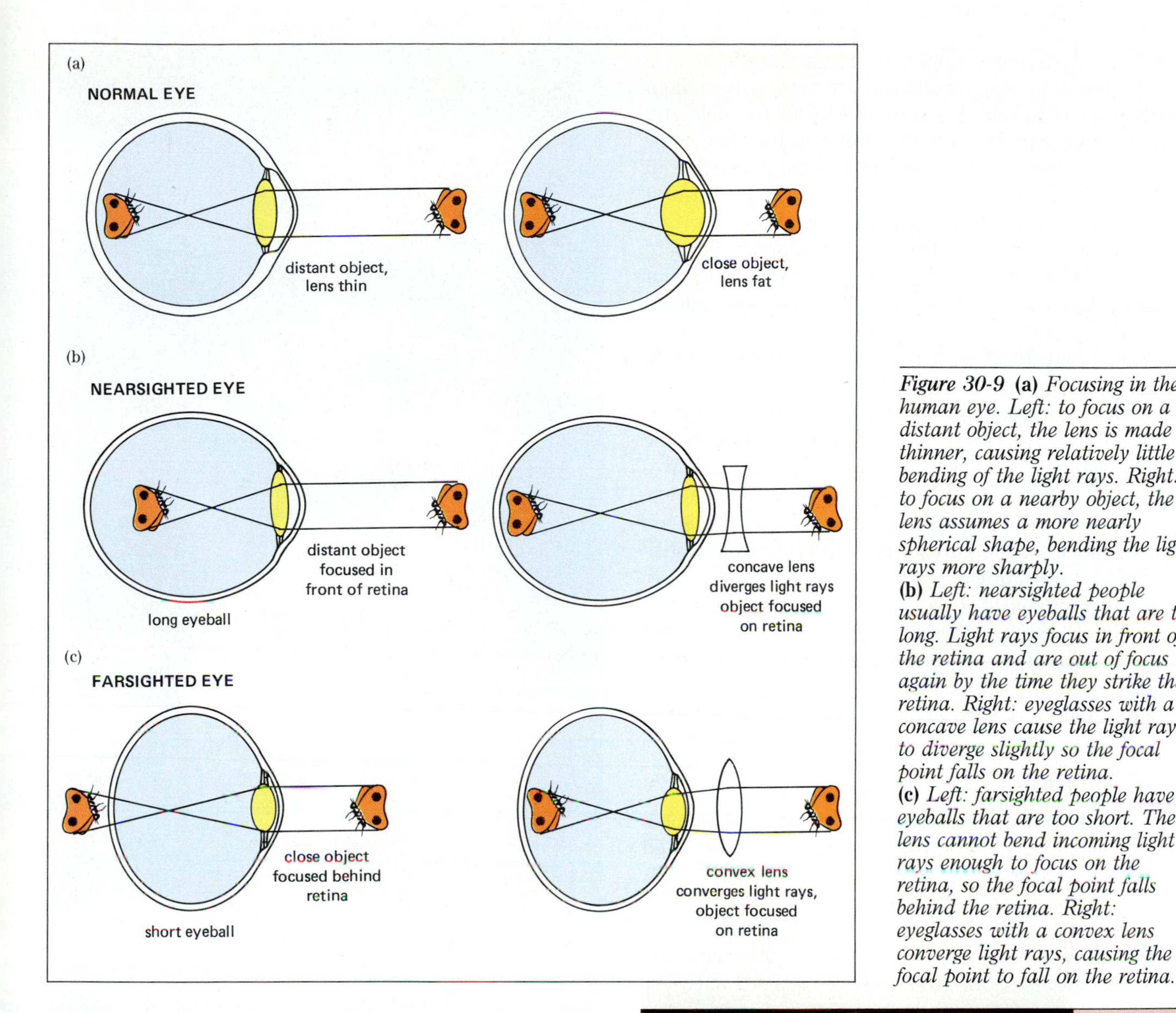

Figure 30-9 **(a)** *Focusing in the human eye. Left: to focus on a distant object, the lens is made thinner, causing relatively little bending of the light rays. Right: to focus on a nearby object, the lens assumes a more nearly spherical shape, bending the light rays more sharply.*
(b) *Left: nearsighted people usually have eyeballs that are too long. Light rays focus in front of the retina and are out of focus again by the time they strike the retina. Right: eyeglasses with a concave lens cause the light rays to diverge slightly so the focal point falls on the retina.*
(c) *Left: farsighted people have eyeballs that are too short. The lens cannot bend incoming light rays enough to focus on the retina, so the focal point falls behind the retina. Right: eyeglasses with a convex lens converge light rays, causing the focal point to fall on the retina.*

the end of the photoreceptors. These membranes form flattened, hollow disks in rods, and are deeply folded in cones, giving them a large surface area bearing the photopigments (Fig. 30-8b).

The 125 million rods are scattered more or less uniformly over the retina, and are marvelously sensitive to light. As shown in Fig. 30-8b, rods have much deeper stacks of pigment-bearing membrane than cones do, and consequently are much more light sensitive than cones. Vision in dim light is almost entirely due to rods. Unlike cones, rods do not distinguish colors, so in the moonlight our world becomes black and white.

The 5 million cones are found throughout the retina, but are concentrated in one small area, the **fovea** (Fig. 30-8a). The fovea, which is where the lens focuses images most sharply, consists entirely of densely packed cones. The fovea appears as a depression on the retina because the layers of neurons covering the photoreceptors are pushed aside here, while still retaining their synaptic connections. Thus light reaches the receptors of the fovea

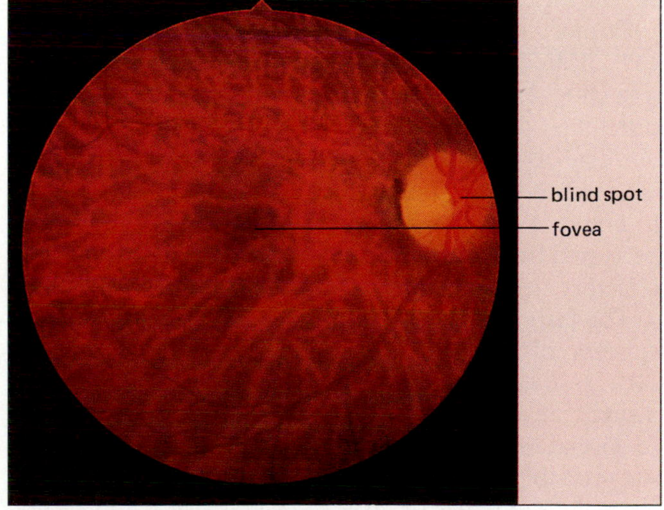

Figure 30-10 A photograph of the human retina, taken through the cornea and lens of a living person. Blood vessels supply oxygen and nutrients. The blind spot (optic disk) and fovea are visible.

with less interference. Cones, having a smaller area of photopigment-bearing membrane, are not nearly as light sensitive as rods and are used mostly in the daylight. Human cones can distinguish colors because they come in three varieties, each containing a visual pigment with a different type of opsin. Each type of cone is most strongly stimulated by a particular color, either red, green, or blue. The brain distinguishes color according to the relative intensity of stimulation of different cones. For example, the sensation of yellow is caused by roughly equal stimulation of red and green cones.

About 5 percent of all males lack normal color vision. Although described as "color blind," they are actually only color deficient. The most common abnormality is red-green color deficiency, due to a recessive allele on the X chromosome (see Chapter 12) that codes for a defective opsin in the red cones. The altered red photopigment has about the same light-absorbing properties as the green photopigment, so the affected person has trouble distinguishing red from green.

Not all animals have both rods and cones. Animals active almost entirely during the day (certain lizards, for example) may have all-cone retinas, while night-active animals (such as the ferret) or those dwelling in dimly lit habitats (such as deep-sea fishes) often have mostly rods.

Chemical Senses

Virtually all animals have **chemoreceptors** that sample the chemical composition of the environment. Through chemical senses, animals find food, avoid poisonous materials, and may locate homes or find mates (Fig. 30-11). Terrestrial vertebrates have two separate chemical senses: one for airborne molecules, called smell or **olfaction,** and one for chemicals dissolved in an aqueous medium (water or saliva), called **taste.**

In most vertebrates, receptors for olfaction are nerve cells located in tissues lining the back of the nasal cavity (Fig. 30-12). These sensory neurons have hairlike dendrites that protrude into the cavity and sample incoming air. The olfactory neurons are specialized, bearing different receptor molecules on their hairs. Each responds only to one or a few types of chemicals. Individual odors are perceived by the brain as patterns of activity in particular olfactory neurons.

The taste buds, located on the tongue, consist of small clusters of taste receptors and supporting cells (Fig. 30-13). Taste receptor cells protrude microvilli through a small pore. Dissolved chemicals enter the pore and bind to special receptor molecules on the microvilli. Although we probably perceive hundreds of distinct tastes, there are only four types of taste receptors: sweet, sour, salty, and bitter. The great variety of tastes is produced in two ways. First, a particular substance may stimulate two or more receptor types to different degrees and taste, for example, "salty–bitter." More important, material being tasted usually also gives off molecules into the air inside the mouth. These odor molecules diffuse to the olfactory receptors (remember, the mouth and nasal passages are connected). The fact that what we call "taste" is mostly

Figure 30-11 *Male moths* **(a)** *find females* **(b)** *not by sight, but by following airborne scents (pheromones) released by the females. These odors are sensed by receptors on the male's huge antennae, whose enormous surface area maximizes the chances of detecting the female scent.* **(c)** *When dogs meet they usually sniff each other about the base of the tail. Scent glands there seem to broadcast information about sex (both type and interest in) and status, and influence what behaviors follow.*

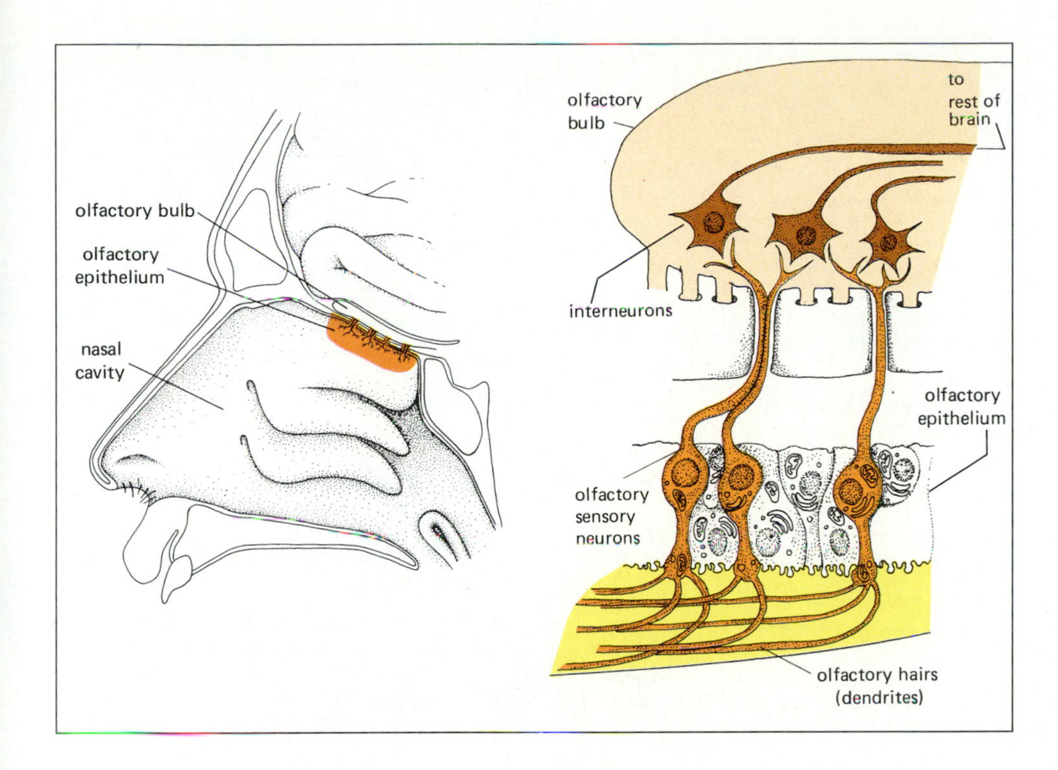

Figure 30-12 The receptors for olfaction in humans are neurons that bear hairlike projections protruding into the nasal cavity. The projections are embedded in a mucus layer, in which odor molecules dissolve before contacting the receptors.

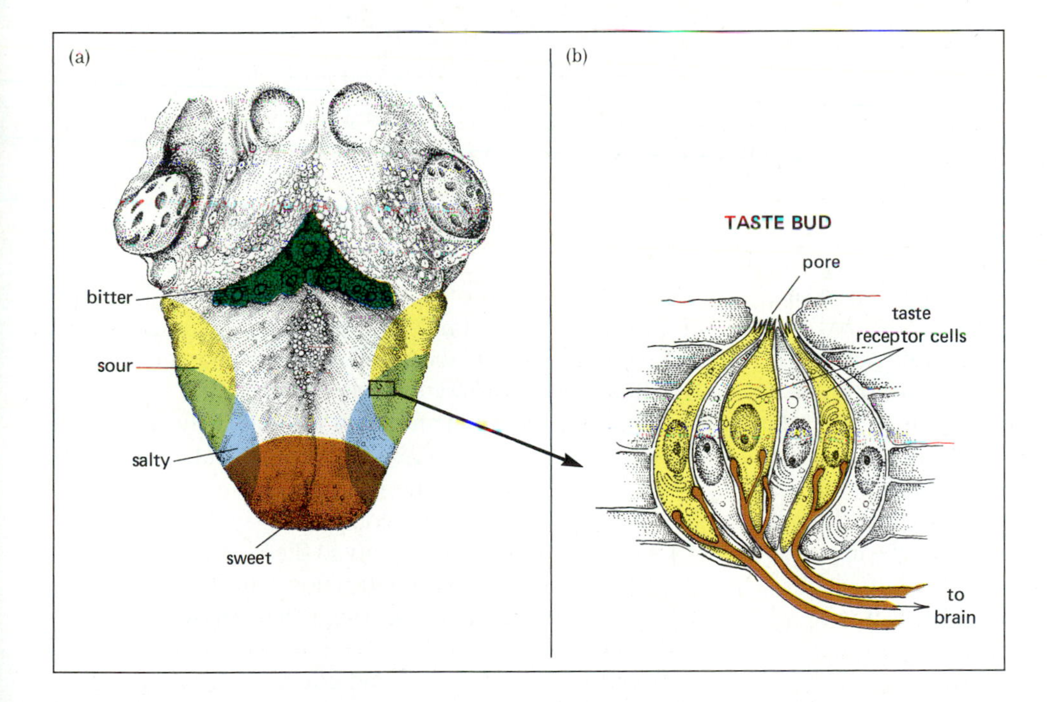

Figure 30-13 **(a)** *The human tongue bears numerous bumps in which are clustered masses of taste buds.* **(b)** *Each taste bud consists of supporting cells and several taste receptors, whose microvilli bind the tasty molecules. Axons from neurons in the brain receive synapses from the receptor cells and transmit messages to the brain.*

due to our sense of smell is clearly illustrated by the blandness of normally tasty foods when a cold plugs up our nasal passages.

Pain

Pain is an unusual sense in that it is somewhat nonspecific. Whether you burn, cut, or crush a fingertip, you will feel pain. This gives us a clue to the nature of pain perception: most pain is produced by tissue damage, regardless of the cause. Over the last few years, researchers have found that pain perception is actually a special kind of chemical sense (Fig. 30-14). When cells are broken open by a cut or a burn, for example, their contents flow into the extracellular fluid and blood. The cell contents include enzymes that convert certain blood proteins into a chemical called **bradykinin.** Pain receptors, which are dendrites of specific sensory neurons, have receptor molecules for bradykinin. Binding of bradykinin to these receptors results in action potentials that are interpreted as

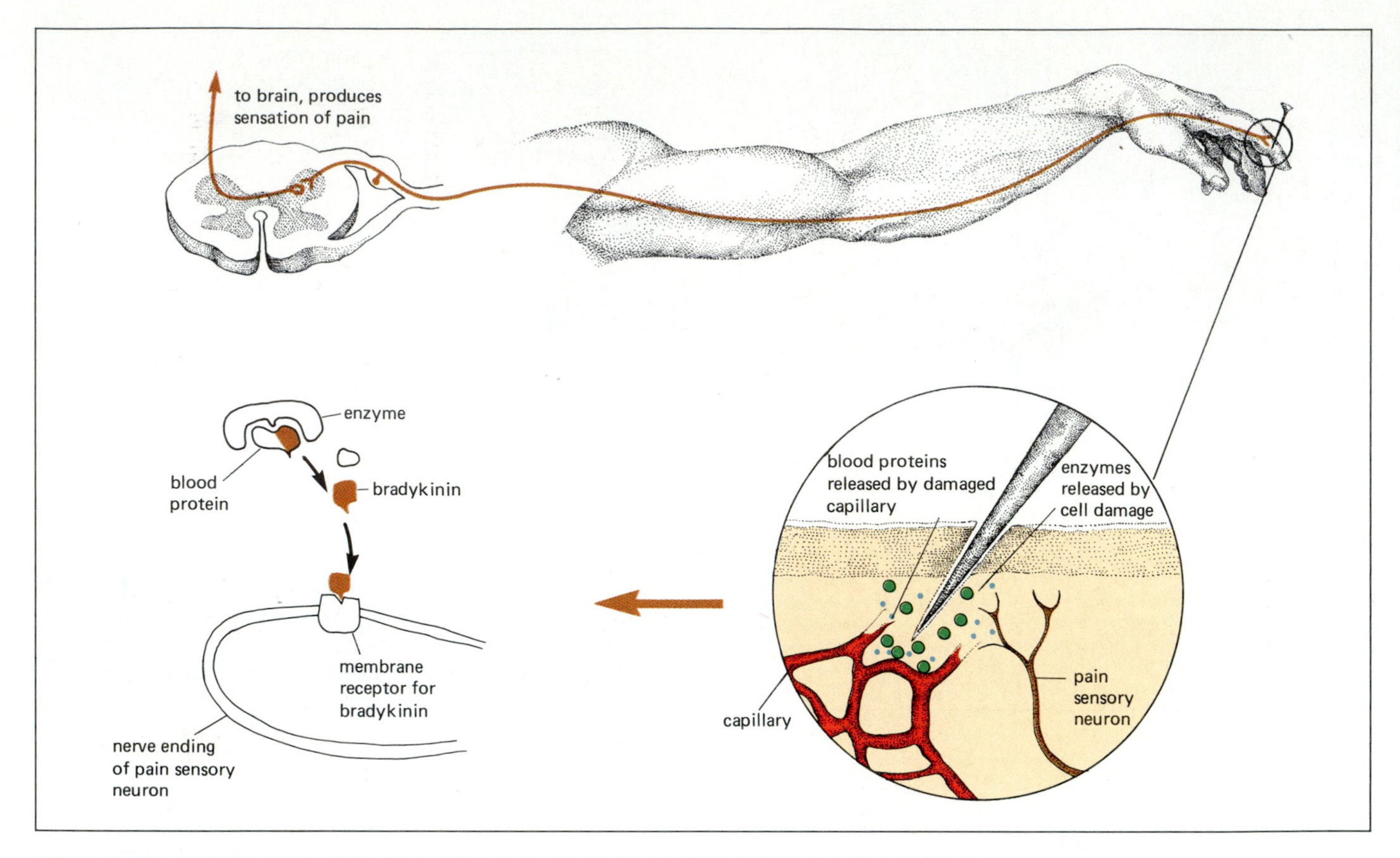

Figure 30-14 Pain perception is a specialized chemical sense. An injury damages both cells and blood vessels. The cells release enzymes that convert certain blood proteins into bradykinin. Bradykinin stimulates pain-sensitive neurons by binding with membrane receptor molecules.

pain by the brain. Since each part of the body has a separate set of pain neurons that provide input to particular brain cells, the brain knows where the pain is occurring.

Drugs that provide pain relief, such as morphine or demerol, block synapses in the pain pathways of the brain or spinal cord. In ways that we are just beginning to understand, the brain can modulate its perception of pain through its own narcotic-like **endorphins** (see Chapter 29). In critical situations, such as combat or during escape from a fire, endorphins may allow us to function by blocking our perception of pain until the emergency is over.

ACTION

Sensory receptors perceive the external and internal environment and communicate information to the brain, which determines a course of action. The role of the muscles and skeleton, discussed in this section, is to implement the decisions of the brain.

Muscle Structure

Many animals cells are capable of some sort of movement (see Chapter 5). Cellular motion is often based on the relative movements of two types of protein strands, **actin** and **myosin.** In animals, the evolution of muscles has been a remarkable elaboration of this preexisting system. Animals have a variety of muscle types, each specialized to perform a particular function. Mammals have evolved three distinct types of muscle: cardiac, smooth, and skeletal. All work on the same basic principles, but differ in function, appearance, and control.

Cardiac muscle is found only in the heart. Cardiac muscle can initiate its own contractions; in fact, vertebrate hearts will often keep on beating when removed from the body. The cardiac muscle cells must contract almost simultaneously to squeeze blood out of the heart chambers efficiently. To accomplish this, specialized porous membranes connect adjacent cells, allowing electrical potentials to travel from one cell to the next, synchronizing their contractions.

Smooth muscles surround blood vessels and most hollow organs, such as the uterus, bladder, and digestive tract. They usually produce either sustained contractions (such as constriction of the arterioles to elevate blood pressure during times of stress) or slow, wavelike contractions (such as the peristaltic waves that move food along the digestive tract). Like cardiac muscle, smooth muscle cells are often directly connected to one another,

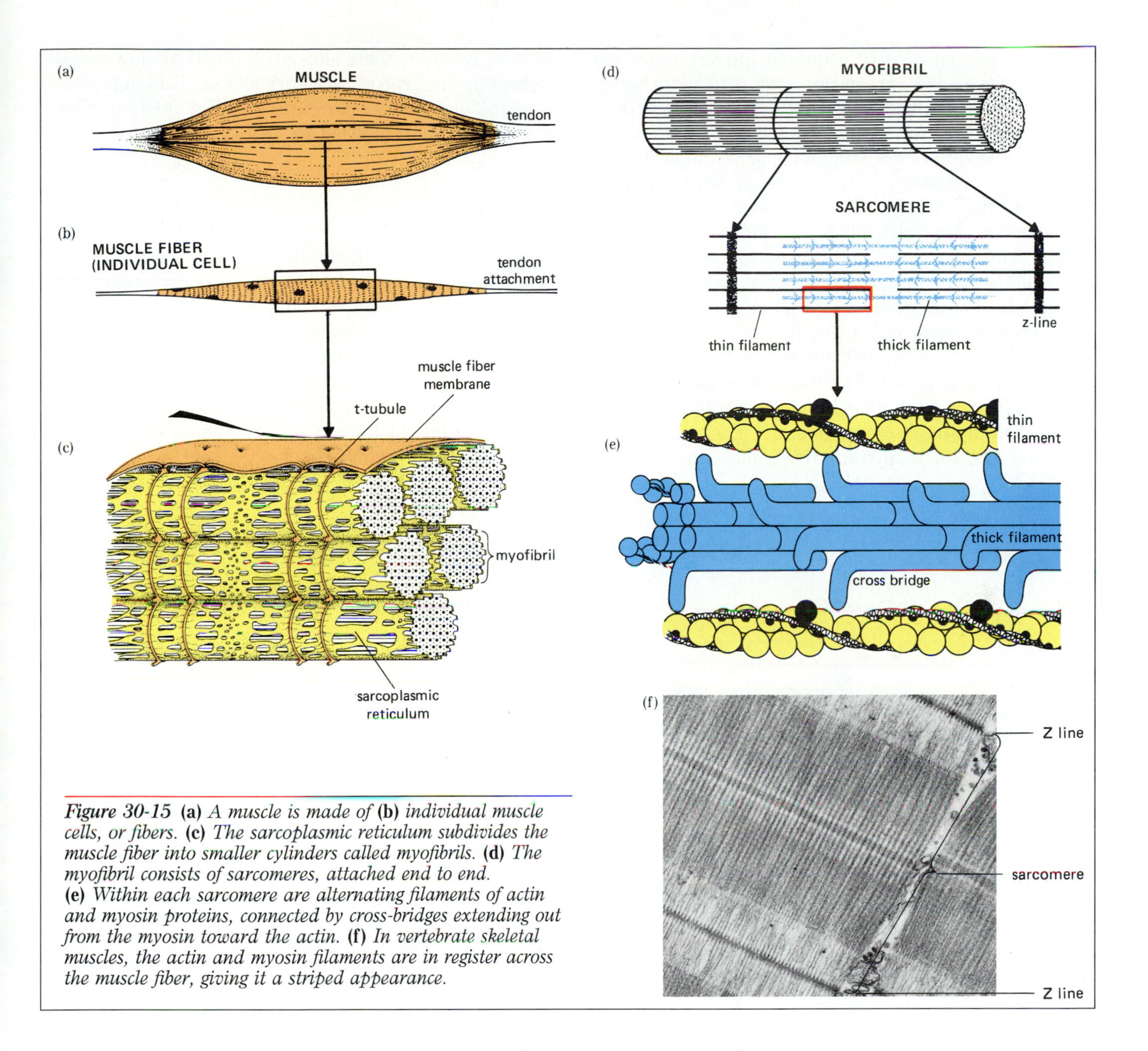

Figure 30-15 **(a)** *A muscle is made of* **(b)** *individual muscle cells, or fibers.* **(c)** *The sarcoplasmic reticulum subdivides the muscle fiber into smaller cylinders called myofibrils.* **(d)** *The myofibril consists of sarcomeres, attached end to end.* **(e)** *Within each sarcomere are alternating filaments of actin and myosin proteins, connected by cross-bridges extending out from the myosin toward the actin.* **(f)** *In vertebrate skeletal muscles, the actin and myosin filaments are in register across the muscle fiber, giving it a striped appearance.*

allowing synchronized contraction. Smooth muscle contraction may be activated by stretch, by hormones, by nervous signals, or by some combination of these stimuli.

Skeletal muscle, also called **striated muscle** because of its striped appearance under the microscope, is used to move the skeleton. Skeletal muscles are under the direct control of the nervous system and can produce contractions ranging from quick twitches (as in blinking) to powerful, sustained tension (as in carrying an armload of textbooks). Our discussion of muscle function will focus on skeletal muscles.

ANATOMY OF SKELETAL MUSCLE. Individual muscle cells, or **muscle fibers,** are among the largest cells in the human body. Ranging from 10 to 100 micrometers in diameter, most muscle fibers run the entire length of the muscle, which may be as much as 30 centimeters in a human thigh (Fig. 30-15a,b). Each muscle fiber in turn contains many individual contractile subunits, the **myofibrils,** extending from one end of the fiber to the other. Each cylindrical myofibril is surrounded by membranes called the **sarcoplasmic reticulum** (Fig. 30-15c). Like the endoplasmic reticulum from which it is derived, the sarcoplasmic reticulum is a series of double sheets of membrane enclosing a space. Deep indentations of the cell membrane, called **T tubules,** extend down into the muscle fiber, passing very close to portions of the sarcoplasmic reticulum. This arrangement of T tubules and sarcoplasmic reticulum (whose spaces contain a high concentration of calcium ions) is crucial to the control of muscle contraction, described later.

Within each myofibril, there is a beautifully precise arrangement of filaments of actin and myosin, organized into subunits called **sarcomeres** (Fig. 30-15d,f). Sarcomeres are attached end to end throughout the length of the myofibril; their junction points are called **Z-lines.** Attached to the Z-lines are strands composed of actin plus two accessory proteins. These three proteins form the **thin filaments** (Fig. 30-15d,e). Suspended between the thin filaments are **thick filaments** composed of myosin protein. The thick and thin filaments are lined up in all the myofibrils, giving the cell its striped appearance (Fig. 30-15d,e,f). The strands of myosin extend small arms, called **cross-bridges,** contacting the actin filaments (Fig. 30-15e). The complex structure of the thin filament is crucial to the regulation of muscle contraction. The actin protein is formed from a double chain of subunits resembling a twisted double strand of pearls. Each subunit has a binding site for a myosin cross-bridge. In a relaxed muscle, however, these sites are covered by thin strands formed by one type of accessory protein, held in place by a second globular accessory protein (Fig. 30-15e). These accessory proteins prevent the myosin cross-bridges from attaching.

The Sliding Filament Model of Muscle Movement

When a muscle contracts, the accessory proteins of the thin filament are moved aside, exposing the binding sites on the actin. As soon as the sites are exposed, myosin cross-bridges attach. Using energy from splitting ATP, the cross-bridges repeatedly bend, release, and reattach farther along, much like a sailor pulling in an anchor line hand over hand (Fig. 30-16a). The thin filaments are pulled past the thick filaments, shortening the sarcomere and contracting the muscle (Fig. 30-16b).

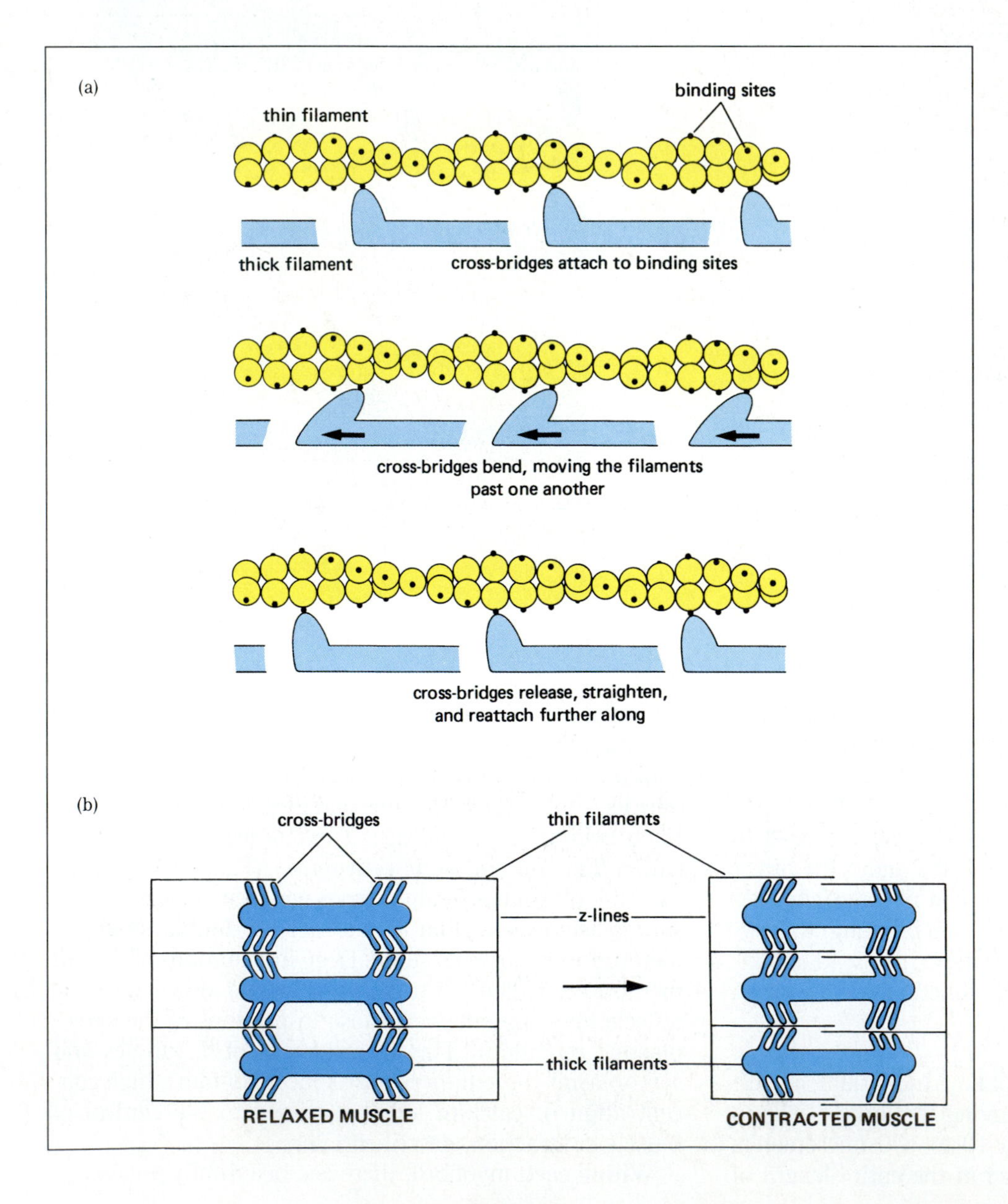

Figure 30-16 Muscle movement. **(a)** *The cross-bridges connecting actin to myosin swivel as if on hinges, pulling the actin filaments (which are attached to the ends of the sarcomere) toward the middle. Repeated cycles of attachment, swivelling, release, and reattachment result in muscle contraction.* **(b)** *Muscle contraction causes the thick and thin filaments to slide past one another, shortening the individual sarcomeres and hence the muscle cell.*

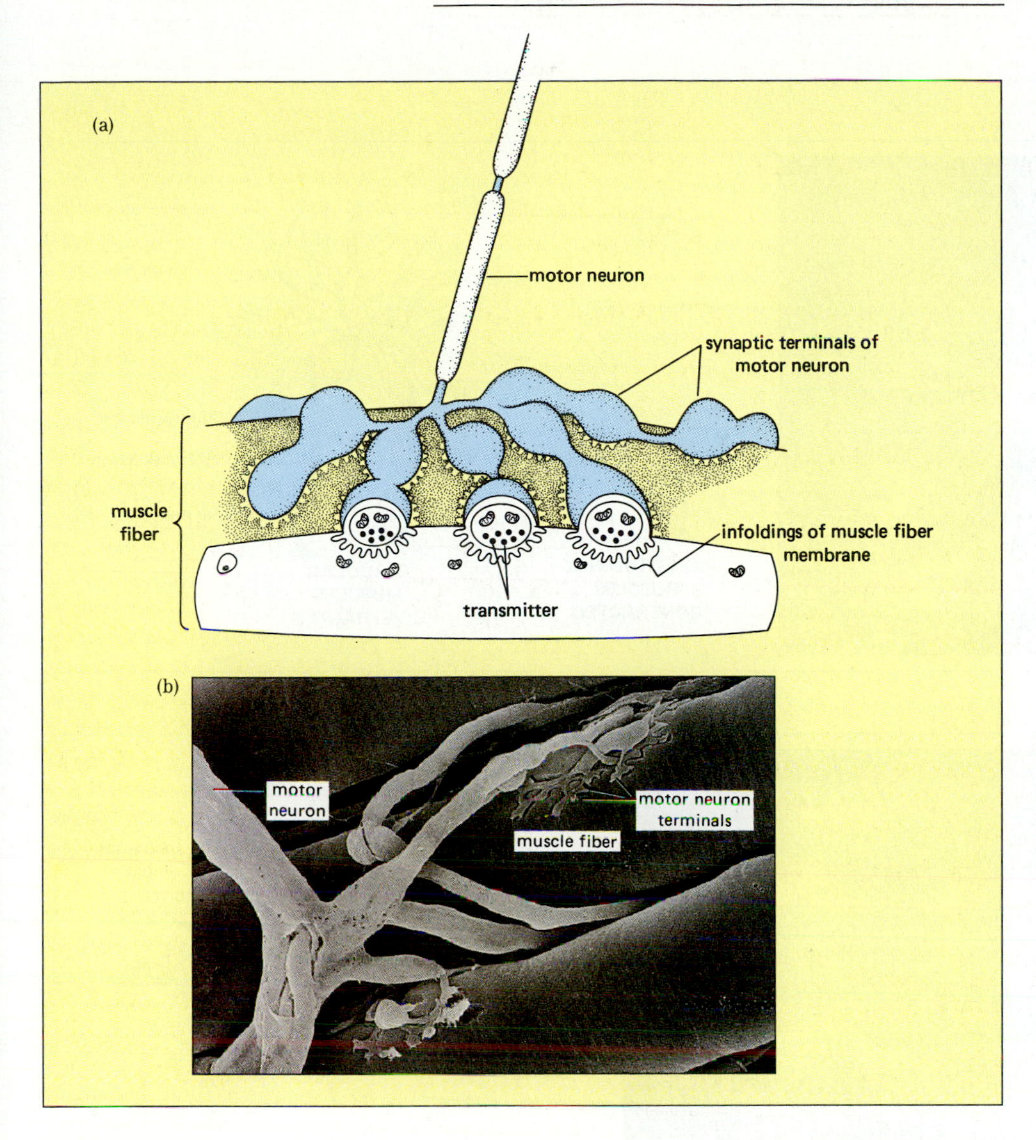

Figure 30-17 The neuromuscular junction. **(a)** *Diagram of a neuromuscular junction in cross section. Action potentials in the motor neuron cause transmitter release from the synaptic terminals. Transmitter binds to receptors on the muscle cell membrane, which is folded beneath the terminal to allow more surface area for receptors.* **(b)** *A scanning electron micrograph shows motor neuron terminals synapsing on muscle fibers.*

Muscle Control

Muscle contracts when the binding sites on actin are exposed, and relaxes when the binding sites are covered by accessory thin filament proteins. But what regulates the position of these accessory proteins? The answer is the concentration of calcium ions around the filaments, which in turn is under the control of the nervous system.

In many respects, skeletal muscles are much like neurons. They have resting potentials, they are stimulated by synaptic contact with neurons, and they produce action potentials when stimulated above threshold. Skeletal muscles contract as a result of their own action potentials, which are caused by synaptic stimulation from motor neurons. The cell bodies of the motor neurons (most of which reside in the spinal cord) send axons out the spinal nerves to the muscles, where they form synapses on the muscle fibers (Fig. 30-17). These **neuromuscular junctions** work much like any other synapse: when an action potential reaches the synapse, the motor neuron releases a neurotransmitter (acetylcholine) that diffuses across the synaptic gap to receptors on the muscle fiber membrane, producing an excitatory postsynaptic potential. Neuromuscular junctions are exceedingly strong, and every action potential in a motor neuron causes a large enough excitatory synaptic potential to evoke a muscle action potential (Fig. 30-18).

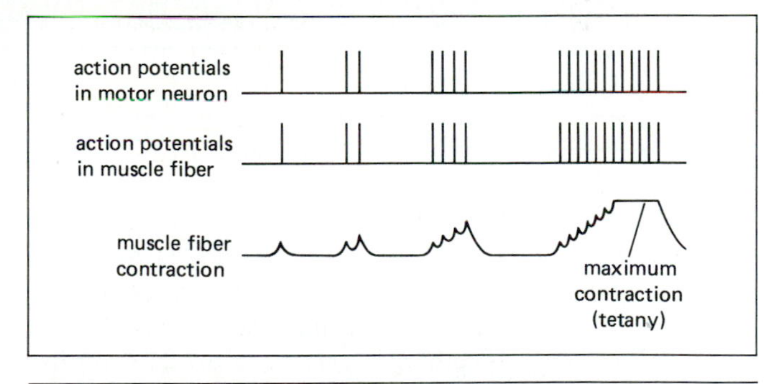

Figure 30-18 The strength of contraction of a single muscle fiber depends on how fast it is stimulated by its motor neuron. Each individual muscle action potential causes only a partial contraction. Up to a point, more rapid action potentials cause the individual contractions to summate, producing larger overall contractions of the muscle. Continued rapid firing of motor neurons gives rise to tetany: sustained, maximal contraction of the muscle.

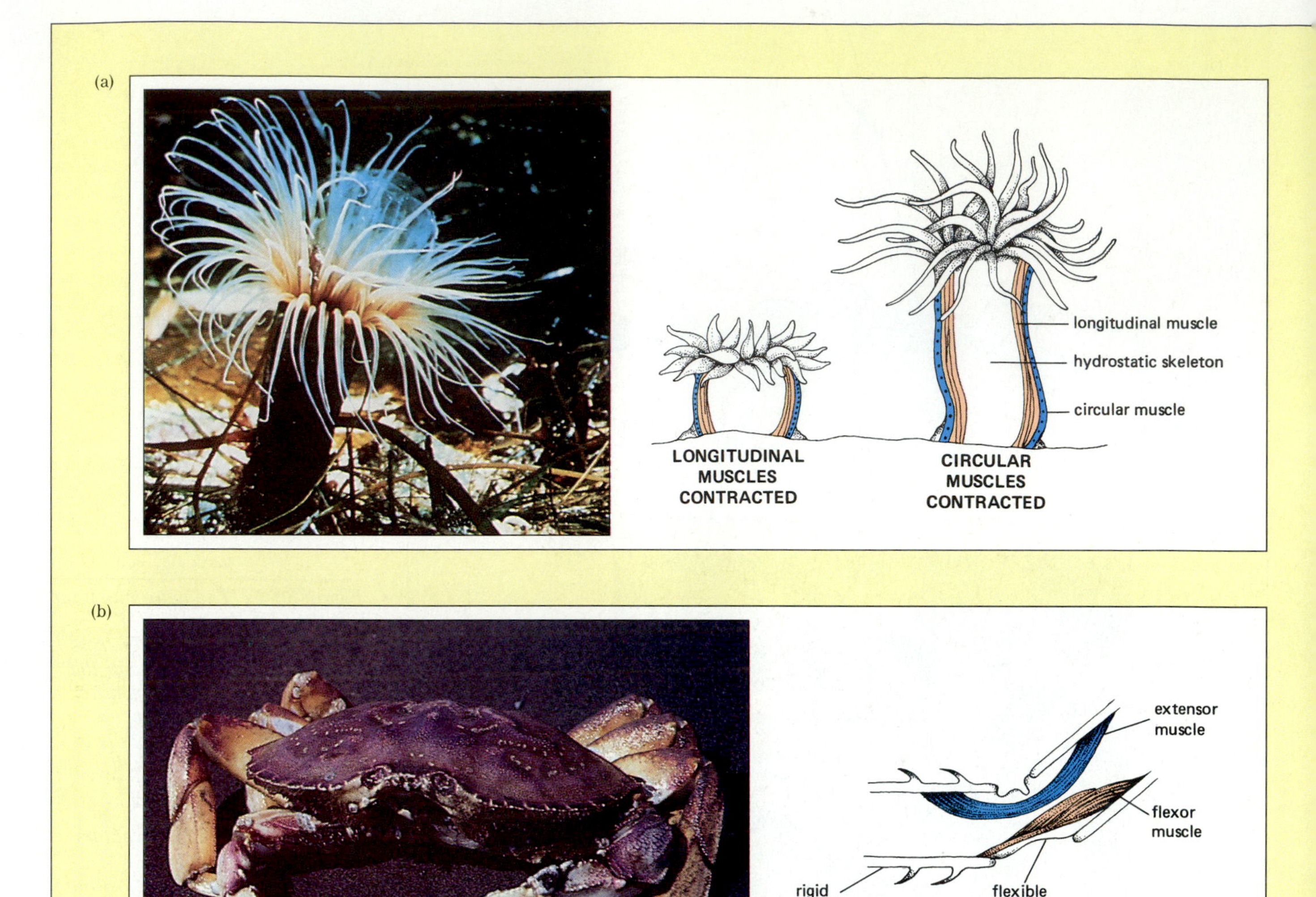

The muscle action potential invades the interior of the muscle cell by passing down the T tubules (see Fig. 30-15c). The change in potential causes the sarcoplasmic reticulum to release calcium ions into the myofibrils. Calcium ions bind to the globular accessory proteins on the thin filament, altering their shape and causing them to pull the thin-stranded accessory proteins off the binding sites on the actin. This allows the myosin cross-bridges to attach to the actin, initiating contraction. As soon as the action potential fades away, active transport proteins in the sarcoplasmic reticulum membrane pump the calcium back inside the reticulum. The accessory proteins move back into place on the thin filaments, preventing further cross-bridge binding and ending the contraction.

If each action potential in a motor neuron elicits an action potential in a muscle fiber, causing all its sarcomeres to contract, how does the nervous system control the strength and degree of muscle contraction? *The degree of muscle contraction depends on the number of muscle fibers stimulated and the frequency of action potentials in each fiber.*

Most motor neurons innervate more than one muscle fiber. The group of fibers on which a single motor neuron synapses is called a **motor unit.** The number of muscle fibers in a motor unit varies from muscle to muscle. Large muscles used for gross movement, such as those of the thigh or buttocks, may have hundreds of muscle fibers in each motor unit. In muscles used for fine control of small body parts, such as those of the lips, eyes, and tongue, only a few muscle cells may be innervated by each motor neuron. Therefore, when a single motor neuron fires an action potential, it may cause contraction of a few muscle cells or of many, depending on the size of the motor unit. In any case, one action potential does not cause a muscle

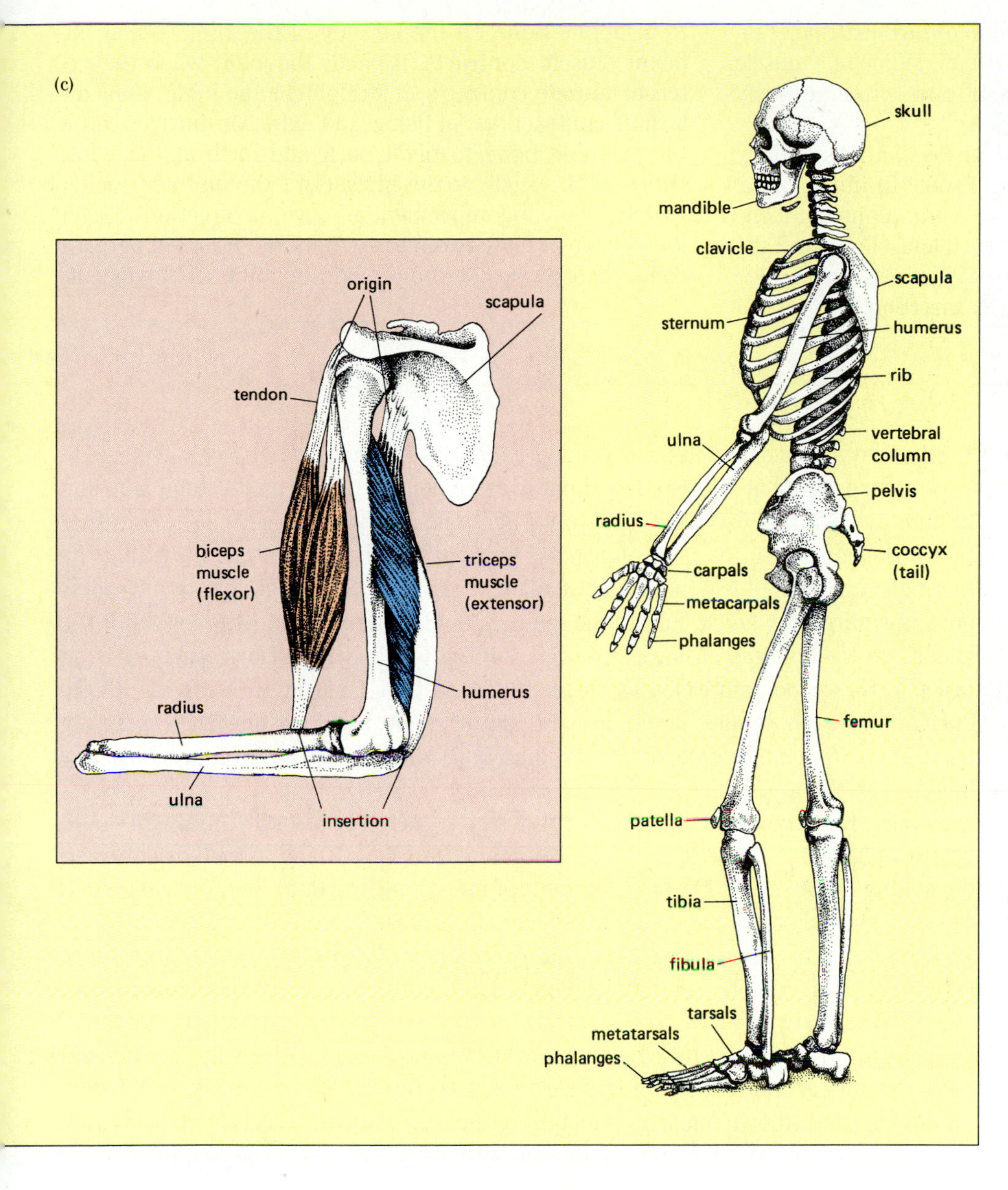

Figure 30-19 Body movement in animals is based on a support system, the skeleton, and antagonistically arranged muscles.
(a) Hydrostatic skeletons: *The skeleton of worms, many molluscs, and cnidarians such as this anemone is essentially a fluid-filled tube with soft walls. Two layers of muscles are arranged perpendicularly to one another; a layer of circular muscles forms a band around the circumference of the tube and a layer of longitudinal muscles runs lengthwise. Since fluids are incompressible, if the circular muscles contract, the animal will become long and thin; if the longitudinal muscles contract, it will become short and fat.*
(b) Exoskeletons: *Arthropods have armorlike skeletons on the outsides of their bodies. Joints allow movement, produced by pairs of muscles spanning the joint.*
(c) Endoskeletons: *Humans and other vertebrates possess rigid internal skeletons. Pairs of antagonistic muscles are attached on either side of joints. When the extensor muscle contracts, it straightens the joint, while contraction of the flexor muscle bends the joint.*

cell to contract fully. After an action potential, the concentration of calcium around the filaments is high for just a short time, and only a few cross-bridge movement cycles can occur before the calcium is pumped away. How much the muscle cells of a given motor unit contract will depend on how fast the motor neurons are firing: faster firing produces stronger contractions, since the contractions produced by each action potential add to one another. This buildup of contraction caused by rapid firing is called **summation** (Fig. 30-18).

Body Movement

The ultimate function of motor neurons firing and muscle fibers contracting is, of course, to move parts of the body in appropriate ways. If muscle contraction is to create useful movement rather than just a quivering lump of protoplasm, the muscles must have an underlying structure on which to act.

Animals are given shape by a framework, the **skeleton.** There are three general types of skeleton found in the animal kingdom: **hydrostatic skeletons,** consisting of water contained within a cylinder, **exoskeletons,** relatively rigid armor surrounding the animal, and **endoskeletons,** hard internal frameworks (Fig. 30-19). While animals with hydrostatic skeletons have a somewhat fluid shape (pun intended!), those with exoskeletons and endoskeletons have a more rigid configuration. Flexible joints between segments of the skeleton allow movement.

Vertebrate skeletal muscles are attached to the skeleton and usually move it when they contract. We have described a mechanism for muscle *contraction,* but not *extension.* This was not an oversight: *muscles can only actively contract.* To extend, the muscle must be pulled out again. Most muscles are arranged in *antagonistic pairs* on opposite sides of a joint (Fig. 30-19b,c). When one muscle contracts, it moves the skeleton and simultaneously stretches out the opposing muscle.

Many joints, such as the elbow or knee, are essentially hinges, movable in only two dimensions. In hinge joints, pairs of muscles lie in roughly the same plane as of the joint (Fig. 30-19c). One end of each muscle, called the **origin,** is fixed to a relatively immovable bone on one side of the joint, while the other end, the **insertion,** is attached to a mobile bone on the far side of the joint. When the **flexor muscle** contracts, it bends the joint; when the **extensor muscle** contracts, it straightens the joint. Thus alternate contractions of flexor and extensor muscles cause the movable bone to pivot back and forth at the joint. Other joints, such as the hip, are of the ball-and-socket variety, allowing movement in several directions. Such joints have at least two pairs of muscles, oriented perpendicularly to each other, to provide flexibility of movement.

REFLECTIONS ON DESIGN IN EVOLUTION

Sense organs and muscles are striking examples of evolution at work. Marvel at the startling agility of Mikhail Baryshnikov as he bounds off the ballet stage, appears to hang poised in midair, and lands once again as if he descended willingly, rather than in obedience to the laws of gravity. Or consider the prairie falcon, riding the thermals hundreds of feet in the sky, scouring the plains below for sight of a careless ground squirrel for lunch. Surely the muscles of Baryshnikov and the vision of the falcon are the epitome of flawless design, superbly suited to perform their allotted tasks. Yet if we look more closely, we see that evolution, far from working by design, operates through chance mutations and natural selection. Despite their seeming perfection, from an engineering viewpoint both of these structures are basically flawed.

Ponder, for a moment, how you might design a "perfect muscle." Surely, it would be most efficient if a muscle could, on command, forcibly extend as well as contract, depending on the need. The same amount of muscle mass would then provide much more strength. For example, a ballet dancer could spring into the air by contracting the muscles in front of his thigh, while simultaneously forcefully extending those behind. Or perhaps one set could be eliminated entirely, greatly reducing weight. We might have to raise the stage ceiling to accommodate the leaps such a modified Baryshnikov could perform! However, evolution never stumbled across an arrangement for dual-action muscles of this sort: whether in jellyfish or worms, in lobsters or humans, all known muscles can only contract. Natural selection has favored mutations for optimal placement of muscles around joints, but it has been restricted to using a contract-only design.

The vertebrate eye is another case of evolution "making do" with mutations that happened to occur. From the point of view of visual acuity, the vertebrate retina is built backwards; the receptors not only face away from the light, but in most parts of the retina, they are also screened by several layers of neurons (see Fig. 30-8). As a result, only about 10 percent of the light entering the eye actually reaches the photopigments of the rods and cones. As you know from personal experience, you cannot perceive colors at night, because the light is too dim to activate the cones in your retina. If they were at the front of the retina instead of the back, perhaps we could have color vision, and sharper acuity as well, in much dimmer light.

One might argue that, with muscles, there are principles that we have not yet grasped which would make dual-action muscles less effective than the contract-only type. But "receptor-side out" retinas are clearly not impossible, since some molluscs have them (Fig. 30-7). An important principle of evolution illustrated by the examples of muscles and retinas is that natural selection can only preserve and promote structures that happen to arise through mutation. If the right mutation never occurs, the "perfect" structure will never occur, although its design might seem obvious to a human engineer.

SUMMARY OF KEY CONCEPTS

Perception

In response to external stimuli, sensory receptor cells produce a graded receptor potential. In sensory neurons, this may cause action potentials whose frequency is proportional to the size of the receptor potential. Alternatively, some receptor cells release transmitter onto a postsynaptic neuron in amounts proportional to the size of the receptor potential. This results in action potentials in the postsynaptic cell whose frequency reflects the intensity of the stimulus. The most common receptor types respond either to mechanical deformation (touch, stretch, hearing, gravity) or to stimuli that influence receptor molecules in the receptor cell membrane (taste, odor, light, and damaging stimuli causing pain).

The Major Senses

The major senses include mechanical senses, hearing, vision, chemical senses, and pain.

Mechanical senses: The skin contains dendrites of many different types of sensory neurons, each sensitive to a different mechanical stimulus, such as pressure, touch, or vibration. Joints and muscles have similar receptors, sensitive to stretch, that inform the brain about the position of the body. Many internal organs have stretch receptors that signal fullness.

Hearing: Hearing is a modified mechanical sense specialized for reception of vibrations of the air or water. In the mammalian ear, air vibrates a set of membranes in the inner ear. The membrane vibrations bend hairs, producing a receptor potential in the hair cells that causes action potentials in the axons of the auditory nerve to the brain. The inner ear also contains the semicircular canals, which detect gravity and acceleration.

Vision: All eyes have photoreceptor cells that contain a pigment which, upon absorption of light, causes a receptor potential. The vertebrate eye contains two types of photoreceptor, rods and cones. Rods are more light sensitive than cones and provide vision in dim light. Cones occur in three types, each sensitive to a different wavelength of light, and provide color vision.

Chemical senses: Terrestrial vertebrates detect chemicals in the external environment either by olfaction (for airborne sources of chemicals) or taste (chemicals dissolved in saliva). Each olfactory or taste receptor cell type responds to only one or a few specific types of molecules, allowing discrimination among tastes and odors.

Pain: Pain is a type of chemical sense, in which sensory neurons respond to chemicals produced by damage to cells in the body.

Action

Vertebrates have three types of muscle cells: smooth (surrounding hollow organs such as arteries and digestive tract), cardiac (forming heart muscle), and skeletal (moving the skeleton).

All muscle contraction is based on movements of two proteins, actin and myosin, within the individual muscle cells (also called fibers). Actin and myosin filaments alternate in the sarcomere within the myofibril, which is a subunit of the muscle fiber. During contraction, myosin filaments pull actin filaments closer together. Since the actin filaments are attached to the ends of the sarcomere, when they are brought closer together, the cell contracts.

During contraction, projections from the myosin filament, called cross-bridges, move back and forth, grabbing onto the actin filaments and pulling the actins along. Energy for cross-bridge movement, and hence for muscle contraction, is provided by ATP.

Skeletal muscles contract under the command of the nervous system. Motor neurons form synapses on muscle cells. Action potentials in motor neurons cause large postsynaptic potentials in the muscle cell, which always exceed threshold and trigger an action potential in the muscle cell. The muscle action potential initiates a chain of events culminating in contraction of the cell. The force and distance of contraction of a whole muscle are controlled by two factors: (1) the number of muscle cells contracting at the same time, and (2) the frequency of action potentials in muscle cells, which determines the amount of contraction of the individual cell.

Muscles move bodies via their attachments to skeletons. Three types of skeletons in the animal kingdom are hydrostatic skeletons, exoskeletons, and endoskeletons. In each type, muscles are usually arranged in antagonistic pairs capable of moving the skeletal part in opposite directions.

GLOSSARY

Actin (ak′-tin) one of the major proteins of muscle, whose interactions with myosin produce contractions; found in the thin filaments of the muscle fiber. *See also* Myosin.

Aqueous humor (ā′-kwē-us) clear, watery fluid between the cornea and lens of the eye.

Auditory nerve (aw′-dih-tory) the nerve leading from the mammalian cochlea to the brain, carrying information about sound.

Basilar membrane (bas′-eh-lar) a membrane in the cochlea that bears hair cells which respond to the vibrations produced by sound.

Bradykinin (brā-dē-kīn′-in) a chemical formed during tissue damage that binds to receptor molecules on pain nerve endings, giving rise to the sensation of pain.

Camera eye the type of eye found in vertebrates and molluscs, in which a lens focuses an image on a sheet of light-sensitive receptor cells (the retina).

Cardiac muscle (kar′-dē-ak) specialized muscle of the heart, able to initiate its own contraction independent of the nervous system.

Choroid (kōr′-ōyd) a layer of tissue behind the retina that contains blood vessels and pigment that absorbs stray light.

Cochlea (kōk′-le-uh) a coiled, bony, fluid-filled tube found in the mammalian inner ear, which contains receptors (hair cells) producing the sense of hearing.

Compound eye a type of eye found in arthropods, composed of numerous independent subunits, called ommatidia. Each ommatidium apparently contributes a single piece of a mosaic-like image perceived by the animal. *See also* Ommatidium.

Cone a cone-shaped photoreceptor cell in the vertebrate retina, not as sensitive to light as the rods. The three types of cones are most sensitive to different colors of light, and provide color vision. *See also* Rod.

Cornea (kōr′-nē-uh) the clear outer covering of the eye in front of the pupil and iris.

Cross-bridge in muscles, an extension of myosin that binds to and pulls on actin to produce contraction of the muscles.

Endoskeleton (en′-dō-skel′-uh-tun) a rigid internal skeleton with flexible joints to allow for movement.

Exoskeleton (ex′ō-skel′-uh-tun) a rigid external skeleton with flexible joints to allow for movement.

Eyespot a simple, lensless eye found in various invertebrates, including flatworms and jellyfish. Eyespots provide information about light versus dark, and sometimes the direction of light, but cannot form an image.

Fovea (fō′-vē-uh) the central region of the vertebrate retina, upon which images are focused. The fovea contains closely packed cones (about 150,000 per square millimeter).

Ganglion cell (gang′-lē-un) a cell type comprising the outer layer of the vertebrate retina. Its axons comprise the optic nerve.

Hair cell the receptor cell type found in the inner ear (cochlea and semicircular canals). Hair cells bear hairlike projections. Bending of the hairs between two membranes causes the receptor potential.

Hydrostatic skeleton (hī-drō-sta′-tic) a skeleton composed of fluid contained within a flexible, usually tubular, covering.

Inner ear the innermost part of the mammalian ear, composed of the bony, fluid-filled tubes of the cochlea and semicircular canals.

Insertion the site of attachment of a muscle to the relatively movable bone on one side of a joint.

Iris the pigmented part of the vertebrate eye, surrounding the central hole of the pupil.

Joint a flexible region between two rigid units of an exoskeleton or endoskeleton, to allow for movement between the units.

Lens a clear object that bends light rays; in eyes, a flexible or movable structure used to focus light on a layer of photoreceptor cells.

Middle ear part of the mammalian ear composed of the tympanic membrane and three bones (hammer, anvil, stirrup) that transmit vibrations from the auditory canal to the oval window.

Motor unit a single motor neuron and all the muscle fibers (cells) on which it synapses.

Muscle fiber an individual muscle cell.

Myofibril (mī′-ō-fī′bril) a cylindrical subunit of a muscle cell, surrounded by sarcoplasmic reticulum.

Myosin (mī′-ō-sin) one of the major proteins of muscle that interacts with actin to produce contraction; found in the thick filaments of the muscle fiber. *See also* actin.

Neuromuscular junction the synapse formed between a motor neuron and a muscle fiber.

Olfaction (ōl-fak′-shun) a chemical sense, the sense of smell; in terrestrial vertebrates, the result of detection of airborne molecules.

Ommatidium (ōm-ma-tid′-ē-um) an individual light-sensitive subunit of a compound eye. Each ommatidium consists of a lens and several (usually eight) receptor cells.

Optic disk (op′-tik) the area of the retina at which the axons of the ganglion cells merge to form the optic nerve; the blind spot of the retina.

Optic nerve (op′-tik) the nerve leading from the eye to the brain, carrying visual information.

Organ of Corti (kōr′-tē) part of the central canal of the cochlea, consisting of the basilar membrane, its ciliated receptor cells, and the tectorial membrane.

Origin the site of attachment of a muscle to the relatively stationary bone on one side of a joint.

Outer ear the outermost part of the mammalian ear, including the external ear and auditory canal leading to the tympanic membrane.

Oval window the membrane-covered entrance to the inner ear.

Proprioception (prō′-prē-ō-cep-shun) a sense that monitors the position of the parts of the body and their direction of movement.

Receptor (1) a cell that responds to an environmental stimulus (chemicals, sound, light, pH, etc.) by changing its electrical potential; (2) a protein molecule in a cell membrane that binds to another molecule (hormone, neurotransmitter, odorous compound, etc.) triggering metabolic or electrical changes in a cell.

Receptor potential a graded electrical potential change in a receptor cell produced in response to reception of an environmental stimulus (chemicals, sound, light, pH, heat, cold, etc.) The receptor potential is proportional to the intensity of the stimulus.

Retina (ret′-in-a) a sheet of tissue at the rear of camera-type eyes, composed of photoreceptor cells plus associated nerve cells that refine the photoreceptor information and transmit it to the optic nerve.

Rod a rod-shaped photoreceptor cell in the vertebrate retina, sensitive to dim light, but not involved in color vision. *See also* Cone.

Round window the membrane-covered opening in the cochlea, located below the oval window.

Sarcomere (sark′-ō-mēr) the unit of contraction of a muscle fiber; a subunit of the myofibril, consisting of actin and myosin filaments and bounded by Z-lines.

Sarcoplasmic reticulum (sark-ō-plas′-mik re-tik′-ū-lum) specialized endoplasmic reticulum found in muscle cells. The sarcoplasmic reticulum stores calcium ions and releases them into the interior of the muscle cell to initiate contraction.

Sclera (sklāra) a tough white connective tissue layer that covers the outside of the eyeball and forms the white of the eye.

Semicircular canal one of three fluid-filled semicircular tubes of the inner ear which function in the detection of rotational movements of the head.

Sensory receptor a cell specialized to respond to particular internal or external environmental stimuli by producing an electrical potential.

Skeletal muscle also called striated or voluntary muscle; the type of muscle that is attached to and moves the skeleton.

Skeleton a supporting structure for the body, upon which muscles act to change the body configuration.

Smooth muscle type of muscle found around hollow organs, such as the digestive tract, bladder, and blood vessels, normally not under voluntary control.

Taste a chemical sense; in mammals, perceptions of sweet, sour, bitter, or salt produced by stimulation of receptors on the tongue.

Tectorial membrane (tek-tōr′-ē-ul) one of the membranes of the cochlea, in which the hairs of the hair cells are embedded. During sound reception, movement of the basilar membrane relative to the tectorial membrane bends the cilia.

T tubules indentations of the muscle cell membrane that conduct the action potential inside the cell.

Tympanic membrane (tim-pan′-ik) the eardrum; a membrane stretched across the opening of the ear, which transmits vibration of sound waves to bones of the middle ear.

Vitreous humor (vit′-rē-us) a clear jellylike substance that fills the large chamber of the eye between the lens and retina.

Z-lines fibrous protein structures to which the thin filaments of skeletal muscle are attached, forming the boundaries of sarcomeres.

STUDY QUESTIONS

1. Describe the series of events leading from an environmental stimulus to an action potential in a sensory neuron.
2. Describe the anatomy of two types of receptor cell responsive to mechanical deformation; one found in the skin, the other in the inner ear.
3. How are chemicals detected by receptor cells? How could a body distinguish among many different chemicals?
4. For any sense, describe how the brain receives information about the type of stimulus and about stimulus intensity.
5. What are proprioceptors? For what types of behavior are they important?
6. Describe the structure and function of the various parts of the human ear. Do this by tracing a sound wave from the air outside the ear to the cells causing action potentials in the auditory nerve.
7. How does the structure of the inner ear allow for the perception of pitch?
8. Diagram the overall structure of the human eye. Label the cornea, iris, lens, sclera, retina, and choroid. Describe the function of each structure.
9. How does the lens change shape to allow focusing of faraway objects? What defect makes focusing on faraway objects impossible, and what is this condition called? What type of lens can be used to correct it, and why?
10. List the similarities and differences between rods and cones.
11. Distinguish between taste and olfaction.
12. Diagram the anatomy of a muscle and its various subunits, starting with the whole muscle and working down to the level of actin and myosin filaments. Label your drawings, and briefly define each part.
13. Describe the process of skeletal muscle contraction, beginning with an action potential in a motor neuron and ending with the relaxation of the muscle. Your answer should include the following words: neuromuscular junction, T tubule, sarcoplasmic reticulum, calcium, thin filaments, binding sites, thick filaments, sarcomere, and active transport.
14. What are the three types of skeletons found in animals? For one of these, describe how the muscles are arranged around the skeleton and how contractions of the muscles result in movement of the skeleton.

SUGGESTED READINGS

Cohen, C. "The Protein Switch of Muscle Contraction," *Scientific American,* November 1975 (Offprint No. 1329). Cohen describes the interaction of calcium ions and the accessory proteins of the thin filament, which together control the contraction of muscles.

Eckert, R. Randall, D., and Augustine, G. *Animal Physiology,* 3rd ed. New York: W. H. Freeman and Company, Publishers, 1988. The best explanations of how muscles and senses work.

Horridge, G. A. "The Compound Eye of Insects." *Scientific American,* July 1977. Compound eyes provide fairly poor vision compared to our camera eyes, yet insects navigate, find flowers, prey, and mates, and even migrate long distances.

Merton, P. A. "How We Control the Contraction of Our Muscles." *Scientific American,* May 1972. Useful muscle activity requires precise simultaneous adjustment of the contractions of many muscles in the body.

Parker, D. E. "The Vestibular Apparatus." *Scientific American,* November 1980 (Offprint No. 1484). Complete perception of gravity and acceleration requires not only the utricle, saccule, and semicircular canals, but also input from other senses.

Ratliff, F. "Contour and Contrast." *Scientific American,* June 1972. Interactions among retinal cells provides us with more visual contrast than actually exists in the light entering our eyes.

Regan, D., Beverley, K., and Cynader, M. "The Visual Perception of Motion in Depth." *Scientific American,* July 1979 (Offprint No. 586). Binocular vision provides us with two views of the world, which the brain fuses into a coherent three-dimensional representation.

31
Animal Reproduction

The word "reproduction" may bring to mind images of courtship, cute babies, and cuddly kittens. However, from an evolutionary perspective, romance and the universal appeal of babies are frills that have evolved only because they further the real point: to pass on one's genes to another generation, in a sense cheating death and achieving immortality through one's offspring.

From this viewpoint, an animal's life can be divided into three stages. First, it is born or hatched from an egg, and grows to sexual maturity. Second, it gathers the resources needed to reproduce, which may include stores of food, impressive strength or weaponry, or a territory. Finally, it finds a mate (if necessary) and reproduces, which may include caring for its offspring until they can fend for themselves. The marvelous adaptations that we have discussed in the last few chapters, such as sophisticated sensory equipment or complex digestive systems, have evolved through millennia of mutation and natural selection because they have allowed successful reproduction by the animals that possess them. Reproduction is the key to the continued existence of the species.

REPRODUCTIVE STRATEGIES

Animals reproduce either sexually or asexually. As you learned in Chapter 8, in **sexual reproduction** an animal produces haploid gametes through meiosis. Two gametes, usually from separate parents, fuse to form a diploid offspring. Since an offspring receives genes from two parents, it is genetically different from either. In most forms of **asexual reproduction,** on the other hand, a single animal produces offspring through repeated mitosis of cells in some part of its body. Therefore, the offspring are genetically identical to the parent. We humans reproduce sexually, and we tend to regard sexual reproduction as the normal, best way to do it. On a firmer biological footing, by bringing together genes from two different parental organisms, sexual reproduction allows for new gene combinations that may enhance the survival and reproduction of the offspring. Nevertheless, asexual reproduction is more efficient, since there is no need to find a mate, court, and fend off rivals, and no waste of thousands, millions, or even billions of sperm and eggs that never unite to form an offspring. Not surprisingly, a number of animals reproduce asexually, at least some of the time. Let's begin, then, with a brief survey of asexual reproduction among animals.

Asexual Reproduction

Regeneration from body fragments is a potential form of reproduction in some animals, such as sea stars (Fig. 31-1a). If sea stars are cut up, fragments that contain part of the central disk can regrow the rest of the star. (In an effort to reduce predation on their "herd," oyster "ranchers" used to catch sea stars, hack them to pieces, and throw the parts back into the sea. Much to their dismay, this merely resulted in more stars than ever, as the fragments regenerated entire animals.) A few brittle stars routinely reproduce in a similar fashion, by splitting apart across the disk, with each half regenerating a complete animal. Despite these asexual capabilities, sea stars usually reproduce sexually, casting huge numbers of sperm and eggs into the sea.

Some animals reproduce by **fission** to produce two new adults. A few corals can divide longitudinally to produce two smaller but complete individuals (Fig. 31-1b). Some flatworms and annelids divide transversely and regenerate the missing parts (Fig. 31-1c). This, of course, means that the "tail half" of the animal must regenerate the head, including the brain!

Many sponges and coelenterates, such as *Hydra* and some anemones, reproduce by **budding** (Fig. 31-2). A miniature version of the animal (a **bud**) grows directly on the body of the adult, drawing upon its parent for nourishment. When it has grown large enough, the bud breaks off and becomes independent.

Figure 31-1 *Methods of asexual reproduction.*
(a) *Many sea stars can regenerate new individuals from fragments if the fragment includes part of the central disk.*
(b) *Some corals divide longitudinally, producing two thin but complete new individuals.*
(c) *Certain flatworms divide transversely. At first, each offspring is missing half the adult body, but these are regrown from "pseudoembryonic" cells near the broken edge.*

Figure 31-2 *The offspring of some cnidarians, such as the anemone shown here, grow as buds upon the body of the parent. When sufficiently developed, the buds break off and assume independent existence.*

Finally, the females of some animal species can reproduce by a process known as **parthenogenesis.** Haploid egg cells are produced in the ovary by meiosis. However, an egg then develops into an adult without being fertilized. In some animals, parthenogenetically produced offspring remain haploid. Male honeybees, for example, are haploid, developing from unfertilized eggs; their diploid sisters develop from fertilized eggs (Fig. 31-3). On the other hand, some fish, amphibians, and reptiles regain the diploid number of chromosomes in parthenogenetically produced offspring. This is accomplished by duplicating all the chromosomes either before or after meiosis. The resulting offspring are all females. Some species of fish, including relatives of the mollies and platies found in tropical fish stores, and some lizards, such as the whiptail, have done away with males completely. Their populations consist entirely of parthenogenetically reproducing females (Fig. 31-4). Still other animals, such as the aphid, can reproduce either sexually or parthenogenetically, depending on environmental factors such as the season of the year or the availability of food (Fig. 31-5).

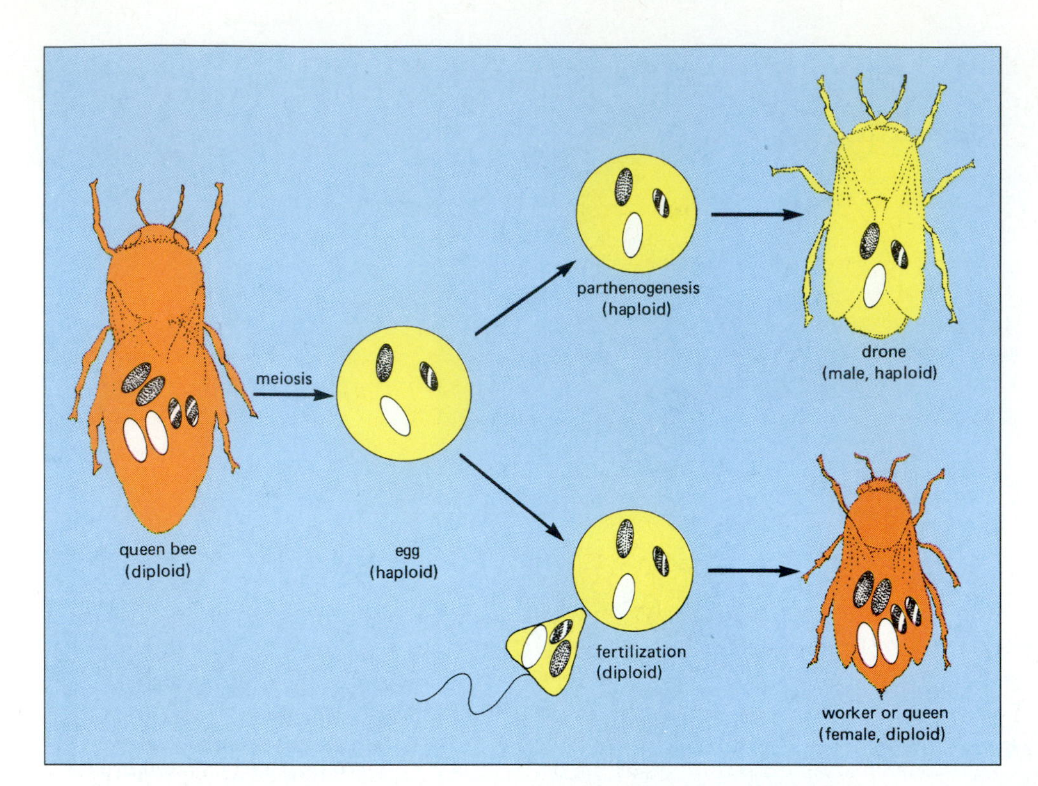

Figure 31-3 Queen honeybees produce haploid eggs by meiosis. The eggs then take one of two developmental paths. If unfertilized, the eggs develop parthenogenetically into haploid drones (males). Fertilized eggs develop into females, either workers or (rarely) new queens. (Only three chromosome pairs are shown for clarity.)

Figure 31-4 Several species of fish and reptiles, such as the whiptail lizard shown here, consist entirely of parthenogenetically reproducing females.

Figure 31-5 A female aphid gives live birth. In spring and early summer, when food is abundant, aphid females reproduce parthenogenetically. In fact, the development of the ovaries proceeds so rapidly that females are born pregnant! In fall, reproduction becomes sexual, as the females mate with males. Aphids have thus evolved the ability to exploit the advantages of asexual reproduction (rapid population growth during times of abundant food, no energy spent in seeking a mate, no wasted gametes) and sexual reproduction (genetic recombination).

Sexual Reproduction

In animals, sexual reproduction occurs when a haploid sperm fertilizes a haploid egg, generating a diploid offspring. In most animal species, males and females are separate individuals. These species are termed **dioecious** (Greek for "two houses"). The sexes are defined by the type of gamete that each produces. Females produce **eggs,** which are large, nonmotile cells containing substantial food reserves. Males produce small, motile **sperm** that have almost no cytoplasm and hence no food reserves. In **monoecious** ("one house") species, such as earthworms and many snails, single individuals produce both sperm and eggs. Such individuals are commonly called **hermaphrodites,** after Hermaphroditos, a male Greek god whose body was merged with that of a female water nymph. Some hermaphrodites can fertilize their eggs with their own sperm if necessary. These animals, including tapeworms and many pond snails, are relatively immobile and may find themselves isolated from other members of their species. Obviously, the ability to fertilize oneself is advantageous under these circumstances.

For dioecious species and for hermaphrodites that cannot self-fertilize, successful reproduction requires that sperm and eggs from different animals be brought together for fertilization. This is accomplished in a variety of ways, depending on the mobility of the animals and on whether they breed in water or on land.

External fertilization. In **external fertilization,** the parents release sperm and eggs into water, through which the sperm swim to reach an egg. This procedure, called **spawning,** is obviously restricted to animals that breed in water. Since sperm and egg are relatively short-lived, spawning animals must synchronize their reproductive behaviors, both *temporally* (male and female spawn at the same time) and *spatially* (male and female spawn in the same place). Animals employ a combination of environmental cues, pheromones, and behaviors to synchronize spawning.

Most spawning animals rely on environmental cues to some extent. Breeding usually occurs only during certain seasons of the year, but more precise synchrony is required to coordinate the actual release of sperm and egg. Grunion, fish that inhabit southern California coastal waters, time their strange reproductive rituals by the season, time of day, and phases of the moon (Fig. 31-6a). On nights of the highest tides in fall (which occur during a full moon), they swim up onto the beach. Writhing

Figure 31-6 Some animals use environmental cues to synchronize spawning.
(a) *At the highest tides of fall, grunion swarm ashore on the few undeveloped beaches left in southern California. The fish burrow slightly into the sand and release sperm and eggs. The eggs hatch in the warm sand and develop over the following two weeks. When the next "high-high" tide comes, the juveniles wash out of the sand back into the ocean.*
(b) *Along the Great Barrier Reef of Australia, thousands of corals spawn simultaneously, creating this "upside-down blizzard" effect. The inset photo shows a package of sperm and eggs erupting from a spawning coral.*

Figure 31-7 Violent courtship rituals among Siamese fighting fish (Betta splendens) *ensure fertilization of the female's eggs, as male and female curl about one another, releasing sperm and eggs together. The male retrieves the eggs as they fall, spits them into his bubble-nest, and cares for the offspring during their first few weeks of life.*

masses of males and females release their gametes into the wet sand, and then swim back out to sea on the next wave. Many corals of Australia's Great Barrier Reef also synchronize spawning by the phase of the moon. On the fourth or fifth night after the full moons of November and December, all the corals of a particular species on an entire reef release a blizzard of sperm and eggs into the water (Fig. 31-6b).

Other animals communicate their sexual readiness to one another by releasing pheromones into the water. A pheromone is a chemical that is released from the body of one animal and affects the behavior of a second animal. Pheromones synchronize spawning in many sessile or sluggish invertebrates such as mussels and sea stars. Usually, when a female is ready to spawn, she releases eggs and a pheromone into the water. Nearby males, detecting the mating pheromone, quickly release millions of sperm.

Temporal synchrony alone does not guarantee efficient reproduction. Corals, sea stars, and mussels all waste enormous quantities of sperm and eggs because they are released too far apart. In mobile animals, both temporal and spatial synchrony can be ensured by mating behaviors. Most fish, for example, have some sort of courtship ritual in which the male and female come very close together and release their gametes in the same place and at the same time (Fig. 31-7). Frogs and toads carry this one step farther, by assuming a characteristic mating pose called **amplexus** (Fig. 31-8). At the edges of ponds and lakes, the male mounts upon the back of the female and prods her in the side. This stimulates her to release eggs, which he immediately fertilizes.

INTERNAL FERTILIZATION. In **internal fertilization,** sperm are taken into the body of the female, where fertilization occurs. This has two advantages. First, sperm are provided with a direct fluid path to reach the eggs. In terrestrial environments, this path can only be guaranteed inside the body of the female. Second, even in aquatic environments, internal fertilization improves the likelihood that most eggs will be fertilized, since the sperm are not left to thrash about in a large volume of water to find the eggs.

Internal fertilization usually occurs by **copulation,** in which the penis of the male is inserted into the body of the female and releases sperm (Fig. 31-9). In a variation of internal fertilization, males of some animals package their sperm in a container called a spermatophore (Greek for "sperm carrier"). Males of some species of mites and scorpions simply drop the spermatophore on the ground. If found by a female, she then fertilizes herself by inserting the spermatophore into her reproductive cavity. The male squid is somewhat more careful with his spermatophore, which he picks up with a tentacle and inserts into the female. In either case, the sperm are then liberated inside the female's reproductive tract.

Just because sperm are deposited in the body of the female does not guarantee fertilization. Fertilization can

Figure 31-8 Amplexus in golden toads, as the smaller male rides atop the female and stimulates her to release eggs. A glance at the large eggs surrounded in a transparent jelly coat suggests why the eggs of frogs and toads are so popular for studies of embryonic development.

Figure 31-9 Internal fertilization is essential for reproduction on land and is found in insects, reptiles, birds, and mammals.

occur only if an egg is mature and released into the female reproductive tract during the limited time when sperm are present. In most mammals, copulation occurs only at certain seasons of the year, or when the female signals readiness to mate, which often coincides with ovulation. Copulation itself triggers ovulation in a few animals, such as rabbits. An alternative strategy, employed by many snails, is to store sperm for days, weeks, or even months, thus assuring a supply of sperm whenever eggs are ready.

MAMMALIAN REPRODUCTION

In mammals, male and female reproductive systems are found in separate individuals. Many mammals reproduce only during certain seasons of the year, and consequently produce sperm and eggs only at that time. Human reproduction is similar to that of other mammals, except for a loss of seasonality. Men produce sperm more-or-less continuously, while women ovulate about once a month. Our discussion of mammalian reproduction will concentrate on humans.

The Male Reproductive Tract

The male reproductive tract consists of the paired **gonads** where sperm are produced, and accessory structures that store the sperm, produce secretions that activate and nourish them, and finally conduct them to the inside of the female reproductive tract (Fig. 31-10 and Table 31-1).

TESTIS. The male gonads, the **testes** (singular **testis**), produce both sperm and male sex hormones. The testes are located in the **scrotum,** a pouch that hangs outside the main body cavity. This location keeps the testes about 4°C cooler than the core of the body and provides the optimal temperature for sperm development. (Tight jeans may look sexy, but they push the scrotum up against the body, raising the temperature of the testes. Some researchers think that this may reduce sperm counts and hence reduce fertility. This is not, however, a reliable means of birth control!) Coiled, hollow **seminiferous tubules,** in which sperm are produced, nearly fill each testis. In the spaces between the tubules are the **interstitial cells,** which synthesize the male sex hormone testosterone.

Just inside the wall of each seminiferous tubule lie the diploid germ cells, or **spermatogonia,** from which all the sperm will eventually arise, and the much larger **Sertoli cells** (Fig. 31-10c and d). Spermatogonia can take one of two developmental paths. First, they may undergo mitosis, thereby providing a steady supply of new spermatogonia throughout life. Second, they may undergo **spermatogenesis,** that is, meiosis and differentiation into sperm (Fig. 31-11). Spermatogenesis begins with growth and differentiation of spermatogonia into **primary spermatocytes.** These are large diploid cells that are developmentally committed to spermatogenesis. The primary spermatocytes then go through meiosis (see Chapter 8). At the end of meiosis I, each primary spermatocyte gives rise to two haploid **secondary spermatocytes.** Each secondary spermatocyte divides again during meiosis II to produce two **spermatids,** for a total of four spermatids per primary spermatocyte. Spermatids undergo radical rearrangements of their cellular components as they differentiate into sperm.

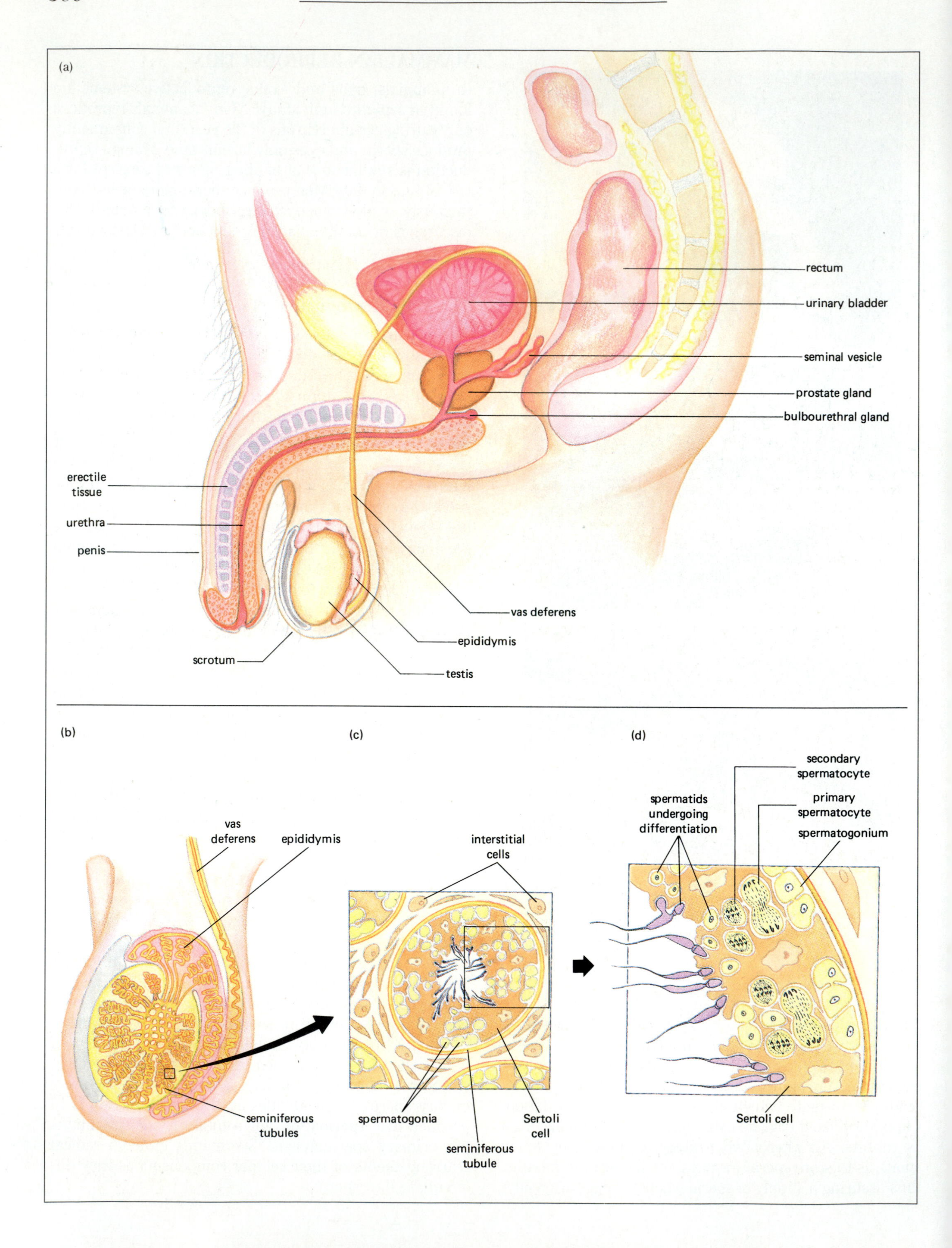
(a)
rectum
urinary bladder
seminal vesicle
prostate gland
bulbourethral gland
erectile tissue
urethra
penis
vas deferens
epididymis
scrotum
testis
(b)
vas deferens
epididymis
seminiferous tubules
(c)
interstitial cells
spermatogonia
seminiferous tubule
Sertoli cell
(d)
secondary spermatocyte
spermatids undergoing differentiation
primary spermatocyte
spermatogonium
Sertoli cell

Figure 31-10 (at left) **(a)** *The human male reproductive tract. The male gonads, the testes, hang beneath the abdominal cavity in the scrotum. Sperm pass from the seminiferous tubules of a testis to the epididymis, and thence through the vas deferens and urethra to the tip of the penis. Along the way, fluids are added from three sets of glands, the paired seminal vesicles and bulbourethral glands and the unpaired prostate gland.* **(b)** *A longitudinal section of the testis, showing the location of the seminiferous tubules, epididymis, and vas deferens.* **(c)** *A cross section of the seminiferous tubules. The walls of the seminiferous tubules are lined with spermatogonia and Sertoli cells protruding into the lumen of the tubule.* **(d)** *As spermatogonia undergo meiosis, the daughter cells move inward, embedded in invaginations of the Sertoli cells. There they differentiate into sperm, drawing upon the Sertoli cells for nourishment. Mature sperm are finally freed into the lumen of the tubules for transport to the penis. Testosterone is produced by the interstitial cells found in the spaces between tubules.*

Table 31-1 **Structures and Functions of the Human Male Reproductive Tract**

Structure	Type of Organ	Function
Testis	Gonad	Produces sperm and testosterone
Epididymis and vas deferens	Duct	Stores sperm; conducts sperm from testes to penis
Urethra	Duct	Conducts semen from vas deferens and urine from urinary bladder to the tip of the penis
Penis	External "appendage"	Deposits sperm in female reproductive tract
Seminal vesicles	Gland	Secrete fluids that contain fructose (energy source) and prostaglandins (possibly cause "upward" contractions of vagina, uterus, and oviducts, assisting sperm transport to oviducts); fluids may wash sperm out of ducts of male reproductive tract into vagina
Prostate	Gland	Secretes fluids that are basic (neutralize acidity of vagina) and contain factors that enhance sperm motility
Bulbourethral glands	Gland	Secrete mucus (may lubricate penis in vagina)

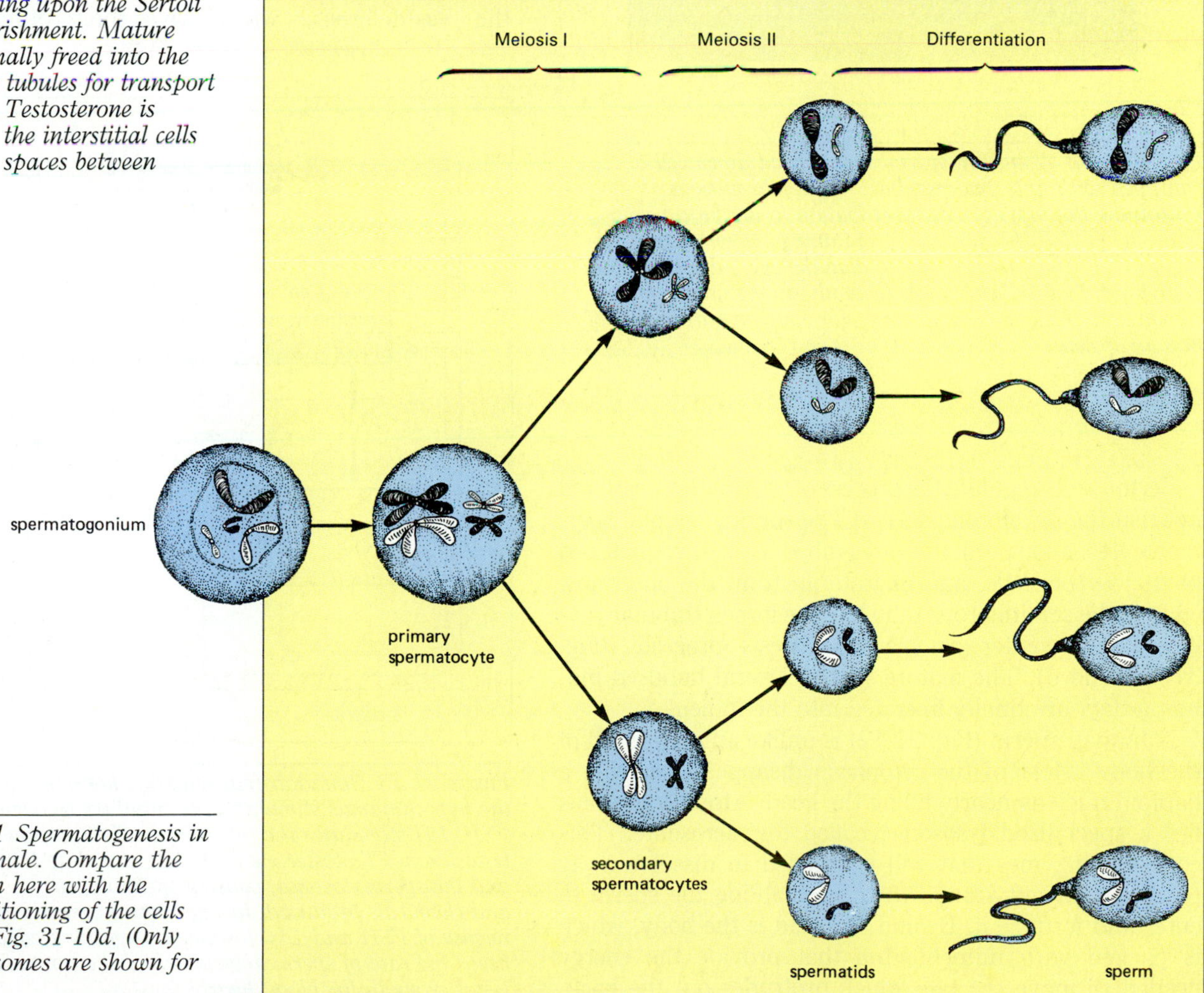

Figure 31-11 Spermatogenesis in the human male. Compare the events shown here with the physical positioning of the cells depicted in Fig. 31-10d. (Only four chromosomes are shown for clarity.)

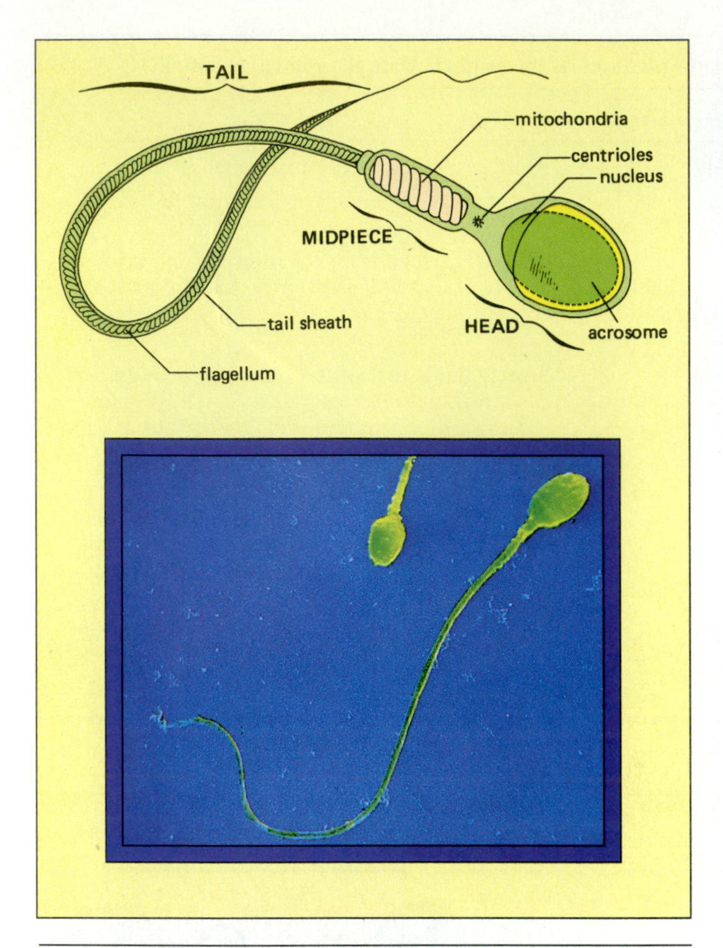

Figure 31-12 A mature sperm is a stripped-down cell equipped with only the essentials: a haploid nucleus containing the male genetic contribution to the future zygote, a lysosome (called the acrosome) containing enzymes that will digest away the barriers surrounding the egg, mitochondria for energy production, and a tail (actually a long flagellum) for locomotion. The photo is a false-color electron micrograph of human sperm.

Sertoli cells regulate the process of spermatogenesis and nourish the developing sperm. The spermatogonia, spermatocytes, and spermatids are embedded in invaginations of the Sertoli cells, and migrate up from the outermost edge of the seminiferous tubule to the lumen (tubular cavity) at the center as spermatogenesis proceeds (Fig. 31-10c and d). The mature sperm, several hundred million a day, are finally liberated into the lumen.

A human sperm (Fig. 31-12) is unlike any other cell of the body. Most of the cytoplasm disappears, leaving a haploid nucleus nearly filling the head. Atop the nucleus lies a specialized lysosome, called the **acrosome.** This contains enzymes that will be needed to dissolve away protective layers around the egg, enabling the sperm to enter and fertilize it. Behind the head is the body, which is packed with mitochondria that provide the energy needed to move the tail, which protrudes out the back. Sculling movements of the tail, really a long flagellum, propel the sperm along inside the female reproductive tract.

Spermatogenesis begins in puberty as a result of the interplay of **luteinizing hormone (LH)** and **follicle-stimulating hormone (FSH)** from the anterior pituitary, and testosterone from the testes themselves (Fig. 31-13). Luteinizing hormone stimulates the interstitial cells to produce testosterone. The combination of testosterone and follicle-stimulating hormone stimulates the Sertoli cells and spermatogonia, causing spermatogenesis.

Testosterone also stimulates the development of secondary sexual characteristics, maintains sexual drive, and is required for successful intercourse. Sperm, however, are not involved in these functions. Therefore, if one could suppress FSH release (blocking spermatogenesis) but not LH release (thereby allowing continued testosterone production), a man would be infertile but not impotent. Efforts are under way to develop a drug to do just that, as a male birth control chemical.

Accessory structures. The seminiferous tubules merge together, ultimately forming a single large tube, the **epididymis** (Fig. 31-10). The epididymis leads to the **vas deferens,** which leaves the scrotum and enters

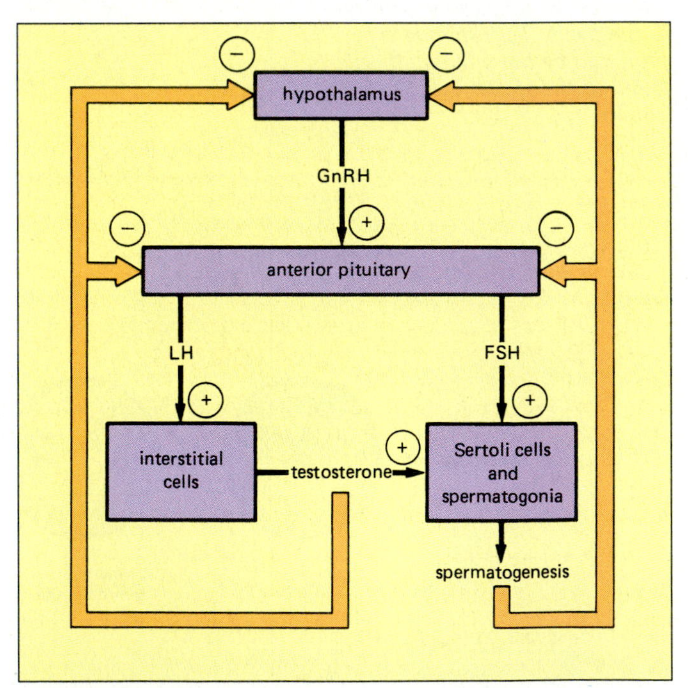

Figure 31-13 Gonadotropin releasing hormones (GnRH) from the hypothalamus stimulates the pituitary to release LH and FSH. LH stimulates the interstitial cells to produce testosterone. Testosterone and FSH stimulate the Sertoli cells and the spermatogonia, causing spermatogenesis. Testosterone and chemicals produced during spermatogenesis inhibit further release of FSH and LH, forming a negative feedback loop that keeps the rate of spermatogenesis and the concentration of testosterone in the blood nearly constant.

the abdominal cavity. Most of the hundreds of millions of sperm produced each day are stored in the vas deferens and epididymis. The vas deferens joins the **urethra,** leading from the bladder to the tip of the penis. This final common path is time-shared by sperm (during ejaculation) and urine (during urination).

The fluid ejaculated from the penis, called **semen,** consists of sperm mixed with secretions from three glands that empty into the vas deferens or urethra: the **seminal vesicles,** the **prostate gland,** and the **bulbourethral gland.** The secretions activate swimming by the sperm, provide energy for swimming, and neutralize the acidic fluids of the vagina (Table 31-1).

The Female Reproductive Tract

The female reproductive tract is almost entirely contained within the abdominal cavity (Fig. 31-14 and Table 31-2).

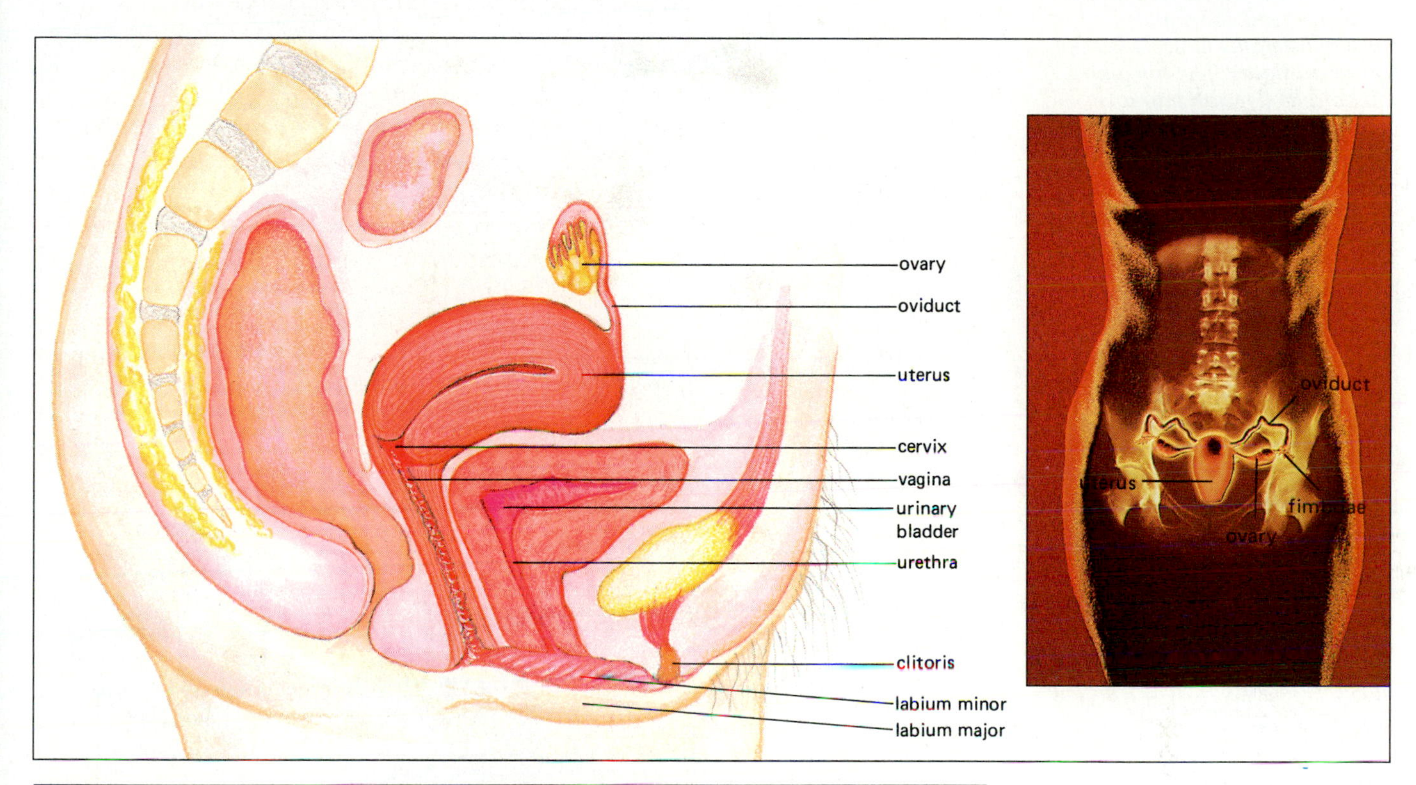

Figure 31-14 The human female reproductive tract, in side view (left) and frontal view (right). Eggs are produced in the ovaries and swept into the oviduct. A male deposits sperm in the vagina, from which they move up through the cervix and uterus into the oviduct. Sperm and egg usually meet in the oviduct, where fertilization occurs. The fertilized egg attaches to the lining of the uterus, where the embryo develops. The external genitalia, consisting of inner and outer labia and the clitoris, are highly sensitive tissues, stimulation of which may lead to orgasm.

Table 31-2 **Structures and Functions of Human Female Reproductive Tract**

Structure	Type of Organ	Function
Ovary	Gonad	Produces eggs, estrogen, and progesterone
Fimbria	Mouth of duct	Cilia sweep egg into oviduct
Oviduct	Duct	Conducts egg to uterus; site of fertilization
Uterus	Chamber	Site of development of fetus
Cervix	Connective tissue ring	Closes off lower end of uterus, supports fetus and prevents foreign material from entering uterus
Vagina	Large "duct"	Receptacle for semen; birth canal

Figure 31-15 *Oogenesis in human females.*
(a) *The development of follicles in an ovary, portrayed in a time sequence going counterclockwise from the lower left. A primary oocyte begins development within a follicle. The follicle grows, providing both hormones and nourishment for the enlarging oocyte. At ovulation, the secondary oocyte bursts through the ovary wall, surrounded by some follicle cells (now called the corona radiata). The remaining follicle cells develop into a secretory organ, the corpus luteum. If fertilization does not occur, the corpus luteum degenerates after a few days.*
(b) *The cellular stages of oogenesis. The oogonium enlarges to form the primary oocyte. At meiosis I, almost all of the cytoplasm is included in one daughter cell, the secondary oocyte. The other daughter cell is a small polar body that contains chromosomes but little cytoplasm. At meiosis II, almost all of the cytoplasm of the secondary oocyte is included in the egg, and a second small polar body discards the remaining "extra" chromosomes. The first polar body sometimes also undergoes the second meiotic division. In humans, meiosis II does not occur unless the egg is fertilized.*

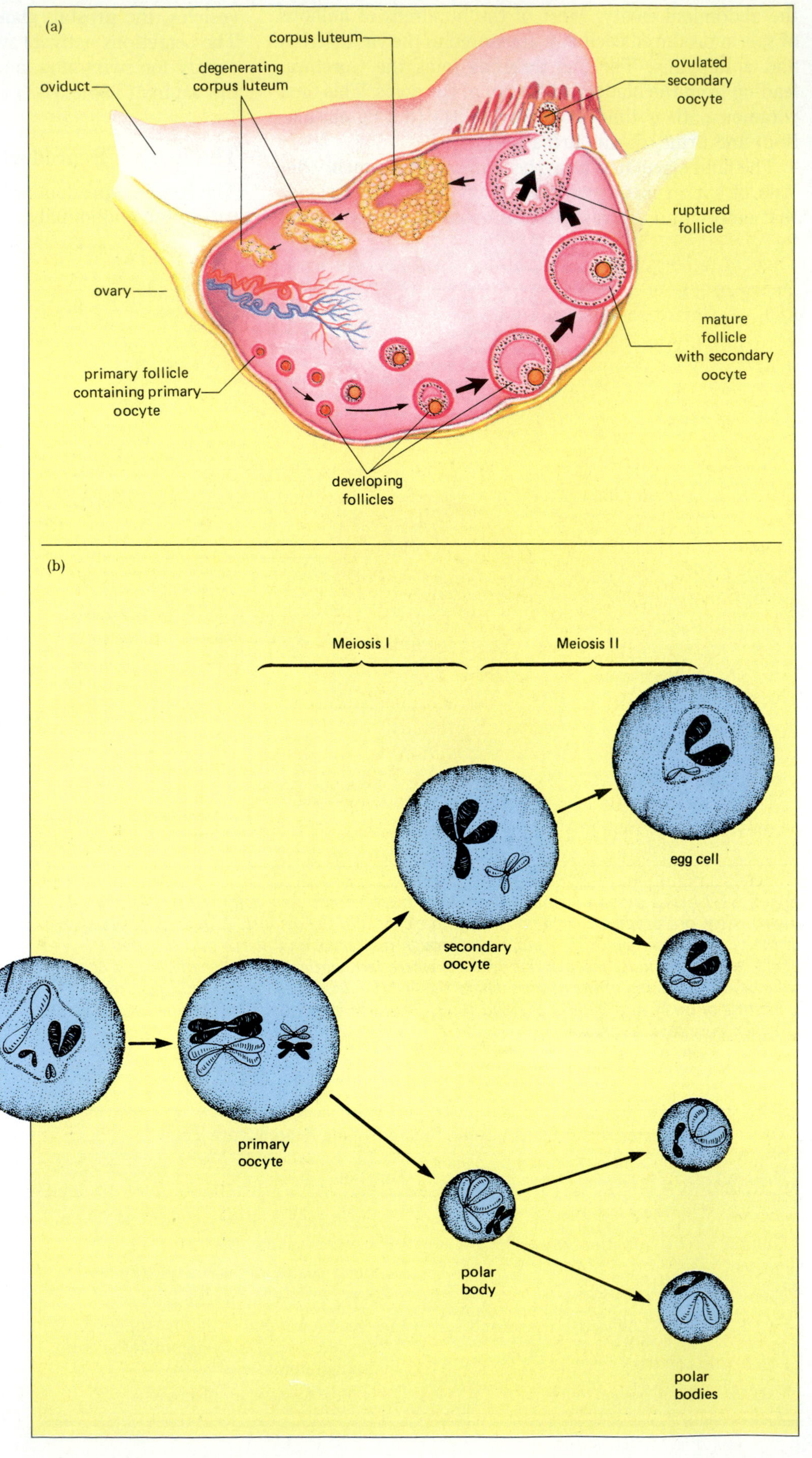

It consists of paired gonads, the **ovaries,** and accessory structures that accept sperm, conduct the sperm to the egg, and nourish the developing embryo.

OVARY. While she is still a fetus in her mother's womb, primordial egg cells, or **oogonia,** form within a woman's developing ovary. The oogonia divide by mitosis and then grow into **primary oocytes.** No oogonia remain after her third month of fetal development, and no new ones form during the rest of her life. Still during the fetal stage, all the primary oocytes begin meiosis but then halt during prophase I. At birth, the ovaries contain about 2 million primary oocytes; many die each day, until at puberty (usually 11 to 14 years of age) only about 400,000 remain. Since only a few oocytes resume meiosis during each month of a woman's reproductive span (from puberty to menopause at age 45 to 55), there is no shortage of oocytes.

Surrounding each oocyte is a layer of much smaller cells that both nourish the developing oocyte and secrete female sex hormones. Together, the oocyte and these accessory cells comprise a **follicle** (Fig. 31-15a). Each month during a woman's reproductive years, pituitary hormones stimulate development in a dozen or more follicles, although usually only one completely matures. The primary oocyte completes the first meiotic division to become a single **secondary oocyte** and a **polar body,** which is little more than a discarded set of chromosomes (Fig. 31-15b). Meanwhile, the small cells of the follicle multiply and secrete estrogen. As it matures, the follicle grows, eventually erupting through the surface of the ovary and releasing the secondary oocyte (Fig. 31-16). The second meiotic division occurs, not in the ovary, but in the oviduct, and then only if the secondary oocyte is fertilized. For convenience, we will refer to the ovulated secondary oocyte as the "egg."

Some of the follicle cells leave with the egg, but most remain behind in the ovary. These cells enlarge and become glandular, forming the **corpus luteum,** which secretes estrogen and a second hormone, progesterone. If fertilization does not occur, the corpus luteum degenerates a few days later.

ACCESSORY STRUCTURES. Each ovary is adjacent to, but not continuous with, an **oviduct** (often called the uterine tube or Fallopian tube in humans). The open end of the oviduct is fringed with ciliated "fingers" called **fimbriae** that nearly surround the ovary. The cilia create a current that sweeps the egg into the mouth of the oviduct. Fertilization usually occurs in the oviduct, and the new zygote gently tumbles down the oviduct into the pear-shaped **uterus,** or womb, in which it will develop for the next nine months. The wall of the uterus has two layers, which correspond to its dual functions of nourishment and childbirth. The inner lining, or **endometrium,** is richly supplied with blood vessels, forming the mother's part of the **placenta,** the structure that transfers oxygen, carbon dioxide, nutrients, and wastes between fetus and mother. The outer muscular **myometrium** contracts strongly during delivery, expelling the infant out into the world.

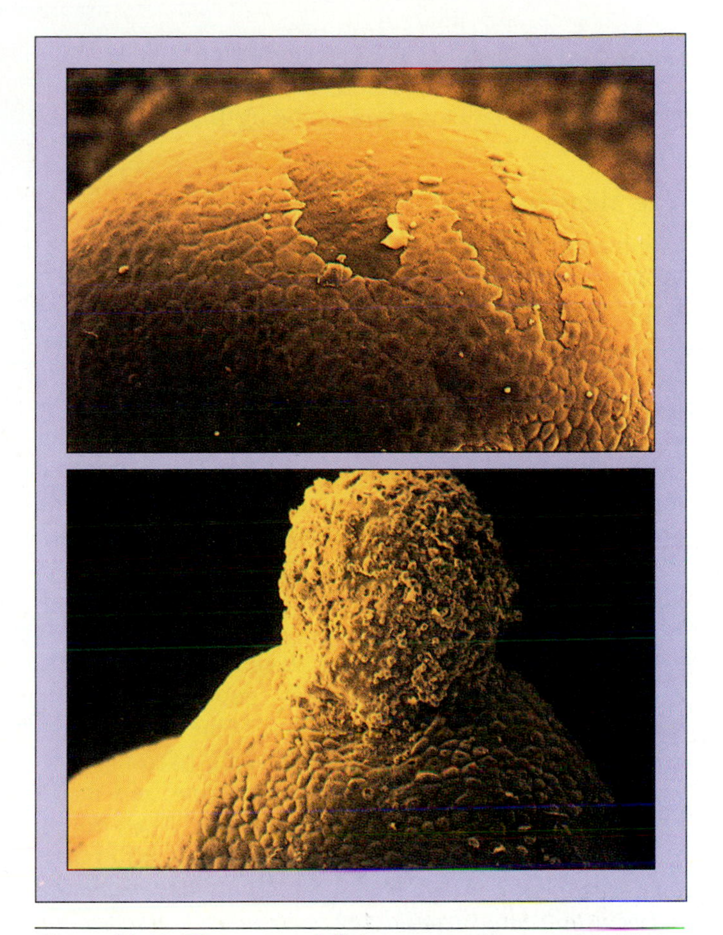

Figure 31-16 The mature follicle grows so large, and is filled with so much fluid, that it moves to the surface of the ovary and literally bursts through the ovary wall like a miniature volcano.

Developing follicles secrete estrogen, which stimulates the endometrium to grow an extensive network of blood vessels and nutrient-producing glands. After ovulation, estrogen and progesterone released by the corpus luteum promote continued growth of the endometrium. Thus, if an egg is fertilized, it encounters a rich environment for growth. If the egg is not fertilized, however, the corpus luteum disintegrates, estrogen and progesterone levels fall, and the endometrium disintegrates as well. The myometrium contracts, causing menstrual cramps and squeezing out the remnants of endometrium. The resulting flow of tissue and blood is called **menstruation** (from the Greek "mensis," meaning month).

The outer end of the uterus is nearly closed off by a ring of connective tissue, the **cervix.** The cervix holds the developing baby in the uterus, expanding only at the onset of labor to permit passage of the child. Beyond the cervix lies the **vagina,** which opens to the outside of the body. The vagina serves both as the receptacle for the penis during intercourse and as the birth canal.

The Menstrual Cycle

The human male produces sperm continuously, an adaptation selected during evolution since it permits fertilizaton of any willing female at any time, thereby increasing the number of potential offspring that a man can father. A woman, however, must carry the developing embryo, nourishing it for about nine months before it can survive outside the womb. Accordingly, it would be fruitless for her to ovulate unless her reproductive tract were prepared for pregnancy. The human female reproductive system goes through a complex **menstrual cycle,** in which hormonal interactions among the hypothalamus, pituitary gland, and ovary coordinate ovulation and the development of the uterus.

You will recall from Chapter 28 that hormone release by the anterior pituitary gland is controlled by neurosecretory cells in the hypothalamus. Some of these neurosecretory cells produce **gonadotropin-releasing hormone (GnRH),** which stimulates endocrine cells in the anterior pituitary to release FSH and LH. *A key to understanding the menstrual cycle is that these neurosecretory cells spontaneously release GnRH all the time, unless actively prevented from doing so by other hormones, notably progesterone.* We will begin our discussion of the menstrual cycle with the spontaneous release of GnRH.

GnRH stimulates the anterior pituitary to release FSH and LH (step 1 in Fig. 31-17). FSH and LH circulate in the bloodstream to the ovaries and initiate the development of several follicles. The follicle cells surrounding the developing oocyte are stimulated by FSH and LH to secrete estrogen. Under the combined influences of FSH, LH, and estrogen, the follicles grow during the next two weeks (step 2). Simultaneously, the primary oocyte within each follicle enlarges, storing both food and regulatory substances (mostly proteins and messenger RNA) that will be needed by the fertilized egg during early development (see Chapter 32). For reasons that are not completely understood, only one, or rarely two, follicles complete development each month. As the maturing follicle enlarges, it secretes ever greater amounts of estrogen (step 3). This estrogen has three effects. First, it promotes the continued development of the follicle itself and the primary oocyte contained within it. Second, it stimulates growth of the endometrium of the uterus. Third, high levels of estrogen stimulate both the hypothalamus and pituitary, resulting in a surge of LH and FSH at about the twelfth day of the cycle (step 4).

The function of the peak in FSH concentration is not known, but the surge of LH has three important consequences: (1) it triggers the resumption of meiosis I in the oocyte, resulting in the formation of the secondary oocyte and the first polar body; (2) it causes the final explosive growth of the follicle, culminating in ovulation (step 5); and (3) it transforms the remnants of the follicle that remain in the ovary into the corpus luteum.

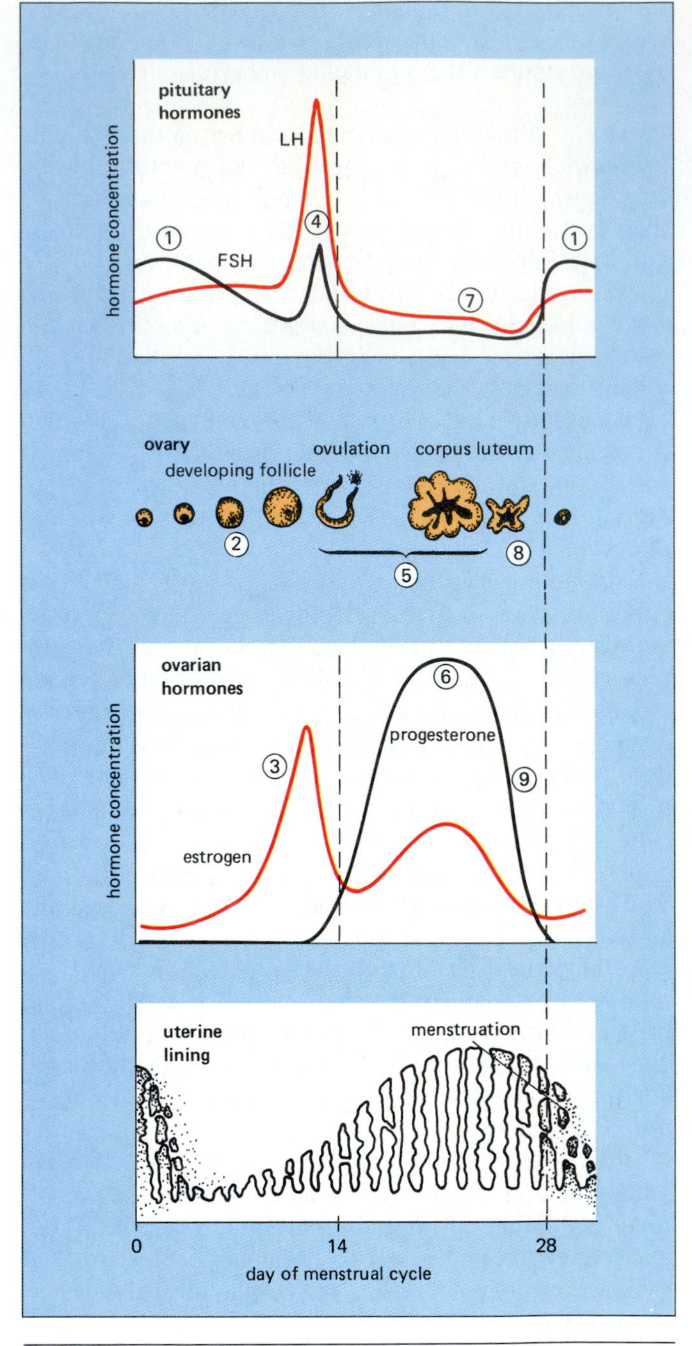

Figure 31-17 The menstrual cycle is generated by interactions among the hormones of the hypothalamus, anterior pituitary gland, and ovaries. The hormonal changes in turn drive cyclic changes in the uterine lining. The numbers on the graphs refer to the hormonal interactions discussed in the text. They zigzag back and forth among pituitary hormones, ovarian structures, and ovarian hormones, as each of these exerts effects on the others.

The corpus luteum secretes both estrogen and progesterone (step 6). The combination of these hormones inhibits the hypothalamus and pituitary, shutting down the release of FSH and LH (step 7), which prevents the development of any more follicles. Simultaneously, estrogen and progesterone stimulate further growth of the endometrium, which eventually becomes about 5 millimeters thick. (Note that the effects of progesterone and estrogen depend on the target organ. Progesterone *stimulates* the endometrium, but *inhibits* hormone release from the hypothalamus and pituitary.)

In menstrual cycles in which pregnancy does not occur, the corpus luteum essentially commits suicide. The corpus luteum survives only while it is stimulated by LH (or by a similar hormone released by the developing embryo, as we describe below). However, progesterone secreted by the corpus luteum shuts off LH secretion, so the corpus luteum dies around the twenty-first day of the cycle (step 8). With the corpus luteum gone, estrogen and progesterone levels plummet (step 9). Deprived of stimulation by estrogen and progesterone, the endometrium of the uterus also dies, sloughing off as the menstrual flow beginning about the twenty-seventh or twenty-eighth day of the cycle. Simultaneously, the reduced progesterone level no longer inhibits the hypothalamus and pituitary, so spontaneous release of FSH and LH resumes (step 1). This initiates development of a new set of follicles, starting the cycle over again.

You may have noticed that the menstrual cycle is an exception to the general rule of hormonal control systems: that negative feedback regulates the concentration of hormones and the functioning of the body at relatively even levels. The reason for this is that *the menstrual cycle includes both positive and negative feedback.* During the first half of the cycle, FSH and LH stimulate estrogen production. High levels of estrogen *stimulate* the midcycle surge of FSH and LH release (positive feedback). During the second half of the cycle, estrogen and progesterone together *inhibit* the release of FSH and LH (negative feedback). The early positive feedback causes hormone concentrations to reach high levels, while the later negative feedback shuts the system down again unless pregnancy intervenes.

Copulation and Fertilization

As terrestrial mammals, humans employ internal fertilization to deposit the sperm in the moist environment of the female reproductive tract. To do this, the penis is inserted into the vagina, where sperm are released during ejaculation. The sperm swim upward in the female reproductive tract, from the vagina through the opening of the cervix into the uterus, and on up into the oviducts. If the female has ovulated within the last day or so, the sperm will meet an egg in one of the oviducts. A single lucky sperm may succeed in fertilizing it, starting the development of a new human being.

Copulation. The male role in copulation begins with erection of the penis. Before erection, the penis is flaccid, because the arterioles supplying it are constricted, allowing little blood flow (Fig. 31-18a). Under the dual influences of psychological and physical stimulation, the arterioles dilate and blood flows into vascular spaces within the penis. As these swell, they squeeze off the veins that drain the penis (Fig. 31-18b). Pressure builds up, causing an erection. After insertion into the vagina, movements of male or female or both further stimulate touch receptors on the penis, triggering ejaculation. Ejaculation occurs when muscles encircling the epididymis, vas defer-

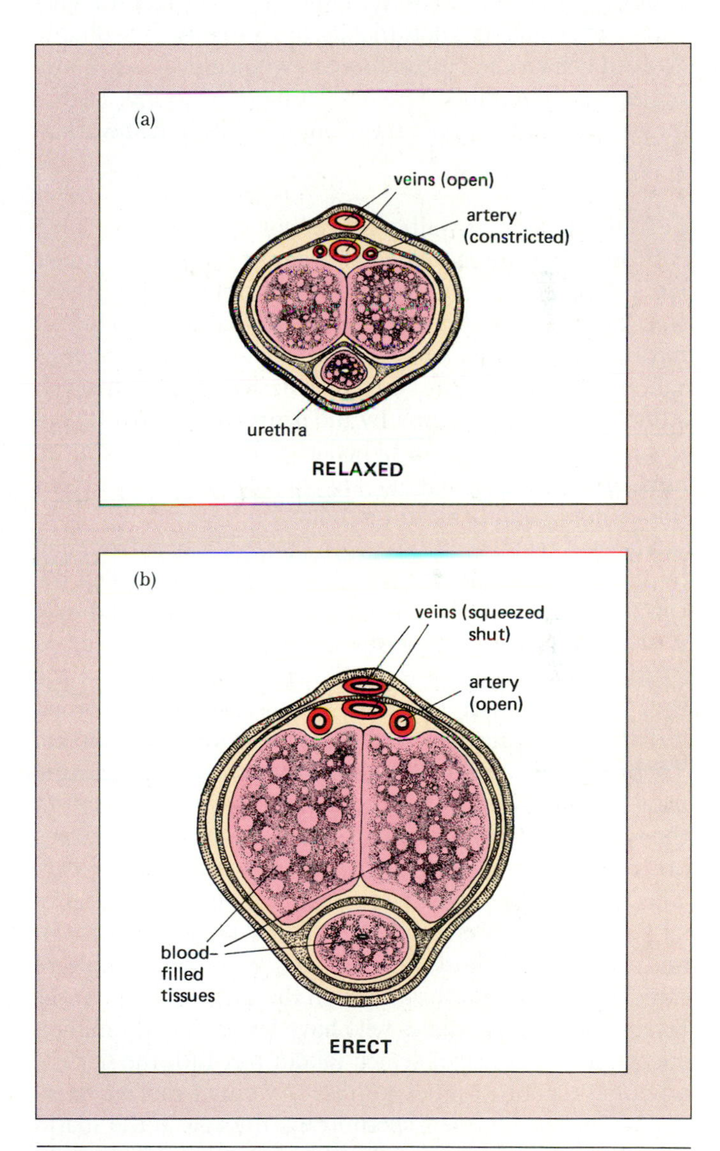

Figure 31-18 *Changes in blood flow within the penis cause erection.*
(a) *Normally, smooth muscles encircling the arteries leading into the penis are contracted, limiting blood flow.*
(b) *During sexual excitement, these muscles relax, and blood flows into spaces within the penis. The swelling penis squeezes off the veins leaving the penis, raising the hydrostatic pressure and causing the penis to become elongated and rigid.*

ens, and urethra contract, forcing semen out of the penis and into the vagina. The average ejaculation consists of 3 or 4 milliliters of semen, containing 300 to 400 million sperm. Ejaculation coincides with orgasm, a feeling of intense pleasure and release.

Similar changes occur in the female. Sexual excitement causes increased blood flow to the vagina and external genitalia, including the vulva and clitoris (Fig. 31-14). The clitoris, which is embryologically similar to the penis, becomes erect. Stimulation by the penis of the male often, but not always, results in female orgasm, a series of rhythmic contractions of the vagina and uterus accompanied by sensations of pleasure and release. Although enjoyable, female orgasm is not necessary for fertilization. (As students of biology, we should note that the pleasures of sex probably evolved because ancestral humans who enjoyed sex more, copulated more, and therefore left more offspring. Sexual pleasure is another example of natural selection at work.)

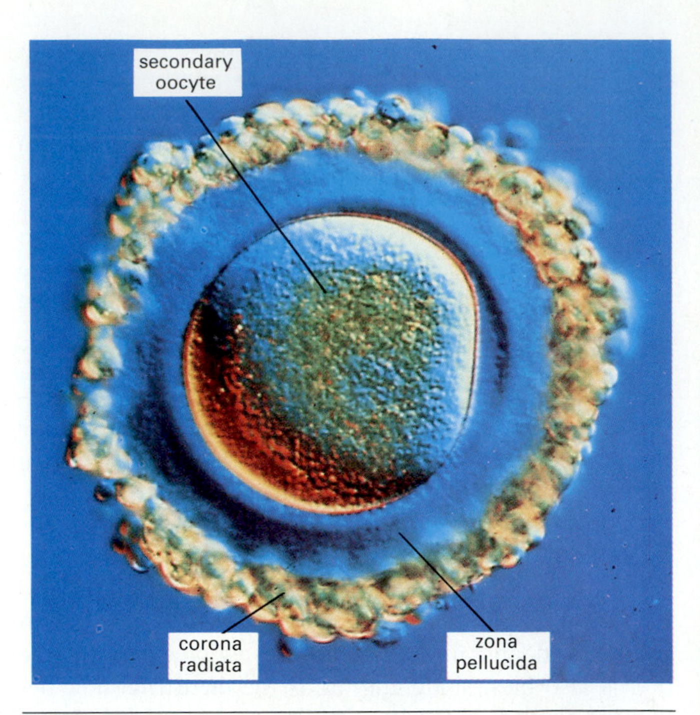

Figure 31-19 A human secondary oocyte shortly after ovulation. Sperm must digest their way through the small follicular cells of the corona radiata and the clear zona pellucida to reach the oocyte itself.

FERTILIZATION. Neither sperm nor egg lives very long. An egg may remain viable for a day, while sperm, under ideal conditions, may live for two. Therefore, fertilization can succeed only if copulation occurs within a couple of days before or after ovulation. In some animals, such as rabbits, synchrony is assured by reflex ovulation, in which stimulation of the vagina by the penis triggers ovulation. In humans, things seem to be left more to chance, but in light of our rapidly burgeoning population, in most cases this is not a handicap to reproduction.

You will recall that the egg leaves the ovary surrounded by follicle cells (Fig. 31-19). These cells, now called the **corona radiata,** form a barrier between sperm and egg. A second barrier, the membrane of the **zona pellucida** ("clear area"), lies between the corona radiata and the egg. In the oviduct, scores of sperm encircle the corona radiata, each one releasing enzymes from its acrosome. These enzymes weaken both the corona radiata and the zona pellucida, allowing the sperm to wriggle through to the egg. If there aren't enough sperm, not enough enzymes are released, and none of the sperm will reach the egg. This may be the selective pressure for the ejaculation of so many sperm. Perhaps 1 in 100,000 reach the oviduct, and 1 in 20 of those find the egg, so only a couple of hundred sperm join the attack on the surrounding barrier layers. In humans, males who have fewer than 20 million sperm per milliliter of semen (about one-fifth the normal amount) usually cannot fertilize a woman during intercourse because too few sperm reach the egg. If the sperm are otherwise normal, such men can father children by artificial insemination, in which a large amount of semen is injected directly into the oviduct.

One sperm finally contacts the surface of the egg. The cell membranes of egg and sperm fuse, and the sperm head passes into the cytoplasm of the egg. Penetration by the sperm triggers two vital changes in the egg. First, vesicles near the surface of the egg release chemicals into the zona pellucida, reinforcing it and preventing further sperm from entering the egg. Second, the egg undergoes its second meiotic division, producing a haploid gamete at last. Fertilization occurs as the haploid nuclei of sperm and egg fuse, forming a diploid nucleus that contains all the genes of a new human being.

Pregnancy

The zygote begins to divide, all the while drifting down the oviduct to the uterus. By about a week after fertilization, the zygote has developed into a double ball of cells, the **blastocyst** (Fig. 31-20). The **inner cell mass** will become the embryo itself, while the sticky outer ball will adhere to the uterus and burrow into the endometrium, a process termed **implantation.** Blood from ruptured uterine vessels plus glycogen secreted by endometrial glands nourish the growing embryo. The details of embryonic development are presented in Chapter 32.

In the usual menstrual cycle (Fig. 31-17), the corpus luteum disintegrates about a week after ovulation. This removes the source of estrogen and progesterone needed to sustain the endometrium, resulting in menstruation. The embryo itself prevents these changes from occurring during pregnancy. Shortly after the outer cells of the blastocyst begin to implant in the endometrium, they start secreting an LH-like hormone called **chorionic gonadotropin (CG).** This hormone travels in the bloodstream to

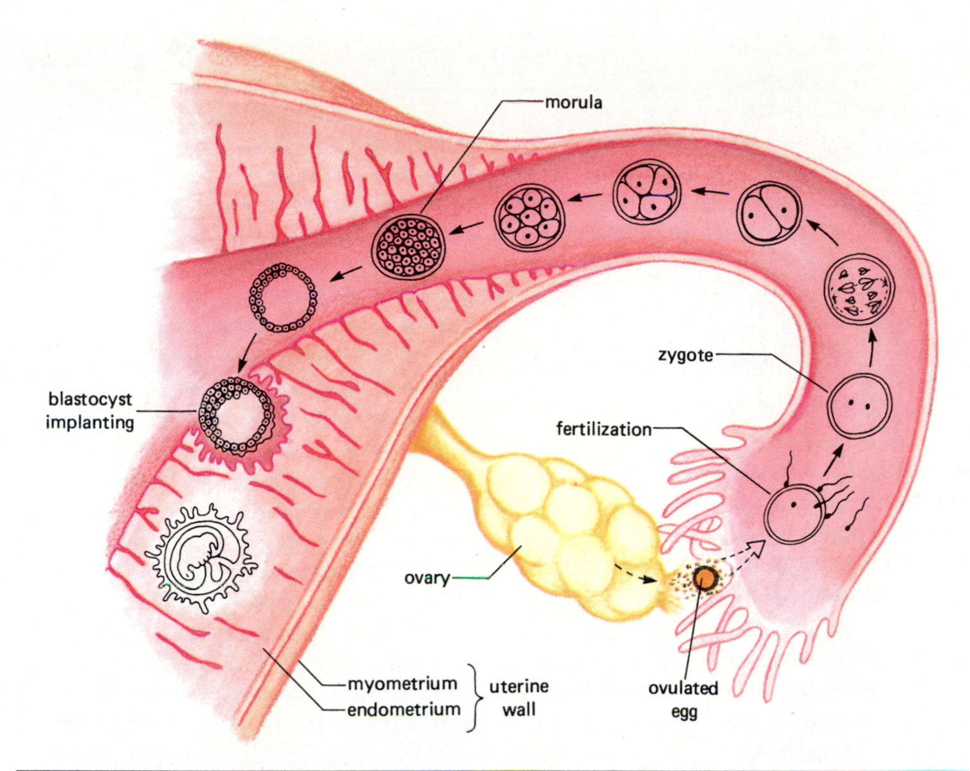

Figure 31-20 *The first few days of pregnancy. The egg is fertilized in the oviduct and slowly travels down to the uterus. Along the way, a few cell divisions occur, until a hollow blastocyst is formed. The inner cell mass will form the embryo proper, while the surrounding cells will adhere to the uterine endometrium, burrow in, and begin forming the placenta.*

the ovary, where it prevents degeneration of the corpus luteum. The corpus luteum continues to secrete estrogen and progesterone, and the uterine lining continues to grow, nourishing the embryo. So much CG is released by the embryo that it is excreted by the mother in her urine; most tests for pregnancy are assays for CG in maternal urine or blood.

Obtaining nutrients directly from the nearby cells of the endometrium will suffice only for the first week or two of embryonic growth. During this time, the placenta begins to form, composed of interlocking tissues of the embryo and the endometrium. Through the placenta, the embryo will receive nutrients and oxygen, and dispose of wastes into the maternal circulation (see Chapter 32).

Delivery

Near the end of the ninth month, give or take a few weeks, the process of birth begins. It is thought that the infant itself somehow signals "readiness for birth," perhaps via a hormone, but no one knows for certain what the signal is. However they are triggered, the immediate causes of delivery seem to be a complex interplay among prostaglandins released by the uterus and the fetal membranes, stretching of the uterus and cervix as the baby grows and is delivered, and oxytocin released by the posterior pituitary gland.

Unlike skeletal muscles, uterine muscles can contract spontaneously, and these contractions are enhanced by stretching. As the baby grows, it weighs more and fills up the uterus, stretching the uterine muscles. These occasionally contract weeks before delivery. In apparent response to the fetal "signal," prostaglandins are released both by the uterus and the membranes surrounding the fetus. These hormones cause intense contractions to begin, signaling the onset of labor. As the contractions proceed, the baby's head pushes against the cervix, making it dilate. Stretch receptors in the walls of the cervix send signals to the hypothalamus, triggering oxytocin release. Under the dual stimulation of prostaglandin and oxytocin, the uterus continues to contract, finally expelling the baby. After a brief rest, uterine contractions resume, so that the uterus shrinks remarkably. During these contractions, the placenta is sheared off from the uterus and is expelled through the vagina as the "afterbirth."

Further prostaglandin release in the umbilical cord causes muscular contraction around the fetal blood vessels in the cord, shutting off blood flow. (Tying off the cord is standard practice but not usually necessary; if it were, dogs, cats, elephants, and all the rest of the mammals could not survive birth!) Although still intimately dependent on its parents for survival, a new human being has been born.

On Limiting Fertility

Successful reproduction is essential if any species, including our own, is to endure. During most of human evolution, mortality was high. Children were especially vulnerable, dying from malnutrition, accidents, disease, predators, and conflicts. Therefore, natural selection favored people with strong sexual drives, who would copulate often and produce many children. Evolution could not foresee the development of modern society, with efficient agriculture, effective hygiene, and quality medical care. With a few tragic exceptions, people today enjoy low infant mortality and a life span triple that of ancient times. Parents need not have a dozen children to ensure that a few survive to adulthood. Today, most people wish to determine the timing and/or limit the number of children they have. Some are concerned about overpopulation (see Chapter 35). Others wish to postpone starting a family until college is finished or a career is established. Still others feel unprepared for the responsibilities of parenthood, or simply want to avoid embarrassment or inconvenience. One way to prevent childbirth, of course, is to abstain from intercourse. However, human genes cannot change as rapidly as cultures do: today's humans have sexual drives appropriate to prehistoric times, but live under vastly different circumstances.

Historically, it has not been easy to separate the acts of copulation and conception. Primitive cultures tried such inventive, if bizarre, techniques as swallowing froth from the mouth of a camel or using vaginal suppositories of crocodile dung. Even 50 years ago there were no reliable methods of birth control. Over the last 20 years, however, several effective techniques have been developed for preventing pregnancy (Table E31-1). As you will see, however, the "perfect" method—completely safe, effective, effortless, and reversible—has not yet been found.

Permanent Contraception

Probably the safest, most effective, and in the long run, most effortless method of contraception is **sterilization,** in which the pathways through which sperm or egg must travel are interrupted. Worldwide, over 100 million people have chosen sterilization as their preferred method of birth control. In men, the vas deferens leading from each testis may be severed in an operation called a **vasectomy** (Fig. E31-1a). Sperm are still produced, but they cannot reach the penis during ejaculation. Aside from the inconvenience of surgery (usually requiring only local anaesthetics), vasectomy has no known physical side effects on health or sexual performance. The similar, but slightly more complex, operation of **tubal ligation** renders a woman infertile by cutting her oviducts (Fig. E31-1b). Ovulation still occurs, but sperm cannot travel to the egg, nor can the egg reach the uterus. Occasionally, tubal ligation leads to irregularity of menstrual periods and increased menstrual flow. Sterilization has the important advantage that no further thought need be given to contraception. However, although in a few instances a surgeon can reconnect the vas deferens or oviducts, sterilization is usually permanent.

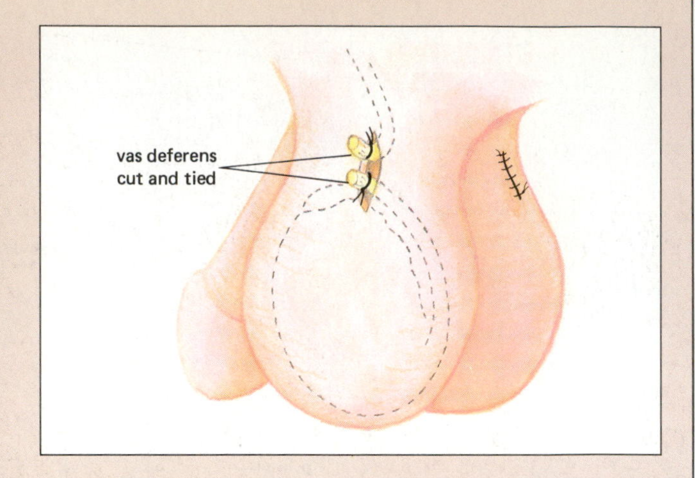

Figure E31-1a

Temporary Contraception

People who may wish to have children sometime in the future must employ temporary methods of contraception. These techniques fall into three general categories: preventing ovulation, preventing sperm and egg from meeting when ovulation does occur, and preventing implantation of a fertilized egg in the uterus.

As you learned earlier in this chapter, during a normal menstrual cycle ovulation is triggered by a midcycle surge of LH; later, LH secretion is inhibited by estrogen and progesterone released by the corpus luteum. An obvious way to prevent ovulation is to provide a continuing supply of estrogen and progesterone, thereby suppressing LH release. Estrogen and progesterone (usually in synthetic form) are the components of **birth control pills.** The Pill is extremely effective, but must be taken daily, usually for 21 days each menstrual period. Unfortunately, the hormones in the Pill induce a mild form of some of the same side effects as pregnancy: water retention, soreness of the breasts, nausea, and especially in women over 40, an increased risk of blood clots. These side effects are extremely rare in women taking pills that have very low doses of estrogen, and for women under 30, the Pill is probably one of the best contraceptive techniques.

On Limiting Fertility

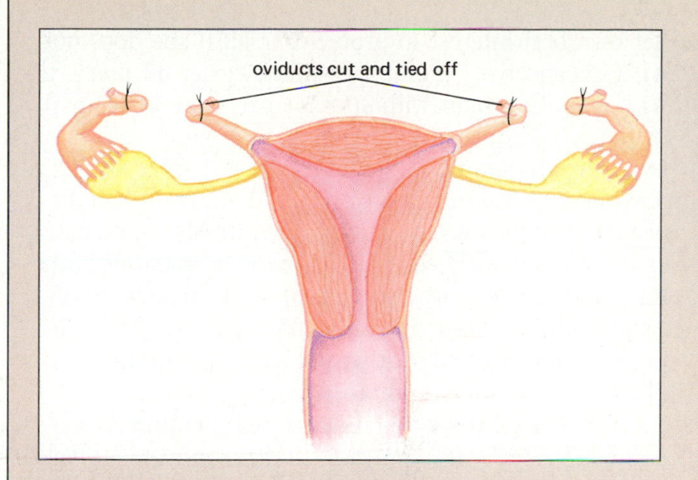

Figure E31-1b

There are several effective ways to prevent the encounter of sperm and egg. One is the **diaphragm,** a rubber cap that fits snugly over the cervix, preventing sperm from entering the uterus. In conjunction with a spermicide, diaphragms are very effective and have no known side effects. Recently, the **contraceptive sponge** has been introduced. A soft, spermicide-impregnated plug is inserted in the vagina up against the cervical opening, physically blocking off the opening and killing any sperm that get near. Alternatively, a **condom** may be worn over the penis of the male, preventing sperm from being deposited in the vagina. Diaphragms, sponges, and especially condoms must be applied shortly before intercourse, often at a time when the participants would rather be thinking about something else. Furthermore, if the diaphragm or condom happens to have even a small hole, sperm may still enter the uterus and fertilize the egg. (It seems to be a singular irony of nature that a man ejaculating 50 million sperm will often be infertile, but if just a few drops of semen escape past a diaphragm or condom, pregnancy may occasionally result.)

Other, less effective procedures include **spermicides alone, withdrawal** (removal of the penis from the vagina just before ejaculation), and **douching** (washing sperm out of the vagina, hopefully before they have had a chance to enter the uterus). Spermicides have some contraceptive effect, but withdrawal and douching are essentially useless. A final method of preventing fertilization is **rhythm:** abstinence from intercourse during the ovulatory period of the menstrual cycle. Although completely safe and theoretically effective, in practice rhythm usually has a high failure rate, due to variability in the exact timing and duration of the menstrual period from month to month, lack of discipline on the part of the users, and inaccuracies in determining the menstrual cycle. Since a slight rise in body temperature usually coincides with ovulation, the incidence of pregnancies in women using the rhythm method can be reduced if the body temperature is recorded daily and sexual activity is regulated accordingly.

Even if an egg is fertilized, pregnancy will not occur unless the zygote implants in the uterus. The **intrauterine device (IUD)** is a small copper or plastic loop, squiggle, or shield that is inserted into the uterus and remains there more or less indefinitely. Although highly effective (if it stays in place), IUDs seem to work by irritating the uterine lining so that it cannot receive the embryo. Nor surprisingly, this sometimes causes discomfort and uterine infection, and occasionally the IUD even punctures the uterine wall. A second method of preventing implantation is the "morning after" pill, which contains a massive dose of estrogen. For some women, preventing implantation has the major drawback that it is, in effect, an extremely early abortion.

Abortion

When contraception fails, the pregnancy may be terminated by an **abortion.** Abortion has medical, financial, and ethical drawbacks. Abortion is much more dangerous to a woman's health than most of the contraceptive techniques described above. All abortions, even those performed during the first trimester of pregnancy, are also quite expensive. Finally, of overriding importance to many women is the matter of killing a human fetus. Science can provide information regarding the progress of fetal development during pregnancy, but cannot provide judgments about when a fetus becomes a "person," legally and ethically, or about the relative merits of fetal versus maternal rights. Therefore, whether abortion involves destruction of a person, and if so, whether the fetus's rights should override those of the woman, are likely to remain extremely controversial issues.

Future Contraceptive Methods

Further advances in contraception are on the way. One possibility is a removable plug for the oviduct. A thin tube is inserted through the vagina and uterus into the mouth of the oviduct, and a stream of liquid silicone rubber is squirted into the oviduct. The rubber gels, plugging up the oviduct. Although it is too early to predict a success rate, the plug may be removable if the woman later decides that she wants to become pregnant. There are also once-a-month contraceptive

On Limiting Fertility

pills under development, most of which prevent the uterus from becoming fully prepared for implantation. Recently, an inventor patented a battery-powered contraceptive. Inserted into the cervix, it generates a weak electrical field that immobilizes sperm. Finally, a hormone-impregnated, silicone rubber block might be inserted beneath a woman's skin, slowly releasing hormones over months or even a year, providing the same contraceptive effects as the Pill. Many of these products are currently undergoing clinical trials of safety and effectiveness. Some may be available by the time you read this book.

You may have noticed that most of the contraceptive techniques are directed at the woman, not the man. There are several reasons for this, not all of them chauvinistic. First, although both sexual partners are equally responsible for conception, it is the woman who becomes pregnant and who bears the major burden of childbirth and postnatal care. Therefore, it is in her own best interest to protect herself if she does not wish to conceive. Second, it is much easier to interfere with ovulation than with sperm formation. A woman ovulates only once a month, ovulation can be blocked by relatively simple and safe hormonal intervention, and there is little evidence that ovulation itself influences a woman's sexual drives. In contrast, men create sperm continuously, and testosterone is essential both for sperm formation and sexual performance; early "male pills" caused not only infertility but also impotence. Third, methods that prevent implantation can obviously be applied only to women.

Nevertheless, there is a major research effort in several countries to develop male contraceptives equivalent to the Pill. A promising contraceptive is a daily dose of testosterone and a modified form of gonadotropin releasing hormone, which together seem to block sperm production without affecting sexual performance. Clinical trials are now underway.

Table E31-1 **Birth Control Techniques**

Method	Mechanism of Action	Side Effects	Failure Rate (pregnancies per 100 women per year)
Vasectomy	Sever vas deferens, so sperm cannot reach penis	None known	0
Tubal ligation	Sever oviduct, so egg cannot reach uterus	Occasional menstrual irregularity	0
Birth control pill	Prevent ovulation	Blood clotting, breast soreness, water retention, nausea, vitamin deficiencies	0–5
Diaphragm (with spermicide)	Block opening of cervix, kill sperm	Spermicide may irritate vaginal lining	3–20
Contraceptive sponge	Block opening of cervix, kill sperm	Spermicide may irritate vaginal lining	? (prob. low)
Condom	Cover penis, prevent sperm from entering vagina	Reduced sensation in penis	3–20
Spermicide alone	Kill sperm	Spermicide may irritate vaginal lining	3–30
Withdrawal	Remove penis from vagina before ejaculation	None	10–30
Douching	Wash sperm out of vagina before they can pass through cervical opening	Douching solution may irritate vaginal lining	20–40
Rhythm	Avoid intercourse during probable time of ovulation	None	5(?)–30
Intrauterine device	Prevent implantation of fertilized egg	May irritate or perforate uterine lining	1–5
"Morning after" pill	Prevent implantation of fertilized egg	Breast soreness, water retention, nausea, blood clotting, cancer(?)	? (prob. low)

SUMMARY OF KEY CONCEPTS

Reproductive Strategies

Animals reproduce either sexually or asexually. Sexual reproduction involves the union of haploid gametes, usually from two separate parents, which produces an offspring that is genetically different from either parent. In asexual reproduction, offspring are usually genetically identical to the parent. Asexual reproduction may occur by regeneration, fission, budding, or parthenogenesis.

Among animals that engage in sexual reproduction, the female is defined as the parent that produces large, nonmotile eggs, while the male produces small, motile sperm. Animals may be either monoecious, in which a single animal produces both sperm and eggs, or dioecious, in which a single animal produces one type of gamete. The union of sperm and egg, called fertilization, may occur outside the bodies of the animals (external fertilization) or inside the body of the female (internal fertilization). External fertilization must occur in water, so that the sperm can swim to meet the egg. Most internal fertilization is through copulation, in which the male deposits sperm directly into the female reproductive tract, which provides the moist environment needed for sperm survival and swimming.

Mammalian Reproduction

The human male reproductive tract consists of paired testes that produce sperm and testosterone, and accessory structures that conduct the sperm to the female's reproductive tract and contribute fluids, nutrients, and activating factors required for sperm motility. In human males, spermatogenesis and testosterone production are stimulated by FSH and LH, secreted by the anterior pituitary. Spermatogenesis and testosterone production are nearly continuous, beginning at puberty, and lasting until death.

The human female reproductive tract consists of paired ovaries that produce eggs and the hormones estrogen and progesterone, and accessory structures that conduct sperm to the egg and receive and nourish the embryo during prenatal development. In human females, oogenesis, hormone production, and development of the lining of the uterus vary in a monthly menstrual cycle. The cycle is controlled by hormones from the hypothalamus (gonadotropin releasing factor), anterior pituitary (FSH and LH), and ovaries (estrogen and progesterone).

During copulation, the male inserts his penis into the female's vagina and ejaculates semen. The sperm move through the vagina and uterus into the oviduct, where fertilization usually takes place. The unfertilized egg is surrounded by two barriers, the corona radiata and the zona pellucida. Enzymes released from the acrosomes at the tips of sperm digest away these layers, permitting sperm to reach the egg. Only one sperm enters the egg and fertilizes it.

The fertilized egg undergoes a few cell divisions in the oviduct and then implants in the uterine lining. Implantation and subsequent release of chorionic gonadotropin by the embryo eliminate further menstrual cycles until the pregnancy is over.

GLOSSARY

Acrosome (ak′-rō-sōm) an enzyme-containing vesicle located at the tip of an animal sperm.

Amplexus (am-plecks′-us) a form of external fertilization found in amphibians, in which the male holds the female during spawning and releases his sperm directly onto her eggs.

Asexual reproduction. reproduction not involving the union of genetic material from two different organisms. Usually, asexual reproduction is accomplished by mitosis, producing genetically identical copies of the parent organism.

Blastocyst (blas′-tō-sist) an early stage of human embryonic development, consisting of a fluid-filled ball with walls one cell layer thick, enclosing a mass of cells attached to its inner surface.

Bud in animals, a small copy of an adult that develops on the body of the parent; eventually breaks off and becomes independent.

Budding asexual reproduction by growth of a miniature copy of the adult animal on the body of the parent. The bud breaks off to begin independent existence.

Bulbourethral gland (bul-bō-ū-rē′-thrul) in male mammals, a gland that secretes a basic, mucus-containing fluid that forms part of the semen.

Cervix (ser′-vicks) a ring of connective tissue at the outer end of the uterus, leading into the vagina.

Chorionic gonadotropin a hormone secreted by the chorion (one of the fetal membranes), which maintains the integrity of the corpus luteum during early pregnancy.

Corona radiata (ka-rō′-na rā-dē-a′-ta) the layer of cells surrounding an egg after ovulation.

Corpus luteum (kor′-pus lū′-tē-um) in the mammalian ovary, a structure derived from the follicle after ovulation, which secretes the hormones estrogen and progesterone.

Dioecious (dī-ē′-shus) pertaining to organisms in which male and female gametes are produced by separate individuals.

Egg the haploid female gamete, usually large and nonmotile, containing food reserves for the developing embryo and regionally localized gene-regulating substances that direct early development.

Endometrium (en-dō-mē′-trē-um) the nutritive inner lining of the uterus.

Epididymis (e-pi-di′-dē-mus) tubes that connect with and receive sperm from the seminiferous tubules of the testis.

Estrogen in vertebrates, a female sex hormone produced by follicle cells of the ovary, which stimulates follicle development, oogenesis, development of secondary sex characteristics, and growth of the uterine lining.

External fertilization union of sperm and egg outside the body of either parental organism.

Fimbria (fim′-brē-a; pl. fimbriae) in female mammals, the ciliated, fingerlike projections of the oviduct that sweep the ovulated egg from the ovary into the oviduct.

Fission asexual reproduction by dividing the body into two smaller, complete organisms.

Follicle in the ovary of female mammals, the oocyte and its surrounding accessory cells.

Follicle-stimulating hormone a hormone produced by the anterior pituitary gland that stimulates spermatogenesis in males and development of the follicle in females.

Hermaphrodite (her-ma′-frō-dīt) an organism that produces both male and female gametes.

Implantation the process whereby the early embryo embeds itself within the lining of the uterus.

Internal fertilization union of sperm and egg inside the body of the female.

Interstitial cells (in-ter-sti′-shul) in the vertebrate testis, testosterone-producing cells located between the seminiferous tubules.

Luteinizing hormone a hormone produced by the anterior pituitary gland that stimulates testosterone production in males and development of the follicle, ovulation, and production of the corpus luteum in females.

Menstrual cycle in females of some primate species, the roughly 28-day cycle of development and degeneration of the lining of the uterus, accompanied by discharge of tissue and blood through the vagina.

Menstruation in females of some primate species, the monthly discharge of uterine tissue and blood from the vagina.

Monoecious (mon-ē′-shus) pertaining to organisms in which male and female gametes are produced in the same individual.

Myometrium (mī-ō-mē′-trē-um) the muscular outer layer of the uterus.

Oogonium (ō-ō-gō′-nē-um) a diploid cell in female animals that gives rise to a primary oocyte.

Ovary the gonad of female animals.

Oviduct in mammals, the tube leading from the ovary to the uterus.

Parthenogenesis (par-the-nō-gen′-i-sis) a specialization of sexual reproduction, in which an egg undergoes development without fertilization.

Placenta in mammals, a structure formed of both embryonic and maternal tissues, which serves to exchange nutrients and wastes between embryo and mother.

Polar body in oogenesis, a small cell containing a nucleus but virtually no cytoplasm.

Primary oocyte (ō′-ō-sīt) a large diploid cell, derived from the oogonium by growth and differentiation, which undergoes meiosis to produce the egg.

Primary spermatocyte (sper-ma′-tō-sīt) a diploid cell, derived from the spermatogonium by growth and differentiation, which undergoes meiosis to produce four sperm.

Progesterone (prō-ge′-ster-ōn) a hormone produced by the corpus luteum that promotes development of the uterine lining.

Prostate gland (prō′-stāt) a gland that produces part of the fluid component of semen. The prostate fluid is basic and contains a chemical that activates sperm movement.

Regeneration (1) regrowth of a body part after loss or damage; (2) asexual reproduction by regrowth of an entire body from a fragment.

Scrotum (skrō′-tum) the pouch of skin containing the testes of male mammals.

Secondary oocyte (ō′-ō-sīt) a large haploid cell derived by meiosis I from the diploid primary oocyte.

Secondary spermatocyte (sper-ma′-tō-sīt) a haploid cell derived by meiosis I from the diploid primary spermatocyte.

Semen the sperm-containing fluid produced by the male reproductive tract.

Seminal vesicle in male mammals, a gland that produces a basic, fructose-containing fluid that forms part of the semen.

Seminiferous tubules (sem-i-ni′-fer-us) a series of tubes in the vertebrate testis in which sperm are produced.

Sertoli cell a large cell in the seminiferous tubule that regulates spermatogenesis and nourishes the developing sperm.

Sexual reproduction a form of reproduction in which genetic material from two parental organisms is combined in the offspring. Usually, two haploid gametes fuse to form a diploid zygote.

Spawning a method of external fertilization in which male and female parents shed gametes into the water, and sperm must swim through the water to reach the eggs.

Sperm the haploid male gamete, usually small, motile, and containing little cytoplasm.

Spermatid a haploid cell derived from the secondary spermatocyte by meiosis II. The mature sperm is derived from the spermatid by differentiation.

Spermatogenesis the formation of sperm.

Spermatogonium (pl. spermatogonia) a diploid cell lining the walls of the seminiferous tubules that gives rise to a primary spermatocyte.

Testis (pl. testes) the gonad of male animals.

Testosterone in vertebrates, a hormone produced by the interstitial cells of the testis; stimulates spermatogenesis and the development of male secondary sex characteristics.

Urethra (ū-rē′-thra) the tube leading from the urinary bladder to the outside of the body; in males, the urethra also receives sperm from the vas deferens and conducts both sperm and urine (at different times) to the tip of the penis.

Uterus in female mammals, the part of the reproductive tract that houses the embryo during pregnancy.

Vagina the passageway leading from the outside of the body to the cervix of the uterus.

Vas deferens (vas de′-fer-ens) the tube connecting the epididymis of the testis with the urethra.

Zona pellicida (pel-ū′-si-da) a clear, noncellular layer between the corona radiata and the egg.

Zygote the fertilized egg.

STUDY QUESTIONS

1. Distinguish between sexual and asexual reproduction, and list at least one advantage of each.
2. Describe three types of asexual reproduction and give an example of an animal showing each type.
3. Describe external fertilization, spawning, and amplexus.
4. How do animals using external fertilization synchronize the release of eggs and sperm?
5. Why is internal fertilization essential for terrestrial animals?

6. List the structures, in order, through which a sperm passes on its way from the seminiferous tubules of the testis to the oviduct of the female.
7. Name the three accessory glands of the male reproductive tract. What are the functions of the secretions they produce?
8. Draw the structure of a mature human sperm, and list the functions of each part.
9. What is the corpus luteum? From what structure in the ovary is it derived? What determines its survival after ovulation?
10. Diagram the menstrual cycle and describe the interactions among hormones produced by the pituitary gland and ovaries that produce the cycle.
11. Diagram the structure of the ovulated egg. How do the sperm penetrate the barriers surrounding the egg?
12. What event early in pregnancy prevents the degeneration of the corpus luteum?

SUGGESTED READINGS

Gold, M. "The Baby Makers.' " *Science 85,* April 1985. *In vitro* fertilization is the only way that some otherwise infertile couples can conceive. Gold describes the various techniques fertility clinics use to promote pregnancy in their patients.

Gordon, M. S., Bartholomew, G. A., Grinnell, A. D., Jorgensen, C. B., and White, F. N. *Animal Physiology,* 4th ed. New York: Macmillan Publishing Company, 1982. Reproduction in animals is incredibly varied and interesting. Some of the reproductive strategies evolved by nonhuman animals are briefly discussed in this textbook.

Grobstein, C. "External Human Fertilization." *Scientific American,* June 1979. The technique of *in vitro* fertilization raises both ethical and legal questions.

Vander, A.J., and Sherman, J. H. *Human Physiology,* 4th ed. New York: McGraw-Hill Book Company, 1985. An understandable presentation of human reproductive physiology, with an emphasis on experimental data.

32
Animal Development

Out of millions of contestants, a single sperm fuses with the egg. The two haploid nuclei—one from the egg, contributed by the mother, and one from the sperm, contributed by the father—fuse to create a diploid cell, the beginning of the new generation. How does this single cell give rise to the trillions of cells of the adult body? It cannot be a simple matter of cell division after cell division, for that would merely give rise to a massive lump of identical cells. Rather, as the cells divide, they also must **differentiate;** that is, they must specialize to become particular cell types, such as liver, brain, or muscle.

How can one cell become different from other cells, when all are descended from the same fertilized egg? How do all the organs of the adult body come to be positioned in the correct locations, and connect up with one another in the proper way? What governs the onset of puberty and reproductive behavior? Why do animals age? Is death inevitable? These questions inspire the study of development, the process by which an organism proceeds from fertilized egg through adulthood to eventual death. In this chapter we examine animal development, especially the early embryonic stages, and discuss some of the mechanisms that control it. Finally, we describe the development of the human embryo.

DIFFERENTIATION

How do cells become differentiated from one another during development? Since the characteristics of each cell are ultimately determined by its genes, one possibility might be that differentiation results from a progressive loss of genes. By this scenario, the zygote would contain all the genes needed to direct the construction of the whole organism, and each differentiated cell would lose those genes that are not needed for its particular function in the body. Pancreas islet cells, for example, might not possess the genes for hair proteins, while hair follicle cells might have lost the genes for insulin production. In an ingenious experiment, J. B. Gurdon showed that *gene loss cannot be the mechanism of differentiation.* Gurdon implanted nuclei from intestinal cells of tadpoles of the African clawed frog, *Xenopus,* into unfertilized eggs whose own nuclei had been destroyed (Fig. 32-1). Although the operation was very difficult and most of the "patients" died, some of the eggs with intestinal nuclei developed into adult frogs. These experiments demonstrated that *differentiated cells contain all the genetic information needed for the development of the entire organism.*

If differentiated cells contain a full complement of genes, then *differentiation must occur because of differential use of genes—that is, which genes are transcribed to messenger RNA and translated into proteins must differ among cell types.*

Gene Regulation During Development

In Chapter 11 we discussed some of the mechanisms that control gene transcription. Although geneticists are just beginning to unravel the details of gene regulation, the principles are straightforward. Cellular materials, usually proteins or proteins combined with activating substances such as steroid hormones, travel to the nucleus and bind to the chromosomes. These proteins then block transcription of certain genes or promote the transcription of other genes. Which genes are transcribed largely determines the shape, structure, and activity of the cell.

There are two major sources of information that control gene usage and hence direct cell differentiation: (1) substances positioned in specific locations of the egg cytoplasm, and (2) messages received from other cells. During oogenesis, gene-regulating substances come to be positioned in specific places in the egg cytoplasm (Fig. 32-2a). (How this is accomplished is not known.) The fertilized egg then divides in particular orientations; its daughter cells, therefore, receive different gene-regulating substances (Fig. 32-2b). Thus *the developmental fate of a daughter cell is determined by the part of the egg cytoplasm it receives, and hence by the gene-regulating substances it inherits.* During later embryonic development

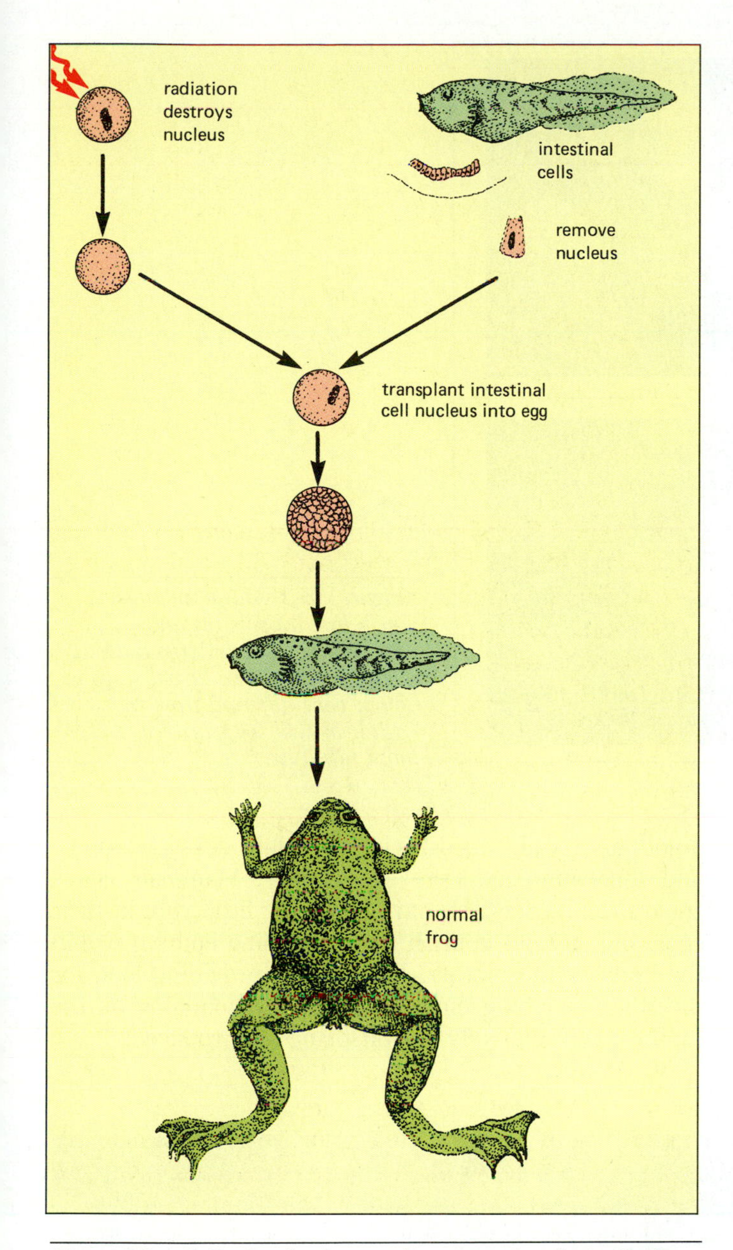

Figure 32-1 *J. B. Gurdon's experiment proving that cells do not lose genes as they differentiate. Gurdon destroyed the DNA of unfertilized frog eggs, and then transplanted nuclei of intestinal cells from a tadpole into the eggs. The resulting egg cells developed into normal tadpoles and eventually adult frogs.*

and continuing throughout adult life, cells constantly receive chemical messages, including nutrients, hormones, and neurotransmitters, from other cells of the body. *These chemical messages can alter the transcription of genes and the activity of enzymes within a cell.* As we proceed through this chapter, we will describe evidence of these two processes at work during development.

EGGS, DEVELOPMENT, AND ANIMAL LIFE HISTORIES

All animal eggs contain food reserves of lipid- and protein-rich **yolk.** This is crucial to the early development of the embryo. Since a zygote has no mouth or digestive tract, its ability to acquire food from its environment is limited. The yolk provides nourishment for the early embryo until it develops into a form that can obtain food from outside sources. The amount of yolk in an egg corresponds closely with the life history of the animal.

Animal development usually proceeds down one of two paths: direct or indirect development. In **indirect development,** the juvenile animal that hatches from the egg differs significantly from the adult, as a caterpillar differs from a butterfly. Indirect development occurs in most of the invertebrates, including insects and echinoderms, and a few vertebrates, notably the amphibians. Animals with indirect development typically produce huge numbers of eggs, each with only a small amount of yolk. The yolk nourishes the developing embryo during a rapid transformation into a small, sexually immature feeding stage called a **larva** (Fig. 32-3). In many animals, the larvae feed on different organisms than the adults do (Fig. 32-3). Eventually, the larvae undergo a revolution in body form, or **metamorphosis,** and become sexually mature adults. Although people tend to regard the adult form as the "real animal" and larvae as "preparatory stages," in some animals, especially insects, most of the life cycle is spent as a larva. The adult may live for only a few days, reproducing frantically and in some cases not even eating.

Other animals, including such diverse groups as reptiles, birds, mammals, and land snails, show **direct de-**

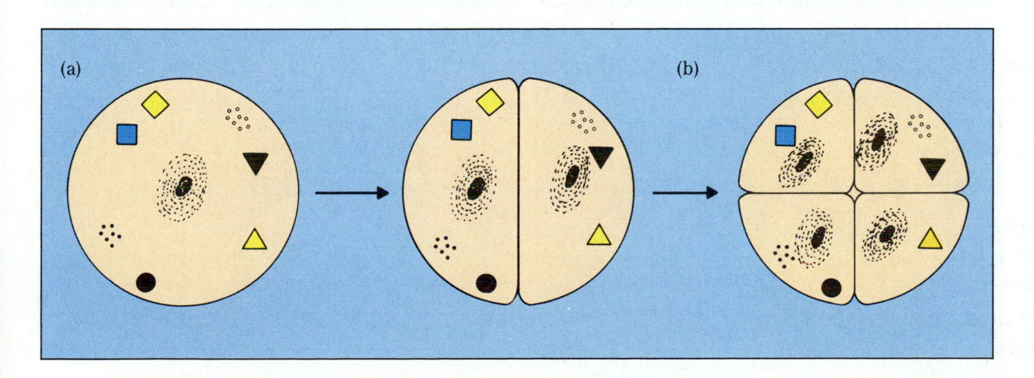

Figure 32-2 *Inheritance of gene-regulating substances during division of a fertilized egg.* **(a)** *Different gene-regulating substances (represented by the various symbols) are positioned in particular places in the egg cytoplasm during oogenesis.* **(b)** *As the egg divides, these materials remain in about the same positions, so that the daughter cells inherit different substances.*

Figure 32-3 Many animals, including limpets **(a)** *and dragonflies* **(b)**, *undergo indirect development. The larval stage is often very different from the adult in size, appearance, diet, and life-style.*

velopment, in which the newborn animal is a sexually immature, miniature version of the adult (Fig. 32-4). These juveniles are typically much larger than larvae, and consequently need much more nourishment before emerging out into the world. Two strategies have evolved to meet the embryo's food requirement. Reptiles, birds, and land snails produce large eggs containing prodigious amounts of yolk: an ostrich egg weighs several pounds, and dinosaur eggs were even larger. Mammals, some snakes, and a few fish have relatively little yolk in their eggs, but instead nourish the developing embryo within the body of the mother. Either way, providing food for directly developing embryos is quite demanding on the mother, and relatively few offspring are produced.

Figure 32-4 The offspring of animals with direct development closely resemble their parents from the moment of birth, except of course in size. Lizards **(a)**, *land snails* **(b)**, *and birds* **(c)** *hatch from large, yolk-filled eggs, while mammalian mothers* **(d)** *nourish their young within their bodies for weeks or months. The availability of food supplies during early development permits the offspring to be born as miniature versions of the adult, bypassing a larval feeding stage.*

PRINCIPLES OF ANIMAL DEVELOPMENT

Animal development (particularly in the vertebrates) may be loosely divided into several stages. Cleavage, gastrulation, organogenesis, and growth occur during embryonic life (Fig. 32-5). Almost all of the organs that will be present in the adult are formed during embryonic development. After birth, the animal undergoes further growth, achieves sexual maturity and reproduces, ages, and finally dies. Let's briefly look at each of these events.

Cleavage: Distributing Gene-Regulating Substances

As you know, an egg is a very large cell. *Development begins with* **cleavage** *of the fertilized egg, which reduces the cell size and distributes gene-regulating substances to the daughter cells.* Although most cell divisions proceed in a three-step cycle (divide, grow, duplicate genetic material, then divide again), embryonic cells skip the growth phase during cleavage. Consequently, as cleavage progresses, the available cytoplasm is split up into ever-smaller cells, finally forming a solid ball of cells (the **morula**) that is still about the same size as the zygote. Then a cavity opens in the midst of the morula, so that the cells become the outer covering of a hollow ball, the **blastula.** To a considerable extent, the pattern of cleavage is controlled by the amount of yolk, because yolk hinders cytokinesis (cytoplasmic division) during cell division. The almost yolkless eggs of sea urchins divide symmetrically, but eggs with extremely large yolks, such as a hen's egg, do not even divide all the way through. Nevertheless, a hollow blastula is always produced, although in reptiles and birds the blastula is a flat disk rather than a sphere.

During cleavage, many of the gene-regulating substances laid down in the egg stay in the same positions. Therefore, different gene-regulating materials become incorporated into different daughter cells (see Figure 32-2). This was discovered decades ago in experiments on frog eggs (Fig. 32-6). The unfertilized frog egg has pale yolk on the "bottom" or vegetal pole and pigmented cytoplasm on the "top" or animal pole. At fertilization, some of the pigment shifts toward the animal pole, leaving behind a **gray crescent** of intermediate pigmentation. The first cleavage division of frog eggs normally passes through the center of the gray crescent, so that each daughter cell receives roughly half of the crescent (Fig. 32-6a). If the two cells are gently separated, each will develop into a normal tadpole. Embryologists can force the first division to occur in a plane rotated 90°, so that one daughter cell receives the entire gray crescent (Fig. 32-6b). If the cells are then separated, the cell with the gray crescent still develops into a tadpole, while the one without any crescent material merely forms a lump of cells that soon die. Clearly, gene-regulating substances (presently unidentified) in the gray crescent region are required for normal development.

Gastrulation

In the next step of development, a dimple called the **blastopore** forms on one side of the blastula. Cells migrate in

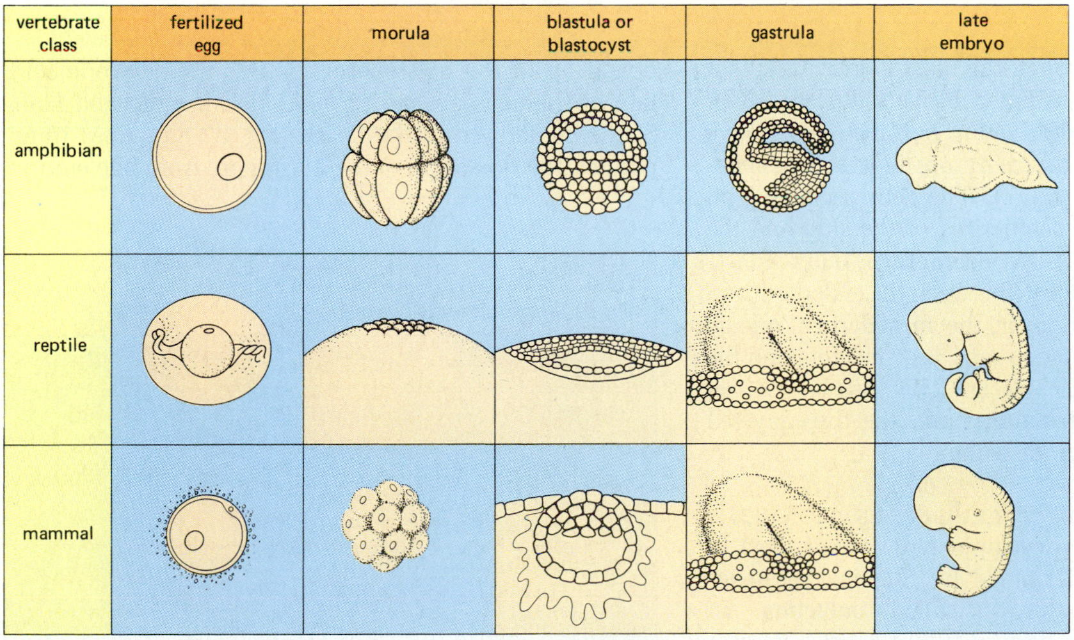

vertebrate class	fertilized egg	morula	blastula or blastocyst	gastrula	late embryo
amphibian					
reptile					
mammal					

Figure 32-5 Stages in the development of amphibians, reptiles, and mammals. Use these illustrations as a guide to accompany the descriptions to follow in the text.

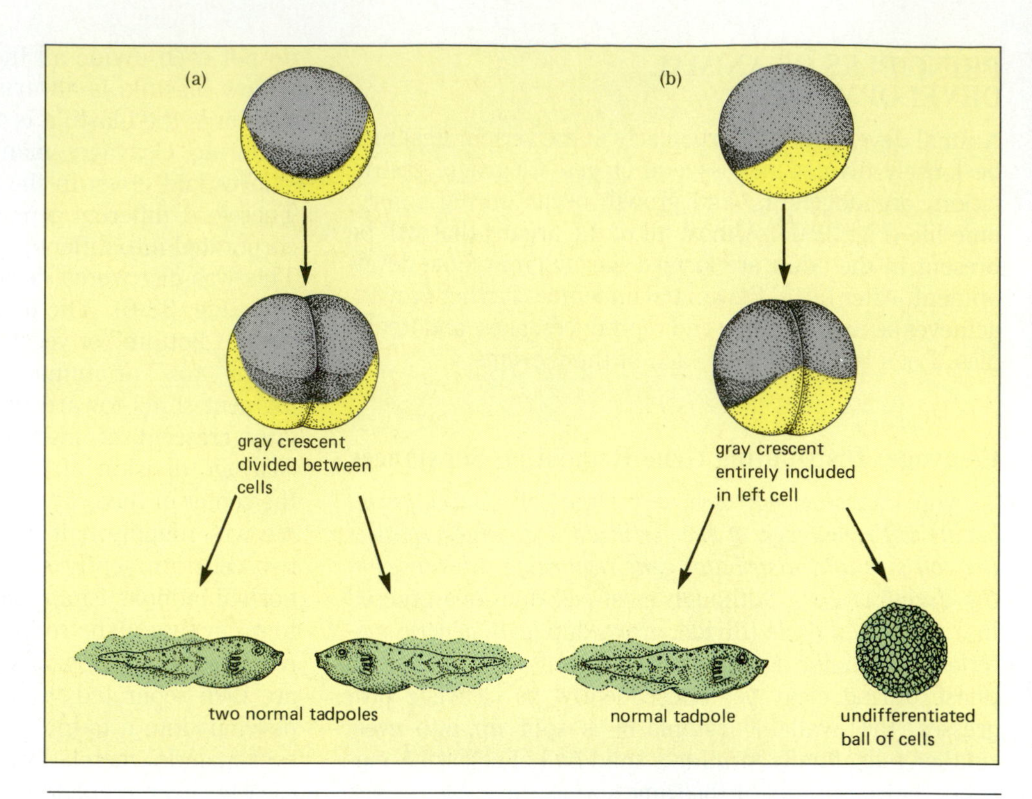

Figure 32-6 Cleavage distributes gene-regulating substances differentially. In frog eggs, a pigmented region known as the gray crescent contains substances needed for normal embryonic growth and differentiation. **(a)** *Normally, the first cleavage division cuts neatly through the center of the gray crescent. If the two daughter cells are separated in the lab, each cell contains gray crescent material, and each can subsequently develop into a normal tadpole.* **(b)** *Embryologists can experimentally force the first cleavage division to miss the gray crescent, so that one daughter cell receives the entire gray crescent, while the other receives none. The cell with the crescent material develops normally, but the cell lacking crescent substance cannot.*

a continuous sheet in through the blastopore, much as if you punched in an underinflated basketball (Fig. 32-7). The enlarging dimple is destined to become the digestive tract; the cells that line its cavity are now called **endoderm** (Greek for "inner skin"). The cells remaining on the outside will form the epidermis of the skin and the nervous system, and are called **ectoderm** ("outer skin"). Meanwhile, some cells migrate between the endoderm and ectoderm, forming a third layer, the **mesoderm** ("middle skin"). Mesoderm gives rise to muscles, skeleton, and the circulatory system (Table 32-1). This process of cell movement is called **gastrulation,** and the three-layered embryo that results is the **gastrula.**

Influences from the cellular environment. During gastrulation, the developmental fate of most of the embryo's cells is determined by chemical messages received from other cells, a process called **induction.** In amphibian embryos, cells derived from the gray crescent region of the zygote form the site of dimpling as the blastula is transformed into the gastrula. This area (called the dorsal lip of the blastopore) controls the developmental fate of the cells around it, as Hilde Mangold and Hans Spemann showed in the 1920s (Fig. 32-8a). They transplanted the dorsal lip of the blastopore from one embryo

Table 32-1 **Derivation of Adult Tissues from Embryonic Cell Layers**

Embryonic Layer	Adult Tissue
Ectoderm	Epidermis of skin; lining of mouth and nose; hair; glands of skin (sweat, sebaceous, and mammary glands); nervous system; lens of eye; inner ear
Mesoderm	Dermis of skin; muscle, skeleton; circulatory system; gonads; kidneys; outer layers of digestive and respiratory tracts
Endoderm	Lining of digestive and respiratory tracts; liver; pancreas

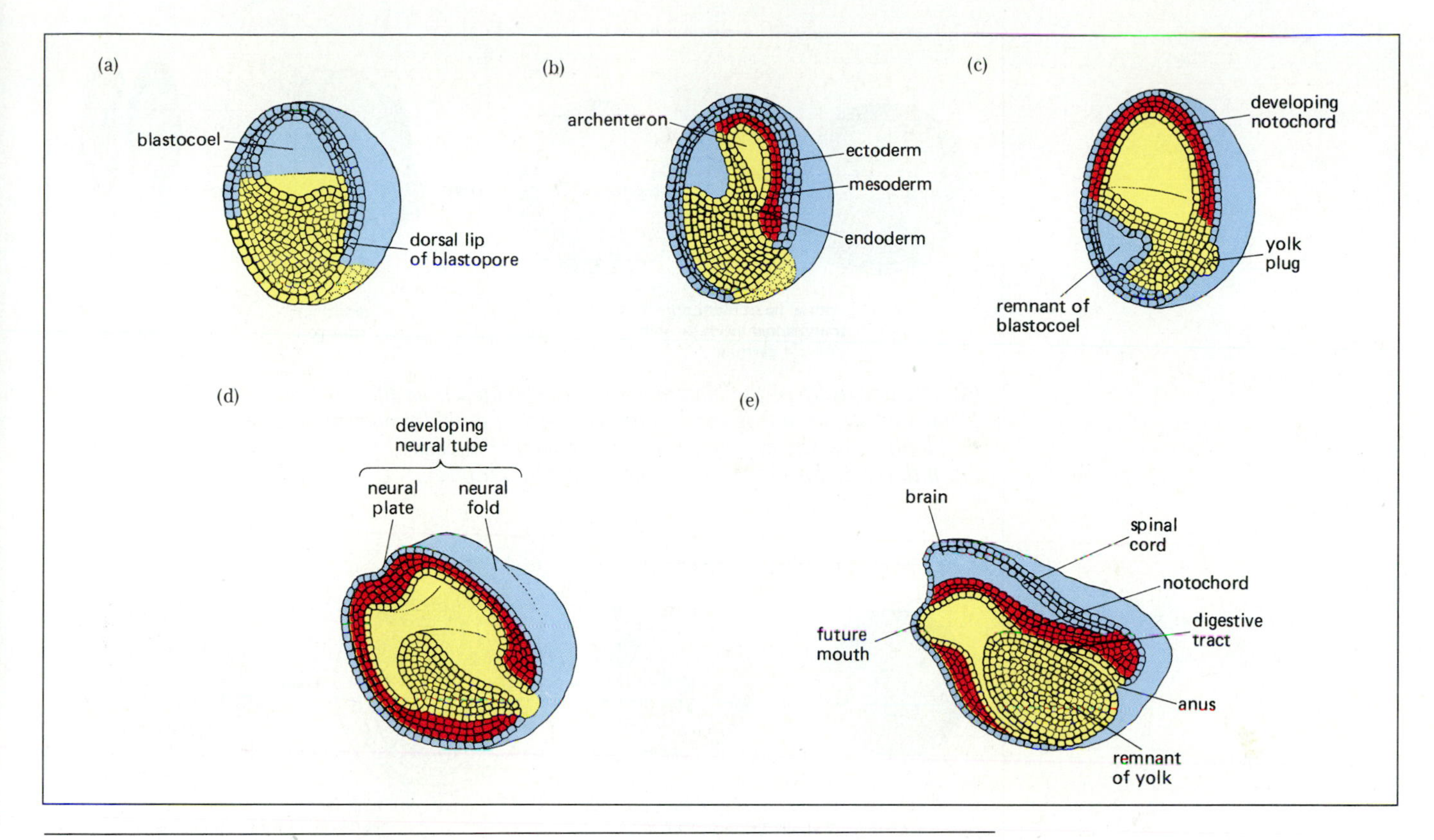

Figure 32-7 *Gastrulation in the frog.*
(a) *The blastula just before gastrulation. The blastopore is the site at which gastrulation will begin.*
(b) *Cells from the surface migrate into the interior of the blastula through the blastopore. These cells will form the endoderm and mesoderm layers of the gastrula, while the cells remaining on the surface form ectoderm. The endoderm encloses the primitive gut, the archenteron.*
(c) *Mesoderm cells differentiate into the notochord and muscle masses.*
(d) *The notochord induces ectoderm cells lying directly above it to form the neural tube.*
(e) *The archenteron will break through at the end of the embryo opposite the blastopore, forming a tubular digestive tract. The blastopore is the future anus, while the second opening becomes the mouth. The neural tube enlarges and differentiates into brain and spinal cord.*

to another. The transplanted dorsal lip then induced the nearby cells of the host to form a second embryo, showing that *dorsal lip tissue controls differentiation in the surrounding cells.* One can also perform the reverse experiment, transplanting cells from regions of the gastrula other than the dorsal lip of the blastopore (Fig. 32-8b). These cells give rise to tissues appropriate for the region into which they were transplanted rather than tissues appropriate for the region from which they were taken. These experiments clearly show that gene usage within differentiating cells (nonblastopore cells) can be regulated by substances originating in other cells (the dorsal lip of the blastopore).

Organogenesis: Developing Adult Structures

Gradually, ectoderm, mesoderm, and endoderm rearrange themselves into the organs characteristic of the animal species. This process, called **organogenesis,** also usually occurs by induction. This was demonstrated for eye formation in frog embryos by Warren Lewis (Fig. 32-9). During normal development, the brain develops a pair of swellings, the optic cups, which grow out toward the epidermis of the head. Where an optic cup touches the epidermis, cells of the epidermis divide, invaginate, and form the future lens. Lewis found that if he transplanted an optic cup into the tail of the embryo, a lens would form from the overlying epidermis of the tail, while no lens would form on the side of the head from which the optic cup had been removed. Therefore, the optic cup induces formation of the lens from epidermal tissue.

Maturation and Aging

Development doesn't stop once an animal is born. Animals continue to change throughout their lives.

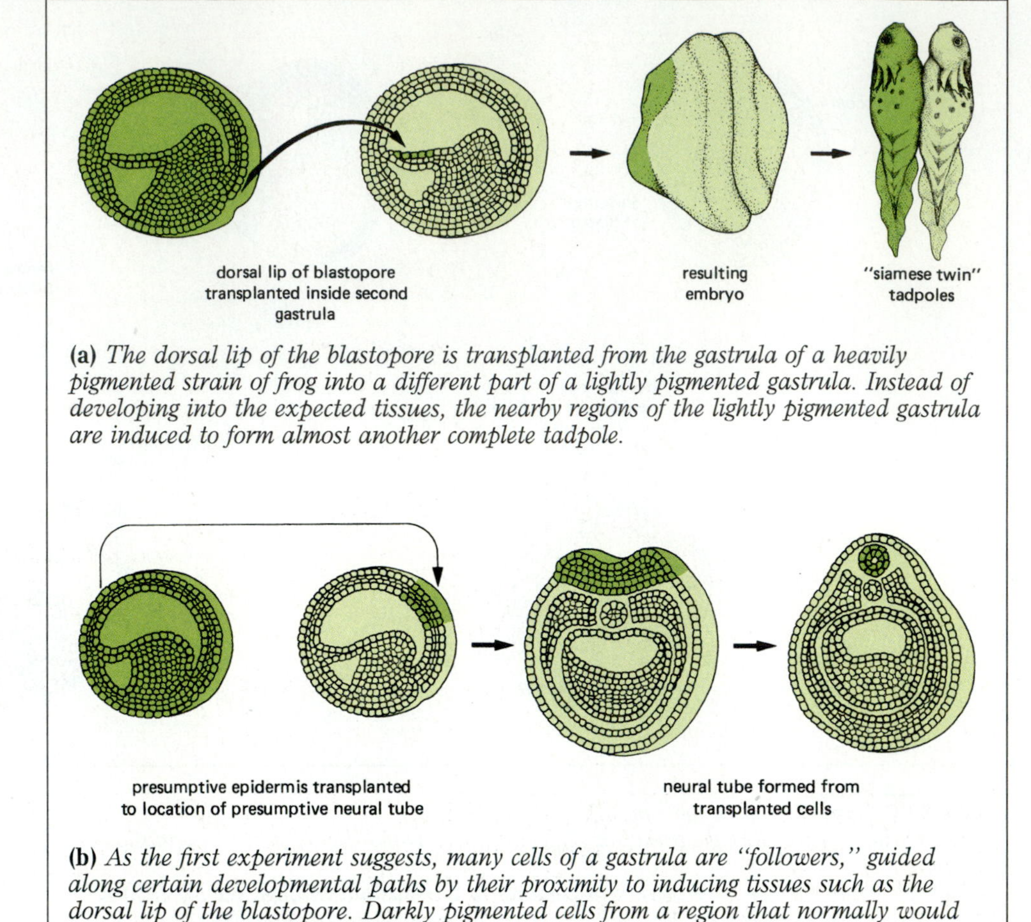

(a) *The dorsal lip of the blastopore is transplanted from the gastrula of a heavily pigmented strain of frog into a different part of a lightly pigmented gastrula. Instead of developing into the expected tissues, the nearby regions of the lightly pigmented gastrula are induced to form almost another complete tadpole.*

(b) *As the first experiment suggests, many cells of a gastrula are "followers," guided along certain developmental paths by their proximity to inducing tissues such as the dorsal lip of the blastopore. Darkly pigmented cells from a region that normally would become skin tissue are transplanted to other sites on pale host gastrulas. Their developmental fate depends not on the region from which they were taken, but on the region into which they were transplanted. They can be induced to form tissues as diverse as brain or muscle.*

Figure 32-8 *Induction as a means of differentiation in frog embryos.*

MATURATION. Animals become sexually mature at an age that is determined by both genes and environment. Many animals undergo months to years of growth and development before they can become sexually mature. This developmental process is largely regulated by environmental stimuli. Songbirds, for example, almost always become sexually mature in the spring, stimulated by the increasingly long days. Internal and perhaps social factors also influence maturation in many species. The age of puberty among women, for instance, has dropped substantially during the last few centuries. This is due in part to improved nutrition, but social stimulation may also be involved.

AGING. No animal is static. Even at the peak of its prowess, the dominant lion in a pride constantly changes. The cells lining his stomach, for example, are exposed to protein-digesting enzymes and strong acids, and consequently are short-lived. They are replaced about every two weeks. On the other hand, most brain cells do not reproduce, and the lion was born with all the nerve cells he will ever have. Even in the carefully regulated environment of his central nervous system, some nerve cells die each day, never to be replaced. Many other cells function less efficiently, or divide more slowly, as he ages. Why do these cells deteriorate and die? Is death, for cells and for entire organisms, a programmed part of life?

For some cells, death is indeed programmed to occur at a precise time during development. At least two mechanisms seem to be at work in different tissues. First, *some cells die during development unless they receive a "survival signal."* Embryonic vertebrates, for example, have far more motor neurons in their spinal cords than adult animals do. Motor neurons are programmed to die unless they successfully innervate a skeletal muscle, which releases a chemical that prevents the death of its own motor

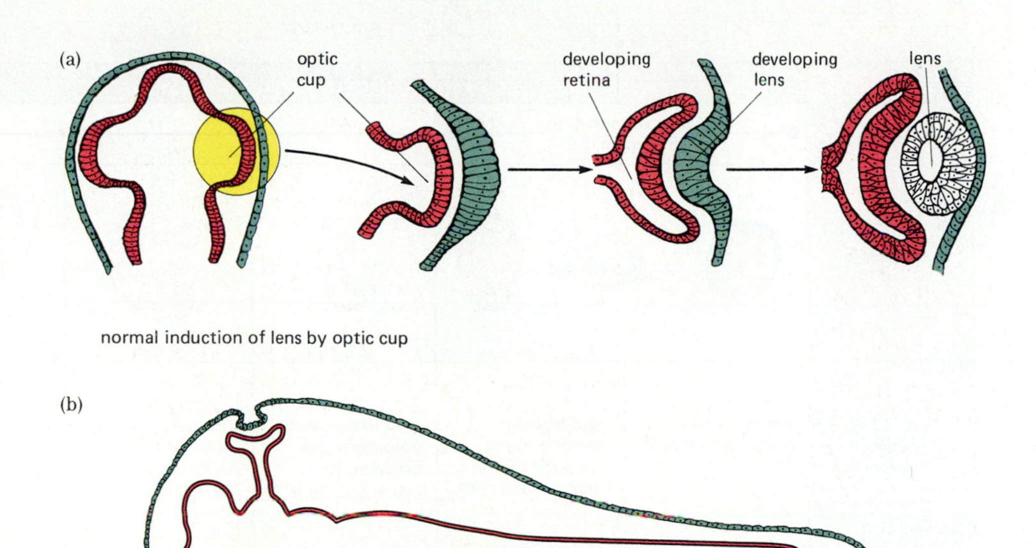

Figure 32-9 *Warren Lewis showed that lens formation is induced by the presence of presumptive retinal cells growing out from the embryonic nervous system.*
(a) *Normally, the optic cup grows out from the brain and ectoderm cells form a lens where the cup touches the skin of the head.*
(b) *If an optic cup is removed from the head and transplanted beneath the skin of the tail, no lens forms in the head, but one does form from the skin of the tail. Skin cells anywhere can generate a lens if induced to do so by the presence of an optic cup.*

neuron. For other cells, the situation is just the reverse: *some cells live unless they receive a "death signal" from other cells of the developing animal.* Many embryonic structures disappear during development, including tails in tadpoles and webbing between the fingers and toes in people. In tadpoles, thyroid hormone (which triggers metamorphosis into a frog) stimulates cells in the tail to synthesize enzymes that digest the tail away. If the thyroid gland is surgically removed, the tail is never resorbed.

Even if a cell survives the drastic changes of embryonic life, it still has a finite life span. If cells are removed from normal tissues and grown in culture in the laboratory, they divide a few times, then stop, and eventually die. This seems to imply that a cell lineage (a parent cell and all its offspring for all generations) has a built-in maximum life span. Recent research has suggested that this maximum life span varies from species to species, and that longevity depends on the ability of the cell to repair damage to its DNA. Longer-lived cells and longer-lived animals are better at repairing damaged DNA. Nevertheless, all normal cells seem to die eventually.

Not all cells, however, are "normal." Cancer cells survive and reproduce indefinitely in cell culture or in the body (which is what makes cancer so catastrophic). Some lines of cancer cells have been reproducing in culture every week or two for decades. This suggests that, whatever the mechanism that regulates the life span of a cell, it can be bypassed. Can we discover a method of defusing the self-destruct mechanism while retaining proper controls over the cells in our bodies? No one knows.

HUMAN DEVELOPMENT

Human development is controlled by the same mechanisms that control the development of other animals. In fact, our development strongly reflects our evolutionary heritage, a fact that we shall emphasize in the brief dis-

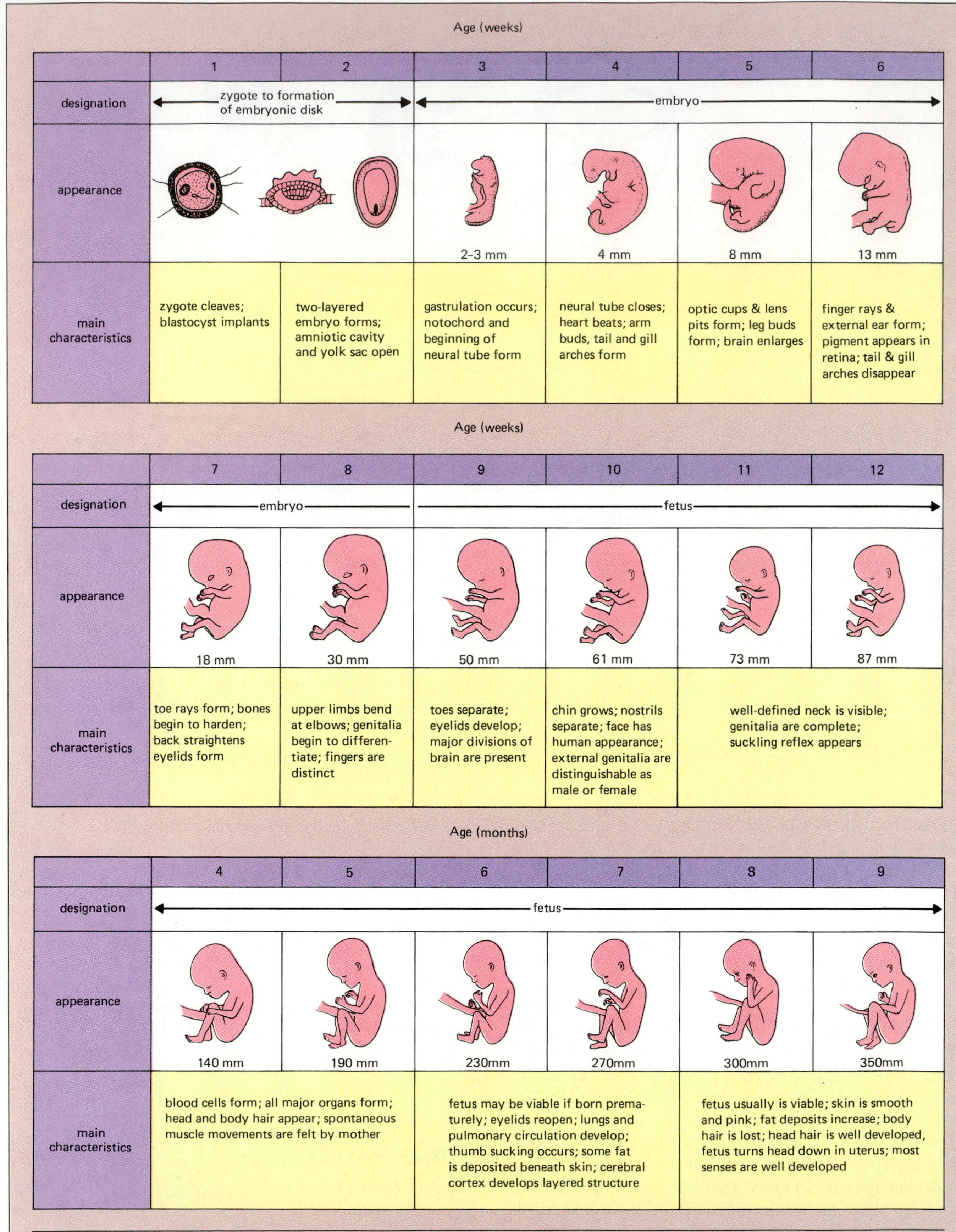

Figure 32-10 *A calendar of human embryonic development, from fertilized egg to birth.*

cussion to follow. Figure 32-10 summarizes the stages of human embryonic development. You may want to refer back to this figure as we go along, and, perhaps years from now, if you have a child of your own.

The First Two Months

As you know, mammals evolved from reptiles, which lay large, yolky eggs (Chapter 18). The massive yolk of reptile eggs prevents division of the entire egg during cleavage and later development. Instead of forming a round ball of cells, as sea urchins and amphibians do, the blastulas of reptiles and their evolutionary descendants, the birds, are confined to a flat disk of cells perched atop the yolk (Fig. 32-5). As an adaptation to embryonic development on land, the embryo produces four membranes, the **amnion, chorion, yolk sac,** and **allantois,** which enclose the embryo within a watery environment, exchange gases with the air, and sequester yolk and wastes (Fig. 32-5). Although mammalian eggs have no yolk to speak of, much of the reptilian genetic program for development still persists, including the initial formation of a disk-shaped embryo and the four embryonic membranes (Fig. 32-11b). The table in Fig. 32-11 compares the structures and functions of the extraembryonic membranes in reptiles and mammals.

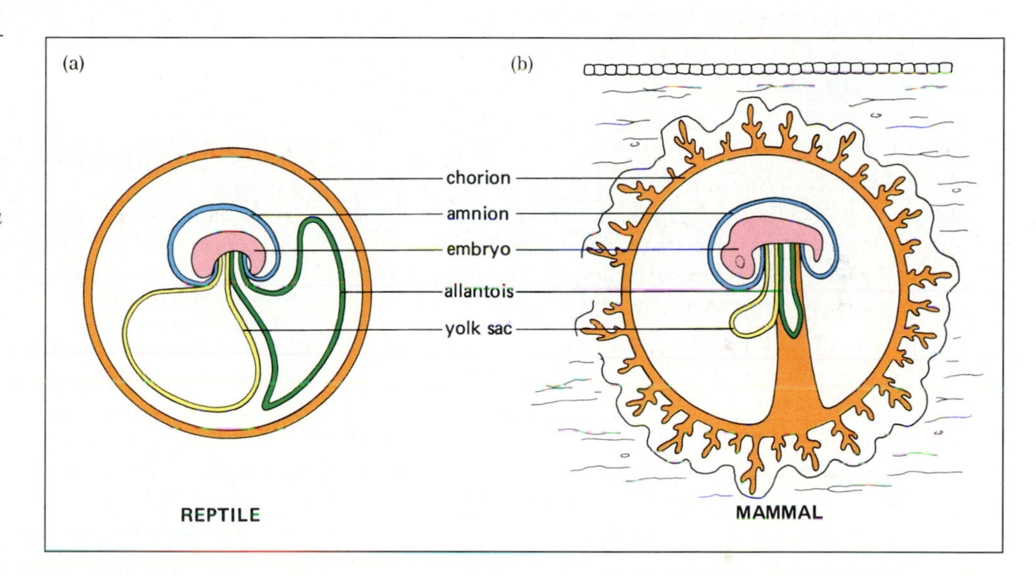

Figure 32-11 **(a)** *Reptiles were the first vertebrates whose embryos developed on land rather than in aquatic habitats. The extraembryonic membranes were an essential adaptation to development on land, providing a "private pond" in which the embryo develops, respiratory surfaces to acquire oxygen and get rid of carbon dioxide, and sacs in which to store food and wastes.* **(b)** *Mammalian embryos, including human embryos, retain these membranes, in modified form, to perform the same vital functions.*

Vertebrate Embryonic Membranes

	Reptilian Embryo		Mammalian Embryo	
Membrane	**Structure**	**Function**	**Structure**	**Function**
Amnion	Sac surrounding embryo	Encloses embryo in fluid	Sac surrounding embryo	Encloses embryo in fluid
Chorion	Membrane lining inside of shell	Acts as respiratory surface; regulates exchange of gases and water between embryo and air	Fetal contribution to placenta	Provides surface for exchange of gases, nutrients, and wastes between embryo and mother
Allantois	Sac connected to embryonic urinary tract; capillary-rich membrane lining inside of chorion, with blood vessels connecting to embryonic circulation	Stores wastes (esp. urine); acts as respiratory surface	Provides blood vessels of umbilical cord	Carries blood between embryo and placenta
Yolk sac	Membrane surrounding yolk	Contains yolk as food; digests yolk and transfers nutrients to embryo; forms part of digestive tract	"Empty" membranous sac	Forms part of digestive tract

How Are Complex Body Shapes Induced?

Developmental biologists have known for decades that embryos develop according to a complex scheme arising partly from gene-regulating substances laid down during oogenesis and partly from interactions among groups of cells as the embryo grows. Early in this century, Lewis, Mangold, and Spemann discovered the phenomenon of induction, whereby certain groups of cells can cause other cells to differentiate into specific tissues, such as the nervous system or the lens of an eye (see Fig. 32-9). But the question remains: *how does induction work?*

A likely hypothesis is that induction must be caused by chemicals that are released by inducer cells, diffuse to other cells, and direct differentiation in the recipient cells. Some inducing chemicals have recently been identified, but the overall process is still poorly understood. After all, induction cannot be as simple as flooding a group of cells with a chemical and expecting those cells to develop into, for instance, a brain.

Throughout the history of biology, difficult problems have often been approached by looking for a system in which fundamentally similar processes operate in a simpler way: in genetics, for example, you are familiar with Mendel's peas and Morgan's fruit flies. Simple systems for studying developmental processes are the patterns found in the wings of butterflies and moths (Fig. E32-1). Butterfly wings are essentially two-dimensional structures, covered with beautiful patterns of colored scales. By studying these two-dimensional patterns, we may discover principles of development that may also apply to the more complex three-dimensional patterns of vertebrate embryos.

Let's look at one of the most striking, yet simplest, of wing patterns, the eyespot. Most eyespots consist of a set of concentric circles, closely resembling an eye with a central "light reflection" patch of white, surrounded in order by a dark "iris," the "white of the eye," and even rather round "eyelids." How does such a pattern arise?

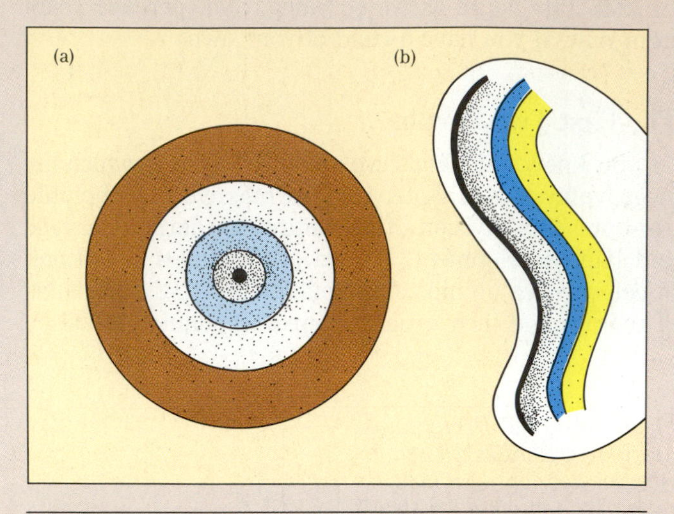

Figure E32-2 *Mechanisms of formation of butterfly wing patterns.*
(a) *Circular eyespots probably develop in response to a gradient of inducing chemical (stippling) released from cells in the center of the spot. Different concentrations of inducer chemical direct synthesis of different pigments in surrounding cells.*
(b) *Wing bars develop in response to gradients of chemical inducers released by a row of cells at the edge of the wing.*

(a)

(b)

Figure E32-1 **(a)** *Each scale of a moth or butterfly wing is a single color. The precise positioning of the colored scales builds up the beautiful pattern on the wings.* **(b)** *The Polyphemus moth shows two of the most common wing patterns: spots (often strikingly realistic eyespots such as the ones on the hindwings) and rows of parallel stripes.*

How Are Complex Body Shapes Induced?

As a wing develops in a butterfly pupa, the center of each eyespot probably induces color patterns in surrounding cells. A tiny group of cells, or perhaps even a single cell, releases a chemical that diffuses evenly in all directions (Fig. E32-2a). *As the chemical encounters nearby cells, it causes them to synthesize certain pigment molecules, depending on its concentration.* A high concentration may stimulate synthesis of white pigment for the central "light reflection." Slightly lower concentrations may induce synthesis of black "iris" pigments, and so on out to the border of the eyespot. Each cell in the wing produces a particular amount of a specific pigment, according to the concentration of inducing chemical to which it is exposed.

Most butterfly wing patterns, of course, are not simple circular eyespots, but similar developmental rules may nevertheless apply. For example, bars are probably formed by a gradient of another inducing chemical released from cells along the edge of the wing and diffusing out, causing stripes of pigment formation as it goes (Fig. E32-2b). Two gradients, from a spotlike inducer in the middle of the wing and a strip-shaped inducer at the edge, generate ellipses, fans, and V-shaped wing patterns. By understanding how chemical gradients interact in the development of butterfly wings, we not only appreciate these insects better, but also learn fundamental principles that may govern our own development.

A human egg is normally fertilized in a woman's oviduct and undergoes a few cleavage divisions on its way to the uterus. Just before implantation in the uterus, the embryo, now called a **blastocyst** (the mammalian version of a blastula), consists of a thin-walled, hollow ball with a thicker **inner cell mass** on one side (see Fig. 32-5). The thin outer wall becomes the **chorion,** and will form the embryonic contribution to the placenta, while the inner cell mass develops into the embryo and the three other embryonic membranes.

After implantation, the inner cell mass grows and splits, forming two fluid-filled sacs separated by a double layer of cells called the **embryonic disk** (Figs. 32-12 and 32-13a). One sac, bounded by the **amnion,** forms the amniotic cavity. The amnion eventually grows around the embryo, enclosing it in what one author has called its "private aquarium," providing the watery environment needed by all animal embryos. The **yolk sac,** homologous to the yolk sac of reptiles and birds, forms the second cavity, although in humans it contains no yolk. At this stage, the embryonic disk consists of a layer of ectoderm cells (on the side facing the amniotic cavity) and a layer of endoderm cells (on the side facing the yolk sac).

Gastrulation begins about the fourteenth day after fertilization (Fig. 32-13b). The endoderm and ectoderm split apart slightly, and a slit, the **primitive streak** (analogous to the blastopore), appears in the center of the ectoderm. Ectoderm cells migrate through the primitive streak into the interior of the embryo, forming mesoderm. One of the earliest mesoderm structures to develop is the notochord, a supporting rod found at some stage in all chordates (Fig. 32-13c).

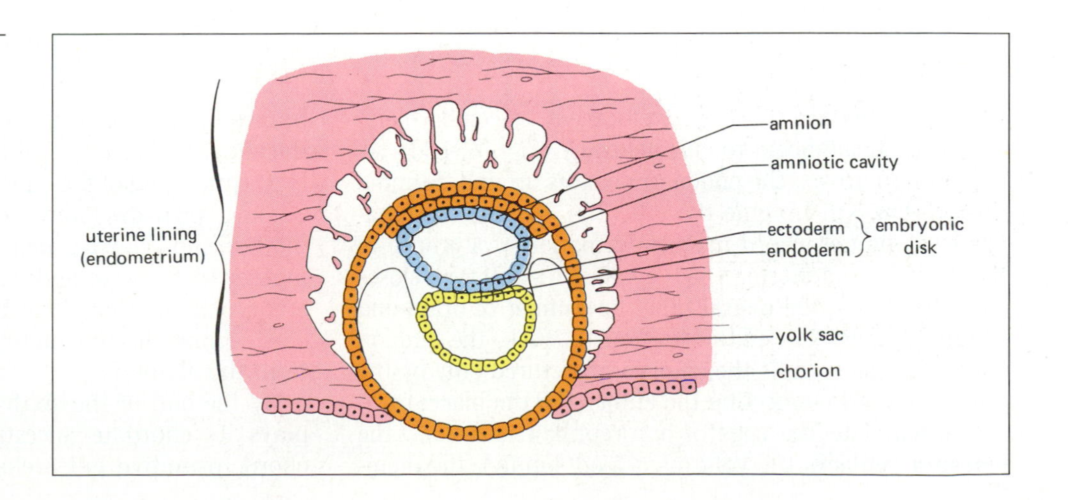

Figure 32-12 A human embryo shortly after implantation. The inner cell mass of the blastocyst develops into the amnion, yolk sac, and embryonic disk.

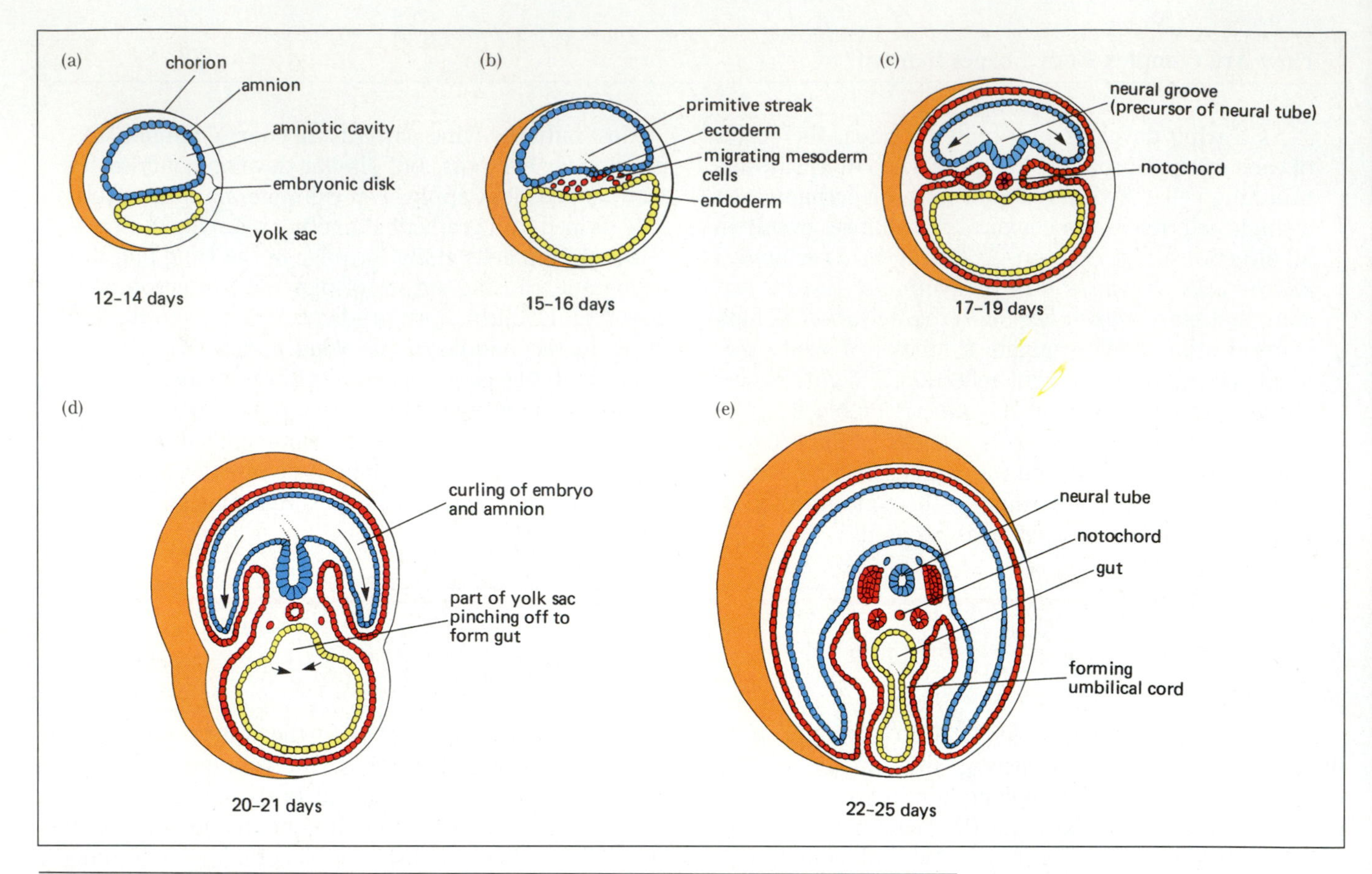

Figure 32-13 Early human development. **(a)** *Shortly after implantation, the inner cell mass separates, forming the amnion, the yolk sac, and the embryonic disk.* **(b)** *The two-layered embryonic disk soon splits open, and the primitive streak develops in the ectoderm. Ectoderm cells migrate in, forming mesoderm.* **(c)** *Some mesoderm cells form the notochord, which induces development of the neural tube, forerunner of the brain and spinal cord.* **(d)** *During the third and fourth weeks of development, the embryo curls around ventrally, forming a tubelike embryo typical of vertebrates.* **(e)** *As the embryo curls, part of the (empty) yolk sac pinches to form the gut. The amnion curls with the embryo, eventually completely enclosing it, except where the yolk sac and allantoic blood vessels extend through as the beginnings of the umbilical cord.*

During the third week of development, the embryo and its amniotic sac begin to curl ventrally (Fig. 32-13d). As the embryo grows, the endoderm curves around, forming a tube that will become the gut (Fig. 32-13e). Simultaneously, the notochord induces formation of a groove in the overlying ectoderm, which invaginates and then closes over to become the **neural tube,** forerunner of brain and spinal cord. By the end of the fourth week, the amnion completely surrounds the embryo, punctured only by the umbilical cord connecting the embryo to the placenta.

Compared to the eggs of our reptilian ancestors, the **placenta** replaces the yolk (as a food supply), the membranes lining the inside of the shell (for respiration), and the allantois (for waste elimination). The placenta, formed by the mingling of the chorion with the endometrium (Fig. 32-14), provides for exchange of materials between mother and embryo: mainly nutrients and oxygen from mother to embryo, and wastes and carbon dioxide from embryo to mother. The blood vessels of the umbilical cord, connecting the embryo to the placenta, are derived from the allantois.

By the end of the sixth week, the embryo clearly displays its chordate ancestry, having developed a notochord, primitive gill arches, and a prominent tail (Fig.

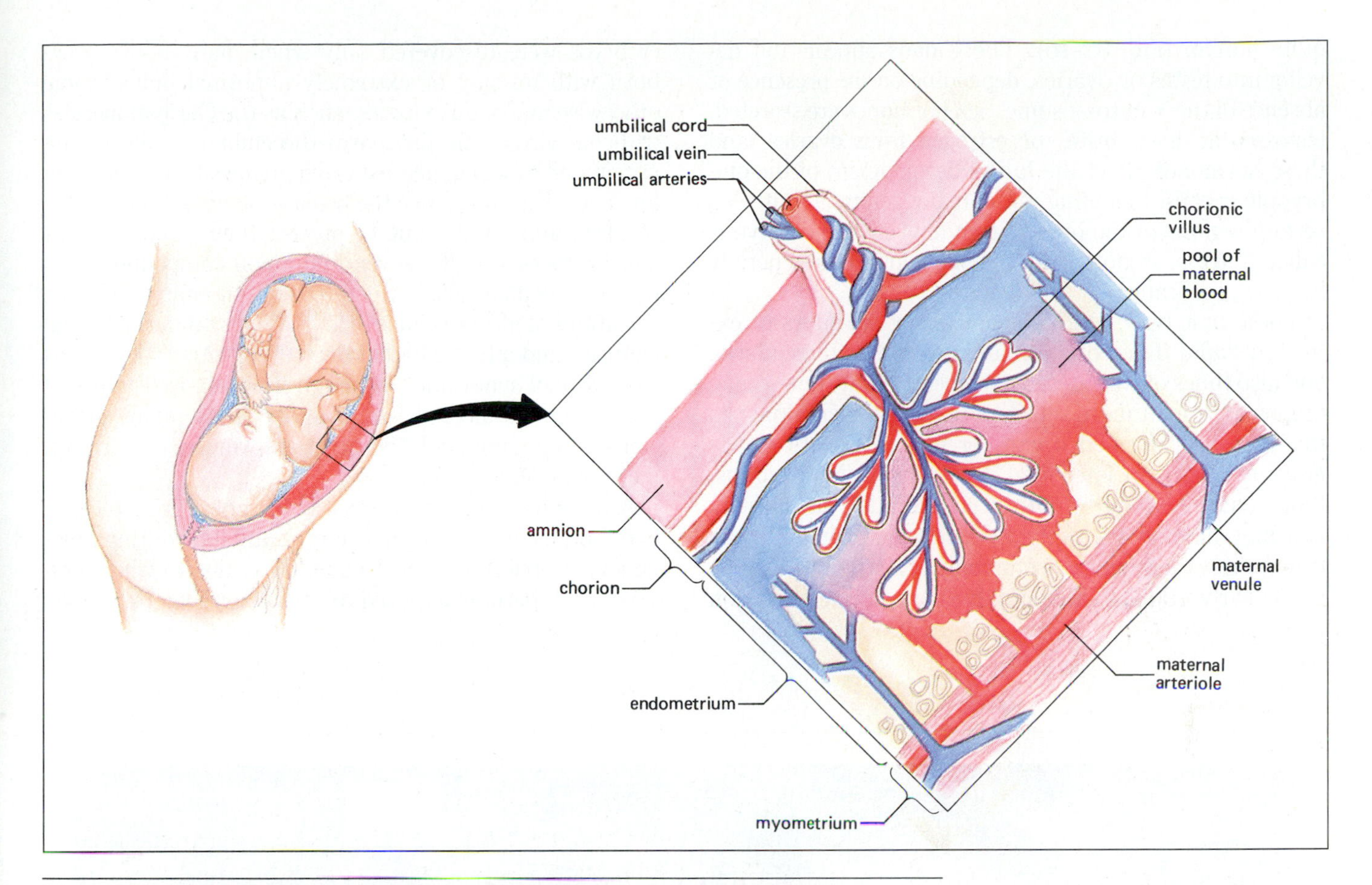

Figure 32-14 The placenta is formed from both the chorion of the embryo and the endometrium of the mother. Capillaries of the endometrium break down, releasing blood to form pools within the placenta. Meanwhile, the chorion develops projections (the chorionic villi) extending into these pools of maternal blood. Blood vessels from the umbilical cord branch extensively within the villi. The resulting structure separates the maternal and fetal blood supplies, while generating a large surface area for diffusion of oxygen, carbon dioxide, nutrients, and wastes between the fetal capillaries and the maternal blood pools.

32-15). These, of course, disappear as gestation continues. The embryo already has the rudimentary beginnings of the eyes, a beating heart, and traces of fingers on its tiny hands. Especially notable at this stage is the rapid growth of the brain, which is nearly as large as the rest of the body. Many of the structures of the adult brain are already recognizable.

As the second month draws to an end, nearly all the major organs have formed, and the embryo begins to look

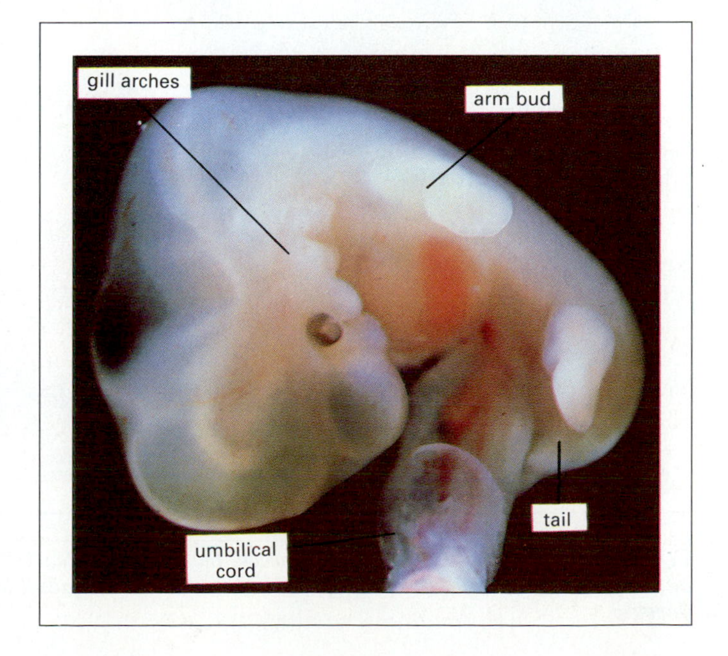

Figure 32-15 At the end of the sixth week, the human embryo is about half head. The feet and hands have begun to develop fingers, and the tail is receding.

quite human (Fig. 32-16). The gonads appear and develop into testes or ovaries, depending on the presence or absence of the Y chromosome. Sex hormones are secreted, testosterone from testes or estrogen from ovaries, and these hormones affect the future development of the embryonic organs, including not only the genitalia, but even certain regions of the brain. From now on, the embryo is called a **fetus,** denoting that it has taken on a generally human appearance.

These first two months of pregnancy are times of extremely rapid differentiation and growth for the embryo, and also times of considerable danger. Rapidly developing organs are much more sensitive to environmental insults than are fully formed organs or even the more slowly growing organs of later fetal development. Many drugs, though harmless to the mother, pass through the placenta and may damage the embryo. A notable example is the tranquilizer thalidomide, commonly prescribed in Europe in the early 1960s. Thalidomide's devastating effects on embryos were discovered only when many babies were born with missing or extremely abnormal limbs. Some otherwise innocuous viruses, such as the German measles (rubella) virus, can also harm the embryo. Children are vaccinated in school against German measles not so much for their own sake, since the disease is very minor in both children and adults, but to protect their unborn sisters and brothers. Finally, several common compounds, particularly alcohol, are extremely toxic to embryos. Large quantities of alcohol cause brain damage, abnormal facial features, and a host of other defects collectively known as fetal alcohol syndrome. Although the precise tolerance of the embryo to alcohol is still under study, many physicians now recommend that pregnant women avoid drinking *any* alcohol at all.

Although the embryo is most vulnerable during the first two months, ill effects can occur at any time during pregnancy. Probably the most common toxins to which embryos may become exposed are those in cigarette smoke.

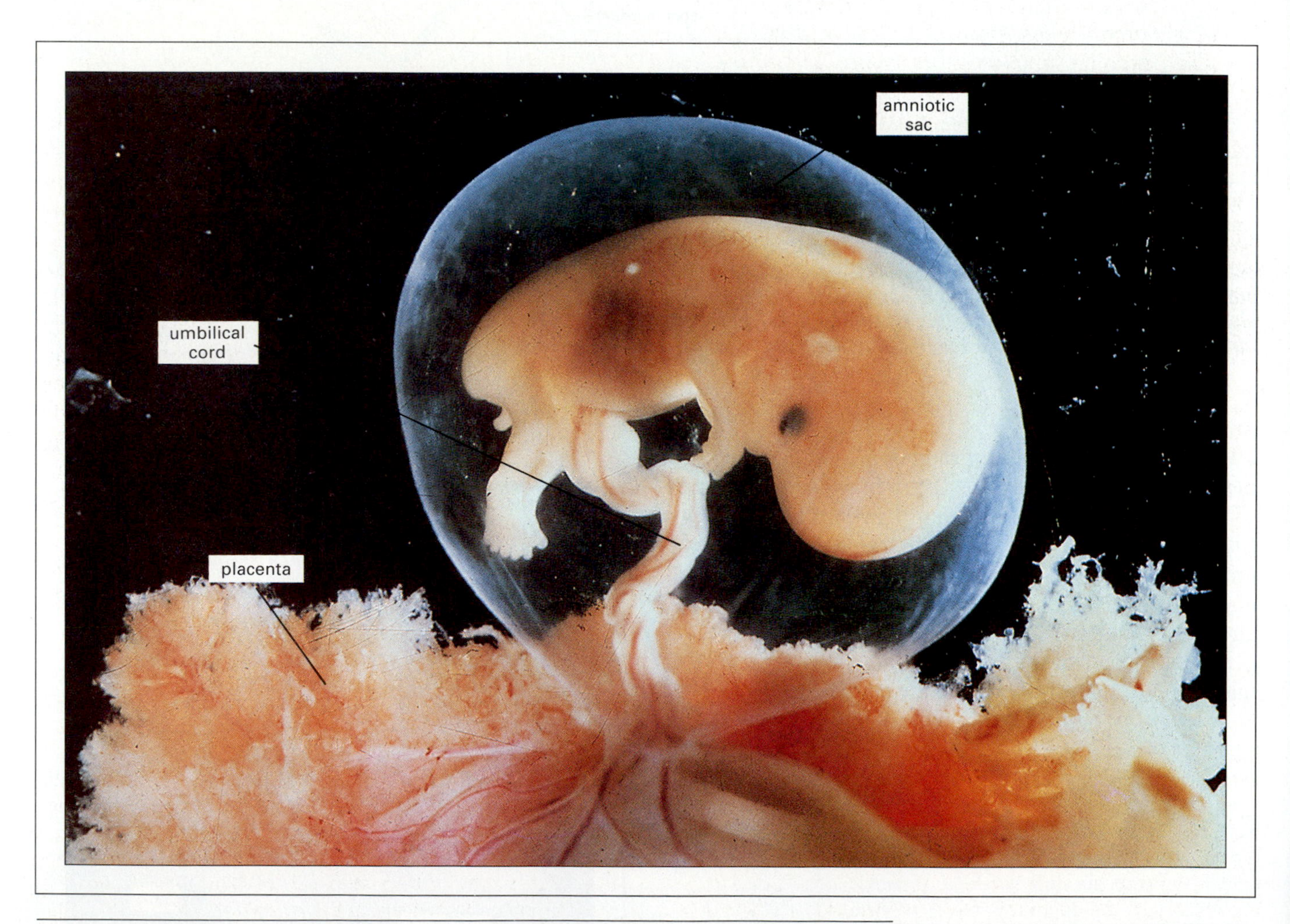

Figure 32-16 *At the end of the eighth week, the embryo is clearly human in appearance and is now termed a fetus. Most of the major organs of the adult body have begun to develop.*

Since about a quarter of all women smoke, and are usually so addicted that they cannot readily stop, even during pregnancy, almost a million human embryos are exposed to these poisons every year in the United States alone. On the average, babies born to smokers are smaller, are less well developed, and have a higher incidence of various abnormalities.

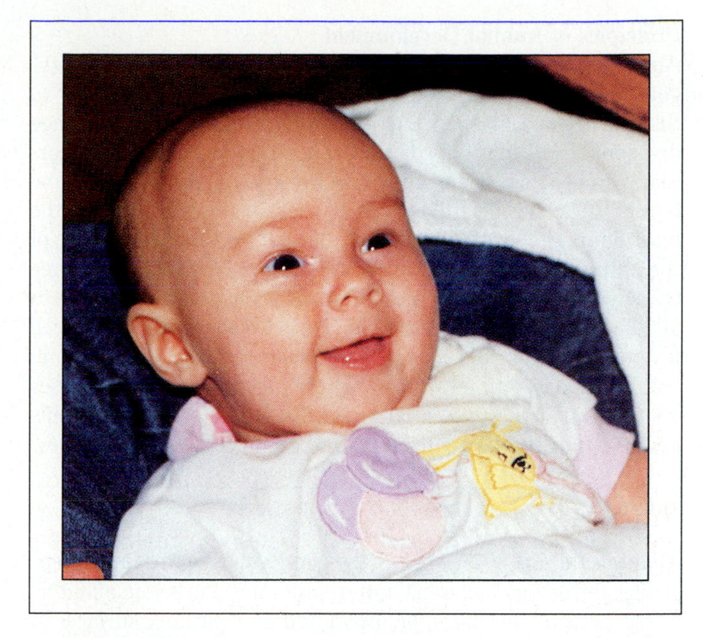

Figure 32-17 On her own.

The Last Seven Months

The fetus continues to grow and develop for another seven months. Although the rest of the body is "catching up" with the head in size, the brain continues to develop rapidly, and the head remains disproportionately large. Nearly every nerve cell ever formed during the entire human life span develops during embryonic life. As the brain and spinal cord grow, they begin to generate noticeable behaviors. As early as the third month of pregnancy, the fetus begins to move about and respond to stimuli. Some instinctive behaviors appear, such as sucking, which will have obvious importance soon after birth (see Fig. 34-22). Structures that the fetus will need when it emerges from the womb—lungs, stomach, intestine, kidneys—enlarge and become functional, although they will not be used until after birth.

Fetuses seven months or older can usually survive outside the womb, but larger and more mature fetuses have a much greater chance of survival. Evolutionarily speaking, it appears that human fetuses stay in the womb as long as possible, but after about nine months of development, the head is so large that it can barely fit out through the mother's pelvis. Even then, the skull must be compressed from a sphere almost into a cylinder to pass through the birth canal.

Birth

Normally, during the last weeks of pregnancy the baby becomes positioned head downward in the uterus, with the crown of the skull resting against (and being held up by) the cervix. No one knows precisely what stimulus initiates the delivery of the baby, although most researchers believe that the fetus itself somehow signals its readiness. The weight of the baby, combined with contractions of the uterus, pushes the head against the cervix, forcing the cervix to dilate. The baby is usually delivered head first. The infant is in for a rude awakening. Formerly, all the world was soft, fluid cushioned, and warm. Nutrients and oxygen were provided free of charge through the placenta from the circulatory system of the mother, and wastes were discharged that way as well. Suddenly, the baby must obtain oxygen and eliminate carbon dioxide by breathing. It must regulate its own body temperature. Although still provided by its mother, food must be obtained by active suckling. Considering that they have just experienced what by all odds is a change for the worse, babies are often remarkably cheerful (Fig. 32-17)!

SUMMARY OF KEY CONCEPTS

Differentiation

All the cells of an animal body contain a full set of genetic information, yet each cell is specialized for a particular function. During development, cells differentiate by stimulating and repressing transcription of specific genes. Gene transcription is regulated in two ways: (1) The egg cytoplasm contains gene-regulating substances that are incorporated into specific daughter cells during the first few cleavage divisions. The developmental fate of the daughter cells is determined by which substances they receive. (2) Later in development, certain cells produce chemical messages that induce other cells to differentiate into particular cell types.

Eggs, Development, and Animal Life Histories

Animals undergo either direct or indirect development. In indirect development, eggs (usually with relatively little yolk) hatch into larval feeding stages that later metamorphose into adults with body forms notably different from that of the larvae. In direct development, the newborn animal is sexually immature but otherwise resembles a small adult. Animals with direct development usually either have large, yolk-filled eggs, or nourish the developing embryo within the mother's body.

Principles of Animal Development

Animal development occurs in several stages. *Cleavage:* the zygote undergoes cell divisions with little intervening growth, so that the egg cytoplasm is partitioned into smaller cells. Cleavage divisions result in the formation of a solid ball of cells, the morula. A cavity then opens up within the morula, forming a hollow ball of cells, the blastula. *Gastrulation:* a dimple forms in the blastula, and cells migrate from the surface into the interior of the ball, eventually forming a three-layered gastrula. The three cell layers of ectoderm, mesoderm, and endoderm give rise to all the adult tissues (see Table 32-1). *Organogenesis:* the cell layers of the gastrula form organs characteristic of the animal species. *Growth and maturation:* the juvenile animal increases in size and achieves sexual maturity. *Aging and death:* cells begin to function less efficiently, and eventually the animal dies.

Human Development

The human developmental program clearly reflects our reptilian ancestry. Reptilian embryos develop four extraembryonic membranes, the chorion, allantois, amnion, and yolk sac, that function in gas exchange, waste storage, provision of the watery environment needed for development, and storage of yolk. These membranes are retained by the human embryo but have been modified to suit development inside the mother's womb. Human embryonic development follows the same principles as the development of other animals. The stages of human development are summarized in Fig. 32-10.

GLOSSARY

Allantois (al-an-tō′-is) one of the embryonic membranes of reptiles, birds, and mammals. In reptiles and birds, the allantois serves as a waste-storage organ and a respiratory surface. In mammals, the allantois forms most of the umbilical cord.

Amnion (am′-nē-on) one of the embryonic membranes of reptiles, birds, and mammals, enclosing a fluid-filled cavity that envelops the embryo.

Blastocyst in mammalian embryonic development, a hollow ball of cells formed at the end of cleavage.

Blastopore the site at which a blastula invaginates to form a gastrula.

Blastula in animals, the embryonic stage attained at the end of cleavage, in which the embryo usually consists of a hollow ball with a wall one or several cell layers thick.

Chorion (kor′-ē-on) the outermost embryonic membrane in reptiles, birds, and mammals. In birds and reptiles, the chorion functions mostly in gas exchange. In mammals, the chorion forms most of the embryonic part of the placenta.

Cleavage the early cell divisions of embryos, in which little or no growth occurs between divisions.

Differentiation the process whereby relatively unspecialized cells, especially of embryos, become specialized into particular tissue types.

Direct development a developmental pathway in which the offspring is born as a miniature version of the adult and does not radically change its body form as it grows and matures.

Ectoderm (ek′-tō-derm) the outermost embryonic tissue layer, which gives rise to structures such as hair, the epidermis of the skin, and the nervous system.

Embryo the early stage of development of animals, while the juvenile animal is still contained in the egg or in the mother's body. In mammals, this term usually refers to the early stages in which the developing animal does not yet resemble adults of the species.

Embryonic disk in human embryonic development, the flat, two-layered group of cells derived from the inner cell mass of the blastocyst, which will develop into the embryo proper.

Endoderm (en′-dō-derm) the innermost embryonic tissue layer, which gives rise to structures such as the lining of the digestive and respiratory tracts.

Fetus the later stages of embryonic development, when the developing animal has come to resemble the adult of the species.

Gastrula (gas′-trū-la) in animal development, a three-layered embryo with ectoderm, mesoderm, and endoderm cell layers. The endoderm layer usually encloses the primitive gut.

Gastrulation (gas-trū-lā′-shun) the process whereby a blastula develops into a gastrula.

Indirect development a developmental pathway in which a free-living offspring goes through radical changes in body form as it matures.

Induction the process by which a group of cells causes other cells to differentiate into a specific tissue type.

Inner cell mass in human embryonic development, the cluster of cells on one side of the blastocyst, which will develop into the embryo.

Larva an immature form of an animal with indirect development, often much different in body form from the adult.

Mesoderm (mes′-ō-derm) the middle embryonic tissue layer, lying between the endoderm and ectoderm, and usually the last to develop. Mesoderm gives rise to structures such as muscle and skeleton.

Metamorphosis (met-a-mor′-fō-sis) in animals with indirect development, a radical change in body form from one larval stage to another or from larva to adult.

Morula (mor′-ū-la) in animals, an embryonic stage during cleavage, when the embryo consists of a solid ball of cells.

Organogenesis (or-gan-ō-jen′-i-sis) the process by which the germ layers of the gastrula develop into organs.

Placenta in mammals, a structure formed partly from the uterine lining and partly from the embryonic membranes, especially the chorion; functions in gas, nutrient, and waste exchange between embryonic and maternal circulatory systems.

Primitive streak in reptiles, birds, and mammals, the region of the ectoderm of the two-layered embryonic disk through which cells migrate to form mesoderm.

Yolk protein or lipid-rich substances contained in eggs as food for the developing embryo.

Yolk sac one of the embryonic membranes of reptile, bird, and mammalian embryos. In birds and reptiles, the yolk sac is a membrane surrounding the yolk in the egg. In mammals, the yolk sac is empty, but forms part of the umbilical cord and the gut.

STUDY QUESTIONS

1. Define differentiation. How do cells differentiate; that is, how is it that adult cells express some but not all the genes of the fertilized egg?
2. Describe the two processes of differentiation during development: the influence of the egg cytoplasm and induction by other cells.
3. Distinguish between direct and indirect development, and give examples of each.
4. What is yolk? How does it influence cleavage?
5. Name two structures derived from each of the three embryonic germ layers—endoderm, ectoderm, and mesoderm.
6. What is gastrulation? Describe gastrulation in frogs and humans.
7. Describe the process of induction and give two examples.
8. How does cell death contribute to development?
9. List the four embryonic membranes and the function of each in a reptile and a human embryo.

SUGGESTED READINGS

Beaconsfield, P., Birdwood, G., and Beaconsfield R. "The Placenta." *Scientific American,* August 1980 (Offprint No. 1478). The placenta is one of the most remarkable of mammalian structures, allowing internal development of offspring within the mother.

Cooke, J. "The Early Embryo and the Formation of Body Pattern." *American Scientist,* January–February 1988. Although the developmental patterns of worms, fruit flies, and humans are superficially very different, they are probably based on common mechanisms.

DeRobertis, E. M., and Gurdon, J. B. "Gene Transplantation and the Analysis of Development." *Scientific American,* December 1979. Genes can be injected into amphibian eggs, allowing direct observation of the effects of specific gene products on development.

Hall, S. S. "The Fate of the Egg." *Science 85,* November 1985. A look at some of the attempts to understand development in everything from fruit flies to people.

Ingelman-Sundberg, A. *A Child Is Born.* New York: Dell Publishing Company, 1966. Simple, clear explanations of the stages of human development, with magnificent photographs by Lennart Nilsson.

Nijhout, H. F. "The Color Patterns of Butterflies and Moths." *Scientific American,* November 1981. The structure and development of wing patterns is a fascinating study for its own sake, as well as for what it may tell us about developmental principles.

Saunders, J. W., Jr. *Developmental Biology.* New York: Macmillan Publishing Company, 1982. Excellent descriptions and illustrations of developmental processes on both the morphological and molecular levels.

33
Animal Behavior I. Principles of Individual Behavior

As you hike through a forest, roll a damp, decaying log to reveal the hidden pillbug, slug, and salamander. Listen for the scolding of squirrel and bird as you pass through their territories. Consider that the existence of these life forms today is the result of eons of successful reproduction. Reproduction requires that the animal live to maturity, find a mate, and gather the resources needed to produce offspring. These activities demand a lifetime of behavior finely tuned to the animal's environment, both physical and social. Anatomy, physiology, genetic programming, and learning all must mesh properly to produce the right action at the right time.

Historically, diverse disciplines have contributed to the science of animal behavior. Ethologists, for example, study behavior as it occurs under natural conditions in a wide variety of organisms. An important principle of ethology is that behavior has evolved as an adaptive trait through natural selection. Thus ethologists view behavior as a genetically programmed response to specific stimuli in the natural environment. Beginning with the observational and intuitive approach of Konrad Lorenz, whose investigations spanned the first half of this century (Frontispiece), ethology was molded into an experimental science by Niko Tinbergen and Karl von Frisch, whose works are mentioned later in this and the following chapter.

Comparative psychologists, in contrast, treat animal behavior as a laboratory science. The evolutionary aspects and adaptive value of behavior are minimized. These investigators generally use a few types of laboratory animals (such as rats and pigeons) under highly controlled conditions. Learning and its mechanisms are the prime focus of their studies. Their ultimate goal is to determine underlying principles that may be applicable to humans.

"As I watched the geese, it appeared to me as little short of a miracle that a hard, matter-of-fact scientist should have been able to establish a real friendship with wild, free-living animals, and the realization of this fact made me strangely happy. It made me feel as though man's expulsion from the Garden of Eden had thereby lost some of its bitterness."

From *King Solomon's Ring* by Konrad Lorenz

Behavioral neurobiologists attempt to decipher the mechanisms by which the brain directs behavior. Advances in technology that make the inner workings of the brain more accessible have contributed to the rapid growth and success of this endeavor. This discipline incorporates elements of psychology, neuroanatomy, brain biochemistry, and the properties of individual neurons. Some investigators use invertebrates to provide insight into how relatively simple nervous systems direct behavior (see the essay, "Of Squids and Snails and Sea Hares," Chapter 29).

Recently, some prominent investigators of animal behavior have adopted a highly theoretical, evolutionary approach. Their studies focus on the adaptive value of broad classes of behavior that do not appear to enhance the survival of the individual. For example, an animal showing altruism may endanger itself to protect others in its social group. Since these individuals are more likely to be killed, why aren't the genes directing this behavior eliminated from the gene pool of the population? This question is addressed in the following chapter (see the essay, "Altruism and the Selfish Gene") and in the essay on kin selection in Chapter 14.

In this chapter we introduce basic concepts of animal behavior, emphasizing individual behavior. Its examples are drawn largely (but not exclusively) from the work of ethologists. These early studies provide insight into the fascinating diversity of animal behavior, its adaptiveness, and its causes. In Chapter 34 we explore animal communication systems and their use in social behavior.

(a)

(b)

Figure 33-1 *Sensory abilities determine how each animal perceives its world.* **(a)** *Bats use sonar to locate their prey (here a tree frog), and avoid objects while flying. By emitting high-pitched, rapid bursts of sound and detecting the patterns of sound reflected back, the bat determines the shape, size, and proximity of objects.* **(b)** *The sensitive pit organs of this tropical viper, located below the eyes, can detect the body heat of nearby prey.*

THE GENETIC BASIS OF BEHAVIOR

All behavior has some genetic basis. For example, much behavior consists of responses to stimuli from the environment. These stimuli are received and filtered by the sense organs. The design of these organs, coded by the genes, determines what stimuli the animal perceives. Because sensory capabilities differ dramatically, each species perceives a different world. Bats, for example, produce and respond to sounds far higher in pitch than the human ear can detect (Fig. 33-1a). A bee observing a flower sees patterns of reflected ultraviolet light invisible to the human eye. The pit organs of rattlesnakes can detect the body heat of a nearby mouse (Fig. 33-1b). The tick is exquisitely sensitive to the odor of butyric acid, produced by the skin of mammals. The scent will cause it to drop off its perch, usually landing on the passing animal. As it sits on a leaf waiting for a suitable victim, it probably senses little else. Clearly, behavior is governed by perception.

Sense organs are extensions of the genetically coded nervous system. Besides sending the impulses that directly drive behavior, the nervous system, by its complexity and its specific neural connections, determines how much and what type of learning is possible.

Another influence of the genes on behavior is seen in the radically differing body forms of different species. It is only within these genetically determined limits that behavior can be influenced by the environment through learning.

As we will see in the descriptions that follow, heredity sets very narrow boundaries for behaviors that are largely **innate** or instinctive, and wider boundaries for learned behaviors.

INNATE BEHAVIOR

A fruit fly, placed in a tunnel with light at one end and darkness at the other, will fly toward the light; a hungry newborn human infant, touched on the side of the mouth, will turn her head and attempt to suckle. These are examples of **innate** or instinctive behaviors. Such acts are performed in reasonably complete form the first time an animal of the right age and motivational state encounters a particular stimulus. Innate behaviors are programmed by the genes and passed from one generation to the next. Since innate behavior is performed without learning or prior experience, it tends to be highly stereotyped. Innate behaviors fall into four categories: (1) kineses, (2) taxes, (3) reflexes, and (4) fixed action patterns.

Kineses

A **kinesis** is a behavior in which an organism changes its speed of random movement in response to an environmental stimulus. As a result, it ends up in a favorable environment by stopping when it blunders there by chance, and escapes hostile conditions by speeding up. For example, the pillbug (a land-dwelling crustacean) uses a kinesis to reach the moist areas it needs to survive. As the air surrounding it becomes drier, the pillbug moves faster (but in no particular direction) until it encounters a damper area, where it slows and eventually stops. This kinesis results in congregations of pillbugs under damp leaves and rotting logs.

Taxes

In contrast to a kinesis, a **taxis** is a *directed* movement toward or away from a stimulus. A moth flying toward a light is one example (Fig. 33-2). Many of the behaviors of very simple organisms, including single-celled protists, are taxes. For example, *Euglena* (see Fig. 17-12) is photosynthetic, but its chlorophyll can be damaged by intense light. As you might predict, *Euglena* shows a positive taxis toward dim light, but a negative taxis toward intense light.

Mosquito behavior involves several taxes. Males orient toward the high-pitched whine of the female. Female mosquitos (only females suck blood) show taxes to the warmth, humidity, and carbon dioxide exuded by their prey. Mosquito repellents probably act by blocking the receptors for these stimuli, rendering the insect incapable of sensing and orienting to her victim.

Taxes can be influenced by the internal state of the organism. The grayling butterfly shows a taxis toward bright light, but only when it is pursued. A grayling fleeing a predator flies directly toward the sun, temporarily blinding its pursuer.

Vertebrates may also show taxes, although vertebrate taxes are more difficult to detect because they interact with other behaviors. For example, fish swim upright by orienting their dorsal surface away from the force of gravity and toward light (normally, the water surface). If the gravity-detecting organ is removed, or the light enters the side of a fish tank, the animal becomes disoriented (Fig. 33-3).

Figure 33-2 Moths and other night-flying insects show a positive taxis toward light, resulting in congregations such as this.

Reflexes

In contrast to a taxis or a kinesis, which involves orientation of the entire body, a **reflex** is a movement of a body part, such as blinking an eye or withdrawing the hand from a hot stove. Reflexes are usually stereotyped and rapid, since the conscious brain is not involved. Take the knee-jerk reflex, for example. Tapping the patellar (kneecap) ligament stretches the thigh muscle to which the kneecap is attached, just as would occur if your knee suddenly buckled. A stretch receptor in the muscle carries an emergency message directly to a motor neuron in the spinal cord (Fig. 33-4). This commands the thigh muscle to contract, suddenly straightening the leg. Had the knee actually buckled, this reflex could prevent a dangerous fall. Your brain, which is informed of this activity via the eyes and spinal cord, does not participate in the reflex. Instead, it observes the reflex response to the doctor's hammer tap with detached interest and curiosity.

Fixed Action Patterns

A **fixed action pattern** is a stereotyped and often complex series of movements. If the animal is at the right developmental stage and properly motivated, a fixed action pattern will be performed correctly the first time the appropriate stimulus, called a **releaser,** is presented.

Fixed action patterns may be so complex and so appropriate that they appear learned. How can we differentiate these complex instinctive acts from learned behaviors? Three approaches are suggested below.

PERFORMANCE WITHOUT PRIOR EXPERIENCE. Scientists can demonstrate that a behavior is innate by depriv-

Figure 33-3 Taxes to light and gravity in the fish Crenilabrus. *Fish on the left are normal, while those on the right have had their inner ears removed, eliminating their response to gravity. Arrows show the direction of light. The fish unable to detect gravity orient to the light alone.*

ing the animal of the opportunity to learn it. For example, red squirrels bury extra nuts in the fall for retrieval during the winter. The squirrel carries a nut to a landmark such as the base of a tree, scratches out a hole, shoves the nut in with its nose, and scrapes and pats dirt over it with its forepaws. Red squirrels can be raised from birth in a bare cage on a liquid diet, providing them with no experience with nuts, digging, or burying. Presented with nuts for the first time, such a squirrel (after eating several) will carry one to the corner of its cage, then make covering and patting motions with its forefeet. Nut burying is therefore a fixed action pattern, and the nut serves as a releaser.

Fixed action patterns can also be recognized by their occurrence immediately after birth, before there is any opportunity for learning. The cuckoo bird, for example, lays its egg in the nest of another bird species, to be raised by the unwitting adoptive parent. Immediately after hatching, the cuckoo chick shoves the nest-owner's eggs (or baby birds) out of the nest (Fig. 33-5).

Manipulating the releaser. Since a fixed action pattern is a rigid response to its releaser, scientists can elicit inappropriate behavior by providing the releaser in the wrong context. An example was inadvertently provided by bird banding. Parent birds clean their nests,

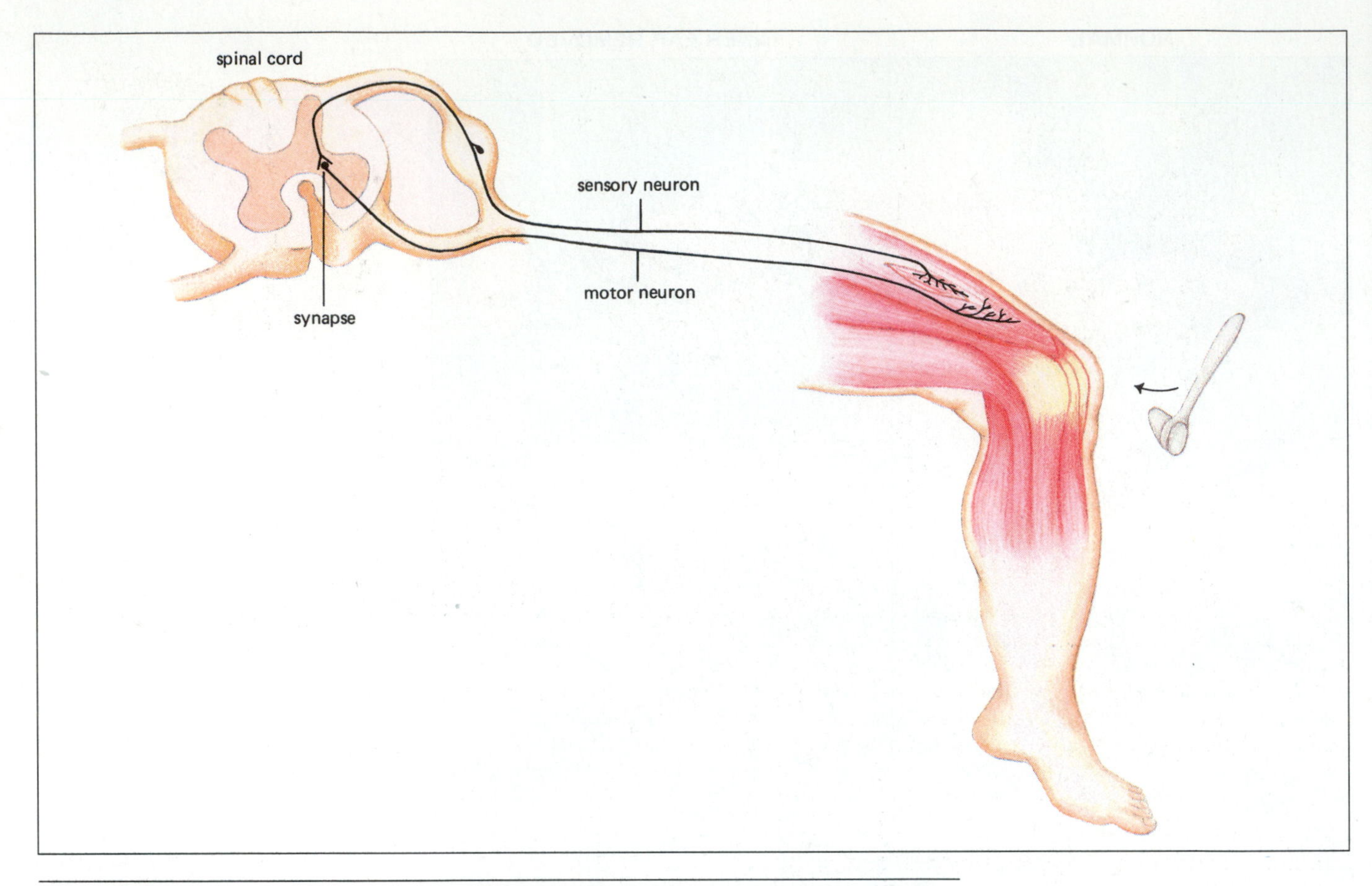

Figure 33-4 *The knee-jerk reflex. Tapping the patellar ligament causes a sudden stretching of the thigh muscle. This activates stretch receptors in the muscle that communicate to the spinal cord. In the cord, the impulse is transmitted directly to a motor neuron, which causes the thigh muscle to contract, jerking the lower leg upward. The brain is not involved in this simple behavior.*

throwing out light-colored objects such as broken eggshells or excrement. Biologists who had placed shiny metal identification bands on the legs of baby birds were dismayed to find the parents attempting to throw the band out of the nest, baby bird and all! The band was a releaser for the fixed action pattern of nest cleaning, regardless of the consequences.

Fixed action patterns are often released only by particular characteristics of the animal toward which they are directed. The female redwinged blackbird signals her readiness to mate by the angle of her tail, which serves as a releaser for mating by the male. He will even attempt to copulate with the stuffed tail of the female, provided that the feathers are angled appropriately (Fig. 33-6a). Konrad Lorenz was attacked by tame and normally friendly jackdaw birds (European relatives of crows) as he walked toward his swimming hole. The limp black bathing trunks he carried released the same behavior as would a dead jackdaw in the clutches of a predator (Fig. 33-6b).

Figure 33-5 *The cuckoo, just hours after it hatches and before its eyes have opened, evicts the eggs of its foster parents.*

Scientists can use models to exaggerate releasers. Such exaggerated releasers, called **supernormal stimuli,** are far more effective than their natural counterparts. By selectively exaggerating specific features of a stimulus, ethologists can determine which features serve as releasers. For

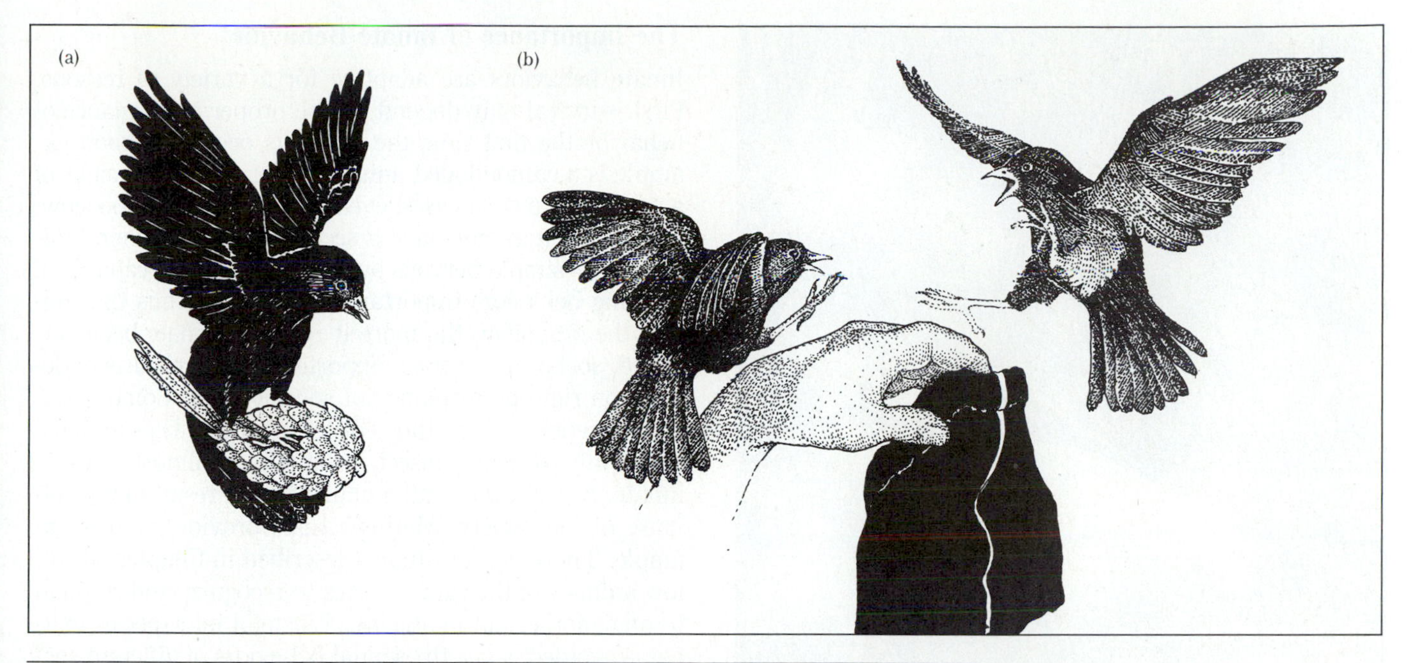

Figure 33-6 **(a)** *A male redwinged blackbird is stimulated to mate by the angle of the female's tail, whether or not she is attached to it.* **(b)** *Konrad Lorenz was attacked by his tame jackdaws as he carried his black bathing trunks. The trunks, resembling a dead jackdaw in the clutches of a predator, acted as a releaser for this defensive behavior.*

example, the herring gull chick, studied by Niko Tinbergen, instinctively pecks at a red spot on the parent gull's beak (Fig. 33-7). This fixed action pattern, which chicks perform immediately after hatching, causes the parent to regurgitate food. Tinbergen found that the releasing features of the bill are its long, thin shape, the red color, and the presence of color contrasts. When he offered chicks a thin red rod with white stripes painted on it, they pecked at it more often than a real beak (Fig. 33-8). Herring gulls,

Figure 33-7 A herring gull chick pecks at the red spot on its mother's bill, causing her to regurgitate food for it. Niko Tinbergen and others have shown that the thin shape of the bill and the contrasting spot serve as releasers for pecking, which occurs immediately after hatching. The pecking then serves as a releaser for food regurgitation by the parent.

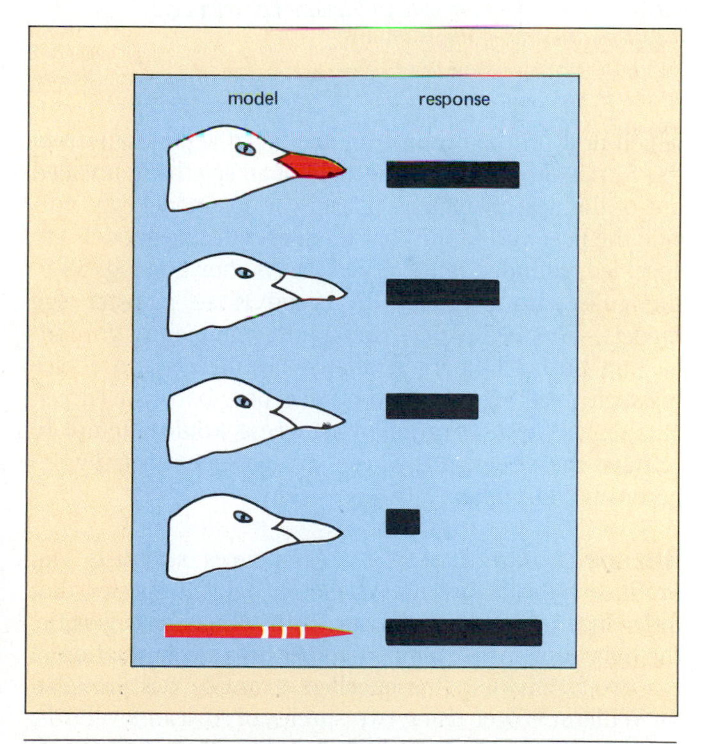

Figure 33-8 Models are used to determine which features of the bill are releasers for pecking behavior. The length of the bar is proportional to the number of pecks delivered to the corresponding model. A brightly contrasting spot improves performance; the color red further increases pecking; but the red knitting needle with contrasting stripes was most effective. Here the long, thin shape, the red color, and the contrast are all exaggerated to produce a supernormal stimulus.

Figure 33-9 A herring gull will ignore her own eggs if offered a supernormal brooding stimulus.

which nest on the ground, show a fixed action pattern of egg retrieval released by the sight of an egg that has rolled out of the nest. Scientists placed various model eggs outside the nest and found that the most effective models are speckled, rounded, and very large. As shown in Fig. 33-9, the gulls also preferred to brood these monster egg models. This behavior is apparently common in ground-nesting birds. Tinbergen offered nesting geese a choice between their own egg and a volleyball, both placed just outside the nest. Invariably, the geese would attempt to retrieve the volleyball, whose size presented a truly supernormal stimulus.

Breeding experiments. Sometimes scientists can produce hybrids by mating closely related species that differ in a particular behavior. If the behavior is genetic, the hybrid offspring may try to perform a combination of the two behaviors. One excellent example was provided by William Dilger using two species of African lovebirds. One species carries nesting material to its nest site in its beak, one piece at a time. A second species tucks several pieces of nesting material into its tail feathers before flying to the nest (Fig. 33-10). Their hybrid offspring tried to place nesting material in their tailfeathers, but positioned it improperly, or failed to release it, apparently trying to carry it in both their tails and their bills.

The Importance of Innate Behavior

Innate behaviors are adaptive for a variety of reasons. First, survival may depend on the proper performance of behavior the first time the stimulus occurs. A good example is a camouflaged animal "freezing" at the sight of a predator. In such cases, even rapid learning is too slow; the animal may not have a second chance. Second, animals with simple nervous systems may not be capable of learning behaviors important to survival. Thus the simpler the organism, the more it relies on innate behavior. Third, social interactions important for survival may depend on rigid performance of specific roles. Complex insect societies such as those of termites and bees are possible only because insect behavior is almost entirely innate. Any show of individuality would result in the collapse of the society. Mating rituals provide another example. These innate rituals (described in Chapter 34) allow animals of the same species to recognize and respond to one another automatically. Wasteful interspecies mating is avoided, since the sexual behaviors of different species are usually incompatible.

LEARNING

The eviction of banded baby birds by their nest-cleaning parents shows that instinctive behavior has its drawbacks, especially when the environment is not entirely predictable. Survival is enhanced if an animal can modify its behavior based on experience. Thus nervous systems have evolved with varying degrees of flexibility. In gen-

Figure 33-10 Lovebirds tearing off strips of nesting material and tucking them among their tailfeathers for transport to the nest site.

eral, the more complex an animal's nervous system, the more its behavior can be modified by experience. This capacity is called **learning.** Here we discuss five categories of learning: imprinting, habituation, conditioning, trial-and-error learning, and insight.

Imprinting

As discussed earlier, learning always occurs within boundaries specified by the animal's genes. This is strikingly illustrated by the unique form of learning called **imprinting.** Imprinting refers to a strong association learned during a particular stage in an animal's life. The process is best known in birds such as geese, ducks, and chickens. These birds learn to follow the animal or object that they most frequently encounter during a sensitive period (for mallard ducks this is about 13 to 16 hours after hatching). In nature, the mother is the object of imprinting. In the laboratory, however, these birds can be forced to imprint on a toy train or other moving object, although if given a choice, they select a duck. Konrad Lorenz coined the word "imprinting" in the 1930s, and demonstrated that the object of imprinting does not have to be appropriate (see the Frontispiece). During sexual imprinting, the young bird learns the characteristics of its future mate. It occurs separately from the following response, often slightly later. During imprinting, the animal is actually learning a releaser (the object followed and later courted) for a fixed action pattern (following, and later, courtship) during a limited period in its development, the sensitive period. Parents may also imprint on their offspring. Herring gulls imprint on their young during the first two days after hatching. During this two-day period, they will accept foster young placed in their nest. Thereafter, the parents will attack and kill baby chicks that are not their own.

Habituation

A more common form of simple learning is **habituation,** defined as a decline in response to a harmless, repeated stimulus. The ability to habituate guards an animal against wasting its energy and attention on irrelevant stimuli. This form of learning is shown by the simplest animals, and has even been demonstrated in the single-celled members of the kingdom Protista. For example, the protist *Stentor* contracts to touch (Fig. 33-11a), but gradually stops retracting if touching is continued. The sea anemone, which also lacks a brain, shows a similar response to touch (Fig. 33-11b). The ability to habituate is clearly adaptive. In this example, if the anemone contracted every time it was brushed by a strand of waving seaweed, it would waste a great deal of energy, and its retracted posture would prevent it from snaring food. Humans habituate to a variety of stimuli: city dwellers to nighttime traffic sounds, and country dwellers to choruses of crickets and tree frogs. Each may initially find the other's habitat unbearably noisy at night.

Classical and Operant Conditioning

A more complex form of learning, most commonly seen in the laboratory, is called conditioning. During **classical**

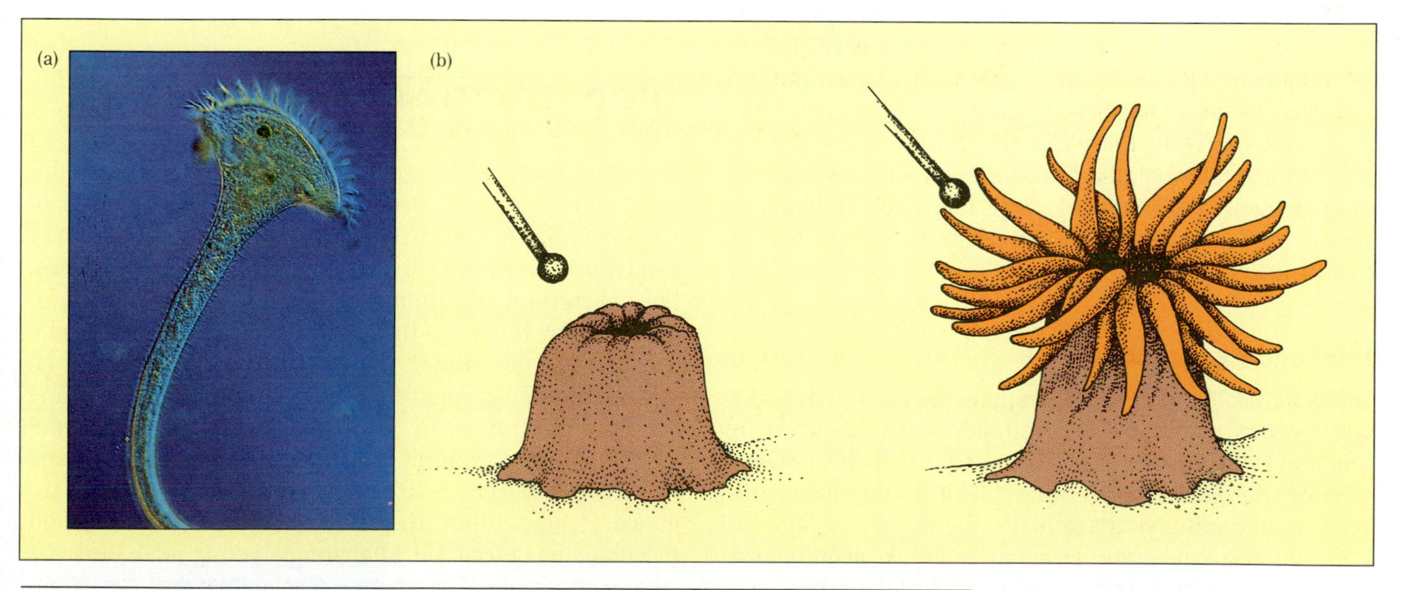

Figure 33-11 **(a)** *The protist* Stentor *is a complex single cell that can show habituation, a simple form of learning.* **(b)** *Habituation in the sea anemone. (Left) The animal is touched for the first time, causing withdrawal. (Right) After many touches, the anemone habituates to this harmless stimulus. Learning occurs even though the anemone possesses only a simple network of neurons and lacks a brain.*

conditioning an animal learns to perform a response (normally caused by one stimulus) to a new stimulus. Ivan Pavlov, a Russian physiologist, is sometimes called the "father of conditioning." Early in the twentieth century, he performed his famous experiment on dogs. Pavlov first placed dried meat powder into a dog's mouth, causing reflex salivation. Then he rang a bell just before the meat powder was presented. After several repetitions, the dog would salivate to the bell alone. The dog who leaps up at the sight of her leash or salivates to the sound of a can opener has been classically conditioned, however unintentionally. This form of learning can also be demonstrated in simple organisms. For example, flatworms can learn to associate a flash of light with an electric shock, which causes them to contract. After several exposures to light paired with shock, they contract to the light alone.

During **operant conditioning** an animal learns to perform a behavior (such as pushing a lever or pecking a button) to receive a reward or avoid punishment. This technique is most closely associated with the American comparative psychologist B. F. Skinner. Skinner designed the "Skinner box," in which an animal (often a white rat or pigeon) is isolated and allowed to train itself. The box might contain a lever that ejects a food pellet when pressed. The animal in its explorations inevitably bumps the lever and is rewarded. Soon it will be pressing the lever repeatedly for food. Skinner boxes can be quite complex. A pigeon might learn that one lever will provide food only when a green light is shining, but when a red light is on, a specific spot on the wall must be pecked to avoid a shock. The Coast Guard has used operant conditioning to train pigeons for sea search-and-rescue missions. The birds are trained to peck a button when they see orange, the color used for lifejackets worldwide. Three trained pigeons are strapped in a clear plexiglass pod on the underside of a helicopter, each with its own button, and each facing in a different direction. When one spots an orange object in the water, it pecks its button, thus activating an alarm and a direction indicator in the pilot's compartment. Pigeons spot more lifejackets bobbing in the sea and make fewer mistakes than do the best human observers.

Trial-and-Error Learning

In the natural environment, animals are faced with naturally occurring rewards and punishments and learn by **trial and error.** (Operant conditioning is actually a special form of trial-and-error learning which occurs under artificial and carefully controlled laboratory conditions.) Through trial-and-error learning, animals acquire new and appropriate responses to stimuli through experience. Much learning of this type occurs during play and exploratory behavior in animals with complex nervous systems (see the essay, "Puzzled About Play"). Trial-and-error learning makes a major contribution to the behavior of young children (not to mention adults). It is how a child learns which foods taste good or bad, that a stove may be hot, and not to pull the cat's tail.

In some instances, trial-and-error learning can modify the releaser for innate behavior, thus making it more adaptive. For example, to the hungry toad, any flying insect is a releaser for feeding. Although this usually results in a meal, occasionally the toad captures a mouthful of trouble: a bee. Having its tongue stung results in trial-and-error learning that takes only a single experience (Fig. 33-12). Learning modifies the releaser to exclude bees and even other insects resembling them.

Insight

The most complex form of learning is called **insight** or reasoning. Insight involves manipulating concepts in the mind to arrive at adaptive behavior. It can be considered a kind of mental trial-and-error learning. Insight arises from the ability to remember a variety of past experiences and to apply these lessons in creative ways to new situations. This ability is demonstrated by animals which can solve a detour problem the first time it is presented (Fig. 33-13a). In 1917, William Kohler showed that a hungry chimp would stack boxes to reach a banana suspended from the ceiling, without any training (Fig. 33-13b). This type of mental problem solving was believed limited to very intelligent animals such as primates. Recently, however, similar insight was demonstrated in the pigeon by R. Epstein and associates at Harvard. As investigators learn more about how to design appropriate experiments, we may find that insight is much more common than originally suspected.

THE INSTINCT TO LEARN AND THE LEARNING OF INSTINCTS

You have probably concluded by now that neither "innate" nor "learned" adequately describes the behavior of any given organism. Especially in an adult animal, every behavior is an intimate mixture of the two. In some cases, the nature and the timing of learning is so rigidly programmed by the genes that we might use the term "innate learning" to describe it.

An example of innate learning is imprinting: learning rigidly programmed to occur at a certain developmental stage. A human example that resembles imprinting occurs in young children, who are far better at acquiring language at age 3 than they will be as adults. More rigid mechanisms determine what song a young male bird will sing as an adult. The white-crowned sparrow, for example, must hear its species' song between the tenth and fiftieth day after hatching or it will not be able to sing properly as an adult, no matter how much coaching it receives. The young bird does not sing at this stage of

Fig. 33.12 Trial and Error learning

(a)

(a) *A naive toad is presented with a bee.*

(b)

(b) *While trying to eat the bee, the toad is stung painfully on the tongue.*

(c)

(c) *Presented with a harmless robber fly, which resembles a bee, the toad cringes.*

(d)

(d) *The toad is presented with a dragonfly.*

(e)

(e) *The toad immediately eats the dragonfly, demonstrating that the learned aversion is specific to bees and insects resembling them.*

Puzzled About Play

One of the most appealing aspects of animals behavior, and one of the most puzzling to the investigator, is play. Play has been observed in many birds and in most mammals, from cows to lions to mice. Pygmy hippopotami push one another, shake and toss their heads, splash in the water, and pirouette on their hind legs. Otters delight in elaborate acrobatics. Bottlenose dolphins balance fish on their snouts, throw objects between their flukes, and carry them in their mouths while swimming. Even baby vampire bats have been observed chasing, wrestling, and slapping each other with their wings.

Play varies enormously. One form of play involves a single animal manipulating an object, like a cat with a ball of yarn or the dolphin with a fish. In another example, a vervet monkey was observed hopping on and off a lilac-colored feather. Play may also be social. Often young of the same species play together (Fig. E33-1), but parents may join them, as may members of other species (cats and dogs from the same household often play together). Social play often includes chasing, fleeing, wrestling, kicking, gentle biting, and grabbing of the playmate.

What are the general characteristics of play? (1) Play seems to lack any clear goal. (2) Animals play when there are no other immediate demands on their time and attention. Play is abandoned in favor of escaping from danger, feeding, and courtship. (3) Play seems to involve feelings of pleasure. (4) Young animals play more frequently than adults. (5) Play often involves movements borrowed from other behaviors (attacking, fleeing, stalking, etc.) (6) Play uses considerable energy. (7) Play is potentially dangerous. Young humans and other animals are frequently injured, at least slightly, during play. In addition, play may distract the animal from the presence of danger while making it conspicuous to predators. The question (at least to scientists) is: Why do they do it?

Probably the most intuitively reasonable theory of play is the "practice theory," first suggested by K. Groos in 1898. He suggested that play allows young animals to gain experience in a variety of behaviors that they will use as adults. By performing these acts repeatedly in a nonserious context, the animal gains skills that will later be important in, for example, hunting, fleeing, or social interactions. This theory has received no good experimental support to date, nor has it been disproven, largely due to the difficulty of experimentation. Although a few studies have shown that play-deprived animals can still hunt, one cannot rule out the possibility that they would perform more efficiently had play been permitted.

A somewhat different interpretation is that play is a mechanism of building cardiovascular fitness, coordination, and endurance. Thus when the young animals leave the care of their parents, play has physically pre-

Figure E33-1 Young red foxes (left) and siberian tigers (right) at play.

Puzzled About Play

pared them to cope with the serious demands of adult life. This was proposed by A. Brownlee in 1954 and today is considered among the most plausible explanations. Physiologically, young organisms respond more readily to vigorous exercise than do adults. For example, studies in rats, rabbits, and humans have shown that in response to equal amounts of exercise, young organisms show a greater increase in heart size and greater increase in ability to extract oxygen from blood than do adults. This would help explain why play is important in young animals. The stimulation of pleasure centers by play is adaptive because it encourages young animals (including humans) to engage in what otherwise would be hard work, "just for the fun of it."

The exercise theory does not account for the monkey playing with a feather or children playing quietly with dolls or blocks. It also fails to explain the strong social component of many forms of play. Clearly, play serves a variety of needs in the developing animal. Manipulating objects and interacting with other individuals may serve as "exercise" for the developing nervous system. Animals raised in impoverished environments where their experiences were limited were found to have more poorly developed neural connections than those in a diverse environment. In play, animals seek out and create novel situations. By quietly manipulating objects, an animal may learn new skills. William Kohler, who studied chimp behavior, offered his chimp Sultan a pair of sticks that could be linked together to form a longer stick. Only the linked sticks would allow Sultan to reach a suspended banana. When presented with the two sticks and the banana, Sultan tried repeatedly to reach the banana with the longer of the two sticks, then finally gave up. Only later, quietly playing with the sticks, did he happen to link them together. He then immediately used his new tool to obtain the food. Considerable trial-and-error learning occurs during play, both social and solitary. Blocks and bikes teach lessons in physics, particularly concerning gravity.

In short, although play uses considerable energy and can be dangerous, it is nearly universal in mammals. One can only conclude that play is a highly adaptive behavior, probably for a combination of the reasons suggested above. But since this is an area that science has yet to fully explain, we can still say that we, and they, do it because it's fun!

development, but stores the song in memory for retrieval several months later.

Adult bees use a type of innate learning to allow them to locate their hive after a foraging trip. Ethologists discovered that bees memorize the location of their hive with respect to certain landmarks only on their first flight of the day. If the hive is moved even a short distance later in the day, the bees will have trouble finding it. If a beekeeper wishes to move the hive, he must do it during the night. The bees then memorize the new location on their first morning flight and return to it effortlessly thereafter.

We have seen how the genes influence learning, but learning may also modify innate behavior. Recall the example of the herring gull chicks pecking their parent's beak for food. Immediately after hatching, the chicks prefer a striped knitting needle to the real beak. Within a few days, however, the chicks learn enough about the appearance of their parents that they begin pecking more frequently at models more closely resembling the parents. After one week, young gulls can distinguish models of their own species from models of a closely related species.

Habituation is another example of learning that can fine-tune an organism's innate responses to environmental stimuli. Baby chicks instinctively crouch whenever a large bird flies overhead, making themselves less visible to predators such as hawks. Early investigators were impressed to see young birds crouching when a hawk flew over, while ignoring harmless birds such as geese. They hypothesized that crouching was released only by the very specific shape of predatory birds. Further research revealed that inexperienced baby birds instinctively crouch when anything flies overhead. Harmless birds, such as geese, appear in the sky so frequently that the chicks quickly habituate to them. The novel shape of a hawk, however, still elicits instinctive crouching. Thus learning modifies the innate response, making it more adaptive.

Trial-and-error learning can also result in more appropriate responses to releasers, as in the case of our bee-catching toad (see Fig. 33-12). The naive toad instinctively snaps at all flying insects of appropriate size, but from painful experience learns to make certain exceptions! Trial-and-error learning also gradually modifies the innate tendency of the human 2-year-old to explore everything with hands and mouth.

Figure 33-13 **(a)** *A dog, lacking insight in this situation, fails a detour problem on the first try. It will eventually learn by trial and error.* **(b)** *Insight in a chimpanzee. Unable to reach the bananas, the chimp stacks boxes beneath them to extend its reach.*

REFLECTIONS ON ETHOLOGY

Why study animal behavior? The more carefully one observes animals, the more fascinating they become. Whether it is the dog next door, a salmon thrashing upstream to spawn, or a honeybee sipping nectar from a flower, a knowledge of ethological principles can enhance our appreciation of other animals and thereby enrich our own lives.

Ethology also has its practical side, for example in agriculture. Understanding the reproductive and other social behavior of domestic animals helps us raise them efficiently in large numbers. Knowledge of the communication system of certain insect pests is allowing the development of natural controls to increase agricultural yields while reducing our use of chemical pesticides.

Humans are extending their influence over the face of the Earth, altering it radically in the process. We are destroying the habitat and interfering with the social interactions necessary for the reproduction of thousands of different species. Many of these will become extinct unless a special effort is made to maintain conditions in which they can thrive and reproduce. For example, we have placed dams in many of the rivers in which salmon swim upstream to lay their eggs. To allow the fish to bypass the dams, people constructed "fish ladders" around the dams, only to find that the fish do not use them. Ethological studies showed that salmon tend to swim into the strongest current, leading them into the outflow from the dam rather than into the ladders. This has led to the design of better ladders and in some cases to netting and trucking the salmon around the dams. Without detailed knowledge of the salmon's homing behavior, there would probably be no salmon in the Columbia River today. In attempting to preserve endangered species from the rapidly disappearing tropical rain forests, knowledge of behavior is crucial. For example, we must learn how much territory each species requires to forage and reproduce successfully. This will allow us to establish a minimum size for preserves.

Another practical reason to study animal behavior is for the insight it can provide into the workings of the nervous system. Reflexes in vertebrates and fixed action patterns such as feeding and swimming in invertebrates are used to probe the mechanisms by which nerve cells interact to produce behavior. Inadequate understanding of behavior can lead to inappropriate experiments, slowing our progress in understanding the nervous system.

Finally, the techniques of detached observation necessary to understand other animals can yield valuable insights into human behavior. One of the greatest contributions of ethology has been the realization that all animals, including humans, are constrained in their behavior. Our options are extremely broad, but they are still circumscribed by the structure of our minds and bodies. The study of animal behavior can help us to decide what questions to ask about human behavior, and how to ask them, in order to learn more about ourselves. Human behavior will be explored further in the following chapter.

SUMMARY OF KEY CONCEPTS

The Genetic Basis of Behavior

All behavior has some genetic basis. Both the complexity of the nervous system that directs the behavior and the physical structure of the body that performs the behavior are genetically determined. Some types of learning, such as imprinting, are rigidly constrained by the genes. The relative contributions of heredity and learning to behavior vary between animal species, and between behaviors within the individual.

Innate Behavior

Innate, or instinctive, behaviors have several possible advantages over learned behaviors. First, they allow the animal to respond properly the first time it encounters a stimulus. Second, they allow animals with simple nervous systems to perform behaviors too complex for them to learn. Third, innate behavior causes members of the same species to perform and respond correctly during courtship and mating.

Innate behaviors fall into four categories. Animals showing kineses orient by varying the speed of essentially random movements, stopping when they encounter favorable conditions. In contrast, taxes are directed movements toward or away from specific stimuli. Reflexes are rapid movements of part of the body. Some, such as the knee-jerk reflex, are directed by the spinal cord and do not involve higher brain centers.

A fixed action pattern is a complex innate behavior elicited by a specific stimulus called a releaser. Learning can sometimes modify the releasers for fixed action patterns. Fixed action patterns can be distinguished from learned behaviors (1) if they are performed without prior experience, (2) if manipulating the releaser can result in inappropriate behavior, and (3) if a genetic basis can be determined through breeding experiments.

Learning

Learning is the ability to make changes in behavior as a result of experience. Learning is especially adaptive in changing and unpredictable environments and may modify innate behavior to make it more appropriate. Forms of learning include imprinting, habituation, conditioning, trial and error, and insight.

Imprinting occurs during a genetically programmed sensitive period. It often involves attachment between parent and offspring, or learning the features of a future mate. Habituation is the decline in response to a harmless stimulus which is repeated frequently. It often modifies innate escape or defensive responses. During classical conditioning, an animal learns to make a reflexive response, such as withdrawal or salivation, to a stimulus that did not originally elicit that response. During operant conditioning, an animal learns to make a new response, such as pressing a button, to obtain a reward or avoid punishment. Trial-and-error learning may modify innate behavior or may produce new behavior as a result of rewards and punishments provided by the environment. Insight can be considered a form of mental trial-and-error learning. An animal showing insight makes a new and adaptive response to an unfamiliar situation.

The Instinct to Learn and the Learning of Instincts

In animals with complex nervous systems, learning and instinct interact to produce adaptive behavior. Certain types of learning occur instinctively, during a rigidly defined time span. Imprinting, song acquisition in some birds, and location of the hive by bee foragers are examples.

Instinctive responses are often relatively unselective. Examples include baby gulls pecking at long thin objects, toads feeding on all flying insects, and chicks crouching whenever large birds fly overhead. Learning allows animals to modify these innate responses so that they occur only to appropriate stimuli.

GLOSSARY

Classical conditioning a training procedure in which an animal learns to make a reflexive response (such as salivation) to a new stimulus which did not elicit that response originally (such as a sound). This is accomplished by pairing a stimulus which elicits the response automatically (in this case, food) with the new stimulus.

Ethology (ē-thol′-ō-gē) the study of animal behavior under natural or near-natural conditions.

Fixed action pattern stereotyped, rather complex behavior which is genetically programmed (innate); often triggered by a stimulus called a releaser.

Habituation (heh-bich-ū-ā′-shun) simple learning characterized by a decline in response to a harmless, repeated stimulus.

Imprinting (im′-prin-ting) the process by which an animal forms an association with another animal or object in the environment during a sensitive period.

Innate (in-nāt′) inborn; instinctive; determined by the genetic makeup of the individual.

Insight a complex form of learning in which the solution to a problem is reached through reasoning.

Instinctive innate; inborn; determined by the genetic makeup of the individual.

Kinesis (kin-ne-sis) an innate process by which an organism achieves an orientation to a stimulus by altering its speed of movement in response to the stimulus.

Learning an adaptive change in behavior as a result of experience.

Operant conditioning a laboratory training procedure in which an animal learns to make a response (such as pressing a lever) through reward or punishment.

Reflex a simple, stereotyped movement of part of the body that occurs automatically in response to a stimulus.

Releaser a stimulus that triggers a fixed action pattern.

Supernormal stimulus a stimulus that exaggerates crucial elements of the releaser, making it more effective than the normal releaser.

Taxis (tax′-is) innate movement of an organism toward or away from a stimulus such as heat, light, or gravity.

Trial-and-error learning process by which adaptive responses are learned through rewards or punishments provided by the environment.

STUDY QUESTIONS

1. Contrast the approach to animal behavior of B. F. Skinner and N. Tinbergen. Name the behavioral "school" of each and describe an important contribution of each scientist.
2. List four types of innate behavior and provide an example of each.
3. Distinguish between taxes and kineses.
4. What is a stimulus called that elicits a fixed action pattern? Give three examples.
5. Assume that you observe two male fish of the same species fighting whenever they approach one another. Design a series of experiments to determine whether this behavior is innate or learned.

6. Assume that your experimental results in question 5 support the hypothesis that fighting in these fish is a fixed action pattern. Design a series of experiments to identify the releaser for this behavior.
7. Compare classical conditioning, operant conditioning, and trial-and-error learning. Give an example of each.
8. Compare insight with trial-and-error learning.
9. Give examples in which habituation and trial-and-error learning make an innate response more appropriate.
10. Explain why imprinting could be described as genetically controlled learning.

SUGGESTED READINGS

Alcock, J. *Animal Behavior,* 3rd ed. Sunderland, Mass.: Sinauer Associates, Inc., 1983. An introduction to animal behavior with an evolutionary approach.

Benzer, S. "Genetic Dissection of Behavior." *Scientific American,* December 1973. Mutant fruit flies are used to study the genetic bases of behavior.

Boycott, B. B. "Learning in the Octopus." *Scientific American,* March 1965. Long- and short-term memory are investigated using the learning ability of the octopus.

Grier, J. W. *Biology of Animal Behavior.* St. Louis: Times Mirror/Mosby, 1984. An up-to-date, complete, and clear presentation of all aspects of animal behavior.

Gould, J. L., and Marler, P. "Learning by Instinct." *Scientific American,* January 1987. What behaviors animals can learn is often influenced by their genes.

Lorenz, K. *King Solomon's Ring: New Light on Animal Ways.* New York: Thomas Y. Crowell, 1952. Beautifully written and full of interesting anecdotes, provides important insights into early ethology.

Tinbergen, N. *The Study of Instinct.* 1951. Introduction to the work of an experimental ethologist.

34
Animal Behavior II. Principles of Social Behavior

Your picnic is complete—ants have discovered the bread crumbs and are advancing single file, then returning by the same path, booty held high. The activity seems efficient and purposeful. Its apparent intelligence is surprising in an animal with a relatively simple nervous system. The ant "bread line" passes over a fallen leaf. When you lift the leaf, immediately the scene changes from one of purposeful intelligence to one of blind chaos! Ants reaching the point where the leaf used to lie abruptly stop their march, searching back and forth. Ants coming and going mill in apparent confusion until the leaf-sized gap is crossed by chance and the march resumed. You have witnessed complex social behavior in a simple organism. Removing the leaf disrupted the communication system—a chemical trail laid by successful foragers—on which this behavior is rigidly based.

COMMUNICATION

Social behavior is exhibited to some degree by all but the simplest organisms. The basis of all social behavior is communication, and the ultimate goal is survival and reproduction. **Communication** may be defined as the production of a signal by one organism that causes another to change its behavior in a way beneficial to one or both. Although communication can cross species lines, most occurs between members of the same species. Because members of the same species have the same needs, it is these individuals who compete most directly with one another for food, space, and mates. Communication must resolve the conflicts that result from these competitive interactions while minimizing harmful encounters.

The mechanisms by which animals communicate are astonishingly diverse and utilize all the senses. In this chapter we discuss basic types of communication, consider some of the mechanisms for resolving competitive interactions, and illustrate some complex, cooperative societies.

Visual Communication

Animals with well-developed eyes, from insects to mammals, use vision to communicate. Visual signals may be **active,** in which a specific movement or posture conveys a message (Fig. 34-1). Alternatively, they may be **passive.** In passive visual signals, the size, shape, or color of the animal conveys important information, often concerning its reproductive state. Active and passive signals may be combined, as illustrated by the lizard in Figure 34-2. Like all forms of communication, visual signals have both advantages and problems. On the plus side, they are instantaneous. The rapidity with which they can be sent and received allows them to impart a great deal of information in a short time. Visual signals may be graded, allowing the animal to convey the intensity of its motivational state (Fig. 34-3). Visual communication is quiet and unlikely to alert distant predators. A negative feature is that the signaler makes itself conspicuous to nearby predators. Visual signals are limited to close-range communication and are not effective in the dark (fireflies excepted), through murky water, or in dense vegetation.

Figure 34-1 Drawings by Darwin, a perceptive student of animal behavior, showing aggressive and submissive displays of dogs. **(a)** *Upright stance, erect hair, ears, and tail, and a direct stare combine to make the aggressor appear formidable indeed.* **(b)** *Notice how these displays are reversed in the submissive pose.*

Figure 34-2 Aggressive display of the South American Anolis lizard. Raising his head high in the air, he extends a brilliantly colored throat pouch, a warning to others to keep their distance.

Communication by Sound

The use of sound overcomes many of the deficiencies of visual displays. Although sound can alert predators, an animal may call from the safety of a burrow, or the camouflage of a dense thicket, thus remaining inconspicuous. Sound is almost instantaneous and can convey a variety of messages. It can be transmitted through darkness, dense vegetation, and water. If the animal is sufficiently energetic, its call may carry much farther than the eye can see. For example, the howls of a wolf pack can be heard for miles on a still night. It is believed that the low, resonant signal of the humpback whale may be detected by other whales hundreds of miles away. Like visual signals, sound can be graded in intensity, thus communicating information about the motivation of the signaler. For example, the louder a bird sings, the more capable

Figure 34-3 The wolf signals increasing aggression by gradually lowering the head, ruffling the fur on its neck and along its back, facing the opponent with a direct stare, and exposing its fangs.

he is of defending a territory. From the whine of the mosquito to the intricacy of bird song to the virtuoso performance of human language, sound is one of the most important media of communication.

Communication by Chemicals

Chemical substances produced by an individual that influence the behavior of others of its species are called **pheromones.** Chemicals may carry messages over long distances, and, unlike sound, take very little energy to produce. Pheromones may not even be detected by other species, including predators, who might be attracted to visual or auditory displays. Like a signpost, a pheromone persists over time and can convey a message after the animal has departed. As anyone who has walked a dog can attest, dog urine carries such a chemical message. Wolf packs, hunting over areas up to 1000 square kilometers, mark the boundaries of their travels with pheromones in urine, thus warning other packs of their presence.

Generally, fewer different messages are communicated with chemicals than with sight or sound, since the animal must have the ability to synthesize and respond to a different chemical for each message. Thus pheromone signals are unlikely to possess the variety and gradation of auditory or visual signals.

Pheromones may act in one of two ways. **Releaser pheromones** cause an immediate, overt behavior in the animal detecting them. They convey messages such as "this area is mine" or "I am ready to mate now." The ants mentioned in the opening paragraph were responding to a releaser pheromone laid down in a trail by successful foragers. Its message: "food is this way." **Primer pheromones** cause a physiological change in the animal detecting them. They may enter through the nose or they may actually be eaten by the receiver of the message. Most primer pheromones affect the reproductive state of the receiver. They are crucial to the maintenance of complex insect societies such as those of termites, ants, and bees. In the honeybee, the queen produces a primer pheromone, called **queen substance,** that is eaten by her hive mates and prevents other females in the hive from becoming sexually mature. Primer pheromones are found in mammals such as mice. In certain mice the urine of mature males contains a primer pheromone that influences female reproductive hormones. This pheromone initiates the first estrus of a newly mature female. It will also cause a female mouse newly pregnant by another male to abort her litter and become sexually receptive to the new male. There is indirect evidence (discussed later) that primer pheromones may even influence human reproductive cycles.

Some pheromones have been chemically analyzed and synthesized in the laboratory. The sex attractant pheromones of agricultural pests such as the Japanese beetle and gypsy moth have been synthesized. They can be used to lure these insects into traps, or sprayed in an infested area to confuse and disrupt mating. The advantages of pest control using pheromones are enormous. Unlike pesticides, which kill beneficial as well as harmful organisms, pheromones are specific to one species and harmless to others. In contrast to pesticides, insects will never become resistant to the attraction of a pheromone. Hopefully, as our understanding increases, so will our use of such natural biological controls.

Communication by Touch

Physical contact between individuals is used in several ways, one of the most important being to establish and maintain social bonds among group members. Primates, including humans, are "contact species" in which a variety of gestures, including kissing, nuzzling, patting, petting, and grooming, play an important social function (Fig. 34-4a). In wolves and dogs, a greeting ceremony involves mutual licking, sniffing, and gentle nipping around the mouth. The bond between parent and offspring is often cemented by close physical contact, and

(a)

(b)

Figure 34-4 **(a)** *A female olive baboon grooms a male. Grooming not only reinforces social relationships but also removes debris and parasites from the fur.* **(b)** *Touch is also important in sexual communication. These land snails* (Helix) *engage in courtship behavior which will culminate in mating.*

sexual activity is frequently preceded by ritualized contact (Fig. 34-4b).

Much communication cannot be categorized simply as sight, sound, scent or touch. Many social interactions use a combination of these as illustrated by the waggle dance of the honeybee (see Fig. 34-18). Nearly all social interactions, however, can be classified according to their functions in the lives of animals. In the following sections we discuss communication used as animals compete for limited resources, reproduce, and cooperate in complex societies.

Figure 34-5 The aggressive display of the male fighting fish includes elevating the fins and flaring the gill covers, thus making the body appear larger.

COMPETITION FOR RESOURCES

Aggression

One of the most obvious manifestations of competition for resources such as food, space, or mates is aggressive behavior between members of the same species. Although the expression "survival of the fittest" evokes images of the strongest animal emerging triumphantly from among the dead bodies of its competitors, in reality most aggressive encounters between members of the same species are rather harmless. Natural selection has favored the evolution of symbolic displays or rituals for resolving conflicts. One reason may be that in a serious fight, even the victorious animal may be injured. Serious fighters are therefore less likely to survive and propagate their genes. Aggressive displays, in contrast, allow the competitors to assess each other and acknowledge a winner on the basis of size, strength, and motivation rather than on wounds inflicted. Visual aggressive displays exhibit the weapons, such as fangs and claws, and often include behaviors designed to make the animal appear larger. These include standing upright to increase apparent height, erecting the fur, feathers, ears, fins, or the entire body (see Figs. 34-1a, 34-2, 34-3, and 34-5). The displays may be accompanied by intimidating sounds (growls, croaks, roars, chirps) whose intensity can also be a factor in deciding the winner. Fighting is usually a last resort when displays fail to resolve the dispute.

In addition to visual and vocal displays of strength, many animal species engage in ritualized combat. Deadly weapons may clash harmlessly or may not be used at all (Fig. 34-6). Frequently, these encounters involve shoving rather than slashing. Again, the strength and motivation of the combatants is determined and the loser slinks away in a submissive posture that minimizes the size of its body (see Fig. 34-1b).

Figure 34-6 Ritualized combat of male impalas, a type of African antelope. The deadly horns, which could gore a predator, clash harmlessly. Eventually one impala, sensing greater vigor in his opponent, will retreat unharmed.

Dominance Hierarchies

Aggressive interactions use a great deal of energy, may cause injury, and can disrupt other important tasks, such as finding food, watching for predators, courting a mate, or raising young. Thus many animals have evolved ways to resolve competition that minimize aggression. One example is the **dominance hierarchy,** in which each animal establishes a rank that determines its social status. Domestic chickens, after a period of squabbling, sort themselves into a reasonably stable "pecking order." Thereafter, when competition for food occurs, all hens defer to the dominant bird, all but she give way to the second, and so on. Conflict is minimized because each bird knows its place. Dominance in male bighorn sheep can be recognized by their horn size (Fig. 34-7). Wolf packs are organized so that each sex has a dominant or "alpha" individual to whom all others are subordinate. Although aggressive encounters occur frequently during the establishing of a hierarchy, they are subsequently reduced by the submission of subordinate individuals. The dominant individuals obtain most access to the resources needed for reproduction, including food, space, and mates.

Figure 34-7 The dominance hierarchy of the male bighorn sheep is signaled by the size of the horns; these rams increase in status from right to left. These backward-curving horns are clearly not designed to inflict injury and are used in ritualized combat.

Territoriality

Territoriality is the defense of an area where important resources are located. The defended resources may include places to mate, raise young, feed, or store food. Territorial animals generally restrict some or all of their activities to the defended area and advertise their presence there. Territories may be defended by males, females, a mated pair, or by entire social groups (as in the case of nest defense by social insects). However, territorial behavior is most often seen in adult males, and territories are usually defended against members of the same species, those who compete most directly for the resources being protected. Territories are as diverse as the animals defending them. They include the patch of algae carefully cultivated by a male fish as a spawning site, a hole in the sand used as a home by a crab, a tree where a woodpecker stores acorns (Fig. 34-8), or an area of forest providing food for a squirrel.

Acquiring and defending a territory requires considerable time and energy, yet territoriality is seen in diverse animals, including worms, arthropods, fish, birds, and mammals. This striking example of convergent evolution suggests that territoriality provides some important advantages. Although the benefits depend on the species and the type of territory it defends, some broad generalizations are possible. First (as with dominance hierarchies), once a territory is established through aggressive interactions, relative peace prevails as boundaries are recognized and respected. The saying "good fences make good neighbors" also applies to nonhuman territories. One reason for this respect is that an animal is highly motivated to defend its territory, and will often defeat larger, stronger animals if they attempt to invade it. Conversely, an animal outside its territory is much less secure and more easily defeated. This principle was demonstrated by Niko Tinbergen using the stickleback fish (Fig. 34-9).

Second, territoriality in some species limits the population, helping keep it within the limits set by the available resources. For example, the Scottish red grouse defends feeding and nesting territories. The size of these territories, and thus the number of individuals in a given area, was found by scientists to vary from year to year. The variations coincided with the availability of food. In years when food was abundant, territories were smaller, allowing higher population density. In lean years, territories were larger, allowing fewer pairs to breed. The effect is to maintain the population at a level that the particular habitat can support.

Territories are advertised through sight, sound, and smell. If the territory is small enough, the owner's mere presence, reinforced by aggressive displays at intruders, may be sufficient defense. In mammals, when the owner cannot always be present it may scent-mark the boundaries using pheromones. Male rabbits use pheromones secreted by chin and anal glands to mark their territories. Hamsters rub the areas around their dens with secretions from a special gland on their flanks.

Vocal displays are a common form of territorial advertisement. Male sea lions defend a strip of beach by swimming up and down in front of it, calling continuously. Male crickets produce a specific pattern of chirps to warn other males away from their burrows. Birdsong is a strik-

Figure 34-8 The acorn woodpecker excavates acorn-sized holes in dead trees, stuffing them with green acorns for dining during the lean winter months. He defends the trees vigorously against other acorn woodpeckers and against acorn-eating birds of other species, such as jays.

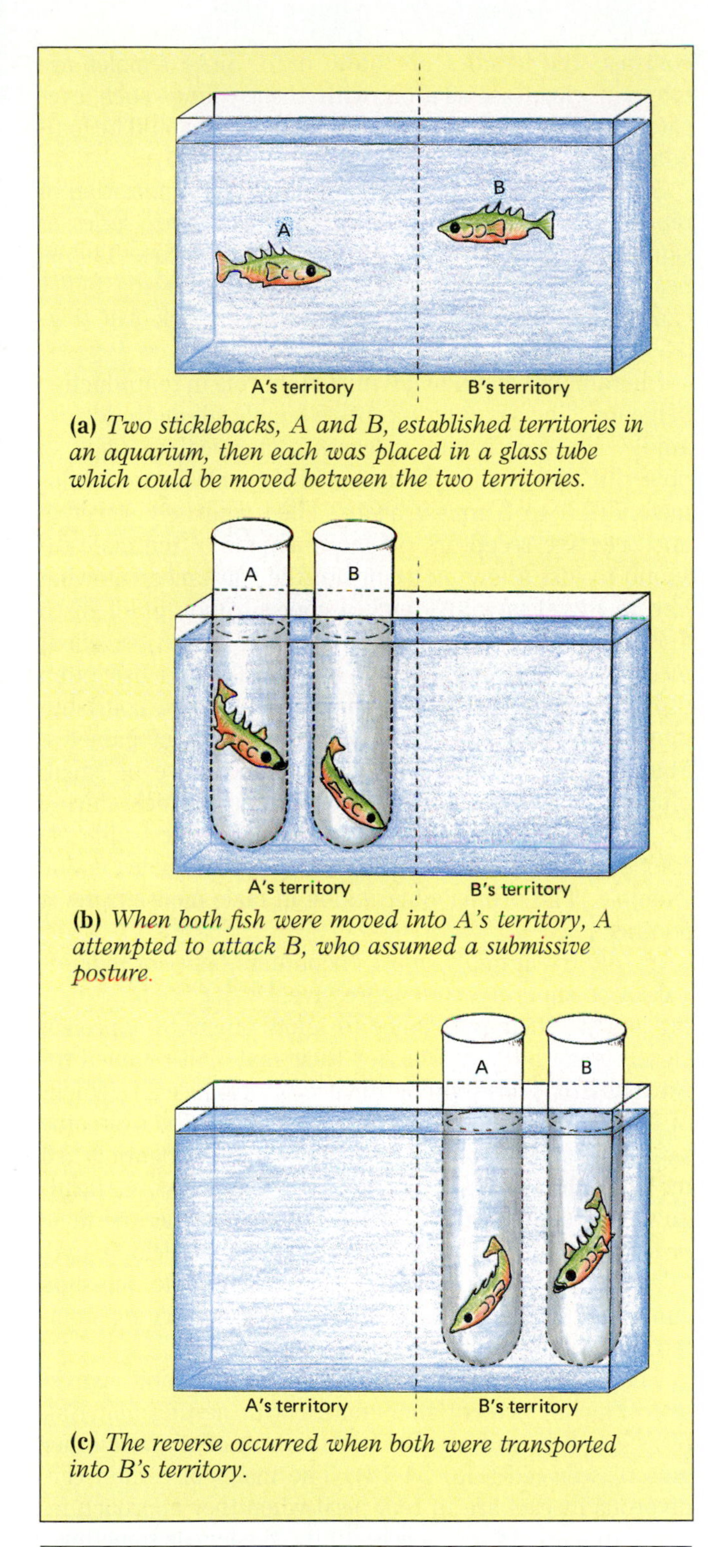

(a) *Two sticklebacks, A and B, established territories in an aquarium, then each was placed in a glass tube which could be moved between the two territories.*

(b) *When both fish were moved into A's territory, A attempted to attack B, who assumed a submissive posture.*

(c) *The reverse occurred when both were transported into B's territory.*

Figure 34-9 *Niko Tinbergen's experiment demonstrating the effect of territory ownership on aggressive motivation.*

ing example. The cheerful melody of the male meadowlark is part of an aggressive display, warning other males to steer clear of his territory (Fig. 34-10). When songbirds migrate into an area in the spring to nest and breed, males move to unoccupied areas and begin singing vigorously.

Figure 34-10 *A male meadowlark announces ownership of his territory to all listeners.*

This is accompanied by much jockeying for position as the most aggressive compete for prize nesting and feeding areas. The loudest singer generally defends the largest territory and is most successful at driving away intruders.

REPRODUCTION

Successful reproduction requires that several criteria be satisfied. Animals must identify one another as members of the same species, as members of the opposite sex, and as being sexually receptive. Many animals resist the close approach of another individual; this must be overcome before mating can occur. Some animals, such as frogs and many fish, must release eggs and sperm at precisely the same moment for fertilization to occur. The need to fulfill all these requirements has resulted in exceedingly complex, diverse, and fascinating courtship behavior.

Individuals who mate with members of other species, or members of the same sex, waste considerable energy and do not pass on their genes. Thus animals have evolved elaborate ways to communicate their species and sex, often using vocalizations. The raucous nighttime chirping that can keep campers awake is probably a chorus of male tree frogs, each singing a species-specific song. Male grasshoppers and crickets advertise species and sex by their calls, as does the female mosquito with her high-pitched whine. Male birds use song to attract a mate as well as defend a territory. For example, the male bellbird of Australia uses its deafening call to defend large terri-

Figure 34-11 **(a)** *The extravagant tail of the male peacock is displayed during courtship. This oversized tail hampers flight and increases the male's vulnerability to predators. It probably evolved as females consistently selected the most flashy birds as mates, preferring this exaggerated releaser to a more practical tail.* **(b)** *The male frigate bird of the Galápagos Islands inflates a scarlet throat pouch to attract passing females.*

tories and attract females from great distances. The females fly from one territory to another, alighting near the male in his tree. The male, beak gaping, leans directly over the flinching female and utters an ear-shattering note. It is believed that the female endures this to compare volumes of the various males, choosing the loudest (who would also be the best defender of a territory) as a mate. In other species, male and female birds join in elaborate duets that help to synchronize reproductive readiness and cement the bond between them.

Many species court using visual displays. The firefly, for example, flashes a message identifying its sex and species. Male fence lizards bob their heads in a species-specific rhythm, and females distinguish and prefer the rhythm of their own species. The tail of the male peacock and the scarlet throat of the male frigate bird serve as flashy advertisements of sex and species (Fig. 34-11). In contrast, the females are quite drab. Since females are often in close association with their young, such eye-catching (and predator-attracting) displays would be maladaptive.

Species and sex recognition and the synchronization of reproductive behavior often require a complex series of signals, both active and passive, by both sexes. This is beautifully illustrated by the complex underwater ballet executed by the male and female stickleback fish (Fig. 34-12).

Pheromones can play an important role in reproductive behavior. The sexually receptive female silkmoth, for example, sits quietly and releases a chemical message so powerful that it may be detected by males 4 to 5 kilometers (2.5 to 3 miles) away. The exquisitely sensitive and selective receptors on the antennae of the male respond to just a few molecules of the substance, allowing him to travel upwind along a concentration gradient to find the female (Fig. 34-13). Water is an excellent medium for dispersing chemical signals, and fish often use a combination of pheromones and elaborate courtship movements to ensure synchronous release of gametes. Mammals, with their highly developed sense of smell, often rely on pheromones released by the female during her fertile periods to attract males. The irresistible attraction of a female dog in heat to nearby males is one example. The primer pheromone in male mouse urine is another.

In solitary animals, most encounters between individuals are competitive and aggressive. During the brief mating season, their reluctance to allow others to approach closely must be overcome for their mate but retained toward others. These conflicting needs introduce an element of tension into sexual encounters that may be overcome by submissive signals. The female Siamese fighting fish mollifies the aggressive male with a submissive, head-down posture. In several species of birds, either the male or female defuses aggressive impulses by mimicking juvenile behavior such as begging. Courting male hamsters emit high-pitched cries like baby hamsters, eliciting a maternal response from the female.

The presentation of gifts also seems to inhibit aggression. Finches present their mates with nesting material, terns give fish, and flightless cormorants offer one another gifts of seaweed (Fig. 34-14). The males of certain carnivorous insects are in real peril when they approach females to mate. In one species of fly, the female sometimes devours the male when he makes sexual advances. The males of a closely related species have solved the problem by presenting the female with a dead insect. While she is busily eating, the male mates and runs. In another species, the male gains extra time by wrapping the insect in a silken web that she is forced to unwrap. A third species merely wraps up a fragment of insect.

This group of related fly species has provided insight into the evolution of complex and seemingly meaningless

a) A male, inconspicuously colored, leaves the school of males and females to establish a breeding territory.

b) As his belly takes on the red color of the breeding male, he displays aggressively at other red-bellied males, exposing his red underside.

c) Having established a territory, the male begins nest construction by digging a shallow pit which he will fill with bits of algae cemented together by a sticky secretion from his kidneys.

d) After tunnelling through the nest to make a hole, his back begins to take on the blue courting color which makes him attractive to females.

e) An egg-carrying female displays her enlarged belly to him by assuming a head-up posture. Her swollen belly, and his courting colors are passive visual displays.

f) He leads her to the nest using a zig-zag dance.

g) After she enters, he stimulates her to release eggs by prodding at the base of her tail.

h) He enters the nest as she leaves and deposits sperm to fertilize the eggs.

Figure 34-12 *Courtship of the three-spined stickleback.*

Figure 34-13 The antennae of the male silkmoth are plume-like structures specialized to detect the female's sex pheromone.

Figure 34-15 **(a)** *A male balloon fly carrying a silken balloon.* **(b)** *He presents it to the female prior to mating.*

Figure 34-14 A flightless cormorant from the Galápagos Islands returns to its nest bearing a "gift" of seaweed for its mate, who will aggressively snatch it away. A bird who fails to bring a gift will be driven away by its mate. The gift appears to allow the nesting bird to take out its aggressive impulses harmlessly.

courtship rituals. One species feeds on plants, so mating is no longer a death-defying act for the male. Observers were mystified to see these males hovering in the air carrying empty silken sacs that they presented to females prior to mating (Fig. 34-15). Study of the related carnivorous species has led to the hypothesis that the behavior first evolved in carnivorous ancestors. Although the insect's diet changed, the presentation of a gift was retained as a courtship ritual.

COOPERATION

Social interactions have evolved because they help the individual and its offspring survive. From this perspective, "opposite" behaviors such as aggression and cooperation are actually alternative means of accomplishing the same end.

The degree to which animals cooperate varies significantly from one species to the next. Many are basically solitary, such as the mountain lion and grizzly bear. Interactions between adults are restricted to brief aggressive encounters and mating. Some animals cooperate based on changing needs. For example, the coyote is solitary when food is abundant, but hunts in packs when food becomes scarce, as in the winter. Some animals form loose social groupings such as herds of musk oxen (Fig. 34-16), pods of dolphins, schools of fish, and flocks of birds. These groups are better able to deter predators and find food. At the far end of the social spectrum are a few highly integrated cooperative societies found primarily among the insects and mammals. We now focus on a few examples of these highly complex societies.

Figure 34-16 Cooperation in loosely organized social groups. A herd of musk oxen functions as a unit when threatened by predators such as wolves. Males form a circle, horns pointed outward, around the females and young.

Insect Societies

The most rigidly organized, most complex societies (humans excepted) are found among the social insects. In these communities, the individual is a cog in an intricate, smoothly running machine. It could no more function on its own than could a spark plug ripped from an engine. Social insects are born into one of several castes within the society. These castes are groups of similar individuals genetically programmed to perform a specific function.

HONEYBEES. Honeybees emerge from their larval stage into one of three major preordained roles. One role is that of queen. Only one queen is tolerated in a hive at any time. Her functions are to produce eggs (up to 1000 per day for a lifetime of 5 to 10 years) and to regulate the lives of the workers. Male bees, called drones, serve merely as sex objects for the queen. Lured by her sex pheromones, drones mate with the queen during her first week of life, perhaps as many as 15 times. This relatively brief orgy supplies her with sperm which will last a lifetime, enough to fertilize over 3 million eggs. Their sexual chore accomplished, the drones become superfluous and are eventually driven out of the hive or killed. The hive is run by the third class of bees, sterile female workers. The tasks of the worker are determined by her age and conditions in the colony (Fig. 34-17). The newly emerged worker starts as a waitress, carrying food such as honey and pollen to the queen, other workers, and developing larvae. As she matures, special glands begin wax production, and she becomes a builder, constructing perfectly hexagonal cells of wax where the queen will deposit her eggs and the larvae will develop. She will take a shift as maid, cleaning the hive and removing the dead, and as a guard, protecting the hive against intruders. Her final role in life is that of a forager gathering pollen and nectar, food for the hive. She will spend nearly half of her two-month life in this role. Acting as a forager scout, she will seek new and rich sources of nectar, and having found one will return to the hive and communicate its location to other foragers using the **waggle dance,** an elegant form of symbolic communication (Fig. 34-18).

Pheromones play a major role in regulating the lives of social insects. In the honeybee, drones are drawn irresistibly to the siren call of the queen's sex pheromone (queen substance), which she releases like a vapor trail during her mating flights. Back at the hive, she maintains her position as the only fertile female using the same substance (now acting as a primer pheromone). The queen substance is licked off her body and passed among all the workers, rendering them sterile. The queen's presence and health are signaled by her continuing production of queen substance; a decrease in production (which occurs normally in the spring) alters the behavior of the workers. Almost immediately they begin building extra large "royal cells" and feeding the larvae that develop in them a special glandular secretion known as "royal jelly." This unique food alters the development of the growing larvae so that, instead of a worker, a new queen emerges from the royal cell. The old queen will then leave the hive, taking a swarm of workers with her to establish residence elsewhere. If more than one new queen emerges, a battle to the death ensues, with the victorious queen taking over the hive.

Vertebrate Societies

Vertebrates possess far more complex nervous systems than insects, and one might therefore expect vertebrate

(a)

(b)

(c)

Figure 34-17 *Some stages in the life a worker bee.* **(a)** *Workers crowd around the queen (center), feeding her and licking the pheromone called "queen substance" from her body.* **(b)** *Workers construct hexagonal cells made of wax to enlarge the honeycomb. The wax is secreted by a gland in the abdomen, passed to the mouth and chewed to a workable consistency.* **(c)** *A forager collects pollen and nectar from a flower. Note the yellow pollen baskets on her legs.*

(a) *The wagging run is symbolic flight, and other foragers crowding around the dancer (identified here by her pollen baskets) in the dark hive hear the buzzing, feel the movements of her abdomen, and translate this into an approximate distance. The smell of the flowers on her body gives them an additional clue as to what to search for, but this is not all they learn.*

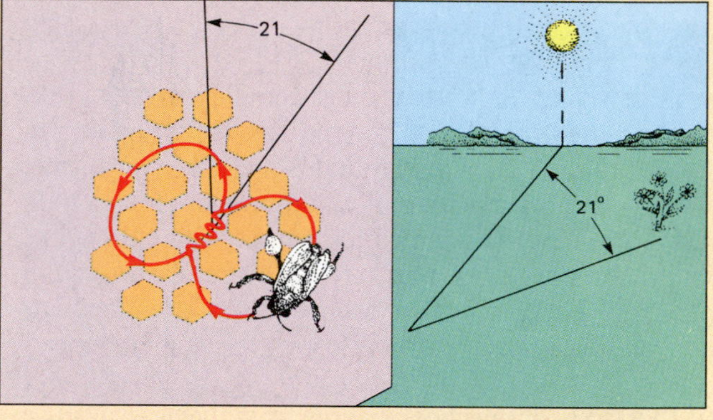

(b) *The forager also tells them in which direction to fly. The dance is normally performed on a vertical wall of the hive, and the angle that the straight run deviates from the vertical represents the angle between the sun and the flowers. In the dark hive, straight up symbolizes the sun, regardless of its actual location in the sky.*

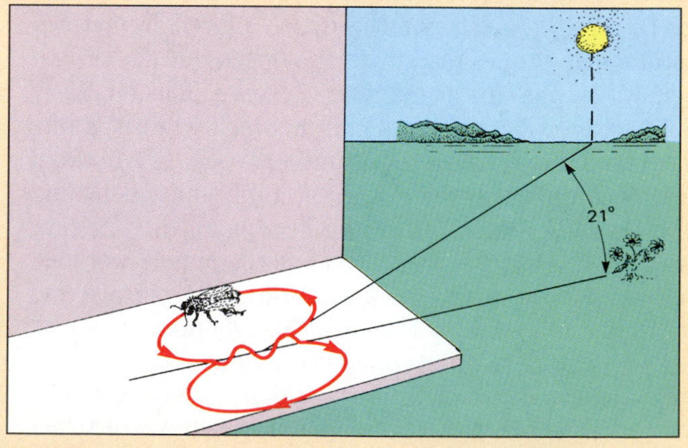

(c) *If the dance is performed on a horizontal surface outside, the straight run is aimed directly at the flowers. Both the performance of the waggle dance and the bees' ability to interpret it correctly are instinctive and can be performed correctly the first time by bees raised in isolation.*

Figure 34-18 *Symbolic communication in the honeybee. A forager, having discovered a distant source of nectar, returns to the hive and communicates its location to other workers using the waggle dance, first described by Karl von Frisch. The informative part of the dance is called the "straight run," during which the bee moves in a straight line while shaking her abdomen back and forth and making a buzzing sound with her wings. At the end of the straight run, the bee makes a semicircle that returns her to the starting point, repeats the run, and circles back in the opposite direction, performing the cycle over and over. The speed with which the straight run is performed is directly related to the distance of the nectar source from the hive. If the nectar is located 300 meters away, the run will last about 0.5 second; a 500-meter distance is communicated by a 1-second run, and so on.*

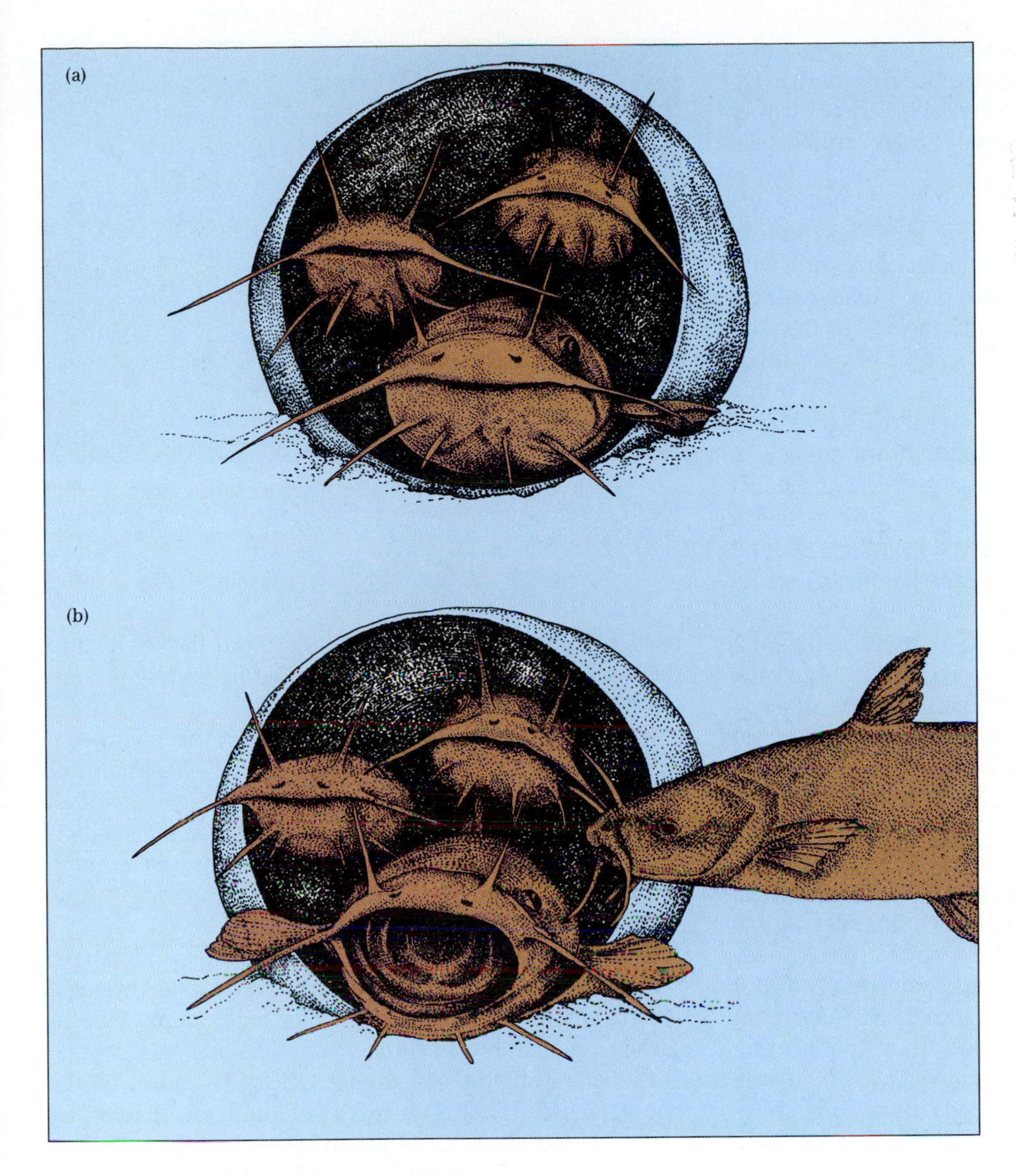

Figure 34-19 Three bullhead catfish occupy a section of pipe in a laboratory tank. The largest of the three fish is recognized by the other two as dominant. He is usually the exclusive occupant of the pipe, which is part of his territory. In **(a)**, *however, he has allowed two subordinate fish to seek refuge in the pipe after detecting the odor of an unfamiliar bullhead. The approach of the strange bullhead elicits an aggressive response from the dominant fish, as is shown by its gaping mouth display* **(b)**. *The tankmates are content to allow the dominant fish to attack the intruder, while they remain safely in the shelter.*

societies to be proportionately more complex. With the exception of human society, however, this is not the case. Perhaps because the vertebrate brain *is* more complex, vertebrate societies tend to be simpler than those of the social insects such as honeybees, army ants, and termites. Each individual is unique, and in vertebrates this uniqueness is enhanced because they exhibit more flexible learned behavior. Although much social behavior has an innate component, vertebrates show a great deal more flexibility (and thus unpredictability) and less of the robotic precision that makes complex insect societies possible.

BULLHEAD CATFISH. The social interactions of the bullhead, described by John Todd of the Woods Hole Oceanographic Institution in Massachusetts, provide a fascinating illustration of a relatively simple vertebrate in which complex social interactions are based almost entirely on pheromones. Todd observed these nocturnal fish in large aquaria in the laboratory. He discovered that when a group of them were housed together, territories were staked out, and a dominance hierarchy was established with the dominant fish defending the largest and best-protected area of the tank. Contests between tank mates consisted of aggressive displays with open mouths, alternately approaching and retreating. Once a fish became dominant, its aggressive displays caused the subordinate fish to flee. Actual violence occurred only when a stranger was introduced into a tank with an established hierarchy. In this case the established group also exhibited cooperative behavior. The dominant fish allowed others to take refuge in his protected territory, then ventured forth to engage the intruder in combat (Fig. 34-19). When the newcomer was defeated and the danger past, the dominant fish chased the others back out of his territory.

Todd discovered that blinding the bullheads did not

cause any appreciable change in their social interactions. When their sense of smell was temporarily destroyed, however, the fish acted like permanent strangers. Neither territories nor dominance hierarchies were established, and aggressive behavior was violent, continuing for weeks until their sense of smell returned. Bullheads clearly recognize one another as individuals by scent and have a long memory. A pair of young bullheads in Todd's laboratory were severely thrashed by a large bullhead who jumped from a neighboring tank into theirs. For the next four months, whenever Todd put a sample of water from the "bully" bullhead's tank into theirs, the two young bullheads ignored their territories and hid together in a protected part of the tank until the odor disappeared. The scent of other fish did not affect them this way. The status of an individual is communicated by scent, as is a change in status. If a dominant fish is removed from his tank and later returned, usually both his territory and his status are remembered and respected by his tankmates. However, if he is removed and subjected to defeat in the tank of a more aggressive fish, his pheromones are somehow altered. Upon return to his home tank, he will be attacked by his former subordinates, who smell the change caused by his defeat.

Under certain circumstances in the wild, or in the laboratory when a large number of newly caught fish are placed in the same tank, bullheads may form a dense and peaceful community lacking territories or dominance hierarchies. Todd established such a community in one tank, and placed a pair of aggressive rival fish in an adjacent tank. When "community water" was pumped continuously from the community tank into the adjacent tank containing the aggressive bullheads, they too became peaceful, only resuming their fighting when the flow was stopped. Under the crowded conditions of the community tank, an antiaggression pheromone is apparently produced, minimizing conflict.

PRAIRIE DOGS. The prairie dog "town" of the American western plains is a relatively complex mammalian society. Black-tailed prairie dogs, which are actually large rodents, live in towns of up to 1000 individuals, but the unit of social behavior is a small group known as the coterie. This usually consists of about 10 individuals, including a few adult males and females and some juveniles and younger pups. These share the same burrows and recognize one another as members of the same coterie. Together they defend a small territory around their burrow, advertising their ownership by barks delivered with such vigor that the animals sometimes fall over backward. The members of a coterie change from month to month as individuals are born, die, or emigrate, but the boundaries of the territory remain the same, passed on by learning from one generation to the next. Social ties are strengthened by a great deal of contact. When these social rodents encounter one another, they greet with an opened-mouth "kiss" (Fig. 34-20), and if they belong to the same coterie, this may be followed by mutual grooming of the fur.

Figure 34-20 *A greeting kiss is exchanged when two prairie dogs meet.*

Living in exposed areas of the plains, prairie dogs benefit from community life, where many eyes can watch for danger. When a predator such as a golden eagle is spotted by an alert "watch dog," it warns nearby animals with a distinctive yipping bark which is quickly passed through the town.

HAMADRYAS BABOONS. Among nonhuman mammals, the primates, our closest relatives, are among the most social. African Hamadryas baboons, for example, exhibit social organization on three levels. The first level, which is the smallest unit of baboon society, is known as the one-male unit. This is an extended family consisting of a male and his "harem," a group consisting of two to five females and their young. The male, who is twice the females' size, dominates them using slaps, bites, or aggressive stares. He leads the group from place to place and, like a protective father or a jealous husband, intervenes if any of his followers attempt to associate with strangers, fight among themselves, or stray too far from him. A young male baboon begins forming his unit by adopting or kidnapping juvenile females from their families. They remain with him permanently, becoming his mates when they reach sexual maturity. This harem system inevitably results in an excess of males, which gather together in all-male "bachelor units." Food gathering is performed independently by the various units in the baboon troop, probably because the acacia trees that supply food are too widely spaced to allow the entire troop to feed together. At the end of each day, several one-male units congregate to form a band in which they travel to water holes and sleeping cliffs. At the large sleeping cliffs, several bands may gather to form the largest social grouping, the troop, which can include up to 750 individuals. When aggressive

Figure 34-21 Threat display of a male baboon shows formidable weapons indeed. Despite potentially lethal fangs (seen only in males), aggressive encounters between baboons rarely result in injury.

interactions occur, for example when strange bands meet, the males are responsible for defending the band. Filming of these encounters has revealed that they consist largely of elaborate bluffs, much like the fights on TV westerns. The formidable fangs are prominently displayed and the hands are used in a slapping motion, but actual contact is infrequent (Fig. 34-21). By exposing his neck to his opponent (a submissive gesture), a defeated male quickly ends the dispute.

HUMAN ETHOLOGY

Students of animal behavior, observing a specific activity, seek its basis in the animal's genetic makeup and past experience and attempt to determine how the behavior helps the animal survive and reproduce. The behavior of all nonhuman animals is assumed to be influenced to some extent by heredity. Even if the behavior is acquired primarily through learning, human observers do not hesitate to postulate an inherited tendency to learn that type of behavior. This, however, is not always true of observers of human behavior, some of whom argue vigorously that our acts are uniquely unaffected by our genes and by the pressures of our environment over evolutionary history. Human behavior, they say, is entirely a product of learning. Opposing this viewpoint are the sociobiologists and human behavioral geneticists, who believe that many human tendencies have a genetic basis. Because it is new as a rigorous discipline, and because it deals with broad tendencies in large groups of diverse individuals, human ethology is still relatively speculative and inexact. To demonstrate that human behavior does have some innate basis, scientists have attempted to isolate the genetic components of human behavior. In the following sections we review some of these studies.

Studies of Young Children

One way to minimize the effects of learning is to observe the behavior of very young infants. Rhythmic movement of a baby's head in search of the mother's breast is a human fixed action pattern which may be observed during the first days after birth. Suckling, which can even be observed in the human fetus, is equally instinctive (Fig. 34-22). Other fixed action patterns seen in newborns or premature infants include walking movements when the body is supported and grasping with the hands and feet. Another example is smiling, which may occur soon after birth. Initially, smiling can be released by almost any object looming over the newborn. Before an infant is 2 months old, an exaggerated releaser (see Chapter 33) may be constructed consisting of two dark, eye-sized spots on a light background. This elicits smiling even more successfully than an accurate representation of a human face. As the child matures, learning and further development of the nervous system interact to limit the response to increasingly accurate representations of a face.

Another way to minimize the effects of learning is to

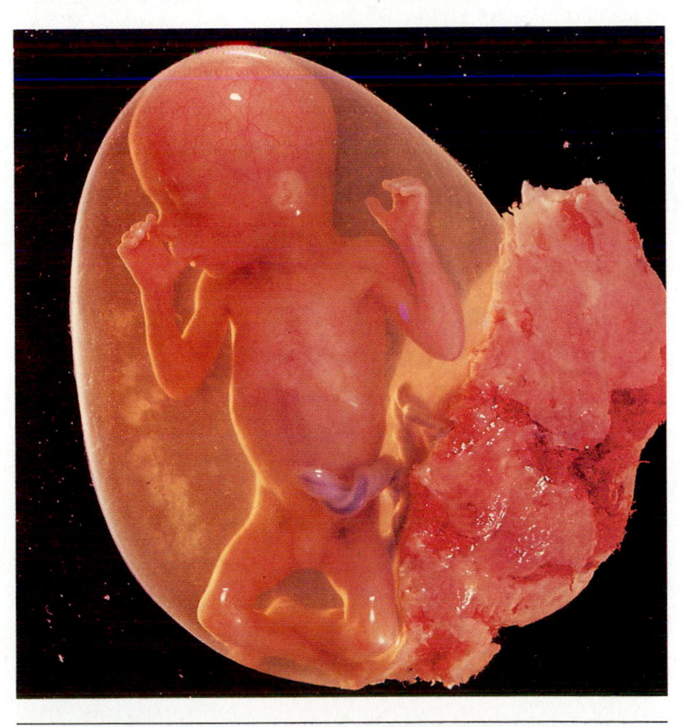

Figure 34-22 Thumb sucking is a difficult habit to discourage in young children, since suckling on appropriately sized objects is an instinctive, food-seeking behavior. This fetus (which is about 8 inches in length) sucks its thumb at the gestational age of 4½ months.

observe children who are blind, deaf, or both, and have thus been unable to learn through sight and sound. Without ever having seen or heard them, these children produce normal smiles and laughter, and expressions of frustration and anger.

Exaggerating Human Releasers

A behavior probably has an instinctive component if the stimulus that causes it (the releaser) can be exaggerated beyond the bounds of reality and elicit an even stronger response. The eyespots that cause young infants to smile are one example. In turn, the smile of an infant, along with certain characteristic baby features, may release protective feelings in adults. These features include a relatively large head with a domed forehead, chubby cheeks, small nose, short arms and legs, and a small, rounded body. Even 3-year-old children respond to these features with "mothering" behavior. The marketplace has exploited the releasing aspects of these features by exaggerating them in baby animals and people and using their innate appeal to sell dolls, posters, calendars, and cards (Fig. 34-23).

One human signal with an innate physiological basis is the involuntary enlargement of the pupil of the eye when viewing something pleasant, be it a loved one or a hot fudge sundae. We also react to this signal in others. To test this, male subjects were shown identical photographs of smiling women which had been retouched to enlarge or contract the pupils. They overwhelmingly preferred those with large pupils, although none were consciously aware of the pupil size or able to pinpoint the reason for their reaction. The positive response of men to enlarged pupils has long been recognized. In the middle ages women artificially enlarged their pupils by using the drug belladonna (meaning "beautiful woman" in Italian). We react positively to someone who gazes at us with dilated pupils (although our recognition of this feature is subconscious), since it implies interest and attraction.

Figure 34-23 *We instinctively respond to certain features associated with infants and very young children. These are sometimes exaggerated to produce supernormal stimuli, as in these dolls.*

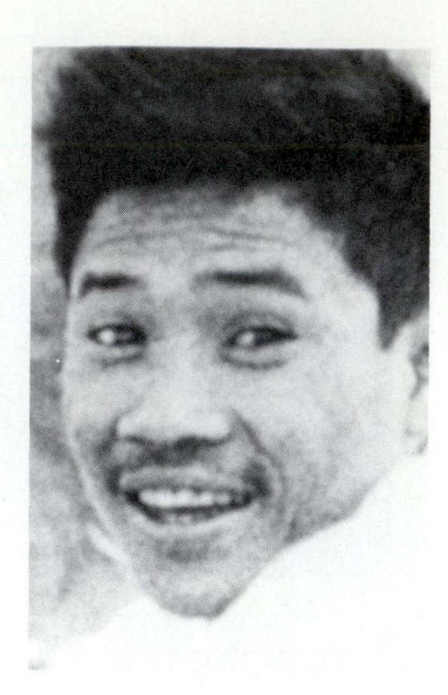

Figure 34-24 *Gestures that have similar meanings in diverse and isolated cultures may be evidence of a common biological heritage. Here motion pictures freeze the "eye-flash greeting" (in which the eyes are widely opened and the eyebrows rapidly elevated) in a person from New Guinea (left) and Bali (right). Watch for this possibly innate response in yourself when you encounter a friend.*

Comparative Cultural Studies

Another way to study the instinctive bases of adult human behavior is to compare simple acts performed by people from isolated and diverse cultures. This comparative approach, pioneered by the ethologist Eibl-Eibesfeldt, has revealed several gestures which seem to form a universal, and therefore probably innate, human language. Such gestures include a variety of facial expressions for pleasure, rage, and disdain, and movements such as the "eye flash" and a hand upraised in greeting (Fig. 34-24).

Human Pheromones?

Humans may have unconscious responses to pheromones. Our sense of smell is poorly developed compared to that of many other mammals, and the role of odor as a means of human communication is largely unknown. However, an interesting study by Martha McClintock of Harvard provided indirect evidence that primer pheromones may influence female reproductive physiology. She studied the menstrual cycles of 135 women living in a college dormitory and found that the menstrual cycles of roommates and close friends became significantly more synchronous over a six-month period. The cycles of women randomly chosen from the dormitory did not. In a further study, she divided the women into groups based on how often during an average week they associated with men. She found that the women who reported spending time with men less than three times weekly had significantly longer cycles than those who saw men more frequently. Although far from conclusive, these findings are tantalizing and call for further investigation of the role of both primer and releaser pheromones in human behavior.

Studies of Twins

By studying identical and fraternal twins, investigators can come as close as possible to controlled breeding experiments in humans. Fraternal twins arise from two individual eggs and are no more similar genetically to each other than they are to other siblings. However, they are exactly the same age and share a very similar environment. Identical twins, arising from a single fertilized egg, have identical genes. The most fascinating twin findings are based on anecdotal observations of identical twins separated soon after birth, reared in different environments, and reunited for the first time as adults. They have been found to share nearly identical taste in jewelry, clothing, humor, food, and names for children and pets. Personal idiosyncrasies such as giggling, nail biting, drinking patterns, hypochondria, and mild phobias may be shared by these unacquainted twins. More rigorous studies are also supporting the heritability of many human behavioral traits. These have documented a significant genetic component for traits such as activity level, alcoholism, sociability, anxiety, intelligence, dominance, and even political attitudes. Based on tests designed to measure many aspects of personality, identical twins are about twice as similar in personality as fraternal twins. Further, identical twins reared apart were found to be almost as similar in personality as those reared together, indicating that the differences in their environments had little influence on their personality development.

The field of human behavioral genetics is controversial, since it challenges the long-held belief that environment is the most important determinant of human behavior. Investigators have progressed past the nature–nurture debate in their investigations of nonhuman animals. As discussed in Chapter 33, we recognize that much behavior has some genetic basis and that complex behavior often combines elements of both learned and innate behavior. In the case of our own behavior, the debate continues.

Human ethology is not yet recognized as a rigorous science, and it will always be hampered because humans can neither view themselves with detached objectivity, nor treat each other as laboratory animals. Despite this, there is much to be learned about the interaction of learning and innate tendencies in people. Such information could be usefully applied to such diverse areas as childrearing, education, understanding interpersonal relationships, and possibly even using synthetic pheromones to regulate reproductive physiology.

SUMMARY OF KEY CONCEPTS

Communication

Communication, an action by one animal that alters the behavior of another, is the basis of all social behavior. It allows animals of the same species to interact effectively in their quest for mates, food, shelter, and other resources. Animals communicate through visual signals, sound, chemicals (pheromones), and touch.

Visual communication is quiet, and can convey subtle, rapidly changing information. Visual signals may be active (body movements) or passive (body shape and color).

Sound communication can also carry a wide range of rapidly changing information, and is effective where vision is impossible. Although sound may attract predators, the animal may remain hidden while communicating.

Chemical signals take the form of primer or releaser pheromones. Primer pheromones alter the physiological state of the recipient, while releaser pheromones influence the recipient's behavior. Pheromones are only detected by others of the same species. Releaser pheromones linger after the sender has departed, conveying a message over time.

Physical contact reinforces social bonds and helps synchronize mating in a variety of animals, from mammals to molluscs.

Competition for Resources

Although competitive interactions are often resolved through aggression, serious injuries are rare. Most aggressive encounters are settled using displays that communicate the motivation, size, and strength of the combatants.

Some species establish dominance hierarchies that minimize aggression and regulate access to resources. Based on initial aggressive encounters, each animal acquires a status in which it defers to more dominant individuals and dominates subordinates. When resources are limited, dominant animals obtain the largest share and are more likely to reproduce.

Territoriality, a behavior in which animals defend areas where important resources are located, also allocates resources and minimizes aggressive encounters. In general, territory boundaries are respected, and the best-adapted individuals are able to defend the richest territories.

Reproduction

Successful reproduction requires that animals recognize the species, sex, and sexual receptivity of potential mates. In some cases they must also overcome a resistance to close approach by another individual. These requirements have resulted in the evolution of a wide variety of sexual displays that utilize all possible forms of communication.

Cooperation

Some species form cooperative societies. The most rigid and highly organized are those of the social insects such as the honeybee, where the members follow rigidly prescribed roles throughout life. These roles are maintained through both genetic programming and the influence of primer pheromones. Nonhuman vertebrates also form complex, but less rigid societies, such as are found among bullhead catfish, prairie dogs, and baboons.

Human Ethology

The degree to which human behavior is genetically influenced is highly controversial. Because humans cannot be treated as laboratory animals, and because learning plays a major role in nearly all human behavior, investigators must rely on observations made under special circumstances. These include studies of newborn infants, observation of responses to exaggerated stimuli, comparative cultural studies, correlations between certain behaviors and physiology (which suggest a role for pheromones), and studies of identical and fraternal twins. Evidence is accruing that our genetic heritage plays a role in personality, intelligence, simple universal gestures, and our responses to certain stimuli.

GLOSSARY

Active visual signal a movement or posture that communicates information.

Aggression antagonistic behavior, usually between members of the same species, often resulting from competition for resources.

Communication the act of producing a signal that causes another animal, usually of the same species, to modify its behavior in a way beneficial to one or both of the participants.

Dominance hierarchy a social arrangement in which animals, usually through aggressive interactions, establish a rank for some or all of the members of the social unit.

Pheromone (făr′-uh-mōn) a chemical produced by an organism that alters the behavior or physiological state of another of the same species.

Primer pheromone a chemical produced by an organism that alters the physiological state of another of the same species.

Queen substance a chemical produced by a queen bee that can act as both a primer and a releaser pheromone.

Releaser pheromone a chemical produced by one organism that alters the behavior of another of the same species.

Territoriality the defense of an area in which important resources are located.

Waggle dance symbolic communication used by honeybee foragers to communicate the location of a food source to their hivemates.

STUDY QUESTIONS

1. Many aggressive encounters are relatively harmless. In what way does this promote the "survival of the fittest"?
2. List four senses through which animals communicate. After each, present both advantages and disadvantages of that form of communication. Also give one example of each.
3. Distinguish between passive and active visual signals, providing an example of each. Which can convey the most rapidly changing information?
4. What are graded visual signals, and what is their purpose?
5. Define and distinguish between primer and releaser pheromones. Give an example of each.
6. Describe and give an example of a dominance hierarchy. What role does it play in social behavior? Give a human parallel and describe its role in human society. Are the two roles similar? Why or why not?
7. Repeat the exercise in Question 6 for territorial behavior.
8. A songbird will ignore a squirrel in its territory, but act aggressively toward a member of its own species. Explain why.
9. Describe some ways in which animals advertise species and sex during reproductive behavior. Include examples using sight, sound, and chemicals. From an evolutionary standpoint, why is such communication important?
10. What type of animal tends to form the most complex societies, and why?
11. A forager honeybee has just discovered flowers directly below the projection of the sun on the horizon (0 degrees from the sun). Compare the orientation of her waggle dance if performed on a horizontal platform outside the hive to that performed on a vertical wall inside the hive.
12. Describe one of the experiments that reveal the importance of chemical communication in bullhead catfish. Suggest why visual communication is not highly developed in this species.
13. How do male Hamadryas baboons acquire females for their harems? Can you suggest why such a system might evolve?

SUGGESTED READINGS

Gould, C. G. "Out of the Mouths of Beasts," *Science 83,* April 1983. Animal communication discussed in an interesting and informative way.

Grier, J. W. *Biology of Animal Behavior.* St. Louis: Times Mirror/Mosby, 1984. A thorough and clearly written text emphasizing the evolution of behavior.

Holldobler, B. "The Ant—Her World and Ways." *National Geographic,* June 1984. Describes the intricate world of ant society in interesting prose and beautiful paintings.

Macdonald, D., and Brown, R. "The Smell of Success." *New Scientist,* May 1985. Describes the amazing diversity of mammalian pheromones.

Pennisi, E. "Not Just Another Pretty Face." *Discover,* March 1986. Describes the newly investigated society of the naked mole rat, a vertebrate whose social behavior resembles that of some social insects.

The drab male bowerbird uses brightly colored objects to attract a female.

Unit V
Ecology

Chapter 35
Population Growth and Regulation

Chapter 36
Community Interactions

Chapter 37
Principles of Ecosystem Structure and Function

Chapter 38
The Earth's Diverse Ecosystems

35
Population Growth and Regulation

There is no exception to the rule that every organic being naturally increases at so high a rate that, if not destroyed, the earth would soon be covered by the progeny of a single pair.

Charles Darwin in *On the Origin of Species.*

INTRODUCTION TO ECOLOGY

In eastern Colorado lies a remnant of shortgrass prairie, dominated by buffalo grass and blue grama grass. In spring and summer it is ablaze with wildflowers: paintbrush, vetch, sunflower, and bladderpod. Prairie dogs stand upright by their burrows, alerting the "town" with vigorous high-pitched cries as a single hawk soars lazily overhead. The prairie is an **ecosystem,** a complex, interrelated network of living organisms and their nonliving surroundings. An ecosystem can be as small as a puddle or as large as an ocean. Within our prairie ecosystem, the wildflowers, prairie dogs, the grass on which they feed, the hawk that preys on them, and the myriad microscopic organisms that keep the soil fertile, constitute a **community.** The community, in turn, is composed of **populations,** each consisting of all the members of a particular species, be they hawks, grasshoppers, buffalo grass, or bacteria.

Just to the north, a farm was abandoned nearly two years ago. The community here is quite different, with Russian thistle, pigweed, amaranth, and cheatgrass invading the new habitat. A "For Sale" sign in the field foreshadows the high-density housing that will soon displace both farmland and prairie. The first human residents will be delighted to see a hawk, a rare and magnificent bird of prey, practically in their backyards. But as the prairie dog town is bulldozed, this large predator will also disappear.

How has the prairie (in contrast to the failed farm) sustained itself for centuries without artificial fertilizer or irrigation? Why are predators such as the hawk rare relative to their prey? What keeps prairie dogs from overpopulating their habitat and starving? What happens when two organisms compete for the same resources? Why is the abandoned farm community different from that of the untouched prairie? What will the new community look like in 40 years, left to itself? Why has the human population continued to increase, while other populations remain stable or decline in the face of human expansion? These are the questions of **ecology,** the science dealing with the interrelationships among living things and their environment. The environment includes the nonliving components of soil, water, and weather, called the **abiotic** portion, and a **biotic** component, including all forms of life within the ecosystem. Ecology is a tremendously diverse, complex, and relatively young scientific discipline.

In preceding chapters we have studied the anatomy, physiology, and behavior of individual organisms. The science of ecology begins at the next level: the population. From this starting point we will proceed to increasing levels of complexity, first to communities and the interactions within them, and finally to entire ecosystems (Fig. 35-1).

POPULATION GROWTH

Studies of ecosystems undisturbed by humans show that many populations tend to remain relatively stable over time. Yet we are vividly aware from the human example that populations can readily increase. Let's first examine

Figure 35-1 Levels in the study of ecology: **(a)** *a population of fish,* **(b)** *a kelp-forest community, and* **(c)** *the ocean ecosystem.*

how and why populations grow, then look at the factors that normally control this growth.

Three factors determine whether and how much the size of a population changes: births, deaths, and migration. Organisms join a population through birth or **immigration** (migration in) and leave it through death or **emigration** (migration out). A population remains stable if, on the average, as many individuals leave as join. Population growth occurs when the number of births plus immigrants exceeds the number of deaths plus emigrants. Populations decline when the reverse occurs. A simple equation for the change in population size is:

$$(\text{births} - \text{deaths}) + (\text{immigrants} - \text{emigrants}) = \text{population change}$$

These numbers are often expressed per thousand individuals per year. For example, if during a given year, a town of 50,000 experienced 1000 births (20 for every 1000 people), 500 deaths (10 per 1000), 100 immigrants (2 per 1000), and 50 emigrants (1 per 1000), its population change would be

$$(20 \text{ births}/1000/\text{yr} - 10 \text{ deaths}/1000/\text{yr}) + (2 \text{ immigrants}/1000/\text{yr} - 1 \text{ emigrant}/1000 \text{ yr}) = 11 \text{ people}/1000/\text{yr}$$

This can also be expressed as a growth rate of 1.1 percent.

In many natural populations, organisms moving in and out contribute relatively little to population change, leaving birth and death rates as the primary factors influencing population growth.

The ultimate size of any population (discounting migration) is the result of a balance between two major opposing factors. The first is **biotic potential,** or the maximum rate at which the population could increase, assuming ideal conditions allowing a maximum birth rate and minimum death rate. Opposing this potential for growth are limits set by the living and nonliving environment. These limits include the availability of food and space, competition with other organisms, and interactions among species such as predation and parasitism. Collectively, these limits are called **environmental resistance.** Environmental resistance can both decrease the birth rate and increase the death rate. *The interaction between biotic potential and environmental resistance usually results in a balance between population size and available resources.* To understand how populations grow and how their size is regulated, we must examine each of these forces in more detail.

Biotic Potential: Exponential Growth

Changes in population size (ignoring migration) are functions of the birth rate, the death rate, and the number of individuals in the original population. Rates of change in populations may be expressed as changes per individual per unit time. For example, the birth rate may be expressed as the number of births per individual per unit time.

The *rate of growth* (r) of a population is determined by subtracting the death rate (d) from the birth rate (b):

$$r = b - d$$

growth rate = birth rate − death rate

To determine the number of individuals added to a population of size N in a given time period, the growth rate (r) is multiplied by the original population size (N):

$$\text{population growth} = rN$$

For example, the annual growth rate of a population of 10,000 in which 1500 births and 500 deaths occur yearly can be calculated as follows:

$$r = \text{birth rate} - \text{death rate}$$

$$r = 1500/10{,}000 - 500/10{,}000 = 0.10 \text{ or}$$

$$r = 0.15 - 0.05 = 0.10$$

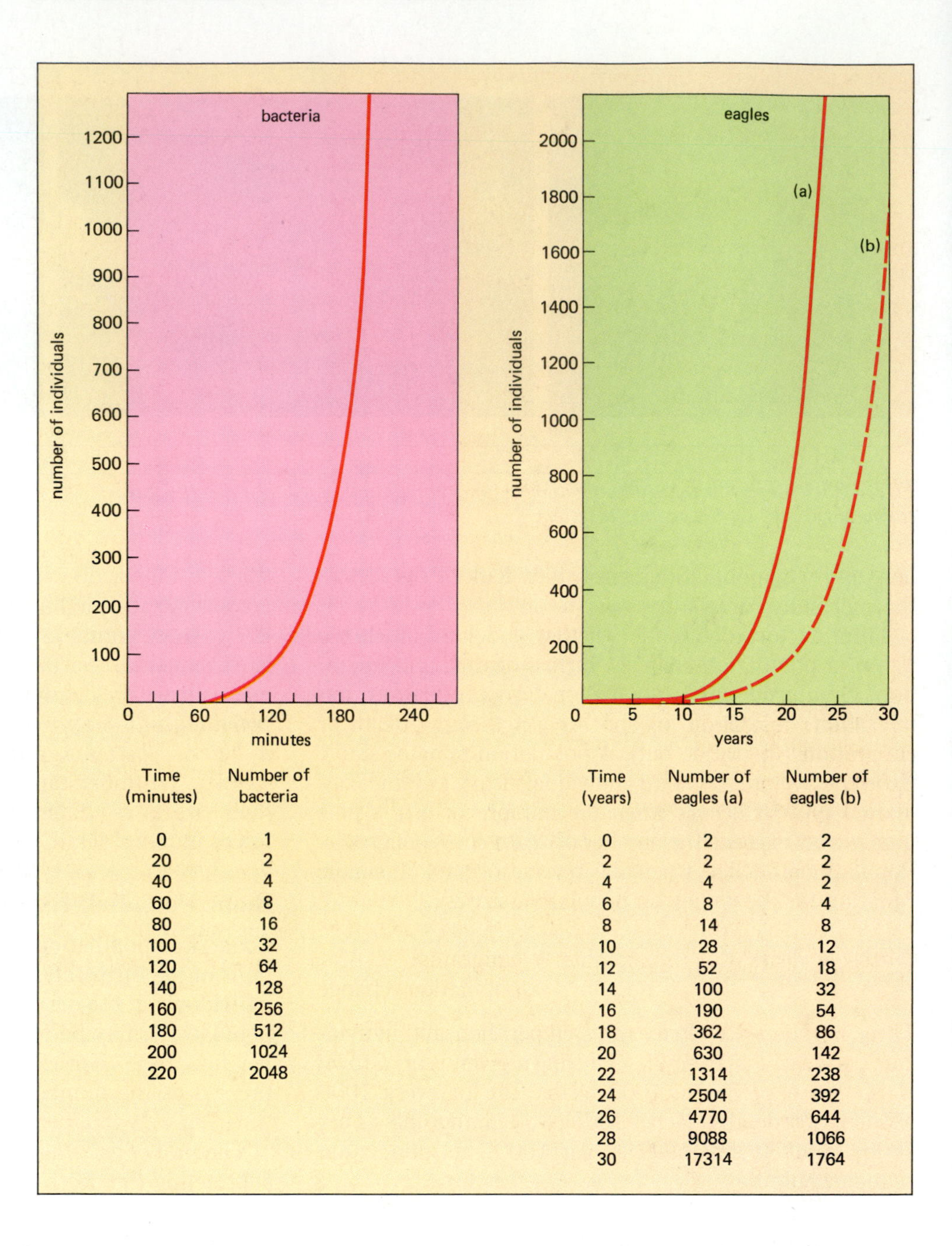

Time (minutes)	Number of bacteria
0	1
20	2
40	4
60	8
80	16
100	32
120	64
140	128
160	256
180	512
200	1024
220	2048

Time (years)	Number of eagles (a)	Number of eagles (b)
0	2	2
2	2	2
4	4	2
6	8	4
8	14	8
10	28	12
12	52	18
14	100	32
16	190	54
18	362	86
20	630	142
22	1314	238
24	2504	392
26	4770	644
28	9088	1066
30	17314	1764

Figure 35-2 *Exponential growth curves all share a similar shape; the major difference is the time scale.*
Left: The growth of a population of bacteria, assuming one individual to start with and a doubling time of 20 minutes.
Right: **(a)** *Growth of an eagle population, starting with a single pair of hatchlings, and assuming that the age at first reproduction is 4 years;* **(b)** *eagle population growth assuming age at first reproduction is 6 years. Notice that after 26 years, those that began reproducing at 4 years have seven times the population as those that began reproducing at 6 years.*

Population growth (rN) equals $0.10 \times 10{,}000 = 1000$.

If this rate of increase persists, then the following year, r must be multiplied by a larger population size ($N + rN = 11{,}000$), resulting in an increase of 1,100 individuals, which in turn is added to N, and so on. This is **exponential growth.** During exponential growth, the population grows (during a given time period) by a fixed percentage of its size at the beginning of that time period. Thus an increasing number of individuals is added to the population during each succeeding time period, causing population size to grow at an ever-accelerating pace. Births will exceed deaths if, on the average, each individual produces more than one surviving offspring during its lifetime. This causes an accelerating increase in population size.

Although the number of offspring produced by an individual each year varies from millions for an oyster to one or fewer for a human, each organism, whether working alone or as part of a sexually reproducing pair, has the potential to replace itself manyfold during its lifetime. This capacity, called biotic potential, has evolved because it helps assure that at least one offspring survives to bear its own young. Several factors influence biotic potential. These include (1) the age at which the organism first reproduces, (2) the frequency with which reproduction occurs, (3) the average number of offspring produced each time, (4) the length of the reproductive life span of the organism, and (5) the death rate of individuals under ideal conditions. Examples in which these factors differ will be used to illustrate the concept of exponential growth (Fig. 35-2).

The bacterium *Staphylococcus* is a normally harmless resident in and on the human body. But in an ideal culture medium such as warm custard, each bacterial cell can divide every 20 minutes, doubling the population three times each hour. (The by-products of the bacteria's metabolism can result in serious food poisoning under these conditions.) The biotic potential of bacteria is so great that, were nutrients unlimited, the offspring of a single bacterium could cover the earth over 7 feet deep within 48 hours! In contrast, the golden eagle is a relatively long-lived, rather slowly reproducing species. Let's assume that the golden eagle can live 30 years, reaches sexual maturity at 4 years, and that each pair of eagles produces two offspring per year for the remaining 26 years. Figure 35-2 compares the potential population growth of eagles to that of bacteria, assuming no deaths occur in either population during the time graphed. Notice that the shapes of the curves are virtually identical. Although the time scale differs, population sizes eventually become astronomical. Figure 35-2 also shows what happens if eagle reproduction begins at 6 years instead of 4. Exponential growth still occurs, but the time required to reach a particular size is increased considerably. This has important implications for the human population: delayed childbearing significantly slows population growth. If each woman has only three children, but has them in her early teens, the population will grow much faster than if women each have *five* children, but begin having them at age 30!

So far we have looked only at birthrates. Even under ideal conditions, however, some mortality occurs. To illustrate the effect of differing death rates, three bacterial populations are compared in Figure 35-3: one in which no deaths occur, one in which 10 percent of the population dies between each division, and one in which 25 percent of the population dies between each division. Again, the shapes of the curves are the same. In each case the population eventually approaches infinite size; only the time required to reach any given population size differs.

In nature, exponential growth curves are observed only under special circumstances, and only for limited periods. For example, in short-lived insects such as houseflies, abundant food allows exponential growth through the spring and summer. In the fall, hard frosts cause a population "crash." Such **boom-and-bust cycles** are typical of short-lived species, whose maximum population size is determined by an environmental variable such as temperature or rainfall (Fig. 35-4). For longer-lived species, populations tend to become relatively stable with minor fluctuations in response to environmental variables such as weather and food availability. However, these populations may show temporary exponential growth under certain circumstances. For example, if a population becomes established in a new area with abundant food and few predators and competitors, it may grow exponentially. This has happened repeatedly when people have introduced foreign or "exotic" species into ecosystems, often with tragic results (see the essay, "Growth Without

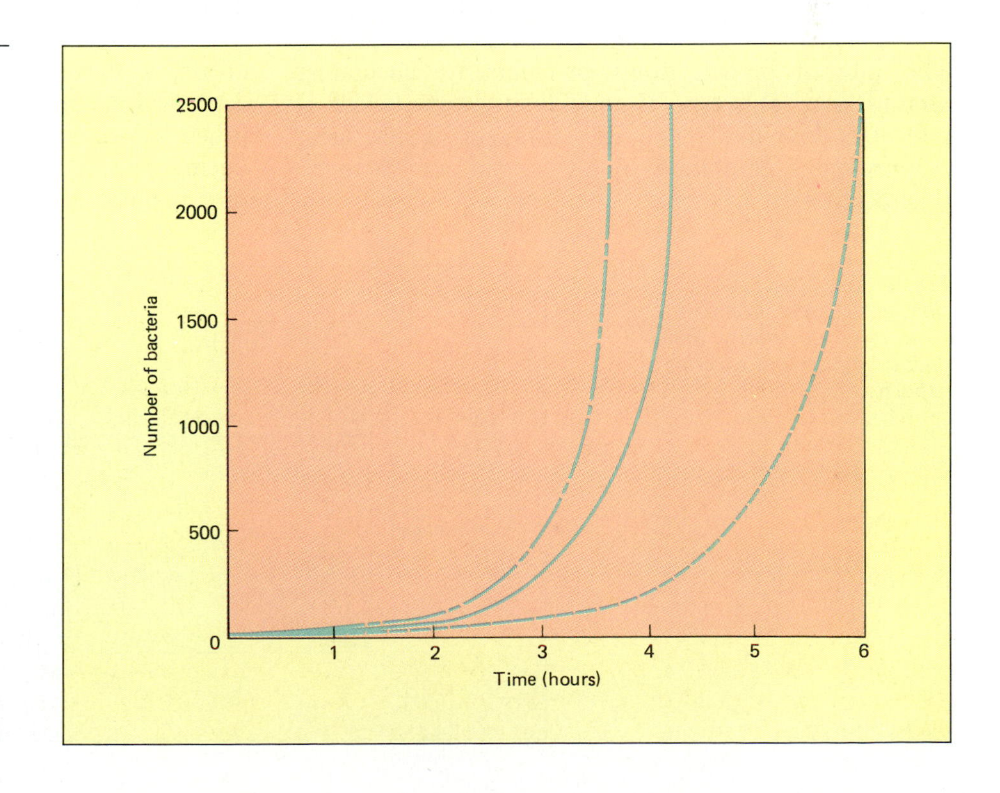

Figure 35-3 The effect of differing death rates on population growth. Left: A bacterial population doubles every 20 minutes without any deaths. Middle: The same population, assuming 10 percent of the population dies between each doubling. Right: In this case, 25 percent die between each doubling.

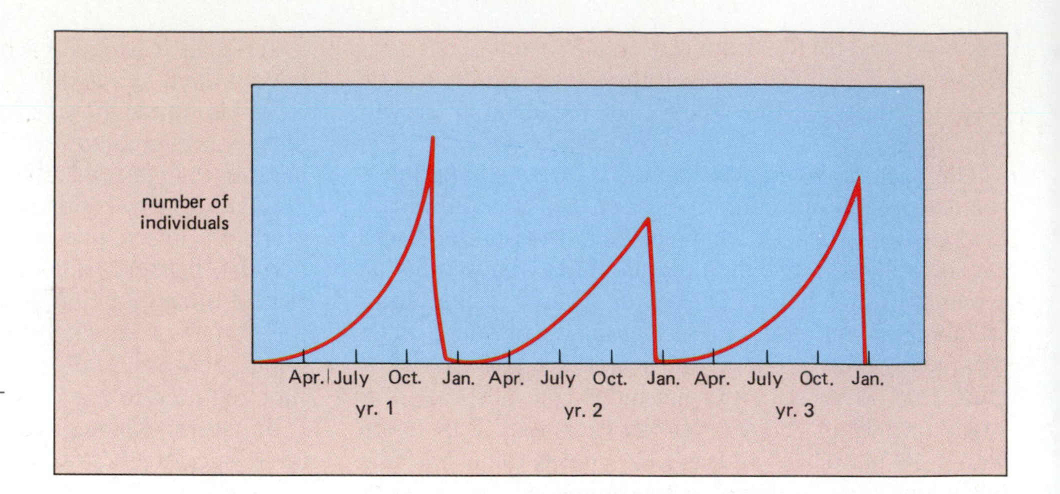

Figure 35-4 Many insects and annual plants show a boom-and-bust cycle in population size.

Resistance: The Problem of Introduced Species"). Temporary exponential growth may also occur in a population initially kept in check by a disease organism or predator, if that limiting factor is suddenly removed. As you will learn in the next section, all exponential growth curves must eventually either flatten or crash.

Environmental Resistance: Limits to Growth

Exponential growth carries with it the seeds of its own destruction. As individuals join the population, competition for resources intensifies. Predators may increase in number or they may make this abundant prey a larger part of their diet. Parasites and diseases spread more readily due to crowding and weakness caused by lack of food or social stress. Consequently, after a period of exponential growth, populations tend to stabilize at or below the maximum size that the environment can sustain. The rate of growth drops precipitously, fluctuating around zero. This type of population growth, which is typical of long-lived organisms colonizing a new area, is represented graphically by a "sigmoid" or "S-curve" (Fig. 35-5).

CARRYING CAPACITY. Populations may stabilize at a level called the **carrying capacity** of the ecosystem. The carrying capacity is the maximum number of organisms that an area can support on a sustained basis. It is determined primarily by the availability of two types of resources: a **nonrenewable resource**—space, and **renewable resources**—nutrients, water, and light. If space requirements are exceeded, animals may emigrate, but often to less suitable areas where their death rate will be higher. Reproduction will decline, since animals may not find adequate breeding sites or the seeds of plants may not reach a suitable place to germinate. If demands on renewable resources such as food, water, and light (the energy source for plants) are too high, organisms will starve. Excess demands may damage ecosystems, reducing their carrying capacity. The result is a population decline until the ecosystem recovers, or a permanently reduced population. For example, overgrazing by cattle on dry western grasslands has given sagebrush (which cattle will not eat) a competitive advantage. Once established, sagebrush thrives, replacing edible grasses and reducing the carrying capacity of the land for cattle. In nature, populations are maintained at or *below* the carrying

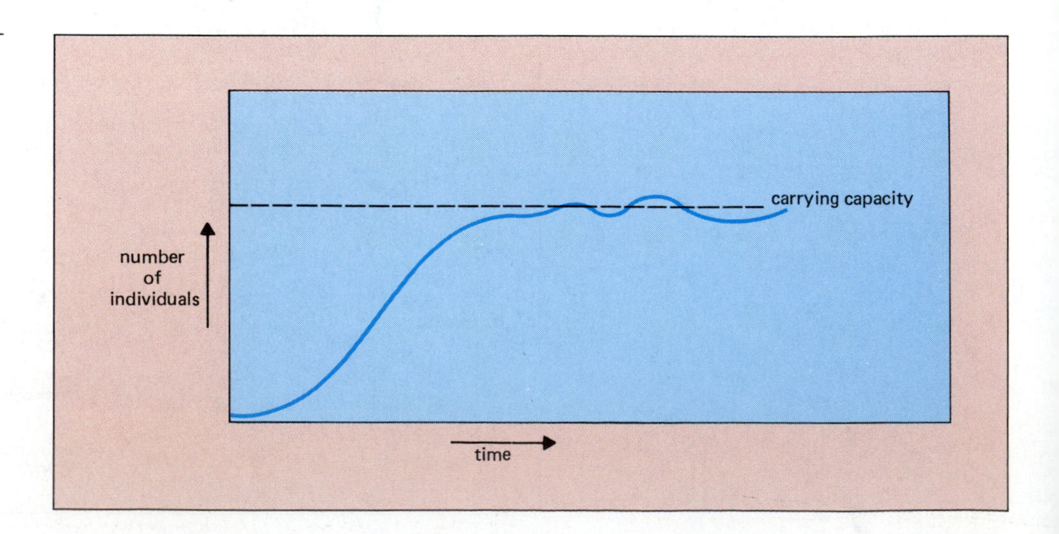

Figure 35-5 The S-curve, showing a population first growing exponentially, then fluctuating around carrying capacity.

capacity of their environment by environmental resistance. Factors of environmental resistance may be classified into two broad categories: **density dependent** and **density independent.**

DENSITY-INDEPENDENT LIMITS TO GROWTH. Density-independent factors limit populations regardless of their size. Perhaps the most important density-independent factor is weather. For example, many insects and annual plant populations are limited in size by the number of individuals that can be produced before the first hard freeze. Weather is largely responsible for the boom-and-bust population curves described above. Such populations typically do not reach carrying capacity, since density-independent factors intervene first. Human activities can also limit the growth of natural populations, in ways that are independent of population density. Pesticides and pollutants may cause drastic declines in natural populations, as does habitat destruction for farms, roads, and housing developments.

Figure 35-6 *Grey wolves have brought down an elk, who may have been weakened by old age or parasites. Predators will often switch to the most abundant prey, limiting prey populations.*

DENSITY-DEPENDENT LIMITS TO GROWTH. Organisms that live several years have evolved various mechanisms to compensate for seasonal changes, thus circumventing density-independent population checks. Many mammals, for example, develop thick coats and store fat for the winter; some also hibernate. Other animals, including many birds, migrate long distances to find food and a hospitable climate. Plants may survive the rigors of winter by entering a period of dormancy, dropping their leaves and drastically slowing their metabolic activities.

By far the most important elements of environmental resistance for these long-lived species are density-dependent factors: those which become increasingly effective as population density increases, thus exerting a negative feedback effect on population size. Density-dependent factors include community interactions such as **predation** and **parasitism** as well as **competition** within the species or with members of other species. These are covered in more detail in Chapter 36.

Predation (Fig. 35-6) becomes increasingly important as population densities increase because individuals are more likely to be encountered by predators. Experiments have shown that some predators eat a variety of prey, depending on what is most abundant and easiest to find. For example, coyotes may switch from eating mice to ground squirrels when the mouse population declines, incidentally allowing the mouse population to recover. As illustrated in Chapter 36, in some cases predators maintain populations well below carrying capacity. In other cases, they may feed on the excess in a population that is beginning to exceed its carrying capacity. In such cases, predators may subsist on prey made vulnerable because they lack adequate food or shelter.

Parasitism (Fig. 35-7) is also density dependent; most parasites have limited motility and therefore spread more readily between individuals at high population densities. For example, parasitic wasps attack the webworm moth in New Brunswick, Canada. J. D. Tothill, a researcher investigating this interaction, found almost no parasite-related deaths during years when the webworm population was very low, and increasing deaths from parasitism as the population increased.

Although enemies to their victims, both predators and parasites can have beneficial effects on the prey population as a whole. As you will learn in the following chapter, predators, parasites, and their prey coevolve. Parasites and predators destroy the least fit of the prey, leaving the better-adapted prey to reproduce. The result is usually a balance in which the prey population is regulated but not eliminated. However, when a predator or parasite is introduced into an area where it did not evolve, the local prey species are often not adapted for it, and the consequences can be disastrous. The conquest of much of the globe by Europeans can be traced partly to the disease organisms they carried with them as they invaded continents whose populations had no previous exposure or resistance to these pathogens. Smallpox, for example, imported by Europeans, ravaged the native population of Hawaii, the Amerindians of Argentina, and the Aborigines of Australia. Equally destructive to natural ecosystems is the case of species introduced into new areas where they have no natural predators or parasites.

Because the resources that determine carrying capacity are limited, any use of these by one individual limits their availability to another. Thus competition, both **interspecific** (between species) and **intraspecific** (between members of the same species), limits population size. Interspecific competition is covered in more detail in Chapter 36. Because the needs of members of the same species for water and nutrients, shelter, breeding sites, light, and

Growth Without Resistance: The Problem of Introduced Species

"Under a spreading chestnut tree, the village smithy stands. . . ." As schoolchildren of the 1950s memorized this well-loved poem, the last of the majestic chestnuts that once dominated parts of the eastern deciduous forest were dying, victims of an imported fungus (Fig. E35-1). The chestnut blight was accidentally brought to the United States from China on some young Chinese chestnut trees planted in the New York Zoological Park in 1895. The Chinese chestnut species, having evolved with the fungus, was hardly affected by it, but the fungus was lethal to the unadapted American species. With little resistance from its new environment, the fungus spread rapidly, killing nearly the entire chestnut population of the United States in 50 years.

The chestnut blight fungus is just one of thousands of species called "exotics": plants, animals, fungi, or microorganisms transported out of their natural habitat into a new area. Over half the insect pests in some areas are imports. Gypsy moths, which destroy 10 million acres of trees per year in the United States, were brought to Massachusetts in the 1860s by a scientist hoping to breed a better silkworm. Dime-sized freshwater clams from Asia caused the shutdown of a nuclear power plant in Arkansas until they could be removed from its cooling pipes and water lines at a cost of $15 million. Wild boars from Europe were introduced deliberately into North Carolina in 1912, where they now damage Great Smoky Mountains National Park. Exotics have been introduced by bird fanciers, such as those who released the first few pairs of English sparrows in 1899. Today, the phenomenal spread of this species is causing the decline of our native bluebird, with whom they compete for nesting sites. The

(a)

(b)

Figure E35-1 **(a)** *The American chestnut was a majestic tree, prized for shade, lumber, and the flavorful nuts it produced.* **(b)** *Today, all that survive are blighted sprouts from the surviving roots of the former giants.*

(a)

(b)

Figure 35-7 **(a)** *The pine bark beetle, an insect parasite of ponderosa pines, tunnels destructively under the bark.* **(b)** *A stand of pines devastated by this pest. Dense stands of similar-age trees, such as regrow after clear-cutting, are particularly susceptible, since competition for water and nutrients weakens the individuals. Thus this parasite helps regulate tree populations in a density-dependent manner.*

Growth Without Resistance: The Problem of Introduced Species

destructive starling was imported from England in 1890 by a New York bird fancier and Shakespeare buff whose goal was to bring to this country all the birds mentioned by Shakespeare.

Exotics have been introduced to "improve" the environment. In 1876, a Japanese vine called kudzu was introduced in southern states as an ornamental, shading the porches of southern mansions from the hot sun. Through the 1940s, kudzu was widely planted along streams and roadsides to control erosion. But by the mid-1950s, it became clear that a menace had been unleashed. No native insect or disease could control this plant, and, although the leaves are killed by frost, the root system remains, spreading each year. Its growth rate is phenomenal. Kudzu can overgrow forests, killing trees and underbrush (Fig. E35-2), and engulf entire houses while their owners are away on vacation.

Figure E35-2 The Japanese vine kudzu will rapidly cover entire trees and houses.

The water hyacinth is another import that has taken over its new habitat. Introduced from South America, this beautiful but uncontrollable floating plant now clogs nearly 2 million acres of lakes and waterways (Fig. E35-3), displacing natural vegetation. Florida, Louisiana, and Texas, the states most plagued by the import, currently spend over $11 million each year to restrict its growth.

Exotics have one thing in common: they face little environmental resistance in their new habitats. No predators or diseases control them, and their new prey may have few defenses against them. Native populations may be outcompeted. The result is often unchecked population growth until the ecosystem is irrevocably altered.

Figure E35-3 The beauty that became a beast, water hyacinths, originally from South America, today infest waterways in our southern states.

other resources are almost identical, intraspecific competition is more intense. Organisms have evolved several ways to deal with this. Some, including most plants and many insects, engage in **scramble competition,** a kind of free-for-all with resources as the prize. For example, when a plant disperses its seeds in a small area, hundreds may germinate. However, as they grow, the larger ones begin to shade the smaller, those with the most extensive roots absorb most of the water, and the weaker individuals eventually wither and die.

Many animals (and even a few plants) have evolved **contest competition,** which helps regulate population size and reduce direct competition. Contest competition consists of social or chemical interactions used to limit access to important resources. One such behavior is **territoriality** (Fig. 35-8; for further discussion, see Chapter 34). Territorial species—such as wolves, many fish, rabbits, and songbirds—defend an area containing important resources such as food or nesting sites. When the population begins to exceed the available resources, only the best-adapted individuals are able to defend adequate territories. Those without territories often do not reproduce and are also easy prey. The creosote bush, which secretes a chemical into the ground that prevents germination of seeds nearby, could be considered a territorial plant (see Fig. 35-10b).

Dominance hierarchies or pecking orders are another form of contest competition observed in many social animals. High-ranking individuals have first access to food, breeding sites, and mates. When resources are limited,

Figure 35-8 Contest competition is illustrated by this grasshopper mouse, who defends his territory by high-pitched howls.

only dominant individuals obtain what they need to reproduce successfully. Thus dominance hierarchies can limit population size.

As population densities increase and competition becomes more intense, some animals react by emigrating. Large numbers leave their homes to colonize new areas, and many, sometimes most, die in the quest. The massive movements of lemmings, which sometimes end in suicidal marches into the sea, may be attempts to migrate in response to overcrowding. Migrating swarms of locusts plague the African continent, stripping all vegetation in their path (Fig. 35-9).

In laboratory studies, overcrowding of small mammals—such as mice, rats, and voles—causes **social stress.** This results in a decrease in the size of reproductive organs, reduced reproductive rate, slower growth, reduced resistance to disease, and cannibalism of young, all of which would reduce population size. In the wild, however, population densities rarely reach the levels achieved in the laboratory, so these findings may have limited applicability to natural populations.

Figure 35-9 Emigration in response to overcrowding reduces local populations. Locust swarms provide a dramatic example.

PATTERNS IN POPULATIONS

Patterns in Space

Organisms may live in flocks, herds, pairs, or as solitary individuals, or they may cluster around resources such as water holes. Distribution may vary with time, changing with the breeding season, for example. Ecologists recognize three major types of spatial distribution: **aggregated, uniform,** and **random** (Fig. 35-10).

AGGREGATED DISTRIBUTION. The most common pattern of distribution is aggregated, in which members of the population tend to live in groups. Many species form family or social groupings, such as elephant herds, wolf packs, prides of lions, flocks of birds, or schools of fish (Fig. 35-1a). What are the advantages of aggregation? Flocks provide many eyes on the lookout for localized food, such as a tree full of fruit. Schooling fish avoid predation by confusing the predator with myriad flashing bodies darting in all directions. Some species form temporary aggregations for mating (Fig. 35-10a). Other plant or animal populations cluster, not for social reasons, but because resources such as nutrients, shelter, or water are localized. Cottonwood trees, for example, grow along streams and rivers in grasslands. Animals also aggregate for water, as in the dry savanna of Africa.

UNIFORM DISTRIBUTION. Uniformly distributed organisms maintain a relatively constant distance between individuals. Spacing is often the result of the defense of scarce resources, such as breeding sites, nutrients, or water. This distribution occurs most frequently among animals that defend territories. Male Galápagos iguanas, which were randomly distributed as they basked along the rocky shore during the winter, establish evenly spaced breeding territories later in the year. Shorebirds are also often found in evenly spaced nests, just out of reach of one another. Other territorial species, such as the tawny owl, mate for life and continuously occupy well-defined, relatively uniformly spaced territories (for breeding animals, the spacing refers to pairs, not individuals). Desert plants growing in poor soil with limited water, such as the creosote bush (Fig. 35-10b), have chemical spacing mechanisms that assure adequate resources for each individual.

RANDOM DISTRIBUTION. Random distribution is the least common. Here individuals do not form social groups, the resources they need are more or less equally available throughout the area they inhabit, and resources are not scarce enough to require territorial spacing. Trees and other plants in rain forests come close to being randomly distributed (Fig. 35-10c), as do the marine iguanas taking advantage of the abundant warmth and sunlight of the rocky shore during their nonbreeding season. There are probably no vertebrate species that maintain random

Figure 35-10 Population patterns in space. **(a)** *Aggregated, as illustrated by this gathering of ladybugs;* **(b)** *uniform, seen in the spacing of these desert creosote bushes; and* **(c)** *the random distribution of trees in a rain forest.*

distribution throughout the year because they must breed, a behavior that makes social interaction inevitable.

Patterns in Time

PATTERNS OF SURVIVORSHIP. Population patterns can be considered from the perspective of time as well as space. Over time, populations show characteristic patterns of deaths or (more optimistically) survivorship. These patterns, called **survivorship curves,** are revealed when the number of individuals of each age is graphed against time. Three different types of survivorship curve, **convex, constant,** and **concave,** are shown in Figure 35-11. Populations with convex survivorship curves have relatively low infant mortality, and most individuals survive to old age. This curve is characteristic of humans and many other large animals, such as Dall mountain sheep. These species produce relatively few offspring, which are protected by the parents. Species with constant survivorship curves have an equal chance of dying at any time during their life span. This phenomenon is seen in the American robin, the gull, and laboratory populations of organisms that reproduce asexually, such as hydra and bacteria. The concave curve is characteristic of organisms producing large numbers of offspring that are left to compete on their own. Mortality is very high among the offspring, but those that reach adulthood have a good chance of surviving to old age. Most invertebrates, most plants, and many fish exhibit concave survivorship curves. In some populations of black-tailed deer, 75 percent of the population dies within the first 10 percent of its life span, giving this mammalian population a concave curve as well.

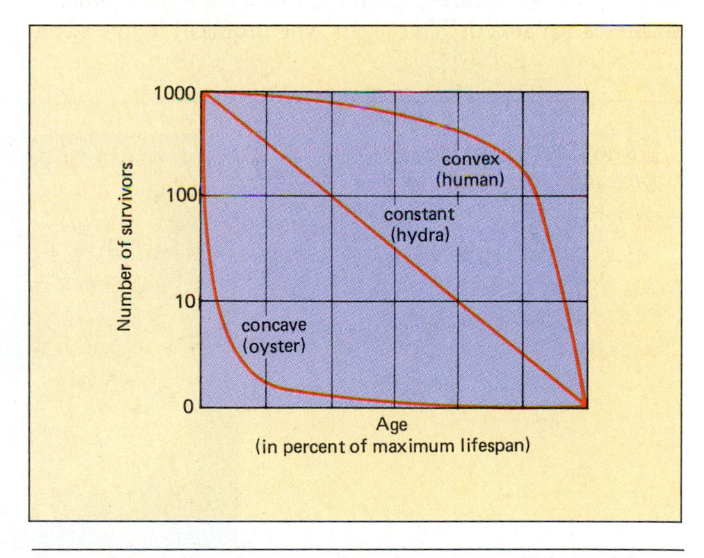

Figure 35-11 Population patterns in time are illustrated by these three types of survivorship curve.

THE HUMAN POPULATION

Your first child received free schooling and free medical care, and you were offered longer maternity leave and a larger pension if you signed a promise to have no more. Pregnant with your second child, you face fines equivalent to all the benefits provided to the first, up to a full year's pay. Each night, delegates come to your home to tell you of the harm you are doing to society. Your neighbors shun you, and the collective pressure finally drives you to the hospital and a state-sponsored abortion. A distant future scenario from a pessimistic sci-fi novel? Hardly. Similar policies were instituted in China. With a population of over 1 billion and an increase of over 15 million people in 1983 alone, the government took drastic measures. China's coercive policies have been eased somewhat, and birth rates are rising again. What is China's demographic future? What of the rest of the world? Let's examine human population in light of what we know of exponential growth and carrying capacity.

World Population Growth

Look now at the growth of the human population graphed in Fig. 35-12 and compare it to Fig. 35-2. The time span is different, but our growth curve is exponential. It took over 1 million years for world population to reach 1 billion, the second billion was added in 100 years, the third in 30, the fourth in 15, and in 1986, the 5-billionth human joined our planet, only 11 years after the 4 billion mark was reached. The sixth will probably come even faster. World population currently grows by over 89 million yearly, *another million people added every 4 days.* The number will be higher as you read this. Why hasn't environmental resistance put an end to exponential growth? What is the carrying capacity of the world for humans?

Like all populations, ours has encountered environmental resistance, but unlike other populations, we have responded to resistance by overcoming it rather than reaching a balance with it. As a result, the human population has grown exponentially for an unprecedented time span; to accommodate our growing numbers, we have altered the face of the globe. Human population growth has been spurred by a series of "revolutions" which conquered environmental resistance and increased the Earth's carrying capacity for people.

Primitive people produced a **cultural revolution** when they discovered fire, invented tools and weapons, built shelters, and designed protective clothing. Weapons meant an increased food supply; clothing and shelter increased the habitable areas of the globe, and population grew as they increased the earth's carrying capacity for humans. Starting about 8000 B.C., the **agricultural revolution,** in which the raising of domesticated crops and animals supplanted hunting and gathering, provided people with a much more dependable food supply. This further increased the carrying capacity. Increased food re-

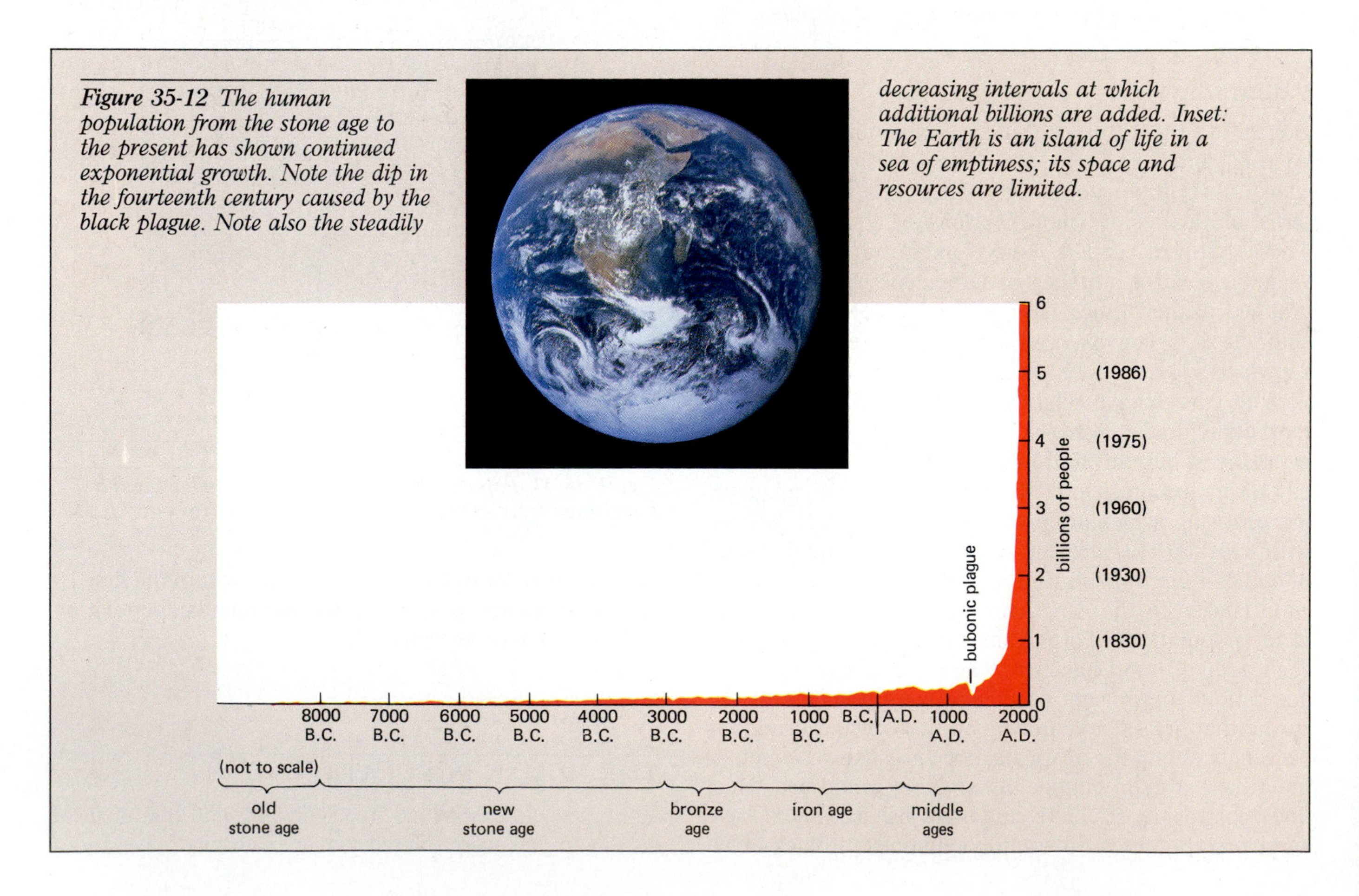

Figure 35-12 The human population from the stone age to the present has shown continued exponential growth. Note the dip in the fourteenth century caused by the black plague. Note also the steadily decreasing intervals at which additional billions are added. Inset: The Earth is an island of life in a sea of emptiness; its space and resources are limited.

sulted in increased longevity and a longer reproductive span, but a high death rate from disease still restricted the rate of growth. Population growth continued slowly for thousands of years until the **industrial-medical revolution** began in England in the mid-eighteenth century, spreading through Europe and North America in the nineteenth century. Medical advances dramatically decreased the death rate by reducing environmental resistance from disease. These advances included the discovery of bacteria and their role in infection, leading to control of bacterial disease through improved sanitation and the use of antibiotics; and the discovery of viruses, leading to the development of vaccines for diseases such as smallpox. The revolution continues today as research proceeds on vaccines against such major killers as schistosomiasis, malaria, and AIDS, as well as sophisticated medical procedures such as coronary bypass operations and organ transplants.

In developed countries, such as those of western Europe, the "industrial-medical" revolution resulted in an initial rise in population due to decreased deaths, but a decline in birthrate followed. This decline can be attributed to many factors, including better education, increased availability of contraceptives, a shift to a primarily urban life-style, and more career options for women. In developed countries such as Switzerland, Sweden, Austria, Germany, and England, populations have more or less stabilized, but these countries are home to only about 4 percent of the world's population.

In less-developed countries, such as most of those in Central and South America, Africa, and Asia, medical advances have decreased death rates and increased life span, but a major decline in birthrate has not occurred. These countries have not experienced the increase in wealth which was partly responsible for the decline in birthrate of developed countries. Children serve as a form of social security in third-world nations since they may be the only support for parents in their old age. In agricultural societies, children are an important source of labor. In extreme poverty, children may be the parents' major source of pride. Social traditions offer prestige to the man who fathers and the women who bears many children. For example, in Nigeria, when women who had already given birth to *nine* or more children were asked if they wanted to stop having children, only 16 percent said yes. Lack of education and lack of contraceptives further impede progress in curbing population growth. Thus birthrates remain high, while death rates have been lowered dramatically by medical advances. Of the 6.3 billion people projected for the year 2000, 5 billion will reside in less-developed countries. The prospects for population stabilization in the near future are nil, barring major catastrophies that might dramatically increase deaths. The reason can be seen clearly by looking at the age structures of the less-developed countries and comparing them to countries with stable populations.

Age Structure and Population Growth

Age-structure diagrams graphically illustrate the number of males and females of various ages comprising the population. All age structure diagrams come to a peak at the top, since relatively few people live into their nineties. The shape of the rest of the diagram, however, shows whether the population is expanding, stable, or shrinking (Fig. 35-13). If the numbers of children (age 0 to 14) exceed the numbers of reproducing individuals (age 15 to 45), the population is growing and the diagram resembles a pyramid. Stable populations have achieved replacement-level fertility, and the number of children is about equal to the reproducing adults. In shrinking populations, there are fewer children than reproducing adults and the figure is constricted at the base. Figure 35-14 shows the average age structures of the populations of less-developed and developed countries. The outermost boundaries represent the projected population structure for the year 2000, the inner ones are for 1984. Each graph has been divided into three parts to show individuals who are pre-reproductive (0 to 14 years old), reproductive (15 to 44 years), and postreproductive (45 and older). In 1988, the less-developed countries (Asia, Africa, India, and South and Central America) had an average annual growth rate of 2.1 percent, and the developed countries (United States, Europe, USSR) showed an average annual growth rate of 0.6 percent. In the developed countries, projections for the year 2000 are only slightly higher than present levels. In contrast, each year in the less-developed countries, increasing numbers of people enter their reproductive years and give birth to an ever-increasing base of infants. Even if these countries were to reach **replacement-level fertility** immediately (i.e., if people of reproductive age had only sufficient children to replace themselves), population growth would continue for decades because of built-in momentum caused by increasing numbers of people reaching reproductive age.

Notice, however, that predictions for the year 2000 show a structure very similar to the present one, the only difference being far more people in all age classes. Although several of these countries have recently initiated family-planning programs, parents in less-developed countries will probably continue to have larger than replacement-level families. Ironically, population growth in

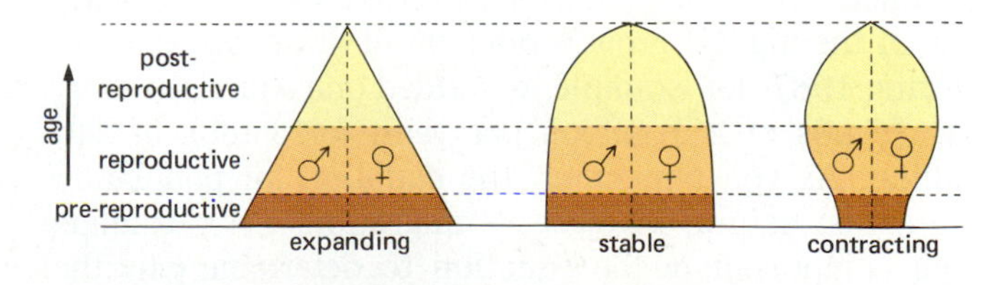

Figure 35-13 Idealized age structure diagrams for expanding, stable, and contracting populations.

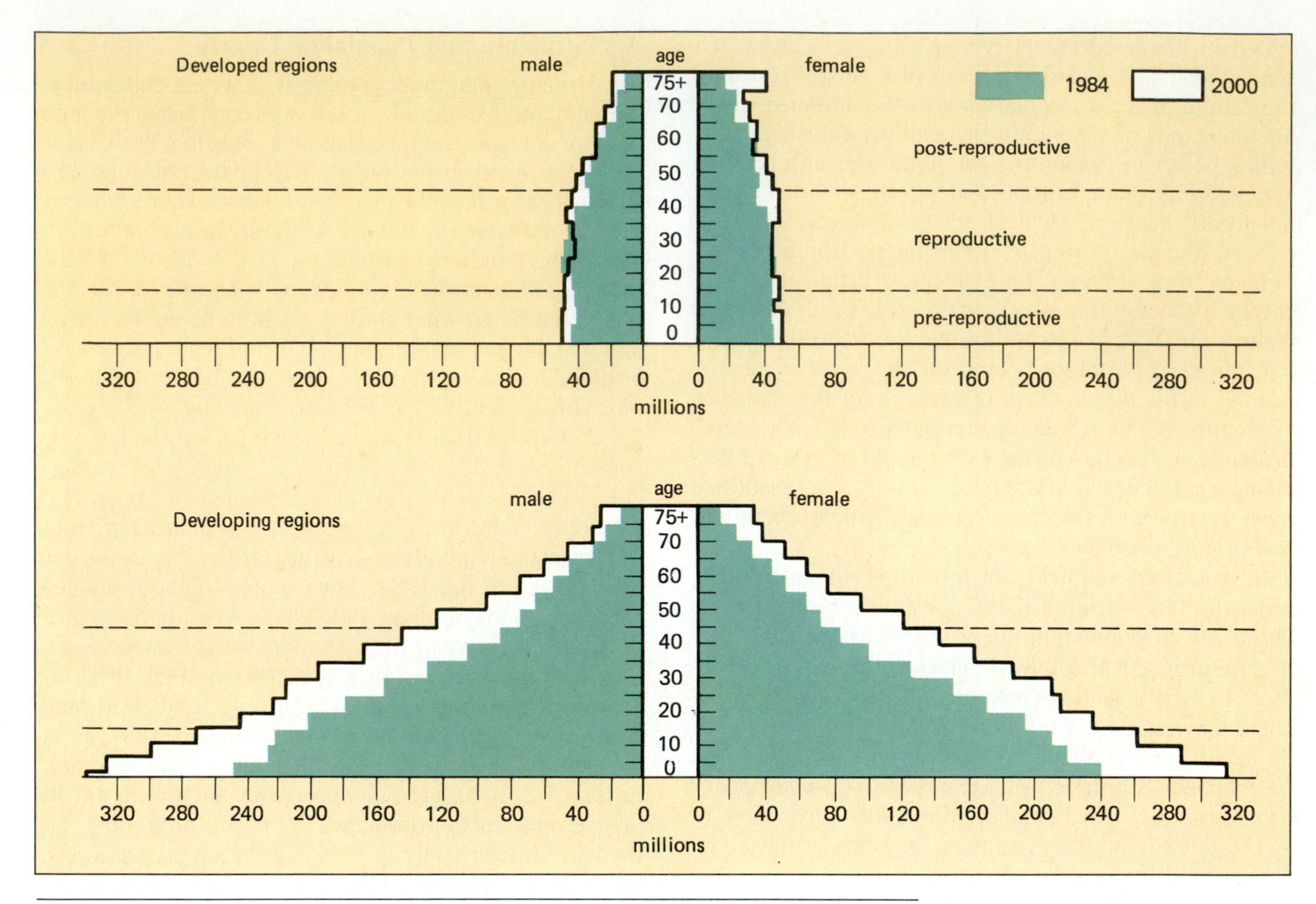

Figure 35-14 Age-structure diagrams for (top) developed countries and (bottom) less-developed countries.

these countries is helping to perpetuate the poverty and ignorance that in turn tends to sustain high birthrates. The relationship of income and education to birthrate has been documented in the United States. Here, women who do not complete high school have twice as many children as those with more than 4 years of college. Women whose family income averages less than $10,000 have twice as many children as those whose income is $35,000 and higher.

Population Growth in the United States

As shown in Fig. 35-15, the U.S. population is growing exponentially. In fact, at over 1 percent annually, we have one of the highest growth rates of all developed nations. During 1987, for example, we added one American every 14 seconds or 2.3 million per year, for a total of 245 million. As you remember, the equation for population change is: natural increase + migration. Let's examine each component of the equation to determine why the United States is growing so rapidly.

If each American woman had 2.1 children, we would have replacement-level fertility (RLF). RLF is slightly higher than 2 since the parents must replace both themselves and the children who die before reaching maturity. The U.S. birthrate is actually about 1.8 children per woman, *below* RLF. Why then do we continue to grow? Two factors are contributing to the rapid growth of the U.S. population: immigration and the past "baby boom."

Part of our current growth rate is a legacy of our recent past. Parents of the late 1940s through the 1960s had larger-than-replacement-level families (Fig. 35-16), resulting in a "baby boom," and a momentum in population growth that has not yet subsided. For example, in 1980 over a million more women entered their reproductive years than did in 1970. Even though these women are averaging fewer than 2.1 children each, because there are more women having children, our population swells.

A second crucial component of the U.S. population equation is immigration, which contributes more to population growth here than in any other nation in the world. Legal immigration in the first half of the 1980s averaged over 430,000 per year, causing about 25 percent of our total growth. Illegal immigration by its very nature is im-

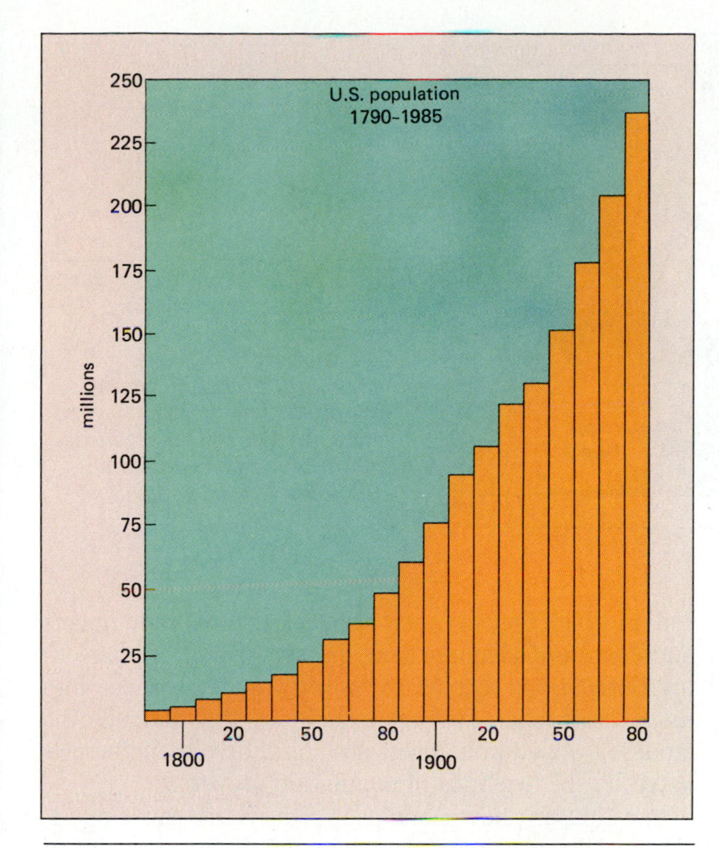

Figure 35-15 Population growth in the United States since 1790.

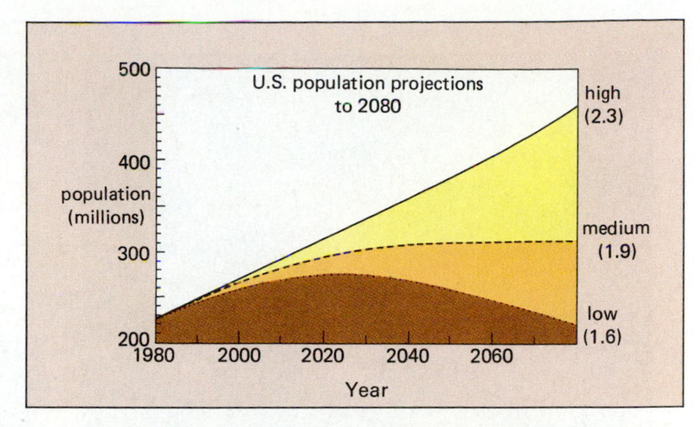

Figure 35-17 Three population scenarios for the United States based on differing fertility rates. [Redrawn from Westoff, C. F. (1986). Science *234: 554–559.]*

possible to measure. In a single year from mid-1982 to mid-1983, over 1 million illegal immigrants were deported; estimates of those who remained range from 100,000 to 1 million. The impact of recent legislation that allows amnesty for many illegal immigrants and imposes hiring sanctions cannot yet be assessed. Even using conservative estimates, legal and illegal immigration to the United States is responsible for 30 to 40 percent of our annual population growth.

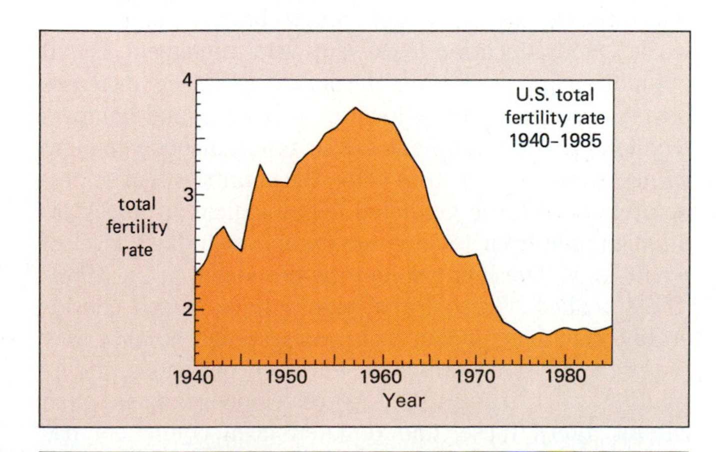

Figure 35-16 Average number of children per U.S. female, 1940–1985. [Redrawn from C. F. Westoff (1986), Science *234: 554–559.]*

According to Census Bureau projections, the U.S. population will not stabilize for decades. Figure 35-17 shows three scenarios, each based on the assumption of continued legal immigration at its present level of about 450,000 per year, but varying fertility rates. Even if fertility drops immediately to an average of 1.6 births per woman, the population will continue to grow for the next 50 years to 275 million before beginning a gradual decline. At 1.9 births per woman (a birthrate that is still below RLF), the population would stabilize in 70 years at about 315 million, sustained by the steady influx of immigrants. But what about the world? When and how will human numbers ever stabilize? How many people can the Earth support?

World Population and Carrying Capacity

A glance at the age structure of developing countries where most of the world's population resides shows a tremendous momentum for continued growth. World population in the year 2000 is predicted to be 6.3 billion, and the population will then be growing at a rate of 100 million annually. A medium-level UN projection is that the human population may stabilize in the year 2110 at 10.5 billion. Can the Earth support over twice its current population?

Earlier we defined carrying capacity as the maximum population that could be indefinitely sustained. This requires that the ecosystem not be damaged in ways that lower its ability to provide necessary resources. By this definition we may have already exceeded the Earth's carrying capacity for people. Each year it is estimated that an area of once-productive land the size of Maine is being turned into desert through overgrazing and deforestation, especially in less-developed countries. The Earth's desert

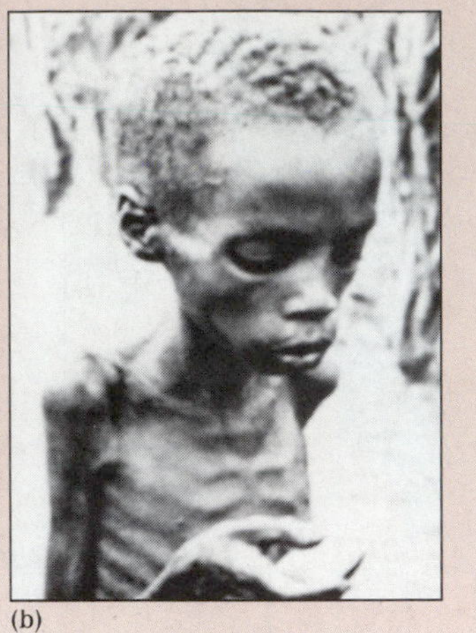

Figure 35-18 **(a)** *Desertification. Human activities, including overgrazing livestock, deforestation, and poor agricultural practices, convert once-productive land into barren desert.* **(b)** *The loss of productive land, when combined with an expanding human population, can lead to tragedy.*

area is projected to increase by 20 percent by the year 2000 as a result of human activities. In a world where between 500 million and 1 billion people are chronically undernourished, a UN survey estimated that 20 percent of the world's farmland is being degraded through erosion and mismanagement (Fig. 35-18). The UN's Food and Agricultural Organization reports that Africa's food production per capita has dropped by 20 percent since 1960, and predicts another 30 percent drop in the next 25 years. Each year the United States is losing 1 million acres of farmland to urban sprawl. World wood production per capita has declined by 9 percent or more since peaking in the mid-1960s. The demand for wood in underdeveloped countries is far outstripping production, and forested areas are being reduced by an area the size of Cuba each year. This in turn causes erosion of precious topsoil, runoff of much-needed fresh water, and the spread of deserts. The fish harvest per capita has fallen by 13 percent or more since 1970, even though expenditures on fishing fleets and fish farming have steadily increased. The destruction of tropical rain forests may exterminate 1 million species of animals and plants by the year 2000, due to the loss of their habitat. These are clear indications that our present population, at its present level of technology, is already "overgrazing" the world ecosystem and decreasing its ability to support all forms of life, including people.

Mass starvation in recent years, such as the 1985 and 1987 famines in Ethiopia, has not occurred because the Earth lacks sufficient food for its population, but because resources are unequally distributed among, and even within, countries. Each natural population must live within the carrying capacity of the ecosystem it occupies. In contrast, human populations often expand well beyond the ability of their local ecosystem to support them, relying on wealth derived from other sources (such as minerals) to allow them to import the food they need. When human populations exceed the carrying capacity of the soil on which they live, and also lack the wealth to import and distribute adequate food, disaster strikes. A burgeoning population, coerced by poverty and ignorance into using destructive farming techniques that damage the land, is trapped in a downward spiral driven and perpetuated by the demands of population growth.

In estimating how many people the Earth can support, we must keep in mind that humans desire more out of existence than a minimum caloric intake each day. It has been estimated that for everyone on Earth to live as Americans do, world population would have to be *reduced* to 500 million, *about one-tenth the present population.* For all to achieve a standard of living similar to that in Europe, where the life-style is less extravagant and untouched wilderness is almost nonexistent, *the present population would have to be reduced by half.* Merely "redistributing the wealth" is obviously not an adequate answer.

There are some who predict that technological advances will continue to increase the Earth's carrying capacity for humans into the indefinite future. In evaluating this forecast, keep in mind that technology takes time, wealth, and education to develop and implement. Rapid population growth in less-developed countries increases poverty, overloads the educational system, and hampers technological development, while simultaneously making its need more urgent. Ironically, the countries that so desperately need these predicted technological advances are often unable even to take advantage of today's level of technology. The high-yield crops developed in the 1960s (the so-called 'Green Revolution") have helped considerably, but their impact has been lessened because they require expensive equipment, fertilizer, pesticides, and irrigation water to cultivate. Less-developed countries often cannot afford these, and donated farm equipment has been left to rust in the fields for lack of fuel as well as the parts and know-how to maintain it.

Hope for the future lies in using the intelligence that

has allowed us to overcome environmental resistance to see signs of overgrazing and act before we have irrevocably damaged our world ecosystem. Human population *will* stop its exponential growth. Either we will voluntarily reduce our birth rate, or various forces of environmental resistance will increase our death rate; the choice is ours. Facing the problem of how to limit births is politically and emotionally difficult, but continued failure to do so will be disastrous. Our dignity and intelligence, and our role as self-appointed stewards of life on Earth, demand that we make the decision ourselves before we have irrevocably decreased the Earth's ability to support all life, including our own.

SUMMARY OF KEY CONCEPTS

Introduction to Ecology

Ecology is the study of the interrelationships between organisms and their environment. Ecology may be approached at the level of the population, the community with its complex interactions, or an entire ecosystem, including communities and their abiotic environment.

Population Growth

Individuals join populations through births or immigration and leave through death or emigration. Thus population change = (births − deaths) + (immigrants − emigrants). The ultimate size of a stable population is the result of interactions between biotic potential (the maximum possible growth rate) and environmental resistance, which limits population growth.

Biotic Potential

All organisms have the biotic potential to more than replace themselves over their lifetime, resulting in population growth. Populations tend to grow exponentially, since in a growing population, the number of reproducing individuals is constantly increasing. Ignoring migration, population growth rate (r) can be expressed by the equation

$$r = b - d$$

where b is the birthrate and d is the death rate. The actual number of individuals added to the population is $r \times N$, where N is the population size. During exponential growth, increasing numbers of individuals are added during each successive period. Factors influencing the maximum rate of growth of a given population include the age at first reproduction, frequency of reproduction, number of offspring produced each time, and the death rate under ideal conditions. Populations cannot continue to grow exponentially indefinitely; they either stabilize or undergo periodic oscillations as a result of environmental resistance.

Environmental Resistance

Environmental resistance restrains population growth by increasing the death rate or reducing the reproductive rate. The maximum size at which a population may be sustained indefinitely by an ecosystem is the carrying capacity, determined by limited resources such as space, nutrients, and light.

Populations are maintained at or below carrying capacity by density-independent forms of environmental resistance such as weather, and density-dependent forms, including predation, parasitism, and competition, both intraspecific and interspecific. Intraspecific competition results in the allocation of resources directly, through scramble competition, or indirectly through chemical interactions or social behaviors, collectively called contest competition. Emigration is another possible outcome of competition.

Patterns in Populations

Populations are spatially distributed according to their physical and social requirements and can be classified as aggregated, uniformly distributed, or randomly distributed. Aggregations may occur for social reasons or around limited resources. Uniform distribution is usually the result of territorial spacing. Random distribution is rare, occurring only when individuals do not interact socially and when resources are abundant and evenly distributed.

Populations show specific survivorship curves that express the likelihood of survival at a given age. Convex curves are characteristic of long-lived species with few offspring that receive parental care. Species with constant curves have an equal chance of dying at any age. Concave curves are typical of organisms such as plants and invertebrates that produce numerous offspring, most of which die.

The Human Population

The human population has exhibited exponential growth for an unprecedented time by overcoming environmental resistance and increasing the Earth's carrying capacity for people. This has been accomplished by the use of tools, agriculture, industry, and medical advances.

Age-structure diagrams depict numbers of males and females in various age groups comprising a population. Expanding populations have pyramidal age structures, stable populations show rather straight-sided figures, while decreasing populations are illustrated by figures that are constricted at the base.

Today most of the world's people live in less-developed countries with rapidly growing populations, where a variety of social and cultural conditions encourage large families. The United States is among the fastest growing of the developed countries, largely due to high immigration rates and the postwar "baby boom."

The carrying capacity of the Earth for humans is unknown, but with a population of over 5 billion, resources are already too limited for all to be supported at a living standard equivalent to that of Europe. A steady decline in productive land, wood, and fish harvests indicates that we are damaging our world ecosystem and decreasing its ability to sustain us.

GLOSSARY

Abiotic (ā-bī-ah′-tik) Nonliving.

Age structure the distribution of males and females in a population according to age categories; often represented graphically.

Aggregated distribution characteristic of populations in which individuals are clustered into groups. These may be social or based on the need for a localized resource.

Biotic (bī-ah′-tik) living.

Biotic potential the most rapid potential growth rate of a population, assuming a maximum birthrate and minimum death rate.

Boom and bust a population cycle characterized by rapid exponential growth followed by a sudden major decline in population size, seen in seasonal species and some populations of small rodents, such as lemmings.

Carrying capacity the maximum population size that an ecosystem can maintain on a sustained basis. Determined primarily by the availability of space, nutrients, water, and light.

Community all the interacting populations within an ecosystem.

Competition interaction that occurs between individuals when both attempt to utilize a resource (e.g., food or space) that is limited relative to the demand for it.

Contest competition a mechanism for resolving intraspecific competition using social or chemical interactions.

Density dependent description of any factor that limits population size more effectively as the population density increases.

Density independent description of any factor such as freezing weather that limits a sensitive population without regard to its size.

Dominance hierarchy a system of social ranks within a population that determines which individuals obtain first access to limited resources.

Ecology (ē-kol′-uh-gē) the study of the interrelationships of organisms with each other and with their nonliving environment.

Ecosystem (ē′kō-sis-tem) all the organisms and their nonliving environment within a defined area.

Emigration (em-uh-grā′shun) movement of individuals out of an area.

Environmental resistance any factor that tends to counteract biotic potential, limiting population size.

Exponential growth a continuously accelerating increase in population size.

Immigration (im-uh-grā′-shun) movement of individuals into an area.

Interspecific competition competition between individuals of different species.

Intraspecific competition competition between individuals of the same species.

Parasitism (pa′-ra-sit-ism) the process of feeding on a larger organism without killing it immediately or directly.

Population a group of interbreeding organisms (organisms of the same species) within an ecosystem.

Predation (pre-dā′-shun) the act of killing and eating another living organism.

Random distribution spacing in which the probability of finding an individual is equal in all parts of an area.

Replacement-level fertility the average birthrate at which a reproducing population exactly replaces itself during its lifetime.

Scramble competition direct interactions between individuals attempting to acquire the same limited resource.

S-curve the growth curve that describes a population introduced into a new area. It consists of an initial period when numbers remain relatively low, followed by a period of exponential growth, followed by decreasing growth rate, and finally, relative stability.

Survivorship curve a curve resulting when the number of individuals in a population is graphed against their age, usually expressed as a percentage of their maximum life span.

Territoriality a behavior in which individuals defend an area in which important resources are located from others of their species.

Uniform distribution a relatively regular spacing of individuals within a population, often as a result of territorial behavior.

STUDY QUESTIONS

1. Calculate the change in the population of a city of 100,000 in which (during a 1-year period) there were 5000 births, 3000 deaths, 1100 immigrants, and 100 emigrants. Express your answer both as a percent growth rate and as change/1000/yr.
2. Define *biotic potential*. Explain natural selection in terms of biotic potential and environmental resistance.
3. Draw the growth curve of a population before it encounters significant environmental resistance. What is the name of this type of growth, and what is its distinguishing characteristic?
4. Distinguish between density-independent and density-dependent forms of environmental resistance.
5. Describe two major forms of intraspecific competition, and give an example of each.
6. Describe (or draw a graph illustrating) what is likely to happen to a population that far exceeds the carrying capacity of its ecosystem. Explain your answer.
7. List three density-dependent forms of environmental resistance and explain why each is density-dependent.
8. Distinguish between populations showing concave and convex survivorship curves. Which is characteristic of Americans, and why?
9. List the human activities that have expanded the Earth's carrying capacity for people. Then list those that have lowered environmental resistance. Can we continue expanding carrying capacity and lowering environmental resistance indefinitely? Explain.
10. Given that the U.S. birthrate is currently below replacement-level fertility, why is our population growing?

11. The United States has a long history of accepting large numbers of immigrants. Discuss the implications of immigration for population stabilization.
12. What factors encourage rapid population growth in less-developed countries? What will it take to change this?
13. Contrast age structure in rapidly growing versus stable human populations. Why is there a momentum in population growth built into a rapidly growing population?

SUGGESTED READINGS

Brown, L. R., and others. *State of the World.* New York: W. W. Norton & Company, Inc., 1988. Annually updated collection of articles concerning global resources, pollution, and population.

Chiras, D. D. *Environmental Science.* 2nd ed. Menlo Park, Calif.: The Benjamin-Cummings Publishing Co., 1988. Thorough, readable coverage of the impact of humans on ecosystems.

Ehrlich, P. R., and Erlich, A. *Extinction.* New York: Random House, Inc. 1981. The causes and the consequences of the disappearance of species.

Ehrlich, P.R., and Roughgarden, J. *The Science of Ecology.* New York: Macmillan Publishing Company, 1987. A readable, complete introductory ecology text, with good coverage of populations.

Hillfry, E. *Ecology 2000. The Changing Face of Earth.* New York: Beaufort Books, Inc., 1980. Engaging reading on the major environmental problems facing the world.

Myers, J. H., and Krebs, C. J. "Population Cycles in Rodents." *Scientific America,* June 1974. Natural populations of small rodents show periodic three- to four-year cycles in population size.

Population Today and *Population Bulletin.* Two regular publications of the Population Reference Bureau, Inc. 777 14th St. NW, Suite 800, Washington, DC 20005.

Westoff, C. F. "Fertility in the United States." *Science* 234:554–559 (1986). Excellent summary of past, current, and projected fertility.

36 Community Interactions

A jumping spider stalks a snowberry fly, approaching with catlike stealth. Suddenly spotting the spider, the fly spreads its wings, brings them forward slightly, and moves back and forth in a jerky dance. The spider hesitates, then flees, leaving the fly untouched. Careful examination of the wing markings of the fly show an uncanny resemblance to the legs of a spider (Fig. 36-1). Further, the jerky movements of the threatened fly resemble those a spider makes when driving another spider from its territory. Natural selection has finely tuned both the behavior and the appearance of the fly to avoid predation by jumping spiders. This mimicry of a predator by its prey is only one of the myriad forms of community interaction that occur in ecosystems.

An ecological **community** consists of all the interacting populations within an ecosystem. Typically, populations within communities have coevolved. During **coevolution,** different species act as agents of natural selection on one another. For example, predators and parasites limit their prey populations without eliminating them. Animals that are preyed upon have evolved elaborate defenses that help them survive. Herbivores have digestive specializations that allow them to eat the local plants. The plants in turn grow rapidly or defend themselves by chemical or physical means, keeping one step ahead of their predators.

Community interactions fall into three major categories: **competition,** in which populations compete for limited resources; **predation,** in which one organism kills and eats another; and **symbiosis,** in which two species live together in close association over an extended time. Symbiotic interactions may be further classified according to how they affect each of the species involved (Table 36-1).

Long-established communities are generally complex and self-sustaining, with interacting populations that remain relatively stable. Communities originate, however, over long periods of gradual change. During these periods, one community gives way to another in a process called **succession.**

In this chapter we focus on the variety of community interactions in stable ecosystems, and the community changes that occur during succession.

Figure 36-1 In response to the approach of a jumping spider (top), the snowberry fly spreads its wings, revealing a pattern resembling spider legs (bottom). The fly enhances the effect by performing a jerky, side-to-side dance that resembles the leg-waving display of another jumping spider defending its territory.

INTERSPECIFIC COMPETITION AND THE ECOLOGICAL NICHE

Although the word "niche" calls to mind a small space or cubbyhole, in ecology it means much more. Each species occupies an **ecological niche** that defines all aspects of its way of life. In addition to the organism's physical

Table 36-1 **Interactions Among Organisms**

Type of Interaction	Effect on Organism A[a]	Effect on Organism B[a]
Competition between A and B	−	−
Predation by A on B	+	−
Symbiosis		
Parasitism by A on B	+	−
Commensalism of A with B	+	0
Mutualism between A and B	+	+

[a]+, benefits; −, harms; 0, neutral or no effect.

home in an ecosystem, its niche includes its food, behavior, predators, and all the physical environmental factors (temperature, rainfall, soil type, degree of shade or sunlight, etc.) necessary for its survival. In short, the niche defines a particular species' role in the community, and for each species the niche is unique.

Just as no two organisms can occupy the same physical space at the same time, no two species can inhabit the same ecological niche. This important ecological rule, often called the **competitive exclusion principle,** was formulated in 1934 by G. F. Gause. If two species with the same or very similar niches are placed together, inevitably one will outcompete the other, and the less well adapted of the two will eventually die out. Gause demonstrated this by forcing two different species of the protist *Paramecium* to occupy almost identical niches under laboratory conditions. Separately, each population thrived; together, one always eliminated the other (Fig. 36-2).

Competition is proportional to the amount of niche overlap; the more overlap, the more intense the competition. Competition intensifies as populations of competing species increase in size. Thus interspecific competition is a density-dependent form of environmental resistance, as described in Chapter 35. In Fig. 36-2, notice that the extreme niche overlap was tolerated by the two species when their populations were small and resources were adequate for all. As the populations grew, competition intensified, eventually eliminating one species.

In natural communities, coevolution has assured that the niches of different species never overlap entirely. For example, the ecologist R. MacArthur tested Gause's principle on five species of North American warbler, all of which hunt for insects in the same type of spruce trees. Although the feeding niches of these birds appear to overlap considerably, MacArthur found that each species concentrated its search in specific areas of the tree, thus minimizing niche overlap and thereby reducing competition. Natural selection favors individuals within each competing population whose feeding preferences differ the most from its competitors, thus reducing niche overlap.

Although coevolution tends to minimize niche overlap, closely related species still compete directly for limited resources. This may restrict the size and distribution of the competing populations. For example, the barnacle *Chthamalus* shares the rocky shores of Scotland with another genus of barnacle, *Balanus,* and their niches overlap considerably (Fig. 36-3). Joseph Connell found that *Chthamalus* dominates the upper shore and *Balanus* dominates the lower. When he removed *Balanus,* however, *Chthamalus* spread downward, covering part of the area its competitor had once inhabited. Similarly, removal of *Chthamalus* allowed *Balanus* to spread upward. Under natural conditions, the distribution of the two species results from the interaction of several factors, including direct competition for space. Where the habitat is appropriate for both species, *Balanus* conquers, since it is larger and faster growing. *Chthamalus* tolerates dryness better than *Balanus,* however. On the upper shore, where only high tides submerge the animals, *Chthamalus* out-

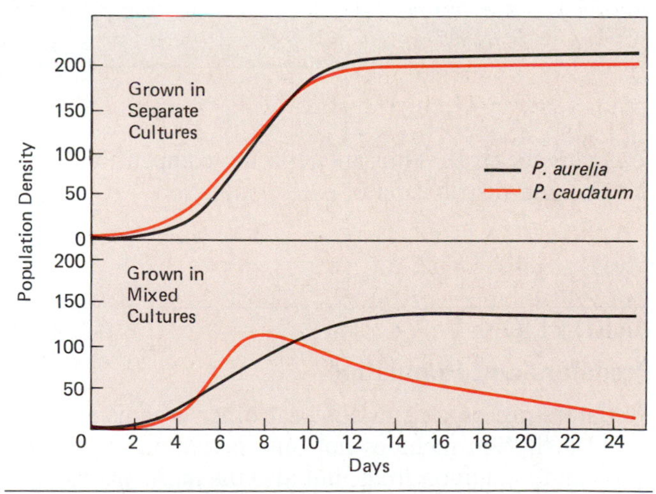

Figure 36-2 *(Top) Raised separately with a constant food supply, both* Paramecium aurelia *and* Paramecium caudatum *show the S-curve typical of a population that grows, then stabilizes. (Bottom) Raised together and forced to occupy the same niche,* P. aurelia *outcompetes* P. caudatum. *(Modified from G. F. Gause,* The Struggle for Existence, *Williams & Wilkins, Baltimore, 1934.)*

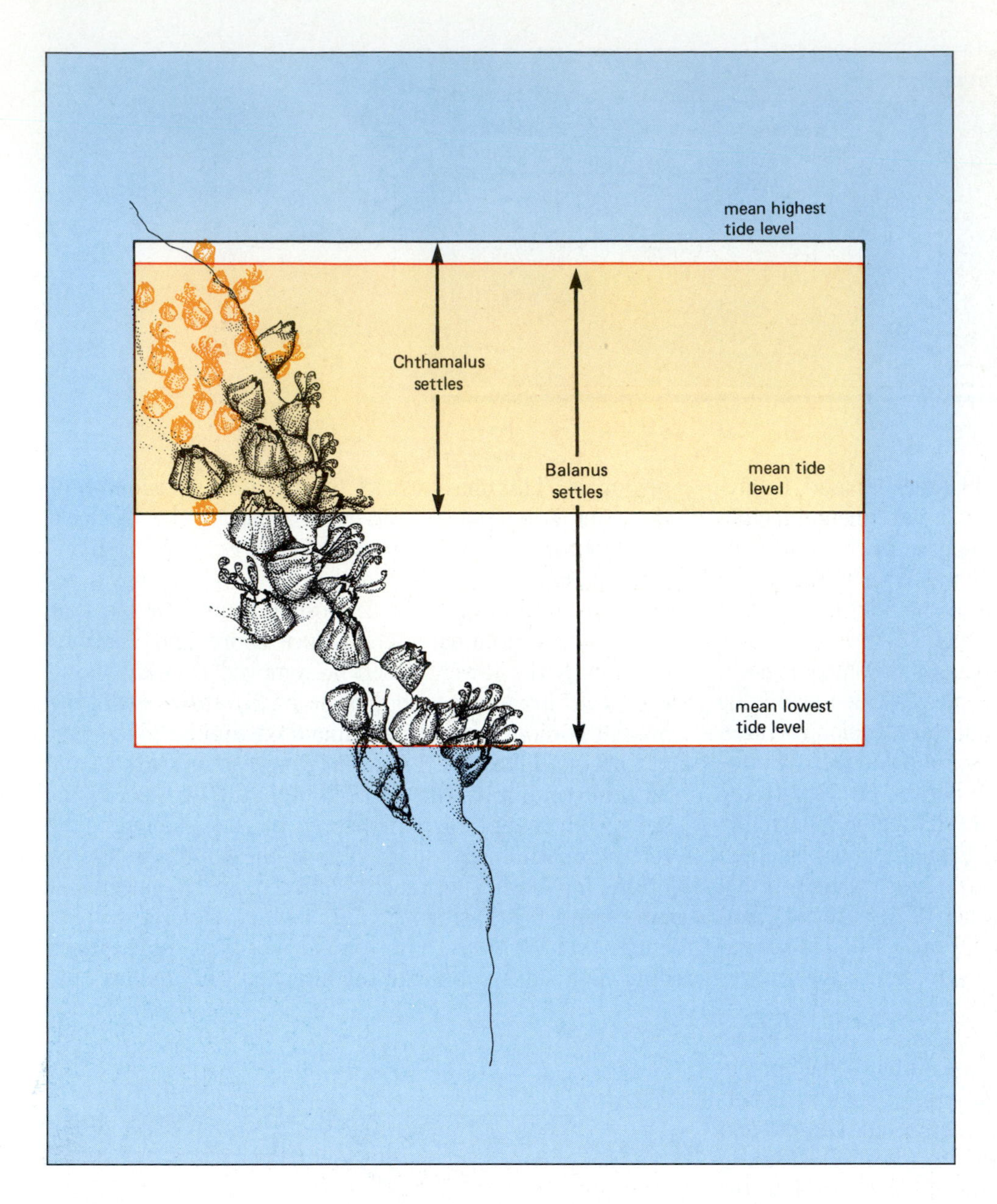

Figure 36-3 Barnacle larvae are dispersed by ocean currents and settle on the rocky shore, where they attach permanently. Although larval settlement of these two species overlaps extensively, adult Balanus *dominate except on the drier upper levels, where desiccation-resistant* Chthamalus *outcompete them.*

competes its rival. Thus interspecific competition limits the size and distribution of each population.

PREDATION

Predators and Population

Predators are organisms that eat other organisms. This broad definition includes not only predation of lions on zebra and bats upon frogs, but also the more prosaic predation of antelope on sagebrush and the exotic predation of sundew upon insect (Fig. 36-4). Predators are usually either larger than their prey or hunt collectively, as wolves do when bringing down a moose. Predators are also less abundant than their prey, for reasons discussed in Chapter 36.

The influence of predators on prey populations varies considerably. Some predators feed on vulnerable prey that have exceeded their carrying capacity; thus predators may maintain prey populations near their optimal maximum density. In other cases, predators may keep prey populations well below the carrying capacity. The best documented cases of predators maintaining prey populations below carrying capacity are those in which exotic species (see the essay in Chapter 35) have been controlled by imported predators. For example, the prickly pear cactus was introduced into Australia and spread uncontrollably, overrunning millions of acres of valuable pasture and range land. In the late 1920s, the cactus moth, a predator of the prickly pear, was imported from Argentina. Within a few years, the cacti were largely destroyed, and the moth continues to keep the pest controlled (Fig. 36-5).

Figure 36-4 Forms of predation. **(a)** *Lions of East Africa hunt cooperatively to bring down large herbivores, including zebra.* **(b)** *A few tropical bats feed on frogs or fish. Predatory bats home in on the mating calls of this tree frog.* **(c)** *Herbivores, such as these pronghorn antelope in Wyoming, are predators of sagebrush.* **(d)** *A predatory plant, the sundew attracts and binds insects with its glistening sticky knobs. The insects provide supplemental nitrogen for the plant.*

Predation may also help maintain a larger number of species in an ecosystem. R. T. Paine of the University of Washington removed predatory seastars from sections of rocky shore along the U.S. Pacific coast. In the absence of this predator, mussels on which it fed became so plentiful that they outcompeted other species. In this case, removing predators caused a reduction in the total number of species in the ecosystem.

Prey populations that fluctuate can have a major impact on their predators. In the arctic, lemming populations show dramatic population cycles. Predators that specialize on lemmings, such as the snowy owl and the arctic fox, regulate the number of offspring they produce according to the abundance of lemmings. The snowy owl may produce up to 13 chicks when lemmings are abundant, but not reproduce at all in years when they are scarce.

Predator–prey interactions may cause cyclic fluctuations of both populations when the predator depends almost exclusively on a single type of prey. In the laboratory, for example, when bean weevils were provided with a steady source of food and maintained with the predatory braconid wasp, both populations fluctuated, but slightly out of synchrony (Fig. 36-6). The wasps lay their eggs on weevil larvae, and the newly hatched wasps feed on the larvae. A large prey population assures a high survival rate for wasp offspring, increasing the predator population. But under intense predation pressure, the weevil population plummets, reducing the survival of the next generation of predators. The reduced wasp population then allows an increase in weevils, and so on. In the wild, such fluctuations are almost impossible to demonstrate because so many other factors affect the populations involved. Wild prey populations, rather than being controlled exclusively by their predators, are often strongly influenced by weather or changes in the availability of their food. Predators, in turn, may switch to other prey when one becomes scarce.

Coevolution of Predators and Prey

To survive, predators must feed and prey must avoid becoming food. These populations thus exert intense selec-

Figure 36-5 (left) A pasture in Queensland, Australia, is blanketed by the imported prickly pear cactus, whose spread was unchecked by predators. (right) The same site three years after the introduction of a predatory cactus moth appropriately named Cactoblastis cactorum. *(Courtesy of the Department of Lands, Queensland, Australia.)*

tive pressure on one another, resulting in coevolution. As prey become more difficult to catch, predators must become more adept at hunting. A complex coevolutionary adaptation is described in the essay, "Ploy and Counterploy." Coevolution has endowed the cheetah with speed and camouflage spots, and its zebra prey with speed and camouflage stripes. It has produced the keen eyesight of the hawk and the warning call of the prairie dog, the stealth of the jumping spider, and the remarkable spider mimicry of the fly it stalks. In this section we examine some of the results of predator–prey coevolution in more detail.

COEVOLUTION OF HERBIVORES AND THEIR PLANT PREY. Plants have evolved a variety of adaptations that deter predators. Many, such as the milkweed, synthesize toxic and distasteful chemicals. Animals rapidly learn not to eat foods that make them sick, so milkweeds and other toxic plants suffer little browsing. Consequently, such plants are often very abundant, and any animal immune to the plant poisons enjoys a bountiful food supply. As plants evolved toxic chemicals for defense, certain insects evolved increasingly efficient ways to detoxify or even store the chemicals. The result is that nearly every toxic plant is eaten by at least one species of insect. For example, monarch butterfly caterpillars consume the toxic milkweed. The caterpillars not only tolerate the milkweed poison, they store it in their tissues as a defense against their own predators.

Grasses have evolved tough silicon (glassy) substances in their blades, deterring all but those with strong, grinding teeth and powerful jaws. Thus grazing animals have come under selective pressure for longer, harder teeth. During the evolution of the horse (see Chapter 13), striking changes occurred in their teeth. As grasses evolved tougher blades that reduce predation, horses evolved longer teeth with thicker enamel coatings that resist wear.

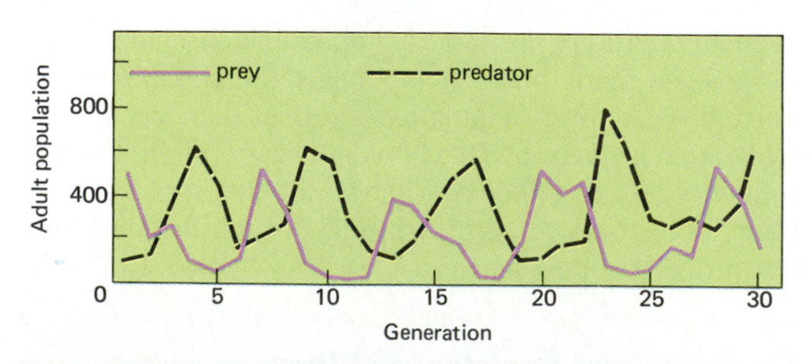

Figure 36-6 Out-of-phase fluctuations in laboratory populations of the azuli bean beetle and its braconid wasp predator.

PROTECTIVE COLORATION. An old maxim of detective novels is that the best hiding place is right out in plain sight. Both predators and prey have evolved colors and patterns that render them inconspicuous even in plain sight. Often, these **camouflaged** animals resemble their general surroundings (Fig. 36-7). But some closely resemble specific (but uninteresting) objects such as leaves, twigs, or even bird droppings (Fig. 36-8). For camouflaged animals, behavior is just as important as coloration. Resembling a bird dropping does an animal little good if it moves about, advertising that it is alive and edible. Thus camouflaged animals tend to remain motionless rather than fleeing their predators.

Some animals have evolved very differently, exhibiting

Ploy and Counterploy

The intricate evolutionary adaptations of predators and prey are beautifully illustrated by the relationship of the assassin bug and the southwestern desert camphor weed, studied by Thomas Eisner of Cornell and his collaborators. The camphor weed exudes a sticky, noxious resin from its leaves that discourages herbivorous predators. The assassin bug (Fig. E36-1) is undeterred by the camphor "glue." The female assassin bug collects the substance and smears it on her abdomen, where it coats her eggs as they emerge, making them undesirable to predators. But the story is not yet over. Soon after hatching, young assassin bugs laboriously scrape the "glue" from their discarded eggshells and transfer it to their forelegs, where it will aid in the capture of their own prey. In this complex association, the defenses evolved by a plant are utilized by the assassin bug both defensively against its own predators and offensively against its prey.

Figure E36-1 A female assassin bug scrapes noxious resin from a camphor plant. Her warning coloration alerts potential predators of her painful sting.

Figure 36-7 Camouflage renders potential prey inconspicuous. **(a)** *The flounder's flat shape and its behavior (flipping sand over its back, then lying perfectly still) help the fish resemble a sandy bottom. The flounder can also alter its color and pattern somewhat to resemble its surroundings. The leeches attached to it are not deceived, since they locate their host by scent.* **(b)** *A cryptic sphingid moth from Trinidad blends perfectly with tree bark.* **(c)** *The leafy sea dragon (an Australian "sea horse" fish) bears leafy projections that allow it to disappear in a bed of seaweed.*

bright **warning coloration** (Figs. 36-9 and E36-1). These animals are usually distasteful and often extremely poisonous. Since poisoning your predator is small consolation if you are already dead, the bright colors declare: "eat me at your own risk." After a single unpleasant experience, predators get the message and avoid these conspicuous prey.

In some instances, different species of dangerous animals have evolved similar warning coloration, for example, the conspicuous stripes on bees, hornets, and yel-

Figure 36-8 Resembling uninteresting parts of the environment allows some animals to avoid predation. **(a)** *A Peruvian katydid perfectly mimics a dead leaf,* **(b)** *a moth from Panama is indistinguishable from a chunk of bark, and* **(c)** *a moth on a Missouri mulberry closely resembles a bird dropping.*

Figure 36-9 Warning coloration. **(a)** *The tropical lionfish, with its venomous spines, and* **(b)** *the South American poison arrow frog, with its poisonous skin, each advertise their unpalatability with bright and contrasting color patterns.*

lowjackets. A common pattern of coloration results in faster learning by predators, and thus less predation on all similar-appearing species.

Once warning coloration evolved, there arose a selective advantage for tasty animals to resemble poisonous ones. For example, predators rapidly learn to avoid the poisonous orange-and-black monarch butterfly. Similar colors reduce predation on the nonpoisonous viceroy butterfly, which has evolved to closely resemble the monarch (Fig. 36-10). This is an example of **mimicry,** in which one (usually tasty) animal species called a **mimic** evolves to resemble a poisonous or distasteful species, the **model.**

Although less common than mimicry by prey species, a few predators have evolved **aggressive mimicry.** Resembling a harmless animal, such predators provide real-life examples of the classic wolf in sheep's clothing (Fig. 36-11).

Certain prey species utilize still another mimicry device, **startle coloration.** Several insects, including certain moths (Fig. 36-12) and caterpillars, have independently evolved spots of color that closely resemble the eyes of a much larger, and possibly dangerous, animal. These animals are usually camouflaged overall and keep the eye-spots hidden. If a predator gets too close, suddenly the prey flashes the eyespots. The predator is startled and momentarily draws back. The prey quickly covers the eye-spots and flees, often escaping to a new hiding place before the predator recovers. A more sophisticated vari-

Figure 36-10 *Mimicry of the warning coloration of the poisonous monarch butterfly (top) by the harmless viceroy (bottom). The monarch stores poisons from milkweeds on which its caterpillar feeds. Birds learn after one taste to avoid the monarch and will also avoid the tasty but nearly indistinguishable viceroy.*

Figure 36-11 **(a)** *Some fish, such as this marine wrasse, obtain food by eating parasites off the bodies of larger fish, a mutualistic relationship called a "cleaning symbiosis." Fish welcome the approach of this predator of parasites.* **(b)** *The saber-toothed blenny has evolved to closely resemble the cleaner wrasse. Fish that allow the blenny to approach usually get a chunk bitten out of them—hence the term "aggressive mimicry."*

Figure 36-12 **(a)** *The peacock moth from Trinidad is well camouflaged, but should a predator approach too closely, it suddenly opens its wings to reveal spots resembling large eyes* **(b)**. *This startles the predator, giving the moth a chance to flee.* **(c)** *Would-be predators of this caterpillar larva of the swallowtail butterfly are deterred by its close resemblance to a snake. Note that the caterpillar's head is the "snake's nose."*

ation on the theme of prey that mimic predators was recently discovered among several related fly species which mimic territorial displays of their spider predators as described early in this chapter (Fig. 36-1).

SYMBIOSIS

Symbiosis, which literally means "living together," is defined as a close interaction of organisms of different species for an extended time. Considered in its broadest sense, symbiosis includes parasitism, commensalism, and mutualism. Although one species always benefits, the second species may be harmed, not affected, or benefited (Table 36-1).

As we have seen, predators eat other organisms, often killing them. **Parasites** also live at the expense of others, but in more subtle and less immediately fatal ways. Although it is sometimes difficult to distinguish clearly between a predator and a parasite, in general parasites are much smaller and more numerous than their prey (also called hosts). Parasites actually live in or on their hosts and do not immediately kill them. Familiar parasites include tapeworms, fleas, and numerous disease-causing microorganisms. The variety of infectious bacteria and viruses and the impressive precision of the immune system that counters their attacks are evidence of the powerful forces of coevolution between parasites and hosts.

Commensalism occurs when the relationship between two species benefits one without affecting the other. Commensalism occurs between large, herd-forming mammals and certain birds. For example, as herds of bison graze, they disturb numerous insects dwelling in the grass. Birds follow the bison, eating the insects that fly up to avoid being trampled. The bison are not affected by the activity of the birds. Birds and the trees in which they nest are also involved in a commensal relationship. The birds obtain shelter and protection while the tree is usually not affected.

When two organisms interact so that both benefit, the relationship is called **mutualism.** The interactions among flowering plants and various pollinators were discussed in Chapter 22. Although these partners are not truly living together, their association benefits both and is often considered a form of mutualism. Mutualistic associations occur in the digestive tracts of cows and termites, where protists and bacteria find food and shelter while helping their hosts extract nutrients, and in our own intestines, where bacteria synthesize certain vitamins. The nitrogen-fixing bacteria inhabiting special chambers on the roots of legume plants are another important example. These bacteria obtain food and shelter from the plant and in return trap nitrogen in a form the plant can utilize. Some mutualistic partners have coevolved to the extent that neither can survive alone. A noteworthy example is the ant–acacia mutualism described in the essay, "Acacia and Ant—Partners for Life." Among vertebrates, mutualistic interactions are relatively rare and typically are less intimate and extended, as in the relationship of the cleaner wrasse (Fig. 36-11a) and the fish it cleans.

SUCCESSION: COMMUNITY CHANGES OVER TIME

In a mature terrestrial ecosystem, the populations comprising the community interact with one another and with their nonliving environment in intricate ways. But this tangled web of life did not spring fully formed from bare rock or naked soil; rather it emerged in stages over a long period, a process called succession. **Succession** is a change in a community and its nonliving environment over time. It is a kind of "community relay" in which assemblages of plants and animals replace one another in a sequence that is at least somewhat predictable.

Succession occurs under a variety of circumstances but is most easily observed in terrestrial and freshwater ecosystems. Freshwater ponds and lakes tend to undergo a series of changes that transform them first into marshes and eventually to dry land. Shifting sand dunes are stabilized by creeping plants and may eventually support a forest. Volcanic eruptions may, as in the case of Mt. St. Helens, wipe out previously existing ecosystems, or they may create new islands which are soon colonized. Forest fires create a nutrient-rich environment which encourages rapid invasion of new life (Fig. 36-13).

The precise changes occurring during succession are as diverse as the environments in which succession occurs, but certain general stages can be recognized. In each case, succession is begun by a few hardy invaders called **pioneers** and ends with a diverse and relatively stable **climax community.** As the community progresses from the pioneers to the climax, the organisms gradually alter the nonliving environment. Ironically, these changes favor competitors, which displace the existing populations. The climax community differs from earlier successional stages because it no longer alters the environment. The climax community will persist unless external forces (such as a gradual change in climate or human activities) alter it.

During succession there are certain general trends in ecosystem structure:

1. The soil increases in depth and in its content of organic material.
2. The overall productivity (the amount of organic material produced in a given area over a given time) increases.
3. The number of different species increases, as does the number of interactions within the community.
4. Longer-lived species come to dominate the ecosystem,

Acacia and Ant—Partners for Life

Daniel Janzen of the University of Pennsylvania, then a doctoral student, was walking down a road in Veracruz, Mexico, when he saw a flying beetle alight on a thorny tree, only to be driven off by an ant. Further observation revealed that the tree, a bull's-horn acacia, was covered with ants. A large ant colony of the genus *Pseudomyrmex* made its home inside the enlarged thorns of the plant, whose soft pulpy interiors are easily excavated to provide secure shelter (Fig. E36-2).

To determine how important the ants are to the tree, Janzen began stripping the thorns by hand until he found and removed the thorn housing the ant queen, thus destroying the colony. He later turned to more efficient but dangerous methods, eliminating all the ants on a large stand of acacias with the insecticide parathion. The acacias were unharmed by the poison, Janzen became ill from it, and the ants were all killed. Within a year of the spraying, the recovered Janzen found nearly all the trees dead, consumed by insects and other herbivores, and shaded out by competing plants. The ground surrounding the trees, which the ants normally kept neatly pruned, was completely overgrown. The trees were apparently dependent on their resident ants for survival.

Figure E36-2 A hole in the enlarged thorn of the bull's horn acacia provides shelter for members of the ant colony. The ant entering the thorn is carrying a food capsule produced by the acacia. As the ant colony grows, more thorns are invaded.

Figure E36-3 Yellow, protein-rich capsules are produced at the tips of certain acacia leaves. These provide food for the resident ants.

Wondering if the ants could survive off the tree, Janzen painstakingly peeled the ant-inhabited thorns off 100 acacia trees, suffering multiple stings in the process. He housed each ant colony in a jar provided with local nonacacia vegetation and insects for food. The colonies all starved. Close inspection of the acacia revealed swollen structures filled with sweet syrup at the base of the leaves and protein-rich capsules on the leaf tips (Fig. E36-3). Together, these provide a balanced diet for the ants.

Janzen's experiments strongly suggest that this species of ant and acacia have an obligatory mutualistic relationship, that is, that neither can survive without the other. Of course, further observations were required to confirm this. The fact that the ants starved in Janzen's jars did not rule out that they might survive successfully elsewhere, but, in fact, this species of ant is never found living independently. Similarly, the bull's-horn acacia is never found without its resident ant colony. Thus a chance observation followed by careful research led to the discovery of an important mutualistic association.

and the rate at which populations replace one another slows. The presence of longer-lived species combined with more complex community interactions results in a community that is more stable and resistant to change.

5. In the climax community, the total weight of living organisms reaches a maximum, and the species present no longer alter the ecosystem in ways that encourage the growth of their competitors.

Figure 36-13 Areas ripe for succession. **(a)** *Hawaiian lava field and* **(b)** *sand dunes are areas where primary succession will occur. The aftermath of a forest fire* **(c)** *provides the basis for secondary succession.*

Primary and Secondary Succession

Succession takes two major forms: primary and secondary. During **primary succession,** an ecosystem is forged from bare rock, sand, or a clear glacial pool where there is no trace of a previous community. The formation of an ecosystem "from scratch" is a process often requiring thousands or even tens of thousands of years. **Secondary succession** occurs after an existing ecosystem is disturbed, as in the case of a forest fire or an abandoned farm field. It happens much more rapidly because the previous community has left its mark in the form of soil and seeds. Succession in an abandoned farm field in the southeastern United States can reach its climax after two centuries. In Fig. 36-14 and in the examples below, we examine these processes in more detail.

PRIMARY SUCCESSION ON BARE ROCK. Figure 36-15 illustrates primary succession as studied on Isle Royale, an island in Lake Michigan. Bare rock, such as that exposed by a retreating glacier, begins to liberate nutrients by weathering. Cracks form as the rock alternately freezes and thaws, contracting and expanding. For lichens (symbiotic associations of fungi and algae), the weathered rock provides a place to attach where there are no competitors and plenty of sunlight. Lichens can photosynthesize, and they obtain minerals by dissolving some of the rock with an acid they secrete. As the lichens spread over the rock, drought-resistant, sun-loving mosses begin growing in the cracks. Fortified by nutrients liberated by the lichens, the moss forms a dense mat that traps dust, tiny rock particles, and bits of organic debris. The death of some of the moss adds to a growing nutrient base, while the moss mat itself acts as a sponge, trapping moisture. Within the moss, seeds of larger plants germinate. Eventually, their bodies contribute to a growing layer of soil. As woody shrubs such as blueberry and juniper take advantage of the newly formed soil, the moss and lichens may be shaded out and buried by decaying leaves and vegetation. Eventually, trees such as jack pine and blue spruce root in the deeper crevices, and the sun-loving shrubs are shaded out. Within the depths of the forest, shade-tolerant seedlings of taller or faster-growing trees, such as balsam fir, paper birch, and white spruce, thrive. Eventually these overtower and replace the original trees, which are intolerant of shade. After a thousand years or more, a tall climax forest thrives on what was once bare rock.

SECONDARY SUCCESSION ON AN ABANDONED FIELD. Figure 36-16 illustrates succession on an abandoned southeastern farm. The pioneers are fast-growing annual weeds such as crabgrass, ragweed, and sorrel, which root in the rich soil already present and thrive in direct sunlight. A few years later, perennial newcomers such as asters and goldenrod, broomsedge grass, and woody shrubs such as blackberry invade. These proliferate and dominate for the next few decades. Eventually, they are replaced by pines and fast-growing deciduous trees such as

Figure 36-14 *Succession in a small freshwater lake in the Rocky Mountains is the result of an influx of material from outside the ecosystem.* **(a)** *Streams feeding the lake carry sediments rich in organic material and nutrients from the land.* **(b)** *Deposited in the still waters of the lake, these sediments gradually make the lake shallower and provide anchorage for lily pads and marsh grasses.* **(c)** *Over the years, the decomposing vegetation further fills the lakebed, transforming it first to a swamp, and finally to dry meadow.*

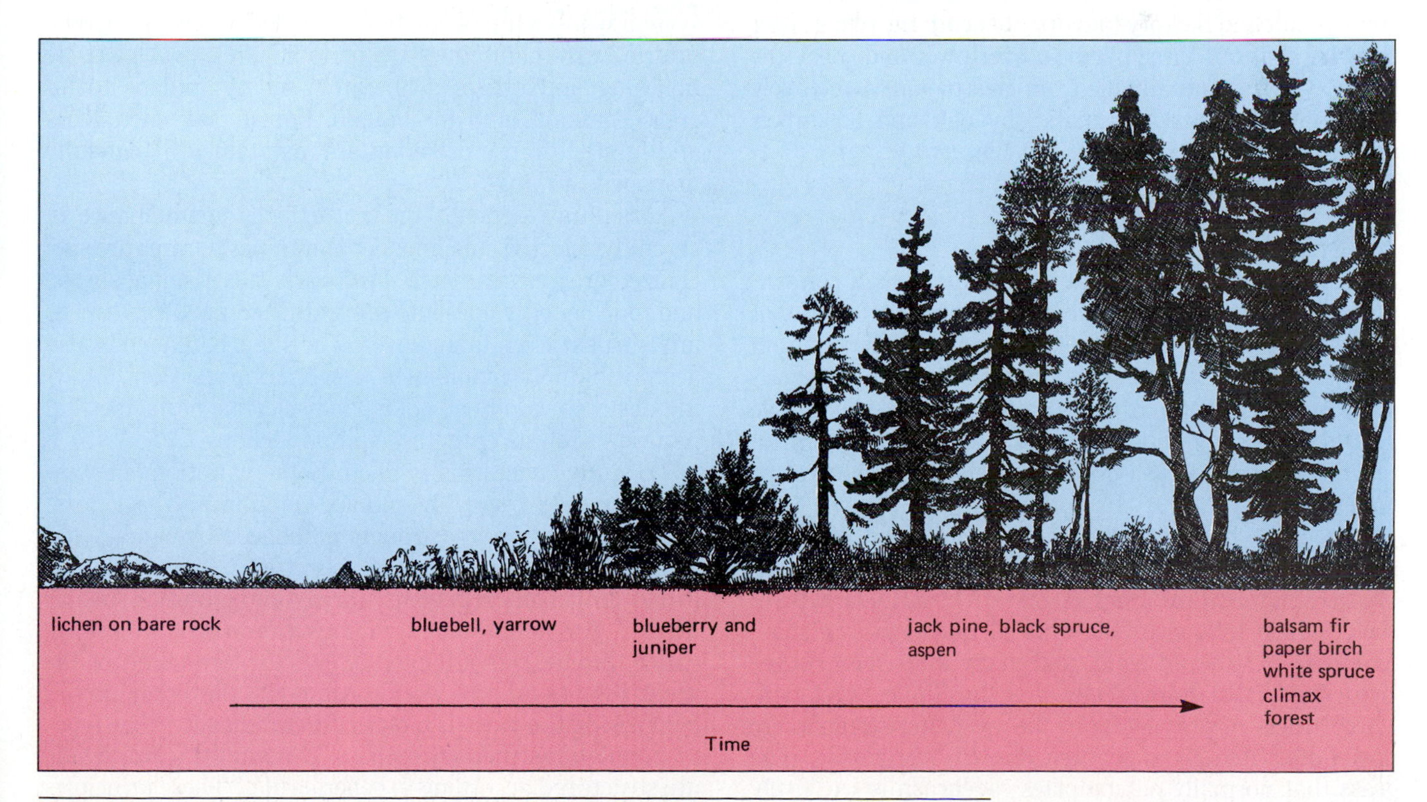

Figure 36-15 *Primary succession on bare rock in upper Michigan.*

Figure 36-16 Secondary succession on a plowed, abandoned farm field.

tulip and sweetgum, which sprout from windblown seeds. These become prominant after about 25 years, and a pine forest dominates the field for the rest of the first century. Meanwhile, shade-resistant, slow-growing hardwoods such as oak and hickory take root beneath the pines. After the first century, these begin to overtower and shade the pines, which eventually die from lack of sun. A relatively stable climax forest dominated by oak and hickory is present by the end of the second century.

The Climax Community

Succession ends with a relatively stable climax community. In your travels, you have undoubtedly noticed that the type of climax community varies dramatically from one area to the next. For example, Colorado has shortgrass prairie as a climax on its eastern plains, spruce–pine forests in its mountains, tundra on their uppermost reaches, and sagebrush-dominated climax in its western valleys.

The exact nature of the climax community is determined by numerous geological and climatic variables, including temperature, rainfall, elevation, latitude, type of rock (which influences the type of nutrients available), exposure to sun and wind, and many more. Human activities may also dramatically alter the climax vegetation. Large tracts of grasslands in the west, for example, are now dominated by sagebrush due to overgrazing. The grass that normally outcompetes sagebrush is selectively eaten by cattle, allowing the sagebrush to prosper.

Many ecosystems are not allowed to reach the climax stage but are maintained in an earlier stage called a **subclimax.** The tallgrass prairie that once covered northern Missouri and Illinois is actually a subclimax of an ecosystem whose climax community is deciduous forest. The prairie was maintained by periodic fires, some set by lightning and others deliberately set by Indians to increase grazing land for buffalo. Forest now encroaches and limited prairie preserves are maintained by carefully managed burning.

Agriculture depends on the artificial maintenance of carefully selected subclimax communities. Grains are specialized grasses characteristic of early successional stages, and much energy goes into preventing competitors (weeds and shrubs) from taking over. The suburban lawn is also a painstakingly maintained subclimax ecosystem. Mowing destroys woody invaders, and selective herbicides kill pioneers such as crabgrass and dandelions.

To study succession is to study the variations in communities over *time.* The climax communities that form during succession are strongly influenced by climate and geology: the distribution of ecosystems in *space.* Over broad geographical regions, climax communities such as deserts, grasslands, and deciduous forests occur; these are called **biomes.** In Chapter 38 we explore some of the great biomes of the world. Although the communities comprising the various biomes differ radically in the types of populations they encompass, communities worldwide are structured according to general rules. These principles of ecosystem structure are described in Chapter 37.

SUMMARY OF KEY CONCEPTS

Interspecific Competition and the Ecological Niche
The ecological niche defines all aspects of a species' interactions with its living and nonliving environment. Interspecific competition occurs when the niches of two populations within a community overlap, with the amount of competition proportional to the amount of niche overlap. In natural communities, interspecific competition limits both population size and distribution. Populations within natural communities have evolved to avoid excessive niche overlap. However, when forced to occupy the same niche under laboratory conditions, one species always outcompetes the other. This observation led G. F. Gause in 1934 to formulate the competitive exclusion principle: no two species can occupy the same niche at the same time.

Predation
Predators eat other organisms. Predators help maintain prey populations at or below carrying capacity, and prey abundance can affect the size of predator populations. By limiting prey population density, predators in some cases help maintain a higher diversity of species in an ecosystem.

Predators and prey act as strong agents of selection on one another. For example, plants that are preyed upon have evolved elaborate defenses ranging from poisons to thorns to overall toughness. These defenses, in turn, have selected for predators that can detoxify poisons, ignore thorns, and grind down tough grasses. Animals have evolved a variety of protective colorations that render them either inconspicuous or startling to their predators. Some prey have become poisonous, and these often exhibit warning coloration so that they are readily recognized and avoided. Prey that have evolved to resemble poisonous animals are called mimics.

Symbiosis
Symbiotic relationships involve two different species that interact closely over an extended period. Symbiosis includes parasitism, in which the parasite feeds on a larger, less abundant host, usually harming it but not killing it immediately. In commensal symbiotic relationships, one species benefits, often by finding food more easily in the presence of another species, which is not affected by the association. Mutualism occurs when both symbiotic species benefit.

Succession
A directional change in the types of populations inhabiting an ecosystem over time is called succession. Primary succession, which may take thousands of years, occurs where no remnant of a previous community exists. Primary succession could occur on bare rock scoured by a glacier or cooled from molten lava, a sand dune, or in a newly formed glacial lake. Secondary succession occurs much more rapidly, since it builds on the remains of a disrupted community, such as the aftermath of a fire, or a plowed and abandoned field. During succession, pioneer organisms invade and alter the environment by their presence. Community changes occur as one group of organisms, particularly plants, alters the environment in ways that encourage growth of their competitors. These replace the first colonizers and are subsequently replaced by others. Some ecosystems, including tallgrass prairie and farm fields, are maintained in relatively early stages of succession by periodic disruptions. Uninterrupted succession ends with a relatively stable group of organisms called the climax community.

GLOSSARY

Aggressive mimicry (mim'ik-rē) the evolution of a predatory organism to resemble a harmless one, thus gaining access to its prey.

Biome (bī'-ōm) a general type of ecosystem occupying extensive geographical areas, characterized by similar plant communities: for example, deserts.

Camouflaged (cam'a-flaged) a term used to describe organisms that resemble their environment, often both in color and pattern.

Climax community a relatively stable community of plants that terminates succession.

Coevolution (kō'ev-ol-oo'shun) the process by which two interacting species act as agents of natural selection on one another over evolutionary time.

Commensalism (kum-en'-sal-ism) a symbiotic relationship between two species in which one benefits while the other is neither harmed nor benefited.

Competitive exclusion principle the concept that no two species can simultaneously and continuously occupy the same ecological niche.

Exotic a species introduced into an ecosystem where it did not evolve.

Mimic (mim'-ik) an organism that has evolved to resemble another.

Model an organism that is mimicked by another.

Mutualism (mū'-chū-al-ism) a symbiotic relationship in which both participating species benefit.

Niche (nitch) the role of a particular species within an ecosystem, including all aspects of its interaction with the biotic and abiotic environment.

Parasitism a symbiotic relationship in which one organism (usually smaller and more numerous than its host) benefits by feeding on the other, usually without killing it immediately.

Pioneer an organism that is among the first to colonize an ecosystem undergoing succession.

Primary succession succession that occurs in an environment, such as bare rock, in which no remnants of a previous community are present.

Secondary succession succession that occurs after an existing community is destroyed, in an environment modified by that community: for example, after a forest fire.

Startle coloration a color pattern (often resembling large eyes) that can be displayed suddenly by a prey organism when approached by a predator.

Subclimax a stage of succession occurring before the climax community is reached, maintained by regular disturbance: for example, tallgrass prairie maintained by regular fires.

Succession (suk-se′-shun) the directional change in the community structure of an ecosystem over time. Community changes alter the ecosystem in ways that favor competitors, until a stable climax community is reached.

Symbiosis (sim′-bī-ō′sis) a close association between different species usually over an extended period. *Symbiosis* as used in this book includes mutualism, commensalism, and parasitism.

STUDY QUESTIONS

1. Define an ecological community, and list three important types of community interactions.
2. An ecologist visiting an island finds two very closely related species of birds, one of which has a slightly larger bill than the other. Interpret this finding with respect to the competitive exclusion principle and the ecological niche, defining both terms.
3. Describe three very different examples of coevolution. In each case, describe the selection pressure that has resulted in the evolutionary adaptations that have occurred.
4. List three different types of symbiosis; define and provide an example of each.
5. What type of succession would occur on a clear cut in a national forest, and why?
6. List two subclimax and two climax communities. How do they differ?
7. Define *succession* and explain why it occurs.

SUGGESTED READINGS

Batten, M. (1984). "The Ant and the Acacia." *Science 84,* April 1984, pp. 59–67. Fascinating description of the work of Daniel Janzen.

Brower, L. P. "Ecological Chemistry." *Scientific American,* 1969, pp. 22–29. Insects feeding on toxic plants become unpalatable to bird predators.

Daniels, P. "How Flowers Seduce the Bugs and the Bees." *International Wildlife,* November/December 1984, pp. 4–11. Beautifully illustrated description of plants and their mutualistic pollinators.

Ehrlich, P. R., and Roughgarden, J. *The Science of Ecology.* New York: Macmillan Publishing Company, 1987. An excellent introductory textbook.

Hopson, J. "Wildlife Poker." *Outside,* December/January 1980, pp. 17–19. The dangers of gambling with exotic species.

Mech, L. D. "How Delicate Is the Balance of Nature?" *National Wildlife,* February/March 1985, pp. 54–59. Wolves and their prey in Minnesota, by one of the leading researchers in wolf ecology.

37 The Structure and Function of Ecosystems

In Chapter 7 you learned how energy is trapped by photosynthesis, released by cellular respiration, and used to construct the complex molecules of life. Here we will relate some of these basic principles to the functioning of ecosystems. All the activity of life is powered by the energy of sunlight, from the leaping of the jackrabbit to the active transport of molecules through a cell membrane. Each time this energy is used, some of it is lost as heat. But while solar energy continuously bombards the Earth and is continuously lost as heat, nutrients remain. They may change in form and distribution but they do not leave the world ecosystem and must be recycled continuously. *The basic laws of ecosystem function are that energy flows through ecosystems, while nutrients cycle and recycle* (Fig. 37-1). These laws shape the complex interactions within living communities.

THE FLOW OF ENERGY

Primary Productivity

Ninety-three million miles away, the sun fuses hydrogen into helium, releasing tremendous quantities of energy. A tiny fraction of this energy reaches the Earth in the form of electromagnetic waves. Of the energy that reaches Earth, much is reflected by the atmosphere, clouds, and the Earth's surface. Still more is absorbed by the Earth and its atmosphere as heat, leaving only about 1 percent to power all life on Earth. Of this 1 percent, green plants capture 3 percent or less. The teeming life on Earth is thus supported by less than 0.03 percent of the energy reaching it from the sun. But how does this energy enter the biological community? *The energy that powers ecosystems enters through the process of photosynthesis.* In photosynthesis (performed by plants, plantlike protists, and cyanobacteria), pigments such as chlorophyll absorb certain wavelengths of sunlight. This solar energy is used to combine carbon dioxide and water into sugar, a compound that stores energy in chemical bonds. Some of this energy is used to power other chemical reactions, converting sugars into starches, cellulose, fats, vitamins, pigments, and proteins (Fig. 37-2). Photosynthetic organisms are called **autotrophs** (Greek for "self-feeders") or **producers,** since they produce food not only for themselves but for nearly all other life as well. The organisms that rely on the high-energy molecules made by autotrophs are called **heterotrophs** (Greek, "other feeders") or **consumers.**

The amount of life an ecosystem can support is determined by the energy captured by the producers. The energy that photosynthetic organisms make available to other members of the community is called **net primary productivity.** Net primary productivity can be measured in units of energy (calories) or as the dry weight of organic material per unit area per year. The productivity of an ecosystem is influenced by many environmental variables, such as the amount of nutrients available to the autotrophs, the amount of sunlight reaching them, the availability of water, and temperature. In the desert, for example, lack of water limits productivity, while in the open ocean, light and nutrients are limited. When resources are abundant, as in estuaries or tropical rain forests, productivity is high. Some average productivities for a variety of ecosystems are presented in Fig. 37-3.

Trophic Levels, Food Chains, and Food Webs

Living things are frequently classified according to their role in the flow of energy through communities. Energy flows through communities from the photosynthetic producers through several levels of consumers. Each category of organism is called a **trophic level** (Greek, "feeding level"). The **producers,** from redwood trees to cyanobac-

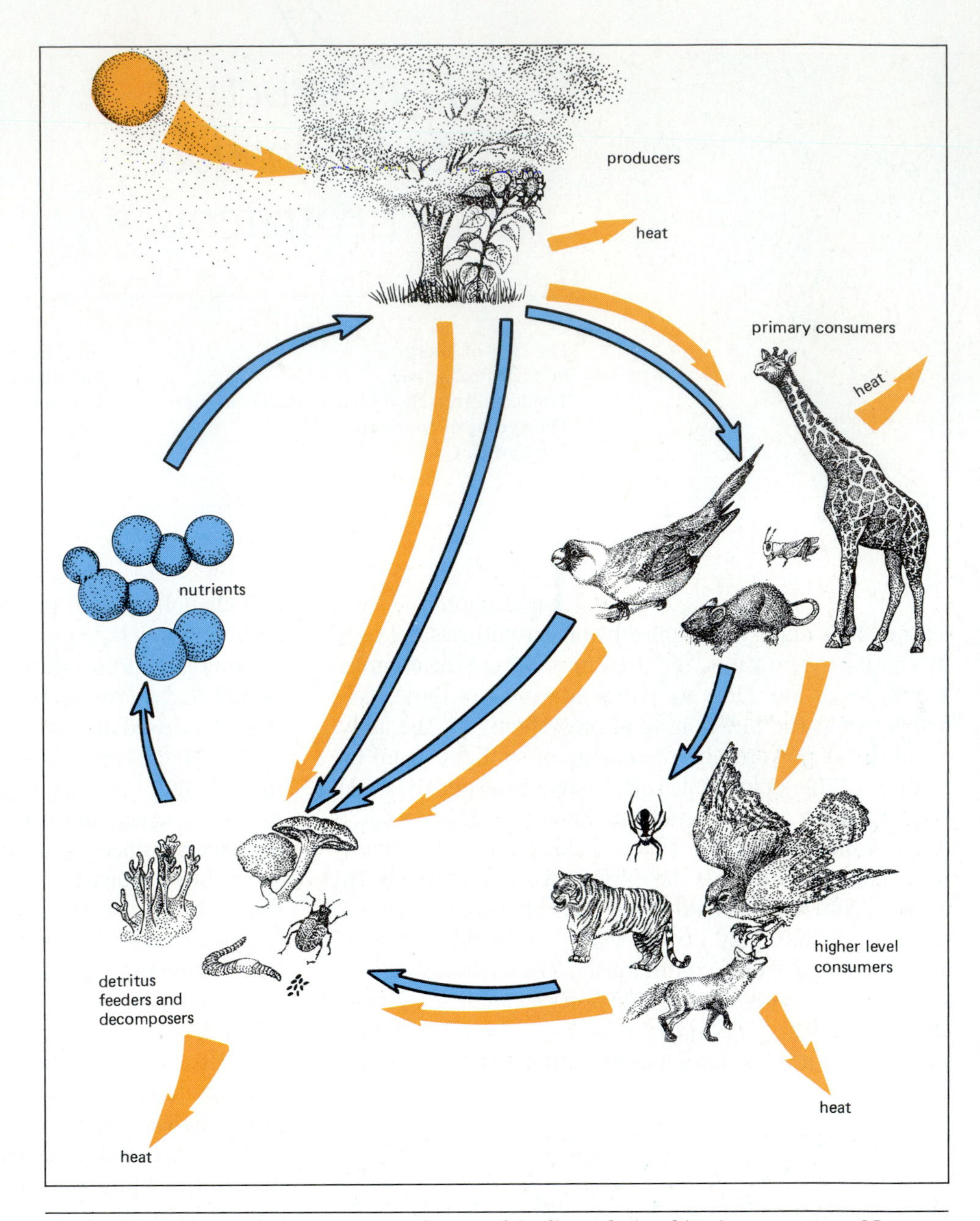

Figure 37-1 *Energy flow, nutrient cycling, and feeding relationships in ecosystems: Note that nutrients (blue) neither enter nor leave the cycle, while energy (orange) is lost at each level as heat and must be constantly replenished by sunlight.*

teria, form the first or lowest trophic level, obtaining their energy directly from sunlight. Certain organisms have evolved to feed directly and exclusively on producers, the most abundant living energy source in the ecosystem. These are the **herbivores,** ranging from grasshoppers to giraffes. As **primary consumers,** they form the second trophic level. **Carnivores** such as the spider, hawk, and wolf are flesh eaters, feeding primarily on herbivores, and are called **secondary consumers.** Some carnivores occasionally eat other carnivores, and when doing so they form the fourth trophic level: **tertiary consumers.**

To illustrate the feeding relationships in an ecosystem, it is common to identify a representative of each trophic level that eats the representative of the level below it. This linear feeding relationship is called a **food chain.** As illustrated in Figure 37-4, different ecosystems have radically different food chains.

Energy transfer through trophic levels. A basic law of thermodynamics is that energy use is never completely efficient. For example, as your car converts the energy stored in gasoline to the energy of movement, about 75 percent is immediately lost as heat. So too in

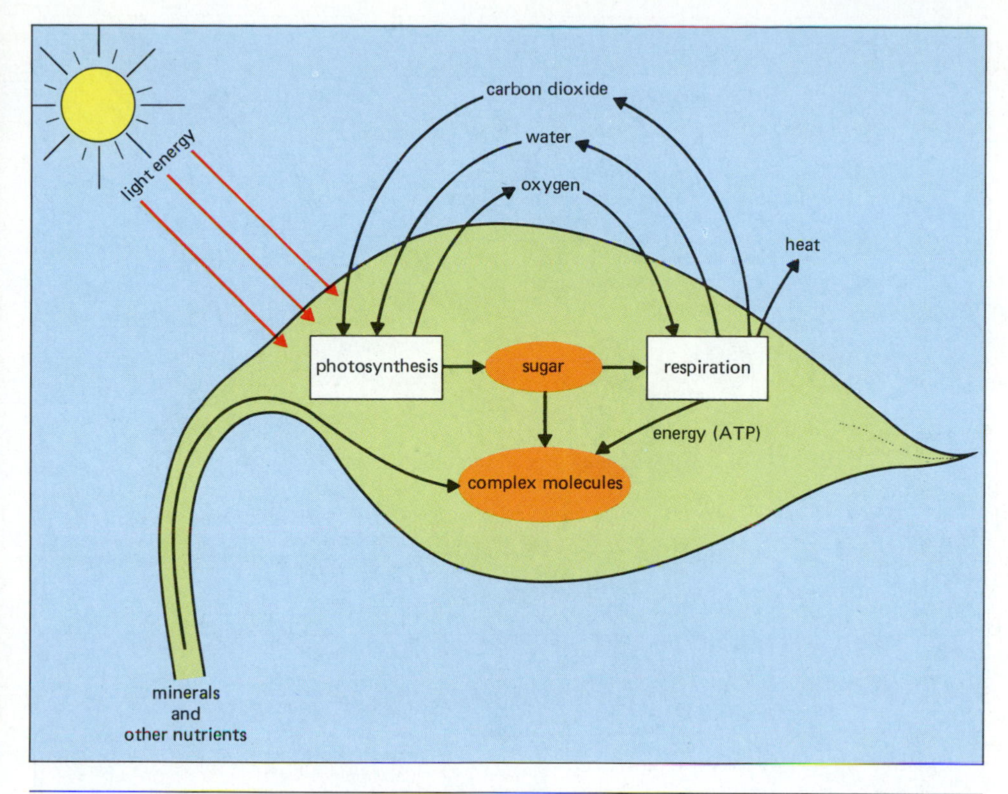

Figure 37-2 *During photosynthesis, plants capture the energy of sunlight and store it in sugar and other high-energy carbohydrates synthesized from carbon dioxide and water. Oxygen is released as a by-product. Some of the energy stored in sugar is liberated during respiration and used in the construction of more complex molecules. In the process, oxygen is used and some energy is lost as heat.*

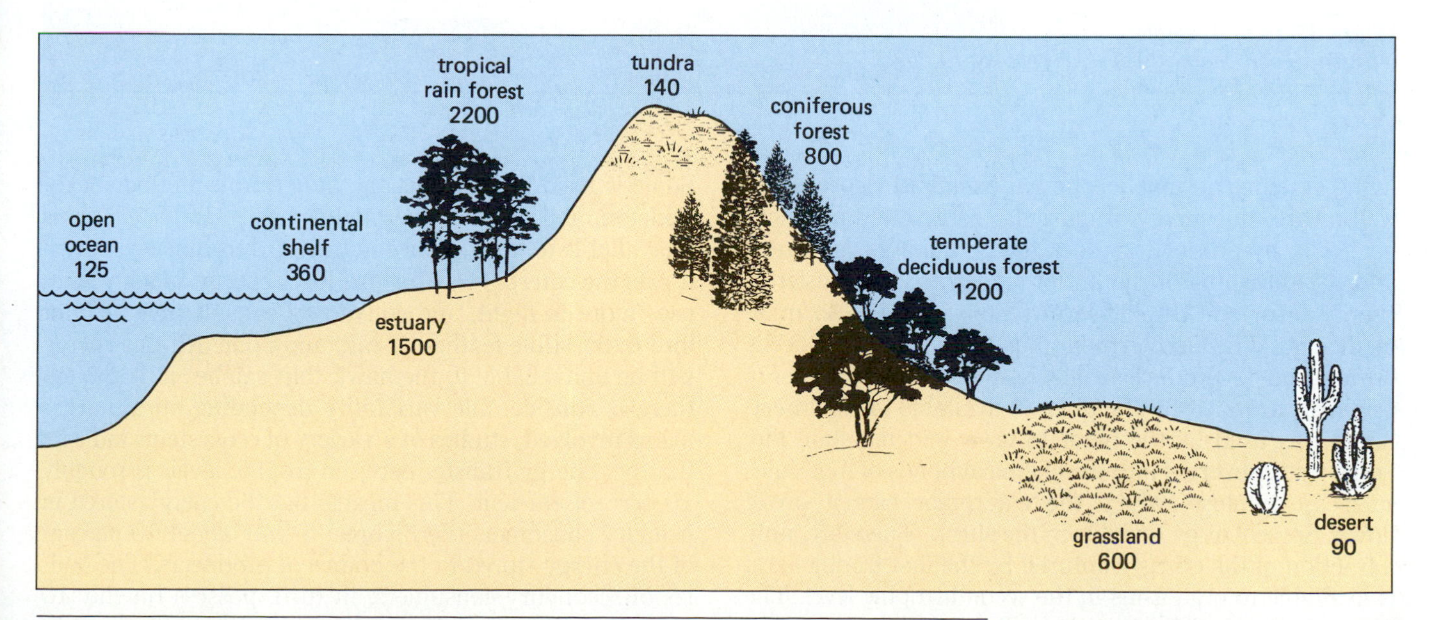

Figure 37-3 *Average primary productivity in grams of organic material per square meter per year of some terrestrial and aquatic ecosystems.*

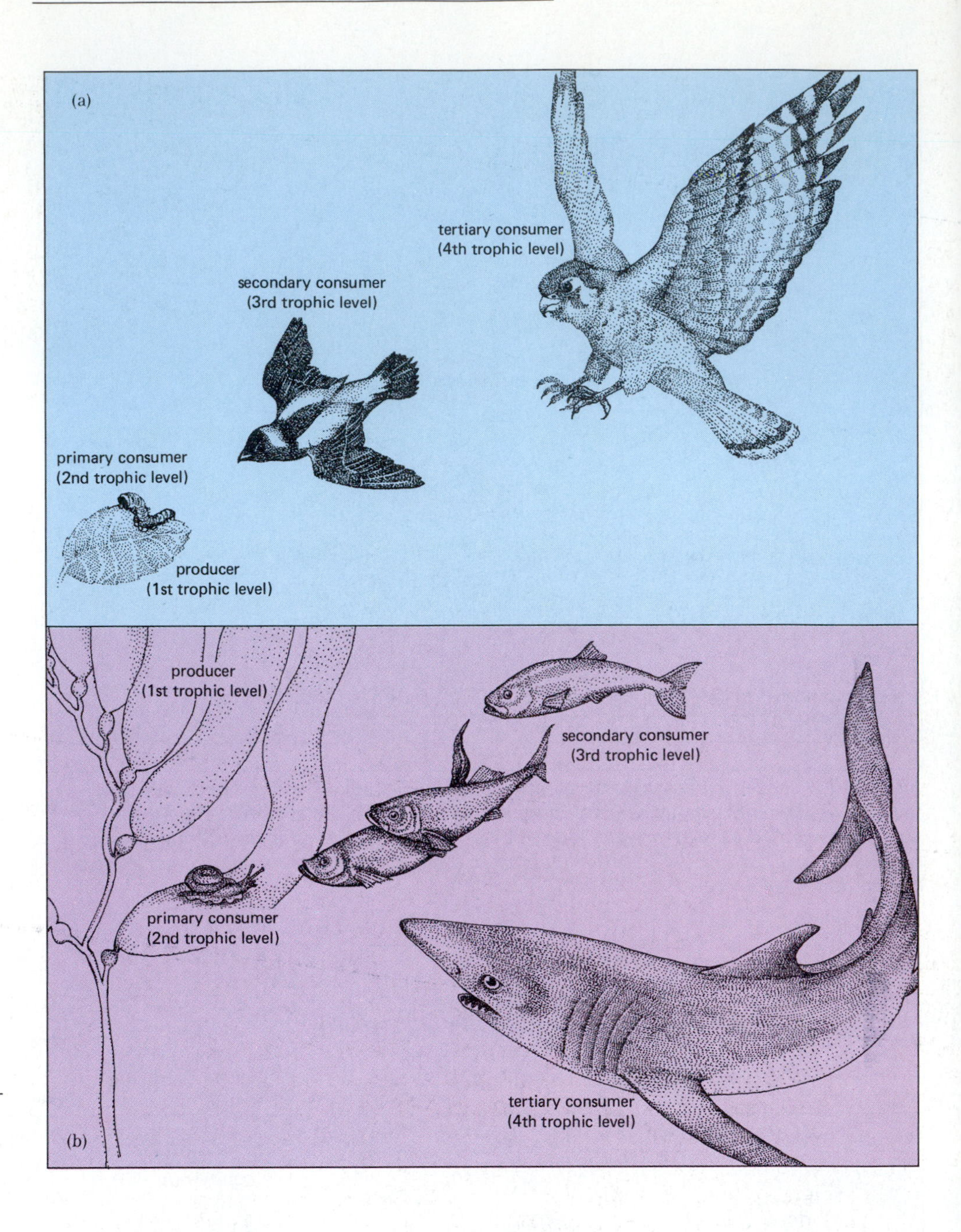

Figure 37-4 **(a)** *A simple terrestrial food chain.* **(b)** *A simple marine food chain.*

living systems; as anyone who has exercised vigorously is well aware, the energy of muscular contraction produces heat as a by-product, as does the germination of a seed and the thrashing of the tail of a sperm. The transfer of energy from one trophic level to the next is also quite inefficient. When a caterpillar (a primary consumer) feeds on a shrub (a producer), only some of the solar energy originally trapped by the plant is available to the insect. Some was used by the plant to grow and maintain life. Some was converted into the chemical bonds of molecules such as cellulose, which the caterpillar cannot break down. Some energy remains in the shrub. Therefore, only a fraction of the energy captured by the first trophic level is available to organisms in the second trophic level. The energy consumed by the caterpillar in turn is partially used to power crawling and the gnashing of mouthparts. Some is used to construct the indigestible chitinous exoskeleton, and much is liberated as heat. All this energy is unavailable to the songbird in the third trophic level when it eats the caterpillar. The bird loses energy as body heat, uses more in flight, and converts a considerable amount into indigestible feathers, beak, and bone. All this energy will be unavailable to the hawk that catches it. Although there is considerable variability depending on the organisms involved, studies of a variety of ecosystems indicate that net energy transfer between trophic levels is roughly 10 percent efficient. This means that the energy stored in primary consumers (herbivores) is only about 10 percent of the energy stored in the bodies of producers. The bodies of secondary consumers, in turn, possess roughly 10 percent of the energy stored in primary consumers. In other words, for every 100 calories of solar energy cap-

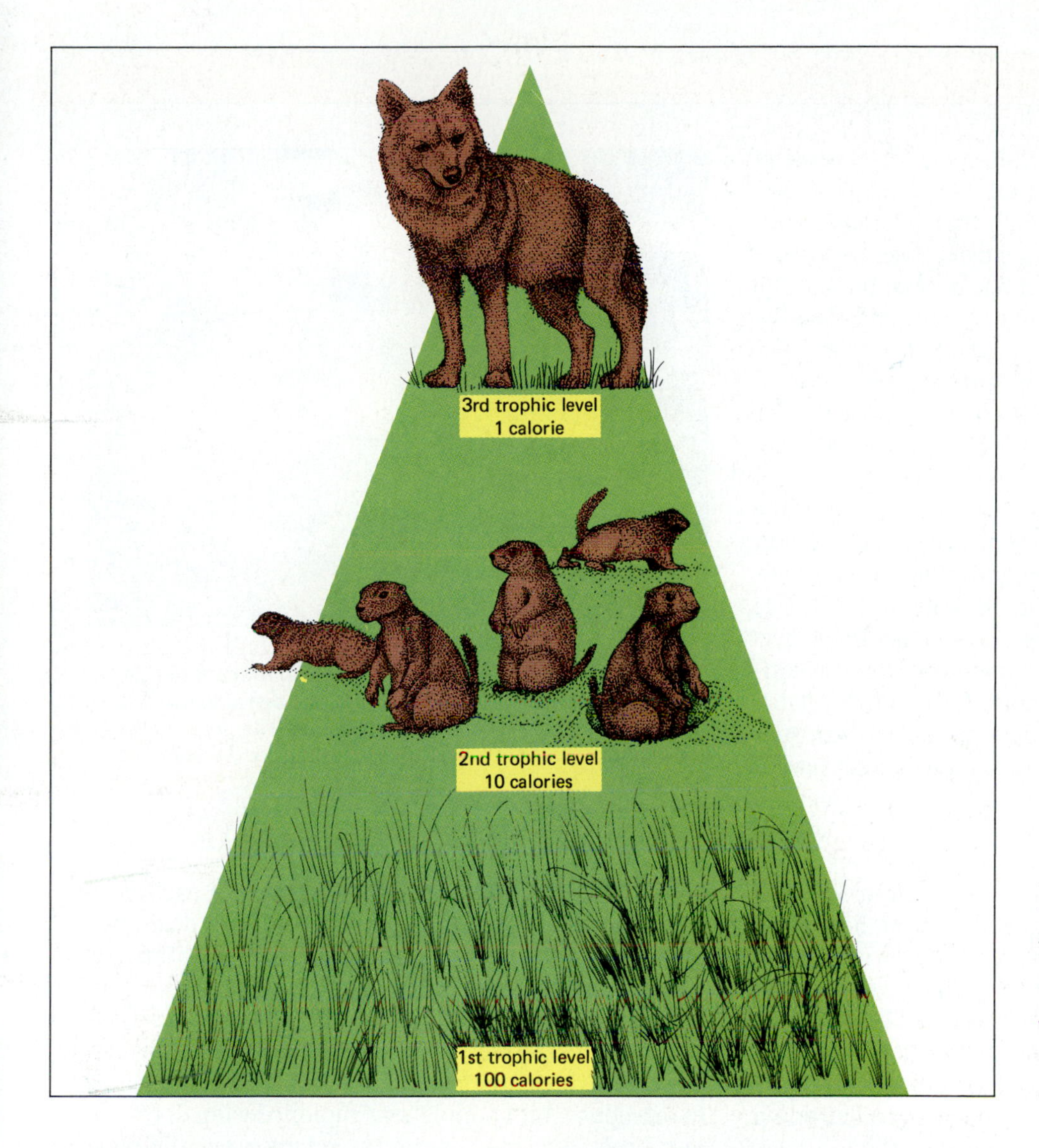

Figure 37-5 *A productivity pyramid. Each successive trophic level stores only about 10 percent of the energy found in the preceding level. Thus for every 100 calories stored in producers, there is likely to be only 1 calorie stored in secondary consumers.*

tured by grass, only about 10 calories are converted into herbivores, and only 1 calorie into carnivores. An **ecological pyramid** illustrates these relationships graphically (Fig. 37-5).

What does this mean for ecosystems? If you wander through an undisturbed ecosystem, you will notice that the predominant organisms are plants; these have the most energy available to them because they trap it directly from sunlight. The most abundant animals will be those feeding on plants, while carnivores will always be relatively rare. The inefficiency of energy transfer also has important implications for human food production. The lower the trophic level we occupy, the more food energy will be available to us; far more people can be fed on grain than on meat.

An unfortunate side effect of the inefficiency of energy transfer, coupled with human production of toxic chemicals, is the phenomenon of **biological magnification.** This is the concentration of certain persistent toxic chemicals in the bodies of carnivores, including ourselves (see the essay, "Biological Magnification").

FOOD WEBS. Natural communities rarely contain well-defined groups of primary, secondary, and tertiary consumers. A **food web** describes the actual feeding relationships within a given community much more accurately than does a food chain (Fig. 37-6). Some animals, such as raccoons, bears, rats, and humans, are omnivores, acting at different times as primary, secondary, and occasionally tertiary consumers. Many carnivores will eat either herbivores or other carnivores, thus acting either as secondary or tertiary consumers. An owl, for instance, is a secondary consumer when it eats a mouse, but a tertiary consumer when it eats a shrew, which feeds on insects. If the shrew ate a carnivorous insect, it would be a tertiary consumer, and the owl that fed on it would then be a quaternary (fourth-level) consumer. A carnivorous plant such as the sundew, when digesting a spider, can tangle the web hopelessly by serving simultaneously as a producer and a secondary consumer!

DETRITUS FEEDERS AND DECOMPOSERS. Among the most important strands in the food web are the decom-

Biological Magnification

In the 1950s and 1960s, wildlife biologists witnessed an alarming decline in populations of several predatory birds, especially fish eaters such as bald eagles, cormorants, ospreys, and brown pelicans. These top predators are never abundant, and the decline pushed the brown pelican close to extinction. Suspicion fell on the insecticide DDT, but although the ecosystems supporting these birds had been sprayed, the levels of chemical used were never high enough to account for the toxic effects. When bodies of these predators were analyzed, they were found to have concentrations of insecticide in their bodies up to *1 million* times the amount that had been added to the water. This led to the discovery of **biological magnification** and the subsequent ban of DDT by the United States in 1973. The explanation for biological magnification lies in the pyramid of energy and certain properties of these **chlorinated hydrocarbon** insecticides. DDT and related compounds have two properties that make them subject to biological magnification: (1) they do not readily break down into harmless substances, and (2) they are fat soluble but not water soluble. Thus they are not excreted in the watery urine but accumulate in the fat of animals. Since the transfer of energy from lower to higher trophic levels is extremely inefficient, herbivores must eat large quantities of plant material (which may have been sprayed with DDT), carnivores must eat many herbivores, and so on. Since DDT is not excreted, the predator gets the accumulated dosage from all its prey over a long period of time. Thus DDT reaches its highest level in top predators, as illustrated by the following example.

For many years, DDT was sprayed on the marshes of Long Island to control mosquitos. Instead of being washed out to sea and diluted, the toxic residues accumulated on detritus and built up in detritus feeders such as shrimp. Since fish such as minnows must eat many shrimp each day, they accumulated all the DDT from thousands of shrimp during their lifetimes. Predatory fish such as needlefish and pickerel consume many minnows daily, acquiring all the DDT which each minnow had accumulated and storing it in fat. Cormorants and merganser ducks, feeding on a variety of fish, received massive doses of DDT with each meal (see Table E37-1). DDT in sufficiently high doses will kill adult birds directly; at sublethal concentrations its effects are less direct but equally disastrous. DDT interferes with the deposition of calcium in eggshells, resulting in extremely fragile eggs that are frequently crushed by the parents in the nest (Fig. E37-1). Unable to reproduce, bird populations rapidly declined. Predatory birds, which are never abundant, were the most seriously affected. Many are still endangered. Today, the eggshells of peregrine falcons still average 14 to 18 percent thinner than they were in 1947, before the introduction of DDT. This serves as dramatic testimony to the persistence of this pesticide in the food web.

Figure E37-1 *DDT interferes with calcium deposition in the eggs of predatory birds, who receive high doses due to their position in the food chain. The resulting fragile eggs are often accidentally crushed in the nest by the brooding parent.*

Although DDT is the best-known example, a number of substances bioaccumulate, including dioxin, other chlorinated hydrocarbons, mercury, and some radioactive compounds. When the Atomic Energy Commission's Hanford plant in eastern Washington

posers and detritus feeders By liberating nutrients for reuse, they form a vital link in the nutrient cycles of ecosystems. The **decomposers** are primarily fungi and bacteria, which digest food outside their bodies, absorb the nutrients they need, and free the remaining nutrients into the soil or water. The **detritus feeders** are an army of small and often unnoticed animals and protists that live on the refuse of life: molted exoskeletons, fallen leaves, wastes, and dead bodies. The network of detritus feeders is extremely complex, including earthworms, mites, protists, centipedes, some insects and crustaceans, nematode worms, and even a few large vertebrates such

Biological Magnification

released trace quantities of radioactive phosphorus into the Columbia River in the 1950s, geese nesting downstream later laid eggs in which radioactive phosphorus was concentrated 2 million-fold over the amount in the river. Mercury from ocean pollution reached such high levels in swordfish (top predators) in the 1960s that the sale of swordfish meat was banned temporarily.

Understanding the properties of pollutants and the workings of food webs has become increasingly important to prevent widespread loss of wildlife and human health hazards. Humans have considerable cause for concern, because we feed high on the food chain. When we eat tuna, for example, we are tertiary or even quaternary consumers. In addition, our unusually long life span provides more time for substances stored in the body to accumulate to toxic levels. For example, prior to its ban, DDT in nursing mothers' milk sometimes exceeded the safe limit for human consumption.

Table E37-1 **Biological Magnification of DDT on Long Island**

Found In:	DDT (parts per million)
Water	0.00005
Plankton	0.04
Sheepshead minnow	0.94
Pickerel	1.33
Needlefish	2.07
Merganser duck	22.8
Cormorant	26.4

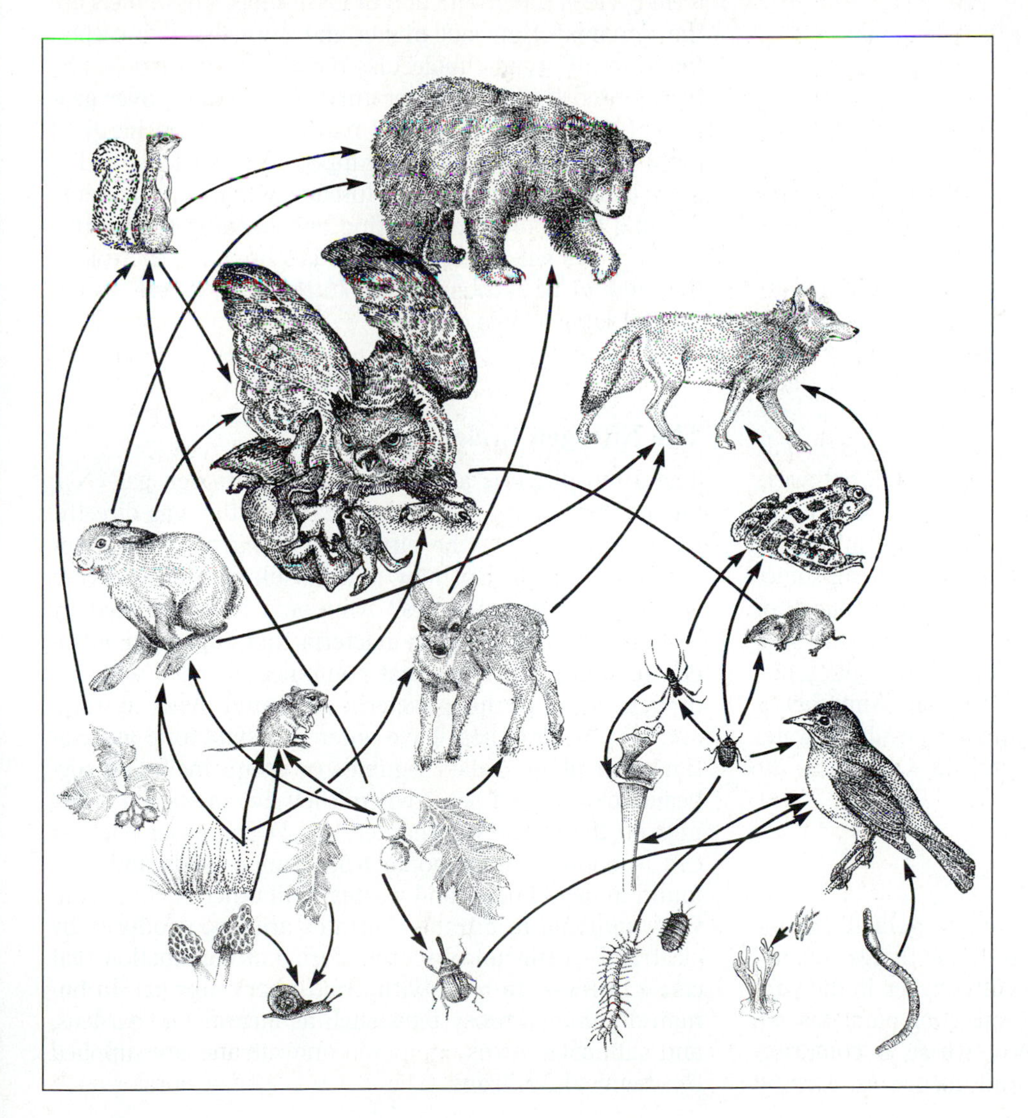

Figure 37-6 *A simple terrestrial food web.*

as vultures. These organisms consume dead organic matter, extract some of the energy stored within it, and excrete it in a still further decomposed state. Their excretory products serve as food for other detritus feeders and decomposers, until most of the stored energy has been utilized. The once-living substance is reduced to simple molecules such as carbon dioxide and water that return to the atmosphere, and minerals and organic acids that return to the soil. In some ecosystems, such as deciduous forests, more energy passes through the detritus feeders and decomposers than through the primary, secondary, and tertiary consumers. This inconspicuous portion of the food web is absolutely essential to life on Earth. If the detritus feeders and decomposers were to disappear suddenly, communities would gradually be smothered by accumulated wastes and dead bodies. The nutrients stored in these bodies would be unavailable and the soil would become poorer and poorer until plant life could no longer be sustained. With plants eliminated, energy would cease to enter the community and the higher trophic levels would disappear as well.

THE CYCLING OF NUTRIENTS

In contrast to the energy of sunlight, nutrients do not flow down onto the Earth in a steady stream from above. For practical purposes, the same pool of nutrients has been supporting life for over 3 billion years. Nutrients are all the chemical building blocks of life. Some, called **macronutrients,** are required in large quantities. These include water, carbon, hydrogen, oxygen, nitrogen, phosphorus, and calcium. **Micronutrients,** including zinc, molybdenum, iron, selenium, and iodine, are required only in trace quantities. **Nutrient cycles** describe the pathways these substances trace as they move from the living to the nonliving portions of ecosystems and back again to living tissues. Food webs depict the movement of a substance through living organisms. However, the major source, or **reservoir,** of important nutrients is generally in the nonliving environment. For example, the reservoir of carbon is the atmosphere and oceans, the reservoir of nitrogen is the atmosphere, and the major supply of available water is found in the oceans. As examples, we will outline the cycles of carbon and nitrogen.

The Carbon Cycle

Chains of carbon atoms form the framework of all organic molecules. The carbon available to living things is found most abundantly in combination with oxygen in the form of carbon dioxide (CO_2). This gaseous compound has two major reservoirs: the atmosphere, where it comprises 0.033 percent of the total gases, and the oceans, where it is dissolved in seawater. The oceans help regulate atmospheric CO_2 by dissolving and releasing it, but the interactions between the oceanic and atmospheric reservoirs are complex and still poorly understood.

Carbon enters food webs through producers, which trap CO_2 during photosynthesis. Some CO_2 is returned to the atmosphere through cellular respiration, while some that is incorporated into the plant body is later passed to herbivores. They, in turn, respire some of it and incorporate some into their tissues. All living things are eventually consumed by predators, detritus feeders, and decomposers, and ultimately most carbon is returned to the atmosphere as CO_2 (Fig. 37-7).

Some carbon cycles much more slowly. For example, molluscs extract carbon dioxide dissolved in water and combine it with calcium to form calcium carbonate ($CaCO_3$), from which they construct their shells. Shells of dead molluscs collect in undersea deposits and may eventually be converted to limestone. Limestone may dissolve gradually as it is exposed to water, making the carbon available to living organisms once more. Another long-term cycle is the production of fossil fuels. Fossil fuels are the remains of ancient plants and animals. The carbon found in all organic molecules remains in these deposits, transmuted by high temperature and pressure over geological time to coal, oil, and natural gas. The energy of prehistoric sunlight is also trapped in fossil fuels and is released by combustion. Human activities, including burning fossil fuels and cutting and burning the Earth's great forests where much carbon is stored, are increasing the amount of carbon dioxide in the atmosphere, as described later in this chapter.

The Nitrogen Cycle

The atmosphere is about 79 percent nitrogen gas (N_2), but neither plants nor animals can use this gas directly. Instead, plants must be supplied with nitrates (NO_3^-) or ammonia (NH_3). But how is atmospheric nitrogen converted to these molecules? Ammonia is synthesized by certain bacteria and cyanobacteria that engage in **nitrogen fixation,** a process that combines nitrogen with hydrogen. Some of these bacteria are found living in water and soil, while others have entered a symbiotic association with plants called legumes (a group including soybeans, clover, and peas) where they live in special swellings on the roots (see Chapter 17). Decomposer bacteria can also produce ammonia from amino acids and urea found in dead bodies and wastes. Still other bacteria convert ammonia to nitrates. Nitrates are also produced by electrical storms and by other forms of combustion that cause nitrogen to react with atmospheric oxygen. In human-dominated ecosystems such as farm fields, gardens, and suburban lawns, ammonia and nitrates are supplied by chemical fertilizers.

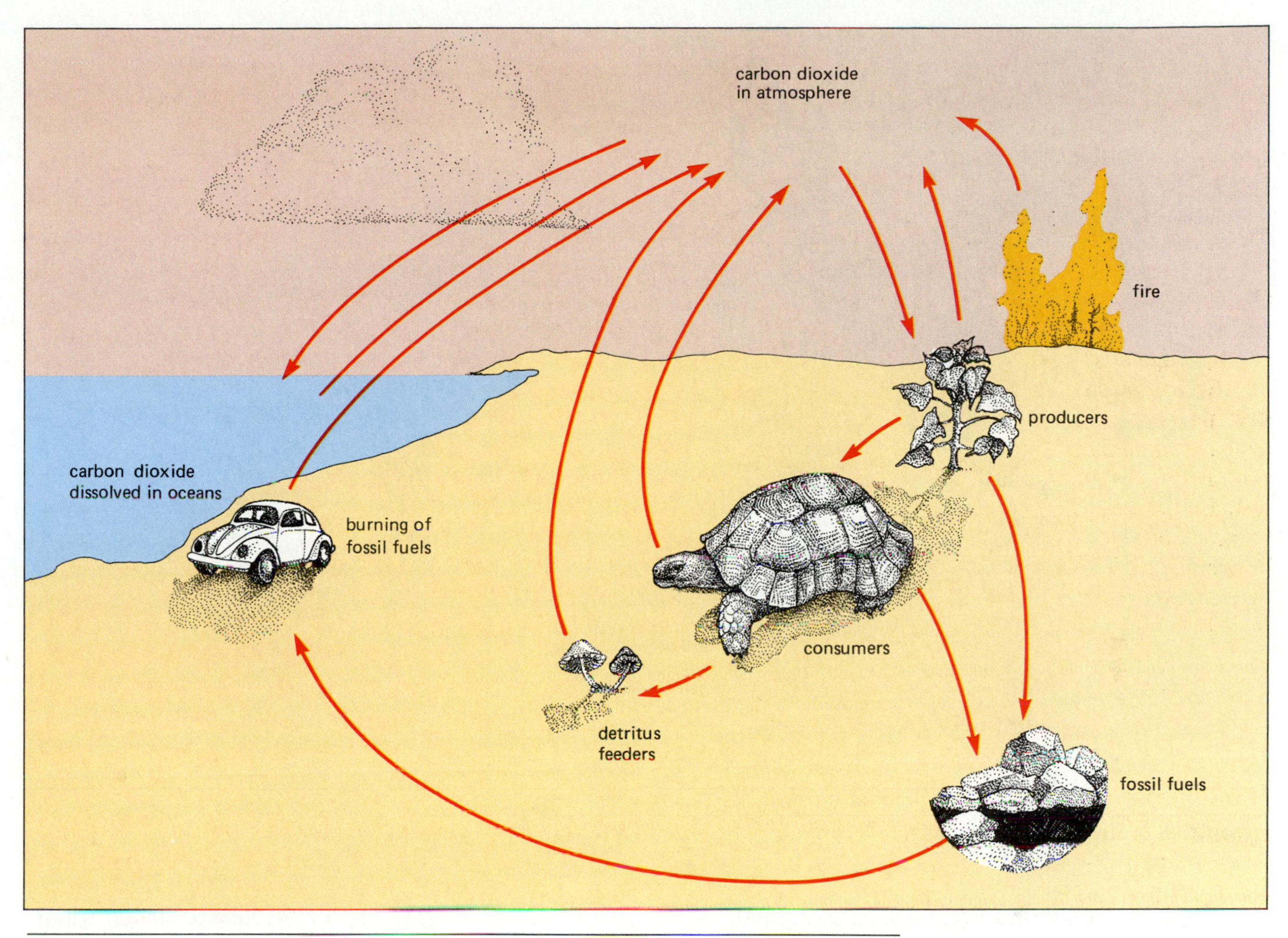

Figure 37-7 A simplified carbon cycle.

Plants incorporate the nitrogen from ammonia and nitrates into amino acids, proteins, nucleic acids, and vitamins. These nitrogen-containing molecules from the plant are eventually consumed, either by primary consumers, detritus feeders, or decomposers. As it is passed through the food web, some of the nitrogen is liberated in wastes and dead bodies, which decomposer bacteria convert back to nitrates and ammonia. The cycle is balanced by a continuous return of nitrogen to the atmosphere by denitrifying bacteria. These residents of mud, bogs, and estuaries break down nitrates, releasing nitrogen gas (Fig. 37-8).

POLLUTION AND NUTRIENT CYCLES

Pollutants are substances that contaminate the air, water, or soil, interfering with the natural cycles in communities. Substances that can act as pollutants fall into three major categories: (1) Materials that are found naturally in ecosystems, but are produced in excess by human activities, including carbon dioxide, carbon monoxide, sulfur dioxide, nitrogen oxides, animal waste products, and sediments from the land that enter waterways; (2) substances that occur naturally in the Earth but are not normally free in the environment, such as arsenic, radioactive minerals, lead, mercury, oil, and asbestos; and (3) synthetic compounds invented by humans, including pesticides, chlorfluorocarbons, and dioxin. These pollutants disrupt ecosystems in two ways: the first class of pollutants are potential nutrients, but when produced in excess, they flood the nutrient cycles with more material than organisms can process and assimilate. The other two classes of pollutants are substances that are foreign to ecosystems. They are poisons to many organisms, which have not evolved the ability to use them or break them down.

Flooding the Nutrient Cycles

In decomposing bodies, bacteria release sulfur dioxide, but no harmful effects are observed. In contrast, power

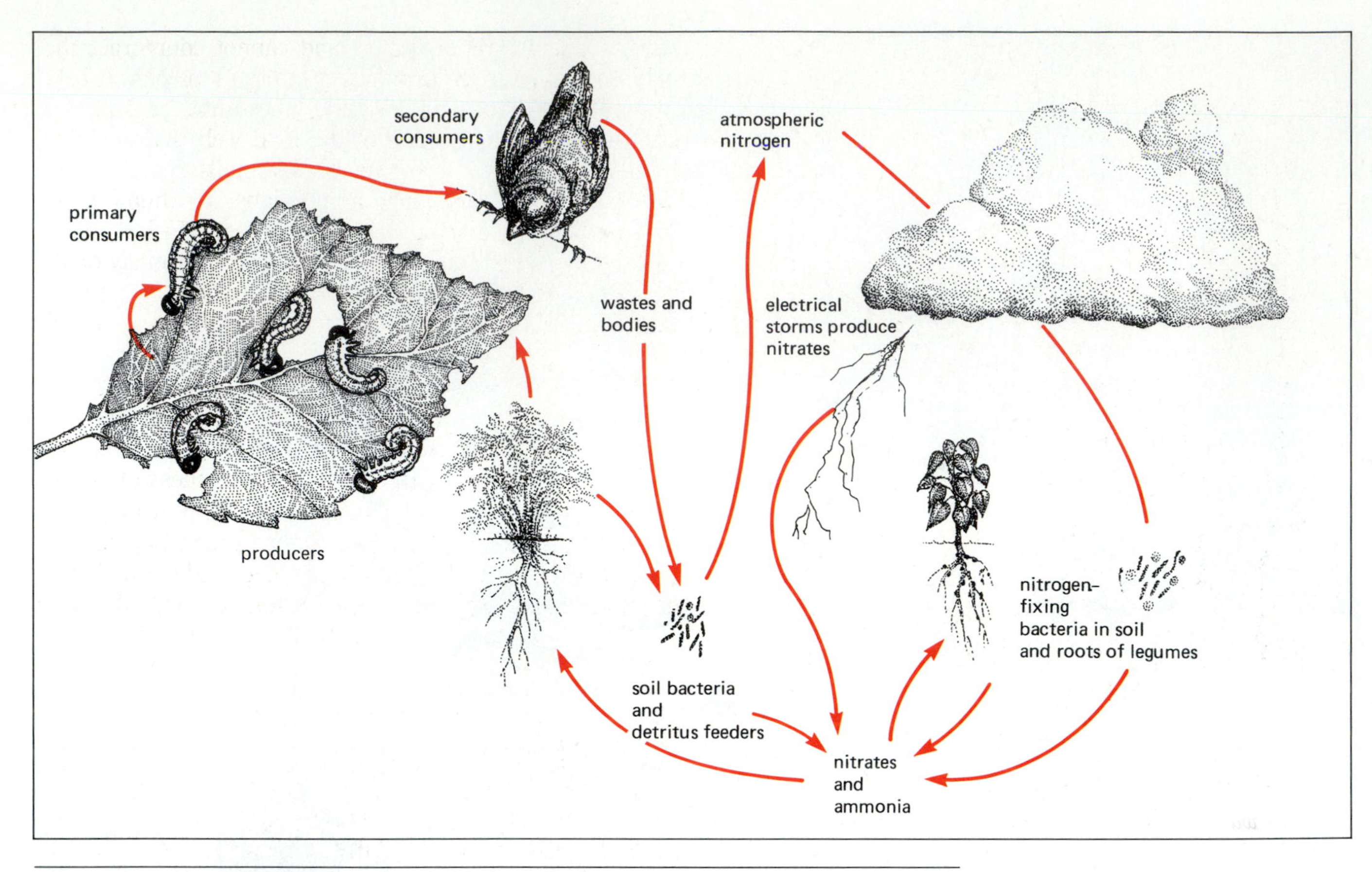

Figure 37-8 *A simplified nitrogen cycle.*

plants spew out sulfur dioxide, and in New England, Canada, and Scandinavia, thousands of lakes become lifeless, victims of acid rain. Countless millions of fish and invertebrates excrete their waste products into a lake, yet plants and animals thrive there and people seek it out for swimming. But when a sewage plant begins releasing treated human wastes into the lake, fish die, the water becomes thick with cyanobacteria and begins to smell, and swimming is banned. As these examples illustrate, some of the most serious pollutants are actually found naturally in ecosystems but are produced in excess by human activities. This excess comes from industrial processes, from enterprises that concentrate the substances locally, or from human populations which are simply too dense for the ecosystem in which they are living. Let's look at a few examples in more detail.

EUTROPHICATION OF LAKES AND STREAMS. A clear lake or stream has water low in nutrients, which limits the growth of microscopic organisms. Water discharged from a sewage plant, even if the solid wastes have been removed, is rich in dissolved nutrients such as phosphates and nitrates, derived both from the organic material and from laundry detergents. Excessively enriched water also washes off fertilized farm fields or feed lots where the manure of thousands of cattle has accumulated. The addition of these nutrients to a clear lake or stream allows a sudden, explosive growth of algae, photosynthetic protists, and cyanobacteria. The process of adding nutrients to a body of water is called **eutrophication** (Greek, "feeding well," in this case *too* well!) The producers cloud the water and form a scum on the surface, blocking sunlight from the submerged plants. The organisms die faster than they are consumed, and their bodies are decomposed by bacteria. Decomposition rapidly depletes the oxygen dissolved in the water. Deprived of oxygen, fish and invertebrates such as crayfish, snails, and insect larvae die. As their bodies fuel more growth of decomposer bacteria, further oxygen depletion occurs (Fig. 37-9). Even without oxygen, certain bacteria can still survive. These further foul the water by producing noxious-smelling gases such as hydrogen sulfide ("rotten egg gas"). Although it is full of life, the lake is considered "dead" by people. It smells dead, most of the trophic levels, including the fish, have been eliminated, and the community is now dominated by bacteria and microscopic algae.

ACID RAIN. Each year, the United States alone discharges about 30 million tons of sulfur dioxide into the atmosphere, two-thirds of it from power plants burning coal or oil (Fig. 37-10). The rest is largely a by-product of industrial boilers, smelters, and refineries. Humans

Figure 37-9 Fish kills such as this result from O_2 depletion of a body of water, often caused by an influx of sewage or other organic material.

contribute 90 percent of the sulfur dioxide in the atmosphere. Twenty-five million tons of nitrogen oxides are released by the United States each year, 40 percent from transportation sources such as autos, trucks, and planes, 30 percent by power plants, and most of the rest by industrial sources.

Both sulfur and nitrogen are crucial to life; for example, both are found in amino acids. But despite their importance as nutrients, beginning in the late 1960s these substances were identified as the cause of an insidious environmental threat: acid rain. Combined with water vapor in the atmosphere, nitrogen dioxide is converted to nitric acid and sulfur dioxide to sulfuric acid. Days later, and often hundreds or thousands of miles from the source, the corrosive acids fall, damaging trees and crops and rendering lakes lifeless.

Although sources of sulfur dioxide are widely distributed, they tend to be concentrated in the upper Ohio valley, Indiana, and Illinois, where high-sulfur coal is burned in old power plants with few emission controls. As chemical reactions in the atmosphere produce sulfuric acid, winds carry it toward New England, which is particularly vulnerable. Some areas of the country have rock rich in calcium carbonate, a natural buffer for acidity (and a major ingredient in some remedies sold for "acid indigestion"). In New England, however, hard rock such as granite and basalt predominates, and cannot counteract the acid rain. In the Adirondack mountains of New York, acid rain has rendered over 200 lakes unfit for fish. But before the fish die, much of the food web that sustains them has also been destroyed. Clams, snails, crayfish, and insect larvae die first, then amphibians, and finally fish. The result is a crystal clear lake, beautiful—but dead. Although the northeast has been the most seriously damaged, it is not the only vulnerable area; about 75 percent of the United States is at least moderately sensitive to acid rain.

Acid rain is not only a threat to lakes. Recent studies under laboratory conditions have shown that it also interferes with the growth and yield of many farm crops. Acid water percolating through the soil leaches out essential nutrients such as calcium and potassium. It may also kill decomposer microorganisms, thus preventing the return of nutrients to the soil. Plants, deprived of nutrients, become weak and more susceptible to infection and insect

Figure 37-10 (Top) Power plants burning high-sulfur coal with few emission controls are a major source of atmospheric sulfur dioxide, the prime contributor to acid rain. (Bottom) A forest in the White Mountains of New England shows the effects of acid rain.

attack. High in the Green Mountains of Vermont, scientists have witnessed an alarming 50 percent dieback among red spruces, 47 percent among beeches, and 32 percent among sugar maples since 1965. The snow, rain, and heavy fog that cloaks these mountaintops is highly acidic. In this fragile and stressful environment, acid precipitation may well be tipping the balance against these forests. In the famed Black Forest of Germany, the devastation is even greater, with over half the trees affected. Many of the firs have turned yellow or brown, trunks have gnarled and roots have shrunk, and plagues of insects and fungal diseases are attacking the weakened trees. It is almost impossible to prove the source of the problems, but German officials are convinced that acid rain from the burning of high-sulfur coal in conjunction with other pollutants is the cause. Vigorous pollution control programs have been started, but many fear that the forests have been damaged irrevocably. In 1984, scientists confirmed that the growth rate of eastern U.S. forests had slowed dramatically over the past several decades. Germany's problems may foreshadow similar disasters in U.S. forests.

Acid rain not only harms animals and plants directly, it also increases their exposure to toxic metals, which are relatively inert in natural ecosystems. Metals, including aluminum, lead, nickel, mercury, cadmium, and copper, are far more soluble in acidified water. Plant growth may be inhibited by metals such as aluminum. Other metals, taken up by plants along with acidified groundwater, may threaten animals higher on the food chain. In Sweden, the kidneys and livers of moose grazing on plants grown in acidified water sometimes contain enough cadmium to be lethal to a person eating them. In acid lakes and streams, aluminum leached from the surrounding rock and soil causes mucus to accumulate in the gills of fish, suffocating them. Drinking water in some households has been found dangerously contaminated with lead dissolved by acid water from lead pipes. Massachusetts spends $1 million annually to add chemicals to Boston's water supply to counteract the acidity. Fish in acidified water have been found to have dangerous levels of mercury in their bodies. The combination of acid and mercury causes the formation of methylmercury, an extremely toxic compound that is subject to biological magnification as it is passed through food chains.

Figure 37-11 This marble statue shows the corrosive effects of acid rain.

Acid rain is corroding our bridges, eating away at our buildings and monuments, and destroying ancient and priceless works of art at a cost estimated at $2 billion annually in the United States alone (Fig. 37-11). The east wall of our capitol building is pitted with small craters and looks as if it has been hit by shrapnel. In Baltimore, the famous statue "The Thinker" by Rodin has been moved indoors to protect it from the corrosive rain. The stone and statuary of ancient Rome cannot be protected and are steadily being dissolved. The damage to both natural ecosystems and property caused by this insidious form of pollution is incalculable.

Carbon dioxide and climate. Between 345 and 280 million years ago, under the unique conditions of the Carboniferous period, huge quantities of carbon were diverted from the carbon cycle when the bodies of plants and animals were buried in sediments, escaping decomposition. There, heat and pressure over time converted their bodies to fossil fuels such as coal, oil, and natural gas. Since the industrial revolution, modern cultures have increasingly relied on the energy stored in these fuels. As we burn them in our power plants, factories, and cars, we release carbon dioxide into the atmosphere.

A second important source of added CO_2 is global deforestation, the cutting of millions of acres of forests each year. This is occurring principally in the tropics, where rain forests are rapidly being eliminated in an attempt to increase agricultural land. Much of the carbon stored in these massive trees eventually returns to the atmosphere after they are cut, through either burning or decomposition. Burning fossil fuels and deforestation have caused a 13 percent increase in the carbon dioxide content of the atmosphere since 1860, from 0.029 percent to 0.033 percent. Recent studies estimate that deforestation accounts for about 30 percent of that increase. Projections indicate that the present CO_2 concentration of the atmosphere could be doubled within the next century.

So what? The projected levels will probably stimulate plant growth. However, carbon dioxide has an important property which makes its buildup a cause for concern: it traps heat in the atmosphere. Atmospheric CO_2 acts something like the glass in a greenhouse, allowing energy in the form of sunlight to enter but absorbing and holding that energy once it has been converted to heat. This "greenhouse effect" could cause a rise of about 4°C by the middle of the next century. Although this may not seem like much, the comings and goings of ice ages over the past million years have involved temperature changes of only 6° to 7° C.

As the icecap and glaciers melt in response to atmospheric heating, sea level will rise, with estimates ranging from inches to 2 feet by the year 2000. A more serious potential consequence of the warming trend is a shift in the global distribution of temperature and rainfall. Even small temperature changes can dramatically alter the paths of major air currents. This would change precipitation patterns in unpredictable ways. Some land which is currently cultivated only with the help of irrigation might become too dry for agriculture. Other areas might receive more rain. Agricultural disruption could be disastrous for human populations on the brink of starvation. But the changes will be slow, and their magnitude is highly speculative. Humans might have ample time to adapt and adjust to the changing climate. This uncertainty makes it very difficult for countries to take the drastic measures necessary to halt the trend. One thing is certain: for the first time in the Earth's 5 billion year history, a single species is progressively altering its climate. Your grandchildren will experience firsthand the consequences that experts can only guess at today.

Poisoning the Nutrient Cycles

Human industrial activities, especially mining and manufacturing, have exposed natural communities to a wide variety of "foreign" substances. These include arsenic, lead, asbestos, mercury, aluminum, oil, and radioactive substances, which are normally buried deep underground, and chemicals never before encountered on Earth, such as DDT and other pesticides, and dioxin. Because these substances are new to ecosystems, most organisms are unable to use or detoxify them. In some cases, such as with dioxin, DDT, certain radioactive substances, and mercury, the toxic effects are subject to biological magnification. Aluminum, lead, and mercury are more likely to enter living organisms as a result of acid rain. Let's examine a few of these foreign substances in more detail.

TOXIC MINERALS: LEAD AND ASBESTOS. Lead enters communities through human activities such as lead refining, the burning of fossil fuels in which lead is present as contamination, and primarily from burning leaded gasoline (although leaded gas is gradually being phased out in the United States). Up until the early 1960s, household paints contained up to 40 percent lead, and many children were poisoned by consuming flakes of the paints as it peeled from walls. In old houses, especially in inner cities, this is still a threat. Lead interferes with blood cell production and attacks the central nervous system. Young children are especially vulnerable because their nervous systems are still developing. Symptoms of lead poisoning include weakness of extremities, slowed reflexes, anemia, and nausea. However, in young children, blood lead levels far too low to cause these symptoms have been correlated with learning disabilities and hyperactivity.

Asbestos, a mineral with fire- and heat-resistant properties, has been widely used as insulation and fireproofing. It is currently used in brake linings because it is one of the few materials able to withstand the high temperatures generated in brakes. Since early in the century, evidence has mounted that the microscopic fibers that comprise asbestos become deadly in the human lung. Two fatal diseases, asbestosis (an emphysema-like lung condition) and mesothelioma (a fatal cancer of the membranes surrounding the lungs), have been linked exclusively to asbestos exposure. In contrast to some poisons, which become health hazards only with prolonged exposure, cancer can be triggered by brief contact with airborne asbestos fibers. Even the families of asbestos workers are at risk, presumably from exposure to the fibers carried home on the workers' clothing.

NEW SYNTHETIC COMPOUNDS: PESTICIDES AND DIOXIN. In the 1940s, the properties of DDT seemed close to miraculous. It was cheap and far more effective than other insecticides against nearly all insects. In less-developed countries, especially those of the tropics, it saved millions of lives by killing the mosquitos that spread malaria. Increased crop yields resulting from DDT's destruction of insect pests saved millions more from starvation. DDT is long-lasting, so a single application keeps killing. The discoverer was awarded the Nobel Prize, and people looked forward to a new age of freedom from insect pests. Little did they realize that this indiscriminate pesticide was unraveling the complex web of life. For example, in the mid-1950s, as part of a malaria control program, the World Health Organization sprayed DDT on the island of Borneo. Although the spraying successfully controlled the malaria epidemic, there were some unexpected side effects. Many insects inhabiting thatched-roofed houses dropped dead, to the delight of the islanders. However, a caterpillar that fed on the thatch was relatively unaffected, but the wasp that preyed on it was destroyed. Delight turned to dismay as thatched roofs collapsed, eaten by the uncontrolled caterpillars. Gecko lizards that ate the poisoned insects built up high concentrations of DDT in their bodies. Both they, and the village cats that ate the geckos, died of DDT poisoning. With the cats eliminated, the rat population exploded and the village was threatened with an outbreak of plague. This was avoided only by airlifting new cats to the villages.

In the 1960s, when many insects had become immune to DDT and our predatory bird populations were severely depleted, we began to realize that the ecological price was too high. In 1973 the chemical was banned in the United States, although it is still used in many less-developed countries. Several compounds of similar chemical structure (called chlorinated hydrocarbons) were also used as pesticides, including aldrin, dieldrin, chlordane, and heptachlor. By 1975, all of these had been banned as well.

In general, the pesticides that have replaced the chlorinated hydrocarbons are less persistent. However, they are often more toxic to humans, birds, and other vertebrates. Like chlorinated hydrocarbons, they kill helpful predatory insects along with the pests, sometimes allowing a surge in the pest population. In addition, our use of pesticides has increased dramatically according to EPA estimates, from 540 million pounds per year in 1964 to over a billion pounds per year today.

In our quest to destroy insects, we have inadvertently exposed ourselves to another hazard, dioxin. This is among the deadliest compounds known, at least to laboratory animals (1 gram will kill 50 million mice). It causes cancer and birth defects in laboratory animals (and possibly in humans as well), and exposed people have suffered skin ailments, liver damage, and nervous system disorders. Dioxin is not manufactured deliberately, but is a by-product of the production of certain pesticides, herbicides, and wood preservatives. Dioxin made the headlines in 1982 when the town of Times Beach, Missouri, was found to be heavily contaminated *11 years* after its streets had been doused with dioxin-contaminated oil to keep down the dust. In 1983, the U.S. government evacuated the town and purchased the homes of its residents at a cost of $36 million (Fig. 37-12). Dioxin has also been found at Love Canal near Niagara Falls, New York, where tons of chemical wastes were buried. A total of 250,000 Vietnam veterans have filed suit against the manufacturers of Agent Orange, a dioxin-contaminated herbicide sprayed heavily over the jungles of Vietnam as a defoliant. Although the veterans have suffered health problems ranging from cancer to birth defects in their children, hard scientific evidence linking their disabilities to dioxin is lacking and may be impossible to obtain. In the United States we are still using over a million pounds of herbicides (Silvex and 2,4,5-T), which are contaminated with traces of dioxin, on our crops and grazing land each year. As the lesson of Times Beach illustrates, dioxin is not broken down by decomposer organisms in the soil. Although it can be destroyed by very high temperature burning (this was done with leftover Agent Orange), for practical purposes, dioxin in soil is there to stay. Since it is stored in fat, it is subject to biological magnification. No one knows what the long-term effect on humans of low-level exposure will be.

Figure 37-12 *Times Beach, Missouri, 1983.*

The synthetic chemicals discussed here were chosen because they are extreme examples whose harm has been clearly demonstrated. The fact that these particular chemicals are no longer widely used in the United States should not make us feel complacent. Each year, hundreds of new compounds are introduced into the environment. Years from now, some of these will be the basis for sections such as this about human "advances" gone amok. We are, however, becoming increasingly aware of the interrelatedness of living communities and human dependence on them. Farmers are more wary of indiscriminate use of pesticides, and in some cases, biological controls such as natural insect predators or insect diseases have reduced the need for these poisons.

SUMMARY OF KEY CONCEPTS

Ecosystems are sustained by a continuous flow of energy from sunlight and a constant recycling of nutrients.

The Flow of Energy

Primary Productivity Energy enters the biotic portion of ecosystems when it is harnessed by autotrophs during photosynthesis. Primary productivity is the amount of energy that autotrophs store in organic material over a given period of time. It is extremely high in ecosystems such as estuaries and tropical rain forests, where light, nutrients, appropriate temperature, and adequate water are available, and low in deserts and the open ocean, where some of these conditions are restricted.

Trophic Levels, Food Chains, and Food Webs Trophic levels describe feeding relationships in ecosystems. Autotrophs are the producers, the lowest trophic level. Herbivores occupy the second level as primary consumers. Carnivores act as secondary consumers when they prey on herbivores, or tertiary or higher-level consumers when they eat other carnivores.

In general, only about 10 percent of the energy captured by organisms at one trophic level is converted to the bodies of organisms in the next highest level. The higher the trophic level, the less energy is available to sustain it. As a result, plants are more abundant than herbivores, and herbivores more common than carnivores. The storage of energy at each trophic level is illustrated graphically as an ecological pyramid.

Feeding relationships in which each trophic level is represented by one organism are called food chains. In natural ecosystems, feeding relationships are far more complex and are described as food webs.

Detritus feeders and decomposers, which digest dead bodies and wastes, free nutrients for recycling.

The Cycling of Nutrients

A nutrient cycle depicts the movement of a particular nutrient from its reservoir (usually in the abiotic portion of the ecosystem) through the food web and back to its reservoir, where it is again available to the producers.

The Carbon Cycle The reservoir for carbon is CO_2 gas, found in atmosphere and dissolved in the oceans. Carbon enters the producers through photosynthesis. From the autotrophs it is passed through the food web and released to the atmosphere as CO_2 during cellular respiration.

The Nitrogen Cycle Although the atmosphere is mainly nitrogen gas, producers are unable to use nitrogen in this form. First, it must be converted to ammonia or nitrate. Nitrogen gas is converted to ammonia by nitrogen-fixing bacteria in soil or in root nodules of legumes. Other soil bacteria convert ammonia to nitrates. Humans synthesize ammonia and nitrates from nitrogen gas for fertilizer. Lightning and other forms of combustion produce nitrates. Nitrogen passes from producers to consumers and is returned to the environment through excretion and the activities of detritus feeders and decomposers. Some release nitrates, some ammonia, and others convert nitrates back to nitrogen gas.

Pollution and Nutrient Cycles

Pollution occurs when human activities produce more nutrients than the natural cycles of a local ecosystem can absorb. It also occurs when humans release into the environment chemicals which are foreign and toxic to the natural community.

Flooding the Nutrient Cycles Overproduction of substances natural to ecosystems can disrupt their normal function. One example is the release of large quantities of human and livestock waste into bodies of water, causing eutrophication. Another is the excessive release of carbon dioxide into the atmosphere through burning fossil fuels and deforestation, which threatens a dramatic change in climate through the greenhouse effect. A third is the production of large quantities of oxides of sulfur and nitrogen through burning fossil fuels, leading to acid rain.

Poisoning the Nutrient Cycles Mining and industrial processes have exposed natural ecosystems to substances such as lead, asbestos, and a wide array of new synthetic compounds, such as pesticides and dioxin. These pose a threat to human health, as well as disrupting natural ecosystems.

GLOSSARY

Autotroph (aut′-ō-trōf) a photosynthetic organism.

Biological magnification the increasing accumulation of a toxic substance in increasingly high trophic levels.

Bloom a dense growth of algae and cyanobacteria that can occur in response to an influx of nutrients into an aquatic ecosystem.

Carnivore (kar′-neh-vōr) literally "meat eater," a predatory organism feeding on other heterotrophs.

Chlorinated hydrocarbon an organic compound that includes a chain of carbon atoms some of which have chlorine atoms bonded to them.

Decomposers a group of decay organisms, mainly fungi and bacteria. These digest organic material by secreting digestive enzymes into the environment. In the process they liberate nutrients into the environment.

Denitrifying bacteria (dē-nī′-treh-fī-ing) bacteria that break down nitrates, releasing nitrogen gas to the atmosphere.

Detritus feeders (de-trī′-tus) a diverse assemblage of organisms ranging from worms to vultures which live off the wastes and dead remains of other organisms.

Ecological pyramid a graphical representation of the energy contained in succeeding trophic levels, with maximum energy at the base (primary producers) and steadily diminishing amounts at higher levels.

Eutrophication (ū′-trif-i-kā′-shun) the process by which a body of water becomes enriched with nutrients and supports increasingly dense growths of photosynthetic and decomposer organisms. There is frequently oxygen depletion due to decomposition of plant material, and a progressive accumulation of organic material.

Food chain an illustration of feeding relationships in an ecosystem using a single representative from each of the trophic levels.

Food web a relatively accurate representation of the complex feeding relationships within an ecosystem, including many organisms at various trophic levels, with many of the consumers occupying more than one level simultaneously.

Legumes (leg′-yūm) a group of plants (including alfalfa, clover, and soybeans) which harbor colonies of nitrogen-fixing bacteria in swellings or nodules on their roots.

Macronutrients molecules used in relatively large quantities in the metabolic activities of organisms.

Micronutrients molecules required by organisms in trace quantities.

Net primary productivity the energy stored in the primary producers of an ecosystem over a given period.

Nitrogen fixation the process of converting atmospheric nitrogen into ammonia.

Primary consumer an organism that feeds on producers; an herbivore.

Producer a photosynthetic organism.

Reservoir the major source of any particular nutrient in an ecosystem, usually in the abiotic portion.

Secondary consumer an organism that feeds on primary consumers; a carnivore.

STUDY QUESTIONS

1. What are the two basic laws of ecosystem function concerning energy and nutrients?
2. What is an autotroph? What trophic level does it occupy, and what is its importance in ecosystems?
3. Define primary productivity. Would you predict higher productivity in a farm pond or an alpine lake? Defend your answer.
4. List the first three trophic levels. Among the consumers, which are most abundant? Relate your answer to the "10 percent law."
5. How do food chains and food webs differ? Which is the most accurate representation of actual feeding relationships in ecosystems?
6. Define detritus feeders and decomposers and explain their importance in ecosystems.
7. Trace the movement of carbon from its reservoir, through the biotic community, and back to the reservoir. How have human activities altered the carbon cycle, and what are the implications for future climate?
8. Define the greenhouse effect and explain how it occurs.
9. How does nitrogen get from the air to a plant?
10. Explain how the following nutrients can become pollutants, and what their effects are: CO_2, nitrates, treated sewage.
11. Define and give an example of biological magnification. In what trophic level are the problems worst, and why?
12. Describe the health effects of asbestos and lead. How does each of these enter the environment?
13. What is dioxin, how are people exposed to it, and what are its effects?

SUGGESTED READINGS

Chiras, D. D. *Environment Science: A Framework for Decision Making.* Menlo Park, Calif.: The Benjamin-Cummings Publishing Co., 1988. Discusses ecosystem structure and human impacts on the biosphere in an authoritative, readable style.

Cloud, P., and Gibor, A. "The Oxygen Cycle." *Scientific American,* September 1970. Traces the movement of oxygen through the living and nonliving parts of ecosystems.

Ehrlich, P. R. *The Machinery of Nature.* New York: Simon and Schuster, 1986. An eloquent description of the workings of the natural world, written for the educated layperson.

Ehrlich, P. R., and Roughgarden, J. *The Science of Ecology.* New York: Macmillan Publishing Company, 1987. An excellent introductory ecology textbook.

Gosz, J. R., Holmes, R. T., Likens, G. E., and Bormann, F. H. "The Flow of Energy in a Forest Ecosystem." *Scientific American,* March 1978. Gives experimental data about a specific forest ecosystem, relating it to the general principles of energy flow through ecosystems.

Hutchinson, G. E. "The Biosphere," *Scientific American,* September 1970 (Offprint No 11). Overview of the earth's thin film of life and how it is supported by the flow of energy and cycling of nutrients.

May, R. M. "The Evolution of Ecological Systems." *Scientific American,* September 1978 (Offprint No 1404). Coevolution of species is examined in light of their interrelationships within complex food webs in specific ecosystems.

Peakall, D. B. "Pesticides and the Reproduction of Birds." *Scientific American,* April 1970. How chlorinated hydrocarbon insecticides interfere with the reproduction of predatory birds.

Penman, H. L. "The Water Cycle." *Scientific American,* September 1970. Traces the movement of water through the biosphere.

Woodwell, G. M. "Toxic Substances and Ecological Cycles." *Scientific American,* March 1967 (Offprint No 1066). How bioaccumulation enhances the danger of certain poisons.

38
The Earth's Diverse Ecosystems

In preceding chapters we discussed the dynamics of population growth, the flow of energy, the cycling of nutrients, and the basic structure of communities. The communities in which these processes occur are extraordinarily diverse, yet there are clear patterns within this diversity. For example, deciduous forests can be recognized on any continent of the world, although their tree composition will vary. Similarly, deserts, grasslands, coniferous forests, coral reefs, and estuaries all have distinguishing features that identify them no matter where they are found or what particular species inhabit them. The reason we are able to identify these communities is that they are dominated by organisms (on land the dominant organisms are always plants) that are specifically adapted for particular environmental conditions. The unique attributes of desert plants, for example, are adaptations to heat and drought, while the ability to shed their leaves in winter to conserve water distinguishes the trees of deciduous forests. The distribution of communities on Earth, then, is determined by the environment. Different types of environment vary in the relative abundance of the four basic resources that provide the requirements of life: nutrients, energy, water, and temperatures appropriate to metabolic reactions.

Figure 38-1 *Old Faithful geyser in Yellowstone National Park. Earthquakes have disrupted the formerly regular cycles of gushing water and steam.* Inset: *The water flowing from a geyser field can be over 40°C but supports a thriving population of heat-tolerant algae.*

THE REQUIREMENTS OF LIFE

From the lichens on bare rock to the thermophilic (Greek, "heat-loving") algae in the hot springs of Yellowstone (Fig. 38-1), to the bacteria thriving under the pressure-cooker conditions of the deep-sea vent, the Earth fairly teems with life. Life is so ubiquitous on Earth because the fundamental requirements of living things are very few: nutrients from which to construct living tissue, energy to power that construction, liquid water to serve as a solvent for metabolic reactions, and a reasonable temperature in which to carry out these processes. Although these four basic requirements for life are simply met, the requirements of individual organisms may be very demanding. The bacteria from deep-sea vents stop reproucing at temperatures below 100°C. It is very easy to kill a cactus with too much moisture or a fern with too little. The distribution of organisms, then, is limited and defined by the abiotic environment. The location of the ecosystem on Earth determines the availability of water, the range of temperature, and the amount of sunlight it receives.

Thus communities vary according to their location on the Earth, and consist of organisms which are specially adapted to those particular conditions.

LIFE ON LAND

Life on land is restricted in its distribution by the availability of water and appropriate temperatures. Terrestrial ecosystems receive plenty of light, even on an overcast day, while the rocks and soil provide a constant source of nutrients. The available water, however, is limited and very unevenly distributed, both in place and in time. Terrestrial organisms must be adapted to obtain water when it is available, and conserve it when scarce. As with water, favorable temperatures are also very unevenly distributed in place and time. At the south pole, even in summer, the average temperature is well below freezing, and, not surprisingly, life is scarce. Other places, such as Siberia, have favorable temperatures only during certain seasons of the year, while the tropics have a uniformly warm, moist climate.

The distribution of temperatures on the Earth is fairly predictable based on latitude (Fig. 38-2). Near the equator, sunlight hits the surface nearly at a right angle, and the amount does not vary much during the year. Therefore, the weather tends to be constantly warm, with little seasonal variation. Farther north or south, the surface is oblique to the sun's rays, so that a given amount of sunlight is spread over a larger area, resulting in lower overall temperatures. Because the Earth tilts on its axis as it revolves around the sun, these higher latitudes also experience considerable variation in the directness of sunlight as the year progresses, resulting in pronounced seasons.

The amount of rainfall a given area receives is much less predictable than temperature, but there is still a general pattern. Due to global airflows related to warming and cooling of large air masses by the sun, land near the equator tends to receive generous rainfall, distributed quite evenly throughout the year. Between 20° and 30° north and south of the equator, there is generally little rainfall, and we find the Earth's great deserts, such as Africa's Sahara and Kalahari and those of Australia, Saudi Arabia, and Mexico. Farther toward the poles, rainfall increases, only to decrease once again at the poles, so that there is almost no precipitation at the north or south pole. (There are continuous glaciers in the Antarctic not because it snows a lot, but because what little snow does fall never melts.) Large land masses, however, complicate these overall rainfall patterns. For example, mountain ranges such as the Sierra Nevada and Rocky Mountains wring the rain out of the westerly winds, leaving deserts in the **rain shadow** on their eastern sides, even at latitudes far north of 30°.

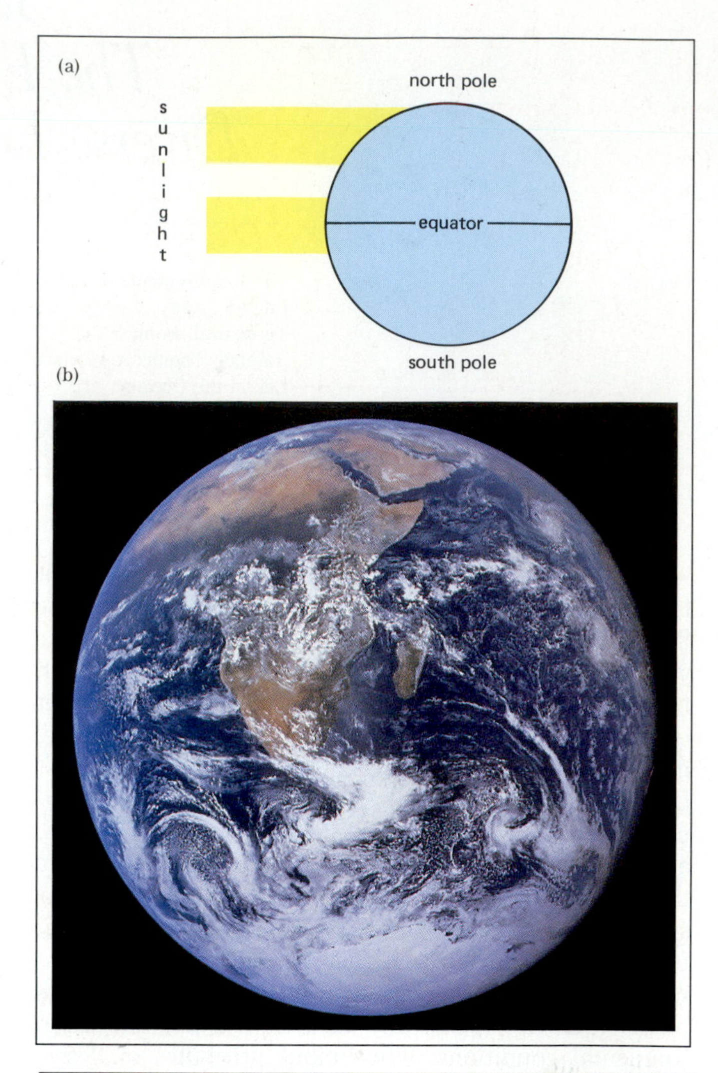

Figure 38-2 The distribution of temperature and rainfall over the surface of the Earth. **(a)** *Near the equator, sunlight falls perpendicularly onto the surface, while toward the poles a beam of the same size covers a much larger surface area. The energy of sunlight thus is greater near the equator, resulting in a temperature gradient with latitude.* **(b)** *The distribution of rainfall is more irregular than that of temperature, but is still fairly predictable. In this photograph taken by* Apollo 11 *astronauts, note the lack of clouds over the Sahara and Arabian deserts near 30° north and the South African desert near 30° south. Along the equator, there are heavy clouds over the rain forests of the Congo in western Africa and thinner clouds above the east African savanna. Temperate Europe to the north is also largely obscured by clouds.*

Terrestrial Biomes

Terrestrial communities are dominated and defined by their plant life. Since plants cannot escape from drought, sun, or winter weather, they tend to be very precisely adapted to the living conditions in a particular locale. If you travel around the world, you will be struck by the

Figure 38-3 Convergent evolution in response to similar environments has molded the bodies of **(a)** *American cacti and* **(b)** *South African euphorbs into nearly identical shapes, although they are in completely different families.*

great diversity of plant assemblages, but you will also notice that although the species may differ, very similar groups of plants are found wherever a particular climate exists. For example, although they are in separate families, the cacti of the American Mojave desert are strikingly similar to the euphorbia of South Africa; their spines and fat bodies are clearly adaptations for water conservation (Fig. 38-3). Similarly, the different species of plants in the arctic tundra and the alpine tundra of the Rockies show growth patterns clearly recognizable as adaptations to a cold, dry, windy climate.

Large land areas with similar environmental conditions and characteristic types of plants are called **biomes** (Fig. 38-4) and are usually named after the major vegetation found there. The predominant vegetation of each biome is determined by the complex interplay of rainfall and temperature (Fig. 38-5). In addition to the total amount of rainfall and the overall yearly average temperature, the variability of rain and temperature over the year also determines which plants can grow in an area.

TROPICAL FORESTS. Large areas of South America and Africa lie along the equator. Here the temperature averages between 25° and 30°C with little variation, and rainfall ranges from 250 to 400 centimeters yearly. These evenly warm, evenly moist conditions combine to create the most diverse biome on Earth, the **tropical rain forest,** dominated by huge broadleaf evergreen trees (Fig. 38-6). Tropical rain forests typically are stratified, with several layers of vegetation. The tallest giants reach 50 meters, towering above the rest of the forest. Below there is a fairly continuous canopy of treetops at about 30 to 40 meters. Still closer to the forest floor is often another layer of shorter trees. Huge woody vines, often 100 meters or more in length, grow up the trees, reaching the sunlight far above. These layers of vegetation block out virtually all of the sun, and plants that live in the dim green light that filters through to the forest floor often have enormous leaves to trap the little available energy.

Due to the relative scarcity of edible plant material close to the ground, much of the animal life in tropical rain forests is arboreal (living in the trees), including numerous birds, monkeys, and insects. Competition for the nutrients that do reach the ground is intense, both among animals and plants. Even such unlikely sources of food as the droppings of monkeys are in great demand, in this case by dung beetles that feed and lay their eggs on droppings. When ecologists attempted to collect droppings of the South American howler monkeys to find out what the monkeys had been eating, they found themselves in a race with the beetles, hundreds of which would arrive within minutes after a dropping hit the ground!

Almost as soon as bacteria or fungi release any nutrients from dead plants or animals into the soil, rain forest trees and vines absorb the nutrients. This is one of the reasons why, despite the teeming vegetation, agriculture is very risky and usually destructive in rain forests. Virtually all of the nutrients in a rain forest are tied up in the vegetation, so if the land is cleared, and especially if the trees are removed, the soil is very infertile. Further, even if the nutrients are released by burning the vegetation, the heavy year-round rainfall quickly dissolves and erodes them away, leaving the soil infertile after a few seasons of cultivation. The exposed soil, which is rich in

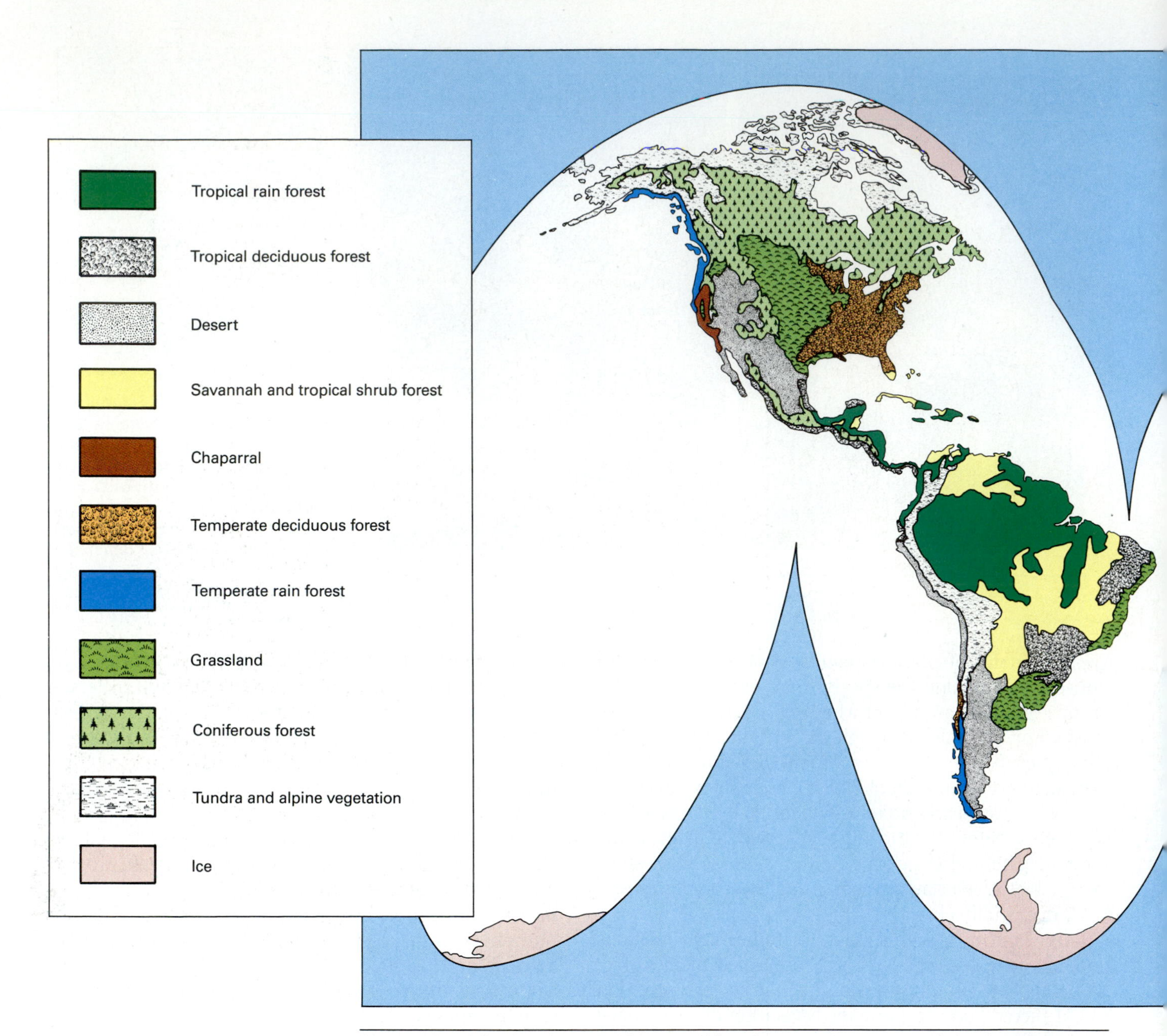

Figure 38-4 *The distribution of biomes. Although mountain ranges and the sheer size of the continents complicate their pattern, note the overall consistencies. Tundra and taiga always occur in the northernmost parts of the northern hemisphere, while the deserts of Mexico, the Sahara, Saudi Arabia, South Africa, and Australia are located around 20 to 30° north and south latitude.*

iron and aluminum, then takes on an impenetrable, brick-like quality as it bakes in the tropical sun. As a result, secondary succession on cleared rain forest land is slowed significantly, and the lush tropical growth is, for practical human purposes, irrevocably lost.

Despite this, rain forests are being felled for lumber and agriculture at a rate of about 110,000 square kilometers per year. About 25 percent of the original rain forests are now gone, and at current rates of cutting, all the forests not specifically protected in parks could disappear within the next hundred years. Although rain forests cover only 6 percent of the Earth's total land area, they are believed to be home to about two-thirds of the world's species. Currently protected areas are less than is cut in one year. Ecologists estimate that loss of all the unprotected forest would result in massive extinctions, the like of which have not occurred since the massive dinosaur extinctions of the Late Cretaceous period over 65 million years ago.

Slightly farther away from the equator, the rainfall is not nearly as constant, and there are pronounced wet and

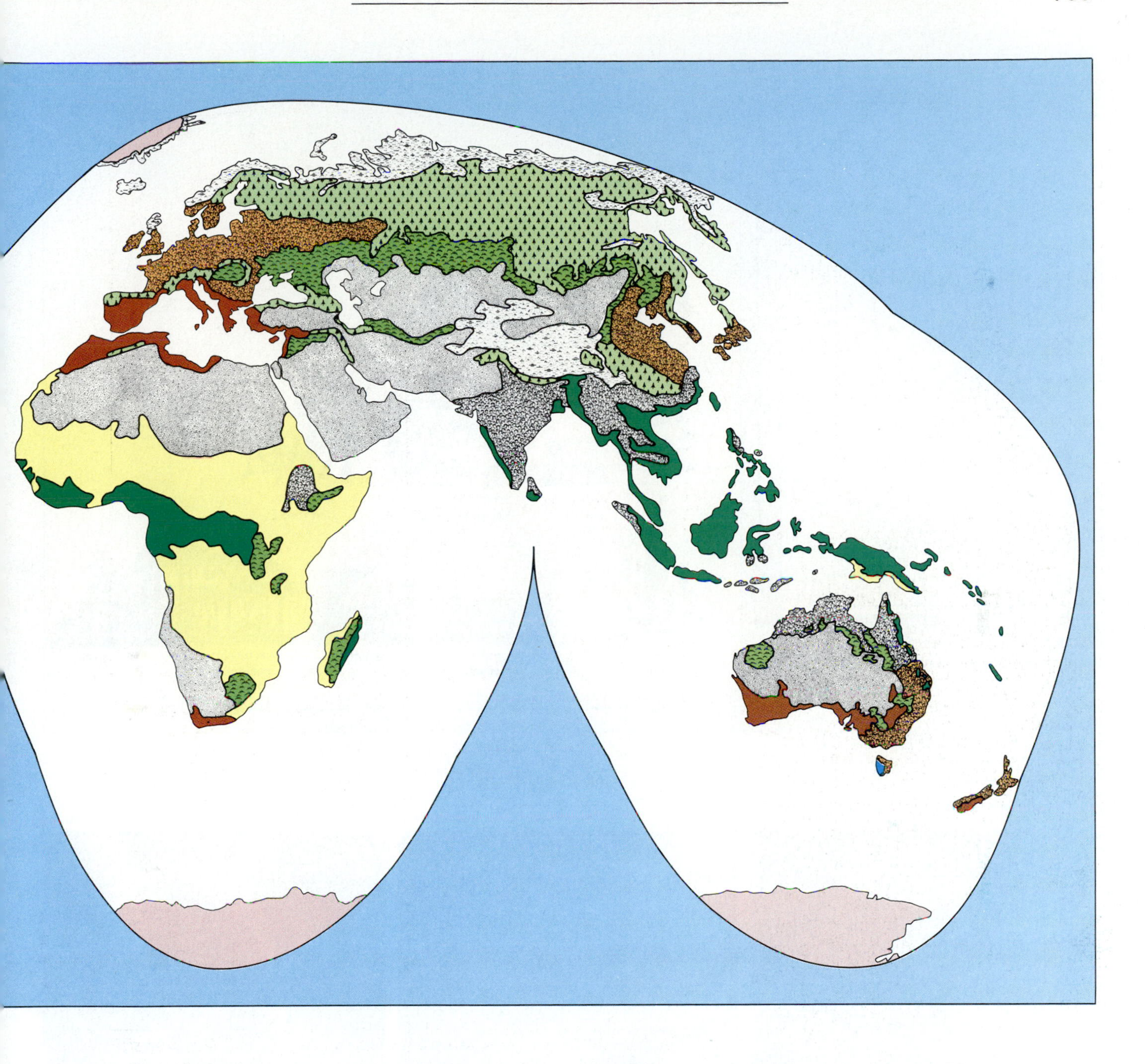

dry seasons. In these areas, for example India and much of southeast Asia, **tropical deciduous forests** grow. During the dry season, the trees cannot get enough water from the soil to compensate for evaporation from their leaves. As a result, the plants have adapted to the dry season by shedding their leaves, thereby minimizing water loss. If the rains fail to return on schedule, the trees merely wait it out until times are better.

SAVANNA. Along the edges of the tropical deciduous forest, the trees gradually become more widely spaced, with grasses growing beneath them. Eventually grasses become the dominant vegetation, with only scattered trees and thorny scrub forests here and there; this is the **savanna** (Fig. 38-7). Savanna grasslands typically have a rainy season during which virtually all of the year's precipitation falls (30 centimeters or less). When the dry season arrives, it comes with a vengeance. No rain may fall for months, and the soil becomes hard, dry, and dusty. Grasses are well adapted to this type of climate, growing very rapidly during the rainy season, and dying back to drought-resistant roots during the dry times. Only a few specialized trees such as the thorny acacia or the water-storing baobab can survive the devastating savanna dry seasons. In areas in which the dry season becomes even more pronounced, virtually no trees at all can grow, and the savanna imperceptibly grades into tropical grassland.

The African savanna has probably the most diverse and impressive array of large mammals on Earth, including numerous herbivores such as antelope, wildebeest, buf-

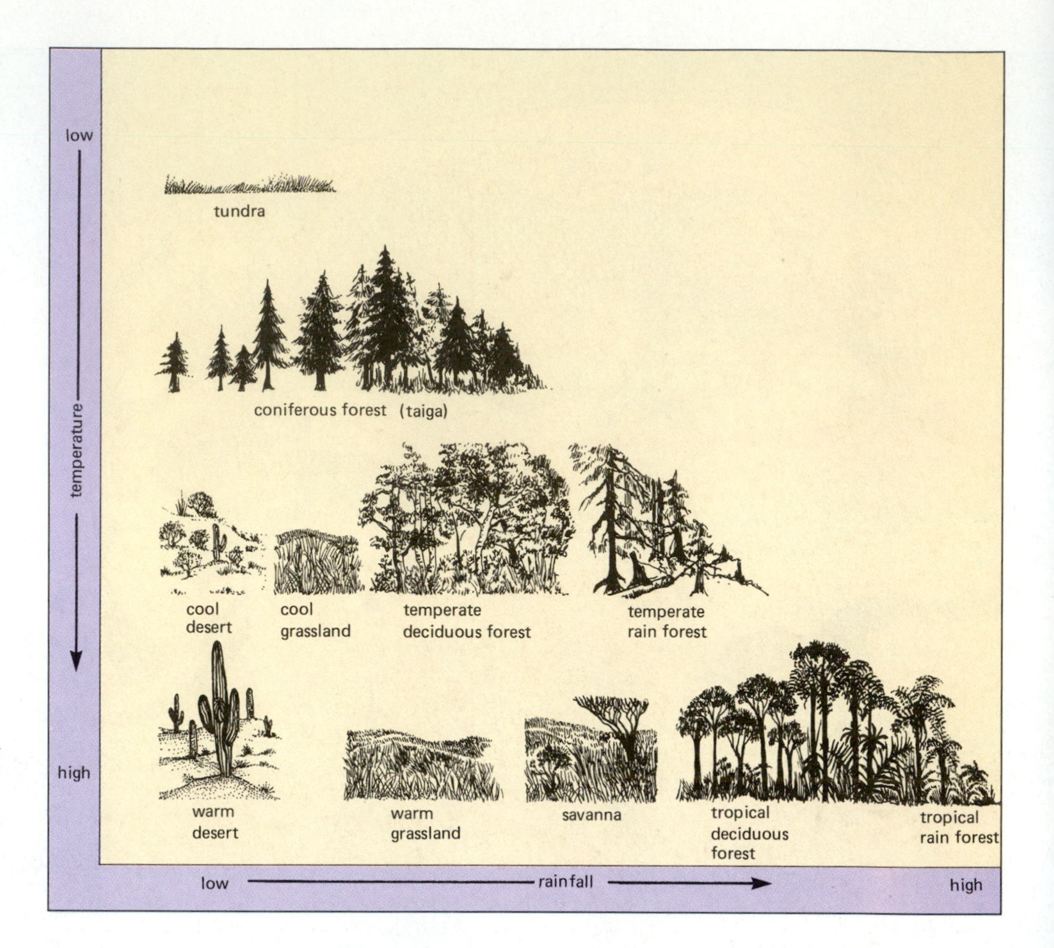

Figure 38-5 The distribution of biomes on land depends on temperature and rainfall, which together determine the available soil moisture needed for plant growth.

Figure 38-6 Reaching for the light in the dense tropical rain forest, **(a)** *vines climb tree trunks, while* **(b)** *epiphytes cling to the bark of trees, high above the dim forest floor.* **(c)** *Since most food is found high in the trees, so are the animals, as illustrated by this howler monkey.*

Figure 38-7 On the African savanna, **(a)** *zebra and* **(b)** *cape buffalo graze fearlessly near resting lions. The herds of herbivores provide food for the greatest assortment of large carnivores on the planet, including lions, leopards, cheetahs, hyenas, and wild dogs. Hunting cooperatively, female lions can bring down a buffalo many times their individual weight.* **(c)** *The abandoned carcass is a feast for scavenging hyenas and vultures.*

falo, and giraffes, and carnivores such as the lion, leopard, hyena, and wild dog. The abundant grasses which make the savanna a suitable habitat for so much wildlife also make it suitable for grazing domestic cattle. As the human population of East Africa increases, so does the pressure of cattle grazing upon the savanna. Fences increasingly disrupt the migration of the great herds of herbivores in search of food and water. Ecologists have discovered that the native herbivores are much more efficient at converting grass into meat than are cattle. Perhaps the future African savanna may see herds of domesticated antelope.

Deserts. Even drought-resistant grasses need at least 25 to 50 centimeters of rain a year, depending on its seasonal distribution and the average temperature. When less rain than this falls, many grasses fail and **desert** biomes occur (Fig. 38-8). These biomes are found on every continent, often around 20° to 30° north and south latitude, and also in the rain shadows of major mountain ranges. As with all biomes, the designation "desert" includes a variety of environments. On one extreme are certain areas of the Sahara or Chile, where it virtually never rains and there is no vegetation at all. The more common deserts, however, are characterized by an intermittent vegetation and large areas of bare ground. The plants are often spaced evenly, as if by hand (Fig. 38-8a). Frequently, the perennial plants are bushes or cacti with large, shallow root systems. The shallow roots quickly soak up the soil moisture after the infrequent desert storms, while the rest of the plant is typically covered with a waterproof, waxy coating to prevent evaporation of precious water. Water is stored in the thick bodies of cacti and other succulents. The spines of cacti are leaves modified for protection and water conservation, since they present almost no surface area for evaporation. In many deserts, all the rain falls in just a few storms, and specialized annual wildflowers take advantage of the brief period of moisture to race through germination, growth, flowering, and seed production in a month or less (Fig. 38-9).

The animals of the deserts, like the plants, are specially adapted to survive on little water. Most deserts appear to

Figure 38-8 Distinctive deserts of western North America. **(a)** *Throughout much of Utah and Nevada, the Great Basin Desert presents a monotonous landscape of widely spaced shrubs, such as sagebrush and greasewood. These shrubs often secrete a growth inhibitor from their roots, preventing germination of nearby plants and thus reducing competition for water.* **(b)** *The Sonoran Desert in southern Arizona and northern Mexico is typified by the saguaro cactus, prickly pear, and the spindly ocotillo, whose red flowers are pollinated by hummingbirds. The body of the tail saquaro is pleated and can expand after a rain to store water.* **(c)** *The bobcat is an elusive denizen of the Sonoran Desert.*

be completely devoid of animal life during the day, because the animals seek relief from the sun and heat in cool underground burrows. After dark, when the deserts cool down considerably, reptiles such as horned lizards and snakes emerge to feed, as do mammals such as the kangaroo rat, and birds such as the burrowing owl. Most of the smaller animals survive without ever drinking at all, deriving all the water they need from their food and that produced during cellular respiration in their tissues. Larger animals such as desert bighorn sheep are dependent on permanent water holes during the driest times of the year.

Figure 38-9 The Mojave desert in southern California in spring is carpeted with desert dandelions. Through much of the year, annual wildflower seeds lie dormant waiting for the spring rains to fall.

Figure 38-10 The chaparral biome is limited primarily to coastal mountains in dry regions, such as the San Gabriel Mountains in southern California. Chaparral is maintained by frequent fires set by summer lightning. Although the tops of the plants may be burned off, the roots send up new sprouts the next spring.

CHAPARRAL. In many coastal regions which border on deserts, such as southern California and much of the Mediterranean, a unique type of vegetation grows called **chaparral.** The annual rainfall in these regions is similar to that of a desert, but the proximity of the sea provides a slightly longer rainy season in the winter and frequent fogs even during the spring and fall which reduce evaporation. Chaparral consists of small trees or large bushes with thick waxy or fuzzy evergreen leaves that conserve water. These shrubs are also able to withstand the frequent summer fires started by lightning (Fig. 38-10).

As is often the case with deserts, grasslands, and forests, the extent of chaparral vegetation is influenced by human activities. About 400 B.C., the Greek philosopher Plato wrote that "there are mountains in Attica which can now keep nothing but bees, but which were clothed, not so very long ago, with timber suitable for roofing very large buildings." The original Greek forests were cut down not only for ceiling beams but also for the great Athenian naval fleets. Subsequently, heavy grazing by goats prevented regrowth of the forests. Only the chaparral plants which could survive in the hot sun and were unpalatable or poisonous to goats regrew. The forests have never returned.

GRASSLANDS. In the temperate regions of North America, deserts occur in the rain shadows east of the mountain ranges, such as the Sierra Nevada, Cascades, and Rockies. Farther east, as the rainfall gradually increases, the deserts begin to support more and more grasses, giving way to the prairies of the Midwest. These **grassland** biomes usually have a continuous cover of grass and virtually no trees at all except along the rivers. From the tallgrass prairies of Iowa and Illinois (Fig. 38-11) to the shortgrass prairies of eastern Colorado, Wyoming, and Montana (Fig. 38-12), the North American grassland once stretched across almost half of the continent.

Water and fire are the crucial factors in the competition between grasses and trees. The hot dry summers and frequent droughts of the shortgrass prairies can be tolerated by grass but are fatal to trees. In the more eastern prairies, trees can grow but historically were destroyed by frequent fires, often set by the Indians to maintain grazing land for the bison. Although the tops of the grasses are killed by fire, their root systems usually survive; trees, on the other hand, are killed outright. The resulting grasslands of North America once supported huge herds of bison, as many as 60 million in the early nineteenth century. Grasses growing and decomposing for thousands of years produced perhaps the most fertile soil in the world. With the elimination of the bison and the development of plows

Figure 38-11 Tallgrass prairie in Missouri. In the central United States, moisture-bearing winds out of the Gulf of Mexico produce summer rains, allowing a lush growth of tall grasses and wildflowers such as these coneflowers (inset).

Figure 38-12 **(a)** *The lands east of the Rocky Mountains receive relatively little rainfall and shortgrass prairie results, characterized by low-growing bunch grasses such as buffalo grass and grama grass.* **(b)** *Wild pronghorn antelope,* **(c)** *protected bison, and* **(d)** *prairie dogs still graze the prairie.*

that could break the dense turf, the former prairie has become the breadbasket of North America.

In the western shortgrass prairie, cattle and sheep have replaced the bison and pronghorn antelope. As a result of their overgrazing the grasses, the boundary between the cool deserts and the grassland has often been altered in favor of desert plants. Much of the sagebrush desert of the American west is actually the result of overgrazing shortgrass prairie (Fig. 38-13). Cattle prefer grass to sagebrush, so heavy grazing destroys the grass. This leaves moisture in the soil that the grass would have absorbed, encouraging the growth of the woody sagebrush. Since the cattle will not eat the sagebrush, it soon becomes the dominant vegetation; thus the prairie grasses are replaced by plants characteristic of cool desert.

Figure 38-13 Sagebrush desert or shortgrass prairie? Biomes are not only influenced by temperature, rainfall, and soil, but also by human activities. The shortgrass prairie field on the right has been overgrazed by cattle, causing the grasses to be replaced by sagebrush.

TEMPERATE FORESTS. At the eastern edge of the North American grasslands, the prairie merges into the **temperate deciduous forest** biome (Fig. 38-14). There is

Figure 38-14 **(a)** *The temperate deciduous forest of the eastern United States, in which the whitetail deer (inset) is the largest herbivore.* **(b)** *In spring, a profusion of woodland wildflowers bloom briefly before the trees leaf out.*

higher precipitation than in the prairie biome (75 to 150 centimeters), and in particular there is more summer rainfall. The soil therefore retains enough moisture for trees to grow, and the resulting forest shades out grasses. In contrast to the tropical forests, the temperate deciduous forest biome has cold winters, usually with at least several hard frosts and often long periods of subfreezing weather. Winter in this biome has an effect on the trees similar to that of the dry season in the tropical deciduous forests: during periods of subfreezing temperatures, liquid water is not available to the trees. To reduce evaporation when water is in short supply, the trees drop their leaves in the fall and leaf out again in the spring when liquid water becomes available. During the brief time in spring when the ground has thawed but the trees have not yet blocked off all the sunlight, numerous wildflowers grace the forest floor.

Insects and other arthropods are numerous and conspicuous in deciduous forests. The decaying leaf litter on the forest floor also provides food and habitat for bacteria, earthworms, fungi and small plants, and many arthropods feed on these or on each other. A variety of vertebrates, including mice, shrews, squirrels, raccoons, and many species of birds dwell in the deciduous forests. Large mammals such as black bear, deer, wolves, bobcats, and mountain lions were formerly abundant, but the predators have been largely exterminated by humans.

In North America, most of the temperate forests are found in the eastern third of the United States. However, on the Pacific coast, from the lowlands of the Olympic peninsula in Washington to southeast Alaska, is the **temperate rain forest** biome (Fig. 38-15). As in the tropical rain forest, there is no shortage of liquid water year round. This is due to two factors. First, there is a tremendous amount of rain. The Hoh River rain forest in Olympic National Park receives over 400 centimeters of rain each year, over 60 centimeters in the month of December alone. Second, the moderating influence of the Pacific Ocean prevents severe frost from occurring along the coast, so the ground seldom freezes.

The abundance of water means that the trees have no need to shed their leaves in the fall, and almost all the trees are evergreens. In contrast to the broadleaf evergreens of the tropics, the temperate rain forests are dominated by conifers. The trees are green year round, and the ground and often the trunks of the trees are covered with mosses and ferns. As in the tropical rain forests, so little light reaches the forest floor that tree seedlings usually cannot become established. Whenever one of the forest giants falls, however, it opens up a patch of light, and new seedlings quickly sprout, often right atop the fallen log. This produces a "nurse log" shown in Fig. 38-15b.

Figure 38-15 **(a)** *The Hoh River temperate rain forest in Olympic National Park. The coniferous trees do not block off the light as effectively as broadleaf trees do, and so there are ferns, mosses, and wildflowers growing in the pale green light of the forest floor (inset).* **(b)** *The dead feed the living, as new trees grow from the decay of this fallen giant.*

TAIGA. North of the prairies and temperate forests, the **northern coniferous forest,** also called the **taiga** (Fig. 38-16), stretches horizontally across the entire continent of North America, mainly in southern Canada. Conditions in the taiga are harsher than those in the temperate deciduous forest. In the taiga, the winters are longer and colder and the growing season is shorter. The few months of warm weather are too short to allow trees the luxury of regrowing leaves in the spring. As a result, the taiga is populated almost entirely by evergreen coniferous trees with small, waxy needles. The waxy coating and small surface area of the needles prevents water loss by evaporation during the cold months, so the leaves can stay on the trees year round. Thus the tree can grow slowly for much of the year and is instantly ready to take advantage of good growing conditions when spring arrives.

Because of its harsh climate, the diversity of life in the taiga is much lower than in many other biomes. Vast stretches of central Alaska, for example, are covered by a somber forest composed almost exlusively of black spruce and an occasional birch. Due to the remoteness of much of the taiga and the severity of the climate, a greater percentage of the taiga remains in primeval condition than any other North American biome except the tundra. Large mammals such as the wood bison, grizzly bear, moose, and wolf, mostly exterminated from the southern regions of their original range, still roam the taiga, along with smaller animals such as wolverine, marten, foxes, snowshoe hare, and deer.

TUNDRA. The last biome encountered before reaching the polar ice caps is the arctic **tundra,** a vast treeless region bordering the Arctic Ocean (Fig. 38-17). Conditions in the tundra are severe. Winter temperatures in the arctic tundra often reach to -40°C or below, winds howl at 50 to 100 kilometers per hour, and precipitation averages 25 centimeters or less per year, making this a freezing desert. Even during the summer, the temperatures can drop to freezing, and the growing season may last only a

Figure 38-16 **(a)** *The taiga. The small needles and pyramidal shape of conifers allows them to shed heavy snows.* **(b)** *Winter is not only a challenge for the trees but for animals such as this snowshoe hare and the lynx that preys on it. They face diminished food supply but increased energy requirements during subfreezing weather.*

Figure 38-17 Life on the tundra is adapted to cold. **(a)** *Plants such as dwarf willows and perennial wildflowers grow low to the ground, escaping the chilling tundra wind. Tundra animals such as caribou* **(b)** *and arctic foxes* **(c)** *can regulate blood flow in their legs, keeping them just warm enough to prevent frostbite, while preserving precious body heat for the brain and vital organs.*

few weeks before a freeze-up recurs. Somewhat less cold but similar conditions produce alpine tundra on mountaintops above treeline. The cold climate results in **permafrost,** a permanently frozen layer of soil often no more than half a meter below the surface. As a result, when summer thaws come, the melted snow and ice cannot soak into the ground, and the tundra becomes a huge marsh. Trees cannot survive in the tundra for several reasons, one of which is that the permafrost limits root growth to the topmost foot or so of soil, which turns to cold soup in summer. The high winds would easily topple a tree if it did not drown first.

Nevertheless, the tundra supports a surprising abundance and variety of life. The ground is carpeted with small perennial flowers and dwarf willows no more than few inches tall, and often with a large lichen called reindeer moss, a favorite food of caribou. The standing pools provide superb mosquito habitat, and tundra summers produce a veritable blizzard of mosquitos. The mosquitos and other insects provide food for numerous birds, most of which migrate long distances to nest and raise their young during the brief summer feast. The tundra vegetation supports lemmings (a type of rodent), which are eaten by wolves, snowy owls, arctic foxes, and even grizzly bears.

The tundra is perhaps the most fragile of all the biomes because of its short growing season. A 4-inch-high willow may have a 3-inch-diameter trunk and be 50 years old. Human activities in the tundra leave scars that persist for centuries. Fortunately for the tundra inhabitants, the impact of civilization is localized around oil drilling sites, pipelines, mines, and military bases.

Rainfall, Temperature, and Vegetation

Terrestrial biomes are greatly influenced by both temperature and rainfall, whose effects interact. Temperature strongly influences the effectiveness of rainfall in providing soil moisture for plants and standing water for animals to drink. The hotter it is, the more rapidly water evaporates, both from the ground and from plants. As a result of this interaction of temperature with rainfall (and to a lesser extent, the distribution of rain throughout the year), places that receive almost exactly the same rainfall can have startlingly different vegetation, all the way from desert to taiga. Take a trip with us from southern Arizona to northern Alaska as we visit ecosystems that receive around 30 centimeters (about 12.5 inches) of rain annually.

The Sonoran Desert near Tucson, Arizona (Fig. 38-3b) receives 28 centimeters of rain yearly, with an average annual temperature of 20° C. The landscape is dominated by giant saguaro cactus and low-growing, drought-resistant bushes. Fifteen hundred kilometers north, in eastern Montana, rainfall is about the same (31 centimeters), but we have passed into the shortgrass prairie biome (Fig. 38-12), largely because the average temperature is much lower, about 7°C.

Central Alaska receives the same annual rainfall (about 28 centimeters), yet is covered with taiga forest (Fig. 38-16a). Due to the low annual temperature (about −4°C), permafrost underlies much of the ground. During the summer thaw, the taiga earns its Russian name "swamp forest," although its rainfall is the same as the Sonoran desert.

LIFE IN THE OCEANS

Of the four essentials for life, liquid water is available in abundance in the Earth's oceans. Except at the polar icecaps, the temperature is also benign, at least compared to the land, never falling below about −2°C or climbing much above 30°C. *The major factors that limit the quantity and type of life in the oceans are energy and nutrients.*

Energy for life is provided by sunlight, either directly in the case of plants or indirectly via the plants for most other life forms. Water, no matter how clear it may appear to be in a glass, is actually quite opaque to light. Even in the purest and clearest water, the intensity of light decreases rapidly with depth, so that at a depth of 100 or 200 meters there is little left, and therefore no energy to power photosynthesis. If the water is at all cloudy, for example due to suspended silt or microorganisms, the depth of light penetration is greatly reduced. The depth of water in which the light is strong enough to support photosynthesis is called the **photic zone;** below this, the only energy available comes from the excrement and bodies of other organisms that sink or swim down.

The oceans are also usually short of nutrients, or to be more precise, most of the nutrients in the oceans are at or near the bottom, where there is no light. Nutrients dissolved in the water of the photic zone are constantly being taken up and incorporated into the bodies of living organisms. When these organisms die, some are eaten or decay while still in the photic zone, but some inevitably sink into the depths, or **benthic zone.** Thus nutrients constantly rain down from the photic zone into the benthic zone. If no new nutrients entered the photic zone, life would soon cease to exist there, and deprived of the rain of nutrients from the photic zone, life in the benthic zone would also disappear.

Fortunately, there are two sources of new nutrients: the land, from which rivers constantly remove nutrients and carry them to the oceans, and **upwelling,** where cold, nutrient-laden water from the ocean depths is brought to the surface. This often occurs along coastlines, as in California, Peru, and West Africa, where prevailing winds displace surface water, causing it to be replaced by water from below. Upwellings also occur around Antarctica.

Not surprisingly, the major concentrations of life in the oceans are found where abundant light is combined with a source of nutrients. This happens most often in shallow coastal waters, including coral reefs.

Coastal Waters

The most abundant life in the oceans is found in a narrow strip surrounding the Earth's landmasses, where the water is shallow and a steady flow of nutrients is washed off the land. Coastal waters include the **intertidal zone,** the area that is alternately covered and uncovered by water with the rising and falling of the tides, and the **nearshore zone,** bays, salt marshes, estuaries, and shallow subtidal areas (Fig. 38-18). Here is the only part of the ocean where large plants can grow, anchored to the bottom. In addition, the abundance of nutrients and sunlight promotes the growth of a veritable soup of photo-

Figure 38-18 The variety of environments at the meeting place of land and sea. **(a)** *A salt marsh in the eastern United States. Expanses of shallow water fringed by marsh grass* (Spartina) *provide excellent habitat and breeding grounds for many marine organisms and shorebirds.* **(b)** *Although the shifting sands present a challenge to life, grasses stabilize them, and (inset) animals such as this* Emerita *crab burrow in the sandy intertidal zone.* **(c)** *A rocky intertidal shore in Oregon, where animals and algae grip the rock against the pounding waves and resist drying during low tide. Inset: A sea star and anemone are surrounded by barnacles and mussels in a shallow tidepool.* **(d)** *Towering kelp sway through the clear water off southern California, providing the basis for a diverse community of invertebrates, fishes, and an occasional sea otter (inset).*

Climb Every Mountain: Death Valley to the High Sierra

In the lower 48 states, the hottest place is Death Valley, at 280 feet below sea level, and the highest place is Mount Whitney, at 14,495 feet above sea level. These two places are not more than 150 kilometers apart in south-central California, yet they have enormously different climates and vegetation. Why? Because going up a mountain is similar in its effect on temperature to going north toward the pole: it gets colder. The vegetation at increasing elevations reflects this cooling trend, so we can see samples of many of the northern biomes in a single day just by hiking up a mountain. In this photo essay, we will travel from the incredibly barren landscapes of Death Valley, through the northern Mojave Desert, swing around to the gentle western slopes of the Sierra Nevada, and climb up the highest reaches of these mountains that John Muir called the Range of Light. Along the way, refer back to the photographs of the major biomes for comparison.

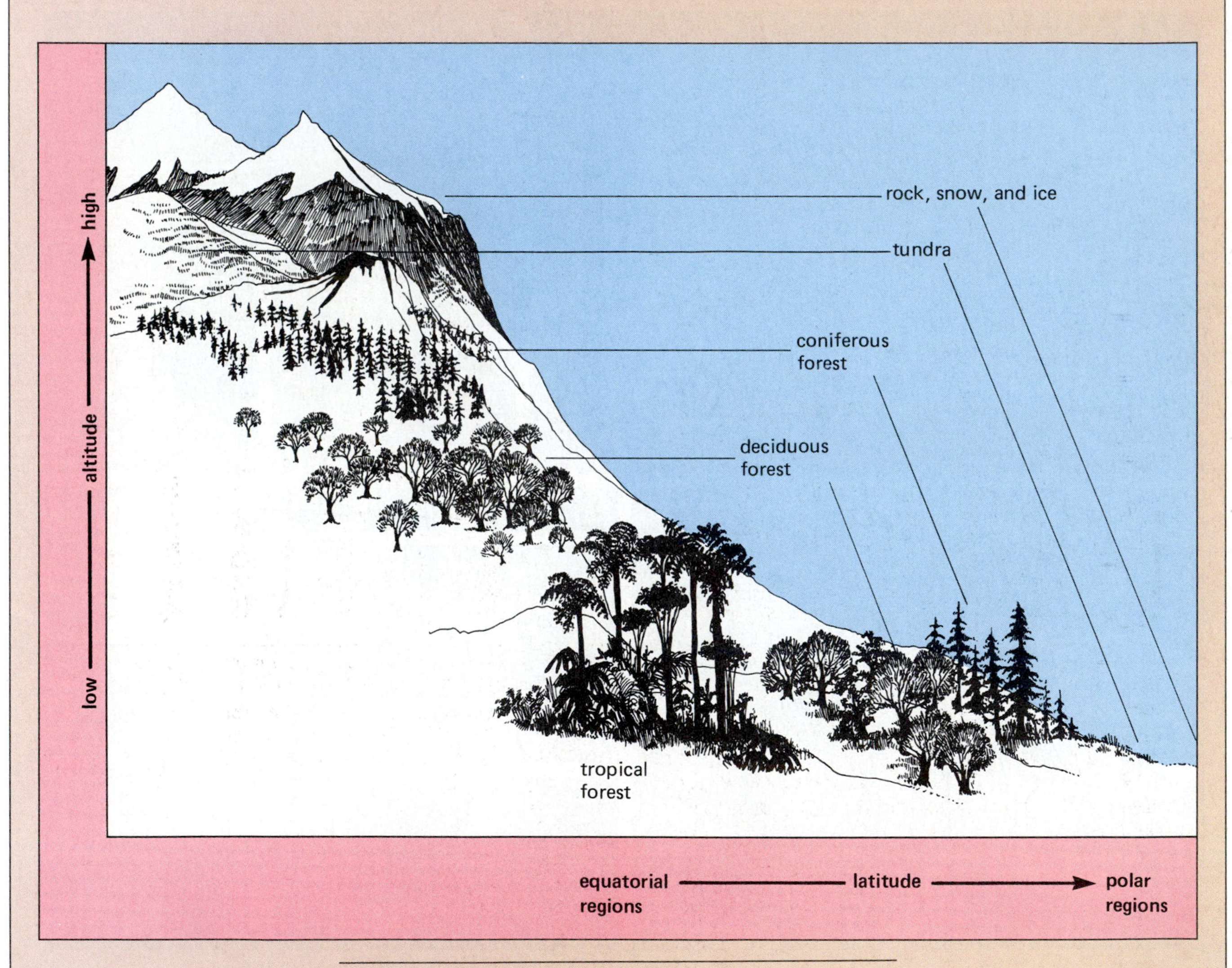

E38-1 Climbing a mountain in some ways is like going north: in both cases, increasingly cool temperatures produce a similar series of biomes.

Climb Every Mountain: Death Valley to the High Sierra

E38-2 The Devil's Golf Course in Death Valley. During the last glacial epoch, Death Valley was filled with water. Today, the valley floor is completely dry. The Devil's Golf Course is encrusted with lumps of salt slowly leaching up from the prehistoric lake bed.

E38-3 Joshua trees are the most striking plant in the Mojave Desert. These relatives of the yucca only produce their huge waxy flower heads in the spring following an unusually wet winter.

E38-4 When the Spaniards came to California in the eighteenth century, the Great Central Valley was a lush grassland. The grasses grew green during the warm winter rains and then turned crisp and brown in the fierce summer sun. The western foothills of the Sierra Nevada, shown here, are almost a North American savanna. The Central Valley grasses sweep up to the feet of the Sierra, where they mingle with live oak and piñon pine.

E38-5 As we climb up into the mountains, at elevations of about 5,000 to 10,000 feet we find a distinctive Sierra Nevada version of the taiga. At the lower elevations, yellow and Jeffrey pines and white fir predominate. Farther up, we may encounter entire forests of lodgepole pines.

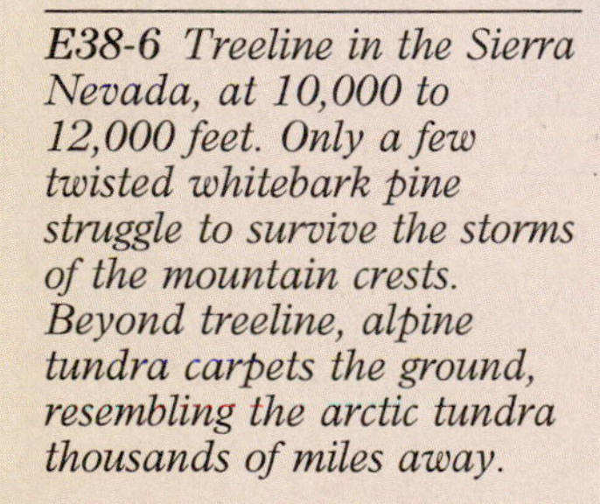

E38-6 Treeline in the Sierra Nevada, at 10,000 to 12,000 feet. Only a few twisted whitebark pine struggle to survive the storms of the mountain crests. Beyond treeline, alpine tundra carpets the ground, resembling the arctic tundra thousands of miles away.

The highest peaks of the Sierra often bear their own ice caps, the southernmost glaciers in North America; the Coness glacier is shown here. Routine hard freezes prevent snowmelt and allow the formation of permanent glaciers along the crest at 12,000 to 14,000 feet.

synthetic protists called **phytoplankton** (Greek, "drifting plants"). Associated with these plants and protists are animals from nearly every phylum: annelid worms, sea anemones, jellyfish, sea urchins, starfish, mussels, snails, fish, and sea otters, to name a few. A large number and variety of organisms live permanently in coastal waters, but many that spend most of their lives in the open ocean come into the coastal waters to reproduce. Bays, salt marshes, and estuaries in particular are the breeding grounds for a wide variety of organisms, such as crabs, shrimp, and a host of fish, including most of our commercially important species.

Coastal regions are of great importance not only to the organisms that live or breed there, but also to humans who wish to use them for food sources, recreation, mineral and oil extraction, or living places. These many uses are often incompatible with one another, and conflict often arises over which use should predominate in a particular area. As populations increase in our coastal states and resources such as oil become increasingly scarce, the conflict between preservation of our coastal wetlands as wildlife and animal habitat and development of these areas for housing, marinas, and energy extraction will become increasingly intense. Since much of the life of the entire ocean is inextricably dependent on the well-being of the coastal waters, it is imperative that we protect these fragile and vital areas.

Coral Reefs

Coral reefs are actually sheltered shallow-water environments created by animals and plants. In warm tropical waters, with just the right combination of bottom depth, wave action, and nutrients, corals and specialized algae build reefs from their own skeletons of calcium carbonate. These reefs provide an anchoring place for many other algae and bottom-dwelling animals, and shelter and food for the most diverse assemblage of invertebrates and fish to be found in the oceans (Fig. 38-19).

Coral reefs are extremely sensitive to certain types of disturbance, especially silt eroding from nearby land. Many corals harbor photosynthetic protists in their tissues and derive a significant portion of their energy from the products of photosynthesis. As silt clouds the water, light is diminished and photosynthesis reduced, hampering growth of the corals. The reef may eventually become buried in mud, the corals smothered, and the entire marvelous community of diverse organisms destroyed. Several reefs near Honolulu have been lost to siltation due to erosion from construction, roadways, and poor land

Figure 38-19 **(a)** *Coral reefs are composed of the skeletons of certain algae and corals and provide habitat for a variety of extravagantly colored creatures. These include the sponges in this photograph,* **(b)** *fish, such as the queen angelfish, shown here with a red sponge and an anemone in the background, and* **(c)** *invertebrates, such as the blue ring octopus, shown on a yellow coral on Australia's Great Barrier reef. This tiny (6 inches, fully extended) octopus is one of the world's most poisonous animals, possessing enough venom to kill 10 adult men.*

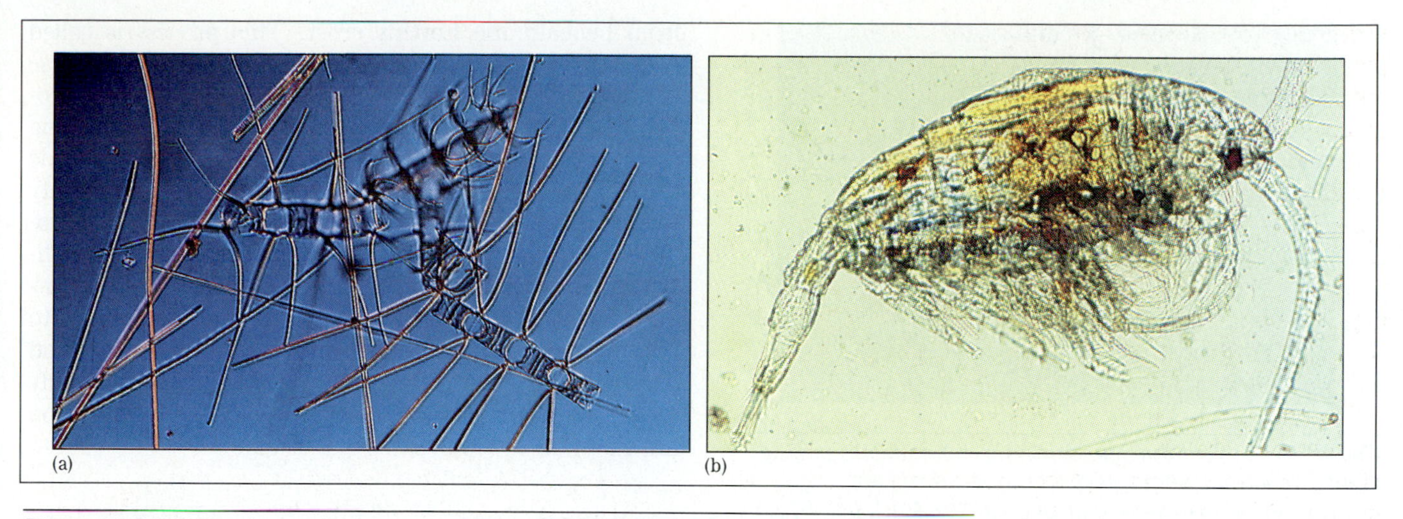

Figure 38-20 **(a)** *In the open ocean, the photosynthetic phytoplankton are the producers, on which most life depends.* **(b)** *They are eaten by zooplankton, represented by this microscopic crustacean. The spiny projections on these planktonic creatures help keep them from sinking below the photic zone.*

management. The reef inhabitants have been replaced by large numbers of sediment-feeding invertebrates such as sea cucumbers.

The Open Ocean

Beyond the coastal regions lie vast areas of the ocean in which the bottom is too deep to allow plants to anchor and still receive enough light to grow. Most life in the open ocean is limited to the upper photic zone, in which the life forms are **pelagic,** that is, free-swimming or floating for their entire lives. The base upon which life exists in the open sea is the **plankton** (Greek, "to drift"), a collection of free-floating, often microscopic protists (phytoplankton) and animals (zooplankton) (Fig. 38-20). The photosynthetic phytoplankton (mainly diatoms and dinoflagellates) are the ultimate food source for the other inhabitants of the open ocean. Most of the planktonic animals are minute crustaceans, relatives of crabs and lobsters, that feed on the phytoplankton or on each other. In their turn, they serve as food for larger invertebrates, small fish, and marine mammals such as the great blue whale.

One challenge faced by denizens of the open ocean is how to stay afloat in the photic zone, where sunlight and food are abundant. Many members of the planktonic community have elaborate flotation devices such as oil droplets in their cells or long projections to slow down their rate of sinking (Fig. 38-20). Most fish have swimbladders that can be filled with gas to regulate their buoyancy. Some animals, and even some of the phytoplankton such as the dinoflagellates (see Chapter 17), actively swim to maintain their position in the photic zone. Some small crustaceans migrate to the surface at night to feed, sinking into the dark depths during daylight, thus avoiding visual predators such as fish.

The amount of pelagic life varies tremendously from place to place in the ocean. The crucial factors regulating the density of life are nutrients, especially phosphorus and nitrogen. Except in upwelling regions, phytoplankton quickly incorporate all the available nutrients into their cells until a shortage of one or more nutrients develops. As the plants and animals die and sink out of the photic zone, the open sea becomes increasingly devoid of nutrients, and hence of life. The blue clarity of tropical waters is a result of a lack of nutrients, which limits the concentration of plankton in the water. The blue color is due to the refraction of light by pure water. Nutrient-rich waters that support a large plankton community are greenish and relatively murky, as sunlight is scattered by the microscopic bodies of myriad protists and animals.

Below the photic zone, in most cases the only available energy comes from the excrement and dead bodies raining down from above. Nevertheless, a surprising quantity and variety of life exists in the benthic zone, including fishes of bizarre shape, worms, sea cucumbers, sea stars, and molluscs.

In 1977, a new and unusual source of nutrients, forming the basis of a spectacular undersea community, was discovered in the benthos. Geologists exploring the Galápagos Rift (an area of the Pacific floor where plates forming the Earth's crust are separating) found vents spewing superheated water black with sulfur and minerals. Surrounding these vents was a rich community of pink fish, blind crabs, mussels, tube worms, and sea anemones, some reaching enormous size (Fig. 38-21). In this unique ecosystem, sulfur bacteria serve as the primary producers, harvesting the energy stored in hydrogen sulfide spewed

Figure 38-21 *Vent communities in the ocean depths include giant 12-foot-tall worms, red with oxygen-trapping hemoglobin. These worms have no digestive tract, but are host to sulfur bacteria, which provide the worms with energy by oxidizing hydrogen sulfide.*

from beneath the Earth's crust. This process is called chemosynthesis, and replaces photosynthesis in these depths, 2500 meters below the surface. The bacteria proliferate in the warm water surrounding the vents, covering nearby rocks with thick, matlike colonies. These provide the food base on which the animals of the vent community thrive. In the laboratory, some rift vent bacteria reproduce rapidly at 250°C under enormous pressure, conditions that would instantly destroy all other life forms known, but stop reproducing if the water is "cooled" to boiling temperature. The world still holds wonders and mysteries for those who seek them, and we have only begun to explore the versatility and diversity of life on Earth.

REFLECTIONS: HUMANS AND ECOSYSTEMS

The expanding human population has left relatively few ecosystems undisturbed. Our impacts on natural ecosystems are so diverse and wide ranging that they far exceed the scope of this book. However, we can identify some general characteristics of ecosystems dominated by humans and contrast them to the characteristics of undisturbed ecosystems. Below are listed six characteristic differences and a few ideas for minimizing them.

First, ecosystems dominated by people tend to be simpler, that is, to have fewer species and fewer community interactions, than undisturbed ecosystems. Although a city street bustles with life and apparent complexity, count the number of different species you encounter in an average block and compare it to the number you encounter while hiking in a wilderness for a similar distance. As humans enter an ecosystem, animals in the highest trophic levels are the first to go. Carnivores are always relatively rare, and their specialized needs are most easily disrupted. Many big carnivores, such as the wolf and mountain lion, require large undisturbed hunting territories. Humans often selectively destroy large predators, believing them a threat to people or their livestock. As a result, even in relatively undisturbed areas they have often been eliminated. Grizzlies and wolves no longer roam the Rocky Mountains. On the western shortgrass prairie, prairie dog towns fall as human towns and ranches rise (Fig. 38-22). The black-footed ferret, a predator of prairie dogs, faces possible extinction. Farm fields have been deliberately simplified from the original prairie biome to eliminate competition and predation and allow the maximum productivity of a single crop species. Nowhere is the contrast between human and natural ecosystems greater than in the tropical rain forests, whose unparalleled diversity is replaced by failed attempts at farming.

There is probably no practical way to restore to human ecosystems the great diversity found in undisturbed areas, nor is this always desirable from a human standpoint. Agriculture, for example, demands a simplified ecosystem (although not all biomes are suited to it). Cities concentrate human activities and culture and may lessen the human impact on the surrounding countryside. However, as we recognize the benefits of artificially simplified ecosystems, we must be aware of the need to preserve intact as many natural communities as possible. The undisturbed forest traps and purifies water and can reduce air pollution. Swamps and estuaries, when

Figure 38-22 *This prairie dog town has since been bulldozed under for another Denver suburb.*

not dredged or filled, contain a wealth of detritus feeders and decomposers that purify water. Coastal wetlands are breeding places for millions of birds and spawning sites for a majority of our commercially important fish and crustacean species. In undisturbed, diverse ecosystems we find aesthetic pleasure, as well as a storehouse of species whose commercial or medicinal values are not yet recognized. For example, nearly half of the medicines in use today were originally discovered in plants, and humans have examined only a small fraction of existing plants for possible medical uses.

Second, whereas natural ecosystems run on sunlight, human ecosystems have become dependent on nonrenewable energy from fossil fuels. From the suburbanite pouring gas into his lawn mower to the farmer driving a combine, managing a simplified ecosystem is an energy-intensive proposition. Energy must be expended to oppose the tendency of the natural system to restore complexity. Fertilizers also require large amounts of energy to produce, and farms must be heavily fertilized because nutrient cycles have been disrupted.

To counteract this trend, some farmers are returning to organic farming. Plant and animal waste and natural nutrient cycles are used to maintain soil fertility. Alternating legume crops such as soybeans and alfalfa helps maintain nitrogen in the soil. Mulching can help retain fertility, reduce water loss, and control weeds. The use of natural insect predators and limited pesticide spraying at critical times during a pest's life cycle can dramatically reduce the need for poisons. Organic farms can be as productive as more conventional farms while using 15 to 50 percent less energy per quantity of food produced.

In our homes and commercial structures, better insulation and increased use of solar heat can result in dramatic energy savings. By increasing our use of renewable energy sources such as wind and especially sunlight, we can conserve fossil fuels, dramatically reduce pollution, and move human ecosystems a step closer to those which occur naturally.

Third, natural ecosystems recycle nutrients, while human ecosystems tend to lose nutrients. Walk around a suburban neighborhood on trash-pickup day. Grass clippings and leaves are packed in plastic bags to be hauled away. To compensate for this loss, suburban lawns and gardens are often heavily fertilized. A similiar trend has occurred in modern farming. The exposed soil is eroded away by wind and water, removing crucial nutrients and requiring large inputs of fertilizer to replace them. Runoff from the field, carrying fertile topsoil and artificial fertilizers, may enter and pollute nearby rivers, streams, and lakes. Pesticides may kill detritus

Figure 38-23 Contour planting on sloping fields minimizes erosion.

feeders and decomposers, further disrupting natural nutrient cycles. Thus, while fertile soil accumulates in many natural ecosystems, it tends to be lost in those dominated by humans. Some 3 billion tons of topsoil are eroded from farms each year in the United States. The Mississippi River alone carries off 40 tons per hour.

Again, organic farming can help reverse this trend, and we can apply the same principles in our own lawns and gardens. Farmers can counteract erosion by using contour planting, in which row crops are oriented so they retard the flow of water instead of funneling it down a slope (Fig. 38-23). Row crops, which tend to allow erosion, can be alternated in strips with dense, soil-catching crops such as wheat. Planting rows of trees as windbreaks helps prevent soil loss from blowing wind, as well as creating a more diverse ecosystem and a nesting place for insect-eating birds. Farm fields need not be plowed in the fall and left unplanted to erode during the winter; in fact, an increasing number of farmers are planting crops in the stubble from the previous season, reducing erosion.

Fourth, natural ecosystems tend to store water and purify it through biological processes, whereas human ecosystems tend to pollute water and shed it rapidly. A thundershower strikes a forest, an adjacent city, and a farm. The rich soil of the forest sponges up the water, which gradually percolates into the ground, filtered by the soil and purified by decomposers that break down organic contaminants. In the nearby city, water pours from sidewalks, rooftops, and streets, picking up soot, silt, oil, lead, and refuse. It races down gutters into storm sewers, and a weakly toxic soup gushes into the nearest stream or river. Farm runoff carries priceless topsoil, expensive fertilizer, and animal manure into rivers and lakes, where these resources become pollutants.

Preventing erosion will simultaneously conserve water and reduce water pollution from farm runoff. Manure from cattle feedlots, which is a significant source of both groundwater and surface water pollution, could be placed on fields, where it would restore needed nutrients.

Although water will continue to run off our cities, the pollutants it carries can be reduced by minimizing our reliance on fossil fuels. We must eliminate leaded gasoline and tighten standards for emissions from diesel and gasoline engines and smokestacks. Efficient rapid-transit systems will reduce pollution and the massive frustration caused by congested traffic. Insulation will reduce power consumption in our homes and offices, as will increased reliance on solar heat.

Fifth, simple human ecosystems such as farms tend to be extremely unstable. Natural ecosystems have many species and tend to remain stable over time. Herbivorous insects are controlled by natural predators, including birds, shrews, and other insects. Insect populations are also limited since their preferred plants are interspersed among many others rather than growing in a pure stand, as on a farm. Although both farm pests and their natural predators are exposed to pesticides, the pests often develop resistance to the poison, while their predators are killed. A resistant pest, its predators destroyed and its favorite food surrounding it, will reproduce rapidly and devastate a crop. In these simplified communities, the introduction of an exotic species (see Chapter 36) is often disastrous, since most of its potential predators and competitors have been removed.

Simplified human ecosystems are also more vulnerable to unfavorable environmental conditions—a hailstorm, too much or too little rain, or an early or late freeze.

Farmers can counteract this trend by planting smaller fields with a wider variety of crops. Alternating crops not only helps maintain soil fertility, but also helps prevent the proliferation of disease and insect pests that are specialized for a particular crop. Populations of corn borers, for example, will die off during years when the field is planted in alfalfa. Use of biological controls such as natural insect predators and insect diseases can reduce reliance on pesticides.

Finally, human ecosystems are characterized by continuously growing populations, whereas nonhuman populations in natural ecosystems are relatively stable. As our population expands, the spread of human-dominated ecosystems presents a growing threat to the diversity of species and to the delicate balance that has evolved over the 3-billion-year history of life on Earth.

In summary, natural ecosystems tend to be complex, stable, and self-sustaining, powered by solar energy and nourished by recycling nutrients. They provide diverse habitats for wildlife, tend to purify contaminants through the action of decomposers, and build up nutrient-rich soil. Modern human ecosystems are relatively simple and are sustained by large inputs of energy from fossil fuels. They tend to minimize wildlife habitat, contaminate soil and water, and lose nutrients and fertile soil. These problems are compounded by continued population growth, which is causing expansion of human-dominated ecosystems at the expense of undisturbed ones.

As we have pointed out, human ecosystems do not have to be as disruptive and alien to the operation of natural ecosystems as we have allowed them to become. Through understanding, education, and commitment, appropriate use of technology, and stabilization of our population, we can reverse many of these destructive trends.

SUMMARY OF KEY CONCEPTS

The Requirements of Life

The requirements for life on Earth include nutrients, energy, liquid water, and a reasonable temperature. The differences in the form and abundance of living things in various locations on Earth are largely attributable to differences in these four factors.

Life on Land

On land, the crucial limiting factors are temperature and liquid water. Large regions of the continents which have similar climates will have similar vegetation, determined by the interaction of temperature and rainfall. These regions are called biomes. Biomes are influenced by the temperature and rainfall patterns, which vary with latitude. The equatorial tropical rain forests have a consistently warm, moist climate, while from about 20° to 30° north and south latitude, the Earth's great deserts are found. Farther north, grasslands and temperate deciduous forests predominate, giving way to the taiga or northern coniferous forest. The last stronghold of life before the polar icecap is the tundra. Continental topography also influences terrestrial biomes. In North America, deserts are often found in the eastern rain shadows of mountain ranges. Since temperature decreases both with increasing latitude and increasing elevation, the vegetation changes observed going up a mountain are similar to those seen while traveling from the equator toward the poles.

Tropical forests are warm and wet, dominated by huge broadleaf evergreen trees. Most nutrients are tied up in vegetation, and most animal life is arboreal. These areas, which are home to the most diverse assemblage of plants and animals on Earth, are rapidly being cut for agriculture, although the soil is extremely poor.

The African savanna is an extensive grassland with pronounced wet and dry seasons. It is home to the world's most diverse and extensive herds of large mammals.

Deserts, hot and dry, are found primarily between 20° and 30° of latitude. Plants are intermittent and have adaptations to conserve water. Animals tend to be small and nocturnal, also adapted to drought.

Chaparral exists in desertlike conditions which are moderated by their proximity to a coastline. This allows small trees and bushes to thrive.

Grasslands, concentrated in the centers of continents, have a continuous grass cover and few trees. They produce the world's richest soils, and have largely been converted to agriculture.

Temperate deciduous forests, whose broadleaf trees drop their leaves in winter to conserve moisture, dominate the eastern half of the United States. The wet temperate rainforests, dominated by evergreens, are found on the North Pacific coast of the United States.

The taiga or northern coniferous forest, covering much of the northern United States and southern Canada, is dominated by conifers whose small waxy needles are adapted for water conservation and year-round photosynthesis.

The tundra is a frozen desert where permafrost prevents the growth of trees, and bushes remain stunted. Nonetheless, a diverse array of animal life and perennial plants flourishes in this fragile biome.

Life in the Oceans

Energy and nutrients are usually limiting factors in the distribution of life in the oceans. Energy for life ultimately comes from the sun, and most nutrients are either washed off the land or brought up from the depths in areas of upwelling. Thus most oceanic life is found in shallow water, where sunlight can penetrate, and is concentrated near the continents and near upwelling currents, where nutrients are most plentiful.

Coastal waters, consisting of the intertidal zone and the nearshore zone, contain the most abundant life. Producers consist of aquatic plants anchored to the bottom and photosynthetic protists called phytoplankton.

Coral reefs are confined to warm, shallow seas. The calcium carbonate reefs secreted by corals form a complex habitat supporting the most diverse undersea ecosystem.

In the open ocean, most life is found in the photic zone, where light supports phytoplankton. In the lower benthic zone, life is supported by nutrients which rain down from the photic zone. Specialized vent communities, supported by bacteria, thrive at great depths in the superheated waters where the Earth's crustal plates are separating.

Humans and Ecosystems

Natural ecosystems tend to be complex, stable, and self-sustaining, powered by solar energy and nourished by recycled nutrients. They provide diverse habitats for wildlife and they may build up rich soil. In contrast, human ecosystems in developed countries are relatively simple and are maintained using large inputs of fossil fuel. Wildlife habitat is minimized, soil, water, and air are contaminated, and nutrients and fertile soil are lost. Human population growth compounds these problems. Solutions lie in the stabilization of population, use of more solar energy, fewer poisons, and more natural controls and nutrients.

GLOSSARY

Benthic zone (ben′-thik) the ocean depths, into which not enough sunlight can penetrate to support photosynthesis.

Biome (bī′-ōm) a terrestrial region with a characteristic climate and vegetation. Regions with similar climates will have similar vegetation and belong to the same biome (grassland, desert) even though widely separated.

Chaparral (shap-eh-rel′) a temperate coastal biome with hot dry summers and cool, somewhat rainy winters with frequent fogs. The typical vegetation consists of small trees or large bushes which are drought and fire resistant.

Climax community a relatively stable and predictable assemblage of organisms, the endpoint of succession.

Desert a biome in which potential evaporation greatly exceeds rainfall. Perennial vegetation is widely spaced over the ground and has drought-resistant features such as waxy or spiny leaves.

Grassland a biome characterized by a relatively continuous ground cover of grasses, with few or no trees except along watercourses. Typically, grassland biomes have a prolonged dry season and frequent fires, both of which favor grasses over trees.

Intertidal zone the area along the edges of land masses which is alternately flooded and left bare by the rising and falling of the tides.

Nearshore zone shallow-water areas of the ocean just below the low-tide line. Due to the abundance of nutrients and strong sunlight, the nearshore often has the greatest concentration of life in the ocean.

Northern coniferous forest a northern biome dominated by conifers; taiga.

Pelagic (pel-a′-jik) referring to organisms that spend their lives in open water, floating or swimming.

Permafrost soil layer a short distance below the surface of the tundra which stays frozen year round.

Photic zone (fō′-tik) the surface waters of the ocean, into which sunlight penetrates well enough for plants to grow.

Phytoplankton (fī′-tō-plank-ton) microscopic photosynthetic protists that form the basis of ocean food webs.

Plankton (plank′-ton) microscopic pelagic protists and animals found in the photic zone of the oceans, whose movement is determined primarily by movement of the water. Phytoplankton are the main producers in the open ocean.

Rain shadow an area of low rainfall on the side of a mountain facing away from the prevailing winds.

Savanna a transition biome between the tropical deciduous forest and tropical grassland, characterized by a nearly continuous cover of grasses with occasional drought-resistant trees. The savanna climate includes such a severe dry season that most trees cannot grow.

Taiga (tī′-ga) the cold coniferous forest biome, characterized by needleleaf evergreen trees. The waxy needles of the trees reduce evaporative water loss during winter.

Temperate deciduous forest a biome having cold winters but a fairly long, moist growing season. The dominant vegetation is broadleaf deciduous trees which shed their leaves in winter as a protection against water loss.

Temperate rain forest a biome of very limited distribution, along certain coasts where the climate is fairly mild year round and rainfall is very high, characterized by coniferous evergreen trees, often of enormous size.

Tropical deciduous forest a broadleaf deciduous forest biome found in areas with a warm climate year round but a pronounced wet and dry season. The trees drop their leaves in the dry season.

Tropical rain forest a broadleaf evergreen forest biome found in the tropics, where rainfall is adequate and temperatures are warm year round. The tropical rain forest has the greatest diversity of both plants and animals to be found anywhere on land.

Tundra a treeless biome characterized by permafrost and an extremely short growing season. Most of the plants are low-growing perennial shrubs, grasses, and wildflowers.

Upwelling movement of deep, nutrient-rich water to the surface.

Zooplankton (zō′-plank-ton) a diverse assemblage of small pelagic animals dominated by crustaceans.

STUDY QUESTIONS

1. What are the four major requirements for life? Which two are most often limiting in terrestrial ecosystems? In ocean ecosystems?
2. Explain why traveling up a mountain takes you through biomes similar to those you would encounter traveling north for a long distance.
3. Where are the nutrients of the tropical forest biome concentrated?
4. Explain two different undesirable effects of agriculture in the tropical rain forest biome.
5. Why is life in the tropical rain forest concentrated high above the ground?
6. List some adaptations of (a) desert plants and (b) desert animals to heat and drought.
7. On what biome does much of the world's agriculture occur?
8. How are trees of the taiga adapted to a lack of water and a short growing season?
9. Where is life in the oceans most abundant, and why?
10. What is the most diverse undersea biome?
11. Distinguish between the photic and benthic zone. How do organisms in the photic zone obtain nutrients? How are nutrients obtained in the benthic zone?
12. What unusual primary producer forms the basis for rift-vent communities?
13. List the terrestrial biomes and some dominant features of each.
14. How do deciduous and coniferous biomes differ?
15. What single environmental factor best explains why there is shortgrass prairie in Colorado, tallgrass prairie in Illinois, and deciduous forest in Ohio?
16. Where are the world's largest populations of large herbivores and carnivores found?
17. Compare and contrast human-dominated with undisturbed ecosystems. List the six general differences discussed in this chapter and explain how each can be minimized. Add some ideas of your own as well.

SUGGESTED READINGS

Bell, R. H. V. "A Grazing Ecosystem in the Serengeti." *Scientific American,* July 1971. Describes the great migrations of the herbivores of the African savanna.

Brownlee, S. "Bizarre Beasts of the Abyss." *Discover,* July 1984. Beautiful photography and informative narrative about the rift-vent communities.

Clapham, W. B., Jr. *Natural Ecosystems.* New York: Macmillan Publishing Company, 1983. A readable introductory textbook.

Goreau, T. F., Goreau, N. I., and Goreau, T. J. "Corals and Coral Reefs." *Scientific American,* August 1979. Discusses the ecology of the great reefs built by tiny polyps harboring symbiotic protists.

Horn, H. S. "Forest Succession." *Scientific American,* May 1975 (Offprint No. 1321). Describes succession in eastern deciduous forests.

Isaacs, J. D. "The Nature of Oceanic Life." *Scientific American,* September 1969 (Offprint No. 884). Describes the oceanic food chain with its microscopic primary producers.

Richards, P. W. "The Tropical Rain Forest." *Scientific American,* December 1973 (Offprint No. 1286). Describes this reservoir of genetic diversity and its rapid destruction.

Glossary

Abiotic (ā-bī-ah′-tik) Nonliving.

Abscisic acid (ab-sis′-ik) a plant hormone that is generally inhibitory, enforcing dormancy in seeds and buds, and closing stomata.

Abscission (ab-si′-shun) separation of leaves, flowers, or fruits from a stem, due to formation of a weakened layer of cells at the site of attachment to the stem.

Abscission layer a layer of thin-walled cells at the base of the petiole of a leaf, flower, or fruit, the usual site of separation from the stem.

Absorption the movement of nutrients into cells.

Accessory pigments colored molecules other than chlorophyll that absorb light energy and pass it to chlorophyll.

Acid a substance that releases hydrogen ions (H^+) into solution; a solution with a pH of less than 7.

Acrosome (ak′-rō-sōm) an enzyme-containing vesicle located at the tip of an animal sperm.

Actin (ak′-tin) one of the major proteins of muscle, whose interactions with myosin produce contractions; found in the thin filaments of the muscle fiber. *See also* Myosin.

Action potential a rapid change from a negative to a positive electrical potential in a nerve cell. This signal travels along an axon without change in size.

Activation energy in a chemical reaction, the energy needed to force the electron clouds of reactants together, prior to the formation of products.

Active site a group of amino acids on a protein that binds specific molecules and/or performs specific functions; the region of an enzyme molecule that binds substrates and performs the catalytic function of the enzyme.

Active transport the movement of molecules across a cell membrane through the use of cellular energy, usually against a concentration gradient.

Active visual signal a movement or posture that communicates information.

Adaptation a specific structure, physiological mechanism, or behavior that promotes the survival and/or reproduction of an organism in a particular environment; also the process of acquiring such characteristics.

Adaptive radiation extensive speciation occurring among related populations as a result of adaptation to a wide variety of habitats.

Adenosine triphosphate (a-den′-ō-sēn trī-fos′-fāt; ATP) a molecule composed of ribose sugar, adenine, and three phosphate groups. The last two phosphate groups are attached by "high-energy bonds" that require considerable energy to form and release that energy again when broken. ATP serves as the major energy carrier in cells.

Adrenal gland an endocrine gland consisting of an outer cortex and inner medulla. The cortex secretes steroid hormones that regulate metabolism and salt balance. The medulla secretes adrenalin and noradrenalin.

Aerobic using oxygen.

Age structure the distribution of males and females in a population according to age categories; often represented graphically.

Agglutination (a-glū-tin-ā′-shun) clumping of foreign substances or microbes, caused by binding with antibodies.

Aggregated distribution characteristic of populations in which individuals are clustered into groups. These may be social or based on the need for a localized resource.

Aggression antagonistic behavior, usually between members of the same species, often resulting from competition for resources.

Aggressive mimicry (mim′ik-rē) the evolution of a predatory organism to resemble a harmless one, thus gaining access to its prey.

Algae (al′-gē; sing. alga) a general term for simple aquatic plants lacking vascular tissue.

Allantois (al-an-tō′-is) one of the embryonic membranes of reptiles, birds, and mammals. In reptiles and birds, the allantois serves as a waste-storage organ and a respiratory surface. In mammals, the allantois forms most of the umbilical cord.

Allele (al-ēl′) one of several alternative forms of a particular gene, usually giving rise to a characteristic form of phenotype (e.g., purple or white flower color).

Allele frequency for any given gene, the relative proportion of each allele of that gene found in a population.

Allergy an inflammatory response produced by the body in response to invasion by foreign materials, such as pollen, which are themselves harmless.

Allosteric inhibition (al-ō-ster′-ik) enzyme regulation in which an inhibitor molecule binds to an enzyme at a site away from the active site, changing the shape or charge of the active site, so that it can no longer bind substrate molecules.

Alternation of generations a life cycle typical of plants in which a diploid sporophyte (spore-producing) generation alternates with a haploid gametophyte (gamete-producing) generation.

Altruism a behavior that benefits another organism, usually at some risk to the altruistic organism.

Alveolus (al-vē′-o-lus; pl. alveoli) a tiny air sac within the lungs surrounded by capillaries where gas exchange with the blood occurs.

Amino acid the individual subunit of which proteins are made, composed of a central carbon atom to which is bonded an amino group ($—NH_2$), a carboxylic acid group ($—COOH$), a hydrogen atom, and a variable group of atoms denoted by the letter R.

Ammonia a highly toxic nitrogen-containing waste product of amino acid breakdown which is converted to urea in the mammalian liver.

Amniocentesis (am-nē-ō-sen-tē′-sis) a procedure for sampling the amniotic fluid surrounding a fetus. Various tests may be performed on the fluid and the fetal cells suspended in it to provide information on the developmental and genetic state of the fetus.

Amnion (am′-nē-on) one of the embryonic membranes of reptiles, birds, and mammals, enclosing a fluid-filled cavity that envelops the embryo.

Amniotic egg (am-nē-ot′-ik) the egg of reptiles and birds. It contains an amnion that encloses the embryo in a watery environment; this allows the egg to be laid on dry land.

Amplexus (am-plecks′-us) a form of external fertilization found in amphibians, in which the male holds the female during spawning and releases his sperm directly onto her eggs.

Amygdala (am-ig′-da-la) part of the forebrain of vertebrates, involved in control of emotions and instinctive behaviors.

Amylase (am′-ē-lās) an enzyme that catalyzes the breakdown of starch, found in saliva and pancreatic secretions.

Anaerobic (an-er-ō′-bik): capable of living and obtaining energy in the absence of oxygen.

Analogous structures structures that have similar functions and superficial appearance but very different anatomy, such as the wings of insects and bats. The similarities are due to similar selective pressures.

Anaphase (an′-a-fāz) the stage of mitosis and meiosis II in which the sister chromatids of each chromosome separate from one another and are moved to opposite poles of the cell. In meiosis I, the stage in which homologous chromosomes are separated.

Angina pectoris (an-jī′-na pek-tōr′-is) chest pain associated with reduced blood flow to the heart muscle caused by obstruction of coronary arteries.

Angiosperm (an′-jē-ō-sperm) a flowering plant (division Angiospermophyta); produces seeds within a ripened ovary.

Anterior (an-tēr′-ē-ur) the front, forward, or head end of an animal.

Antheridium (an-ther-id′-ē-um) a structure in which male sex cells are produced, found in the bryophytes and certain seedless vascular plants.

Anther (an′-ther) the uppermost part of the stamen in which pollen develops.

Antibody a protein produced by cells of the immune system which combines with a specific antigen and usually facilitates its destruction.

Anticodon a sequence of three nucleotides in transfer RNA that is complementary to the three nucleotides of a codon of messenger RNA.

Antidiuretic hormone (an-tē-dī-ūr-et′-ik) also called ADH; a hormone produced by the hypothalamus and released by the posterior pituitary gland into the bloodstream. It acts on the nephron of the kidney and causes more water to be reabsorbed into the bloodstream.

Antigen a complex molecule, usually protein or polysaccharide, that stimulates the production of a specific antibody.

Apical dominance (āp-i-kul) the phenomenon whereby a growing shoot tip inhibits the sprouting of lateral buds.

Apical meristem (āp′-i-kul mer′-i-stem) the cluster of meristematic cells found at the tip of a shoot or root (or one of their branches).

Aqueous humor (ā′-kwē-us) clear, watery fluid between the cornea and lens of the eye.

Archegonium (ar-ke-gō′-nē-um) structure in which female sex cells are produced, found in the bryophytes and certain seedless vascular plants.

Arteriole (ar-tēr′-ē-ōl) a small artery that empties into capillaries. Contraction of the arteriole regulates blood flow to various parts of the body.

Artery (ar′-tur-ē) a vessel with muscular, elastic walls that conducts blood away from the heart.

Asexual reproduction. reproduction not involving the union of haploid gametes. Usually, asexual reproduction is accomplished by mitosis, producing genetically identical copies of the parent organism.

Association neuron in nervous circuits, a nerve cell that is postsynaptic to a sensory neuron and presynaptic to a motor neuron. In actual circuits, there may be many association neurons between individual sensory and motor neurons.

Aster during cell division in animals and some protists, a star-shaped array of microtubules extending in all directions outward from the centrioles.

Atherosclerosis (ath′-er-ō-skler-ō′-sis) a disease characterized by obstruction of arteries by cholesterol deposits and thickening of the arterial walls.

Atom the smallest particle of an element that retains the properties of the element; composed of a central nucleus containing protons and neutrons, and outer orbitals containing electrons.

Atomic number the number of protons in the nuclei of all atoms of a particular element.

Atrioventricular node (ā′-trē-ō-ven-trik′-ū-lar nōd) a specialized mass of muscle at the base of the right atrium through which the electrical activity initiated in the SA node is transmitted to the ventricles.

Atrium (ā′-trē-um) a chamber of the heart that receives venous blood and passes it to a ventricle.

Auditory nerve (aw′-dih-tory) the nerve leading from the mammalian cochlea to the brain, carrying information about sound.

Autoimmune disease a disorder in which the immune system produces antibodies against the body's own cells.

Autonomic nervous system part of the peripheral nervous system of vertebrates that innervates mostly glands and internal organs and produces largely involuntary responses.

Autosome (aw′-tō-sōm) a chromosome found in homologous pairs in both males and females, and which does not bear the genes determining sex.

Autotroph (aw′-tō-trōf) an organism that can manufacture all its high-energy organic molecules (e.g., sugars, proteins) from simple inorganic molecules (such as carbon dioxide, water, and minerals), using a nonliving energy source (usually sunlight); an organism that does not have to consume organic molecules as food.

Auxin (awk′-sin) a plant hormone that influences many plant functions, including phototropism, apical dominance, and root branching. Auxin generally stimulates cell elongation and, in some cases, cell division and differentiation.

Axon a long process of a nerve cell, usually extending from the cell body to synaptic endings on other nerve cells or on muscles.

B cell a type of lymphocyte that secretes antibodies into the circulatory system in response to stimulation by antigens.

Bacillus (buh-sil′-us; pl. bacilli) a rod-shaped bacterium.

Bacterial conjugation: the exchange of genetic material between bacteria.

Bacteriophage (bak-tēr′-ē-ō-fāj) a virus that infects bacteria.

Bacterium (bak-tēr′-ē-um; pl. bacteria): an organism consisting of a single prokaryotic cell surrounded by a complex polysaccharide coat.

Bark the outer layer of a woody stem, consisting of cork cells, cork cambium, and phloem.

Barr body an inactive X chromosome found in somatic cells of mammals that have at least two X chromosomes (usually females). The Barr body usually appears as a dark spot in the nucleus.

Basal body the organelle, structurally identical to a centriole, that gives rise to the microtubules of cilia and flagella.

Base (1) a substance that releases hydroxide ions (OH^-) into a solution or that is capable of combining with and neutralizing hydrogen ions, producing a solution with a pH greater than 7; (2) in molecular genetics, one of the nitrogen-containing, single- or double-ringed structures that distinguish one nucleotide from another.

Base-pairing rule the rule that only complementary bases can pair during DNA replication or RNA synthesis from DNA.

Basilar membrane (bas′-eh-lar) a membrane in the cochlea that bears hair cells which respond to the vibrations produced by sound.

Behavioral isolation lack of mating between species of animals that differ substantially in courtship and mating rituals.

Benthic zone (ben′-thik) the ocean depths, where sunlight is insufficient to support photosynthesis.

Bilateral symmetry body plan in which only a single plane drawn through the central axis will divide the body into mirror-image halves.

Bile (bīl′) a liquid secretion of the liver stored in the gallbladder and released into the small intestine during digestion. Its role is to emulsify or disperse lipids into small particles on which lipid-digesting enzymes may act.

Biological clock a metabolic timekeeping mechanism found in most organisms, whereby the organism measures the approximate length of a (24-hour) day even without external environmental cues such as light and dark.

Biological magnification the increasing accumulation of a toxic substance in increasingly high trophic levels.

Biome (bī′-ōm) a general type of ecosystem occupying an extensive geographical area. The same biome (desert, grassland) in different parts of the world will have similar climate and similar vegetation.

Biosphere (bī′-ō-sfēr) that part of the Earth inhabited by living organisms; includes both the living and nonliving components.

Biotic (bī-ah′-tik) living.

Biotic potential the most rapid potential growth rate of a population, assuming a maximum birthrate and minimum death rate.

Bladder a muscular storage organ for urine.

Blade the flat part of a leaf.

Blastocyst (blas′-tō-sist) an early stage of mammalian embryonic development, consisting of a fluid-filled ball with walls one cell layer thick, enclosing a mass of cells attached to its inner surface.

Blastopore the site at which a blastula invaginates to form a gastrula.

Blastula in animals, the embryonic stage attained at the end of cleavage, in which the embryo usually consists of a hollow ball with a wall one or several cell layers thick.

Bloom a dense growth of algae and cyanobacteria that can occur in response to an influx of nutrients into an aquatic ecosystem.

Book lungs thin layers of tissue resembling pages in a book, enclosed in a chamber and used as a respiratory organ by certain types of arachnids.

Boom and bust a population cycle characterized by rapid exponential growth followed by a sudden major decline in population size, seen in seasonal species and some populations of small rodents, such as lemmings.

Bowman's capsule the portion of the nephron in which blood filtrate is collected from the glomerulus.

Bradykinin (brā-dē-kīn′-in) a chemical formed during tissue damage that binds to receptor molecules on pain nerve endings, giving rise to the sensation of pain.

Brain the part of the central nervous system of vertebrates enclosed within the skull.

Bronchiole (bron′-kē-ōl) a narrow tube formed by repeated branching of the bronchi, which conducts air into the alveoli.

Bronchus (bron′-kus) a tube that conducts air from the trachea to each lung.

Bryophyte (brī′-ō-fīt) a division of simple nonvascular plants including mosses and liverworts.

Bud in plants, an embryonic shoot, usually very short and consisting of an apical meristem with several leaf primordia; in animals, a small copy of an adult that develops on the body of the parent; eventually breaks off and becomes independent.

Budding a form of asexual reproduction in which the adult produces miniature versions of itself that drop off and assume independent existence.

Bulbourethral gland (bul-bō-ū-rē′-thrul) in male mammals, a gland that secretes a basic, mucus-containing fluid that forms part of the semen.

C_3 cycle the cyclic series of reactions whereby carbon dioxide is fixed into carbohydrates during the light-independent reactions of photosynthesis. Also called the Calvin–Benson cycle.

C_4 pathway the series of reactions in certain plants that fixes carbon dioxide into organic acids for later use in the C_3 cycle of photosynthesis.

Calorie the amount of energy required to raise the temperature of 1 gram of water 1°C. When capitalized (i.e., Calorie), the amount of energy required to raise the temperature of 1000 grams of water 1°C., i.e., 1000 calories. The energy content of foods is measured in Calories.

Calvin-Benson cycle *see* C_3 cycle.

Cambium (kam′-bē-um) a lateral meristem that causes secondary growth of woody plant stems and roots. *See also* Cork cambium; Vascular cambium.

Camera eye the type of eye found in vertebrates and molluscs, in which a lens focuses an image on a sheet of light-sensitive receptor cells (the retina).

Camouflaged (cam′a-flaged) a term used to describe organisms that resemble their environment, often both in color and pattern.

Cancer a disease in which some of the body's cells grow without control.

Carbohydrate a compound composed of carbon, hydrogen, and oxygen, with the chemical formula $(CH_2O)_n$; includes monosaccharides, disaccharides, and polysaccharides (starches, glycogen, and cellulose).

Carbon fixation the initial steps in the C_3 cycle, in which carbon dioxide reacts with ribulose bisphosphate to form a stable organic molecule.

Cardiac muscle (kar'-dē-ak) specialized muscle of the heart, able to initiate its own contraction independent of the nervous system.

Carnivore (kar'-ni-vōr) literally "meat eater," a predatory organism feeding on other heterotrophs.

Carnivorous (kar-niv'-e-rus): feeding on the bodies of other animals.

Carotenoid (ka-rot'-en-oyd) a family of pigments, usually yellow, orange, or red, found in chloroplasts of plants and serving as accessory light-gathering molecules in thylakoid photosystems.

Carpel (kar'pel) the female reproductive structure of a flower, composed of stigma, style, and ovary.

Carrier (1) an individual who is heterozygous for a recessive condition. Carriers display the dominant phenotype but can pass on their recessive allele to their offspring; (2) a protein in a cell membrane that binds specific molecules and facilitates their transport across the membrane.

Carrying capacity the maximum population size that an ecosystem can maintain on a sustained basis. Determined primarily by the availability of space, nutrients, water, and light.

Cartilage (kart'-lij): flexible, translucent tissue that serves as the forerunner of bone during embryonic development in most vertebrates. In the class Chondrichthyes, cartilage is retained and forms the entire skeleton.

Casparian strip (kas-par'-ē-an) a waxy, waterproof band in the cell walls between endodermal cells in a root, which prevents the movement of water and minerals in and out of the vascular cylinder via extracellular space.

Catalyst (cat'-a-list) a substance that speeds up a chemical reaction without itself being permanently changed in the process. Catalysts lower the activation energy of a reaction.

Catastrophism the hypothesis that the Earth has experienced a series of geological catastrophes, much like Noah's flood, probably imposed by a supernatural being.

Causality the scientific principle that natural events occur as a result of preceding natural causes.

Cell the basic unit of life, consisting of an outer cell membrane surrounding cytoplasm and genetic material.

Cell body part of a nerve cell in which most of the common cellular organelles are located. Also often a site of integration of inputs to the nerve cell.

Cell cycle the sequence of events in the life of a cell, from one division to the next.

Cell division in eukaryotes, the process of reproduction of single cells, usually into two identical daughter cells, by mitosis accompanied by cytokinesis.

Cell-mediated immunity an immune response in which foreign cells or substances are destroyed by contact with T cells.

Cell membrane the outer membrane of a cell, composed of a bilayer of phospholipids in which proteins are embedded.

Cell plate in plant cell division, a series of vesicles that fuse to form the new cell membranes and cell wall separating the daughter cells.

Cell theory a theory stating that all living things are composed of cells, cells are the functional units of living things, and all cells arise from preexisting cells.

Cell wall a layer of material, usually made up of cellulose or cellulose-like materials, found outside the cell membrane of plants, fungi, bacteria, and plantlike protists.

Cellular respiration the oxygen-requiring reactions occurring in mitochondria that break down the end products of glycolysis into carbon dioxide and water, while capturing large amounts of energy as ATP.

Cellulose an insoluble carbohydrate composed of glucose subunits; forms the cell wall of plants.

Central nervous system in vertebrates, the brain and spinal cord.

Central vacuole a large, membrane-bound organelle, containing mostly water and dissolved substances, that occupies most of the volume of mature plant cells.

Centriole (sen'-trē-ōl) in animal cells, a microtubule-containing structure found at the microtubule organizing center and the base of each cilium and flagellum. Gives rise to the microtubules of cilia and flagella, and may be involved in spindle formation during cell division.

Centromere (sen'-trō-mēr) the region of a replicated chromosome at which the sister chromatids are held together.

Cerebellum (ser-uh-bel'-um) part of the hindbrain of vertebrates, concerned with coordination of motor activities.

Cerebral cortex (ser-ē'-brel kōr'-tex) a thin layer of neurons on the surface of the vertebrate cerebrum, in which most neural processing and coordination of activity occurs.

Cerebral hemisphere one of two nearly symmetrical halves of the cerebrum, connected by a broad band of axons, the corpus callosum.

Cerebrum (ser-ē'-brum) Part of the forebrain of vertebrates concerned with sensory processing, direction of motor output, and coordination of most bodily activities. The cerebrum consists of two nearly symmetrical halves (the hemispheres) connected by a broad band of axons, the corpus callosum.

Cervix (ser'-vicks) a ring of connective tissue at the outer end of the uterus, leading into the vagina.

Chaparral (shap-eh-rel') a temperate coastal biome with hot dry summers and cool, somewhat rainy winters with frequent fogs. The typical vegetation consists of small trees or large bushes which are drought and fire resistant.

Chemical bond the force of attraction between neighboring atoms that holds them together in a molecule.

Chemical equilibrium the condition in which the "forward" reaction of reactants to products proceeds at the same rate as the "backward" reaction from products to reactants, so that no net change in chemical composition occurs.

Chemoreceptors (kē'-mō-rē-sep-ters): sensory cells specialized for detecting chemicals, such as food.

Chemosynthetic (kēm'-ō-sin-the-tic): capable of oxidizing inorganic molecules to obtain energy.

Chiasma (kī-as'-ma; pl. chiasmata) during prophase I of meiosis, a point at which a chromatid of one chromosome crosses with a chromatid of the homologous chromosome. Exchange of chromosomal material between chromosomes takes place at a chiasma.

Chitin (kī'-tin) a compound found in the cell walls of fungi and the exoskeletons of arthropods, composed of chains of nitrogen-containing, modified glucose molecules.

Chlorinated hydrocarbon an organic compound that includes a chain of carbon atoms some of which have chlorine atoms bonded to them.

Chlorophyll (klor'-ō-fil) a green pigment found in chloroplasts that captures light energy during photosynthesis.

Chloroplast (klor′-ō-plast) the organelle of plants and plantlike protists that is the site of photosynthesis; surrounded by a double membrane and containing an extensive internal membrane system bearing chlorophyll.

Cholecystokinin (kō′-lē-sis-tō-kī′-nin) a digestive hormone produced by the small intestine that stimulates release of pancreatic enzymes.

Chorion (kor′-ē-on) the outermost embryonic membrane in reptiles, birds, and mammals. In birds and reptiles, the chorion functions mostly in gas exchange. In mammals, the chorion forms most of the embryonic part of the placenta.

Chorionic gonadotropin a hormone secreted by the chorion (one of the fetal membranes), which maintains the integrity of the corpus luteum during early pregnancy.

Choroid (kor′-ōyd) a layer of tissue behind the retina that contains blood vessels and pigment that absorbs stray light.

Chromatid (krō′-ma-tid) one of the two identical strands of DNA and protein forming a replicated chromosome. The two sister chromatids are joined at the centromere.

Chromatin (krō′-ma-tin) the complex of DNA and proteins that makes up eukaryotic chromosomes.

Chromosome (krō′-mō-sōme) in eukaryotes, a linear strand composed of DNA and protein, found in the nucleus of a cell, containing the genes; in prokaryotes, a circular strand composed solely of DNA.

Chyme (kīme) an acidic, souplike mixture of partially digested food, water, and digestive secretions that is released from the stomach into the small intestine.

Ciliate (sil′-ē-et): a category of protozoan characterized by cilia and a complex unicellular structure, including harpoon-like organelles called trichocysts. Members of the genus *Paramecium* are well-known ciliates.

Cilium (sil′-ē-um; pl. cilia): A short, hairlike organelle projecting through the cell membrane, containing microtubules in a 9 + 2 arrangement; usually numerous and engaged in coordinated beating which moves a cell through a fluid environment or moves the fluid over the surface of the cell.

Circadian rhythm (sir-kā′-dē-un) an event that recurs with a period of about 24 hours, even in the absence of environmental cues.

Citric acid cycle a cyclic series of reactions in which the pyruvic acid produced by glycolysis is broken down to CO_2, accompanied by the formation of ATP and electron carriers. Occurs in the matrix of mitochondria.

Classical conditioning a training procedure in which an animal learns to make a reflexive response (such as salivation) to a new stimulus which did not elicit that response originally (such as a sound). This is accomplished by pairing a stimulus which elicits the response automatically (in this case, food) with the new stimulus.

Cleavage the early cell divisions of embryos, in which little or no growth occurs between divisions.

Climax community a relatively stable and predictable assemblage of organisms, the endpoint of succession.

Closed circulatory system a type of circulatory system in which the blood is always enclosed in the heart and vessels.

Coccus (ka′-kus; pl. cocci) a spherical bacterium.

Cochlea (kōk′-le-uh) a coiled, bony, fluid-filled tube found in the mammalian inner ear, which contains receptors (hair cells) producing the sense of hearing.

Codominance the relation between two alleles of a gene such that both alleles are phenotypically expressed in heterozygous individuals.

Codon a sequence of three nucleotides of messenger RNA that specifies a particular amino acid to be incorporated into a protein. Certain codons also signal the beginning and end of protein synthesis.

Coelom (sē′-lōm) a space or cavity within the body separating the body wall from the inner organs.

Coenzyme (kō-en′-zīm) an organic molecule that assists enzymes in their actions.

Coevolution the evolution of adaptations in two species due to their extensive interactions with one another, so that each species acts as a major force of natural selection upon the other.

Cohesion the tendency of a substance to hold together.

Coleoptile (kō-lē-op′til) a protective sheath surrounding the shoot in monocot seeds.

Collar cells specialized cells lining the inside channels of sponges. Flagella extend from a sievelike collar, creating a water current that draws microscopic organisms through the collar to be trapped.

Commensalism (kum-en′-sal-ism) a symbiotic relationship between two species in which one benefits while the other is neither harmed nor benefited.

Communication the act of producing a signal that causes another animal, usually of the same species, to modify its behavior in a way beneficial to one or both of the participants.

Community two or more populations of different species living and interacting in the same area.

Companion cell a cell adjacent to a sieve-tube element in phloem, involved in control and nutrition of the sieve-tube element.

Competition interaction that occurs between individuals when both attempt to utilize a resource (e.g., food or space) that is limited relative to the demand for it.

Competitive exclusion principle the concept that no two species can simultaneously and continuously occupy the same ecological niche.

Competitive inhibition in enzyme-catalyzed reactions, a condition in which two molecules (at least one a substrate for the enzyme) compete for entry into the active site of the enzyme, thus slowing down the rate of reaction.

Complementary referring to a nucleotide that can pair with another nucleotide via hydrogen bonding.

Complement a group of blood-borne proteins that participate in the destruction of foreign cells to which antibodies have bound.

Complement reactions interactions among foreign cells, antibodies, and complement proteins, resulting in the destruction of the foreign cells.

Complete flower a flower that has all four floral parts (sepals, petals, stamens, and carpels).

Compound eye a type of eye found in arthropods, composed of numerous independent subunits, called ommatidia. Each ommatidium apparently contributes a single piece of a mosaic-like image perceived by the animal. *See also* Ommatidium.

Compound a substance composed of two or more elements that can be broken into its constituent elements by chemical means.

Concentration gradient a difference in concentration of a substance between two parts of a system or across a barrier such as a cell membrane.

Condensation a chemical reaction in which two molecules are joined by a covalent bond, with the simultaneous removal of a hydrogen from one molecule and a hydroxyl group from the other, forming water.

Cone a cone-shaped photoreceptor cell in the vertebrate retina, not as sensitive to light as the rods. The three types of cones are most sensitive to different colors of light, and provide color vision. *See also* Rod.

Conifer (kon′-eh-fer) a class of tracheophyte that reproduces using cones and retains its leaves throughout the year.

Contest competition a mechanism for resolving intraspecific competition using social or chemical interactions.

Contractile vacuole a membrane-bound organelle found in certain protists that takes up water from the cell, contracts, and expels the water outside the cell via a pore in the cell membrane.

Convergence a condition in which a large number of nerve cells provide input to a smaller number of cells.

Convergent evolution the independent evolution of similar structures among unrelated organisms, due to similar selective pressures.

Cork cambium a lateral meristem in woody roots and stems that gives rise to cork cells.

Cork cell a protective cell of the bark of woody stems and roots; at maturity, cork cells are dead, with thick, waterproofed cell walls.

Cornea (kōr′-nē-uh) the clear outer covering of the eye in front of the pupil and iris.

Corona radiata (ka-rō′-na rā-dē-a′-ta) the layer of cells surrounding an egg after ovulation.

Corpus callosum (kōr′pus kal-ō′-sum) the tract of axons that connect the two cerebral hemispheres of vertebrates.

Corpus luteum (kor′-pus lū′-tē-um) in the mammalian ovary, a structure derived from the follicle after ovulation, which secretes the hormones estrogen and progesterone.

Cortex the part of a primary root or stem located between the epidermis and the vascular cylinder.

Cotyledon (kot-ul-ē′don) also called a seed leaf; a leaflike structure within a seed that absorbs food from the endosperm and transfers it to the growing embryo.

Coupled reactions a pair of reactions, one exergonic and one endergonic, that are linked together so that the energy produced by the exergonic reaction provides the energy needed to drive the endergonic reaction.

Covalent bond (kō-vā′-lent) a chemical bond between atoms in which electrons are shared.

Creationism the hypothesis that all species of organisms on Earth were created in essentially their present form by a supernatural Being, and that significant modification of those species, specifically their transformation into new species, cannot occur through natural processes.

Crista (kris′-ta; pl. cristae) a fold in the inner membrane of a mitochondrion.

Crop an organ found in both earthworms and birds in which ingested food is stored temporarily before passing to the gizzard, where it is pulverized.

Crossing over the exchange of corresponding segments of the chromatids of two homologous chromosomes during meiosis.

Cross-bridge in muscles, an extension of myosin that binds to and pulls on actin to produce contraction of the muscles.

Cross-fertilization union of sperm and egg from two different individuals of the same species.

Cultural evolution changes in the behavior of a population of animals, especially humans, by learning behaviors acquired by members of previous generations.

Cuticle (kū′-ti-kul) a waxy or fatty coating on the exposed epidermal cells of many land plants, which aids in the retention of water.

Cyanobacteria photosynthetic prokaryotic cells, utilizing chlorophyll and releasing oxygen as a photosynthetic by-product, sometimes called "blue-green algae."

Cyclic AMP a cyclic nucleotide formed within many target cells as a result of the reception of modified amino acid or protein hormones, and which causes metabolic changes in the cell; often called a second messenger.

Cyclic nucleotide (sik′-lik nū′-klē-ō-tīd) a nucleotide in which the phosphate group is bonded to the sugar at two points, forming a ring. Cyclic nucleotides serve as intracellular messengers.

Cyst (sist) an encapsulated resting stage in the life cycle of certain invertebrates, such as parasitic flatworms and roundworms.

Cytokinesis (sī-tō-ki-nē′-sis) division of the cytoplasm and organelles into two daughter cells during cell division. Usually cytokinesis occurs during telophase of mitosis or meiosis.

Cytokinin (sī-tō-kī′-nin) a plant hormone that promotes cell division, fruit growth, and sprouting of lateral buds, and prevents leaf aging and leaf drop.

Cytoplasm (sī′-tō-plazm) the material contained within the cell membrane but outside the nucleus.

Cytoskeleton a network of protein fibers in the cytoplasm that gives shape to a cell, holds and moves organelles, and is often involved in cell movement.

Cytosol (sī′-tō-sol) the fluid part of the cytoplasm.

Day-neutral plant a plant in which flowering occurs under a wide range of daylengths.

Decomposers a group of decay organisms, mainly fungi and bacteria. These digest organic material by secreting digestive enzymes into the environment. In the process they liberate nutrients into the environment.

Degeneracy the property of the genetic code whereby several codons may specify the same amino acid.

Deletion a mutation in which one or more nucleotides are removed from a gene.

Dendrite (den′-drīt) the site of signal input to a nerve cell, usually takes the form of branched fibers located close to the cell body.

Denitrifying bacteria (dē-nī′-treh-fī-ing) bacteria that break down nitrates, releasing nitrogen gas to the atmosphere.

Density dependent description of any factor that limits population size more effectively as the population density increases.

Density independent description of any factor such as freezing weather that limits a sensitive population without regard to its size.

Deoxyribonucleic acid (dē-ox-ē-rī-bō-nū-klā′-ik; DNA) a molecule composed of deoxyribose nucleotides; the genetic information of all living cells.

Desert a biome in which potential evaporation greatly exceeds rainfall. Perennial vegetation is widely spaced over the ground and has drought-resistant features such as waxy or spiny leaves.

Desmosome (dez′-mō-sōm) a strong cell-to-cell junction that functions in attaching cells to one another.

Detritus feeders (de-trī′-tus) a diverse assemblage of organisms ranging from worms to vultures which live off the wastes and dead remains of other organisms.

Diaphragm (dī′uh-fram) a dome-shaped muscle forming the floor of the chest cavity. Contraction of this muscle pulls it downward, enlarging the cavity and causing air to be drawn into the lungs.

Diatom (dī′-e-tom) A category of protist that includes photosynthetic forms with two-part glassy outer coverings which separate when the cell divides. Diatoms are important primary producers in fresh and salt water.

Dicotyledon (dī′-kot-ul-ēd′-un) a class of angiosperm whose embryo has two cotyledons, or seed leaves.

Differential permeability the property of a membrane by which some substances can permeate more readily than other substances.

Differentiated cell a mature cell specialized for a specific function; in plants, differentiated cells usually do not divide.

Differentiation the process whereby relatively unspecialized cells, especially of embryos, become specialized into particular tissue types.

Diffusion the net movement of particles from a region of high concentration to a region of low concentration, driven by the concentration gradient.

Digestion the process by which food is physically and chemically broken down into molecules that can be absorbed by cells.

Dihybrid cross a breeding experiment involving parents that differ in two distinct, genetically determined traits.

Dinoflagellate (dī-nō-fla′-gel-et) A category of protist that includes photosynthetic forms in which two flagella project through armorlike plates. Abundant in oceans, these sometimes reproduce rapidly, causing "red tides."

Dioecious (dī-ē′-shus) pertaining to organisms in which male and female gametes are produced by separate individuals.

Diploid (dip′-loyd) referring to a cell with pairs of homologous chromosomes.

Direct development a developmental pathway in which the offspring is born as a miniature version of the adult and does not radically change its body form as it grows and matures.

Directional selection a type of natural selection in which one extreme phenotype is favored over all others.

Disaccharide (dī-sak′-a-rīd) a carbohydrate formed by the covalent bonding of two monosaccharides.

Disruptive selection a type of natural selection in which both extreme phenotypes are favored over the average phenotype.

Divergence a condition in which a small number of nerve cells provide input to a larger number of cells.

Division: a taxonomic category in botany; the equivalent of an animal phylum.

DNA polymerase an enzyme that covalently bonds DNA nucleotides together into a continuous strand, using a preexisting DNA strand as a template. DNA polymerase catalyzes the duplication of the DNA of chromosomes during interphase prior to mitosis and meiosis.

Dominance hierarchy a social arrangement in which animals, usually through aggressive interactions, establish a rank for some or all of the members of the social unit. High-ranking individuals obtain first access to limited resources.

Dominant an allele that can determine the phenotype of heterozygotes completely, so that they are indistinguishable from individuals homozygous for the allele. In the heterozygotes, the expression of the other (recessive) allele is completely masked.

Dormancy a state in which an organism does not grow or develop; usually marked by lowered metabolic activity and resistance to adverse environmental conditions.

Dorsal (dōr′-sul) the top, back, or uppermost surface of an animal oriented with its head forward.

Dorsal root ganglion a ganglion located on the dorsal (sensory) branch of each spinal nerve, containing the cell bodies of sensory neurons.

Double bond a covalent bond that occurs when two atoms share two pairs of electrons.

Double fertilization in flowering plants, a phenomenon in which two sperm nuclei fuse with the nuclei of two cells of the female gametophyte. One sperm fuses with the egg to form the zygote, while the second sperm nucleus fuses with the two haploid nuclei of the primary endosperm cell to form a triploid endosperm cell.

Down syndrome a genetic disorder caused by the presence of three copies of chromosome 21. Common characteristics include mental retardation, abnormally shaped eyelids, a small mouth with protruding tongue, short fingers, heart defects, and unusual susceptibility to infectious diseases.

Ecological isolation lack of mating between organisms belonging to different populations that occupy distinct habitats within the same general area.

Ecological pyramid a graphical representation of the energy contained in succeeding trophic levels, with maximum energy at the base (primary producers) and steadily diminishing amounts at higher levels.

Ecology (ē-kol′-uh-gē) the study of the interrelationships of organisms with each other and with their nonliving environment.

Ecosystem (ē′kō-sis-tem) (1) all the organisms and their nonliving environment within a defined area. (2) one or more communities together with their nonliving surroundings.

Ectoderm (ek′-tō-derm) the outermost embryonic tissue layer, which gives rise to structures such as hair, the epidermis of the skin, and the nervous system.

Effector (ē-fek′-tōr) a part of the body (usually a muscle or gland) that carries out responses as directed by the nervous system.

Egg the haploid female gamete, usually large and nonmotile, containing food reserves for the developing embryo and regionally localized gene-regulating substances that direct early development.

Electron a subatomic particle, found in the orbitals outside the nucleus of an atom, bearing a unit of negative charge and very little mass.

Electron carrier a molecule that can reversibly gain and lose electrons. Electron carriers generally accept high-energy electrons produced during an exergonic reaction and donate the electrons to acceptor molecules that use the energy to drive endergonic reactions.

Electron shell all the electron orbitals at a given distance from the nucleus of an atom.

Electron transport system a series of molecules found in the inner membrane of mitochondria and the thylakoid membranes of chloroplasts, which extract energy from electrons, which is used to generate ATP or other energetic molecules.

Element a substance that cannot be broken down to a simpler substance by ordinary chemical means.

Embryo an early stage of development; in mammals, this term usually refers to the early stages in which the developing animal does not yet resemble adults of the species.

Embryo sac the haploid female gametophyte of flowering plants.

Embryonic disk in human embryonic development, the flat, two-layered group of cells derived from the inner cell mass of the blastocyst, which will develop into the embryo proper.

Emigration (em-uh-grā'shun) movement of individuals out of an area.

Emphysema (em-fuh-sē'-muh) a condition in which the alveoli become brittle and rupture, causing decreased area for gas exchange.

Endergonic (en-der-gon'-ik) pertaining to a chemical reaction that requires an input of energy to proceed; an "uphill" reaction.

Endocrine gland a ductless, hormone-producing gland that releases its secretions into the extracellular fluid within the body, from which the secretions diffuse into nearby capillaries.

Endocytosis (en-dō-sī-tō'-sis) the movement of material into a cell by a process in which the cell membrane engulfs extracellular material, forming membrane-bound sacs that enter the cell interior.

Endoderm (en'-dō-derm) the innermost embryonic tissue layer, which gives rise to structures such as the lining of the digestive and respiratory tracts.

Endodermis (en-dō-der'-mis) the innermost layer of cells of the cortex of a root.

Endometrium (en-dō-mē'-trē-um) the nutritive inner lining of the uterus.

Endoplasmic reticulum (en-dō-plaz'-mik re-tik'-ū-lum) a system of membranous channels within eukaryotic cells; the site of most protein and lipid biosynthesis.

Endorphin (en-dōr'-fin) one of a group of peptides in the vertebrate brain that mimics some of the actions of opiates. Endorphins reduce the sensation of pain.

Endoskeleton a supportive structure within the body; an internal skeleton. It may be nonliving, as in echinoderms and sponges, or living, as in vertebrates.

Endosperm (en'-dō-sperm) the stored food within a seed used to nourish the developing plant embryo.

Endosymbiosis the hypothesis that certain organelles, especially chloroplasts and mitochondria, evolved from bacteria captured by ancient predatory prokaryotic cells.

End product inhibition in enzyme-mediated chemical reactions, the condition in which the product of a reaction inhibits one or more of the enzymes involved in synthesizing the product.

Energy the ability to do work.

Energy carrier a molecule that stores energy in "high-energy" chemical bonds and releases the energy again to drive coupled endergonic reactions. ATP is the most common energy carrier in cells.

Energy level a particular amount of energy characteristic of a given electron shell in an atom.

Enhancer in eukaryotes, a stretch of DNA that influences the access of RNA polymerase to the promoter region of a structural gene.

Entropy (en'-trō-pē) a measure of the amount of randomness and disorder in a system.

Environmental resistance any factor that tends to counteract biotic potential, limiting population size.

Enzyme a protein molecule that speeds up specific chemical reactions but which is not itself used up or permanently altered; a protein catalyst.

Epicotyl (ep'-ē-kot-ul) the part of the embryonic shoot located between the tip of the shoot and the attachment point of the cotyledons.

Epidermis (ep-i-der'-mis) the outermost layer of cells of a leaf, young root, or young stem; in mammals, the outermost layers of the skin.

Epididymis (e-pi-di'-dē-mus) tubes that connect with and receive sperm from the seminiferous tubules of the testis.

Epiglottis (ep-eh-gla'-tis) a flap of cartilage in the lower pharynx that covers the opening to the larynx during swallowing. This directs the food down the esophagus.

Equilibrium population a population in which allele frequencies do not change from generation to generation.

Erythrocytes (ē-rith'-rō-sītes) red blood cells active in oxygen transport, which contain the red pigment hemoglobin.

Esophagus (eh-sof'-eh-gus) a muscular passageway connecting the pharynx to the next chamber of the digestive tract, the stomach in humans and other mammals.

Estrogen in vertebrates, a female sex hormone produced by follicle cells of the ovary, which stimulates follicle development, oogenesis, development of secondary sex characteristics, and growth of the uterine lining.

Ethology (ē-thol'-ō-gē) the study of animal behavior under natural or near-natural conditions.

Ethylene a plant hormone that promotes ripening of fruits, and leaf and fruit drop.

Euglenoid (ū'-gle-noid) a category of protist characterized by one or more whiplike flagella used for locomotion and a photoreceptor for detecting light. Euglenoids are photosynthetic, but some are capable of heterotrophic nutrition if deprived of chlorophyll.

Eukaryotic (ū-kar-ē-ot'-ik) referring to cells characteristic of organisms of the kingdoms Protista, Fungi, Plantae, and Animalia. Eukaryotic cells have their genetic material enclosed within a membrane-bound nucleus and contain other membrane-bound organelles.

Eutrophication (ū'-trif-i-kā'-shun) the process by which a body of water becomes enriched with nutrients and supports increasingly dense growths of photosynthetic and decomposer organisms. There is frequently oxygen depletion due to decomposition of plant material, and a progressive accumulation of organic material.

Evolution the descent of modern organisms from preexisting life forms; strictly speaking, any change in the overall genetic composition of a population of organisms from one generation to the next.

Excitatory synapse a synapse between two nerve cells in which the resting potential of the postsynaptic cell becomes less negative due to the activity of the presynaptic cell.

Exergonic (ex-er-gon′-ik) pertaining to a chemical reaction that releases energy; a "downhill" reaction.
Exocrine gland a gland that releases its secretions into ducts that lead to the outside of the body or into the digestive tract.
Exocytosis (ex-ō-sī-tō′-sis) the movement of material out of a cell by a process whereby intracellular material is enclosed within a membrane-bound sac that moves to the cell membrane and fuses with it, releasing the material outside the cell.
Exon a segment of DNA in a eukaryotic gene that codes for amino acids in a protein.
Exoskeleton (ex′ō-skel′-uh-tun) a rigid external skeleton with flexible joints to allow for movement.
Exotic a species introduced into an ecosystem where it did not evolve.
Expiration (ex-per-ā′-shun) the act of exhaling, which results from relaxation of the respiratory muscles.
Exponential growth a continuously accelerating increase in population size.
External fertilization union of sperm and egg outside the body of either parental organism.
Extinction the death of all members of a species.
Extracellular (ex-tra-sel′-ya-ler) occurring outside cells.
Extracellular digestion the physical and chemical breakdown of food that occurs in a digestive cavity.
Eyespot a simple, lensless eye found in various invertebrates, including flatworms and jellyfish. Eyespots provide information about light versus dark, and sometimes the direction of light, but cannot form an image.
Facilitated diffusion diffusion of molecules across a cell membrane, assisted by protein pores or carriers embedded in the cell membrane.
Fat a lipid composed of three saturated fatty acids covalently bonded to glycerol; fats are solid at room temperature.
Fatty acid an organic molecule composed of a long chain of carbon atoms, with a carboxylic acid (—COOH) group at one end. Fatty acids may be saturated (all single bonds between the carbon atoms) or unsaturated (one or more double bonds between carbon atoms).
Fermentation anaerobic reactions that convert the pyruvic acid produced by glycolysis into lactic acid or alcohol and CO_2.
Fertilization the fusion of male and female haploid gametes to form a diploid zygote.
Fetus in animals, the later stages of embryonic development, when the developing animal has come to resemble the adult of the species.
Fibrin (fī′-brin) a clotting protein formed in the blood in response to a wound. Fibrin binds with other fibrin molecules and provides a matrix around which a blood clot forms.
Fibrous root system a root system characterized by many roots of approximately the same diameter arising from the base of the stem.
Fimbria (fim′-brē-a; pl. fimbriae) in female mammals, the ciliated, fingerlike projections of the oviduct that sweep the ovulated egg from the ovary into the oviduct.
Fission asexual reproduction by dividing the body into two smaller, complete organisms.
Fixed action pattern stereotyped, rather complex behavior which is genetically programmed (innate); often triggered by a stimulus called a releaser.
Flagellum (fla-jel′-um; pl. flagella) a long, hairlike extension of the cell membrane. In eukaryotic cells, contains microtubules arranged in a 9 + 2 pattern. Movement of flagella propels some cells through fluid media.
Flame cells cells in flatworms specialized for excretion and fluid regulation. They enclose a small chamber full of beating cilia, whose flickering appearance gives them their name.
Flower the reproductive structure of an angiosperm.
Fluid mosaic a model of cell membranes. According to this model, the membrane is composed of a double layer of phospholipids in which a variety of proteins are embedded. The lipid bilayer is a somewhat fluid matrix that allows movement of proteins within it.
Follicle in the ovary of female mammals, the oocyte and its surrounding accessory cells.
Follicle-stimulating hormone a hormone produced by the anterior pituitary gland that stimulates spermatogenesis in males and development of the follicle in females.
Food chain an illustration of feeding relationships in an ecosystem using a single representative from each of the trophic levels.
Food vacuole a membrane-bound sac within a cell in which food is enclosed. Digestive enzymes are released into the vacuole and intracellular digestion occurs here.
Food web a relatively accurate representation of the complex feeding relationships within an ecosystem, including many organisms at various trophic levels, with many of the consumers occupying more than one level simultaneously.
Fossil the remains of an organism, usually preserved in rock. Fossils include petrified bones or wood; shells; impressions of body forms such as feathers, skin, or leaves; and markings made by organisms such as footprints.
Founder effect a type of genetic drift in which an isolated population founded by a small number of individuals may develop allele frequencies that are very different from those of the parent population, because of chance inclusion of disproportionate numbers of certain alleles in the founders.
Fovea (fō′-vē-uh) the central region of the vertebrate retina, upon which images are focused. The fovea contains closely packed cones (about 150,000 per square millimeter).
Free-living: not parasitic.
Fruit in angiosperms, the ripened ovary (plus, in some cases, other parts of the flower).
Fruiting body: a reproductive structure of fungi in which spores are formed.
Functional group one of several groups of atoms commonly found in organic molecules, including hydrogen, hydroxyl, amino, carboxyl, and phosphate groups.
Gallbladder a small sac adjacent to the liver in which the bile secreted by the liver is stored. Bile is released from the gallbladder via the bile duct to the small intestine.
Gamete (gam′-ēt) a haploid sex cell formed in sexually reproducing organisms.
Gametic incompatibility the inability of sperm from one species to fertilize eggs of another species.
Gametophyte (ga-mēt′-ō-fīt) the multicellular haploid stage in the life cycle of plants; produces haploid sex cells by mitosis.
Ganglion (gang′-lē-un) a collection of nerve cells.
Ganglion cell (gang′-lē-un) a cell type comprising the outer layer of the vertebrate retina. Its axons comprise the optic nerve.
Gap junction a type of cell-to-cell junction in animals, in which channels connect the interiors of adjacent cells.

Gastrovascular cavity a chamber that has both digestive and circulatory functions, found in simple invertebrates. A single opening serves as both mouth and anus, while the chamber provides direct access of nutrients to the surrounding cells.

Gastrula (gas′-trū-la) in animal development, a three-layered embryo with ectoderm, mesoderm, and endoderm cell layers. The endoderm layer usually encloses the primitive gut.

Gastrulation (gas-trū-lā′-shun) the process whereby a blastula develops into a gastrula.

Gene a unit of heredity containing the information for a particular characteristic. A gene is a segment of DNA located at a particular place on a chromosome.

Gene flow the movement of alleles from one population to another due to migration of individual organisms.

Gene pool for a single gene, the total of all the alleles of that gene that occur in a population; the total gene pool is the total of all alleles of all genes in the population.

Generative cell in flowering plants, one of the haploid cells of a pollen grain. The generative cell undergoes mitosis to form two sperm cells.

Genetic code the collection of codons of mRNA, each of which directs the incorporation of a particular amino acid into a protein during protein synthesis.

Genetic drift a change in the allele frequencies of a small population purely by chance.

Genetic recombination assembling a new combination of preexisting genes through crossing over.

Genotype (jēn′-ō-tīp) the genetic composition of an organism; the actual alleles of each gene carried by the organism.

Genus (jē′-nis) a taxonomic category consisting of very closely related species.

Geographical isolation the separation of two populations by a physical barrier.

Germination the growth and development of a seed, spore, or pollen grain.

Gibberellin (jib-er-el′-in) a plant hormone that stimulates seed germination, fruit development, and cell division and elongation.

Gills in aquatic animals, a branched tissue richly supplied with capillaries around which water is circulated for gas exchange.

Gizzard a muscular organ found in earthworms and birds in which food is mechanically broken down prior to chemical digestion.

Glomerulus (glō-mer′-ū-lus) a dense network of thin-walled capillaries located within the Bowman's capsule of each nephron. Here blood pressure forces water and dissolved nutrients through capillary walls into the nephron.

Glucose the most common monosaccharide, with the molecular formula $C_6H_{12}O_6$. Most polysaccharides, including cellulose, starch, and glycogen, are made of glucose subunits covalently bonded together.

Glycerol (glis′-er-ol) a three-carbon alcohol to which fatty acids are covalently bonded to make fats and oils.

Glycogen (glī′-kō-gen) a polysaccharide composed of branched chains of glucose subunits, used as a carbohydrate storage molecule in animals.

Glycolysis (glī-kol′-i-sis) anaerobic reactions carried out in the cytosol that break glucose down into two molecules of pyruvic acid, producing two ATP molecules. Glycolysis does not require oxygen, but can proceed when oxygen is present.

Glycoprotein a protein to which a carbohydrate is attached.

Golgi complex (gōl′-jē) a stack of membranous sacs found in most eukaryotic cells, which is the site of processing and separation of membrane components and secretory materials.

Gonorrhea (gon-a-rē′-uh) a sexually transmitted bacterial infection of the reproductive organs. Untreated gonorrhea may result in sterility.

Graded potential in a nerve cell, an electrical response due to sensory input or synaptic input from another nerve cell. Graded potentials may be positive or negative and vary in amplitude with the strength of stimulation.

Gradualism: the hypothesis that species arise through slow, gradual changes over hundreds of thousands, even millions, of years.

Granum (gra′-num; pl. grana) in chloroplasts, a stack of thylakoids.

Grassland a biome characterized by a relatively continuous ground cover of grasses, with few or no trees except along watercourses. Typically, grassland biomes have a prolonged dry season and frequent fires, both of which favor grasses over trees.

Gravitropism growth with respect to the direction of gravity.

Guard cell one of a pair of specialized epidermal cells surrounding the central opening of a stoma of a leaf, which regulates the size of the opening.

Gymnosperm (jim′-nō-sperm) a vascular plant that produces seeds not enclosed in an ovary.

Habituation (heh-bich-ū-ā′-shun) simple learning characterized by a decline in response to a harmless, repeated stimulus.

Hair cell the receptor cell type found in the inner ear (cochlea and semicircular canals). Hair cells bear hairlike projections. Bending of the hairs between two membranes causes the receptor potential.

Halophile (hā′-lō-fīl) a salt-loving organism.

Haploid (hap′-loyd) referring to a cell that has only one member of each pair of homologous chromosomes.

Heat of fusion the energy that must be removed from a substance to transform it from a liquid into a solid at a constant temperature.

Heat of vaporization the energy that must be supplied to a substance to transform it from a liquid into a gas at a constant temperature.

Helix (hē′-licks) a corkscrew-shaped object, as if a wire were wrapped around a cylinder.

Helper T cell a type of T cell that aids other immune cells to recognize and act against antigens.

Hemocoel (hē′-mō-sēl): a blood cavity within the bodies of certain invertebrates in which blood bathes tissues directly. A hemocoel is part of an open circulatory system.

Hemoglobin (hē′mō-glō-bin) an iron-containing protein that gives red blood cells their color. Hemoglobin binds to oxygen in the lungs and releases it to the tissues.

Hemophilia a recessive, sex-linked disease in which the blood fails to clot normally.

Hermaphrodite (her-ma′-frō-dīt) an organism that produces both male and female gametes.

Heterotroph (het′-er-ō-trōf) an organism that cannot use inanimate energy sources (such as sunlight) to synthesize all its energy-rich organic molecules; hence heterotrophs must acquire energy-rich organic molecules manufactured by other living organisms as food.

Heterozygote (het-er-ō-zī′-gōt) an organism carrying two different alleles of the gene in question; sometimes called a hybrid.

Hippocampus (hip-ō-cam′-pus) part of the forebrain of vertebrates, important in motivation, emotion, and especially learning.

Histamine a substance released by certain cells in response to tissue damage and invasion of the body by foreign substances. Histamine promotes dilation of arterioles and leakiness of capillaries, and triggers some of the events of the inflammatory response.

Homeostasis (hō-mē-ō-stā′-sis) the process of maintaining a relatively constant internal environment in the face of variations in the external environment.

Homologous structures structures that may differ in function but that have similar anatomy, presumably because of descent from common ancestors.

Homologue (hō′-mō-log) a chromosome that is similar in appearance and genetic information to another chromosome with which it pairs during meiosis. Also called homologous chromosome.

Homozygote (hō-mō-zī′-gōt) an organism carrying two copies of the same allele of the gene in question; also called a true-breeding organism.

Hormone a chemical synthesized by one group of cells and carried in the bloodstream to other cells, whose activity is influenced by reception of the hormone.

Humoral immunity an immune response in which foreign substances are inactivated or destroyed by antibodies circulating in the blood.

Hybrid an organism that is the offspring of parents differing in at least one genetically determined characteristic; also used to refer to the offspring of parents of different species.

Hybrid infertility reduced fertility (often complete sterility) in hybrid offspring of two different species.

Hybrid inviability the failure of a hybrid offspring of two different species to survive to maturity.

Hydrogen bond the weak attraction between a hydrogen atom bearing a partial positive charge (due to polar covalent bonding with another atom) and another atom, usually oxygen or nitrogen, bearing a partial negative charge. Hydrogen bonds may form between atoms of a single molecule or of different molecules.

Hydrolysis (hī-drol′-i-sis) the chemical reaction that breaks a covalent bond through the addition of hydrogen to the atom forming one side of the original bond, and a hydroxyl group to the atom on the other side.

Hydrophilic (hī-drō-fil′-ik) pertaining to a substance that is attracted to and usually dissolves in water. Such molecules either are ions or have polar parts.

Hydrophobic (hī-drō-fō′-bik) pertaining to a substance that is insoluble in water; usually uncharged and lacking any polar parts.

Hydrostatic skeleton (hī-drō-stat′-ik) the use of fluid contained in body compartments to provide support for the body and mass against which muscles can contract.

Hypertension arterial blood pressure that is chronically elevated above the normal level.

Hypertonic (hī-per-ton′-ik) referring to a solution that has a higher concentration of dissolved particles (and therefore a lower water concentration) than a cell.

Hypha (hī′-pha; pl. hyphae) threadlike structure consisting of elongated cells, often with many nucleii. The body of a fungus is composed of numerous hyphae.

Hypocotyl (hī′-pō-kot-ul) the part of the embryonic shoot located between the attachment point of the cotyledons and the root.

Hypothalamus (hī-pō-thal′-uh-mus) part of the forebrain of vertebrates, located just below the thalamus, involved in regulation of hormonal activities (especially of the pituitary gland), and many behaviors such as feeding, drinking, sex, aggression, and fear responses, largely through activation of the autonomic nervous system.

Hypothesis (hī-poth′-e-sis) in science, a supposition based on previous observations, which is offered as an explanation for an event, and used as the basis for further observations or experiments.

Hypotonic (hī-pō-ton′-ik) referring to a solution that has a lower concentration of dissolved particles (and therefore a higher water concentration) than a cell.

Immigration (im-uh-grā′-shun) movement of individuals into an area.

Immune deficiency disease a disorder in which the immune system is incapable of responding properly to invading disease organisms.

Immune response a specific response by the immune system to invasion of the body by a particular foreign substance or microorganism, characterized by recognition of the foreign material by immune cells and its subsequent destruction by antibodies or cellular attack.

Implantation the process whereby the early embryo embeds itself within the lining of the uterus.

Imprinting (im′-prin-ting) the process by which an animal forms an association with another animal or object in the environment during a sensitive period.

Inactivation a process whereby certain chromosomes or parts of chromosomes are converted into a dense mass, preventing transcription.

Incomplete dominance a pattern of inheritance in which heterozygotes have a phenotype intermediate between those of the two homozygotes.

Incomplete flower a flower that is missing one of the four floral parts (sepals, petals, stamens, or carpels).

Indirect development a developmental pathway in which a free-living offspring goes through radical changes in body form as it matures.

Induced fit a theory of enzyme activity proposing that binding of substrates to an enzyme active site changes the shape or charge of both the substrates and the active site.

Induction the process by which a group of cells causes other cells to differentiate into a specific tissue type.

Inflammatory response a nonspecific, local response to injury to the body, characterized by phagocytosis of foreign substances and tissue debris by white blood cells, and "walling off" of the injury site by clotting of fluids escaping from nearby blood vessels.

Inheritance the transmission of inborn characteristics from parent to offspring.

Inheritance of acquired characteristics the hypothesis that organisms' bodies change during their lifetimes by use and disuse, and that these changes are inherited by their offspring.

Inhibiting hormone a hormone secreted by the hypothalamus that inhibits the release of specific hormones from the anterior pituitary gland.

Inhibitory synapse a synapse between two nerve cells in which the resting potential of the postsynaptic cell becomes more negative as a result of the activity of the presynaptic cell.

Innate (in-nāt′) inborn; instinctive; determined by the genetic makeup of the individual.

Inner cell mass in human embryonic development, the cluster of cells on one side of the blastocyst, which will develop into the embryo.

Inner ear the innermost part of the mammalian ear, composed of the bony, fluid-filled tubes of the cochlea and semicircular canals.

Inorganic referring to any molecule that does not contain both carbon and hydrogen.

Insertion (1) the site of attachment of a muscle to the relatively movable bone on one side of a joint; (2) a mutation in which one or more nucleotides are inserted within a gene.

Insight a complex form of learning in which the solution to a problem is reached through reasoning.

Inspiration the act of inhaling air into the lungs by enlarging the chest cavity.

Instinctive innate; inborn; determined by the genetic makeup of the individual.

Integration in nerve cells, the process of adding up electrical signals from sensory inputs or other nerve cells, to determine the overall electrical activity of the nerve cell.

Integument (in-teg′-ū-ment) the layers of the ovule immediately surrounding the embryo sac; develops into the seed coat.

Intensity the strength of stimulation or response.

Intermediate filament part of the cytoskeleton of eukaryotic cells, probably functioning mainly for support.

Intermembrane compartment the fluid-filled space between the inner and outer membranes of a mitochondrion.

Internal fertilization union of sperm and egg inside the body of the female.

Internode the part of a stem between two nodes.

Interphase the stage of the cell cycle between cell divisions. During interphase, chromosomes are replicated, and other cell functions occur, such as growth, movement, and acquisition of nutrients.

Interspecific competition competition between individuals of different species.

Interstitial cells (in-ter-sti′-shul) in the vertebrate testis, testosterone-producing cells located between the seminiferous tubules.

Interstitial fluid (in-ter-sti′-shul) fluid similar in composition to plasma (except lacking large proteins) which surrounds the body cells. This fluid leaks from the capillaries and acts as a medium of exchange between the body cells and the capillaries.

Intertidal zone the area along the edges of land masses which is alternately flooded and left bare by the rising and falling of the tides.

Intracellular (in-tra-sel′-ya-ler) occurring within individual cells.

Intracellular digestion the chemical breakdown of food, which occurs within single cells.

Intraspecific competition competition between individuals of the same species.

Intron a segment of DNA in a eukaryotic gene that does not code for amino acids in a protein.

Invertebrate (in-vert′-uh-bret) a category of animals that never possess a vertebral column.

Ion (ī′-on) an atom or molecule that has either an excess of electrons (and hence is negatively charged) or has lost electrons (and is positively charged).

Ionic bond a chemical bond formed by the electrical attraction between positively and negatively charged ions.

Iris the pigmented part of the vertebrate eye, surrounding the central hole of the pupil.

Isotonic (ī-sō-ton′-ik) referring to a solution that has the same concentration of dissolved particles (and therefore the same water concentration) as a cell.

Joint a flexible region between two rigid units of an exoskeleton or endoskeleton, to allow for movement between the units.

Killer T cell a type of T cell that directly destroys foreign cells upon contacting them.

Kinesis (kin-ne-sis) an innate process by which an organism achieves an orientation to a stimulus by altering its speed of movement in response to the stimulus.

Kinetic energy (kin-et′-ik) the energy of movement.

Kingdom the most inclusive category in the classification of organisms. The five kingdoms are Monera, Protista, Fungi, Plantae, and Animalia.

Kin selection selection favoring a certain allele because of benefits accruing to relatives bearing the same allele.

Klinefelter's syndrome a set of characteristics typically found in individuals who have two X chromosomes and one Y chromosome. These individuals are phenotypically males, but sterile, and have several female-like traits, including narrow shoulders, broad hips, and partial breast development.

Krebs cycle the citric acid cycle (in honor of Hans Krebs, who discovered many of its biochemical details).

Lactose (lak′-tōs) a disaccharide composed of glucose and galactose; found in mammalian milk.

Larva an immature form of an animal with indirect development, often much different in body form from the adult.

Larynx (lār′-inx) that portion of the air passage between the pharynx and the trachea. The larynx contains the vocal cords.

Lateral bud a bud located at a node of a stem, usually in the crotch between the stem and the petiole of the leaf found at the same node.

Lateral meristem also called cambium; a meristematic tissue in dicot stems and roots, usually found between the xylem and phloem (vascular cambium) and just outside the phloem (cork cambium).

Leaf an outgrowth of a stem, usually flattened and photosynthetic.

Leaf primordium (prī-mor′-dē-um) the outgrowth of a shoot that develops into a leaf.

Learning an adaptive change in behavior as a result of experience.

Legumes (leg′-yūm) a group of plants (including alfalfa, clover, and soybeans) which harbor colonies of nitrogen-fixing bacteria in swellings or nodules on their roots.

Lens a clear object that bends light rays; in eyes, a flexible or movable structure used to focus light on a layer of photoreceptor cells.

Leukocyte (loo′-kō-sīt) any of the white blood cells circulating in the blood.

Lichen (lī′-ken): a symbiotic association between an alga or cyanobacterium and a fungus, resulting in a composite organism.

Light-dependent reactions the first step of photosynthesis, in which the energy of light is captured as ATP and NADPH; occurs in thylakoids of chloroplasts.

Light-harvesting complex in photosystems, the assembly of pigment molecules (chlorophyll and often carotenoids or phycocyanins) that absorb light energy and transfer the energy to electrons.

Light-independent reactions the second stage of photosynthesis, in which the energy obtained by the light-dependent reactions is used to fix carbon dioxide into carbohydrates; occurs in the stroma of chloroplasts.

Limbic system a diverse group of brain structures, mostly in the lower forebrain, including the thalamus, hypothalamus, amygdala, hippocampus, and parts of the cerebrum, involved in emotion, motivation, and learning.

Linkage the inheritance of certain genes as a group because they are parts of the same chromosome. Linked genes do not show independent assortment.

Lipase (lī′-pāse) an enzyme that catalyzes the breakdown of lipids, such as fats.

Lipid (li′pid) one of a number of water-insoluble organic molecules, containing large regions composed solely of carbon and hydrogen. Lipids include oils, fats, waxes, phospholipids, and steroids.

Long-day plant a plant that will flower only if the length of daylight is greater than some species-specific duration.

Loop of Henle (hen′-lē) a specialized portion of the tubule of the nephron in birds and mammals which creates an osmotic concentration gradient in the fluid immediately surrounding it. This in turn allows the production of urine more osmotically concentrated than blood plasma.

Luteinizing hormone a hormone produced by the anterior pituitary gland that stimulates testosterone production in males and development of the follicle, ovulation, and production of the corpus luteum in females.

Lymph (limpf) a fluid resembling blood plasma that collects in special lymph vessels and eventually returns to the bloodstream.

Lymph nodes small structures that act as filters for lymph. These contain both lymphocytes and macrophages, which inactivate foreign bodies such as bacteria.

Lymphatic system a system consisting of lymph vessels, lymph capillaries, lymph nodes, and the thymus and spleen. The system helps protect the body against infection, absorbs fats, and returns excess fluid and small proteins to the blood circulatory system.

Lymphocyte (lim′-fō-sīt) white blood cell type important in the immune response.

Lysosome (lī′-sō-sōm) a membrane-bound organelle containing intracellular digestive enzymes.

Macronutrient a nutrient needed in relatively large quantities (often defined as composing more than 1 percent of an organism's body).

Macrophage (mak′-rō-faj) a cell derived from white blood cells called monocytes, whose function is to consume foreign particles including bacteria.

Maltose (mal′-tōs) a disaccharide composed of two glucose molecules.

Mammary glands (mam′-uh-rē) milk-producing organs used by female mammals to nourish their young.

Mantle (man′-tul) an extension of the body wall in certain invertebrates, such as molluscs. It may secrete a shell, protect the gills, and, as in cephalopods, aid in locomotion.

Marsupial (mar-sū′-pē-ul) a type of mammal whose young are born at an extremely immature stage and undergo further development in a pouch while they remain attached to a mammary gland. Includes kangaroos, opossums, and koalas.

Matrix the fluid contained within the inner membrane of a mitochondrion.

Matter the material of which the universe is made.

Mechanical incompatibility the inability of male and female animals to mate because their genitalia do not fit together properly.

Medulla (med-oo′-la) part of the hindbrain of vertebrates that controls automatic activities such as breathing, swallowing, and heartbeat.

Medusa (meh-dū′-suh) a bell-shaped, often free-swimming stage in the life cycle of many cnidarians. Jellyfish are one example.

Megakaryocyte (meg-a-kār′-ē-ō-sīt) a large cell type that remains in the bone marrow, pinching off pieces of itself. These cytoplasmic fragments enter the circulation as platelets.

Megaspore a haploid cell formed by meiosis from a diploid megaspore mother cell. Through mitosis and differentiation, the megaspore develops into the female gametophyte.

Megaspore mother cell a diploid cell contained within the ovule of a flowering plant, which undergoes meiosis to produce four haploid megaspores.

Meiosis (mī-ō′-sis) a type of cell division found in eukaryotic organisms, in which a diploid cell divides twice to produce four haploid cells.

Membrane in cells, a thin sheet of lipids and proteins that surrounds the cell or its organelles, separating them from their surroundings.

Memory cell a long-lived descendant of a B or T cell that has been activated by contact with antigen. Memory cells are a reservoir of cells that rapidly respond to reexposure to the same antigen.

Menstrual cycle in females of some primate species, the roughly 28-day cycle of development and degeneration of the lining of the uterus, accompanied by discharge of tissue and blood through the vagina.

Menstruation in females of some primate species, the monthly discharge of uterine tissue and blood from the vagina.

Meristem cell (mer′-i-stem) an undifferentiated cell that remains capable of cell division throughout the life of a plant.

Mesoderm (mes′-ō-derm) the middle embryonic tissue layer, lying between the endoderm and ectoderm, and usually the last to develop. Mesoderm gives rise to structures such as muscle and skeleton.

Mesophyll (mez′-ō-fil) cells located between the epidermal layers of a leaf.

Messenger RNA (mRNA) a strand of RNA, complementary to DNA, that conveys the genetic information in DNA to the ribosomes to be used during protein synthesis. Sequences of three nucleotides (codons) in mRNA specify particular amino acids to be incorporated into a protein.

Metamorphosis (met-a-mor′-fō-sis) in animals with indirect development, a radical change in body form from one larval stage to another or from larva to adult.

Metaphase (met′-a-fāz) a stage in mitosis and meiosis in which the nuclear membrane has completely disappeared, the chromosomes are lined up along the equator of the cell, and each chromosome is attached to microtubules of the fully formed spindle.

Methanogen (me-than′-ō-gen): a type of anaerobic archaebacterium capable of converting carbon dioxide to methane.

Microfilament part of the cytoskeleton of eukaryotic cells, composed of the proteins actin and (sometimes) myosin; functions in movement of cell organelles and in locomotion by pseudopodia.

Micronutrient a nutrient needed in relatively small quantities (often defined as composing less than 0.1 percent of an organism's body).

Microsphere a small, hollow sphere formed from proteins or proteins complexed with other compounds.

Microspore a haploid cell formed by meiosis from a microspore mother cell. Through mitosis and differentiation, the microspore develops into the male gametophyte.

Microspore mother cell a diploid cell contained within an anther of a flowering plant, which undergoes meiosis to produce four microspores.

Microtubule a hollow cylindrical strand found in eukaryotic cells, composed of the protein tubulin; part of the cytoskeleton used in movement of cell organelles, cell growth, and construction of cilia and flagella.

Microtubule organizing center a region of a eukaryotic cell at which tubulin is assembled into microtubules.

Microvilli (mī-krō-vi′-lī) a series of folded projections of the cell membrane which increase its surface area.

Middle ear part of the mammalian ear composed of the tympanic membrane and three bones (malleus, incus, stapes) that transmit vibrations from the auditory canal to the oval window.

Migration in population genetics, the flow of genes between populations.

Mimic (mim′-ik) an organism that has evolved to resemble another.

Mineral an inorganic substance, found in rocks or soil.

Mitochondrion (mī-tō-kon′-drē-un) an organelle bounded by two membranes that is the site of the reactions of aerobic metabolism.

Mitosis (mī-tō′-sis) a type of nuclear division found in eukaryotic cells. Chromosomes are duplicated during interphase before mitosis. During mitosis, one copy of each chromosome moves into each of two daughter nuclei. The daughter nuclei are therefore genetically identical to each other.

Model an organism that is mimicked by another.

Molecule a particle composed of one or more atoms, held together by chemical bonds. A molecule is the smallest particle of a compound that displays all the properties of that compound.

Molt to shed an external body covering, such as an exoskeleton, skin, feathers, or fur.

Monera (mō′-ne-ra) a taxonomic kingdom consisting of unicellular prokaryotic organisms, including bacteria, archaebacteria, and cyanobacteria.

Monocotyledon (mahn′-eh-kot-ul-ēd′-un): a class of angiosperm plant in which the embryo has one cotyledon, or seed leaf.

Monoecious (mon-ē′-shus) pertaining to organisms in which male and female gametes are produced in the same individual.

Monohybrid cross a breeding experiment in which the parents differ in only one genetically determined trait.

Monosaccharide (mo-nō-sak′-a-rīd) the basic molecular unit of all carbohydrates, composed of a backbone of carbon atoms to which are bonded hydrogen and hydroxyl groups.

Morula (mor′-ū-la) in animals, an embryonic stage during cleavage, when the embryo consists of a solid ball of cells.

Motor neuron a neuron that carries information from the central nervous system and stimulates effector organs such as muscles or glands.

Motor unit a single motor neuron and all the muscle fibers (cells) on which it synapses.

Mucous membrane the lining of the inside of the respiratory and digestive tracts.

Muscle fiber an individual muscle cell.

Mutation a change in the base sequence of DNA; usually refers to a genetic change that is significant enough to change the appearance or function of the organism.

Mutualism (mū′-chū-al-ism) a symbiotic relationship in which both participating species benefit.

Mycelium (mī-sēl′-ē-um) the body of a fungus, consisting of a mass of hyphae.

Mycorrhiza (mī-kō-rī′za; pl. mycorrhizae) a symbiotic relationship between a fungus, often a basidiomycete or an ascomycete, and the roots of a vascular plant.

Myelin (mī′-eh-lin) a wrapping of lipid-rich membranes of Schwann cells around the axon of a vertebrate nerve cell. Myelin increases the speed of conduction of action potentials.

Myofibril (mī′-ō-fī′bril) a cylindrical subunit of a muscle cell, surrounded by sarcoplasmic reticulum.

Myometrium (mī-ō-mē′-trē-um) the muscular outer layer of the uterus.

Myosin (mī′-ō-sin) one of the major proteins of muscle that interacts with actin to produce contraction; found in the thick filaments of the muscle fiber. *See also* actin.

Natural selection the differential survival and/or reproduction of organisms due to environmental forces, resulting in the preservation of favorable adaptations. Usually, natural selection refers specifically to differential survival or reproduction based on genetic differences among individuals.

Nearshore zone shallow-water areas of the ocean just below the low-tide line. Due to the abundance of nutrients and strong sunlight, the nearshore often has the greatest concentration of life in the ocean.

Negative feedback a type of control mechanism in which the output of a system causes actions that suppress further output.

Nematocyst (nēm-āt′-ō-sist) a specialized cell found in cnidarians which, when disturbed, ejects a sticky or poisoned thread. Used by cnidarians to trap and sting their prey.

Nephridium (nef-rid′-ē-um) a type of excretory organ found in earthworms, molluscs, and certain other invertebrates. A nephridium somewhat resembles a single vertebrate nephron.

Nephron (nef′-ron) the functional unit of the kidney, where blood is filtered and urine formed.

Nerve a bundle of axons of nerve cells, bound together in a sheath.

Nerve cord also called the spinal cord of vertebrates, a hollow nervous structure lying along the dorsal side of the body of chordates.

Nerve net a loosely coordinated network of neurons.
Net primary productivity the energy stored in the primary producers of an ecosystem over a given period.
Neurohormone a chemical synthesized by a specialized nerve cell (called a neurosecretory cell) and secreted into the bloodstream as a hormone.
Neuromuscular junction the synapse formed between a motor neuron and a muscle fiber.
Neuron (nur′-on) a single nerve cell.
Neurosecretory cell a specialized nerve cell that synthesizes and releases hormones.
Neurotransmitter a chemical released by a presynaptic cell at a synapse, which binds to receptors on the postsynaptic cell, causing changes in the electrical potential of the second cell.
Neutralization the process of covering up or inactivating a toxic substance with antibody.
Neutral mutation a mutation (change in DNA sequence) that has little or no phenotypic effect.
Neutron a subatomic particle found in the nuclei of atoms, bearing no charge and having mass approximately equal to that of a proton.
Niche (nitch) the role of a particular species within an ecosystem, including all aspects of its interaction with the biotic and abiotic environment.
Nitrogen fixation the process of converting atmospheric nitrogen (N_2) to ammonium (NH_4^+).
Nitrogen-fixing possessing the ability to remove nitrogen from the atmosphere and combine it with hydrogen to produce ammonium.
Node a region of a stem at which leaves and lateral buds are located.
Node of Ranvier (nōd of ron′-vē-ā) an interruption of the myelin on a vertebrate myelinated axon, at which action potentials are generated.
Nondisjunction an error in meiosis in which chromosomes fail to segregate properly into the daughter cells.
Nonpolar covalent bond a covalent bond with equal sharing of electrons.
Northern coniferous forest a northern biome dominated by conifers; taiga.
Notochord (nōt′-ō-kōrd) a stiff but somewhat flexible, supportive rod found in all members of the phylum Chordata at some stage of development.
Nuclear envelope the double membrane system surrounding the nucleus of eukaryotic cells. The outer membrane is often continuous with the endoplasmic reticulum.
Nucleic acid (nū-klā′-ik) an organic molecule composed of nucleotide subunits. The two common types of nucleic acids are ribonucleic acids (abbreviated RNA) and deoxyribonucleic acids (DNA).
Nucleoid (nū′-klē-oyd) the location of the genetic material in prokaryotic cells; not membrane enclosed.
Nucleolus (nū-klē′-ō-lus) the region of the eukaryotic nucleus engaged in ribosome synthesis, consisting of the genes encoding ribosomal RNA, newly synthesized ribosomal RNA, and ribosomal proteins.
Nucleotide (nū′-klē-ō-tīd) an organic molecule composed of a phosphate group, a five-carbon monosaccharide (ribose or deoxyribose), and a nitrogen-containing base. The individual subunit of which nucleic acids are composed.
Nucleus (1) the central region of an atom, composed of protons and neutrons. (2) the membrane-bound organelle of eukaryotic cells that contains the cell's genetic material.
Nutrients the atoms and molecules that living organisms need to acquire in their diets.
Oil a lipid composed of three fatty acids, some of which are unsaturated, covalently bonded to a molecule of glycerol. Oils are liquid at room temperature.
Olfaction (ōl-fak′-shun) a chemical sense, the sense of smell; in terrestrial vertebrates, the result of detection of airborne molecules.
Ommatidium (ōm-ma-tid′-ē-um) an individual light-sensitive subunit of a compound eye. Each ommatidium consists of a lens and several (usually eight) receptor cells.
Oncogene a gene that, when transcribed, causes a cell to become cancerous.
One-gene, one-protein hypothesis the proposition that each gene encodes the information for the synthesis of a specific protein.
Oogonium (ō-ō-gō′-nē-um) a diploid cell in female animals that gives rise to a primary oocyte.
Open circulatory system a type of circulatory system in arthropods and molluscs in which the blood is pumped through an open space (the hemocoel), where it bathes the internal organs directly.
Operant conditioning a laboratory training procedure in which an animal learns to make a response (such as pressing a lever) through reward or punishment.
Operator in prokaryotes, a segment of DNA that controls access of RNA polymerase to the promoter.
Operon a unit of organization of prokaryotic chromosomes, in which several genes that specify related functions (e.g., enzymes in the same biosynthetic pathway) are grouped together on the chromosome, are transcribed at the same time, and are regulated together.
Opioid (ōp′-ē-ōyd) a group of peptides found in the vertebrate brain that mimic some of the actions of opiates (such as opium, heroin, morphine). Besides analgesia (pain relief), opioids seem to be involved in many behaviors, including emotion, learning, and the control of appetite.
Optic disk (op′-tik) the area of the retina at which the axons of the ganglion cells merge to form the optic nerve; the blind spot of the retina.
Optic nerve (op′-tik) the nerve leading from the eye to the brain, carrying visual information.
Orbital the region of an atom, outside the nucleus, in which an electron is likely to be found.
Organ a structure of an organism (e.g., intestine), usually composed of several tissue types, that is organized into a functional unit.
Organ of Corti (kōr′-tē) part of the central canal of the cochlea, consisting of the basilar membrane, its ciliated receptor cells, and the tectorial membrane.
Organ system two or more organs working in together in the execution of a specific bodily function (e.g., digestive tract).
Organelle (or-ga-nel′) a structure found in the cytoplasm of eukaryotic cells that performs a specific function; sometimes used to refer specifically to membrane-bound structures such as the nucleus or endoplasmic reticulum.
Organic referring to a molecule that contains both carbon and hydrogen.
Organism (or′-ga-niz-em) an individual living thing.

Organogenesis (or-gan-ō-jen′-i-sis) the process by which the germ layers of the gastrula develop into organs.

Origin the site of attachment of a muscle to the relatively stationary bone on one side of a joint.

Osculum (os′-kya-lum) relatively large opening in the sponge body through which water is expelled.

Osmosis (oz-mō′-sis) the diffusion of water across a differentially permeable membrane, usually down a concentration gradient of free water molecules. Water moves into the solution that has a lower water concentration from the solution with the higher water concentration.

Osmotic pressure a measure of the tendency of water to move from a solution with a lower concentration of water into one with a higher concentration of water; the physical pressure which must be applied to a solution to prevent water movement into it from pure water.

Outer ear the outermost part of the mammalian ear, including the external ear and auditory canal leading to the tympanic membrane.

Oval window the membrane-covered entrance to the inner ear.

Ovary (1) in flowering plants, a structure at the base of the carpel containing one or more ovules; develops into the fruit. (2) the gonad of female animals.

Oviduct in mammals, the tube leading from the ovary to the uterus.

Ovule a structure within the ovary of a flower, inside which the female gametophyte develops. After fertilization, the ovule develops into the seed.

Palisade cells mesophyll cells just beneath the upper epidermis of a leaf; usually elongated perpendicularly to the epidermis.

Pancreas (pan′-krē-as) a combined exocrine and endocrine gland located in the abdominal cavity. Its endocrine parts secrete the hormones insulin and glucagon, which regulate glucose concentrations in the blood. Its exocrine parts secrete enzymes for fat, carbohydrate, and protein digestion into the small intestine.

Parasitic (par-uh-sit′-ik) living in or on the body of another organism, and causing it harm as a result.

Parasitism a symbiotic relationship in which one organism (usually smaller and more numerous than its host) benefits by feeding on the other, usually without killing it immediately.

Parasympathetic nervous system the division of the autonomic nervous system that produces largely involuntary responses related to maintenance of normal body functions, such as digestion.

Parathyroid a set of four small endocrine glands embedded in the surface of the thyroid gland that produce parathormone, which (with calcitonin from the thyroid) regulates calcium ion concentration in the blood.

Parthenogenesis (par-the-nō-gen′-i-sis) a mode of reproduction, in which an egg undergoes development without fertilization.

Passive transport movement of materials across a cell membrane down a gradient of concentration, pressure, or electrical charge, and not using cellular energy.

Pathogen (path′-ō-gen) an organism capable of producing disease.

Pedigree a diagram showing genetic relationships among a set of individuals, usually with respect to a specific genetic trait.

Pelagic (pel-a′-jik) referring to organisms that spend their lives in open water, floating or swimming.

Peptide (pep′-tīd) a chain composed of two or more amino acids linked together by peptide bonds.

Peptidoglycan (pep-tid-ō-glī′-can): Material found in prokaryotic cell walls consisting of chains of sugars cross-linked by short chains of amino acids.

Pericycle (per′-i-sī-kul) the outermost layer of cells of the vascular cylinder of roots.

Peripheral nervous system in vertebrates, that part of the nervous system located outside the brain and spinal cord, consisting of the nerves leading to and from the brain and spinal cord, and the ganglia of the autonomic nervous system.

Permafrost soil layer a short distance below the surface of the tundra which stays frozen year round.

Permeate (per′-mē-āt) to pass through, as through pores in a membrane.

Petal part of a flower, often flat and brightly colored, serving to attract potential animal pollinators.

Petiole (pet′-ē-ōl) the stalk that connects the blade of a leaf to the stem.

Phagocytosis (fā-gō-sī-tō′-sis) a type of endocytosis in which extensions of a cell membrane engulf extracellular particles and transport them into the interior of the cell.

pH a scale with values from 0 to 14, used for measuring the relative acidity of a solution. At pH 7 a solution is neutral, pH 0 to 7 is acidic, and pH 7 to 14 is basic. Each unit on the pH scale represents a tenfold change in the concentration of hydrogen ions.

Pharynx (fār′-inx) (1) a chamber at the back of the mouth shared by the digestive and respiratory systems; (2) a portion of the digestive system between the mouth and the esophagus. In flatworms, it is developed as an extensible, muscular organ.

Phenotype (fēn′-ō-tīp) the physical properties of an organism. Phenotype can be defined as outward appearance (e.g., flower color), behavior, or in molecular terms (e.g., ABO glycoproteins on red blood cells).

Phenylketonuria (fe-nul-kē-tō-nū′-rē-a) a recessive disease in which the enzyme that catalyzes the conversion of the amino acid phenylalanine to tyrosine is faulty.

Pheromone (fār′-uh-mōn) a chemical produced by an organism that alters the behavior or physiological state of another organism of the same species.

Phloem (flō′-um) in plants, the vascular tissue that conducts a concentrated sugar solution.

Phospholipid (fos-fō-li′-pid) a lipid consisting of glycerol to which two fatty acids and one phosphate group are bonded. The phosphate group bears another group of atoms, often containing nitrogen, and usually either polar or bearing an electrical charge.

Photic zone (fō′-tik) the surface waters of the ocean, into which sunlight penetrates well enough for plants to grow.

Photon (fō′-ton) the smallest unit of light.

Photoreceptors (fō′-tō-rē-sep-ters) sensory cells specialized for detecting light and differences in light intensity.

Photosynthesis the series of chemical reactions in which the energy of light is used to synthesize high-energy organic molecules, usually carbohydrates, from low-energy inorganic molecules, usually carbon dioxide and water.

Photosystem in thylakoid membranes, a light-harvesting complex and its associated electron transport system.

Phototropism growth with respect to the direction of light.

Phycocyanin (fī-kō-sī′-a-nin) a bluish pigment found in the membranes of chloroplasts and used as an accessory light-gathering molecule in thylakoid photosystems.

Phytochrome (fī′-tō-krōm) a light-sensitive plant pigment that mediates many plant responses to light, including flowering, stem elongation, and seed germination.

Phytoplankton (fī′-to-plank-ten) a general term describing photosynthetic protists that are abundant in marine and freshwater environments.

Pinocytosis (pī-nō-sī-tō′-sis) a type of endocytosis in which part of the cell membrane, when contacted by appropriate extracellular substances, forms a vesicle that enters the cytoplasm, carrying materials into the cell.

Pioneer an organism that is among the first to colonize an ecosystem undergoing succession.

Pith cells at the center of a root or stem.

Pituitary an endocrine gland located at the base of the brain that produces several hormones, many of which influence the activity of other glands.

Placenta in mammals, a structure formed partly from the uterine lining and partly from the embryonic membranes, especially the chorion; functions in gas, nutrient, and waste exchange between embryonic and maternal circulatory systems.

Plankton (plank′-ton) microscopic pelagic protists and animals found in the photic zone of the oceans, whose movement is determined primarily by movement of the water. Phytoplankton are the main producers in the open ocean.

Plasma the fluid, noncellular portion of the blood.

Plasma cell an antibody-secreting descendant of a B cell.

Plasmodesma (plaz-mō-dez′-ma; pl. plasmodesmata) a cell-to-cell junction in plants that connects the interiors of adjacent cells.

Plastid (plas′-tid) in plant cells, an organelle bounded by two membranes that may be involved in photosynthesis (chloroplasts), pigment storage, or food storage.

Platelets (plāt′-lets) cell fragments formed from megakaryocytes in bone marrow. Platelets, which lack nuclei, circulate in the blood and play a role in blood clotting.

Pleated sheet a type of secondary structure of a protein. In a pleated sheet, protein chains lie side by side, held to one another by hydrogen bonds.

Polar body in oogenesis, a small cell containing a nucleus but virtually no cytoplasm.

Polar covalent bond a covalent bond with unequal sharing of electrons, so that one atom is relatively negative, while the other is relatively positive.

Pollen the male gametophyte of gymnosperms and angiosperms.

Pollination in angiosperms, the deposition of pollen onto the stigma of a flower of the same species; in gymnosperms, the deposition of pollen within the pollen chamber of a female cone of the same species.

Polygenic inheritance a pattern of inheritance in which the interactions of two or more genes determine phenotype.

Polyp (pol′-ip) the sedentary, vase-shaped stage in the life cycle of many cnidarians. Hydra and sea anemones are examples.

Polyploid (pol′-ē-ployd) having more than two homologous chromosomes of each type.

Polysaccharide (pol-ē-sak′-a-rīd) a large carbohydrate molecule composed of branched or unbranched chains of repeating monosaccharide subunits, usually glucose or modified glucose molecules. Polysaccharides include starches, cellulose, and glycogen.

Population a group of individuals of the same species, found in the same time and place, and actually or potentially interbreeding.

Population bottleneck a form of genetic drift in which a population becomes extremely small, which may lead to differences in allele frequencies as compared to other populations of the species, and to a loss in genetic variability.

Posterior (pos-tēr′-ē-ur) the tail, hindmost, or rear end of an animal.

Postmating isolating mechanism any mechanism that prevents organisms of two different populations, once mating has occurred, from producing vigorous, fertile offspring.

Postsynaptic referring to the nerve cell at a synapse which changes its electrical potential in response to a chemical (the neurotransmitter) released by another (presynaptic) cell.

Potassium channel a hollow-cored protein that spans a nerve cell membrane, forming a pore, and which only allows potassium ions to flow through the pore.

Potential energy "stored" energy, usually chemical energy or energy of position within a gravitational field.

Preadaptation: a feature evolved under one set of environmental conditions that, purely by chance, helps an organism to adapt to new environmental conditions.

Prebiotic evolution: evolution before life existed; especially abiotic synthesis of organic molecules.

Predation (pre-dā′-shun) the act of killing and eating another living organism.

Premating isolating mechanism any mechanism that prevents organisms of two different populations from mating.

Presynaptic referring to a nerve cell that releases a chemical (the neurotransmitter) at a synapse, which causes changes in the electrical activity of another (postsynaptic) cell.

Primary consumer an organism that feeds on producers; an herbivore.

Primary growth growth in length and development of initial structures of plant roots and shoots, due to cell division of apical meristems and differentiation of the daughter cells.

Primary oocyte (ō′-ō-sīt) a large diploid cell, derived from the oogonium by growth and differentiation, which undergoes meiosis to produce the egg.

Primary phloem phloem produced from an apical meristem.

Primary root the first root that develops from a seed.

Primary spermatocyte (sper-ma′-tō-sīt) a diploid cell, derived from the spermatogonium by growth and differentiation, which undergoes meiosis to produce four sperm.

Primary structure the amino acid sequence of a protein.

Primary succession succession that occurs in an environment, such as bare rock, in which no remnants of a previous community are present.

Primary xylem xylem produced from an apical meristem.

Primer pheromone a chemical produced by an organism that alters the physiological state of another of the same species.

Primitive streak in reptiles, birds, and mammals, the region of the ectoderm of the two-layered embryonic disk through which cells migrate to form mesoderm.

Producer a photosynthetic organism.

Product an atom or molecule resulting from a chemical reaction.

Progesterone (prō-ge′-ster-ōn) a hormone produced by the corpus luteum that promotes development of the uterine lining.

Prokaryotic (prō-kar-ē-ot′-ik) referring to cells of the kingdom Monera. Prokaryotic cells do not have their genetic material enclosed within a membrane-bound nucleus and also lack other membrane-bound organelles.

Promoter a specific sequence of DNA to which RNA polymerase binds, initiating gene transcription.

Prophase (prō′-fāz) the first stage of mitosis or meiosis, in which the chromosomes first become visible in the light microscope as thickened, condensed threads, the nuclear membrane disintegrates, and the spindle forms. In meiosis I, the homologous chromosomes pair up and exchange parts at chiasmata.

Proprioception (prō′-prē-ō-cep-shun) a sense that monitors the position of the parts of the body and their direction of movement.

Prostaglandin (pro-sta-glan′-din) a family of modified fatty acid hormones manufactured by many cells of the body.

Prostate gland (prō′-stāt) a gland that produces part of the fluid component of semen. The prostate fluid is basic and contains a chemical that activates sperm movement.

Protease (prō′-tē-ās) an enzyme that digests proteins.

Protein an organic molecule composed of one or more chains of amino acids.

Protista (prō-tis′-tuh) a taxonomic kingdom including unicellular, eukaryotic organisms.

Proton a subatomic particle found in the nuclei of atoms, bearing a unit of positive charge and a relatively large mass roughly equal to the mass of the neutron.

Protozoan (prō-te-zō′-an; pl. protozoa) a nonphotosynthetic or animal-like protist.

Pseudopod (sū′-dō-pod) a temporary extension of the cell membrane used for locomotion or phagocytosis in certain cells such as the protist *Amoeba* or white blood cells of vertebrates.

Pulmonary circulation the pathway of blood from the right ventricle through the pulmonary artery to the lung capillaries and back to the left atrium through the pulmonary vein.

Punctuated equilibrium the hypothesis that the characteristics of species are constant for long periods of time (e.g., millions of years) and that new species arise rapidly over a few thousand years.

Purine a nitrogen-containing nucleic acid base consisting of fused six- and five-sided rings. The common purines in nucleic acids are adenine and guanine.

Pyrimidine a nitrogen-containing nucleic acid base consisting of a single six-sided ring. The common pyrimidines in nucleic acids are cytosine, thymine, and uracil.

Quaternary structure (kwat′-er-nā-rē) the complex three-dimensional structure of a protein that is composed of more than one peptide chain.

Queen substance a chemical produced by a queen bee that can act as both a primer and a releaser pheromone.

Radial symmetry a body plan in which many planes drawn along a central axis will divide the body into approximately mirror-image halves. Cnidarians and many adult echinoderms show radial symmetry.

Radioactive pertaining to an atom with an unstable nucleus that disintegrates spontaneously with the emission of radiation.

Radula (ra′-dū-luh) a ribbon of tissue in the mouth of gastropod molluscs that bears numerous teeth on its outer surface and is used to scrape and drag food into the mouth.

Rain shadow an area of low rainfall on the side of a mountain facing away from the prevailing winds.

Random distribution spacing in which the probability of finding an individual is equal in all parts of an area.

Reactant an atom or molecule that is used up in a chemical reaction to form a product.

Reaction center in the light-harvesting complex of a photosystem, the chlorophyll molecule to which light energy is transferred by the antenna pigments. The captured energy ejects an electron from the reaction center chlorophyll, and the electron is transferred to the linked electron transport system.

Receptor (1) a cell that responds to an environmental stimulus (chemicals, sound, light, pH, etc.) by changing its electrical potential; (2) a protein molecule in a cell membrane that binds to another molecule (hormone, neurotransmitter, odorous compound, etc.) triggering metabolic or electrical changes in a cell.

Receptor potential a graded electrical potential change in a receptor cell produced in response to reception of a stimulus from the external or internal environment (chemicals, sound, light, pH, heat, cold, etc.). The receptor potential is proportional to the intensity of the stimulus.

Recessive an allele expressed only in homozygotes and which is completely masked in heterozygotes.

Recombinant DNA a laboratory technique of inserting foreign DNA into an organism, producing new gene combinations not previously existing in nature.

Reflex a simple, automatic, usually unconscious behavior performed by part of the body in response to a stimulus.

Regeneration (1) regrowth of a body part after loss or damage; (2) asexual reproduction by regrowth of an entire body from a fragment.

Regulatory gene a gene that controls the timing or rate of transcription of other genes.

Releaser a stimulus that triggers a fixed action pattern.

Releaser pheromone a chemical produced by one organism that alters the behavior of another of the same species.

Releasing hormone a hormone secreted by the hypothalamus that causes the release of specific hormones by the anterior pituitary gland.

Replacement-level fertility the average birthrate at which a reproducing population exactly replaces itself during its lifetime.

Reproductive isolation the failure of organisms of one population to breed successfully with members of another population; may be due to premating or postmating isolating mechanisms.

Reservoir the major source of any particular nutrient in an ecosystem, usually in the abiotic portion.

Respiratory center a location in the brainstem that sends rhythmic bursts of nerve impulses to the respiratory muscles, resulting in breathing.

Resting potential an electrical potential found in unstimulated nerve cells; the inside of the cell is negatively charged.

Reticular activating formation (reh-tik'-ū-lar) a diffuse network of neurons extending from the hindbrain, through the midbrain, and into the lower reaches of the forebrain, involved in filtering sensory input and regulating what information is relayed to higher centers in the cerebrum for further attention.

Retina (ret'-in-a) a sheet of tissue at the rear of camera-type eyes, composed of photoreceptor cells plus associated nerve cells that refine the photoreceptor information and transmit it to the optic nerve.

Rhizome (rī'-zōm) an underground stem, usually horizontal and functioning in food storage.

Ribonucleic acid (rī-bō-nū-klā'-ik; RNA) a single-stranded nucleic acid molecule composed of nucleotides, each of which consists of a phosphate group, the sugar ribose, and one of the bases adenine, cytosine, guanine, or uracil.

Ribosome (rī'-bō-sōm) an organelle consisting of two subunits, each composed of ribosomal RNA and protein. Ribosomes are the site of protein synthesis, in which the sequence of nucleotides of messenger RNA is translated into the sequence of amino acids in a protein.

Ribozyme an RNA molecule that can catalyze certain chemical reactions, especially those involved in synthesis and processing of RNA itself.

RNA polymerase an enzyme that catalyzes the covalent bonding of free RNA nucleotides into a continuous strand, using RNA nucleotides that are complementary to those of a strand of DNA.

Rod a rod-shaped photoreceptor cell in the vertebrate retina, sensitive to dim light, but not involved in color vision. *See also* Cone.

Root the part of a plant, usually below ground, that anchors the plant in the soil, absorbs and transports water and minerals, stores food, and produces certain hormones.

Root cap a cluster of cells at the tip of a growing root, derived from the apical meristem. The root cap protects the growing tip from damage as it burrows through the soil.

Root hair a fine projection from the epidermal cells of a young root.

Rough endoplasmic reticulum endoplasmic reticulum lined on the outside with ribosomes.

Round window the membrane-covered opening in the cochlea, located below the oval window.

Runner a horizontally growing stem that may develop new plants at nodes that touch the soil.

Sarcodine (sar-kō'-dīn) a category of nonphotosynthetic protist (protozoa) characterized by the ability to form pseudopodia. Some, such as amoebae, are naked, while others have elaborate shells.

Sarcomere (sark'-ō-mēr) the unit of contraction of a muscle fiber; a subunit of the myofibril, consisting of actin and myosin filaments and bounded by Z-lines.

Sarcoplasmic reticulum (sark-ō-plas'-mik re-tik'-ū-lum) specialized endoplasmic reticulum found in muscle cells. The sarcoplasmic reticulum stores calcium ions and releases them into the interior of the muscle cell to initiate contraction.

Saturated referring to a fatty acid with as many hydrogen atoms as possible bonded to the carbon backbone; a fatty acid with no double bonds in its carbon backbone.

Savanna a transition biome between the tropical deciduous forest and tropical grassland, characterized by a nearly continuous cover of grasses with occasional drought-resistant trees. The savanna climate includes such a severe dry season that most trees cannot grow.

Schwann cell a cell that forms the myelin coating of vertebrate axons by wrapping its cell membrane many times around the axon. *See also* Myelin.

Sclera (sklāra) a tough white connective tissue layer that covers the outside of the eyeball and forms the white of the eye.

Scramble competition direct interactions between individuals attempting to acquire the same limited resource.

Scrotum (skrō'-tum) the pouch of skin containing the testes of male mammals.

S-curve the growth curve that describes a population introduced into a new area. It consists of an initial period when numbers remain relatively low, followed by a period of exponential growth, followed by decreasing growth rate, and finally, relative stability.

Second messenger a term applied to intracellular chemicals, such as cyclic AMP, that are synthesized or released within a cell in response to the binding of a hormone or neurotransmitter (the first messenger) to receptors on the cell surface. Second messengers bring about specific changes in the metabolism of the cell.

Secondary consumer an organism that feeds on primary consumers; a carnivore.

Secondary growth growth in diameter of a stem or root due to cell divisions in lateral meristems.

Secondary oocyte (ō'-ō-sīt) a large haploid cell derived by meiosis I from the diploid primary oocyte.

Secondary phloem phloem produced from cells arising at the outside of the vascular cambium.

Secondary spermatocyte (sper-ma'-tō-sīt) a haploid cell derived by meiosis I from the diploid primary spermatocyte.

Secondary structure a repeated, regular structure assumed by protein chains, held together by hydrogen bonds; usually either a helix or a pleated sheet.

Secondary succession succession that occurs after an existing community is destroyed, in an environment modified by that community, for example, after a forest fire.

Secondary xylem xylem produced from cells arising at the inside of the vascular cambium.

Seed the reproductive stage of gymnosperms and angiosperms, usually including an embryonic plant and a food reserve, enclosed within a resistant outer covering.

Seed coat the outermost covering of a seed, formed from the integuments of the ovule.

Segmentation (seg-men-tā'-shun) division of the body into repeated, often similar units.

Segmentation movements asynchronous contractions of the small intestine which result in mixing of the partially digested food and digestive enzymes. The movements also bring nutrients into contact with the absorptive intestinal wall.

Self-fertilization union of sperm and egg from the same individual.

Semen the sperm-containing fluid produced by the male reproduction tract.

Semicircular canals three fluid-filled semicircular tubes of the inner ear which function in the detection of rotational movements of the head.

Semiconservative replication the process of replication of the DNA double helix; the two DNA strands separate, and each is used as a template for the synthesis of a complementary DNA strand. Each daughter double helix therefore consists of one parental strand and one new strand.

Seminal vesicle in male mammals, a gland that produces a basic, fructose-containing fluid that forms part of the semen.

Seminiferous tubules (sem-i-ni′-fer-us) a series of tubes in the vertebrate testis in which sperm are produced.

Senescence in plants, a specific aging process, often including deterioration and dropping of leaves and flowers.

Sensory neuron a nerve cell that carries information about internal or external environmental conditions to the central nervous system.

Sensory receptor a cell specialized to respond to particular internal or external environmental stimuli by producing an electrical potential.

Sepal (sē′-pul) one of the protective outer coverings of a flower bud, often opening into green, leaflike structures when the flower blooms.

Sertoli cell a large cell in the seminiferous tubule that regulates spermatogenesis and nourishes the developing sperm.

Sessile (ses′-ul) not free to move about, usually permanently attached to a surface.

Sex chromosome one of the pair of chromosomes that differ between the sexes and usually determine the sex of an individual; for example, human females have similar sex chromosomes (XX) while males have dissimilar ones (XY).

Sex-influenced inheritance a mode of inheritance in which traits of a nonsexual nature are more common in one sex than in the other, often due to differing levels of sex hormones.

Sex linkage a pattern of inheritance characteristic of genes located on one type of sex chromosome (e.g., X) and not found on the other type (e.g., Y).

Sex-linked inheritance inheritance of traits controlled by genes carried on the X chromosome. Females show the dominant trait unless they are homozygous recessive, whereas males will express whatever allele is found on their single X chromosome.

Sexual recombination the formation of new combinations of alleles due to inheritance of chromosomes from two different parental organisms during sexual reproduction.

Sexual reproduction a form of reproduction in which genetic material from two parental organisms is combined. In eukaryotes, two haploid gametes fuse to form a diploid zygote.

Sexual selection a type of natural selection in which the choice of mates by one sex is the selective agent.

Shoot all the parts of a vascular plant exclusive of the root. Usually above ground, consisting of stem, leaves, and reproductive structures.

Short-day plant a plant that will flower only if the length of daylight is shorter than some species-specific duration.

Sickle-cell anemia a recessive disease caused by a single amino acid substitution in the hemoglobin molecule. Sickle-cell hemoglobin molecules tend to cluster together in long chains, distorting the shape of red blood cells and causing them to break and clog the capillaries.

Sieve plate a part of the cell wall between two sieve-tube elements in phloem, with large pores interconnecting the cytoplasm of the elements.

Sieve tube in phloem, a tube made from sieve-tube elements, which transports sugar solutions.

Sieve-tube element one of the cells of a sieve tube.

Simple diffusion diffusion of water, dissolved gases, or lipid-soluble molecules through the phospholipid bilayer of a cell membrane.

Single bond a covalent bond that occurs when two atoms share a single pair of electrons.

Sinoatrial node (sī′-nō-āt′-rē-ul nōd) also called the SA node, this is a small mass of specialized muscle in the wall of the right atrium. It generates electrical signals rhythmically and spontaneously and serves as the heart's pacemaker.

Skeletal muscle also called striated or voluntary muscle; the type of muscle that is attached to and moves the skeleton.

Skeleton a supporting structure for the body, upon which muscles act to change the body configuration.

Smooth endoplasmic reticulum endoplasmic reticulum without ribosomes.

Smooth muscle type of muscle found around hollow organs, such as the digestive tract, bladder, and blood vessels, normally not under voluntary control.

Sodium channel a protein that spans a nerve cell membrane, forming a pore, and which allows only sodium ions to flow through the pore.

Sodium-potassium pump an enzyme-like protein in nerve cell membranes which actively transports potassium ions into the cell and sodium ions out of the cell.

Solvent a liquid that is capable of dissolving (uniformly dispersing) other substances in itself.

Spawning a method of external fertilization in which male and female parents shed gametes into the water, and sperm must swim through the water to reach the eggs.

Species (spē′-sēs) the sum of all the populations of organisms that are potentially capable of interbreeding under natural conditions and that are reproductively isolated from other populations.

Specific heat the amount of energy required to raise the temperature of 1 gram of a substance 1°C.

Sperm the haploid male gamete, usually small, motile, and containing little cytoplasm.

Spermatid a haploid cell derived from the secondary spermatocyte by meiosis II. The mature sperm is derived from the spermatid by differentiation.

Spermatogenesis the formation of sperm.

Spermatogonium (pl. spermatogonia) a diploid cell lining the walls of the seminiferous tubules that gives rise to a primary spermatocyte.

Spinal cord part of the central nervous system of vertebrates, extending from the base of the brain to the hips, protected by the vertebrae of the spine; contains the cell bodies of motor neurons innervating skeletal muscles, the circuitry for some simple reflex behaviors, and axons communicating with the brain.

Spindle a football-shaped array of microtubules that moves the chromosomes to opposite poles of a cell during anaphase of meiosis and mitosis.

Spindle apparatus a structure found in dividing eukaryotic cells, composed of microtubules, which appears to guide the movement of chromosomes.

Spiracles (spī′-re-kul) openings in the abdominal segments of insects through which air enters the tracheae.

Spongy layer irregularly shaped cells located just above the lower epidermis of a leaf.

Spontaneous generation the proposal that living organisms can arise from nonliving matter.

Spore (spōr) a resistant or resting structure that disperses readily and withstands unfavorable environmental conditions.

Sporophyte (spōr′-ō-fīt) the diploid form of a plant that produces haploid, asexual spores through meiosis.

Sporozoan (spōr-ō-zō′-en) a category of parasitic protist. Sporozoans have complex life cycles often involving more than one host, and are named for their ability to form infectious spores. A well-known member (genus *Plasmodium*) causes malaria.

Stabilizing selection a type of natural selection in which those organisms displaying extreme phenotypes of any sort are selected against.

Stamen (stā′-men) the male reproductive structure of a flower, consisting of a filament and an anther in which pollen grains develop.

Starch a polysaccharide composed of branched or unbranched chains of glucose molecules, used by plants as a carbohydrate storage molecule.

Startle coloration a color pattern (often resembling large eyes) that can be displayed suddenly by a prey organism when approached by a predator.

Start codon a codon in messenger RNA that signals the beginning of protein synthesis on a ribosome.

Stem the normally vertical, aboveground part of a plant body that bears leaves.

Steroid a lipid composed of four fused rings of carbon atoms to which functional groups are attached.

Stigma (stig′-ma) the pollen-capturing tip of a carpel.

Stoma (pl. stomata; stō′-ma) adjustable openings in plant leaves surrounded by guard cells. Most gas exchange between leaves and the air occurs through the stomata.

Stop codon a codon in messenger RNA that stops protein synthesis and causes the completed protein chain to be released from the ribosome.

Stroke an interruption of blood flow to part of the brain, caused by the rupture of an artery or the blocking of an artery by a blood clot. Loss of blood supply leads to rapid death of the area of the brain affected.

Stroma (strō′-ma) the semifluid material of chloroplasts, in which the membranous grana are embedded.

Structural gene a gene that codes for a protein used by the cell for purposes other than gene regulation. Structural genes code for enzymes or for proteins that are structural parts of a cell.

Style a stalk connecting the stigma of a carpel with the ovary at its base.

Subatomic particles the particles of which atoms are made: electrons, protons, and neutrons.

Subclimax a stage of succession occurring before the climax community is reached, maintained by regular disturbance: for example, tallgrass prairie maintained by regular fires.

Substrate the atoms or molecules that are the reactants for an enzyme-catalyzed chemical reaction.

Succession (suk-se′-shun) the directional change in the community structure of an ecosystem over time. Community changes alter the ecosystem in ways that favor competitors, until a stable climax community is reached.

Sucrose a disaccharide composed of glucose and fructose.

Supernormal stimulus a stimulus that exaggerates crucial elements of the releaser, making it more effective than the normal releaser.

Suppressor T cell a type of T cell that depresses the response of other immune cells to foreign antigens.

Surface tension the property of a liquid to resist penetration by objects at its interface with the air, due to cohesion between molecules of the liquid.

Survivorship curve a curve resulting when the number of individuals in a population is graphed against their age, usually expressed as a percentage of their maximum life span.

Symbiosis (sim′-bī-ō′sis) a close association between different species usually over an extended period. *Symbiosis* as used in this book includes mutualism, commensalism, and parasitism.

Sympathetic nervous system the division of the autonomic nervous system that produces largely involuntary responses that prepare the body for stressful situations.

Synapse (sin′-apz) the site of communication between nerve cells. One cell (presynaptic) usually releases a chemical (the neurotransmitter) which changes the electrical potential of the second (postsynaptic) cell.

Synaptic potential at a synapse between two nerve cells, the electrical change occurring in the postsynaptic cell as a result of reception of neurotransmitter released by the presynaptic cell.

Synaptic terminal the terminal branches of an axon, where the axon forms a synapse, usually with the dendrites of the second nerve cell.

Syphilis (si′-ful-is) a sexually transmitted bacterial infection of the reproductive organs which, if untreated, can damage the nervous and circulatory systems.

Systemic circulation the pathway of the blood from the left ventricle through the aorta and other large arteries to the capillaries of the body tissues and back through the large veins to the right atrium.

T cell a type of lymphocyte that recognizes and destroys specific foreign cells or substances, or that regulates other cells of the immune system.

T tubules indentations of the muscle cell membrane that conduct the action potential inside the cell.

Taiga (tī′-ga) the cold coniferous forest biome, characterized by needleleaf evergreen trees. The waxy needles of the trees reduce evaporative water loss during winter.

Taproot a relatively thick, vertically oriented root that develops from the primary root of dicots.

Target cell a cell upon which a particular hormone exerts its effect.

Taste a chemical sense; in mammals, perceptions of sweet, sour, bitter, or salt produced by stimulation of receptors on the tongue.

Taxis (tax′-is) innate movement of an organism toward or away from a stimulus such as heat, light, or gravity.

Taxonomy (tax-on′-uh-mē) the science by which organisms are classified into hierarchically arranged categories that reflect their evolutionary relationships.

Tay–Sachs disease a recessive disease caused by a deficiency in enzymes regulating lipid metabolism in the brain.

Tectorial membrane (tek-tōr′-ē-ul) one of the membranes of the cochlea, in which the hairs of the hair cells are embedded. During sound reception, movement of the basilar membrane relative to the tectorial membrane bends the cilia.

Telophase (tel′-o-fāz) the last stage of meiosis and mitosis, in which a nuclear membrane re-forms around each new daughter nucleus, the spindle disappears, and the chromosomes relax from their condensed form.

Temperate deciduous forest a biome having cold winters but a fairly long, moist growing season. The dominant vegetation is broadleaf deciduous trees, which shed their leaves in winter as a protection against water loss.

Temperate rain forest a biome of very limited distribution, along certain coasts where the climate is fairly mild year round and rainfall is very high, characterized by coniferous evergreen trees, often of enormous size.

Temporal isolation the inability of organisms to mate if they have significantly different breeding seasons.

Temporal lobe part of a cerebral hemisphere of the human brain, involved in recall of learned events.

Tendril a slender outgrowth of a stem that coils about external objects and supports the stem; usually a modified leaf or branch.

Tentacle (ten'-te-kul) an elongate, extensible projection of the body of cnidarians and cephalopod molluscs that may be used for grasping, stinging, and immobilizing prey, and locomotion.

Terminal bud the bud at the extreme end of a stem or branch.

Territoriality the defense of an area in which important resources are located.

Tertiary structure (ter'-shē-ār-ē) the complex three-dimensional structure of a single peptide chain. The tertiary structure is held in place by disulfide bonds between cysteine amino acids, by attraction and repulsion among amino acid side groups, and by interaction between the cellular environment (water or lipids) and the amino acid side groups of the protein.

Test cross a breeding experiment in which an individual showing the dominant phenotype is mated with an individual that is homozygous recessive for the same gene. The ratio of offspring with dominant versus recessive phenotypes can be used to determine the genotype of the phenotypically dominant individual.

Testis (pl. testes) the gonad of male animals.

Testosterone in vertebrates, a hormone produced by the interstitial cells of the testis; stimulates spermatogenesis and the development of male secondary sex characteristics.

Thalamus part of the forebrain, the thalamus serves as a relay network between other parts of the nervous system and the cerebrum.

Theory in science, an explanation for natural events that is based on a large number of observations and is in accord with scientific principles, especially causality.

Thermoacidophile (ther-mō-a-sid'-eh-fīl) a form of archaebacterium that thrives in hot, acidic environments.

Thorn a hard, pointed outgrowth of a stem; usually a modified branch.

Threshold the electrical potential (less negative than the resting potential) at which an action potential is initiated.

Thylakoid (thī'-la-koyd) a disk-shaped, membranous sac found in chloroplasts, the membranes of which contain the photosystems and the ATP-synthesizing enzymes used in the light-dependent reactions of photosynthesis.

Thyroid an endocrine gland, located in front of the larynx in the neck, that secretes the hormones thyroxine (affecting metabolic rate) and calcitonin (regulating calcium ion concentration in the blood).

Tight junction a type of cell-to-cell junction in animals that prevents the movement of materials through the spaces between cells.

Tissue a group of (usually similar) cells that together carry out a specific function.

Trachea (trā'-kē-uh) a rigid but flexible tube supported by rings of cartilage, which conducts air between the larynx and the bronchi.

Tracheae (trā'-kē) elaborately branching tubes that ramify through the bodies of insects and carry air close to each body cell. Air enters the tracheae through openings called spiracles.

Tracheid (trā'-kē-id) an elongated xylem cell with tapering ends containing pits in the cell walls; forms tubes that transport water.

Transcription the synthesis of an RNA molecule from a DNA template.

Transfer RNA (tRNA) a type of RNA that (1) binds to a specific amino acid and (2) bears a set of three nucleotides (the anticodon) complementary to the mRNA codon for that amino acid. Transfer RNA carries its amino acid to a ribosome during protein synthesis, recognizes a codon of messenger RNA, and positions its amino acid for incorporation into the growing protein chain.

Transformation a method of genetic recombination whereby DNA from one bacterium (usually released after the death of the bacterium) becomes incorporated into the DNA of another, living, bacterium.

Translation the process whereby the sequence of nucleotides of messenger RNA is converted into the sequence of amino acids of a protein.

Transpiration evaporation of water from a leaf.

Trial-and-error learning process by which adaptive responses are learned through rewards or punishments provided by the environment.

Trichocyst (trik'-eh-sist) a stinging organelle of protists.

Triple bond a covalent bond that occurs when two atoms share three pairs of electrons.

Trisomy 21 *see* Down syndrome.

Trisomy X a condition of females who have three X chromosomes instead of the normal two. Most of these women are phenotypically normal, and are fertile.

Tropical deciduous forest a broadleaf deciduous forest biome found in areas with a warm climate year round but a pronounced wet and dry season. The trees drop their leaves in the dry season.

Tropical rain forest a broadleaf evergreen forest biome found in the tropics, where rainfall is adequate and temperatures are warm year round. The tropical rain forest has the greatest diversity of both plants and animals to be found anywhere on land.

True-breeding pertaining to an individual all of whose offspring produced through self-fertilization are identical to the parental type. True-breeding individuals are homozygous for the trait in question.

Tube cell the outermost cell of a pollen grain; the tube cell digests a tube through the tissues of the carpel, ultimately penetrating into the female gametophyte.

Tube feet cylindrical extensions of the water-vascular system of echinoderms, used for locomotion, grasping food, and respiration.

Tubule (tūb'-ūle) the tubular portion of the nephron. It includes a proximal portion, the loop of Henle, and a distal portion. Urine is formed from the blood filtrate as it passes through the tubule.

Tundra a treeless biome characterized by permafrost and an extremely short growing season. Most of the plants are low-growing perennial shrubs, grasses, and wildflowers.

Turgor pressure pressure developed within a cell (especially the central vacuole of plant cells) as a result of osmotic water entry.

Turner's syndrome a set of characteristics typical of a woman with only one X chromosome. These women are sterile, failing to develop normal ovaries. They also tend to be very short, and to lack normal female secondary sexual characteristics.

Tympanic membrane (tim-pan′-ik) the eardrum; a membrane stretched across the opening of the ear, which transmits vibration of sound waves to bones of the middle ear.

Uniformitarianism the hypothesis that the Earth developed gradually through natural forces similar to those at work today.

Uniform distribution a relatively regular spacing of individuals within a population, often as a result of territorial behavior.

Unsaturated referring to a fatty acid with fewer than the maximum number of hydrogen atoms bonded to its carbon backbone; a fatty acid with one or more double bonds in its carbon backbone.

Upwelling movement of deep, nutrient-rich water to the surface.

Urea (ū-rē′-uh) a water-soluble, nitrogen-containing waste product of amino acid breakdown which is one of the principal components of mammalian urine.

Ureter (ū′-re-tur) a tube that conducts urine from each kidney to the bladder.

Urethra (ū-rē′-thra) the tube leading from the urinary bladder to the outside of the body; in males, the urethra also receives sperm from the vas deferens and conducts both sperm and urine (at different times) to the tip of the penis.

Uric acid (ūr′-ik acid) a nitrogen-containing waste product of amino acid breakdown which is a relatively insoluble white crystal. Uric acid is excreted by birds, reptiles, and insects.

Uterus in female mammals, the part of the reproductive tract that houses the embryo during pregnancy.

Vaccine a material injected into the body that contains antigens characteristic of a particular disease organism, and that stimulates an immune response.

Vacuole (vak′-ū-ōl) a large vesicle consisting of a single membrane enclosing a space. *See* Central vacuole; Food vacuole; Contractile vacuole.

Vagina the passageway leading from the outside of the body to the cervix of the uterus.

Variable a condition, particularly in a scientific experiment, that is subject to change.

Vascular (vas′-ku-ler) possessing or composed of conducting tissue or vessels.

Vascular bundle (vas′-kū-lar) a strand of xylem and phloem found in a leaf; commonly called a vein.

Vascular cambium a lateral meristem located between the xylem and phloem of a woody root or stem.

Vascular cylinder the centrally located conducting tissue of a young root, consisting of primary xylem and phloem.

Vas deferens (vas de′-fer-ens) the tube connecting the epididymis of the testis with the urethra.

Vein a large-diameter, thin-walled vessel that carries blood from venules back to the heart.

Ventral (ven′-trul) the lower, or underside of an animal whose head is oriented forward.

Ventricle (ven′-trē-kul) the lower muscular chamber on each side of the heart, which pumps blood out through the arteries. The right ventricle sends blood to the lungs, and the left to the rest of the body.

Venule (ven′-yul) a narrow vessel with thin walls that carries blood from capillaries to veins.

Vesicle (ves′-i-kul) a small membrane-bound sac within the cytoplasm of a cell.

Vessel a tube of xylem composed of vertically stacked vessel elements, with perforated or missing end walls, leaving a continuous, uninterrupted hollow cylinder.

Vessel element one of the cells of a xylem vessel; elongated, dead at maturity, with thick lateral cell walls for support but with end walls either lacking entirely or heavily perforated.

Vestigial structures (ves-tij′-ē-ul) structures with no known function, but which are homologous to functional structures in related organisms.

Villus (vi-lus) projections of the wall of the small intestine which increase its absorptive surface area.

Virus (vī′-rus) a noncellular parasitic particle consisting of a protein coat surrounding a strand of genetic material. Viruses can multiply only within the cells of living organisms.

Vitamin any one of a group of diverse chemicals that must be present in trace amounts in the diet to maintain health. Vitamins are used by the body in conjunction with enzymes in a variety of metabolic reactions; a few are involved in growth and differentiation.

Vitreous humor (vit′-rē-us) a clear jellylike substance that fills the large chamber of the eye between the lens and retina.

Waggle dance symbolic communication used by honeybee foragers to communicate the location of a food source to their hivemates.

Water-vascular system a system in echinoderms consisting of a series of canals through which seawater is conducted and used to inflate tube feet for locomotion, grasping food, and respiration.

Wax a lipid composed of fatty acids covalently bonded to long-chain alcohols.

Xylem (zī′-lum) a conducting tissue of vascular plants that transports water and minerals from root to shoot.

Yolk protein or lipid-rich substances contained in eggs as food for the developing embryo.

Yolk sac one of the embryonic membranes of reptile, bird, and mammalian embryos. In birds and reptiles, the yolk sac is a membrane surrounding the yolk in the egg. In mammals, the yolk sac is empty, but forms part of the umbilical cord and the gut.

Z-lines fibrous protein structures to which the thin filaments of skeletal muscle are attached, forming the boundaries of sarcomeres.

Zona pellucida (pel-ū′-si-da) a clear, noncellular layer between the corona radiata and the egg.

Zooflagellate (zō-ō-fla′-gel-et) a category of nonphotosynthetic protist that move using flagella.

Zooplankton (zō′-plank-ton) a diverse assemblage of small pelagic animals dominated by crustaceans.

Zygospore (zī′-gō-spōr) a fungal spore surrounded by a thick, resistant wall, which forms from a diploid zygote.

Zygote (zī′-gōt) (1) in sexual reproduction, a diploid cell formed by the fusion of two haploid cells; (2) the fertilized egg.

Photo Acknowledgments

xi	Don W. Fawcett/Science Source/Photo Researchers
xii	Doug Allan/Oxford Scientific Films/Animals, Animals
xiii	Chip Clark/Smithsonian Institution
xiv	Jane Burton/Bruce Coleman
xvii	Philip Green

Chapter 1

1-1	NASA
1-2a	Dr. Jeremy Burgess/Science Photo Library/Science Source/Photo Researchers
1-2b	Sven-Olof Lindblad/Photo Researchers
1-2c	Kim Taylor/Bruce Coleman
1-4	E. H. Newcomb/BPS
1-5	Jeff Apoian/Photo Researchers
1-6	Tom McHugh/Photo Researchers

Chapter 1 Gallery

p. 6(top left)	CRNI/Science Photo Library/Science Source/Photo Researchers
p. 6 (bottom left)	Eric V. Gravé/Science Source/Photo Researchers
p. 6(top right)	James L. Castner
p. 6 (bottom right)	Catalina Marine Science Center
p. 7(left)	Dwight R. Kuhn
p. 7(top right)	Audesirk
p. 7 (center right)	Audesirk
p. 8(top left)	Dr. Ronald L. Shimek
p. 8 (bottom left)	Dr. Ronald L. Shimek
p. 8(top right)	Sea Studios, Inc./Peter Arnold
p. 8 (bottom right)	J. W. Porter, Univ. of Georgia/BPS
p. 9(top left)	P. J. Bryant, Univ. of California, Irvine/BPS
p. 9 (bottom left)	Nancy Sefton/Photo Researchers
p. 9(top right)	William E. Townsend, Jr./Photo Researchers
p. 9 (right center)	Gregory G. Dimijian/Photo Researchers
p. 9 (bottom right)	Michael Fogden/Oxford Scientific Films/Animals Animals
1-8	John Durham/Science Photo Library/Science Source/Photo Researchers
1-10	Catalina Marine Science Center
1-11	Tom Algire/Tom Stack & Associates
1-12	Carolina Biological Supply Company
1-13	Chip Clark/Smithsonian Institution
1-17	Audesirk

Unit I Opener Don W. Fawcett/Science Source/Photo Researchers

Chapter 2

2-1	Andy Levin
2-11a	Michael P. Gadomski/National Audubon Society/Photo Researchers
2-11b	Audesirk

Chapter 3

3-4a	Eric V. Gravé/Science Source/Photo Researchers
3-4b	Don W. Fawcett/Science Source/Photo Researchers
3-5b	J. R. Waaland, Univ. of Washington/BPS
3-5c	Audesirk
3-6b	Richard Kolar/Animals Animals
3-10a	Robert Peargy/Animals Animals
3-10b	Russ Kinne/National Audubon Society/Photo Researchers
3-10c	Floyd Schrock/Tom Stack & Associates

Chapter 4

4-1	Omikron/Science Source/Photo Researchers
4-2	Dr. Stanley L. Erlandsen
E4-1a-e	M. I. Walker/Science Source/Photo Researchers
E4-1f	Courtesy, R. D. Allen, Univ. of Hawaii, Honolulu
E4-2	Don W. Fawcett/Science Source/Photo Researchers
E4-3	Philips Industries, The Netherlands
E4-4	Manfred Kage/Peter Arnold
E4-5	Dr. K. Tanaka
4-4	Dr. F. G. Leedale/Biophoto Associates/Science Source/Photo Researchers
4-10a-c	Dr. Joseph Kurantsin-Mills
4-13	Eric V. Gravé/Photo Researchers
4-14	Dr. K. Hausmann/Peter Arnold
4-15	Keith R. Porter, Laboratory of Cell Biology, Univ. of Maryland
4-16	L. Andrew Staehelin
4-17	Drs. Don W. Fawcett, K. Hama, D. Albertini/Science Source/Photo Researchers
4-18	E. H. Newcomb and W. P. Wergin, Univ. of Wisconsin, Madison/BPS
4-19	Michael Giannechini/Photo Researchers

Chapter 5

5-1a	J. J. Cardamone, Jr., Univ. of Pittsburgh/BPS
5-1b	M. W. Steer and E. H. Newcomb, Univ. of Wisconsin, Madison/BPS
E5-1a	E. H. Newcomb and T. D. Pugh, Univ. of Wisconsin, Madison/BPS
E5-1b	Omikron/Science Source/Photo Researchers
E5-2	Drs. Don W. Fawcett and Kanaseki/Science Source/Photo Researchers
5-4	Keith R. Porter, Laboratory of Cell Biology, Univ. of Maryland
5-5	E. H. Newcomb and W. P. Wergin, Univ. of Wisconsin, Madison/BPS
5-6	K. G. Murti/Visuals Unlimited
5-7a	K. G. Murti/Visuals Unlimited
5-7b	Drs. Don W. Fawcett and D. Friend/Science Source/Photo Researchers
5-8	Gary W. Grimes/Taurus Photos
5-9	Zdenek Hruban, Univ. of Chicago
5-11	Biophoto Associates/Science Source/Photo Researchers
5-12a,b	Thomas Eisner, Cornell University
5-14	J. J. Cardamone, Jr., Univ. of Pittsburgh/BPS
5-15	W. Rosenberg, Iona College/BPS
5-16	Carolina Biological Supply Company
5-17a	Miles Scientific
5-17b	Keith R. Porter, Laboratory of Cell Biology, Univ. of Maryland
5-19	Keith R. Porter, Laboratory of Cell Biology, Univ. of Maryland
5-20 top	W. L. Dentler, Univ. of Kansas/BPS
5-20 bottom	Don W. Fawcett/Science Source/Photo Researchers

Chapter 6

6-1	Alan Carey/Photo Researchers

Chapter 7

7-1	Yann Arthus-Bertrand/Peter Arnold
7-15a	Spuk Ltd./Focus West

7-15b Jonathan Watts/Science Photo Library/Photo Researchers

Unit II Opener Doug Allan/Oxford Scientific Films/Animals Animals

Chapter 8

8-3 Dr. Gopal Murti/Science Photo Library/Science Source/Photo Researchers
8-4 CNRI/Science Photo Library/Science Source/Photo Researchers
8-5 R. Knauft, from Biology Media/Science Source/Photo Researchers
8-6 Audesirk
8-7 Andrew Bajer
8-8a Keith R. Porter, Laboratory of Cell Biology, Univ. of Maryland
8-8b Carolina Biological Supply Company
8-9a T. E. Schroeder, Univ. of Washington/BPS
8-10a B. A. Palevitz and E. H. Newcomb, Univ. of Wisconsin, Madison/BPS
8-11a Biophoto Associates/Science Source/Photo Researchers
8-11b Carolina Biological Supply Company
8-11 Audesirk
8-13 James Kezer, Univ. of Oregon

Chapter 9

9-1a,b Art Resource
9-2 Archiv/Science Source/Photo Researchers
9-8 Darwin Dale/Photo Researchers
9-10 Carolina Biological Supply Company
9-12a James Kezer, Univ. of Oregon

Chapter 10

10-1 NIH/Science Source/Photo Researchers
10-4 Tom Broker, Univ. of Rochester Medical Center
10-7 Rosalind Franklin/Science Source/Photo Researchers
E10-1 Omikron/Science Source/Photo Researchers

Chapter 11

11-14 Science Source/Photo Researchers
11-15a,b March of Dimes
11-16 Audesirk
11-17b Prof. Stanley Cohen/Science Photo Library/Science Source/Photo Researchers

Chapter 12

12-3a Marlene and Bob Lippman
12-4 Audesirk
12-5a Omikron/Science Source/Photo Researchers
12-5b Science Source/Photo Researchers
12-6a Visuals Unlimited
12-6b Dr. Richard Lindenberg
12-8 John D. Cunningham/Visuals Unlimited
12-11 Elizabeth Crews

Unit III Opener Chip Clark/Smithsonian Institution

Chapter 13

13-1 Art Resource
13-4 Audesirk
13-6a Alex Kerstitch
13-6b Chip Clark/Smithsonian Institution
13-6c American Museum of Natural History
13-9b R. Ake Norberg
13-9c D. Cavagnaro/Peter Arnold
E13-1 The Bettmann Archive
E13-3a,b Carolina Biological Supply Company
13-14 (turtle) G. B. Ruff/Animals Animals
13-14 (mouse) Dan McCoy/Rainbow
13-14 (human) Petit Format/Nestle/Science Source/Photo Researchers
13-14 (chicken) Carolina Biological Supply Company
13-16a Z. Leszczynski/Animals Animals
13-16b Robert Percy/Animals Animals
13-16c Margot Conte/Animals Animals
13-16d Cooke/Photo Researchers
13-17a,b M. W. F. Tweedie/Bruce Coleman
13-18 Audesirk

Chapter 14

14-1 Frans Lanting
14-5a Jeff Foott
14-5b David Burney, Lida Pigott Burney
14-6 Gregory Dimijian/Photo Researchers
14-7b Michael Cunningham
14-8 William Ervin
14-9 Audesirk
14-10a M. P. L. Fogden/Bruce Coleman
14-10b Carol Hughes/Bruce Coleman
14-13 (c) Manuel A. Rodriguez
14-15 Michael McCoy/Photo Researchers
14-16 Jeff Lepore/Photo Researchers
E14-1 John W. Fitzpatrick and Glen E. Woolfenden
14-17a,b Pat and Tom Leeson/Photo Researchers
14-18 Jack E. Williams, Univ. of California, Davis
14-19 James A. Kern

Chapter 15

15-1 Science Source/Photo Researchers
15-5 Science Source/Photo Researchers
15-7 H. Kawakami and N. Kawakami. *Journal of Protozoology* 25: 217–225, 1978.
15-10a Field Museum of Natural History
15-10b Alex Kerstitch
15-10c Carolina Biological Supply Company
15-10d Ronald L. Shimek
15-11 Peabody Museum of Natural History, Yale University. Painted by Rudolf F. Zallinger
15-13 D. Lyons/Photo Researchers
15-14 Peabody Museum of Natural History, Yale University. Painted by Rudolf F. Zallinger
15-15a,c Tom McHugh/Photo Researchers
15-15b Martha E. Reeves/Photo Researchers
15-18 Chip Clark/Smithsonian Institution
15-19 American Museum of Natural History

Chapter 16

16-1a Audesirk
16-1b H. Oscar/Visuals Unlimited
16-1c Tom McHugh/Photo Researchers

Chapter 17

17-1b C. McClaren/Visuals Unlimited
17-2b Lee Simon/Stammers/Science Photo Library/Science Source/Photo Researchers
17-3a,b Z. Skobe/BPS
17-3c P. W. Johnson and J. McN. Sieburth, Univ. of Rhode Island/BPS
17-4a CNRI/Science Photo Library/Science Source/Photo Researchers
17-4b Prof. L. Caro/Science Source/Photo Researchers
17-5 T. J. Beveridge, Univ. of Guelph/BPS
17-6 David Parker/Science Photo Library/Science Source/Photo Researchers
17-7a,b C. P. Vance/Visuals Unlimited
E17-1a,b Eli Lilly
17-8a N. J. Lang, Univ. of California, Davis/BPS
17-8b Biophoto Associates/Science Source/Photo Researchers
17-9a Carolina Biological Supply Company
17-9b Eric Gravé/Science Source/Photo Researchers
17-10 Micrograph by Elaine P. Simons, Univ. of British Columbia, Dept. of Oceanography
17-11 Jan Hinsch/Science Photo Library/Science Source/Photo Researchers
17-13 Carolina Biological Supply Company
17-14 Stanley L. Erlandsen and D. E. Feely

17-15a,b M. I. Walker/Science Source/Photo Researchers
17-15c Eric Gravé/Science Source/Photo Researchers
17-17a,b Biophoto Associates/Science Source/Photo Researchers
17-17c Gary W. Grimes and S. W. L'Hernault/Taurus Photos

Chapter 18

18-2a (top) Catalina Marine Science Center
18-2a (bottom) Ronald L. Shimek
18-3a Alex Kerstitch
18-3b Catalina Marine Science Center
18-3c Ronald L. Shimek
18-3d John D. Cunningham/Visuals Unlimited
18-7 Carolina Biological Supply Company
18-10 T. E. Adams/Peter Arnold
18-11a Carolina Biological Supply Company
18-11b Dr. Howard Schiang, DVM
18-14a Dave Woodward/Tom Stack & Associates
18-14b J. H. Robinson/Science Source/Photo Researchers
18-15 Audesirk
18-16 Audesirk
18-18 Dr. William Stark
18-20a Carolina Biological Supply Company
18-20b P. J. Bryant, Univ. of California, Irvine/BPS
18-20c Stephen Dalton/Photo Researchers
18-21a Carolina Biological Supply Company
18-21b John D. Cunningham/Visuals Unlimited
18-21c Audesirk
18-22a,c Carolina Biological Supply Company
18-22b P. J. Bryant, Univ. of California, Irvine/BPS
18-22d Catalina Marine Science Center
18-23a H. D. Jones, Univ. of Manchester
18-23b Alex Kerstitch
18-24a Fred Bavendam/Peter Arnold
18-24b Audesirk
18-25a Jeff Rotman
18-25b S. K. Webster, Monterey Bay Aquarium/BPS
18-25c Alex Kerstitch
18-26a,b Catalina Marine Science Center
18-26c Chris Newbert/Bruce Coleman
18-27 Michael Male/Photo Researchers
18-29 John Giannicchi/Science Source/Photo Researchers
18-30d Catalina Marine Science Center
18-31a David Hall/Photo Researchers
18-31b Jeff Rotman
18-32a Peter David/Planet Earth Pictures
18-32b Mike Neumann/Photo Researchers
18-32c S. K. Webster, Monterey Bay Aquarium/BPS
18-32d Peter Scoones/Planet Earth Pictures
18-33a Breck P. Kent/Animals Animals
18-33b Dwight R. Kuhn/Bruce Coleman
18-33c Cosmos Blank/National Audubon Society/Photo Researchers
18-34a John D. Cunningham/Visuals Unlimited
18-34b R. K. Burnard/BPS
18-35b Carolina Biological Supply Company
18-36a Walter E. Harvey/National Audubon Society/Photo Researchers
18-36b Carolina Biological Supply Company
18-36c Jen and Des Bartlett/Bruce Coleman
18-37 Carolina Biological Supply Company
18-38a Flip Nicklin & Associates
18-38b Jonathan Watts/Science Photo Library/Photo Researchers
18-38c David Burney, Lida Pigott Burney
18-39a Jean-Paul Ferrero/Auscape
18-39b David C. Fritts/Animals Animals

Chapter 19

19-1a J. R. Waaland, Univ. of Washington/BPS
19-2a,b Carolina Biological Supply Company
19-3a W. K. Fletcher/Photo Researchers
19-3b John D. Cunningham/Visuals Unlimited
19-4 Audesirk
E19-1 Doisneau-Rapho/Photo Researchers
E19-2 G. L. Barron, Univ. of Guelph/BPS
E19-3 Nancy Allin and G. L. Barron, Univ. of Guelph
19-5a New York Botanical Gardens
19-5b Audesirk
19-5c A. Davies/Bruce Coleman
19-5d Carolina Biological Supply Company
19-7 John D. Cunningham/Visuals Unlimited
19-8a-c Audesirk
19-9a P. W. Grace/Science Source/Photo Researchers
19-9b Carolina Biological Supply Company
19-11 Catalina Marine Science Center
19-12a Audesirk
19-12b Catalina Marine Science Center
19-12c Robert Evans/Peter Arnold
19-13a Carolina Biological Supply Company
19-13b M. I. Walker/Science Source/Photo Researchers
19-14 Carolina Biological Supply Company
19-15b Carolina Biological Supply Company
19-16a Dwight R. Kuhn
19-16b Carolina Biological Supply Company
19-16c Audesirk
19-17b Audesirk
19-19a Audesirk
19-19b A-Z Collection/Photo Researchers
19-21a,b Audesirk
19-22a-c Audesirk
19-23 Audesirk

Unit IV Opener Jane Burton/Bruce Coleman

Chapter 20

20-2a R. Humbert/BPS
20-2b Audesirk
20-2c Tom McHugh/Photo Researchers
20-3a Audesirk
20-3b John D. Cunningham/Visuals Unlimited
20-4 Audesirk

Chapter 21

21-1 Catalina Marine Science Center
21-5a Lynwood M. Chace/Photo Researchers
21-5b Dwight R. Kuhn
21-6(1) Ed Reschka/Peter Arnold
21-6(2) E. R. Degginger/Earth Scenes
21-8 J. R. Waaland, Univ. of Washington/BPS
21-10a Gene Cox/Bruce Coleman
21-10b J. R. Waaland, Univ. of Washington/BPS
21-10c Carolina Biological Supply Company
21-11 J. N. A. Lott, McMaster University/BPS
21-12 Dr. G. F. Leedale/Biophoto Associates/Science Source/Photo Researchers
21-16a Carolina Biological Supply Company
21-16b John D. Cunningham/Visuals Unlimited
21-17 John D. Cunningham/Visuals Unlimited
21-19 Audesirk
21-22a Biophoto Associates/Science Source/Photo Researchers
21-22b John A. Menge, Dept. of Plant Pathology, Univ. of California, Riverside
21-23b Dwight R. Kuhn
21-23c E. H. Newcomb and S. R. Tandon, Univ. of Wisconsin, Madison/BPS
E21-2 James L. Castner
21-25a M. I. Walker/Science Source/Photo Researchers
21-25b Dr. Jeremy Burgess/Science Photo Library/Science Source/Photo Researchers
E21-4 Biophoto Associates/Science Source/Photo Researchers
E21-5 Audesirk
21-27 Dr. Julian I. Schroeder, Lewis Research Center, UCLA School of Medicine
21-28a,b Dr. Martin H. Zimmermann
21-30 Audesirk
21-31 Bill O'Connor/Peter Arnold

21-32a Audesirk
21-32b Carolina Biological Supply Company
21-33a John D. Cunningham/Visuals Unlimited
21-33b,c,e Audesirk
21-33d Dwight R. Kuhn
21-34a Oxford Scientific Films/Earth Scenes
21-34b Nuridsany and Perrennou/Photo Researchers

Chapter 22

22-1 Kevin Scharer/Peter Arnold
22-3 Terry Domico/Earth Scenes
22-4b James L. Castner
22-5 Audesirk
22-6 Audesirk
22-7 James L. Castner
22-8a,b Thomas Eisner
22-9 Audesirk
22-10a Merlin Tuttle
22-10b Oxford Scientific Films/Animals Animals
22-11 Bob and Clara Calhoun/Bruce Coleman
22-12 Dr. J. A. L. Cooke/ Oxford Scientific Films/Animals Animals
22-13a,b Audesirk
22-14a-c Carolina Biological Supply Company
22-15a J. N. A. Lott, McMaster University/BPS
22-15b P. Wiese/Visuals Unlimited
22-16a-c Carolina Biological Supply Company
22-17 Carolina Biological Supply Company
22-19b,c Carolina Biological Supply Company
E22-1a,b Audesirk
E22-1c Carolina Biological Supply Company
E22-2 Audesirk
E22-3 Dwight R. Kuhn
E22-4 W. J. Weber/Visuals Unlimited
22-21 Audesirk
22-22a George Holton/ Photo Researchers

Chapter 23

23-8a,b Audesirk
E23-1,2 Carolina Biological Supply Company
23-10 Dwight R. Kuhn
23-11a Richard Parker/ National Audubon Society/Photo Researchers
23-11b John Shaw/Tom Stack & Associates

Chapter 24

Frontispiece Marty Stouffer Productions/Animals Animals
24-2a Dwight R. Kuhn/ DRK
E24-1 Audesirk

Chapter 25

E25-2 C. Allan Morgan
E25-3 M. P. Kahl/Photo Researchers

Chapter 26

26-8 Lennart Nilsson
E26-1 Audesirk
26-9 D. Phillips/Visuals Unlimited
26-10 Bill Longcore/ Science Source/ Photo Researchers
26-12 W. J. Johnson/ Visuals Unlimited
26-13 Manfred Kage/Peter Arnold
26-16 John D. Cunningham/Visuals Unlimited
26-18a P. J. Bryant, Univ. of California, Irvine/ BPS
26-18b E. D. Brodie, Jr., Univ. of Texas, Arlington/BPS
26-21a (top) Dr. E. R. Degginger/ Bruce Coleman
26-21a (bottom) John Shaw/Bruce Coleman
26-21b Gary Milburn/Tom Stack & Associates
26-22b H. R. Duncker, Dept. of Anatomy, Univ. of Giessen, West Germany
E26-3 Bill Travis, M. D./ National Cancer Institute
E26-4 O. Auerbach/Visuals Unlimited

Chapter 27

27-6 (both) David M. Phillips/ Visuals Unlimited
27-7 (both) Andrejus Liepins/ Science Photo Library/Science Source/Photo Researchers
E27-1 CNRI/Science Photo Library/Science Source/Photo Researchers
E27-2 Dr. Matthew A. Gonda, Ph.D., Head, Laboratory of Cell and Molecular Structure/National Cancer Institute

Chapter 28

E28-1b K. B. Raper
E28-2,3 D. W. Francis
28-9 Biophoto Associates/ Science Source/ Photo Researchers
28-11a L. Orci, M. D., Universite de Geneve, Institut d'Histologie et d'Embryologie
E28-4 (all) P. J. Bryant, Univ. of California, Irvine/ BPS

Chapter 29

Frontispiece Art Resource
29-2a *Journal of Physiology*
29-2b Carolina Biological Supply Company
29-12a Richard Humbert/ BPS
29-12b Audesirk
29-12c Ronald L. Shimek
E29-1 (both) Audesirk

Chapter 30

E30-2 Johnny Johnson/ DRK
30-4a,b Joseph E. Hawkins, D. Sc., Kresge Hearing Research Institute, Univ. of Michigan Medical School
30-6 Dr. William Stark
30-7 David Doubilet
30-10 Audesirk
30-11a (1) Oxford Scientific Films/Animals Animals
30-11a (2) D. R. Specter/ Animals Animals
30-11b Audesirk
30-15f Franzini-Armstrong/ Science Source/ Photo Researchers
30-17b Don W. Fawcett/ Science Source/ Photo Researchers
30-19a Catalina Marine Science Center
30-19b Audesirk

Chapter 31

31-2 F. Stuart Westmorland/Tom Stack & Associates
31-4 Z. Leszczynski/ Animals Animals
31-5 P. J. Bryant, Univ. of California, Irvine/ BPS
31-6a Tom McHugh/Photo Researchers
31-6b Audesirk
31-6b (inset) Peter Harrison
31-7 (c) Anne and Jacques Six
31-8 Michael Fogden/ Animals Animals
31-9 (top) Audesirk
31-9 (middle) Tom McHugh/Photo Researchers
31-9 (bottom) Robert Hernandez/ Photo Researchers
31-12(2) CNRI/Science Photo Library/Science Source/Photo Researchers
31-14 (right) Bill Longcore/ Science Source/ Photo Researchers
31-16 (both) G. G. Martin/Visuals Unlimited
31-19 Lennart Nilsson

Chapter 32

32-3a,b(1) Oxford Scientific Films/Animals Animals
32-3b(2) Dwight R. Kuhn
32-4a Z. Leszczynski/ Animals Animals
32-4b Oxford Scientific Films/Animals Animals
32-4c Tom Bledsoe/Photo Researchers
32-4d William Ervin
E32-1a,b Carolina Biological Supply Company
32-15 Lennart Nilsson
32-16 Lennart Nilsson
32-17 Audesirk

Chapter 33

Frontispiece Thomas McEnvoy, Life Magazine, (c) 1955, Time, Inc.
33-1a Merlin Tuttle/Photo Researchers
33-1b Michael Fogden/ Bruce Coleman
33-2 G. I. Bernard/Oxford Scientific Films/ Animals Animals

33-5 Oxford Scientific Films/Animals Animals
33-7 Louis Darling
33-9 Thomas McEnvoy, Life Magazine, (c) 1955, Time, Inc.
33-10 William Vandivert
33-11a Manfred Kage/Peter Arnold
33-12 (all) Lee Boltin Picture Library
E33-1(1) Stephen J. Krasemann/DRK
E33-1(2) Gary Milburn/Tom Stack & Associates

Chapter 34

34-2 Richard K. Laval/Animals Animals
34-4a William Ervin/Comstock
34-4b Hans Pfletsinger/Peter Arnold
34-5 (c) Anne and Jacques Six
34-6 William Ervin
34-7 William Ervin
34-8 John D. Cunningham/Visuals Unlimited
34-10 Harold Hoffman/Photo Researchers
34-11a Audesirk
34-11b John D. Cunningham/Visuals Unlimited
34-13 P. J. Bryant, Univ. of California, Irvine/BPS
34-14 Jen and Des Bartlett/Bruce Coleman
34-17a,b Carolina Biological Supply Company
34-17c Audesirk
34-18a Carolina Biological Supply Company
34-20 William Ervin
34-21 William Ervin
34-22 Lennart Nilsson
34-23 Audesirk
34-34 Dr. I. Eibl-Eibesfeldt

Unit V Philip Green

Chapter 35

35-1a Larry Lipsky/Tom Stack & Associates
35-1b Bill Curtsinger/Photo Researchers
35-1c Catalina Marine Science Center
35-6 Tom McHugh/Photo Researchers
E35-1a,b USDA Forest Service
35-7a,b USDA Forest Service
E35-2 Chuck Pratt/Bruce Coleman
E35-3 U. S. Army
35-8 Thomas A. Wiewandt
35-9 Gianni Tortoli/Photo Researchers
35-10a-c Audesirk
35-12 NASA
35-18a Tim McCabe/Photo Researchers
35-18b H. Oscar/Visuals Unlimited

Chapter 36

36-1 Dr. Monica Mather, Simon Fraser University
36-4a David Burney, Lida Pigott Burney
36-4b M. D. Tuttle/Photo Researchers
36-4c Audesirk
36-4d (both) Carolina Biological Supply Company
36-5a,b Courtesy of G. Diatloff, Dept. of Lands, Queensland, Australia
E36-1 Thomas Eisner
36-7a Ronald L. Shimek
36-7b James L. Castner
36-7c David Doubilet
36-8a,b James L. Castner
36-8c Audesirk
36-9a Carl Roessler/Bruce Coleman
36-9b James L. Castner
36-10 (both) R. Humbert/BPS
36-11a,b Z. Leszczynski/Animals Animals
36-12a-c James L. Castner
E36-2 Carol Himes/Bruce Coleman
E36-3 Gilbert Grant/Photo Researchers
36-13a Greg Vaughn/Tom Stack & Associates
36-13b Stephen J. Krasemann/DRK
36-13c Keith Gunnar/Bruce Coleman
36-14a-c Audesirk

Chapter 37

37-9 Terry Domico/Earth Images
37-10a Henry M. Mayer/Photo Researchers
37-10b Tom Bean/DRK
37-11 Ray Ellis/Photo Researchers
37-12 Owen Franken/Sygma

Chapter 38

38-1, and (inset) Audesirk
38-2b NASA
38-3a,b Audesirk
38-6a,c Tom McHugh/Photo Researchers
38-6b B. J. Miller/BPS
38-7a Frans Lanting
38-7b,c David Burney, Lida Pigott Burney
38-8a-c Audesirk
38-9 Audesirk
38-10 Brian Parker/Tom Stack & Associates
38-11 Frank Staub
38-11 (inset) R. K. Burnard/BPS
38-12a,b Audesirk
38-12c John D. Cunningham/Visuals Unlimited
38-12d Frank Staub
38-13 Audesirk
38-14a, and (inset) Carolina Biological Supply Company
38-14b A. C. Twomey/Photo Researchers
38-15a, (inset), b Audesirk
38-16a Audesirk
38-16b Marty Stouffer Productions/Animals Animals
38-17a,b Robert Hernandez/Photo Researchers
38-17c James Simon/Photo Researchers
38-18a Townsend P. Dickenson/Photo Researchers
3-18b,c and (inset) Audesirk
38-18b (inset) R. Lennon/Ocean Researchers Films
38-18d Catalina Marine Science Center
38-18d (inset) Stan Wayman/Photo Researchers
E38-2 through 5 Audesirk
E38-6 Bob and Clara Calhoun/Bruce Coleman
38-19a-c Alex Kerstitch
38-20a Manfred Kage/Peter Arnold
38-20b Fred Bavendam/Peter Arnold
38-21 Dr. Frederick Grassle
38-22 Audesirk
38-23 Budd Titlow/Tom Stack & Associates

Index